SUPERCOLLIDER 5

SUPERCOLLIDER 5

Edited by

Phyllis Hale

Superconducting Supercollider Laboratory
Dallas, Texas

Springer Science+Business Media, LLC

Library of Congress Cataloging-in-Publication Data

Supercollider 5 / edited by Phyllis Hale.
      p.   cm.
   "Proceedings of the Fifth International Industrial Symposium on
the Super Collider, held May 6-8, 1993, in San Francisco,
California"--T.p. verso.
   Includes bibliographical references and index.
   ISBN 978-1-4613-6036-0   ISBN 978-1-4615-2439-7 (eBook)
   DOI 10.1007/978-1-4615-2439-7
   1. Superconducting Super Collider--Congresses.   I. Hale, Phyllis.
II. International Industrial Symposium on the Supercollider (5th :
1993 : San Francisco, Calif.)   III. Title: Supercollider five.
QC787.P7S8668   1994
539.7'3--dc20   94-3824
                                                                  CIP

Proceedings of the Fifth International Industrial Symposium on The Super Collider,
held May 6–8, 1993, in San Francisco, California

ISBN 978-1-4613-6036-0

© 1994 Springer Science+Business Media New York
Originally published by Plenum Press, New York in 1994

Preface

The fifth annual International Symposium on the Super Collider was a great success. Over 700 participants from around the country and the world gathered on May 6-8, 1993, in San Francisco to mark the progress of the SSC, to discuss current issues, and to chart a course of action for the continued development of our understanding of basic subatomic matter.

Together, the American public, academic communities, private sectors and governments from around the world have embarked on a project critical to maintaining our nation's preeminence as the world's leader in basic scientific research and the practical application of scientific knowledge. America has long maintained a commitment to investing in our nation's future. The Super Collider represents an essential next step in the direction of scope of human knowledge. The theme of the conference reflects these important goals: "SSC-Focusing the World on Next Generation Science."

The challenge for us today is to spread the message of the importance of investing in America's future. This is our task, and the task of supporters of the Super Collider throughout the nation. Without employing all of our energies, our nation will miss an historic opportunity to ensure America's scientific technological and economic leadership in the years ahead as we enter the next millennium.

Many exciting parallel and plenary sessions were planned in order to educate the participants on the great scientific activities and technological developments that are happening at the Super Collider. The quality and diversity of this year's sessions and speakers allowed the IISSC to meet its goals of promoting a fruitful exchange of ideas among the SSC community.

This year's plenary session speakers included:

> Mr. Charles Anderson, Air Products and Chemicals, Inc.
> Mr. Joseph R. Cipriano, DOE Project Manager for the SSC
> Dr. Charles R. Perry, Chairman of the Texas National Research Laboratory Commission
> Dr. Roy Schwitters, Director of SSCL
> Dr. Wolfgang K.H. Panofsky, Director Emeritus of Stanford Linear Accelerator Center
> Dr. George F. Smoot, Lawrence Berkeley Laboratory
> Dr. Nicholas P. Samios, Director of Brookhaven National Laboratory
> Dr. John Toll, President of URA
> Mr. Robert W. Galvin, Chairman of the Executive Committee of Motorola, Inc.

Congressional Panel:
        Honorable Jim Chapman
        Honorable Ron Packard
        Honorable Martin Frost
Dr. Raphael Kasper, Associate Director of SSC Laboratory
Dr. William Happer, Director of Energy Research, U.S. Department of Energy
Kasuke Takahashi, Deputy Director, KEK, Japan
Fang Shounxian, Director, BEPC, Peoples Republic of China
Vladimir Kadyshevskiy, Director, Joint Institute of Nuclear Research, Dubna,
        Russia

The IISSC presented its annual award to Congressman Tom Bevill of Alabama, Chairman of the House Appropriations Subcommittee on Energy and Water. As a senior Member of the powerful House Appropriations Committee, our award recipient is responsible for legislation which funds all the Nations major energy research projects. Our award recipient has been a staunch supporter of the Super Collider since its inception. His support for the project has been instrumental in moving the project forward and his efforts deserve our recognition.

Another group which deserves special recognition is this year's program committee. It was chaired by Ms. Regina Borchard, from Martin Marietta. Her committee consisted of the following members for the IISSC Board of Directors:

    Mr. Charles Anderson
    Dr. Paul Ayres
    Mr. Coby Chase
    Dr. William R. Frisken
    Mr. Henry Gandy
    Dr. Eric Gregory
    Dr. Hiromi Hirabayashi
    Dr. Thomas Kirk
    Mr. Ted Kozman
    Dr. Uriel Nauenberg
    Dr. Satoshi Ozaki
    Dr. Clyde Taylor
    Mr. Ken Wilson

The program committee's effort was also supplemented by numerous other associates from government, industry, and national laboratories.

For the fifth consecutive year, Pamela Patterson served as conference manager. Year after year, she manages a successful conference, despite the large size and scope of the IISSC. We appreciate and recognize her time and commitment to the SSC and to the conference.

The following companies, organizations, societies, and agencies assisted in producing IISSC 1993. Those listed below provided support in many ways, including financial, inkind, volunteers, or combinations of all three.

    ABB Technology Company
    AccSys Technology, Inc.
    AFA Industries
    Air Products and Chemicals, Inc.
    Allegheny Ludlum Corporation
    Ansaldo Componenti S.p.A.

Babcock & Wilcox
Bechtel National, Inc.
Brookhaven National Laboratory
Brown & Root
C. Itoh Pipe & Tube, Inc.
Copper and Brass Sales
CVI Incorporated
E. I. du Pont de Nemours & Company
Furukawa Electric
General Dynamics Space Systems Division
Grumman Space Systems
Handy & Harman
H.C. Starck Inc.
Helicoflex Company
Hitachi, Ltd.
Hudson International Conductors
IGC Advanced Superconductors, Inc.
Intermagnetics General Corporation
Ishikawajima-Harima Heavy Industries (IHI)
KEK National Laboratory for High Energy Physics
Lawrence Berkeley Laboratory
Lockheed Engineering & Sciences Company
Lotepro Corporation
Marotta Scientific Controls
Martin Marietta Corporation
Minnesota Valley Engineering, Inc.
Morrison Knudsen Corporation
New England Electric Wire Corporation
Noell, Inc.
Nippon Steel USA, Inc.
Oak Ridge National Laboratory
Oxford Superconducting Technology
Parsons Brinckerhoff Quade & Douglas, Inc.
SSC Laboratory
Siemens AG
Teledyne Japan K.K.
Teledyne Wah Chang Albany
Tempel Steel Company
Texas National Research Laboratory Commission
The PB/MK Team
The Sakura Bank
Toshiba Corporation
University of Colorado
U.S. Department of Energy
Westinghouse Electric Corporation
York University

The 1993 International Industrial Symposium on the Super Collider is a private non-profit corporation, whose Board of Directors serves as volunteers. The IISSC exists to promote the SSC program by sponsoring an annual symposium. Attendees can exchange ideas and learn about the possibilities created by the SSC program in furthering science, technology, and education. The 1993 Board was composed of the following members from industry, universities, government, and national laboratories.

Member	Affiliation
Mr. Charles Anderson	Air Products and Chemicals, Inc.
Dr. Paul Ayres	Babcock & Wilcox
Mr. Robert Baldi	General Dynamics Space Systems
Ms. Regina Borchard	Martin Marietta Energy Systems
Mr. Coby Chase	SSC Laboratory
Mr. Tony Favale	Grumman Space Systems
Dr. William Frisken	York University, Canada
Mr. Henry Gandy	Texas National Research Laboratory Commission
Mr. Paul Gilbert	The PB/MK Team
Dr. Eric Gregory	IGC Advanced Superconductors, Inc.
Ms. Phyllis Hale	SSC Laboratory
Dr. Hiromi Hirabayashi	KEK National Laboratory, Japan
Mr. Andrew Jarabak	Westinghouse Electric Corporation
Dr. Thomas Kirk	SSC Laboratory
Mr. Ted Kozman	SSC Laboratory
Mr. T. Scott Kreilick	Hudson International Conductors
Dr. Uriel Nauenberg	University of Colorado
Mr. Ronald Naventi	Bechtel National, Inc.
Mr. John Nonte	Lockheed Engineering & Sciences Company
Dr. Satoshi Ozaki	Brookhaven National Laboratory
Mr. James Richardson-Gonzales	SSC Laboratory
Dr. Giuseppe Scarfi	ANSALDO Componenti S.p.A.
Mr. Mark Sisinyak	Brown & Root, Inc.
Dr. Clyde Taylor	Lawrence Berkeley Laboratory
Mr. Kuniyasu Toga	Hitachi, Ltd.
Mr. Kenneth Wilson	CVI Incorporated

The Board of Directors met in San Francisco on May 5, 1993, prior to the start of the symposium, to review preparations for the 1993 symposium and to elect new officers and members of the IISSC, Inc. for the coming year.

President and Conference Chairman	Ken Wilson
Vice President	Eric Gregory
Secretary	Scott Kreilick
Treasurer	Paul Ayres
Administrative Officer	Phyllis Hale
Program Committee Chairman	Ted Kozman

To assure the success of the SSC in the coming years, the Board of Directors urges all the attendees to renew their commitment to this premier research program. This symposium is an excellent forum in which to learn the great news about the Super Collider.

Remember, it is our responsibility as industry, scientific and academic leaders to educate the American people, our elected officials and the world about the importance of the Super Collider.

On behalf of the IISSC Board of Directors, we thank all those who attended and contributed to IISSC 1993.

Charles E. Anderson

Charles E. Anderson
Chairman
IISSC 1993

Acknowledgments

There are many individuals who contributed to this manuscript. Several deserve special recognition. They are:

Chuck Anderson, President and Meeting Chairman for providing insight and leadership.

Pam Patterson, Conference Manager who, once again, proved to be the ultimate professional in sucessful meeting organization.

Regina "Gi Gi" Borchard, Program Chair for pulling together a program with excellent technical content.

Valerie Kelly who, along with the SSC Laboratory's Technical Information group, did all those little (and not so little) things which made the editor's job look easy. It should be noted that Valerie has been consistently associated with the IISSC publication. In addition to her technical expertise, she brings a beautiful dedication to this work.

Phyllis

Contents

DATA ANALYSIS AND COMPUTING II

Chairman: Irwin Gaines
Fermi National Accelerator Laboratory

SUPERCONDUCTORS I

Chairman: Donald W. Capone II
Superconducting Super Collider Laboratory

4. TECHNICAL POSTER SESSION I

Chairman: Roger W. Coombes
Superconducting Super Collider Laboratory

5. PARALLEL TECHNICAL SESSIONS III

DETECTORS III

Chairman: George P. Yost
Superconducting Super Collider Laboratory

ACCELERATORS III

Chairman: Y. H. Cai
Superconducting Super Collider Laboratory

MAGNETS III—THERMAL PERFORMANCE

Chairman: Peter O. Mazur
Fermilab National Accelerator Laboratory

CRYOGENICS III

Chairman: Robert L. Powell
Process Systems International, Inc.

6. NEXT GENERATION SCIENCE

Chairman: Wolfgang K. H. Panofsky
Stanford Linear Accelerator Center

7. TECHNICAL POSTER SESSION II

Chairman: Roger W. Coombes
Superconducting Super Collider Laboratory

8. PARALLEL TECHNICAL SESSIONS IV

DETECTORS IV

Chairman: Michael Marx
Superconducting Super Collider Laboratory

ACCELERATORS IV—RF AND OTHER TOPICS

Chairman: Jerry M. Watson
Superconducting Super Collider Laboratory

MAGNETS IV—MEASUREMENT TECHNIQUES

Chairman: Joe Muratore
Brookhaven National Laboratory

MAGNETS V

Chairman: Robert J. Malnar
Superconducting Super Collider Laboratory

CONVENTIONAL CONSTRUCTION II

Chairman: Paul H. Gilbert
The PB/MK Team

DATA ANALYSIS AND COMPUTING IV

Chairman: Pekka Sinervo
University of Toronto

DETECTORS V

Chairman: Tony A. Gabriel
Oak Ridge National Laboratory

9. WORLD CONVERGING ON SSC (INTERNATIONAL PARTICIPATION)

Chairman: Raphael Kasper
Superconducting Super Collider Laboratory

1 Partners Meeting Milestones

SSC Status

IT'S THE SCIENCE STUPID

Joseph Cipriano

Project Director
Department of Energy
2550 Beckleymeade Avenue
Dallas, Texas 75237-3997

Few publicly funded programs have been subjected to more scrutiny, suffered from more misinformation, and remained more misunderstood than the Superconducting Super Collider.

I want to set the record straight on several issues and discuss my assessment, as project director, of the status of the SSC project and its most probable future course.

First of all, the SSC is no longer just a concept - it is emerging from the Texas prairies, from within the Austin Chalk, and from the spotless manufacturing lines throughout the United States and the laboratories of the world in the form of high technology machines.

Just one month ago the SSC achieved "first light" with the successful acceleration of hydrogen ions through the first stage of the SSC's linear accelerator.

This significant first step promises that we are a mere two years away from a fully operational linac. A linear accelerator which can be used for cancer research and therapy as well as its primary role as part of the SSC's injector complex.

Less than one year ago, in August 1992, and six weeks ahead of schedule, a string of full scale collider dipole superconducting magnets, spool pieces, and quadrupole magnet were successfully tested at full operational current.

To date 20 prototype collider dipole magnets and six quadrupole magnets have been built by American industry and tests show that they exceed all the planned requirements.

The first phase of qualification for eight superconducting cable manufacturers has been successfully completed with the production of 3.5 metric tons of cable by each company.

Over thirty eight miles of the collider tunnel are now under contract and more than three miles are already completed. New records have been set for best day, week, and month for a hardrock tunnel boring machine by an SSC contractor, Gilbert/Shea, using a Robbin's tunnel boring machine.

More then one million square feet of laboratory, industrial, and office buildings are now complete and largely occupied by over 2,400 full and part-time laboratory personnel.

Rights to nearly all of the 16,000 acres of land required for the project have been acquired. The state of Texas must be congratulated for its unwavering support in acquiring this land for the facility and providing promised funding.

Design will be complete for almost 75% of the conventional construction and 25% of the construction will be complete by the end of FY 93.

Two large detector collaborations have been assembled with over 1600 participants to build the worlds most sophisticated particle detectors.

$140 million in contract awards have been made to small and disadvantaged businesses.

Over 50,000 students have been reached by SSC education programs and over 7,000 direct jobs have been created.

These are all significant accomplishments - accomplishments that are real - not subject to argument, or misinterpretation. While the politics of the SSC has been debated we have been building it.

Now, let me clearly address some "bad press," also referred to as "Donaldson's Discourses."

Myth #1: the SSC project is over budget and the key to federal deficit reduction

The truth is that the SSC project execution is tracking the baseline estimates well. We do not yet have enough data to accurately predict the final outcome from trends and so, even though the trends happen to be good today, I am going to resist the temptation that the opponents couldn't resist when the trends happened to be down and simply say preliminary trends are encouraging.

The much discussed General Accounting Office report of its audit of the SSC's cost and schedule was, I believe, misinterpreted by some and misused by others.

Here are the facts:

It is a fact that expectations for every aspect of the SSC are high. We have not met all those expectations and have been criticized for these shortcomings. I accept the criticisim as a challenge. It is a challenge I believe we can meet.

In spite of these shortcomings...

To date all seven of the major, scheduled, SSC major programmatic milestones have been met.

As of January only 3.5 percent of the $850 million contingency dollars of the program had been committed even though the project was 17 percent completed.

Current projections, based on actually completed activities, suggest that the project can be completed within baseline cost and schedule if the funding were available.

Mr. Gilbert is not going to overrun his construction budget. Are you Paul?

Furthermore, the SSC does not significantly impact the deficit since it represents less than one percent of the total Federal R&D Budget and all the money appropriated is pumped into the economy, creating jobs, and strengthening new industries.

Finally we do not want to spend our time talking about the cost of the SSC, we want to talk about its benefits.

Myth #2: the SSC is a "Texas pork barrel project"

First of all, Texans don't eat pork - they eat beef - as any BBQ aficionado would tell you.

4

The SSC is not pork science. It has had a most rigorous series of peer reviews, all of which have concluded it is good high priority science. The highest priority in high energy physics.

Second, Texans are paying for the privilege of having the SSC in their state.

$1 billion plus all the land, relocations, highway expenses, and other inconveniences that go along with major construction.

Finally, of over $951 million in contracts awarded to date over 61% has been awarded to corporations and institutions located in 48 states other than Texas.

The SSC is a national project.

Another significant feature of contracts awarded by the project is that the the SSC project has demonstrated its understanding of the importance of the small business community to science and the economic development of the nation.

To date nearly 14% of all contracting by the SSC project has been placed to small and disadvantaged business.

It is important that these businesses be a part of this program and that we take full advantage of the opportunity we have to strengthen America's business base.

"Spin-offs" have also been a much misunderstood aspect of the SSC project.

We are not building the SSC for spinoffs, but spinoffs are a natural result of any high technology construction project and we should not be embarrased to talk about them.

Spinoffs reduce the real cost of the program to the taxpayers by providing early payback.

The SSC program has already contributed the following to the nation:

Over $500 million to defense conversion;

Using OMB's baseline of 55 jobs per million dollars of public funding over 100,000 direct and secondary jobs have been supported by SSC funding to date with the number swelling to nearly one-half million by project completion;

SSC supported development of superconducting cable has resulted in a ten-fold reduction in its cost, and a 30% increase in performance. As the cost continues to come down, the applications will grow in number.

A state-of-the-art parallel processing system has been developed at the SSC laboratory that is both powerful and economical.

This system of integrated workstations has the computational power equivalent to 30 conventional mainframe computers, but at one-seventh the cost;

Over 50,000 students in 26 states have been to SSC designed programs aimed at stimulating their interest in math and science education;

As the SSC moves into full scale production of superconducting magnets, the injector complex comes on-line,

Superconducting technology will be fully industrialized through development of fully automated manufacturing lines for the over 10,000 large-scale magnets required by the SSC and through an additional ten-fold reduction in the cost of superconducting cable.

These advances will support the development of superconducting magnetic energy storage systems which offer the promise of greater efficiency in the generation and transmission of electric power and measurable beneficial effects on the global environment due to decreases in the use of fossil fuels.

A cancer research and therapy center, directed at the treatment of certain kinds of cancer located deep within the body, will become operational.

And finally, I am convinced that just as the discoveries which arose from research on particle accelerators over the last 75 years have resulted in many of the products and processes we depend on today, once operational, the SSC will unveil fundamental secrets of the universe - secrets that will not only advance scientific understanding of nature but expand the limits of technologies that will fuel economic growth in the latter half of the 21st century.

As a result of these exciting accomplishments and benefits of the SSC my heart is full of hope and confidence for the future.

I believe that we have been our own worse enemy.

We have failed to articulate the benefits of the project in clear and understandable English.

We have failed to move as rapidly as we should have in establishing the management and reporting systems which provide objective evidence of the control we have over the project.

We must do a better job in explaining the benefits to the average citizen and this burden must primarily be borne by those of you in industry and universities that are involved in the project.

We must all be ever vigilant against the waste of resources, be they financial, personnel, or material. The measure is not just what is lawful, but what is wise and prudent.

We must keep the ultimate scientific goals of the project in mind but along the way we must be alert for opportunities to "exploit" developments for the good of the private sector.

If we can accomplish these goals I am confident that the American people will recognize and embrace the benefits of the SSC.

In science and technology, as in all aspects of our nation's history, Americans have consistently sought the roles of pioneers and leaders.

America has always been eager to search out new frontiers and define the unknown.

A string of rapid advances in science and technology have merged in such a way that permits for the first time in history a unique opportunity to develop the largest most sophisticated scientific instrument the world has ever known.

The SSC will enable scientists to explore the smallest parts of the atom and the fundamental forces of our universe.

It will spawn dramatic breakthroughs in physics, in technology, and in medicine.

It will stimulate education, industry, and the economy.

Ultimately, it will improve the quality of life for all mankind.

Our national pursuit of this opportunity not only continues our successful leadership tradition, but also is a sound investment in America's future.

This, ladies and gentlemen, is our message.

Spread the word!

2 Parallel Technical Sessions I

In Depth Review of Major SSC Programs

BUSINESS OPPORTUNITIES IN THE SSCL GLOBAL CONTROL SYSTEM

J. W. Heefner, R. G. Bork, and D. P. Gurd

Superconducting Super Collider Laboratory[*]
Accelerator Systems Division/Controls Department, MS-4002
2550 Beckleymeade Avenue
Dallas, TX 75237-3997

INTRODUCTION

The SSC consists of a series of six accelerators: an injector complex made up of a LINAC, a Low Energy Booster (LEB), a Medium Energy Booster (MEB), and High Energy Booster (HEB), as well as two storage rings together known as the "Collider." Although controls requirements and issues related to this complex of accelerators are not inherently different from those of other large accelerator laboratories, some special problems result from the large number of control points involved (greater than 500,000), the great distances between the various components (up to 100 Km), and the very high reliability required (0.986 availability for the global control system).

The SSCL will employ an integrated control system, which at the highest level is called the Global Accelerator Control System (GACS). The GACS is defined as that system which provides the infrastructure and environment in which all systems directly affecting the accelerated beam are integrated, operated and controlled. The GACS is divided into two primary subsystems, Beam Controls and Process Controls. Beam Controls are those required to operate systems which directly support beam acceleration, steering, focusing and diagnostics. The precisely timed and synchronized, high speed, high quality Beam Controls are required only during beam operations. In contrast, the Process Controls operate independently, update at slower rates and do not require precise timing or synchronization.

BEAM CONTROL SYSTEM

The SSCL Beam Control System (BCS) must interface with various accelerator subsystems including RF, Magnet Power Systems, Quench Protection Systems, Beam Instrumentation, Beam Synchronization Systems, Beam Abort/Permit Systems, and Personnel Safety Systems. The BCS will provide all of the Operator Stations, Engineering Development Stations, VME crates, VXI crates, commercially available I/O modules, communications modules, and Control software necessary for accelerator control. Table 1 shows the anticipated quantities of the equipment types described below.

---

[*] Operated by the Universities Research Association, Inc., for the U.S. Department of Energy under Contract No. DE-AC35-89ER40486.

Table 1. Anticipated quantities of equipment.

Machine/ Equip.	LINAC	LEB	MEB	HEB	Collider	Main Control Room
Workstations	3	6	8	10	10	12
File Servers	2	2	2	4	20	
X Terminals				2	10	36
Console Racks				2	10	12
VME/VXI Crates	60	100	100	296	1480	
VME/VXI Processors	60	100	100	296	1480	
Rack Power Supplies	15	25	25	100	400	
I/O modules	Numerous	Numerous	Numerous	Numerous	Numerous	
ADM	2	6	8	34	200	1
Fiber	1.23 Km	2.01 Km	5.41 Km	15.79 Km	96.68 Km	
Routers	1	1	1	2	10	1

Control Room Hardware

Operator stations will be UNIX workstations with at least 32 MByte of main memory, 2 to 6 GByte of disk storage, one keyboard, mouse, 3 to 4 displays (for control room) each and rated at 50-100 SPECmarks. In addition, the workstations will be required to support audio alarms and messages.

Control system file servers will be UNIX computers with at least 64 MByte of main memory, 4 to 10 GByte of disk space, one keyboard, mouse, one display each and rated at 50-100 SPECmarks.

X-terminals will be used as additional status displays in the Main Control Room and in each sector control room. It is also planned that a large video wall will be located in the Main Control Room. This video wall will be made up of multiple displays and will be used to display status information that is useful to operators and experimenters in the control room.

Peripherals such as printers, and disk drives will be procured for use in each sector control room and the Main Control Room.

Console racks will be 5 bay low profile racks. A table top will be attached to the console and will have space for an operator station keyboard, and mouse or trackball, and operator logbooks.

Communications Equipment

The SSCL will install a high speed telecommunications infrastructure to transmit the massive quantities of data over the geographically large site area. Network cable right of ways will be placed in the floor of the various accelerators. The backbone of the system is a redundant fiber Synchronous Optical Network (SONET).

Add-Drop-Multiplexers (ADM) will be used to interface the low speed T1, and OC-1 links with the main fiber backbone. ADM's will have various low speed connections ranging from T1 to OC-1 and high speed connections ranging from OC-3 to OC-48.

The fiber cable used in the communications system will be single mode, in 24 and 48 fiber cable bundles. The cable will meet or exceed NEC 770, NFPA 262-1985 and Bellcore Standard TR-20. The jacket will be Kevlar reinforced. Connectors will be FC/PC types. Small quantities (less than 5Km) of multi-mode cable will be used. Multi-mode connectors will be ST or SMA.

Routers will be used to interface standard networks such as Ethernet, or FDDI to the fiber backbone. T1 and OC-1 interface modules will be used to interface VME and VXI processors to the communications infrastructure.

Test equipment such as Optical Time Domain Reflectometers (OTDR) and fiber network analyzers will be purchased commercially and used to maintain and test the communications infrastructure.

Front End Equipment

Front End systems are those system that directly interface the GACS with accelerator subsystems such as RF, Magnet Power Supplies, and Beam Instrumentation Systems. The SSCL has chosen EPICS (Experimental Physics and Industrial Control System) as the software that will run in Front End System processors. The primary hardware supported by EPICS is VME and VXI based, but some industrial I/O busses will be supported.

VME and VXI crates will be procured and located in control racks in the above ground buildings of the LINAC, LEB, MEB, HEB and Collider. In addition, crates will be located in each niche of the HEB and Collider. These crates will meet all minimum requirements of the VME and VXI specifications, respectively, plus additional specifications imposed by the SSCL on power supply regulation, minimum crate cooling requirements, cable routing, and physical size limitations.

VME and VXI processors will be procured for Front End Systems. These processors will meet the minimum performance specifications of Motorola 68040 processors and be capable of running EPICS.

VME and VXI I/O modules will be procured. These modules will be used to control and monitor the various systems to which the GACS must interface. Typical module and signal types include: 24 VDC inputs, 24 VDC Outputs, TTL Inputs, TTL Outputs, Contact Outputs, RTD's, Thermocouples, Silicon Diodes, Helium liquid level probes, 4-20 mA inputs, 4-20 mA Outputs, Analog Voltage Output (typically 0-5, +/- 5, 0-10, +/- 10 VDC), LVDT's, Strain Gauge, Analog Voltage Input (typically 0-5, +/- 5, 0-10, +/- 10 VDC).

Power conditioning equipment such as rack mounted UPS's, isolation transformers, filter transformers, and DC power supplies will be procured for Front End System Racks. Interconnection devices such as terminal strips, terminal blocks and connectors will be used within racks and to connect to various accelerator subsystems and devices.

Commercial Software

Artificial Intelligence Supervisory Control packages and predictive failure analysis and simulation software may be procured for use within the GACS.

PROCESS CONTROL SYSTEM

The SSCL Process Control System (PCS) is divided into 15 separate, but linked systems that correspond to each of the 10 Collider sectors, 2 HEB sectors, the MEB, LEB, and LINAC. Specifically, the PCS includes the controls for HEB and Collider cryogenics refrigerators, transfer lines and spools pieces, and LEB, MEB, HEB and Collider vacuum, LCW, ICW, and Instrument Air systems. It has been estimated that the total number of I/O points covered by the PCS is in excess of 150,000 points.

Commercial vendors will be asked to supply all necessary components in the appropriate quantities for the SSCL PCS. Each Process Control System consists of Operator Stations, Engineering Development Stations, control processors, input and output modules, racks, power supplies, configuration software, and run-time system software. In addition, there are components that will reside in the SSCL Process Control Main Control Room. It is envisioned that these components will include Operator Stations and Engineering Development Stations and all necessary communications equipment to tie the Main Control Room to each sector PCS. These stations may be separate from the Beam Controls Operator and Engineering Development Stations.

The operator stations will contain all the hardware and software necessary to provide a Man-Machine Interface (MMI) and will consist of CRT displays, UNIX computers, keyboards, mouse and software to provide a variety of MMI displays.

The Engineering Development Stations will contain all hardware and software required to develop process control applications. The hardware will include UNIX computers, keyboards and a mouse. The configuration tools available within the system will allow the process engineer to configure a system and applications using ladder logic, function blocks, sequential function charts, or script. In addition the system will include an Applications Programmer's Interface (API) that will contain software routines and templates required to develop new interfaces and applications processes based on the C and C++ programming languages.

I/O modules and signal conditioning devices will be required in each niche of the Collider (approximately 200), each niche of the HEB (24), in each Sector Refrigerator (12), and in numerous locations in LEB and MEB. The number of I/O points in each niche of the Collider and HEB has been estimated to be 500 points, split approximately 60/40 binary to analog, respectively.

STATUS AND SCHEDULE

Beam Control System

Equipment for the BCS of each machine and the Main Control Room will be procured as follows:

Table 2. BCS Equipment Procurements.

Machine	Fiscal Year
LINAC	1993
LEB	1994
MEB	1995
HEB	1997
Collider	1997
Main Control Room	1994

Process Control System

It is currently planned that the SSCL Process Control System will be procured in multiple steps. The first part of the process will be the procurement of the first Collider sector's worth of equipment. This procurement will be a two step procurement. In the first step, the SSCL Controls Department will issue a Request for Proposal in which vendors will be asked to provide technical proposals for the first sector of the Process Control System. These proposals will be evaluated by the SSCL and the top three proposals will be selected for the second step of the procurement. In this step, the three successful vendors will be asked to provide evaluation systems that meet the minimum requirements set forth in the Prime Item Development Specification and Statement of Work. The SSCL will then evaluate each system in accordance with predefined evaluation criteria. An RFP for the delivery of the first sector's worth of equipment will then be issued. Award of the contract will be based on the results of the SSCL evaluation and responses to the RFP. A preliminary schedule of events for this first procurement is shown below.

Table 3. PCS Procurement Schedule.

CBD Announcement	April 1993
RFI Issued	May 1993
First RFP Issued	July 1993
Contract Award	November 1993
Evaluation Systems delivered	December 1993
RFP for sector equipment Issued	March 1994
Contract Award	May 1994
Equipment Delivery Complete	October 1994

Additional PCS sector procurements will be as follows:

Table 4. PCS Sector Procurement Schedule.

Fiscal Year	Sectors Procured
1994	2
1995	4
1996	6
1997	3

EQUIPMENT ACQUISITION PLANS FOR THE SSCL MAGNET EXCITATION POWER SYSTEMS

Russell A. Winje

Superconducting Super Collider Laboratory*
2550 Beckleymeade Avenue
Dallas, TX 75237-3997

INTRODUCTION

The particle beam production capability of the Superconducting Super Collider Laboratory (SSCL) consists of the injector accelerator systems and the collider accelerator systems. The basic parameters of these accelerators are given in Table 1.

Table 1. Technical Parameters of the Injector and Collider Accelerators.

PARAMETER	INJECTORS				COL
	LINAC	LEB	MEB	HEB	
Length/Circumference	150 m	540 m	3960 m	10 890 m	87 120 m
Injection Momentum	N/A	1.2 GeV/c	12 GeV/c	200 GeV/c	2 TeV/c
Maximum Momentum	1.2 GeV/c	12 GeV/c	200 GeV/c	2 TeV/c	20 TeV/c
Cycle Time	100 ms	100 ms	8 sec	515 sec	Continuous

The Linac is a linear accelerator composed of an H⁻ ion source, the radio frequency quadrupole accelerator, the drift tube linac accelerator and the cavity coupled linac. The Low Energy Booster (LEB) is the first of three injector synchrotrons which collectively raise the momentum of the proton beam from 1.2 GeV/c to 2 TeV/c. Both the LEB and the Medium Energy Booster (MEB) have resistive magnets in their lattice while the High Energy Booster (HEB) has superconducting magnets. The HEB is a bi-directional accelerator in which protons are first accelerated in one direction (CW) filling one CW collider (COL) ring followed by operation in the reverse direction (CCW) filling the other collider ring. As noted above, there are two collider rings with superconducting magnets, each operating independently. During experimental times, the proton beams in the two accelerators are brought to the same orbit, causing collisions between protons.

One of the key subsystems of all of the accelerators are the magnet excitation power systems. These power systems, along with auxiliary support subsystems, are provided by the Electrical Engineering Department in the Accelerator Systems Division (ASD/EE). The power systems provide well regulated and repeatable current to dipole, quadrupole and other multipole magnets in the accelerators and their interconnecting beam lines. These magnets may be individual magnets such as those of the pulsed power and correction element power supplies or may be strings of magnets which are typical of the ring magnet and beam line power systems.

The output current from the power systems may be: continuous steady state as used in the Linac; ramped (programmed) as in the LEB, MEB, HEB and COL synchrotrons and some of the beam lines; pulsed as in the injection and extraction kicker subsystems; or may be half sine wave as in the pulsed septum and Lambertson magnets. In all cases the output current is required to be

* Operated by the Universities Research Association, Inc., for the U.S. Department of Energy under Contract No. DE-AC35-89ER40486.

continuously adjustable over nearly the full range of current (i.e. from a few percent up to 100% of the maximum output current).

Two unique electrical subsystems also provided by the ASD/EE, while technically not magnet excitation power systems, are the quench protection subsystem and the energy extraction subsystem. This equipment is required for protecting the superconducting magnets in the HEB and COL from being destroyed when they unexpectedly transition from the superconducting state to the normal (resistive) state with high levels of current flowing in their coils (quenching). This situation is made worse with strings of superconducting magnets when energy from other magnets in the string can feed into the quenched magnet. For economic reasons, superconducting magnets are not made with sufficient copper conductor in their coils to support full current conduction when the current leaves the superconductor filaments and flows instead in the copper surrounding the superconductor. The quench protection subsystem detects this transition condition and applies stored electrical energy to heaters imbedded in the superconducting coils to drive the entire coil normal, distributing the magnetic energy through the entire coil volume. At that time, bypass diodes, in parallel connection to the magnets, become forward biased diverting the current from the coil and thereby protecting the coil. The energy extraction subsystem protects the bypass circuit from damage by inserting resistors in the circuit reducing the current in the magnet coils and absorbing the magnetic energy stored in the magnet string.

What follows is a brief description of the major electrical technical equipment used in the accelerators and the present laboratory plans for the acquisition of the equipment.

PULSED POWER SUPPLIES AND KICKER MAGNETS

A number of pulsed power supplies and specialized fast rise and fall time magnets are needed for various injection and extraction operations of the several accelerators. Table 2 lists some of the critical features of the kicker magnet subsystems.

Table 2. Kicker System Critical Performance Criteria.

Accelerator System	Rise Time μ sec	Fall Time μ sec	Pulse Length μ sec	Peak Current Amperes
LEB extract	0.08	N/A	2.5	1400
MEB inject	0.06	2	2.0	1400
MEB extract	2	N/A	15	2000
MEB abort	2	N/A	15	2000
HEB inject	0.4	1.7	13	3200
HEB extract	1.7	N/A	35	6000
HEB abort	1.7	N/A	35	10000
COL inject	1.7	4	35	3500
COL abort	4	N/A	300	15000

These magnet and power supply subsystems will be developed by the technical staff of the laboratory in accordance with the system level technical specifications to the extent ranging from preparing detailed fabrication designs to preparing detailed technical performance specifications. In the case where detailed designs are being prepared, for example the LEB extraction magnets, the laboratory will develop the detailed magnet fabrication and assembly drawings, contract for the fabrication and assembly of the hardware and take delivery of the completed magnets. Acceptance will be by testing and inspection. For example, the subcontractor will measure the leak rate of the vacuum envelope and will perform high potential testing of the coil assembly.

In those cases, such as the modulator and charging power supplies, where equipment is acquired through performance specifications, the laboratory staff will prepare detailed technical specifications for the equipment with the fabrication and assembly design being prepared by the subcontractor, subject to the review and approval of the laboratory technical staff. As part of the work, the subcontractor will fabricate and assemble the modulators. Acceptance of the completed equipment will be based on tests and inspection.

Table 3 lists the main components which the laboratory plans to purchase in support of the pulsed power work and the calendar year in which the equipment is expected to be delivered. These components will be purchased both by performance specifications and by fabrication and assembly documentation.

Table 3. Pulsed Power Major Components.

Component	CY 93	CY 94	CY 95	CY 96	CY 97
HV Cable RG 220	20 000 ft	20 000 ft	100 000 ft		
Modulator & Charging PS Assembly					
MEB extract			12		
HEB inject				18	
HEB abort				20	
HEB extract				20	
COL inject					20
COL abort				40	20
Ferrite Core 1x6x6 inch	200	200			
Ferrite Core Magnet Assembly Gap: 5x8x500 mm				12	18
Laminated Iron Core Magnet Assembly 5x5x1500 mm				80	40

CORRECTION ELEMENT POWER SUPPLIES

The principle function of the correction element power supplies is to provide the accelerator operators the capability to correct the orbit of the proton beam due to misalignment errors in the main dipole and quadrupole magnets and for betatron tune correction. Although these errors may be small, the accumulated effect is to reduce the brightness of the beam which in turn reduces the luminosity of the colliding beams or the production capability of the laboratory. Table 4 shows the correction element power supplies and the calendar year in which the order is expected to be delivered. Acquisition of this equipment will be through detailed performance specifications.

Table 4. Correction Element Power Supplies.

Accelerator/Power Supply	CY 93	CY 94	CY 95	CY 96	CY 97	CY 98	CY 99
LEB							
4 Quadrant							
70 V, 25 A		90					
130 V, 15 A		76					
130 V, 25 A		48					
190 V, 100 A		18					
70 V, 3.5 A		12					
MEB							
4 Quadrant							
100 V, 15 A			250				
Unipolar							
800 V, 600 A			2				
HEB							
4 quadrant							
30 V, 100 A					254	200	
300 V, 100 A					8		
COLLIDER							
4 quadrant							
30 V, 100 A					700	800	700
200 V, 100 A					100	100	

RING MAGNET POWER SUPPLIES

The ring magnet power systems furnish the excitation power to the main dipole and quadrupole magnets in the four synchrotrons. In the LEB, HEB and COL, both dipoles and quadrupoles magnets are powered in long series-connected magnet strings. In the MEB, the dipoles and the two quadrupole circuits are independently powered. The main parameters of the ring magnet power system circuits are given in Table 5. The COL circuit is subdivided into 10 independent circuits per ring to minimize the recovery time from magnet quenches.

Table 5. Ring Magnet Power System Specifications.

PARAMETER	LEB	MEB D	MEB QF	MEB QD	HEB	COL
Maximum Current	4000 A	5200 A	4500 A	4500 A	7000 A	7000 A
Cycle Time	0.1 s	8.5 s	8.5 s	8.5 s	250 s	cont
System dI/dt	125 kA/s	1690 A/s	1500 A/s	1500 A/s	70 A/s	4 A/s
Power Supplies	3	8	1	1	4	20
Pulse Number	24	24	24	24	24	12
Total Ripple Voltage	<1%	<1%	<1%	<1%	<1%	<1%
Ripple Current	50 ppm	50 ppm	50 ppm	50 ppm	10 ppm	1 ppm

The design of the ring magnet power systems are based line commutated power conversion systems operating as current regulated power supplies. The design emphasizes common hardware in all power systems with variations between the injector power supplies and the collider power supplies. In all cases except the LEB and the MEB QF and QD power circuits, one of the power supplies will operate in the current regulated mode with the remainder in the voltage regulated mode. In these cases, the power supplies are operated in the voltage regulated mode imbedded in the current regulation circuit.

The major components of the system will be acquired through performance specifications prepared by the SSCL according to the delivery schedule in Table 6. Equipment design and fabrication/assembly will be by the subcontractor.

Table 6. Acquisition Schedule for the Ring Magnet Power Supplies.

COMPONENT	CY 93	CY 94	CY 95	CY 96	CY 97	CY 98	CY 99
15 kV SWITCHGEAR		4	12		5		
15 kV RECT XMFR		6	20	4			
5 kV SWITCHGEAR					10	6	4
5 kV RECT XMFR					10	6	4
POWER CONVERTER		6	20	4	12	20	12
THYRISTOR ASSY		36	120	24	48	120	72
OUTPUT FILTER		2	10	4	9	8	7
CURRENT XDUCR		1	3		12	20	9
AC HARMONIC FILT		1	3		1	5	5

At the present time, the technical requirements have been determined for all of the technical components of the system and the prerequisite procurement documents have been written for much of the equipment. Most of the equipment is of conventional design and construction. However a few components are special and require some explanation. The 15 and 5 kV rectifier transformers are multiphase units that are designed for both three phase bridge type circuits (used in the Injector synchrotrons) and for dual six phase star circuits as used in the COL. The transformers are characterized by tight tolerances on impedance, phase angle and turns ratio. The power converter equipment contains thyristor gate drive equipment and the interconnecting ac and dc busses. In the case of the HEB and COL power converters, those units also contain interphase transformers to assist in minimizing unbalanced current and voltage ripple. The thyristor assemblies support three thyristors and the necessary voltage snubbers. These units are water cooled and are housed in the power converter assemblies. The output filters are required to reduce the output voltage ripple to the level needed to meet the magnet circuit current ripple requirements.

There are two types of current transducers. One type IS the conventional current magnitude devices that measure the output current of the power circuit. Three current ranges are planned: 4000 A, 5200 A and 7000 A. The transducter has a D/A converter for the digital reference signal, an analog difference circuit with a gain of stage 32 and an A/D to produce a digital difference signal for the current regulation system. The AC harmonic filter will be acquired for both 15 kV and 5 kV utilization voltages. These units are the band pass type and are required to reduce harmonic voltages on the AC system to acceptable levels and to improve the power factor.

In all cases, the performance requirements of the equipment will be by the SSCL staff. The subcontractors will be responsible for the detailed design and fabrication. Acceptance will be by performance testing and inspection.

QUENCH PROTECTION EQUIPMENT

As explained in the introduction, the quench protection equipment is designed to protect the superconducting magnets from damage during quench operations.

The QPS procurement strategy is for the SSCL to procure non-critical items under performance specifications, procure commercial off-the-shelf assemblies under source control drawings, and design and produce fabrication drawings for procurement of assemblies or chassis that are critical to magnet protection, in particular those having intricate functionality not normally seen in industry. Use of a system integrator to assemble and comprehensively check out fully configured QPS systems (above ground) is under consideration. Test requirements for procured items will be defined in detail by the SSCL. Testing of major components will be the responsibility of the contractor. Embedded software will be designed and produced by SSCL. The quantities of major components required for the QPS are listed in Table 7 along with the planned delivery schedule.

Table 7. Quench Protection Equipment Acquisition.

COMPONENT	CY 93	CY 94	CY 95	CY 96	CY 97	CY 98	CY 99
Heater Firing Unit	180	360	360	360	360	360	240
Volt to Freq Convert	40	65	65	65	65	65	45
Quench Prot Monitor	40	65	65	65	65	65	45
Bypass Diode Assy	160	320	320	320	320	320	170
Uninterruptible Power Supplies	40	65	65	65	65	65	45

ENERGY EXTRACTION EQUIPMENT

The Energy Extraction equipment consists of 7000 A dc circuit breakers and temperature dependent resistors capable of absorbing up to 1000 MJ of energy stored in the HEB and COL superconducting magnets. The circuit breaker is controlled from the QPS. On the initiation of a quench in one of the magnets, the circuit breaker opens and switches the current into the resistor. At the same time, the power supply is turned off and the current begins to decay with a 36 second time constant. Table 8 lists the major equipment required for the energy extraction system and the calendar years for delivery. The requirements for this equipment will be detailed by the SSCL with design, fabrication and assembly by the subcontractor. Acceptance is by inspection and performance testing.

Table 8. Energy Extraction Equipment Delivery Schedule.

COMPONENT	CY 93	CY 94	CY 95	CY 96	CY 97	CY 98	CY 99
DC CRKT BRKRS			10	33	33	20	
RESIST ASSY			10	33	33	20	

BEAM LINE POWER SUPPLIES

A number of current regulated power supplies are required for the beam transfer lines interconnecting the various accelerators and the test beam facility. These power supplies are mainly programmed types for the purpose of energy conservation with a few being continuous duty types. These power supplies will be free standing units and will have current regulation requirements of 1000 ppm and better. Table 9 lists the major units with ratings greater than 200 kW to be purchased and the year of delivery. Additional units less than 200 kW will also be purchased.

Table 9. Beam Line Power Supplies Delivery Schedule.

COMPONENT	CY 93	CY 94	CY 95	CY 96	CY 97	CY 98	CY 99
100 V, 3000 A				8			
100 V, 5000 A					8		
250 V, 1000 A					4		
500 V, 700 A				5			
300 V, 2500 A					2		
125 V, 6000 A						6	
170 V, 2000 A						2	
65 V, 3000 A			1				
200 V, 1000 A			12	13			

All of the beam line power supplies will be purchased to technical specifications prepared by the SSCL. The subcontractor will provide the design and will fabricate and assembly the equipment. Acceptance will be by performance testing and inspection.

CONTROL AND REGULATION EQUIPMENT

Most of the equipment described above will be complete with regulation and control hardware ready to be installed and connected to the control system. A serial fiber optic interface for connecting the power supplies to the host control system is being developed at the SSCL. The specifications for this hardware will be included in the power supply technical specifications and will provide the interface protocol and functional diagrams. The detailed design will be performed by the subcontractor.

For other equipment such as the ring magnet power supply, the control and regulation equipment will be designed by the SSCL and will be acquired by purchasing components at the module level. Some special equipment will be required which will be designed by the SSCL. The design of this hardware is based on digital regulation and control schemes. Single board computer hardware and operating system software with a VME back plane will be purchased. Detailed application software will be written by the SSCL.

SCHEDULE

At this time, extension of the collider completion schedule is being considered by the DOE and laboratory management. However, to provide a consistent schedule, the delivery dates given above are based on the present integrated project schedule with the completion of the project in 1999.

CONCLUSIONS

A comprehensive listing of forthcoming procurement activities for special technical electrical equipment for the SSCL has been given. Some high level technical specifications have been provided to assist readers in evaluating the scope of these acquisitions.

BEAM INSTRUMENTATION AND PRECISION TIMING EQUIPMENT
FOR THE SSC ACCELERATOR COMPLEX

Robert C. Webber

Beam Instrumentation Department, Accelerator Systems Division
Superconducting Super Collider Laboratory[*]
2550 Beckleymeade Avenue
Dallas, Texas 75237

INTRODUCTION

This article is intended to provide an introduction to the beam instrumentation systems and the precision timing system of the SSC Laboratory. The Beam Instrumentation Department of the SSC Accelerator Systems Division is responsible for the design of these systems for the Collider rings and the injector accelerators. Status and plans are briefly described.

The signal sensors and low level signal processing electronics of the beam instrumentation systems are the eyes and ears into the world of the proton beam. Situated between that invisible beam and the accelerator operator, they monitor and measure the behavior and performance of the otherwise intangible protons as they are guided and accelerated through the Linac and the injector synchrotrons, and finally stored in the Collider rings.

The precision timing system provides the vital time references required to synchronize critical accelerator hardware functions, beam manipulations, and data acquisition over the geographical extent of the Supercollider complex. Several fundamental timing signals are distributed over dedicated fiber optic networks. Special programmable electronic modules process the timing information to provide precisely timed outputs to control the many accelerator systems and devices.

BEAM INSTRUMENTATION

Monitoring and controlling the beam of protons through acceleration and storage requires measurement of a variety of beam parameters. Additionally, peak accelerator performance is achieved only with a thorough understanding of the "as built" machine. The most sensitive probe of the behavior of the accelerator magnet and RF systems is the beam itself, and the instrumentation systems provide the tools for measuring beam response. Parameters to be measured and the respective systems to be designed include: the quantity of protons - beam intensity monitors; transverse location of the beam within the evacuated beam tube - beam position monitors (BPMs); transverse beam size and distribution - profile monitors; loss of beam particles due to scraping, misalignment, or instabilities - beam loss monitors (BLMs); and temporal distribution of particles in the beam - wideband current monitors.

Acquiring measurements requires interaction with the beam particles themselves or their associated electromagnetic fields. As a result, most beam instrumentation systems include sensors which must be an integral part of the beamline vacuum system. Monitors that interact directly with the particles nearly all need mechanical actuators to allow remote insertion and withdrawal from the beam path. Electromagnetically coupled monitors must have coupling loops or electrodes within the beam vacuum or a vacuum tight insulating "window" to the fields associated with the beam. Nearly all instrumentation devices require

---

* Operated by the Universities Research Association, Inc., for the U. S. Department of Energy under Contract No. DE-AC35-89ER40486.

vacuum sealing electrical signal feedthroughs. In some cases multi-pin connectors are appropriate and in many other cases coaxial feedthroughs with controlled high frequency performance are demanded. Beam loss monitors, sensitive to radiation penetrating the beam tube walls, are the exceptional devices which generally do not interface to the vacuum system. In all cases, the sensors must provide a long useful lifetime in the radiation environment of the beamlines. Front-end low level processing of the instrumentation signals is typically done in the tunnel niches in the HEB and Collider rings and in surface level equipment buildings for the conventional injector machines.

Intensity monitors and wideband current monitors are an assortment of electrical current transformers designed to measure microamperes to hundreds of milliamperes with bandwidths ranging from DC to several gigahertz. A beam position monitor consists of one or two pair of electrodes in the vacuum, electronics to determine the relative amplitude difference between signals from opposing electrodes over a large dynamic range, and necessary interconnecting cabling as much as several hundred meters long. Transverse beam size and profile measurements, which will be among the most difficult and challenging, utilize a variety of techniques ranging from thin wires or grids to synchrotron light, ionization, and transverse probe beams of electrons or neutral ions. Accurate determination of beam sizes as small as 100 microns must be made in the high energy machines. Beam loss monitors are typically ionization chambers or solid state radiation detectors located outside but near the beam tube.

Many parameters, including beam current, profile, and temporal distribution, are adequately determined by measurements made at only one or a small number of points around each ring. Since each machine has different beam aperture requirements, beamline space constraints, and/or measurement range and resolution requirements, devices to determine these parameters are most often one of a kind mechanical designs. These instruments frequently are complex, precision assemblies with positioners or actuators designed for remote control. Commercial concerns with the capability to economically design and fabricate such custom assemblies would find the most interest in these types of beam instrumentation components.

Beam position and beam loss measurements are required to be made at many locations, more or less regularly spaced around each ring. These are the instrumentation systems that would be attractive to businesses interested in quantity production items. The sensor mechanics and the electronics for each of these types of systems number in the three thousand range.

Beam Position Monitor Systems

The position monitor sensors (usually just called BPMs) are approximately 20 cm long and 10 cm diameter in all rings. They consist of one or two pair of electrodes in the vacuum around the beam line, acting as antennae to the beam. The BPMs for each accelerator have unique aperture requirements and unique external mechanical interfaces and constraints. In the LEB and MEB, they are room temperature devices. In the HEB and Collider they are located in the cryogenic spool pieces and must operate at 4° K. The total number of BPMs is approximately 2650; of these, 2250 are 'cold' and 400 are 'warm'. Mechanical reliability of the 'cold' BPMs is critical, since failure requiring repair or replacement means days of downtime for warm-up, repair, and cool-down.

The electrical center, the axis on which the beam must travel to impart equal signal to opposing electrodes, of BPMs installed in each ring must ultimately be known relative to the ideal nominal beam line to an rms. accuracy of better than 1.0 mm in the LEB and 0.2 mm in the Collider. Positional stability over lifetime and temperature is important. This must be accomplished by an economical combination of manufacturing tolerance, electrical measurement, mounting, and installation survey.

Vacuum tight coaxial electrical feedthroughs pass the signals through the BPM body. In the superconducting machines, these feedthroughs must survive numerous temperature cycles and perform reliably for 25 years. Special coaxial cables exhibiting good high frequency transmission characteristics, low heat conductance, and high radiation resistance are required to carry the signals from the cold BPMs through the insulating vacuum to the cryostat wall. Another feedthrough is used there to bring the signal finally out to atmosphere. Over 17,000 high quality coaxial feedthroughs and 8,000 of the special cables are required.

Coaxial cables of RG8 or flexible solid copper shielded type shall carry the signals from the BPM in the radiation environment to the signal processing electronics in the niches of the superconducting machines or to the surface buildings in the warm machines. Over 600 kilometers of such cable is required.

Signals from the BPMs will be a train of sharp bipolar pulses approximately one nanosecond long with a 16 nanosecond repetition period. Each pulse train may range from a burst of as few as 10-20 pulses to a continuous stream depending on the distribution of particles in the machine. The pulse amplitude may vary from less than 0.1 volt to greater than 20 volts in proportion to the instantaneous beam current. Over this range of operating conditions, the processing electronics must extract beam position information by determining the normalized amplitude difference of signals from opposing electrodes. The signal processing bandwidth must be several megahertz. We anticipate this electronics to be constructed in VXI or VME format with analog signal processing, video digitizing, and memory on a single board. About 2,650 channels of such electronics required. Reliability of the electronics is of concern since much of it will be housed in the tunnel niches which are not normally accessible. However, in most instances, there is considerable redundancy provided by the large number of channels.

The mechanical BPMs for the Linac and LEB have been designed in-house and are in prototype fabrication. Within the coming several months we expect to solicit bids for production. MEB BPM design is scheduled to begin soon. To date, a few prototype Collider BPMs (HEB units will be quite similar or perhaps identical) have been fabricated and a current design is complete. That design has been provided to the spool piece contractors, Martin Marietta and Westinghouse, to produce 'build to print' units to include in their first prototype spools. A performance spec is in the final stages of preparation to contract for engineering services to improve manufacturability and reduce cost of the current design. A decision has yet to be made as to whether production BPMs will be supplied to the spool vendors as government furnished equipment or be contracted directly as a part of the spool production. Development contracts for 'cold' feedthrough, cryostat wall feedthrough, and 'warm' feedthrough designs are in place. A performance specification for the special coaxial cables used inside the cryostats has been written and will soon be released for solicitation of proposals. The BPM mechanical budget is approximately $6 million dollars.

The development, design, and qualification work on the Collider BPM, feedthroughs and cables must be completed this year in order to stay in tune with the 1994 spool piece production schedule. Preliminary development and design of the BPM electronics is underway in-house at this time. In parallel, a specification is being written for the electronics to obtain assistance in areas of manufacturability and reliability. The BPM electronics budget is also about $6 million dollars.

Beam Loss Monitor Systems

The beam loss monitor system to be provided by the Beam Instrumentation Department shall be used for accelerator commissioning, operational, and diagnostics purposes. It does not form any part of the personnel radiation safety system, though it does play an important role in accelerator technical equipment protection. The system will be capable of sensing and locating small beam losses for the purposes of monitoring loss patterns, identifying potential aperture restrictions or obstacles in the beam tube, tuning the machine for the cleanest possible operation, and triggering a controlled abort of the beam when the potential for losses that may lead to equipment damage or high irradiation of accelerator components is sensed. By providing a monitor to reduce operational beam losses within the tunnel to a low level, residual radiation induced into accelerator components that require access for maintenance can be minimized.

The beam loss monitors (BLMs) are expected to be gas filled ionization or proportional chambers or perhaps solid state radiation detectors. They are distributed more or less uniformly around each accelerator, mounted to the tunnel walls, magnet stands or similar location. The BLMs must exhibit properties including radiation resistance, high and stable sensitivity to radiation signals, low noise or dark current, wide dynamic range, low maintenance and high reliability. Approximately 3500 BLMs and associated electronics channels are required.

Development and design work has begun for both the sensors and the electronics. It is likely that a small number of BLM development contracts will be placed within the next two years. Assessment of the suitability of gas filled chamber type detectors as used at Fermilab and Brookhaven National Laboratory is now underway. Investigation is also ongoing into the possible use of mercuric iodide and other materials for a solid state detector.

BLM electronics must be suitable to process small signal currents expected from the sensors over a dynamic range of 10,000:1. For purposes of accelerator equipment protection from large beam losses, a response time on the order of ten microseconds is required . Each channel must have sufficient memory to provide a postmortem record for analysis purposes of beam aborts. As with the BPMs, we anticipate the electronics to be constructed in VXI or VME format with analog processing, digitization, and memory on a single board. Reliability of the electronics and sensors is of special concern since failure to trigger necessary beam aborts may result in severe magnet damage and each abort caused by false alarms will result in hours of unacceptable machine downtime.

The schedule for these devices is somewhat relaxed as compared to the BPM mechanics since the BLMs are expected to be relatively fragile components and among the last items to be installed in the tunnels. Significant quantities are not expected to be needed before mid 1995. The budget for BLMs including sensors and electronics is over $2 million dollars.

PRECISION TIMING SYSTEM

Operation of the SSC accelerator complex requires synchronization between many different systems and devices with widely varying time precision over a large geographical region. The magnetic cycles of the different accelerators must play out relative to each other with millisecond fidelity; transfer of beam from one synchrotron to the next requires 100 picosecond accuracy between the RF systems of the respective rings as well as nanosecond accuracy of fast pulsed kicker magnets. The Precision Timing System will be composed of a fiber optic distribution system for the required time base signals, a means of encoding resets or markers onto the time bases, and an electronic receiver or clock module to keep time per the time base, to respond to the resets, and to generate outputs as programmed via the Global Accelerator Control System.

The main timing references for the accelerators are the utility company power line, which is highest frequency external interface to the accelerators, and the RF or revolution frequencies of the beam particles as they orbit around each synchrotron. The nominal 60 Hz power line provides a suitable time base for the synchronization of the power supplies and magnet cycles. The revolution frequencies, and multiples thereof, provide necessary time bases for beam control and manipulation operations. The revolution frequency of particles in any one synchrotron is a function of the machine circumference and the time varying parameters of beam energy and magnetic field strength. At any given instant, there is no direct correlation between the frequencies of any two synchrotrons; yet within relatively narrow windows of time, transfer of beam from one machine to the next must be made under stringent conditions of frequency and phase match. A mechanical analogy might be a multigear transmission in which synchronization must be achieved at the time of shifting, not simply to mesh the gears but also to align a specific tooth on one gear with another specific tooth on the next. In addition, the geographical extent of the accelerator systems requires careful attention to and control of timing signal time-of-flight delays.

A global time base with information related to the power line frequency and machine cycle resets will be distributed throughout the entire accelerator complex. In addition, each synchrotron will have distributed around it the respective RF time base with the revolution frequency information contained. This RF time base is also required at the RF control point of the next lower energy machine to facilitate the beam transfer synchronization. The RF time bases are nominally 60 MHz with some variation from one synchrotron to another; the LEB RF frequency for instance sweeps from 47 to 60 MHz during that machine's 50 millisecond acceleration cycle. All distribution systems must provide a high degree of point-to-point signal transit time determinacy, requiring a dedicated purpose system. Active feedback control of signal transit delays is expected to be required along some segments of the distribution networks to compensate for temperature and aging effects. The respective time bases must be made available at each niche or surface equipment building for each accelerator. Studies are underway to determine suitable reset or marker coding schemes to understand the impact on timing system jitter and noise.

A general purpose clock module, dubbed the Universal Timing Module (UTM), is being designed to serve as a receiver and decoder of the time base signals. It needs to consist of 60 MHz counters, control logic, and an interface to the control system. The control system provides a 'schedule' which programs the outputs of the UTM. The outputs shall be TTL and/or ECL pulses or gates for direct timing control of hardware devices and data acquisition systems. Somewhere in the range of 2,000 UTMs are expected to be required. Prototype modules, likely to be less than full-featured, are now well into the design stage for Linac applications where they are needed by the end of this year. Larger quantities demanding near full performance are required for the LEB in early to mid 1995. VXI format is planned for the initial boards.

The budgeted value of the entire precision timing system including fiber optic systems and electronics is approximately $7 million dollars.

SUMMARY

Considerable work on many components for the beam instrumentation and precision timing systems is underway with industrial participation on levels ranging from systems engineering, to detailed design, to prototype and soon production fabrication. Much effort yet remains to establish final requirements definitions for many aspects of the systems before detailed design and production contracts are ready to be placed. Specifications for beam instrumentation equipment often converge only late in the accelerator design stage. This is because the instrumentation is frequently subjected to spatial limitations imposed by less flexible beamline components such as magnets and RF systems, and also because many instrumentation performance requirements are intimately dependent on planned and desired accelerator operational flexibility which the accelerator designers and physicists are understandably reluctant limit early in the design.

To abide by the current schedule calling for Collider operation before the year 2000, initial contracts for BPM fabrication and electronics design will need to be in place by the end 1993. The total budgeted cost for design, procurement, fabrication and test of all beam instrumentation and precision timing systems for the entire injector and Collider complex is approximately $50 million dollars.

AN OVERVIEW OF THE COLLIDER DIPOLE MAGNET PROGRAM

R.W. Baldi, L.R. Patterson, and G. Salvador

General Dynamics Space Systems Division
Space Magnetics
P.O. Box 85990
San Diego, CA   92186

ABSTRACT

"World Class" design, engineering, and planning have resulted in successful execution of each phase of the Collider Dipole Magnet (CDM) Program. To date, every key milestone and objective have been achieved. In partnership with the SSCL, the General Dynamics team, stationed at the Fermi National Accelerator Laboratory, built seven industry demonstration magnets. These magnets were built on schedule and successfully performed all objectives in the Accelerator System String Test (ASST) which was conducted in July of 1992. Prototype magnet production was started on schedule in October of 1992 and is well underway at General Dynamics' Hammond, Louisiana facility. This paper describes the key elements of this successful program and details the activities completed to date.

INTRODUCTION

The Collider Dipole Magnet (CDM) program has been carefully structured around a phased developmental approach in order to achieve the overall program goal of developing an industrial capability for the rate production of dipole magnets. In October of 1990, General Dynamics was selected by the Super Collider Laboratory (SSCL) as the Leader contractor for the Collider Dipole Magnet (CDM) developmental program. Westinghouse Electric Corporation was selected as the Follower contractor. Under the terms of the Annex I Leader contract which began in May of 1991, GDSS is responsible for producing a final production design for the CDM which is capable of meeting all Collider performance and operational specifications, proving out the rate production capability of magnet manufacturing tooling, and manufacturing magnets (4 model, 12 prototype, 35 preproduction, and 251 Low Rate Initial Production). It is planned that the Annex I program will lead to a follow-on rate production effort under which nearly 8,000 magnets will be produced. Production rates approaching 10 units per day are required in order to meet current program delivery schedules.   As the follower, Westinghouse will produce 35 preproduction and 251 Low Rate Initial Production magnets. The overall Collider Dipole Magnet program schedule is shown in Figure 1.

PRODUCT DEVELOPMENT AND INDUSTRIAL DEMONSTRATION

The product development phase of the CDM Program involved the completion of several major tasks. At the commencement of the Annex I program, activities were undertaken by the SSCL in order to disseminate existing magnet technical design and development information developed by the DOE National Laboratories. A technology transfer program was conducted by the SSC Laboratory through which General Dynamics and Westinghouse Electric Corporation technical personnel participated in a detailed technical review of the existing Fermi National Accelerator Laboratory (FNAL) and Brookhaven National Laboratory (BNL) designs for the Collider Dipole magnet. In order to support the Congressionally mandated requirement to demonstrate that industry could successfully manufacture dipole magnets, an industrial demonstration program was conducted by the SSCL at FNAL. In April of 1991 General Dynamics relocated a team of personnel to Fermilab for the purpose of manufacturing seven (7) full-length CDMs. All magnets built by GD were successfully cold tested by March of 1992. Subsequent testing of these CDMs in the Accelerator System String Test (ASST) proved fully successful. Westinghouse similarly manufactured five (5) industrial demonstration CDMs at BNL. Figure 2. shows a completed CDM as built by General Dynamics.

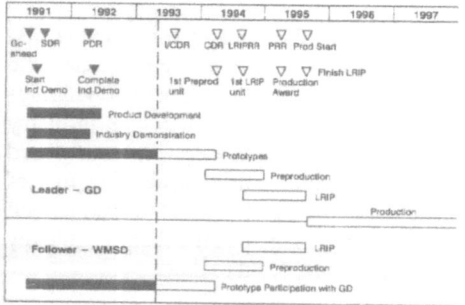

Figure 1. The CDM Program Schedule

Figure 2. Industrial Demonstration CDM built by General Dynamics

Concurrent with the technology transfer and industrial demonstration programs, General Dynamics began the production design of the CDM in San Diego. GD's design approach was to use the Fermi CDM design as the baseline and modify it where necessary to incorporate producibility enhancements thus making it suitable for high rate production. Input from the technology transfer and lessons learned from the industrial demonstration program were fed into the design process. Producibility enhancements developed by GD through independent research and development (IRAD) were also assessed for their inclusion into the final design. During the design process, General Dynamics utilized an active Integrated Product Development Team (IPDT) approach. In conjunction with the CDM magnet final design, GD initiated final design of the production tooling with its selected suppliers. The major tools required for magnet build (coil winder, curing press, collaring and weld press) were the focus of initial design and development activities. Minor tools were introduced into the design process as major tool designs matured. In August of 1991 General Dynamics successfully completed the Systems Design Review (SDR) with the SSCL. The purpose of the SDR was to review and finalize the CDM Prime Item Development Specification (PIDS) and to obtain SSCL approval to proceed with final design activities. SDR also approved the start of initial long-lead procurement (materials and tooling) activities to support prototype magnet build. Magnet and tooling final design continued throughout the remainder of 1991. In March of 1992 General Dynamics successfully completed the Preliminary Design Review (PDR) for the CDM with SSCL. The purpose of the PDR was to review the final design of the CDM as completed by GD and the design of all major and minor tooling. In addition, other topics reviewed during the PDR were GD's plan for activation of the Hammond Manufacturing Facility, the magnet master test plan, magnet systems engineering program, and magnet quality assurance program. The approval by SSCL of a successful PDR provided GD with go-ahead to proceed with procurement of long lead materials for the preproduction magnets. In August of 1992 General Dynamics completed release of 100 percent of the drawings for the CDM design.

HAMMOND FACILITY ACTIVATION

GD began implementation of its Hammond, Louisiana manufacturing facility modification and activation program after SDR. After development of specifications and plans for the facility activation, GD awarded an architectural and engineering contract in November of 1991 which was followed by the award of a construction management contract in March of 1992. Construction began immediately following this award. Major tasks completed during the construction upgrade included the installation of laboratory quality lighting, installation of a new temperature and humidity control system, construction of tooling foundations and separation walls, preparation and surface treatment to interior walls and floors, upgrading of utilities services, and construction of a new adjoining structure designed to house coil curing tooling. Another significant part of the Hammond Facility activation program was the design and manufacture of the magnet cryogenic test facility. Air Products and Chemicals Inc., under subcontract to General Dynamics, completed the design for the liquid helium refrigeration plant and cold test stand system. Construction of these systems began in October of 1992. Currently, the liquid helium refrigeration plant and cryogenic test stands have been completed and preliminary acceptance testing is underway. As the focus of the CDM program began to shift from product development to readiness for the start of prototype magnet production, key members of General Dynamics team (located in San Diego) began to transition to Hammond. Transition to Hammond of program personnel was completed in September of 1992.

MODEL AND PROTOTYPE MAGNET MANUFACTURING

The purpose of building both model (1.8 meter long cold masses) and full-length (15.2 meter) prototype magnets is to prove out the magnetic design of the CDM and assess its performance against the Collider performance specification. In order to support model magnet manufacturing, General Dynamics relocated personnel to the SSCL's Magnet Development Laboratory (MDL) in April of 1992. Model magnet assembly was performed by GD at the MDL and at present, the first, second, and third model magnets have been completed. The fourth model magnet will be used to demonstrate the use of low cost injection moldable coil end part material. All model magnets will be subjected to a detailed test program which includes verifying quench performance and magnetic field quality.

In support of prototype magnet production, General Dynamics began the installation and proofing of major tooling at the Hammond Facility in August of 1992. Figure 3. shows major tool installation in the Hammond Facility. Production start of the first prototype magnet began in October of 1992. Prototype magnets will be manufactured in three lots (A, B, and C) and all units will undergo cold testing. Each lot is being built to achieve specific goals. Lot A (4 units) will be used to assess the magnetic performance of the GD design.

Lot B (3 units) will be used to further proof out the coil manufacturing process. Lot C (5 units) will be magnets built with the next iteration of the design (in terms of magnetic cross section and end design enhancements based on test data). One unit from lot C will be subjected to accelerated life testing. The use of three lots in the prototype phase permits design and process iterations to mature the production design. Currently, the first three prototype CDMs are under construction.

Figure 3. Activation of the Hammond Facility includes the installation of major and minor magnet production tooling.

CONCLUSION

The SSCL's Collider Dipole Magnet development program is progressing on-schedule. General Dynamics has successfully completed the final production design of the CDM and released 100 percent of the final production drawings. Seven (7) industrial demonstration magnets have been built and tested at Fermilab and subsequently performed as designed in the Accelerator System String Test. The activation of General Dynamics' Hammond Facility is well underway with a current staff of over 175 personnel in place. Air Products and Chemicals, Inc. have completed construction of the magnet cryogenic test facility and are presently testing the system. GD has completed 3 of 4 model magnets, of which the first will begin cold testing in May of 1993. Full length prototype magnet production is underway with the first prototype scheduled to begin cold test in August of 1993. In terms of subcontracting, GD is meeting and/or exceeding all established targets for contracting with small disadvantaged business suppliers. Overall, General Dynamics has achieved all contract milestones (5) and Contract Data Requirements List (CDRL) items (over 250) on schedule to the SSCL. We fully expect to continue to meet all contractual milestones.

ACKNOWLEDGMENT

The work described herein was accomplished under contract to the Universities Research Association in support of the Superconducting Super Collider project for the U.S. Department of Energy.

DESIGN OF SSC COLLIDER STRUCTURES

James E. Monsees

The PB/MK Team
5510 S. Westmoreland Road
Dallas, TX 75237

INTRODUCTION

We would like to set the record straight! To date, underground construction contracts on the SSC main ring have been bid at a savings of $77 million dollars or 33 percent <u>below</u> the baseline cost estimate. Sorry, Sam Donaldson, you should check your facts more closely.

The SSC is the largest single underground project ever built anywhere in the world. When completed we will have approximately 70 miles of tunnels, 60 shafts, two huge underground experiment halls -- each the size of a football stadium -- and numerous other structures, each of which would be considered a major facility on any other project.

COMPUTER MODELING

Because of the generally massive and uniform geologic deposits that are the home of SSC structures in Ellis County, we have used state-of-the-art computer modeling extensively as a design tool to model the insitu conditions, the excavation stages, and the final composite structure of man-made elements and geology. This work involved the largest three-dimensional finite element model ever used for an underground project. You may remember that the geology for the main collider tunnels breaks down approximately as shown in the following Table 1:

Table 1. Summary of Geology.

Geologic Material	Unconfined Compressive Strength	Percent of Tunnel	Overstress at Depth
Austin Chalk	2,000-3,000 psi	56	No
Taylor Marl	400-600 psi	32	Yes
Eagle Ford Shale	200-400 psi	12	Yes

Supercollider 5, Edited by P. Hale
Plenum Press, New York, 1994

FIELD OBSERVATIONS

We have designed the underground structures using the observational approach -- constantly feeding measurements and observations from the field back into the design procedure. By doing this we have been able to modify the designs of later structures to effect major savings in their construction costs. In effect, we have used early structures as full-scale field tests to improve our understanding of the geologic materials and how they behave under construction and long term loadings.

Exploratory Shaft

We first used the observational approach on the exploratory shaft (ES). This shaft is 250 ft. (82m) deep, 16 ft. (5m) and 10 ft. (3m) in diameter, and has a test adit (stub tunnel) at the contact between the Austin Chalk (AC) and Eagle Ford Shale (EFS). We did not expect the Austin Chalk to be overstressed at depth but the EFS was expected to be overstressed. Thus, we had to answer three basic questions:
- Will the EFS stabilize if we apply a reasonable support load (and develop an elastic-plastic zone around openings)?
- Will the EFS undergo a long-term creep behavior that will in time require the support system to resist something approaching the in situ stress condition?
- Will the EFS experience an elastic-clastic behavior where the stresses cause the EFS to break up around the opening as it did in San Antonio?

Results from the 12 instrument systems in the exploratory shaft and adit and observations of rock behavior in the field have enabled us to supply the following answers to the questions asked above.
- When continuous reasonable support, such as shotcrete, is applied to the EFS within about two hours, the material generally follows elastic-plastic behavior.
- The support required for stability is then far below (10 to 20 percent of) the in situ stress condition; creep does not appear to be a concern.
- If support is not installed continuously or in a timely manner, the material begins to follow an elastic-clastic behavior (it begins to break up). Uncontrolled, we believe this behavior would lead to instability.

Magnet Delivery Shaft

Observations in the 30 ft by 60 ft magnet delivery shaft (MDS) have further confirmed the results gained from the exploratory shaft. Results can be summarized as follows:
- Radial movements in the chalk are purely elastic and generally less than 0.1 in. for this large structure, even at a depth of 200 ft.
- Radial movements in shale are approximately 0.75 in. or 7.5 times those in the chalk. Noting that the moduli of the two materials differ by a factor of six, these movements would be expected to be approximately 0.6 in. The remaining difference, we believe, is readily explained by the fact that the model does not incorporate strain softening.

Based on these results, we have decided that the portion of the magnet shaft in chalk will not receive a final lining of reinforced concrete, **resulting in a savings of approximately $300,000**. We also decided to continue to observe the behavior of that portion of the magnet shaft in shale. If that behavior continues as it has been to date we will likely leave out additional concrete.

Starter Tunnels

From the instrumentation in the starter tunnels, we have observed (1) that radial deformations of approximately 0.75 in. are being measured, (2) these deformations are close to those predicted for the shale, and (3) that the deformations stabilize and remain so as soon as full closure of the support is completed. This latter observation dramatically proves the old axiom, "To stabilize the opening, one must complete the circle."

COST SAVINGS

Drawing upon the above analyses, and especially the field observations and measurements, we have come to the following conclusions:
- For circular (or near circular) structures, the modeling and field observations are both showing small deformations
- For circular (or near circular) structures, stability has, to date, been obtained by the initial supports of pattern rock dowels and shotcrete in shale, and occasional rock dowels in chalk. Thus, we have deleted final cast-in-place concrete linings in all starter tunnels, adits, and circular shafts for **an estimated total savings of over $2 million**.

DESIGN COMPLETION

At the time of this writing, we have awarded contracts for 37.8 miles (61 km) or 70 percent of the collider main ring. By the end of the calendar year, we will have 48.6 miles (78 km) or 90 percent of the ring under contract. Contrary to some reports in the press, **these contracts have been well below the Baseline Cost Estimate**. Table 2 summarizes those contracts and shows their actual costs. Note that the collider tunnels to date are $77 million or 33 percent below the Baseline Cost Estimates.

Table 2. Summary - Tunnel Contracts.

Contract	Rock Type	Bid Date	Length (Miles)	Tunnel Type Support	Bid Amount	Baseline Cost Estimate	Cost Difference
N15-N20 (CCU A-610)	EFS	12/91	2.7	TBM - Precast Segments	$25	$22	+3
N20-N25 (CCU A-611)	EFS/ AC	03/92	2.7	TBM - Precast Segments	$16	$19	-3
N25-N40 (CCU A-650)	AC	05/92	8.1	TBM - Dowels	$28	$47	-19
N40-N55 (CCU A-670)	AC	07/92	8.1	TBM - Dowels	$31	$47	-16
S40-S55 (CCU A-740)	AC	10/92	8.1	TBM - Dowels	$28	$54	-26
S25-S40 (CCU A-720)	Marl	04/93	8.1	TBM - Precast Segments	$35	$51	-16
T O T A L							$77 Million

As currently scheduled, the remainder of the main ring contracts and the contract for the High Energy Booster will be bid by mid-1995, at which time we will be able to give a report on the bid costs of these elements as well.

CONCLUSION

Underground conventional construction for the main collider is meeting established schedules and running $77 million or 33 percent below the Baseline Cost Estimate. Thus, assuming that budget constraints do not push construction too far into future years, it is apparent that conventional construction of underground collider facilities will come in well below the budget.

CONVENTIONAL CONSTRUCTION SURFACE FACILITIES DESIGN PROGRESS

Robert H. Janowski, Jr.

Design Division Manager
The PB/MK Team
5510 S. Westmoreland Road
Dallas, Texas 75237

Jack H. Clifton

Requirements & Planning Group Leader
Conventional Construction Division
Superconducting Super Collider Laboratory*
2550 Beckleymeade Avenue
Dallas, Texas 75237-3997

COST AND SCHEDULE SUMMARY

The engineering design of the surface facilities for the SSC is a large and diverse task totaling 27 primary projects budgeted at $299,000,000. Design is either completed or in progress on 25 of the 27 projects, with design completion at 83 percent. Construction contract values on these projects range from $200,000 to $46, 700,000. These 25 projects are on schedule, supporting the technical operation of the laboratory, and are overall within five percent of budget. The big success story of the past two years on the surface has been the fast-track design/construction of the accelerator systems string test facility, with a phased occupancy in October 1991. By September 1992, the Superconducting Super Collider Laboratory (SSCL) had successfully proven the operation of the cold magnets string, and had done it several weeks ahead of the first critical technical schedule milestone for the project.

HOW BIG ARE THE SURFACE FACILITIES?

Surface facility requirements for the SSC are massive as illustrated by the examples in Table 1.

---

* Operated by the Universities Research Association, Inc., for the U. S. Department of Energy under Contract No. DE-AC35-89ER40486.

Table 1. Illustration of Scale of Surface Facilities.

- Total construction site: 16,000 acres, the size of Kalamazoo, Michigan
- Total power requirement: 225 megawatts of electricity, enough to serve the entire state of Vermont or the city of Dallas
- 15 cooling ponds: combined total surface area of 268 acres, total heat rejection capacity of 146 megawatts
- Population employed at the facility : over 2000
- Total earthwork volume: over 3,000,000 cubic yards
- Cut-and-cover tunnels in the injector area: 5900 linear feet total
- Bored tunnels: 20,600 linear feet total
- Small diameter or micro-tunnels: 2,980 feet total[1]
- 19.5 miles of roads: included as part of the SSC construction
- 30 miles of state roads and 13 miles of county roads: being upgraded by the Texas National Research Laboratory Corporation
- 102 buildings, totaling 1,500,000 square feet
- Other surface facility systems include: parking lots, telecommunications, ductwork, natural gas, potable water, raw water, sanitary sewers, sewage treatment, cooling tower, and HVAC systems
- For the total SSC, both surface and underground, over 100 separate design/construction packages on the SSC conventional construction: total value in excess of $1 billion, each contains some surface and infrastructure design work

---

[1]Tunnels near the surface at the LINAC, LEB, MEB, and Test Beams are included with the surface facilities.

DESIGN STATUS SUMMARY

A summary of the status of design engineering for the conventional surface facilities is shown in Table 2 and discussed here. Exclusive of two major projects scheduled for the future years (the West Campus and Permanent Power) the surface facility design is 83 percent complete. In total, with the future campus and power included, the surface facility design is 45 percent complete.

N15 Site

Clustered at N15 are the magnet delivery laboratory (MDL), magnet test laboratory (MTL), Accelerator System String Test (ASST), collider access shafts, and cooling pond. The N15 site design is 100 percent complete, and construction is 100 percent complete with two exceptions: the roads, which will be done this July, and the N15 warehouse, which is scheduled for occupancy in 1994.

Injector Area

The Linear Accelerator (LINAC) and the Low Energy Booster (LEB) are both designed and under construction. The LINAC will be ready for occupancy this July, and the LEB is on schedule for occupancy in April 1994.

The Medium Energy Booster (MEB) has been designed, bid at $46,700,000 ($2,000,000 below the baseline cost estimate), and the award of the construction contract is currently awaiting the approval of the Secretary of Energy.

The Test Beam area is currently 90 percent complete in design and is scheduled to go out for bid in June 1993, with a construction start in August.

Table 2. Status of Conventional Construction.

Facility	Value $Millions	Design % Comp.	Construction Start	SSCL Occupancy
Design and Construction Complete				
MDL	10.0	100	10/90	10/91
ASST	7.0	100	1/91	10/91
N15 Cooling Tower	0.2	100	11/91	9/92
N15 Sewage Treatmt. Plant	1.4	100	3/91	10/92
MTL	5.8	100	1/92	10/92
N15 Roads	5.5	100	10/91	7/93
Design Complete / Construction in Progress				
LINAC	5.0	100	5/92	7/93
LEB	8.6	100	2/92	4/94
Design Complete / Construction in Future				
N15 Warehouse	6.0	100	-	1/94
MEB	46.7	100	5/93	11/94
Injector Cool Pond	3.5	100	5/93	11/94
Design In Progress				
East Complex Utilities	5.2	5	10/93	10/94
East Complex Roads	4.7	60	12/92	11/94
Test Beam Area	17.2	90	8/93	1/95
West Side Power Dist.	26.6	50	2/94	2/95
West I.R. Pond	3.8	50	2/94	2/95
IR-5, IR-8 Cool Ponds	4.8	30	7/94	10/95
Permanent Power	95.0	0	9/94	1/97
West Campus	42.6	0	5/95	1/97
TOTAL	299.6			
Design Completed	134.9	45%		

East Complex

Surface facilities including the utilities, roads, and IR-5 and IR-8 cooling ponds are all currently in design, with completion ranging from 5 percent to 60 percent.

West Campus

The West Interaction Region Pond is 50 percent complete in design and scheduled for bid later this year. The West Campus, which is currently planned to be sited around the pond, includes 500,000 square feet of facilities, including a 1000-seat auditorium, meeting facilities, library, cafeteria, computer center, laboratory space, and the SSCL Operations Control Building which will monitor and control the injection accelerators, test beams, the collider and the experimental detector systems. Campus design is scheduled to start in May of 1994.

FUTURE YEARS

The future surface design activity will center primarily on the permanent electrical power for the SSCL. Current plans are for the Department of Energy Project Office to procure the system using a request-for-proposal (RFP) for design/build. The $95,000,000 project is scheduled for RFP release in June 1993 with design starting in January 1994 and facility completion in January 1997. The system includes two 345kV/69kV substations, a 69 kV distribution system in the collider tunnel, ten 69kV/12.47kV substations around the ring, and a 12.47 kV distribution system.

LABORATORY TECHNICAL SERVICES PROVIDES BUSINESS OPPORTUNITIES FOR SUPERVISORY CONTROL AND DATA ACQUISITION SYSTEMS

Wayne Ballard

Operations Group Supervisor
Superconducting Super Collider Laboratory*
2550 Beckleymeade Ave.
Dallas, Texas 75237-3997

INTRODUCTION

At this time, I would like to present some additional information about what I consider are some really great opportunities for the business community to participate with us in developing the greatest scientific project in the history of mankind.

Facility Engineering Services is part of Laboratory Technical Services. As part of this group, we have the responsibility to direct the construction of interim facilities, scientific labs, production process, cooling towers, cooling ponds and the operation and control of SSC Laboratory conventional support systems. These operations and controls will be accomplished through the employment of a Supervisory Control and Data Acquisition system (SCADA).

DISCUSSION

From a master control room located at Central Facility, the operating status of all environmental control equipment within the SSC Laboratory will be monitored and building environments controlled. Deviations from normal values will be quickly detected, isolated and corrective action taken. Vital power distribution system characteristics will be observed, analyzed and deviations from normal accounted for and corrections made. When new campuses and/or facilities are added to the SSC Laboratory, their related system monitors and controls will be integrated into and made a part of the Supervisory Control and Data Acquisitions System.

Transmitting electronic input signals from multiple vendors, multiple control systems and from multiple physical locations across a sophisticated and highly active computer network into one automated and integrated system monitored and controlled from one central location is a technically complex task.

---

* Operated by the Universities Research Association, Inc., for the U. S. Department of Energy under Contract No. DE-AC35-89ER40486.

The complexity of this project makes it necessary for the SSCL to select experienced vendors that can demonstrate both an understanding of the functions of energy management and the capability to apply current information systems technology to the design of centralized and integrated non-proprietary controls for geographically dispersed multiple vendor environmental systems.

AREAS OF OPPORTUNITIES

- Supervisory Control and Data Acquisition Systems
- Integration of multi-vendor proprietary platforms
- Chemically controlled water treatment and analysis systems with read/write capability
- Ion controlled water treatment and analysis systems with read/write capability
- State of the art, flow transmitters, temperature transmitters and pressure transmitters
- Remotely controlled by computer lighting systems utilizing solid state switching
- Water quality and quantity monitoring systems
- Air quality monitoring systems
- Thermal monitoring systems
- Vibration analysis

These are but a few of the possibilities in which the Business Community can participate. We in Facility Engineering Services are more than happy to work with each of you and to assist you in both the immediate and long term needs of the Laboratory Technical Services as we work to support the Superconducting Super Collider.

INDUSTRY PARTICIPATION IN THE SSCL INSTALLATION PROGRAM

Frank Spinos

Superconducting Super Collider Laboratory*
2550 Beckleymeade Ave.
Dallas, TX 75237

INDUSTRIAL PARTICIPATION

The uncertainties of funding levels or cancellation makes involvement in controversial programs a hazardous business for any industrial organization and effectively eliminates small corporations from the competition.

Corporate Executive Board decisions to allocate bid and proposal money are usually made with the understanding that the corporate investment would be returned along with some profit in the out years. Their major concern is the real possibility that additional funds will be required to support the bid team beyond the original projected date of contract awards. Experience has shown that awards are usually one to two years later than the projected award dates.

American industry has complained that the National Laboratories have not made an attempt to transfer state of the art technology to the industrial sector. It has always been industry's position that both the laboratories and industry would be better served if the scientific breakthroughs were accomplished as a partnership. The SSC program is concentrating on using industry at the outset of each facet of the program to make the best use of the technical talent resident in both sectors.

EXAMPLES OF NATIONAL LABORATORY INDUSTRIALIZATION

Portrayed below are two large programs conducted at separate National Laboratories in the past decades. They are examples of two distinctly opposite approaches to bringing a major high energy physics device on line within the cost and schedule goals of the Department of Energy (DOE).

Fermi National Accelerator Laboratory (FNAL) is an example of a program where industry was brought on board to fill voids in the personnel and construction expertise. In general, industry was not involved in design recommendations or solutions.

FNAL has been in operation for over twenty years during which time the concentration has been on the building and operation of the Tevatron, one of the worlds major accelerators. The majority of the technical development, design and manufacture of the various accelerator components were accomplished by FNAL staff personnel. Generally the direction and approval of the manufacturing techniques, procurement policies, installation practices and commissioning were provided by the physicists responsible for the function of the machine.

---

* Operated by the Universities Research Association, Inc., for the U. S. Department of Energy under Contract No. DE-AC35-89ER40486.

The fusion experiment at the Princeton Plasma Physics Laboratory (PPPL), started in the early 1970's, is an example of a program where the transfer of the developing technology to American industry was a specific aim of the program.

Industry participation was evident from the beginning of the program when many industrial teams bid on the design and construction of the fusion reactor. The DOE chose to award and control the conventional construction of the laboratory office building and the cells for the machine and the utilities. PPPL retained the fabrication of the large poloidal coils and certain of the vacuum components since there were already staffed coil and vacuum shops on the Princeton campus. Physics requirements and design performance parameters were provided by the PPPL staff. Oversight of the program was maintained by the PPPL and DOE management on site.

The benefit to the program became obvious as various problems were encountered. The resources of the major subcontractors became available immediately as a problem surfaced. Expertise in properties of materials, structural test laboratories and geodesy were easily accessed as the need arose without the need for additional procurement actions.

SSCL INSTALLATION PROGRAM PHILOSOPHY

The Universities Research Association (URA) determined that a cooperation with industry similar to that of other recent large projects like the TFTR experience would be the key to a successful program and that only major corporations would have the depth of technical skills, financial resources and facilities to manage a program of the scope and complexity of the SSC installation program.

The three compelling reasons for selecting only major industrial organizations were as follows: First is the varied resource pool required of the selected organizations. Skills ranging from program management to warehouse control will be required. The availability of large numbers of experienced personnel in each skill category will be especially important. Secondly, there will not be enough time for the selected subcontractor to hire, train and develop a task team drawing from the open market from the time of the award to the beginning of critical path activity. Finally, the ability of the subcontractor to fund the costs of a major proposal and the initial costs of getting started.

The definition of a major subcontractor as the prime to the URA was never intended as a barrier to companies of lesser size. Specific instructions were provided to the bidders that defined the goals of the URA as they applied to small, disadvantaged businesses as well as minority and women owned businesses. Teaming with other small or medium sized businesses was also encouraged. The result, as will be evident in further discussion, was the receipt of proposals in which a number of industrial teams were formed.

Implementation of the Solicitation

The Commerce Business Daily notice was posted late in 1991 inviting industrial contractors to participate in the installation program. Approximately forty industrial concerns attended the bidder's conference.

Immediately following the publication of the announcement the SSCL received visits from the corporate heads of many major companies. It was surprising that companies whose existing contracts were counted in billions of dollars would be committed to winning the installation subcontract whose final value would be a fraction of what was their business norm. Discussions that revolved around that point revealed that the business climate was such that winning a billion dollar award was no assurance that the program would remain viable. In fact, they cited examples of programs where millions of bid and proposal dollars were spent winning a particular award which was canceled prior to the recovery of any of the expended dollars. Although industry felt that the jeopardy of fiscal year funding placed them in a poor position, it was necessary to actively pursue contracts of this type and size.

At the time of the briefing, the superconducting magnet procurements had been awarded to General Dynamics, Westinghouse and Babcock and Wilcox. The magnet technology transfer program had been started at FNAL and at Brookhaven National Laboratory where a number of superconducting dipole magnets were under construction by the magnet subcontractor personnel. The SSCL was confident that there was sufficient definition of the configuration of the Collider Storage Ring Arc sections to permit a full description of the installation requirements. The attendees were given a full briefing of the arc section requirements as well as defined in the Statement of Work that accompanied the solicitation.

The program was further expanded shortly after the briefing to include all of the installation activities for the Linear Accelerator (LINAC), Low Energy Booster Ring, Medium Energy Booster Ring, High Energy Booster Ring and the remaining areas of the Collider Storage Ring even though some of the ring configurations were not fully defined.

The bidders were instructed to provide proposals, based on the information available, to include the approach envisioned for the management of the program, the transportation and handling systems suggested and the existing planning programs that could be applied to the program.

The solicitation had indicated that the SSCL would award two to four study subcontract awards of $500,000 each during which the subcontractors would receive a thorough technology transfer. The terms of the solicitation required that each of the subcontractors place a full time representative at the SSCL who would interact on a daily basis with the SSCL Installation Manager as well as with each other. The goal was to develop certain key specifications and installation concepts that would be included in the second phase definition.

Subcontract Activity

Four proposals were received in answer to the solicitation. Three were accepted by the SSCL based on evaluation criteria scoring. The selected subcontractors were: Bechtel International Corporation, Brown and Root Corporation and Martin Marietta Corporation. Subtier contractors to the three prime subcontractors noted above included General Dynamics, Westinghouse, Lockheed, Science Applications International, Belding, the University of Texas at El Paso and FATA Automation.

A program kick off meeting was held to establish the ground rules that would govern the interaction between the URA Management and the representatives of the selected subcontractors. The rules were designed to ensure that all three subcontractors would be provided equal access to all available information and still guard the proprietary data of each.

The essence of the working arrangement was the daily interaction between the subcontractor representatives at the SSCL and the URA Installation Manager. The representatives of each of the subcontractors were invited to all meetings that contained information relative to the installation program. There were times when one or another were unable to attend. We did not attempt to hold a separate briefing to guard against the possibility that additional information might have been passed on to a single subcontractor.

The three subcontractors came to an agreement that individual meetings were to be permitted when discussions of a proprietary nature were desired. It was agreed to rely on the veracity of the URA Installation Manager to ensure that only proprietary information was discussed in the private meetings.

Technical meetings were arranged to permit the various subcontractor associates to participate. Many meetings had twenty or more subcontractor representatives in attendance; question and answer sessions were always held en mass. Documentation was provided to each subcontractor using transmittal sheets which were signed as evidence that the documents had been received. Records of the meeting attendees and the receipt of documents have been retained on file.

Working Relationships

The relationship that was established between the three subcontractor representatives and the lead personnel at the SSCL soon matured to a point where the group was able to differentiate between those items that needed to be established for the good of the program and those that could be retained by each subcontractor as discriminators in the competition.

The complex issue of determining the most cost effective tunnel transportation system was addressed in many common meetings. The group was able to establish the type of power to be used to drive the tunnel transport system, the methods of conveyances needed for transporting both components and personnel and the tunnel floor finish requirements. These decisions permitted the conventional construction contracts to be awarded with confidence that the installation requirements had been thoroughly investigated. Technical information meetings were held on a very intensive schedule over a four month period after which the subcontractors began the formalization of their individual approaches to the installation effort.

Phase I Report / Presentation

The fifth month was devoted to the preparation of the final subcontractor reports and presentations to the SSCL and DOE personnel. The requirement for the presentation was initiated to provide each subcontractor the opportunity to describe their preferred installation process. The flaw in that thinking was that each subcontractor had reservations about divulging all of their program developments for fear that the information would be of use to their competition.

Those reservations resulted in presentations that gave the impression that the propositions were not well thought out when in reality the missing information was intentionally left out. Evaluations of all three subcontractors' presentations were essentially the same; ranging from poor to barely acceptable. There were some bright spots in each of the presentations and reports. Whether intentional or the fact that a particular subject was of particular interest, each report contained at least one item that was well addressed. This led to the implementation of a "Bridge Task" program which was designed to further refine the specifics of the installation program. These refinements were implemented to provide ongoing funded tasks for the subcontractors to amplify the requirements of specific technical systems.

It became obvious that there would be some delay in the issuance of the second phase solicitation for the actual installation of the components. We selected specific tasks to be accomplished by each of the three Phase I subcontractors. The tasks assigned were for the further development of databases, a more specific set of requirements for the tunnel transportation system, activity to prepare cable definition, a specific plan for receiving and handling materials being received from overseas suppliers and the design of a magnet stand for the Medium Energy Booster dipole magnets.

The formal reports and presentations that were scheduled at the end of a two month performance period were given to the URA, the DOE and the other two installation subcontractors. This resulted in a further leveling of the competitive field since considerable more detail was included in each bridge task. The results of the bridge tasks were incorporated into the Phase II Installation Solicitation that was to be released for bid in the first quarter of 1993.

DOE Policy Changes

Additional perturbations to the planned issue date for the Request for Proposal (RFP) resulted from the changes in the procurement regulations as they apply to subcontracts in excess of twenty five million dollars. The Acquisition Plan, the Solicitation and the recommendation for the subcontract award must now be approved by the Secretary of Energy. This policy will, in all likelihood, add time to the procurement process.

Response to the Solicitation

One of the major advantages of industrial participation in the preparation of the solicitation was that we were able to reduce the response time for the proposal. It is our intention to provide a formal draft to the three bidders at the same time that the document is sent to the Secretary of Energy. We expect that the DOE approval will take thirty days. Once approved, the released RFP will be provided to the bidders with a thirty day response time stipulation.

3 Parallel Technical Sessions II

Part 2 Technical Sessions II

FRONT END SIGNAL PROCESSING ELECTRONICS FOR THE SDC STRAW TRACKING SYSTEM

F.M. Newcomer, S.Tedja, R. Van Berg, J. Van der Spiegel
and H.H. Williams

University of Pennsylvania, Philadelphia, PA 19104

INTRODUCTION

The SDC tracker poses significant competing requirements for short measurement time ($\approx$5ns), good double pulse resolution ($\approx$ 20ns), low power ($\approx$15mW/channel), low operational threshold ($\approx$1fC) and good radiation resistance. Borrowing from existing signal processing techniques, we have designed a custom integrated circuit to address these signal processing issues on a single monolithic substrate. The ASD-8 contains eight channels of amplifier, shaper and discriminator. It was intended as a prototype to study the feasibility of including amplifier, detector tail compensation and comparator functions on the same chip. Threshold control with lockout for each channel and a programmable current output allows easy interfacing to a variety of triggering and time measurement devices. Tests with fabricated parts indicate excellent yield and stable operation with less than 3% crosstalk between adjacent channels.

THE SDC STRAW TRACKER

The 140,000 channel straw tracking system being designed for the SDC detector at the SSC consists of axial layers of $4mm$ diameter proportional drift tubes, straws, varying in length from 2.4 to 4m. High occupancy, up to 10% for inner layer straws at full SSC luminosity, makes it important to achieve good double pulse resolution[3].

When a charged particle traverses a gas filled straw, it leaves an ionization trail of 4 to 5 ion clusters per mm [2]. Ionized electrons are attracted by a positive potential on the wire. As they enter the high field gradient region near the wire, large numbers secondary ionizations occur giving rise to a detectable signal.

Electrons from these secondary ionizations move to the wire with very little change in potential leaving a cloud of positive ions behind. These ions induce a current on the wire as they move toward the cathode. The time dependence of this current pulse is of the form $1/(t + t_0)$[1] and lasts for about $25\mu s$. For good double pulse resolution, this long signal *tail* must be eliminated in the signal processing electronics.

For good spatial resolution and efficiency near the edges of the straw tube, it is important to be able to detect a single ionization cluster. Noise in the electronics is a critical issue, since the straw gas gain has a limited range of adjustment for stable performance and the signal collection time is limited by the double pulse resolution requirement. To ensure that the position resolution is not degraded by the electronics, the discriminator is required to have sub-nanosecond timing accuracy. The resultant high bandwidth, low threshold, readout is susceptible to RF sources and requires careful treatment of its inputs. By mounting the electronics on the detector, both signal to noise losses and pickup can be minimized. This adds the burden of increased power dissipation inside detector cavity making low power design an additional priority.

TECHNOLOGY

Bipolar technology was chosen since it offers low intrinsic noise, high speeds and the

Supercollider 5, Edited by P. Hale
Plenum Press, New York, 1994

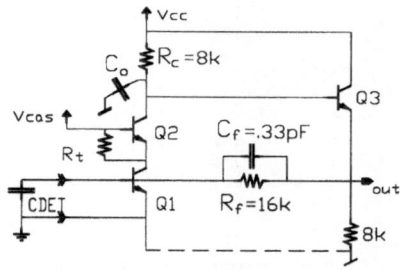

Figure 1. ASD Preamp.

highest transistor gain, g_m, per unit of quiescent current of any commercially available technology. This combination of properties makes it ideal for fast, low power and low noise designs. The Tektronix SHPi analog bipolar process was chosen for the ASD-8. This high speed, low capacitance process offers excellent device matching and well characterized SPICE models. Two micron layout rules for the two metal layers allow close packing of devices. Active devices include: vertical npns, lateral pnps and jfet transistors. Passive components include: active base, p+ and nichrome resistors, and MOS capacitors.

THE CIRCUIT

The ASD-8 consists of 8 identical preamplifier, shaper and discriminator channels with differential inputs and outputs and individual threshold control. The preamplifier converts charge at the input into a voltage output minimizing the noise added to the signal to the extent that the technology permits. The shaper eliminates the preamplifier tail utilizing the well known pole-zero cancellation technique and bandwidth limits the remaining signal using three equivalent integrations. Two additional pole-zero networks serve to reject the characteristic detector tail. The shaped signal is then DC coupled into a two stage timing discriminator. A single channel consumes about $18mW$ broken down as follows: $3.5mW$ in the dual preamplifiers, $4mW$ in the shaper, $7mW$ in the discriminator and $3mW$ minimum to power the output driver section. A thorough discussion of this circuit can be found in [5]

Preamplifier

Each ASD channel is implemented with two fully functional preamplifiers. On chip, this pseudo-differential configuration provides good DC matching at the input to the shaper and good common mode noise rejection. Off chip, it allows the board designer to exploit differential pickup rejection techniques. The three transistor, cascoded common emitter circuit is shown Figure 1. Q1 is implemented as a parallel pair of transistors to facilitate a symmetric layout of the two preamplifiers. The size was selected as a good tradeoff between low base resistance for good noise performance and reasonable current density for radiation hardness. Based on our measurements of SHPi transistors exposed to 0.6×10^{14} neutrons per square centimeter, we expect a closely matched loss of no more than 30% in HFE for the input transistor in each preamp after ten years of SSC operation.

Most of the impulse charge is integrated onto C_f resulting in an output voltage of $\approx Q/C_f$. The extent to which the amplifier falls short of this depends most directly on the total capacitance at the input and the open loop gain, $g_m R_c$. In this circuit, the current in Q1 is $550\mu A$ and R_c is fixed at $8K\Omega$. The resultant open loop gain is 170.

Shaper

The main features of the shaper circuit are shown in Figure 2. The preamplifier tail, due to $R_f C_f$ is cancelled by a differentiating network in the emitters of the first shaping stage. A short integration, $\approx 2ns$ is achieved at the collector node of each of the three stages, resulting in a $(RC)^3$ shaping. The choice of three stages is based on the need for good double pulse resolution without compromise to noise performance. With fewer stages, the bandwidth of the output would need to be significantly higher to maintain the same double pulse resolution.

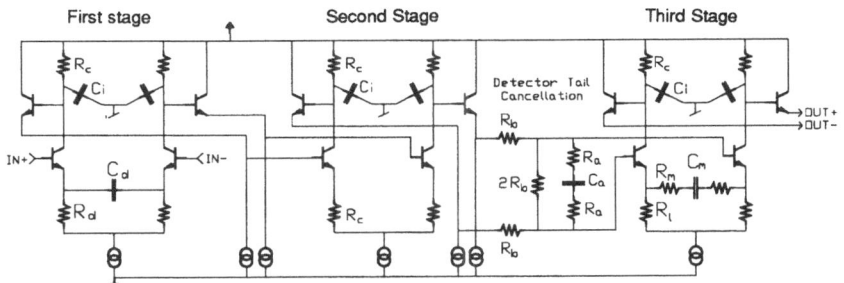

First stage   Second Stage   Third Stage

Figure 2. ASD Shaper.

Table 1. Discriminator Specifications

Power Dissipation	$7mW$
Time Slewing	Less than 1ns/decade of overdrive
Internal offset	$1mV$ (.05fC) or less
Threshold Range	± 10fC
Threshold Uniformity	Better than 10% of setting (chip to chip)

Using a method developed at Harvard [7], the ASD employs two networks, each with a pole and a zero, to cancel the signal tail due to the positive ion motion. In one network, the pole is higher than the zero, in the other, the zero is higher than the pole. The output of the shaper connects directly to the discriminator. SPICE calculations indicate that the measurement time, t_m, at the shaper output is approximately 6ns, and that the DC gain of the shaper is 2.2.

Discriminator

Important features of the two stage, differential discriminator are given in Table 2. This circuit has been previously described in [6]

FABRICATION AND MEASUREMENTS

The ASD-8 along with a silicon strip amplifier designed at UC Santa Cruz, was fabricated in a six wafer lot. Each wafer contained more than 100, 2.7 X $4.3mm$, ASD-8 sites.

We have measured more than 100 ASD-8 chips and find the yield for packaged die is about 80%. Approximately 15% of the chips have one bad channel and the majority of the remaining 5% have all bad channels.

The offset voltage of a sample of 100 working channels was measured at the shaper output. The RMS value for these offsets was $2.5mV$, or about 0.18fC referred to the input.

Temperature sensitivity was studied by applying a $3cm$ square brass heating block to the top of the packaged ASD-8. The temperature of the ceramic package was monitored and the threshold adjusted for 50% efficiency for a 2fC input pulse. Over a range of $40°C$ the effective threshold was observed to change by $30mV$, corresponding to a shift of 0.12fC in effective input threshold.

The noise performance was measured using data taken by an HP54111D digitizing scope. Output noise was measured by sampling the signal at the monitor output at a one gigahertz rate for a period of 5μus. The RMS value of this data was then related to the equivalent input noise charge using the pulser calibrated gain. The measured noise is linear with capacitance, as expected, and found to be 78 electrons/pF + 850 electrons.

In Figure 3 the analog response of the ASD-8 connected to a 4m long $4mm$ diameter straw is shown before and after tail cancellation. These measurements were made at Duke University with an Fe^{55} source. The nearly symmetric, 20ns FWHM, plot on the left indicates that the desired signal processing objectives have been successfully implemented in the ASD-8.

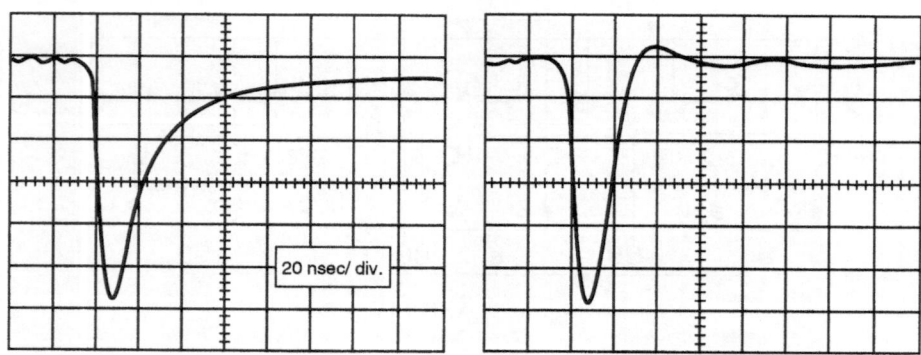

Figure 3. The plots above, provided by Duke University, show the ASD-8 response, in arbitrary units of amplitude, when connected to a 4m X $4mm$ straw tube illuminated using a Fe^{55} source. The analog response is shown before (left) and after (right) detector tail cancellation.

CONCLUSION

ASD-8's are currently being used with prototype straw systems at Colorado, Duke, Indiana and Tokyo Metropolitan Universities. While improved versions are being planned, the first pass success of this, rather complex design is a tribute to the accuracy the SPICE models, modeling tools and layout technique.

References

[1] R. A. Boie, A.T. Hrisoho and P. Rehak, *Signal Shaping and Tail Cancellation For Gas Proportional Detectors* NIM, A192 (1982) 365-374.

[2] J. Fischer, A. Hrisoho, V. Radeka and P. Rehak, *Proportional Chambers for Very High Counting Rates Based on Gas Mixtures of CF_4* , NIM, A238 (1985) 249-264.

[3] G.Hanson et.al., *Wire Chamber Requirements and Tracking Simulation Studies for Tracking Systems at the SSC*, NIM, Vol. A283(1989) 735-743.

[4] V. Radeka, *Low Noise Techniques in Detectors*, Annual Reviews of Nuclear and Particle Science, Vol. 38 (1988).

[5] F.M. Newcomer, S.Tedja, R. Van Berg, J. Van der Spiegel and H.H. Williams, *A Fast, Low Power, Amplifier-Shaper-Discriminator for High Rate Tracking Systems, IEEE Trans. Nucl. Sci.*, TBP August 1993.

[6] F.M. Newcomer and S. Tedja, *A Low Power Timing Discriminator for SSC Applications*, Symposium on Detector Research and Development for the SSC, 1990: Fort Worth, Tex.

[7] John Oliver, *A Tail Cancellation Circuit for Pulses From Wire Proportional Chambers*, SDC note(TBA)

A CMOS TIME TO DIGITAL CONVERTER WITH ANALOG MEMORY
FOR HIGH ENERGY PHYSICS PARTICLE DETECTORS

E. J. Gerds,[1] J. Van der Spiegel,[1] H. H. Williams,[2] and R. Van Berg [2]

[1]Electrical Engineering Dept., [2]Physics Dept.
University of Pennsylvania
200 South 33rd St., Rm 308
Philadelphia, PA 19104

ABSTRACT

A data driven TDC (Time to Digital Converter) has been designed and fabricated in HP's 1.2 μm nwell CMOS process. The circuit was designed to work with the straw tube electronics of the Superconducting Supercollider (SSC), where we wish to measure the arrival time of electrons at a sense wire. The TCCAMU (Time to Charge Converter with an Analog Memory Unit) measures the time between an edge of the system clock and the leading edge of an asynchronous signal, and then gives a digital output representing that time measurement. Analog data sparsification occurs before the digitization with the help of an analog Level 1 / Level 2 storage system; Level 1 to Level 2 data transfers are virtual, in the sense that one swaps capacitor addresses instead of moving charge.

Two separate fabrication runs resulted in chips that have ~108 ps / LSB resolution for any particular storage location. The measurement range is 8 - 24 ns, but adding digital logic to count the reference clock will extend the range to ~ 1 second.

I. INTRODUCTION

There are a variety of different TDC circuit implementations throughout the literature. One such example is the TVC / TAC (Time to Voltage Converter / Time to Amplitude Converter) discussed by Stevens [1,2]. Basically, the time interval being measured is converted to a voltage by integrating a constant current onto a capacitor; the voltage could then be digitized. This TVC approach produced a system with sub nanosecond resolution. This scheme does depend on capacitor matching, however, as well as the matching of the currents from channel to channel.

Another possible TDC architecture called the TMC (Time Memory Cell) has recently been proposed by Arai [3] as an all digital alternative. Basically, a direct time to digital conversion is achieved by recording input signals to memory cells in 1 ns intervals. With the use of feedback stabilized delay elements, it was possible to obtain ~ 1 ns resolution in time measurements.

In this paper, we will discuss the TCCAMU (Time to Charge Converter with Analog Memory Unit) [4]. The advantage of converting time to charge, and then digitizing the charge is the relative insensitivity to mismatches in capacitors or integrating currents. However, we still achieve sub nanosecond resolution. This scheme is sometimes referred to as 'dual slope integration'.

The requirements for an SSC straw tube TDC are rather stringent. We need a resolution of < 0.5 ns along with a double pulse resolution of 20 - 30 ns. Because there will be ~ 140,000 straws in a detector, the VLSI chip must use a minimum of power and area. The slope of the TDC input vs. output curve must also match well from channel to channel so as to minimize the number of calibration constants. Finally, the TDC must be capable of storing data on chip for at least 1 μs until a readout trigger can be generated to accomplish data sparsification.

II. OPERATION OF THE TCCAMU CHIP

The TCCAMU chip has 2 major inputs - the Clock and Discriminator signals (see figure 1). The 16 ns system Clock represents the beam crossing rate for the SSC, and the Discriminator pulse represents the time that a cluster of electrons has hit a particular sense wire in the straw tube drift chamber. The purpose of this TDC is to measure the time between a Clock edge and the Discriminator edge, and to supply a digital output representing that interval.

When a Discriminator input arrives, one of the width generators (even or odd) produces an output pulse (signals Qe / Qo as in figure 1) of width equal to the time interval being measured. During the width generator pulse, a selected Level 1 storage capacitor in the analog memory will integrate a current 'I'. The resulting charge is simply $Q = I t$, where 't' is the integration time. Although the measurement time range is 16 ns, the actual charge integration occurs for 8 - 24 ns to avoid any nonlinearity near zero integration time.

The Discriminator input also drives the Delay Generator, which produces an output

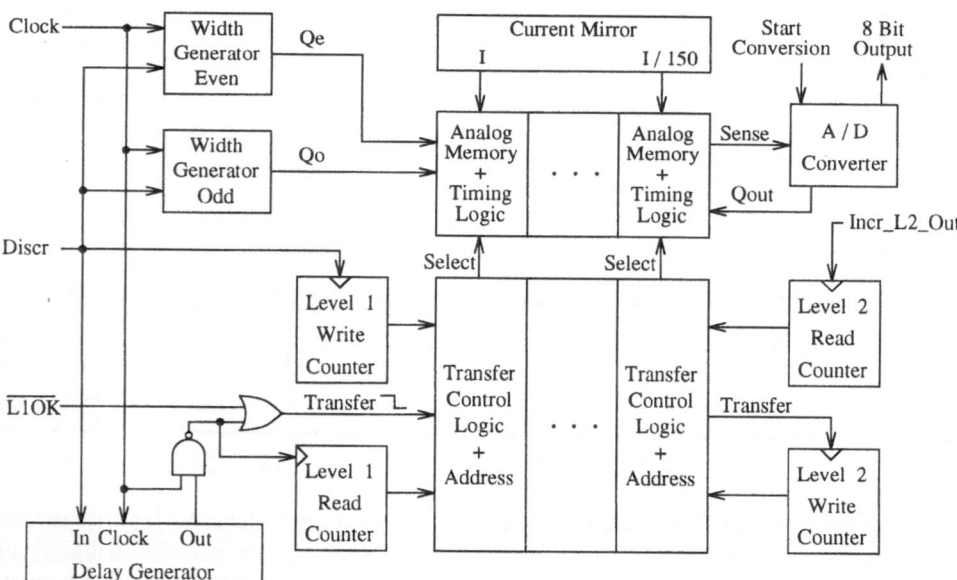

Figure 1. A simplified block diagram of the TCCAMU architecture.

pulse 1 μs after the input. During that delay time, a readout trigger system would decide whether to accept or reject the data associated with that Discriminator input.

If the Level 1 datum were to be accepted, then a 'L1OK' pulse would be supplied to the TCCAMU. This in turn would cause the datum to be transferred from Level 1 to Level 2. Note, however, that we have a *virtual* transfer and not a *physical* one. We accomplish this virtual transfer simply by swapping addresses between the 2 appropriate Transfer Control blocks. This scheme permits a transfer in 16 ns to avoid deadtime, and does so without the use of high power op amps otherwise needed for a physical charge transfer.

Once the charge datum has reached the Level 2 virtual buffer, one can use the Incr_-L2_Out input to the TCCAMU to skip a particular Level 2 location, hence rejecting a particular datum. If that datum is accepted while in Level 2, however, then the charge would be digitized into an 8 bit word. Digitization is accomplished by subtracting the charge off the capacitor using a *ratioed* current of 'I/150'. During a maximum conversion time of ~ 4 μs, a counter in the A / D converter keeps track of how many clock cycles (16 ns period) are needed to remove all the charge from the analog location. This scheme thus would be independent of storage capacitor variations, as well as variations in 'I' from channel to channel.

Each storage capacitor in the analog memory has a Level 1 or a Level 2 address associated with it. The addresses are stored in the Transfer Control Logic blocks shown in figure 1. There are 12 such Control blocks (for 8 Level 1 + 4 Level 2 locations), each of which has associated with it one physical capacitor in the analog memory. Writing to a Level 1 location, for example, is accomplished with a Level 1 Write counter. The counter output might be a '3', meaning that the particular Transfer Control block that contains a Level 1 address of '3' would activate its own storage capacitor for writing.The same scheme applies to the Level 1 Read, Level 2 Write, and Level 2 Read counters.

III. MEASUREMENTS AND RESULTS

Testing of the TCCAMU chips was done at SSC Laboratory using an HP82000 chip tester. The input Clock was given a 16 ns period. The delay of the leading edge of the input Discriminator pulse relative to a particular Clock edge was systematically varied from 0 to 16 ns for all the measurements. The digital outputs for each Discriminator input were then recorded for the purpose of analysis.

The resulting data points were used to draw a best fit straight line through the data. A plot of the input delay vs. digital output is shown in figure 2 for one particular chip. The slopes of the 6 curves (for 6 different Level 1 analog storage locations) shown vary from 107.8 ps / LSB to 108.4 ps / LSB. Note that the curves in figure 2 are periodic due to the periodicity of the input Clock.

The slopes are virtually identical for all the chips that were tested. Four chips in the first fabrication run were fully tested and had slopes varying from 107.2 - 108.1 ps / LSB. Five chips from a second fabrication run were fully tested and had slopes varying from 107.4 - 108.4 ps / LSB. The slopes thus seem to match fairly well regardless of mismatches in fabrication parameters, integration currents, or integration capacitors.

Based on figure 2, it can also be seen that the entire 16 ns range is measurable by the TCCAMU. Also, the low to high output transition appears quite abrupt, as expected. Incrementing the delay by as little as 50 ps at that transition still gave the exact same expected results. With a little extra digital logic, it will also be possible to count a stable reference clock so that the measurement range is extended to ~ 1 second.

Another important result is the integral nonlinearity INL (difference between the measured data and the best fit straight line). For all analog memory locations tested on chips from the second fabrication run, we have a worst case INL < +/- 216 ps and rms INL< 90 ps. The results for all capacitors tested on each of the chips were similar both quantitatively and qual-

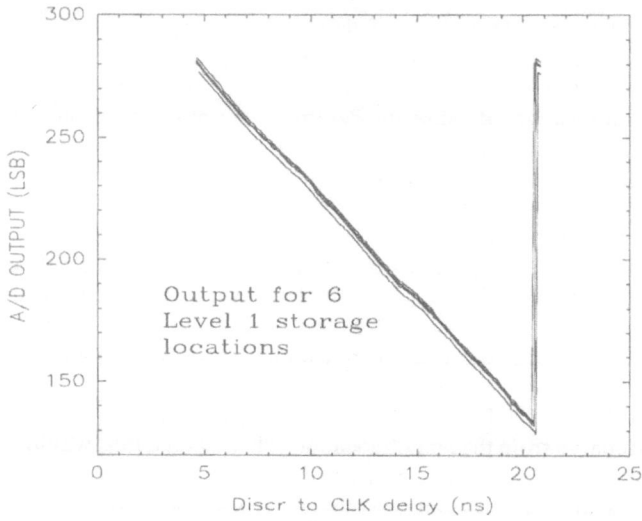

Figure 2. Input delay vs. A/D output for the TCCAMU

itatively. Plots of the INL for various capacitors were so similar that we suspect that most of the INL is due to fixed pattern noise.

The worst case differential nonlinearity DNL is < +/- 108 ps; there are no missing codes. Since the DNL and INL are not independent, we believe that the pattern noise also affects the DNL.

The power dissipations of the chips in the first and second runs are 8.3 mW and 7.2 mW, respectively. The TCCAMU has ~ 10,000 transistors and an area (excluding pads and output drivers) of 2.0 mm x 2.2 mm.

IV. CONCLUSION

A time to digital converter using dual slope integration has been successfully designed and tested. The TCCAMU has 8 Level 1 and 4 Level 2 analog memory locations for data sparsification. It accomplishes a virtual Level 1 to Level 2 datum transfer using an address swapping scheme. The timing resolution is ~ 108 ps/LSB for all storage locations, with INL < +/- 216 ps and rms INL < 90 ps on any chip. The DNL is < +/- 108 ps. Future versions of the TCCAMU with larger analog memories and better INL are being planned.

REFERENCES

1. A. Stevens, "A time-to-voltage converter with analog memory for colliding beam detectors," *IEEE JSSC*, pp. 1748 - 1752, December 1989.
2. A. Stevens, "A fast low-power time-to-voltage converter for high luminosity collider detectors, *IEEE Trans. Nucl. Sci.*, vol. 36, pp. 517 - 521, 1989.
3. Y. Arai, "A CMOS four-channel x 1K time memory LSI with 1-ns/b resolution," *IEEE JSSC*, pp. 359 - 364, March 1992.
4. L. Callewaert, "Front end and signal processing electronics for detectors at high luminosity colliders," *IEEE Trans. Nucl. Sci.*, vol. 36, pp. 446 - 457, February 1989.
5. Dan Porat, "Review of sub-nanosecond time-interval measurements," *IEEE Trans. Nucl. Sci. pp. 36 - 51, 1973*.

A COMPACT FRONT-END ELECTRONICS MODULE FOR THE SDC STRAWTUBE OUTER TRACKER

M.S. Emery[1], G.T. Alley[1], R.M. Leitch[2], R.A. Maples[1], and W. Holmes[1]

[1]Instrumentation and Controls Division
[2]Engineering Division
Oak Ridge National Laboratory
Post Office Box 2008
Oak Ridge, TN 37831

INTRODUCTION

The challenges of building a detector for the Superconducting Super Collider have been talked about for the last several years. Those challenges are proving to be real and in some cases tougher than expected as prototype subsystem and component development continues within the different collaborations. Not to be daunted, engineers and scientists are using ingenuity and novel designs to meet the challenges. One such area has been in the development of the outer tracker readout electronics for the Solenoidal Detector Collaboration (SDC) detector. The tracker has over 100,000 channels and is composed of strawtubes that are 4 mm in diameter and 4 meters long. The sheer number of channels and small-diameter tubes require a very high density packaging scheme with critical attendant concerns, including power consumption, cooling, and crosstalk. This paper describes the novel approach taken to solve some of these challenges.

REQUIREMENTS

The strawtube outer tracker design consists of bundles, or modules, of straws arranged in a trapezoidal pattern. There are two types of straw modules -- stereo modules with 159 channels and trigger modules with 212 channels. Because of the large number of channels present in the straw tracker and several other systems used in the Solenoidal Detector,[1] space for front-end electronics is at a premium. A detailed description of the strawtube tracker can be found in the SDC Modular Straw Outer Tracking System Conceptual Design Report.[2]

Many issues must be considered in the design of high-density, high-speed circuits. Among these are crosstalk and system stability, packaging, and cooling. To get the best performance possible from the system, minimum lead length between the detector and front-end electronics is desirable. Short lead lengths will reduce the possibility of crosstalk and pickup of other noise sources and improve the system stability.

The straws also require a 2000-V anode bias voltage, which must be decoupled from the low-voltage electronics. Besides the obvious catastrophic effects of high voltage breakdown on a particular straw channel, the high voltage could arc from one channel to another in a domino effect because of the close spacing of the components.

Figure 1. Photograph of front-end electronics module designed at ORNL for strawtube tracker.

DESCRIPTION OF MODULE

The trapezoidal cross-section of the electronics module, shown in Figure 1, matches that of the strawtube stereo module. Connection to the strawtube module is made at the near end, and output signal cabling is through the rear. This design utilizes the University of Pennsylvania designed Amplifier-Shaper-Discriminator[3] (ASD8) integrated circuit. Time constants for the shaping stages and tail cancellation circuit and sensitivity of the charge-sensitive preamp stage were selected specifically for these strawtubes.

The three horizontal circuit boards hold surface-mounted components on both top and bottom surfaces. One of the three boards is six layer, consisting of two ground planes, two high-voltage routes, and top and bottom surfaces. The other two boards are four layer, with the ground plane and high-voltage routing merged together. All three boards utilize blind vias for interconnection between layers.

High voltage (HV) is distributed to the straws for each circuit board in four banks. Figure 2 shows the distribution of the high voltage. Any scheme to match the electronics density to the same footprint of the straw module requires HV capacitors to block the dc voltage from the electronics while passing the charge signal on to the ASD8 chip. No commercially available chip capacitors were identified that had a sufficiently small cross-sectional area. Consequently, a custom chip capacitor was designed for this application by KD Components Corporation located in Carson City, Nevada. The cross section was reduced to 2.5 x 2.5 mm, at the expense of making the device longer (i.e., 7.6 mm).

A number of low-voltage control lines and power supply lines are required for proper operation of the ASD chip. Although they are not shown in Figure 2, these lines provide threshold levels for the discriminator function, programming for the output current amplitude, and power supply lines. A complete description of the ASD chip requirements and specifications may be found in Reference 2.

Very conservative design rules, consistent with MIL-STD-275D,[4] were used for one of the three component-carrying boards. A minimum of 80-mil spacing was maintained between any trace carrying high voltage and a low-voltage trace on a layer. This would provide for arc-free operation on a clean surface with voltages well in excess of the maximum operating voltage of 2500 V. These relatively large spacings, along with the desire for a continuous ground plane, forced the use of six layers with blind vias.

However, much of the high voltage was handled on traces that were in interior layers. Therefore, it was felt that more aggressive spacing could be used on interior layers, and at the suggestion of the PC board vendor, a four-layer design was used for the other two boards. This change was made at the expense of slicing up the ground plane layer and inserting the high-voltage buses.

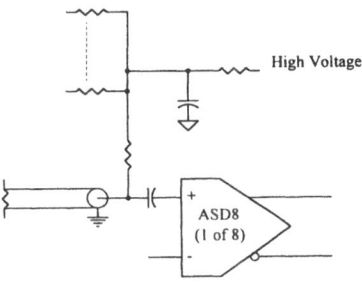

High Voltage

ASD8
(1 of 8)

Figure 2. Simplified circuit schematic of strawtube readout electronics.

THERMAL ANALYSIS AND COOLING

Cooling the front-end electronics is necessary to ensure reliable operation and to maximize operating lifetime. In addition, it is important to maintain a certain level of thermal stability for structural purposes to maintain critical tracking alignment requirements. The cooling system should also be designed to mitigate the potential fire hazard caused by supply voltages that are present.

A test was conducted for a 159-channel stereo module to determine the amount of flow needed to maintain the electronics at 50°C or below. The cooling system consisted of a hose supplying air to an aluminum cover that directed the flow of the air over the electronics. This test used room temperature air (23°C), but because of the potential fire hazard, the actual design will use nitrogen at a temperature of 10-15°C. Since air and nitrogen have similar properties, the results of this test are valid for nitrogen.

The power dissipation of the 159-channel module was measured as 8 W (50 mW per channel). A series of tests was conducted at various orientations with and without active cooling. The test without cooling was made so that the characteristics of the powered boards could be studied and compared to the counterpart cases when active cooling was applied. The various orientations were necessary to determine whether certain geometries significantly affected the heating of the electronics.

Thermocouples were mounted on the electronics boards in 12 different locations. The locations were chosen so that the maximum temperature and maximum gradient in a board could be measured. The temperature of the electronics was allowed to stabilize for both the heating and cooling portions of each test cycle to provide the time constants for the system. For the purposes of this test, the time constant is defined as the time required to reach two-thirds of the maximum steady-state temperature.

Several conclusions can be drawn from these tests. First, active cooling is required since the temperature of the electronics (70°C) exceeded the desired 5°C when no cooling was used. Secondly, the test results indicate that a flow of 0.25 L/s (~0.5 cfm) is sufficient to maintain the electronics boards at the desired temperature of 50°C or less with a maximum temperature gradient in the boards of approximately 10°C. At this flow rate, the exit air temperature is 10 to 11°C warmer than the inlet air temperature. Finally, the orientation of the electronics module does not appear to be a significant factor.

A thermal math model of the electronics was generated using the thermal analysis program SINDA85.[5] This model was correlated with the test results by increasing the published value of thermal conductivity of the circuit board material, FR4. This change was necessary since the circuit board material contained more copper than standard FR4 circuit board material.

A comparison of the model results and thermocouple number 10 from the test with cooling is shown in Figure 3. The maximum error was 18.7% in the cooling phase of the cycle. However, the average error was 0.5%, indicating that the correlation is very good and that SINDA85 will be a useful tool for designing the cooling system of the straw tracker.

PERFORMANCE

All exposed high-voltage sections of the circuit boards are covered with a conformal coating. This reduces the likelihood of any flashovers or arcing and improves long-term reliability by keeping dirt and moisture from forming paths for leakage currents. Successful tests to 2500 V have been made; routine operation is around 2000 V.

An output signal for an input signal of 0.5 fC is shown in Figure 4. The output load is 100 Ω driven through 18 in. of 50-mil spaced ribbon cable. The differential output pair is shown as two separate traces on the oscilloscope photograph. Vertical sensitivity is 50 mV/div; horizontal sensitivity is 50 ns/div.

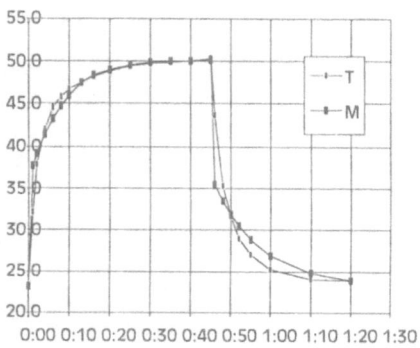

Figure 3. Comparison of measured temperature with Thermal Math Model.

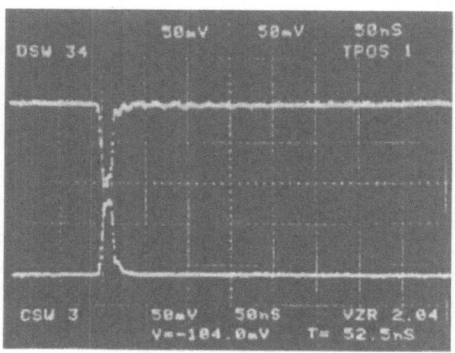

Figure 4. Differential output signals from electronics module. Input signal is 0.5 fC.

FUTURE IMPROVEMENTS

Several improvements and additions are planned for this basic design. A test signal input for diagnostics and calibration must be provided for the actual tracking system. Although most of the input protection for the ASD chips is built into the monolithic chip, survivability in the case of a sudden short in a straw can be improved by the addition of a small value resistor placed in series with the input line. When chosen carefully, the resistor will have only a negligible effect on the equivalent input noise level. These additions, among other changes, will be included in a unit designed to match a strawtube module for the trigger layer.

CONCLUSIONS

A compact and densely packed front-end electronics module has been designed and fabricated using high-speed, high-gain integrated circuits for use in the strawtube tracker section of the SDC detector. Tests have shown the electronics package to withstand operating voltages of 2000 V with a safety margin of at least 500 V. Stable operation of the ASD8 chips has been demonstrated. Long-term reliability can be enhanced since low operating temperatures can be maintained with only 0.5 cfm of air flow. This concept shows considerable promise for use with the final strawtube tracker design.

ACKNOWLEDGMENTS

The authors wish to thank Brig Williams, Mitch Newcomer, Rick Van Berg, and Nandor Dressnandt of the University of Pennsylvania for their suggestions for improving this package design. Also thanks to ORNL technician Peggy Johnson for the tedious work in assembling prototypes used for testing.

REFERENCES

1. "SDC Technical Design Report," SDC-92-201, SSCL-SR-1215 (April 1, 1992).
2. "SDC Modular Straw Outer Tracking System Conceptual Design Report," SDC-91-00125 (Nov. 7, 1991).
3. Newcomer, et al., "A Fast, Low Power, Amplifier-Shaper-Discriminator for High Rate Straw Tracking Systems," presented at IEEE Nuclear Science Symposium (1992).
4. "MIL-STD-275D Printed Wiring for Electronic Equipment," 4.5.1.4 Conductor Spacing (Oct. 5, 1981).
5. "SINDA85 - Systems Improved Numerical Differencing Analyzer," Developed under NASA contract NAS9-18411 by Martin Marietta Astronautics Group, Martin Marietta, Denver, Colorado (1985).

OPTOELECTRONIC TECHNOLOGIES FOR DETECTOR READOUT AT THE SSC

Max Buttinger[1], John Davies[1], Peter Duthie[1], Nick Green[1], David Hall[1], and Andrew Moseley representing the CERN RD-23 Project Collaboration[2]

[1] GEC-Marconi Ltd., Stanmore, Middlesex HA7 4LY, England
[2] CERN RD-23 Project: supported by Birmingham/CERN/EPFL Lausanne /Marconi Defence Systems/Imperial College/Lund/Oxford/RAL

INTRODUCTION

The high density of particle detectors in Hadron Collider systems, such as the SSC, requires a compact and low power dissipation readout system[123]. We present the results of the development and application of two optical modulation technologies to the optical fibre transfer of analog or digital signals from the detector front end. These are the Mach-Zehnder interferometer in lithium niobate and the novel multiquantum well reflective modulator in indium phosphide. The trade off between architecture flexibility with performance and reliability is analysed. A key to the success of an optical readout system for the SSC is in the packaging of the devices for production. Here the economics of application to the central tracking detectors is discussed.

ARCHITECTURES

Mach Zehnder Modulator System

Guided wave devices on LiNbO$_3$ are used in this application with the laser source and receivers outside the detector and only the modulator placed next to the front end electronics. The integration of an optical splitter and many parallel modulator devices enables the technology to be viable at acceptable cost as packaging and assembly costs currently dominate. In the current programme a 16 channel demonstrator device has been designed and fabricated. It consists of an integrated polariser at the input, followed by a 1:16 optical splitter and 16 identical modulators in parallel (total 40mm x 4mm) . The splitter occupies about two thirds of the chip area. Two alternative splitter layouts have since been modelled and tested, and the more compact device shown to have improved performance [4].

The waveguide devices are polarisation sensitive requiring the polarisation to be set correctly for optimum operation. There are several methods available: polarisation maintaining fibre (PMF), actively scrambled launch polarisation or active polarisation control. For the demonstration system, PMF was chosen but the provision of the on-chip polariser allows other techniques to be used at a later date.

To maintain the noise performance of the system it is important to reduce the optical reflections. A simple butt-joint between glass fibre and lithium niobate yields a return loss of 14 dB which can be easily increased to 50 dB with angled junctions. An angled input fibre was used with a normal, non-angled, 16 output fibre array. This gave a return loss of 14 dB plus twice the device insertion loss. The output fibre array could also be angled, if greater return loss were required.

The V_π and insertion loss (including 12dB splitter loss) for each of the 16 channels has been measured Figure 1. The modulators have a mean V_π of 4.07V with a standard deviation of 0.1V, and mean excess loss per channel of 9.2dB with standard deviation of 0.8dB. An extra 2dB for connectors will apply. The interchannel electrical crosstalk has been measured at -46dB.

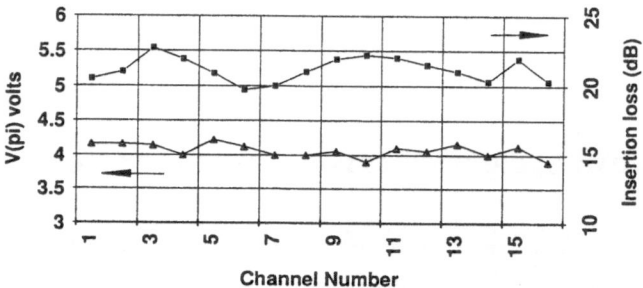

Figure 1. Results for 16-Channel Modulator.

Maximum 1% linearity is achieved for a 2V monopolar signal by setting the modulator bias at 0.17 V/V_π using laser trimmed resistors. This provides a possible 50% modulation depth. Biasing using resistors dissipates heat. therfore non heat dissipation methods such as laser ablation of waveguides will ultimately be used. Operating the modulator at below its quadrature point reduces receiver shot noise whilst increasing optical absorption, the reverse occurs when operating above the quadrature point. The choice will depend on the the acceptable level of optical power dissipation and desired dynamic range.

The modulator has a capacitance of 10pF, and therefore requires 10^4 to 10^5 charge amplification between Si detector and modulator, for full 50% modulation. The heat dissipated from charge amplification requires trading off with that of optical power to achieve optimum performance.

For 100:1 dynamic range a minimum power of -20dBm is required at the receiver (responsivity 0.8A/W, transimpedance amplifier of 14KΩ) and a laser with RIN<-135 dB/Hz. A commercial diode-pumped Nd:YAG (P_o=10dBm, RIN<-160 dB/Hz), is more than capable of running 10 such systems.

It has been shown that $LiNbO_3$ has negligible reduction in performance for total doses of 10^{14}neutrons/cm^2 and 1Mrad.

Modulators in III-V Technology

M-Z modulators based on III-V technology could also be used in this application. The III-V technology also offers a number of other approaches such as guided wave absorption based devices and reduced capacitance. Performance is generally enhanced through the use of quantum well structures, or Braquetts (Braqwets). Recently vertical cavity devices operating in a reflective mode have been developed, which offer several advantages over the waveguide devices [5]. The very small size (comparable to LED chips) and parallelism is useful for the eventual integration of many modulators to the particle detector and processing electronics.

The modulator is an asymmetric Fabry-Perot [5] device based on a vertical optical cavity formed between a metal mirror and multi layer stack, in between the mirror and stack is the multi quantum well absorber. Previous devices have been fabricated on both GaAs and InP substrates for operation around 850 nm and 1600 nm respectively, and have demonstrated modulation up to 20 GHz [6]. The present design has been fabricated for operation at 1.53 μm allowing access to current lower cost commercial telecoms laser devices. It is optimised for low insertion loss and maximum linearity for analogue or digital applications.

InP has a technological advantage over GaAs as a result of the transparent substrate. This enables electrical and optical access from opposite sides of the structure. The InP Fabry-Perot structure is sensitive to the wafer uniformity and the absolute thickness of each layer in the structure, thus requiring tight control of the tolerances throughout design and fabrication. The demonstrator packaged devices were produced each using 4-channels with V-groove alignment of fibres and flip-chip solder bumping, (a well known technique in electronics), to provide the device mountings and electrical connections. This technology can also be used to provide the optical alignment to the device, since it has a large active area (~30µm) and is polarisation insensitive. This can lead to designs for low cost integration of the optical front-end.

Using a reflective device requires the optical source and the receiver to be coupled to a single fibre. It is also possible to use a single laser source to supply several channels via a low cost optical splitter, such as polymer based devices. An alternative solution is to use optoelectronic integrated circuits (OEIC) containing an array of basic units, where a basic unit consists of a laser, photodiode and 3 dB coupler.

The first MQW demonstrator devices have a reflectivity of within 3% of the desired 28% limited by variations in fabrication and absorption. The total insertion loss per channel is 27.5dB made up of 4.5dB modulator insertion loss, 3 dB fibre interfacing loss, 12 dB (1x16) splitter loss, 6 dB (1x2) splitter loss (double path), and 2dB connector loss.
Bias at -9V will allow a few % linearity and 10% modulation depth to be achieved with a 4V monopolar swing. Therefore for 100:1 dynamic range, 50 µW is required at the receiver. A major advantage of the MQW is its low capacitance (0.5pF). It therfore requires a low charge amplification of $<10^3$ after the Si tracker strip. This is in contrast to charge amplification in LEDs of 10^5 and LiNbO$_3$ of 10^4.

The absorption band edge becomes sharper at lower temperatures thereby making them suitable for cryogenic calorimeters. MQW devices have also shown no appreciable reduction in performance after exposure to 10^{14} neutrons/cm^2.

Power dissipation in the MQW device is almost entirely due to the optical power incident on the device. e.g. for 100µW in the input fibre, 81µW is due to the photo induced current (device responsivity of 0.2 A/W), 36µW due to material absorption of the light and a further 37µW due to losses in the fibre coupling, total 154 µW. This is virtually constant independent of data rate. This level of dissipation is slightly higher than LiNbO$_3$ but less than for LEDs, at the expected occupancy.

FUTURE

The LiNbO$_3$ M-Z structure will be scaled up to offer more channels per chip, lower capacitance and bias setting with no power dissipation. Packaging technologies such as flip chip techniques on Opto Silicon Hybrids will ensure large channel capacity (64 to 128), high volume and low cost units.

The MQW modulator will similarly be improved with a greater modulation depth and reduced pigtailing losses. An increase in the number of channels per chip from 4 to between 16 and 64, with solder bumping for both chip placement and fibre alignment will occur. Likely numbers per 3" wafer amount to 10,000 plus with likely yields higher than LEDs. Source and receiver units will be fabricated as OEICs. This provides space saving, high channel capacity, low cost units. Numbers of devices on a 3" wafer can be >1000.

Programmes such as RACE 2010 and RACE 2001 are developing 1 nm, 2 nm and 4 nm wavelength spacing for wave division multiplexing (WDM) as well as electronic switches for 2.5 Gbps packet data transmission. Novel integrated WDM M-Z and MQW devices are also under development. This opens up a whole new area of channel multiplexing possibilities for reducing: front-end electronics, power consumption, fibre count, radiation damage, and hopefully costs on a timescale commensurate with both the SSC and LHC experiments, Figure 2. The 2.5 Gbps drivers and receivers will undoubtedly find application as Asynchronous Transfer Mode hardware supporting detector back-end data transfer.

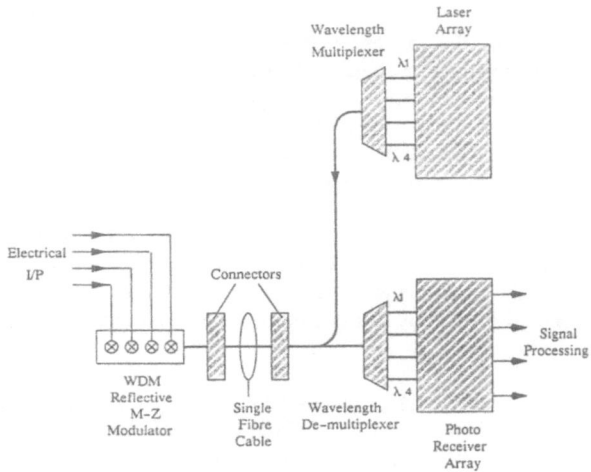

Figure 2. Possible WDM architecture for reducing fibre count.

CONCLUSION

For current and future front end electronic readouts, externally modulated links have been shown to have lower power consumption and dissipation, and lower charge amplification requirement than directly modulated links.

The application of two different technologies for external modulation has been discussed with reference to initial device results and proposed future developments. Analogue data transfer requires less electronics than digital readout, resulting in lower mass and power consumption at the front end, due to the absence of the ADC.

ACKNOWLEDGEMENTS

This work was supported by Marconi Defence Systems Ltd. and CERN Programme RD23.

REFERENCES

1.   SDC "Technical Design Report" SDC-92-201, SSCL-SR-1215 1/4/92.
2.   N. Ellis, S. Cittolin and L. Mapelli "Signal Processing, Triggers and Data Acquisition." CERN/ECP90-17 21/12/90.
3.   S. Gadomski, G. Hall, T.Hogh, P. Jalocha, E. Nygard and P. Weilhammer "The Deconvolution Method of Fast Pulse Shaping at Hadron Colliders" CERN-PPE/92-94 31/1/92. (Submitted to Nuclear instruments and Methods A).
4.   T.P. Young, A.J.S. Aitken and D.D. Hall "Design and Initial evaluation of a compact optical splitter and 16-Channel modulator array on Ti:LiNbO$_3$" ECIO'93, p.10-10.
5.   A.J. Moseley, J. Thompson, N.Q. Kearley, D.J.Robbins, M.J. Goodwin "Low Voltage InGaAs/InP Multiple Quantum Well Reflective Fabry-Perot Modulator." Electron. Lett. vol 26 pp.913-4 1990.
6.   C.C. Barron, M. Whitehead, K.K. Law, J.W. Scott, M.E. Heimbuch and L.A. Coulden, "K-band operation of symmetric Fabry-Perot Modulators" IEEE J.Lightwave Tech, LT-9 pp1639-1645 (1991).
7.   Cashmore, Brooks, Martin, Nickerson "Use of low-speed Fibre Optic Links For The SDC Si-Tracker." Dept. Physics, Oxford University. July '92.
8.   N.P.Green, M.R.Buttinger, P.Duthie. D.D.Hall, A.J.Moseley. "Optoelectronic Modulator Technology For Analogue Data Transfer" ECIO '93.
9.   CERN, "Optoelectronic Analogue Signal Transfer for LHC Detectors", CERN/DRDC/91-41, Now project RD-23.

A DATA COLLECTION SYSTEM
FOR THE SDC STRAW TUBE TRACKER

A. Hölscher,[1] K. Ragan,[2] P. Sinervo,[1] G. Stairs,[1] and K. Strahl[2]

[1] University of Toronto, Toronto, ON
[2] McGill University, Montreal, PQ
    Institute for Particle Physics *, Canada

INTRODUCTION

The Solenoid Detector Collaboration (SDC) has proposed to construct and operate a large, general-purpose, solenoidal detector that combines excellent charged particle detection, a hermetic, scintillator-based calorimeter and a large muon identification system.[1] A principle component of the SDC detector is the outer tracker that provides precise momentum measurement of charged particles from a radius of 85 cm to 165 cm from the proton beam. The outer tracker front-end electronics capture and digitize the data it produces and a data collection circuit (DCC) collects this data and passes it to the SDC data acquisition (DAQ) system. We describe in this article the specifications and design of the outer tracker DCC.

OUTER TRACKER DATA COLLECTION CIRCUIT

The outer tracker consists of 137,000 straw tube chambers that provide timing information used to measure the position of charged particles. The analog signals are processed using an amplifier-shaper-discriminator circuit[2] and are digitized by a time memory chip[3] (TMC). The digitized data is held by the TMC and a level 2 buffer (L2B) until a decision to either reject or accept the data is made by the level 1 and level 2 (L2) trigger system. Upon a L2 trigger accept signal, the DCC acquires the relevant data from the TMC/L2B's, formats it into data packets, and passes these packets to the DAQ system. Each TMC/L2B instruments four straw tubes, eight TMC/L2B s are arranged on a small PC board (Time Memory Assembly or TMA), and either five or seven of these boards are mounted in a small housing known as a

* Supported by the Natural Sciences and Engineering Research Council of Canada and the Texas National Research Laboratory Commission.

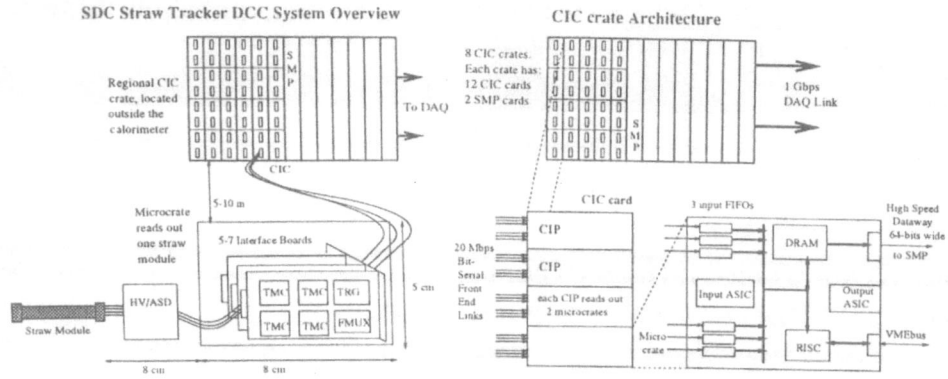

Figure 1. The data collection circuit described in the text. Each FMUX reads out eight TMC/L2B chips, and either five or seven FMUX's are connected together and read out 156 or 212 straw tubes, respectively. The CIC is located in an SDC standard DAQ crate.

microcrate. Each microcrate instruments the 212 (156) straws of an axial (stereo) module of the outer tracker.

Our proposed DCC system[4] consists of two levels: i) an application-specific integrated circuit known as the Front-End Multiplexor (FMUX) located on each TMA, and ii) a set of high-performance, low-cost microprocessors (Crate Interface Processors or CIP's) that collect the data from the microcrates. Each FMUX collects data from eight TMC/L2B chips for a given interaction. Once all the data has been collected, the FMUX transmits this data to a CIP located on a Crate Interface Card (CIC) that reside off the detector. Each CIC has four CIP's that process the data from eight microcrates. A total of 96 CIC cards are required to collect the data for the outer tracker. The CIC cards are housed in SDC Standard DAQ crates, each of which are expected to carry 12 CIC's. The CIC format and compress the data for each interaction and then transfer the data to the DAQ system via a Scheduler/Monitor Processor (SMP). We illustrate this system in Fig. 1.

DCC Design Specifications

We have performed detailed Verilog[5] simulation studies in order to estimate the DCC buffering and bandwidth requirements.[4] Although other studies have been made of data collection and DAQ systems for SSC detectors,[6] they have either been generic in nature or focused on devices that present different data collection issues (such as the SDC calorimeter). We took into account the physics correlation of data, multiple interactions in the same beam crossing and crossings adjacent in time to the trigger, and the effects of finite buffer sizes and bandwidths. These studies show that the occupancy of a single straw tube could be as large as 0.10 hits/interaction at the design SSC luminosity of 10^{33} cm^{-2}s^{-1}. The DCC will therefore read out an maximum average of 0.4 hits/event from each TMC/L2B. The rate of interactions accepted for read out (the L2 rate) are expected to be of order 10^3 Hz, but we have conservatively assumed a maximum rate of 10^4 Hz. We estimate that it will require 3 bytes to encode each datum, so an average maximum bandwidth of $\sim$ 10 Kbytes/second between each TMC/L2B and FMUX is required. Our simulation studies have shown that a maximum burst bandwidth of $\sim$ 100 Kbytes/second is sufficient.

The DCC will read out anywhere from zero to ~ 10 data from each TMC/L2B for a specific interaction. A data-driven DCC is therefore attractive as it minimizes the amount of buffer space and bandwidth, and it decentralizes the control of the system. However, it is not possible to anticipate the total amount of buffer space required at each stage in the system, so conservative estimates must be made of the necessary buffer space in order for the system to function with high data collection efficiency. In any event, the system must be able to recover gracefully when one or more buffers fill completely.

The preliminary specifications for the SDC DAQ system require that data be collected in channel and event order. Channel ordering of the data can be ensured by reading out the TMC/L2B's in order after the data for an interaction is available. This would require the DCC to begin data collection only when the data for the first channels are available. It implies that data for a subsequent trigger must be held until all of the TMC/L2B chips have been read out for the current trigger. However, we assume that data can be passed to the DAQ system regardless of event order as long as the data packets contain the information necessary to identify the interaction the data is associated with.

The DCC must be highly reliable, as many of its components will be installed in areas that will be infrequently serviced. The system must also be fault-tolerant, so that a single component failure does not significantly impact the overall performance of the system.

Features of Proposed DCC

Our proposed DCC operates in a data-driven mode where data is read out of each TMC/L2B whenever it is available by the FMUX and stored in a FIFO buffer. The first FMUX in a microcrate waits until all eight of its TMC/L2B chips have signalled the data for the next trigger has been transferred to the FMUX buffers. The FMUX then transmits the data from its buffer to the CIC and signals the next FMUX in the microcrate. This continues until the entire microcrate is read out. The CIC formats the data from each microcrate as it appears in an input buffer. When all data has been collected from the microcrates serviced by a CIC card, the CIC transfers the data packet to the DAQ system. Our studies have shown this system to be able to handle the dataflow requirements of the outer tracker.

The interface between each TMC/L2B and FMUX consists of 2 control lines and 4 data signals. The relatively modest bandwidth requirement allows us to clock the data into the FMUX at a rate of ~ 2 MHz, which reduces the digital noise that could affect the performance of the nearby analogue electronics. Three redundant, bit-serial lines are used to transfer the data from the FMUX to the CIC using a group encoding scheme that is robust and self-synchronizing.

Most of the monitoring functions in the DCC are implemented in the CIC. We have determined that currently-available microprocessors are sufficiently powerful to be able to validate each datum as the CIP creates the data packets. This allows the CIP to check for data integrity and to log any exceptions encountered. The CIP can also keep statistical information such as the amount of data read out and the number of exceptional conditions encountered, and provide that information upon request.

We expect the reliability of the DCC components to be high given the level of integration used in the custom digital circuits and the use of commercial components for the CIC. We therefore believe the largest potential source of failure will be interconnections between components. Our DCC design reduces the exposure to such failures in two ways. A failure of a single TMC/L2B will only affect the four channels readout by that chip and not compromise the readout of the 7 other TMC/L2B's on the same TMA (this is not necessarily true if all TMC/L2B's were read out using a single bus. Data transmission from the FMUX's can take place over any one of the three bit-serial data paths connected to the microcrate (the datapath in use is configurable). This reduces the impact of a failure in a specific FMUX; only data from the 32 channels read out by that FMUX would be compromised.

We are continuing a more detailed fault analysis in order to pinpoint other aspects of the design that could cause significant loss of system capability.

Prototype Development and Conclusions

We are currently engaged in an R & D program to build a DCC prototype system. A prototype FMUX with virutally all of the functionality described above is currently being manufactured by Northern Telecom. This prototype will be able to read out two TMC/L2B's and has a buffer depth of 8 hits for each TMC/L2B. This design will subsequently be scaled up for eight TMC/L2B chips and a 16-deep data buffer.

We have developed a conceptual design for the CIC and expect to have a detailed layout of a prototype board completed by fall 1993. This prototype will be implemented on a 6U standard VME card and will be designed to readout a single set of FMUX's.

ACKNOWLEDGEMENTS

We acknowledge the Natural Sciences and Engineering Research Council of Canada and the Texas National Research Laboratory Commission for their support of this research, and the Canadian Microelectronics Corporation for their assistance in the ASIC design and fabrication.

REFERENCES

1. The SDC Collaboration, "SDC Technical Design Report," SSCL-SR-1215 (1992).
2. F. Newcomer, R. Van Berg, H. Williams, S. Tedja, and J. Van Der Speigel, "Front End Signal Processing Electronics for the SDC Straw Tracking System," these proceedings.
3. Y. Arai, T. Matsumara and K. Endo, "A CMOS four channel × 1K time memory LSI with 1-ns/b Resolution," *IEEE Journal of Solid State Circuits* 27:359 (1992).
4. A. Hölscher, K. Ragan, P. Sinervo, G. Stairs and K. Strahl, "Data Collection Circuit for the SDC Straw Tube Tracker," UTPT-92-15 (1992).
5. The Verilog simulation and modeling language is available from Cadence Design.
6. See, for example, A. W. Booth *et al.*, SSCL-SR-1148, June 1992, and E. Hughes *et al.*, "Modeling and Simulation of the SDC Data Collection Chip," IEEE Trans. Nucl. Sci. April 1992.

THE ACCELERATOR SYSTEMS STRING TEST:

OBJECTIVES, PROBLEMS, RESULTS, ANECDOTES, AND REFLECTIONS

Tom Dombeck

Superconducting Super Collider Laboratory*
2550 Beckleymeade Ave.
Dallas, TX 75237

INTRODUCTION

The Accelerator System String Test (ASST) was a congressionally mandated milestone for the Superconducting Super Collider Laboratory (SSCL) to demonstrate powered operation of a half-cell of industrially-fabricated collider magnets before mass production could begin. In this paper I describe the installation of the string components and the results from the full-powered tests of the magnet string, consisting of five dipole magnets having an aperture of 50 mm and a single quadrupole having an aperture of 40 mm. Power and cryogenic connections were made to the string through spool pieces that were prototypes for SSC operations. The string was cooled to cryogenic temperatures in early July, 1992 and power tests were performed at progressively higher currents up to the nominal SSC operating point near 6500 Amperes.

ASST COMPONENTS

The construction of the ASST buildings and installation of critical equipment was completed in a little over one year's time in a rural area of Texas rangeland that was the first SSCL site to be developed. The cryogenic refrigerator began installation in March and was commissioned in April of 1992. Magnet and spool piece installation in the string took place over the period from February to early June. Installation and commissioning of the power supply, quench protection, and controls equipment was completed in late June. By the beginning of July, the string had been cooled to cryogenic temperatures, the final safety reviews had taken place, and the final stage of commissioning the magnet quench protection system leading to the power tests of the string had begun. Power tests were performed at progressively higher currents over a three week period leading to the test at the nominal SSC operating point of 6500 Amperes in mid-August. The completion of the ASST milestone is a continuation of string tests dating back over 15 years as shown in Table 1.

*Operated by the Universities Research Association, Inc., for the U.S. Department of Energy under Contract No. DE-AC35-89ER40486.

Table 1. Significant Superconducting Magnet String Tests that Have Been Performed:

ESCAR (LBL)

November 1977 to June 1978
Twelve 1-m Long Dipoles

B12 (FNAL)

1976 to 1981
One 3-m Long Dipole, Up To Sixteen 7-m Long Dipoles

A SECTOR (FNAL)

January 1982 to June 1982
One-Eighth of the Tevatron Ring

CBA/ISABELLE (BNL)

February 1982 to May 1983
Six 4.4-m Long Dipoles and Two 1.4-m Long Quadrupoles

HERA (DESY)

1988
Three 9-m Long Dipoles and Two 2-m Long Quadrupoles

RHIC (BNL)

1990
Two 9-m Long Dipoles and Two 1-m Long Quadrupoles

SUPER COLLIDER (FNAL)

1990 to 1991
Two to Five 17-m Long Dipoles (40-mm Aperture)

SUPER COLLIDER (SSCL)

1992
Five 15-m Long Dipoles (50-mm Aperture) and
One 5.2-m Long Quadrupole (40-mm Aperture)

The half-cell of magnets comprising the ASST is the basic subunit of the SSC collider. A general description of its cryogenic and electrical operation is discussed in ref. [1]. The ASST string was designed to be as close as possible to a prototype half-cell for the collider in order to test the general features of magnet interconnection to determine cryogenic heat loads under different operating conditions, and to demonstrate the safe operation of the string under normal powered conditions and during magnet quenches. The dipole magnets were fabricated by the General Dynamics Corporation using tooling in place at Fermi National Laboratory (FNAL). The quadrupole magnet was fabricated at Lawrence Berkeley Laboratory (LBL) and placed in its cryostat at the SSCL.

The string was assembled in the front half of a 200-m long enclosure building with associated electronics installed in a niche area off to the side of the string and in a controls trailer outside the enclosure. The refrigerator was installed in a building attached to the enclosure with its compressor and storage dewars mounted outside on a concrete pad. The power supply and its associated equipment were installed in a room about 80 m away from the string. The basic parameters for the ASST components are summarized in Table 2.

Table 2. Parameters of the ASST Magnets and Equipment.

Dipole Magnet

Inductance	75 mH
Aperture	50 mm
Length	15 m
Operating Field	6.5 T
Nominal Operating Current	6506 A
Stored Energy	1.6 MJ
Quench Heater Energy	3.5 kJ/heater pair
MIITs Limit on Inner Coils [2]	20
MIITs Limit on Outer Coils [2]	15
Mass	13.6 metric tons
Anticipated Cold Mass Heat Leak	363 mW

Quadrupole Magnet

Inductance	7 mH
Aperture	40 mm
Length	5.2 m
Operating Field	206 T/m
Nominal Operating Current	6506 A
Stored Energy	0.15 MJ
Quench Heater Energy	1.2 kJ/heater pair
MIITs Limit on Inner Coils [2]	12
MIITs Limit on Outer Coils [2]	10
Mass	1.32 metric tons

Spool Pieces

Mass Flow on Power Leads	2 g/s

Refrigerator

Mass Flow	54 g/s
Liquefaction Rate	150 l/hr

Power Supply

Maximum Voltage	10 V
Maximum Current	8,000 A
Maximum Ramp Rate	6 A/s
Dump Time Constant-all 6 Magnets	24 s
Dump Time Constant-3 Dipole Magnets	13 s

RESULTS OF THE POWER TESTS ON THE ASST

A detailed account of the operating principles of the string components and the cryogenic data and the results of the many power tests and magnet quenches induced in the string are presented in ref. 3. Throughout the program to bring the string progressively to higher currents we did not observe any natural quenches in the magnets, though we induced quenches at various current levels to study the response of the quench protection (QPM) and cryogenics systems. We also studied quench propagation from one magnet to another. For instance in a quench at 5500 A induced in the quadrupole magnet, the quench was isolated to the last three magnets in the string. However in a test at 4500 A where the quench was induced in the third dipole magnet, all six of the magnets in the string quenched in what appears to have been the result of thermal propagation of the quench from the first three magnets to the last three magnets.

The quench at 5500 A was initiated in one of the pairs of strip heaters in the quadrupole magnet and was quickly detected by the QPM and activated the firing of all the heaters in the fourth and fifth dipole magnets as well as the other set of heaters in the quadrupole magnet. The power supply was turned off and the dump switch opened placing the dump resistor into the circuit to extract the current still flowing in the first three dipole magnets in the string that remained superconducting throughout the whole quench sequence. The string current exhibited an exponential decrease with a 12 s time constant, consistent with only half of the magnets and therefore one half of the inductance of the entire string as opposed to the 24 s time constant expected for a dump of the whole string. The maximum number of MIITs generated in the dipole magnets was 7.5. The MIITs in the quadrupole magnet was 4.7. This was well below the estimated cable limits given in Table 2 for these magnets.

No matter in which magnet in the back half of the string we initiated the quench, the largest voltages were generated in the fifth dipole (DCA316). We also observed the highest voltages in the third dipole magnet (DCA319) if the quench was initiated anywhere in the first half of the string. We believe this was due to the smaller Residual Resistivity Ratio of the copper used in the outer coil cables in these two magnets that was about one-half the value of the cable used elsewhere in the string. The smaller ratio implies a higher resistance in the copper that carries the current after the superconducting strands in the cable go normal in a quench and therefore higher resistive voltages will be generated.

The pressures generated in the string in the 5500 A magnet quench reached a maximum of about 0.83 MPa occurring as a pressure spike just after the quench. It was observed in the interconnect region between the fourth and fifth dipoles and was caused by the rapid heating and expelling of the helium vapor from the magnet coils where the quenches occurred. This pressure measured about twice that seen in single magnet tests at FNAL because of the rush of hot gas from the quenching dipoles on either side of the interconnect in the string.

The highest cold mass temperatures reached above 9 K and appeared in the quadrupole and the fifth dipole magnets immediately after the quench. This reflects the flow of hot gas down the string toward the quench valve in the end spool piece. After exiting the string, this hot helium gas was diverted to the refrigerator storage dewar. Immediately the refrigerator began to recool the string requiring about 3.5 hrs to stabilize the string temperature to that where current could be run again. Much of this recovery process was controlled manually by the refrigerator operators.

In mid-August the string was ramped up at 4 A/s to 6000 A and the current held steady for many minutes. Then the string was ramped up at 2 A/s to a current of 6520 A, a little above the ASST milestone goal of 6500 A. Again the current was held constant for a number of minutes and then the string was ramped down at 4 A/s. There were no quenches during this sequence and the string components performed as designed.

REFERENCES

1. J. R. Sanford and D. M. Matthews (eds.), *Site-Specific Conceptual Design of the Superconducting Super Collider*, Report Number SSCL-SR-1056, July, 1992.
2. J. Nonte (ed.), *Proc. of the Third International Industrial Symposium on the Super Collider* (Plenum Press, New York,1991).
3. W. Burgett et al., Full-Power test of a string of magnets comprising a half-cell of the Superconducting Super Collider, Report Number SSCL-Preprint-162, October, 1992 (Accepted for publication in *Particle Accelerators*).

THE ACCELERATOR SYSTEMS STRING TEST PROGRAM

P. Kraushaar

Superconducting Super Collider Laboratory*
2550 Beckleymeade Avenue, MS 6000
Dallas, Texas 75237-3997

INTRODUCTION

The Accelerator Systems String Test (ASST) complex is located at the N15 site in Ellis County Texas. The complex was initially constructed to demonstrate the operation of a standard half-cell of the Collider machine lattice using prototypical superconducting magnets. The ASST half cell consisted of five 50 mm aperture dipoles, one 40-mm aperture quadrupole and three spool pieces. This demonstration was a Congressionally-mandated milestone for the Superconducting Super Collider Laboratory (SSCL) and was scheduled for completion by October 1, 1992. The milestone test (Run 1) was completed with the successful powering of the half cell to 6500 A in mid-August, 1992.[1]

ASST PROGRAM MANAGEMENT

The completion of the milestone marked a transition point for the ASST management structure. The task force organization that was used for the milestone effort was focused on accomplishing a single task. When that task was completed, the organization was dissolved and the ASST entered a new phase. The planning for this transition was started well in advance of the completion of the milestone. The new organization had to focus on the operation of a test facility rather than accomplishing a single task as before. The philosophy that drove the new organizational structure adopted for the ASST was simple. The SSCL required a facility where technical components could be integrated into collider prototypical systems and subsystems for testing and operation. Without this capability, the SSCL could not meet the quality assurance goals stated for the laboratory by test verification of the superconducting accelerator's [the Collider and High Energy Booster (HEB)] level 3B specifications.[2,3] In addition to providing a test bed for systems and personnel training, the ASST test program needed to provide for testing critical component parameters that could not be verified in single component testing. One example of this would be the heat leak to the 4 K cryogenic circuit in the superconducting magnets. Another would be the response of spool piece components to magnet quenches.

---

* Operated by the Universities Research Association, Inc., for the U. S. Department of Energy under Contract No. DE-AC35-89ER40486.

The Collider Machine Group was given the responsibility for the program management and the initial operation of the ASST. The ASST Test Group was formed from members of the Collider Machine Group and the Project Management Office (PMO) who participated in the milestone effort, and the support personnel from the ASST Task Force. The principle positions within the test group are the ASST Program Manager, the Program Physicist, Program Engineer, Program Analyst and the Systems Engineer. These people are assisted by six support staff members. The established reporting structure has the members of the test group reporting to the program manager, with the program manager reporting to the Collider Machine Leader who reports to the Deputy Project Manager for Superconducting Accelerators.

The ASST is an operating facility at the SSCL and as such, requires test technicians and operators to function. This type of personnel cannot be provided by the Collider Machine Group or PMO in general. The laboratory has recently established the Accelerator Operations Department (AO) under the Deputy Project Manager for Accelerator Operations within PMO. This department will ultimately be responsible for the operation of the Collider and the injector accelerators once they are commissioned by the machine groups. In the interim, AO provides the safety and operations oversight for the ASST. Eventually, AO will provide the test technicians and operators for the string test facility along with the operations management function. Until AO has the staff to provide these positions, the ASST Test Group will rely on the technical divisions, Magnet Systems Division (MSD), and Accelerator Systems Division (ASD), for technician support. Current staffing requests call for a core group of six test technicians to support the installation and checkout of string components. This group would then function as string operators during the testing phase of operations. This staffing requirement is now being filled by technicians from the Cryogenics, Mechanical Engineering, and Electrical Engineering Departments of ASD.

Two committees were established to assist the ASST Test Group with the program and operational aspects of the string test facility. The first was the ASST Safety Review Committee. This committee is composed of members of the SSCL not directly involved in the day-to-day operations of the string test. They are responsible for reviewing changes in component configurations or operating conditions at the string test with respect to compliance with the ASST Safety Analysis Report (SAR) and general safety requirement of the SSCL. Prior to commissioning the string for operation, the committee reviews any changes that have taken place since the last test run. If the committee is satisfied with the configuration and proposed operating envelope, a recommendation is made to the Deputy Project Manager for Accelerator Operations to grant the test group a permit to operate. During the commission phase, conditional permits to operate are often given prior to the general one.

The second committee established was the Program Steering Committee. This group is chaired by the ASST Program Manager and is composed of representatives from the superconducting machine groups in PMO, and the technical departments within ASD and MSD. The charge to this committee was to establish the test program direction by the review of test requests submitted to the ASST Test Group and the assignment priorities to those requests based on technical merit and the SSCL project requirements. The testing schedule, run configurations and integrated test plans are also reviewed and approved by the program steering committee. This group reports to the Deputy Project Manager for Superconducting Accelerators.

THE TESTING PROGRAM

In order to facilitate the submission and evaluation of test requests for the ASST, the test group developed a test request procedure and forms.4 During the period from August, 1992 through April, 1993, over 50 test requests were received by the test group. In the beginning, the type of testing that was appropriate for the ASST was not well understood by the user community at the laboratory. Of the first 20 requests received, the steering committee rejected five of them. The rejected proposals generally involved measurements which could better be performed on single component test stands. Most test requests are now approved for integration into the testing program. The majority of the test requests come from the technical divisions and not the machine groups. ASD accounts for 50% of the submitted requests, followed by PMO with 28% and finally, MSD with 22%. However, test requests from PMO are largely generated by the ASST Test Group and are usually system level tests. These requests generally define the basic string configuration for a run, the testing priorities and require the most resources to support.

Roughly, 80% of the test requests fall into two categories of testing. These catagories are the measurement of the thermal and electrical characteristics of the string system and components. The thermal tests typically focus on determining the heat leak to the various cryogenic circuits that exist in the dipole and quadrupole magnets, the spool pieces and the interconnect region between string components. The heat leak budget for these components is well defined and the capacity of the collider cryogenic system is based on those values. If the accelerator components fail to meet the defined budget, the capacity of the present design of the refrigerator system will be inadequate. During Run 2, which took place from November 1992 through January 1993, several weeks of testing were devoted to thermal measurements. During this period, heat leak measurements were made on the dipole magnets and the SPR spool. Measurements were also made of the temperature distribution of the heat shields in the interconnect region. The results of the heat leak measurements on the half-cell configuration acquired in Runs 1 and 2 indicate that the heat leak to the 4 K circuit in the dipoles is significantly over budget. These results are fully discussed in Reference 5.

Test requests for electrical measurements generally involve the magnet power system (CECAR), the quench protection system (QPS) and the response of string components to a magnet quench. During Run 2, tests were conducted on the energy dump time constant, the energy dump switch, the SPR spool bypass lead, the power supply regulation, the quench detection threshold, the down ramp rate response of the magnets, and the quench response of the string to quenches induced by spot and quench heaters. The principle results from the electrical tests for Run 2 are reported in Reference 6.

The initial full cell testing program (Run 3) at the ASST will start in June 1993 and last for approximately five months. The priorities for this run as determined by the ASST Program Steering Committee are first, the thermal measurements of the string using a highly instrumented dipole magnet (DCA323) and SPR spool piece; second, the measurement of the quench response of a full cell at 4.25 K and third, the testing of two prototype SPR spools designed and fabricated by vendors. The initial five month period of testing will address the first two priorities for the full cell program. The third priority, the testing of the two vendor spools in the string, will require an additional four to five months of effort.

FUTURE PROGRAM PLANS

The test program at the ASST has thus far used magnets designed by the laboratory and fabricated at either Fermi National Accelerator Laboratory (FNAL) or at Brookhaven National Laboratory (BNL) by industry. The data collected from these tests will be fed back into the design process of the industrial magnet vendors. The Collider Machine Group still has the responsibility to verify that the magnet designs produced by industry meet the Collider Level 3B Specifications. Planning is in progress for a full cell test of industry designed and built collider dipole and quadrupole magnets. This test would take place in the first half of FY95. The goals of this string test would be to verify the thermal and electrical performance of the magnets and spools, to test the magnet stand design and installation techniques, and to provide a training exercise for the installation contractor selected to install the Collider components. This string test would be the final opportunity to test these components and procedures before the underground installation is started for the Early Cryo Loop Test.

The Early Cryo Loop Test is another M1 milestone for the Collider project and is currently scheduled for March 1996. This is a string test of four cells of the collider lattice installed in the tunnel starting at the N15 and proceeding northward. Planning for this test is in progress. The ASST Test Group will be responsible for the test plan and management of this effort. Since the Early Cryo Loop Test will be underground in the collider tunnel, the ASST surface facility will be available for other testing programs. The HEB Machine Group is in the process of defining such a test program. Currently, the HEB string test effort is being planned to start in FY96 and last for approximately one year. Given the shorter lengths of magnets, the HEB string test could be several cells in length.

SUMMARY

The SSCL has established a testing facility at the ASST complex dedicated to providing the scientific and engineering staff of the laboratory with a facility in which to test superconducting accelerator systems, subsystems and components. Two data-taking test runs have been completed using laboratory designed Collider prototype magnets in a half-cell configuration. A third testing period is scheduled for the summer of 1993 using a full cell of these prototype magnets. Future plans include the testing of a full-cell string of Collider magnets based on the industrial design. The ASST testing program for the Collider will culminate with the Early Cryo Loop Test, a SSCL M1 milestone tentatively scheduled for March, 1996. The HEB Machine Group has formulated plans to conduct HEB magnet string tests at the ASST beginning in FY96.

REFERENCES

1. W. Burgett et al., "Full-Power Test of a String of Magnets Comprising a Half-Cell of the Superconducting Super Collider," SSCL-Preprint-162 (1992), submitted for publication in "Particle Accelerators".
2. Project Management Plan for the Superconducting Super Collider, SSCL Document No. P40-000021.
3. Element Specification-Collider Accelerator Arc Sections, SSCL Document No. E10-000027.
4. Accelerator Systems String Test Request Procedure, SSCL Document No. E10-0000103.
5. W. Burgett et al., "Cryogenic Characteristics of the SSC Accelerator Systems String Test (ASST)," Proceedings of the Fifth Annual 1993 International Industrial Symposium on the Super Collider.
6. W. Burgett et al., "Electrical Performance Characteristics of the SSC Accelerator Systems String Test," Proceedings of the Fifth Annual 1993 International Industrial Symposium on the Super Collider.

THE ASST CRYOGENICS

G. T. Mulholland

Superconducting Super Collider Laboratory[*]
2550 Beckleymeade Ave.
Dallas, TX 75237-3997

INTRODUCTION

The ASST magnet string cryogenic refrigeration requirements were planned around an original design and a much smaller, backup, He refrigeration system. The ASST schedule required that the backup, or Plan B, helium refrigerator provide and meet all the requirements of the milestone test. The Plan B design, layout, sub-system commissioning tests, and the performance schedule will be provided. The magnet string cryogenic system pump and purge, cooldown and warmup, central and multiple shield cooling, temperature control and subcooling, and recooler operating experience are reported.

The ASST cryogenic system static performance and the dynamic provisions, response, and recovery to magnet quenches will be described.

SYSTEM DESCRIPTION

The Plan B backup system is a basic CCI Cryogenics model 600 Helium Refrigerator/Liquefier (R/L) outfitted, modified and augmented to provide the cooling for the Accelerator Systems String Test (ASST) half cell string test program.[1] The basic 600-W R/L was modified to allow for series operation of the low temperature expander. Additional equipment included an auxiliary cold box containing a precooler, subcooler, cold compressor, and a boil-off shielded 5,000-liter helium dewar. Custom, 50 g/s supercritical helium feed and return, helium recooler vapor return, 20-K shield return, and quench helium recovery lines and controls provisions were made to support the full energy ASST testing program.[2]

The system includes a 70 g/s Sullair, CM25, 17.5 bar, compressor, special oil removal package, two temperature level, single piston, expander main cold box, cold compressor equipped auxiliary cold box, dewar, associated transfer lines, special in-line quench valving,

---

[*]Operated by the Universities Research Association, Inc., for the U. S. Department of Energy under Contract No. DE-AC35-89ER40486.

200 m^3 warm gas storage, and a set of Moore 351 and 352 electronic controllers. The SSC designed and provided the custom VJ system to provide the string connection and its isolation.[3] The equipment layout was constrained to fit in the magnet "laydown" area, located to avoid the Plan A distribution box, and the auxiliary cold box and the storage/recovery helium dewar coupled as closely as practical to the magnet string feed spool. The string and refrigerator were independently operable, after a brief connection installation period.

COMMISSIONING, ACCEPTANCE TESTS

The installed equipment was commissioned and acceptance tested using a calorimeter internal to the precooler and an outlet-to-return VJ jumper to measure refrigeration, and a system gas makeup measurement to measure liquefaction to the 5,000 liter dewar. The measured values were 550 W and 135 l/hr. These values include the load of the auxiliary cold box (refrigeration), dewar and associated lines (liquefaction), but do not include the ASST connecting lines.

ASST CONDITIONING

The 100–m long magnet string was helium pumped and purged to a few ppm of air components, but the water content could not be reduced below approximately (ca.) 50 ppm. In the interests of the schedule the refrigerator cleaned up this situation. This proved successful and constant monitoring of GN_2 and water measured $\leq$ 7 ppm throughout all the subsequent running.

COOLDOWN

The string was cooled down twice from room temperature and once from nitrogen temperature. The total time to reach the string operating temperature was ca. six days in each major cooldown. The cooldowns were preceded by the string and auxiliary cold box nitrogen shield cooldowns.

The nitrogen temperature cooldown rate was to be limited to 20 g/s He flow to keep the gas below the fluidization velocity of the charcoal Neon Adsorber at 80 K.[4] The cooldown rate was later measured by energy considerations to be closer to 25 g/s. The ca. 1.3 magnet long, 80–300 K, temperature wave front moved through the string in ca. 60 hrs. Note that the lower the cooldown flow, the greater the magnet temperature gradient. The 25 g/s temperature gradient ($\approx$ m$^{-0.2}$) is 1.31 times that expected for the 100 g/s collider specified nitrogen temperature cooldown flow[5], i.e., the slower cooldown is more stressful.

The low temperature expander flow can be operated in series with the load or recycled to the low pressure side. The choice implies a flow fixed by the engine speed and inlet conditions, and an outlet temperature fixed by the engine inlet temperature, or an adjustable outlet temperature. We chose the adjustable outlet temperature in the initial ASST cooldown. The expander cooldown was operated to approximately halve the source temperature with each successive cooling wave, after the first was set to 50 K. The expanders can deliver 2 kW down to an outlet temperature of ca. 7 K. The string expander cooled to operating temperatures in ca. 40 hours.

The first cooldown was conducted at string pressures low enough to allow two phase conditions in the string. The onset of significant two phase quantities and liquid helium surges down the inclined (1 m/200 m) string, led to periodic, widely fluctuating, return flows that strained the compressor pressure control response and range. Pressurization above the critical pressure when the low temperature expander was reconfigured in the series arrangement eliminated the variations.

20 K, 80 K SHIELDS

The 20-K shield flow is provided through a JT valve from the source supercritical helium line at the far end. The flow was fixed by a manual valve and varied from 1–2 g/s dependent on the mode of operation and detailed shield temperature requirements.

The 80 K shield flow was nominally two phase nitrogen, but could be run on saturated nitrogen vapor to support shield heat leak measurements.[6] The saturated vapor generator used could not provide low flow string source temperature values below ca. 90 K. It will be replaced for subsequent runs with a significantly improved version.

SUBCOOLED TEMPERATURE OPERATION

The cold compressor in the auxiliary cold box is a motor driven piston expander with compressor suction and discharge check valves. Its suction is arranged to reduce the return pressure on the subcooler vessel and the shell side of the string feed spool and spool piece recoolers. The subcooler pressure is directly reduced by the cold compressor, while the recooler is effective only if filled to an operating level with liquid helium. The ASST string has been cooled to a vapor pressure temperature of 3. 8 K, 0.63 bar.

RECOOLER OPERATION

The recooler system functioned as expected at a load of 54 W once the liquid level was established.[7] The nominal 100 W duty of the recooler was not tested.

LOAD SUMMARY

The 4 K equivalent load for the first assembly of the ASST as measured or estimated is listed in Table 1. The 20 K shield was provided as a liquefaction load (ambient temperature return) at a considerable refrigeration load to allow accurate room temperature flow measurement. Note the close approach of the total to the full Plan B capacity.

Table 1. The 6500 A load distribution summary of the first assembly.

• Connecting Transfer lines, in-line valves	30 W	
• (5) Dipoles,(1) quad magnet and		
(3)Spools (2 with a 7 kA power lead pair)	102 W	
• 20 K shield line (2 g/s @4 K)	200 W	
• (4) 120 scfh (air) lead flows, (1.6 g/s@4 K)	160 W	
• SCHe return line	10 W	est.
• He Vapor return line	10 W	est.
• He Dewar and lines	10 W	est.
Total	522 W	est.

QUENCH RESPONSE

The refrigerator quench response is initiated by a quench signal and, in default, by the rise of magnet pressure above 5 bar. The magnet return is isolated from the refrigerator LP side, the 20-K return is isolated from the refrigerator, and the 20 K/quench line is opened to the He dewar through a 2", cooled, VJ line within a second of the receipt of the quench signal. The quench driven He is shunted to the dewar, and the refrigerator LP side is isolated from the high pressures associated with the magnet quench. Each control function is paralleled by one or more reliefs to assure system safety in the event of a controls malfunction. The Plan B refrigerator quench control scheme functioned flawlessly throughout the 34 quench program.

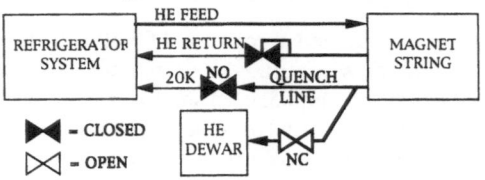

Figure 1. He Quench Recovery Arrangement.

Quench recovery requires a) recovering and refrigerating the contents of the dewar, b) cooling the magnet string, and c) refilling the string operations.

CONCLUSION

The Plan B backup R/L provided the necessary equipment capability, capacity, reliability, and schedule to allow the completion of the ASST half cell, full current, milestone test ahead of schedule. It served to test concepts and understandings, and supported a significant quench study program up to, and including, 6,500 A full string quenches. Although lacking the capacity to make rapid recoveries from quenches, and falling short of containing all the quench helium above ca. 5500 A quenches because of the limited dewar size, it successfully filled an important role scheduled for a unit eight times larger. The complete dedication of Richard Ahlman, Harry Carter, the Plan B crew, and a willing and resourceful supplier, CCI Cryogenics, are directly responsible.

Plan B R/L has since been retired from the N15 site and will return, as the Phoenix from the ashes, as the helium refrigeration source at the SSCL Central Facility for the test stands of the Spool Piece Test Facility in July.

REFERENCES

1. T. Dombeck, et al., "Cryogenic and Power Testing of 50-mm Aperture Dipole Magnets at the SSCL," DESY Conference, August, 1992.
2. W. Burgett, et al., "Full-Power Tests of a String of Magnets Comprising a Half Cell of the SSCL," SSCL Preprint 162 (1992), accepted for publication by Particle Accelerators.
3. G. Hanowski, "Plan B-ASST Transfer Line Design," SSCL Tech Note, March, 1993.
4. A. Winters, CCI Cryogenics, Private Communication, May 11,1992.
5. A. Yucel, "Effect of helium mass flow on cooldown," SSCL Memo, January 8,1991.
6. W. Burgett, et al., "Cryogenic Characteristics of the SSC ASST," IISSC paper, May, 1993.
7. A. Gulperi, A. Scheidemantle, "Recooler Test Data Analysis-ASST," Draft, January 11,1993.

ELECTRICAL SYSTEMS FOR THE
ACCELERATOR SYSTEMS STRING TEST

Gerry Tool

Gerry Tool & Co., Inc.
Sister Bay, WI 54234-9605

ABSTRACT

The Accelerator Systems String Test (ASST) is a facility for development and test of the many magnet associated systems required in the Collider and High Energy Booster (HEB) accelerators. This paper describes the magnet electrical system and the powering and quench protection systems initially installed in the ASST for Phase I, the congressionally mandated milestone.

The string of 5 dipoles, 1 quadrupole and 3 spools and the other systems were installed in the ASST enclosure in the spring of 1992, cooled down in early July 1992, and powered in late July and early August. The nominal Collider operating current of 6500 A was achieved on 14 August 1992 after operation at lower current levels successfully tested the operation of the magnet string, its power system and quench protection system, including intentional quenches in all of the magnets.

The electrical systems, the installation, checkout and operating history of the tests leading to completion of this key milestone for the SSC Laboratory are described.

INTRODUCTION

The ASST was designated as a vehicle for the accomplishment of a milestone mandated by Congress to demonstrate powered operation, in a colliderlike configuration, of a half cell of industrially produced prototype magnets. Successful completion of this milestone has allowed mass production of magnets and tunnel boring to commence.

Creation of this facility and operation of the systems to achieve the milestone represented one of the first major efforts of the Laboratory to move from the conceptual design, cost estimating and scheduling phase into the design, manufacture, installation and operation of a complex technical system. Several papers presented at this conference describe different aspects of this activity, and a comprehensive test report has been published[1].

SYSTEM CONFIGURATION

The ASST consisted of five Dipoles, one Quadrupole, three Spool Pieces and external systems as shown in the schematic, Figure 1, emulating a half cell of the Collider lattice.

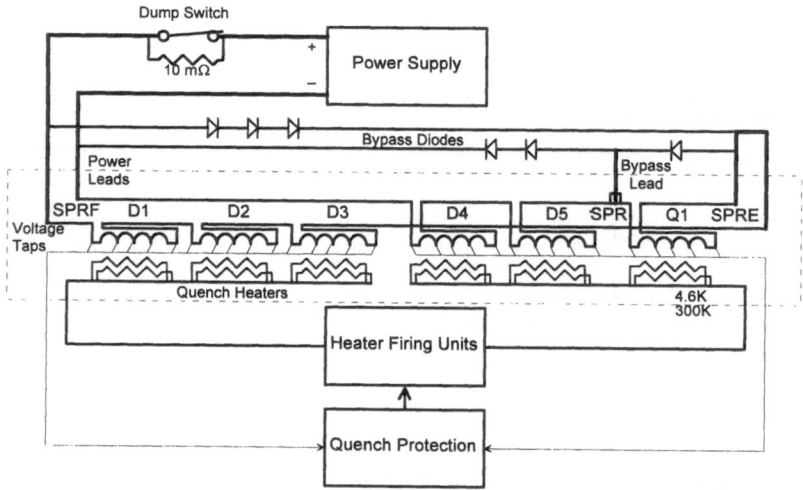

Figure 1. Simplified Electrical Schematic of ASST Phase I Configuration

The magnets were wired as two independent quarter cells to allow the quench detection response to propagate quenches into only half of the magnets, making observations of quench coupling between the two quarter cells due to electrical signal misinterpretation and cryogenic flow possible. Initial operation uncovered configuration faults that caused detection interpretation coupling, and higher energy quenches exhibited cryogenic coupling from an initial quench in the upstream quarter cell to the adjacent quarter cell.

The 40 mm aperture quadrupole is constructed of lower capacity superconducting cable then the 50 mm dipoles, so it requires an independent bypass to prevent energy stored in the adjacent dipoles from dissipating in the quad if it should be the initially quenching magnet. This feature required that the SPR spool which would normally be on the right side of the quad in a Collider half cell be placed between the adjacent dipole and the quad for this test. Major parameters of the configuration are given in Table 1.

Table 1. Major Parameters of ASST Electrical Systems

Magnet Inductance	380 mH Total
Individual Dipole	75 mH
Individual Quadrupole	7 mH
Maximum Operating Current	6500 A (8000 A P.S.)
Maximum Ramp Rate	± 4 A/s
Power Supply Voltage	10/20/40 V
Power Supply Configuration	12 Pulse SCR, 2 Quadrant, Passive Filter
Current Regulation	.01% of Full Scale
Energy Extraction Resistance	10 mΩ
Time Constant	38 s nominal (no quench)
Energy	8 MJ @ 6500 A
Voltage	65 V @ 6500 A

The Quench Protection System is an active system modeled after the Tevatron, using microprocessor based quench detection and control of redundant Heater Firing Units, passive external Bypass Diodes, and coordinated control of the Power Supply and Energy Extraction systems.

INSTALLATION, CHECKOUT AND COMMISSIONING

Individual component testing was very extensive, incorporating formal quality assurance and testing procedures and documentation, as did each step of the installation and system integration. All components attached to the main bus system were acceptance hipotted to 2 kV. Each interconnection of major components was followed by hipotting and configuration checking before proceeding to the next interconnect. The Power Supply, Bus, Energy Extraction and Bypass Diode systems were tested at full current before connection to the magnets.

After cooldown of the magnet string, the total system was hipotted, with breakdown occurring at less than 1000 V, an unexpected result. Tests identified the problem to be caused by poorly potted voltage tap connector pins on the Spool Pieces. Since each of these signal wires has a series current limiting 200 Ω resistor attached to the tap point inside the magnet, this breakdown did not threaten the magnets themselves, so a decision was made to proceed with power testing, with the operating procedure to include analysis of the voltage to ground after each quench event to verify that the operating levels were less than our hipot test capability.

A full dry run of all Quench Protection System functions was conducted prior to powering the magnet system. Initial powered operation was limited to currents of 500 and 1000 A during which the Energy Extraction system was used to check that all monitor signals were functioning and calibrated properly. At a current level of 2 kA, each magnet heater system was individually tested to verify that it would independently protect its magnet by purposely inducing twelve quenches in the magnet string and studying the system response after each quench. This testing uncovered several system configuration problems which were fixed prior to higher energy operation.

OPERATION

Each day of system operation was preceded by formal checklist completion, including analysis of the response to a simulated quench. After the 2 kA Quench Protection System verification, the operating current was raised in 500 A steps. At each new current, one or more quenches were induced by firing a heater in a selected magnet. The voltage to ground profile for the string was characterized and projected to the next operating level before proceeding to ensure that the hipot level would not be exceeded during the next run. This series of tests identified one magnet, D3 (DCA319), as the "worst case" magnet, from a voltage to ground viewpoint, in which to induce a quench[2]. Further analysis indicated that this magnet was one of the least likely to spontaneously quench as we raised the operating current, so after encountering a voltage to ground maximum of 714 V while quenching this magnet at 4500 A, we subsequently performed our test quenches on the quadrupole, which single magnet test data indicated was the most likely magnet to spontaneously quench as we approached the target operating current of 6500 A.

Successful quenching of the string at 5000 A and 5500 A provided confidence that the goal of 6500 A could be achieved without danger to the magnets from a spontaneous quench that might occur. Table 2 summarizes the important test quench parameters.

Table 2. ASST Phase I Test Quench Summary

Magnet-Htr	Date	I (kA)	T_q (ms)	Miits Dipole	Miits Quad	|VTG| max	Comments
D1-1	Jul 29	2	167	4.7		45	Both Buses (signal)
D2-1	Jul 31	2	167	6.6		29	
D5-1	Jul 31	2	167	4.6		42	Both Buses (signal)
D3-2	Jul 31	2	183	4.6		53	Both Buses (signal)
D4-1	Aug 1	2	167	6.6		24	
Q1-1	Aug 3	2	233	6.1	4.4	33	
D1-2	Aug 6	2	150	6.8		30	
D4-2	Aug 6	2.5	133	5.9	3.2	44	
D2-2	Aug 6	2.5	150	6	2.1	61	Both Buses (signal)
D5-2	Aug 7	2.5	133	5.7	2.9	97	
D3-1	Aug 7	2.5	150	5.8	2.5	122	Both Buses (signal)
Q1-2	Aug 7	2.5	217	4.7	2.9	61	
D5-2	Aug 8	3	133	6.7	4.1	138	
D1-1	Aug 8	3	133	7.3		86	
D3-1	Aug 10	3.5	117	7.4		357	
D5-2	Aug 10	3.5	117	7.9	4.2	211	
D3-1	Aug 11	4	117	8	3.8	531	Both Buses (cryo)
D3-1	Aug 12	4	117	8	4	522	Both Buses (cryo)
D3-1	Aug 12	4.5	100	8.7	5.3	714	Both Buses (cryo)
Q1-1	Aug 13	5	167	7.5	4.4	296	
Q1-1	Aug 13	5.5	150	8.2	4.7	337	

The milestone effort came to a successful conclusion, six weeks ahead of schedule, on the morning of August 14, 1992 when the current was raised to 6000 A at a rate of 4 A/s, held there for several minutes while joint resistance data was acquired, and then raised at 2 A/s to a level of 6520 A. The current was held at this level for several minutes and then returned to zero at -4 A/s. This initial run thouroughly proved successful operation of the magnet power and quench protection system design concepts and provided extensive system performance data in addition to achieving the congressional milestone.

POST MILESTONE OPERATION

Subsequently, the defective electrical connectors were replaced after warming the system up. The string was again cooled down and operated in a comprehensive Collider related power testing program, including high field quench events initiated with both quench protection heaters and spot heaters from October, 1992 through January, 1993. The ASST is currently undergoing extension of its configuration to include a full cell of magnets and Collider prototype Spool Pieces, and is scheduled to resume test operation in June, 1993. In future years, runs emphasizing High Energy Booster systems tests are planned.

REFERENCES

1. T. Dombeck, et al. "Full-Power Test of a String of Magnets Comprising a Half-Cell of the Superconducting Super Collider", SSCL-Preprint-162, October, 1992.
2. W. Robinson, et al. "Electrical Performance Characteristics of the SSC Accelerator System String Test", Paper IV-61 of this conference.

DATA ACQUISITION AND CONTROLS FOR THE SSCL ACCELERATOR SYSTEMS STRING TEST PHASE I

R. Bork, M. Christiansen, E. Faught, K. Goetze, D. Haenni, S. Lee,
D. Murray, J. Wang, E. Williams, M. Wylie, and J. Zatopek

Superconducting Super Collider Laboratory*
Accelerator Systems Division/Controls Department MS-4002
2550 Beckleymeade Ave.
Dallas, TX 75237

INTRODUCTION

The Accelerator Systems String Test (ASST) was a major milestone in SSCL R&D. Phase I demonstrated that the smallest repetitive sequence of bending magnets (a half cell) could be installed, leak checked, cooled to liquid helium temperatures, energized, and safely quenched. To support this activity, a control system had to be developed to operate and monitor cryogenic systems in the string, along with LCW and vacuum systems. Also, since this is a test facility, the magnet systems were heavily instrumented to provide sufficient data to confirm that design requirements were met, that the system was operating safely and as expected, and allow further design studies necessary for the construction of the SSC. This required the design and implementation of an acquisition system capable of collecting relatively large amounts of data, at various data rates, and presenting this data both to operations personnel and to a database for off-line analysis. In this paper, we describe the design, implementation, and operation of the data acquisition system and controls employed on the ASST.

SYSTEM REQUIREMENTS SUMMARY

The string components included a feed spool piece/recooler (HSPRF), five dipoles, a spool piece recooler (SPR), one quadrupole, and an end spool piece (HSPE). Instrumentation from these components included 120 cryogenic temperature sensors, 20 cryogenic pressure sensors, 24 vacuum gauges, 2 strain gauges, 3 linear position sensors, 32 accelerometers, 3 flow sensors, and 2 level sensors. Control devices included 2 immersion heaters, 6 on/off valves, 2 variable position valves, and 9 mass flow controllers. String electrical monitoring included 5 current transductors, 50 differential coil voltage taps, and 3 system voltages. The ASST specification dictated requirements for log data acquisition, transient data acquisition, and process controls to support this mix of instrumentation and

---

*Operated by the Universities Research Association, Inc., for the U.S. Department of Energy under Contract No. DE-AC35-89ER40486.

control devices. The interface configuration of the instrumentation had to be flexible to support the special requirements of individual test requests to be performed on the ASST.

The transient data acquisition system presented a challenge with channel count and per channel sample size. A maximum of 256 transient event recording channels were required for the half cell configuration of the ASST. Synchronous recording of all transient channels is initiated on a triggered event, such as a signal from the magnet quench protection system. Sampling rate and sampling time requirements varied from 2000 samples per second for 30 seconds to 10 samples per second for 30 minutes. High sampling rates were required for the magnet power system, the string electrical signals, and the accelerometers. Medium sampling rates were required for cryogenic pressure and valve actuator signals. Lower rates were used for string temperatures.

Data Acquisition Electronics

To meet high accuracy measurement requirements, a sensor excitation and signal conditioning subsystem was designed and fabricated for the critical instrumentation. This subsystem is based on an 8-channel excitation and conditioning board packaged as a shielded 9U Eurocard module. An integrated backplane on a 19 inch rack mountable crate accommodates eight of the 8-channel modules and a single computer interface card. All analog, digital, and DC power supply I/O is accomplished through a crate backplane with rear access connectors. The front panels of the 8-channel modules have analog test points, excitation polarity switches, and gain setting displays for each channel.

Each channel has a fully independent excitation source. The excitation circuitry features computer controlled polarity switching, front panel polarity override switch, DIP switch programmable excitation values, and a 1K ohm 0.01% precision reference resistor. The reference voltage developed across the precision resistor is conditioned by a programmable gain instrumentation amplifier stage and routed to the crate backplane independently for each channel.

The signal conditioning for a sensor voltage input consists of an independent programmable gain instrumentation amplifier stage. The amplifier output for each channel is routed to the crate backplane. The amplifier gain setting is displayed using front panel LED and is also available for readout by the computer interface card.

The sensor signals from the excitation and conditioning subsystem are cabled to 64 channel VXI relay multiplexers. Pre-conditioned signals from flow controllers, vacuum process controllers, pressure sensors, and other devices are connected directly to VXI relay multiplexer inputs. A 5 1/2 digit VXI voltmeter module measures the voltage from the selected multiplexer channel.

A second set of multiplexers and a second voltmeter are provided for the conditioned excitation reference voltage outputs cabled from the excitation and conditioning subsystem. During sampling of a high accuracy channel, the conditioned sensor voltage is connected to one voltmeter and the conditioned excitation reference voltage is connected to the other voltmeter. The two voltmeters are armed and the VXI instrument trigger is used to initiate a simultaneous reading of the sensor voltage and reference. The excitation source for the high accuracy channel is then switched to the opposite polarity by the VXI controller, and after some settling time a second pair of simultaneous voltage readings are taken. The positive and negative voltage readings taken from the sensor and the excitation reference are algebraically manipulated to eliminate offset voltages from thermoelectric effects, the instrumentation amplifier stage, and the voltmeter. A precise value for sensor resistance is then calculated from these results.

The ASST transient data acquisition requirements were met by the joint development of a specification for a VME module by the SSC Controls Department and Analytek, Ltd. of San Jose. Analytek then designed and built the modules (designated model 2032LC) for use in the ASST acquisition system. Eight of these 32-channel modules were connected to a single transient acquisition controller managing the configuration of the digitizers and the retrieval of sampled data at the completion of post-event sampling.

The VME and VXI crates associated with data acquisition are interconnected via ethernet to UNIX workstations. These computers act as file servers, database servers, man-machine interfaces, and data analysis stations.

Data Acquisition Software

Periodic data logging is performed at a sample rate of five minutes. This data is used in both the real-time monitoring of the condition of the string and for off-line data analysis of heat leak measurements and other slow processes.

Transient data recording captures the conditions and state changes of the string during a quench event or during a magnet current dump. In contrast to the periodic data logging, transient data is acquired at much higher sample rates for the same set of transducers, as well as additional transducers which record string voltages and currents developed during a quench event. Sample rates are assigned to individual transducers to meet specific instrumentation and data analysis requirements. Transient sample rates can be as high as 10kHz, and several thousand samples can be recorded for a given transducer during an event.

Data acquired for both the log and transient data acquisition is managed by and made available from a commercial Relational Database Management System (DBMS). Oracle was the DBMS selected for this purpose. The Application Programmers Interface of ORACLE was an important component in the implementation of our distributed Database access routines.

All Database access uses a library of routines written to provide application software distributed access to the information maintained by the database. Implemented using Remote Procedure Calls (RPC), these routines provide a set of function calls which allow configuration, log and transient data to be stored and retrieved from the database server. The calls were developed specifically for the data tables and data schema maintained by the DBMS.

These routines allow read and write access to channel configuration data, logged data, and transient data related to individual quenches. For each channel, the configuration database contains such channel attributes as its hardware address, its sampling rate, and engineering unit conversion data. The application software which runs in the real-time embedded systems uses the channel configuration information to set up the hardware and drive the real-time acquisition process.

The process of engineering unit conversion is implemented using an embedded RPN expression interpreter. This interpreter provides all standard math functions, as well as standard polynomial and Cheveychev polynomial evaluation of the Nth degree.

There were two major applications written to support the acquisition and recording of data for the log and transient data channels. The first is an application which manages the log data hardware and channels. This application, running in a Hewlett Packard Model 300 VXI system, configures and manages the multiplexers, digital multi-meters, parallel I/O cards, and counter card which comprise the hardware supporting log data acquisition. On request, this application reads the configuration of the log data channels from the DBMS for use in data acquisition, conversion and display.

The second major application of the data acquisition components of the system is the software which manages the transient data recorders. This software runs in a mode similar to the log data acquisition process described above, retrieving channel configuration information from the database, configuring and arming the Analytek recorder cards, and causing the data recorded in a triggered card to be written out to the database.

Data Analysis Software

The second major portion of this system are those components which are used by the operators and analyst to monitor and perform data analysis tasks on the logged data. It includes the DBMS described above; human interface which allows the user to select from the thousand or more channels maintained by the system some set on which they wish to operate; a trending and plotting package, and a commercial data analysis package (N!Power).

PROCESS CONTROLS SYSTEM

String cryogenics process control variables included power lead temperatures, recooler levels, shield temperatures, and valve positions. String vacuum process controls monitor 12 Convectron gauges and 12 cold cathode gauges. The LCW system included a cooling tower, a heat exchanger, a closed loop cooling water system, a deionization loop, and a backup water loop. LCW controls were programmed for supply pressure, temperature, and resistivity and for a full monitoring and alarm system with interlocks to the magnet power supply.

The string cryogenics, string vacuum, and LCW process control systems were implemented using TI (Siemens) model 545 programmable logic controllers (PLCs) and 505 series I/O modules. A separate PLC was programmed for each function. The string vacuum and LCW systems were configured in single 16-slot crates. The string cryogenics system required a second I/O crate with a remote base controller (RBC) to accommodate the number of I/O points involved. A TISTAR Model 20 system was used as a common operator interface for all three functions.

CONSTRUCTION, COMMISSIONING AND INITIAL OPERATION OF 2400W

REFRIGERATOR AND COLD TEST STAND FOR CDM TESTING

John D. Dubbs[1] and Ken Kreinbrink[2]

[1]Project Manager, Air Products and Chemicals, Inc.
Allentown, PA

[2]Manager Systems Engineering, CVI Inc.
Columbus, OH

INTRODUCTION

Air Products and CVI collaborated to design, construct and commission a refrigerator, test stands and integrated control system for the General Dynamics Collider Dipole Magnet Cold Test Facility (CTF) in Hammond , LA. The original project schedule required the cold test facility to be operational within 17 months of the notice to proceed. Midway through the project, changes in General Dynamics magnet testing requirements necessitated doubling the plant capacity, but the on stream date for the initial capacity increment could not be relaxed. The Air Products/CVI team had to adapt the project execution strategy to mitigate the schedule impact of the expansion in a cost effective manner without impacting system functionality, quality or safety.

An equally challenging aspect of the job was that the (CTF) was being designed while several major systems that would interface with the CTF were being engineered. General Dynamics, Air Products and CVI had to work very closely to manage the interface issues. The teams efforts were very successful. The Hammond refrigerator/liquifier was started up on schedule. The first two test stands are currently being commissioned and will be on stream just six weeks later than the pre-expansion schedule target and all four test stands will be operational in time to support General Dynamics magnet testing requirements.

PROJECT BACKGROUND

General Dynamics contracted with the Air Products/CVI team to supply a 6 g/sec LHe/600W integrated liquifier /refrigerator to support the testing of collider dipole magnets. Subsequently, two test stands and a test stand control system were added to the scope. When the decision was made to manufacture and test special dipole magnets at the Hammond facility, the refrigerator/liquifier capacity was increased to 12 g/sec LHe/1200W and an additional two test stands were added. This expansion included minimal changes to warm end equipment and modifications to the coldbox. A liquid helium circulation pump box and two additional test stands were fabricated.

The CTF interfaced with several other systems being designed concurrently with the CTF. The CTF design team had to anticipate interface requirements and react to changes necessitated by the design evolution of other systems while limiting impact of the CTF capacity upgrade on those systems. The primary areas of concern were: (1) the interface with the magnet shuttle delivery system which drove the test stand magnet handling design and dictated test stand/refrigerator location; (2) effective two way communication between the test stand control system and the Test Program Control Computer (which controls all magnet production tools) and the quench protection system; (3) the shop area available for the test stand equipment including DC power supplies and associated equipment. Ultimately to better control the detailed interface issues involved, installation of General Dynamics' supplied DC power supply equipment and other miscellaneous test stand area wiring was added to the Air Products CTF installation scope.

The milestones for the plant expansion and the DC power installation are compared against the original project schedule in Figure 1.

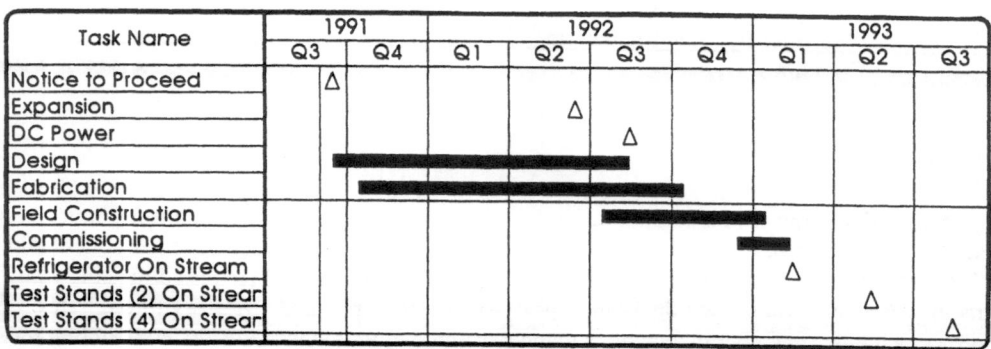

Task Name	1991		1992				1993		
	Q3	Q4	Q1	Q2	Q3	Q4	Q1	Q2	Q3
Notice to Proceed	Δ								
Expansion				Δ					
DC Power					Δ				
Design		▬▬▬▬▬▬▬▬▬▬							
Fabrication		▬▬▬▬▬▬▬▬▬▬▬▬							
Field Construction						▬▬▬▬▬			
Commissioning							▬▬		
Refrigerator On Stream							Δ		
Test Stands (2) On Stream								Δ	
Test Stands (4) On Stream									Δ

Figure 1. Hammond CTF Project Schedule

The schedule challenge is immediately apparent in that the decision to expand the facility was made nearly at the end of design and well into the fabrication. A staged approach to the installation was chosen. Installation of the full capacity refrigerator and two test stands would take place per the original schedule with the new equipment (pump box and additional two test stands) following in six months. To meet the refrigerator on stream date it was necessary to carefully analyze the impact of the expansion on design, fabrication, construction and commissioning. It was also necessary to carefully analyze the project schedule to determine the interdependencies between all activities to determine how more tasks could be accomplished in parallel to compress the schedule.

DESIGN

The key objectives of the design strategy were to: (1) minimize changes to long lead equipment items; (2) plan to start-up the refrigerator and two test stands with the later addition of two more stands and a helium circulation pump box; (3) ensure that the refrigerator commissioning could proceed in parallel with test stand installation and; (4) install the pump box and additional test stands with minimal impact on ongoing magnet testing.. The first objective was met when General Dynamics worked with URA to relax the delivery pressure requirement for magnet clean-up, cool-down, warm-up, and purge gas. This allowed the refrigerator capacity to be increased with no changes to the compression system and with a manageable impact on coldbox design. The second objective was necessitated by equipment delivery constraints. The "warm" plant areas were least affected by the expansion and could be delivered according to the original schedule. The coldbox required modification and would slip at least one month from the original target. The pump box was a completely new addition to the scope which could not be designed and fabricated in time to install for initial start-up. It and the additional test stands would be delivered six months later than the originally targeted on stream date. The third objective was included since the test stand area was most at risk for schedule delay due to the large number of systems integration issues affecting the design. The final objective was imposed since the General Dynamics magnet testing program would be underway and interruptions would have serious effects on those program goals.

Hazards and operability reviews were carried out with these objectives in mind. Appropriate modifications were made to the piping, instrumentation and structural systems to allow safe operation of the plant subsystems in stages while construction work continued.

FABRICATION

Since brazed aluminum heat exchangers, coldbox vessels and the coldbox head had already been fabricated, the design team was challenged to add two new exchangers and incorporate larger expanders in the cold box without exceeding the original vacuum can envelope. The modification had to be designed in such a way as to allow coldbox assembly to proceed almost to completion while waiting for the new exchangers. This was successfully accomplished and the coldbox shipped within six weeks of the original pre-expansion schedule. The delay in turbo-expander delivery caused by changing frame sizes was offset by installing the turbine casings in the field rather than in the fabrication shop. This increased the risk of encountering leaks internal to the coldbox vacuum can after expander installation, but was felt to be an acceptable risk/schedule tradeoff.

CONSTRUCTION

After award of any installation subcontract, a detailed resource loaded construction schedule was developed with the successful contractor. The schedule reflected equipment delivery constraints, constraints imposed by coordination with other construction work and the requirements of the commissioning schedule. This schedule became the basis for the weekly earned value monitoring of construction progress. Deviations from the baseline could be seen very early and corrective action taken to minimize impact upon successor activities.

Since key components (coldbox, test stand equipment) were to be delivered late in the construction phase, it was important that field installation effort be minimized. Wherever practical, pipe and structural steel were prefabricated in advance of the start of mechanical construction. Deliveries of large equipment items and prefab components were carefully coordinated with the mechanical construction activity. In almost all cases equipment was set directly from the truck to the foundation most within a one week period at the start of the mechanical construction.

Due to the small site size there was little opportunity for overlap of civil work and equipment installation. It was important to get the foundation and underground work done prior to the start of other construction activity. This was accomplished in eight weeks despite the interruption forced by Hurricane Andrew and continued abnormally rainy weather throughout the fall.

It was necessary to have a high degree of overlap between the mechanical subcontractor and the instrument/electrical subcontractor to meet the on stream target. This required considerable coordination of installation tasks but was a manageable problem. Electrical construction started only one month after mechanical and they finished almost simultaneously in each plant area. Construction safety was of particular concern under these conditions. Air Products practice is to only allow construction contractors with superior safety records to bid on subcontracts. All construction personnel attend safety training and compliance with all safety procedures is a precondition of continued access to the job site.

COMMISSIONING

It was recognized that there would be considerable overlap in construction and commissioning. To insure efficient coordination, the commissioning schedule was developed as a precursor to finalizing construction plans. A major goal of the Air Products construction supervisor and installation subcontractors was to not only to meet the final completion date for the entire facility but to consistently meet the intermediate milestone completion dates required to keep the commissioning effort on track. The operations plant manager was brought onboard early in the construction phase to assist in assuring that plant subsystems were completed as necessary to support checkout and startup.

Air Products pulled experienced operators and technicians from existing commercial helium liquefaction facilities to support the commissioning effort and operator training. The plant operating training was held just prior to the start of commissioning. The new operators then participated in the check out and initial operation of the plant equipment. The interaction with experienced staff and involvement in plant trouble shooting greatly enhanced the effectiveness of operator training.

To ensure a safe transition from installation to operation, an operational readiness inspection was held for each plant area prior to the start of commissioning. Once operation of any plant subsystems had commenced, remaining installation activity was controlled via a safety work permit system. The installation subcontractors were required to outline the work scope to be performed and any lockouts, tagouts or other precautions necessary for sign-off by the plant operator prior to each shift. The result was that there was only one minor OSHA recordable safety incident at the Hammond site and that was unrelated to start-up activities.

SUMMARY

The integrated management of construction and commissioning resulted in the refrigerator/liquefier being on stream on the pre-expansion target date. The refrigerator /liquefier is fully operational. The test stands and test stand control system are currently being commissioned and will be operational within six weeks of the original target date despite the difficult interface issues that were resolved during the design. The pump box and additional test stands are in fabrication and will be installed six months after initial refrigerator startup as per the plant expansion scheduled. That the refrigerator startup and operation were unaffected by the delay in the test stand area or the absence of the pump box is a reflection of the careful planning that took place.

The successful execution of the Hammond CTF project demonstrates how close teamwork between the General Dynamics, Air Products and CVI allowed a fast and cost effective response to changing customer needs. The importance of considering commissioning and startup requirements in the design and construction approach were borne out as was the necessity for proactive schedule management throughout all phases of the project.

COLD TEST FACILITY FOR 1.8 M SUPERCONDUCTING
MODEL MAGNETS AT THE SSCL

A. LaBarge, R. Althaus, R. Bird, J. Baron, J. Chagnon,
M. Deak, M. Scott, V. Vasilyev, and G. Williamson

Superconducting Super Collider Laboratory[*]
2550 Beckleymeade Ave.
Dallas, TX 75237

INTRODUCTION

A new facility has been constructed to measure the characteristic features of superconducting model magnets and cable at cryogenic temperatures—a function which supports the design and development process for building full-scale accelerator magnets. There are multiple systems operating in concert to test the model magnets, namely: cryogenic, magnet power, data acquisition and system control.

A typical model magnet test includes the following items: (1) warm measurements of magnet coils, strain gauges and voltage taps; (2) hipot testing of insulation integrity; (3) cooling with liquid nitrogen and then liquid helium; (4) measuring quench current and magnetic field; (5) magnet warm-up.

While the magnet is being cooled to 4.22 K, the mechanical stress is monitored through strain gauges. Current is then ramped into the magnet until it reaches some maximum value and the magnet transitions from the superconducting state to the normal state. Normal-zone propagation is monitored using voltage taps on the magnet coils during this process, thus indicating where the transition began. The current ramp is usually repeated until a plateau current is reached, where the magnet has mechanically settled.

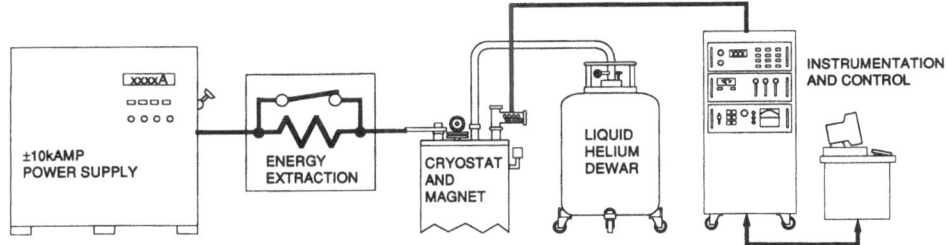

Figure 1. A simplified block diagram of the model magnet testing facility, illustrating current flow through the energy extract dump resistor, LHe supplied through portable 500 L dewars and the data acquisition and control system.

Many variations on the current ramping sequence are used to study different phenomena associated with magnet performance, *e.g.* magnetization hysteresis, eddy current losses, cryogenic stability, etc.

A warm bore cryostat with a rotating coil is inserted in the magnet to measure field strength and homogeneity. These types of measurements yield multipole and current versus field data.

---

[*]Operated by the Universities Research Association, Inc., for the U.S. Department of Energy under Contract No. DE-AC35-89ER40486.

CRYOGENIC SYSTEM

The testing vessel is an open mouth helium dewar designed for vertical testing of short dipole and quadrupole superconducting magnets. The nominal clear ID of the dewar is 711 mm except at the lambda plate support channel where the ID is 664 mm. The working depth below the Lambda plate is approximately 2400 mm and the overall height of the dewar to the top-plate is 3529 mm. The associated top-plate assembly includes a subcooler, Lambda plate and HE II heat exchanger to permit magnet operation over the temperature range from 4.5 K down to 1.8 K at 1.0 to 1.3 bar. The dewar is equipped with a liquid nitrogen-cooled shield to reduce the heat leak to liquid helium. The liquid nitrogen system terminates in an external keepful which automatically controls venting without release of liquid.

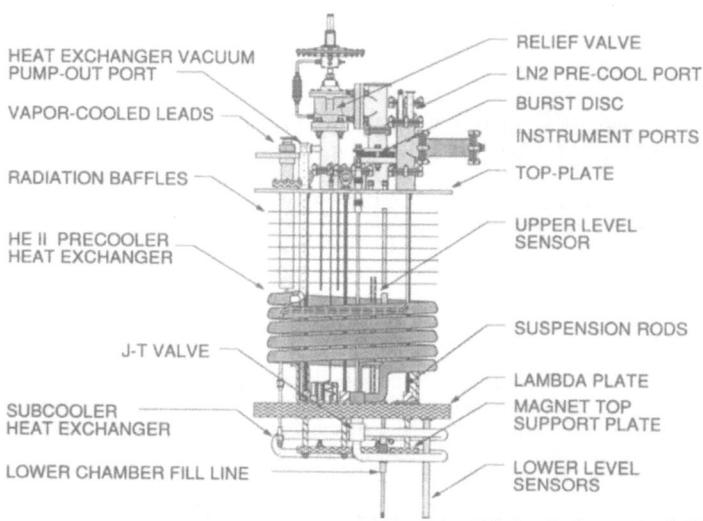

Figure 2. The top-plate assembly contains all of the active elements of the cryostat and provides an interface for the cryogenics and instrumentation. The magnet under test is suspended below the Lambda plate.

Additional features of the cryostat include a manual Joule-Thomson valve to control flow from the subcooler to the HE II heat exchanger, a vacuum pumping line for 10 to 12 Torr vapor from the superfluid heat exchanger, two superconducting liquid level gauges above and two below the Lambda plate, a Lambda plate relief device, and a capillary line for pressurization of the HE II volume.

POWER SYSTEM

A Dynapower four quadrant ±10 kA power supply is used to energize the model magnet. It is series connected to the dump resistor which provides 1 MJ of energy extraction. The ±10 kA power supply has two six (6) pulse secondary silicon controlled rectifier (SCR) cycloconverters in series operating in a circulating current mode. The passive Prague filter with split-leg choke and active *ripple bucker* together provide excellent noise performance with a measured voltage ripple of less than 3.0 mV. Three primary taps are available on the transformer to configure the output voltage to ±6 V, ±12 V, and ±24 V. The current is reduced to ±5 kA at the ±24 V setting.

Power supply stability and regulation is maintained by an ultra-stable zero flux current transducer (ZFCT) with temperature controlled burden resistor, digital-to-analog converter and summing amplifier. The ZFCT accounts for the 2 ppm stability and 50 ppm absolute accuracy.

Operating wave forms for either voltage or current are downloaded to the power supply's embedded controller through an IEEE-488 interface. This wave form is executed as a piece-wise linear curve. Each set-point is described by a final current or voltage value and an associated ramp rate. Access to the digital-to-analog converter is also possible for arbitrary wave form generation.

One rail of the power supply output is connected to the energy extraction circuit as illustrated below in figure 3. Ordinary operation of energizing the magnet bypasses the dump resistor by allowing current to flow through the active parallel SCR.

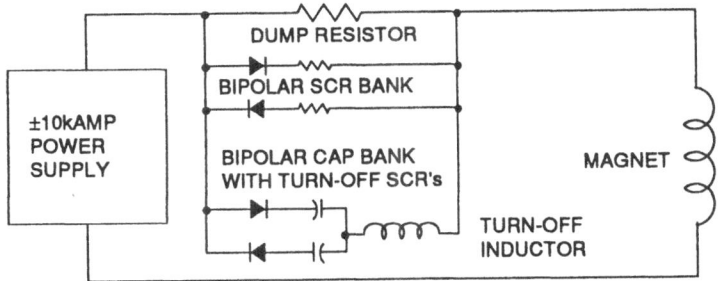

Figure 3. The ±10 kA power supply is series connected to the magnet under test through an energy extraction dump resistor. Two sets of SCR's and capacitor banks are utilized for bipolar operation.

When the magnet transitions to normal state, the active SCRs are disabled by removing the gate drive signal. The SCRs will stop conducting when current stops flowing through the junction and the gate charge is removed. This is accomplished by discharging the associated capacitor bank into the magnet, which turns-off the active SCRs, thus allowing the energy stored in the magnet to dissipate through the dump resistor. During energy extraction the power supply reference is set to zero which causes the output to invert and assist the extraction. The power supply output may also be clamped by a set of bypass thyristors.

DATA ACQUISITION AND CONTROL

The data acquisition and control system is comprised of the following instruments: (1) 256 channel voltage tap, (2) 96 channel strain gauge, (3) power supply, (4) cryogenic, (5) magnetic measurement. Each of these instruments is connected, via ethernet, to a SPARCstation 2 control console. Ethernet provides the principal data and control path for communication with each instrument. Data synchronization is handled at each instrument through an on-board time stamp module which is connected through a fiber optic cable to the master time stamp controller in the power supply instrument. Sampled data is stored locally at each instrument until the experimental run is finished. At this time the data is transferred to the SPARCserver 690MP for archive to the main database, Sybase.

Control software at each instrument consists of VxWorks, a real-time operating system, and application code for data acquisition and control. Network sockets are used to communicate back to the control console with data and status. At the control console, application software is running under SunOS UNIX with DataViews as the graphical user interface. PV~Wave is used to plot results to a postscript printer.

Voltage Tap Instrument

The Analogic Data Acquisition Instrument (DAI) is a 256 channel transient signal recorder with a wide dynamic range and 13-bit resolution. Each channel is a low noise, floating, autoranging analog-to-digital converter which can digitize signals from ±5 µV to greater than ±1000 V. A 16-bit word consisting of a two's complement 13-bit mantissa and a 3-bit exponent is produced for each sample. The digitizer sampling rate ranges from 1 kHz to 100 kHz and each channel has 1 Mbyte of buffer space allocated, thus 5.12 seconds of data may be recorded at 100 kHz. Each exponent value corresponds to seven (7) gain ranges, the full-scale ranges are as follows: (1) 9.76 mV, (2) 78.1 mV, (3) 0.625 V, (4) 5 V, (5) 40 V, (6) 320 V, (7) 2560 V. The resolution at the lowest gain range is 2.3 µV and the system bandwidth is approximately -3 dB @ 25 kHz with a noise density less than 30 nV/(Hz)$^{1/2}$.

The DAI contains a precision voltage source, which is used for channel calibration before each experimental run. The calibration coefficients are stored in each of the digitizers and used in real-time to compensate for channel offsets. Each set of 32 channels is funneled through a digital signal processor consisting of two (2) TMS320C30s, each capable of 33 million floating point operations per second.

Strain Gauge Instrument

The strain gauge instrument provides 96 channels of precision 4-wire resistance measurements with continuous sampling at approximately 1 Hz. The illustration in Figure 4 outlines a typical set of seven strain gauge channels and their interconnection to the instrument. A 16-bit isolated analog-to-digital converter is used as a precision current source for a series of seven active and compensating strain gauges. The center of the current loop is a Vishay resistor with a temperature drift of 0.3 ppm/°C. This sense resistor is sampled along with the strain gauges as an accurate measure of the excitation current. Thermal offsets present in the interconnection cable, relays and FET multiplexers are significantly reduced by switching the source current and voltmeter terminals.

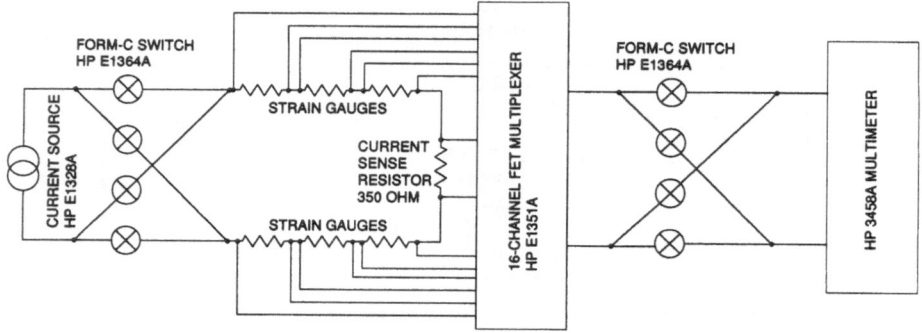

Figure 4. A functional diagram of the strain gauge instrument depicting the strain gauge current loop with the central current sense resistor.

Additionally, a precision resistor (traceable to NIST) is measured by each channel during a calibration cycle and the offsets are used for subsequent channel compensation.

Power Supply and Cryostat Instruments

The power supply instrument is primarily responsible for accurately measuring the magnet voltage and current in addition to downloading wave forms to the ±10 kA power supply. The HP 3458A 81/2 digit multimeter is used to measure magnet current from the ZFCT in the power supply, an additional meter is used to measure magnet voltage. Safety interlocks, data synchronization, and energy extraction are also managed from this instrument.

The cryostat instrument handles all temperature, pressure and mass flow measurements, including liquid level monitoring and a weighing pad for portable 500 L LHe dewars. The Lake Shore model 820 cryogenic thermometer is used to monitor carbon-glass and platinum temperature sensors. The MKS model 147B flow and pressure controller is used to monitor and control cryostat pressure and mass flow through the vapor-cooled current leads.

RESULTS

Each magnet test is archived to an on-line database. Once the experimental run is complete, the data is collected from each instrument and stored in the database. Setup configuration, such as channel assignment and strain gauge calibration coefficients, are also stored. The database easily handles multiple user requests so data collection and analysis may operate concurrently. The database conforms to the ANSI standard for structured query language (SQL) transactions.

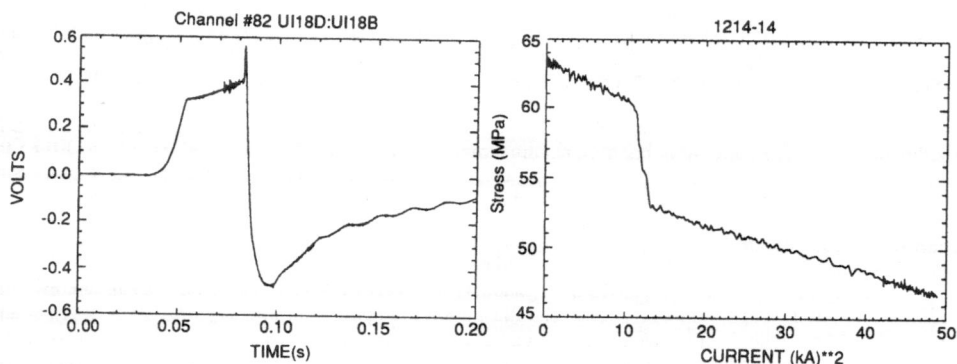

Figure 5. A typical voltage tap and strain gauge plot from a *ramp to quench* run on a model magnet. These plots were generated directly from the magnet database.

PERFORMANCE OF THE MAGCOOL-SUBCOOLER CRYOGENIC
SYSTEM AFTER SSC QUADRUPOLE QUENCHES*

K. C. Wu

RHIC Project
Brookhaven National Laboratory
Upton, New York 11973-5000

INTRODUCTION

The subcooler assembly installed in the MAGCOOL magnet test area at Brookhaven National Laboratory has been used for testing SSC dipoles, quadrupoles and a spool piece since 1989. A detailed description of the system, its steady state capacity and the performance after quenches of a 50 mm SSC dipole were given[1,2]. Subsequent studies on low current quenches of the SSC dipoles[3] and quenches of the RHIC dipoles[4] were also carried out. In this paper, the performance of the subcooler after quenches of the SSC quadrupole QCC404 is presented. Pressures, temperatures and flow rates in the magnet cooling loop after magnet quenches are given as a function of time. The cooling rates and total energy removed by cooling during quench recovery have been calculated for quench currents between 2000 and 7952 amperes. Because the inductance of the quadrupole is about one tenth that of a SSC dipole, the stored energy released is small and the impact on the system is mild. The cooling loop pressure never exceeds 12 atmospheres and the cryogenic system recovers in less than 15 minutes. As in all past studies, the peak pressure and temperature in the magnet cooling loop are linearly proportional to the energy released during a quench and excellent agreement between the total cooling provided and the magnetic stored energy is found.

SYSTEM DESCRIPTION

The MAGCOOL subcooler assembly is designed to provide supercritical cooling to the superconducting magnet. Figure 1 shows the flow schematic for the subcooler and the cooling loop. A circulating compressor is used for closed loop circulation of single phase helium which delivers cooling from the subcooler helium pot to the magnet. When the magnet quenches, the magnetic stored energy is deposited as heat in the magnet cooling loop. The pressure and temperature in the loop first increase and then return to their original values after the magnet is cooled again to the test temperature prior to a quench. Typical operating conditions prior to a magnet quench are also given in Figure 1 in which temperatures are shown without units and are in Kelvin. The volume of the circulating loop piping is about 200 liters and there is a warm line of 10 liters volume connecting the loop to the warm relief valve. It is believed the 10 liters warm volume reduces the peak pressure by some amount compared to a system without a warm volume, but has little effect on other properties.

*Work performed under contract with the U.S. Department of Energy.

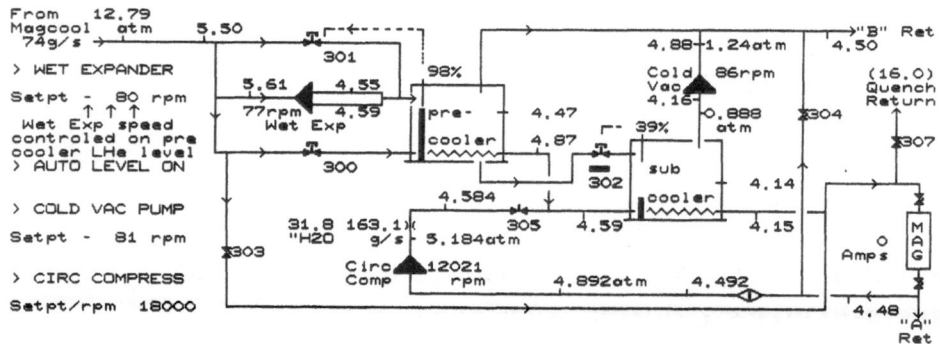

Figure 1. Flow schematic and operating conditions prior to a magnet quench

PRESSURE, TEMPERATURE AND FLOW RATE AFTER QUENCH

The following results were obtained from quenches of the SSC QCC404 quadrupole in MAGCOOL Test Stand B. The magnet was maintained at 4.3 K prior to a quench. A total of ten quenches with quench currents between 2000 and 7953 amperes were investigated. The 2000, 4000 and 5000 ampere quenches were initiated by a strip heater located on the magnet coil. The other quenches were natural quenches occurring between 7265 and 7953 amperes. The corresponding magnetic stored energies varied from 15 to 240 kilo-joules. The results for the 4000, 5000, 7396 and 7952 ampere quenches are presented graphically below.

The loop pressures as a function of time after a quench are given in Figure 2. As seen, the loop pressure first increases after a quench due to the heat released from the magnet to the cooling loop. As the system is cooled to the condition prior to the quench, the pressure decreases. The peak loop pressure increases linearly with the amount of energy released as shown in Figure 3. The time at which these peak pressures occur also increases with the energy released. In all cases, the loop pressure returned to the original pressure before the quench in less than 4 minutes.

Temperatures recorded at the return line in the subcooler assembly after quenches are given in Figure 4. The return temperature increases initially as heat is carried to the subcooler. The temperature reaches a peak value before returning to test condition. The return temperature is higher for higher current quenches whereas the supply temperatures is essentially the same for all cases.

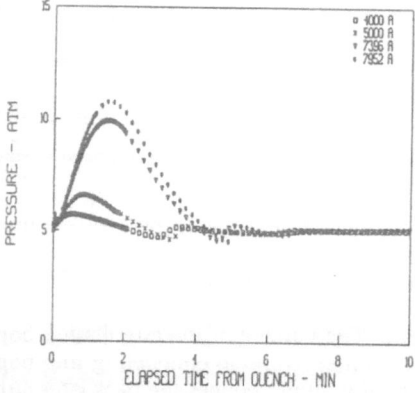

Figure 2. Loop pressures after quenches

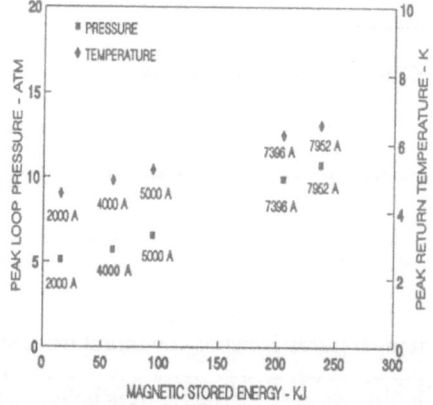

Figure 3. Peak loop pressure and temperature versus stored energy

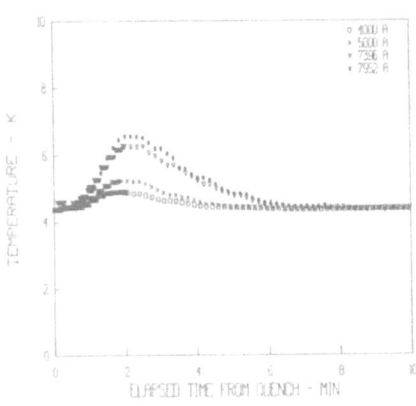

Figure 4. Return temperatures after quenches

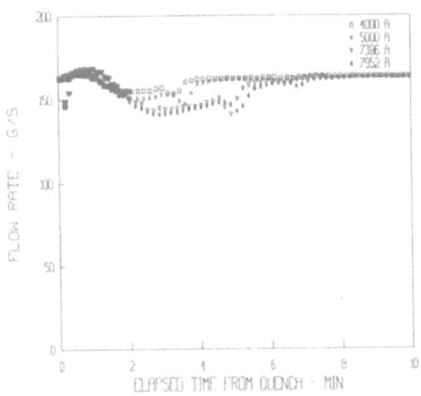

Figure 5. Mass flow rates after quenches

The peak temperatures recorded are proportional to the magnetic stored energy as shown in Figure 3. The time at which the peak temperatures were recorded depends primarily on the circulation of helium and was approximately 2 minutes after each quench. The recovery time for temperatures in the circulating loop is less than 10 minutes.

The helium flow through the magnet as a function of time is given in Figure 5. As can be seen, the mass flow rate only varies slightly because the energy released into the helium loop is not large. However a large energy release does cause a larger perturbation in flow rate.

COOLING RATE AND TOTAL COOLING

The apparent cooling rate applied to the magnet is defined as the difference in the enthalpy flux between the helium in the supply and the return lines. The net cooling rate for quench recovery equals the apparent cooling rate minus the background heat load. Because the system which started at a test condition is cooled to the original condition after a quench, the integration of the net cooling rate represents the total amount of cooling provided for quench recovery and should be equal to the stored energy released by the magnet. The apparent cooling rates during quench recovery are given in Figure 6. The total net cooling provided during quench recovery for each of the four quenches is given in Figure 7. The cooling provided increases with time and reaches a plateau when the loop is cooled to conditions existing prior to the quench.

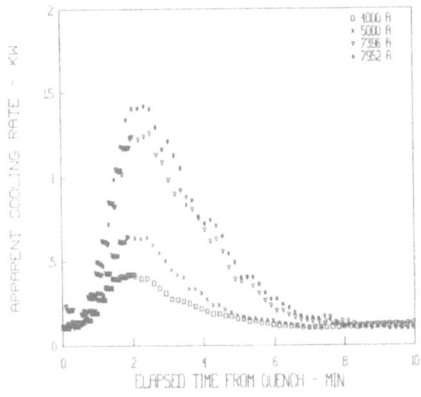

Figure 6. Apparent cooling rates

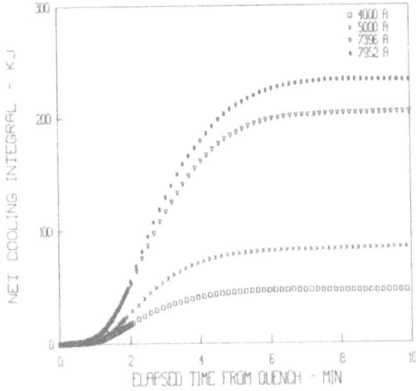

Figure 7. Total net cooling provided

SUMMARY

Key parameters and the corresponding stored energies for the ten quenches studied including the four quenches given above are summarized in Table 1. As can be seen, the total net cooling provided agrees with the magnetic stored energy very well for quench currents above 4000 amperes. For the 2000 ampere quenches, accuracy of the measurements is not good because the energy released is simply too small.

Table 1. Results from quench of QCC-404 quadrupole*

Quench Current	Peak Pres.	Time to Peak Pres.	Peak Return Temp.	Time to Peak Return Temp.	Max. Cooling Rate	Total Net Cooling Provided	Magnetic Stored Energy $1/2\,L\,I^2$	Ratio of Cooling to $1/2\,L\,I^2$
ampere	atm	sec	K	sec	kw	kj	kj	
2000	5.14	N.A.	4.52	N.A.	0.18	8.4	15.0	0.56
4000	5.74	N.A.	4.91	N.A.	0.42	52.4	60.0	0.87
5000	6.61	52	5.24	110	0.65	92.3	93.8	0.98
7265	9.64	92	6.16	132	1.18	201.3	197.9	1.02
7396	9.92	90	6.27	115	1.26	210.7	205.1	1.03
7715	10.07	90	6.51	131	1.38	222.2	223.2	1.00
7749	10.20	91	6.54	122	1.39	242.1	225.2	1.08
7773	10.18	93	6.53	133	1.39	244.1	226.6	1.08
7953	10.81	92	6.59	133	1.40	239.6	237.2	1.01
7952	10.78	90	6.57	120	1.42	239.4	237.1	1.01

*0.0075 Henries is used for the inductance of the quadrupole.

CONCLUSION

These results from testing a SSC quadrupole and those from earlier tests of SSC and RHIC dipoles provide useful information and a method for analyzing other superconducting magnet systems. The thermal characteristics of the MAGCOOL system after a magnet quench is primarily determined by the magnetic stored energy released into the helium cooling loop. As in all past tests, the peak loop pressure and peak return temperature are found to be linearly proportional to the magnetic stored energy. The agreement between the total net cooling provided and the magnetic stored energy has been confirmed. The energy releases for a quadrupole quench are less than 250 kilo-joules. The peak loop pressure observed is less than 12 atms and no helium was vented from the cooling loop. The system recovers in less than 15 minutes.

ACKNOWLEDGEMENT

The author would like to thank D. Zantopp for developing the data acquisition software, P. Radusewicz for providing information on the quadrupole, A. Prodell for providing helpful comments and S. Agnetti for preparing this paper.

REFERENCES

1. K. C. Wu, D. P. Brown, J. H. Sondericker, Y. Farah, D. Zantopp and A. Nicolletti, Subcooler assembly for SSC single magnet test program, in "Advances in Cryogenic Engineering", 37A:763, Plenum Press, New York (1991)
2. K. C. Wu, Performance of the MAGCOOL-subcooler cryogenic system after 50 mm SSC dipole quenches, in "Supercollider 4", p.483, Plenum Press, New York (1992)
3. K. C. Wu, Peak loop pressure and temperature and comparison of total cooling provided to the energy released after low current quenches of SSC dipoles in MAGCOOL cryogenic system, to be presented in the CEC-ICMC conference, July 12-16, 1993, Albuquerque, N.M.
4. K. C. Wu, Thermal characteristics of the MAGCOOL cryogenic system after quenches of RHIC dipoles, to be presented in the CEC-ICMC conference, July 12-16, 1993, Albuquerque, N.M.

A HIGH EFFICIENT 12KW HELIUM REFRIGERATOR FOR THE LEP 200 PROJECT AT CERN

B. Chromec,[1] W.K. Erdt,[2] D. Güsewell,[2] K. Löhlein,[1] A. Meier,[1]
A.-E. Senn,[1] T. Saugy,[1] N.O. Solheim,[2] U. Wagner, [1] G. Winkler,[2]
and B. Ziegler,[1] et al.

[1] LINDE KRYOTECHNIK AG, CH-8422 Pfungen, Switzerland
[2] CERN, Div. AT, CH-1211 Geneva, Switzerland

ABSTRACT

CERN has ordered helium refrigeration / liquefier plants for the LEP200 project in 1991 with an equivalent refrigeration capacity of 12kW at 4.5 K. The cold equipment of these plants is divided into two parts with a cut on a temperature level of roughly 20 K. One bigger coldbox with temperatures between ambient and 20 K is installed at ground level, whereas the smaller box between 20 and 4.5 K is placed in the underground LEP tunnel and is therefore limited regarding its constructional size. The boxes are interconnected by a four stream transferline system mainly vertically arranged in machine access shafts of 90 to 140 m depth. The helium refrigeration plant delivered by LINDE KRYOTECHNIK AG combines an extremely compact construction especially of the lower coldbox with a remarkably good cycle efficiency. This paper describes the thermodynamic process, the construction of the plant and the control concept of the system.

In addition, the cold boxes are already prepared for a later expansion to 18kW refrigeration capacity without need of changing internal equipment.

INTRODUCTION

During the coming years the European Laboratory for Physics of Elementary Particles (CERN) near Geneva will increase the particle energy of its 26.7 km long underground electron positron collider LEP from 55 GeV to about 90 GeV. This will be achieved in the upgrade project known as LEP-200, by progressively installing superconducting cavities for acceleration. To cool these strings of superconducting cavities, four helium plants with a nominal capacity of 12 kW at 4.5 K each are presently being commissioned at the even LEP interaction points. Two of these plants are delivered by LINDE KRYOTECHNIK AG.

PLANT DESCRIPTION

The cryogenic equipment described in this paper consists of a compressor group, two cold boxes and an interconnecting transfer line.

The compressors and the large upper cold box (UCB) are located at ground level. The smaller lower cold box (LCB) is installed in the LEP tunnel below the surface to overcome static pressure effects at low temperature. The depth of the tunnel is 90 meters at point 6 and 140 meters at point 4. The interconnecting transferline (ITL) between UCB and LCB operates at a temperature level of 20 K which results in a good balances of static and dynamic pressure loss of the 1.1 bar return flow.

The specified cooling capacity and the expected power input of the plant is given in Table 1. The coldboxes are prepared to enable a capacity increase to 18 kW equivalent at 4.5K for covering future demands of the LEP 200 project without major changes at the cold boxes. This means, that all heat exchangers, lines and valves are sized to cover the increased capacity. In addition spaces for two more turboexpanders are foreseen.

Table 1.   Specified cooling capacity and expected power input

refrigeration capacity at 4.5 K	10000	W
liquefaction capacity at 1.3 bar / 4.5 K	13	g/s
refrigeration capacity at < 80 K	6700	W
total power input at terminals	2600	kW

REFRIGERATION PROCESS

Helium gas is compressed by a set of five screw compressors, three of them working as boosters in parallel at a suction pressure of 1.01 bar and an outlet pressure of roughly 4 bar. Two are operated in parallel as second stage machines with a discharge pressure of 20 bar. (Fig. 1) Each compression stage is followed by a bulk oil separator and a gas cooler. A fine oil separation, consisting of a three-stage coalescing filter and a charcoal adsorber, reduces oil contamination to less than 10 ppb (mass).

In the cold box refrigeration is achieved by a total of 7 turbines and 13 heat exchangers, arranged in four blocks, as shown in a simplified flowsheet in Fig. 2 and the corresponding T-s diagram in Fig. 3. A first loop, consisting of 4 turbines in series, expanding helium from 19 bar to 1.2 bar, provides precooling of the JT-stream to roughly 20 K. A part of the JT-stream is extracted at 29 K, used to provide shield cooling and returned to the turbine cycle upsteam of the second turbine. Below 20 K the JT-stream is passed through the ITL to the LCB and then fed to two parallel turbines[1], operated at inlet temperatures of 17 K and 9 K, respectively. After passing through heatexchanger E4B and the last turbine which operates as a wet expander, it is available as subcooled liquid at 1.5 bar. Liquid nitrogen cooling is not used.

COMPRESSORS

The five compressors used are of the lubricated screw compressor type, produced by Stal, Norrköping, Sweden. All machines are directly coupled to the electric motors. They feature hydraulically operated slide valve, rotor balancing piston and gliding ring shaft seal. Their rotors are axially supported by angular contact ball bearings and radially by pressure lubricated slide bearings.

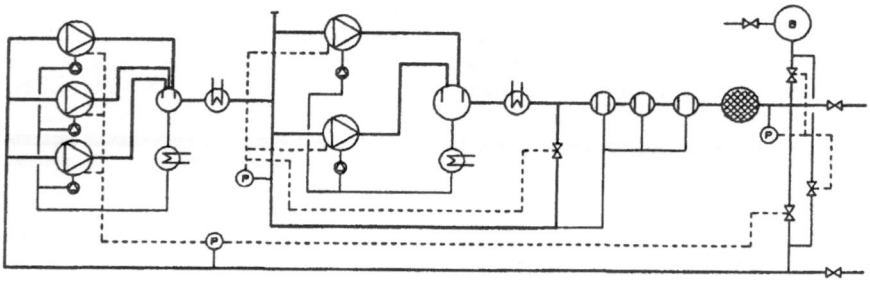

Figure 1.   Flowsheet of the compressors

The booster compressors S93 are equipped with a separate external oil pump, whilst the high stage machines S75 have integrated oil pumps directly coupled to the rotor. The three booster compressors share one common bulk oil separator, oil cooler and gas cooler. The whole system was preassembled on a total of eight skids.

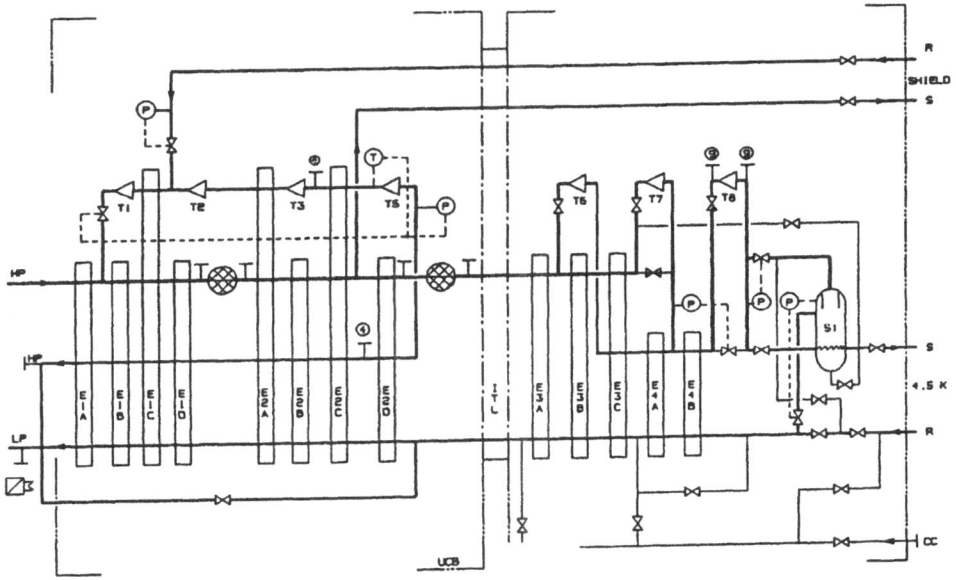

Figure 2.   Flowsheet of the cold box

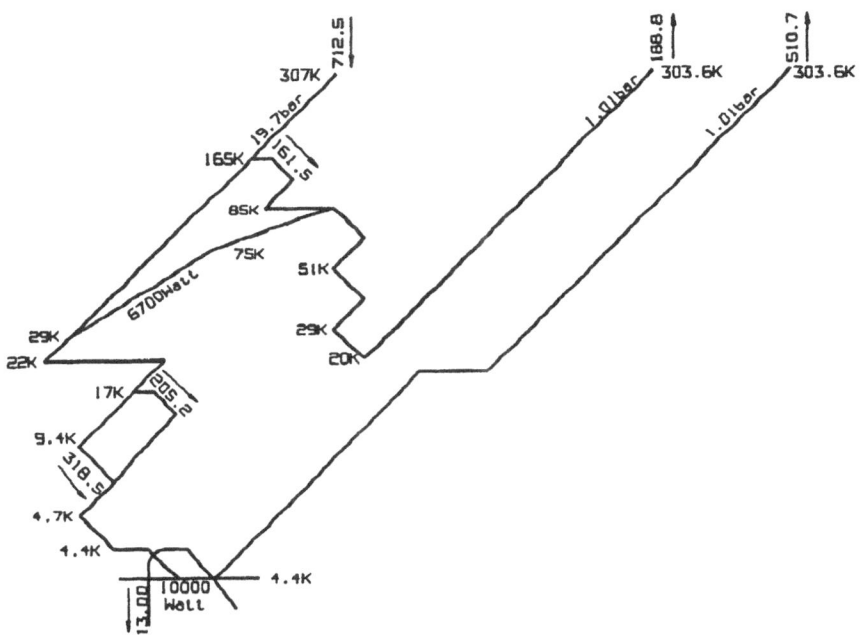

Figure 3.   T-s diagram

COLD BOX

The cold box design is very much dominated by space limitations both in the access shaft and in the tunnel. The maximum dimensions are:

vertical shaft:			
	length	6.0	m
	width	2.0	m
	max. acceptable weight	12.5	tons

On a very early stage it was agreed to divide the cold box in two units. The UCB has nearly no size limitations and incorporates the first eight heat exchangers in two blocks, the adsorbers at 80K and 20 K level and the upper four turbines. The weight of the first heat exchanger block only is 8.3 tons. To increase the capacity to 18 kW provisions for a fifth turbine are made. The LCB is built of two tubes 1.9 m in diameter, one horizontal and one vertical to be lowered to the tunnel through the shaft in vertical position.

The very compact design of the LCB required special attention during fabrication. The fabrication sequence and the corresponding quality checks, such as radiography of welded joints and leak testing, had to be carefully planned, since access to the internal parts became more and more limited during progress on cold box assembly. Fig. 4 shows a view on the cold boxes.

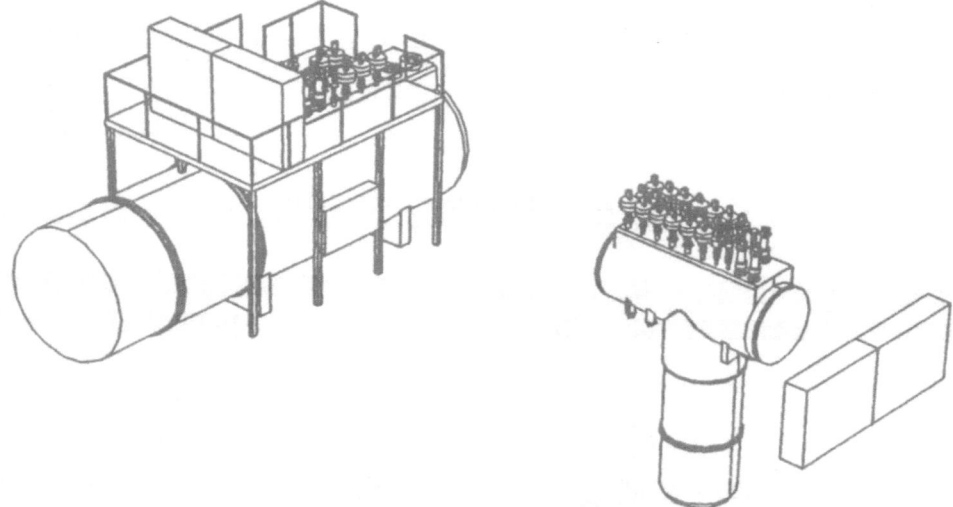

Figure 4.   CAD view on the cold boxes

CONTROL SYSTEM

The 12 kW helium refrigerator will be run by a team of operators together with several plants and systems built by other suppliers. For reasons of simplified operation and maintenance all these systems will be equipped by CERN with the same microprocessor based industrial process control system. The established CERN standard is the Master System supplied by ABB. It provides all local control functions required and - via one of CERN's data transmission networks - remote operation from the central cryogenics control room up to 10 km away from the individual plants.

A so called Master Piece, an independent intelligent local unit, is provided both for the compressor system and the cold box, each with its own battery backed 24 VDC supply system. Separate units, called Master View, are placed in the control room for process visualisation and operator / plant interface. The operation of the cryoplant will, in general, not be affected by failures of the communication network or the display system.

To limit potential damages of machinery, in particular compressors and turbines are equipped with an additional basic protection using hard wired interlock. It is intended to operate the plant fully automatic and without intervention or even presence of operators.

COMMISSIONING

Compressors

Compressors were first started in October 1992 and had successfully completed a 200 hour test run in December 1992. Massflow and efficiency measurements, however, were slightly below the guaranteed design values.

Table 2.   Design and operating data of the compressors

		total	
		design	measured
suction temperature	[K]	303	294
suction pressure outlet	[bar]	1.01	1.01
delivery pressure inlet	[bar]	20.0	20.0
massflow	[g/s]	705	704 (-3.1%)
electrical power input	[kW]	2600	2558
isothermal efficiency	[%]	51	50 (-2%)

Cold box

The cold box was first cooled down to LHe temperature on March 21st 93 and has passed a 100 h run at full capacity from April 4th to 8th 93. During the 100h test run a capacity test was performed, which results are listed in table 3. The cold box is presently prepared to run the final acceptance tests.

Table 3. Operating data of CERN's 12 kW plant for LEP Point 6

		total	
		design	measured
refrigeration at 4.5 K	[W]	10000	10020
liquefaction at 4.5 K	[g/s]	13.0	12.6
refrigeration at < 75 K	[W]	6700	6700
hp massflow	[g/s]	712.5	712.3
hp pressure	[bar]	19.7	19.8
suction pressure	[bar]	1.01	1.01
suction temperature	[K]	304	295
cold box efficiency	[%]	59.7	61.4

CONTROL CONCEPT

Compressors

The process control at the compressors is done by three pressure control loops which are arranged to keep the plant at a good overall efficiency (s. Fig. 1).

The booster suction pressure is mainly controlled by the Hp/Lp bypass. Depending on the bypass position, the slide valve positions of the three boosters are increased or decreased in split range. If very low capacity is demanded by the cold box the third booster is automatically stopped and started if needed again.

The second stage suction is controlled by the stage bypass and the slide valves of the hp machines in split range. The setpoint of the controller is set to a low value to maintain the slide valves at full power position for most of the load cases.

The delivery pressure to the cold box is controlled by the load and unload valves to the helium buffer.

Cold Box

The process control of the cold box is arranged to allow the plant easy adaption to the cold loads (s. Fig. 2).

The four upper turbines operating between 165 K and 20 K, are controlled according to the outlet temperature of T5 during design operation and by the inlet pressure of T5 during cool down operation. (Turbine T4 is not yet installed, as it is needed only for the 18 kW expansion of the cold box.) The shield cooling flow, which is added to the turbine flow out of T1 is controlled by the shield return pressure. The setpoint for the shield return pressure is set according to the delivery pressure.

Turbine T6, T7 and T8 are not controlled, their inlet valves are just ramped open. The J-T-flow is controlled by three valves. One is keeping the outlet pressure of T6 and T7 at about 6.4 bar, one controls the outlet pressure of T8 and the subcooler pressure at 1.5 bar, and the third controls the pressure of the phase separator.

REFERENCES

1   A. Kündig, Method and apparatus for liquefying a low boiling gas, US Patent Number: 4.606.744

THE STRUCTURAL ANALYSIS OF THE GENERAL DYNAMICS SSC
BASELINE DESIGN COLLIDER DIPOLE MAGNET

Gregory Mehle and John Wohlwend

General Dynamics Space Systems Division
Space Magnetics
PO Box 85990
San Diego, CA 92186

ABSTRACT

This paper presents the methodology for the structural analysis of the 13m and 15m Collider Dipole Magnet (CDM) coldmass. A geometric nonlinear Finite Element solution was used to determine the deflection characteristics, and stress state of the typical two-dimensional magnet cross-section through all assembly operations, and for magnet energization at 4K.

INTRODUCTION

The typical mechanical cross-section of the SSC CDM coldmass is shown in Figure 1. The coldmass is composed of the superconducting coils, laminated collars, laminated yokes, and the helium containment vessel. The function of the coldmass structure is to control the location and displacements of the superconducting coils during magnet operation. Therefore, all structural components (coils, collars, yokes, and vessel) must be in predictable contact throughout the CDM operating cycle. GDSS is utilizing numerical methods (Finite Element Method) to predict the structural (stress and displacement) behavior of the magnet structural system.

The CDM magnet has inner and outer coils made up of two grades of NbTi cabled conductor. These conductors are wrapped with a polyimide film insulation. The inner coils have nineteen turns spaced at three locations with copper wedges; the outer coils have twenty-six turns spaced at one location with a copper wedge. The stiffness of the conductor pack greatly affects the stresses in the coil, and the deflections of the coil and surrounding structure.

The coils are preloaded and restrained by fine-blanked Nitronic 40 steel collars. The collars accurately locate the coils during fabrication, and react coil loads in concert with the yokes and vessel during coil energization. The geometries of the laminated collars and tapered keys (that join the upper and lower collars together) are tailored to minimize the ovalization that occurs when the coil is preloaded during the collaring operation.

Laminated low-carbon steel yokes surround the CDM collars. In addition to their magnetic function, the yokes restrain collar lateral and axial displacements by providing additional restraint to collar. The yokes must be in contact with the collars to provide structural restraint. The collar/yoke contact is introduced during the build-up of the magnet coldmass. Welding of the 304LN CRES helium containment vessel creates circumferential prestress which restrains the yokes, and completes the coil support structure.

The objective of the finite element analysis is:
— Predict the stresses and deformations of the magnet structure
— Determine collar ovalization from the coil preloading
— Predict material nonlinearities in the high stress regions in the collars
— Determine yoke and containment vessel preloading during CDM assembly
— Develop collar/yoke interface pressures during magnet cooldown and energize
— Predict the horizontal mid-plane closure of the yokes after magnet cooldown

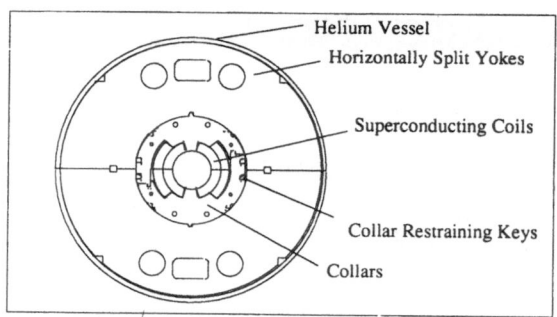

Figure 1. CDM Coldmass Components

FINITE ELEMENT ANALYSIS APPROACH

The approach for the CDM coldmass analysis was the development of a Finite Element Method (FEM) solution for predicting the deformations, and stress state of the coldmass components. FEM models were developed for each assembly phase for the coldmass. A model of the coils and collars was used to represent the collaring and post-keyed condition of the coils. Yokes and helium shell geometry was added for predicting the yoking, post-skinned, cooled-down and energized condition of the of the magnet coldmass.

All analyses assume 1/4-model symmetry, and a (two-dimensional) Plane Stress condition for all the components within the cross-section. Plane stress implies all out-of-plane stresses are zero, and out-of-plane deformations have no influence the stresses or deflections of the cross-section. The components are modeled with four node quadrilateral elements. The collars, restraining keys, yokes, and coil wedges use isotropic material properties; the coils are cylindrically orthotropic. The conductor properties were developed through SSCL and GDSS test programs. Contact regions between the components are modeled with nonlinear formulated gap elements to eliminate misconstrained deformations of the structure.

The boundary conditions for the FEM model components are shown in Figure 2. Anti-symmetric and symmetric boundary conditions for the collars are applied through multi-point constraint (MPC) equations. MPC equations allow dependent nodal degrees of freedom to be set equal or opposite to identified (independent) nodal degrees of freedom. General symmetry boundary conditions for the shell, yoke and coils are applied through single point constraint (SPC) equations for specific nodal degrees of freedom.

The coil preloads from collaring and keying are obtained from an applied force at the coil midplane region of the model. MPC equations are applied to the coil midplane nodes to maintain a symmetric boundary condition at the region. The derived deformation at the inner and outer coil midplane is then used as an nodal offset value which is applied within all subsequent analyses. In these subsequent analyses coil preload is obtained by utilizing SPC equations to displace the offset coil midplanes to the geometric midplane of the model. The collared coil pole face target prestress values are 69 Mpa for the inner coil; 55 Mpa for the outer coil.

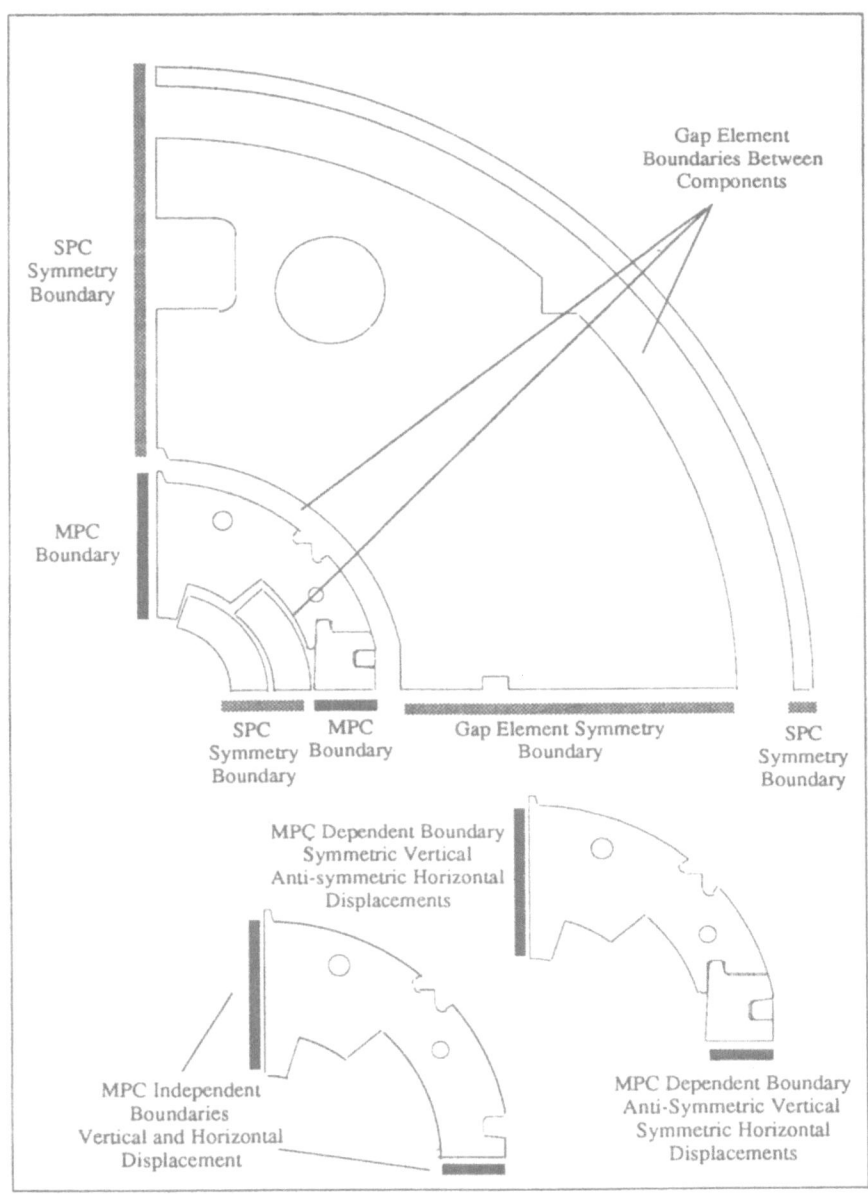

Figure 2. Quarter symmetry FEM model anti-symmetric and symmetric boundary conditions

MSC-NASTRAN is used for all analyses. The solution is geometrically nonlinear which is a Newton-Raphson based (virtual work) iterative method. Although there are small deformations in the FEM model, the geometric nonlinear approach is required for stiffness matrix reformulation for gap element convergence at the contact regions of the model.

All major analyses of the CDM cross-section have been completed. However, the evolution of the numerical model is continuing. The work is focusing on improving the resolution of the FEM models to better represent the as-fabricated condition of the components.

A specific area of investigation is the condition of generalized plane strain of the coils after magnet cooldown. The basic assumption within all completed analyses is the condition of plane stress. From collaring through skinning, the applied coil preload causes axial elongation of the coils due to the poisson effect. Since the coils are not restrained from growing axially there is no effect to coil deflections or stresses within the 2-dimensional analysis cross-section. The plane stress assumption is valid in this situation.

The coldmass assembly acts as a single axial structure due to the mechanical and frictional interactions between the coils and collars; collars and yokes; and yokes and shell. The single axial structure has a thermal expansion that is driven primarily by the yokes. The yokes have a large surface area, a high Young's modulus, and the low coefficient of thermal expansion.

The coils have a high coefficient of thermal expansion and a low axial direction Young's modulus. The coils are constrained axially from the mechanical/frictional interactions with the other coldmass components. As cooldown occurs the yoke constraint on the coil induces a tensile axial strain in the coils. This condition changes the effective the cross-sectional stress state of the coils. This is considered to be a Generalized Plane Strain condition— known values of out-of-plane strain. Analyzing the cooled down and energized state of the coldmass under a generalized plane strain condition will change the coil preload values as compared to the plane stress condition.

CONCLUSIONS

This paper has presented the basic analytical assumptions, and FEM modelling methodology used for the 13m and 15m CDM. Data from the model magnet program and the prototype magnet program is the mechanism used to validate the analyses.

Measured vertical and horizontal deflections of the GDSS configuration collars where taken after the coil collaring/keying operations on the CDM model magnets. The as-built (diameter based) deflections are approximately .65 mm (.026 inch) vertical, and -.25 mm (-.01 inch) horizontal. The predicted values from the CDM FEM analysis model is .62 mm vertical, -.41 horizontal at consistent coil pole face stress values.

The validation of all analyses will continue as more model and prototype CDM test information is developed. Based upon the existing information, the resolution of the FEM analyses allows an accurate predictive tool for the program.

ACKNOWLEDGMENTS

The work described herein is being accomplished under contract to the Universities Research Association in support of the Superconducting Super Collider project for the U.S. Department of Energy. Special thanks to Chuck Gibson for presenting this paper at the IISSC Conference.

REFERENCES

1. Turner, J., "Mechanical Analysis of the W6733H Cross-Section," MD-TA-143, SSCL, July 1990

2. Turner, J., "Results from ANSYS Studies on the DCX201 Cross-Section," MD-TA-200, SSCL, December 1991

HIGH ENERGY BOOSTER QUADRUPOLE COLD MASS DEVELOPMENT AND INDUSTRIALIZATION PROGRAM

G. Ducos,[1] J. Giacometti,[1] P. Giovannoni,[1] D. Leboeuf,[1]
F. Le Coz,[1] C. Lyraud,[1] C. Michez,[1] J. F. Millot,[1] D. Orrell,[2]
J. Perot,[1] J. M. Rifflet,[1] J. C. Toussaint,[1] J. Turner,[2]
P. Vedrine,[1] and J. Cortella[1]

[1] CEA SACLAY DAPNIA/STCM*
91191 Gif sur Yvette Cedex France
[2] SSCL Representatives working at Saclay

INTRODUCTION

The department DAPNIA of the CEA Saclay has been involved in High Energy Physics for several decades, working on projects such as Detectors, Superconducting magnets (STCM), Thermonuclear Fusion machine (TORE SUPRA [1]) and accelerator magnets. Considerable research and development effort have gone into the design and production of quadrupole magnets for HERA [2], and, over the last two years, for LHC [3] .

In January 1992 a subcontract was placed between URA and the CEA Saclay in France : from the SSC technical specification , the CEA Saclay has to design , study , fabricate , test the prototypes and develop all the production processes, as well as the tooling required to build and test in US Industry the High Energy Booster (HEB) Arc and special Quadrupole Cold Masses . This paper presents the overall program and the status of the work after 16 months .

HIGH ENERGY BOOSTER QUADRUPOLE COLD MASS OVERALL PROGRAM

The overall program is to be carried out within 41 months. 2 magnets are to be studied and produced as prototypes as well, and another is to be fully instrumented . For all magnets the complete documentation (Technical specifications, Process specifications,Travellers, Control , Tests ...) is to be produced for the technology transfer to US Industry. One set of all the special machines will be designed and produced at Saclay. They will conform to the US regulation (UL NEC OSHA) as well as the European standard.

The magnets to be studied are listed in table 1.

---

* DAPNIA (Departement d'Astrophysique, de Physique des Particules , de Physique Nucléaire, et d'Instrumentation Associée)
STCM (Service Technique de Cryogènie et de Magnétisme)

TABLE 1 List of magnet to be designed and prototypes produced at Saclay

QUADRUPOLE	Length in M	Quantity for HEB	Prototypes built at Saclay	Tooling
Arc HQCM	1,6	274	YES	YES
Arc HQCM fully instrumented	1,6	1	YES	NO
SPECIAL HQCM 1	0,902	16	NO	NO
SPECIAL HQCM 2	1,173	4	NO	NO
SPECIAL HQCM 3	2,258	8	NO	NO
SPECIAL HQCM 4	3,347	16	YES	YES

Main Milestones : (see planning)

Contract award	01 January	92
Preliminary Design Requirement Review	05 February	92
Preliminary Design Review	15 May	92
Critical Design Review	30 June	93
ARC QUADRUPOLE PROTOTYPE	October	94
ARC QUADRUPOLE INSTRUMENTED	March	95
HQM4 SPECIAL QUADRUPOLE	April	95
Build to Print Package Review Arc quad	July	94
Completion	July	95

PLANNING

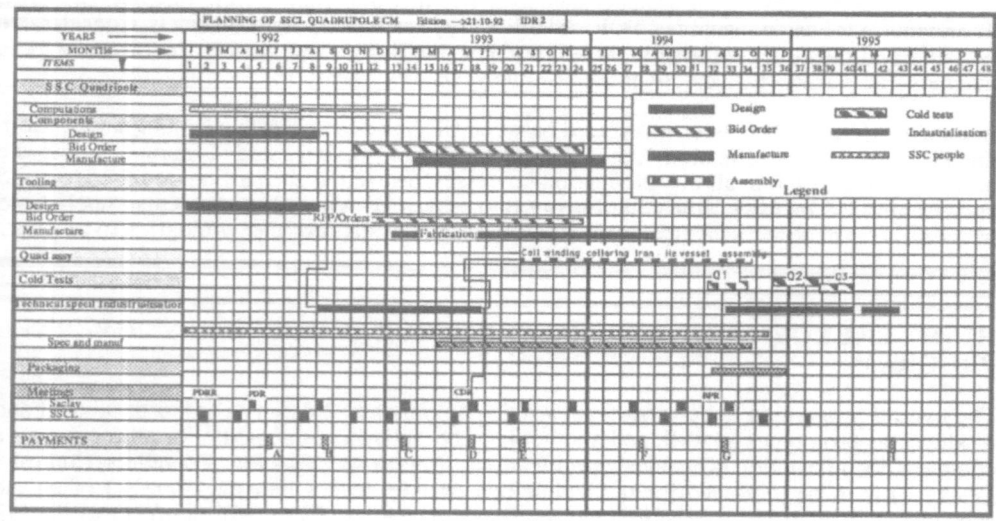

HEB COLD MASS TIME SCHEDULE

QUADRUPOLE COLD MASS DESIGN

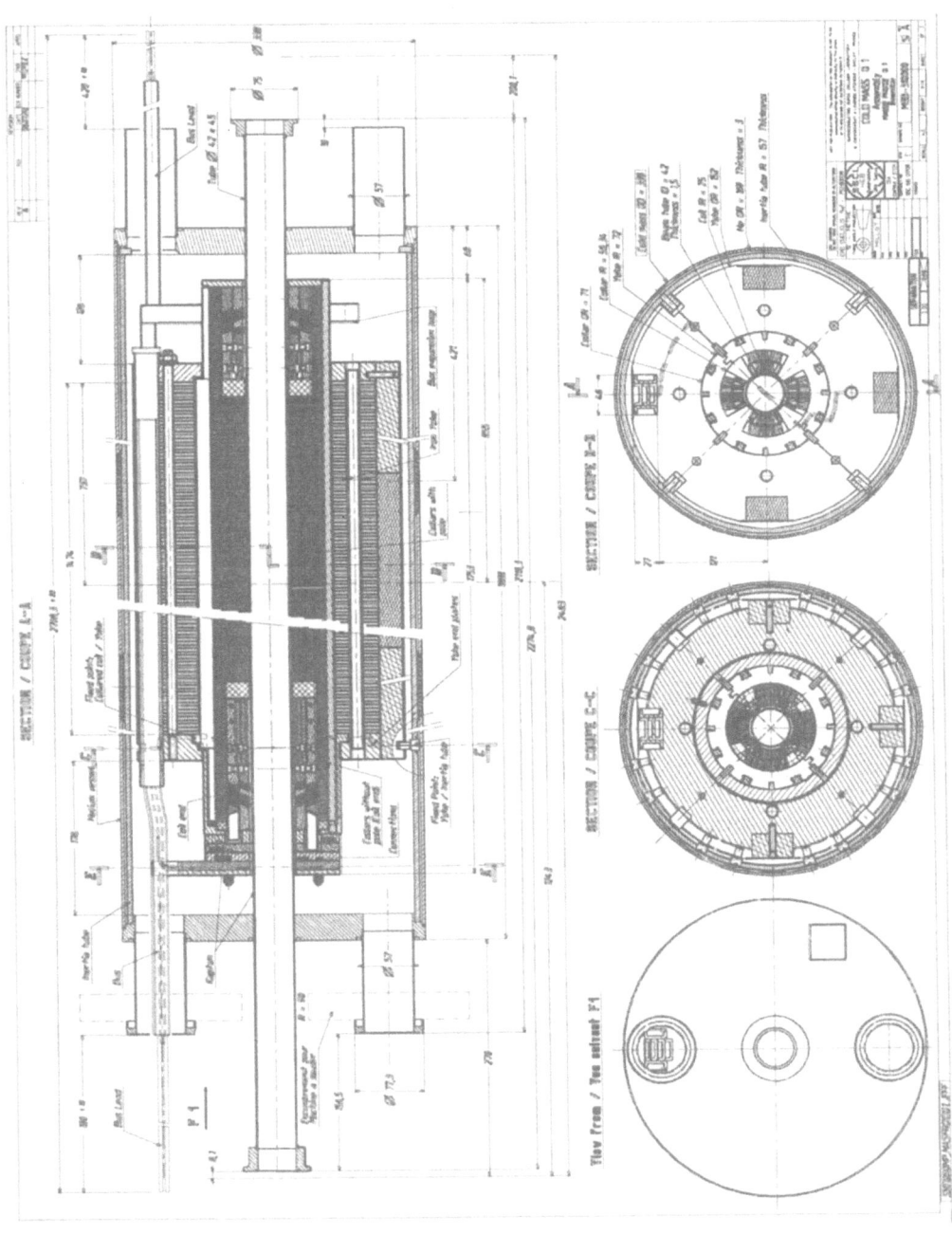

STATUS AFTER 16 MONTHS

Computation and Design

The Conceptual Design Report is complete. All mechanical, thermal, electromagnetic calculations have been carried out . All the technical specification requirements are fulfilled. The most difficult one is the RMS multipoles which are not easy to compute, however taking into account the HERA scaled values (1), it seems possible to remain within the specification.

The Cold Mass Design is complete. An original feature of Saclay's Quadrupole Design is to fabricate most of the components without machining. Fine Blanking and Cold Drawing are used to insure a very good reproducibility. As all the Lorentz coil forces are self-contained in the collars, it is not necessary to use a welded clamping tube in order to prestress them. The magnet is self-aligned when fitting the collared coil and the yoke assembly in the Inertia Tube around which the helium vessel , complying with the ASME regulation, is fitted.

The conductor is ready to be insulated, the angular wedges, end parts, ground insulation, collars and tapered keys for collaring are ready to be ordered.

Tooling and special production Machines

The curing molds are designed and ordered, while the curing press, Heating Unit Controller and Coil Young Modulus Measurement Machine are under construction. All other tooling or machines are under study.

Test facilities

Saclay is now equipped with a vertical cryostat capable of testing cold mass magnets in 4,2K (up to 3,5K) boiling helium bath.

TEST CRYOSTAT CHARACTERISTICS

Diameter	Cold mass Length
860 mm	6000 mm

The power supply available is ± 10 000 A and 4 Volts giving the possibility to program the SSC bipolar cycle with an accuracy of 10^{-4} .

Industrialization

The tooling requirements have been calculated in order to satisfy the production rate demanded by the SSC. The process specification and travelers are written for all procedures concerning the coil manufacture and assembly.

CONCLUSION

The Preliminary Design Requirement Review and the Preliminary Design Review were held successfully in 1992 . The Critical Design Review will occur at Saclay in June 93 and should allow the laboratory to order the components and start the first magnet assembly. The first HQM prototype should be ready for cold tests by August 94 .

REFERENCES

(1) TORE SUPRA DSM DRFC CEN Cadarache 13100 France
(2) J Perot, JM Rifflet " Measurement data taken during the industrial fabrication of the HERA superconducting quadrupoles "
(3) LHC CERN Geneva Suisse

SSC QUADRUPOLE MAGNET PERFORMANCE AT LBL

A. F. Lietzke, P. Barale, R. Benjegerdes, S. Caspi, J. Cortella, D. Dell'Orco, W. Gilbert, M.I. Green, K. Mirk, C. Peters, R. Scanlan, C.E. Taylor, and A. Wandesforde

Superconducting Magnet Group
Lawrence Berkeley Laboratory
Berkeley, CA 94720

INTRODUCTION

LBL contracted to design, contract, and test four short (1m) models and six full-size (5m) models of the Superconducting Super Collider (SSC) main-ring 5 meter focusing quadrupole magnet (211 Tesla/meter). The training performance of these magnets are herein summarized.

MAGNET CONSTRUCTION

Each magnet contained eight coils arranged in a two-layer "cos 2θ" pattern around a circular (40mm) bore (Figure 1). Coil construction and magnet assembly details can be found elsewhere [1,2,6]. In summary, all coils were wound under tension with 30-strand NbTi cable, and insulated with Kapton and epoxy-impregnated fiber-glass cloth. The coils were compressed to final size and heated enough to polymerize the epoxy. The resulting rigid coils were assembled onto an alignment mandrel and compressed by several quadrupole-symmetric, interlocking aluminum or S.S. collar-plate assemblies. The collar-packs were aligned by four collar tabs which fit into key-way grooves machined into the iron "yoke-blocks". Structural analysis of this collaring system was done by D. Dell'Orco [3]. Each iron yoke-block was aligned by inserting keys through openings in the shell. An end-to-end twist of less than 3m radians was achieved for all 5m magnets. The end-regions, outside the iron flux-return, were compressed azimuthally by an aluminum collet during the collaring procedure (QSC401 & 402, used S.S. C-shells.) Each end was immobilized and compressed axially before testing. Construction extended over a period of two years. Changes were made to the magnet end-region clamping and immobilization systems, collaring material, collaring pressure, collaring mandrel length and hardness, collar/yoke friction, conductor manufacturer and curing & collaring procedures.

TEST PROCEDURES AND RESULTS

The test procedures are discussed in detail elsewhere [6]. Magnets were tested in a horizontal boiling helium (1Atm) cryostat. The magnetic, strain-gage and training responses to thermal cycles were measured. The quadrupole gradient, and relative multipole purity were determined from Fourier analysis of the rotating coil signals [4,6]. Magnetic and strain-gage measurements were taken on-the-fly. The voltage-tap data was analyzed to determine quench-origin and propagation characteristics. Quench-training proceeded at 4.3K until a plateau was achieved or subcooling (2.5K) was used to accelerate the training process. The early short (1m) magnets were also trained at 1.8K (10kA) to help identify potential weak areas. The MIITs [6] were calculated to compare various magnet protection methods.

Ramp-rate sensitivity (tested from 1A/s to 10kA/s) varied from magnet to magnet (Figure 2). The low ramp-rate sensitive magnets had either been collared more than once (QSC401, QCC405) or used a new conductor surface treatment (QCC406). These magnets also exhibited higher maximum MIITs [6] (<9 MIITs, unprotected).

Figure 3 shows the 4.3K, 16A/s training behavior of these magnets. All training quenches started in the pole-turn, usually in the inner coil at the start of training, but moving to the outer coil (as expected [1] when the plateau was achieved. Long (5 meter) magnets (Figure 4a) trained as fast as short (1 meter) magnets [6].

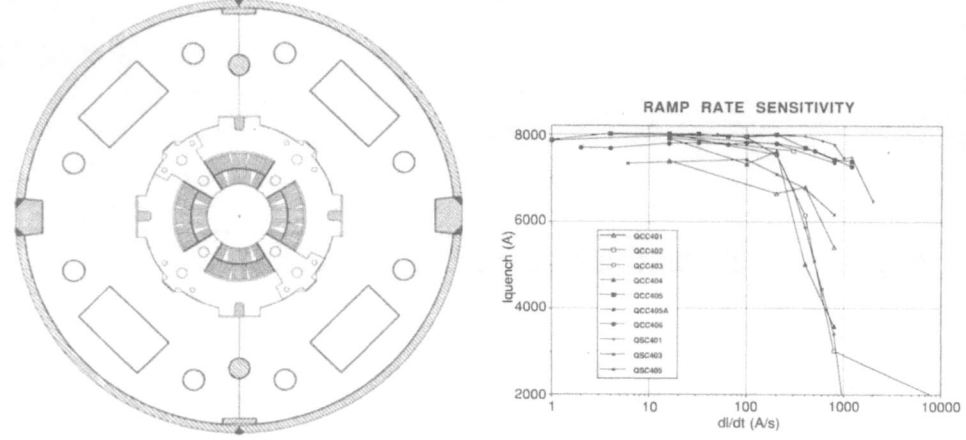

Figure 1. Collared Coil cross-section: coils, pole-pieces, collars, keys and yoke alignment tabs

Figure 2. Ramp-rate Dependence of Plateau Current (all Magnets)

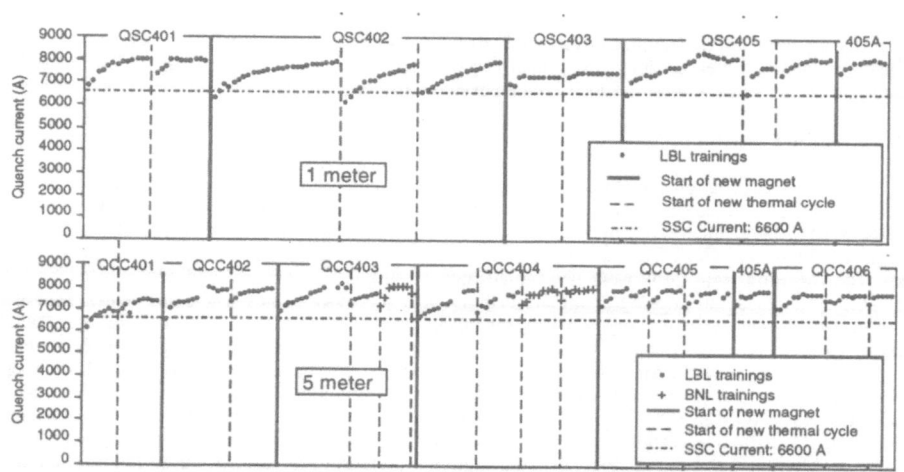

Figure 3. Quadrupole Training Records (16A/s, 4.3K)

Performance improved substantially between the first and last magnets. The last two magnets required no training below 7170A. Later magnets generally started at a higher current (except QCC 404 and QCC 406) and required fewer quenches to achieve and maintain plateau levels. Magnets that had lower plateau currents required fewer quenches to attain plateau: QSC403, QCC406. Sub-cooling (<2.5K) reduced the number of quenches needed to reach the plateau current.

All magnets had to be retrained (Figure 4b) after being warmed to room temperature. Subsequent thermal cycles usually started higher and trained faster (some more dramatically than others). Clamping the collared coil firmly in the yoke did not significantly change the retraining (compare QCC405A with QCC405, Figure 4b).

Quench-origins (Q-O's): All training quenches originated in the pole turn (shown in Figures 5a,b, relative to voltage taps, strain-gages, collar-pack boundaries, hard-mandrel edges, and the inter-layer ramp-splice).

Each magnet showed a characteristic pattern of quench-origins (Q-O's). QCC404 (Figure 5a) revealed a band of Q-O's near the middle, and QCC406 (Figure 5b) suffered most of its Q-O's near its ends. Axial nonuniformities were usually preserved over thermal cycles but usually with a change in quadrant. Many Q-O's were at, or near, collar-pack boundaries. Strain-gage-triggered quenches (x) were used to estimate

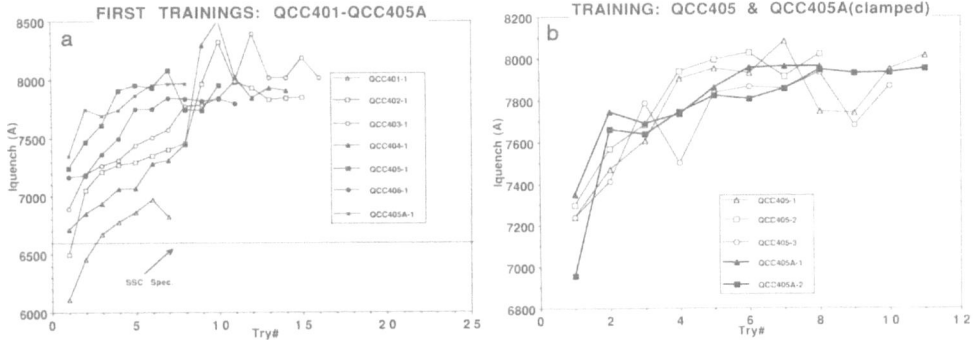

Figure 4a. 1st Trainings: 5m Quadrupoles (4.3K,16A/s)

Figure 4b. Re-Training after clamping & thermal cycling (4.3K,16A/s)

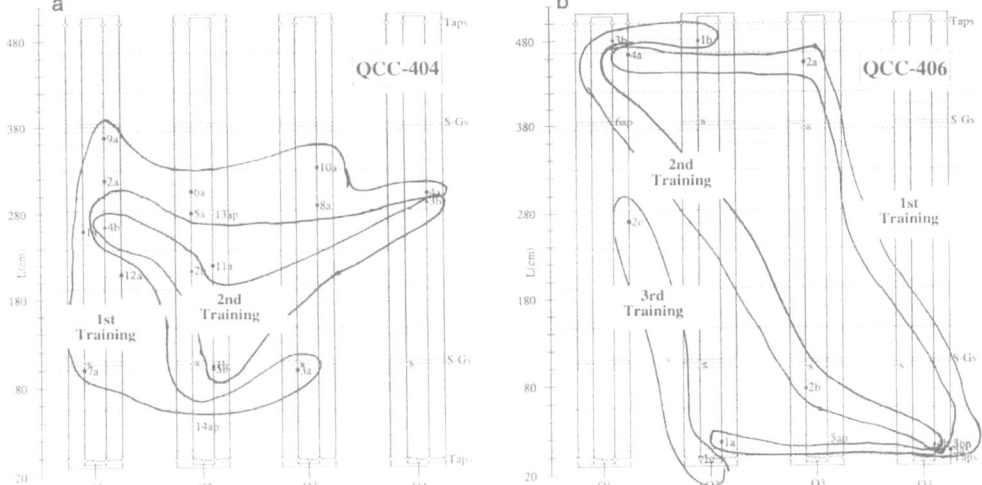

Figure 5a. Quench-Origins: QCC404 Trainings

Figure 5b. Quench-Origins: QCC 406 Trainings

systematic and random errors. A magnet rarely repeated any particular quench-origin. Exceptions QSC401 and QSC402 were later discovered to have end-clamp design oversights [6], and QCC406 is believed to have had a broken strand (right corner, Figure 5b).

Strain-gage response: After training, most magnets produced a linear response to the Lorentz load [6]. In spite of considerable variations in the initial stress (correlated to the measured coil sizes), substantial pole pressures usually remained up to 8kA. QCC406, however, suffered an unexpected amount of creep. Some locations were unloaded above 6kA [6], yet no obvious effect on training was observed. Short models were also tested to 10kA @ 1.8K with no lack of trainability. Ramping into virgin territory (I>5kA, Figure 6a) revealed a steeper unloading rate (i.e., lower coil stiffness). This "mechanical training" characteristic repeated on subsequent thermal cycles and is suspected to be related to the manner in which these magnets also "forgot" their training. While ramping to 1st quench, an anomalous stiffening was often observed prior to quenching (Figure 6b). The quench-origin was often located near the load cell which exhibited the largest anomaly. Magnetic multipole purity was very good [7].

111

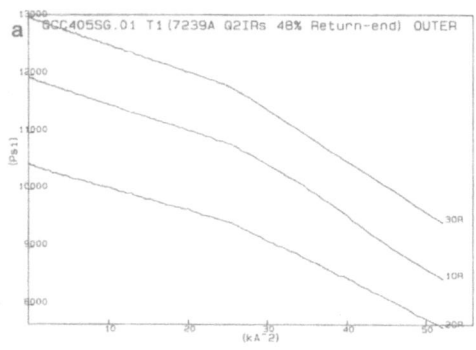

Figure 6a. Strain-Gage: Entering Virgin Territory

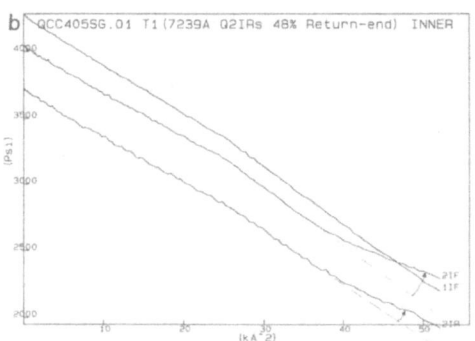

Figure 6b. Strain-Gage: Near 1st Quench

DISCUSSION AND SUMMARY

The importance of adequate coil clamping was dramatically supposed by the improved training after the end-region cold-clamping pressure was retained (magnets after QSC402, where some repetitive quench-origins were eliminated). How much pressure is enough, is questionable in view of QCC406's unloading, yet second best training.

The issue of optimum yoke-collar friction is also unclear. QSC405A (re-collared, welded and clamped in the yoke) trained the fastest of all magnets. QCC404, on the other hand (lightly clamped), trained much more slowly than magnets built to slide (QCC403, 405, 406); and QCC405A (clamped QCC405) showed no improvement at all.

The larger compliance observed during the first excitation, suggests that the cable/collar system has some freedom to deform under Lorentz loading. Stiffening observed just prior to quenching suggests that some parts of the system resist this deformation, and may constitute a source of stored energy capable of triggering quenches. Lack of memory retention after thermal recycling suggests the existence of restoring forces that are strong enough to restore a somewhat "virgin" state during some part of the thermal cycle. Movement of the quench origins to new locations implies that the magnet is "reset" to a different "virgin" condition.

The MIITs/ramp-rate sensitivity correlation is evidence that dI/dt-triggered quenching can decrease the cable temperature.

Ten pre-prototype magnets including six 5m models were designed, constructed and tested at LBL for proof-of-principle demonstrations of the design proposed to meet SSC operational requirements. Except for modest training above the anticipated SSC operating point, the magnets performed very well and proved to be self-protecting. Some design flaws were identified and corrected. The last two 1m models and all the 5m models have been reinstalled in cryostats at the SSC Laboratory, retested and used to achieve various milestones in their program.

REFERENCES

[1] C. E. Taylor, et al; *IEEE Trans. on Mag.*, vol. 27, No. 2, March 1991.
[2] S. Caspi, "The 40mm SSC Arc Quadrupole – Magnetic Design," SC-MAG-314, LBID-1677, November 1990.
[3] D. Dell'Orco, "Finite Element Analysis of the QC Quadrupole Magnet for the SSC," LBL-29600, October 1989.
[4] M. I. Green et al., "Measurements of Magnetization Multipoles in Four Centimeter Quadrupoles for the SSC," Cryogenic Engr. Conf., Huntsville, AL, June 11–14, 1991.
[5] J. M. Cortella, et al., "Mech. Prop. of 5m SSC Quadrupoles," Proc. of ASC, Chicago, Aug. 1992, SC-MAG 401.
[6] A. F. Lietzke, et al., "Quadrupole Magnets for the SSC," Proc. of ASC, Chicago, August 1992, SC-MAG 400.
[7] P. J. Barale, et al., "Magnetic Meas. of 5 Meter Quadrupoles at LBL, (paper IV-C-2, this conference).

Work supported by the Director, Office of Energy Research, Office of High Energy and Nuclear Physics, High Energy Physics Division, U.S. Department of Energy, under Contract No. DE-AC03-76SF00098.

FABRICATION AND AS-BUILT DESIGN OF THE 50 MM APERTURE SSC MODEL QUADRUPOLE MAGNETS

N. Hassan, D. Albone, C. Arden, D. Bailey, D. Bein, D. Block,
K. Couzens, S. Dwyer, R. England, A. Fluhmann, A. Jaisle, R. Jayakumar,
S. Krishnamurthy, E. McGuire, R. Mihelic, F. Nobrega, S. Phillips,
G. Snitchler, S. Smith, S. Stromberg, E. Vrsansky, R. Wood, and R. Zeigler

Superconducting Super Collider Laboratory*
2550 Beckleymeade Avenue
Dallas, TX 75237-3997

INTRODUCTION

Three 50 mm aperture model quadrupole magnets, designated as QSE-101 through 103, have been built at SSC Laboratory, two of which have have been tested at 4.25 K. The original concepts for the magnet and tooling design as well as assembly procedures have been evaluated in this program. Throughout this process, several magnet components were modified to improve the performance, reliability, and manufacturing of the magnet. The original magnet assembly procedures and design specific tooling were also evaluated and modified to accommodate fabrication of the model magnet. Fabrication and redesign of the model quadrupole magnet has resulted in good magnet performance as well as a more reliable, cost effective, and more production oriented magnet and tooling design. The new concepts which have emerged and the lessons learned from this effort are being applied to the design of magnets and tooling for the SSC Interaction Region quadrupoles as well as other SSC magnets which may be fabricated at the SSC. In this paper a discussion of the magnet and tooling design modifications, assembly processes, and the lessons learned during assembly and evaluation of QSE-101 through 103 is presented.

QSE-101 THROUGH 103

A brief discussion of design modifications, assembly processes, and lessons learned during fabrication of QSE-101 through 103 is presented in this section. Major parameters and cross section of the magnet are presented in Table 1 and Figure 1, respectively.

Traditionally, SSC inner and outer coils are wound and cured separately and are then spliced together. Since both inner and outer coils are made of the same cable, the splice between inner and outer coils has been eliminated in QSE magnets. Prior to coil winding, the total conductor for an inner and outer coil is divided on two spools and the inner/outer coil ramp is formed. In the coil winding process the inner coil is wound and cured on the winding mandrel. An intercoil spacer, made of G-11CR and Kapton[1], is placed on the inner coil which serves as the winding mandrel for the outer coil. The outer coil is wound and cured as an assembly with the intercoil spacer and the inner coil. The major challenge for winding these spliceless coils involves handling of the outer cable spool during the winding

* Operated by the Universities Research Association, Inc., for the U. S. Department of Energy under Contract No. DE-AC35-89ER40486.

of inner coils. In this process, the spool of the outer cable is suspended above the winding table and is allowed to rotate with the inner coil turns[2].

The inner radius of the QSE inner coil is 0.08 mm larger than that for the 50 mm aperture dipole magnet built at the SSC. The dipole coil winding mandrel was used to wind the QSE coils, lined with a 0.08 mm thick layer of Mylar[3]. This approach, which has been successfully implemented, saved the QSE program a considerable amount of time and money that would otherwise be required for fabrication of a new coil winding mandrel.

Table 1. Major magnet parameters.

Coil Inner Diameter	49.70 mm
Coil Outer Diameter	98.72 mm
Intercoil Spacing	0.55 mm
Coil End-to-End Length	1209 mm
Coil End Length	95 mm
Coil Straight Length	1019 mm
Inner/Outer Pole Angle	31.39 °
Inner Layer	-----
Number of turns	11
Conductor blocks	8, 3
Number of wedges	1
Outer Layer	-----
Number of turns	16
Conductor blocks	13, 3
Number of wedges	1
Collar Outer Diameter	140.40 mm
Yoke Outer Diameter	266.70 mm
Cold Mass Length	1.4 m
Cold Mass Diameter	276.5 mm

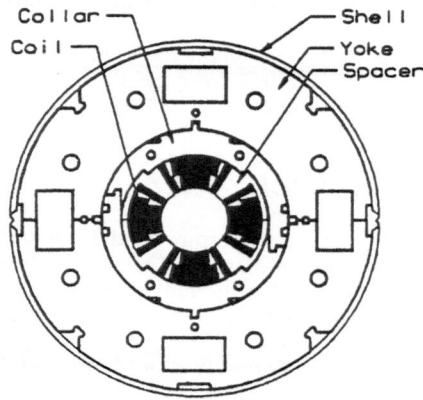

Figure 1. QSE cold mass cross section.

Inspection of QSE practice coils revealed separation of conductor blocks from adjacent coil end parts. This separation was more evident in the spacers and the fillers. The coil end parts had to be re-machined and re-molded to optimize the winding and to remove the improper turn placement after curing[2]. Particularly, the lead end filler part was reshaped using a new technique of shaping and molding with prepregs. The inspection of practice coils also revealed that the ramp between the inner and outer coils was too short. The lead end key, and the ramp forming process were modified to improve the configuration of the ramp. The short length of the ramp also resulted in a "dip" in the first turn of both inner and outer coils which are located at either side of the ramp. Experiments have indicated that increasing the length of the ramp solves this problem.

The cable used for winding QSE coils is a 36 strand cable, which is also used for winding SSC dipole magnet outer coils. This is a "left hand lay" cable. It has been determined that the left hand lay cable winds best clockwise and the right hand lay cable winds best counter clockwise[4]. Since the inner and outer coils are wound in opposite directions, one of them has to be wound in the "unfavorable direction". That is, the bend around the ends tightens the pitch of the cable which causes the strand to pop out of place. Typically, it is preferred to wind the inner coil in the favorable direction to avoid "strand popping". This is an important consideration, particularly since the tighter turn radius of the inner coil makes it more prone to the dangers of popped strands. The QSE inner coil is designed for counter clockwise winding, which is the unfavorable direction. A trial and error process was required to determine an acceptable cable tension that was also suitable for manipulating the cable around the ends to insure the integrity of the cable. The problem with popped strands which appeared in the practice coils was resolved as the cable tension was adjusted and the coil winding technicians gained experience in handling the cable for this particular design.

The QSE coil azimuthal size was measured in the "deviation from the master" manner[2]. The inner coil is measured alone, after it is cured. Following the final cure of the inner/outer coil pair, the coils are measured as a pair. Ground insulation was added to the coil pair to duplicated the ground insulation in the magnet. Azimuthal coil size measurements for the practice magnet, QSE-101, and 102 indicated the inner coils to be 0.05 mm smaller and the inner/outer coil pairs to be 0.18 mm larger than the design size. The coil size deviation from design target values was due to deviation of coil curing tooling from the design. The coil curing approach uses the "fixed volume mold cavity" concept. Since all tooling parts have a fixed geometry, the coil azimuthal size is determined by the available volume between the sizing bars and the keys, which occupy the pole region, unless, shimming or another mechanism is used for adjusting the coil size. In an attempt to bring the outer coil size closer to nominal, a 0.20 mm thick shim was placed on top of the sizing bars for curing QSE-103 outer coils. The result was a reduction of 0.10 mm in coil size. Similar behavior has been observed in other magnet programs. Further coil size reduction was not attempted as indications of excessive stress appeared on coil end parts. The coil stress indicated by strain gage collar pack was considerably higher than the design target values for QSE-101 and 102. The coil stress in QSE-103 was very near the target values indicating the effectiveness of the attempt to reduce the coil size in 103. All three magnets used the same size collaring shims.

The ground insulation on coil ends was also modified to remove difficulties in assembly. The stepped cutouts in the end insulation were eliminated. Instead, parts of the straight section insulation were extended in to the ends and additional pieces of insulation were inserted in the ends. The ground insulation modifications still kept the insulation seams on successive layers in a staggered arrangement.

The spacer pack assembly design included using roll-pins to stack the laminations. These roll-pins were inserted into the holes of the stacked and aligned laminations using a tabletop arbor press. The longest available roll-pin for the specified holes in the spacer lamination was 30 mm long. This configuration, which was used for QSE-101, was determined to be ineffective and time consuming. QSE-102 and 103 used 160 mm long tubes with swaged ends for assembly of the spacer packs. This modification has considerably improved the assembly process. The strain gage blocks which have the same profile as the space laminations were 0.20 mm longer than the coil, causing concerns about interference at the mid-plane between top and bottom assemblies. The gage blocks were modified for QSE-102 and 103 by removing the excess material.

The collar laminations were about 0.1 mm smaller in the horizontal diameter and about 0.15 to 0.25 mm smaller on the vertical diameter due to fabrication error[2]. The collaring keys were about 0.08 mm smaller than the collar keyways, also due to fabrication error. Normally the smaller diameter of the collars would have caused a great concern. However, the vertical collar deflection after keys were inserted was considerably more than predicted by the finite element analysis that was performed in the design phase. The predicted and observed deflections of the collared coil are listed in Table 2. Limitations of the finite element model were determined to be the source of discrepancy between the predicted and observed collar behavior. The observed horizontal deflection is identical to that observed in dipole magnets. This was expected, despite predictions of the finite element model, since QSE magnets use dipole style collar laminations. The vertical collar deflection, however, is considerably larger than that observed in dipole magnets. The collar design has been modified for future 50 mm quadrupole magnets.

Table 2. Predicted and observed collar deflections of QSE-101 through 103

---	Horizontal	Vertical
Predicted Collar Deflection, (mm)	0.11	0.12
Observed Collar Deflection, (mm)	-0.03	0.30

Due to use of separate pole spacers, QSE coils can not be assembled in the same manner used for assembly of conventional dipole and quadrupole coils. The QSE coils are assembled on a plastic mandrel which is removed after collar packs have been installed. In the coil assembly process the tooling holds the coils and the insulation as they are rolled over to insulate each quadrant[2]. This process is followed by installation of end ground insulation and collaring shoes. The spacer packs are installed and held in place by the tooling as the assembly is rotated to install spacer packs in each quadrant. Collar packs are the last components to be installed prior to keying of the collars, which is done in an identical manner to that used for keying of dipole magnet collars. The tapered collaring keys of the QSE magnet were inserted using a "square key" method. An approximately 50 % spring-back loss in coil stress has been observed when the vertical pressure from the press is removed.

Both the low carbon steel laminations and the stainless steel yoke laminations were stamped from the same die. Stamping and fineblanking experts agree that stainless steel laminations produced by a die that is made for stamping low carbon steel laminations are very likely to be distorted. Inspection of the QSE stainless steel yoke laminations revealed that the inner diameter of the stainless steel laminations was about 0.05 mm larger that the design size. This however, is thought to have been compensated for by the large vertical deflection of collar laminations. The stainless steel yoke laminations, used in the cold mass end yoke modules were to be epoxied to each other to prevent buckling of these laminations under press and welding loads. This process was not employed. Instead, the Stainless steel laminations were welded together after they were stacked and aligned in a fixture. In the welding process, filler material was not used and weld beads were made in locations were interference with adjacent or mating parts could not be caused.

The shell used in 40 mm dipole magnet was also used in QSE magnets. The arc length of the shell was reduced to make it compatible with the QSE alignment bars which are welded to the shells. The splice contour on the first splice plate which contains one of the three coil pair-to-coil pair splices, had to be modified to increase the hard bend radius, since the small bend radius caused the cable to collapse in first few attempts to form the cable.

ACKNOWLEDGEMENTS

The authors would like to express their appreciation to the SSC Magnet Systems Division staff, particularly the technical staff of the Magnet Development Laboratory for their dedicated efforts to make the 50 mm quadrupole program a significant accomplishment for SSC Laboratory.

REFERENCES

1. Kapton is a registered trademark of DuPont, Inc.
2. R. Jayakumar, et al, 50 mm quadrupole model magnet, QSE-101, MD-ENG-92-004, (1992).
3. Mylar is a registered trademark of DuPont, Inc.
4. G. Jochen, private communication.

BUILDING THE REPOSITORIES TO SERVE

DeLynden Lersch

Superconducting Super Collider Laboratory*
2550 Beckleymeade Ave., MS 1074
Dallas, TX 75237-3997

The project to design and build the Superconducting Super Collider (SSC) Laboratory also includes the exciting opportunity to implement client/server information systems. Lab technologists were eager to take advantage of the cost savings inherent in the open systems and a distributed, client server environment and, at the same time, conscious of the need to provide secure repositories for sensitive data as well as a schedule sensitive acquisition strategy for mission critical software.

During the first year of project activity, micro-based project management and business support systems were acquired and implemented to support a small study project of less than 400 people allocating contracts of less than $1 million. The transition to modern business systems capable of supporting more than 10,000 participants (world wide) who would be researching and developing the new technologies that would support the world's largest scientific instrument, a 42 Tevatron, superconducting, super collider became a mission critical event.

This paper will present the SSC Laboratory's strategy to balance our commitment to open systems, structured query language (SQL) standards and our success with acquiring commercial off the shelf software (COTS) to support our immediate goals. Included will be an outline of the vital roles played by other labs (Livermore, CERN, Brookhaven, Fermi and others) and a discussion of future collaboration potentials to leverage the information activities of all Department of Energy (DOE) funded labs.

INTRODUCTION

The Superconducting Super Collider (SSC) Laboratory was established to design, build, maintain, and operate the Superconducting Super Collider, a high energy subatomic particle accelerator to be used in basic scientific research to learn more about the fundamental nature of matter and energy. When completed in 1999, the Super Collider will be the most powerful subatomic particle accelerator in the world.

Knowledge about the basic particles and forces of nature has reached a critical point in the advance of science and civilization. Our understanding of the complexities of our universe has dramatically increased during the past 20 years, but puzzling questions still exist about how the universe is constructed and how it behaves and the relationship of matter to energy. The scientists of the Super Collider probe for answers to these questions and challenge our understanding of the fundamental particles and forces of nature.

The SSC Lab is being developed and managed for the U.S. Department of Energy by the Universities Research Associates, Inc., a non-profit institution established by 79 major research universities in the U.S. and Canada.

The project combines many elements of education on the grand scale: collaborationists from more than 100 colleges and universities, a strong commitment to improve American math and science education that includes talented primary and secondary teachers from all 50 states, and

---

* Operated by the Universities Research Association, Inc., for the U. S. Department of Energy under Contract No. DE-AC35-89ER40486.

exciting programs to interest and challenge young (potential) scientists from all economic levels. Other important products of the project's advancements include new cancer treatment techniques and highly competitive commercial advancements in the use of massively parallel computing tools.

At the same time, an important scientific instrument design and a major construction project will dominate the culture of the Lab for the next few years and those projects require support and excellent management and control systems at every stage. One of the decisive events of the project has been its transition from the older, primarily desktop, platforms for business and project control systems to state-of-the-art, robust business platforms.

While dozens of magazine articles today tout the cost benefits/challenges of "downsizing," none counsel strategies and tactics for "upsizing," i.e. the establishment or retrofit of the repositories necessary to review and manage important tasks. In 1992, the Super Collider initiated its overhaul of its mission-critical management information systems.

ARCHITECTING THE FUTURE

In the initial project schedule, the Lab was to have replaced its early, small project management and business support systems acquired in 1991 to support its early feasibility studies. At that time, major systems design and testing was to begin, large equipment acquisition and installation development would be underway; in short: systems capable of supporting a $300 million/annum project would be required. Intensely conscious that the decisions made during the next few months would mark the project's information systems architecture for years, the selection teams of 1990 and 1991 evaluated many architectures: hierarchical, relational, object oriented. These early selection teams had a real desire to implement object oriented technologies. Unfortunately, the commercial application of object oriented database (OODB) was still immature: few commercial packages were available in 1991 to support mainstream business functions. Worse, many vendors in this area were new to the market, their stability and commercial viability characteristics unknown. Relational database systems and more traditional hierarchical systems were available but they would have to be carefully evaluated, and their integration strategies plotted and approved before they could be considered.

After months of thoughtful debate, the Lab agreed upon a mixed strategy of SQL-based, commercial-off-the-shelf-software (COTS) whenever possible. Selection teams and support groups were formed to support the Laboratory's diverse computing and network communications needs (see figure 1).

The first project undertaken was to establish the core of the major business systems. In the winter of 1991, the Superconducting Super Collider Lab began efforts to develop or obtain an on-line, highly integrated, laboratory procurement system that would handle the 15,000 purchase requisitions and 10,800 purchase orders that would be processed and placed among a pool of more than 10,000 vendors each year beginning in FY93. This mission critical upgrade effort turned into a project that would ultimately result in the overhaul of all the Lab's basic business systems. This project became "P.A.R.I.S.," the Lab's Purchasing, Accounting, and Requisition Information System.

One initial thought was to attempt to integrate the newer system with the existing accounts payable (AP), general ledger (GL) inventory, and manufacturing resource planning systems and broaden support for receiving and encumbrance tracking functions. However, early analysis confirmed that the ramp-up in purchasing workload was, of course, MATCHED by a ramp up in accounts payable workload. Therefore, it became apparent during the initial analysis that in order to improve the entire acquisition process, the project should be expanded to include a new accounts payable software package.

The needs analysis also identified critical functional areas and users to represent those critical functional areas to be included on both the software selection and implementation committee. Active, sometimes full time, participation of these users turned out to be a major strength of the project.

Requests for Information (RFI's) were sent to over 25 software vendors. As the results from the RFI's were reviewed, it was apparent that replacement of the existing General Ledger systems should be included in the P.A.R.I.S. project. A General Ledger RFI was subsequently prepared and sent to the original vendors. In addition, twelve additional vendors were then invited to participate. Although the initial assumption within the Super Collider applications teams was that no one package would satisfy all three groups, it became apparent that a one vendor solution would help ensure integration.

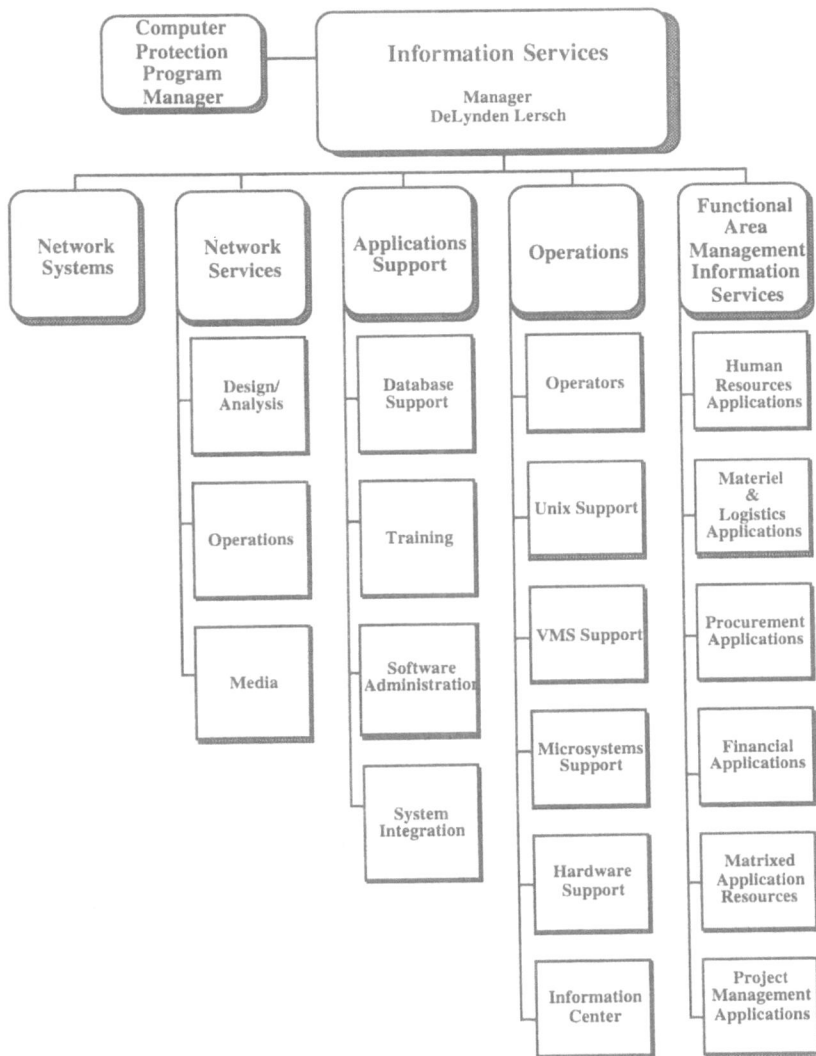

Figure 1. Organizational Chart of Information Services (3/2/93).

The responses to a combined selection package of nearly 400 requirements clearly indicated that while many of the vendors had integrated Procurement, AP, and GL applications, few had incorporated government accounting and on-line procurement requirements into relational database systems. Procurement and Financial teams agreed on the value and priorities of requirements, most of which were met by several vendors. Finally, two requirements eliminated many of the contenders: the winning package would have both online procurement tracking and government fund accounting. Only one package claimed to meet all the other requirements and supported both online procurement tracking and government fund accounting: Oracle Government Financials.

LEAD, FOLLOW OR GET OUT OF THE WAY

Mission Critical is an overused phrase these days, but in this instance, it was accurate. The Lab had to migrate from the small, primarily PC-based systems that had served well during the early design and feasibility stages of the project onto systems that could manage, rapidly and accurately report the plans and progress of international scientific collaborations and a $350 Million a year major construction project. The Lab's new Integrated Planning and Scheduling System would be using Work Package Breakdown Structure (WBS) package identifications that that were 12 characters in length (the Lab's existing General Ledger package allowed only 6 characters). In addition, the ability to accurately track sources and uses of funds, types of costs and labor categories absolutely had to be in place for FY93 to drive the new Integrated Project Schedule and all the new DOE reporting requirements. That meant the SSC Lab had 10 months in which to do a 27 month job.

That was the bad news. The good news was nearly everyone on selection committees understood that and committed to get the job done.

Three key factors were necessary to support the schedule:

Organized DOE support
Prompt assistance from other laboratories
Experienced, professional contractor assistance.

OUTLINE THE PLAN

Fortunately, the Lab had kept its DOE reviewers briefed on all developments. Detailed plans for a fast track implementation were developed and approved. Eight contractors would be called in to give support in both the implementation of the new software and to maintain the existing systems in order to free permanent labor to train and supervise the implementation. Other laboratories supplied temporary database support and draft procedures.

More than 150 users, administrative support professionals, analysts, technicians and programmers worked days, nights and weekends through the summer and fall to set up the logic, check, electronically transfer and recheck and verify the information associated with older non-integrated systems. New procedures were written; teams and processes were integrated; and problems were identified and resolved. "The systems have been operational since late October and we are very pleased with the user support and acceptance," says Tony Reed one of the two PARIS project managers. Sharon Valenzula, the Oracle System Application Administrator adds, "We have over 450 active users at present, 200 of whom are usually on line."

P.A.R.I.S. IN THE FALL

The software was stable and the contractors who came to help were resourceful and very professional. The challenge for the conversion turned out to be the data itself: years of keeping separate lists of data, unintegrated, on small personal computers meant hundreds of exceptions among invoice numbers, purchase orders, old Work Breakdown Structures and new Work Breakdown Structures. All in all, the costs associated with more than $1.5 Billion and 75,000 data entries had to be converted, reviewed and confirmed. The effort took more than 100 days, with teams working holidays and weekends.

THE SSC LAB IN THE SPRING

The Lab has officially converted now to the new software. Not all functions and reports within the new system are running in production. (Budget data is still being collected and carried in the Integrated Project Schedule. That data should be released for use by the new financial and procurement systems in April.) The conversion caused huge variances in all the monthly (in current period) Cost Performance Reports for the first quarter. More than 300 people worked on the project, most without overtime or any extraordinary compensation.

CURRENT STATUS

As of March 15, 1993, all software is functional (although the budgeting family of reports is awaiting approval of official budget data before reporting is instituted). All of the pre-FY93 data has been transferred, validated and integrated. Approximately 2 weeks of work remains to "catch up" with the data input neglected, by necessity, during the implementation. (Accounts Payable actions were conducted manually for much of the first 90 days of operation.) Requisition time from inception to buyer release has dropped from an average of 20 days to 10.

Most of the next few months will be spent developing commitment and financial reporting systems: our users have discovered a treasure trove of data is now available and have submitted more than 80 programs, primarily reports to be programmed.

Would we do it again, 100 hour work weeks and all? Yeeeeessssss, but...

LESSONS LEARNED

Not all policies and procedures governing laboratory business had been reviewed and approved and institutionalized prior to software acquisition and implementation. Some work groups, quite properly, seized upon the implementation of new software as an opportunity to also implement additional business control/integration. Procurement in particular had just finished a review of the purchasing practices of other laboratories and was able to bring in "best of breed" business practices as part of the implementation.

THE IMPORTANCE OF OTHER LABORATORIES/DOE PROJECTS

Special thanks must go to other institutes that shared knowledge, sent personnel and encouragement to the SSC Lab:

LLNL	– Jean Deir	NREL
– Bob Zanetel	– Gordon Jurgenson	– Paul Dragseth
– George Beck	– Eric Davies	– Mary Meinecke
– Rick Locatelli		
– Brad Calderon	Brookhaven National Lab	Wichita Airport Authority
– Mary Orr	– Ed Gallagher	– Karlis R. Otannkis

FUTURE COLLABORATIONS

This support has not waned with the successful implementation of the first Lab-wide business systems. At the time of the submission of this paper, we are currently reviewing the Environmental Health & Safety software developed by Fermi Lab and many of the logistics support packages used by CERN.

FUTURE PROJECT EFFORTS

These inter-Lab cooperative efforts encompass only a portion of the Lab projects during 1993 and 1994. During that time the Super Collider will be initiating/implementing whole families of applications:

- Overhaul of the Lab's Human Resources Information Systems
- Additions to the Lab's Financial, Procurement, Project Management Business Systems
- Major Subcontract Tracking System
- Fiscal Year Planning System
- Laboratory Funds Management System
- Directorate Executive Information Systems & Director-Net Project
- Lab-wide Calendering Support
- Standardized Facilities Information Systems & Geographic Information Systems (interfacing with the Lab's existing CAD deliverables)
- Upgrades to the Lab's Electronic mail systems
- Upgrades to the Lab's Document Management
- Testing of Massively Parallel Systems
- Continued development of the challenging Geodesy Systems
- Upgrades to the Lab's Warehouse System
- Integration of many of the systems to support Equipment Installation
- Equipment Systems Support and Tracking
- Automated Data Processing Short Range Planning, Information Resource Management,
  - Enhancements to the Lab's Library Systems
  - Implementation of ES&H Systems

We hope many of you who offer solutions in these areas will contact us to discuss them, register for our bidder lists and help us implement the open architecture, distributed systems for the next 10 years.

CONSTRUCTION PHASE APPLICATIONS OF THE SSC GEOGRAPHIC INFORMATION SYSTEM

A. Oslin, R. Brees, and S. Sipes

Superconducting Super Collider Laboratory[*]
2550 Beckleymeade Ave.
Dallas, TX 75237

INTRODUCTION

A Geographic Information System (GIS) was recognized as a tool for facility planning and public relations coordination early during SSC construction. Since October 1991, the SSC GIS group has been compiling regional, planning, engineering and surveyed information about surface and sub-surface features in the project area. Descriptive information (attributes) are stored in a relational database and associated with the mapped features. GIS provides a support service for many people working on the project by coordinating geographic database development between SSCL, DOE, PB/MK, Texas National Research Laboratory Commission (TNRLC), Ellis County, and the North Central Texas Council of Governments. Several construction phase applications have been identified and are in various stages of development, including GIS input to: site development planning; facility construction tracking; Land Information System (LIS); Real Property Inventory System (RPIS); building and structure naming; construction access routing; environmental permitting and annual impact reporting; chemical and hazardous material inventory system; emergency preparedness planning and E-911 system interface; land use management; and natural and cultural resource baseline characterization and monitoring.

GIS GENERAL DESCRIPTION

The SSC GIS is a computer-based system that provides a development, management and analysis environment for geographically positioned features integrated with descriptive information (attributes) in a relational database. GIS stores surface and sub-surface information about the project area as a series of coverages each representing a different theme or topic of information. The concept of coverage implies that an "intelligent" or topological structure exists between the geographic elements representing mapped features and database attributes associated with the features. Questions can be asked concerning the

---

[*] Operated by the Universities Research Association, Inc., for the U. S. Department of Energy under Contract No. DE-AC35-89ER40486.

relationship between features and/or attributes. Maps and reports can be created to illustrate or summarize the results. GIS is an analysis tool where spatial relationships between various map features are identified then used to resolve conflict, evaluate impact and help manage project resources.

The system is being designed to encompass information that many users across multiple disciplines require for effective planning, design, engineering, construction, environmental monitoring and facility management. User needs will define the ultimate goals and informational content of the GIS and these needs vary from discipline to discipline. The SSC GIS will evolve to accommodate applications during all phases of the project's life. The GIS will integrate with other SSCL Technical Systems for applications that require geographic and related attribute data. In particular, site development planning, facility and real property management, environmental compliance and monitoring, radiation modeling, emergency preparedness and management, and materials management are identified as areas where technical systems integrate with GIS.

Intergraph, Arc/Info and ERDAS software are being used to develop two forms of geographic data; the vector model that represents map features as topologically structured points, lines or boundaries (polygons) and the raster model that represents information as images of pixels (picture elements) or grid cells. Attribute or descriptive data is to be maintained in an Oracle relational database with ties to the geographic data models. Digital data of potential use in the development of the SSC GIS include: Oracle and other relational database fields and records; ASCII data, File Maker Pro and spreadsheet files; CADD 2D and 3D drawings; point, line and polygon topological coverages; elevation and 3D models; digital orthophotography and satellite imagery; video and scanned photography; and scanned documents. These data types can be related to each other by a common geodetic network that provides NAD 83 Texas State Plane Coordinates to geographically tie the information together.

GIS application development is a process whereby the relationship between user needs, business functions, data sources and levels of accuracy are defined, then materialized into a system using documented standards. Several user groups from scientific, engineering and environmental disciplines have been identified. It is apparent that the SSC GIS must be multi-purpose and accommodate a wide variety of applications at different levels of map accuracy. Accuracy requirements have been grouped into six general categories; the survey level Land Information System (LIS), Engineering (1:480 to 1:2,400), Vicinity (1:4,800 to 1:9,600), Planning (1:12,000 to 1:48,000), Regional (1:100,000 and 1:250,000), and U.S./World Atlas (1:2,000,000). GIS provides digital base maps that can be used to manage attribute and other map information at these various levels of scale and accuracy.

CONSTRUCTION PHASE GIS APPLICATIONS

Site development planning is an ongoing process at the SSC. Now that GIS data is accessible, the system is being used in the design process. Potential conflicts can be identified when planning surface facilities or predicting surface impacts of underground structures. The GIS can overlay existing natural and man-made features as well as the proposed structure to "flag" interferences early in the design process. The GIS can also be used to calculate a rated plan for facility siting by indexing physical, environmental and other constraints. The system is used to evaluate design and alternatives so that better options can be forwarded through the design process. GIS is being utilized in this manner as the SSCL revises the Site Development Plan. GIS brings up-to-date facility and environmental resource information to the planning table so that master planning can progress efficiently.

Standard project maps have been prepared to locate facility sites, buildings and structures. The GIS Section coordinates with construction Task Leaders to ensure that building and structure numbers and names are consistent with the SSCL Standard Naming Convention. Buildings and structures are added to the composite database no sooner than the 90% design submittals. The location of buildings and structures is converted from the original CADD drawing files and processed into the GIS buildings and structures coverage, where structures are associated with descriptive information in the relational database. The geographic position of buildings and structures is updated throughout the remainder of the design process.

As-built footprints of buildings and structures will be entered into the database as they become available. However, there is a significant amount of time between when a building or structure is completed and when the as-built information is reflected in the contractor's CADD drawings. In this interim period, aerial photographs will be geo-referenced to satellite imagery and digital orthophotography, then used to temporarily map facility features. This expedites the visualization of major design changes and allows the database to track the location of temporary structures such as trailers and contractor lay down areas. GIS provides the location information to the Real Property Inventory System (RPIS), used by DOE to track real property.

A dynamic map has been developed by the GIS group to track contract awards, general construction activity and progress of the Tunnel Boring Machines. Information on construction status is collected at the beginning of each week and used to update the Collider Construction Progress Map. Copies of the map are posted bi-weekly at several locations throughout the Laboratory. A subset of this mapped information is illustrated in Figure 1.

The GIS is used to assist in the management of SSCL land. During the land acquisition process Universal Field Services, Inc. under the direction of TNRLC, compiled all surveyed parcels into a composite land base for the project. The GIS parcels coverage was processed from the composite land base. The coverage is used to track land ownership status for construction access and land use maintenance. Even though parcel boundaries within the site are no longer indicative of land ownership; historical, archeological, environmental, and land use management information is tied to the geographic features representing expired parcels. Utility easements, Right-of-Way boundaries, and survey markers will be extracted from the electronic plats for the land use management applications.

Applications are being developed for the SSCL Environmental Safety and Health (ES&H) Division and ES&H Oversight Department as well. Map and location information from the system is being used for permitting and other environmental compliance and mitigation issues. The GIS is working with ES&H groups to establish the environmental baseline to which changes will be measured and monitored. GIS coverages and spatial analysis capabilities are being utilized in emergency preparedness planning and management. The ES&H chemical inventory database will also interface to the GIS for location information and spatial analysis as required for material management.

Priority GIS construction phase tasks include: 1) development of all LIS and engineering coverages for use in construction, facility planning and land use management activities; 2) survey system integration for as-built and other coordinate data visualization; and 3) production of thematic coverages such as the Ellis County street network, construction access routes, geotech and soil information, hydrology, land use - land cover, political boundaries, utilities, and environmental resource data for planning and construction applications. GIS operation and maintenance is scheduled to begin Fall 1993 for critical applications such as land use and real property management, building and structures naming, site development planning, emergency preparedness, and survey information system interface. Additional GIS applications will be defined as end user interfaces are developed for these critical projects. Ultimately, database query, cartographic system interface and menu options will be developed so that efficient links to other technical systems are provided for SSCL facility management.

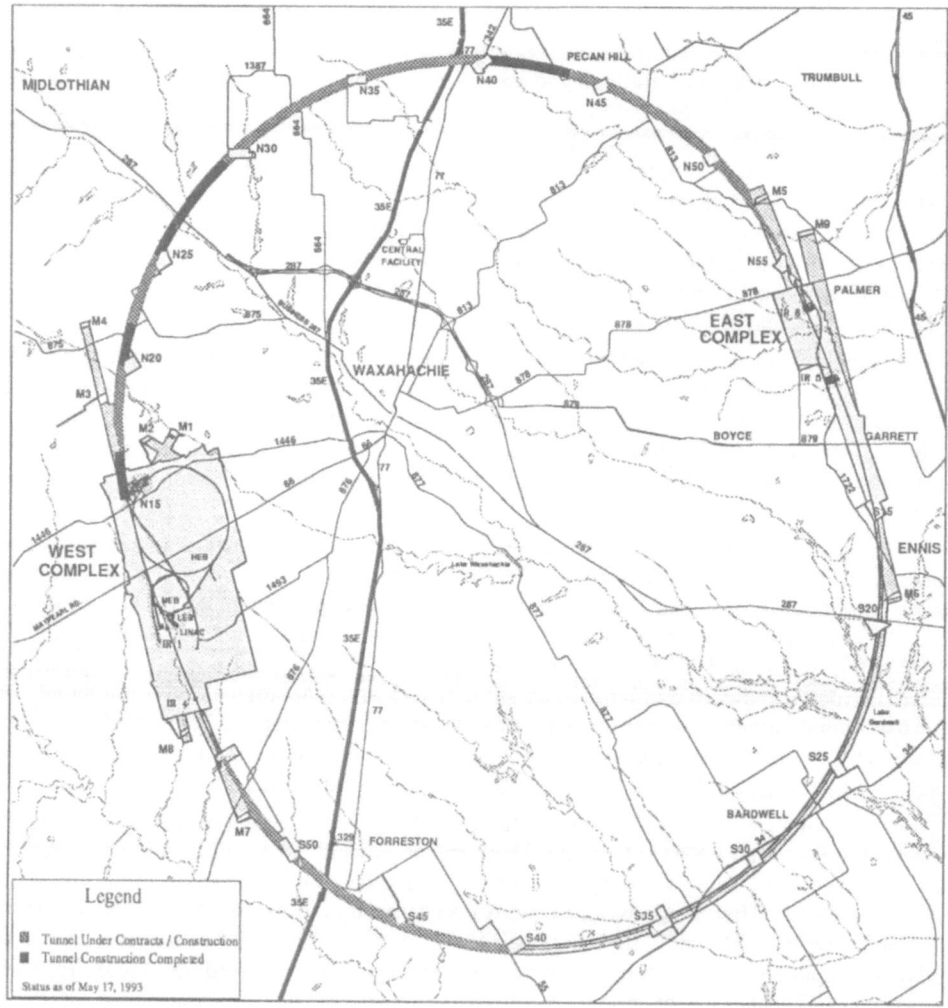

Figure 1. Information subset from the Collider Construction Progress Map showing weekly advancement of Tunnel Boring Machines.

FACILITY MODELING OF WEST UTILITY STRAIGHT TUNNEL SECTION

Martin W. Butalla,[1] Brett Parker,[2] Dale Orth,[2] and Amanda Elioff[3]

[1]Conventional Construction Division
[2]Program Management Organization
 Superconducting Super Collider Laboratory*
 2550 Beckleymeade Ave.
 Dallas, TX 75237

[3]The PB/MK Team
 5510 S. Westmoreland Road
 Dallas, Texas 75237

ABSTRACT

The Project Management Organization (PMO) and the Conventional Construction Division (CCD) of the Superconducting Super Collider Laboratory (SSCL) used the computer as a tool to build a repository for both physical and attribute information, in graphical and textual forms, related to vital technical and conventional components of the West Utility Straight Section of the SSC main tunnel. CCD and the PMO realized the benefits of computer intelligent modeling during conceptual design of the technical component configurations in this tunnel section. Benefits included interference resolutions, time savings, visual problem solving, underground space conceptualization, redefinition of multiple "what-if" concept scenarios, and the preparation of design requirements.

INTRODUCTION

The 87 km circumference SSC main tunnel will house the liquid-helium-temperature superconducting magnets (cold magnets). In a few places around the ring, these cold magnets will be omitted to provide space for warm components. All of the technical components necessary for getting the beam into and out of the main collider accelerator, as well as all of the radio frequency (rf) power equipment needed to accelerate the beam, are concentrated in a single warm region, collider's West Utility Straight Section. Although superconducting bending magnets are omitted from this region, there are isolated cold quadrupole magnets located in this region to keep the beam well focused, thereby necessitating a cryogenic bypass line through the entire region.

---

* Operated by the Universities Research Association, Inc., for the U. S. Department of Energy under Contract No. DE-AC35-89ER40486.

This single utility region will contain a grouping of the most complex technical equipment (most still under conceptual design) within the collider system: warm and cold magnets, pulsed kicker magnets, beam scrapers, collimators, beam instrumentation and damping equipment, low-conductivity water (LCW) and power supply hardware, cryogenic bypass components, and rf klystrons and waveguides. In addition, complex underground construction features in this region include rf galleries, painter magnet galleries, two external beam abort line tunnels extending from the main tunnel to the beam backstop facilities, a portion of the High Energy Booster (HEB) tunnel, the transfer tunnels from the HEB, and eight shafts. To add to this already complex tunnel configuration, two of the eight shafts will have connecting adits to both the main collider tunnel and the HEB. The transfer and abort lines cut across all of the other subsystems in this region, thereby inducing a tremendous coordination effort.

PURPOSE AND SCOPE

During the summer of 1991, PMO and CCD began a study to develop an accurate electronic model of the technical and conventional components of this complex region. The primary purpose of the study was to provide a mechanism through which members of the technical design teams could develop individual specific requirements for this area as the design of their technical components and systems progressed. A secondary purpose was for the joint development of the design requirements for the conventional facilities that would house and support the technical systems.

The model included primary equipment, space allocations for piping, cabling, and ducting, and pertinent architectural/structural features. In addition secondary components were added to the model, including supporting cryogenic equipment and piping, and access galleries. Future additions to the model would include electrical and mechanical support equipment. The model represented approximately 1200 m of main collider tunnel centering around the intersection of the HEB injector and abort beams. Each abort beam is contained within tunnels that are approximately 1500-m long extending northward and southward from the center of the straight section to the backstops.

TECHNICAL ACCELERATOR COMPONENT MODELING

The coordination necessary for the design, construction, and integration of technical systems in this region generated concern. The limited, confined tunnel and niche spaces could have potentially interfered with the installation and/or operation of technical systems and components. Interferences discovered too late during design could have seriously jeopardized cost and schedule. Early design considerations were communicated to all design groups as early as possible so that resolutions could be tested and proved across all systems. Design considerations included component installation and maintenance, and whether components needed to be removed or serviced in place.

During the summer of 1991, people and computer equipment were scarce throughout the SSCL. Although the SSCL was beginning to acquire resources to meet its forecasted manpower and equipment needs, modeling needed to begin quickly to maximize communication during the conceptual evolution of the design. In spite of the drastic lack of resources (yet with hopes that the model would generate a fully-functioning information system), Construction Systems Associates (CSA) of Marietta, Georgia was contracted to begin a seed modeling effort.

CSA had provided modeling services for the world engineering community, focusing on the highly-complex nuclear power industry. CSA was able to provide the computer and personnel resources necessary to initiate the modeling effort. As a outside force, CSA was able to extract technical information from the large number of engineers and physicists at the

SSCL, incorporate the information, and present the results to the SSCL in the form of a computer model. Once the iterative design process was established, PMO, which had organized an in-house modeling team, continued the effort when CSA's work was completed. The model has continued to advance to what it is today, a three-dimensional electronic representation, with database intelligence and to scaleable accuracy, of all major components and assemblies in the collider West Utility Straight Section.

During model development physical and spatial interferences were detected, discussed, resolved, and tested between magnets and other major components. Abort beam line interferences were detected as beam line paths traveled through intricate components. Most potential major technical configuration problems were resolved in a time frame that permitted the conventional facilities design requirements document (DRD) to be completed within three months of the initial baseline schedule for delivery to the Architect-Engineer/Construction Manager (A-E/CM). Future phases of this work will include: component stand and interference studies, installation sequencing, operation and maintenance studies, and radiation shielding studies.

CONVENTIONAL FACILITIES DESIGN

The DRD for the collider tunnel contracts included a written description of the facilities and the items to be designed, applicable standards, and two dimensional drawings of the facilities showing minimum/maximum dimensions of the underground openings and future equipment layouts for tunnel, shaft, and adit cross-sections. For the basic tunnel design of the West Utility Straight Section, the DRD also included the electronic model. Conventional facilities requirements and construction contracts for the main collider tunnel have been separated into basic and finish contracts so that basic tunnel construction (excavation and initial support) may begin as early as possible with finish work completed by a follow-on contractor. This breakup of contracts allows additional time for the completion of finish design work that includes tunnel cooling, ventilation, and electrical systems. The ability to include future systems in the model to establish overall space requirements allowed timely development of the basic DRD.

During development of the basic DRD, equipment layout was arranged electronically to establish required tunnel cross-sections. For example, the studies concluded that the West Utility Straight Section would require a 17 foot minimum tunnel diameter rather than the 14 foot diameter used in other portions of the collider tunnel. The additional area was needed to accommodate the scraper shielding, beam abort and tune-up beam line magnets, 34 inch cryogenics bypass line (weighing 450 kg/m), transport vehicles, and installation and maintenance lanes, among other components. Through use of the model, it was also found that the tunnel width at the abort tunnel transition could be shortened if the abort lines followed the inside of the abort tunnel. Further refinements in the tunnel centerline alignment relative to the beam line are being studied during the Title I (30 percent) design phase currently in progress.

The West Utility Straight Section requirements for the basic tunnel design and construction are unique. The most challenging feature of this section involving geotechnical and structural engineering analyses is the HEB-to-collider tunnel transfer region. The HEB tunnel is situated vertically over the main collider within the West Utility Straight. The two tunnels are connected via transfer tunnels, each approximately 250 meters long, inclined at 1.6 degrees. The HEB tunnels will be excavated entirely within the Austin Chalk (AC) Formation, a stable limestone, while the collider tunnel will be excavated in the relatively weak Eagle Ford Shale (EFS) formation. The two main tunnels lines will be separated by approximately 32 feet of rock, which will contain the AC/EFS contact, a zone of complex geology. Several large open galleries along and parallel to the collider tunnels will also require unique geotechnical analyses and construction methods.

The typical main collider tunnel in the straight section is expected to be excavated by tunnel boring machine and to be supported by a precast concrete segmental lining. At the tunnel transitions to the transfer and abort tunnels, as well as to adits and niches, specialized excavation and support techniques will be required for the EFS. These techniques will likely use sequential excavation with multiple drifts, rock bolts, and shotcrete. Significant additional design and construction costs associated with revisions in space requirements will be reduced through studies with the model.

The basic tunnel DRD issued for the West Utility Straight Section included the electronic model (in Intergraph format) for use as the basis of Title I design. The visual presentation of the model allowed an early, better understanding of the tunnel intersection and space requirements for planning and coordinating this tunneling system, 200 feet below the ground submerged in structurally weak rock. The model aided a team of geotechnical, structural, and construction engineers in reaching a workable solution concerning tunnel stability during excavation, tunnel construction sequence, and tunnel lining procedures.

The model will be continuously updated as new tunnel configurations and technical systems are incorporated. Future uses of the model will include establishing the finish contract DRDs, developing magnet support and installation methods in the transfer region, and verifying construction document cross-sections and alignments.

CONCLUSION

Communication of technical systems and components occurring early in the conceptual design phases of the technical systems resulted in significant cost and time savings for the conceptual design and development of design requirements of the West Utility Straight Section. The initial contract with CSA was successful in that it identified to management a real need for modeling. Management, in turn, responded with the necessary in-house resources to address and continue the modeling task. The groups responsible for further developing the model quickly rendered an initial working model for communication among the design groups.

The A-E/CM's continued use of the model during design is also proving the value of this unique design tool for minimizing cost and time resources associated with the conventional facility design and construction.

ACKNOWLEDGMENTS

The authors wish to thank all of the following individuals for assistance in the preparation of the design of the West Utility Tunnel Section: all members of the CCD Collider West Tunnel Occupation Worry-List Network (COWTOWN) Working Group, Tony Massing, David Silveira, Tracy Lundin, Richard Molenaar, and members of The PB/MK Team, the A-E/CM for the SSC Project. We would also like to acknowledge Jon Ives and Rainer Meinke, who had the vision to support this design effort.

SSCL DD2 DRIVER PORT AND PROJECT UPDATE

Steven L. Mestad

Physics Research Division
Computing Department MS-2003
Superconducting Super Collider Laboratory*
2550 Beckleymeade Ave.
Dallas, TX 75237
stevem@diehard.ssc.gov

ABSTRACT

A paper previously published in the 1992 ICHEP proceedings [ref. 1] outlined the SSC's need for a high speed, high capacity tape drive to store detector data. Also described were stages and lessons learned while developing a custom device driver for the Ampex DD2 tape drive on a Silicon Graphics 4D/310. This paper updates the work on the SGI driver and describes the efforts in porting the driver to a Sun Microsystems 670 server.

INTRODUCTION

The SSCL's interest in the DD2 tape drive is driven by the predicted data storage requirements of the lab's detectors. The detectors are expected to generate data at 100 megabyte/second thereby producing several petabytes of data per year. The detector collaborations require keeping a year's worth of data on line for analysis and comparisons to simulation data generated independently. These requirements for large storage with high transfer rates clearly call for something other than the typical tape or optical disk drive.

The DD2 tape drive is a 19 millimeter digital video tape drive produced by AMPEX. AMPEX has adapted their video tape drive, used in the professional broadcasting industry, for high speed high density storage of digital computer data. The DD2 tape drive is a promising device for meeting the SSCL's requirements. Currently it is available with a 15 megabyte/second transfer rate and the capability of storing 25 gigabytes on a $60 tape. A four drive tape robot system can store over 6 terabytes with access to any byte averaging less than 30 seconds. The costs for such a robot system is comparable to a 1 terabyte 3480 based tape silo system.

The DD2 mass storage project at the SSCL started by integrating a DD2 tape drive and a VME IPI-3 controller, made by Control Data, with a Silicon Graphics computer system. This required writing a custom device driver for the tape drive. Currently the driver is being ported to a Sun 670 computer system.

---

*Operated by the Universities Research Association, Inc., for the U.S. Department of Energy under Contract No. DE-AC35-89ER40486.

PROJECT STATUS

The SGI driver has undergone further debugging and testing since the ICHEP paper was written. Error reporting and recovery has been enhanced and the front panel message display is now supported. Proper operation of multiple controllers and drives has been verified. The best data transfer rate achieved is 13.8 megabytes per second with over 10 megabytes per second sustained. Further performance tuning is still needed. Also, further work is required to allow proper operation on multiple cpu SGI systems.

Currently the focus has been on porting the driver to a Sun 670 system in order to support the PASS project. PASS (Petabyte Access Storage Solution) is experimenting with various data storage methods to find an optimal strategy for storing lab data [ref. 2].

PROJECT STAGES

Porting the SGI driver to the Sun presented a few problems right from the start. Since a skeletal driver containing all the required functionality could not be found, setting up the Sun specific code was a problem. Although Sun's manuals give driver examples [ref. 3, 4], these often did not address problems presented by our driver. In addition, SunOS 4.1.3 does not include support for partitioned DAT drives so it was not possible to map several DD2 drive operations into SunOS supported operations as was done on the SGI. Moving these operations into device specific commands means the MT(1) utility will not support several required DD2 drive commands so a custom MT type utility needs to be written. SunOS also did not include a device specific ioctl(2) so the mtio.h include file also required modifications.

In porting to SunOS, it was expected that the tape drive specific portions of the driver would be largely unaffected and that the SGI specific portions would need large modifications. These changes would include restructuring the initialization and setup portions of the driver to fit Sun conventions as well as switching over to Sun specific driver data structures. There are also kernel calls available on the SGI which do not have a comparable call under SunOS. These calls must be replaced by functionally equivalent code.

Once the structural changes were roughed out and the code compiled, the debugging began. After getting past several problems related to acquiring configuration information, the first device specific problems arose. After solving a system lockup problem due to incorrect backplane jumper settings [ref. 5], the interface card failed its self test with a DMA error. This problem could have several different causes. The most unusual clue however was the VME bus analyzer did not register any attempts to access memory.

The interface vendor, CDC, directed inquiries to the board designers. After using a diagnostic port on the interface board to verify the DMA maps were setup properly and the Sun memory was indeed accessible from the VME card, it was determined that the device driver was requesting block transfer mode on the VME bus. Once this mode was turned off, the board initialized correctly. Though the interface will work without block transfer, performance will be degraded. Information from Sun about their VME bus support [ref. 6] is forthcoming and should help determine if block transfer mode is supported. At the time this paper was written, the Sun driver is not completely debugged. There are problems synchronizing execution of the driver. The driver is passing commands to the tape drive and accepting asynchronous interrupts; therefore the command block assembler, interrupt handler and response block parser are all operating. Hopefully the driver can be made functional in the time between the paper deadline and the IISSC presentation.

MORE LESSONS

One of the lessons learned was that finding a good skeletal driver to work from can save large amounts of time. Piecing together the correct approach from several incomplete driver fragments using trial and error can be time consuming and frustrating. Debugging can also be difficult as most problems have several causes and finding the correct solution from an assortment of possibilities involves a large amount of investigation.

A previously learned lesson, the importance of vendor documentation, also came up again. The standard documentation set shipped with our previous Sun equipment contains no information on making use of the Sun Open Boot PROM used in newer SPARC systems such as the 670. This information is contained in Sun's Sbus Developers Toolkit [ref. 7]. Sun's manual describing their VME implementation was unknown until after inquiring about support of block transfer mode. Also, SGI issued an update to their device driver manual. This was discovered upon learning that a better method of using semaphores and spinlocks was covered in a manual section we did not have [ref. 8, 9]. However, don't discard old manuals as newer ones may not contain necessary information. Sun's older debugging manual [ref. 10] contains information on using ADB(1) for kernel debugging. This topic is not present in a newer debugging manual which concentrates on newer debugging tools.

The previously learned lesson on the importance of test equipment was once again demonstrated by the VME bus analyzer. Because the bus trace showed no attempt at accessing memory, the possibility of incorrectly setup DMA maps or a bad initialization address could not be causing the problem experienced. Such problems would be the more likely explanation of a DMA selftest failure but the trace showed that something very unusual was taking place.

A new lesson, the importance of maintaining current hardware revisions, was demonstrated by the arrival of the dual port hardware. Once installed, it became apparent something was wrong when trying to use both controllers in a single system. Instead of behaving like 2 separate drives on 2 separate controllers as expected, the second interface would fail to initialize properly. Investigation showed the new Rhino board, an IPI interface in the drive itself, was a newer revision than the original one. Replacing the original Rhino with an upgrade helped but did not solve the problem. Further investigation found that our tape drive contained hardware which was seriously obsolete. This would be unusual for a one year old product, but as this tape drive is still in development, the situation is not unheard of. After the drive was upgraded to current hardware and firmware levels, the dual port hardware operated as expected.

The ICHEP paper also touched upon the importance of keeping an archive of previously working versions of code to allow comparisons to the current code when it has stopped working. While this practice has been helpful during software development, it also turned out vital for resuming work after several hardware changes. After installing a backplane upgrade and multiple cpu board in our SGI system, the driver failed to work properly. This was not terribly surprising as it was known the semaphores are not properly guarding critical sections. However, it was surprising when the code continued to fail after swapping the single cpu board back into the system. Along with the SGI cpu upgrade, AMPEX had performed a major upgrade on the tape drive and the effects of this work were still unknown. A modified version of the driver supplied by AMPEX would run but less than one half the speed of our original code. On the possibility that building a multi cpu kernel changed the configuration, SGI sent a copy of the default kernel configuration code. However, kernels built with it showed the same problems.

At this point it was time to try old drivers to see if the current version was defective. The oldest working archived driver also no longer ran. We then loaded the kernel configuration code from an old system backup and built a kernel with the oldest driver version. It worked and ran as fast as expected. We then tried running the most recent driver archived with all three sets of kernel code. It ran properly and as fast as expected with all sets of kernel code. Further investigation is needed to understand why the latest version of the driver no longer works and why the oldest version failed unless built from the old backup files. Analysis of why AMPEX's code is slower than our code may be helpful in understanding which modifications will improve performance. Determining which configuration files are automatically changed by a multiple cpu kernel build might shed some light on why the SGI driver fails on a multiple cpu system when locked to a single cpu although it will run properly on a single cpu system.

FUTURE DEVELOPMENT

As outlined in the ICHEP paper, there are several areas for future work on the DD2 project. Because of the concentration on the Sun port, little recent work has been done on the SGI driver. The SGI driver needs further tuning to enhance throughput. Modifications are also required to permit proper operation on multiple cpu systems. Essentially this work involves ensuring the critical sections of driver code are protected by semaphores to prevent simultaneous update problems in the driver data structures. However, to ensure good performance of the tape drive and computer system, semaphore and spin locks must protect the smallest amount of code possible. The locks must also be implemented carefully to avoid the possibility of deadlock. The Sun driver, once fully functional, will require further changes as well. Plans are to upgrade the 670 system to Solaris 2.1 which will require major changes to the driver. This involves porting the driver yet again as SunOS is based on a BSD kernel whereas Solaris is based on a System V kernel.

In order to make use of the dual port hardware, the driver will need some modifications to support it. In particular, the proper handling of tape drive resets, current tape position, and the setting of block sizes will need attention. Access control and IPI port allegiance also needs support; probably through some sort of drive request daemon.

The solution to the problem of how to feed a 15 megabyte per second tape drive has been decided on. It is expected that the computing systems used for data analysis will have sufficiently large amounts of memory; enough to hold several entire data sets. Therefore the data will be written and read directly to memory and striped disk arrays will not be needed.

Other projects under consideration include acquiring a D2 tape robot so that a base of experience and support code can be developed. Integration of D2 into the PDSF for tertiary storage is also desirable due to the existing need for large amounts of fast storage [ref. 11]. The 8mm drives currently in use are too slow and their reliability disappointing.

NOTES

UNIX, SGI, IRIX, SUN, SUNOS, and possibly other terms in this document are registered trademarks of USL, Silicon Graphics, Sun Microsystems, and other various companies.

REFERENCES

1. "SSCL DD2 Mass Storage Project", Steve Mestad, *ICHEP Proceedings*, 1992.
2. "The PASS Project: Database Computing for the SSC", E. May, et. al., *Proposal to the DOE High Performance Computing and Communication Initiative*, March 1992.
3. "Writing Device Drivers", Sun Microsystems part no. 800-3851-10.
4. "Writing SBus Device Drivers", Sun Microsystems part no. 800-5322-10.
5. "SPARCsystem 670MP", Sun Microsystems part no. 825-1390-01.
6. "SPARC 600MP VME Implementation Guide", Sun Microsystems part no. 800-6738-10.
7. "Open Boot PROM Toolkit User's Guide", Sun Microsystems part no. 800-5279-10.
8. "Writing Device Drivers for Silicon Graphics Computer Systems", Version 1.0, SGI document number 007-0911-010, 1989.
9. "IRIX Device Driver Programming Guide", Silicon Graphic document number 007-0911-020.
10. "Debugging Tools", Sun Microsystems part no. 800-3849-10.
11. "SSCL-PDSF Data Management System", Jeffrey L. Allen, *ICHEP Proceedings*, 1992.

SISSY: AN EXAMPLE OF A MULTI-THREADED, NETWORKED, OBJECT-ORIENTED DATABASED APPLICATION

B. Scipioni, D. Liu, and T. Song

Physics Computing Department
Superconducting Super Collider Laboratory*
2550 Beckleymeade Avenue
Dallas, TX 75237

ABSTRACT

The Systems Integration Support SYstem (SISSY) is presented and its capabilities and techniques are discussed. It is a fully automated data collection and analysis system supporting the SSCL's systems analysis activities as they relate to the Physics Detector and Simulation Facility (PDSF).[1, 2] SISSY itself is a paradigm of effective computing on the PDSF. It uses home–grown code (C++), network programming (RPC, SNMP), relational (SYBASE) and object–oriented (ObjectStore) DBMSs, UNIX operating system services (IRIX threads, cron, system utilities, shell scripts, etc.), and third party software applications (NetCentral Station, Wingz, DataLink) all of which act together as a single application to monitor and analyze the PDSF.

INTRODUCTION

Using networks of computers as a single system, that is, to perform tasks in a parallel or distributed fashion, has created a need for support services similar to those found on single computer systems. Systems administrators, systems integrators, developers and users all have need for certain information concerning the functioning and performance of a system of computers. SISSY is an automated tool which provides this information in a timely way and which relies upon UNIX standards in a heterogeneous environment.

MOTIVATION

There are two reasons for the choice of architecture for SISSY. First, it was desired to easily develop software which was maintainable, high performance, and makes maximum use of commercial applications. This was expected to result in fast development time and very low

*Operated by the Universities Research Association, Inc., for the U.S. Department of Energy under Contract No. DE-AC35-89ER40486.

manpower for development and maintenance. Second, one of the goals of the computing group is to promulgate and infuse modern software technology and engineering techniques into the HEP computing community at the SSCL. SISSY is a vehicle for demonstrating these techniques.

REQUIREMENTS

The PDSF,[1] Figure 1, is in its second phase of operations with Phase III in procurement and Phase I having been a prototype and testing ground for hardware and software integration techniques. It was mandatory, then, that all aspects of subsystem performance be monitored so systems analysis activities could provide directions for future computer acquisitions. This is an ongoing process with overall and detailed system performance and utilization reviewed on a weekly basis. The data required for this analysis effort includes average and peak utilization of all CPU, network, disk and tape systems throughout the PDSF. These hardware systems represent the bulk of the monetary resource invested and, therefore, a measure of effectiveness and efficiency of the facility is required. This analysis also points out hot spots and bottlenecks to guide future architectural changes and additions. In addition, both aggregate and individual user statistics need to be provided to the main groups of users, in particular the detector collaborations, so they may plan their current and future computing activities. The following list represents current requirements on weekly reporting, but it is only a part of the data collected and filtered:[2]

- % CPU utilization daily on a 24 hour basis, and weekly average for each cluster of compute servers, each data server, and each support computer (database and console multiplexers).

- Maximum and average data rates (KB/s) for each FDDI network every hour throughout the week. Maximum data rate (Kb/s) and packet rate (packets/s) separately for input and output for each computer and router network interface throughout the PDSF.

- Weekly average disk and tape storage in use, available and change from previous week.

- Hourly traces of the total number of distinct users with processes on the PDSF and within individual computing clusters.

- Total weekly CPU utilization (in minutes) consumed by each user, on each architecture and grouped by user organization. A list of all processes with more than one CPU minute for each user.

- Total CPU utilization for the week (in minutes) on each architecture broken down by user organization and location (in/out of the lab).

DESIGN

There is a particular software design strategy implicit in the hardware architecture of the Physics and Detector Simulation Facility (PDSF) at the Superconducting Super Collider Laboratory (SSCL).[2] In particular, the many nodes which comprise the compute servers on the PDSF are intended, by design, to be able to both compute and perform I/O to network, disk and tape independently, or in parallel. In addition, the data servers act as a collection point for files common to applications running in parallel on the clusters of nodes, and as launch points for jobs executing on the compute servers. For this reason Symmetric Multi–Processing (SMP) is required for the data servers. Any application which takes maximum advantage of this architecture will be both multi–threaded and networked. In the case of SISSY there is a collector process on each data server, which is a multi–threaded object. Each thread executes a (non–blocking) Remote Procedure Call (RPC) to collect data from one of the compute servers of the associated cluster, as shown in Figure 2. In this way the RPCs can be used as a parallel network program, and all nodes in the system can be polled simultaneously rather than visited in series. This results in both higher data consistency and an increase in performance by more

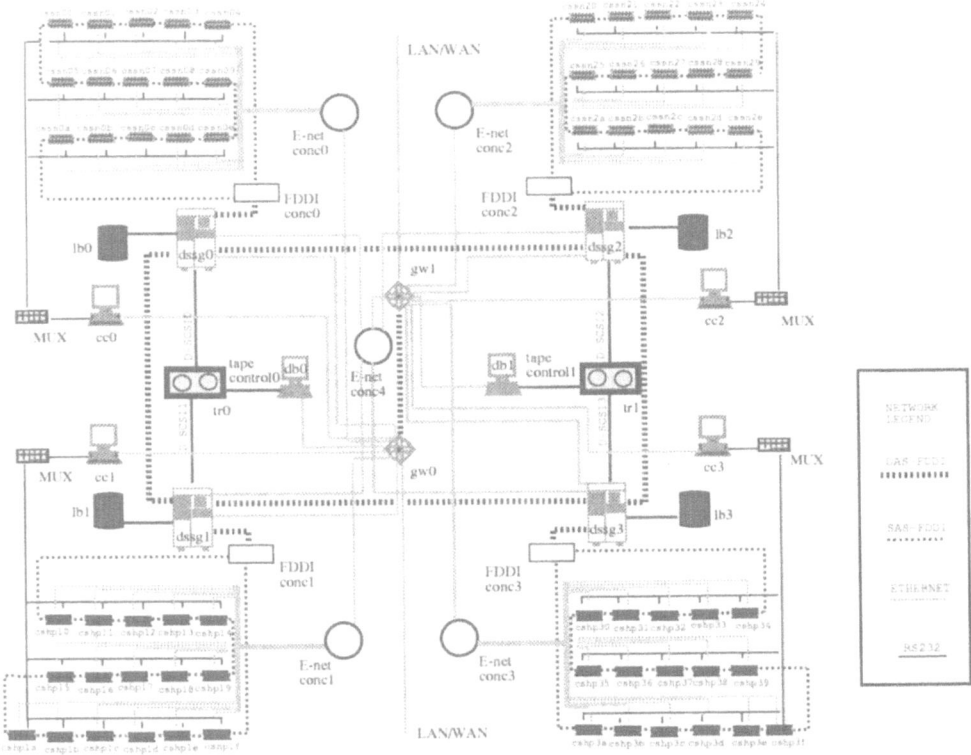

Figure 1. Physics Detector Simulation Facility (PDSF) Architecture.

than an order of magnitude. The polling schedule and type of information requested is maintained in a database and queried by the collectors in order to determine when to execute data collection. This schedule can be changed asynchronously, and take immediate effect by a database update. Each compute server throughout the PDSF collects the requested information in parallel, usually by executing a UNIX command, then filters and packages the result in SQL and ships it back to appropriate collector. Here at the collector, along with information from the other nodes in the cluster, it is inserted over the network into the database. Scripts can be executed at any time to query the latest information in the database, but normally this is done weekly by an automated procedure. Every Thursday morning a process awakens and executes all the database scripts, feeds the resulting tables to a spreadsheet application running in batch mode and automatically pipes the output viewgraphs and tables to a color printer for presentation.

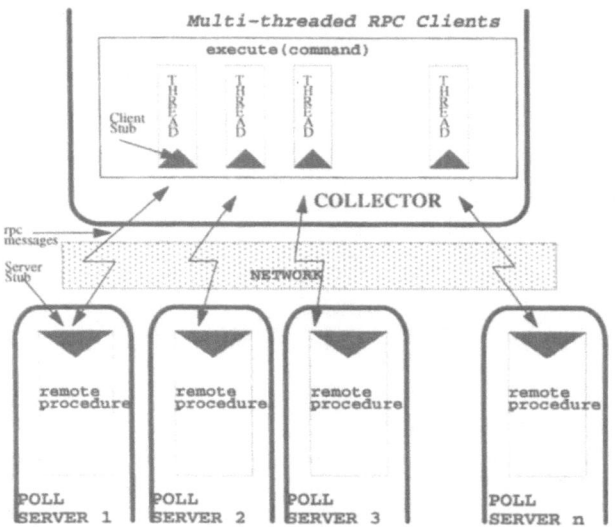

Figure 2. Threaded RPCs.

IMPLEMENTATION

All software written for SISSY is written in C++. The polling data are kept in Object Store object oriented database management system (OO–DBMS). An OO–DBMS is used because the polling process is a continuous loop through the collector objects. So the procedural logic of the queries and the repetitive nature of the polling benefit greatly from the features of the OO–DBMS. This ensures that the appropriate data persists in the program's data structures, reducing traffic and simplifying code. The actual systems data, however, is returned as SQL statements and inserted into two R–DBMS (Sybase) on two database workstations. The data entities were designed using both Software Through Pictures (IDE) and ERdraw (LBL), and the SQL create table statements (and design) were then generated automatically. The collector programs are instances of the same C++ class with all the functional details of the RPCs hidden away in the object methods. Each collector object invokes a public threaded method which then calls the RPC by invoking a private method. The threads mechanisms is provided through the SGI m_* library routines. These require the shared data to be global to the threaded function, which is natural for data which are static object members. There are only two tricks to get this to work correctly. First the RPC clients must be created in series because of the way they use operating system resources. This goes fast and is put into the class constructor. Second, the threaded function must be of static storage class and visibility with its shared data global to the module. This allows the same copy of the method to be shared by all objects of the class and inhibits the compiler from putting in an extra pointer to the method and thus confusing the threads library routines. Some of the network data is collected by NetCentral Station network management software through the SNMP daemons executing on all computers and routers in the systems. NetCentral is implemented on top of Sybase which makes for a seamless integration of commercial and homegrown tools residing on a single database. The report generation is provided by WINGZ using its hyperscript batch capability and then printed on a Tektronix Phaser II color printer.

FAULT TOLERANCE AND UPGRADE

Currently higher fault tolerance and a natural language interface are being incorporated into SISSY. Since there are two database computers, either can pick up the data collection of the other if it is down.This up/down information on each node in the system will be maintained in the database by a program which attempts a remote execution. Since each computer has a minimum of two network interfaces (Ethernet and FDDI) if a data server is down severing the FFDI connectivity another data server can pick up the task over Ethernet. Design is complete and implementation underway. Also the natural language interface development has just begun using Natural Language from Natural Language, Inc. as a conversational English database interface.

CONCLUSION

SISSY is a portable, automated, high performance, systems analysis and maintenance tool based upon homegrown and commercial software, which is low in manpower requirements and provides the necessary information to successfully operate a parallel, distributed, computing environment. It is also a successful example of simple software development techniques which may be useful in production physics code running on the PDSF and other HEP computing facilities.

REFERENCES

1. G. Chartrand, L. Cormell, R. Hahn, D. Jacobson, H. Johnstad, P. Leibold, M. Marquez, B. Ramsey, L. Roberts, B. Scipioni, N. Shivapuja, G. Yost, "Physics and detector simulation facility specifications," SSCL–275, Attachment A, July (1990).
2. B. Scipioni, "Physics and detector simulation (PDSF) architecture/utilization", IISSC5(these proceedings).

DISTRIBUTED COMPUTING AT THE SSCL

Laird R. Cormell and Randle C. White

Superconducting Super Collider Laboratory*
Computing Department MS-2003
2550 Beckleymeade Ave.
Dallas, TX 75237-3997
cormell@sscvx1.ssc.gov
rwhite@sscvx1.ssc.gov

ABSTRACT

The rapid increase in the availability of high performance, cost-effective RISC/UNIX work-stations has been both a blessing and a curse. The blessing of having extremely powerful computing engines available on the desk top is well-known to many users. The user has tremendous freedom, flexibility, and control of his environment. That freedom can, however, become the curse of distributed computing. The user must become a system manager to some extent, he must worry about backups, maintenance, upgrades, etc. Traditionally these activities have been the responsibility of a central computing group. The central computing group, however, may find that it can no longer provide all of the traditional services. With the plethora of workstations now found on so many desktops throughout the entire campus or lab, the central computing group may be swamped by support requests. This talk will address several of these computer support and management issues by discussing the approach taken at the Superconducting Super Collider Laboratory (SSCL). In addition, a brief review of the future directions of commercial products for distributed computing and management will be given.

INTRODUCTION

Dramatic changes have occurred throughout the computing industry and its user community since the revolution of distributed desktop computing began in the 1980's. The abundance of cheap, powerful computing engines is driving this revolution at a feverish pace. Participants in the revolution are the high speed networks and distributed peripherals such as laser printers, SCSI disks, and 8-mm tapes that make distributed computing a viable option to the more traditional mainframe-oriented, centralized computing environments. Networked desktop computing has changed forever our computing and working environments. In fact, the sociology of how we do our work has also changed. Some of that change can be traced back to the interactive time-sharing environments of the 70's, of course. But in the 90's environment, a typical user is no longer dependent on the central computing organization to meet his needs. The computer has become another tool in the integrated office/work place of the 90's. Around the corner lie many new integrating tools in the area generally referred to as groupware that should further expand the utilization of computing in the office. These products include integrated phone, voice mail and e-mail services, desktop video conferencing, multi-media applications, etc. While a discussion of groupware is beyond the scope of this paper, planning for the management and support of these products must begin soon. There are many management issues to be addressed, but at this time there does not seem to be that many solutions

---

* Operated by the Universities Research Association, Inc., for the U. S. Department of Energy under Contract No. DE-AC35-89ER40486.

readily available. We will provide examples of the methods employed by the SSCL to address these problems and then discuss some of the efforts under way in the computer industry.

DISTRIBUTED COMPUTING MANAGEMENT ISSUES

We've already alluded to the fact that there are both good and bad aspects of distributed computing environments. There are some simple, perhaps obvious, observations that we can make. First the advantages.

Advantages of Distributed Computing

Some of the advantages of distributed computing include:

- Distributed computing working environment is good. Most people do most of their work in their offices. Efficiency and productivity increase when computing and other resources are available in the office work place.
- Working group clusters are good. Most people work in groups on problems common to that group. A distributed environment that is sub-netted by group can be very effective in providing common shared resources that are group specific. For example, an engineering group may need CAD and finite element analysis packages, but have no need for the CERN Program Library. Client/server architectures can be utilized to their fullest capability with minimal effect on the network backbone.
- Dedicated resources are good. Placing a cheap RISC work-stations on every desktop can virtually eliminate time sharing. A typical user will have his own dedicated resource.
- Independence is good. Individual users and working groups relish the independence they have when in control of their own environments. They can choose configurations that are tailored to their needs rather than the general purpose environments found in centralized mainframe environments.
- The economics are good. First, RISC/UNIX workstations and personal computers, for that matter, are cheap. And as mentioned above, working groups or individuals buy only the applications and tools that meet their needs. They configure the hardware to meet their needs alone, no more. Shared multiple users licensing can keep the cost of software down. Installation and maintenance fees are typically small.

Disadvantages of Distributed Computing

There are likewise a number of disadvantages associated with distributed computing environments.

- Distributed computing working environment is bad. The individual user must become his own system manager to some extent. He must become concerned with issues of installation maintenance, updates, backups, etc. If he is not prepared to perform these functions himself, where does he turn, who provides the support?
- Working group clusters are bad. Individual working groups may chose to handle their own support issues. This can lead to duplication of efforts in several service areas. A further complication arises when different groups develop completely different solutions to the same problem.
- Dedicated resources are bad. When dedicated computing is provided for individual users, there are typically many compute cycles and other resources going to waste.
- Independence is bad. Individual users and working groups often don't have the experience or knowledge to configure their environment to meet their needs in an effective efficient manner.
- The economics are bad. While it is true that RISC/UNIX workstations are cheap, there are often hidden costs. There is a need to build up a support staff familiar with several UNIX (or PC) platforms. There are potentially several areas for duplication of efforts. The purchase of software and services may not be coordinated throughout the site and thereby lead to wasted expenditures.

So where does this leave us? While there are certainly some disadvantages associated with the management of distributed computing environments, most of the arguments come down in favor of such environments. The raw computing power, versatility, and freedom offered by distributed computing at an unrivaled price/performance point makes them extremely attractive. The "pros and cons" of centralized vs. distributed computing environments may be debated at length, but there is no escaping from the fact that the era of distributed computing is upon us.

Management Issues

In the era of distributed computing and heterogeneous environments, diverse systems are networked throughout the high energy physics (HEP) community. Unfortunately, the diversity that lets users choose the system that best meets their needs also creates an administrative nightmare: it may require system administrators to develop different management scheme for each hardware platform. It is possible to integrate large multi-vendor systems when standards can be applied. These standards, however, are not all employed in the same way by every vendor. Most users enjoy the independence offered by RISC/UNIX workstations, but many of them still expect (and sometimes demand) the services normally offered by more centralized operations. System managers must find novel ways to deal with these expectations in multi-vendor configurations; they must address several issues including:

- Training: Support personnel must receive training for several different hardware platforms and operating systems. Users also need training. Clearly UNIX managers must provide training for their personnel, but who trains the users?

- Installation and setup: Workstation vendors are typically operating on small profit margins and may not provide free installation. Users often don't want to pay for the additional installation and setup services. But the typical physicist/user may have little or no experience in system management or much knowledge about UNIX environments. How do UNIX managers provide the support?

- Hardware maintenance: UNIX workstations are inexpensive and fairly reliable, but they still must be maintained throughout their life cycles. There are typically product enhancements, firmware fixes, and upgrades offered by the vendors that extend the useful life of the equipment. While system administrators are familiar with this phenomena, it's not clear that the end-user "keeps up" with the latest improvements or that he knows what to do. Who should provide the maintenance and ensure that the system is current?

- Software maintenance: Likewise for software upgrades, only more so. Bug fixes, patches, and new software versions are released on a regular basis. How does a UNIX manager install these on all platforms, how does he maintain configuration control, who tests new versions for compatibility?

- Network management: In many distributed environments the network is an integral part of computing, or as some have stated: "the network is the computer". In working group clusters files and applications are shared in a client/server arrangement; workstations and X-terminals are mounted to hosts; mail, ftp and other communication services are crucial. The HW and SW issues mentioned above apply to NW management as well. In addition there are questions of how to configure the network to provide optimized access to file servers, printers, etc.

- Distributed services: There are a number of services traditionally provided in centralized configurations that must be dealt with in new and innovative ways. How does a system administrator provide uniform printing, backup, licensing, batch, and tape services that users require?

- Accounting: How does an administrator in a distributed environment collect statistics on system utilization, disk usage, print service, and traffic over the network? At some sites the computer group provides resources on a "charge-back" basis. How should these charges be assessed in a distributed environment?

- Documentation: The usefulness of a system is often determined by the quality and accuracy of its documentation. How can the UNIX manager provide current documentation to all the users when some are located not only on the other side of campus but on the other side of the world?

- Host management: In a distributed environment the system administrator is faced with the difficulty of providing user and group accounts and privileges on a number of different hosts. In addition operations on the system such as shutdown/reboot, naming, addressing etc. must be performed.

Obviously there are a number of challenges to be faced in managing distributed systems. So, how do we cope with distributed computing? In the next section we discuss the approach that has been taken at the SSCL. In the section following that we will identify some new developments from the commercial side that may provide some solutions.

MANAGEMENT OF DISTRIBUTED COMPUTING AT THE SSCL

The SSCL is committed to developing a distributed computing environment.[1] Use of a centrally controlled but otherwise decentralized and distributed environment allows maximum functionality and flexibility of access to shared data, print services, electronic mail, backup systems and nameservice while minimizing network overloads across the primary backbone.

Operations support: The Informations Services Operations Group has 25 employees that provide all levels of support from Help Desk assistance to system programming on all central computing resources at the SSCL. They support five operating systems (UNIX, VMS, MS/DOS, Novell, and MAC-OS.) and approximately 3000 user accounts. The central resources include several VAXclusters, QuickMail servers, Appleshare servers, and Novell servers as well as the UNIX workstation clusters networked together in the Physics Detector Simulation Facility (PDSF). There are approximately 1500 Macintosh systems, 300 PCs and 700 UNIX workstations. Some groups (particularly those in Accelerator Systems Division) provide their own system support for specific functions.

Distributed Servers: At present, management functions and critical applications are primarily performed using centralized server systems. The network impact of centralized data has reduced the system's efficiency somewhat. They plan to place a number of servers in strategic laboratory locations. These server locations are being selected on the basis of user population density and networking considerations. The critical applications and services to be distributed include the various file systems (AFS, NFS, Appleshare, and PC volume services), network applications (nameservice, NTP, and USENET), print services (IMPRINT, Alisashare, and Pathworks), mail gateway services, BITNET routing , and software metering for Macintosh and PC software permitting efficient sharing of infrequently used desktop applications.

Distributed Backups: The goal at the Lab is to provide administratively centralized, unattended backups on most SSCL servers by the end of '92. The volume has grown to greater than 250 GB of routinely backed-up data on UNIX, VMS, Macintosh, and Novell systems. In most cases the data is copied to 8-mm tapes in stacker units from remote systems over the FDDI backbone. System backups are currently performed using native operating system utilities. These utilities all fall short in some areas and none provide tape management or simple user archival capabilities.

Site Licensing: The SSCL is site-licensed for the SUN operating system and is in the process of procuring a site-license for SUN unbundled products. Procedures have been established for the distribution of copies of media and documentation from a centralized source to technical contacts throughout the SSCL. Other software that does not have licensing restrictions is made available through AFS or FTP. Client-server software (particularly that which is run on MACs or PCs) is a difficult problem. The SSCL is currently working on procedures to control and automate distribution and installation of client software through use of tools such as Timbukto for Macintosh.

Distributed Printing: The Operations Group has developed a distributed printing utility, lwprint, that permits access to most of the laser writers on AppleTalk or Ethernet. Print jobs are spooled and queued by a central UNIX print server. Users can select the printer of their choice from an interactive menu. Post script is the default format and several options such as landscape and two pages per sheet are being implemented.[2]

FUTURE DIRECTIONS

Common Open Software Environment (COSE)

In a move towards application and end-user interoperability in a heterogeneous UNIX platform network, a unified model for a common open software environment was announced by Hewlett-Packard, Sun Microsystems, IBM, Univel, Santa Cruz Operation (SCO), and UNIX System Laboratories (USL) at the March, 1993 Uniform conference in San Francisco. The agreement includes specifications for a common desktop environment, networking, graphics, multimedia, object technology, and systems management.

Desktop Environment: The specification for the common desktop environment will be given to X/Open for insertion into the X/Open Portability Guide for interoperability and portability of applications among the various platforms. The specification will base a consistent graphical user interface (GUI) and a consistent applications programming interface (API) on X.11 Windows, OSF's Motif toolkit, SunSoft's ToolTalk, and HP's Encapsulator. Sun agreed to migrate from OpenLook to Motif although current OpenLook applications will still be able to run under the new common desktop environment.

Networking: A heterogeneous distributed workstation network environment with network transparency for applications will be supported on the various platforms through OSF's Distributed Computing Environment (DCE) and Distributed Management Environment (DME) as well as Sun's Open Network Computing (ONC+) environment and Univel's Netware.

Graphics: A common graphics application environment will be supported through the use of the X Consortium's facilities including Xlib/X for 2D pixel graphics, PEXlib/PEX for 2D/3D geometry graphics, and XIElib/XIE for advanced imaging.

Multimedia: A specification for common Distributed Media Services (DMS) and a Desktop Integrated Media Environment (DIME) will be submitted to the Interactive Multimedia Association. The multimedia environment will be an integrated part of the common desktop environment.

Object Technology: The Common Object Request Broker Architecture (COBRA) specification from the Object Management Group (OMG) will be supported for common object management across multiple heterogeneous platforms.

Systems Management: A working group will be set up to promote industry acceptance of specifications on systems management applications including security management, software installation, software license management, data backup and restore, print spooling, and distributed file system management.

SUMMARY AND CONCLUSIONS

High energy physicists are resourceful and creative. Given that, it's not surprising to see: first, that physicists, ever eager to obtain more computing power, have exploited RISC/UNIX systems with a passion; and second, that they have attempted, with some success, to develop integration and management tools to make effective use of these new systems such as those found in the SSCL UNIX environment.

But we should expect, in fact demand, tools and integrated packages from the workstation vendors. These products should be interoperable, coherent, scalable, and affordable. Fortunately, there are several promising efforts underway, by groups such as OSF and COSE, but how long can we wait and how successful will their products be?

REFERENCES

1. Brenda Ramsey, Information Services Department, SSCL, private communication.
2. Dave Somogyi, Information Services Department, SSCL, private communication.

INITIAL RESULTS OF STRAND PRODUCED IN PHASE II OF THE SSCL VENDOR QUALIFICATION PROGRAM

M.J. Erdmann, D. W. Capone II, E.S. Coleman,
B.A. Jones, and J. M. Seuntjens

Superconducting Super Collider Laboratory[*]
2550 Beckleymeade Ave., MS 8000
Dallas, TX 75237-3997

INTRODUCTION

In 1991, the Superconducting Super Collider Laboratory (SSCL) instituted a program to qualify specific superconductor manufacturers for production of cable acceptable for use in both Collider Dipole (CDM) and Quadrupole (CQM) magnets. The SSCL Vendor Qualification Program (VQP) was designed with two Phases. Phase I was divided into two additional phases, IA and IB, which ran concurrently. In Phase IB, each vendor was directed to manufacture roughly 3000 kg of cable using a "baseline" process. The baseline process was agreed to by both the SSCL and the vendor at the beginning of the VQP. In this phase, process control was closely monitored with the use of statistical methods[1] and each vendor was graded based on these results. Phase IA, known as the R&D phase, was developed to allow each vendor an opportunity to optimize and improve on their baseline process in terms of both cost and manufacturability. In this phase, multifilament billets were designed to explore several key variables such as alternate alloy sources, process modifications and improved billet designs. At the end of Phase I, the results from both IA and IB were evaluated at a review between the SSCL and each vendor, and a final Phase II process was generated and fixed using the best results. In Phase II, each vendor is required to manufacture roughly 6000 kg of superconducting cable under a firm fixed price contract which can then be used to create an accurate price estimate for competitive bidding on the full rate production CDM and CQM contracts. Participating vendors include Alsthom Intermagnetics (AISA), Outokumpu (OTU), Oxford Superconducting Technology (OST), Hitachi (HIT), and Furukawa Electric (FEC) for outer cable, and Intermagnetics General Corporation (IGC), Teledyne SC (TSC), and Sumitomo Electric (SEI) for inner cable.

At the end of Phase II, each vendor must meet the minimum requirements outlined in the contract to become a qualified superconducting cable supplier. For one requirement, critical process variables identified by the SSCL Conductor Department at the beginning of the VQP will be evaluated to determine the quality and uniformity of the material produced during Phase II of the program. Variables considered critical are steady state Cu/SC ratio, final wire diameter, critical current and RRR in strand, and keystone angle, width, mid-thickness, twist, pitch and critical current in the cable.

---

* Operated by the Universities Research Association, Inc., for the U. S. Department of Energy under Contract No. DE-AC35-89ER40486.

Preliminary results are reported with emphasis on Phase II strand produced by each of the eight vendors in the VQP. Analysis of the critical electrical and mechanical properties of the strand are presented, with discussions concentrating on the statistical variation and process control of each vendor. The piece length and yield statistics of the final strand are also examined.

STRAND PERFORMANCE

In the early stages of SSCL-type superconductor development, a major obstacle to overcome during the manufacture of both inner and outer strand was piece length. During the Accelerator System String Test (ASST) program, most wire manufacturers participating in the VQP were consistently achieving average piece lengths from 1500 to 4000 meters.[2] For Phase II of the VQP, preliminary data shows that for most vendors, piece lengths have dramatically improved over ASST results for both inner and outer strand as can be seen in the distributions, average piece lengths and mass median piece lengths shown in Figure 1. Seuntjens, et al, has shown that if a mass median piece length of 10,000 m for inner or 15,000 m for outer is attained, strand to cold weld free cable yields of approximately 97% would be expected.[3] The baseline cost estimate target cable yield of 94% would then be feasible assuming losses due to strand breaks, crossovers, etc., do not exceed 3%.

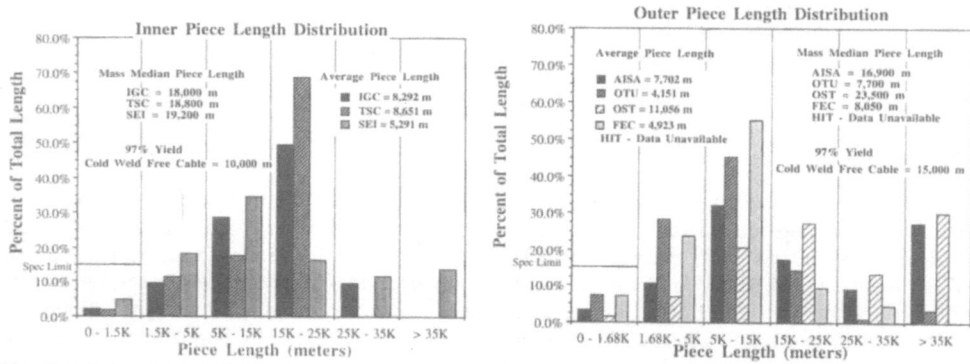

Figure 1. Length distribution, average piece length, and mass median piece length for inner and outer wire manufacturers.

In addition to piece length, multifilament yield has a large impact on the cost of the final cable. SSCL's baseline cost estimate target value which was generated at the beginning of the program assumed a 90% yield for multifilament billets. Preliminary results show that most vendors are making adequate progress towards meeting this value (see Table 1).

Table 1. Percent of Phase II completion, percent yield and coefficient of variation of multifilament billets for inner and outer wire manufacturers.

	VENDOR	% Complete	% YIELD	CV %
OUTER	AISA	100	85.3	2.9
	OTU	100	79.2	3.2
	OST	100	85.1	2.6
	FEC	30	81.3	0.4
	HIT	6	92.1	3.4
INNER	IGC	17	84.0	1.3
	TSC	14	79.0	1.4
	SEI	60	87.7	1.4

In Phase I of the VQP, Cu/SC ratio was the most difficult parameter to control with most manufacturers having performance index's[1] (C_{pk}'s) between 0.5 and 1. Phase II results were similar to Phase I with Cpk's remaining near 1 for most vendors as seen in Figure 2. High Cu/SC ratio variation is an inherent problem in the superconducting wire production process due to unsteady state material created at both ends of the billet during the multifilament extrusion. One method to increase Cpk would be to scrap acceptable material from the ends thereby reducing the standard deviation of the population. This, however, would cause a dramatic reduction in billet yield. To resolve this problem, SSCL will evaluate this parameter during qualification only in the completely steady state region. Figure 2 shows the data including the unsteady end effect material to make comparisons to Phase I meaningful.

A box plot showing the range of I_C for each manufacturer along with the coefficient of variation[1] (CV) for the data are shown in Figure 3. This I_C data was generated by production unit testing which was implemented for Phase II. The production unit test method for I_C requires that the Cu/SC ratio be determined for each piece of wire in a production unit, and I_C tests performed on any three pieces of wire with the average Cu/SC in that production unit. From the figure, you can see that the CV's for all vendors is 2% or less which is similar to the Phase I and ASST[2] results. Based on this, one can conclude that production unit testing gives an accurate interpretation of the average I_C of a production unit without testing every piece of wire. In addition, extracted strand I_C results from Phase IB cable show CV's of less than 2%[4] which confirm that the need for strand mixing in the cable based on I_C's is unnecessary to achieve low CV's in cable.

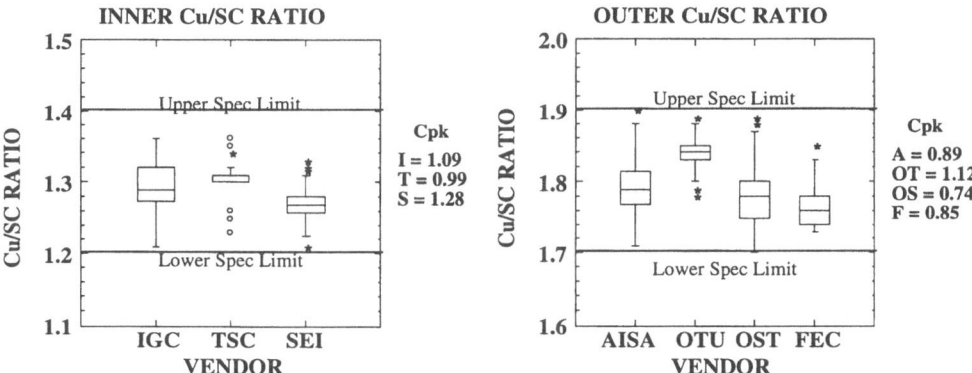

Figure 2. Box plot and performance index of Cu/SC ratio for inner and outer wire manufacturers.

The box plot and C_{pk}'s of the strand diameter data from Phase II is shown in Figure 4. All manufacturers were extremely successful at controlling this parameter with Cpk's well over 1.5. This equates to less than 6 strand diameter nonconformances per 1 million data points.

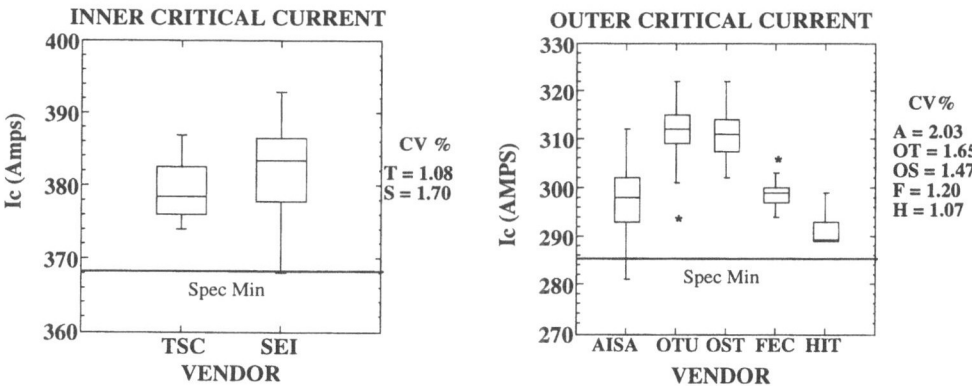

Figure 3. Box plot and coefficient of variation of I_C for inner and outer wire manufacturers.

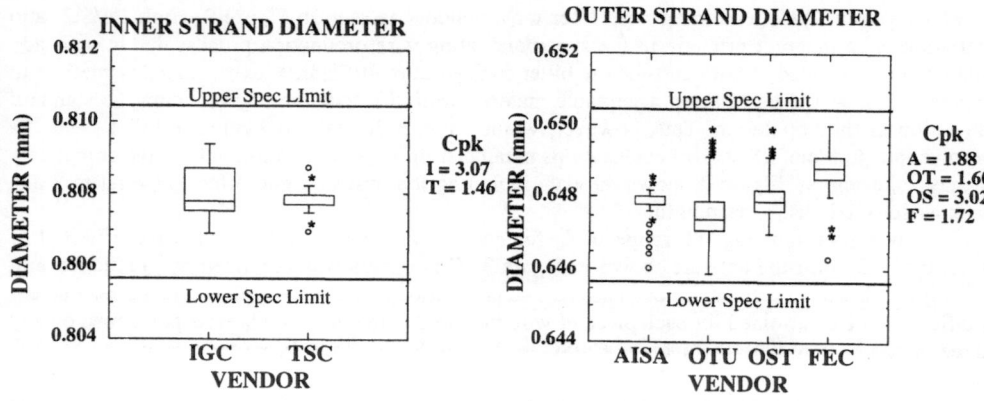

Figure 4. Box plot and performance index of strand diameter for inner and outer wire manufacturers.

SUMMARY

In Phase II of the VQP, each manufacturer was required to produce roughly 6000 kg of cable. Piece length has dramatically improved over ASST results for most vendors making cold weld free cable with strand to cable yields of approximately 97% expected. Most vendors are making adequate progress towards the baseline cost estimate target value of 90% for multifilament billet yield. The Cu/SC results are similar to Phase I resulting in C_{pk}'s of around 1 for most manufacturers. I_c results using production unit testing achieved CV's of 2% or less which is comparable to ASST and Phase I results. This substantiates that production unit testing gives an accurate interpretation of the average I_c of a production unit without testing every piece of wire. The variation of strand diameter is very low with manufacturers achieving C_{pk}'s consistently greater than 1.5.

REFERENCES

1. J.M. Juran. "Juran's Quality Control Handbook Fourth Edition," McGraw-Hill, Inc., New York (1988).

2. D. Christopherson, D.W. Capone II, J.M. Seuntjens, C.R. Hannaford, S. Graham, and D. Pollock. "Summary of the Performance of Superconducting Cable Produced for the Accelerator System String Test Program," IISSC IV: 25-32 (1992).

3. J.M. Seuntjens, D. Christopherson, F.Y. Clark, and D.W. Capone II. "Cold Weld Analysis in SSC Strand and Cable," IISSC IV: 685 - 693 (1992).

4. E.S. Coleman, D.W. Capone II, M.J. Erdmann, B.A. Jones, and J.M. Seuntjens. "Results of Cabling from Phase IB of the SSC Vendor Qualification Program," Paper III-F-3, to appear in Supercollider 5., (1993).

COMPARISONS OF PROCESSES AND PERFORMANCE OF SSC-VQP MATERIAL

J. M. Seuntjens, F. Y. Clark, M. J. Erdmann, E. S. Coleman, and B. A. Jones

Superconducting Super Collider Laboratory[*]
2550 Beckleymeade Avenue
Dallas, TX 75237-3997

INTRODUCTION

The Superconducting Super Collider's (SSC) cable Vendor Qualification Program (VQP) will end in FY 1993. At the time of this writing, all 8 vendors involved in this program have demonstrated capability to fabricate conductor which meets SSC specifications. The magnet vendors have hard choices to make in calendar year 1993 in deciding which cable vendors will make the production cable. It is well accepted that because of requirements of magnet uniformity, that only one vendor will be chosen for dipole Inner cable, one vendor for dipole Outer cable, and one vendor for quadrupole Outer cable. The production quantities are nominally 500, 500, and 200 metric tonnes, respectively. Among the many deciding factors are a technically sound production process, process control, and production quantity capability of each cable vendor. Qualified vendors will have proven their technical process and process control is adequate for production quantities. This paper is part of ongoing effort to provide technical information for the magnet vendor's decision making process. Some of the Phase IB process data is summarized and well as results of a portion of the materials characterization performed at the SSC Laboratory. Key process and final product parameters for each cable vendor are compared without identifying specific vendor's process detail.

VARIATIONS BETWEEN STRAND PROCESSES AND FINAL STRAND PROPERTIES

All 8 vendors in the VQP use a common double extrusion process. A monofilament with 4 % nominal diffusion barrier is assembled, extruded, drawn, cut to hexagonal elements, and stacked into a multifilament billet. The multifilament billet is extruded, rod drawn, given 3-4 heat treatments with intermediate strains, and drawn to final size. At this time, all vendors are using a common NbTi alloy source as well as OFHC copper for their raw materials. Raw material differences have been presented earlier.[1] Since each qualified vendor's material meets SSC specifications, the most relevant technical issue is to identify differences between each vendor's materials.

Key process parameters are recorded for each billet produced in the VQP. This data has been taken from the SSC-VQP database[2] and averaged for each vendor's phase IB production. Table 1 below lists the minimum and maximum average Phase IB values for key parameters in the double extrusion processes. Monofilament processes vary widely. However, the parameters near the end of the process tend to have less variation. All strand processes attempt to provide adequate superconducting properties and maximize mechanical robustness., which explains some of the similarities near the end of the process.

---

*Operated by the Universities Research Association, Inc., for the U.S. Department of Energy under Contract No. DE-AC35-89ER40486.

Characterization of monofilament from the Phase IB processes has been presented earlier.[3] The final strand properties are given in table 2 for each cable vendor. The Ic, Jc, and RRR are as reported from the vendor. The slope was measured at the SSCL between 5.6 T and 7 T for Outer and 7 T and 8 T for Inner. The r/D, final strand tensile strength, and springback measured at SSCL is also presented for comparison. In each case, the average value from Phase IB is reported. The properties of the cables made from this strand are discussed elsewhere in these proceedings.[4]

Table 1. Key strand process parameters and the minimum and maximum Phase IB average values reported in different vendor's processes.

Parameter	Min	Max
Nb barrier fraction (% of filament)	3.8	4.3
Virgin copper RRR	180	380
Mono. billet aspect ratio	2.6	6.3
Mono. extrusion temp. ($^{\circ}$C)	240	810
Mono. extrusion reduction ratio	9.7	92
Mono. restack diameter (mm)	1.8	3.3
Multi. theoretical yield (km)	62	223
Multi. Billet aspect ratio	2.2	5.7
Multi HIP temp. ($^{\circ}$C)	25	580
Multi. extrusion temp. ($^{\circ}$C)	500	650
Multi. extrusion reduction ratio	6.3	17
Last heat treatment temperature ($^{\circ}$C)	375	405
Number of heat treatments	3	4
Heat treatment total time (hours)	160	320
Final drawing strain	3.9	4.6

Table 2. Phase IB averages for final strand properties for each vendor.

Vendor	Ic (A)	Jc (A/mm^2)	Slope (A/mm^2 T)	RRR (final)	r/D	Tensile, final (MPa)	Springback (degrees)
AISA	289	2484	550	187	0.113	849	875
OTU	316	2689	600	209	0.098	856	850
FEC	293	2438	606	108	0.095	875	925
HIT	300	2431	539	157	0.108	860	780
OST	305	2568	550	222	0.131	835	760
IGC	383	1736	530	181	0.087	937	854
TSC	387	1737	596	139	0.096	855	837
SEI	374	1649	511	135	0.092	949	825

The tensile strengths are similar for all Inner and all Outer cable vendors. If we average the similar vendor's numbers we get 914 MPa for Inner (56.5 % copper) and 855 MPa for Outer (64.3 % copper). Smith[5] reports a tensile strength of approximately 60 ksi (414 MPa) for fully cold worked copper. If one places these 3 data points on a common plot of tensile strength vs. % copper, one can predict the tensile strength of 6 μm NbTi filaments. The plot in figure 1 shows good correlation and a predicted NbTi tensile strength of 1595 MPa at the Y intercept. In earlier work sponsored by SSCL, Liu[6] reported that NbTi filaments have a tensile strength of 195 ksi (1334 MPa) at a filament size of 260 μm. This suggests that in these composites with high strain, the rule of mixtures is obeyed. Also included in figure 1 is an average tensile strength for the restack monofilaments of 920 MPa and 665 MPa for bare NbTi and copper clad monofilament, respectively, reported earlier for cases without monofilament anneal.[7] The monofilament has no precipitation and far less cold work and clearly does not fall on the line representing the extreme cold worked condition.

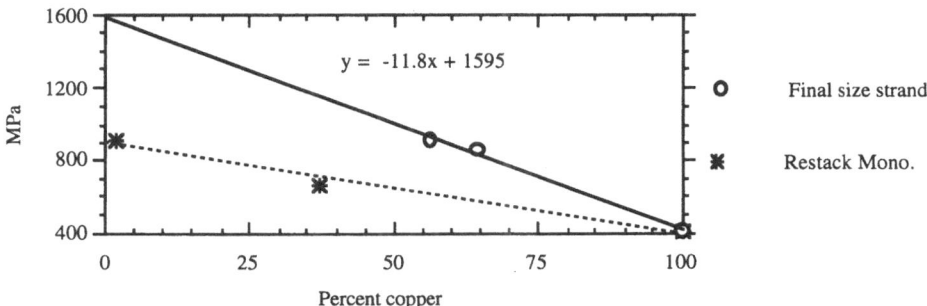

Figure 1. (Upper) Tensile strength vs. % copper for pure copper, the average Inner and the average Outer strand values. A linear extrapolation for 6 μm NbTi filaments is 1595 MPa. (Lower) Tensile strength of bare NbTi and copper clad monofilaments at restack. Bare NbTi is shown slightly off set from the Y axis.

A series of image analysis measurements on a select series of samples from the each vendor's program have been made. The filament roundness and filament area coefficient of variation (CV) have bee measured on 3-5 billet's samples from each vendor. This data is intended to present typical values for the vendor. In any one vendor, the number of samples are too small to yield statistical information. The techniques used in these measurements have been discussed in detail elsewhere.[7]

Figure 2 plots the filament area CV after first and last heat treatment as well as final size. The data after first heat treatment is incomplete at this time. Figure 3 displays the filament roundness for each vendor's material at monofilament restack, previously reported[7], as well as after first heat treatment, after last heat treatment, and final size strand.

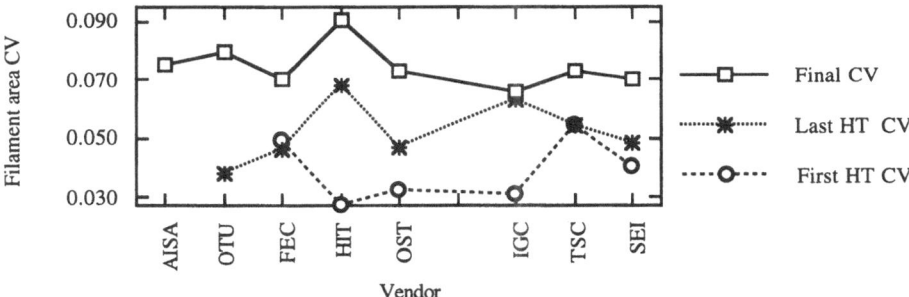

Figure 2. Filament area CV for each vendor after first and last heat treatment as well as final size strand. The data after first heat treatment is incomplete at this time.

DISCUSSION

The Phase IB monofilament billet design and process parameters vary widely. The multifilament billets also vary widely in aspect ratio and theoretical yield. However, their extrusion temperatures and extrusion reduction ratios are considerably closer to each other than the monofilament billets. The variation in the multifilament heat treatment temperature and total heat treatment time are reflected in the 9 % range for Ic between the vendors shown in table 2. The final strain varies by no more than 40 % (approximately 2 drawing die passes) for all of the vendors.

The slope of the Jc vs. field has a range of about 20 % between vendors and does not correlate to other process parameters. The final RRR is on the order of 60 % of the raw copper value. The RRR annealing procedures were not identical for each vendor in Phase I. The r/D of the Inner composites is similar. OST material has significantly different r/D than all other Outer vendors.

The tensile data for Inner and Outer strand, combined with cold worked copper data, indicate that these composites have little deviation in strength from the rule of mixtures. The calculated tensile strength of 6 μm filaments is approximately 1600 MPa. The monofilament data also follow the rule of mixtures, however the cold worked (strain of >3) NbTi tensile strength, in the absence of precipitation heat treatment, is 920 MPa.

The filament area CV is typically 3-4 % after first heat treatment, 4-7 % after the last heat treatment, and 7 to 9 % at final size. The data indicate that the filament "sausaging" occurs gradually throughout the multifilament process. The roundness data show that the filaments distort between the monofilament restack and first heat treatment, falling from 0.82 to 0.72, respectively. The roundness does not appear to further degrade during the heat treating and drawing strain as does the filament area CV. Note that the filaments with the highest degree of roundness also have the highest filament area CV. Distortion of the filaments is most noticeable to the eye and is undesirable because it increases the filament perimeter which reduces the barrier thickness. In the present case of adequate barriers for all vendors, distortion is not a serious problem because it is not generally severe enough to "puncture" the barrier or allow electrical "bridging" of filaments. Sausaging is less noticeable to the eye and is thought to be more troublesome in that it reduces the composite Ic. These result suggest that filament distortion and filament sausaging are separate phenomena.

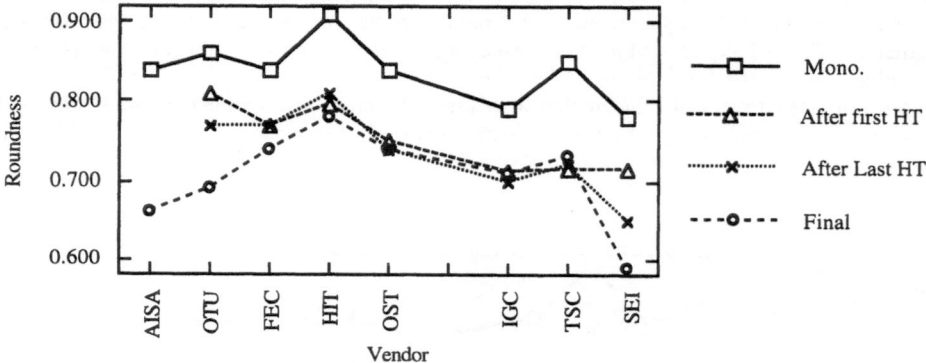

Figure 3. Filament roundness for each vendor at monofilament restack, after first heat treatment, after last heat treatment, and final size strand. A roundness value of unity represents a filament with a circular cross section.

REFERENCES

1). J. M. Seuntjens, V. A. Bardos, D. W. Capone II, F. Y. Clark, E. S. Coleman, M. J. Erdmann, and B. A. Troupe, Supercollider IV, p. 661, (1992).

2). V. A. Bardos, E. S. Coleman, M. J. Erdmann, K. Kozman, D. Little, B. A. Jones, and J. M. Seuntjens, *Databases For Analysis of Superconducting Cable Manufacturing,* paper VII-10 , to appear in Supercollider V., (1993).

3). J. M. Seuntjens, V. A. Bardos, D. W. Capone II, D. Christopherson, F. Y. Clark, E. S. Coleman, M. J. Erdmann, T. J. Headley, B. Jones, and D. K. Washburn, *Analysis of Monofilament and Multifilament samples Obtained From Phase I of the SSCL Vendor Qualification Program,* to appear in *IEEE-Trans. Appl. Superconductivity,* no. 3, March, (1993).

4). E. S. Coleman, D. W. Capone II, M. J. Erdmann, B. A. Jones, and J. M. Seuntjens, *Results of Cabling from Phase IB of the SSC Vendor Qualification Program,* paper III-F-3 , to appear in Supercollider V., (1993).

5). W. F. Smith, *Structure and Properties of Engineering Alloys,* McGraw-Hill, p. 220, (1981).

6). H. Liu, Masters Thesis, Oregon State University, p. 28, (1991).

7). J. M. Seuntjens, F. Y. Clark, T. J. Headley, and N. Y. C. Yang, Applied Superconductivity Conference, paper LOD-3, 1992, to appear in *IEEE-Trans. Appl. Superconductivity,* no. 3, March, (1993).

RESULTS OF CABLING FROM PHASE IB OF THE SSC VENDOR QUALIFICATION PROGRAM

E. S. Coleman, D. W. Capone II, M. J. Erdmann, B. A. Jones, and J. M. Seuntjens

Superconducting Super Collider Laboratory*
2550 Beckleymeade Ave.
Dallas, TX 75237

INTRODUCTION

The Superconducting Technology Department of the Superconducting Super Collider Laboratory (SSCL) is directing the Vendor Qualification Program (VQP) to identify vendors of superconducting material to fabricate 6-micron filament conductor for various magnet programs. Eight suppliers are participating in the qualification program: Intermagnetics General Corporation, Inc. (IGC), Teledyne SC (TSC), and Sumitomo Electric Industries (SEI) are fabricating 30-strand inner cable and Oxford Superconducting Technology (OST), Furukawa Electric Co. (FEC), Hitachi Cable, Ltd. (HIT), Alsthom Intermagnetics, s.a. (AISA), and Outokumpu Copper Superconductors (OTU) are fabricating 36-strand outer cable. Phase I of the VQP included two portions: Phase IA, the R&D portion, and Phase IB, a simulated production quantity processed on a single production process. During Phase IB of the VQP, each of the suppliers was to fabricate approximately 3500 kg of finished cable, meeting the appropriate cable specification. This cable is being used to manufacture the prototype magnets in each of the major magnet programs underway within the Magnet Systems Division. In order to meet the needs of the Collider Quadrupole Magnet program, four of the five outer cable vendors were directed to divert approximately 350 kg of finished strand to the fabrication of 30 strand quadrupole cable.

Each vendor was responsible to direct the cabling activities for his own material. Four vendors (IGC, SEI, FEC, and HIT) chose to perform the cable manufacturing on machines at their own facilities. The other four chose cabling subcontractors; OST and OTU cabled at New England Electric Wire Corp. (NEEW), and AISA and TSC cabled at Superconducting Cabling and Engineering, Inc. (SCE).

CABLE SUMMARY

Cabling Yield: All vendors began Phase IB with the plan to complete the full quantity of 3500 kg. The planned quantity of inner cable was reduced to 3150 kg due to a change of Cu/SC specification at the beginning of the program. The finished amount was a result of both the yield of the wire processing and the cable manufacturing. The results in Table 1 only represent yields of the

*Operated by the Universities Research Association, Inc., for the U.S. Department of Energy under Contract No. DE-AC35-89ER40486.

Table 1. SSC cable data from extracted strand measurements performed at the SSCL. Reported I_c values correspond to testing fields of 7 T for inner and 5.6 T for outer and quadrupole cables, at temperature of 4.22 K.

Vendor	Type	Quantity (meter)	Cabling Yield (%)	Cable I_c (Amp)	Cable I_c CV (%)	Ave Deg. (%)	J_c Slope (A/mm²/T)	Slope CV (%)
IGC	Inner	25,841	78%	11,113	1.66	3.51	567	3.0
SEI	Inner	15,807	76%	10,822	1.45	3.75	539	1.7
TSC	Inner	231	--	11,233	--	4.02	578	--
OST	Outer	28,393	88%	10,538	0.93	3.96	569	1.7
FEC	Outer	26,046	88%	10,564	1.20	0.36	607	0.9
HIT	Outer	14,216	56%	10,179	1.42	6.83	552	1.6
AISA	Outer	26,126	82%	10,399	0.29	1.36	571	0.8
OTU	Outer	18,995	87%	10,968	0.81	3.65	618	1.5
OST	Quad	4,691	--	8,672		5.35	552	
FEC	Quad	4,456	--	8,750	0.99	1.1	597	0.4
AISA	Quad	4,698	--	8,593		2.61	558	
OTU	Quad	4,571	--	9,023	0.62	3.3	598	1.2

strand-to-cable portion of the process. The total length of within-specification strand was compared to the length of the delivered cable multiplied by the number of strands in the cable. The baseline expected yield before beginning the program was 91%, which includes an approximate 4% loss due to the lay pitch of the strand in the cable. Many extra losses occurred during Phase IB, including extra material used for cable startup, crossovers and strand breaks, and mishandling of the cable being wound onto spools.

Critical Current: In Figure 1 the critical current results of the cables from each supplier are compared. The results are from extracted-strand testing done at the SSCL. A specific explanation about the extracted-strand testing can be found in an accompanying paper by Kovachev[1]. The cable-to-cable coefficient of variation (CV, where CV = standard deviation, σ, divided by the mean, $\overline{X}$) within any one supplier is no greater than 1.7%. This variation compares to maximum CV of 1.5% within the cable batches produced for the Accelerator System String Test (ASST) in 1991[2]. Taken as a whole, the entire population of inner cables have an average I_c of 11,035 A, with a variation of 1.99%. CDM outer cables average 10,538 A, with 2.26% variation, and CQM cables showed an average of 8,834 A, with 2.17% variation.

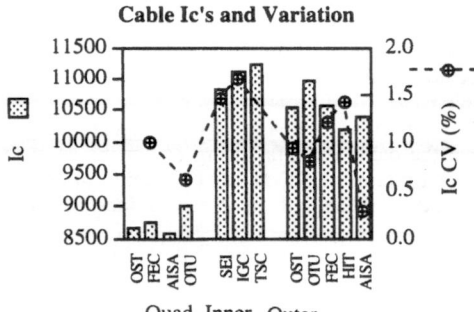

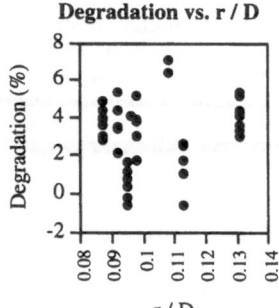

Figure 1. Average Critical Current and Coefficient of Variation for each cable batch. Specified Ic's are 8,160 A 10,500 A, and 9,780 A for Quad, Inner, and Outer cables respectively.

Figure 2. Extracted Strand Degradation factor versus r / D ratio for each vendor's strand.

Degradation: Previous summaries of cable results have indicated that the degradation in the cable has correlated strongly with a parameter of the superconducting strand geometry called "r / D"[3]. This r / D parameter is a ratio of the minimum thickness of the copper jacket around the filament array (r) to the outside diameter of the finished strand (D). Outer-type cables used in the ASST particularly indicated that by using a thicker copper jacket around the filament array, and therefore increasing the r / D ratio, the cabling degradation could be reduced. Figure 2 displays a plot of measured degradation versus r / D for the cables in Phase IB. In this case no correlation can be drawn between the parameters. Cables for previous SSC programs have been electrically tested at Brookhaven National Laboratory, in the full cable state. The procedure for those tests included a mathematical correction for the self field of the cable[3]. When the degradation was calculated, it was then possible to achieve negative degradation factors. In the extracted strand testing procedure, randomly selected strands were extracted from the cable, and tested individually. The average critical currents of the extracted strands were compared to the average critical currents of all the strands in the cable as reported by the manufacturer. Because the extracted strands are tested in the same magnetic conditions as the precabled wires, the reduction in current is a true degradation factor. Because the wire may be further deformed by handling in preparation for testing, this degradation is a worst case estimate of the true degradation.

J_C vs. Field Slope: The J_C slope was calculated using the ratios of the extracted strand J_C's at 7 T and 8 T for inner and 5.6 T and 7 T for outer cables. The average slope has a range of 15% between all of the vendors. Most have a variation near 2%. Observations from Phase I of the VQP program are that while a vendor can consistently produce a certain J_C slope, process variables are so interrelated that it is difficult to adjust this to a target value.

CABLE DIMENSIONS

During Phase IB of the VQP, the ability of each cable supplier to maintain control of the cable dimensions was evaluated by the statistical index called Process Capability Index (C_{pk}). C_{pk} is defined as:

$$C_{pk} = \frac{Min[(\overline{X} - LSL), (USL - \overline{X})]}{3\,\sigma},$$

where $\overline{X}$ is the mean, USL is the Upper Specification Limit, LSL is the Lower Specification Limit, and σ is the standard deviation of the data. The target C_{pk} for the VQP is 1.5, which indicates a process that will have 6.8 defects per million measurements. A typical measurement frequency during cable production is once per 3 meters of cable. A company meeting this benchmark would have one out of tolerance condition every 49,000 meters of cable. Figure 3 shows the resulting C_{pk}'s for each dimension and each type of cable.

Keystone Angle: The actual variation in the keystone angle is approximately the same for all three cable types, but because the tolerance range of 36-strand outer cable is $\pm 0.06^o$ rather than $\pm 0.1^o$, the variation is larger relative to the tolerance band, and the C_{pk}'s are generally lower.

Width: The most noticeable group of results is in Sumitomo's batch of inner cables, where Cpk ranges from 0.8 to 1.5. As the width specification is written as 12.34 +0.05, -0.00 mm, SEI targeted the nominal value of 12.34 mm. If the evaluation had been done with the index C_p, which does not take into account whether the results are centered within the tolerances, SEI would have results from 2.0 to 3.0. In comparison, IGC's results would range from 3.0 to 5.2. The other suppliers of outer and quad cable meet the expectations in nearly every case.

Midthickness: It can be seen that at this point the control of the keystone angle and width dimensions have been accomplished much better than the control of the midthickness. In comparison between the width and midthickness dimensions, both are produced by the same mechanical method using ground rollers, and therefore, the variation of both will be approximately the same. However, the tolerance range for midthickness is only 0.012 mm, one fourth of the range for the width, which is 0.05 mm. Some cases of thickness control have been demonstrated,

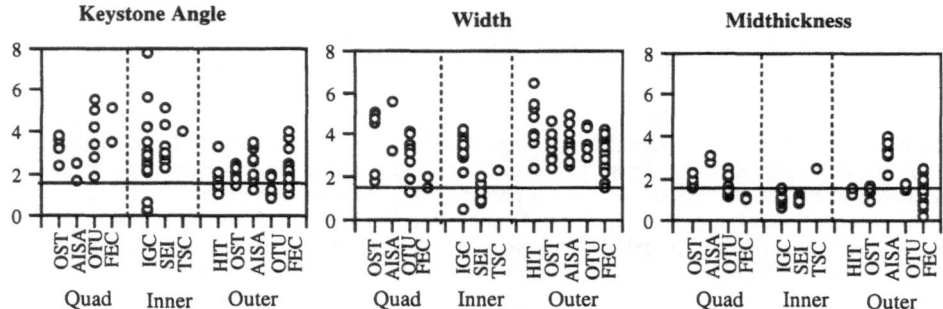

Figure 3. Process Capability Indices (Cpk's) for the dimensionsions of each cable batch, grouped by cable types.

but generally there is still improvement to be made in this area. Companies who maintained the best control generally did so by eliminating eccentric runout of the rollers and bearings, and controlling the temperature of the turkshead unit. During the cable run, heat generated from the strand deformation warms the rollers. It has been universally observed that unless the turkshead rollers are preheated to the same temperature as during the cable run, the diameter of the rollers will increase due to the added heat, and the thickness of the resulting cable will become thinner as the run progresses. Temperature changes in the room environment can affect the thickness in the same way. In order to counter this effect, some vendors have added climate control for the turkshead and/or temperature control methods for the rollers.

SUMMARY

The number of participating vendors in the VQP program is more than in any of the previous cable programs. Electrical results and variations are comparable, but slightly more variations occur in some vendors' cases. Dimensional variations of the cable are being addressed, and in some cases variations are being controlled to the level of the program goals. The Phase IB cable does demonstrate that the SSC cable can be fabricated in uniform production batches. Points for further improvement for Phase II have been identified for each supplier, which range from yield improvement to dimensional control through specific machine parameters.

REFERENCES

1. V. Kovachev, D. W. Capone II, E. S. Coleman, B. A. Jones, J. T. Madison, J. Qin, and J. M. Seuntjens, Extracted Strand I_c Degradation for SSC Cables, Paper III-F-4, to appear in Supercollider 5 (1993).
2. D. Christopherson, D. W. Capone II, J. M. Seuntjens, C. R. Hannaford, S. Graham, and D. Pollock, Summary of the Performance of Superconducting Cable Produced for the Accelerator System String Test Program., IISSC IV: 25-32 (1992).
3. M. Garber, A. K. Ghosh and W. B. Sampson, The Effect of Self Field on the Critical Current Determination of Multifilamentary Superconductors, IEEE Trans. Magn. MAG-25, 1940-44 (1989).

EXTRACTED STRAND I_C AND STRAND I_C DEGRADATION FOR SSC CABLES

V. T. Kovachev, D. W. Capone II, E. S. Coleman,
B. A. Jones, J. T. Madison, J. Qin, and J. M. Seuntjens

Superconducting Super Collider Laboratory[*]
2550 Beckleymeade Avenue
Dallas, TX 75237-3997

INTRODUCTION

The performance testing of the large quantities of superconducting cables for SSC magnets is a daunting challenge, at best. In an effort to reduce the quantity of full cable testing required, an investigation is underway to evaluate the utility of determining the performance of the SSC cables using extracted strand testing. It is believed that the cable performance can be quite accurately determined by measuring the critical current (I_c) on strand samples removed, at random, from the manufactured cable.[1] These strand measurements are used to derive the cable critical current. The measured critical current (I_{ce}) are then compared to the virgin strands (I_{cv}) input into the cable to determine the degradation resulting from the cable fabrication. The advantage of this type of certification is two fold; firstly the manufacturer can certify the performance using existing strand measurement equipment. This allows for a reduction in the lead time between manufacture and delivery of the cable. Secondly, the SSC can perform random sampling, as opposed to 100% cable testing and still maintain adequate visibility into the performance of the cables. The following sections cover the techniques used and the results obtained for measurements of extracted strand I_c (I_{ce}) and field dependence of I_{ce} for more than fifty representative cables. The data for another five cables for which full cable I_c values were previously measured at Brookhaven National Laboratory are also presented and compared with full cable I_c results. The details of the full cable measurements performed at BNL are described elsewhere[2]. The SSCL full cable test facility will soon be in routine operation which will allow for precise verification of the validity of extracted strand testing in the determination of I_c in SSC cables.

EXPERIMENTAL DETAILS

The extracted strands are prepared by cutting a one meter length from each cable sample. Six strands are randomly taken from each cable sample for test. These strands are

---
*Operated by the Universities Research Association, Inc., for the U.S. Department of Energy under Contract No. DE-AC35-89ER40486.

removed so as to not disturb the "zig-zag" imparted by the path of the wires through the cable itself. The sample rig is based on the successful design used at BNL for many years.[3] The sample holder consists of a G-10 barrel with grooves for three strand samples. The geometry of the 38 mm diameter barrel is such that each sample has a pitch of 12.7 mm. This combines to reduce, to minimal levels, the solenoidal contribution to the self-field on the sample during energization. The gage length between voltage taps is 579 mm. The samples are independently energized using three current leads and a single common lead so as to maintain the Lorentz forces inward during the I_C measurements.

During the mounting of these samples no effort is made to retain the original geometry of the wire pitch in the cable. The strands are pulled into the grooves of the sample holder, and straightened considerably, but not completely in the process. This is done to ensure adequate strand tension to avoid training behavior during the measurements. This sample mounting technique increases the possibility of introducing additional damage to the sections of the strand deformed at the edges of the cable. However, as the measurements show, apparently the additional movement introduced by hand is insignificant compared to the deformation imparted by the turkshead rollers during the manufacturing process.

The background field for the measurement is obtained using a 15 Tesla superconducting solenoid magnet with a 64 mm aperture, produced commercially by Oxford Instruments, Inc. The three samples reside, during the measurement, within the 1 % uniformity sphere defined by the magnet geometry.

The I_C measurements are performed using a computerized data acquisition system consisting of two Keithley 182 microvoltmeters, a Hewlett Packard 6464C DC power supply, a Hewlett Packard 6031A DC power supply, and a Dynapower zero flux current transformer. These instruments interface over an IEEE-488 bus. The system is operated using a LabVIEW virtual instrument called ICTEST running on a Macintosh computer with 5 Mb of RAM and a 40 Mb hard drive.[4] Data is collected at a rate of 2-3 Hz and full data reduction occurs within 3-5 seconds after measurement with typical data files of 200-300 data points. Typical noise levels in this system are 100-200 nV. I_C is defined using a $1 \times 10^{-14}\Omega$-m resistivity criterion. For a gage length of 579 mm, the criteria corresponds to 4-6 μV at I_C. Initial runs on the NIST Standard Reference Material,[5] as well as the round robin testing certification procedures implemented for the VQP,[6] have shown compliance with required accuracy and repeatability levels. The repeatability, reproducibility, and accuracy variations are 1.31 %, 0.41 %, and 1.16 %, respectively. The total error of the Ic measurements is 2.23 %.

Six strand I_C measurements for each cable sample are averaged to produce a strand I_C value for the cable. No correction to the solenoid or the self-field of the wire is performed. The solenoid field produced by the strand is negligible (approximately 2.0×10^{-3} T). The maximum self-field for the outer and the inner wire (at 5.6 T and 7 T respectively) is less than 9.0×10^{-2} T. There is no necessity in self-field correction for comparison of virgin wire I_C and extracted strand I_C

The cable critical current is determined by the product of this average I_C and the number of strands in the cable. The average Ic for the virgin case and the extracted strand case is not corrected for self field and geometery effects. In addition to the cable I_c, the degradation of the strands as a result of the deformation during cabling is determined. The degradation is determined as the percent change of the average extracted strand I_C from the weighted average I_C of the virgin wires used in the cable manufacturing. The weighted average I_C of the wires used for the cabling was determined from the map of wire positions within the cable which is used to manufacture the cable. The weighted average I_C is the sum of the measured I_C of a wire times the fractional length of that wire in the cable divided by the number of strands in the cable.

RESULTS

Six cable samples were selected from the archive storage maintained within the SSC Magnet Systems Division Conductor R&D group. Three samples each of Inner and Outer cable were used. The critical current of these cables all had been previously measured at BNL. The cable samples selected, along with the relevant BNL results, are identified in Table 1.

Table 1. Relevant parameters of the BNL measured cables used for this paper.

Cable sample	Vendor ave. virgin Ic (A)	SSCL ave. extracted strand (A)	SSCL cable calculation (A)	BNL cable measurement (A)	SSCL measured degradation (%)	Difference SSCL vs. BNL (%)
Inner						
3-I-00044	368	343	10290	10764	6.80	-4.4
3-I-00054	n/a	341	10286	10904	n/a	-6.0
3-I-00064	n/a	343	10290	10799	n/a	-4.9
Outer						
4-K-00025	318	302	10872	11188	5.1%	-2.9
4-K-00026	319	306	11016	11207	4.0%	-1.7

The average I_C, n (at 5.6T and 7.0T), dI_C/dB, and degradation data for 29 SSC Outer cables are presented in Table 2. The average I_C, n (at 7.0T and 8.0T), dI_C/dB, and degradation data for 17 SSC inner cables are presented in Table 3.

Table 2. Average I_C, dI_C/dB, n, and degradation for extracted Outer strands

Vendor		Ic (5.6 T)	n (5.6T)	Ic (7.0 T)	n (7.0 T)	Degradation (%)	dI_C/dB (A/T)
AISA	Ave.	288.3	37.7	196.0	32.0	1.6	-66.2
	Std dev.	1.1	0.9	0.8	1.3	1.2	0.7
FEC	Ave.	292.1	34.3	190.5	29.8	0.8	-72.7
	Std dev.	2.8	2.8	2.4	1.4	0.8	0.7
HIT	Ave.	282.8	40.1	188.6	34.1	6.0	-67.3
	Std dev.	3.1	2.6	2.1	1.4	1.1	0.9
OTU	Ave.	303.0	42.0	203.6	35.0	3.5	-71.1
	Std dev.	2.7	3.8	1.6	2.4	1.0	1.2
OST	Ave.	292.2	47.1	198.5	37.6	4.1	-67.0
	Std dev.	2.7	6.1	1.2	2.5	0.8	1.3

Table 3. Average I_C, dI_C/dB, n, and degradation for extracted Inner strands

		Ic (7.0 T)	n (7.0 T)	Ic (8.0 T)	n (8.0 T)	Degradation (%)	dI_C/dB (A/T)
IGC	Average	369.58	38.01	243.89	30.48	3.51	-125.70
	Std dev.	5.44	2.58	3.56	3.49	0.70	2.79
SEI	Average	356.89	36.88	237.49	31.80	3.64	-123.71
	Std dev.	9.23	1.01	3.01	3.59	1.07	1.91

DISCUSSION

The BNL value for the cable has a self field correction for their test configuration, according to Garber et. al.[2] They have developed a strand correction of $B_p = B_a + \pi * (10^{-4}) * J * D / (1 + x)$, where B_a is the applied field, J is the critical current density, D is the strand diameter, and x is the copper to superconductor ratio. Using a J slope of 29 % per Tesla, these corrections are 5.5% for Inner and 5.3 % for Outer strand. In addition, they have a field profile calculation for the cable cross section in their test rig which results in a correction about 1 % lower (i.e. 4 %) than the strand self field correction. This comes from figure 7 in the Garber paper for B perpendicular to the cable face and adding to the cable narrow edge. In the BNL case, the gap between the cables in the sample holder is 0.8 mm and the cable type is 40 mm (23 strand) Inner. Therefore, the uncorrected self field values with the SSCL extracted strand measurement are about 4-5 % lower than the self field corrected values of BNL procedure.

The difference between BNL and SSCL critical current values is as predicted by the self field correction. Based on this data and the round robin work performed on strand, it is felt that SSCL and BNL data are in agreement to better than 2 %. It is felt that the agreement are not likely to be better than this value because the strands in the cable are not exposed to a homogeneous field.[6] A single linear self field correction only approximates the situation of the strand in the cable. More in depth consideration of magnetic field and geometry corrections in SSC cables will be made when full cable measurements are made on the cables already characterized by extracted strand testing in this work.

The extracted strand n values are approximately 10-20 % lower than those on the virgin strands reported by the vendors. This reflects the extended superconducting to normal transition and Ic degradation found in the extracted strand.

It is noted that the degradation of the Phase IB cables is generally low, averaging about 3 % for Inner and Outer cables. Cabling result of SSC VQP Phase IB cable is discussed in more detail elsewhere.[7]

ACKNOWLEDGEMENT

The authors thank R. Taliaferro for his technical assistance in these measurements.

REFERENCES

1). S. L. Wipf, Preprint, DESY HERA 90/15, August, (1990).

2). M. Garber and W. Sampson, IEEE Trans. Mag. 25, no. 2, pp. 1940-1944, (1989).

3). M. Garber, private communication.

4). J. M. Seuntjens and E.S. Coleman, SSC publication SSCL-N-796, (1992).

5). National Institute for Standards Testing, Standard Reference Material Now 1457, (1984).

6). M. J. Erdmann, D.W. Capone II, and J. M. Seuntjens, Continuing Results of Systematic Error in Ic Testing, to appear in *IEEE-Trans. Appl. Superconductivity*, no. 3, March, (1993).

7). E. S. Coleman, D. W. Capone II, M. J. Erdmann, B. A. Jones, and J. M. Seuntjens, *Results of Cabling from Phase IB of the SSC Vendor Qualification Program*, paper III-F-3 , to appear in Supercollider 5 (1993).

DISCUSSION OF RESULTS OBTAINED DURING THE PHASE II
PRODUCTION IN THE SSC VENDOR QUALIFICATION PROGRAM

H. Ii,[1] T. Shimada,[1] K. Ogawa,[1] T. Suzuki[1] and M. Ikeda[2]

[1]Nikko Research Laboratory, The Furukawa Electric Co., Ltd.,
500 Kiyotaki, Nikko, Tochigi 321-14, Japan
[2]Superconducting Products Department, The Furukawa Electric Co., Ltd.,
2-6-1 Marunouchi, Chiyoda-ku, Tokyo 100, Japan

INTRODUCTION

The production process of the SSC CDM outer wire and cable was almost established through the successful Phase I production of highly uniform cable of about 4,000 kg in weight[1]. However, some minor problems were found. For example, there was a small margin over the required wire critical current (Ic) because of low critical current density (Jc). And part of the wire did not survive the sharp bend test.[2] In order to solve these problems, wire cross-sectional design was slightly changed in Phase II. This paper describes the R&D results for those improvements, as well as the evaluation of the performance of the wire produced in Phase II.

CROSS-SECTIONAL DESIGN OF WIRE

Filament Spacing

The Ic of the Phase I wire indicated the possibility that part of the wire might have lower Ic than the specified lower limit because of (average - 3 x standard deviation) / (specified lower limit) = 0.98. It meant that the Jc had to be improved for Phase II. Because the experimental results showed that the longer heat treatment time did not contribute to the Jc enhancement[1], the low Jc was thought to be not due to the low α-Ti precipitation density, but due to external factors, such as filament sausaging. Therefore, 100 filaments were measured on cross-sectional area by using an image processing device. Its coefficient of variation (standard deviation / average) was evaluated. The average filament area was 28.05 μm^2 with a standard deviation of 3.51 μm^2 and its coefficient of variation was 12.5 %. This variation was extremely large, indicating that the filament was suffering remarkable sausaging. One of possible causes in copper matrix NbTi superconducting wire is the large filament spacing.[3] The filament spacing of the Phase I wire was 1.1 μm, but changed to 1.0 μm to suppress filament sausaging in Phase II. By reducing the filament spacing, the coefficient of variation of the filament area decreased to 9.3 %. As a result, the Jc of the wire enhanced, as shown in Fig. 1.

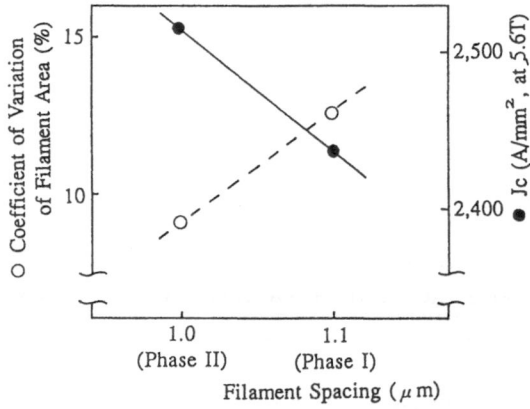

Fig. 1. Variation of filament area and Jc at 5.6 T as a function of filament spacing.

Copper Sheath Thickness of Wire (Ratio of r/D)

Sharp bend test on wire, which is intended to simulate cabling, is one of the wire inspections required by the specification[2]. Part of the Phase I wire did not survive in the test. The copper sheath thickness (r) was assumed to be sensitive in the test. The r/D ratio was evaluated, where D was the wire diameter. The ratio was varied from 0.06 to 0.12. The result is shown in Table 1. It is known from the result that the ratio must be higher than 0.10 in order to prevent the sharp bend failure. On the basis of the above result, the ratio was changed from 0.09 to 0.127 in Phase II. Accordingly, the cross-sectional structure of the wire was changed. The cross sections of the Phase I and Phase II wire are shown in Fig. 2. The Phase II wire was produced in about 25 % required quantity and showed no sharp bend failure.

Table 1. Effect of r/D on the sharp bend failure.

r/D	0.06	0.08	0.10	0.12
Defective Wire on Sharp Bend Test (%) (n=12)	67	25	0	0

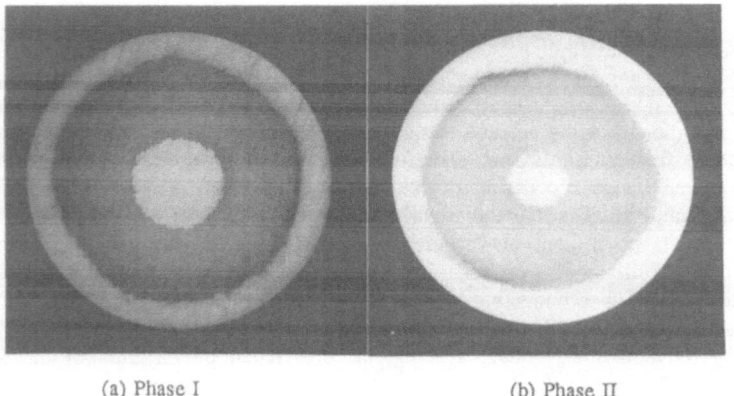

(a) Phase I   (b) Phase II

Fig. 2. Wire cross sections of Phase I and Phase II

The improvement of the sharp bend quality, on the other hand, should suppress the Ic degradation due to cabling. Christopherson et al asserts that there exists a negative correlation between the r/D and the Ic degradation due to cabling.[4] In this view, the Ic degradation of the Phase II wire was compared with that of the Phase I wire. To estimate the Ic degradation, 6 samples taken from the end of each cable (the sample with wire portion before cabling) were used. That is, the Ic at 5.6 T was measured and evaluated before and after cabling on the same wire. The degradation of the Phase I wire was 2.5 % in average and at maximum 4.5 %. That of the Phase II wire was 1.7 % in average and at maximum 3.2 %. Fig. 3 shows the effect on the Ic degradation due to cabling.

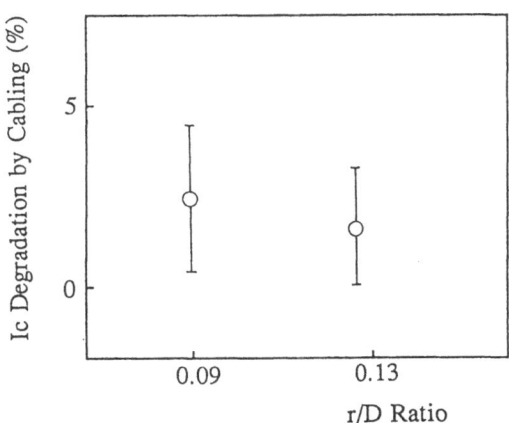

Fig. 3. Effect of increase in r/D on suppression of Ic degradation due to cabling.

RESULTS IN PHASE II

As described above, the cross-sectional structure of the wire was changed for Phase II. Along with the changes of filament spacing and r/D, the fraction of the central copper core in the total sectional area decreased from 9.4 % to 3.6 %. And the variations in some performnce such as the RRR of the copper material and the wire diameter were reduced. There was no other substantial change including processes. Large-scale production using fifteen multifilament billets of 11 inches in diameter was made. The performance of 25 % of the total production quantity of the Phase II wire was evaluated and was compared with the Phase I wire as summarized in Table 2.

The average of in-line mesurements made at a 80 meter interval over the whole length with a laser micrometer during the twisting and drawing process was regarded as a wire diameter for the evaluation. As the coefficient of variation is as low as 0.09 %, the process capability index (Cp) of 1.50 is higher than that in Phase I.

Cp of the Cu/SC ratio is as high as 1.48. Its strict control is necessary because the low ratio leads to the low r/D and the high ratio leads to the low wire Ic.

The Ic and the n value were measured on six samples from each production unit. A production unit consisted of material from a single multifilament billet which underwent identical mechanical and thermal processing. The obtained Ic, Jc and n value exceeded those in Phase I, indicating that change of the filament spacing proved remarkably effective.

Samples for the RRR measurement were annealed at 270 C for 2 hours. Because the RRR of the wire is determined mostly by the RRR of the copper for multifilament billet can and central core, its copper was selected by means of its RRR. As a result, the wire RRR was improved to be the average of 143 and the coefficient of variation of 4.0 %.

Table 2. SSC CDM outer wire performance of Phase I and Phase II

Factor / Performance	Phase II (1,800 kg)					Phase I (4,000 kg)				
	Average	Standard Deviation	Coefficient of Variation (%)	(Average-3xS.D.) /(Spec.Minimum)	Cp	Average	Standard Deviation	Coefficient of Variation (%)	(Average-3xS.D.) /(Spec.Minimum)	Cp
Ic at 5.6T (A)	300.6	3.87	1.3	1.01	-	293.5	4.74	1.6	0.98	-
Jc at 5.6T (A/mm2)	2518.4	25.93	1.0	-	-	2437.6	38.86	1.6	-	-
n value at 5.6T	42.5	1.29	3.0	1.10	-	44.3	3.78	8.5	0.94	-
Wire Diameter(mm)	0.6487	0.0006	0.09	-	1.50	0.6491	0.0009	0.14	-	0.89
Cu/SC Ratio	1.762	0.023	1.3	-	1.48	1.747	0.022	1.3	-	1.47
RRR	142.7	5.76	4.0	1.79	-	108.4	13.82	12.7	0.96	-

DISCUSSIONS

When the filament spacing was reduced, filaments restrained each other, suppressing longitudinal and/or cross-sectional abnormal deformation, such as the filament sausaging. As a result, the Jc at 5.6 T was enhanced by 3.3 %.

The high wire r/D eliminated sharp bend failure while suppressing the Ic degradation due to cabling. The sharp bend failure and the Ic degradation due to cabling are occur when the NbTi filaments located on the outside of the bent wire are exposed to strong tensile stress to cause filament breaks or area reduction in cross sections. This tensile stress grows as the filaments move nearer to the outside, and so the r/D was increased and the filaments were arranged in the area near to the center. In this manner, the above improvement was achieved.

The absolute value and the uniformity of performance of the wire were better than in Phase I, because the cross-sectional structure was changed. The SSC CDM outer wire thus produced well satisfied all the specification requirements.

CONCLUSIONS

(1) Decreasing the filament spacing from 1.1 μm to 1.0 μm suppressed the filament sausaging, resulting in the Jc enhancement by 3.3 %.

(2) The wire cross-sectional structure, the filament arrangement, strongly effected the percent defective of sharp bend test and the Ic degradation due to cabling. Increasing the r/D above 0.10 reduced the percent defective to zero and the degradation ratio was as small as 1.7% when r/D = 0.127.

(3) On the basis of above acquirements, some design changes for the Phase II production, such as the change of the wire cross-sectional structure and the reduction of the tolerances of some performances, improved the wire and cable performance. Namely, the Ic, Jc, n value, wire diameter and RRR were improved both in terms of the absolute value and the uniformity.

REFERENCES

1. K. Susai, H. Ii, K. Ogawa, T. Suzuki, M. Ikeda and S. Shiga, "Critical Current Properties of Fine Filament Superconducting Wires for the SSC", ASC, Chicago, ILL, Aug. 23-28, 1992.
2. SSC Specification SSC-MAG-M-4146 Rev. 2
3. E. Gregory, H. Liu, G. M. Ozeryansky, M. D. Sumption, K. R. Marken Jr. and E. W. Collings, "Experiments to Improve Materials for SSC Magnets", IISSC, New Orleans, LA, Mar. 4-6, 1992.
4. D. Christopherson, D. W. Capone II, J. M. Seuntjens, C. R. Hannaford, S. Graham and D. Pollock, "Summary of the Performance of Superconducting Cable Produced for the Accerelator System String Test Program", IISSC, New Orleans, LA, Mar. 4-6, 1992.

4 Technical Poster Session I

STRUCTURAL AND THERMAL ANALYSIS OF A SOLID-COOLED, LOW ENERGY BOOSTER, RADIO-FREQUENCY-CAVITY TUNER AT THE SUPERCONDUCTING SUPER COLLIDER

R. Ranganathan, A. Propp, B. Dao, and B. Campbell

Mechanical Engineering Department*
Accelerator Systems Division,
SSC Laboratory, Dallas, TX 75237

ABSTRACT

A three-dimensional heat conduction and structural model was developed to analyze and optimize the design of a solid-cooled low energy booster (LEB) radio-frequency (RF) cavity tuner concept. Consideration was given to three cooling options: (i) using beryllium oxide (BeO) disks, (ii) using aluminum nitride (AlN) disks and (iii) using neither BeO nor AlN disks. The results indicate that solid cooling is feasible from thermal and structural viewpoints if a minimum of two BeO disks or four AlN disks are used.

INTRODUCTION

The RF cavity tuner (modeled) consisted of five ferrite disks encased in a copper housing (Figure 1-a). The perpendicular biasing of the ferrites used for tuning the LEB RF cavity results in heat generation in the ferrites, housing and in the BeO (or AlN). A cooling system is needed to remove this heat[1,2] and ensure that the peak ferrite temperature and stress is maintained at safe levels. Therefore, a model was developed to analyze solid-cooling of the tuner. In this concept, the heat generated in the ferrites will be transferred by conduction to an external water jacket either directly through the ferrites (option iii above) or by using annular BeO (or AlN) disks glued between the ferrites (options i and ii). The BeO (or AlN) facilitates heat conduction from the ferrites. Details of this work are documented elsewhere.[2]

ANALYSIS

The temperature and stress distributions were assumed to be steady and three-dimensional. The thermo-structural properties were assumed to vary with temperature. The problem was solved using ANSYS.[3] Due to symmetry, only 1/32 of the tuner was modeled. Thus, the model spans 22.5 degrees in the tangential direction and includes 2 1/2 ferrites (with a symmetry plane passing through the innermost ferrite). Figure 1-b illustrates the computational domain that includes the ferrites, the epoxy, the BeO (or AlN) disks, the housing side walls and the ribs. Cooling water at 35 °C flows through the water jacket.

RESULTS

Quantitative Sensitivity Results (Table 1)

When neither AlN nor BeO were used (rows 1, 4, 7 and 10), the peak stress in the ferrite resulted in a safety factor of less than 2 (compared to its tensile strength of 39 MPa). Further, the same cases also indicated the highest temperatures in the ferrites and the epoxy.

---

*Operated by the Universities Research Association, Inc., for the U.S. Department of Energy under Contract No. DE-AC35-89ER40486.

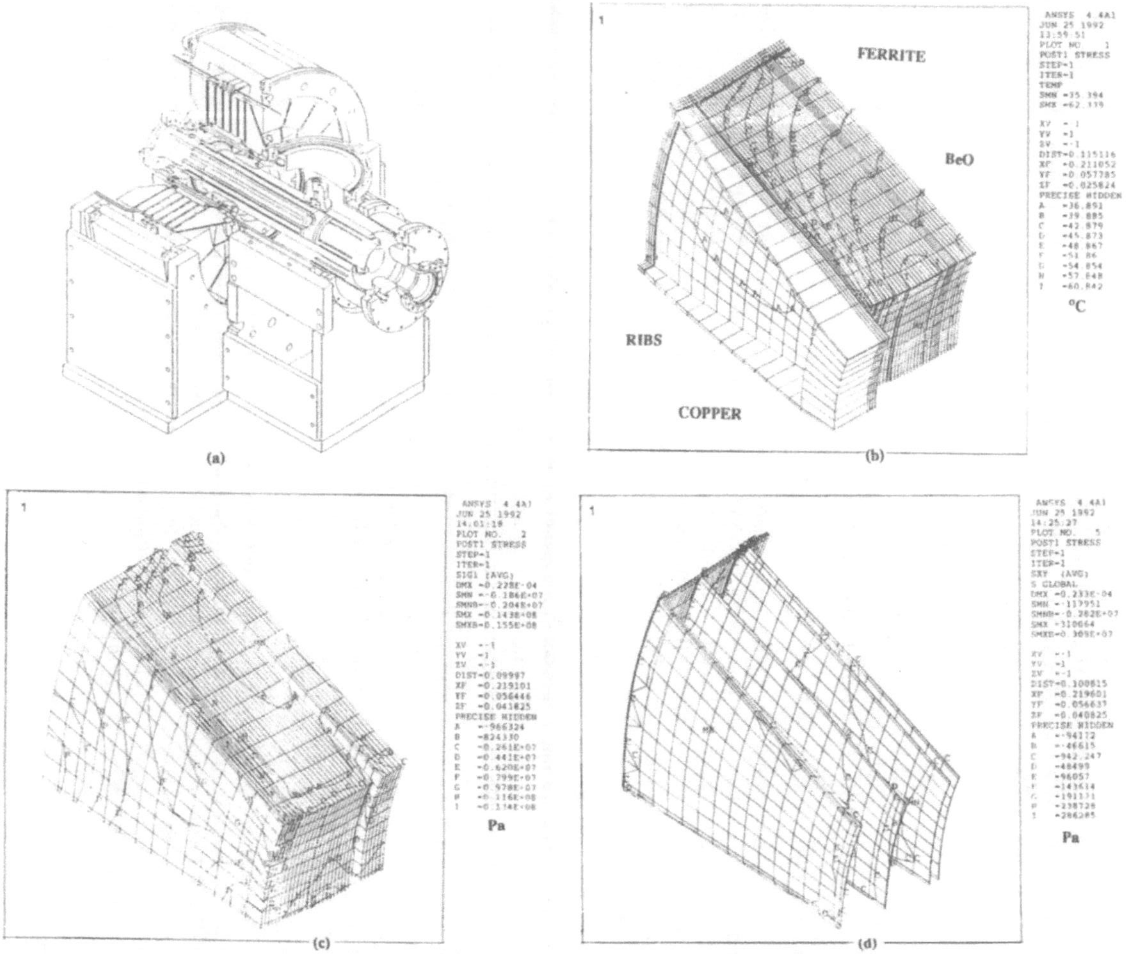

Figure 1. Qualitative results for a BeO cooled tuner.

Table 1. Sensitivity Results.

Row	no. of AlN	no. of BeO	x_g	k_g	T_a	T_b	T_f	T_e	S_a	S_b	S_f
1	-	-	50	0.8	-	-	85	83	-	-	26
2	2	0	50	0.8	63	-	66	66	32	-	15
3	-	2	50	0.8	-	59	62	61	-	8	14
4	-	-	200	0.8	-	-	85	83	-	-	21
5	2	-	200	0.8	63	-	66	66	23	-	14
6	-	2	200	0.8	-	59	63	62	-	8	13
7	-	-	50	11.6	-	-	75	74	-	-	28
8	2	-	50	11.6	56	-	58	56	24	-	28
9	-	2	50	11.6	-	51	54	52	-	6	16
10	-	-	200	11.6	-	-	75	74	-	-	23
11	2	-	200	11.6	55	-	58	55	17	-	16
12	-	2	200	11.6	-	51	54	52	-	7	14
13	4	-	50	0.8	60	-	63	60	26	-	11
14	-	4	50	0.8	-	56	59	56	-	6	10
15	6	-	50	0.8	61	-	63	61	27	-	12
16	-	6	50	0.8	-	57	59	57	-	10	9

Cases using AlN (rows 2, 5, 8, 11, 13 and 15) indicated factors of safety well in excess of 2 in the ferrites and between 1 and 3 in the AlN (compared to its flexural strength of 46 MPa). Further, the peak temperature in the ferrite and epoxy were less than 70°C which is safe compared to the maximum operating temperature of epoxy of 150°C and the curie temperature of ferrite of 200°C. When using BeO (rows 3, 6, 9, 12, 14 and 16), the safety factors were always greater than 2 in the ferrites and about 15 in the BeO (compared to its yield point under tension of 151 MPa). For the same cases the peak temperatures in the ferrite and epoxy were always less than 65°C.

Note, there is an uncertainty in the curie temperature and the tensile strength of ferrite. There is also an uncertainty in the thermal and structural properties of the low conductivity epoxy. Therefore, safety factors for stress (based on tensile strength) greater than 3 may be desirable. Also, note that a higher thermal conductivity of epoxy lowers the peak temperatures and stresses (Table 1). A thicker layer of epoxy lowers the stresses but raises the temperatures slightly. Using four or six BeO (or AlN) disks lowers both the peak stresses and the temperatures (rows 13–16).

Note, for all the cases shown in Table 1, the peak temperature and Von-Mises stress in the copper as well as the maximum shear stress in the epoxy were small compared to their corresponding thermal and structural limiting values. In addition, recent experiments indicate that steady state is reached within 30 minutes of the startup of the tuner and so transient results are not of interest since steady state represents the worst case condition.

Qualitative Results

To obtain a physical feel for the phenomena, the qualitative results for the case when two BeO disks are employed (Table 1, row 3) are shown in Figures 1 (b–d). For each of these figures, the scale on the right gives the magnitudes. The isotherms in the tuner (ferrites, BeO, housing walls, ribs etc.) show that the peak temperatures (symbol I) are located at the inner radius of the ferrite midway between the housing and the BeO as shown in Figure 1-b. The lowest temperatures (symbol A) are present on the copper.

The principal stresses in the ferrites given in Figure 1-c indicate that the peak value is located at the inside radius, near the junction of the ferrite and copper housing side wall. This may explain why a thicker layer of epoxy reduces the peak stresses (Table 1). Figure 1-d shows that the shear stresses in the epoxy are small compared to its lap shear strength of 7 MPa.

SUMMARY

A three-dimensional, structural and thermal model was developed to evaluate a solid-cooled LEB RF cavity tuner. The results were found to be sensitive to: the use of BeO or AlN or neither of them, the number of BeO (or AlN), the thermal conductivity and the thickness of the epoxy. The use of BeO gave the lowest peak temperature and stress in the ferrite.

REFERENCES

1. R. Ranganathan, "Three dimensional numerical analysis of a liquid-cooled LEB RF cavity tuner," AMC-2210001, SSC Laboratory, (1992).

2. R. Ranganathan, "LEB RF cavity tuner solid cooling," AMC-2210005, SSC Laboratory, (1993).

3. ANSYS Engineering Analysis System, Revision 4.4, Swanson Analysis Systems, Inc., Houston, Pennsylvania, (1989).

NOMENCLATURE AND UNITS

k_g thermal conductivity of epoxy, W/ (mC)
S_a peak AlN stress (MPa)
S_b peak BeO stress (MPa)
S_f peak ferrite stress (MPa)
T_a peak AlN temperature (°C)
T_b peak BeO temperature (°C)
T_e peak epoxy temperature (°C)
T_f peak ferrite temperature (°C)
x_g thickness of the epoxy between the ferrites and BeO (or AlN), microns

THE QUALIFICATION AND RELIABILITY VERIFICATION OF A LOW COST
CRYOGENIC SUPPORT POST FOR SSC QUADRUPOLE MAGNETS

M.W. Hiller, R.J. Kunz, G.A. Lehmann, and M.J. Nilles

Babcock & Wilcox - Accelerator & Magnet Systems
Lynchburg, VA   24505-0785

ABSTRACT

A cryogenic support post has been designed and tested for use in the superconducting 5.4 meter long SSC Collider Quadrupole Magnet (CQM). The support post is injection molded from a commercially available glass fiber reinforced thermoplastic and can be mass produced at a lower cost than other available SSC support post configurations.

This paper discusses both the validation of FEA predicted structural performance through component and material testing, and the delineated reliability method used for verifying compliance with apportioned reliability targets using a synthesis of the FEA and test data.

INTRODUCTION

Approximately 1700 CQMs will be produced for the SSC by Babcock & Wilcox (B&W) in Lynchburg, VA. Each CQM contains two support posts within the cryostat to support and maintain alignment of the cold mass and thermal shields relative to the vacuum vessel. Over 3400 support posts of the type illustrated in Figure 1 will be required for test purposes and magnet production over the next 5 years to fulfill CQM contract requirements. A description of the overall CQM cryostat configuration[1] is referenced for the interested reader.

While prototype support posts have been developed at Fermi National Accelerator Lab (FNAL) to meet the requirements of SSC magnets, the high cost of its numerous tightly-toleranced piece parts and intricate assembly procedures are undesirable on mass produced magnets. Cost effective support post alternatives were investigated by B&W for the CQM program, and the post that promised the greatest cost savings was developed by Brookhaven National Lab (BNL) for the RHIC program[2].

The BNL post was injection molded from the glass reinforced Ultem® 2000 series of polyetherimide thermoplastics from the General Electric Company. Ultem® was selected because it exhibits excellent material properties for this application. Though BNL selected the 10% and 20% glass reinforced (by weight) versions of Ultem® (referred to as 2100 and 2200 respectfully), B&W selected Ultem® 2300 (30% glass reinforced) because the additional glass content enhanced the strength, dimensional stability, and creep resistance of the post.

Supercollider 5, Edited by P. Hale
Plenum Press, New York, 1994

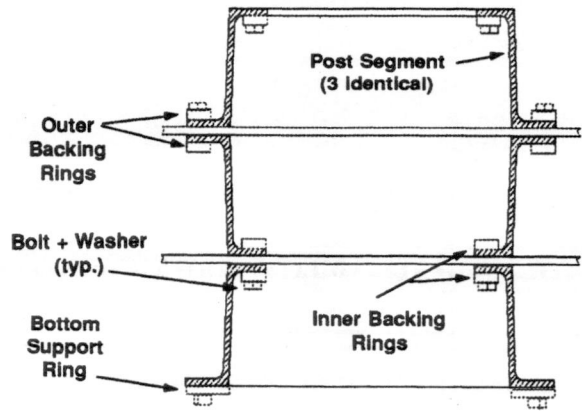

Figure 1. CQM Injection Molded Support Post Cross-Section

The post utilizes the near-net shape injection molding manufacturing technology in order to reduce the part count and complexity of the component while simplifying the cryostating of the magnet; in addition, the B&W post is comprised of three identical segments to further reduce manufacturing and development costs.

DESIGN and ANALYSIS APPROACH

ANSYS FEA software and hand calculations were used to optimize the configuration of the CQM support post to concurrently meet the structural and thermal performance requirements in the CQM Product Specification[3]. The geometry of the post was optimized to reduce stress risers and deleterious effects during molding[4].

MANUFACTURING STATUS

The support post design and drawings were completed in May 1992. A molding vendor was selected and a mold was then fabricated. Support post segments began arriving at B&W by the end of that year. Each molded support post segment requires a total molding cycle of less than 3 minutes. The only secondary manufacturing step is the removal of the diaphragm gate at the top of each segment. Although the molding vendor has the capacity to produce enough post segments to fulfill the entire CQM contract (10,000+ segments) in a matter of months, the current plan at B&W is to procure components in smaller lots.

Support post segments for over 50 Posts have been received by B&W for test and prototype CQM purposes. The dimensions of the segments have been inspected on a coordinate measuring machine, and the demonstrated part-to-part variability is typically less than +/- 0.063 mm (+/-.0025"). The production cost per support post, including backing rings and fasteners, will be less than $300. This compares very favorably to production cost estimates in excess of $700 each for the alternate post designs.

STRUCTURAL LOAD TESTING

Molded support post segments have been tested in compression by B&W and at the National Institute of Science and Technology (NIST). The test results shown in Table 1 demonstrate that the design and FEA methodology developed by B&W for molded support posts accurately predicts delivered performance.

Table 1. Comparison of Compressive Load Test Results to Predicted Performance.

Segment Designation	indicated failure load N (lb.)	FEA correlated Stress MPa
GE data (a)	521091 (117,099) (a)	210 (a)
Proto 1 (b)	519760 (116,800)	211
BW11 (c)	525545 (118,100)	214
BW33 (c)	529105 (118,900)	215
BW14 (c)	534445 (120,100)	217
BW64 (c)	535780 (120,400)	218
BW34 (c)	538895 (121,100)	219
BW12 (c)	539340 (121,200)	219

(a) FEA prediction based on published GE material properties[5]

(b) tested at B&W on 1/13/93

(c) tested at NIST week of 3/18/93

COMPRESSIVE STRENGTH RELIABILITY ANALYSIS

A support post reliability analysis was performed for vertical weight loading using the test data in Table 1.   The probability that the magnet compressive load will exceed the material allowable was determined, and the corresponding reliability of the post under this load was calculated.   The data yields a mean failure stress of 216.03 MPa and a sample standard deviation of 2.97 MPa; this results in a coefficient of variation (c.v.)= .0137 or 1.37%. According to the FEA analysis the steady-state stress due to magnet weight loads is 4.2 MPa.   These values were applied to the standard normal deviate equation shown below which includes a correction factor (non-central t) for the small sample size:

$$[\text{mean failure stress - operational stress}]/[\text{standard deviation*correction factor}] =$$
$$(216.03-4.2)/(2.97*1.225) = 58.22 \tag{1}$$

The reliability analysis methodology illustrated in Figure 2 denotes increasing stress levels along the horizontal axis.   The 1g load is the long term steady-state load case for the magnet.   The 3g vertical load is also shown because it is the maximum vertical "post delivery" load identified in the CQM product specification. Since the resulting peak stress levels do not overlap (nor come near) the material strength distribution, the component demonstrates a high reliability in this load configuration.

The standard normal deviate of 58.22 calculated in Equation 1 corresponds to a "probability of failure" $P(f) << 3\text{x}10^{-138}$, which results in a reliability far in excess of .999999999.   This reliability prediction assumes that the test samples to date (Table 1) are representative of the total production variability.   Though composites typically exhibit higher coefficient of variation (c.v.) and are usually quite sensitive to off nominal production variations, the current status looks very promising.   Component testing is scheduled to obtain test data to extend this reliability analysis to load cases containing lateral loads.

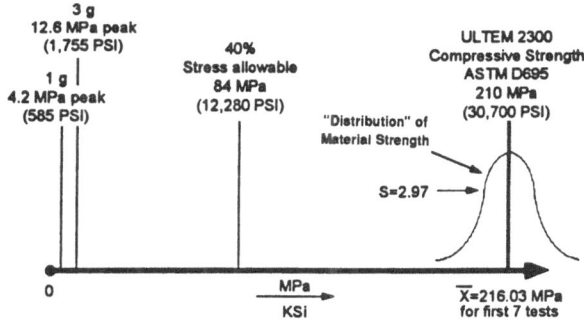

Figure 2. Post Compressive Load Failure Function.

RELIABILITY ANALYSIS and TEST PLANS

While the demonstrated reliability of the post due to the load discussed above is excellent, the post is the principle alignment maintenance component of the magnet so the issue of excessive post creep must be addressed. Testing is currently being performed to characterize the room temperature creep behavior of the post. This data will be used to verify that the post can maintain alignment within a creep allocation of less than 0.05mm (.002") over its 260mm (10.25") height during the 25 year life of the CQM.

Another part of the reliability analysis is the effect of radiation on the post. The current CQM product specification requires that the posts withstand .0020 MGrey (.2 MRad) gamma radiation. The preliminary results of the Ultem® 2300 electron beam irradiation tests being conducted by the SSCL look very promising at this radiation level. In addition, published data[6] from GE for the Ultem® resin report that there is less than 7% degradation in ultimate tensile strength at 500 MRad. As the SSCL test data becomes formally released it will be incorporated into the CQM post reliability analysis.

SUMMARY

The FEA predicted performance of the Post (using published material properties) compares very favorably with test results to date. Engineering practices and assumptions relating to the design and analysis of molded support post have been developed and verified through testing. The test data was incorporated into a reliability analysis to demonstrate the reliability of the component. Additional test plans were described to ascertain material properties and component performance to further verify the adequacy of the design and reliability of the post. The molded support post design has been adopted by other industrial affiliates working on SSC magnet contracts because B&W has proven that it is the lowest cost, proven support post approach currently available.

ACKNOWLEDGMENTS

Work was supported by the University Research Association through SSCL under contract SSC-91-B-01703 with Babcock & Wilcox. Jack Sondericker, Tom Nicol, and Arie Lipski are thanked for blazing the path concerning support post alternatives and for providing valuable data and technical discussions. Dennis Rule (NIST) is acknowledged for providing valued test services and expertise. Thanks go to Dave Mapes (B&W) for procurement support, and to Gary and Pat McCready and Dick Steward of TSI Plastics in Minneapolis, MN for the excellent molding job.

REFERENCES

1) M.W. Hiller et al, SSC Quadrupole Magnet Cryostat Design Alternatives, in: Advances in Cryogenic Engineering, Vol. 37A, Plenum Press, NY (1991).
2) J.H. Sondericker, Alternate Concepts for Structurally Supporting the Cold Mass of a Superconducting Accelerator Magnet, in: IISSC-Supercollider 3, Plenum Press, NY (1991).
3) CQM System Specification, SSCL Doc. No. M80-000007, September 21, 1992.
4) M.W. Hiller and J.A. Waynert, A Cryogenic Support Post for SSC Quadrupole Magnets, in: IISSC-Supercollider 4, Plenum Press, NY (1992).
5) GE Publication ,Ultem Properties Guide, ULT-306G, (1990).
6) GE Publication ,Ultem Resin: Advanced Technology for Reusable Medical Devices, ULT-314A, (1991), pg 8.

DESIGNING, FABRICATING, AND TESTING COST EFFECTIVE STRUCTURAL COMPOSITE FOR THE SSCL MAGNETS

Fred Nobrega

Superconducting Super Collider Laboratory*
2550 Beckleymeade Avenue, MS 1009
Dallas, TX 75237-3397

INTRODUCTION

Particle accelerators like the Superconducting Super Collider (SSC) use super-conducting dipole magnets to bend the particle bunches around the 54-mile ring and super-conducting quadrupole magnets to focus the particles. The heart of these magnets is the superconducting niobium-titanium copper cable which carries extremely high current because the internal resistance is zero at liquid helium temperatures. With these high currents, the magnets generate large magnetic fields on the order of 6.7 Tesla.

The superconducting cable is insulated with a wrap of polyimide film on the first layer and a second layer wrap of either a polyimide film with adhesive or a fiberglass epoxy prepreg. The insulated cable is wound into long coils and cured. All coil materials must withstand temperature extremes from 220°C (428°F) to -269°C (-452°F) at loads as high as 104 MPa (15 ksi). In addition, all magnet components must survive for 25 years with a total radiation dose of 1000 MRad. The parts at the end of a coil are used to support and restrain the conductors during magnet energization, The most common end part materials used to date have been G-10 and G-11 fiberglass and epoxy tubes and laminates in NEMA grades and CR type. Developments in polyimides like bismaleimides, copolymers like the newly developed PT resins and advanced epoxy blends like CTD101 and CTD102 are materials of choice for magnet components because of their radiation resistance. An extensive testing program is currently underway by the SSCL to measure the radiation degradation of these and many other materials.

END PART DESIGN

End part design begins with a 2D magnetic cross section of the coils. Conductor parameters such as width, keystone angle, mid-thickness and other information are used with program BEND. This Fortran program optimizes conductor geometry to minimize

---

* Operated by the Universities Research Association, Inc., for the U. S. Department of Energy under Contract No. DE-AC35-89ER40486.

strain and maximize cable bend radius in the coil end region. The numerical output from the program BEND defines the surfaces that support the conductor. The keys, spacers, fillers and saddles are machined directly from the programs geometric output.

The minimum strength requirements of the end parts have not been fully characterized. For years, G-10 and G-11 type materials have been used successfully in the Tevatron accelerator at Fermi National Laboratory. However, the radiation requirements for the Supercollider are orders of magnitude higher than the Tevatron. This precludes using a generic epoxy resin system.

The ideal end part totally supports the end turns, is flexible enough to accommodate the uncured cables during the winding process, and after curing the coil assembly, meets the strength and radiation requirements. The mechanical and thermal loads on the end parts during coil curing, collaring and magnet operation are high. During winding, the coil is uncompressed and somewhat larger than the final cured size. A "low" flexural modulus is desirable for the saddle and spacer end parts because all end parts are designed to the final cured size of the coil. The end parts must be flexible enough to accommodate the larger uncured coil. The keys have very high loads during curing and therefore a higher modulus is preferred. The flexural problem with G-10CR was temporarily solved at the SSC by adding slots to the saddles and spacers. During magnet assembly and operation, flexural and compressive strength and coefficient of thermal expansion become the important material parameters.

FABRICATION

There are several fabrication methods that can be employed for end parts. They include injection, compression and resin transfer molding of thermoplastic and thermoset materials. Use of metal castings, either coated with organic insulators or anodized, is another viable fabrication technique. An additional requirement for metal end parts is that they must be nonmagnetic. Machining solid end parts from G-10 or G-11 tube has been the standard method at the laboratories due to the small quantities of parts required and the high cost of the other fabrication methods.

Injection Molding

"Injection molding of thermoplastics is a process by which plastic is melted and injected into a mold cavity. Once the melted plastic is in the mold, it cools to a shape that reflectsthe form of the cavity. The resulting form usually is a finishedpart needing no other work before assembly into or used as a finished product."[1]

Due to the high cost of tooling and low part count, injection molded end parts were machined from injection molded tubes at Fermi.[2] Three thermoplastics, Torlon 5030, PEEK (polyetheretherketone) and PEK (polyetherketone) all with 30% glass fiber were investigated. Lack of adhesion between the cured coil and the end part was a problem particularly with Torlon 5030.

Compression Molding

"The most apparent advantage of compression molding of thermosets is the simple system involved. The material is placed in a heated cavity and is pressurized for the required cure time. Tooling costs are inexpensive because of the simplicity. Fillers and reinforcements are random and lead to fair strength properties. Material is not wasted because of the absence of sprues and runners."[3]

Compression molded end parts using filled epoxy systems were also studied by Fermi.[2] The study focused primarily on the manufacturing process as opposed to material selection.

Resin Transfer Molding (RTM)

When producing an intermediate volume of reinforced plastic parts, RTM is a cost effective technique.[4] Low viscosity resin is injected into a mold at low injection pressures. Typically the mold cavity contains a fiber preform with fiber orientation determined by strength requirements. Compound transfer molding is very similar to RTM, however the resin contains chopped fibers and the compound is injected into the mold. For low and moderate part counts, RTM tooling is more cost effective than injection molding.

Unlike injection molding, where the tooling needs to be made of steel because of the high temperatures and pressures, RTM molds can be made from composite materials for low quantity production runs. Mass cast epoxy tooling is being used to evaluate a phenolic triazine resin, Cryorad by Allied Signal. This resin as with other cyanate ester resins has the ability to be B-staged or undercured into adhesives, laminates and end parts.

TEST RESULTS

To demonstrate the feasibility of an undercured, conformable end part, The SSC, Grumman Areospace, Allied Signal, and Cooper Composites pooled their expertise in 1991 to make Cryorad end saddles using the RTM process. A sample outer coil was wound and cured. After a thorough visual inspection and coil measurement, the coil was potted and sectioned. The sectioned parts were inspected for conductor insulation damage, conductor placement and compaction, and for voids and defects in the molded part. No voids were found and conductor placement, compaction and insulation were very good. One of the more important goals was also met. The Cryorad end saddle formed or flowed during the curing cycle into a small gap between the last conductor and the shelf which demonstrates the parts will indeed conform to the conductor group. This is important as a coil made entirely of Cryorad or cyanate ester type end parts would provide total and rigid conductor support.

Cryorad end parts and film adhesive were used in model magnet DSA334 and cold tested in September 1992. Masters made of G-10 were used to fabricate the mold cavities. Structural preforms made of S2 glass and milled fibers were used as fillers. The structural preforms were fabricated on a CNC sewing machine. The end parts used the same RTM manufacturing process as the previous years demonstration part. As before, end parts provided the necessary support for the cable during the winding and curing phase coil manufacturing. The magnet did not show any degradation due to the Cryorad end parts, even though the end saddle was only partially cured to 50% and the winding key was fully cured during coil winding. During the coil cure cycle, the end parts cured further.

Cryorad manufacturing and material costs have been of great concern even though according to Jacobs,[3] the cost of a given RTM tool can be from 2 to 25% of the cost of an injection mold tool. As recent as March 1992, Grumman has lowered its price estimate to below $100 per benchmark part at production quantities. However, when compared to injection molding of thermoplastics at $10 to $25 for the same part, more effort is required to lower the cost of the RTM part. It must be remembered that several of the salient features of the cyanate ester thermoset resins mentioned above are not possible with thermoplastics.

FUTURE PLANS

A coil winding experiment is planned for mid 1993 that uses end parts made of two different materials and processes. One of the materials is an advance epoxy blend and the other is a BMI blend. The manufacturing processes used are bulk compression molding and RTM. Results from this experiment and studies completed to date will be used to determine the material selection and manufacturing process for the BV1C vertical bending dipole magnet.

CONCLUSIONS

There are several cost effective fabrication methods available for end parts. The resin selected will play a major role in the fabrication method chosen. More development and testing is necessary to develop the optimum end part material and manufacturing process.

REFERENCES

1. S. Kirkham, Injection molding thermoplastics, *in:* "Modern Plastics Encyclopedia," McGraw Hill (1990).
2. A. Lipski, et al., Alternate manufacturing processes and materials for the SSC dipole magnet coil end parts, *SSCL-146* March (1992).
3. R. Whitesides, Compression and transfer molding, *in:* "Modern Plastics Encyclopedia," McGraw Hill (1990)
4. K. Jacobs, Resin transfer molding, *in:* "Modern Plastics Encyclopedia," McGraw Hill (1990).

EXPERIMENTAL YOUNG'S MODULUS CALCULATIONS

Y. Chen, R. Jayakumar, and K. Yu

Superconducting Super Collider Laboratory*
2550 Beckleymeade Ave.
Dallas, TX 75237-3997

INTRODUCTION

Coil is a very important magnet component. The turn location and the coil size impact both mechanical and magnetic behavior of the magnet. The Young's modulus plays a significant role in determining the coil location and size. Therefore, Young's modulus study is essential in predicting both the analytical and practical magnet behavior.

To determine the coil Young's modulus, an experiment has been conducted to measure azimuthal sizes of a half quadrant QSE101 inner coil under different loading. All measurements are made at four different positions along an 8-inch long inner coil. Each measurement is repeated three times to determine the reproducibility of the experiment. To ensure the reliability of this experiment, the same measurement is performed twice with a "dummy coil," which is made of G10 and has the same dimension and similar azimuthal Young's modulus as the inner coil. The difference between the G10 azimuthal Young's modulus calculated from the experiments and its known value from the manufacturer will be compared. Much effort has been extended in analyzing the experimental data to obtain a more reliable Young's modulus. Analysis methods include the error analysis method and the least square method.

THE EXPERIMENTS

Experiments are implemented based on the SSC MSD Process Control Specification: "QSE Coil Size Measurements." Step (1): pole stress of a steel inner coil master is increased from 0 to −10.08Mpa, −20.16Mpa, −30.24Mpa, −45.35Mpa and −60.47Mpa[†] (calculated from applied hydraulic load); then the pole stress is decreased through the same path. All corresponding azimuthal sizes measured are recorded. Step (2): The inner coil master is replaced with an 8-inch long half quadrant inner coil, then step (1) is repeated every other 2 inches along the coil. Step (3): Step (2) is repeated twice. Step (4): Step (2) is repeated twice with a G10 tube.

---

* Operated by the Universities Research Association, Inc., for the U. S. Department of Energy under Contract No. DE-AC35-89ER40486.

[†] 60.47Mpa is about 130% designed inner coil pole stress for QSE101 magnet.

YOUNG'S MODULUS CALCULATION

In this experiment, the following equation aids in evaluating the local azimuthal Young's modulus.

$$E_{ij\,up/dn} = \Delta\sigma_j / (\Delta L_{ij\,up/dn} / L) \qquad (1)$$

where, E is a local azimuthal Young's modulus; $\Delta\sigma_j = \sigma_{j+1} - \sigma_j$, and σ_{j+1}, σ_j are consecutive pole stresses; $\Delta L_{ij\,up/dn} = [(SZC_{ij+1} - SZM_{j+1})_{up/dn} + SM_{j+1}] - [(SZC_{ij} - SZM_j)_{up/dn} + SM_j]$, is the change of the azimuthal deflection resulting from the change between two consecutive pole stresses, where SZC and SZM are respectively the azimuthal sizes of the coil/G10 tube and the master, $SM_j = \sigma_j * L / E_{steel}$ is the calculated master deflection in which $E_{steel} = 195,000$ Mpa; L=18.0681 mm (711.35 mil) is the middle arc length of the half quadrant inner coil cross-section designed for QSE101 also. "i" $\in \{1, 2, 3, 4\}$ denotes a position along the coil/G10 tube, "j" denotes a state at a certain pole stress, "up" denotes the branch of increasing pole stress, and "dn" denotes the branch of decreasing pole stress.

Based on equation (1), local Young's moduli for the coil at decreasing pole stress is bigger than that at increasing pole stress. This is primarily because the coil is a composite material; thus, different materials in the coil expand back differently when decreasing the compressing pole stress.[1] Therefore, this hysteresis of the coil deflections in loading and unloading results the hysteresis of its Young's modulus. This phenomena is more obvious at high pole stress. This is because the coil azimuthal size is measured shortly after decreasing pole stress, so that the coil has not been sitting long enough to complete the relaxation for the pole stress drop from the process of increasing pole stress; therefore the deflection change is very small, resulting a jump in Young's modulus. Further study on coil relaxation is necessary in order to obtain realistic coil azimuthal Young's moduli at decreasing pole stress. Thus, at this stage, only the coil azimuthal Young's moduli at increasing pole stress are presented. All the moduli from EQ. (1) are shown in Table 1 and Table 2.

Table 1. Coil Azimuthal E

P (psi)	500	1000	1500	2250
σ (Mpa)	10.08	20.16	30.24	45.35
Position	E (Mpa) - 1st			
1 up	5796	6921	8536	23901
2 up	5363	6311	11203	19555
3 up	5796	6921	10146	19555
4 up	4665	6921	9272	23901
	E (Mpa) - 2nd			
1 up	6040	7660	12506	23901
2 up	5571	9742	11819	21511
3 up	5571	7660	14152	19555
4 up	5571	8577	12506	16547
	E (Mpa) - 3rd			
1 up	5571	6921	14152	19555
2 up	5169	7660	12506	23901
3 up	5571	7660	12506	23901
4 up	6040	6921	12506	19555

P :   pump load
σ :   pole stress

Table 2. G10 Azimuthal E

P (psi)	500	1000	1500	2250
σ (Mpa)	10.08	20.16	30.24	45.35
Position	E (Mpa) - 1st			
1 up	8083	9742	14152	11950
1 dn	8566	12236	12153	14835
2 up	8083	9742	16296	13444
2 dn	9729	13377	13701	13878
3 up	8566	10452	16296	11950
3 dn	9729	13377	14633	14835
4 up	8566	11274	16296	11950
4 dn	10437	16445	13701	13878
	E (Mpa) - 2nd			
1 up	8083	9742	13278	11321
1 dn	8270	12896	14152	16547
2 up	9729	9742	15148	11950
2 dn	13869	10930	15148	14340
3 up	9111	10274	16296	12653
3 dn	10000	12896	15148	14340
4 up	9111	9742	17632	11950
4 dn	9345	14170	15148	13444

From Tables 1 and 2, it is clear that the coil azimuthal Young's modulus increases as the pole stress increases, which means that coil becomes stiffer at high pole stresses than at low pole stresses. The G10 azimuthal Young's modulus does not vary as much as the coil's does, and the hysteresis of its modulus is smaller than the coil's.

A. The Error Analysis For Young's Modulus

The azimuthal size measurement for coil/G10 tube involves several mechanical instruments, which may bring errors to the data recorded and result in errors in calculating the azimuthal Young's modulus from EQ. (1). Therefore, the error bar analysis is performed to predict the range of possible azimuthal Young's moduli. In these measurements, errors that might occur are: 50 psi reading error from the pump pressure meter and 0.05 mil reading error for the size measurement from the LVDT reader. Since EQ. (1) is used to calculate each azimuthal Young's modulus, the error of the calculated Young's modulus can be evaluated as following:

$$\Delta E_{ij\ up/dn} = [\delta(\Delta\sigma)/\Delta\sigma_j + \delta(\Delta L)/\Delta L_{ij\ up/dn}] * E_{ij\ up/dn} \qquad (2)$$

where $\delta(\Delta\sigma) = 100$ psi and $\delta(\Delta L) = 0.1$ mil denote respectively the maximum pump pressure error and size measurement error. The first term represents the Young's modulus error from pump pressure and the second term represents that from size measurement. The results of the error analysis from EQ. (1) and EQ. (2) are representatively shown in Figures 1 through 3, where each plot shows the calculated azimuthal Young's modulus with its error bar versus pole stress at a certain position along the coil/G10 tube. For coil, the higher the pole stress is, the bigger the error bar is. This is because the coil becomes very stiff at high pole stress so that its deflection change due to the pole stress change is very small, thus the error from the second term in EQ.(2) becomes large. The Young's modulus error at low pole stress is dominated by the pump pressure error; while at high pole stress, it is dominated by the coil size measurement error. For G10, the Young's modulus error at different pole stresses is quite uniform simply because its modulus at different pole stresses is nearly constant.

B. The Least Square Analysis For Young's Modulus

From the size data recorded, the deflection curve for G10 is nearly linear (refer to Figure 4); therefore, the least-square method is used for the coil/G10 azimuthal Young's modulus calculation, determining an average modulus. In Figure 4, a linear line is drawn by using the least-square criterion in a plot of relative deflection versus pole stresses. Then the product of the line reciprocal slope and the middle arc length L=18.0681 mm gives a constant modulus. Each relative deflection in each plot is the average relative deflection from repeated measurements. All constant moduli calculated in this way are recorded in Table 3 for both coil and G10 tube.

In Table 3, the hysteresis of the azimuthal Young's modulus is obvious and is bigger for coil than for G10 tube. In this experiment, the G10 azimuthal deflection is more linear than the coil's; therefore, the G10 azimuthal Young's moduli from the least square method are more reliable than the coils'.

Table 3. Azimuthal Young's Modulus from the Least Square Method

Position	1 up	1 dn	2 up	2 dn	3 up	3 dn	4 up	4 dn
Coil E (Mpa)	9854	15364	9995	15125	10043	15338	9624	14628
G10 E (Mpa)	11045	12482	11953	13236	12344	13417	12264	13476

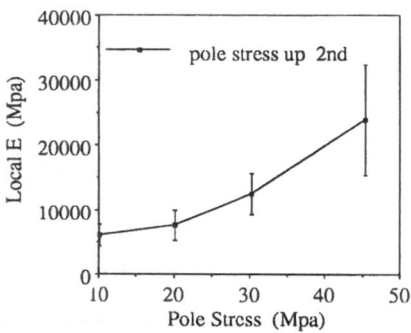

Figure 1. Local Inner Coil Azimuthal Young's Modulus for loading - Position 1

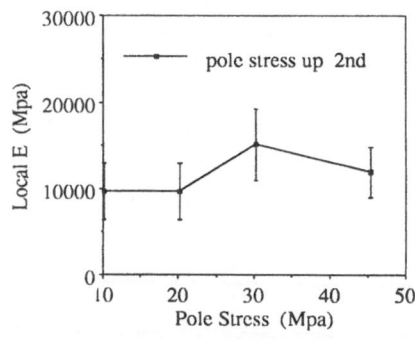

Figure 2. Local G10 Azimuthal Young's Modulus for loading - Position 1

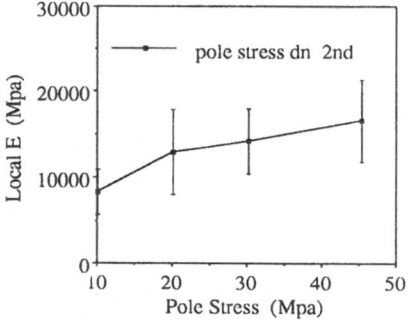

Figure 3. Local G10 Azimuthal Young's Modulus for unloading - Position 1

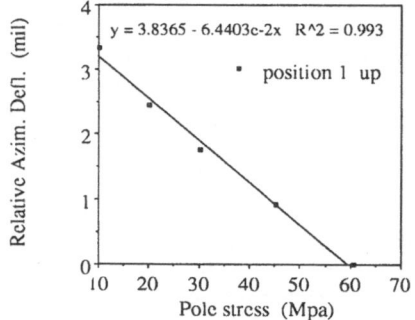

Figure 4. Relative G10 Azimuthal Deflection Vs. Pole Stress for loading G10 - Position 1

CONCLUSION

This paper presents a practical method to evaluate coil's azimuthal Young's modulus during the process of increasing pole stress. The results obtained yield several conclusions. First, coil's azimuthal Young's modulus increases as pole stress increases, in other words, the coil becomes stiffer. Secondly, the calculated G10's azimuthal Young's modulus shown in Tables 9 through 11 are very close to the known value from the manufacturer, which is 14,000 Mpa.[1] This proves that this experiment can be conducted to estimate inner coil's azimuthal Young's modulus. Finally, data from repeated measurements are reproducible, which ensures the reliability of the results.

The coil's azimuthal Young's modulus during the process of decreasing pole stress can be calculated in the same way. However, the coil size measuring technique needs to be improved upon further study of coil relaxation.

REFERENCES

1. Private conversation with S. Krishna.

MECHANICAL ANALYSIS OF THE DSB CROSS-SECTION

Yanping Chen

Superconducting Super Collider Laboratory*
2550 Beckleymeade Ave.
Dallas, TX 75237-3997

INTRODUCTION

This paper presents the preliminary mechanical finite element analysis for the SSCL designed DSB dipole magnet. This SSCL version 50 mm aperture dipole magnet is for the SSCL High Energy Booster with nineteen turns, three wedges for inner coil and twenty six turns, one wedge for outer coil, the round collar is nineteen mm thick, the yoke and the shell are adopted from the design for quadrupole QSE101.

The main purposes of this mechanical study are to ensure that there are no excessive stresses in the cold mass under different loading, to avoid coils unloading from the collar at excitation of 6500 A, to ensure collar-to-yoke, line-to-line fit after welding the shell, and also to ensure the yoke midplane gaps are closed at an operating current of 6500 A.

Therefore, the analyses performed include magnet assembly (collaring) to 69 Mpa azimuthal stress at the inner coil pole and 55 Mpa azimuthal stress at the outer coil pole; shell welding to 207 Mpa azimuthal stress in the shell; magnet cooldown to 4.25 K; and Lorentz excitation at a current of 6500 A.

FINITE ELEMENT MODEL

Analysis is performed on a quarter model since the geometry and loading are symmetrical.

Model Assumption[1]

* The model assumes plane stress analysis
* All materials are linearly elastic, except that plasticity is considered for the collar. The coils are orthotropic in radial and azimuthal directions both for Young's Moduli and thermal coefficient. The Young's Moduli for the inner coil are higher than that for the outer coil due to the stiffness increment caused by curing the inner coil twice.

---

* Operated by the Universities Research Association, Inc., for the U. S. Department of Energy under Contract No. DE-AC35-89ER40486.

* Young's Modulus for each material at 4.25K is higher than that at room temperature
* All contact surfaces are assumed frictionless or bound
* No interferences exist between coil turns and wedges
* The front collar is spot welded with the back collar, so that the displacements of front and back collars in the horizontal midplane are the same radially, but the same magnitude in the opposite direction azimuthally, also similarly in the vertical plane
* Lorentz forces are calculated with infinite permeability iron.
* Half collaring prestress is modeled by specifying coil midplane displacement, another half by specifying an interference between coils and collar at pole planes.

Material Properties[1]

	Elastic Modulus	Thermal Coefficient
Inner Layer Insulated Cable	Room Temperature Eazim = 11,000 Mpa Eradial = 17,600 Mpa Cold Temperature Eazim = 13,200 Mpa Eradial = 18,320 Mpa	αazim = 1.40e-5 (1/K) αradial = 1.53e-5 (1/K)
Outer Layer Insulated Cable	Room Temperature Eazim = 10,000 Mpa Eradial = 16,000 Mpa Cold Temperature Eazim = 12,300 Mpa Eradial = 17,120 Mpa	αazim = 1.40e-5 (1/K) αradial = 1.53e-5 (1/K)
Copper wedge, Brass Shoe, Brass Key	Room Temperature E = 120,000 Mpa Cold Temperature E = 150,000 Mpa	α = 1.11e-5 (1/K)
Kapton	Room Temperature E = 3,000 Mpa Cold Temperature E = 3,810 Mpa	α = 2.00e-5 (1/K)
Intercoil Spacer	Room Temperature E = 10,000 Mpa Cold Temperature E = 12,300 Mpa	α = 1.8e-5 (1/K)
Nitronic 40 Stainless Steel	Room Temperature E = 195,000 Mpa Eplastic = 40,000 Mpa yield stress = 620 Mpa Cold Temperature E = 206,700 Mpa	α = 0.92e-5 (1/K)
Iron Yoke	Room Temperature E = 205,000 Mpa Cold Temperature E = 211,000 Mpa	α = 0.698e-5 (1/K)
Stainless Steel Shell	Room Temperature E = 195,000 Mpa Cold Temperature E = 206,700 Mpa	α = 1.073e-5 (1/K)

RESULT

Table 1. Inner and outer coil average azimuthal stresses; collar horizontal and vertical deflections at pole under various loading.

Loading Condition	Pole Stress (Mpa)		Collar Deflection (mm)	
	Inner	Outer	Horizontal	Vertical
Collaring	69.7	54.8	0.0264	0.0995
Welding to 207 Mpa*	85.9	68.0	0.00655	0.0586
Cooldown to 4.25 K*	56.5	51.2	0.0324	0.0459
Energization to 1000 A*	55.7	50.6	0.0331	0.0454
Energization to 2000 A*	53.3	49.0	0.0353	0.0441
Energization to 3000 A*	49.3	46.2	0.0388	0.0418
Energization to 4000 A*	43.7	42.3	0.0431	0.0387
Energization to 5000 A*	36.7	37.4	0.0486	0.0347
Energization to 6000 A*	28.0	31.4	0.0554	0.0297
Energization to 6500 A*	23.1	28.0	0.0593	0.0268

Table 2. Collar to yoke contact force; midplane yoke to yoke contact force and minimum and maximum gap under various loading.

Loading Condition	Collar-Yoke Contact F (N)		Yoke-Yoke		
	Horizontal	Vertical	Contact F (N)	Max.	Min. Gap (mm)
Welding to 207 Mpa	942	1012	0	0.0488	0.0353
Cooldown to 4.25 K	165	332	1590	0	0
Energization to 1000 A	172	327	1594	0	0
Energization to 2000 A	198	315	1607	0	0
Energization to 3000 A	244	297	1625	0	0
Energization to 4000 A	312	274	1649	0	0
Energization to 5000 A	400	246	1678	0	0
Energization to 6000 A	511	213	1712	0	0
Energization to 6500 A	578	198	1728	0	0

Where deflection "*" is calculated as:

Deflection = R.α.dT - Abs. Dfl.

dT = 293 – 4.25 K

R = collar outer most radius at horizontal and vertical directions

α = thermal coefficient

Abs. Dfl. = absolute deflection of the point from ANSYS analysis

Here, R.α.dT are 0.1998 mm and 0.1838 mm in vertical and horizontal directions.

In this analysis, the stretching stress in the collar near the key way is about 720 Mpa.

For the comparison of this round collar design with an anti-oval collar design, the analysis for the anti-oval collar with 0.1 mm total ovality is performed also. The results show that there is no contact between collar and yoke, which means that collar is free to move inside the yoke, but they are in contact at operating current. Some experts insist on collar to yoke contacting all the time after cooldown in order to reduce the risk of quench due to the movement of the collar inside the yoke, others consider that no contact between the collar and the yoke after cooldown and at low excitation currents is acceptable as long as they contact each other at operating current.

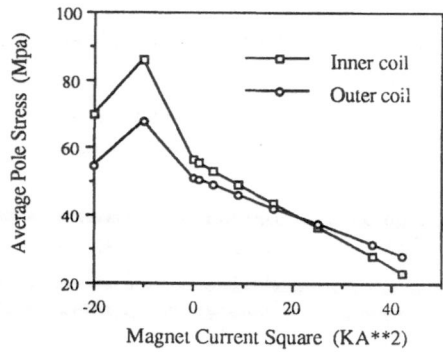

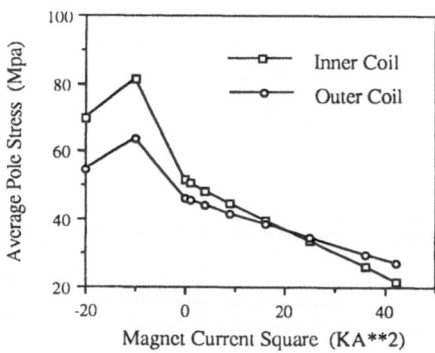

Figure 1. Calculated average azimuthal pole stress pole stress for round collar

Figure 2. Calculated average azimuthal for anti-oval collar

From the data in Table 1, the calculated average azimuthal pole stress for a coil under different loading is shown in Figure 1. The same type of curve for the anti-oval collar is plotted in Figure 2. As can be seen in the figures, the pole stress for the round collar is only several Mpa greater than the corresponding one for the anti-oval collar.

CONCLUSION

From the ANSYS model analysis, it is clear that coil prestresses are far away from unloading at operating current; that yoke and collar are in tight contact after welding and remain in good contact after cooldown and excitation to 6500A; also that the yoke midplanes contact after cooldown and thereafter, although there is no any contact between them after welding.

However the analysis also shows that the tensile stress in the collar near the key way is over yielding stress after collaring, but this would not effect the overall mechanical performance of the magnet, since it is in a highly localized area.

According to this analysis, DSB with round collar would perform similarly to the collider dipole in mechanical aspects.

ACKNOWLEDGMENTS

My thanks to Nick Hassan and J. Turner for their support.

REFERENCES

1. J. Turner, G. Spigo, Y. Chen: Mechanical Analysis of the QSE101 Cross-Section, March, 1992.

188

DESIGN AND ISSUES ASSOCIATED WITH THE HDM ELECTRICAL INSULATION SYSTEM

J. F. Roach,[1] S. K. Singh,[1] O. R. Christianson,[1] D. J. Hall,[1]
and A. G. McConnon[2]

[1]Westinghouse Science and Technology Center
1310 Beulah Road
Pittsburgh, PA 15235

[2]Westinghouse Magnet Systems Division
I-H 35 North and Westinghouse Road
Round Rock, TX 78680

INTRODUCTION

The Westinghouse Electric Corporation (WEC) is under contract to design and build the High Energy Booster dipole magnets (HDM's) for the SSCL through low rate initial production (LRIP). The first phase of the HDM program is the fabrication and test of short 1.8 m HDM model magnets designed by the SSCL. This technology transfer phase is well underway with the delivery of the first WMSD built HDM model magnet, DSB701 to the SSCL and the completion of the test program conducted at the SSCL Superconducting Cable and Magnet Test Laboratory (SMCTL) in April of this year. This paper presents a summary of reverse engineering analyses of the HDM model magnet electrical insulation system performed by the WEC. The electrical stresses in the 2-D magnet cross section are estimated under rated voltage conditions for Hipot tests in air. A transient voltage analysis is presented for the ringer circuit Results of an analysis of quench voltage behavior as a function of protection circuit parameters is also presented. The lumped quench circuit model predicts the terminal and splice voltages, coil resistance, hot spot temperature, and MIITS. Deficiencies in the electrical insulation HDM model magnet design are addressed.

MODEL MAGNET DIELECTRIC STRESS ANALYSIS

The HDM model magnet electrical insulation design for the 2D cross section is shown in Figure 1. The coil superconducting cables and wedges are insulated with the Kapton/epoxy pre-impregnated woven fiberglass system. The ground wall insulation configuration is based on the Fermi National Accelerator Laboratory (FNAL) scheme, except as modified to accomodate the coil-on-coil cure fabrication process. The essential feature of the coil-on-coil design is that the outer coil is wound over the cured inner coil with G-10 CR spacer, Item 15 and an inner coil adhesived Kapton, Item 14 in place. The inner and outer coil winding assembly along with the outer coil adhesived Kapton, Item 7 and the internal ramp splice assembly are then cured together. Kapton sheets, Items 5,6 replace the quench heaters.

An analysis of the insulation design was performed to estimate the dielectric stresses induced under rated voltage requirements in the HEB dipole magnet specification.[1] Table 1 presents the dielectric stress analysis results and includes estimated capacitances used in the transient voltage analysis described below. The minimum insulation ratings in air are specified to be 3 kV coil-to-coil and 5 kV coil-to-ground. The turn-to-turn rating in air is 3 kV for all turns within a given coil.

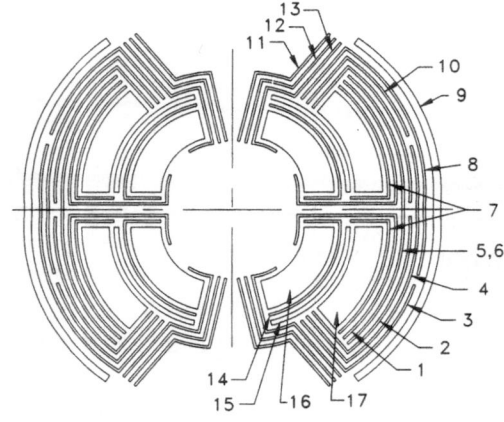

Item		Thk(mm)
Kapton	2-5, 8,10-13	0.13
Kapton	7,14	0.09
Kapton	1,6	0.05
G-10/CR	15	0.39
Brass	9	0.51

All Kapton items LT film except
Items 7 and 14 which use 0.05 mm
HN film with 0.04 mm adhesive

Figure 1. HDM model magnet electrical insulation assembly for SSCL coil on coil cure design.

The enhanced void stresses in Table 1 are for 0.013 mm air voids assumed within the insulation and were calculated from [2] $E = E_{av} \varepsilon / [1 + (\varepsilon - 1) d_v / d]$ where the dielectric constant ε is taken to be 3.5 for Kapton and 5.0 for G-10 CR, d_v is the void size, d the insulation spacing, and E_{av} the average stress.

Table 1. Electrical insulation analysis of the HDM model magnet design.

Stressed Region	Spacing	Minimum Creep Path	Test Stresses		Capacitance
			Average	Void	
	mm	mm	kV/mm	kV/mm	nF/m
Turn to Turn Inner Coil	0.168	NA	17.9	51.8	0.393
Turn to Turn Outer Coil	0.178	NA	16.9	50.7	0.285
Inner Coil to Outer Coil	0.80	13	3.75	17.6	3.58
Upper Coil to Lower Coil	0.42	30	7.14	23.4	1.76
Inner Coil to Collar	1.02	32	4.90	23.5	1.08
Inner Coil to Pole	0.46	20	10.87	36.0	1.55
Outer Coil to Shoe	0.87	13	5.75	19.7	3.58
Outer Coil to Pole	0.85	NA	5.88	20.1	0.81

The expected breakdown stress for 0.013 mm air void is estimated to be 28 kV/mm based on Paschen's curve for air.[3] Therefore, air void partial discharges are expected during full coil Hipot testing at 5 kV between the inner coil and pole ground. Such partial discharges result in test leakage currents. The HDM specification[1] requires the maximum allowable leakage current at rated voltage (during the useful life of the magnet) to be 25 µA. The contribution of trapped gas void discharges to the net leakage current under test conditions is proportional to the void fraction, which is expected to be quite small. The high void stresses between adjacent turns of a coil at the 3 kV rated voltage is not believed to be of real concern since such high turn-to-turn voltages could only be seen under short transient operational conditions. The turn-to-turn stress is never really tested at 3 kV in practice, except for the verification of punch-through capability at the component level, for example in 2-stack tests where 2 kV is normally the applied voltage and mechanical stress at failure is measured.

For creep stress assessment a failure stress of 0.3 kV/mm was assumed, derived from 1/10 of Paschen's air breakdown for gaps greater than 10 mm. With this criteria only the 13 mm creep path from the outer coil to shoe ground is not acceptable for the 5 kV design rating.. This short creep path is considered a design deficiency. It results from a failure to interleave the inner and outer coil to pole creased Kapton as can be seen in Figure 1. A solution is to interleave Kapton Item 13 with the outer coil pole face and Kapton

Item 8. Another potential problem with the coil-on-coil design is that the outer coil inner radius corners are not directly capped with Kapton. This means that any flaw in the cable Kapton/epoxy/fiberglass insulation is exposed and high edge stress more likely to initiate a dielectric failure. A recommendation was made to the SSCL that the HDM model magnet coil to ground Hipot test be limited to 3 kV. This recommendation was accepted by the SSCL and adopted for the HDM model magnet testing.

The insulation ratings in air are essentially based on twice the expected peak operating, upset and test voltages plus 1 kV. This safety margin approach[4] is based on commercial power frequency equipment and its validity for magnets, where transient high voltage only occurs during dump conditions, has not been well established. Higher dielectric safety magins may be required to meet the high reliability requirements in the HEB ring. A better understanding of the correlation between magnet testing in air at rated voltage and leakage current limit and operation under liquid and gaseous helium states is needed. As a general rule, the dielectric strength of helium will be as good as in air for small voids and gaps <1 mm for helium density >15 kg/m[3] based upon air[3] and helium[5] Paschen's curves. The dielectric strength of liquid helium[2,4,6,7] follows Paschen's law below about 15 kg/m[3] and tends to reach a limiting field at > 80 kg/m[3]. Dielectric failures are not expected based upon the inherent strengths of Kapton films and operational helium states. Catastrophic failure is generally due to flaws in film insulation such as pinholes, fabrication damage, and trapped contamination particles, leading to ultimate punch-through and shorts of the insulation system under high compression loadings and cyclic fatigue.

TRANSIENT VOLTAGE ANALYSIS

Circuit simulations to predict transient response of the collared coil were conducted using the PSPICE code from Microsim Corporation. The HDM model magnet was divided into four coils, upper and lower inner/outer coil pairs, each coil treated as an L,R section with distributed series and ground capacitances. The self and mutual inductance matrix was calculated using a specialized version of the Westinghouse finite element code WEMAP. The inner and outer coil resistances were assumed to be 160 mΩ and 255 mΩ, respectively, (measured values are 118 mΩ and 192 mΩ). The estimated distributed capacitances are given in Table 1. Resistive losses in the collar due to changing magnetic field were not taken into account. The effective length of the magnet was taken to be 1.48 m. Figure 2 shows the predicted voltage response for a ringer circuit dumping a 20 μF capacitor charged to 1500 V across the magnet terminals with one terminal grounded. Ringer circuit resistance, stray inductance, stray capacitance as well as lead and splice resistance are not accounted for. The model predicts that 92% of the initial capacitive voltage is dropped across the high voltage side inner coil, and that under long term L,R,C response, 18% of the applied voltage drops across each inner coil and 32% across each outer coil. Simulations were also conducted to determine the effect of a single shorted turn. A shorted turn in the high voltage side inner coil results in a 6 Hz shift in ringing frequency from 482 Hz to 488 Hz. A shorted turn in the ground side outer coil results in a 7 Hz shift to 489 Hz. Therefore, the predicted shorted turn detection sensitivity based on change in ringer circuit frequency is > 1.2 %. The damping shown in Figure 2 is due only to coil resistance. Actual ringer tests performed by WEC on collared HDM model magnets DSB 701 and 702 show terminal voltage is damped out rapidly in about 2-cycles indicating significant AC losses in the collar; the ringing frequency of the collared model magnets was about 525 Hz compared to 482 Hz predicted by the model circuit analysis.

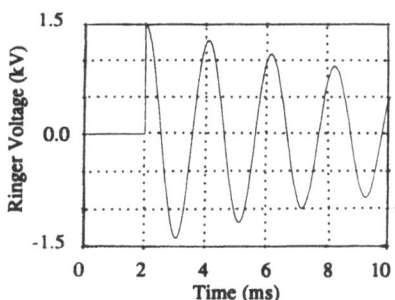

Across model magnet terminals

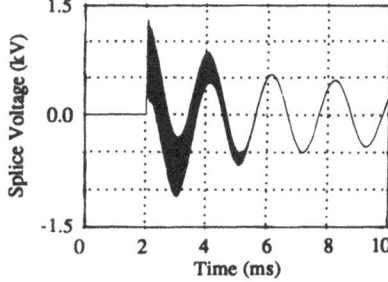

External splice terminal to ground

Figure 2. Predicted ringer circuit voltage response for collared HDM coil assembly.

QUENCH VOLTAGE ANALYSIS

A quench protection circuit analysis was performed to predict quench voltages in the model magnet expected under test conditions at the SMCTL. The magnet terminal, internal splice, and external splice voltages were predicted as well as magnet resistance, hot spot temperature (HST), and MIITS as functions of time for selected dump circuit delay and resistance parameters. The calculations assumed an outer coil quench initiating at the mid-plane and located half way along the axial length. In all cases the peak quench terminal voltage was less than 650 V. Table 2 gives predictions for quench parameters for 34 ms and 150 ms circuit delays with 0 mΩ and 100 mΩ dump resistance. Actual dump circuit values during tests of model magnet DSB 701 at the SMCTL are 45 ms delay and 100 mΩ dump resistor.

Table 2. Quench circuit analysis results for the HDM model magnet.

Dump Resistor mΩ	Circuit Delay ms	Peak Voltage V	Magnet Resistance mΩ	HST K	MIITS
0	34	55	100	112	6.3
0	150	68	115	122	6.7
100	34	630	7	51	2.6
100	150	470	100	110	6.3

In the case of 100 mΩ dump resistor and 34 ms delay, the quench voltage and current decay time is less than 200 ms. The helium density state as a function of time was calculated under these quench conditions.[8] For protection circuit delays of 34 ms and greater, the helium density is greater than 15 kg/m^3 in helium gaps and voids during sensible quench voltage within the magnet. Thus, the local dielectric strength in stressed helium regions is at least as good as under simulated Hipot rated voltage testing in air. No dielectric problems are expected under quench conditions expected for the HDM model magnet.

CONCLUSIONS

The HDM model magnet electrical insulation system was discovered to have a potential creepage path design deficiency during model magnet dielectric stress analysis. A recommendation was made by WEC to reduce Hipot test voltage for assessment of ground insulation integrity to from 5 kV to 3kV and was adopted by the SSCL for the HDM DSB series model magnet testing. This creep path problem is readily solved by minimal design change to permit proper interleaving of inner and outer coil pole ground wall Kapton insulation. Ringer circuit analysis predicts about a 1.2 % increase in ringer frequency for a shorted turn. The ringer circuit analysis did not account for AC losses in the collar which lead to more damping than predicted. The peak quench voltage expected during testing of the model magnet at SMCTL is less than 650 V and no dielectric problems are anticipated under quench protection circuit operation.

REFERENCES

1. SSC HEB Dipole Magnet Specification, M80-000045A, January 15, 1993.
2. J. Hiley and R.S. Dhariwal, Dielectric breakdown in high density helium and in helium impregnated solid dielectrics, Cryogenics 25:334 (1985).
3. T.W. Dakin et al., Breakdown of gases in uniform electric fields, *Electra* 32:61 (1974).
4. S.W. Schwenterly, Design and testing of electrical insulation for superconducting coil, in: "Advances in Cryogenic Engineering," Vol. 33, Plenum Press, New York (1988), p. 271.
5. H. Winkelnkemper et al., Breakdown of gases in uniform electric fields, *Electra* 52:67 (1977).
6. J. Gerhold, Dielectric breakdown of cryogenic gases and liquids, *Cryogenics* 19:57 (1979).
7. J.L. Wu and J.F. Roach, DC dielectric breakdown tests of liquid helium at temperatures1.8K to 4.2K, in: "Supercollider 4," J. Nonte, ed, Plenum Press, New York (1992), p. 731.
8. Private communication, E.F. Daly, Westinghouse STC calculated the helium density as a function of time and quench protection circuit delay.

ANALYSIS, DESIGN, AND TESTING FOR THE GDSS MODEL

MAGNET QUENCH HEATERS

Daniel J. Kinzie and Scott D. Peck

General Dynamics Space Systems Division
Space Magnetics
P.O. Box 85990
San Diego, CA 92186

ABSTRACT

Five quench heater designs have been developed for testing in the GDSS lot A dipole model magnets. Design test features were chosen to quantitatively indicate the quench performance enhancements for various modifications to the heater strip geometry and encapsulation. The quench performance of each heater design has been predicted using previously-developed analysis tools and these predictions will be compared with test data. The conclusions reached on the relative performance benefits gained will be applied to the baseline quench heater design for the SSC collider dipole magnets.

INTRODUCTION

The major design objective for the quench heaters of the SSC collider dipole magnets (CDM's) is the reduction in the amount of heater energy required for coil protection. Considerable analysis has been done to support this objective; heater evaluation including the testing of five designs in the GDSS CDM model magnets will be done to confirm the analysis and design choices. Rapid thermal response and reduced energy requirements for normal zone initiation can be obtained by reducing the resistive volume of the heater and the insulation thicknesses. Previous work [1,2,3] has produced a protection system design in which pairs of encapsulated heater strips are mounted in opposite quadrants of each dipole between the outer conductor layer and the collars and connected in series. A plan view illustrating the heater design is shown in Figure 1. The heaters are fired by the discharge of electrolytic capacitors producing a heater firing unit (HFU) energy pulse with an RC time decay. The performance of the heater when fired is a strong function of the HFU parameters (capacitance(C), voltage (V)), heater geometry and conductor conditions.

Analytic models have been developed in [3] to predict quench performance (i.e. time to normal zone initiation (t_fn), maximum heater and fault temperatures) as a function of system parameters, heater strip geometry, insulation and encapsulation configuration, and conductor conditions). Based on these models, design attributes were chosen for evaluation in the model magnet quench heaters. The important remaining heater features for evaluation include large reductions in active resistive zone volume and the use of alumina-loaded polyimide for heater encapsulation. Differences in cooling mode (pool boiling vs. pressurized liquid helium) cause the quench performance results from the model magnets to

Supercollider 5, Edited by P. Hale
Plenum Press, New York, 1994

not be directly applicable to the prototype CDM performance [2]; however, the results are indicative of the relative benefits of various design modifications. This information will be applied to the final quench heater design development for the full-scale CDM's.

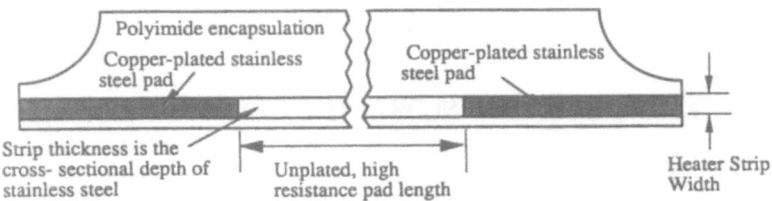

Figure 1. Plan view of a quench heater strip

ANALYSIS TOOLS

The theory and analysis used in predicting quench heater performance were developed in detail in [3]. These analysis methods have been used here to predict quench performance given system and HFU parameters. Only the final pertinent analysis results will be shown again here. By application of an energy balance to the system of the quench heater, the ground plane insulation, the outer conductor layer, and the collar, the following relationship for the heater temperature $T_{htr}(t)$ as a function of time after heater firing can be found:

$$T_{htr}(t) \approx \left(\frac{2h^2E_o}{\tau_e}\right)\left(\frac{R_{Kapo}}{3/\gamma - 2\tau_{to}/\tau_e}\right)\left(\exp\left(\frac{-2t}{\tau_e}\right) - \exp\left(\frac{-3t}{\gamma\tau_{to}}\right)\right) - \frac{2}{3}(T_{kap} - T_{cond})\exp\left(\frac{-3t}{\gamma\tau_{to}}\right) + \frac{T_{cond} + 2T_{Kap}}{3}$$

C , V, I = HFU Capacitance, Voltage and Current
T_{cond} = Conductor sink temperature
E_o = Stored energy of protection system = $0.5CV^2$
h = 1/(# of parallel resistive paths in protection circuit)
g = (# of htr in series per parallel path)/(# of parallel paths)
R_{strip} = heater strip resistance
τ_e = Electrical time constant = $R_{sys}C = GR_{stri}$
$\gamma = R_{Kapc}/(R_{Kapc} + R_{Kapo})$

t = time after firing HFU
$T_{Kapo}(t)$ = Kapton temperature
C_{htr} = Heat capacity of heater strip material
x = lead cable resistance/$2R_{strip}$
G= g+x
R_{Kapo} = thermal resistance of Kapton
τ_{to} = Thermal time constant = $R_{Kapo}C_{htr}$

From this heater temperature history, the heat deposition into the outer conductor layer can be calculated. A quench occurs at t_fn when the heat deposited into the conductor is sufficient to heat the outer conductor beneath the quench heater to the conductor current-sharing temperature. This corresponds to the deposited energy equaling the available enthalpy between T_b and the current-sharing temperature of the conductor, T_{cs}. At 4 K, the vast majority of the available heat capacity is supplied by the helium within the nominal 5% void space of the collared coil. The available helium enthalpy can be calculated for the limiting cases of constant volume (C.V.) and constant pressure (C.P.) heating processes. For the purposes of predicting of the GDSS model quench heater performance, test data from short dipole magnets tested at FNAL is available from which to extract effective enthalpies (DSA 324, 328, 332, and 333) [2]. The effective helium enthalpies determined from the test data should fall between these minimum C.V. and maximum C.P. curves. This effective enthalpy can be backed out from data by modeling the test system and determining the helium enthalpy necessary to match the t_fn data. The best fit curve to the available helium enthalpy data can then be used to predict the performance of the model magnet heater designs. Figure 1 illustrates the curves of effective helium enthalpy as a function of conductor current predicted for 5% cable porosity C.V. and C.P. processes, and for the average. Heater test data was only available for coil currents of 2 kA and 5 kA.

The conductor enthalpy calculated from the test data falls within the predicted envelope and matches expectations well. An unfortunate feature of the enthalpy data from the FNAL-tested short magnets is the scatter of effective conductor enthalpies over a large portion of the range between C.P. and C.V. processes. This can be compared with the correlations found for the DCA 311-322 where the available conductor enthalpy was found to concentrate towards the average enthalpy value or, in the case of the reduced ground plane insulation system, the C.V. enthalpies [3]. Unfortunately, the pool of available short magnet test data is currently too small for a developing pattern to be seen in the different behaviors. Given that, the curve of the average enthalpy between a C.P. and C.V. process provides a reasonable fit to the experimental data for the purposes of this investigation. Calculations assuming C.V. process enthalpies have also been made for comparison.

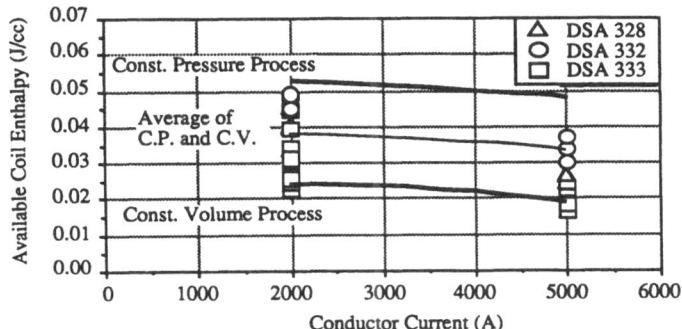

Figure 2. Available conductor enthalpy falls within predicted envelope

MODEL MAGNET DESIGNS AND PREDICTED PERFORMANCE

GDSS will test five heater designs in its four lot A CDM model magnets. Design modifications focus on the reduction of the "active" resistive zone of the quench heater while maintaining high resistance. The GDSS heater designs have reduced strip widths, thickness, and active pad lengths corresponding to reduction in resistive volume by a factor of 4 and 16 over the "baseline" heater. The use of alumina-loaded MT Kapton is also being investigated to determine if the increased thermal conductivity at low temperatures will more than offset the increased heat capacity of the loaded polyimide and result in improved heater performance. The geometries of heater designs built for testing in the model magnets are listed in Table 1. The model magnets will be tested with two different quench heaters in each magnet, thus increasing the amount of direct comparison data obtained.

TABLE 1. Model magnet design attributes

Design	Unplated pad length	Strip thickness	Strip width	Encapsulation
# 1	640 mm	0.0254 mm	12.7 mm	LT Kapton
# 2	320 mm	0.0254 mm	6.35 mm	LT Kapton
# 3	320 mm	0.0254 mm	6.35 mm	MT Kapton
# 4	160 mm	0.0127 mm	6.35 mm	LT Kapton
# 5	480 mm	0.0254 mm	9.5 mm	LT Kapton

1. The baseline design for performance comparison (previous SSCL heater baseline)
2. A 4-fold reduction in resistive volume (50% length and 50% width)
3. A 4-fold reduction in resistive volume (50% length,width) with MT Kapton
4. A 16-fold reduction in resistive volume (25% length, 50% width, thickness)
5. A 44% reduction in resistive volume (75% length and 75% width)

The testing of each heater will determine the HFU voltage necessary to obtain a t_fn of 0.2 s at a current of 2 kA and a t_fn of 0.1 s at a current of 5 kA, all for an HFU RC time-constant of 0.18 s. These standards for measuring performance allow comparison of data with earlier FNAL short dipole magnet test data. The cold resistance of each GDSS model magnet heater (two strips in series) is near 2.26 Ω and the total HFU circuit resistance will be ~4.5 Ω. To obtain an RC time constant of 0.18 s, an HFU capacitance of 40 mF is required. For these parameters, and for both average enthalpy and C.V. enthalpy processes, the HFU voltages have been calculated for the required t_fn at 2 kA and 5 kA. These results are shown in Table 2 along with the predicted reduction in HFU energy for each design relative to design #1. It is the variation in predicted effective helium enthalpy that results in different HFU voltages predicted for the C.V. and average processes. It is expected that the GDSS model magnets will exhibit more parallel quench performance given the parallel construction and testing methods used.

TABLE 2. Predicted model magnet quench heater performance

Design	V_{HFU} (t_fn = 0.1 s @ 5kA)		E_{rqrd}/E_{base}	V_{HFU} (t_fn = 0.2 s @ 2 kA)		E_{rqrd}/E_{base}
	@ΔH_{avg}	@$\Delta H_{C.V.}$		@ΔH_{avg}	@ΔH_{CV}	
#1	70 V	61 V	---	63 V	56 V	---
#2	53 V	47 V	0.59	45 V	40 V	0.51
#3	53 V*	47 V*	0.59*	45 V*	40 V*	0.51*
#4	47 V	42V	0.46	41 V	36 V	0.41
#5	62 V	55 V	0.81	55 V	49 V	0.76

*Overall performance effects of alumina-filled Kapton encapsulation not quantified

Reductions in HFU energies of 60% or possibly more are clearly obtainable. Testing of the first GDSS model magnet will begin in late May 1993. Comparisons between the predicted and actual performance will be made as test data becomes available.

SUMMARY

Five quench heater designs have been developed for testing in the GDSS CDM model magnets. These have been configured to indicate the relative performance advantages for specific modifications in heater strip geometry and encapsulation. Analysis tools were used to estimate the range of HFU voltages and energies required to meet t_fn target levels at specified coil current levels of 2 kA and 5 kA for average values of the coil enthalpy. These predictions will be compared with test data as the testing program progresses.

ACKNOWLEDGMENTS

We would like to thank Chris Haddock of SSCL for all the help that he has provided. The work described herein is being accomplished under contract to the Universities Research Association in support of the Superconducting Super Collider project for the U.S. Department of Energy.

REFERENCES

1. C. Haddock, et al, "SSC Dipole Quench Protection Heater Test Results," Proceedings of the 1991 IEEE PAC, 6-10 May 1991, San Francisco, CA, pp. 2215-2217.
2. Communications with C. Haddock, April 1992 - January 1993.
3. D.J. Kinzie and S.D. Peck, "Analysis of SSC Dipole Quench Behavior as a Function of Quench Protection Heater Configuration," Proceedings of the 1992 Applied Superconductivity Conference, 24-28 August 1992, Chicago, IL.

HIGH ENERGY BOOSTER DIPOLE MODEL MAGNET (HDMM) MARGIN

O.R. Christianson, H.L. Chuboy, M.P. Krefta, and S.K. Singh

Westinghouse Science and Technology Center
1310 Beulah Road
Pittsburgh, PA 15235

INTRODUCTION

The HDMMs are required to be designed with a field margin of greater than 10% and a temperature margin greater than 0.6 K. The field and temperature margins estimated for the HDMM DSB701 exceeds these specifications. For calculation purposes, the operating temperature is taken as 4.25 K. The field margin, found from the load line of the magnet and the critical surface of the superconductor, is about 13% in cables 17 and 19 where the maximum field occurs. The corresponding temperature margin, found from the magnetic field dependent critical current, critical temperature, and the operating temperature of the superconductor, is 0.78 K.

Westinghouse uses Morgan's correlation[1] as a standard and, as an independent verification, a procedure similar to Green's[2] and Lubell's[3] where the critical current follows a linear flux pinning relationship at high magnetic fields, the Kim-Anderson relationship at medium magnetic fields, and a square root of the magnetic field at low magnetic fields.

TABLE 1. HDMM CONDUCTOR SPECIFICATIONS

	Inner	Outer
Strands	30	36
Cu/SC	1.3	1.8
B_ϕ	6.67	
B_{peak}	6.9	5.71
Max Temp	4.25 K	4.25 K
J_c	9662 A @7T,4.2 K	9780 A @5.6T, 4.2K
I_{op}	6640 A	6640 A

CONDUCTOR PERFORMANCE

The wire that was initially specified for use in the HDMM model exceeds the critical current design specifications. See Table 2.

TABLE 2. Critical currents for HDMM inner and outer cables

Inner Cable	SSC-3-I-101		
	Measured:	Specification:	
I_c strand	384.5 A @ 7T	339 A @ 7T	No self field correction
I_c cable	11020 A @ 7T		Self field corrected
Outer Cable	SSC-4-K-00026		
	Measured:	Specification:	
I_c strand	316 A @ 5.6T	286 A @ 5.6T	No self field correction
I_c cable	11207 A @ 5.6T		Self Field corrected

FIELD AND TEMPERATURE MARGIN

The field margin is defined as FM = $(B_q - B_0)/B_0$, where B_0 is the peak field in the conductor when the dipole magnet is excited to a current corresponding to 20 TeV operation of the collider, and B_q is the maximum field in the conductor at the point where the load line intersects the critical current curve. The load line and critical currents for the inner coil of the HDMM are shown in figure 1.

TABLE 3. Margin calculations for the HDMM DSB701.

	Inner Coil	Outer Coil
Field Margin	12.9 %	18.4 %
Critical Temperature	5.052 K	5.428 K
Temperature Margin	0.802	1.178

The magnetic field is different in each cable with the maximum field toward the pole turns, and the magnetic field decreases across the width of a cable. Cables 17 and 19 have the largest magnetic field at close to 7 Tesla, with a field margin about 13%, see figure 2.

The temperature margin is the difference between the critical temperature and the operating temperature of the conductor. The critical temperature is the temperature at which the superconductor is superconducting at the field and current defined by the load line of the magnet. The HDMM meets the required temperature margin of 0.6 K on the inner and outer coils.

MARGIN SENSITIVITIES

Several factors influence margin, including sag in the transfer function, cable critical currents, and sensitivities associated with the calculation of the magnetic field in the conductor region.

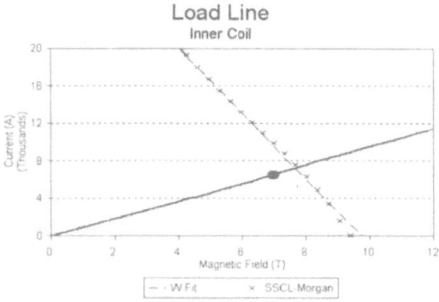

Figure 1. Load line and critical current for the inner coil of the HDMM at 4.25 K.

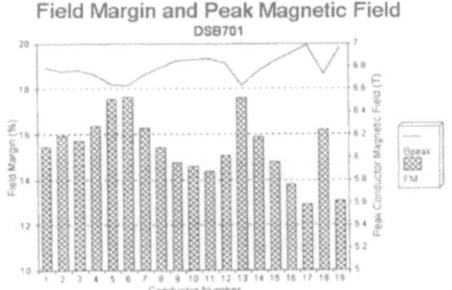

Figure 2. Field margin and peak magnetic field as a function of cable number.

The load line of the magnet deviates from linear due to saturation of the iron in the collar and yokes. This saturation is included in the transfer function for a particular cross section, collar, and yoke design. Comparing with SSCMAG, which does not include sag, with Westinghouse calculations for the DSB701 cross section at 4.222 K using a critical current of 2750 A/mm^2 and 5% degradation gives a 0.2% change in the field margin, see Table 4

Table 4. Field margin comparison between SSCL SSCMAG and the Westinghouse calculation.

	Field Margin
SSCMAG	14.1384%
Westinghouse, no sag	14.1389%
Westinghouse, sag	13.9359%

The field margin is dependent upon the critical current of the superconductor. Measured critical currents are used to estimate the field margin, but measured strand and cable critical currents are not exactly the same. Measured strand critical currents include the self field but do not include any degradation. Measured cable critical currents are corrected for the self field. Cable critical currents are used in the Westinghouse margin calculations.

The magnetic fields are calculated using a finite element code, WEMAP. To facilitate the finite element analysis each cable is represented by current in a rectangle that is compacted azimuthally. This allows the use of elements with good aspect ratios between the cables. This current compaction does not substantially affect the harmonics in the central region, but slightly increases the maximum magnetic field in the conductor region. Discretization of the cable current more closely represents the actual physical geometry, figure 3; and results in a slightly smaller maximum magnetic field, table 5, and increases the field margin by about 1%.

TABLE 5. Calculated field margin using cable and strand representations for current.

	Maximum Magnetic Field	Field Margin
Current Compacted	6.97769 T	13.02%
Discretized Current	6.90407 T	13.99%

Careful examination of the magnetic fields in the conductor region indicate minor field variations from radial distributions, see figure 3. The peak magnetic field occurs on the top of a cable group (the bottom of a wedge) toward, but not at the inner radius. The minimum magnetic field occurs on the bottom of a cable bundle (the top of a wedge), toward, but not at the outer radius.

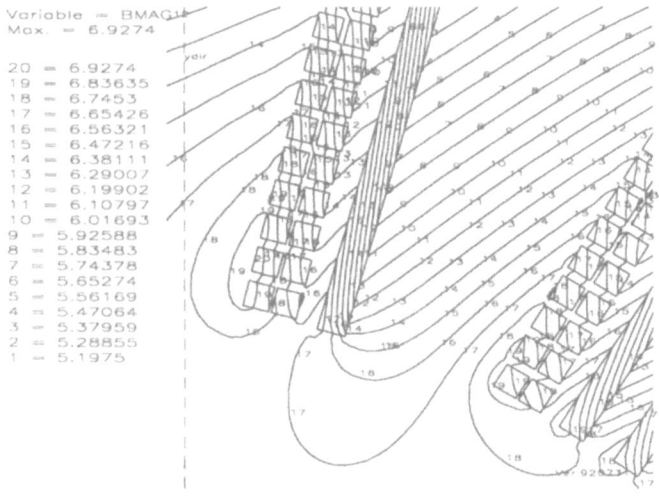

Figure 3. Magnetic field contours where cables 17 and 19 have been discretized.

COMPARISON OF FIELD MARGIN TO TEMPERATURE MARGIN

The temperature margin is found by raising the temperature until the critical current intersects the operating point. The minimum temperature margin for DSB701 is about 0.78 K, see figure 4.

The temperature margin follows the field margin, suggesting a field margin to temperature margin correlation. A crude overlay suggests that a 10% field margin in DSB701 is approximately a 0.7 K temperature margin. One can solve for the quench current, or field, as a function of temperature. Defining the field margin and temperature margin in terms of the quench current yields a relationship between the field margin and temperature margin. This relationship, neglecting sag for simplicity, is graphed in figure 5. The 10% field margin is equivalent to a 0.6 K temperature margin.

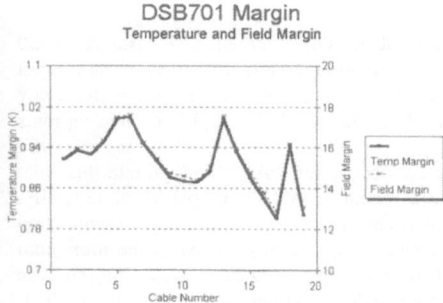

Figure 4. Temperature and field margin by cable illustrating the correlation between temperature and field margin.

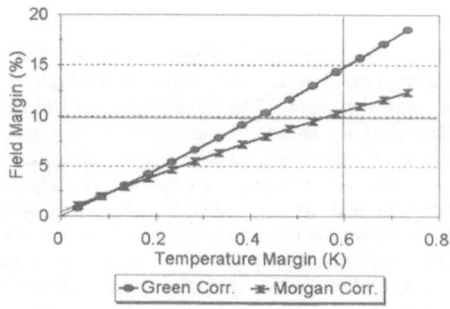

Figure 5. Relationship of field margin to temperature margin for different critical current correlations and where sag is not included in the load line.

CONCLUSIONS

The HDMM design meets the required system specifications of 10% field margin and 0.6 K temperature margin. The margin is sensitive to the critical current value, transfer function and sag, and characterization of the current source.

REFERENCES

1. G. Morgan, "A Comparison of Two Analytic Forms for the Jc(B,T) Surface," SSC Tech. Note. No. 310-1 (SSC-MD-218), (January 6, 1989).
2. M. A. Green, "Calculating the Jc, Bc, Tc, Surface for Niobium Titanium Using a Reduced-State Model," Proc. 1988 Applied Superconductivity Conference, San Francisco, August 21-25, 1988.
3. J. S. Lubell, "Empirical Scaling Formulas for Critical Current and Critical Field for Commercial NbTi," IEEE Trans. on Magnetics, Vol MAG-19, No. 3, p. 754 (May 1983).
4. S. Peck, "Definition of "Margin" for SSC CDM's," General Dynamics 890-91-P793, September 11, 1991.
5. Y. Zhao, "Effect of Persistent Magnetization Currents in SSC Dipole Magnets," SSC MD-TA-218, April 24, 1992.

HIGH ENERGY BOOSTER DIPOLE MODEL MAGNET (HDMM) QUENCH PERFORMANCE

O.R. Christianson, J.F. Roach, and S.K. Singh

Westinghouse Electric Corporation
Science & Technology Center
1310 Beulah Road
Pittsburgh, PA 15235

INTRODUCTION

To ensure that no damage to a HDMM results from a quench, the quench performance of the HDMM DSB701 in the short magnet quench protection circuit is modeled. If the temperature rise is excessive, damage to the Kapton insulation could occur, above 500 K, or the superconducting critical current could degrade, above 850 K. The magnet current, coil hot spot temperature, and magnet resistance assuming adiabatic quench propagation velocities are calculated as a function of time. Voltages developed within the circuit during a quench are also estimated. The fraction of the stored energy deposited in the coil and in the dump resistor is calculated, which determines the amount of LHe vaporized during a quench.

MIITS CALCULATIONS

The HDMM operating point is far from the cryostable condition, joule heating is large, and current decay times are small compared to thermal time constants in the windings, such that an adiabatic one dimensional approximation of the temperature rise during a quench is valid. The adiabatic one dimensional energy balance is

$$\frac{I^2(t) \bullet \rho(T)}{A_{C_u}} = A \bullet C(T) \bullet dT \tag{1}$$

where I is the transport current, t is time, r is the copper resistivity, T is temperature, A_{Cu} is the copper cross-sectional area, A is the conductor cross-sectional area, and C is the average volumetric specific heat. This expression is rearranged separating the temperature dependent quantities from the time dependent quantities,

$$\int_0^t I^2 dt = A_{C_u} A \bullet \int_{T_o}^{T} \frac{C(T)}{\rho(T)} dT.$$ (2)

The final temperature reached during a quench may be found from the value of the right hand integral as a function of temperature. The units are for the left hand integral MIITS, 10^6 Amp^2-sec, and the right hand, 10^6 J/W. The MIITS tabulation for copper and for niobium-titanium used in this work were given by McAshan[1]. The MIITS integrals for the HDM are displayed in Fig. 1. Since the temperature rise in the outer coil is larger than that in the inner coil for a given MIITS, the magnet quench performance is determined by the outer coil.

HDMM QUENCH ELECTRICAL CIRCUIT

The HDMM quench protection circuit, shown in Fig. 2, consists of an HDMM and a dump resitor. After a quench the power supply is 'crow-barred' out of the circuit and the dump resistor is inserted into the circuit. The rate at which the current decays in the magnet is dependent upon the magnet inductance and the dump resistance. The inductance of an HDMM is about 7 millihenries. The resistance of the magnet, which is initially small compared to the dump resistor, is a function of the quench propagation, the delay in inserting the dump resistor into the circuit, and the temperature dependent electrical resistivity.

The quench propagation velocity is large, on the order of 70 m/s. Calculated quench propagation velocity in the adiabatic approximation underestimates the measured propagation velocity, see Fig. 3. Furthermore, the magnetic field varies azimuthally, generally increasing toward the pole turn, and varies across the width of each cable. These effects are included in a quench propagation velocity calculation, shown in Fig. 4. The quench propagation velocity is therefore taken as an input to calculating the normal zone size and resistance. Use of the calculated adiabatic quench propagation allows the propagation velocity to be treated as a function of magnet current.

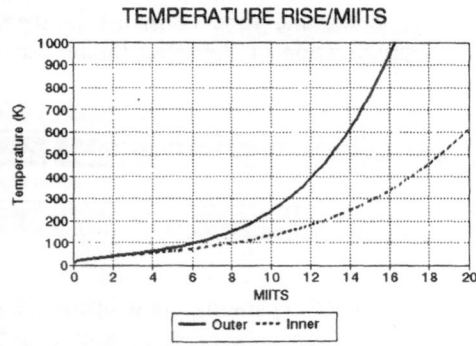

Fig. 1. Temperature versus MIITS for the inner and outer coils in the HDMM.

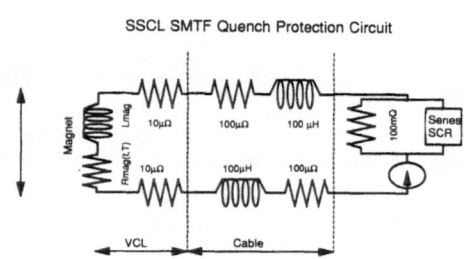

Fig. 2. The HDMM quench protection circuit showing the dump resistor.

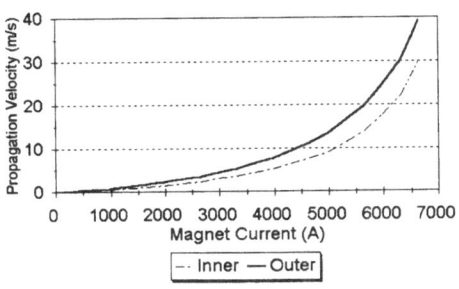

Fig. 3. Calculated adiabatic quench propagation velocities.

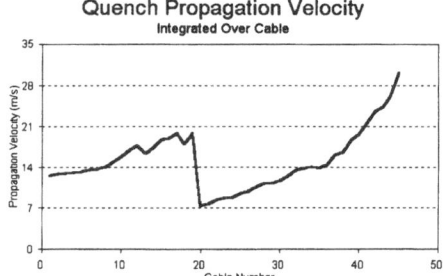

Fig. 4. Calculated quench propagation velocities including the variation of the magnetic field cable by cable and across the width of a cable.

HDMM QUENCH TEMPERATURE EXCURSIONS

The calculation of the temperature rise in an HDMM during a quench using the MIITS integrals is dependent upon calculating the current in the magnet as a function of time. The rate of current decay is dependent upon the resistance of the dump resistor. The current in the magnet is described by

$$I(t) = I_o \exp\left(-\frac{1}{L}\int_0^t ds R(s)\right),$$ (3)

where I_0 is the initial transport current, L is the inductance, R is the resistance which initially is dominated by the dump resistor, and s is an integration variable for time.

The resistance of the normal zone is dependent upon the size of the normal zone. It is assumed that the normal zone shape is a combination of longitudinal and azimuthal extent where the longitudinal extent is given by the quench propagation velocity and the time, and where the azimuthal extent is given by a calculated transverse propagation. Experimental measurements suggest that transverse quench propagation in SSC magnets is more complex than expected from just conduction.[2]

The temperature of the normal zone is found from the MIITS integrals. An numerical procedure is used where the resistance of the magnet is found from the size and temperature of the normal zone in the previous time step. The total resistance, both the dump resistance which is dominant and the coil resistance, then determines how quickly

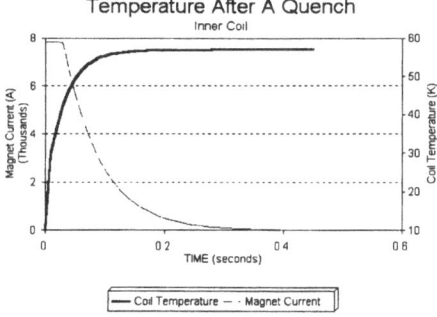

Fig. 5. Projected temperature rise in the HDMM during a quench.

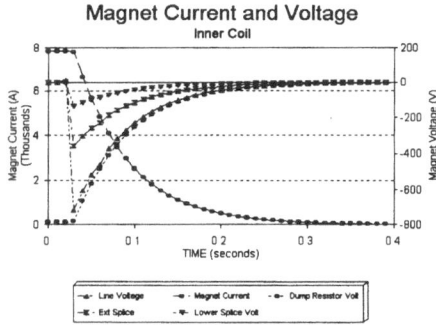

Fig. 6. Voltages in the HDMM during a quench.

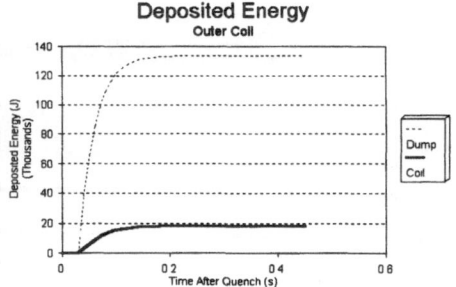

Fig. 7. Energy deposited in the coil and the dump resistor during a quench.

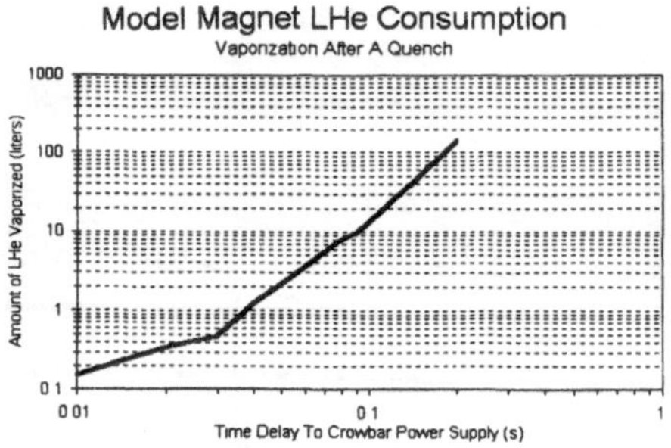

Fig. 8. Amount of cryogen vaporized during a quench while using a dump resistor as a function of the time the dump resistor is inserted into the circuit.

the current decays in the next time step. The number of MIITS increases which results in a temperature rise. Comparison to a measurement on a single dipole magnet[3,4] is good for both the current decay and voltage in the magnet as well as the temperature rise. Voltages during a quench are predicted by using a lumped electrical circuit where the resistances and currents are calculated by this numerical procedure, see Fig. 6.

ENERGY DEPOSITION IN THE DUMP RESISTOR AND COIL DURING A QUENCH

The energy deposited in the coil and dump resistor during a quench are calculated, see Fig. 7. Most of the HDMM's stored energy is deposited in the dump resistor, and only a small amount of the stored energy is deposited in the coil. This reduces the amount cyogen vaporized during a quench, see Fig. 8, compared to the case where no dump resistor is used and all of the stored energy is deposited in the coil with a larger amount of cryogen vaporized.

SUMMARY

The temperature rise in the HDMM is projected using MIITS integrals. The temperature rise is modest such that no damage should result during a quench. Use of a dump resistor minimizes the amount of cryogen vaporized during a quench.

REFERENCES

1. M.S. McAshan, MIITS integrals for copper and for Nb-46.5% Ti, SSC-N-468 (1988).
2. A. Devred, et al., Quench characteristics of full-length SSC R&D dipole magnets, Advances in Cryogenic Engineering, Vol. 35, pages 599-608 (1990).
3. A. Devred, et al., Investigation of heater-induced quenches in a full-length SSC R&D dipole, 11th International Conference on Magnet Technology (MT-11), Vol. 1, pages 91-95 (1990).
4. O. Christianson, Quench temperature excursion predictions for the SSC high energy booster dipole and quadrupole magnets (HDM and HQM), IEEE Transactions on Applied Superconductivity Vol. 3, No. 1, pages 662-665 (1993).

A STUDY OF THE EFFECT OF CABLE INSULATION ON COIL PROPERTIES

Amanda Spindel, Richard Sims, Steve Bastian, and Douglas Pollock

Superconducting Super Collider Laboratory*
2550 Beckleymeade Ave.
Dallas, TX 75237

INTRODUCTION

Stacks of insulated superconducting cable were cured and tested for mechanical properties. Four different polyimide films were investigated in all polyimide insulation schemes. The effect of bottom layer material, top layer material and wrapping scheme on stack size, modulus, and compressive creep are reported.

TEST METHOD

Ten pieces of manually insulated superconducting cable five inches long were stacked with staggered keystone angles and cured for one hour at 150°C. The central 3 inches of the stack were subjected to pressure. The deflection, temperature, and pressure were measured and recorded once a minute during cure. Three stacks of each combination were cured and tested. The stacks were used to model a coil for the purpose of testing insulation schemes.

For the study of film type, bottom layer films used were Kapton H, Apical NP, and Upilex R and top layer films were Kapton LT, Apical NP, and Upilex R. The bottom and top layers were 50% overlap and there was adhesive only on the outside surface of the top layer. In addition, wrapping configuration was studied. For this part of the study, film type was held constant with Kapton H on the bottom layer and Kapton LT on the top. The 50% overlap top layer was compared to a butt wrap top layer and adhesive on both sides of the top layer was compared to adhesive on one side of the top layer. The modified 3P adhesive system was used throughout because it was the only adhesive available on all film types. The same batch of cable was used for all stacks. Fiberglass systems were not investigated due to a shortage of cable.

The modulus and compressive creep were measured with an MTS servohydraulic test machine. Deflection was measured on each side of the stack with extensometers. Modulus data were taken at a load rate of 1 ksi/sec. The modulus was taken to be the slope of the stress strain curve between 8 ksi and 12 ksi. The height at 10 ksi was found by subtracting the deflection of the stack at 10 ksi as measured in the modulus test from the stack height at 0 ksi. Compressive creep data were taken for four hours at 12 ksi and were taken at least

* Operated by the Universities Research Association, Inc., for the U. S. Department of Energy under Contract No. DE-AC35-89ER40486.

24 hours after the modulus measurements to allow for recovery of the stacks. The coefficient of the best fit logarithmic curve was used as the basis of comparison. Results presented in Tables 1 and 2 are the average values for these trials.

RESULTS

Statistical analysis was performed using analysis of variance (ANOVA). After the existence of a statistically significant difference was found with ANOVA, post hoc tests using pair-wise comparison were used to determine which configuration(s) were statistically different.

The wrap configuration had a significant effect on stack height. For both the 0 ksi and 10 ksi stack height measurements, the butt wrapped top layer stacks were, as expected, smaller than the stacks with 50% overlap top layers and the two sided adhesive stacks were only slightly larger than their one sided adhesive counterparts. The two sided adhesive stacks had significantly higher moduli than the one sided glue stacks. There did not appear to be a difference in modulus between the butt wrapped and 50% overlap configurations. Compressive creep was significantly affected by both the wrapping and the adhesive layers. Creep was higher in the 50% overlap than the butt wrapped configurations and in the one sided glue than the two sided glue configurations. The data can be seen in Figure 1.

Both top and bottom layer materials had a significant affect on the stack height at both 0 ksi and 10 ksi. For the 0 ksi height, the stacks with a bottom layer of Upilex R were significantly smaller than the others with no difference between Apical NP and Kapton H. At 10 ksi, the stacks with bottom layer Apical NP were significantly larger than the other stacks, with no difference between Upilex R and Kapton H. At both 0 ksi and 10 ksi, the stacks with top layer Kapton LT were significantly smaller than the others with no difference between Apical NP and Upilex R. Perhaps the Kapton LT was smaller than the nominal 1 mil and that accounts for the top layer effect. Figures 2 and 3 show this data.

The stack modulus depended only on the bottom layer with Upilex R having a significantly higher modulus than the others and no difference between Apical NP and Kapton H, as can be seen in Figure 3. The creep coefficient depended only on top layer material, as can be seen in Figure 2. Stacks with Apical NP as a top layer material suffered significantly less creep than others and no difference between Kapton LT and Upilex R. This is surprising because Kapton LT is widely believed to reduce creep.

CONCLUSIONS

The optimum coil would have the lowest possible creep rate, the most consistent stack height under load, and the highest modulus. It would be recommended that a butt wrap be used on the top layer with glue on both sides, and that Apical NP be used as the top layer material. Because the bottom layer effect on size and modulus were both due to Upilex R, which is no longer available in the United States, further study may be needed to determine an optimum bottom layer material. Determination of optimum glue type and cure cycle will also require further research.

Table 1. Wrap configuration data.

adhesive	wrapping	height (in) @ 0 ksi	height (in) @10 ksi	modulus (ksi)	creep coef. $(x10^{-4})$
1 side	butt	0.636	0.621	794.4	-5.623
1 side	50%	0.659	0.638	804.2	-5.906
2 side	butt	0.637	0.624	839.1	-5.212
2 side	50%	0.661	0.648	851.2	-5.644

Table 2. Film type data.

bottom layer	top layer	height (in) @ 0ksi	height (in) @10ksi	modulus (ksi)	creep coef. (x10^{-4})
Kapton H	Kapton LT	0.659	0.638	804.2	-5.906
Kapton H	Apical NP	0.657	0.642	777.5	-4.816
Kapton H	Upilex R	0.657	0.641	795.6	-5.556
Apical NP	Kapton LT	0.655	0.639	776.1	-5.063
Apical NP	Apical NP	0.661	0.643	869.3	-4.814
Apical NP	Upilex R	0.660	0.643	823.7	-4.906
Upilex R	Kapton LT	0.652	0.638	881.0	-5.954
Upilex R	Apical NP	0.657	0.643	920.8	-4.291
Upilex R	Upilex R	0.655	0.641	911.5	-5.195

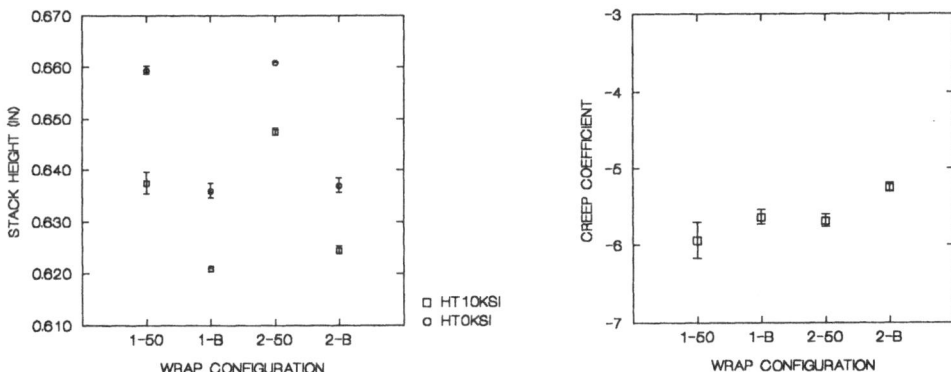

Figure 1. Effect of wrap configuration on stack height and creep coefficient.

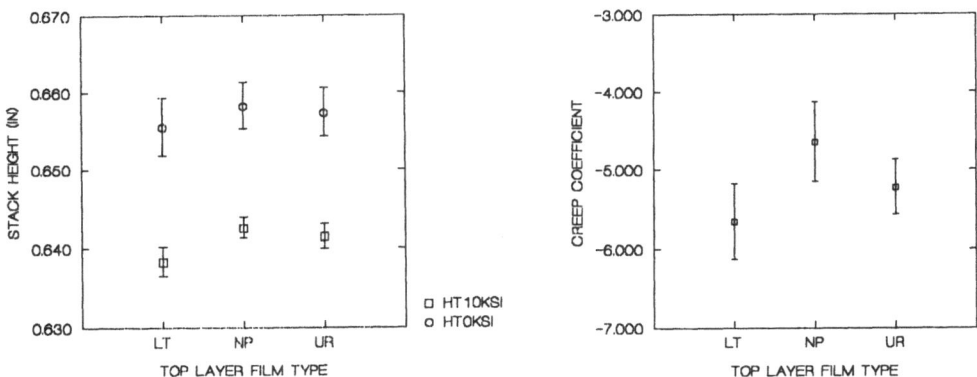

Figure 2. Effect of top layer materials on stack height and creep coefficient.

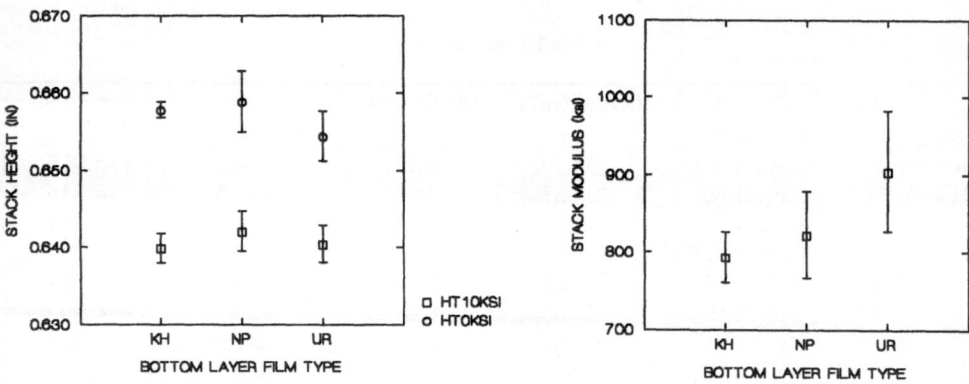

Figure 3. Effect of bottom layer materials on stack height and modulus.

MAGNETIC DESIGN CONSIDERATIONS FOR THE
SSC VERTICAL BENDING (BV1C) MAGNET

V. Venkatraman, C. Goodzeit, R. Jayakumar, F. Nobrega, and G. Snitchler

SSC Laboratory[*]
2550 Beckleymeade Ave.
Dallas, TX 75237

INTRODUCTION

The BV1C magnet is a large aperture, vertical bending magnet to be used to bend proton beams in the interaction region. An aperture larger than 80mm is required. The central field has to be a minimum of 6T with a 10% margin. The lattice requirements for field quality are stringent because two counter beams traverse this magnet off the center axis. This magnet's transfer function sag is specified to match closely the transfer function sag of the low beta quadrupoles. With these specifications in mind, suitable designs for the 2-d magnetic cross-sections have been analyzed.

DESIGN CONSIDERATIONS

A magnetics design study was carried out on hundreds of designs ranging from a 75mm to 100mm aperture dipole magnet two-dimensional cross-section. Both single layer and two layer designs were considered. The design studies yielded input into the possible solutions to beam bending and optics requirements. The design solutions were governed by the harmonic requirements listed in table 1. The resulting conceptual design of the BV1C is an 87mm aperture cross-section which produces 6.4585T at 6714A which is the present baseline high field operating current.

It is useful to review several of the decisions which generated this atypical design. First, this large aperture magnet has an inductance of 167mH for a 16.3 meter magnetic length which represents a large stored energy relative to the collider dipole magnets (CDM). Consequently, this design has potential quench protection issues. Test data from the ASST prototypes suggest that the outer cable for the CDM magnets limited that design from being

* Operated by the Universities Research Association, Inc., for the U.S. Department of Energy under Contract No. DE-AC35-89ER40486.

Table 1. DSC magnet field quality tolerance goal

Normal Terms	Field Quality at 20 TeV [at 1.0 cm, x 10-4] (max. values)	Skew Terms	Field Quality at 20 TeV [at 1.0 cm, x 10-4] (max. values)
b1	0.8280	a1	1.3240
b2	0.2112	a2	0.3504
b3	0.1235	a3	0.2477
b4	0.0727	a4	0.0292
b5	0.0102	a5	0.0102
b6	0.0036	a6	0.0036
b7	0.0013	a7	0.0013
b8	0.0005	a8	0.0005
$b_n; (n \geq 9)$	0.0002	$a_n; (n \geq 9)$	0.0002

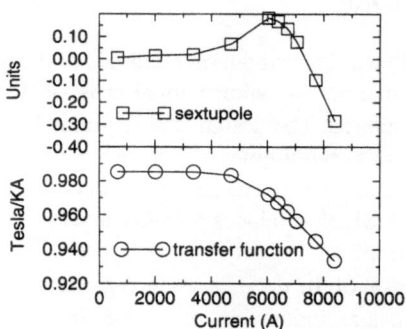

Figure 1. Transfer function and saturable sextupole harmonic.

self-protected. To prevent the requirement of a new cable or even a new strand for the outer cable, the design team elected to use CDM inner cable both in the inner and outer layers of this BV1C design. This allows for a significantly higher MIITS margin during a quench. It does, however, bring into question whether protection heaters should be placed on the inner or outer coil since the design has a very high margin on the outer coil. This issue will be resolved within the framework of the model magnet program.

The second non-standard feature for this design, is that four wedges are used in the cross-section rather than the usual three wedges. This was an added requirement for the sole purpose of increasing the margin of the magnet. Lessons learned from the DSB program suggested that moving Ampere-turns from the pole of the magnet improved the margin of the design. The addition of a wedge near the pole of this design represents a margin increase of approximately 2%. This pole turn wedge did not play a key role in the harmonics optimization. The cross-section is presented in ref 1. The analytic optimization program SSCMAG[2] was used to perform the design study.

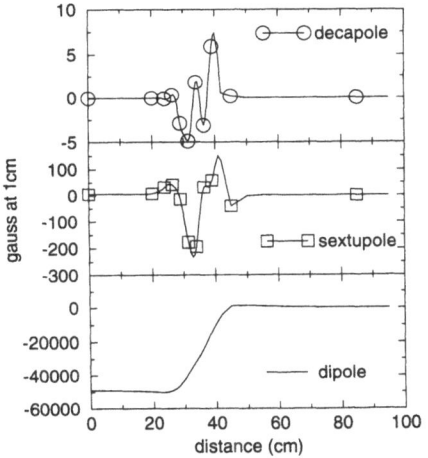

Figure 2. End fields produced by ENDS3D.

The next step in a magnet design is a two dimensional saturable iron finite element calculation. This allows design input into the yoke outer diameter and yoke features. Again, lessons learned from the DSB program are that inner diameter features as well as the standard CDM non-magnetic key play a role in the sextupole saturation profile as a function of current. The stated saturation requirements are to track the CDM within 1.5%. In figure 1, the sextupole and transfer function profiles are presented. Care was taken not to allow the pole alignment key to adversely effect the sextupole performance. When satisfied with the two-dimensional cross-section performance, the next step is to provide a magnetics input into the end-turn design. The design team generated an initial mechanical end part design from the BEND program developed by J. Cook[3]. This program provides input into a program called END3D developed by Lilly, Orrell, and Snitchler[4]. This provides an estimate to the dipole field, sextupole, and decapole harmonics from the end design. The lengths of the end parts are adjusted to lower the end harmonics. In figure 2, the field and harmonics are plotted.

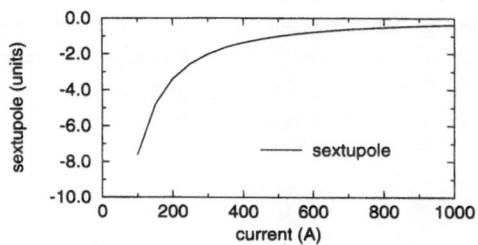

Figure 3. Superconductor magnetization sextupole harmonic.

Other estimates were made to ensure that no major contributor to the field harmonics was ignored. A finite element calculation was performed addressing the collar permeability. Since high-Manganese steel has been selected, there is a very low contribution to the sextupole. No cross-section adjustment was made. Also, the superconductor magnetization sextupole is analyzed using AHARM[2] and are presented in figure 3. The injection sextupole is approximately -.75 units at injection field.

The preliminary BV1C design has been presented. As the detail design progresses, design modifications may be required. A major concern for this design is the amount of pre-stress required to constrain the ampere-turns. Estimates and solutions are presented in ref 1. Special concerns for quench performance remain an open issue. In general, the design parameters meet the requirement guidelines in allowed harmonics, saturable performance, and required field.

REFERENCES

1. F. Nobrega, et. al., "Parameters and Conceptual Design of a Vertical Bending Magnet for the Super Collider," IISSC May 1993.
2. B. Archer, D. Orrell, G. Snitchler, "SSCMAG User's Manual," internal MSD document MD-END-92-A-005.(1992)
3. J. Cook, private communication.
4. Lilly, Orrell, and Snitchler, private communication.

PRELIMINARY RESULTS FROM A STUDY OF COLLAR LAMINATION VARIATION IN SSC PROTOTYPE DIPOLE MAGNETS

R. Gattu, G. M. Brown and D. Pollock

Superconducting Super Collider Laboratory*
2550 Beckleymeade Avenue
Dallas, TX 75237

INTRODUCTION

The collar laminations used in SSC Prototype Collider Dipole Magnets determine the volume within which the magnet coils are constrained after collaring and keying. The uniformity and symmetry of the inside volume of the collars along the length of the magnet may have a significant influence on the *field quality* of the finished assembly. This paper describes an on-going Statistical Quality Control study of collar lamination dimensional variation being performed by SSCL Magnet Systems Division Quality Assurance (MSD QA).

Samples of collars have been measured using a coordinate measuring machine to evaluate manufacturing *process capability* as well as the overall uniformity of the inventory population of collar laminations. The collar data will be used to predict variation in the coil assembly center and radius for inner and outer top-bottom, left-right coil combinations as well as pole angles. Collar results will be combined with azimuthal coil size measurements as part of a manufacturing cause and effect model for predicting axial geometric multipoles based on the observed mechanical variation.

This work focuses on Prototype Collider Dipole Magnet DCA102 currently being built at the SSCL MDL in Waxahachie, Texas. This magnet is being made on the same coil curing and collaring mold cavities that were used for the DCA 300 series magnets built at FNAL in 1991-92 and which were later used in the 1992 Accelerator Systems String Test (ASST). The collars are part of the same procurement used for the DCA300 series magnets.

SAMPLING FROM INVENTORY

Collar laminations have been collected randomly from the SSCL inventory. Ten boxes were selected randomly from inventory for each collar type (left and right hand lamination pairs). From each box of one hundred collars, a sample of ten collars has been collected randomly. Each sample has been identified by a box number and a reference number for measurements.

MEASUREMENT FEATURES

Critical dimensions which influence the coil to collar and collar to yoke interface features have been measured. The important measured features are collar lamination outer radius, lower inner radius, upper inner radius and pole face angles. Each feature has been measured

---

*Operated by the Universities Research Association, Inc., for the U.S. Department of Energy under Contract No. DE-AC35-89ER40486.

at different locations (left to right for radii and top to bottom for pole face angle features) to obtain feature statistics. Each measurement is identified by a ten digit feature number, angular location for radius feature and X Y coordinates for pole face angle features. A total of fifty three individual measurements have been made on each of two hundred collar laminations in the study.

MEASUREMENT PROCEDURE

Collar laminations have been measured using a Sheffield CORDAX 1808M-DCC (Direct Computer Control) Coordinate Measuring Machine (CMM) at the SSCL. MSD QA Inspection & Test Group developed a part program to measure collar laminations on the CMM. Collar laminations are mounted for measurement on a specially designed fixture, which can hold ten laminations for each setup. Initially the outer radius feature is measured at eight different locations and the average value is computed to determine the center for Outer Radius (O4R) feature. From the computed center point for O4R feature, the origin is computed and located as per the drawing[1] requirements and used as a starting point for measurement of additional features. Inner and outer radii features are measured at different horizontal and vertical offsets according to the drawing requirements. Fig 1 illustrates an Inner Left Radius (IL1R) feature measured at four different angular locations.

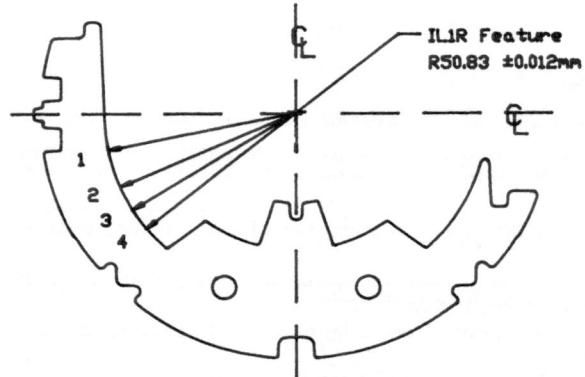

Figure 1. Collider Dipole Magnet Collar Lamination

MEASUREMENT ERROR

Measurement error studies have been conducted to ensure adequate gage and inspection system capability.

Table 1. Machine Repeatability Measurement Analysis

Lamination Feature	σ gage repeatability (mm)	1* σ gage repeatability / Tolerance (mm)	(1* σ gage repeatability / Tolerance)*100 (%)
IL1R01	0.00232	0.193	19.3
IL1R02	0.00092	0.076	7.6
IL1R03	0.00105	0.087	8.7
IL1R04	0.00108	0.090	9.0

To evaluate measurement error, one collar lamination was repeatedly measured for all the features ten times. Four locations for IL1R feature have been selected for comparative study to evaluate the measurement error (σ gage repeatability). Table 1 shows the measurement error (repeatability) values for the four locations IL1R01 - 04. The manufacturer's specified repeatability range is 0.003mm (0.00012"). The drawing tolerance is ± 0.012mm. One σ gage repeatability /tolerance ratio is shown for all four locations measured. Repeatability is typically less than ten percent of the part tolerance.

Current calibration[2] of CMM shows that the accuracy and the repeatability values meet the manufacturer's specifications.

PRELIMINARY DATA ANALYSIS

Measurement data have been obtained for two hundred sampled laminations. Due to space limitations in this paper, only four features (IL1R, Inner lower Radius (I1R), Inner middle Radius (I2R) and O4R) are described for the collars from one box sampled. The feature IL1R is illustrated in figure 1. Graphs and analysis of lamination measurements presented in this paper represent only an initial methodology for describing feature variation. Future studies will analyze the data in detail. Data will be analyzed by two methods, one based on the ideal center (i.e, in accordance with the drawing) and the other based on the best fit center calculated for the part. Both methods will be used to evaluate the shift between the ideal center and the estimated center of the part.

Figure 2 (based on the ideal center measurements) shows variation "within" the lamination and "between" the laminations for the four measurements of the IL1R feature. The two dotted lines in figure 2 represent the drawing lower and upper tolerance limits.

Figure 3 is a box plot for IL1R by measurement. The ideal center data is used for the box plots. The box plot shows an upward trend for the measured position suggesting a systematic deviation in the radius.

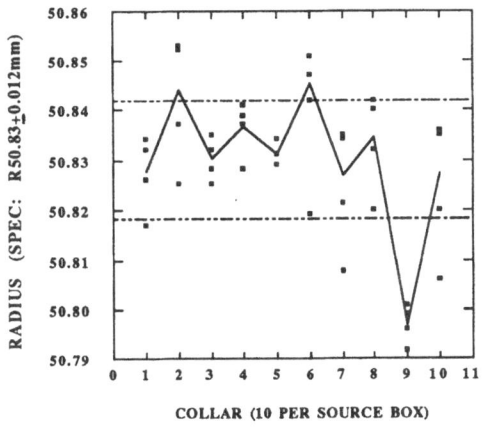

COLLAR (10 PER SOURCE BOX)

Figure 2. Collar Lamination vs Radius Feature Plot

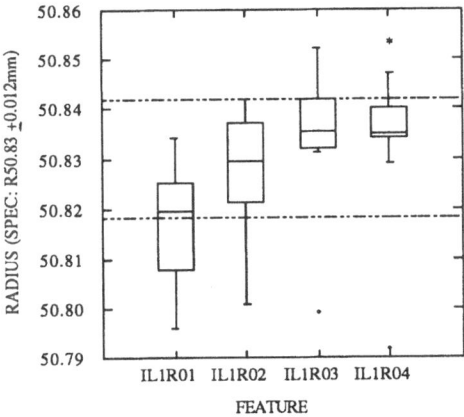

FEATURE

Figure 3. Box Plot for Radius Feature

Figure 4 is a plot based on best fit centers for I1R and I2R features. These features are illustrated in figure 6. Two vertical and horizontal dotted lines in figure 4 represent the drawing tolerance zones for X and Y coordinates for I1R and I2R center points. The enclosed dotted line rectangle is the tolerance zone for the centers of the I1R and I2R features. None of the measured feature centers lie in the drawing specified tolerance zone. These deviations if representative of the inventory population will affect the coil placement in the collared coil. Such deviation in collar geometry is expected to influence field quality[3].

Figure 5 is a plot based on best fit centers for Outside Radius (O4R) feature. The feature is illustrated in figure 6. Best fit center points for O4R feature are lying in the drawing specified tolerance zones for all the laminations except one. Similar comparisons will be studied for other outer surfaces (O4R) to evaluate collar influence on yoke collar interface.

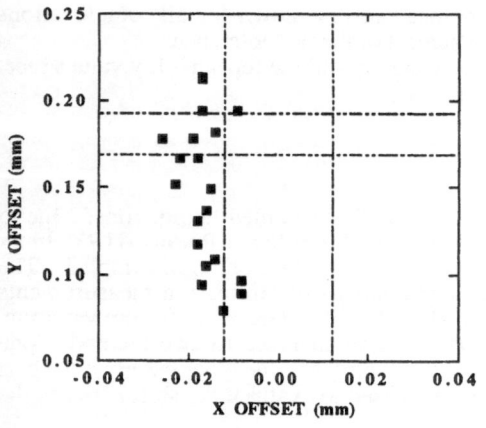

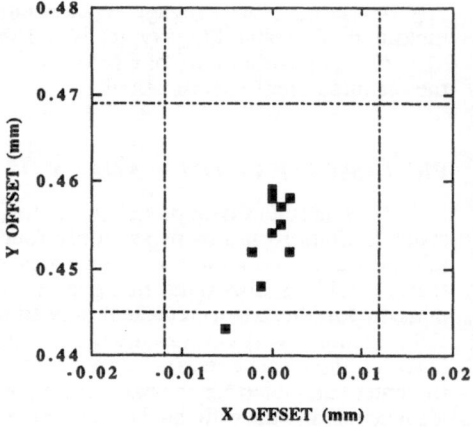

Figure 4. Collar Inner Radius Center (I1R AND I2R) Feature

Figure 5. Collar Outer Radius Center (O4R) Feature

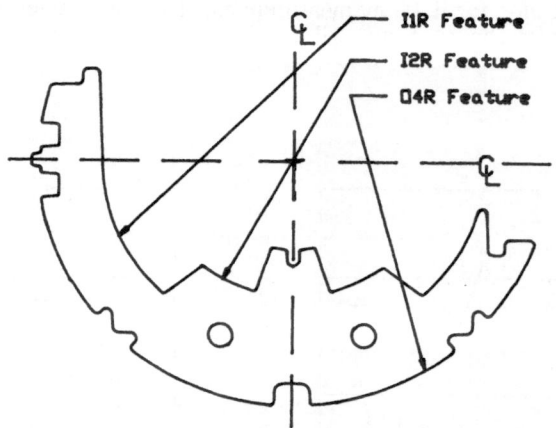

Figure 6. Collar Centers I1R, I2R and O4R Features

CONCLUSIONS

We have described a set of prototype dipole collar lamination measurements taken as part of a statistical quality control study of critical component manufacturing variation. As a result of collecting comprehensive set of measurements from a sample of collars in our inventory, we are now able to study the level of agreement between drawing requirements and manufactured parts in great detail. Collar manufacturing process capability for the source supplier can also be estimated from the compiled results. By mapping the location of the source boxes of sampled laminations, as placed in DCA102, we plan to study the correlation between collar dimensional variation and magnetic field quality.

REFERENCES

1. FNAL Drawing # 0102-ME- 292059, Revision C, SSC 50mm Dipole Collar Lamination For Vertically Split Yoke.
2. CMM Certificate of Calibration, SSCL MSD QA (Inspection & Test Group), January 1993.
3. Document Number MD-MRA-A-93-001, Requirements and Specifications for SSC 50mm Collider Dipole Magnet, Section 2.3.3.6, August 1991.

HDM MODEL MAGNET MECHANICAL BEHAVIOR WITH HIGH
MANGANESE STEEL COLLARS

J.R. Snyder

Westinghouse Science and Technology Center
1310 Beulah Road
Pittsburgh, PA 15235

BACKGROUND

Westinghouse Electric Corporation (WEC) is presently under contract to the SSCL[1] to design, develop, fabricate, and deliver superconducting dipole magnets for the High Energy Booster (HEB). As a first step toward these objectives SSCL supplied a design for short model magnets of 1.8 m in length (DSB). This design was used as a developmental tool for all phases of engineering and fabrication. Mechanical analysis of the HDM (High Energy Booster Dipole Magnets) model magnet design as specified by SSCL was performed with the following objectives: 1) to develop a thorough understanding of the design; 2) to review and verify through analytical and numerical analyses the SSCL model magnet design; 3) to identify any deficiencies that would violate design parameters specified in the HDM Design Requirements Document.

INTRODUCTION

A detailed analysis of the model magnet mechanical behavior was pursued by constructing a quarter section finite element model and solving with the ANSYS finite element code. The model, as shown in Figure 1, consists entirely of 2-D plane stress elements and gap elements placed between the bearing surfaces. All dimensions for constructing the mesh were taken from model magnet drawings provided by SSCL. An endless list of dimensional variations are possible if one considers dimensional tolerancing. For this analysis nominal dimensions were assumed for all components except for deviations in pole face shim size and the amount of stress free interference between the collar and yoke. Analysis runs were made for four conditions through magnet assembly to simulated operation. Conditions imposed at each step were carried through the subsequent steps. First: the coils were collared with an imposed azimuthal prestress of 70 MPa on the inner coil and 56 MPa on the outer coil. Second: The yoke and shell were added with an assumed azimuthal shell tension after welding of 205 MPa. Third: cooldown from room temperature to 4.25 K was simulated. The Fourth and final analysis step was to apply Lorentz forces to the conductors for energization levels up to a

Supercollider 5, Edited by P. Hale
Plenum Press, New York, 1994

field strength of 6.7 T. Figure 2 traces the average pole face stress through the four analysis steps. Other implied assumptions pertinent to the analysis were the specification of all gap elements without friction and the definition of material properties to be linear elastic and isotropic. The coils however were defined with a different elastic modulus in the radial and azimuthal directions.

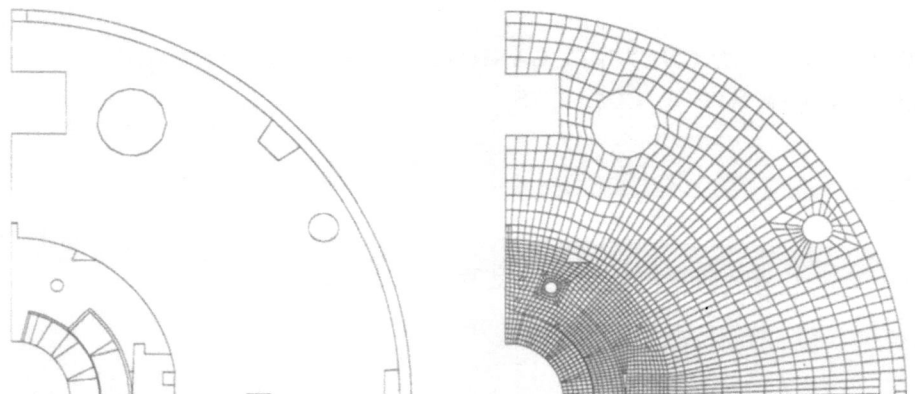

Figure 1. The finite element model is displayed with material boundaries only on the left and the element mesh on the right. The second layer of collar elements is omitted for clarity.

CONSIDERATIONS FOR HIGH-MANGANESE STEEL

Collar materials of Nitronic-40 and High-Manganese steel were both considered for the HEB model magnet program with the High-Manganese being the final selection. The primary mechanical difference in the two materials is the much lower thermal contraction of the High-Manganese steel. With this material the collars will contract less than the enclosing yoke producing an increased collar-yoke interference during cooldown. The analysis showed that if the specified nominal dimensions were used with the collar outer radius equal to the yoke inner radius then the yoke midplane gap would not close after shell welding and would remain open after cooldown and through energization to full field strength. By analyzing cases with varying gaps between the collar and yoke it was determined that the collar outer radius should be dimensioned approximately 40 microns smaller than the inner radius of the yoke in order to achieve both a line contact between the collar and yoke and midplane gap closure. A high degree of sensitivity to dimensional accuracy was observed for achieving both objectives.

STRESSES

Proceeding through the analysis steps previously outlined, the highest average pole face stress in the coils occurred at room temperature after shell welding. As illustrated in Figure 2 the azimuthal stress increased from 70 to 80 MPa on the inner coil and from 54 to 64 MPa on the outer. These magnitudes then dropped to 51 and 55 MPa respectively during the simulated cooldown analysis. No compensation was made to account for relaxation that is known to occur in the conductors and insulation. The computed pole face stress loss due to Lorentz forces followed a linear path when plotted against the square of the imposed current. It is felt that the characteristic nonlinear curves[2] typically displayed when experimental results are plotted would show up analytically if the simulation was performed with orthotropic and nonlinear coil material properties. Large variations in coil pole stress across the face resulted when small deviations in geometry and pole shim size were used. If a larger than nominal shim is required to achieve the desired average stress, the stress level will increase significantly at the inner radius and decrease at the outer radius. Azimuthal stress as high as 140 MPa was calculated at the inner edge of the inner coil when the pole face shim was oversized by 0.150 mm. An undersized pole face shim reversed the stress gradient.

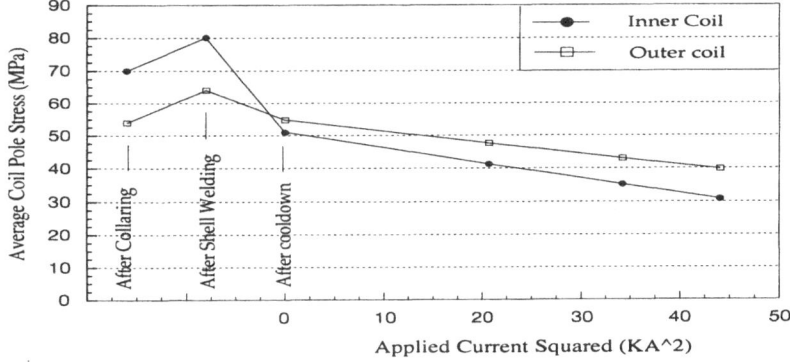

Figure 2. Average pole face stresses calculated for the conductors as the magnet was assembled, cooled to 4.25 K and then subjected to current loadings. A 25 micron stress free gap was specified between the collar and yoke.

With reference to the coil geometry depicted in Figure 1 it is easy to verify that the sharp edge of the outer coil wedge impinges on the second block of the inner coil conductors. When Lorentz forces are applied the sharp rigid point of the wedge creates a localized stress concentration in the insulation. This is not a problem for DSB, however, the repeated loading of the HDM could lead to an eventual breakdown in the insulation over the service life of the magnet. There are two notable areas of high stress that appeared in the collar analysis. One of the areas of high stress occurred in the material around the keyway corners and the other was located at the spot weld. In both cases some yielding in the collar material is indicated however it is limited to small localized areas. With the horizontally split yoke design both of these high stress conditions were greatly reduced when the yoke was added and the shell welded into place. Elevated stress at the spot weld all but disappears with line contact between the collar and yoke. A further reduction in the keyway stress to 10 MPa occurred with cooldown to 4.25 K. When Lorentz forces were applied to the model, stresses around the corners of the key area increase but only to the 60 MPa level. The azimuthal stress in the shell was assumed to be 205 MPa after shell welding. Since friction between the yoke and shell was, for the most part, ignored in this analysis the tensile stress was uniform from the midplane to the vertical axis. Upon cooldown to the operating temperature of 4.25 K the shell stress increased to a uniform 410 MPa. A primary consideration for the HDM shell stress is the possible fatigue failures due to the repeated energization to full load when the HEB is in service. The calculated cycling shell stress was low for all analysis runs with a magnitude dependence on the initial collar and yoke geometry. If excessive interference prevented the yoke midplane gap from closing then the variation in shell stress due to Lorentz forces was in the order of 5 MPa. If contact between the collar and yoke was maintained and the midplane gap remained closed then the cycling stress in the shell was 1 MPa or less.

DISPLACEMENTS

Coil collaring to the design preload values of 70 and 56 MPa resulted in a vertically ovalized coil-collar assembly by displacing the collar 0.118 mm along the vertical axis and 0.005 mm radially outward at the horizontal midplane. After shell welding to 205 MPa the ovalization was reduced 40 percent with constraint of the coil preload being shared by the shell tension and a reduced load in the keys. After cooldown the ovalization is almost entirely removed leaving a near perfect circular shape along the inner radius of the inner coil. A brief investigation into yoke-shell frictional effects[5] displayed a distortion in the yoke causing the outer edge of the yoke halves to contact first. The amount of taper to the yoke midplane gap was easily increased by 0.050 mm by simulating friction between the yoke and shell. Primary objectives of the mechanical design are to maintain prestress in the coils and to constrain the conductors to the idealized position under operating conditions. Assuming that the coil was ideally positioned after shell welding, it was found that after cooldown to 4.25 K the inside radius of the inner coil is within .002 mm of the warm position. The Yoke, which is at least an order of magnitude stiffer than the other components, tends to dominate the displacements as it contracts radially inward. The coil, by design, acts as a spring that maintains radial contact between the coils and collar and in turn between the collar and yoke. The reduction in radial dimension of the coils and collar is almost exactly offset by the inward contraction of the yoke. Conductor displacement due to Lorentz forces was found to be significantly influenced by the dimensional accuracy at the collar-yoke interface. If excessive interference between the collar and yoke prevents the midplane gap from closing then the yoke and conductors will displace 0.032 mm radially at the horizontal axis and contract inward by 0.026 mm along the vertical axis. With a more ideally dimensioned collar and yoke interface the deflections were cut in half to 0.016 mm and 0.014 mm respectively.

CONCLUSIONS

These analytical results are consistent with similar studies reported by other investigators[3][4][5][6] and to a modest extent with experimental results. It is clear that precise dimensional specification is crucial to a successful magnet and the selection of High-Manganese collars requires it's own special configuration.

REFERENCES

1. High Energy Booster Dipole Magnet Program Contract No. SSC-91-B-01705

2. A. Devred,et al., "About the Mechanics of SSC Dipole Magnet Prototypes," SSCL-Preprint-6, Fermilab-Pub 91/330, November 1991.

3. Yanping Chen, "Mechanical Analysis of the DSB corss-section," (Analysis presented to Westinghouse STC, May 1992.) IV-8, 5th IISSC.

4. J. Turner, "Results from recent ANSYS studies on the DSX201 Cross Section," SSC Laboratory Test & Analysis Note MD-TA-200, December 11, 1991.

5. Jim Strait, "Analysis of Yoke-Skin Interaction," Fermilab TS-SSC 90-040, June 28, 1990.

6. B.Wands and M.Chapman, "Finite Element Analysis of Dipole Magnets For The Superconducting Super Collider," SSCL-235, August 1989.

MULTILAYER INSULATION FOR THE INTERCONNECT REGION IN THE ACCELERATOR SYSTEM STRING TEST – A PRACTICAL ENGINEERING APPROACH FOR A NEW SCHEME OF DESIGN AND INSTALLATION BRIDGES

D. Baritchi and A. Jalloh

Superconducting Super Collider Laboratory[*]
2550 Beckleymeade Avenue
Dallas, TX 75237-3997

ABSTRACT

In order to minimize the heat leak in the Accelerator System String Test (ASST) interconnect region, shield bridges and multilayer insulation (MLI) are provided. A sliding joint between shield bridges on adjacent magnets accommodates the contraction that occurs during cooldown. In the original design of the MLI bridges, thermal contraction was provided for by compressing the MLI. During assembly of the interconnect region, it was realized that there was not enough room for the required compression. This resulted in a redesign of the MLI bridges. The new scheme involves splitting and overlapping the MLI. This scheme has worked very well in subsequent assembly of the interconnect region. In this paper, we are going to present the new design scheme. We will also compare this design with the original design and present its advantages.

INTRODUCTION

The objective of the MLI system in the interconnect region is to limit heat leak from thermal radiation and residual gas conduction to the level specified by the design criteria. The MLI must maintain this level for 20 plus years lifetime of the SSC. Essential to meeting design requirements is an insulation system design that addresses transient conditions through high layer density for improved gas conduction shielding, having also enough mass and heat capacity to restrict the effect of thermal transients. At the same time, materials suitable for use in a high radiation environment must be considered. The insulation system must have a mean apparent thermal conductivity of 0.76×10^{-6} W/cm-K in order to meet the design heat load budget. This is achieved by using an MLI system comprised of reflective layers of double aluminized Mylar (DAM) separated by layers of spunbonded polyester (Reemay). The reflective layer consists of flat polyester film (polyethylene tere phtalate PET) aluminized on both sides to a nominal thickness not less than 350 Å. The spacer material consists of randomly-oriented spunbonded polyester fiber mats. The mean apparent thermal conductivity of an MLI comprised of these materials has been measured to be 0.52×10^{-6} W/cm -K.[1-2]

---

* Operated by the Universities Research Association, Inc., for the U. S. Department of Energy under Contract No. DE-AC35-89ER40486.

The MLI system for the ASST 80 K thermal shield consists of two 32 reflective layer blanket assemblies, for a total of 64 reflective layers. The reflective layers are separated by a single layer of Reemay spunbonded polyester. The blankets are designated as the inner 80 K and the outer 80 K blanket, respectively.

The MLI system for the ASST 20 K thermal shield consists of 10 layers of double aluminized Mylar (DAM), each separated by 3 layers of Reemay spunbonded polyester.

DESIGN ANALYSIS

For the ASST dipole the MLI blankets were fabricated using a large diameter winding apparatus.[3] Using this 18 ft diameter apparatus for winding blankets, we are going to get extra material in successive layers in the length and width directions to aid thermal contraction. The first layer of MLI will have the following length:

18 [feet] × 12 [inches] × π = 678.58 [inches].

In order to calculate the length of the last layer of MLI we have to consider the fact that the diameter of the winding apparatus will be increased by the thickness of the MLI. Thus the diameter for the 80 K blanket will be:

18 [feet] + 2 × 5/8 [inches] =217.25 [inches].

The diameter for a 20 K blanket will be calculated as follows:

18 [feet] + 2 × 1/4 [inches] =216.5 [inches].

The length of the last layer of MLI will be:

217.25 [inches] × π = 682.51 [inches] for 80 K blanket,

216.50 [inches] × π = 680.15 [inches]for 20 K blanket.

In order to calculate the extra material due to the increase of the MLI thickness (occurring during winding) we have to divide the maximum diameter of the 80 K MLI by the minimum diameter of the 80 K MLI. We do the same for the 20 K MLI.

682.51 / 678.58 = 1.0057 for the 80 K MLI,

680.15 / 678.58 = 1.0023 for the 20 K MLI.

These numbers show that there is a reserve of 5.7 Mils per inch for thermal contraction for the 80 K MLI and 2.3 Mils per inch for the 20 K MLI.

The distance between two supports posts in a dipole magnet is 125 inches. In order to provide extra material for contraction, the distance between the centers of the holes cut out in the MLI blankets for the posts will be 126.5 inches. From here we are going to get another reserve for thermal contraction.

126.5 [inches] / 125 [inches] = 1.012.

The reserve is 12 Mils per inch for 80 K and also 12 Mils per inch for 20 K. Adding these two numbers gives 17.7 Mils per inch for the 80 K MLI and 14.7 Mils per inch for the 20 K MLI. We are now considering the thermal contraction reserve that occurs at a dipole magnet with 80 K MLI equal to 593 inches length and 20 K MLI equal to 595 inches length.

593 [inches] × 17.7 [Mils / inch] = 10.49 [inches] for 80 K blanket,

595 [inches] × 17.7 [Mils / inch] = 8.75 [inches] for 20 K blanket.

The thermal expansion coefficient for PET (Mylar)[4] is 6.5×10^{-5} K^{-1}.

Next we try to calculate the thermal contraction considering that the above number is a real number for the MLI (in reality the MLI is composed of PET and Reemay). Also let us consider the worst case: a connection between two dipole magnets with an interconnect region with a length of 55 inches for 80 K and 53 inches for 20 K blankets. The formula for thermal contraction is:

Thermal contraction $= \alpha \times l \times \delta T$,   where: α = coefficient of thermal expansion,

δT = change in temperature,

l = length.

The contraction for the 80 K MLI is :

6.5×10^{-5} K$^{-1} \times (593 + 27.5) \times 220$ K = 8.87 [inches].

The contraction for the 20 K MLI is :

6.5×10^{-5} K$^{-1} \times (595 + 26.5) \times 280$ K = 11.3 [inches].

SUMMARY

As we can see the thermal contraction calculated exceeds the reserve provided for the 20 K MLI. For this reason providing extra material for the contraction in the interconnect region for both the 80 K and 20 K blankets has been considered. Besides this consideration the coefficient of the thermal contraction used in the calculations covered only the double aluminized Mylar (DAM). No coefficient of thermal expansion for Reemay spunbonded polyester, the second component of MLI, could be found.

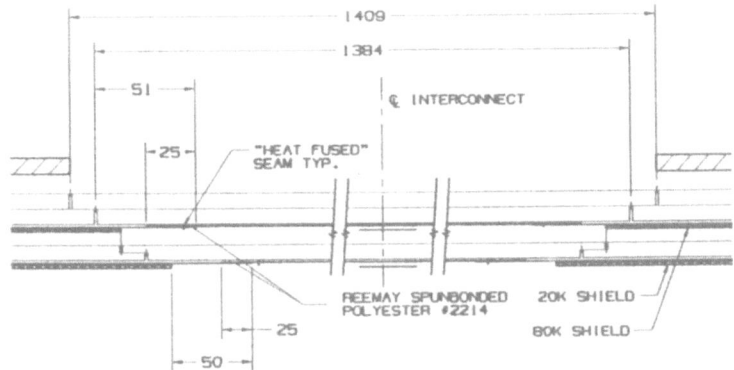

Figure 1.   Interconnect region MLI bridges (The old concept using ultrasonic welding).

In Figure 1 the first concept of the interconnect MLI bridges is shown. In this design we try to compress the MLI. We did not succeed in compressing the MLI and we had to change the conceptual design. The MLI bridge is now split and overlapped. In the first conceptual design we used an ultrasonic welding machine in order to connect magnet MLI with bridge MLI. This ultrasonic welding fuses a heavy-duty cover layer, made from the same material as the spacer (Reemay spunbonded polyester), but five times thicker. This connection can be seen in Figure 1. In the new conceptual design, the connection between magnet MLI and bridge MLI is realized with aluminized tape. With this new scheme the installation of the ASST components is much easier and less expensive. The new conceptual design is shown in Figure 2.

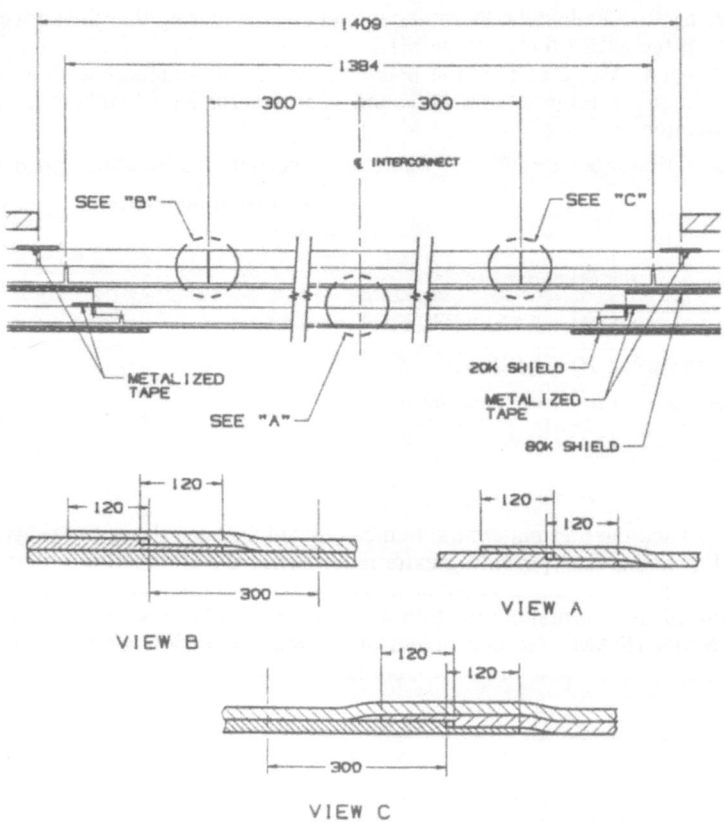

Figure 2. Interconnect region MLI bridges (The new concept with MLI split and overlapped).

REFERENCE

1. W. N. Boroski, J. D. Gonczy and R. C. Nieman, "Thermal Performance Measurement of a 100 Percent Polyester MLI System for the Superconducting Super Collider Part 1: Instrumentation and Experimental Preparation (300K–80 K)," Advance in Cryogenic Engineering, Vol. 35, Plenum Press, New York, 1990, pp. 487–496.
2. W. N. Boroski, J. D. Gonczy and R. C. Nieman, "Thermal Performance Measurement of a 100 Percent Polyester MLI System for the Superconducting Super Collider Part 1: Instrumentation and Experimental Preparation (300K–80 K)," Advance in Cryogenic Engineering, Vol. 35, Plenum Press, New York, 1990, pp. 497–506.
3. J. D. Gonczy, W. N. Boroski, and R. C. Nieman, Multilayer insulation (MLI) in the Superconducting Supercollider – a practical engineering approach to physical parameters governing MLI Thermal performances.
4. U. S. Department of Commerce, Frederick B. Dent, Secretary National Bureau of Standards, Richard W. Roberts, Director Issued September 1973: "A compilation and Evaluation of Mechanical, Thermal, and Electrical Properties of Selected Polymers."

INVESTIGATION OF THE MECHANICAL AND ELECTRICAL PROPERTIES OF SUPERCONDUCTING COILS

T. Saito[1], K. Hara[2], Y. Kojima[2], A. Mori[3], K. Nojima[4], Y. Okamoto[5],
S. Takabayashi[6], T. Tanaka[7], T. Yamagiwa[1], and K. Hosoyama[2]

1 Hitachi Works, Ibaraki, Japan
2 KEK - National Laboratory for High Energy Physics, Tsukuba, Ibaraki, Japan
3 SHINKO Chemical Industrial Co., Ltd. , Fukui, Japan
4 Arisawa Mfg. Co., Niigata, Japan
5 KANEKA Corporation, Shiga, Japan
6 UBE Industrial Co., Ltd., Oosaka, Japan
7 DuPont-TORAY, Tokai, Japan

INTRODUCTION

Measurement of elastic(Young's) modulus of the superconducting coil and electrical punch-through test have been performed at LBL[1] to understand the mechanical and electrical properties of the superconducting coils. We have investigated the elastic modulus of the superconducting coils with six kinds of insulators (made with polyimide-fiberglass-epoxy and all polyimide insulation with epoxy/polyimide adhesive) at room and liquid nitrogen temperatures using samples made of 10 stack of superconducting cables. The samples are cured under varying compression to investigate the curing pressure dependence of Young's modulus of the coils with six kinds of the insulation system. The electrical punch-through test has also performed under compression at room and liquid nitrogen temperatures to investigate electrical integrity of the insulated coils. The tensile strength test of four kinds of polyimide films has been performed at various temperatures (between cryogenic and coil curing temperatures) to understand the mechanical properties of the films.

Table 1. Insulation system and dimensions of the superconducting cables

No.	Materials	thickness [mm] inner layre	outer layre	adhesive type	contents of resin (wt%)	sample cure condition temp. (°C)	time (min.)	pressure (MPa)	dimensions of superconducting wires (for 10 stack) [mm]
1	KAPTON CI	0.025	0.025	—	33	—	—	50,70, 100	
2	UPILEX-S	0.025	0.05	polyi-mide	25	160	60	50,100, 150	
3	UPILEX-R	0.025	0.05	epoxy	26	150	300	50,100, 150	
4	UPILEX-R	0.025	0.05	epoxy	26	150	300	50,100, 150	
5	In:UPILEX-R Out: Fiber Grass	0.025	0.08	epoxy	40	130	90	50,100, 150	
6	Apical NPI	0.025	0.05	epoxy	26	150	300	50,100, 150	

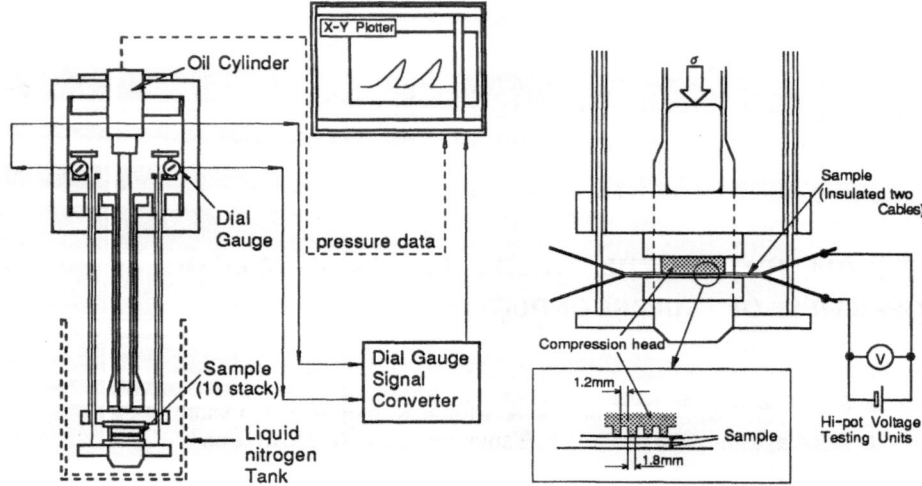

Figure 1. Measurement system for Young's modulus of the 10 stacks.

Figure 2. Measurement set up of electric punch-through test and compression head.

YOUNG'S MODULUS AND PUNCH-THROUGH TEST

The principal parameters of the insulation system and dimensions of the superconducting cables used in the test are tabulated in Table 1. The cables are insulated by helical double 50% overlap winding of 0.025mm thick tape for inner layer and single gap winding of 0.05mm thick pre-impregnated tape for outer. For sample #1 helical double 50% overlap winding of 0.025mm thick pre-impregnated tape is used for outer layer instead of single gap winding. The sample made of 10 stacks of cables are cured at specified temperature listed in Table 1 and under varying compression up to 150MPa. Figure 1 shows a equipment used for measurements of Young's modulus. After three cycles of up and down ramping between 0 and 100MPa., we evaluated Young's modulus of the samples from gradient of up ramping stress-strain curves at 80MPa. Figure 3 shows curing pressure dependence of Young's modulus for the samples at room and liquid nitrogen temperatures. The measured values in Fig.3 are the average of five samples. The values of Young's modulus increase with curing pressure and their dependences seemed to reflect the mechanical property of the insulator tapes. The values of Young's modulus measured at liquid nitrogen temperature are 2 - 6MPa. higher than those at room temperature. The measured values of

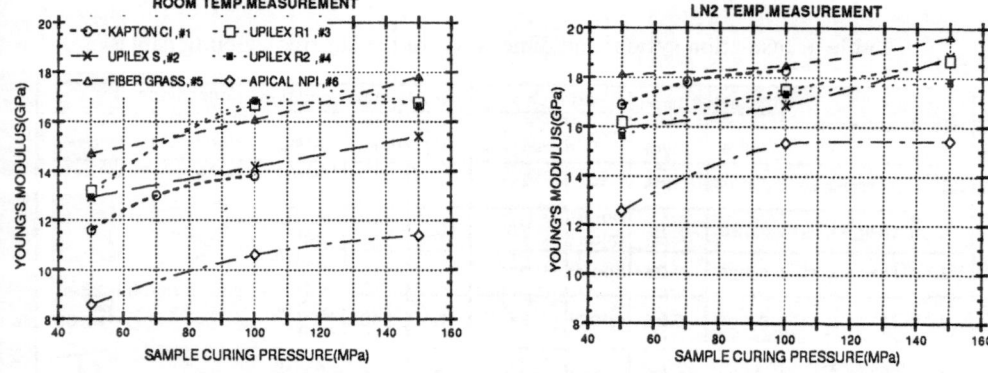

Figure 3. Young's modulus of 10 stack cables.

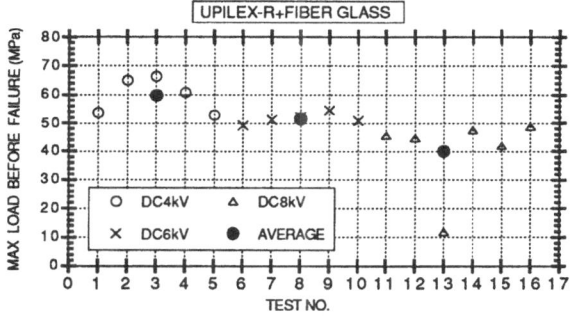

Figure 4. Result of punch-through test of sample #5.

sample #6 shows about 10-20% lower than others. We could explain this by taking account of thickness of the cables.

We performed the electrical punch-through test of samples #1 - #6 in atmospheric environment and liquid nitrogen. Figure 2 shows set up of the test and detail of compression head which simulate the collar laminations. We did not observe the electric breakdown in all samples except for polyimid-fiberglass-epoxy sample #5 under test condition of DC 10kV and compression up to 650MPa. Figure 4 shows the result of electric breakdown test at room temperature for sample #5 .

TENSILE STRENGTH TEST OF POLYIMIDE FILMS

Figure 5 shows a set up of measurement system for tensile strength. This device was specially designed and constructed at KEK to meet requirements for the tensile strength test of insulation tape between cryogenic and curing temperatures. We measured the stress-strain curves of the polyimide films with dimension of 10mm x 80mm under the cross head speed of 12mm/min. between the coil operation temperature of 4.2K and the coil curing temperatures of 473K every 50K. Figure 6 shows stress-strain curves of sample #6 for

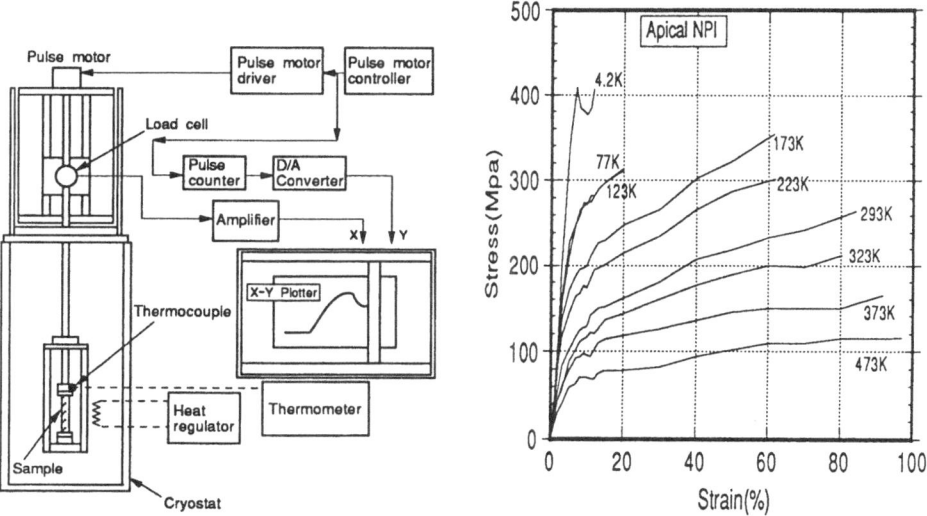

Figure 5. Measurement system for tensile **Figure 6.** Stress - strain curves of sample #6.
strength of insulation tape.

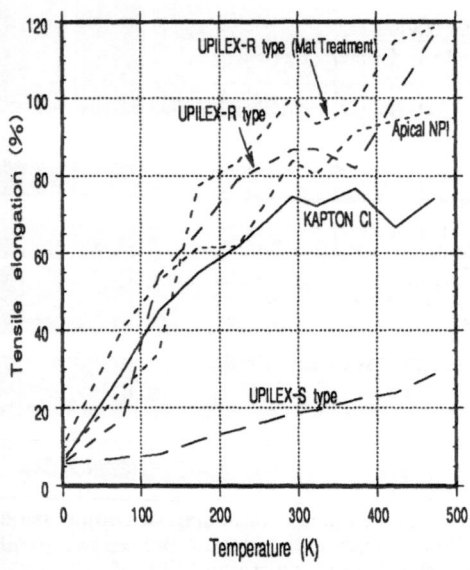

Figure 7. Temperature -Tensile elongation curve.

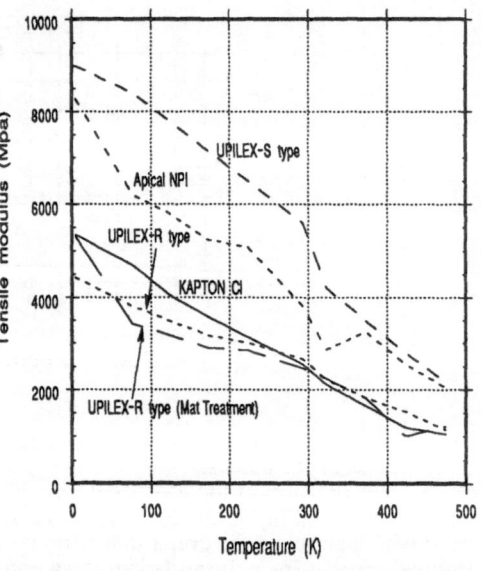

Figure 8. Temperature - Tensile modulus curve.

typical example. Figures 7, 8 show temperature dependence of the tensile elongation and tensile modulus of the tapes, respectively. At liquid helium temperature the tensile elongation of all samples except UPILEX-S® are 5-9% of those at room temperature. The tensile elongation of UPILEX-S®tape is extremely smaller than the others, but this does not change drastically even at liquid helium temperature.

CONCLUSION

We have investigated the mechanical properties of the insulated coils at room and liquid nitrogen temperatures by the measurement of 10 stack of cables with 6 kinds of insulation system which are candidate for SSC magnet insulation system. We also measured temperature dependence of tensile strength and tensile elongation of the insulation tapes used in the insulation system at temperature between 4.2K and 473K to understand the mechanical properties of the coil. Recently we started investigation of irradiation effect on the polyimide films by exposing 2MeV electron beam and stress relaxation of the polyimide insulation films.

ACKNOWLEDGMENTS

The authors express their gratitude to Professor Y. Kimura for his continuous support and encouragement. They also wish to thank to Mr. E. Willen and Dr. R. Coombes for their helpful advice and useful disucussions and Mr. U. Hanai , Mr. T. Onuma and Mr.M. Maeda.

REFERENCES

1. J.Zbasnik and C.Peters,"Conductor Bundle Mechanical Measurements, Coil Size/Prestress Effects and Insulation Test," Proc. SSC MAGNET INDUSTRIALIZATION PROGRAM Phase 1 Technology Orientation Meeting, 1989.

A STATUS REPORT ON THE DEVELOPMENT OF 5-CM APERTURE, 1-M LONG SSC DIPOLE MAGNET AT KEK

K. Hosoyama,[1] K. Hara,[1] N. Higahi,[1] A. Kabe,[1] H. Kawamata,[1] Y. Kojima,[1]
Y. Morita,[1] H. Nakai,[1] A. Terashima,[1] H. Fuse,[2] Y. Imai,[3] H. Morita,[4]
T. Takahashi,[5] A. Yamanishi,[6] T. Shintomi,[1] H. Hirabayashi,[1] and Y. Kimura[1]

[1]KEK-National Laboratory for High Energy Physics,Tsukuba, Ibaraki, Japan
[2]Nippon Kokan K.K., Kawasaki, Japan
[3]Mitsubishi Electric Corporation, Kobe, Japan
[4]Hitachi Work, Hitachi, Ltd., Hitachi, Japan
[5]Mitsubishi Heavy Industries, Ltd., Kobe, Japan
[6]Ishikawajima-Harima Heavy Industries Co., Ltd., Yokohama, Japan

INTRODUCTION

The design and construction of a series of eleven 5-cm aperture, 1-m long SSC model dipole magnets KEK#1-#11 was started in May 1990. We have paid much attention to the design of magnet end part and developed new-type end spacers[1] which have constant perimeter profiles to minimize internal stress of the cables at the coil end. Except for length, these short magnets have the same baseline design and features as the long magnet. We have performed the quench test, ramp rate quench dependence, and AC loss measurement in 1.8K superfluid helium cryostat.[3]

MAGNET DESIGN AND CONSTRUCTION

The basic design of the magnet cross section is W6733[2] which has a two-layer $\cos\theta$ coil geometry with 19 turns in inner coil and 26 turns in the outer. In this design 6 μm NbTi filament cables with 30 strands for inner coil and 36 in the outer are used[1]. The magnets KEK#3, #5 are fabricated with 2.5μm filament conductors to meet the requirement of HEB fast ramp cycles. The cable is insulated with 25μm thick double half wrapped tape and 50μm thick B-stage epoxy impregnated polyimide tapes Upilex-R®. The inner coil is wound on the mandrel and cured at 150 °C for 5 hours in the mold. The outer coil is soldered at the splice part and then wound directory on the cured inner coil and cured under the same curing conditions described above. In the case of the magnets KEK #10 and #11 inner and outer coils are wound and cured individually and then soldered at splice part. The entire coil is clamped firmly in place by a collar stacks constructed of 1.5mm thick YUS130S® stainless steel laminations. Beam type strain gauge transducers are installed in the collar pack to measure coil azimuthal prestress during magnet assembly and testing. The prestress at straight section is controlled to be 90MPa for inner coil and 70MPa for outer coil during the collaring process by using pole brass shims. Horizontally split yoke made of 1.5mm thick low carbon steel laminations is used to enhance the magnetic field and further mechanical stability. The yoke modules at the ends have stainless steel voids to lower the magnetic field there.

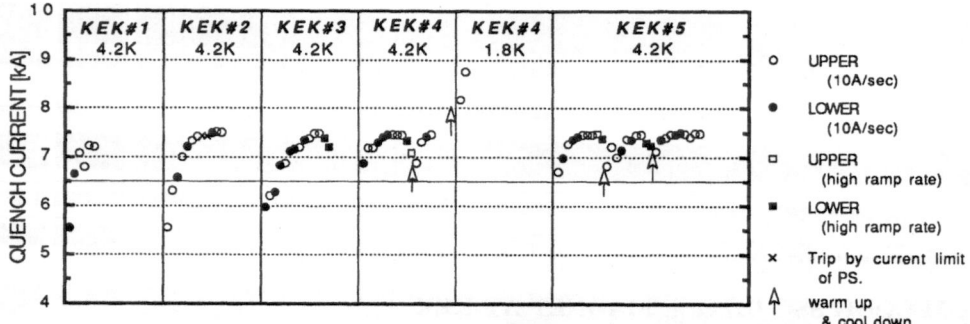

Figure 1. Quench histories of KEK#1 - #5.

EXPERIMENTAL TECHNIQUES

Cryogenic testing of the magnets was performed in vertical cryostats with pool boiling liquid helium at 4.2K and pressurized superfluid helium at 1.8K. The typical test procedure of the magnets consisted of training by ramping up the magnet current at about 10 A/s until a quench was generated and performing this until the quench currents had reached a quench plateau where the current can reach about 7500 A. In order to test the limits of the mechanical performance of the magnet under the stress generated by higher magnetic fields the training tests were also performed in pressurized superfluid helium at 1.7K where the quench current can exceed 9000A. We performed the high ramp rate quench test and AC loss measurement[3] by calorimetric method in superfluid helium at varying ramp rates up to 300 A/s. Quench origin location and velocity of propagation were determined by the resistive signals between a pair of voltage taps surrounding the quench hot spot.

QUENCH TEST AND DESIGN CHANGE

The quench history plots of ten magnets from KEK#1 to #10 are shown in Figs.1,2,3. For the first three magnets of the series the first quench currents are lower than 6000A and it require a few training quenches to reach the design operating current of 6600A and several more quenches to the plateau at 7500A. The quench origins of these low current quenches are located at the splice part due to insufficient clamping force. We have adopted newly designed collars to increase the prestress at the splice part for the series of magnets after #4. The prestress at the splice part is also monitored with the same type strain gauge transducer

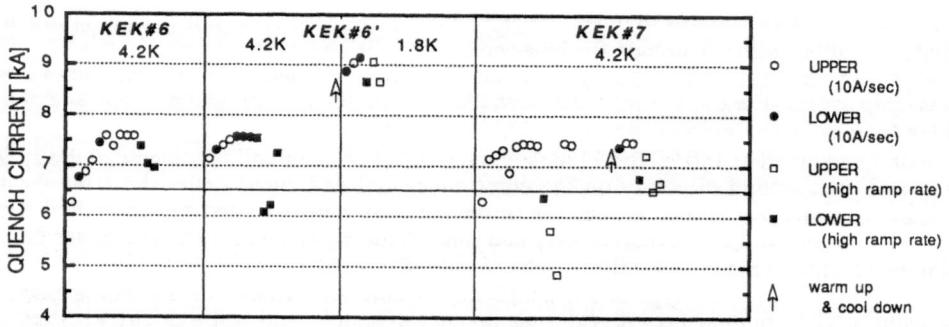

Figure 2. Quench histories of KEK#6 - #7.

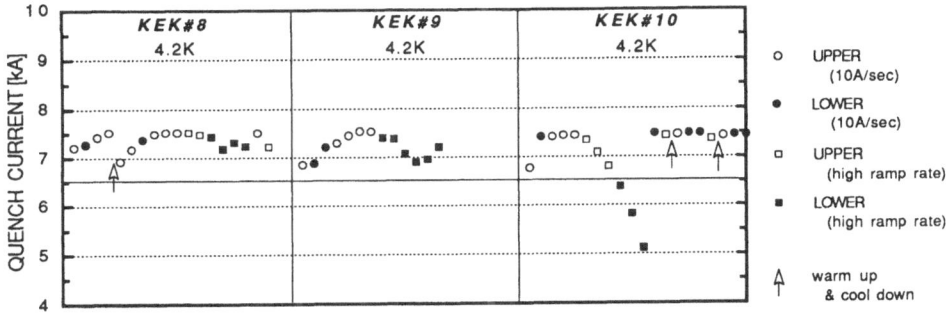

Figure 3. Quench histories of KEK#8 - #10.

used at the straight part, and controlled to be about 80MPa by shim. By this modification we could eliminate the low current quench at the splice part, but several more quenches are required to the plateau. In this case quenches occurred at the starting point of outer coil winding. After the quench test of KEK#6, we disassembled and inspected the weak part of the magnet very carefully and found a pop-up of a strand at the sharp bending corner of the first turn of outer coil. In order to increase the mechanical stability we soldered this part in KEK#6' and #7 (#6' is reassembled magnet of #6). By this modification the quench performance was improved as shown in a quench history plot of KEK#6' in Fig.2. In KEK#8, #9 we adopted a new design of splice to outer coil path which have a smooth curve with constant perimeter surface. Figure 4 shows various designs of the coil end part adopted in our magnets ; (a) for KEK#1 - #3, (b) for KEK#4 - #7, (c) for KEK#8, #9, and (d) for KEK#10,#11. In KEK 8, #9 the bending corner of the first turn of outer coil was not mechanically stabilized by soldering. In the magnets KEK#10, #11 we adopted the splice design developed by BNL, and the quench performance was improved greatly. After the first quench at 7800A this magnet reached the plateau. The quench origin was firstly located at the transition part between end and straight in the outer coil, and then moved to a pole turn straight part after two quenches and remained staying.

RAMP RATE STUDIES

The ramp rate-dependent quench test was performed at varying ramp rates up to 300A/s. The results for eight magnets are summarized in Fig. 5.   The magnets KEK#3, #4,

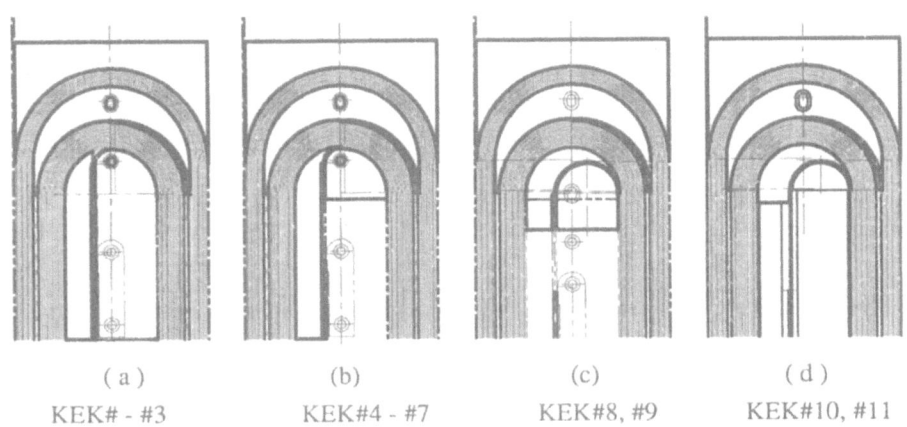

| (a) | (b) | (c) | (d) |
| KEK# - #3 | KEK#4 - #7 | KEK#8, #9 | KEK#10, #11 |

Figure 4. Coil end design.

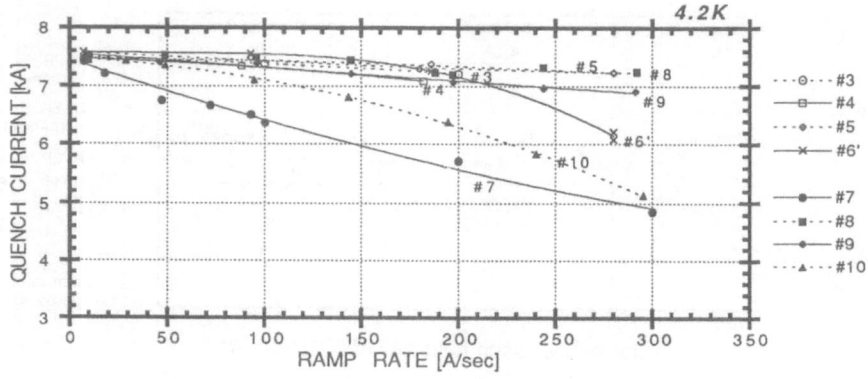

Figure 5. Ramp rate dependence of quench current.

#5, #8, #9 did not show larger ramp rate dependence than those #6', #7, #10 did. In magnets KEK#6' and #7 quenches at low ramp rate were in the ramp section, quenches at higher ramp rate were in the soldered part of bending corner of the outer coil due to the eddy current heating there. The magnet KEK#10 which has the same splice design as DAS207 and DAS208 constructed at BNL showed rather large ramp rate dependence. Quenches at high ramp rate were inside the G-10 box where the cable has been solder-filled to increase the mechanical stability. This ramp rate quench behavior is very similar to those of DSA207 and DSA208.[4]

CONCLUSION

A series of eleven 5-cm aperture, 1-m long SSC model dipole magnets have been constructed and tested at KEK. In the first magnets of the series, quenches at low excitation current are found to originate at the splice part due to insufficient clamping force. We could overcome these quenches by increasing the azimuthal prestress to the splice part with modified collar. The quench characteristics were further improved by adoption of new splice design and the test result was very promising. The test results of ramp rate dependence of quench tell us the importance of further investigation for construction of HEB dipole magnet.

ACKNOWLEDGMENTS

The authors express their gratitude to Professor H.Sugawara for his continuous support and encouragement. They also wish to thank Dr. E.Willen and Dr. R.Coombes for their helpful advice and useful discussions.

REFERENCES

1. K.Hosoyama, et.al., "Fabrication and test of 5-cm aperture 1-m long SSC collider dipole magnet," in Supercollider 3, John Nonte, ed., Plenum Press, New York p615, (1991).
2. R.C. Gupta, S.A. Kahn, and G.H. Morgan, "SSC 50 mm dipole cross section," in : Supercollider 3, John Nonte,ed., Plenum Press, New York and London, p587, (1991).
3. Y.Morita, et.al., "Calorimetric AC loss measurement of 1-m long model dipole magnet by using a 1.8 K cryostat," to be presented at IISSC, San Francisco (1993)
4. J.F. Muratore, et.al., "Construction and test results from 1.8 m-long, 50 mm aperture SSC model dipoles," in Supercollider 4, John Nonte, ed. ,Plenum Press, New York, p559 (1992)

EXPERIMENTS IN DEVELOPMENT OF CRYOGENIC STRUCTURAL MATERIALS FOR ACCELERATOR (SSC) SUPERCONDUCTING MAGNET

Kiyohiko Nohara[1], Shinji Sato[2], Takehito Nomura[2],
Rinzo Tachibana[3], Takeo Higashino[3],
Kagao Okumura[2], and Hiroshi Sasaki[2]

[1]Technical Research Division
[2]Steel Technology Division
[3]Mizushima Works
Kawasaki Steel Corporation
1, Kawasaki-cho, chuoh-ku, Chiba 260, Japan

ABSTRACT

As the candidates of structural materials for a collar and a yoke of accelerator superconducting magnets,both non-magnetic high Mn steel (KHMN) and ferro-magnetic ultra-low C steel(EFE) have been developed.The current investigation was aimed to study 1) fine-blanking and magnet error component with 1.5 and 2.5mm thick KHMN, and 2) compatibility between strength and coercive force with 6 and 1.5mm thick EFE.The positive results were obtained in each subject.

INTRODUCTION

The structural materials of non-magnetic KHMN and ferro-magnetic EFE owe the chemical compositions given in Table1. The former is featured by very high Mn content and the latter by extensively low C content to meet their requirements,respectively. [1-4] Some usabilities of both materials were pursued.

HIGH MN NON-MAGNETIC COLLAR MATERIAL (KHMN)

We tried to measure the flatness of 1.5mm thick collar materials at each process on fine-blanking procedures in collaboration from KEK and Akita Co., Ltd. Figure 1 shows the results of flatness change measurements in head(H),middle(M),and tail(T) of the coil. The measurements were conducted at six different processes. It is noted that the flatness is markedly reduced by tension levelling and reform levelling. The flatness of the initial coil in cross (transverse) to rolling direction is less than 0.20mm, and that of final products is as small as below 0.1mm except for the coil location H where the setting of the operation conditions was likely less optimized.

Photograph 1 shows the micrographs of cross sectional areas after the above mentioned

Table 1. Examples of non-magnetic steel for a collar (high Mn steel, KHMN) and ferro-magnetic steel for a yoke (ultra-low C steel, EFE) in chemical compositions (wt%).

Material		Examples	C	N	Si	Mn	P	S	Al	O	Cr	Ni	V	Ca
Non-magnetic steel (High Mn steel)	KHMN (Collar)	A	0.107	0.1037	0.71	27.75	0.032	0.003	0.020	0.050	7.13	1.01	0.063	0.0043
		B	0.106	0.0998	0.61	28.40	0.033	0.002	0.022	0.056	7.10	1.04	0.063	0.0047
Ferro-magnetic steel (Ultra-low C steel)	EFE (Yoke)	A	0.0011	0.0015	0.003	0.12	0.007	0.0060	0.003	0.0076		—		
		B	0.0010	0.0025	Tr	0.051	0.007	0.0040	0.001	0.0112				

Table 2. Contribution of impurity elements and grain size to yield stress and magnetic characteristics (induction B and coercive force Hc) of (ultra) low carbon steel (EFE).

	Factors	Detailed items	Related actions in actual process
1.	Composition	① Impurity element (C, N, Si, Mn, P, S, Al, O) ①′ Additional minor elements ② Precipitation, segregation ③ Inclusion, oxide	(1) Selection of raw material (2) Steel making/degassing (3) Continuous casting (electro-magnetic stirring)
2.	Micro-structure	① Metallographical texture (Grain size, grain shape) ② Crystallographical texture ③ Internal strain	(1) Slab heat; FET, FDT, CT (2) Continuous annealing (2) Strain annealing (3) Temper rolling, skinpassing

 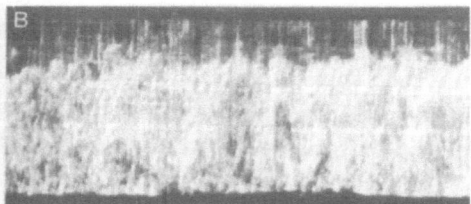

Photo 1. A) Micrographs of cross sectional area of KHMN after fine-blanking or B) conventional stamping.

fine-blanking and conventional stamping. They give a clear-cut difference in the ratio of sheared and fractured surfaces between them. This also means each different amount of residual stresses.

It is possible that 2.5mm thick collar is used instead of 1.5mm for total cost saving. In this case the fine-blanking followed by press-fitting would be desired rather than the conventional stamping. The relation between pull force after fine-blanking / press-fitting and semi-perforation(5mm dia.)height is illustrated in Fig. 2 with interference as a variable. The

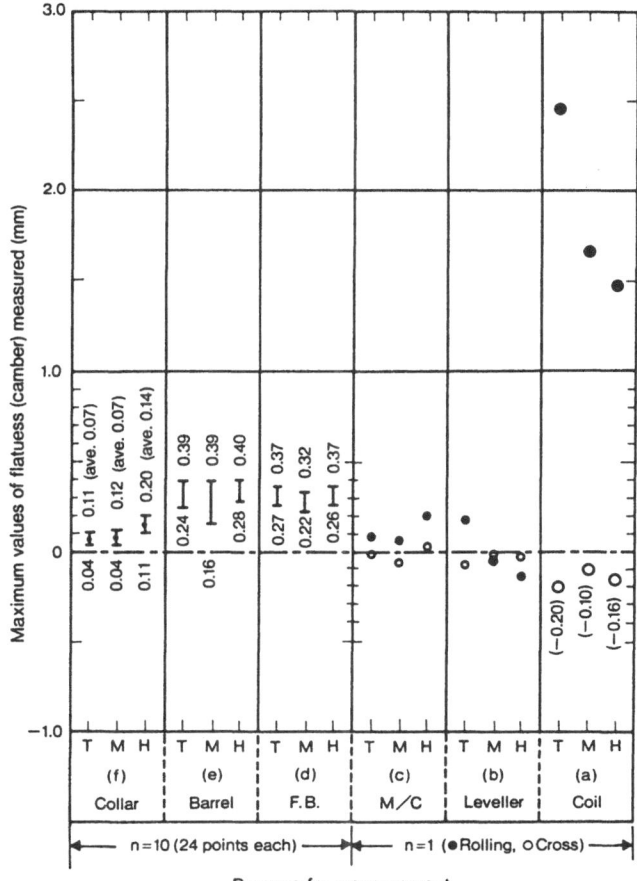

Figure 1. Result of flatness measurement at each process of fine-blanking of a KHMN coil for collar fabrication.

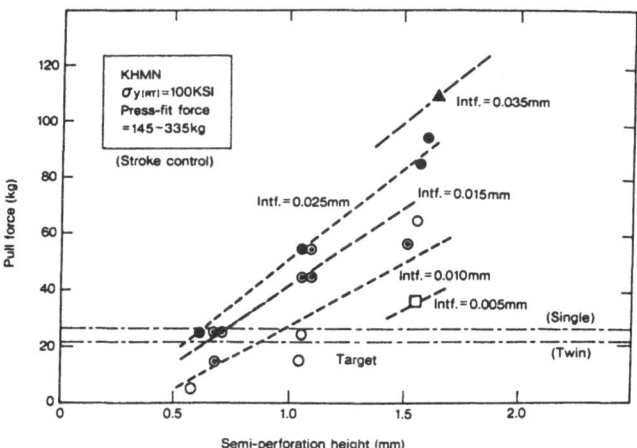

Figure 2. Change of pull force with semiperforation height of 2.5mm thick press-fitted KHMN.

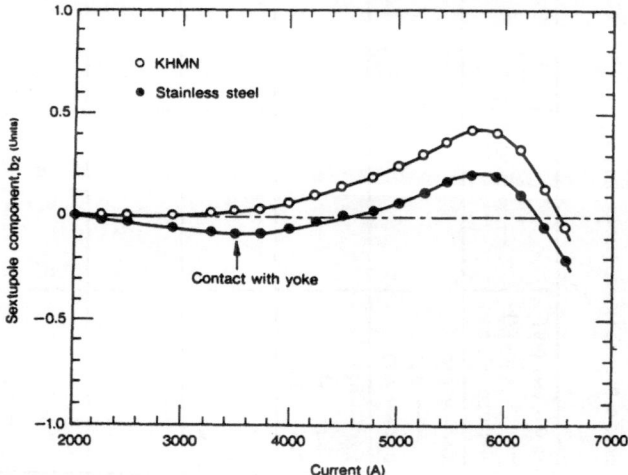

Figure 3. Change of sextupole error component with current in a KHMN collar dipole magnet in comparison with a stainless steel collar magnet to show a smooth increase in the former and a decrease followed by an increase in the latter.

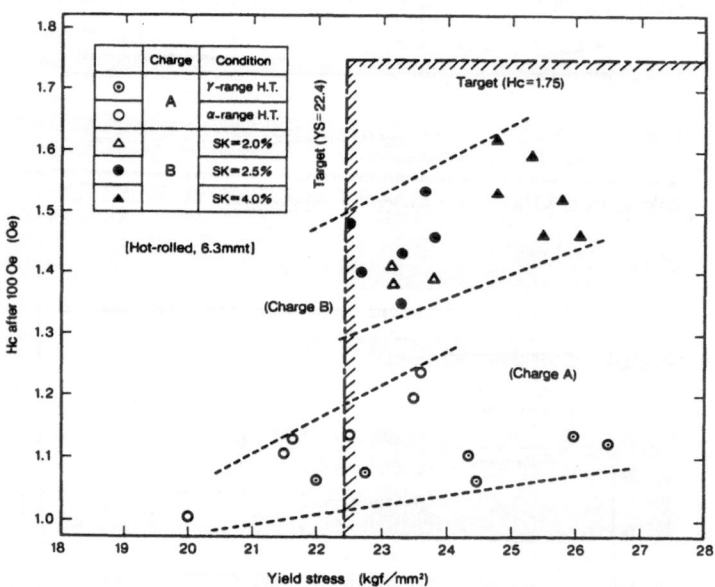

Figure 4. Relation between coercive force (Hc) and yield stress at room temperature of 6.3mm thick hot-rolled EFE fabricated with various conditions described.

linearity can be seen to meet the specified values of pull force. The result is possibly due to the increased amount of contact area between a semi-perforation and a hole.

A short SSC dipole magnet with re-collared KHMN replaced to austenitic stainless steel was employed to the measurement of magnetic performances within a bore tube. The variation tendency of the normal sextupole component(b_2) with current in Fig. 7 was found to be different in KHMN and austenitic stainless steel. While the former varies smoothly,the latter shows an inflexion point at the middle stage of excitation which appears to be related to the contact of a collar with a yoke after the formation of a gap between them during cool down. KHMN can be free from such inflexional behavior thanks to its smaller expansivity.

ULTRA-LOW C FERRO-MAGNETIC YOKE MATERIAL (EFE)

The RHIC project of BNL and the main injector project of FNAL require 6mm thick hot-rolled and 1.5mm thick cold-rolled ultra-low C steel (EFE) , respectively, which shall have both high strength and low coercive force to ensure the reliable tooling and the stable operation.

The yoke material for the RHIC project, 6mm thick, shall be specified by yield strength (σ y) beyond 32KSI, and coercive force (Hc) below 1.75Oe at room temperature each. Figure 4 is Hc- σ y relation with the specified target zone. It can be seen that the increase of σ y invites the degradation of Hc. Nevertheless, the steel as hot-rolled with no annealing followed can meet the current specifications. It is interesting that the experimental plots differ depending on charge A or B (Table 1) in terms of Hc. This is principally caused by N and O contents.

The yoke material for the main injector project,1.5mm thick, shall be specified by hardness (HRB) between 20 and 50, and coercive force (Hc) below 1.0 at room temperature each. Figure 5 is Hc-HRB relation associated with the specified target area. It can be seen that the increase of HRB introduces the worsening of Hc. However, the steel as cold-rolled product can meet the present specifications with not necessarily enough margins. The plots do not so largely differ depending on charge A or B unlike 6mm thick coil.

As stated in the foregoing experiments on EFE, it is noted that the mechanical and magnetic properties are dominated by composition and micro-structure. The detailed influential metallurgical factors and their related manufacturing conditions in actual plant processes are comprehensibly tabulated in Table 2. Making both strength and coercive force get compatible was a current concern to be successfully materialized by the reduction of impurity elements, micro-structural control, and internal temper strain hardening.

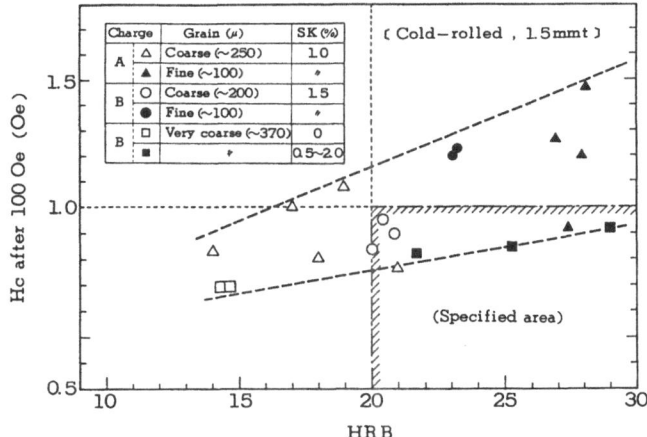

Figure 5. Relation between coercive force (Hc) and the Rockwell hardness (HRB) of 1.5m thick cold-rolled EFE with various conditions mentioned.

CONCLUSIONS

(1) The experiments on usability of non-magnetic high Mn steel (KHMN) for the accelerator magnet collar and the experiments on characteristics of ultra-low C steel (EFE) for the accelerator magnet yoke were carried out.

(2) The fine-blanking of KHMN resulted in the flatness changes in each process, the pull force controllability after press-fitting with semi-perforation height and interference, and the micro-scopic difference in the cross section from the conventional stamping. The sextupole error component of KHMN changed differently from stainless steel during the excitation .

(3) The compatibility of strength and coercive force of both hot-rolled and cold-rolled EFE was able to be met in the actual specifications. This was attained by the suitable control of metallurgical factors such as impurity elements, micro-structure, and internal strain.

---

*The flatness experiment of 1.5mm thick KHMN was carried out in collaboration with KEK and Akita Co.,Ltd.,Japan.

*The press-fit experiment of 2.5mm thick KHMN was performed in cooperation with Amada Metrecs Co.,Ltd.,Japan.

REFERENCES

1. K. Nohara, et al. "Supercollider 2," Plenum Press (1990), p. 765.
2. K. Nohara, et al. "Supercollider 3," Plenum Press (1991), p. 75.
3. K. Nohara, et al. "Supercollider 4," Plenum Press (1992), p. 1055.
4. K. Nohara, et al. "Supercollider 4," Plenum Press (1992), p. 1065.

INDUSTRIAL MANUFACTURING OF MODEL COLLIDER QUADRUPOLE MAGNETS FOR THE SSC PROGRAM

Robert F. Ryan, Jr., John S. Oblinger, and William F. Bensiek

P.O. Box 785
Lynchburg, Virginia 24505

ABSTRACT

In order to fabricate production quadrupole magnets it is first necessary to prove out the magnet design and test all the procedures and equipment to be used in its fabrication. Model magnets are needed to validate basic design parameters and inexpensively demonstrate B&W's ability to fabricate and test functioning magnets. B&W made four (4) short model magnets and is in the process of making four (4) full-length models. Model magnets were fabricated and tested. Interim conclusions are drawn concerning acceptability of the magnet design, tooling and procedures for an industrialized production of the Collider Quadrupole Magnets.

PURPOSE

The purpose of the model magnet program was to prove that B&W could design and fabricate magnets which met the SSCL quench performance goals. The B&W superconducting quadrupole magnet design was to be based on the Lawrence Berkeley Lab (LBL) quadrupole design. B&W was required to verify the LBL design and make changes as they deemed necessary. Early LBL magnet quench behavior was inconsistent. SSCL considered this a significant unacceptable risk and therefore requested B&W to perform a model magnet program to assess magnet quench behavior. B&W proposed making 5 short models to validate basic design and manufacturing concepts. Four (4) long models were then to be made to verify design concepts and demonstrate manufacturing and testing techniques on full length magnets. All this hardware was to be fabricated and tested prior to prototype magnet fabrication.

B&W reviewed the LBL design and concluded that some changes should be made: inter coil splices were eliminated, cold mass utilized a sliding coil to yoke interface, 2 piece collars were used in lieu of collar packs, collar keys were used as sliding interface. The model magnet program was to be used to demonstrate that these changes yielded a fabricable magnet with acceptable quench behavior. To assess performance of the models instrumentation was used to monitor coil prestress, and to monitor end force during cold testing. Voltage taps were used to aid in determination of quench sites during cold testing.

The model program was also to provide B&W an early look at field quality and to further assist in minimizing the costs for materials and processes associated with fabrication of acceptable quadrupole magnets. SSC had agreed to provide magnetic measurement equipment as Lab Furnished Equipment (LFE) to support testing of model magnets. This would give B&W early insight into field quality and another chance to modify the design prior to start of production fabrication.

METHODS

Manufacturing had to test out fabrication processes and procedures on dummy hardware in advance of the actual model magnets. The following methods were used: welds were tested on pipe, trial assembly and disassembly of production hardware were used to confirm fit and alignment. A dummy cold mass was fabricated to test alignment and welding of production quality components.

Assembly and fit of steel laminations, cold mass shells, weld backing strips, expansion flanges, and end rings were tested using the mockup. Using the model magnet routing document procedures and tooling were tested.. The mockup was assembled and compressed in a special press modified for model fabrication. The longitudinal seam was welded by hand as was done on the short models. Upon completion of welding the alignment of the key slot in the·bottom of the laminations were checked using oversize key stock to insure welding did not distort the plane for the key. Expansion flanges (end rings) were welded to test the orbital machine weld machine, planned weld parameters and means of support and alignment of the expansion flange with the cold mass. The dummy cold mass is shown in Figure 1.

All parts were subjected to extensive dimensional and visual inspection to insure the components performed as predicted. Parts were formally released, greatly mitigating risks associated with committing production hardware to untested equipment and procedures. A photographic journal documented the development efforts and has been useful in subsequent production problem solving.

ACCOMPLISHMENTS TO DATE

B&W has completed fabrication and testing of the first three (3) short model magnets. We are currently assembling and testing the fourth. All cold mass fabrication and testing were done at the B&W facility. Collared coils were fabricated at Siemens. Shown in Figure 2 is a graph of the quench curves for the first thermal cycle for QSH-801, 802, and 803. Also shown, is the first thermal cycle for the collared coil for QSH-801 which was tested as a coil before being assembled into a cold mass. The graphs show that in all cases the first quench was above normal operating current and the magnets quickly progressed to plateau. Further detail is available in the reference 1.

Preliminary field quality data on two of the magnets are available at this time. Early indications are that design changes will be required to trim the b_5 multipole.

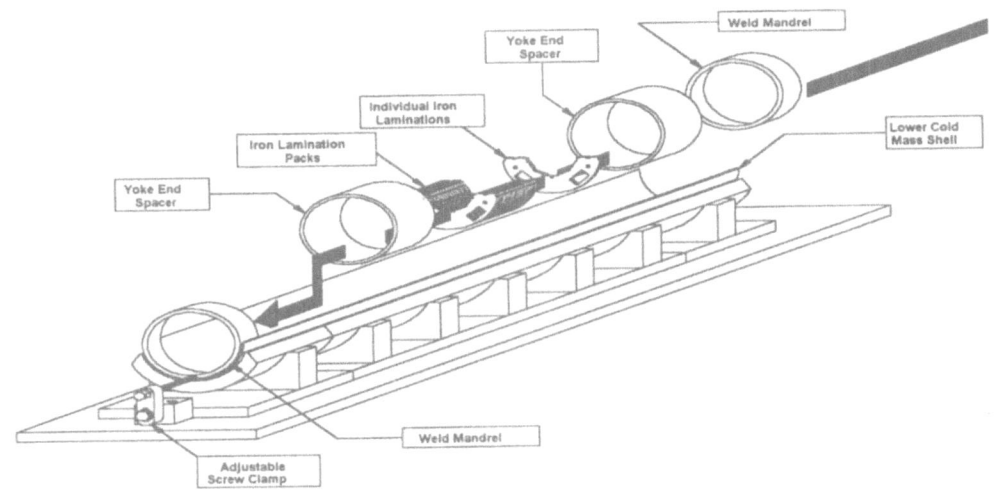

Figure 1. CQM Cold Mass Manufacturing Mockup

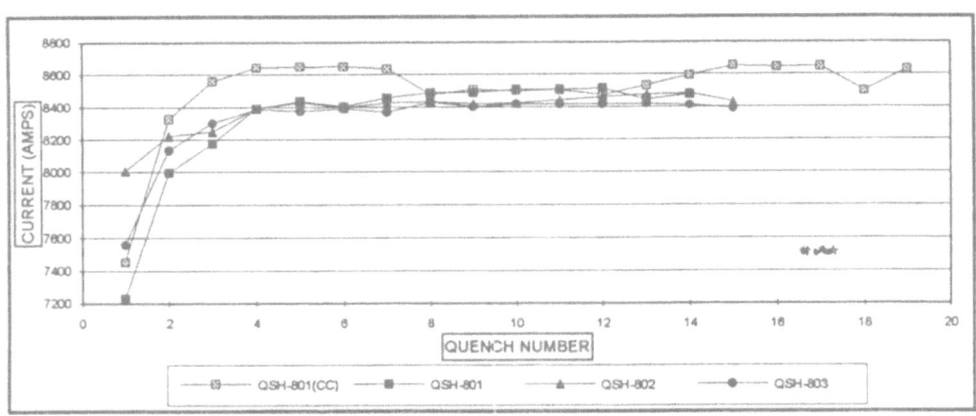

Figure 2. B&W Quadrupole Model Magnet Quench Data

The model program demonstrated the close cooperation necessary between B&W and Siemens and within the various organizations of B&W itself. The model program has demonstrated B&W's capability to utilize orbital machine welding on production components and to maintain registration of all the cold mass parts during the weld process. The model program has also allowed B&W time to develop low cost component designs and to determine lowest cost means of fabrication while developing a low cost vendor base.

OUTLOOK

In the near future we plan to perform field quality inspection on the short models which will provide initial multipole data and to test out the field survey data collection system. B&W will start fabrication of the four (4) long model cold masses in June of '93. We are planning to use production equipment in fabrication of the long models. Length effects such as midplane alignment during welding, interference between the yoke iron and the collared coil due to coil twist, component handling problems, and field gradient will be assessed. We will perform cold mass quench testing and field quality evaluation in our vertical dewar facility. As with all manufacturing results, the data will be used by Design Engineering to iterate the design to better control multipoles and other design parameters. The iterated design will be incorporated in the prototype phase. We will use the SSCL provided alignment system to locate the field centerlines and relate them to external fiducials. QCH-801 will be fully cryostated as a manufacturing trial to test cryostat and vacuum vessel fit-up. Trial vacuum testing may be performed pending equipment availability. Cryostat assembly tooling and procedures will be proof tested.

CONCLUSIONS

The model program has achieved some very significant milestones: B&W demonstrated significant quench margin and consistent quench behavior. B&W demonstrated the ability to manufacture CQM's with a slip fit between the collared coil and the yoke. B&W demonstrated that collared coils could be made at one facility and the cold mass completed at a second facility without magnet degradation. Most importantly we have demonstrated industry's ability to manufacture high quality inexpensive quadrupole magnets.

REFERENCES

1. Rey, C. et all. "Quench Performance of B&W Siemens Short Quadrupole Model Magnets", Unpublished

THE EFFECTS OF THE SSC POWER BUS ON MAGNETIC FIELD HARMONICS

James L. Elliott and Husam F. Gurol

General Dynamics Space Systems Division
Space Magnetics
P. O. Box 85990
San Diego, CA 92186

INTRODUCTION

This paper presents an analysis of the SSC power bus effects on magnetic field harmonics for the collider dipole magnet. The total effects will be approximated by summing the integrated effects calculated in the 2D magnet cross section and the integrated effects of the bus within the 3D interconnect region (between dipole magnets). The "effective" harmonics contributed by each region are a function of the region length, the magnetic length (14.932 m)[1], and the central field in the magnet straight section (6.791 T)[1]. For the purpose of this analysis, the interconnect region is defined to be the length (104.5 cm) between the fully yoked portion of adjacent dipole magnets. This region includes an expansion loop in the bus, a splice used to connect adjacent power bus sections, and a splice used to connect the bus to the inner coil of the magnet.

POWER BUS HARMONICS IN 2D CROSS SECTION

A detailed 360 degree PE2D[2] finite element model was created to calculate the contribution of the bus on field harmonics in the magnet 2D cross section. The model was run with and without the power bus features to isolate the effects of the bus. The skew and normal multipoles were calculated using equations (1) and (2) respectively; $r = 1$ cm and the central field $B_0 = 6.71$ T. The resulting multipoles due to the bus are shown in Table 1.

$$a_n = \frac{10^4}{\pi B_0 r^n} \int_{-\pi}^{\pi} B_y \sin(n\theta)\, d\theta \;\; ; n \geq 1 \quad (1) \qquad b_n = \frac{10^4}{\pi B_0 r^n} \int_{-\pi}^{\pi} B_y \cos(n\theta)\, d\theta \;\; ; n \geq 1 \quad (2)$$

Table 1. Multipoles (in units) produced by power bus in 2D magnet cross section.

a1 =	0.0004		b1 =	-0.0023
a2 =	0.0003		b2 =	0.0002
a3 =	-0.0003		b3 =	-0.0001
a4 =	0.0001		b4 =	0.0001
a5 =	0.0004		b5 =	0.0000
a6 =	-0.0001		b6 =	0.0000
a7 =	0.0000		b7 =	-0.0003
a8 =	0.0002		b8 =	0.0001
a9 =	-0.0002		b9 =	0.0000
a10 =	-0.0002		b10 =	0.0001

POWER BUS HARMONICS IN 3D INTERCONNECT REGION

Model Development

The "clockwise" (CW) configuration of the power bus mechanical design data base was used to construct the magnetics model conductor geometry. I-DEAS[3] was used to convert from Unigraphics[4] "splines" to 8-node brick elements. The I-DEAS data base was converted into an OPERA[2] compatible command input file using the GDSS utility TOSCAX. The bus current used was 6714 A, which produces a central field of 6.791 T in the magnet straight section. The current density for each element was set as a function of the cross section area of the element face through which current enters. The bus model was duplicated, shifted, and reattached to the original model to accurately model the bus splice region. The current density of each element within the bus splice region was reduced by a factor of 2. The conductor elements on one side of the expansion loop were offset by 5.8 cm and the loop elements were incrementally shifted to simulate the effects of thermal contraction on the loop. The conductor model is shown in Figure 1. The axial center of the expansion loop is located at 1536.5 cm. The bus splice occurs between 1584.5 cm and 1595 cm. The vertical planes of the coil splice are centered about 1607.75 cm and 1609.25 cm. The end of the horizontal sections of the coil splice are located at 1626.7 cm.

A full 360 degree 3D model of the magnet end region including the power bus and expansion loop was also created. Harmonics could not be accurately calculated using this model due to insufficient mesh density in the iron yoke. The model was used to calculate Lorentz forces acting upon the bus. Stress analysis performed indicates that the bottom of the expansion loop will deflect 3.6 mm towards the end of the magnet due to the Lorentz load. This deflection was not included in the 3D conductor model.

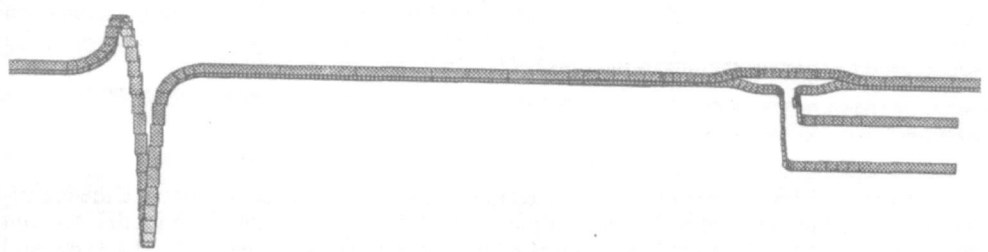

Figure 1. 3D Conductor Model

Harmonics Calculations

For the conductor only model, the OPERA post-processor was used to calculate the skew dipole field, A_0, as well as multipoles a_1-a_{10} and b_1-b_{10} at 0.5 cm increments along a 150 cm axial length (including the 104.5 cm interconnect region). The 2D harmonics formulation (equations 1 and 2) was used with the assumption that $dBz/dz = 0$. A_0 is shown in Figure 2; variations of a1, a2, b1, and b2 versus axial position are shown in Figure 3. Each multipole was numerically integrated over the interconnect region length (from 1527 and 1631.5 cm). The integrated values are listed in Table 2. Although the iron yoke will affect the harmonics locally in the 3D region, the conductor only approximation is believed to be reasonable. Coil only and coil plus iron harmonics calculations were found to be in reasonable agreement[1] as part of the coil end design analysis.

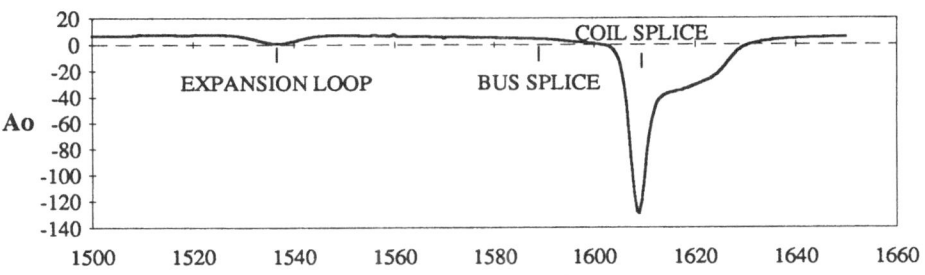

Figure 2. Variation in Ao (Gauss) versus axial position (cm)

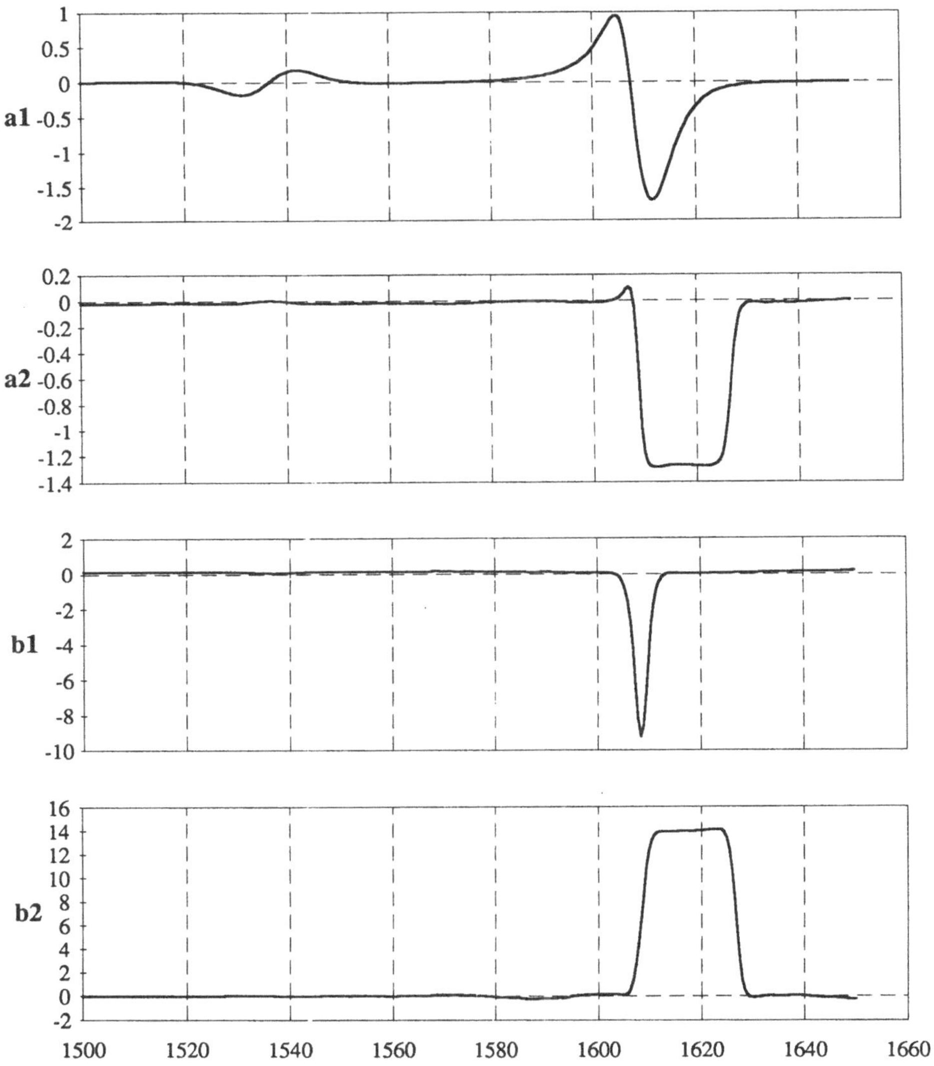

Figure 3. Local multipoles (in units) versus axial position (cm)

Table 2. Integrated multipoles (in unit-cm) over 104.5 cm interconnect region length.

a1 =	-6.6031	b1 =	-20.3272
a2 =	-23.6282	b2 =	251.6503
a3 =	-0.4047	b3 =	-3.0317
a4 =	-4.3707	b4 =	28.4688
a5 =	-0.0712	b5 =	-0.3672
a6 =	-0.7358	b6 =	3.2586
a7 =	-0.0105	b7 =	-0.0823
a8 =	-0.1125	b8 =	0.3827
a9 =	0.0005	b9 =	-0.0342
a10 =	-0.0163	b10 =	0.0710

INTEGRATED HARMONICS INCLUDING 2D AND 3D REGIONS

The total integrated value for each multipole is equal to the integrated 2D value plus the integrated 3D value divided by the magnetic length. The integrated 2D values are calculated by multiplying the constant values shown in Table 1 by the length of the 2D region , which is nominally 1476.63 cm. The integrated 3D values in Table 2 can be used directly . The integrated effects of the SSC power bus on magnetic field harmonics over the collider dipole magnet slot length, based upon the approximations described in this study, are shown in Table 3.

All integrated multipole values are within specification limits for total systematic effects. Multipoles which exceed 10% of the limits are: a1 (10.1%), a2 (48.6%), a4 (28.3%), a6 (12.8%), b1 (39.8%), b2 (21.1%), b4 (24.0%), and b6 (17.0%). The -b1 produced by the CW bus coil splice will be offset by +b1 produced in the CCW bus coil splice. The -a2, -a4, and -a6 produced by the inner coil splice straight sections will be largely offset by +a2, +a4, and +a6 produced by the outer to outer coil splice geometry. The b2, b4, and b6 systematic effects can be offset in the 2D cross section design.

Table 3. Integrated multipoles (in units) normalized by magnetic length. Maximum specification values (for high field) also shown.

a1 =	-0.0040	I0.04I	b1 =	-0.0159	I0.04I
a2 =	-0.0156	I0.032I	b2 =	0.1688	I0.80I
a3 =	-0.0005	I0.026I	b3 =	-0.0021	I0.026I
a4 =	-0.0028	I0.01I	b4 =	0.0192	I0.08I
a5 =	0.0003	I0.005I	b5 =	-0.0002	I0.005I
a6 =	-0.0006	I0.005I	b6 =	0.0022	I0.013I
a7 =	0.0000	I0.005I	b7 =	-0.0004	I0.005I
a8 =	0.0001	I0.005I	b8 =	0.0004	I0.01I
a9 =	-0.0001	I0.005I	b9 =	-0.0001	I0.005I
a10 =	-0.0002	I0.02I	b10 =	0.0001	I0.02I

Acknowledgments

The work described herein is being accomplished under contract to the Universities Research Association in support of the Superconducting Super Collider project for the U. S. Department of Energy.

REFERENCES

1. D.W. Bliss, H.F. Gurol, and M.P. Krefta, "Magnetic and Mechanical Considerations in the Design of the SSC Collider Dipole Magnet End Region", Presented at the Applied Superconductivity Conference, Aug. 23-28, 1992, Chicago, IL.
2. "PE2D" 2D Finite Element Electromagnetics Program, and "OPERA" 3D Pre and Post-Processor for Electromagnetics Analysis, Vector Fields Ltd., Oxford, England.
3. "I-DEAS V", Integrated Design Engineering Analysis Software, Structural Dynamics Research Corp.
4. "Unigraphics", Computer Graphics Program, McDonnell Douglas Corporation.

AN ORDINARY METHOD OF CUTTING A STAINLESS STEEL BEAM
TUBE INSIDE THE SUPERCONDUCTING DIPOLE MAGNET WITHOUT
DISASSEMBLY

M. Deak,[1] A. Emerson,[2] M. Heath,[2] A. LaBarge,[1]
C. Stephens,[2] and V. Vasilyev[1]

[1]Superconducting Super Collider Laboratory*
2550 Beckleymeade Ave.
Dallas, TX 75237-3997

[2]HeathCo Manufacturing
7420 Whitehall Street
Fort Worth, TX 76118

ABSTRACT

The 50-mm superconducting dipole magnet approximately 1.3 m in length was fabricated at LBL to provide a background field up to 7 T in the Cable Test Facility at SSCL. The dipole has a stainless steel beam tube with a 47.88-mm OD and a wall thickness of 2.095 mm which significantly reduced the usefulness of this magnet. There were many unsuccessful attempts to remove this tube at 300 K and 80 K.

The authors have devised a method of an ordinary multi-cutting process inside the magnet without its disassembly. In this article, tooling construction details with regulating cutting depth will be discussed.

INTRODUCTION

The purpose of the SSCL Cable Test Facility is to perform short sample critical current measurements on production lengths of the superconducting cables to be used in SSC magnets. One of the main components in this testing facility is a superconducting dipole magnet that provides a background magnetic field up to 7.5 T.

The dipole D16B-1 was designed in Lawrence Berkeley Laboratory (LBL) and tested in 1990. The coil windings are a two layer, $\cos\theta$ variety with 4 wedges to form a reasonable arc. The inner layer cable consists of 28 strands of 0.808-mm diameter superconducting wire with the Cu:SC ratio of 1.22:1, and the outer layer cable has 36 strands of 0.648-mm diameter with a Cu:SC ratio of 1.66:1. These cables were produced using the same inner and outer edge packing fractions as the SSC cable. In order to maximize the transfer-

---

*Operated by the Universities Research Association, Inc., for the U.S. Department of Energy under Contract No. DE-AC35-89ER40486.

function, the iron yoke was placed close to the winding with a 0.5-mm gap. Room temperature prestress is applied to the coil windings via the 25-mm thick aluminum rings to apply 70 MPa. During magnet cooldown the tension in the aluminum rings increases because the contraction of the iron yoke is less than that of the aluminum and the prestress is maintained on the windings without alteration. A stainless steel skin 3-mm thick surrounds the aluminum rings in order to ensure torsional stiffness and to react with the end loads.

The inner pair of coils were assembled on a 43.7-mm ID bore (beam) tube to protect the winding from mechanical damage and thermal isolation from the quenching sample. The stainless steel beam tube with a 44.45-mm outer diameter × 1.57-mm wall was expanded to an outside diameter of 47.57 mm by pulling a brass plug through the bore. The outer surface of the tube was insulated with 0.025-mm thick Kapton film using 50% overlap, followed by a spiral wrap of nylon monofilament flattened to a size of 0.94 × 2.44 mm for helium flow. The pitch length in the straight section is 25.4 mm and 5 mm on the ends.

LBL proposed to test a stack of SSC cable inside the extrusion holder with the external torque aluminum tube.

In 1992 SSCL was developing a new holder for a two-cable stack inside a high stress rolled G10 rod with 47-mm diameter. For this reason we adopted a solution to remove the beam tube from the dipole D16B-1 without disassembly.

In this article we outline technical information about an ordinary method of cutting the stainless steel beam tube inside the dipole.

DESIGN TOOLING

Several methods of removal had already been attempted by the SSCL including heating, cooling, and pressing. During liquid nitrogen cool down of the tube we observed that the tube contracted only to the lead end of the magnet in accordance with the properties of 304 stainless steel. Below we will explain the behavior of the tube. The authors designed a tool for cutting the tube while in the magnet without disassembly. The tool itself (Figure 1) was made up of three basic components, Cutter Body, Cutter, and Shock Dampening Rod. The cutter was machined from an EDM wire cutter that had four usable cutting edges made of M2 tool steel. In the cutter body there were two adjustable set screws for regulating the depth of the cut, which was set with a micrometer.

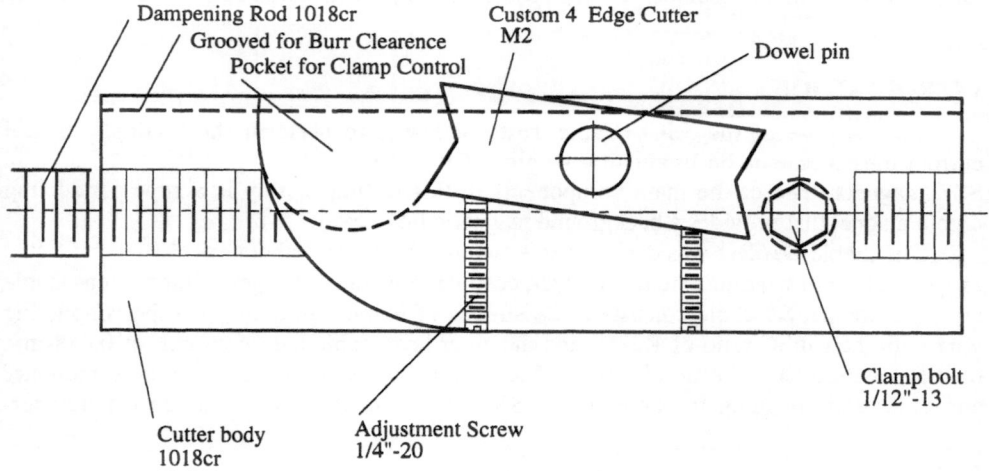

Figure 1. Cross section of tooling (not to scale).

CUTTING PROCESS

This tooling provided the opportunity to cut the tube without damaging the inner surface of the Coil in the Dipole Magnet. The cutting process consists of the following steps: (1) Precision adjustment of cutting tool to desired cutting depth. (2) Connect Dampening Rod to Cutter Body. (3) Insert Tooling through lead end of Magnet. (4) Manually applied dynamical load with hammer weight of 7 kg to Cutter Body to achieve a 12-mm long cut with a cutting depth of 0.13 mm. (5) After each completed pass through the magnet the cutter was repositioned with a new cutting edge. The process was repeated by setting the tool into the same end until the thickness of tube was decreased to 0.2 mm. This allowed the tube to collapse under a small deflection of the groove edge which permitted easy removal from the magnet.

CONCLUSION

After removal of tube, we examined the coil surface and found no damage. The vertical measurement of ID differs from the horizontal by 2.2 mm. This deviation was observed only on the first turn of the upper and lower coils. Shrinking of the tube to the lead end of the magnet can be explained by mechanical contacts between displaced turns and insulated tube.

The magnet was tested at helium temperature by the SSCL, energized with a ramp rate of 20-100 A/sec, up to 7.2 T without a quench. This indicates that the tube was removed without damage to the magnet, and the tube was not a support for the coils.

ACKNOWLEDGMENTS

The authors would like to thank Dr. J. Zbasnik for his consultations and R. Althaus, R. Bird, J. Chagnon, M. Scott, and G. Williamson for their help performing the magnet test.

REFERENCES

1. J. Zbasnik, et al. Short sample testing facility for the Superconducting Super Collider: requirements and development status. *Advances in Cryogenic Engineering*, Vol. 35, Plenum Press, New York, 1990.
2. R. Althaus, et al. The superconducting cable test facility at SSCL to be published in SSCL-Preprint.

VARIATION SIMULATION OF THE CROSS-SECTION
FOR THE SSC DIPOLE MAGNET

Charles Abel, H. Gurol, L.V. Nguyen, and H. van den Bergh

General Dynamics Space Systems Division
Space Magnetics
P. 0. Box 85990
San Diego, CA 92186

INTRODUCTION

This paper reports on the status of using a random coil stack-up approach to analyze the 2D cross-section geometry of the SSC Dipole Magnets. The simulation of the random stack-up of the coil parts due to random variations of parts, tooling, and manufacturing processes will identify those variations that significantly affect the multipoles. This approach is part of the field quality plan that has been implemented to ensure that the RMS specification requirements for the SSC Dipole Magnets are satisfied. This will allow us to adjust our manufacturing processes to satisfy the RMS multipole specifications.

We will describe the features of the simulation program, how various processes are developed for 90° and 360° magnet cross-sections, discuss the magnetic field sensitivity matrix, and provide some sample calculations.

VARIATION SIMULATION ANALYSIS

The tool that is used to simulate the random stack-up of parts in the coil cross-section is called VSA (Variation Simulation Analysis)[1]. VSA uses Monte Carlo simulation to select random tolerance variations from given statistical distributions of part sizes. The simulations are used to predict the amount of variation that can occur in an assembly resulting from specified design tolerances, fixture tolerances, and manufacturing and assembly variations. Additionally, it can determine the location(s) of the predicted variation, the factors that contribute to a given variation, and their relative contribution. Comparative "what if" studies can be quickly evaluated analytically to optimize the process thereby eliminating costly and time consuming "trial and error methods."

The VSA software provides a flexible, concise simulation language providing the necessary capabilities to describe any assembly process. The language also contains the ability to model all standard Geometric Dimensioning and Tolerancing relationships. Included are ten statistical distribution types to describe variation characteristics.

DEVELOPMENT OF 90 DEGREE CROSS-SECTION

The 90 degree model represents one quarter of a full cross-section, and was developed to investigate multipole sensitivities resulting from symmetric variations. All four quadrants then have the same geometrical characteristics. Over 200 individual dimensional variations are simulated, along with three major assembly processes: coil winding, curing, and collaring, discused below.

Statistical Variations Used On The 90 Degree Model

To use VSA as a prediction tool all processes are assumed to be "in control" and have no "special " causes that would skew the data. Statistical Process Control (SPC) defines a "special" cause as a source of variation that is intermittent, unpredictable, and unstable. Geometry variations were assumed to follow normal statistics. Some part dimensions (example: conductor width) have unilateral tolerances producing non-normal distributions. As data is collected, the true range and distribution can be used in the model allowing the critical processes to be monitored during the coil production process.

Detailed Conductor Geometry and Coil Winding

The model was constructed following the build sequence of the SSC Dipole Magnet cold mass. The critical geometry for each conductor and wedge was modeled following the design drawings. Each representation of the conductor cross-section, along with the wedges result in a total of forty-nine components for each cross-section. Geometry representing the polyimide insulation was added to each of the forty-nine components.

In the coil winding process, the geometry for two coil winding mandrels are added to the model, one mandrel each for the inner and outer coils. The insulated conductors and wedges are then "stacked" against the mandrel geometry to simulate the cross-section configuration during the winding process. Variations in the pole angle of the mandrel, and variations within each component being stacked is taken into consideration.

Curing Cycle

To simulate the curing cycle, the required azimuthal coil size was determined from a finite-element NASTRAN stress model. The cured azimuthal coil size was taken to be below the mid-plane line by 0.222 mm for the inner coil, and 0.187 mm for the outer coil. Cured coil sizes that extend below the mid-plane line are needed to meet the required pre-stress pressures in the coils after collaring: 10,000 psi for the inner coil and 8,000 psi for the outer coil. The distribution of variation in the azimuthal coil size was defined by inserting the actual azimuthal coil measurement data taken from DCA311 to DCA319 cured coils in to VSA.

Empirical data was not available to determine the amount of compression for the polyimide insulation and conductor during curing separately. It was decided to compress the insulated conductor as a composite part. The conductor stack is adjusted to simulate compression during curing until it is within 0.0005 mm of the defined azimuthal coil size.

During the curing cycle, it was noted that the conductors and wedges moved outward toward the curing press caul sheet. This required a new "stacking" of the conductors and wedges. Compression of one percent on the outer edge and two percent on the sides of the conductor geometry was simulated during curing. Magnified visual inspection of real cross-section samples of 40 mm dipole magnets also indicated less compression of the inner polyimide insulation than the adjoining sides or the outer surface.

Coil Assembly and Collaring

The coil assembly and collaring processes were simulated by modeling the collar geometry, pole shims, quench heater strips, brass shoes, and the ground plane polyimide insulation tolerances based on existing drawing specifications. The coil geometry was adjusted to simulate the compression seen in the collaring process. The collaring process is similar to the curing process, in that an iteration process is used until the correct compressed coil dimensions are achieved. The coil geometry was compressed to the nominal position with zero mid-plane shift as shown in Figure 1.

To baseline the ability of the model to simulate the correct nominal geometry, the nominal conductor locations were compared with those produced by COP7 [2]. The difference between the nominal coil azimuthal dimension from VSA and COP7 is only 0.0005 mm, which is negligible in that this difference produces negligible multipoles.

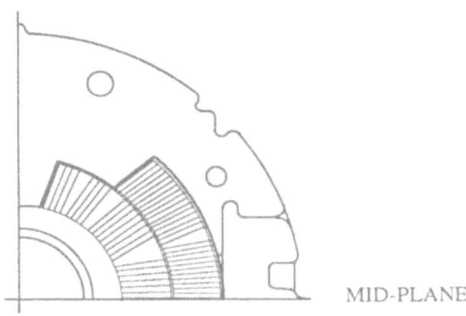

MID-PLANE

Figure 1. Assembled and collared coils -- 90 degree model.

Sensitivity Matrix For Multipole Effects

A sensitivity matrix[3] was developed using 90 degree COP7 and PE2D[4] models to calculate the changes in the allowable multipoles due to conductor displacements in both azimuthal and radial directions. The changes in the multipoles per unit change in azimuthal and radial displacements were computed. The resulting multipoles b_2 through b_{10} due to unit radial and azimuthal displacements in the conductors are stored in a matrix that can be used to calculate the total changes in the multipoles due to any combination of displacements.

The VSA model calculates the centroid for each conductor during each simulation of the cross-section. These centroid values are then multiplied by an array containing the sensitivity matrix values to produce a Δb_n for each conductor. The Δb_n values from all conductors are added to produce the total Δb_n for a given cross-section. The number of times a given Δb_n appears for a given cross-section is counted for each simulation. This produces the expected multipole given the assumed tolerances for the parts, processes and tools. Examples of Δb_2 and Δb_4 are shown in Figures 2 and 3. This simulation allowed all parts within and including the collar to vary. The interpretation of this result depends on the data used to generate it. If the data used is representative of expected variations of part size, curing pressure etc. along the length of the coil, then these distributions would be the expected b_2 and b_4 variations within a given coil. If the input data is representative of coil-to-coil variations, then these are the expected b_2 and b_4 multipoles in 3,000 magnets.

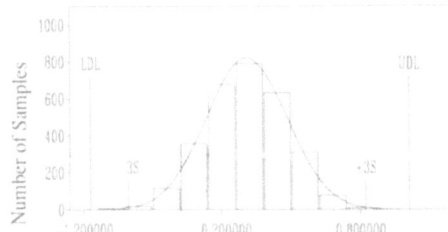

Figure 2. b_2 distribution

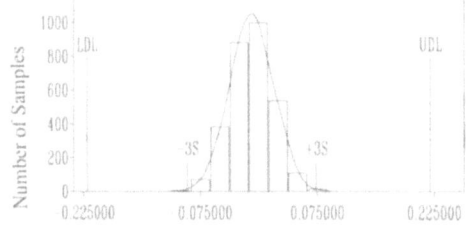

Figure 3. b_4 distribution

These results are summarized in Table 1. Since these results are based on a 90° model, they do not include asymmetric effects, such as mid-plane shift etc. Based on this data, the conclusion is that the b_2 and b_4 RMS specifications can be satisfied.

Multipoles due to a defined systematic error can also be determined using the coupled sensitivity matrix within the VSA model. In fact, all sensitivity multipole studies can be performed very quickly using the VSA model. For example, the change in the multipoles due to .05 mm changes in the pole faces were calculated. The results are shown in Table 2. The resulting multipoles from an azimuthal change of 0.05 mm in the pole faces are more than a factor of 3 within the RMS specifications. This indicates that using the design tolerance of ±.025 mm for the pole angles is reasonable.

Table 1. RMS multipoles

	VSA results	Specification
b2	.3	<1.15
b4	.03	< .22
b6	.004	< .018

Table 2. Multipole result of change in pole faces.

	Δ b2	Δ b4	Δ b6
Inner pole angle	-.41	.05	-.009
Outer pole angle	-.36	-.004	.001

FUTURE WORK AND SUMMARY

A 360 degree model was developed analyze the effects of asymmetrical variations (Figure 4). Each quadrant was modeled to simulate different geometrical characteristics. The 360 degree model has over 800 individual dimensional variations that are simulated per cross-section. We also included non-circular conductor stackup. Collar deflections due to collaring were defined from a NASTRAN finite-element model and were included when defining the collar geometry. The model takes into account variations in insulation thickness, collar ovality effects due to collaring deflections, and inner to outer coil assembly stackup. This model is currently being tested.

We plan on using this approach to analyze data during the prototype development phase to identify those variations that significantly affect the allowable multipoles, and to adjust our manufacturing processes to satisfy the RMS multipole specifications during production.

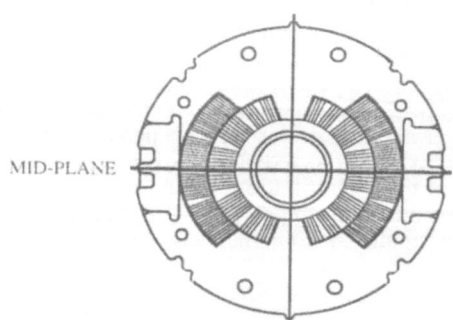

MID-PLANE

Figure 4. 360 degree model

ACKNOWLEDGMENTS

The work described herein is being accomplished under contract to the Universities Research Association in support of the Superconducting Super Collider project for the U.S. Department of Energy.

REFERENCES

1. VSA, Variation Simulation Analysis Software, licensed and trademark of Applied Computer Solutions, St. Clair Shores, MI.

2. COP7, Coil Optimization Program

3. L. V. Nguyen, C. Abel, H. Gurol,"Effects of Manufacturing Errors on the Magnetic Performance of the SSC Dipole Magnets", LN-6, 1992 Applied Superconductivity Conference (ASC92),Chicago, Illinois, August 18-24

4. PE2D, program for the calculation of electromagnetic fields, Vector Fields Limited 24 Bankside, Kidlington Oxford OX5 1JE England.

DESIGN AND TEST OF THE GENERAL DYNAMICS SSC MODEL MAGNETS

C. Gibson,[1] K. Couzens,[2] D. Hoffman,[1] and M. Wadsworth[1]

[1]General Dynamics Space Systems Division
Space Magnetics
P.O. Box 85990
San Diego, CA 92186

[2]Superconducting Super Collider Laboratory
2550 Beckleymeade Avenue
Dallas, TX 75237-3997

INTRODUCTION

The Collider Dipole Magnet (CDM) Program includes four model magnets, labeled DSD101-104, designed and fabricated by General Dynamics in collaboration with the SSCL. The purpose of these 1.8 meter long dipole magnets is to demonstrate the design features of the 15 meter long CDM cold mass. Some of the new features include a redesigned coil cross-section, polyimide cable insulation, Spaulrad/Ultem coil end parts, fine-blanked and semi-perforated collars and yokes, and revised quench heaters. The magnets were fabricated by General Dynamics and SSCL personnel at the SSCL Magnet Development Laboratory in Waxahachie, Texas. This paper discusses the design, fabrication, and test of these model magnets.

MODEL MAGNET DESIGN

The design philosophy used on the CDM Program was to start directly with the design of the Production Magnets. No developmental steps are envisioned, with the exception of one or two iterations on the coil geometry to fine tune the magnetic field quality. The Prototype magnets are therefore, identical to the Production magnets with very few exceptions such as additional instrumentation and variable thickness pole shims.

The CDM model magnets are identical to the Prototype cold mass but are only 1.8 m long instead of 15 m as shown in Figure 1. This includes all key areas such as conductor insulation, coil end parts and collar and yoke laminations. In fact, the majority of the model magnet parts were ordered using Prototype drawings. The only unique model magnet drawings were for those parts which required an interface or length change.

The CDM uses a Rutherford-type cable with a different cable on the inner and outer layers for grading purposes. The inner cable has 30 strands with a Copper to Superconductor ratio of 1.5:1. The outer cable has 36 strands with a Cu:SC ratio of 1.8:1.

The conductor is insulated with two layers of polyimide film both of which are wound 50% overlapped. The first wrap consists of 30 μm thick polyimide film. This film is coated on the outside with XMPI polyimide adhesive which requires a cure temperature of 225°C. The second wrap consists of polyimide film which has been reinforced with alumina for increased punch through resistance. The thickness of this film is 34 μm for the

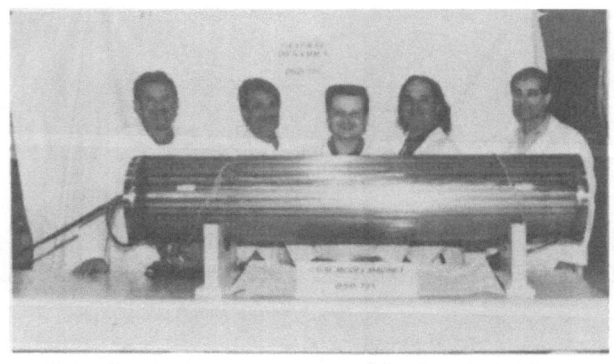

Figure 1. The DSD Model Magnets are 1.8 m long versions of the CDM Prototype cold mass.

inner coil and 32 μm for the outer coil. The second wrap is coated on both sides with XMPI adhesive.

The conductor is wound into coils which have a cosine-theta distribution and has been designated GD_B05. The inner coil has three copper wedges and the outer coil one in order to obtain the correct field. After each coil is wound it is individually cured to the correct geometry.

The coil ends are of a grouped design with the inner coil having four groups and the outer coil two. The coil end parts and several splice parts are made from Spaulrad, a polyimide/glass laminate material,. This material is no longer available and hence the Spaulrad material has been replaced by Ultem 6200, a filled polyetherimide material. The fourth model magnet will use this material for the coil end parts.

After coil winding and curing, the next step is coil assembly. The coils are insulated to a minimum of five kilovolts with formed sheets of 127 μm thick polyimide film. The placement of the sheets is designed to maximize the creep path length. The coil assembly step also includes soldering the inner and outer coils together. This splice is located on the lead end of the magnet and is internal to the coil pairs.

Quench heaters are located on the outside surface of the outer coil. They consist of a copper plated stainless steel strip which has been encapsulated in polyimide film. The quench heaters are energized during a quench to cause more of the coil to go normal, thereby reducing the hot spot peak temperature. A parametric study will be performed on the quench heaters by using five different strip designs in the DSD model magnets.

The coils are restrained by collars which keep them under compressive prestress to limit conductor motion. The collar laminations are fine-blanked from 2.5 mm thick 21-6-9 stainless steel. The collar laminations are joined together using semi-perforations. The collar pairs are assembled into modules nominally 254 mm long. The modules are locked around the coils using tapered phosphor bronze keys.

Iron yokes are placed around the collars to add to and contain the magnetic field. The yokes also transfer load from the coils to the shell in order to distribute the load. The yoke laminations are fine-blanked from 6.0 mm thick low carbon steel and are assembled into modules 100 laminates thick through the use of semi-perforations.

To lower the peak field in the coil ends, the yokes do not extend all the way to the end of the coils. To add support to the end of the collared coil, bronze collets are clamped around the collars. Outside of the collets, thin yoke laminations are used to lower the stray magnetic fields.

A 4.9 mm thick 304LN stainless steel skin is welded around the outside of the yokes. The weld shrinkage helps to close any gap between the upper and lower yokes. The ends of the magnet are contained by 38 mm thick 304LN end plates. Load screws are used to apply axial prestress from the end plates to the coil ends.

The splice between the upper and lower coil halves is made outside of the end plates. This splice is contained inside a pair of Spaulrad splice plates.

All of the design features of the model magnet are identical to the Prototype magnet with the exception of those related to magnet length. The Prototype magnets have additional features not found in the model magnets. Those include beam tube and beam tube supports, end domes, and a cryostat which contains support posts, 20K and 80K shields, multilayer insulation, cryogenic lines and a vacuum vessel.

FABRICATION EXPERIENCE

The fabrication of the CDM Model Magnets was performed at the SSCL MDL facility in Waxahachie, TX. GD provided a liaison engineer, manufacturing engineer and four technicians. These personnel worked with SSCL engineers and technicians in the fabrication of the magnets. The generic MDL tooling, such as the winding machine and collaring press, was utilized. Any modifications to the tooling required for the CDM design was performed by GD.

The model magnets acted as a trail blazer for the CDM Program. Many systems were proofed by the models such as procurement of parts, receiving and inspection, parts fabrication, design and manufacturing processes. The lessons learned from the models have been transferred to the Prototype team in Hammond. Examples of lessons learned are discussed below.

The insulating of the conductor was performed on the SSCL's cable wrapping machine. This machine was not designed to control the registration between the first and second wrap of polyimide film and hence, the specified staggering of wrap joints was not achieved. The GD engineering group decided that this was acceptable for the Model Magnets since there will be a minimum of four layers of polyimide film between turns. The GD cable wrapping machine is designed to control first to second wrap registration.

Coil end parts made from Spaulrad were first tried in the model coils. The end parts fit well but they did require additional slitting to increase their flexibility. Several instances of "popped strands" occurred during winding both in the end and straight sections of the coils. They were fixed by massaging the strands back into place and repairing the insulation if required. No shorts have been found in any of the 16 coils wound to date, which indicates the robustness of the selected coil insulation.

The adhesive on the conductor insulation requires a cure temperature of 225°C. Although coils have been cured at this temperature at Brookhaven National Lab, these were the first high temperature coils cured at the MDL. The curing press was instrumented with thermocouples and large thermal gradients were found. The gradient was reduced by reworking the hydraulic manifolds, adding insulation to the press and increasing dwell times.

Coil azimuthal measurements were performed using the MDL's measuring machine and GD provided standards . The coils azimuthal size varied by about 7-10 μm along the length of the coils. The average coil size was about 381 μm oversize for the inner coils and 305 μm on the outers.

During coil assembly, the CDM ground plane insulation design was validated. This design included the use of subassemblies to reduce fabrication time and the use of heat and/or adhesives to join the insulation components together.

During collaring it was found that the pole shims required were larger than expected. For example on DSD101, the inner coils were oversize by 381 μm and hence it was expected that the pole shims should be 381 μm less than nominal. However, the actual pole shims required were only 51 μm less than nominal. This same phenomena has been observed on other dipole magnets.

After collaring of DSD101, the tapered keys had a tendency to become unengaged. During the collaring operation Teflon lubricant was used to aid in installing the keys. Eliminating the lubricant greatly reduced this problem. The root of the problem is a loose fit between the collar tabs which allows the collars to slowly "scissor" out the keys. For Prototype 8 and on, the tab feature on the collars will be changed to a line-to-line fit to ensure engagement of the keys.

After collaring measurements were taken of the vertical diameter of the collared coil assembly. These measurements are about 360μm larger than analytically predicted. The coil prestresses are about 14 MPa larger than expected and this accounts for some, but not all of the increased vertical collared coil deflection.

The yoking of the models demonstrated the use of the 100 laminate yoke packs. The skin and end plate welding went smoothly and validated the ASME code weld joint designs. The gap between the upper and lower yokes closed to about 127 mm after skin welding. Finite-element analysis shows that this gap will close upon cooldown to 4.2 K.

The fabrication of the model magnets have validated the CDM cold mass design and no show stoppers have been found. As of the 1993 IISSC Conference, the fabrication of DSD101, 102 and 103 have been completed and the magnets are awaiting cold testing. The DSD104 magnet will begin coil wind on May 10, 1993 using the Ultem coil end parts.

MODEL MAGNET TESTING

The CDM Model Magnet test planning was performed jointly by GD and the SSCL. GD provided the Test Objectives[1] and the SSCL was responsible for the Test Plan[2]. The actual testing and the test equipment design were performed by the SSCL.

The in-process Tests are described in detail in the model magnet travelers. These travelers are also used to record the test results and to describe any anomalies and lessons learned. In general, the in-process tests include electrical and dimensional tests performed before, during and after each fabrication step. Examples include lump and pinhole detection during cable wrap, and electrical short checks after winding, curing, collaring and skin welding. Other examples of in-process tests are coil azimuthal measurements after curing and dimensional checks after collaring.

Strain gauge measurements are also an integral part of the in-process tests. The magnets each have a strain gauge collar pack to determine the coil prestress during collaring, skin welding, cooldown and energization. The skins also have strain gauges attached to determine the stress induced in the skin during welding.

Before cold tests the magnets will all undergo warm magnetic tests. These tests will determine the magnetic field quality in terms of multipole values. These measurements will be used to investigate the warm to cold magnetic field correlation. Magnetic measurements will also be taken using the skew quadrupole centering technique developed by GD.[3] In this technique, current flow in the upper half of the magnet is in the opposite direction to the lower half. In the prototype magnets, this technique will be used to determine the magnetic center of the magnet.

The cold tests will be performed at the SSCL's Short Magnet and Cable Test Laboratory as detailed in the DSD Lot A Test Plan.[2] These tests include monitoring the collar pack strain gauges, measuring the residual resistance ratio of the cable, and quench testing. Magnetic measurements will be taken with the magnet hooked up in both the dipole and skew quadrupole configurations. Each model magnet will use a slightly different quench heater to evaluate the quench heater sensitivity. These tests will measure the delay time from when the heater was fired until the normal zone propagation begins and also the amount of energy required to cause the magnet to quench.

After the magnets are cold tested, GD plans to perform additional warms tests. These tests include coil relaxation tests where the prestress as measured by the strain gauge collar packs will be monitored for an extended period of time. Other tests include sectioning both the straight and end sections to determine how close the conductors are to their theoretical locations. Another test will investigate the effect of AC resonance tests on the conductor insulation.

ACKNOWLEDGEMENT

The work described herein is being accomplished under contract to the Universities Research Association in support of the Superconducting Super Collider project for the U.S. Department of Energy.

REFERENCES

[1] C.R. Gibson. "Model Magnet Test Objectives," SSCL Doc #M3A-100014 (1992)

[2] S. Smith, B. Aksel. "DSD Lot A Test Plan," SSCL Doc #M40-000009 (1993)

[3] C.R. Gibson, D.W. Bliss, R.E. Simon, A.K. Jain and P. Wanderer, Locating the magnetic center of the SSC CDM using a temporary quadrupole field, presented at the Applied Superconductivity Conference, Chicago 1992.

MECHANICAL BEHAVIOR OF B&W SHORT QUADRUPOLE MODEL MAGNETS

G. Dun, C. Rey, M. Xu, X. Huang, J. Kelley, J. Savignano,
K. Dixon, B. Cantor, and J. Waynert

Babcock & Wilcox
Accelerator and Magnets Section
P.O. Box 785 MS 80
Lynchburg, VA 24505-0785

INTRODUCTION

B&W planned to build four short quadrupole magnets for the SSC project. The major purpose of building these model magnets is to examine the quench behavior. Good quench performance can be achieved when adequate prestress is applied by the collar to the coils to prevent relative motion between the two. In order to detect and monitor the prestress, a set of eight strain-gage transducers was installed in each model magnet. In this paper, we examine the mechanical behavior of the coils as being detected by these transducers. Three of the short model magnets have been built and tested. To avoid unnecessary repetition and yet retain details, we concentrate on the test results of one magnet -- QSH-802.

STRESS AFTER FIRST COOL DOWN

At room temperature, the average inner coil prestress is 65 MPa and the outer coil 57 MPa. The coils, by their nature of shrinking more than stainless steel collar in the azimuthal direction, were expected to lose prestress after first cool down. Two-dimensional finite element analysis, which took the temperature dependence of Young's moduli into consideration, indicated an approximately 30% decrease in prestress. However, the measurements show, surprisingly, prestress may decrease or increase after cool down. Prestresses on the inner coils have an average decrease of 5 MPa, or 8 %, per gage. On the other hand, a net average increase of 5 MPa, or 9 %, per gage was observed in the outer coils.

The increased prestress after cool-down is not unique to this magnet, neither is it only to B&W measurement equipment. Previously SSCL has tested B&W's QSH-803 in their facility and observed a 5 % increase in prestress in the inner coils and a 13 % rise in the outer coils.

In order to understand this unexpected behavior, the variables entering the calculation of coil prestress were investigated. These variables are the measured resistances and initial resistances of the active and compensating gages, and the coefficients for the cubic equation fitting the coil prestress and beam strain relation. From the measured resistances the mechanical strain was calculated. For the inner coil gages, the decrease of strain as temperature lowers from room temperature (RT) to liquid helium temperature (LHT) is 12.8% of its RT value. For outer coil gages, it is 5.5%. The decrease is compensated by the more than 8% increase in Young's modulus of both beam and coil at low temperature. This means that there is a good possibility that the stress will increase with decreasing temperature.

In the process of investigating our measured resistances, it was found that they increase at a rapid rate with temperature below 20 K (possibly from Kondo effect). We observed the same behavior in the initial resistance measurements at various temperatures supplied by SSCL, in which there is a 0.135 ohms increase (corresponding to 13.5 MPa) per degree K near LHT. This behavior needs to be precisely compensated in order to obtain accurate measurement, if the paired compensating and active gages are not calibrated as a unit (a current practice). It was found that, as temperature lowers, the change in the differences between the outer and inner compensating gages can be as high as 0.07 ohms. If this difference happens between an active gage and its compensating gage, it will be mistakenly counted as mechanical strain and lead to an error of approximately 7 MPa.

The second possible source of error can come from the magnitude of the coefficients for stress-strain relation. The magnitude of the coefficients at LHT will increase from those at RT mainly from the increase of Young's modulus of the beam. The beam was made of stainless steel A-286 with a Young's modulus of 207 GPa at RT and 224 GPa at LHT, an 8% increase. This corresponds to a 8% increase in the coefficient had the stress-strain relation been linear. The measured coefficient for the linear term increases an amount of approximately 12%, a little higher than expected.

The third possible cause can be the shifting in the original values of initial resistances. Since the incident of QSH-803, SSCL has embarked on an investigation of the shifting of initial resistances. Preliminary results indicated that the shifting can be as high as 0.1 ohms , leading to an error of 10 MPa. More experiments are underway for reaching a conclusion.

The fourth possible origin of error is the interaction between the coil, the collar and the transducers, which affects the gage reading, but was not modelled by the stack. One of the concerns is that the thermal differential contraction between the transducer housing and beam may put beam in a multiaxial stress state and thus lead to erroneous readings. Experiments on these effects are yet to be performed.

CURRENT-DEPENDENT STRESS VARIATION

PRE-QUENCH BEHAVIOR

Before quench test, the magnets were subjected to "strain-gage runs". In each run the coils were supplied with current up to either 2000 A, or 4000 A, or 6000 A, and for each supplied current, the strain gages were monitored in 10 even intervals between 0 A and the maximum current. Fig. 1 shows the measured prestress at the outer coils as a function of current squared.

The function is essentially a straight line with a negative slope. This behavior was universally observed in various dipoles[1] and quadrupoles[2] in the SSC project except with different slopes. What it means is that the prestress was chosen high enough such that the unloading behavior of the coil is linear. If the coil is unloaded to the nonlinear region, there is concern of developing gap between the coil/collar pole interface. A small undersize in coil due to manufacturing tolerance will produce significant reduction in prestress. The slope depends on the intensity of magnetic field, typically with the inner coils (slope -0.26 MPa/kA^2) unloading more than the outer coils (slope -0.15 MPa/kA^2), thus form a basis for higher prestress required in the inner coils than in the outer coils .

The zero-current prestress usually decreases after a cycle to a miximum current. The decrease became more sizable at a higher maximum current. A cecrease of 1 to 1.5 MPa was observed after reaching a maximum current of 6000 A. This decrease was possible produced by the Lorentz force, the coil hysteresis, the friction between the coil and the collar. The coil was squeezed azimuthally by the Lorentz force and, after the force disappears, was prevented from returning to its original position by the frictional forces, therefore, leading to a decrease in prestress. Since the decrease is observed to be independent of the level of prestress, this kind of irreversible movement can not be corrected by increasing the prestress.

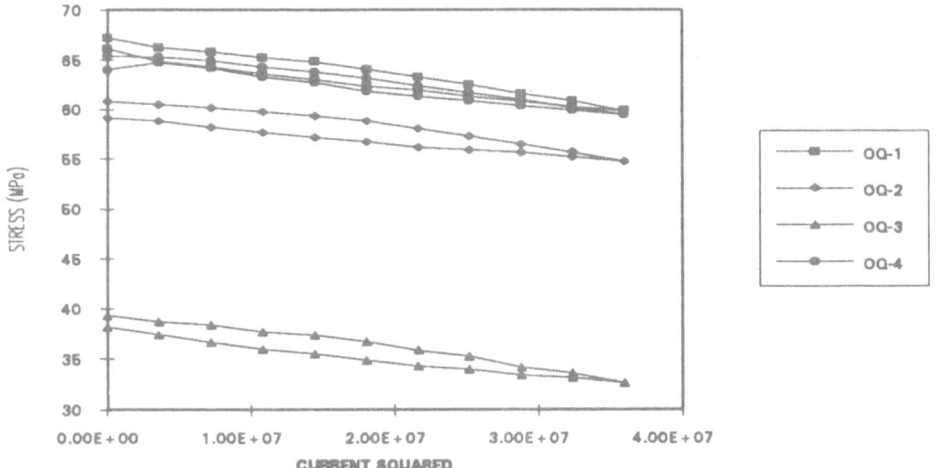

Fig. 1 Pre-quench unloading of the outer coil during energization

POST-QUENCH BEHAVIOR

After training quenches, the prestress at zero current was observed to further decrease from its pre-quench value. Sometimes the magnitude of decrease is up to 15 MPa. However, after a strain-gage run to 8000 A, most of the loss was recovered, leaving a smaller decrease of approximately 5 MPa. This behavior is depicted in Fig. 2 and might be explained as follows: Lorentz forces disappear quickly during a quench, allowing the azimuthally compressed coil to spring back in a fast rate. The external and internal frictions, increasing with the spring back rate, latch the coil. Once the current is gradually ramped up, Lorentz forces allow the release of these frictional forces and therefore the recovery of the prestress. This behavior exists in every following thermal cycle.

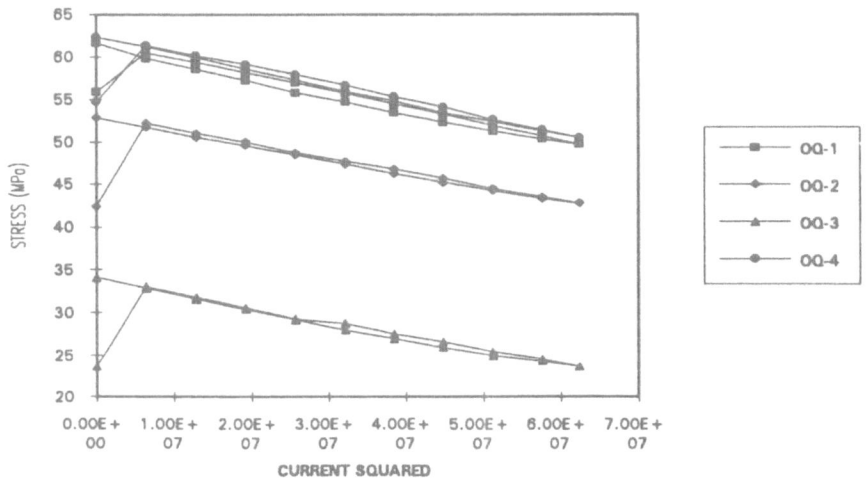

Fig. 2 Post-quench unloading of the outer coil during energization

STRESS AFTER FIRST WARM UP

After the first warm up, the prestress at zero current at room temperature was observed to decrease approximately 6 MPa from its original value, as shown in Fig. 3. This indicates that 80% of the decrease of prestress in a thermal cycle is caused by quenches, about 20% by energization, and very little by the thermal shrinkage. Further thermal cycles add approximately 1 MPa prestress drop for each of the two additional cycles we test

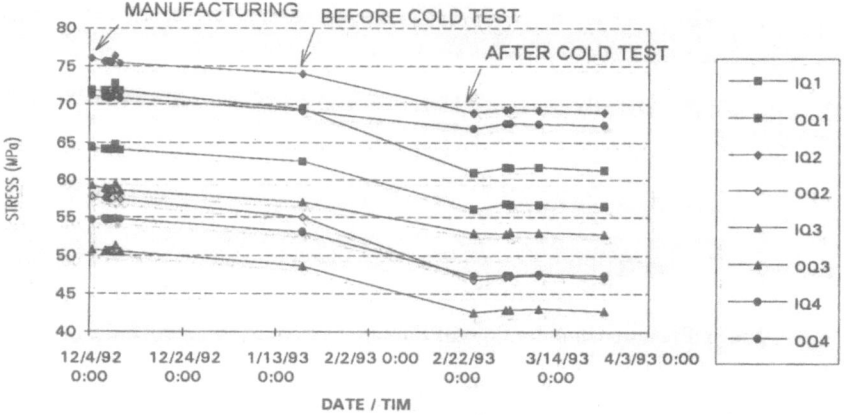

Fig. 3 Prestress before and after first cool down (all RT values)

SUMMARY

A tendency of increase in prestress after cool down was found for B&W short model magnets. It was found that small change in the differential resistances between active and compensating gages from room temperature to liquid helium temperature can contribute to this unusual behavior. Preliminary evidences seem to indicate that the initial resistances of the gages may have already shifted from their original values.

Quenches can contribute up to 80 % of the loss in prestress observed in the thermal cycling, with energization and thermal shrinkage sharing 20%. A thermal cycle produces a loss of approximately 6 MPa. Further two thermal cycles lead to about an additional 1 MPa loss per cycle.

REFERENCES

[1]. A. Devred, et al "About the Mechanics of SSC Dipole Magnet Prototypes", in "The Physics of Particle Accelerators", AIP Conference Proceedings, 1991.

[2]. J. M. Cortella, "Mechanical Analysis of Full-Scale Collider Quadrupole Magnets for the Superconducting Supercollider", Test &Analysis Note MD-TA-212, SSC Laboratory, 1992.

DESIGN AND MAIN PARAMETERS OF THE HIGH ENERGY BOOSTER QUADRUPOLE COLD MASS FOR THE SSC

G. Ducos,[1] J. Giacometti,[1] P. Giovannoni,[1] D. Leboeuf,[1]
F. Le Coz,[1] C. Lyraud,[1] C. Michez,[1] J. F. Millot,[1] D. Orrell,[2]
J. Perot,[1] J. M. Rifflet,[1] J. C. Toussaint,[1] J. Turner,[2] and P. Védrine[1]

[1]CEN Saclay DAPNIA/STCM
91191 Gif-sur-Yvette Cedex France

[2]Superconducting Super Collider Laboratory
2550 Beckleymeade Avenue
Dallas, TX 75237-3997 U.S.A.

INTRODUCTION

CEA/Saclay has a long standing tradition in the area of High Energy Physics, both in equipement and experiments. In the field of Superconducting magnets for accelerators, one can recall the HERA Superconducting quadrupoles[1], and the work presently being made on the LHC Superconducting quadrupoles[2].

The work on the quadrupoles for the SSC HEB described below has greatly benefited from the large experience gained by the Saclay team in the two above mentioned programs made in collaboration with DESY and CERN.

The work has also benefited from the large experience of the SSC on the collider dipole and quadrupole magnets; a common design team (Saclay/SSCL) is working on this project.

The SSC quadrupoles present one new and challenging feature with respect to the HERA and LHC quadrupoles: the ramp rate (60A/S) which increases energy losses in the coils, reduces the safety margin and gives rise to unwanted multipoles.

MAIN CHARACTERISTICS

The magnet is a standard quadrupole design using two layers of Rutherford cable. The cables are arranged in two blocks in the inner layer and one block in the outer layer, for a total of 25 turns. The nominal gradient is 186.5T/m for a current of 6650A. The coil inner diameter is 50mm and the cold mass outer diameter 338mm. The cross-section of the magnet is shown in Figure 1. The magnet is to be built with a number of different length variants. The arc quadrupole has a total length of 1.753 m, while the other length variants are 1.053, 1.325, 2.414, and 3.507 m.

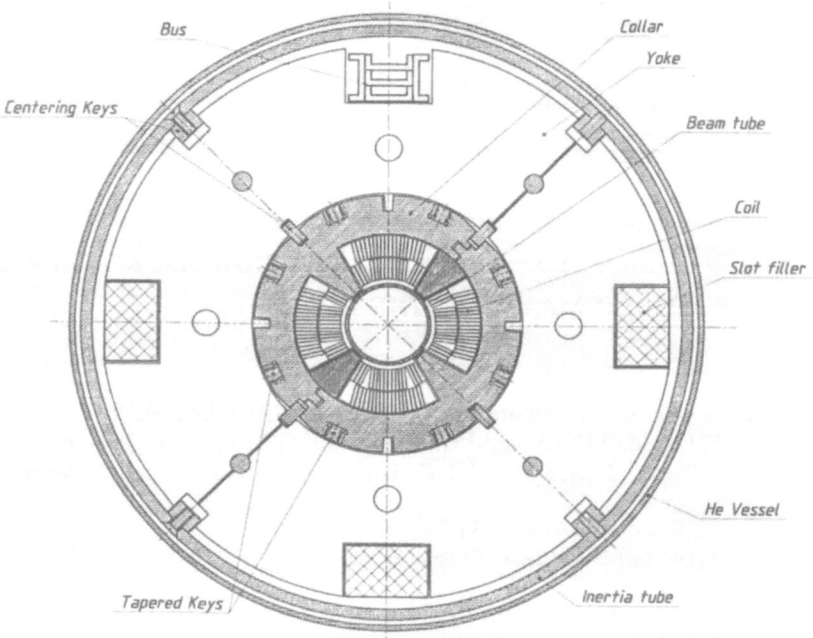

Fig 1. Cross section of the cold mass

DETAIL DESIGN

Coils

The conductor is a keystoned Rutherford cable of the same type as is being used in the outer layer of the collider dipoles. It consists of 36 strands of 0.648mm diameter and a copper to superconducting ratio of 1.8/1. Insulation consists of two layers of 12.5 mm Kapton with 50% overlap, plus a 114 mm thickness of fiberglass epoxy tape. The insulated dimensions, cured and compressed are: (1.22 - 1.43) 11.96mm.

The cross-section was optimized using a combination of analytic and finite element codes to meet the required specifications. Factors influencing the 2-D multipoles were the permeability of the 304 LN beam tube (assumed to be 1.008), permeability of the collar (1.005), persistent currents in the cables, eddy currents in the cables, movement of cables due to Lorentz forces, saturation of the yoke, and the field due to the bus. Table 1 shows the harmonics that we have calculated for injection and full energy. These do not take into account the eddy currents in the cables, which will depend on the as-yet unmeasured interstrand resistance. The eddy current contribution to b5 has been calculated as a function of interstrand resistance[3] and is expected to be in the opposite direction and have similar magnitude to persistent current b5. The gradient has a total sag between injection and full energy of 0.107%, which is due to a combination of Lorentz forces and yoke saturation.

The 3-D harmonics depend on the magnet length variant, but are close to the 2-D harmonics. The end design was optimized by adjusting the straight section length of each coil to cancel the integral b5 through the ends. The ramp splice and the magnet leads were also found to have a significant effect on harmonics, which was minimized by arranging the leads in such a way that the harmonic b5 they create nearly cancels the b5 due to the ramp. There remains a harmonic a1, which can be canceled by rotating the magnet a fixed amount when it is placed in the tunnel. The maximum rotation required is for the shortest magnet and is 1.15 mrad.

Table 1. Calculated Harmonics at r=10mm

Harmonic	2-D inj	2-D full	3-D inj (arc quad)	3-D full (arc quad)	Spec
b5 (units)	-0.346	0.131	-0.324	0.153	0.570
b9 (units)	-0.050	-0.050	-0.050	-0.050	-

Codes used were the SSC programs COP8 (2D optimization), AHARM (2D persistent currents), ENDS3 (3D harmonics), LEAD (lead end design), and the commercially available finite element programs PE2D® (2D), ANSYS® and TOSCA® (3D). Results were also verified (and good agreement found) using the Saclay in-house codes Q3CH (3D harmonics) and CASTEM 2000 (finite element).

Thermal behaviour

As this magnet will be operating at a fairly high ramp rate of 60 A/s, eddy currents will be generated which could substantially lower the magnet's safety margin. For this reason, detailed thermal analyses were made on the magnet cross-section using the .93W/m allowed magnet losses and the specified helium flow. Assuming the inlet helium temperature is 4.25K, the outlet temperature of the helium for the longest quad (3.507m) is 4.43K, and the safety margin of the magnet is reduced from 20% to 15.9%.

Collars

The primary design philosophy for the HEB quadrupole collars was to make the design as similar as possible to the collars used for the HERA and LHC quadrupole magnets in order to minimize risk. Therefore, the collars consist of 180 degree arcs which are stacked around the coils while the coils are held in a vertical position. The collars are then locked together with tapered keys, the prestress being applied entirely by these tapered keys.

After numerous studies of different collar widths and designs with only four keys versus eight keys, the final design chosen was one with eight keys and a width of about 21mm. The calculated peak stress after collaring is about 550 MPa with an average coil prestress of 35 MPa. This necessitates the use of a high strength steel for the collars such as Nitronic 40, which after being properly cold worked has a yield strength of about 630 MPa.

Furthermore, finite element analysis was used to optimize the shape of the key and keyway in order to determine the correct key size and taper angle necessary for the desired coil prestress. Because the collars are not infinitely rigid, they should not be designed to simply fit the desired compressed coil shape. Their elasticity should be compensated for in order to have the correct coil shape and prestress after collaring. In order to accomplish this goal, the key has been slightly oversized with respect to the keyway (0.08mm).

Finally, from the 35 Mpa prestress in the coils after collaring, there remains 23.5 MPa after cooldown, 14 MPa at nominal field and 7 MPa at critical field. The radial collar deflection, for nominal field, is 0.04 mm in horizontal median plane and 0.008mm at 45 degrees.

Yoke

The yoke shall be stacked around the collared coil assembly while the magnet is held in the vertical position. Each yoke lamination consists of a 180° arc so that there are two yoke laminations per layer. On the next layer, the laminations are rotated by 90°. This makes the yoke stacking almost identical to the collar stacking. Keys are placed between the yoke and the collars in order to align the collared coil assembly after yoking. After the keys have been inserted and all of the yoke laminations have been stacked, the assembly is 'squeezed' from all sides in order to align the laminations, and tubes are inserted through holes in the yoke laminations. Next, these tubes are expanded to lock the laminations in place, avoiding any weld.

Inertia Tube and Helium Vessel

The inertia tube is placed around the outside of the yoke to support and align the magnet. It is made from a 6 mm thick stainless steel tube. Very precise holes are placed in the inertia tube, through which pins can travel. These pins also pass through keys which are placed into slots in the yoke, so that the yoke is aligned by the inertia tube.

The entire magnet is then surrounded by a 3.5 mm thick stainless steel shell. This shell serves as a helium containment vessel and as a pressure vessel during a magnet quench. The thickness of the shell was determined by finite element analysis and hand calculations required by the American Society of Mechanical Engineers for pressure vessels.

CONCLUSION

At this time, the main parameters of the magnet are frozen and most of the tooling designed. The conductor provided by SSCL is already available and the main components such as the collar and yoke laminations are being ordered. Some of the tooling such as the winding machine, coil molds and mandrels, and the curing press are either ordered or just about to be ordered. Coil winding should start in the fall of 1993.

Three prototypes will be built: two arc quadrupoles (1.753m long) and one long quadrupole (3.507m long). The first arc quadrupole should be tested in a vertical bath cryostat at Saclay during the summer of 1994. The second arc quadrupole which will be fully instrumented, will be tested at the end of 1994, and the long quadrupole in the spring of 1995.

All toolings and technical documents ("build to print package") must be provided to SSCL in order to allow mass production by US industry in the middle of 1995.

REFERENCES

1. J. Perot, J. M. Rifflet "Measurement data taken during the industrial fabrication of the HERA superconducting quadrupoles". (IISSC) Atlanta, Georgia, USA, 1991

2. J.M. Baze, et al."Design and fabrication of the Prototype Superconducting Quadrupole for the CERN LHC project".MT12 conf. Leningrad, Russia, 1991

3. T. Ogitsu "Dependence of AC Loss and Multipole Coefficients on Cable Eddy Currents". MD-TA-240, Jan. 11, 1993

OPTIMIZING WELDMENTS OF THE CQM BEAM TUBE ASSEMBLY FOR THE SSC

William C. Young and Richard A. Fenolietto

Accelerator and Magnet Systems Department
Naval Nuclear Fuel Division
Babcock and Wilcox Company
PO Box 785, Lynchburg, VA 24505

ABSTRACT

The design constraints on the SSC quadrupole guides the assembly sequence for the beam tube. The sequence requires that the copper plated beam tube be inserted after coil collaring and that flanges be subsequently welded to the beam tube. This creates several challenges in the fabrication process with respect to weldability and copper plating integrity on the tube internal diameter. The results of process development, weld toughness mockups, and other manufacturing efforts will be presented. Data from cryogenic impact testing (Charpy) of XM-11/304LN weldments, and results from copper plate testing are included.

INTRODUCTION

The desired degree of precision in the alignment of the superconducting coils for the quadrupole magnets and prior satisfactory manufacturing experience (on quadrupole magnets for HERA) guided a decision to insert the beam tube after assembly and collaring of the magnet assembly. This processing sequence, in turn, dictates that the beam tube must pass the coil aperture of 40mm. Because of this diametral limitation, the required closure and interface flanges must be installed onto the beam tube after insertion. The beam tube must have a highly conductive and tightly adhered copper coating on the interior surface. The installation of these closure and interface flanges on the copper plated beam tube without deleterious effects on the plating was considered one of the highest risk process points in the envisioned manufacturing process. Although testing continues, preliminary testing indicates that the required flange welding is feasible. The results of that preliminary testing is presented below.

The material for the beam tube was initally dictated by contractual provisions. After that constraint was removed, B&W reviewed alternatives and concluded that the magnetic and cryogenic strength performance of the previously dictated material, XM-11 (21-6-9 stainless steel), was more desireable than the alternatives and should continue to be pursued.

Availability of information which assures the cryogenic toughness of the weldments and HAZ (Heat Affected Zone) was a problem with this material however. To address this concern, an experiment was conducted to evaluate the cryogenic impact toughness of XM-11 and determine if a post weld anneal was necessary for those assembly welds conducted prior to electroplating. The experiment demonstrated XM-11 weld HAZ to have equivalent toughness to that of 304LN grade stainless steel. The results of that testing are also presented below.

CRYOGENIC IMPACT TESTING

Experimental Design

Two stress relieved stainless plates were machined to the configuration shown in figure 1.

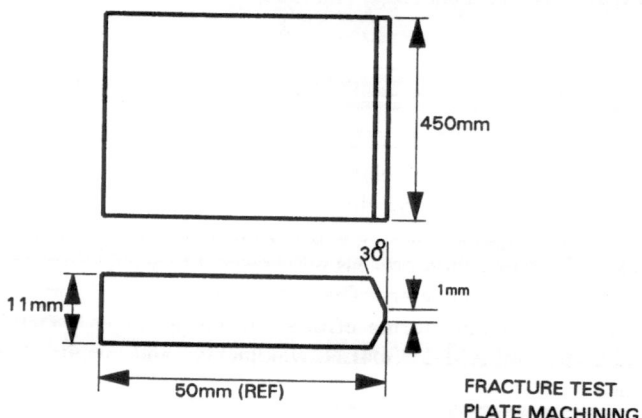

Figure 1. Charpy Specimen Test Plate

One plate was 304LN and the other XM-11. The plates were then machine welded using Gas Tungsten Arc Welding (2 passes on each side of the joint.) with 316L filler wire (Ferrite Number 5.87). After welding the plate was sectioned and one half was annealed for 1/2 hour at 1093°C followed by water quench. Standard size Charpy V-notch specimens were then machined from the as welded and as annealed test pieces. A total of 22 specimens were tested. Testing was conducted at 4K in accordance with ASTM A370. Specimens were transfered from the liquid helium to the test machine within 5 seconds.

Results and Conclusions

The results of impact testing of the XM-11/304LN test specimens are summarized in figure 2. The XM-11 is equivalent in toughness to the 304LN test plate both as welded and after annealling. Because 304LN is used in the as welded condition for most of the cold mass asembly, and because XM-11 weldments appear to be equivalent, the use of XM-11 in the as welded condition for the beam tube is concluded to be satisfactory.

CLOSURE AND END FLANGE WELDING

Experimental Design

B&W has on hand a length of copper coated simulated beam tube. This tube is a vendor sample and, as such, has a questionable pedigree. It is fabricated from XM-11 and has been copper coated by the anticipated production vendor (Fluhman). The coating has not been tested for RRR (Relative Resistivity Ratio) but is believed to be of quality comparable to that which was vended for the HERA program.

4K CHARPY TESTING OF XM-11/304LN WELDMENT

SPECIMEN LOCATION	AS WELDED		WELDED AND ANNEALED	
	IMPACT TOUGHNESS	LATERAL EXPANSION	IMPACT TOUGHNESS	LATERAL EXPANSION
	Nwtn-Mtrs	mm/mm	Nwtn-Mtrs	mm/mm
304LN BASE METAL**	29.2	0.020	107.1	0.050
304LN HAZ	30.3	0.019 *	99.7	0.043
WELD METAL	29.4	0.017 *	118.4	0.042
XM-11 HAZ	33.0	0.018	103.7	0.049
XM-11 BASE METAL**	25.8	0.019	104.4	0.054

*These averages include individual values lower than 0.015
**Base metal represents single specimen others are average of three

Figure 2. Charpy Testing Summary

The production concept which is being pursued utilizes a sleeve over approximately 100mm on the lead end of the beam tube. This sleeve, which is to be installed by welding prior to plating, serves to make the copper coat more remote from the closure and end flange welding. The experiment utilized a sleeve which had been tack welded to the beam tube to simulate this condition. The experiment also utilized a gas jet impinging on the internal surface of the beam tube underneath the weldment. This gas jet served both to keep the copper plating flooded with protective Argon and as a coolant. A closure flange mockup was autogenously fillet welded to the sleeve on both sides (i.e. cold mass and vacuum sides) using GTAW and destructively evaluated.

Results and Conclusions

Figures 3 and 4 show the copper coating remote from and directly under the weld nugget from the weld mockup respectively. There is no gross damage to the copper plating from the welding. It is not observed in the microphotographs, but a small increase in grain size and/or increase in nickel strike diffusion may have occured in the vicinity of the weld. This is considered a satisfactory interim result although evaluation of RRR of production quality plating is considered the real acid test.

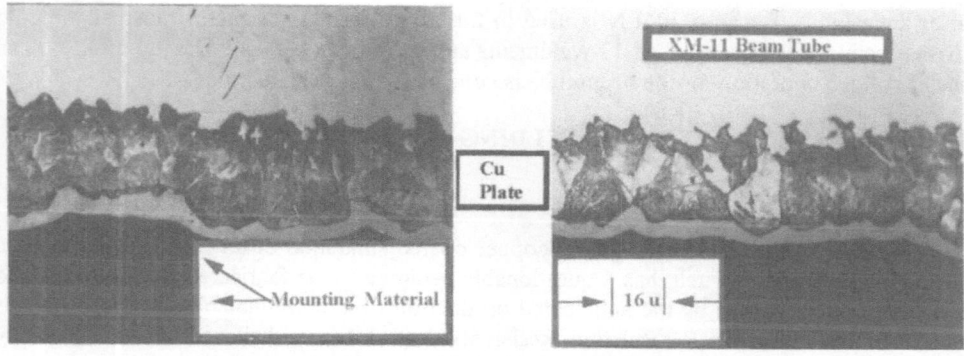

Figure 3. Copper Plating Remote
From Welding

Figure 4. Copper Plating Under
The Weld Nugget

Figure 5 shows the crossection of the weldment which is considered dimensionally and metallurgically satisfactory. There is no sign of grain growth in the XM-11 beam tube under the weld nugget.

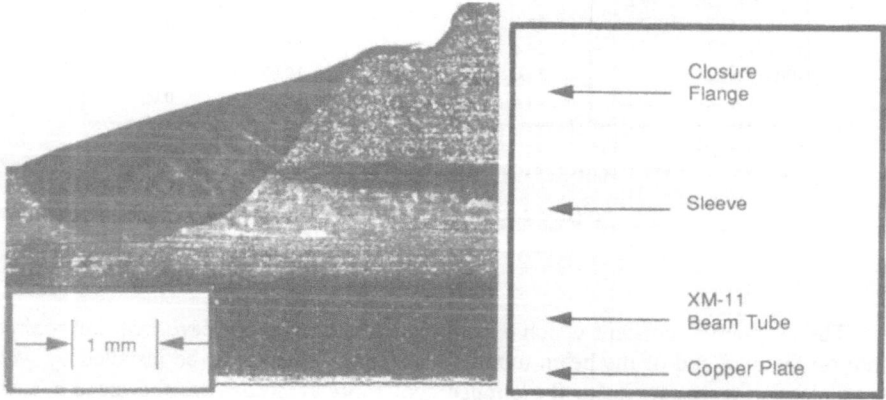

Figure 5. Autogenous GTAW Weld of Closure Flange to Sleeve on Beam Tube

Welding of the closure flange and end flange, which uses a similar weld configuration similar to the closure flange are considered viable with GTAW welding. The gas jet which was used with the mockup should be tried at a higher flow rate (when an appropriate regulator is available. The mockup report herein was welded with approximately 2300lpm., the current equipment capacity.).

EFFECTIVE STRESS OF A 4.2 K BEAM TUBE IN A QUENCHING COLLIDER 50 mm DIPOLE MAGNET FOR THE SSC

K.K. Leung, Q.S. Shu, G. Snitchler, K. Yu, and J. Zbasnik

Magnet Systems Division
Superconducting Super Collider Laboratory*
2550 Beckleymeade Ave.
Dallas, TX 75237-3997

INTRODUCTION

Two mechanical design requirements are defined for the SSC Collider beam tube.[1] First, the vacuum requirement (luminosity lifetime=150 hrs). It requires the design of a pressure boundary within the cold mass vessel to provide a vacuum tunnel for the proton beam and to minimize the synchrotron radiation gas desorbtion with a suitable material. The Collider beam tube design is under an intensive activity to search for a material that will meet the luminosity requirement without a distributed pump or liner.[1] Second is the tube wall's resistivity requirement ($\sigma^*t = 2E5\ \Omega\text{-}1$). For a 4.2 K beam tube,[1] the Cu thickness is 100 µm (RRR=30,6.7 T, σ=2E9Ω-^{1}m-^{1}). The copper yield strength is relatively low in comparison to steel and, therefore, the design of the steel layer is governed by the copper layer yield stress limit. A beam tube subjected to eddy current load in a quenching dipole requires an optimum diameter design to provide maximum aperture and adequate cooling space for the liquid Helium flow to cool the beam tube. This paper presents a mechanical design procedure using an established finite element analysis and modelling method[2] to produce a design with safety, matching the dipole cold mass vessel as designed by the ASME[3] code, and to generate a steel tube wall thickness to ensure the copper coating stress below the yield stress limit in a quenching dipole.

DESIGN CONSIDERATIONS

The SSC beam tube subjected to eddy current force[4,5,6] is designed as two-shell laminate. Material selection maximizes the vacuum and cooling space with acceptable structural safety and vacuum compatibility. Nitronic-40 steel is selected as the outer layer for its structural strength in cryogenic environment, non-magnetic properties and a close thermal expansion match to the copper layer.

*Operated by the Universities Research Association, Inc., for the U.S. Department of Energy under Contract No. DE-AC35-89ER40486.

The steel layer is designed to withstand the thermal and eddy current loads as well as the external pressure from the vaporized liquid Helium. Copper is selected as the inner layer of the beam tube for low electrical resistivity. It minimizes power loss from the image currents induced by the circulating protons on the wall of the beam tube. The copper must also retain high surface finish to shorten the path of the high frequency image current and reduce power loss.

OBJECTIVE OF THE BEAM TUBE EFFECTIVE STRESS ANALYSIS

Effective stress analysis helps produce a high quality design that enhances the magnet cold mass vessel reliability. Beam tube stress in a quenching dipole involves dynamic, magnetic, thermal, and structural loading on non-linear material. The Safety factor of the ASME SEC.VIII, DIV. 1 is employed for the beam tube design. SSC test data is used in computing the beam tube stress including, for example, the magnetic field strength and its rate of change with respect to time in a quenching dipole, and the dynamic amplification factor for the duration and shape of the transient eddy current pressure. The use of "stress ratios," the non dimensional coefficients denoting the fraction of the allowable stress (including the buckling strength by ASME code) for the combined loadings help qualify the beam tube design. Design of the beam tube is determined by its structural integrity, but more importantly to sustain the high surface finish of the copper layer and to ensure the copper stresses within the elastic limit during quench. Predicting the service life of the beam tube with plastic stress in copper layer is a complicated process. Post yield copper layer stress analysis needed information of the cold work substantiated by tests and, therefore, is not considered in the present analysis.

DESIGN LOADINGS, STRESS RATIOS AND ALLOWABLE STRESS

The beam tube loadings except the external buckling load are not included in ASME code[2] (UG-22). Use of ASME U-2 (g) and stress analysis by the finite element method for the following loads on non-linear material are proposed: (a) eddy current and quench helium external pressure, (b) eddy current torque or eddy current lateral bending force, (c) cooldown thermal load on the bimetallic tube, and (d) axial load induced by temperature rise of the beam tube heating by the coils can be reduced with a bellow. The beam tube's high ratio of length to radius of gyration ($l/\gamma > 200$) produces zero axial structural capacity. Beam tube stresses are calculated under loadings (a), (b), (c). Stress ratios are applied to qualify the beam tube design. The sum of the stress ratio is directed to be near 1.0 for a minimum tube wall thickness design. Yield strength of the Nitronic-40 at 4 K is 1185 MPa (172 ksi). The allowable tensile is established as 172/4 = 43 ksi using the ASME factor. For allowable buckling stress, no general theory exists which will apply in all cases. A 46.5-mm OD beam tube with 2-mm N-40 steel wall is calculated to have 131 MPa (19 ksi) allowable buckling stress from ASME UCS-28.3.

EDDY CURRENT PRESSURE EVALUATION

(A) The static equatorial eddy current pressure (see Figures 1 and 2) may be evaluated by finite element codes such ANSYS or EMAS. Code calculated pressure by SSC indicates that the following closed form equation[4] is conservative for beam tube design.

$$P_{Lmax} = B * (dB/dT) * b * t * \sigma$$

P_{Lmax}	= equatorial eddy current pressure.
B	= dipole field strength. (SSC test data[7] see Figure 3)
dB/dt	= field strength change at quench. (SSC test data,[7] see Figure 3)

b	= the "mean" radius of the copper layer.
t	= the layer thickness of the copper.
σ	= the copper electrical conductivity depends on RRR ratio, temperature, magnetic flux density and cold-worked condition.

For a 4.2 K beam tube with ID= 42.3 mm and copper layer = 100 microns. The calculated pressure is 93 psi with its distribution shown in Figure 2.

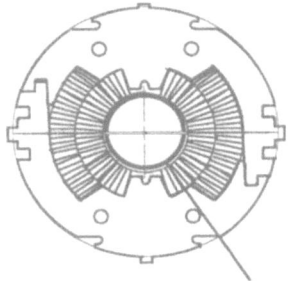

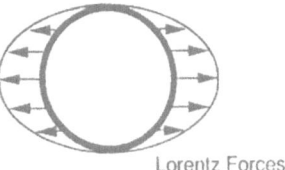

Figure 1. Beam Tube in Dipole. **Figure 2.** Eddy Current Pressure.

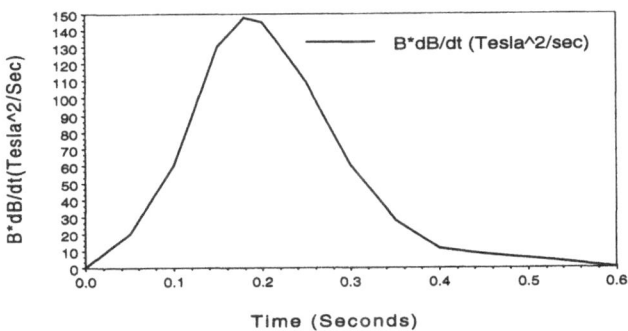

Figure 3. Dipole Field Strength at Quench.[7]

Considering eddy current load on copper[1] RRR=30 The equatorial static eddy current pressure (max. Lorentz pressure) equation is given as

P_{Lmax} = B * (dB / dT) * b* t * σ = * B(tesla) * dB/dT (tesla/sec) *b (m) *t (m)
σ (Ω-1 m-1) (MPa).

Where T= 0.18 sec., B = 6.01 (tesla), dB/dT= 24.52 (tesla/sec)
B*dB/dT= 147.26 T^2/sec,
r = 0.02115 (m),
b = 0.0217
t =1.0-4 m
$\sigma^1_{(6.7, 4.2))}$= 2E9 x (Ωm) $^{-1}$

P_{Lmax} = 1.45E-4 *147.26 T^2/sec* 0.0217 (m)* 1.E-4 (m)
2 E9 (Ω-1*m -1) = 0.64 MPa (93 psi)

STRESS ANALYSIS FOR A 42.3 MM ID COLLIDER DIPOLE MAGNET (CDM) BEAM TUBE WITH 2.0 mm STEEL WALL AND 0.1 mm COPPER LAYER

A sample calculation is included for the proposed SSC CDM beam tube. The CDM beam tube is subjected to the largest eddy current force and its thickness represents a conservative design for all other SSC mgnet systems. The design data for the CDM beam tube design is summarized as follows:

(1) L. He. Pressure = 2 MPa = 290 psi at 0.20 seconds, SSCMSD specification
(2) The static eddy current pressure is 0.64 MPa (93 psi) at 0.18 seconds
(3) Beam tube buckling stress from L. He. = 290*46.5 / (2*2) = 23.2 MPa (3.37 ksi).
     The Stress ratio = 3.37 / 19 (ASME Code Allowable) = 0.17737
(4) Eddy current and cool down stress = 19.45 ksi (From 3D model FEM analysis).
     The stress ratio = 19.45 / 43 (ASME Code safety factor included)) = 0.463
(5) Estimated maximum non-axisymmetrical eddy current pressure = 20 psi,[6] and the tube bending stress is = 28.9 MPa (4.2 ksi). The stress ratio = 4.2/43 (ASME Code safety factor included) = 0.1
(6) The total stress ratio = 0.17737 + 0.463 + 0.1 = 0.74 < 1.0 and is acceptable
(7) The combined copper stress is 34.5 MPa (5.02[8] ksi) <44 MPa (6.4 ksi) yield
(8) The dynamic factor is estimated as 1.1 to 1.3 and is not included in this analysis.

CONCLUSIONS AND DISCUSSIONS

A beam tube with 100 μm copper (RRR-30) bonded to a 2.0 mm layer of stainless steel (Nitronic-40) fulfills the structural requirement based on the SSC Magnet System Division quench pressure specification.[8,9] Reduction of the steel wall thickness is possible because test data[7,8] indicating that the helium pressure is less than the specification limit. Designing a beam tube with a minimum wall should be governed by the copper yield stress limit. Designing the copper for post-yield strength is not acceptable because of the development of micro-cracking in the copper. A bellow is required to reduce the beam tube axial stress.The high slenderness ratio of the beam tube ($l/\gamma > 200$) practically has no axial load capacity. The non-axisymmetrical eddy current loads need to be considered in the bellow and the end joints design. The present beam tube analysis is performed on a 3-D finite element model in a non-linear stress analysis including non-linear material properties and "beam-column" effect. We learn from this analysis that a 2D linear stress analysis produces unacceptable erroneous copper layer stress.

REFERENCES

1. W. C. Turner. "Collider Beam Tube Vacuum," Presentation at SSCL, March 1983.
2. K. K. Leung. "Seismic Stress of Heat Exchanger Supporting Structures," 3rd International Conference on Structural Mechanics in Reactor Technology, London, 1975.
3. American Society of Mechanical Engineers. "ASME Code Sec. VIII, Div. 1," 1992.
4. A. Chao. "More on Copper Coating Considerations," Report, SSC-N-434.
5. K. Y. Ng, et al., "Allowable Stress in The SSC Beam Tube During A Quench," SSCL Report, SSC-168, 1988.
6. K. K. Leung. "Non-axisymmetrical Eddy Current Loads on Beam Tube in Quenching Dipole Magnet," SSCL Report, MD-TA-243.
7. Q. S. Shu. "Status of Tests on RF Surface & Magneto-Resistance," SSC Meeting, March 1993.
8. K. K. Leung. "Design of the SSC Collider 4.2 K Beam Tube," SSC Meeting, March 1993.
9. SSCL Specification, "Doc. No.: M80-000001 Rev: A, Magnet System Specification," SSCL, Jan. 1992.

DYNAMIC ANALYSIS OF SIX-STRUT SUPPORTING SYSTEM FOR ACCELERATOR MAGNET

K. K. Leung

Magnet System Division
Superconducting Super Collider Laboratory*
2550 Beckleymeade Ave.
Dallas, TX 75237

ABSTRACT

A six-strut magnet support system designed by Lawrence Berkeley Laboratory (LBL) is considered as an alternative to the current SSC magnet support system. The LBL designed a six-strut support system based on the kinematics mount concept that is generally used in the optical and the laser communication industries. The six-strut system is defined by six static degrees of freedom that constrain a point in space with no redundant restraint. Adjustment of any strut's length means redefining the translation or rotational degree of freedom of the mounting point and produces the desirable movement of the magnet system. The accurately operated six-strut mounting system used in the Berkeley's Advance Light Source (ALS) magnet support is able to maintain the magnet system structural integrity to survive a 7 earthquake,[1] position the magnet to high tolerances, have a small footprint, simple to operate, and adjust to a micron level of accuracy.

Though finite element simulation has been used for years in safety analysis, such as seismic dynamic response analysis in nuclear reactor and piping supports, in late 1970, it was employed in the dynamic study for a magnet system in Lawrence Berkeley Laboratory in the late eighties.[2] The modeling methodology developed in LBL for the six-strut system design, especially for the critical mounting joint design under dynamic loads, is presented in this paper and may be employed for prospective SSC accelerator magnet supporting system design.

INTRODUCTION

The dynamic response stress results of the critical magnet supporting components, subjected to seismic excitation, is presented in this paper. The components consist of the post, girder base plate, and main supporting bolt.

---

*Operated by the Universities Research Association, Inc., for the U.S. Department of Energy under Contract No. DE-AC35-89ER40486.

This calculation is of interest for three reasons: (1) the loading occurs when the horizontal (x) seismic force combining with the dead load of the girder producing a 30 kips downward force to a strut that is cantilevered out from the center of a steel plate, (2) the main bolt is held by the base plate in supporting the magnet system, (3) the anchor bolt and its tensile load must be compared with a limited number of separate field tests that gives experimentally measured values, of the the ultimate bond strength, of the anchor bolts in the concrete slab.

DESCRIPTION OF THE FINITE ELEMENT MODEL

A three dimensional Finite Element Model (FEM) is employed for LBL's ALS booster ring girder in order to calculate earthquake dynamic response. The booster ring girder is constructed by joining two (2) straight Wide Flange (WF) beams, with the web positioned parallel to the floor, then welding the center of the girder at approximately 10 degrees. A plate is then welded to one side of the top and bottom flange of the WF beam that serves as a top plate to support all the magnets. The steel girder model consists of plate elements with six (6) positions of adjustment in order for plane bending, moment connections and shear connections of load to

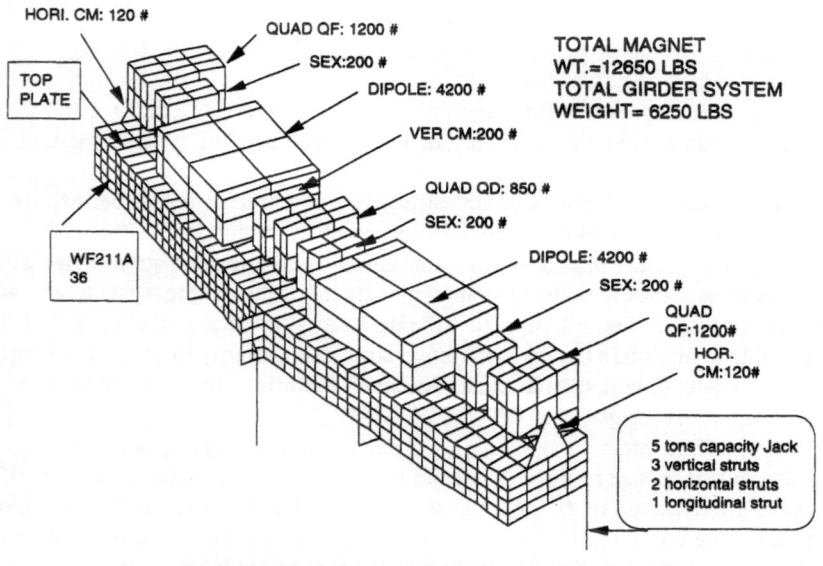

FEM FOR BOOSTER RING GIRDER SEISMIC ANALYSIS

Figure 1. ALS Booster Ring Magnet and Supporting System of LBL

be accomplished. The parameters controlling the booster ring girder system dynamic response, in the FEM are summarized as follows: a) lumped mass magnet locations, b) experimental compliance data of the magnet and girder supporting struts. The correct compliance for struts is critical in dynamic modelling.

Figure 1 presents the FEM of the booster ring magent and the supporting system. Figures 2 through 4 show the posts, the base plate and the main bolt. Figure 5 shows maximum stress in base plate, and Figure 6 shows maximum strut loads.

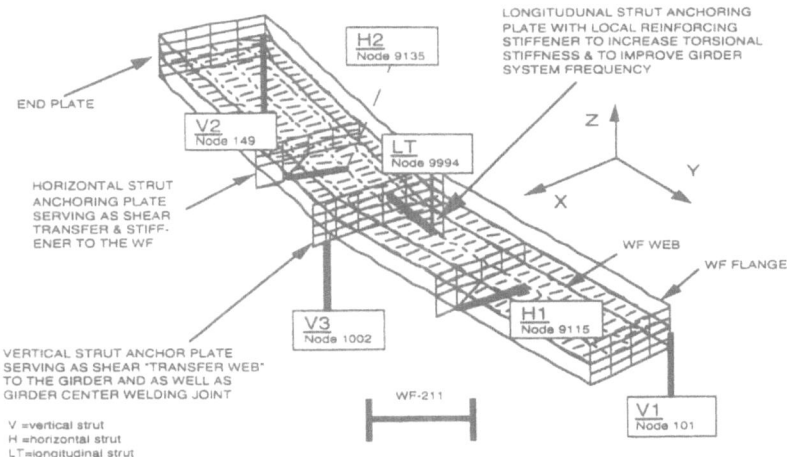

Figure 2. Magnet Support System

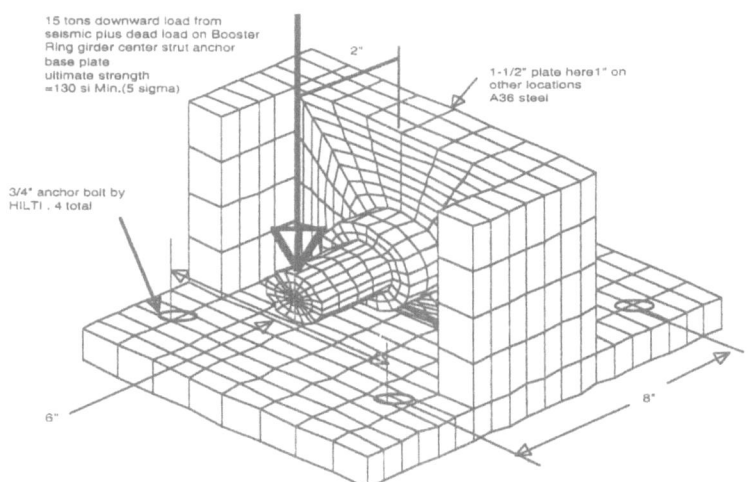

Figure 3. Base Plate

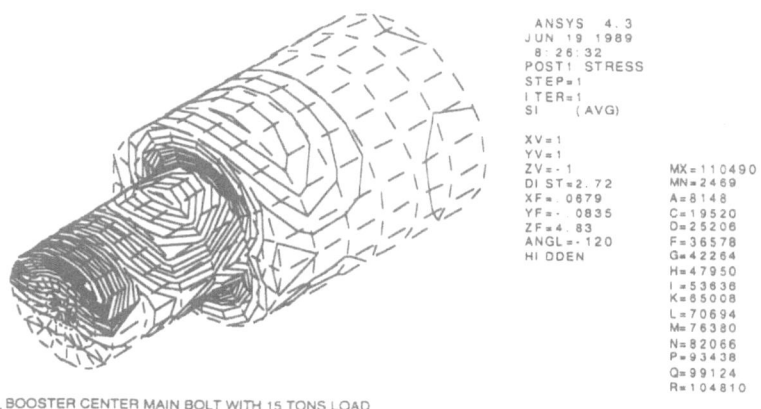

Figure 4. Maximum Stress in the Main Supporting Bolt

277

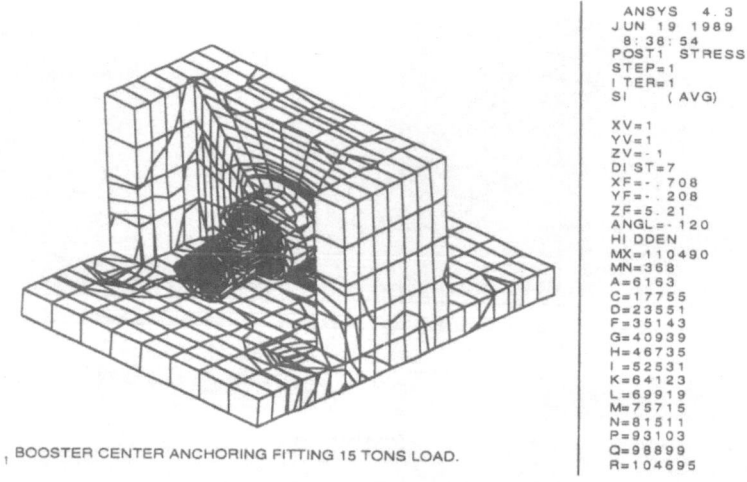

ANSYS   4. 3
JUN 19 1989
8:38:54
POST1 STRESS
STEP=1
I TER=1
SI   (AVG)

XV=1
YV=1
ZV=-1
DIST=7
XF=-.708
YF=-.208
ZF=5.21
ANGL=-120
HIDDEN
MX=110490
MN=368
A=6163
C=17755
D=23551
F=35143
G=40939
H=46735
I =52531
K=64123
L=69919
M=75715
N=81511
P=93103
Q=98899
R=104695

BOOSTER CENTER ANCHORING FITTING 15 TONS LOAD.

Figure 5. Max. Stress in the Base Plate

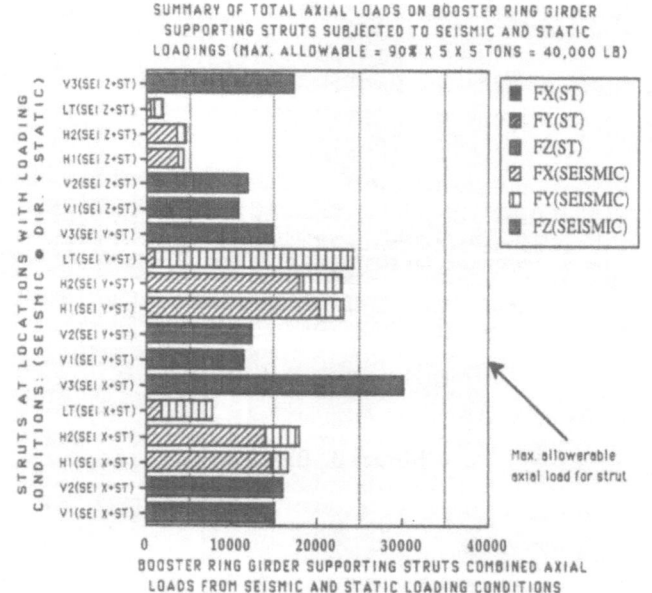

SUMMARY OF TOTAL AXIAL LOADS ON BOOSTER RING GIRDER
SUPPORTING STRUTS SUBJECTED TO SEISMIC AND STATIC
LOADINGS (MAX. ALLOWABLE = 90% X 5 X 5 TONS = 40,000 LB)

Figure 6. Max. Strut Loads

ACKNOWLEDGMENTS

The author wishs to thank A. Paterson, and T. Lauritzen of LBL for their support.

REFERENCES

1. B. A. Bolt. "Strong Seismic Ground Motion for Design Purpose at the Lawrence
   Berkeley Laboratory," Report LBL-17377 (July 1979)
2. K. Leung, A. Paterson. "ALS Seismic Design Procedures," LSME-135, May 17, 1989.

USE OF AUTOMATED TEST EQUIPMENT AND "PAPERLESS" PROCESS CONTROL TO IMPLEMENT EFFICIENT PRODUCTION OF SSC DIPOLE MAGNETS

T. Tobin, R. Fagan, and D. Mitchell

General Dynamics Space Systems Division
Space Magnetics Division
P.O. Box 85990
San Diego CA 92186

ABSTRACT

In an effort to minimize human error and maximize process control and test capabilities during Collider Dipole Magnets (CDM) production, General Dynamics is developing automated test and process control equipment; known as Test & Process Control Modules (TPCM's). When used along with software designed to create "paperless" process control documentation, the system becomes the Test & Process Control System (TPCS). This system simplifies business decisions and eliminates some problems normally associated with process control documentation, while reducing human errors during CDM production. It is also designed to reduce test operator errors normally incurred during test setup and data analysis.

We will present an overview of the TPCS hardware and software being developed at General Dynamics, along with the process control techniques included in TPCS.

INTRODUCTION

TPCS is a system, and as such, the hardware, and software in this system develop in parallel around several fundamental system components.

System components were selected using the following criteria: capability to perform required task, proven technology, cost, reliability, system integration capability, availability. These criteria, all possessing a weighted value relative to each other, form a matrix by which components were judged and either integrated into the system or rejected.

HARDWARE

The host computer for the entire TPCS is a HP 9000 Server built by Hewlett Packard. This computer ties all the individual departments and software to a single data base for concurrent planning, process, and material control.

TPCM hardware design evolved around the need for diversified capabilities at each production process for CDM. As a result, seven different types of TPCM's were identified; each tailored to its task. Some contain computers for process control and test control. Some contain test equipment and power sources. Some contain computers and test equipment.

The heart of most TPCM types is an industrialized rack mounted i386 PC Clone computer built by Texas MicroSystems. Through this computer run the hardware interfaces to tooling, test equipment, as well as the software to control process, test, planning, and material. A super VGA touch screen monitor allows easy user interface in a production environment.

Serial and parallel communications tie the computer to test equipment and tooling. A Local Area Network (LAN) connects the individual TPCM's to the host computer and the rest of the TPCS.

Resistance test data from certain CDM production process points is retrieved by a HP 3457A multimeter built by Hewlett Packard. Data is then passed to the computer for pass fail evaluation and data download to the host computer. Hi-pot test data is taken by a M100DC tester built by ROD-L. Data is then passed to the computer for pass fail evaluation and data download to the host computer. Inductance test data in taken by a HP4278A LCZ meter built by Hewlett Packard.

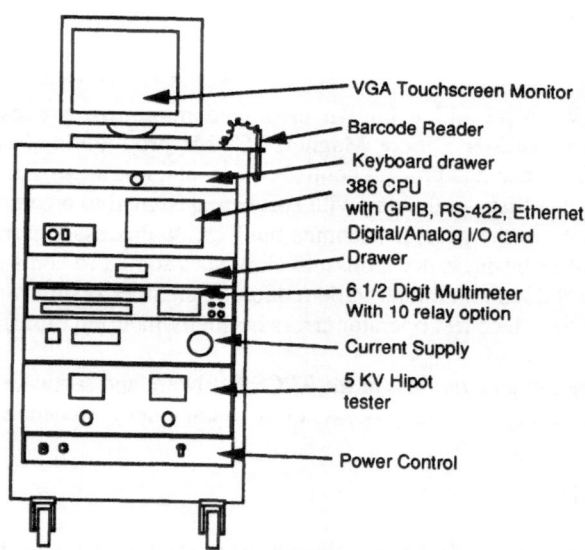

Figure 1. Typical TPCM containing resistance monitoring equipment, hi-pot equipment and process control computer.

SOFTWARE

TPCS software is divided into two areas. The first is software on the host computer used for planning, shop floor control, data gathering, Statistical process control (SPC), configuration management, and budget control. The second is software on the TPCM's located throughout the plant, used for process control, tool interface, automated testing and process flow control.

Host computer software includes the following:

1. Fourth Shift Corp. Fourth Shift Manufacturing Resource Planning (MRPII) for material planning, material ordering, and associated material financial data.
2. Cimflex Teknowledge Corp. Delta for analysis and historical data collection, fault reporting, "as built" information, serialization, SPC, trend analysis and shop floor control.
3. Welcom Software Corp. Open Plan for tracking activities and events during process phases, man power requirements tracking
4. Electronic Data Systems Unigraphics and InfoManager for automatic Engineering drawing configuration management
5. MicroFrame Corp. Control/Award for budget control

TPCM computer software includes the following:

1. Microsoft Windows for a graphical user interface (GUI)
2. Microsoft Word for writing Process Requirements Documents (PRD's) and other process control documents
3. Custom (GDSS) application written in C++ for use under Microsoft Windows for controlling all functions of the TPCM such as automated test processes, PRD's, tooling interfaces, host computer interfaces

PROCESS CONTROL

Process Requirements Documents (PRD's) are the basic interface document between the production technician and the TPCM and the entire TPCS. PRD's provide step by step work and test instructions for the technician and the TPCM. These documents, created in Microsoft Word, are called by the GDSS application and provide "paperless" work instructions for the production technician to build and test CDM's. They also provide instructions for the TPCM to communicate with tooling, test equipment, perform tests, and download and upload information with the host computer. Through the PRD flow command sequences, test data, historical data and process flow commands that update and create the entire TPCS database.

CONCLUSIONS

The TPCS allows a consolidated approach to CDM production. All areas of production simultaneously receive pertinent data critical to the respective departments whether it is material procurement, test, or planning. The shared database in the TPCS allows faster, and more efficient business decisions in all departments. Since the system is automated, it minimizes human induced deviations and errors. The following examples represent some of the advantages of the automated TPCS.

The TPCM test current sources are automatically set to a predetermined level whenever a precision resistance measurement is performed on a CDM coil during production. Since this level in consistently driven by the computer, human tendencies to adjust the current level slightly higher than the recommended level are eliminated. This provides a similar test setup for each magnet tested, thereby making the database created free from this subtle error induced by different personnel performing the same task.

As technicians install parts into a magnet, they enter serial numbers of critical parts into the computer via the bar code wand. Software tracks quantities of non serialized parts consumed with the completion of each PRD. Software then tracks remaining quantities in stock so the material control department can do real time control. This allows "Just In Time" manufacturing techniques to be implemented. In other words, parts enter into the plant just before installation into the magnet, thus eliminating large stockrooms containing parts; some of which are 15 meters long.

Quality engineering overlays test data from several process steps for different magnets. Through SPC they recognize trends that could identify tooling wear or out of tolerance material or processes. Since the database is automated, Quality quickly analyzes data and corrects problems before they cause larger out of tolerance conditions.

ACKNOWLEDGMENTS

Portions of the work described herein are being accomplished under contract to the Universities Research Association in support of the Superconducting Super Collider project for the U.S. Department of Energy.

SYSTEMS ANALYSIS DETERMINING CRITICAL ITEMS, CRITICAL ASSEMBLY PROCESSES, PRIMARY FAILURE MODES AND CORRECTIVE ACTIONS ON ASST MAGNETS

Craig S. Arden

Superconducting Super Collider Laboratory*
2550 Beckleymeade Avenue
Dallas, TX 75237-3997

INTRODUCTION

During the assembly process through the completion of the Accelerator Surface String Test (ASST) phase one test, Magnet Systems Division Reliability Engineering has tracked all the known discrepancies utilizing the Failure Reporting, Analysis and Corrective Action System (FRACAS) and data base.

This paper discusses the critical items, critical assembly processes, primary failure modes and corrective actions (lessons learned) based on actual data for the ASST magnets. The ASST magnets include seven Brookhaven Lab Dipoles (DCA-207 through 213), fourteen Fermi Lab Dipoles (DCA-310 through 323) and five Lawrence Berkeley Lab Quadrupoles (QCC-402 through 406).

Between all the ASST magnets built there were one hundred eighty six (186) class one discrepancies reported out of approximately eleven hundred total discrepancy reports. The class one or critical discrepancies are defined as form, fit, function, safety or reliability problem. Each and every ASST magnet is considered a success, as they all achieved the quench performance requirements and were capable of being incorporated into the string test.

This paper will also discuss some specific magnet discrepancies, including failure cause(s), corrective action and possible open issues.

CRITICAL ITEMS AND ASSEMBLY PROCESSES

As shown in percentage of class one failures, Figure 1, Critical Items List, the top two items of concern are cryostat components. Problems encountered were cryogenic tubes not in the proper position and thermal shields not within the dimensional tolerances causing possible thermal shorts. These problems did not effect the string test, but do effect the heat load specified for the magnet and may increase the work load on the cryogenic system. The third item shown is strain gauge instrumentation, which is important for prototype test data, but has no effect on magnet quench performance and is not planned for production hardware.

Magnet cold mass items, which include the superconducting coils, power bus assembly, and the quench protection heaters are the next critical items on the list. Coil problems primarily include turn-to-turn shorts or coils damaged during the curing process. The primary bus problem was Hi-pot failures due to the bus absorbing moisture as a result using improper soldering-flux. Although this

*Operated by the Universities Research Association, Inc., for the U.S. Department of Energy under Contract No. DE-AC35-89ER40486.

problem was observed early and fixed, subsequent bus problems were due to mechanical fit. The quench protection heaters had some dimples, pin holes in the kapton jacket and some discoloration along its length. Work developing manufacturing processes for the heaters continued with the vendor throughout the ASST magnet assembly phase. The low volume, tight tolerance and the long length of the heaters resulted in some discrepancies. Not all heater problems were due to manufacturing, but assembly processes as well.

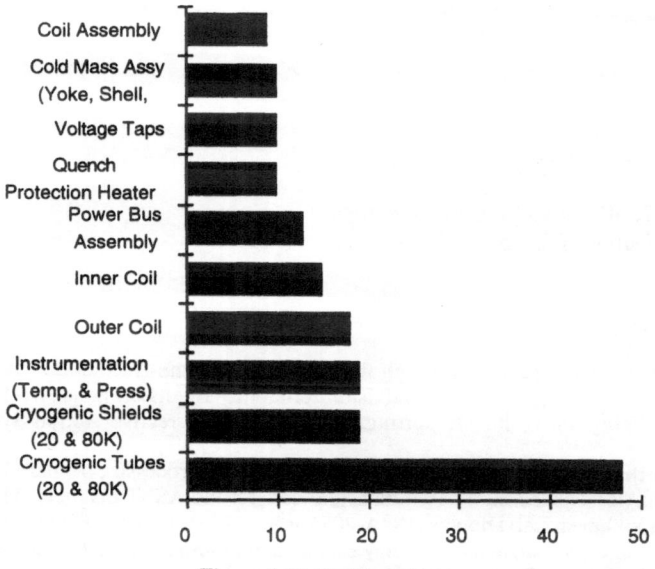

Figure 1. Critical Item List

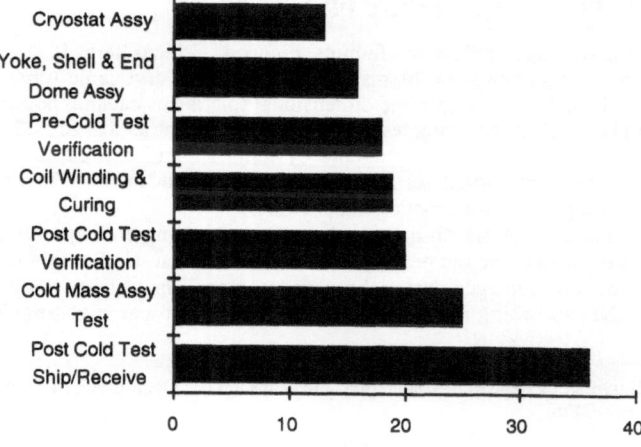

Figure 2. Critical Assembly Processes

The data in Figure 2, Critical Assembly Process, is somewhat in error due to the inclusion of post cold test, shipping and receiving. Not that these checks are important or critical but they are not part of the magnet assembly. Post cold test failures were primarily due to open instrumentation voltage taps or cryogenic tubes out of location. Similarly, shipping and receiving problems included cryogenic tube position and damaged instrumentation.

Figure 2 shows that during primary electrical verification, which includes cold mass assembly test, yoke, shell and end dome assembly and post collar keying test, is where many of the failures were found. Figure 2 also shows that coil winding and curing, which is one of the primary assembly processes, is most critical among the magnet assemblies.

PRIMARY FAILURE MODES AND CAUSES

The top three primary failure modes shown in Figure 3: Primary Failure Modes, dimensional, workmanship, and configuration are basically due to initial assembly start-up and should be discovered in the prototype and preproduction stages. The coil shorts, power bus shorts, quench heater shorts, coil open and coil bonding items are considered more important, requiring assembly verification to detect potential failures.

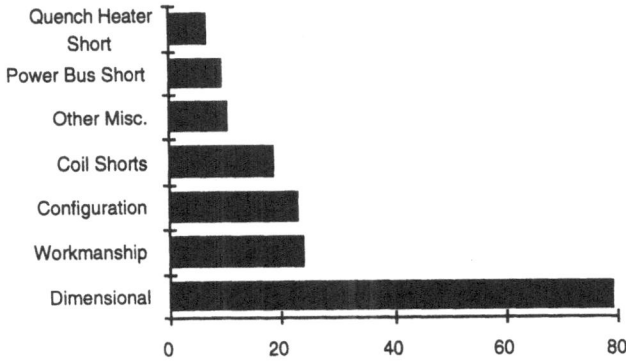

Figure 3. Primary Failure Modes

A fundamental part of the FRACAS procedures is to eliminate the cause of failures. Consequently, seven causes were identified and presented in Figure 4: Design was the primary cause for the failures identified; and points to the cryogenic tube problem as well as the short start-up time from design to magnet assembly. Similarly, processes were being developed along with initial assembly. These causes are basically due to start-up and could be controlled if not eliminated in a mature production environment.

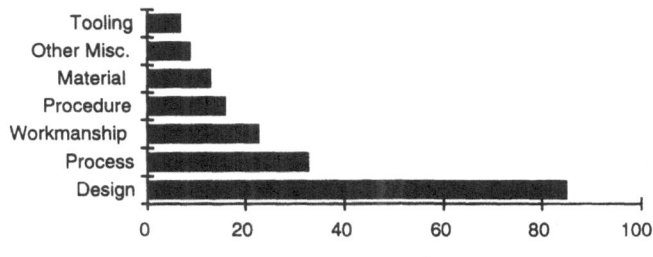

Figure 4. Primary Failure Causes

HARDWARE DISPOSITION AND CORRECTIVE ACTION

Hardware disposition identifies what actions are necessary in order for the magnet to be usable in the ASST. Therefore it is odd that Figure 5 shows "Use as Is" as the primary hardware disposition. This shows that even though some parts may not have been per print, but were more then just OK, and had no effect on the product integrity.

Briefly, the difference between repair and rework, is that repair includes disassembly of an assembly in order to perform the required repair. Similarly, replace is defined as remove and replace the in-doubt component.

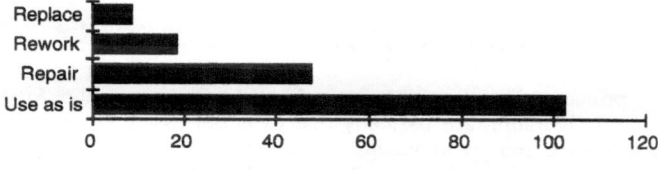

Figure 5. Hardware Disposition

In the long term, it is cost effective to prevent now known problems by incorporating corrective action. Figure 6 contains a list of actions taken or need to be taken for this purpose. "Vendor to Address" means that no effective corrective action was feasible during the brief and schedule driven ASST production and that the problem (primarily cryogenic tube position items) was left for the vendor to analyze and address during the next design generation magnets. "None" or no corrective action appears to be a poor closed loop preventive action. However, if the cause of the problem can not be specifically found, such as in coil turn-to-turn shorts, besides hardware repair, no logical corrective action can be or should be taken.

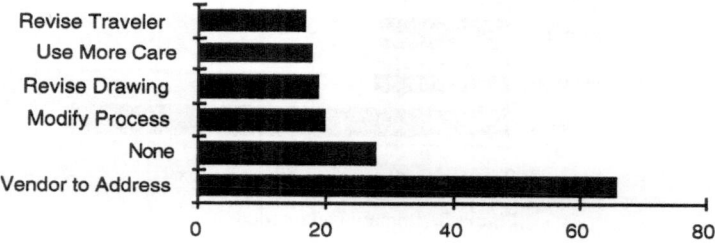

Figure 6. Corrective Actions Taken

SUMMARY

It is true many discrepancies were found, but it is also true that all the discrepancies which were found and corrected, was prior to magnet cold quench performance testing. This proves that with the proper verification and check points the SSCL superconducting magnets, which are critical items, are very achievable to produce and prove to be within or exceed the quench performance specification.

The primary open issue lies in the cryogenic tube position. Basically, the initial interface control attempted to make all items rigid and control tolerances, but this was not achieved. Magnet contractors have been made aware of this problem and have proposed various solutions that should prevent it from occurring on production magnets. These concepts allow final adjustment of pipe ends in late production stages or even in the tunnel, do not increase heat loads, and stay within allowable bellows misalignment requirements.

ACKNOWLEDGEMENTS

Thanks goes to John M. Robinson, SSCL, Reliability Engineering Support, who performed all the FRACAS data input, provided technical input and the data in a well organized manner.

And to James Franciscovich, SSCL and Lockheed Systems Engineer, Systems Engineering Group Manager, who provided technical input and document review.

MEASUREMENT OF COMPLEX SURFACES

Graham M. Brown

Superconducting Super Collider Laboratory*
2550 Beckleymeade Ave.
Dallas, TX 75237

INTRODUCTION

Several of the components used in coil fabrication involve complex surfaces and dimensions that are not well suited to measurements using conventional dimensional measuring equipment. Some relatively simple techniques that are in use in the SSCL Magnet Systems Division (MSD) for incoming inspection will be described, with discussion of their suitability for specific applications.

Components that are submitted for MSD Quality Assurance (QA) dimensional inspection may be divided into two distinct categories; the first category involves components for which there is an approved drawing and for which all nominal dimensions are known; the second category involves parts for which 'reverse engineering' is required, the part is available but there are no available drawings or dimensions. This second category typically occurs during development of coil end parts and coil turn filler parts where it is necessary to manually shape the part and then measure it to develop the information required to prepare a drawing for the part.

Measurement of three-dimensional components is normally performed on a coordinate measuring machine (CMM) and it is important to briefly review here some of the operational characteristics of these machines. A touch sensitive probe with a spherical tip is used to contact the surface of a component and the CMM initially records the position of the center of the spherical tip. It is not possible to directly measure points along an edge. The probe center position is then compensated by the probe radius, to determine the actual location of the contact point. In the case of surfaces that consist of planes, lines and circular or cylindrical features, this compensation can be performed by standard geometric techniques. In the case of complex surfaces, this compensation can only be performed when the probe approaches the surface along the direction of the surface normal. The magnitude of the probe correction depends on the tip diameter and on the deviation of the approach direction from the surface normal. It might appear possible to eliminate the need for compensation by using a 'point' probe. Examinations of typical 'point' probe tips on an optical comparator have shown tip radii of the order of 0.075mm. This dimension is 75% of a typical total profile tolerance for coil end parts.

*Operated by the Universities Research Association, Inc., for the U.S. Department of Energy under Contract No. DE-AC35-89ER40486.

MEASUREMENT OF COMPONENTS WITH DEFINED SURFACES

A typical coil end part is shown in Figure 1. This part is fabricated as the intersection of two ruled surfaces with the inner and outer surfaces of a cylinder. The inner and outer ruled surfaces are each defined by a total of 400 points along the intersections of the ruled surfaces with the surfaces of the cylinder (2 intersections * 2 symmetry about a longitudinal axis * 2 edges * 50 points per edge). Coordinates of these points are available from the computer assisted design (CAD) system, but surface normals are not available. A ruled surface has one of the principal radii of curvature equal to infinity and it is important that this feature be maintained during fabrication. Some care is required in programming numerical control machining systems to generate ruled surfaces; the standard method of determining a tool path by connecting defined points with spline curves and generating a surface from those curves does not usually result in a ruled surface.

The approach used for inspection of components with surfaces defined by ruling lines is as follows:

• straight lines are generated between corresponding ruling points on the inner and outer cylinder surfaces, these lines are the ruling lines used to define the surface.

• point coordinates are calculated at the one-third and two-third length locations along the lines. The selection of these fractions is somewhat arbitrary. Any position sufficiently far from the edge for good data would be acceptable.

• for all ruling lines other than those at the ends of the surface, approximate normals are generated from the vector product of vectors drawn along the diagonals of a panel formed from the ruling lines on either side of the line under consideration. At the panel ends, approximate normals are calculated from the end ruling line and the line adjacent to it. This method can be shown to produce approximate normals that are typically within +- 5 degrees of the actual surface normal and the cosine correction for deviations from the actual surface normal can be neglected.

• a 'vector touch' is performed at each of the calculated points, along the calculated normal to the surface (Actually, to the point on the surface that lies on the line through the nominal point with the calculated normal direction. This method requires operation of the CMM under program control and with small target tolerances).

• the distances between the nominal and actual points can then be calculated and compared with the allowed tolerance. This task is simplified by printing, in addition to the actual measured and nominal data, a table in which the condition of each point is denoted by a one-character symbol that shows the degree of deviation from the nominal value.

A standard inspection program has been written to read nominal point coordinates from a data file and perform the inspection of many different parts based on the information in the file. A standard tip definition file that lists all required probe tip orientations is used with a general calibration program for automatic calibration of all required tips, and to verify in the inspection program that all required tip calibrations have been performed prior to the start of the inspection.

This method is reliable and measurements on metal components typically show repeatability within the nominal CMM specifications (+- 0.005mm). Measurements on glass-reinforced epoxy components typically show repeatability of +-0.02mm. The degradation is primarily attributed to surface irregularities of the composite and to slight variations in the actual probe locations in repeated measurements. Repeatability on measurements of composites can be improved by decreasing the allowed touch probe target tolerances. CMM hardware does impose a lower limit on target tolerance values.

MEASUREMENT OF COMPONENTS WITH ARBITRARY SURFACES

In the case of components with surfaces that do not consist of points with known coordinates (see Figure 2) direct probe compensation is not possible and a more complex

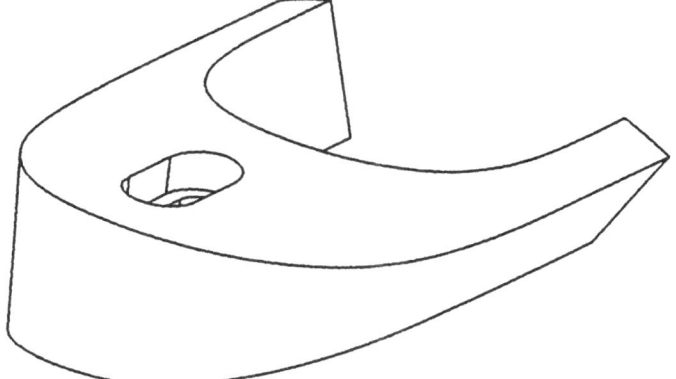

Figure 1. A typical part constructed from ruled surfaces and concentric cylinders

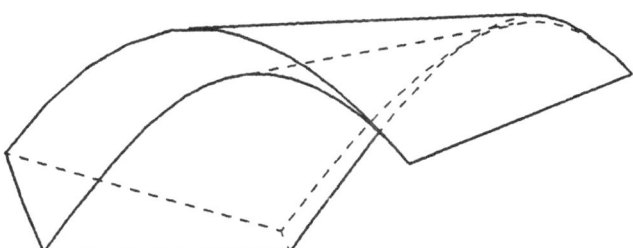

Figure 2. A typical part constructed from arbitrary surfaces

approach is required. If a set of points is measured without probe compensation, the measured values will correspond to points on a surface that is locally parallel to the desired surface and distant from it by one probe radius along the local surface normal. CAD programs with three-dimensional capability allow the translation of the measured surface along local surface normals by a prescribed distance.

One approach that has been used successfully is as follows:
• divide the surface to be measured into a set of adjoining quadrilaterals, with (possibly) a different probe tip orientation for each quadrilateral. The size of each quadrilateral depends on the configuration of the surface and on the available access for the probe tip. The normal to the best-fit plane through the vertices of the quadrilateral is used to determine the direction of approach to the surface and points are measured without probe compensation.
• divide each quadrilateral into geometrically similar subelements, based on a maximum step size along each of the two lines intersecting at a specified boundary corner of the quadrilateral.
• measure points at each of the interior points and along the boundary of each quadrilateral. Measured values are written to a disk file in a format appropriate for data input to a CAD program.
• use the points in this data file to generate a three-dimensional CAD surface that is then translated in the appropriate direction along the local surface normals by one probe radius, to generate a surface that corresponds to the actual measured surface.

It might appear that this would be an efficient method for measuring the ruled surfaces discussed in the previous section. In fact, due to CAD limitations, comparisons between actual and nominal positions become very difficult if this approach is used. Ideally, it requires a general three-dimensional surface fitting routine that supports all six degrees of freedom in determining the best fit between two surfaces, and with the ability to determine distances between two surfaces along the normals to one surface at prescribed points on that surface. The problem is further complicated by the fact that the total component consists of several individual surfaces with discontinuous curvatures at their intersections and the best-fit is required on all surfaces simultaneously.

SUMMARY

Two techniques have been presented for measurement of the complex surfaces typically found in the coil end pieces of superconducting magnets. Applicability criteria have been listed, together with some limitations of existing hardware and software.

SUPPLIER PERFORMANCE EVALUATION AND RATING SYSTEM (SPEARS)

M. Oged, D. G. Warner, and E. Gurbuz

Superconducting Super Collider Laboratory*
2550 Beckleymeade Ave.
Dallas, TX 75237-3997

INTRODUCTION

The SSCL Magnet Quality Assurance Department has implemented a Supplier Performance Evaluation and Rating System (SPEARS) to assess supplier performance throughout the development and production stages of the SSCL program. The main objectives of SPEARS are to promote teamwork and recognize performance. This paper examines the current implementation of SPEARS.

MSD QA supports the development and production of SSC superconducting magnets while implementing the requirements of DOE Order 5700.6C. The MSD QA program is based on the concept of continuous improvement in quality and productivity. The QA program requires that procurement of items and services be controlled to assure conformance to specification. SPEARS has been implemented to meet DOE requirements and to enhance overall confidence in supplier performance. Key elements of SPEARS include supplier evaluation and selection as well as evaluation of furnished quality through source inspection, audit, and receipt inspection. These elements are described in this paper.

THE SELECTION AND EVALUATION PROCESS

Supplier selection and evaluation are accomplished initially by a "pre-award survey" and later through periodic evaluation of performance. The purpose of the pre-award survey is to determine a supplier's ability to: 1) produce items that will meet SSCL specifications, 2) maintain a Quality Assurance program which ensures finished quality, and 3) minimize cost by reducing losses. The pre-award survey is based on MIL-STD-45208 quality assurance requirements for suppliers of magnet and tooling component parts, or MIL-Q-9858 quality assurance requirements for major magnet suppliers. An "Approved Suppliers List" (ASL) is maintained by MSD QA. Only suppliers on the ASL are used for procurement of magnet and tooling items.

---

* Operated by the Universities Research Association, Inc., for the U. S. Department of Energy under Contract No. DE-AC35-89ER40486.

PERIODIC EVALUATION

Supplier quality is periodically evaluated by MSD QA. Supplier "Approved" status may be removed as a result of poor performance. Under SPEARS, each approved supplier is evaluated quarterly. Performance is measured by incoming Quality (defined as the percent of conforming lots received). A three level Quality Rating is assigned depending on the percent of acceptable lots received as described in Table 1. Quality Rating Parameters. The supplier rating becomes part of the ASL. In the future, we plan to expand the evaluation criteria to include "Schedule" as well as frequency of "Waiver/Deviations".

Table 1. Quality Rating Parameters.

Rating	Category	Quality (% acceptance Rate)
1	Outstanding	95-100%
2	Good	90-94%
3	Unsatisfactory	Below 90%

Suppliers having a Quality Rating of 3 receive increased MSD surveillance and may be recommended for restricted use or removal from the ASL. If supplier quality remains less than 90% for one calendar quarter, Procurement Quality Assurance (PQA) will inform the supplier by means of a performance status letter requesting corrective action. If supplier quality remains at less than 90% for a second quarter within one year of the first quarter, the supplier is down graded to "conditional status" with restricted procurement action until corrective actions are complete. If supplier quality remains at less than 90% for a third quarter within one year, the supplier becomes "disapproved" and is removed from the ASL.

Exceptions to or deviations from these guidelines may be accomplished by direction of the MSD Associate Director. Re-qualification is accomplished through a "pre-award survey" as described earlier and by demonstrated evidence of supplier improvements. Consideration is given to product deficiencies not within supplier control. Deviations/Waivers submitted and approved prior to SSCL inspection are not counted against the supplier.

Summary data of supplier performance for four consecutive quarters (January through December, 1992) is provided in Table 2. Supplier Performance Summary (calendar 1992). See Figure 1. SPEARS Summary (Status) and Figure 2. SPEARS Summary (Active Status) for an additional description of recent supplier performance.

Table 2. Supplier Performance Summary (calendar 1992).

	ACTIVITY PERIOD			
	JAN-MAR (Q1)	APL-JUN (Q2)	JUL-SEP (Q3)	OCT-DEC (Q4)
Approved Suppliers	122	132	132	148
Inactive	81	100	110	124
Active*	41	32	22	24
Unsatisfactory	11	9	3	9
% Unsatisfactory	27	28	14	38

* Active Suppliers: Suppliers that had receipt or source inspection performed by MSD on products in each quarter.

Analysis of the data for 1992 shows that out of a total of 148 suppliers, only 61 remained active in all four quarters while 19 remained unsatisfactory through four consecutive quarters. A large percentage of our suppliers appear to be inactive. We are investigating options to improve usefulness of the ASL by maintaining both an "Active" and an "Inactive" Approved Supplier List. Any reduction of our ASL will be accomplished in strict accordance with SSCL MSD QA standards.

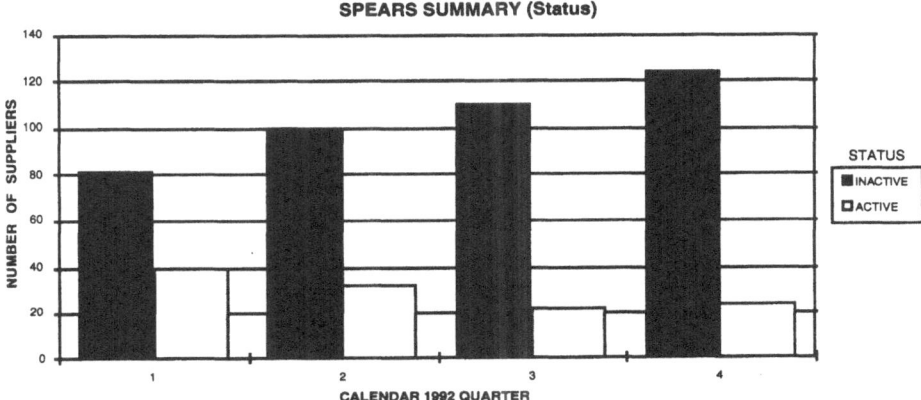

Figure 1. SPEARS Summary (Status) of the Active or Inactive status of MSD suppliers by calendar quarter in 1992.

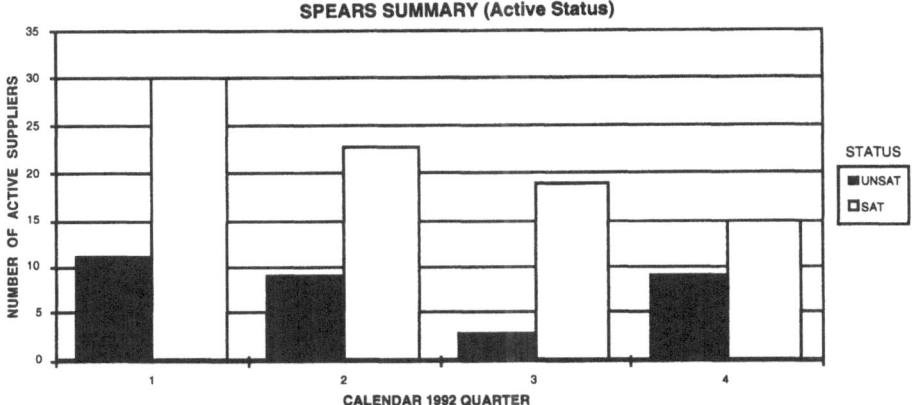

Figure 2. SPEARS Summary (Active Status) of the quality rating as satisfactory or unsatisfactory for active suppliers by calendar quarter in 1992.

SUPPLIER EVALUATIONS

During the evaluation of supplier performance, we attempt to identify discrepancies that contribute to a reduction of quality. These discrepancies are analyzed and their root causes documented. We have found that the majority of discrepancies are caused by inadequate communication of requirements, errors in drawing interpretation, late supplier submittal of Waiver/Deviations, and tolerances on dimensions that are accepted for use-as-is.

CONTINUOUS IMPROVEMENT

In our effort to reduce procurement cost by reducing losses, MSD PQA has implemented the following strategies during 1992:

(1) *Geometric Dimensioning and Tolerancing* (GDT). A GDT course has been offered to SSCL and supplier personnel (at no cost to suppliers). The GDT course was given to help suppliers better understand SSCL drawing requirements. Approximately forty suppliers attended the course.

(2) *Procurement Team* (PT). A PT has been created to work with our critical suppliers to improve procurement quality. The PT includes representatives from MSD QA, Magnet Development (Engineering) and Procurement. The main objective of the PT is to establish and communicate requirements for SSCL procured material. Quality improvements include but are not limited to: adding MSD QA criteria to Purchase Orders, identifying pertinent engineering specifications and drawings, communicating MSD QA inspection plans and/or supplier inspection plans, and the use of SPEARS. These improvements are being developed with supplier input. We are working to establish long term relationships with suppliers of critical parts.

(3) *A Quality Satisfaction Questionnaire* has been implemented to give each MSD engineer the opportunity to communicate parts problems on individual orders.

(4) *Communication*. Suppliers have been encouraged to honestly and fairly inform the SSCL of the quality status of delivered items. We have also made them aware of the impact of delivering non-conforming products. Unintentional discrepancies that could jeopardize a supplier's program are clearly communicated to responsible individuals to assure maintenance of the supplier's quality and ethical reputation.

(5) *Performance Recognition*. Good performance by our suppliers is now recognized with a certificate of commendation. The award provides recognition of achievement and encourages suppliers to improve performance. A supplier with four consecutive quarters of satisfactory quality, 90% or above, receives the certificate from the MSD Associate Director. In addition, we encourage MSD buyers to use the most qualified suppliers. In this way we reward suppliers who consistently produce quality products. This in turn helps minimize total cost to the division.

CONCLUSION

The SPEARS goal is to improve the working relationships between the SSCL and our suppliers; it enables us to identify and resolve quality issues in a timely manner. The SPEARS program can also serve other purposes such as tracking the timeliness of incoming materials and ensuring continuous quality improvement. In addition, the SPEARS program can be used effectively to minimize purchased material cost by reducing quality related losses.

In the future, MSD PQA plans to strengthen communications with our suppliers to provide essential information, performance feed-back, and to identify problem areas as they arise. We want to stimulate corrective action, and will work closely with our suppliers. In this way, we feel that we will continually improve quality and performance while reducing total costs.

QUALITY ASSURANCE PROGRAM IMPLEMENTATION ON
THE COLLIDER QUADRUPOLE MODEL MAGNET CONTRACT

Donald C. Johanson and Jeremy E. VandeBogart

Babcock & Wilcox
Naval Nuclear Fuel Division
P.O. Box 785 Mail Code 80
Lynchburg, VA 24505

INTRODUCTION

Babcock & Wilcox' (B&W) Model Magnet Program was initiated in order to evaluate quench performance of the B&W / Siemens cross section design. The program consists of the manufacture, inspection, warm electrical and quench testing of (four) one meter long cold masses and (four) five meter long cold masses.   The program also includes the capability to modify the design and configuration of the magnet, based upon quench testing results.   Although the program is an added scope of work to be executed ahead of the original schedule, B&W has implemented many of the quality systems that will be required during later contract phases. The rationale for this effort is that if a successful design is to be reproduced, all of the configuration, manufacturing, inspection, and test information must be captured and it must be reliable.   Conversely, if a magnet's performance is substandard, this must be known too, so that it can be corrected in time for fabrication of the next model.

OBJECTIVES AND RELATED QUALITY SYSTEMS

During the Model Magnet program, the B&W Quality Assurance department has set an overall goal of addressing each of the elements listed in the QA Requirements section of the CQM contract. These elements are listed in Table 1.

The primary focus, however, is to control manufacturing, inspection, and test operations or in other words, ensure that all work is performed in an orderly manner and is documented. Therefore, a Component Fabrication Routing (CFR) or shop traveler is issued for each model magnet. The CFR defines the sequence of and instructions for each of these operations as required to assemble, inspect, and test the magnet. This document

Table 1. Collider Quadrupole Magnet contract quality elements.

Management Commitment and Staffing
Magnet Quality Program Requirements
Inspection System
Test and Measuring Equipment and Standards
Subcontractor's - Subcontractor Control
Training
Worksite and Information Control
Failure reporting Analysis, and Corrective Action
Material Review Board
Defect Prevention
Process Control Procedures and Work Instructions
Serial Numbers
Cleanliness controls
Shelf Life and Storage Control
Control of Electronic Data Collection and Non deliverable Software
SSCL - Furnished Material
End Item Documentation Package
Audits
Transition to Production

also serves as an historical record for each magnet produced. CFRs are also used to document receipt inspection operations. An example of a typical CFR page is illustrated in Figure 1. The CFR is supplemented by the data pack which contains all fabrication support data including dimensional inspection and electrical test results.

Controlling the status of piece parts and maintaining proper paperwork so that end item configuration can be established is another important function during the Model Magnet phase. Before any sub component can be installed in the assembly, it must be released by the Quality Assurance department. Typically, this is the final step in the receipt inspection process during which all technical information is systematically reviewed for completeness and adequacy. This activity results in the issuance of a serialized release document which authorizes manufacturing to use the hardware. The release serial number is then recorded in the CFR, thereby linking it to a specific magnet assembly.

It is imperative that a system governing the documentation and proper evaluation of nonconformances is in place during the Model Magnet program, so that Design Engineering is made aware of factors that could potentially effect magnet performance. The Quality Control Deficiency Reporting (QCD) system consists of an engineering level procedure and a form which inspection or test personnel use to document nonconformances on the shop floor. The document is then forwarded to Quality Assurance Engineering for further routing and evaluation by all Engineering disciplines involved including Design, Manufacturing, Reliability, and potentially the customer via a Material Review Board or MRB. The same document incorporates disposition, cause, and corrective action. Upon completion of this process, a copy of the QCD form is inserted into the CFR and becomes part of the magnet end item configuration.

SEQ NO.	OPERATION DESCRIPTION	Reference Document	Rev.
001 PC CM501	Issue routing. Completed by: Date		
002 QA CM211	Perform visual inspection of container and Collared Coil. Evaluate documentation. Completed by: Date:	QAP-700	—
003 QA CM212	Perform electical tests. NOTE: This step may be performed any time between steps 002 and 006. **SSCL CHECKPOINT** **Notify SSCL QA resident for witness and sign-off of this test.** SSCL QA Signature_____ Date_____ (QA Foreman sign if SSCL can not be present) Completed by: Date:	M4D-200012 M4D-200013 M4D-200014 M4D-200016 M4D-200018	— — — — —
004 QA CMTBD	Collect strain gage data. NOTE: This step may be performed any time between steps 002 and 006. Completed by: Date		
005 QA CM213	Dimensionally inspect. NOTE: This step may be performed any time between steps 002 and 006. Completed by: Date	QAP-700	—
006 QA CM214	Audit data. Completed by: Date	QAP-702	—
007 QA CM215	Release. Completed by: Date	QAP-301	—
008 PC CM502	Perform CLOSEJOB/PRODRCPT. Completed by: Date:		

Figure 1. Component Fabrication Routing (CFR) - Collared Coil Receipt Inspection.

OBSTACLES TO QUALITY SYSTEM EFFECTIVENESS

Rigid quality systems require significant effort during implementation (actually generating the necessary procedures and invoking their use on the shop floor). Actually cycling a system consumes considerable resources as well. Nonconformance control, for example, is very time consuming since a variety of disciplines must be involved in the evaluation. All departments must also concur on disposition, cause, and corrective action. However, reaching agreement in these areas is particularly difficult during the development phase since the design is untested and manufacturing capability is being explored and not well established. Furthermore, since Model Magnet quantities are small, piece part order quantities are low and might even be one-of-a-kind due to the developmental nature of the program. In these cases, the value added by identifying and implementing meaningful corrective action (as required by procedure) might not justify expending the required personnel resources.

The manufacturing and inspection processes have been designed with much forethought based upon technology transfer and previous manufacturing experience. Inevitable though, are minor details that could not have been anticipated; a special tool that is needed, a difficult assembly process, or unexpected tolerance stack-up. Since Model Magnets are fabricated in accordance with the CFR, when such obstacles are encountered work stops until engineering changes to instructions or drawings can be implemented. The need for relatively frequent, timely changes has resulted in a modification to change control procedures which allow for performing work to hand written changes in CFRs and Engineering Change Requests attached to drawings.

CONCLUSIONS

As of this date, three short Model Magnet cold masses have been manufactured and have tested with excellent results. The fourth is near the end of the fabrication process and with each successive magnet, the Quality Assurance and Manufacturing departments implement the lessons learned from those previous. The quality systems mentioned are primarily, methods for accomplishing the objectives that were identified as being critical to the success of the Model Magnet program. There are additional benefits to their use however. An obvious benefit of implementing these systems is that Quality Assurance has "test-driven" many procedures and systems that will be carried over into subsequent contract phases. Also, the people who must work within these systems learn them early in the program. For example, step 003 of the CFR of Figure 1 requires notification of an SSCL resident for witness of the electrical test. These "checkpoints" are also inserted in manufacturing CFRs in order to test the process and identify and resolve any complications early on. Still another benefit is that exercising systems such as the QCD system, results in significant interaction between all B&W engineering disciplines, which in itself is constructive.

Overall, implementation of quality systems has been successful; the Quality Assurance department has been audited both internally and by the SSCL with favorable results. The efforts expended during the development stages of fabrication will properly orient those who must work within the limits of the systems, and will allow the Quality Assurance department to focus on improvement of processes at an early stage of production.

MATERIAL HANDLING ADVANCEMENTS FOR COLLIDER DIPOLE MAGNET MANUFACTURING SYSTEMS

C. Patrick Cowan

General Dynamics
Space Systems Division - Space Magnetics
P.O. Box 85990, (MZ: C1-8260)
San Diego, CA 92186-5990

INTRODUCTION

General Dynamics (GD) has developed a manufacturing system for the Collider Dipole Magnet (CDM) Program. Initial tools and equipment have been installed and prototype production is underway in GD's CDM Production Facility in Hammond, Louisiana. A target production capability to build up to ten CDMs a day on a two shift basis drove the initial manufacturing systems design. For example, this rate requirement will require approximately 160,000 kg of material to be moved in and out of the Hammond Facility everyday. These requirements have led to the development of a complex, highly advanced material handling system that supports the manufacturing process. The development of the material handling system has been in work for nearly 6 years and the solutions now being implemented reflect many significant achievements made by numerous GD manufacturing and industrial engineers. This paper will describe the results of these achievements and the process in which they were developed.

Initial Baseline Establishment

The original baseline for manufacturing the CDM was based on tools, methods, and processes in use at both Fermi and Brookhaven National Laboratories during the late 1980's. The material handling methods being used were highly dependent on the use of overhead crane systems and complex handling fixtures. Smaller, easier to handle parts were moved within the facilities on standard 4 wheeled utility carts. This approach was highly effective in the laboratory environment and GD initially established that a similar system would be used for production of the CDM. Plans were then derived for the manufacturing facility using the crane and cart approach and a cost baseline was established. The intention was that by establishing a costed baseline, trade studies investigating new concepts and approaches could then be analyzed. The baseline approach was documented in initial facility layouts and in a station to station material handling flow requirements plan.

Trade Studies

The requirements of manufacturing up to ten CDMs a day drove an internal research and development effort. This effort was conducted by GD manufacturing specialists to analyze all CDM material handling requirements. A matrix was developed that multiplied the total weight of each subassembly times the quantity of each subassembly needed for the product build process. The matrix analysis identified points in the manufacturing process where significant amounts of repetitive handling of large weights occurred in the process. These process steps represented the best opportunities for automation studies. The work stations identified in this analysis were the stations that involved the laminates; yoking and collaring stations.

An automation concept was then developed to incorporate the use of a material and Work In Process (WIP) movement system called the shuttle system. This system "shuttled" the WIP from workstation to workstation where laminates were assembled, thus eliminating the requirement for cranes and overhead lifting. Significant cost savings were realized by eliminating rigging the product in the repetitive operations. The initial shuttle automation concept was then cost traded against the existing baseline and was found to be an overall advantage in reducing the recurring and non-recurring production costs overall.

A new baseline for the manufacturing operation was then established to be a combination of the use of cranes, utility carts, and where cost effective, an automated shuttle movement system for the yoking and collaring stations.

Factory Simulation

The next step in the material handling development process was supported by a computer simulation effort that modeled the entire CDM build process. The simulation analysis established where "bottlenecks" exist. Bottlenecks are steps in the manufacturing process in which capacity is of such short supply that a particular resource used at that step regulates the overall system capacity. The results of the simulation resolved that one such bottleneck in the system was the crane system planned to support the majority of the manufacturing operations. In order to increase overall systems capacity, the bottleneck resources in the process were studied and improvements were made to increase the capacity of the resource, thus increasing the throughput ability of the overall manufacturing system.

In brainstorming sessions held to solve the bottleneck problems of extensive cranes usage, an idea was derived to expand the "shuttle" concept throughout the entire manufacturing process. This approach eliminated the requirements for crane systems used during the WIP build process. This concept was further supported by concerns raised about the environment contamination and particulate generation that cranes can cause. Additionally, the planned use of cranes had significantly increased the product and personnel safety risks. By eliminating the requirement to continuously rig the product, extract the product from the tooling, and move the product overhead, shutting down many ongoing operations, these risks were greatly reduced. New layout plans and cost information was then derived and a new cost trade was conducted to validate the idea. The new expanded shuttle concept was found to be exactly what was required to eliminate the bottleneck in the process while significantly increasing the control and effectiveness of the overall manufacturing system. The baseline was then modified to include the use of the shuttle system throughout the manufacturing facility, eliminating the requirement for cranes for all major WIP movements.

The Shuttle System

The shuttle concept was further developed throughout the product develop phase of the CDM program. An automation tooling vendor was brought on board to complete the design of the shuttle system and the corresponding interface requirements to each of the production and test stations required in the overall manufacturing process. Figure 1 displays the shuttle design currently in use at the CDM manufacturing facility in Hammond, Louisiana. The shuttle is an electrically driven system made up of a common transport device which can automatically go to any preprogrammed address on the factory floor. Each station has unique attachment tooling which can interface with the shuttle and the next assembly station. The shuttle can also circulate attachment tools back to their point of origin. Additional benefits of the shuttle system includes inherent part protection, flow control, reduced banking and inventory, and potential for automatic control.

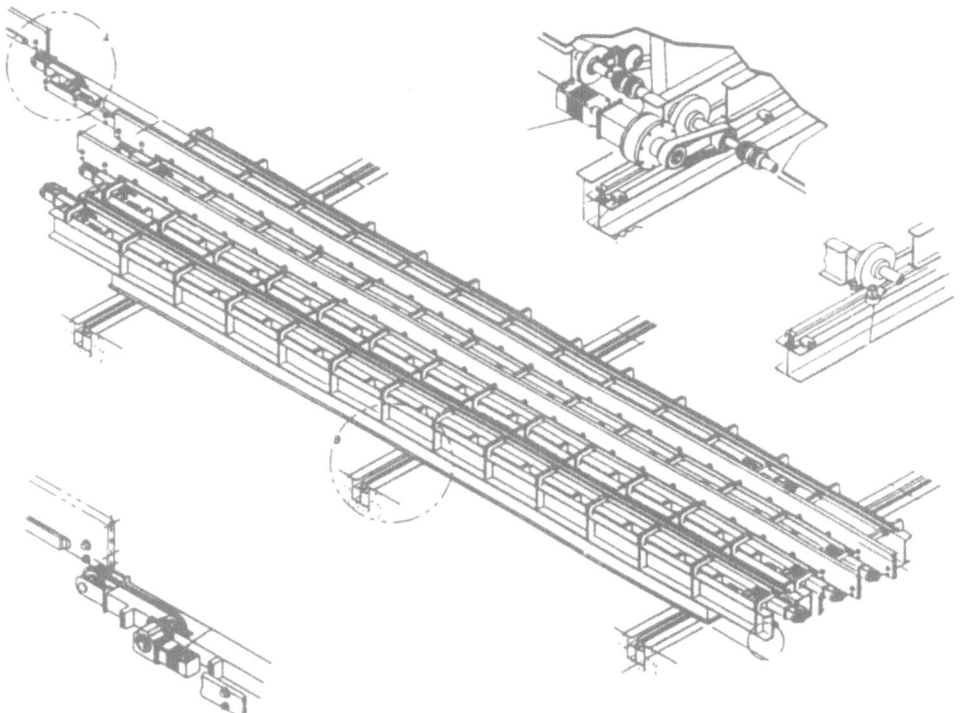

Figure 1. The Shuttle System Design Supports CDM WIP Movement.

PART HANDLING PLANS

An additional aspect of the material handling plans is the movement requirements of the many piece parts and subassemblies that are unloaded from the receiving area and moved to the production stations for use in the build process.

Full Rate Production Plans

Ultimately, CDM parts handling methods will incorporate many "state of the art" material and inventory methods and philosophies. These plans include achieving a "Just In Time" (JIT) system. JIT will reduce the factory space required for materials warehousing and minimize the requirements to handle parts and subassemblies multiple times before assembly into the WIP. JIT methods outline that parts be delivered to the manufacturing facility "just in time" for use in the build process. This approach reduces the inventory waiting to be used by the factory to a minimum. The requirements for achieving a JIT operation are balanced against the total quantities required for each subassembly, the overall production rate requirements, and the capacity of delivery systems provided by the parts and subassembly vendors. Other plans include incorporating customized packaging for each part type and subassembly. Customized packaging involves the use of specially designed containers for the shipment and delivery of parts from the parts vendors. The packaging optimizes ideal quantities to be in each container and dictates that parts be positioned in the final assembly configuration when arriving at the manufacturing facility.. The packaging design also considers the movement off the delivery vehicle to the point of use in the build process. Packaging also will be built for recycling and reuse, minimizing the overall packaging cost over the production program life. To accomplish the overall customized packaging designs, each major part and subassembly will be analyzed and packaging designs are derived considering the following information; end use requirements, part configuration, part vendor concerns and manufacturing cost, part cleanliness requirements, shipping requirements, and the useful life expectancy of the packaging system.

The plans for parts handling, inventory control, and packaging will be implemented over the initial phases of the overall CDM Program. As the production rate requirement increase up to ten magnet a day, cost trades will be updated and aspects of these plans will be implemented at the most optimal timeframe in the program. This approach will keep initial program costs to a minimum and allow for an efficient phase-in of the material handling methodologies.

SUMMARY

The requirements of the overall CDM production program are challenging and represent many opportunities to improve superconducting magnet manufacturing methods. Material handling is a key technological aspect that will be positively impacted by meeting the challenges of the CDM Program. General Dynamics has developed and implemented many significant material handling solutions in the initial CDM production operation and will continue to implement methods and ideas that will ultimately benefit the overall field of superconducting magnet manufacturing technology.

Acknowledgments

Portions of the work described herein is being accomplished under contract to the Universities Research Association in support of the Superconducting Super Collider project for the U.S. Department of Energy.

The paper was made possible by the dedication of many General Dynamics manufacturing and industrial engineers. These people include Ken Miller, Steve Pidcoe, Art Zinzser, Don Paganini, Dick Shertzer, Frank Giuseppini, Shelly Ewing, Earl Powers and Therese Tawoda.

COLLIDER DIPOLE MAGNET COLD TEST FACILITY IMPLEMENTATION

Bradford J. Boulais

General Dynamics Space Systems Division
Space Magnetics
P.O. Box 85990
San Diego, CA 92186-5990

INTRODUCTION

The approach and plans for the facility activation to support the Collider Dipole Magnet (CDM) Cold Test Stands (CTS) and associated equipment will be discussed throughout this paper. Methodologies utilized and results obtained in the layout and planning process represent significant advancements in CTS systems integration for CDM testing.

This paper will discuss the systematic approach to generating design requirements, and the facilitization required to support Cold Test Stand implementation. Discussions will focus on space requirements, material handling, specialized electrical, Heating Ventilating and Air-conditioning (HVAC), and foundations for the CTS and associated equipment.

OVERVIEW OF THE CDM COLD TEST SYSTEMS

Superconducting Super Collider (SSC) Dipole magnets will be tested at the General Dynamics (GD) Hammond, Louisiana CDM production facility under conditions that will simulate a cryogenic/power cycle environment found in the SSC main ring. The cryogenic test system will consist of four (4) Cold Test Stands, and all associated equipment. Two CTS's will have the capability to test both 15M and 13M magnet lengths in an effort to co-utilize existing test stand design by incorporating minor modifications.

Project Management

A GD CTS activation team was formed on 4/1/92 which included a broad range of disciplines (i.e. Design Engineering, Instrumentation, Thermal Control, Facilities etc.) in order to provide design parameters & implement such a complex test system.   Each discipline was chartered to concentrate on the areas and tasks most effecting an efficient execution of the project.

Due to the concurrent engineering of many systems with which the test stands and control systems interface, an extensive dialog was established between GD, GD facility subcontractors, and Air Products to ensure a smooth and continuous evolution of the functional specifications. (Note: Air products was selected to manufacture and install the cryogenics plant and the CTS's)

GD began a series of CTS activation meetings, and a cold test stand activation schedule was developed as a way to track and provide visibility to critical CTS components which included the following:

Factory Data Acquisition System (DAS) - This equipment will obtain strain, deflection and temperature measurements as the CDM magnet is cooled and warmed during cryogenic acceptance testing.

Prototype DAS - This equipment will make high speed (10kHz) measurements of the coil voltages and end temperatures and pressures to determine quench origin, velocity and relative strength.

Minor Tools - This equipment includes a solder cart to solder the power bus leads to the CTS distribution box and all other tools required to provide mechanical connections, as required, to connect the CDM to the CTS.

Magnetic Measurement Equipment (MME) - Is required to make various magnetic field measurements to determine field quality, alignment, and length.

Transport Stub Dock - This tool is simply a raised steel table mounted on casters and is required to bridge the 13' gap between the shuttle clearance aisle and the CTS during magnet insertion or removal. A 13' gap is necessary to allow for placement of the MME at the test stand end cans.

Cold Test Stand - This equipment includes the stands to support the CDM, the vacuum sleeves to provide a leak tight environment for cryogenic testing, and the end can and distribution can which contain the cryogenic and power interface to the CDM during testing operations. The cans also provide the interface for the DAS and Coil Protection System during testing.

End Can Lifting Device - This device will be in the form of a dual hoist gantry crane spanning all four test stands, and will accommodate the 15M and 13M magnet length variation. Its primary function will be to lift the CTS end can from its test position to allow for CDM insertion and removal.

CTS Control System - Is comprised of a VAX computer which will run under the VMS operating system. The software will be capable of 1) operating each test stand and vacuum system, 2) accessing data in the database of current process signals, 3) recording data from the test stand with a resolution of 2 samples per second (Hz), and 4) displaying any measured variable on the system. (Note: the VAX systems interface with Bailey control computers, which control the cryogenics plant operations.

Coil Protection System - This equipment monitors the CDM coil halves for a quench detection, signals the quench protection heaters, shuts off the power supply, and closes the circuit breaker, which allows energy to be routed to dump resistors.

Test Process Control Modules (TPCM's) - This equipment controls the integrated testing of the CDM. All operations are run from integrated work/test instructions and Process Requirements Documents (PRD's). The system also stores all test events in a log file, to automatically record the test history of a CDM.

10KA Power Supplies - This equipment provides the operating current for the CDM during testing. The power supplies are rated for 10,000KA at 40VDC and have a programmable current ramp rate.

The CTS activation schedule identified individual CTS component project milestones through design, procurement, manufacturing, and installation, and showed the critical paths & component interrelationships.

During this process, the facility team members were tasked to 1) Define the facility interface requirements (i.e. power, water, etc.) utilizing CTS design criteria, 2) Perform preliminary & final design of the CTS foundations, and 3) Perform detailed design of all Hammond facility modifications and interfaces required to support installation of the CTS components.

CTS FACILITY IMPLEMENTATION TASKS

Facility Layout Considerations

The first step in implementing the Cold Test area was to determine the overall dimensional characteristics of CTS components and their relative positioning during cold testing. The orientation of the test stands was dictated by the Hammond facility shuttle aisle location. The key dimension, from which all other dimensions were based, was taken from the location of the shuttle clearance aisle. Footprints for the MME transporter system, CTS end can, magnet test stand, etc. were built up from the shuttle clearance aisle to fix the final location of the CTS distribution boxes. Based on CTS distribution box locations, the Air Products (AP) cryogenics facility located outside the West wall of the Hammond plant was designed and located to minimize the piping runs between the AP and the CTS distribution boxes.

Using CTS component footprints, the layout for the Hammond Facility Cold Test area was first validated through the use of a block plan (Figure 1) by considering all aspects of test requirements while utilizing the features of the existing facility as stated above. This posed a special problem given the factory floor space allocated to cold testing was only 5400 SF (36' X 150'). This approach was further supported and validated by using computer animation software called "Interactive Graphics Robot Instruction Package (IGRIP)" by Deneb Robotics Inc. This computer driven system simulated in three dimensions the facility layout and tooling interfaces, and allowed real time simulation of physical movement.

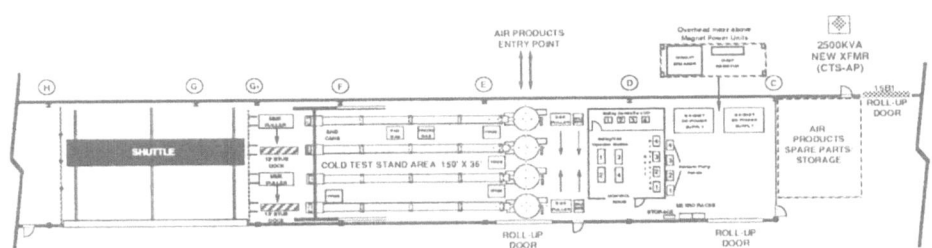

Figure 1.   Cold Test Stand Block Plan

Shuttle Interface & Material Handling Devices

The Hammond facility is utilizing a transfer shuttle system as the major work-in-process material handling device. The shuttle will integrate with all of the Cold Test Stands via a

Transport Stub Dock. The Transport Stub Dock is a device required to bridge a 13' gap between the shuttle clearance aisle and the CTS. (Note: the clearance aisle is required for the MME system) The shuttle will be used to load and unload work from each CTS station, utilizing controlled interfaces.

Two (2) Transport Stub Docks are required to support four (4) CTS's. The stub docks are mounted on castors and move perpendicular to each CTS. Each device will be located relative to each CTS by means of floor mounted locating pins, with final alignment done by using a series of leveling screws.

In order to allow magnet insertion or removal, the CTS's Return End Can must be moved from its test position by means of an End Can lifting device. This device will be in the form of a dual hoist gantry crane spanning both the 15M and 13M magnet length variation.

Utility Requirements

Electrical - A dedicated transformer was installed to support the 10KA DC power supply requirements. This installation required a separate foundation pad for a 2500KVA transformer. The pad measured 15'-6" X 13'-0" and was installed per Louisiana Power & Light (LP&L) drawing #DS-92-285675-027B & C which called for a 6" steel reinforced concrete slab supported by concrete filed auger piles, with an extra 6" concrete pad under the transformer footprint.

HVAC - Specifications for this area include a temperature range of 60-80 degrees (F) with a relative humidity range of 30%-60% and a cleanliness level of 65% (ASRAE - 52-76). This area is serviced by three (3) roof mounted direct expansion HVAC units. Each HVAC unit is controlled by the "Trane Building Management Network" designed to allow the use of personal computers to monitor and interface with "Tracker" panels located throughout the factory. This system allows temperature and humidity trends to be analyzed & stored for historical data control.

CTS Foundation

Foundation requirements were generated based on the design parameters of magnetic measurement equipment (MME System) whose magnetic measurements can be adversely affected by vibratory input to the CTS. This requirement drove the need for vibration isolation. It was determined that one (1) million lbs. of concrete would be needed to support all four CTS's vibration isolation requirements. This foundation measured 70' X 36' and was approximately 36" in depth. The total sub-surface area was steel reinforced with the addition of a 1/2" Unisorb® neoprene vibration isolation pad.

ACKNOWLEDGMENTS

Special thanks to Ron Green for providing the CTS Facility Block Plan, Gary Rowe/Ron Jones/Bob Churchill and the rest of the CTS team members for their technical support.

The work described herein is being accomplished under contract to the Universities Research Association in support of the Superconducting Super Collider project for the U.S. Department of Energy.

A METHODOLOGY TO DESCRIBE PROCESS CONTROL REQUIREMENTS

R. Carcagno and V. Ganni

Superconducting Super Collider Laboratory*
2550 Beckleymeade Ave.
Dallas, TX 75237-3997

ABSTRACT

This paper presents a methodology to describe process control requirements for helium refrigeration plants. The SSC requires a greater level of automation for its refrigeration plants than is common in the cryogenics industry, and traditional methods (e.g., written descriptions) used to describe process control requirements are not sufficient. The methodology presented in this paper employs tabular and graphic representations in addition to written descriptions. The resulting document constitutes a tool for efficient communication among the different people involved in the design, development, operation, and maintenance of the control system. The methodology is not limited to helium refrigeration plants, and can be applied to any process with similar requirements. The paper includes examples.

INTRODUCTION

The Cryogenics Department of the SSC Laboratory has decided to separate the selection and procurement of the computer control system from the procurement of the helium refrigeration plants needed for the project. The scope of supply for the cryogenic equipment vendor includes instrumentation (sensors, actuators) and wiring up to a well-defined interface, but does not include the computer control system needed to run the plant. A separate procurement will provide the computer control equipment and software necessary to run not only the refrigeration plants, but also all the underground cryogenic systems; these constitute 75% of the control system size as measured by the number of input/output points.

An important consequence of the decision to procure the computer control hardware and software separately is the need to define an efficient mechanism for communicating process control requirements between the vendor and the SSC. The methodology used must take into account the design goal of remote, unattended operation of these plants. Unfortunately, although several tools are available (e.g., binary logic diagrams[1] and sequential function charts[2]) there is no standard (such as P&IDs for piping and instrumentation) for defining process control requirements. Depending on the process, some tools work better than others, but in general a combination of tools is needed in order to completely describe the control requirements.

THE METHODOLOGY

In this paper we present a methodology that has been designed to capture and describe process control requirements for the collider helium refrigeration plants. These include monitoring, logic control, regulatory control, sequential control, and supervisory control. The plants must be configured to support the many operating modes of the collider[3], thus some features of batch control are also required.

The process control description document is the basis for the generation of software requirements specifications for the computer control system. The methodology is not limited to helium refrigeration plants, and can be applied to any process with similar requirements.

The methodology is based on a structured, top-down approach where object-oriented concepts are used. It is not concerned with any implementation features such as MANUAL or AUTOMATIC operating modes. The question of implementation is secondary to the process control description.

---

* Operated by the Universities Research Association, Inc., for the U. S. Department of Energy under Contract No. DE-AC35-89ER40486.

The first task in this methodology is to identify the objects and their hierarchical relationships. The system is partitioned into a multilevel structure with each level containing a certain number of functionally modular parts (or objects). The objects should be more or less self-contained from the controls point of view. Ideally, objects communicate with each other exclusively via status words and commands. In practice, some process information (e.g., temperatures, pressures, flows, positions) must be exchanged as well. One of the goals of the object-oriented design is to minimize the amount of process information to be exchanged among objects.

A typical structure includes four levels. At the lowest level (Level 0), base-level objects are identified. Base-level objects are the principal building blocks and deal primarily with *devices* such as sensors, valves (on/off valves, control valves) and motors. Many objects at this level are found in a classic P&ID.

The next level (Level 1) contains a group of Level 1 objects associated with a *module* such as a compressor skid. Typical control functions of this level include regulatory control, sequential control, and process interlocks. Examples of Level 1 control functions are: a compressor skid shutdown sequence triggered by a high-high alarm in the motor winding temperature, and makeup flow regulation to the gas management module.

The next level (Level 2) contains a group of Level 2 objects associated with a *subsystem* such as a compressor group. Typical control functions of this level include supervisory control and equipment scheduling. Examples of Level 2 control functions are: automatic switching of 80 K beds and bed regeneration cycle scheduling, and coordinating compressor skid operation and gas management operation during plant startup.

The next level (Level 3) in the hierarchy of objects contains a group of Level 2 objects associated with a *system* such as a compressor system. Typical control functions of this level include system process management and process coordination with other systems (e.g., the refrigeration system). Examples of Level 3 control functions include configuring the compressor and refrigerator systems for minimum capacity mode.

Once the objects in each level and their relationships with upper level objects have been identified, a functional description of each object is needed. Four major concepts are used in this description: state, transition, function, and status words and commands. Four major tools are used to describe the concepts: the truth table, the sequential function chart, the state and transition diagram, and the cause and effect table.

ELEMENTS OF THE OBJECT FUNCTIONAL DESCRIPTION

An **object state** is associated with a given configuration of object components (e.g., a certain valve lineup is associated with the "ON" state). The major reason for defining a state is to provide a useful reference for the operator or for other objects.

Object states are described by the *truth table*. The truth table describes the status of subordinate control objects for each object state. In a truth table, row labels depict the subordinate objects, and column labels depict the state names. The matrix interstices indicate the subordinate object status. Devices under automatic control are designated with the name of the controlling loop. Analog devices not under automatic control are designated with the constant value at which they are positioned. A dash (-) indicates that the subordinate object status is not relevant to the state definition. Table 1 is an example.

Table 1. Example of truth table: Truth table for object – CMP-11.

SUBORDINATE OBJECT	ON	READY	OFF	LOCAL	SAFE
RMT-SWITCH	TRUE	TRUE	TRUE	FALSE	TRUE
EM-10102	ON	OFF	OFF	-	OFF
FV-10151	OPEN	OPEN	-	-	OPEN
FV-10152	OPEN	OPEN	-	-	OPEN
FV-10155	OPEN	OPEN	-	-	OPEN
FV-10156	OPEN	OPEN	-	-	OPEN
LV-10121-A	LC-10121	LC-101212	-	-	CLOSE
LV-10121-B	LC-10121	LC-101212	-	-	CLOSE
LV-10130	LC-10130	LC-10130	-	-	CLOSE
TIC-10103	AUTO	AUTO	-	AUTO	MANUAL
TY-10103	TIC-10103	TIC-10103	-	TIC-10103	CLOSE

An **object transition** is the sequence of events that must take place in order for the object to change from one state to another (e.g., open/close valves, turn motors on/off, put control loops in manual/auto).

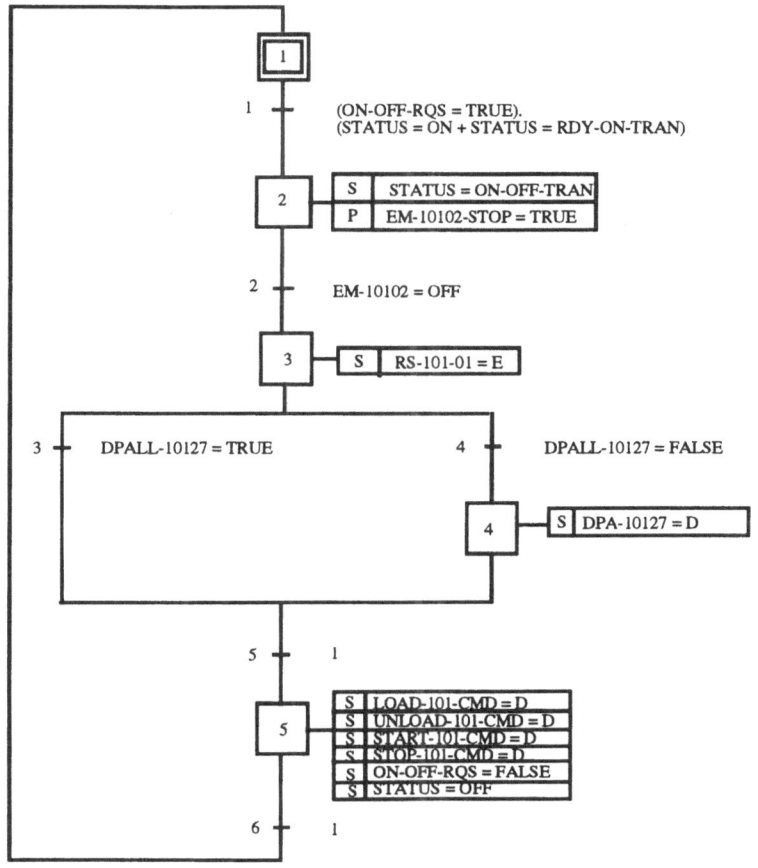

Figure 1. Example of a state and transition diagram.

Object transitions are described by a *sequential function chart* (SFC). An SFC is a graphical description of the transition according to the IEC 848 international standard[2]. Figure 1 is an example.

The relationship between states and transitions is graphically described by the *state and transition diagram* (STD). Figure 2 is an example.

An object function is a piece of logic that the object executes either continually or sequentially in order to perform a certain control task (e.g., interlocks, alarms, PID loops, fill sequence). Functions may be enabled or disabled according to the object state.

Object functions are described by the *cause and effect table*. When functions are too complicated to be described by the cause and effect table, the table includes a reference to another graphic (e.g., binary logic diagrams, SFC) or tabular (e.g., PID table) representation. Table 2 is an example.

Table 2. Example of cause and effect table.

FUNCTION	TYPE	CAUSE			EFFECT		
		OBJECT	TAG	STATE	OBJECT	TAG	STATE
FU-101-03	CM	CMP-11	MMI-101-02	TRUE	CMP-11	ON-OFF-RQS	TRUE
SD-101-02	I	CMP-11	DPALL-10127	TRUE	CMP-11	ON-OFF-RQS	TRUE
SD-101-03	I	CMP-11	TAHH-10102	TRUE	CMP-11	ON-OFF-RQS	TRUE
ON-OFF-TRAN	SFC	CMP-11	ON-OFF-RQS	TRUE	CMP-11	*See Fig. 2*	

Status words and commands are the way objects communicate with each other. The status word is defined according to the object state, transition, or function taking place. There may be more than one status word per object (for example, a gas tank could be ON and PURE). Commands are generated by functions or by the operator. Status words and commands provide the basic linkage among the different tools used in the description.

For a more complete set of examples, see reference 4.

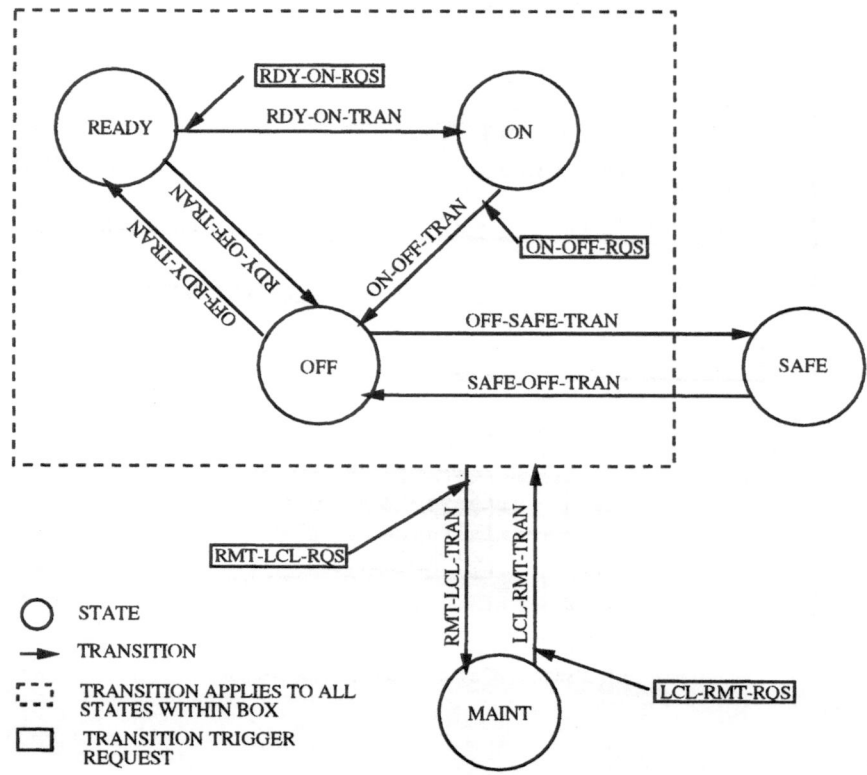

Figure 2. Example of a sequential function chart.

CONCLUSIONS

A well-defined standard for process control requirements description does not exist. In this paper we have presented a methodology designed to capture and describe process control requirements of the type expected in highly automated helium refrigeration plants. It is based on a structured, top-down approach where object-oriented concepts are used. We expect to use this methodology to facilitate the communication of process control requirements among refrigeration system vendors, control system vendors, and the SSCL.

REFERENCES

1. "Binary logic diagrams for process operations," ANSI/ISA S5.2-1976 (R1981). Instrument Society of America (ISA) publication (1981).
2. "Preparation of function charts for control systems," CEI/IEC 848: 1988. International Electrotechnical Commission (IEC) publication (1991).
3. R. Than, S. Abramovich, and V. Ganni, "The SSC cryogenic system design and operating modes," SSC Laboratory, SSC Cryo Note 92-12, October 1992.
4. R. Carcagno, V. Ganni, D. Rivenbark, and B. Shuster, "Process Control Description Example for a Compressor Skid," SSC Laboratory, SSC Cryo Note 92-13, October 1992.

A STUDY ON SECTOR NITROGEN INVENTORY EMERGENCY VENTING

Burt Zhang

Accelerator Systems Division
Superconducting Super Collider Laboratory*
2550 Beckleymeade Avenue
Dallas, Texas 75237

INTRODUCTION

The main ring of the Superconducting Super Collider consists of 10 sectors. Each sector has 4 strings and the total nitrogen inventory in a nominal string (4,320 m) is approximately 8,856 kg (2,894 gal)[1]. There are occasions when part of the collider nitrogen inventory needs to be vented in an emergency fashion. Due to the large quantity of the nitrogen inventory and the high elevation differential between the below-ground collider ring and the above-ground vent stack, it is essential to vent the nitrogen inventory in the vapor phase. Uninsulated dump tanks are provided at all sites to collect and vaporize liquid nitrogen expelled from the strings. Vapor nitrogen is then vented through a vent stack at surface level into the atmosphere. An emergency venting starts when a remote control valve in the nitrogen subcooler box opens. The string nitrogen inventory is then discharged into the dump tank. Around 100 kg nitrogen inventory will be discharged before the string inventory reaches a saturation state at certain location. After that, the static heat leak into the 80 K circuit (16,250 W per string) becomes the dominating driving force to expel the nitrogen inventory. Due to the length of the string, the boil-off bubbles are localized. The mass boil-off rate is translated into a much higher mass flow rate expelled from the string. The string saturation pressure is dynamically balanced by the total pressure drop between the string and the vent stack. The capacity requirement for the nitrogen dump tank reaches the maximum when the dump tank at each site has to handle the nitrogen inventory in two strings, which is approximately 17,712 kg (5,788 gal). This work is intended to analyze the volume capacity requirement for the nitrogen dump tanks for this worst scenario. A computer program is developed to numerically simulate the emergency venting process. The string, dump tank, and venting paths are modeled individually and then linked to impose the constraints.

ANALYSIS

In a venting process, the string pressure is first relieved from nominal operating pressure to a saturation pressure. The static heat leak vaporizes nitrogen at a rate $\dot{M}_v = \dot{Q}_s / L_s$, where $\dot{M}_v$ is the mass rate at which nitrogen is vaporized in the strings, and L_s is the latent heat of nitrogen evaluated at string pressure. Due to the slender geometry of the strings, the vapor generated in time is entrapped along the strings. The mass expelled per unit time $\dot{M}_e$ is much greater than that being vaporized, which is governed by the equation

$$\dot{M}_e = \dot{M}_v \frac{v_g - v_f}{v} \tag{1}$$

here $v = xv_g + (1 - x)v_f$ is the specific volume, x is the quality, and v_g and v_f are the specific

*Operated by Universities Research Association, Inc., for the U.S. Department of Energy under Contract No. DE-AC35-89ER40486

volumes of saturated vapor and liquid respectively. Equation 1 indicates that the driving force that expels the nitrogen inventory in the strings is the difference in specific volume between saturated vapor and saturated liquid. The dump tank collects, temporarily contains, and vaporizes nitrogen expelled from the strings. The venting mass rate from the tank to the vent stack is given by

$$\dot{M}_o = \frac{\dot{Q}_t + \dot{Q}_w}{L_t} + x\dot{M}_v \tag{2}$$

where L_t is the nitrogen latent heat at tank pressure, and x is the quality of nitrogen inventory in the strings. The first term in this equation represents the mass rate being vaporized in the tank, and the last term is the vapor portion of the mass rate into the tank. $\dot{Q}_t$ in Equation 2 represents the heat transfer rate to the bulk nitrogen in the tank from the underground enclosure environment. It has the form

$$\dot{Q}_t = (T_a - T_i)/\sum_{j=1}^{N} R_j, \qquad N = 4 \tag{3}$$

where T_a and T_i are ambient air and bulk nitrogen temperatures respectively. The heat transfer process between the ambient air and the liquid nitrogen in the dump tank includes conduction, convection and radiation modes. The denominator in Equation 3 is the sum of heat transfer resistance which includes the following four resisters

$$\sum_{j=1}^{N} R_j = R_o + R_f + R_w + R_i, \qquad N = 4 \tag{4}$$

where R_o denotes the heat transfer resistance between the ambient air in the tunnel and the outer surface of the frost formation on the dump tank, R_f is the frost conduction resistance, and R_w, and R_i represent the resistances through the tank wall and at the inner surface of the dump tank respectively. The heat transfer between the ambient air in the tunnel and the outer surface of the frost formation is due to free convection and radiation. The resistance is characterized by

$$R_o = \frac{1}{H_c + H_{rad}} \tag{5}$$

where $H_c = N_u k_a/L_c$, L_c is a characteristic length (tank height), k_a is the thermal conductivity of ambient air, and the Nusselt number is evaluated using the formula

$$N_u = \left[.825 + \frac{.387 R_a^{1/6}}{\left[1 + (.492/P_r)^{9/16} \right]^{8/27}} \right]^2 \tag{6}$$

The Prandtl number for air P_r is evaluated at tunnel ambient temperature, and the Rayleigh number $R_a = G_r P_r$ is calculated using

$$G_r = g\beta \frac{(T_a - T_o)L_c^3}{\nu_a^2} \tag{7}$$

$$\beta = \frac{2}{T_a + T_o} \tag{8}$$

T_o here is the temperature at the outer surface of the frost formation, ν_a is the kinematic viscosity of air, and g is the gravitational acceleration. The radiative heat transfer coefficient in Equation 5 is calculated using $H_{rad} = \epsilon\sigma \left(T_a^4 - T_o^4 \right)/(T_a - T_o)$ where ϵ and σ are the emissivity of the frost formation surface and the Stefan-Boltzmann constant respectively. The second resister is due to the conduction resistance of the frost formation. The conduction resistance of the frost formation depends on the venting duration and moisture content of the ambient air in the underground enclosure. A value $R_f = 2 \times 10^{-4}$ K/W is used in this study. The third resister is the stainless steel tank wall. The evaluation of this resister involves an integral of the thermal conductivity of stainless steel over the temperature range between the outer and inner wall surface temperatures T_{wo} and T_{wi}

$$R_w = \frac{(T_{wo} - T_{wi})W}{K} \tag{9}$$

in which K is the thermal conductivity integral for stainless steel over the given temperature range,

$$K = \int_{T_{wi}}^{T_{wo}} k_{ss} dT \tag{10}$$

312

W is the thickness of the tank wall, and k_{ss} is the thermal conductivity of stainless steel, which increases with temperature. This integration process is repeated at each iteration when the temperature range is updated. The fourth resister R_i represents the resistance to convective heat transfer between the inner surface of the tank wall and the contained bulk nitrogen. As the temperature difference $T_{wi} - T_i$ pass the Leidenfrost point and hence the heat transfer process under discussion falls in the film pool boiling regime, the correlation

$$N_u = .64 \left[\frac{g(\rho_f - \rho_g)L'_{N_2}L_c^3}{\nu_g k_g (T_{wi} - T_i)} \right]^{25}$$ (11)

is used to yield the convective heat transfer coefficient with the corrected latent heat given by

$$L'_{N_2} = L_{N_2} + .4C_p(T_{wi} - T_i)$$ (12)

Properties used to calculate the Nusselt number N_u are evaluated at the film temperature $T_f = (T_{wi} + T_i)/2$. The term $\dot{Q}_w$ in Equation 2 represents the heat ejected from the tank walls as the tank is filled and cooled down. It is given by $\dot{Q}_w = C\dot{M}\Delta T_w$, where C is the specific heat of the tank wall material (stainless steel) evaluated at the average wall temperature, $\dot{M}$ is the rate at which the tank wall mass comes in contact with the bulk nitrogen in the dump tank, and ΔT_w is the average temperature change of the tank wall during the initial cool-down. The equations listed in this section are solved to produce two nitrogen mass flow rates. The first is the mass rate expelled from the strings, the second is the one that leaves the tank being vented through the vent stack. Special interest is focused on the process before these two mass rates have the same magnitude when the accumulated liquid in the tank reaches the highest level, which determines the capacity requirement for the dump tank. Thermodynamic properties of nitrogen are generated using a computer subroutine[3]. They are updated at each iteration and time step.

RESULTS

The simulated time history of the venting is presented in Figure 1 in terms of masses in the strings and dump tank. In less than 15 minutes, approximately 7,500 kg (2,450 gal) of nitrogen

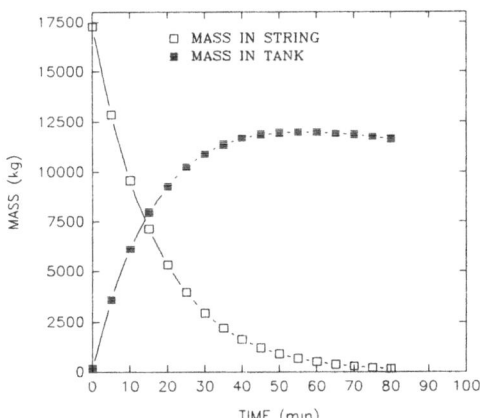

Figure 1. Time history of masses in string and dump tank

will accumulate in the dump tank, which equals to the remaining nitrogen quantity in the strings. The vented mass by this time is slightly over 2,700 kg (882 gal). The accumulated nitrogen mass in the tank reaches a maximum of 12,000 kg (3,921 gal) in less than 60 minutes after the venting started. At this time, over 96% of the string nitrogen inventory has been discharged, and approximately 29% of the inventory has been vented. The mass flow rates into and out the dump tank are compared in Figure 2 as functions of time. The initial mass flow rate into the dump tank is about 17 kg/s which drops rapidly as the string nitrogen reaches a saturation pressure. Further reduction of this mass flow rate is due to the variation of the string nitrogen inventory quality which increases from

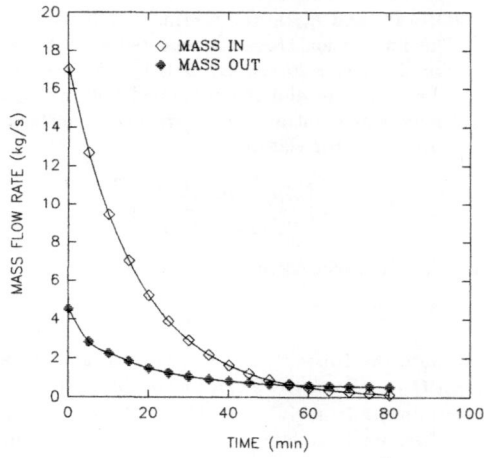

Figure 2. Mass flow rates into and out dump tank as functions of time

0 to over 0.8 within the first 80 minutes of venting. The venting mass flow rate has a starting value of less than 5 kg/s. It drops below 1 kg/s when the dump tank is cooled down and remains almost constant after 40 minutes of venting. The two curves intercept each other in about 55 minutes after the venting started, which corresponds to the moment when the accumulated liquid nitrogen reaches its highest level in the dump tank. After 80 minutes, the remaining mass in the strings falls below 160 kg which is less than 1% of total inventory. This may be considered as the end of the venting process. The subsequent venting of the accumulated nitrogen mass in the dump tank takes approximately 6 hours, which is of less concern.

SUMMARY

The dynamic process of the SSC tunnel nitrogen inventory emergency venting has been numerically simulated. Time histories of both the remaining string nitrogen inventory and the nitrogen accumulation in the dump tank are generated along with the mass flow rates into and out the dump tank. For a predetermined inner diameter of the dump tank, the minimum capacity requirement of the dump tank is determined. It corresponds to the highest liquid nitrogen accumulation level in the dump tank. Heat ejected by the tank wall during the initial flash-off and the subsequent cool-down vaporizes a large portion of the accumulated liquid nitrogen. Since the latent heat of nitrogen slightly decreases with increasing pressure, the string pressure during venting should be kept minimal, which requires a small pressure drop through the venting path. The string pressure will dynamically build up to a level required to overcome the resistance in the venting path. According to the simulation results, the minimum capacity requirement for the dump tank at each site should be 15.14 m³ (4,000 gal).

REFERENCES

1. M. McAshan, S. Abramovich, V. Ganni, A. Scheidemantle, and M. Thirumaleshwar, *84 K Nitrogen System for the SSC*, IISSC, Supercollider 4, J. Nonte, ed., Plenum Press, New York (1992) p. 207.

2. F.P. Incropera and D.P. DeWitt, *Fundamentals of Heat and Mass Transfer*, 2nd Ed., John Wiley & Sons, Inc., New York (1985).

3. V.D. Arp and R.D. McCarty, *GASPAK* user's guide, version 2,2, Cryodata, Niwot, Colorado (1989).

4. R.F. Barron, *Cryogenic Systems*, 2nd Ed., Oxford University Press, Inc., New York (1985).

STRESS ANALYSIS ON SSC CRYOGENIC SHAFT TRANSFER LINE
SUSPENSION SYSTEM

B. Zhang

Accelerator Systems Division
Superconducting Super Collider Laboratory*
2550 Beckleymeade Avenue
Dallas, Texas 75237

INTRODUCTION

The Superconducting Super Collider has a total of twelve refrigeration plants. Each plant requires a cryogenic transfer line to connect the above-ground refrigerator to the below-ground collider main ring. The transfer line consists of seven cryogenic circuits enclosed in a cryostat. It is to be built in a number of pieces (modules) and assembled on-site. Within each transfer line module, the internal elements including the circuit tubes, thermal shield, and multilayer insulation (MLI) blankets are supported by an internal suspension system. The suspension system for a module consists of a longitudinal support and four radial supports. The radial supports restrict any radial movement of the internal elements while allowing longitudinal thermal movement. The longitudinal support anchors all the internal elements to the vacuum jacket. Each type of support consists of three plates for supporting the circuit tubes at three designated temperature levels (4, 20, and 80 K), and two sets of stand-off rods which make joints between the 4 and 20 K plates, and between the 20 and 80 K plates. An isometric view of a radial support and a longitudinal support is shown in Figure 1.

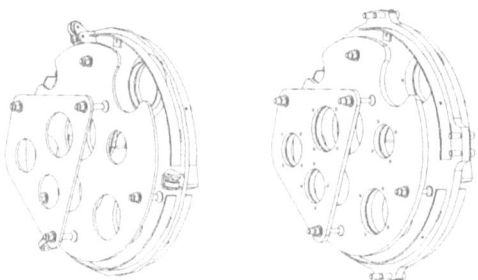

Figure 1. Isometrics of radial and longitudinal supports

The longitudinal support anchors the internal elements to the vacuum jacket through the welds between the vacuum jacket and a metal ring fastened to the 80 K plate. Each transfer line has two 90° bends where straight modules are joined by elbow pieces. There are two internal supports within each elbow piece. The elbow piece supports are similar to the longitudinal supports except they have much higher strength requirements (approximately 10 times). The three stand-off rods connecting the 4 and 20 K plates are strategically located such that an optimal weight distribution

*Operated by Universities Research Association, Inc., for the U.S. Department of Energy under Contract No. DE-AC35-89ER40486

is achieved resulting the least strength requirement for the 20 K plate. The same design principle is applied to locating the rods between the 20 and 80 K plates in order to minimize the strength requirement for the 80 K plate. Due to the geometrical complexity of the internal suspension system design, numerical methods are required for the stress analysis of the suspension system components. The objective of finite element modelling is to investigate the behavior of the support assemblies under specified conditions.

LOADING REQUIREMENTS AND CONSTRAINTS

The loading requirements for the support assemblies are based on the design system requirements and geometrical constraints. The system requirements are characterized by the parameters contained in Table 1. The loads on the radial supports depend on the total weight of the internal

Table 1. Transfer line system characteristics

| Cryogenic circuit | Pressure | | Temperature | Mass flow rate | |
	Design (Bar)	Nominal (Bar)	(K)	Design (g/s)	Nominal (g/s)
4 K helium supply	20	4	4.5	400	400
4 K helium return	20	3.5	4.3	400	100
4 K gas helium	10	1	4.0	400	300
20 K helium supply	20	3	16.5	200	200
20 K helium return	20	2	24.0	200	200
80 K liquid nitrogen	20	10	80.0	400	80
80 K gas nitrogen	10	1	80.0	400	300

elements and the cryogen content (approximately 6,200 N) when the module is installed horizontally. They are negligible when the module is installed vertically. The load on the longitudinal support during normal operation is due to gravity when the module is installed vertically. A 2 g loading is applied to the internal elements for the stress analysis purpose. Expansion joints are provided for the interconnection of the transfer line modules. Therefore, the thermal loading on each circuit tube is neglected. The use of expansion joints (bellows) for the interconnection between an elbow piece and neighboring modules demands the tensile forces in the circuit tubes be isolated within the elbow piece. Therefore, the dominating loading for the elbow support is the hydrostatic force on the circuit tubes at the bend. In addition, loadings due to cryogen momentum change at the bend and thermal contraction are also considered. Detailed loads on the longitudinal and elbow supports are tabulated in Table 2. The loads for the elbow supports are due to hydrostatic forces based on design

Table 2. Loading requirements for longitudinal and elbow supports

| Circuit | Loading (N) | | Coordinates (mm) | |
	Longitudinal Support	Elbow Support	Y	Z
4 K supply	864	8,460	-117.62	2.2
4 K return	864	8,460	-117.62	-112.83
4 K gas	988	7,691	2.79	-55.31
20 K supply	864	8,460	75.64	60.91
20 K return	1,196	15,382	-63.87	139.03
80 K liquid	3,393	3,716	16.15	-228.09
80 K gas	3,615	7,691	203.2	0

pressures. They will be dramatically reduced when the nominal pressures are used. The magnitudes of the actual loads on the elbow supports will vary depending on the finalization of the transfer line design and system operating modes.

MODELING

For the longitudinal support, five work sheets are created for the three plates at 4, 20 and 80 K and two sets of stand-off rods. The plates naturally fall in the plate/shell category, and the rods are considered as 3-dimensional solids. Triangle and quadrilateral elements are used for mesh generation in the plates and rods respectively. No boundary node is assigned to the stand-off rods. Each set of rods is modeled individually and linked to the plates to impose spatial constraints. The meshing statistics and geometry for the longitudinal support are presented in Table 3 and Figure 2 respectively.

Table 3. Statistics for finite element analysis

Component	Total nodes	Nodes w/ B.C.'s	Element type
4 K plate	209	36	triangle
20 K plate	433	24	triangle
80 K plate	472	40	triangle
Rods	1542	0	quadrilateral

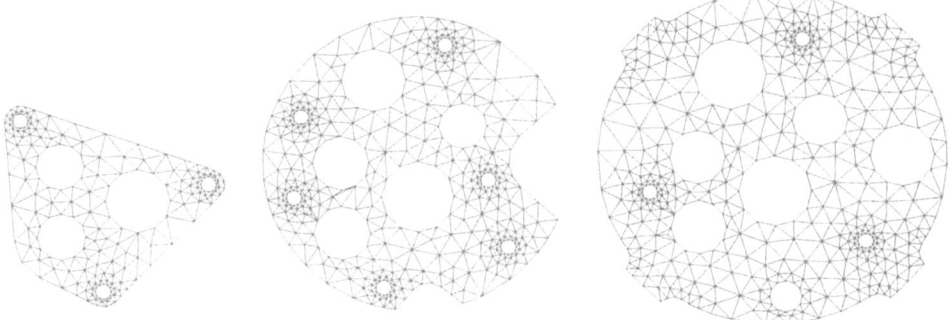

Figure 2. Geometric meshing for finite element analysis

The 4 K plate measures approximately 0.34 × 0.345 m, the 20 K and 80 K plates are circular with outer diameters of 0.482 m and 0.546 m respectively. The geometries of the elbow support plates are the same as those of the longitudinal support plates except thickness (see Table 4) and the number of anchoring locations (four for longitudinal support and eight for elbow support). Young's modulus $E = 3 \times 10^4$ MPa and Poisson's ratio $\mu = 0.2$ are used for the plates, and E=1.93 $\times 10^5$ MPa and $\mu = 0.3$ are used for the rods. The five elements (three plates and two sets of rods) are then decoded individually before they are linked together to impose mutual spatial constraints for the final stress analysis process. The stress analysis is performed using Algor's finite element analysis (FEA) processing system release 9.0.

RESULTS AND DISCUSSIONS

The two models (longitudinal and elbow supports) are processed separately rendering stress and deformation distributions, and other pertinent informations. A screen capture of the stress distribution in the 4 K plate of the longitudinal support is shown in Figure 3 as an example.

The Von Mises stress spectrum is shown in the upper-right corner of the graph in the unit of megapascal. Higher stress is concentrated at locations where the stand-off rods join the 4 K plate as anticipated. The maximum stress in the plate ($\sigma_{max} = 94.5492$ MPa) allows sufficient flexibility in design to provide adequate strength. The stress level in the 20 K plate is relatively low. Since all the loading from the 4 K circuits is transmitted to the 80 K plate through the 20 K plate, only the neighborhood of the stand-off rods connecting the 20 K plate to the 80 K plate needs to be examined. In the 80 K plate, higher stress is shown in the vicinity of the four tabs where the plate is anchored to the vacuum jacket. The stress at the tab areas can be significantly reduced by reducing

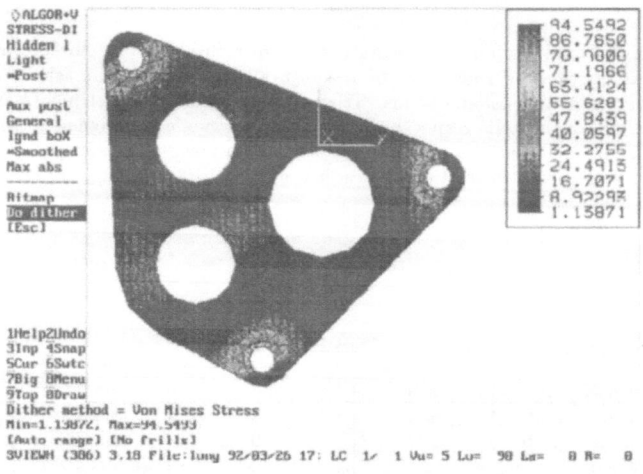

Figure 3. Stress distribution in 4 K plate

Table 4. Results of analysis

Plate Element	σ_{max} (MPa)	e_{max} (mm)	Y (mm)	Z (mm)	Thickness (mm)
4 K longitudinal	95	1.52	-192.5	-117.2	9.53
20 K longitudinal	57	.72	-60	-213	12.7
80 K longitudinal	59	.085	31.2	-6.1	19.05
4 K elbow	169	1.1	-192	-117	19.05
20 K elbow	272	2.73	-157	183	25.4
80 K elbow	105	.99	-19	32	25.4

the inner diameter of the anchor ring. The maximum Von Mises stress and deformation σ_{max} and e_{max}, as well as the coordinates where the maximum deformation occurs are listed in Table 4.

CONCLUSIONS

To predict the strength requirements for the internal suspension system of the SSC cryogenic transfer lines, finite element stress analysis is performed on the longitudinal and elbow supports. The results reveal the behaviors of the construction elements. Computed maximum stress and deformation which occur under complex loading situation serve as reference for material selection and sizing in the preliminary design process. The highly localized stress in the plates may be substantially reduced by improving supporting boundary conditions. Testing apparatus has been designed and fabricated to verify the numerical results. Attention will be focused on the critical elements indicated by the numerical results to ascertain that they have sufficient strength for potential adverse situations.

REFERENCES

1. B. Zhang and V. Ganni, *Conceptual Design of The SSC Cryogenic Transfer Lines*, IISSC, Supercollider 4, J. Nonte, ed., Plenum Press, New York (1992) p. 453.

2. P.R. Smith and T.J. Van Laan. *Piping and Pipe Support Systems*, McGraw-Hill, Inc., New York (1987).

3. *Finite Element Analysis System Processor Reference Manual*, Algor, Inc., Pittsburgh, PA (1989).

EXPERIMENTAL STRENGTH VERIFICATION OF SSC CRYOGENIC TRANSFER LINE INTERNAL SUSPENSION SYSTEM

Kirk Stifle and Burt Zhang

Superconducting Super Collider Laboratory*
2550 Beckleymeade Ave
Dallas, Texas 75237

INTRODUCTION

A transfer line system and support structure has been proposed by Zhang and Stifle[1] and Zhang[2]. The support structure uses three composite (G-10) plates to support the seven cryogenic lines. The cryogenic lines range in temperature from 80 K to 4 K. There are two 80 K nitrogen lines, two 20 K vapor helium lines, and three 4 K helium lines. The 80 K lines are attached to one plate, the 20 K lines are attached to another plate and the 4 K lines are attached to the last plate. Stainless steel rods are used to connect the plates to each other. This arrangement results in essentially three different temperature levels: an 80 K plate, a 20 K plate, and a 4 K plate.

There are three different support structures; an elbow support, a longitudinal support, and a radial support. Figure 1 shows the longitudinal and radial support structures. The main difference between the structures is the thickness of the plates. The elbow support is the thickest because it must carry the largest load and the radial support is the thinnest because it carries the lightest load. There are slight differences in the 80 K plates of each support but the 4 K and 20 K plates remain virtually unchanged except for the thickness. The 80 K plate of the elbow support has more "tabs" which are used to attach it to the vacuum vessel than the 80 K plate of the longitudinal support. The 80 K plate of the radial support is not attached to the vacuum vessel but has rollers which allow it to move as the tubes expand and contract with temperature changes. Only the elbow and longitudinal supports were tested.

EXPERIMENTAL SETUP

The test setup is shown in Figure 2. A complete description of the test setup and test procedure can be found in Stifle and Zhang[3]. The idea was to test the support structure in as mechanically realistic a condition as possible. The 80K plate was bolted to a hollow ring used to simulate the vacuum vessel. A steel plate was placed on top of the ring and 80 K plate. The 4K and 20K plates were placed on top of this steel plate and attached to the 80K plate. The steel plate had holes drilled to allow the support plates to be connected together. Another steel plate was placed on steel rods which were welded to the bottom steel plate. This plate provided a surface for hydraulic jacks which were used to apply the load. Stainless steel inserts were machined and attached to the G-10 plates at the tube locations. Steel rods were connected to the inserts and to the jacks. Pressure gages on the hydraulic jacks were used to measure the applied load.

Strain gages were placed at various locations on the plates. The location of the gages was determined from finite element analysis of the plates. The gages were general purpose gages of Constantan foil with a polyimide film backing. The gages had maximum strain limits of 3% (30,000 μstrain).

* Operated by Universities Research Association, Inc., for the U.S. Department of Energy under Contract No. DE-AC02-89ER40486.

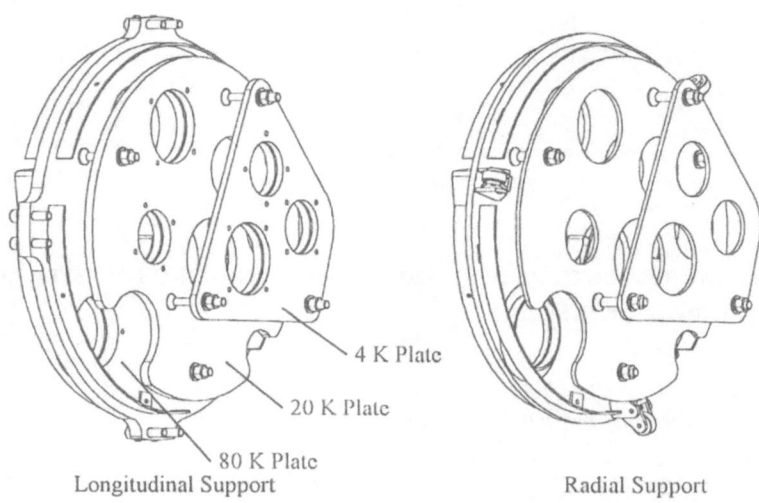

4 K Plate

20 K Plate

80 K Plate

Longitudinal Support

Radial Support

Figure 1. Longitudinal and Radial Supports

The loading on the longitudinal support is due to the weight of the plates, tubes, copper shield, and multilayer insulation blankets. The maximum loading applied to the longitudinal support was three times the design load. The elbow support has the same loading plus the addition of a hydrostatic pressure force from the interconnect bellows. The bellows design has not be finalized and the actual loading cannot be determined until this happens. The actual loading is expected to be in the upper range of the loads tested.

The method of testing was as follows: The jacks were pumped until the desired pressure was achieved on all three plates. Next, the strain from each gage was read from the indicator and manually recorded. The loading on the plates was then increased to the next level and readings taken. The procedure was repeated until the maximum load was reached, after which the pressure was released back to zero. The entire test was then repeated twice more.

RESULTS AND DISCUSSION

For a complete presentation of the data, see Stifle and Zhang[3]. Figure 3 shows typical data for the longitudinal support. The data is for the longitudinal 80 K plate.

The data for all the longitudinal support plates show increasing strain as the loading is increased. The strain for the 80 K plate was almost all compression (negative strain) except for two gages. The two gages were placed on the outside of where the liquid nitrogen lines would be attached and show very large tensile (positive strain) strains. The maximum strain is approximately 1750 μstrain. This gives a stress value of 4375 psi at the SG82 location using a value of 2.5×10^6 for the modulus.

The strains on the 20 K plate were relatively small and should not pose a problem. The strains on the 4 K plate were slightly higher than those on the 20 K plate. However, these values were all much lower than the maximum strains on the 80 K plate.

Figure 4 shows typical data for the elbow support. The data is for the elbow 80 K plate. We believe that the unusual nature of the data is due to the interactions between the plates.

The strain was considerably higher on this support than on the longitudinal support. This was expected because of the increased loading which the elbow support is expected to have. It was anticipated that the majority of the load would be transmitted back to the 80 K plate through the steel connecting rods. The data does not indicate that this happened. The 4 K plate load was not transmitted back to the 80 K plate. This indicates that the 80 K and 20 K plates were too stiff or the 4 K plate was not stiff enough. The 20 K and 80 K plates need to bend more or the 4 K plate needs to bend less to help reduce the strains on the 4 K plate. Otherwise, the 4 K plate will have a much larger stress and will yield before the other plates.

Because no gages placed at the edge of the liquid nitrogen lines, the maximum stress on the 80 K plate may not have been recorded. The maximum measured strain occurred on the 4 K plate and yielded a stress of approximately 7450 psi (2.5×10^6 psi modulus).

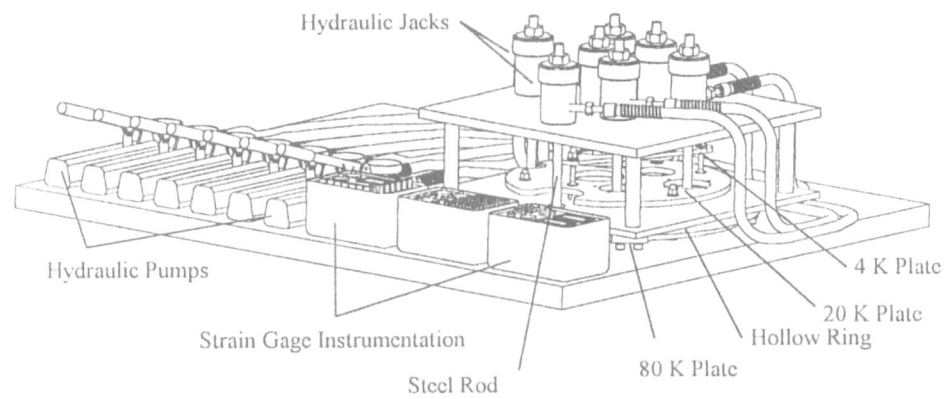

Figure 2. Experimental Setup

The strain gages were placed only on the top surface of the plates. Therefore, we measured axial strain on the top surface of the plates. This number should be compared to the maximum allowable tensile/compressive stress for G-10. A factor of safety of 5 was applied to the ultimate stress to obtain a maximum allowable stress. Using 20% of ultimate gives an allowable stress level of 9000 psi in tension and 10400 psi in compression for the warp direction. The allowable loading is 6000 psi for tension and 8400 psi for compression in the fill direction. The stress on the longitudinal support was below these values. The stress on the elbow support was below these values except for on the 4 K plate where the stress value was larger than the allowable tensile stress in the fill direction.

CONCLUSION

A prototype internal suspension system made from composite (G-10) materials was built and tested. G-10 was chosen as the main support material for convenience and cost and other materials (composite, metal, or combination of composite and metal) could have been used. The parts could have also been made from injection molding or resin transfer molding instead of machined.

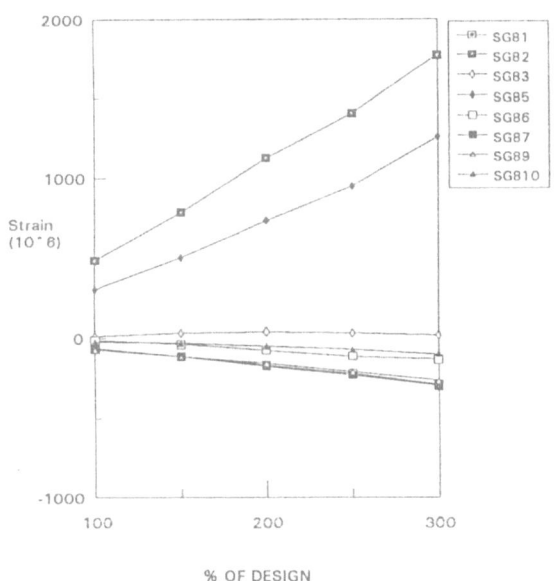

Figure 3. Longitudinal Support 80 K Plate Strain Data

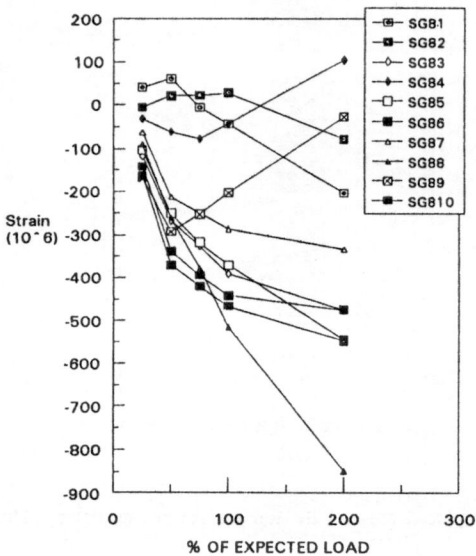

Figure 4. Elbow Support 80 K Plate Strain Data

The prototype suspension system was tested to determine stress levels in the material. We can draw some conclusions from this testing:

1. Neither G-10 support (elbow/longitudinal) yielded when loaded with anticipated loads.

2. The longitudinal support had no problem with the anticipated loads when tested at room temperature..

3. The elbow support stress exceeds 20% of the ultimate stress using a value of 2.5×10^6 psi for Young's modulus. This may cause a problem depending on the final bellows design. However, the plates could be made thicker or another material could be used to reduce the stress.

4. When tested at room temperature, the G-10 structure did not have problems with the applied loading. G-10 or another composite can be considered as options for the support structure if further testing is carried out. This testing should show that the material will not lose strength at cryogenic temperatures over the intended operating life time, i.e. 20 year life.

ACKNOWLEDGMENTS

The authors would like to thank the SSC Cryogenics Shop for help in the set up and assembly of the experimental apparatus.

REFERENCES

1. B. Zhang and K. Stifle, "Preliminary Design of Sector Cryogenic Shaft Transfer Line System", Presented at 5th International Industrial Symposium on the Super Collider (IISSC), San Francisco, CA, May 6-8 (1993).
2. B. Zhang, "Stress Analysis on SSC Cryogenic Transfer Line Internal Suspension System", Presented at 5th International Industrial Symposium on the Super Collider (IISSC), San Francisco, CA, May 6-8 (1993)
3. K. Stifle and B. Zhang, "Development and Testing of a Prototype Internal Suspension System Using Composite Materials for Cryogenic Transfer Lines", SSC Cryo Note 92-16 (1993).

ANALYSIS OF ER STRING TEST THERMALLY INSTRUMENTED INTERCONNECT 80-K MLI BLANKET

E. F. Daly[1] and R. K. Pletzer[2]

[1] Advanced Electromechanical Systems
Westinghouse Science & Technology Center
1310 Beulah Road
Pittsburgh, PA 15235

[2] Magnet Systems Division
Superconducting SuperCollider Laboratory
2550 Beckleymeade Avenue
Suite 125, MS-1001
Dallas, TX 75237

INTRODUCTION

The 40mm CDM string test conducted at Fermi National Accelerator Laboratory (FNAL) served as a springboard for the SSCL Accelerator Systems String Test (ASST) in order to test various operational scenarios, as well as determine electrical and thermal system performance[1]. In support of these efforts, an 80-K Multi Layer Insulation (MLI) blanket in the interconnect region between magnets DD0019 and DD0027 in the 40mm CDM string was instrumented with temperature sensors to obtain the steady state temperature gradient through the blanket after string cool down.

A thermal model of the 80-K blanket assembly was constructed to analyze this temperature gradient, and to estimate the heat flux corresponding to the measured temperature gradients. The results of the analysis predicted a heat flux of 0.363-0.453 W/m^2 based on eleven sets of data. The calculated heat fluxes were 33-46% below the 80-K MLI blanket heat leak budget of 0.676 W/m^2, and are consistent with total heat loads to the 80-K calculated by McInturff et al[1]. The blanket assemblies performed very well when compared with the 80-K static infrared heat leak budget.

EXPERIMENTAL SET-UP

The blanket assemblies consisted of two 32-layer MLI blankets wrapped once each around the 80-K shield bridge in the interconnect region. Each blanket consisted of 32 layers of double aluminized mylar (DAM) reflector material separated by single layers of polyethylene-terapthalate (PET) spacer material.

During cold testing, temperature data was obtained from a series of fourteen cryogenic linear temperature sensors (CLTS). Their low profile and relatively large surface area made them excellent choices for installation within the blanket assemblies. Eleven data sets, each taken over a 24-hour period, were averaged and documented by Augustynowicz[2] of SSC Accelerator Cryogenics. A residual gas analyzer was employed to determine the gas pressure and composition prior to each testing period.

MATHEMATICAL MODEL

A one-dimensional, finite difference thermal model of the 80-K MLI blanket assembly was developed to analyze the temperature distribution using General Dynamics' Convair Thermal Analyzer (CTA) software. Zero-capacitance nodes were used to represent each reflector layer of the blanket assembly and the aluminum shield. The PET spacer and cover layers were not represented with nodes but were accounted for in the thermal resistance formulation for the model. Adjacent reflector layers are connected by a triplet of parallel resistor elements to account for residual gas conduction, thermal radiation and solid conduction through the spacer layers (Fig. 2). This approach is applied between all layer spacings and the shield, and a detailed description is found in Reference [3].

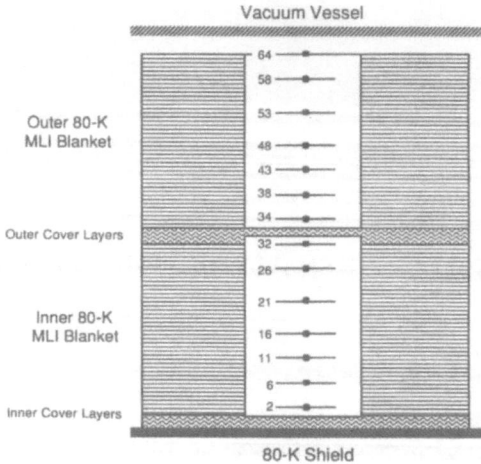

Figure 1. Thermal model of 80-K MLI blanket assembly.

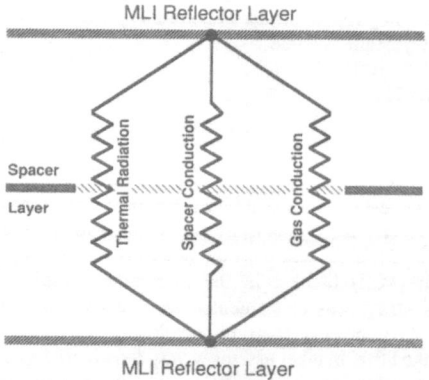

Figure 2. Thermal resistor triplet between DAM reflector layers.

Initially, the effective conductivity of the heavy cover layers were assumed the same as the spacer layers. However, the calculated heat flux across the cover layers were inconsistent with the measured temperature difference from the 80-K shield to layer 2 and from layer 32 to layer 34 (See Fig. 1). Therefore the effective conductivities and emissivities of the heavy cover layers were independently adjusted until the heat flux through the heavy cover layers matched those predicted through adjacent spacer layers to determine the behavior of the cover layers empirically.

ANALYSIS

The eleven sets of steady-state temperature gradient data were analyzed using the following approach. For each set of data, three steady-state simulations were performed.

Case 1: Imposed Temperature Gradient (Qave)

The model temperatures at layer locations corresponding to the sensored layers were fixed at the measured temperatures from those locations. The solver program calculated the remaining reflector layer temperatures and corresponding heat fluxes. Since the heat flux through the blanket must be constant, the values obtained were averaged.

Case 2: Fixed Hot/Cold Sink Temperatures

The model temperatures at only the outermost layer of the outer blanket and the 80-K shield were fixed at the measured temperatures from those locations. The predicted temperatures were compared to the measured values, and the calculated heat flux was compared with the results from the imposed temperature case.

Case 3: Imposed Heat Flux (Qfixed)

The model temperatures at only the 80-K shield was fixed at the measured temperatures, and the average heat flux calculated from the first case was imposed. The predicted temperatures were compared to the measured values, and should closely match the results from the second case.

RESULTS

The effective conductivity of the heavy cover layer pair between the shield and the inner blanket was 2.1 times lower than that of the spacer layer. The effective conductivity of the cover layer between inner and outer blankets was 7 times lower than that of the spacer layers. Effective emissivities of 0.05 were required to obtain acceptable temperature gradients and heat fluxes.

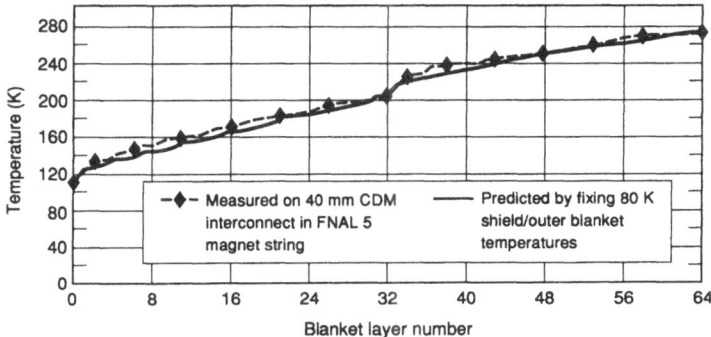

Figure 3. Comparison of measured and predicted temperatures vs. blanket layer number on 10/23/91.

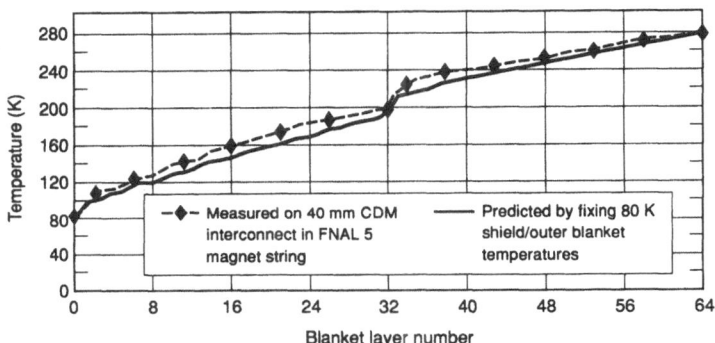

Figure 4. Comparison of measured and predicted temperatures vs. blanket layer number on 10/1/93.

This decrease in effective conductivity is most likely due to contact resistance. The inner cover layer contacts the outer diameter of the 80-K shield more intimately than the outer layer contacts the outer surface of the inner blanket, hence the difference in estimated effective conductivities. The discrepancy in the effective emissivity of the cover layers compared with the room temperature values is attributed to the uncertainty in the reflector layers emissivities used in the analysis.

Temperature profile predictions for the two analysis cases are shown in Figures 3 and 4. Note that the average values apply to the first case, where a particular temperature gradient was imposed through the MLI blankets to replicate the ER string test profiles. The fixed values correspond to the second case, where only the temperatures of the 80-K shield and the outermost surface of the outer blanket were fixed.

Five of the eleven data sets were run with a nominal shield temperature of 80 K, while six were run with a nominal temperature of 105 K. This lead to analysis cases whose difference in temperature across the MLI blankets ranged from roughly 160 K to 195 K.

The predicted temperature gradients appear much more linear through the layers than the corresponding measured temperature gradients. The measured gradients contain some curvature, although small, due to the radiative component of heat transfer. This indicates that the model predicts a larger portion of heat transfer due to conduction than actually existed. Above N=40, the radiative component is minimal, and the gradient becomes linear.

The percent difference in heat flux between cases does not correlate well with the temperature difference across the blanket assembly. Fixed cases were within 9% of the average cases, but the model over predicted the heat leak by about 30 mW. The ranges for the effective resistance were comparable for both case sets, but the values for the fixed cases were lower overall. This is consistent with the idea of a resistance network: as resistance increases, the heat flow decreases for a fixed temperature difference. This implies that the cases analyzed employing fixed-end temperatures stress the conductive mode of heat transfer more than the experimental set-up.

The results of the analysis show that the calculated heat leak, ranging from 0.363-0.453 W/m^2, is below the heat leak budget value of 0.676 W/m^2 for static heat leak to the 80-K cooling system. The range of heat fluxes corresponds to a total of 10.8 to 13.5 W conducted through the 80-K blanket. Heat leak measurements of the ER string test yielded a total heat leak of 23 to 31 W for LN2 flowing at 3.3 g/sec for the interconnect region between magnets DD0019 and DD0027[1].

Heat leak measurements conducted by Boroski on a single 32 layer DAM/PET MLI blanket resulted in a heat leak of 0.648 W/m^2 for a gradient of 299.7 to 79.9 K. For the same temperature difference and twice the blanket thickness, the effective conductivity would be roughly 50%, or 0.324 W/m^2. This value is consistent with data calculated in Table 1.

Table 1. Predicted heat fluxes and effective resistances.

FNAL ER Data Set	Shield/ Outer MLI Temp. Difference (K)	Qave (W/m^2)	Qfixed (W/m^2)	Reff-ave $(K/W/m^2)$	Reff-fixed $(K/W/m^2)$	% Difference (Qf-Qa)/Qa
10/1/91*	194.78	0.453	0.473	430.07	411.54	4.50%
10/23/91	165.22	0.394	0.420	419.13	393.29	6.57%
10/24/91	167.22	0.410	0.426	407.95	393.00	3.81%
10/25/91	167.11	0.391	0.425	427.61	392.92	8.83%
10/26/91*	192.11	0.438	0.467	439.11	411.46	6.72%
10/27/91*	190.50	0.435	0.460	437.53	413.77	5.74%
11/2/91	159.22	0.365	0.397	436.58	400.86	8.91%
11/3/91	159.61	0.363	0.393	439.58	406.13	8.23%
11/4/91	168.78	0.380	0.409	443.81	412.56	7.57%
11/5/91*	185.78	0.416	0.440	446.80	422.61	5.72%
11/6/91*	186.89	0.421	0.443	444.45	421.59	5.42%

CONCLUSIONS

The results show that the DAM/PET 80-K MLI blanket assembly is performing comparable with design estimates, component tests, and system tests. This performance is encouragingly below the budget heat flux.

Differences in the effective conductivities of the cover and spacer layers are most likely due to the particular design and the assembly method. The MLI system relies on intermittent contact and, therefore, on high contact resistance between layers to effect the insulation scheme. This contact resistance is roughly a function of the stress imparted to the blankets during assembly.

The current model predicts a heat leak whose conductive mode is greater than experimental results support. Therefore, cryostat models using this modeling technique will produce conservative estimates of radiative heat leak.

REFERENCES

1. McInturff, A. D., J. G. Weisend II, C. E. Dickey, R. Flora, and D. B. Wallis, Measured control characteristics of the half-cell 40mm aperture magnet string, *Proceedings from the Fourth International Industrial Symposium on the SuperCollider*, March (1992).

2. Augustynowicz, S., Internal SSCL memorandum, December (1992).

3. Daly, E. F. and R. K. Pletzer, Analysis of ER string test thermally instrumented interconnect 80-K MLI blanket", SSCL Tech. Note SSCL-526, May (1992).

LOW CONDUCTIVITY WATER PLANTS-ISSUES OF PROCESS CONTROL

Liana Baritchi and D. R. Haenni

Superconducting Super Collider Laboratory, Dallas, Tx.

ABSTRACT

The SSC laboratory will require a large number of widely distributed low conductivity water (LCW) cooling systems to support accelerator operations. In addition to designing the physical plants, plans must be made for control / information systems and the human organization to run them. Initial considerations in these areas are presented.

INTRODUCTION

Various SSC accelerator components such as resistive magnets, radio frequency equipment, and power supplies will use LCW cooling to reject heat. The cooling water must be deionized to prevent short circuits due to current flow through the water and the erosion of piping components by electrolysis. When cooling is needed for equipment in radiation areas intermediate closed loop systems will be used. Heat from the LCW will be transferred to industrial cooling water, ICW, systems and then to the atmosphere by means of cooling ponds or towers.

A total of about 35,000 gpm of deionized water is needed for the operation of the accelerator systems. This water will be provided from 34 plants located around the complex. The location of these plants is shown in Figure 1. Systems for LINAC, LEB, MEB, HEB, Transfer Lines and Test Beams (TB) will service both above and below ground components. For the Collider only surface components will be water cooled.

The LCW systems contain the equipment, devices, and instruments required to produce, cool and distribute the LCW. For some systems ICW plants will be an integral component of the LCW plants but for others it will be provided from an external source. In addition there will be a centralized regeneration plant for the deionizer beds.

Because of the wide geographical distribution of the LCW plants and a projected small staff to operate and maintain them, each system must operate in a "lights out" or fully automatic mode under control and monitoring from a single central location.

In this paper we discuss the requirements and human organization needed to control and maintain the LCW system.

SYSTEM DESIGN CONSIDERATIONS

The LCW plants are considered as a technical utility and therefore must provide highly reliable continuous service. They will be designed, installed and commissioned individually in time to support their "customers" the accelerator components. The primary requirements for the LCW are for the most part uniform for the whole of the SSC. The systems need to provide a continuous flow of water at a maximum temperature of 90 degrees

*Operated by the Universities Research Association, Inc., for the U.S. Department of Energy under Contract No. DE-AC35-89ER40486.

F and resistivity of 5 MΩ by maintaining a constant differential pressure across supply and return of 80 psid. Environmental considerations require that the return pressure from the LCW must be lower than that of the ICW. Plants diversify in their capacities, level of redundancy, and whether they contain an integral ICW plant.

To develop an LCW system for the SSC one must first define it (decide its boundaries) then determine its requirements and goals. From this, a set of modular tasks can be created which implement the desired LCW system. Conceptually these tasks can be assigned to one of two broad categories depending on if they deal with the movement and transformation of energy / material or information. For implementation these tasks can be handled in one of three ways. They can be carried out by process equipment (plant hardware), information systems (computers, networks, software, etc.), or by humans. This analysis is outlined schematically in Figure 2. The distribution of the tasks for implementation provides insight into the level of automation in the system and the human effort needed to run it. The above generally follows models developed for industrial CIM systems (Williams 1989, 1992).

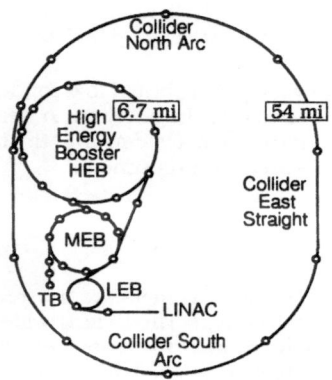

Figure 1. SSC LCW plant locations

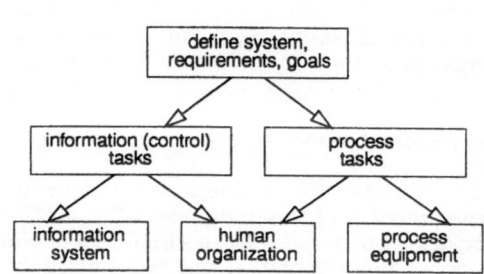

Figure 2. Determining tasks for implementing an LCW system

Qualitatively it is clear that to minimize the long term cost tasks should be implemented with the process equipment and information systems rather than with the human organization. Unfortunately this leads to a higher initial capital cost. Thus the process should (1) be designed to minimize direct manual labor, (2) minimize maintenance labor by maximizing component reliability, providing backup equipment, and selecting components with minimum routine maintenance requirements, and (3) minimize routine operations manpower by providing control automation to the fullest extent possible.

SYSTEM ARCHITECTURE

A suggested architecture for an LCW system "enterprise" to support the SSC is given in Figure 3. It shows process plants, control systems, data management, and major groups in the human organization. This architecture is based on the assumption of an independent LCW unit in the SSC. The ultimate laboratory organization may not reflect this but this exercise helps identify requirements which must be met.

The LCW plants shown at the bottom of the figure are designed in a modular fashion to meet the requirements stated above. Each module carries out a specific system function or regulates a deliverable LCW parameter such as water temperature. The modules necessary to implement ICW systems are also shown but these are only present in those plants with integral ICW control systems. Plants with external ICW would have a communications link with the remote ICW control system.

The information system links the process plants with the human organization. At the bottom this is essentially an industrial process control system. This control system is envisioned as a relatively simple two level hierarchy. On the average each LCW plant is estimated to have 200-300 I/O points and thus should require only a single process controller per site. There may additionally be specialized unit controllers covering for example water quality. Each process controller should be able to support an optional (portable?) local operator interface for plant commissioning, maintenance, and stand alone operation when needed. The second level centralizes control of all the plants and would reside in or near the laboratory central control room. From here an operator should have all control functions available at the local operator interface level. This supervisory level would also link with an information management data base that contained plant historical data, configurational data, and other information relating to the LCW systems. The mandate for data base would be to provide easily accessible, timely, accurate LCW data to those in the organization who need it. This data base forms the core of the upper part of the information system.

As indicated in Figure 3 the process control system and data base would provide links with other SSC control and information systems. This provides the needed integration of LCW systems with the rest of the laboratory.

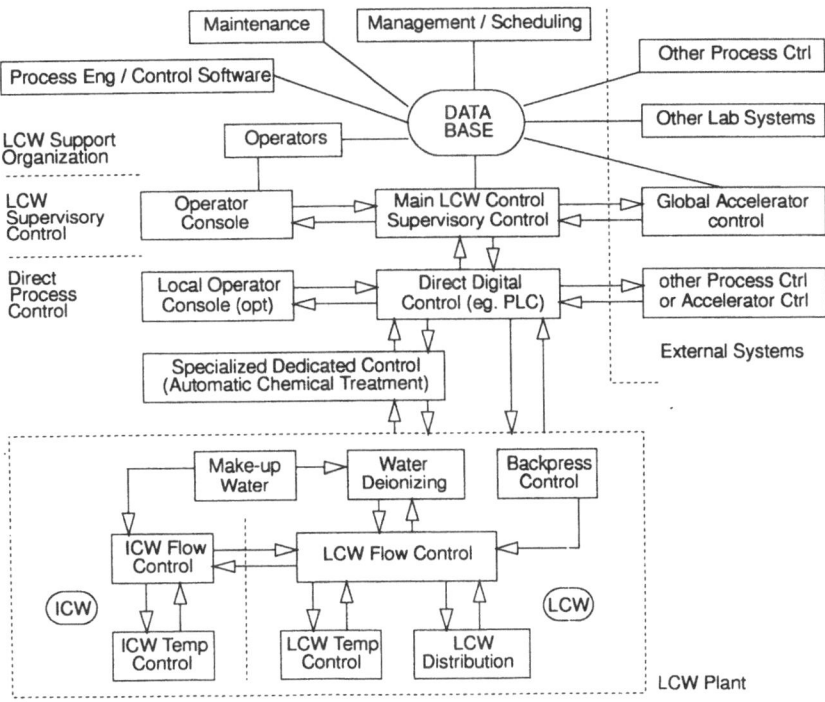

Figure 3. LCW process and the Information management system

At the top of Figure 3. the various functional groups in the LCW systems organization are shown. It may be possible that a given individual could function as a member of more than one group. The groups represent general task areas which must be covered to support the LCW plants and operations. These include maintenance, operators, process engineers, and management / scheduling.

HUMAN ORGANIZATION

Thus far the SSC laboratory has focused on the design and construction of the physical plants and control systems but has not modeled as completely the human efforts involved in running and maintaining them. A few considerations in this area are now presented.

An important question relates to the number of operators needed to run the LCW systems and what are their tasks. Placing a full time operator at each plant (34) is too costly in terms of manpower, thus the plants will be run under remote control from a central location. The minimum number of operators on duty could then be reduced to one. Assuming that an operator delivered six hours of effective work (watching the operator console) during a shift, then each plant would receive ~10.5 minutes of operator attention. Therefore the operation of a plant cannot require careful operator analysis of raw information or long term careful watching of the operation to make adjustments. Obviously to run with a single full time operator the system would need to be fully automated and self diagnosing. In reality is difficult to reach 100% automation and thus an operator is needed to cover a few residual operations. One scenario for a single operator system would be as follows. The operator is responsible for covering these residual control operations for all the 34 systems. This will likely take most of his attention and time. The operator must also provide the first line of fault and problem analysis. When a problem with the system is identified (an alarm or inconsistent operation) the operator would be responsible to determine whether maintenance or process engineering is best suited to handle it. He would then call on someone else to carry out detailed work on the problem. These people in turn can access the information necessary to further analyze the problem through the central data base without having to be in the central control room or at a plant site.

With an enhanced information management system tasks which often fall to operators could be distributed to others in the organization outside of the control room. Process engineers could work on tuning, improving and optimizing control loops. Maintenance personnel could carry out trend analysis on component performance, operation, and failure patterns to identify equipment for repair, recalibration, or replacement before a system failure (which could impact the total accelerator operation) actually occurs. Management could directly generate accurate operations reports and perform other analysis from the information in the system. Some of these activities could in fact be automated by these groups outside operations and the central control room.

Another area where humans interact with the LCW plants is in the replacement and regeneration of deionizer beds and replacement of carbon filters. Should this be done on a time schedule or by monitoring when a bed / filter is actually used up? Carrying this out on a time basis would necessarily assume a worst case water quality and thus cause beds to be regenerated too often. This would prematurely wear out the deionizer resin, increase regeneration costs (material and manpower), and require the construction of a larger (increased capacity) regeneration facility at the SSC. There is also the costs of the human effort necessary to replace and transport the deionizer beds from the LCW plants to the regeneration facility. More efficient use the deionizer beds requires increased plant control and instrumentation. The plant would have to determine when a bed has failed and then have a replacement bed available to automatically switch over. The failed bed could then be replaced and regenerated on a more relaxed schedule (which uses less manpower).

These provide representative issues which must be addressed in the development of LCW systems at the SSC.

REFERENCES

1. Theodore J. Williams. "A Reference Model For Computer Integrated Manufacturing (CIM)" Instrument Society of America, Triangle Park (1989).
2. Theodore J. Williams. "The Purdue Enterprise Reference Architecture" Instrument Society of America, Triangle Park (1992).

ENGINEERED DESIGN OF SSC COOLING PONDS

James E. Bear

Engineering and Design Group
Conventional Construction Division
Superconducting Super Collider Laboratory*
2550 Beckleymeade Avenue
Dallas, Texas 75237

INTRODUCTION

The cooling requirements of the SSC are significant and adequate cooling water systems to meet these requirements are critical to the project's successful operation. The use of adequately designed cooling ponds will provide reliable cooling for operation while also meeting environmental goals of the project to maintain streamflow and flood peaks to preconstruction levels as well as other streamflow and water quality requirements of the Texas Water Commission and the Environmental Protection Agency.

MECHANISMS OF POND SURFACE HEAT TRANSFER

Any body of water can gain or lose heat to the atmosphere and the surrounding earth. However, the transfer of heat between the earth and water is limited to conductive transfer only and is insignificant when compared to that which takes place at the air-water interface. All bodies of water receive natural heat input from short-wave solar radiation and long-wave atmospheric radiation and discharge heat to the atmosphere by back radiation, evaporation, and conduction. The various mechanisms by which heat is exchanged between the water and the atmosphere are depicted on Figure 1.

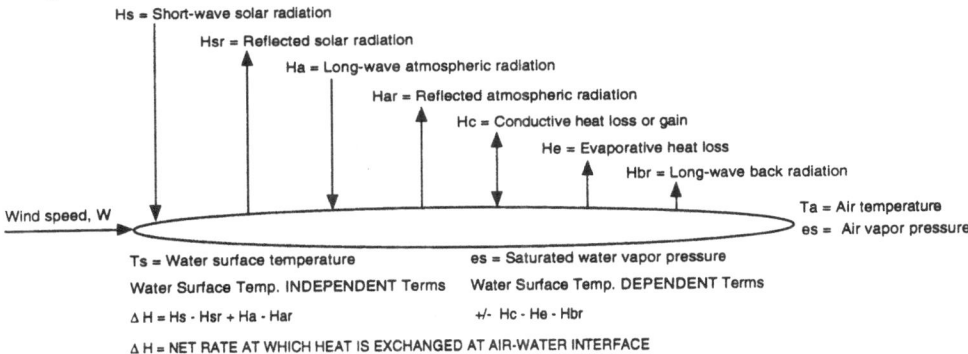

Figure 1. Mechanisms of Heat Transfer Across a Water Surface.

* Operated by the Universities Research Association, Inc., for the U. S. Department of Energy under Contract No. DE-AC35-89ER40486.

The four heat transfer mechanisms that are <u>independent</u> of the surface temperature of the water body are described below:

- The incoming short wave solar radiation, H(s), is the radiant solar energy which passes directly through the atmosphere to the earth from the sun. Its intensity at any location on earth varies with the latitude of the location, time of day, season of the year, and the amount of cloud cover. The amount of short-wave solar radiation reaching the earth's surface can be measured with a pyrheliometer or can be computed based on cloud cover.
- A portion of the incoming solar radiant energy is reflected by the water surface before it can be absorbed by the water. The reflected solar radiation, H(sr), is a function of the sun's altitude and the amount of cloud cover but can be estimated at H(sr) = 0.05xH(s).
- Long-wave atmospheric radiation, H(a), depends primarily on air temperature and humidity and increases as the moisture content of the air increases. It adds the largest amount of heat to a body of water on warm cloudy days when direct solar radiation decreases to zero. This form of radiation is a complex function consisting of many variables and is largely dependent on the distribution of moisture, temperature, ozone, carbon dioxide, and other materials within the atmosphere. It is generally computed by using empirical formulae.
- A portion of the incoming atmospheric radiant energy is also reflected by the water surface before it can be absorbed by the water. The reflected atmospheric radiation is relatively constant at H(ar) = 0.03xH(a).

These four radiation terms - H(s), H(sr), H(a), and H(ar) constitute the net radiation absorbed by the water body and therefore are called total absorbed radiation-- H(R). They can be measured or computed as indicated above and are the four mechanisms <u>independent</u> of the surface temperature of the water body.

The three mechanisms <u>dependent</u> on the surface temperature of the water body are more difficult to determine due to the introduction of another variable. However, these mechanisms, as described below, can be computed using empirical relationships and iterative calculation procedures.

- The conductive heat loss or gain, H(c), is dependent on the temperature difference between the air and the water at the air-water interface. The rate at which heat is conducted between the water and the air is equal to the product of the difference in temperature between the two media and the heat transfer coefficient. Wind speed also affects the rate of heat loss or gain by conduction as it impacts the heat transfer coefficient.
- The evaporative heat loss from a body of water, H(e), can be computed by an empirical relationship dependent on wind speed and the difference between the saturated water vapor pressure of the water surface and the water vapor pressure of the air at the interface.
- Back radiation, H(br), is the ability of a body of water to reject energy to the atmosphere in the form of long-wave radiation since it radiates as an almost perfect black body. The rate at which this heat is lost by a water body is computed from the Stephan-Boltzman fourth power radiation law.

Climatic data requirements used to determine these mechanisms of heat transfer across a water surface include air temperature, relative humidity, wind speed, and solar radiation or cloud cover. The sum of the heat transfer mechanisms at the air-water interface is the net rate at which heat enters or leaves a body of water. For a given set of climatic data, there is a steady-state or thermally balanced condition when the net heat transfer at the air-water interface is zero. The temperature of the water surface for this condition is defined as the natural equilibrium temperature, E.

However, the natural equilibrium temperature is seldom attained in the actual state due to the constantly changing characteristics of the various climatic data. A body of water that has an actual surface water temperature below the natural equilibrium temperature will approach the natural equilibrium temperature by becoming warmer and a body of water having an actual surface water temperature above the natural equilibrium temperature will approach the natural equilibrium temperature by becoming cooler.

Therefore, the rate of heat exchange for a body of water at its air-water interface is proportional to this difference between the actual water surface temperature and the natural equilibrium temperature. For this reason, a body of water receiving an induced heat load that results in an increase in its water surface temperature, has the capacity to dissipate more heat than a body of water without an induced heat load under the same climatic conditions. Thus, the natural equilibrium temperature becomes the reference temperature level for a body of water for a given set of climatic data.

The climatic design requirements for the SSC cooling ponds were established to meet thermal performance requirements 99% of the time or not be exceeded more than 1% of the time in a given year. Since the climatic data requirements for cooling pond design include four climatic variables, summer weather data including all four of these parameters for the past 30 years were evaluated to identify months of high natural equilibrium temperatures.

After the eight highest natural equilibrium temperature months were identified in this monthly analysis, daily computations of natural equilibrium temperatures were made for all eight months. A probability analysis of these daily natural equilibrium temperatures was performed to establish the 1% frequency level. Results of the analysis indicated that the daily natural equilibrium temperature would be between 86° and 87° 1.2 % of the time. Based on this result, the 1% design level was set at 87°F.

THERMAL PERFORMANCE OF POND

The evaluation of the thermal performance of a pond receiving an induced heat load should utilize the time dependent water surface temperature decay equation shown as follows:

$$To = \frac{Ti - E}{e^{\frac{KA}{C_p Q r}}} + E$$

Where: To = Temperature at time t, (°F);
Ti = Initial temperature at t = 0, (°F); E = Equilibrium temp. (°F);
K = Heat exchange coefficient, (BTU/Sq Ft/Day/°F);
A = Surface area, (Sq Ft); r = Density of water, (Lbs/Cu Ft);
C_p = Specific heat of water, (BTU/Lb/°F);
Q = Flow rate (Cu Ft/Day)

While this basic temperature decay equation defines the pond's ability to gain or lose heat through the air-water interface, the complete thermal performance analysis must be based on the pond's ability to maximize surface heat loss as well as its hydrodynamic ability to dampen out thermal transients. Surface heat loss is governed largely by the extent of the pond surface area, but is also a function of the amount of mixing that takes place within the pond. The magnitude of a thermal transient resulting from fluctuating meteorology or induced heat loading is a function of the hydrothermal structure of the pond and its flow circulation patterns which are dependent on location and flow characteristics of intake and discharge facilities as well as any other flow control devices.

The use of computer models to evaluate the thermal performance of power plant cooling water systems has been studied extensively by the Electric Power Research Institute. One of their principal investigators has been Dr. Eric Adams of the R. M. Parsons Laboratory for Water Resources and Hydrodynamics at the Massachusetts Institute of Technology. Dr. Adams was retained as a specialized consultant to review the A-E/CM's methodology of pond design and to assist them in using the MITEMP thermal performance model developed at MIT by Dr. Adams and his colleagues.

The MITEMP model uses a pond classification scheme that includes thermal loading, composite pond number, and horizontal aspect ratio. The SSC cooling ponds are being designed to be classified as the "Shallow Cooling Pond Model with Longitudinal Dispersion." The efficiency of this type of pond is largely dependent on its surface area mobilization and proper pond design should include adequate baffle dikes to guide the circulation of cooling water throughout the pond as well as strategic locations of pond inlet and discharge facilities. Location of the pond inlet to promote surface discharge for the warmer return water and location of the pond discharge to promote subsurface intake of the cooler supply water will also enhance the efficiency of the pond's performance.

The SSC cooling pond design criteria require that an inlet water temperature of 90°F be provided for equipment heat exchangers 99% of the operating time of the machine. This design criteria requires that ponds be designed for a temperature excess of 3°F above the 99% natural equilibrium temperature of 87°F for that condition. Heat exchanger inlet (pond discharge) temperatures were also computed for mean monthly conditions and the temperature excess varies from 2° to 5.9°F for July and January, respectively when the pond discharge temperature varied from 85 to 49.2°F.

THE MEB/INJECTOR POND

The layout of the first SSC pond, the MEB/Injector Pond includes a central baffle dike to guide the circulation path of the cooling water throughout its surface area, and the design will also include some excavation of the pond bottom to eliminate large shallow areas. The proposed design was selected with the aid of costing studies to select the least costly of the alternatives evaluated. The cooling pond water distribution system will serve heat exchangers in the various pump rooms for the LINAC, LEB, MEB, and Test Beam. The pond will provide for 35 MW of cooling per design requirements of not exceeding the 90°F pond outflow temperature, and will also provide for 40 MW of cooling about 95% of the time as well as 47 MW of cooling 90 % of the time without exceeding the 90° F temperature limit.

The MEB/Injector Pond will also provide for flood flow attenuation in accordance with environmental goals of the project to provide stormwater detention in project development areas so as to maintain streamflow and flood peaks to pre-construction levels. The effects of the MEB/Injector Pond storage will reduce the 100-year peak flood flow rates from about 1390 cubic feet per second to about 410 cubic feet per second in the unnamed tributary of Chambers Creek downstream of the pond. The volume of flood water will not be reduced but the peak flood flow will be dampened for a longer period of time. This capability of the MEB/Injector Pond will minimize erosion and help maintain more desirable durations of flow rates downstream.

All water consumed in the cooling process by natural and induced evaporation, as well as any miscellaneous leakage and/or additional requirements for water quality maintenance will be replaced by purchased water from downstream water supply reservoirs. All local inflow to the pond must be released so it will not impact downstream water rights, the majority of which are held by the owner of the downstream reservoirs. Therefore, the pond will be maintained at the normal pool level through water purchases as required and any local inflow will pass on downstream.

CONCLUSIONS

For many years, utilities in Texas have been successfully using power plant cooling ponds and reservoirs to provide reliable cooling water systems while meeting other environmental requirements for these type projects. The SSC can also take advantage of these same qualities with the proper design development, construction, and operation controls of the SSC cooling ponds to insure the successful operation of this important project.

REFERENCES

1. E. E. Adams, A. L. Godbey, D. R. F. Harleman, and K. R. Helfrich, "Evaluation of Models for Predicting Evaporative Water Loss In Cooling Impoundments," Electric Power Research Institute, Research Project CS-2325, Massachusetts Institute of Technology, Cambridge (1982).
2. J. E. Edinger and J. C. Geyer, "Heat Exchange in the Environment," Cooling Water Studies for the Edison Electric Institute, Research Project RP-49, John Hopkins University, Baltimore (1965).

AN ENGINEERING STUDY AND CONCEPT DESIGN OF THE SUPERVISORY CONTROL AND DATA ACQUISITION (SCADA) FOR CONVENTIONAL SYSTEMS

Leonard S. Norman

Engineering and Design Group
Conventional Construction Division
Superconducting Super Collider Laboratory*
2550 Beckleymeade Avenue
Dallas, Texas 75237

ABSTRACT

The study objective was to evaluate several conventional equipment SCADA system architectural concepts and to recommend an approach for development. Each of the concepts given consideration had to satisfy the Superconducting Super Collider (SSC) conventional equipment SCADA application requirements and the evaluation process determined which approach represented the best technical and most cost effective solution to the system requirements. Based on the results of the concept evaluation process, a personal computer based approach was recommended for the SSC conventional equipment SCADA application. Block diagrams and budgetary cost estimate for this approach were developed with specific recommendations with respect to the conventional equipment SCADA system architecture and development process.

INTRODUCTION

This paper discusses the study used to evaluate and recommend a system architecture and concept design for a Supervisory Control and Data Acquisition (SCADA) system for conventional systems of the SSC. Conventional systems are defined as all utility systems such as power distribution, potable water, gas, and HVAC systems. Technical systems such as the low conductivity water, cryogenics and accelerator control systems have separate control systems.

The study evaluated several conventional SCADA system architectural and concept designs and recommended an approach for development. Each concept under consideration during the study had to satisfy SSC conventional systems SCADA application requirements. The evaluation process determined which approach presented the best technical benefits and most cost effective solution to system requirements.

The technical systems will be supported by a high speed networked control system with operations from a central location. The conventional systems SCADA will also have

---

*Operated by the Universities Research Association, Inc., for the U. S. Department of Energy under Contract No. DE-AC35-89ER40486.

centralized monitoring and control with an interface for data exchange between technical and conventional systems.

SYSTEM REQUIREMENTS

The SSC is a large and complicated facility. The collider tunnel ring, will be over 54 miles in circumference. There will be 10 service areas (spaced approximately 5.4 miles apart) that will be for power distribution, technical systems equipment, ventilation, and surface support equipment to the collider ring. Total input/output (I/O) point count for the collider ring has been estimated for conventional systems requirements at 5,000 points.

The injector facilities will consist of a Linear Accelerator (LINAC), a Low Energy Booster (LEB) ring, a Medium Energy Booster (MEB) ring, a High Energy Booster (HEB) ring, and four experiment sites located in areas designated as the east and west campuses. The total I/O point count for these facilities will add about another 5,000 points.

The SCADA system has to provide immediate access to status and control data on all utilities, including alarms and problem conditions, with facilities to provide reports of trends, utility usage, maintenance requirements, and archiving.

Flexibility and-state-of-the-art equipment will be required in the system hardware and software, with the ability to adopt upgrades in the future. The system must be able to include:
- A control system that fits into the construction schedule, which reaches completion in the year 1999 or later.
- A system that runs in parallel with the high speed technical network with minimal involvement from the technical operations center.
- The most cost effective system that satisfies all conventional SCADA operational requirements and maintenance requirements.
- A technology that will carry conventional systems SCADA past 1999 and into the foreseeable future.

STUDY OBJECTIVE AND RESULTS

The study objective was to evaluate conventional SCADA system architectural concepts and to recommend an approach for development. Each of the concepts under consideration must satisfy the SSC conventional equipment SCADA application requirements. The evaluation process had to determine which approach represented the best technical benefit and most cost effective solution to the system requirements.

The information gathered during the study was used to formulate four conventional equipment SCADA system architectural concepts. Each of the concepts considered in the evaluation process was a system configured around commercially available equipment technologies. They were configured to support 30 SSC local sites with 750 data parameters being produced at each site. The actual local site estimate is: collider service areas (10 locations), experimental sites (four locations), main power substation (two locations), and various sites in the injector and campus (four locations), for a total of 20 sites with an I/O count of 10,000 points.

ABBREVIATED CONCEPT DESCRIPTION:

- CONCEPT A - An off-the-shelf distributed control system (such as Westinghouse, or Foxboro) supporting SCADA master units at each SSC local site.
- CONCEPT B - Utilized off-the-shelf open architecture computers (such as IBM, DEC, or HP) as the SCADA host and site computers at each local site.
- CONCEPT C - Utilized off-the-shelf VMEbus crate computers at the central control area and at each local site. One of the VMEbus crates at each local site would act as a site master system.
- CONCEPT D - Utilized off-the-shelf IBM format personal computers as the SCADA host and site computers at each local site.

Concept evaluations were performed using a weighted comparison technique. The weighted evaluation method yields numerical values representing objective and subjective desirability. The objective desirability measures the concept life cycle costs. The subjective desirability measures are the key issues identified in the system requirements.

Based on the results of the concept evaluation process, Concept D was recommended, with additional specific recommendations with respect to the system architecture and development process as follows:

- Hardware should be standard state-of-the-art personal computers.
- Software should be UNIX based or the same software as used by technical systems controls.
- The recommended network for interconnection of site master computers and the host computer is an Ethernet bus conforming to IEEE 802.3 standards.[1]
- A single source for equipment should be specified.
- Acceptable interfaces for technical systems control interconnection with SCADA master computers should be specified.

CONCEPT DESIGN

Concept D, as amended, is shown as a block diagram in Figure 1. The concept SCADA system includes PCs in the central control area and PCs at each of the remote sites. Shared memory will be used at each location for the hot standby and graceful system degradation capabilities needed to support desired system reliability. SCADA system software is based on a QNX (PC version of UNIX) operating system. These choices, coupled with rich market availability of PC hardware/software, will satisfy the concerns of system flexibility and the potential for future upgrades.

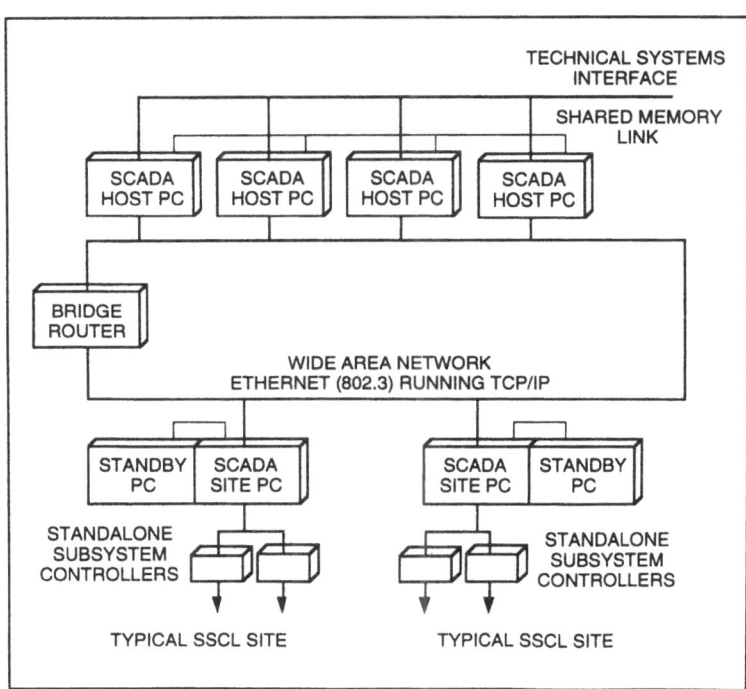

Figure 1. Recommended concept design.

The communications network between the central control area host PCs and the site PCs will be an Ethernet bus conforming to IEEE 802.3 and other related standards. Ethernet is widely supported in the marketplace. Open standards are available for communications (TCP/IP) on an Ethernet network. The medium bandwidth, as specified by the standard, is 10-MB-per-second on an 18-MHz channel. The Ethernet bus is a broadcast type network that can be implemented in segments. Though Ethernet does not use tokens to sequence through a logical ring, routers are used for addressing nodes to permit central computer polling of the remote sites.

CONCLUSIONS

The study and concept design demonstrated that a system can be developed to incorporate complex design and construction schedules. As site design details are established, it is anticipated that the conventional SCADA system will undergo changes in system requirements definition. The need to develop interim central control capabilities may be required. To ensure an engineered rather than evolutionary system design, the SCADA system concept design must be implemented as a standardized system with standard equipment configurations.

ACKNOWLEDGMENTS

The author wishes to recognize and thank James E. Whiteford of Sverdrup Technology Inc. Tullahoma, Tennessee, the Senior Engineer in the development of the study and concept design.

REFERENCES

1. IEEE. 1990. "Carrier Sense Multiple Access/Collision Detection (CSMA/CD)" *Standard* 802.3 The Institute for Electrical and Electronic Engineers (IEEE) 802 Standards Committee, New York, NY.

CONTROL OF THE SSC ACCELERATOR SYSTEM STRING TEST CONFIGURATION

L. Cromer, L. Maxson, D. Pan, P. Ravishankar and V. Saladin

Superconducting Super Collider Laboratory[*]
2550 Beckleymeade Ave.
Dallas, Texas 75237-3997

INTRODUCTION

The Accelerator Systems String Test (ASST) program is a research and development effort located at the SSCL N15 Site. Its purpose is to support the evaluation of key subsystems required in the Superconducting Super Collider (SSC)[1]. A procedure has been generated to allow SSCL scientists and engineers to request testing at the facility[2] and, if approved by the ASST Steering Committee, perform their tests as schedule time and priorities permit.

Implementation of approved test requests is the responsibility of the ASST test group. Since the ASST magnet string is highly instrumented, it is critical that the test group keep accurate records of the configuration. Failure to do so would result in program chaos and the recording of incorrect test data.

THE SYSTEM INTEGRATION APPROACH

Prior to and during the ASST string configuration setup process, the ASST test group generates and maintains a complete set of documents and drawings to support system integration for a test run. To date, two runs have been completed. Each run required approximately 200 system integration drawings and 3 databases with nearly 1800 line entries to define them. Hard copies of the drawings and databases were kept in a "Configuration Notebook" located in the ASST system control room for reference by configuration setup personnel, maintenance personnel and system operators. Also, a master Configuration Notebook was maintained and eight copies distributed and maintained by the ASST test group. Although this paper approach is traditionally used to define and communicate how subsystems would be integrated, it was not recognized when the program started that the traditional approach would be inadequate for an R&D facility, because of its slow turn around time. To correct the inadequacies of the approach, the test group decided to develop a computer based system to support system integration.

---

[*] Operated by the Universities Research Association, Inc., for the U.S. Department of Energy under Contract No. DE-AC35-89ER40486.

The ASST test group held a planning session, with key SSCL engineers to define an approach that would meet the needs of the ASST program. During this planning session, the following basic requirements for the approach were identified:

1. The approach should be simple to use by all requiring system integration information.
2. The database and drawings generated to support system integration should both be accessible by the general user. Redlining of the drawings and database should be allowed so users could make changes as they were required. This redlining function should include the capability of notes attachment, so that the rationale for implementing certain changes could be recorded for future reference. Redlining should not alter the released drawings or database.
3. The database and drawings should be accessible from anywhere within the SSC Lab, with the appropriate hardware and software.
4. A system database/drawing "manager" would be required to oversee the approach. This manager should be the focal point for all data going into the drawings and database as they were created, and the focal point for all changes to them once they were released. The system manager should be automatically notified when drawings had been redlined or notes generated. The manager would be required to process proposed changes and, if approved, ensure the released drawings within the system were updated.

A goal was set to implement this approach by June 1, 1993.

DEFINITION OF THE ASST CONFIGURATION DATABASE/DRAWING SYSTEM

The main goals of the ASST configuration database/drawing system are:

1. To aid in the development of the configuration by focusing the efforts of engineers involved in the ASST system integration.
2. To support the physical integration of subsystems into the ASST facility. (The database and drawings contained within the database/drawing system represent the plan to be followed by subcontractors, lab technicians and engineers involved in the configuration setup.)
3. To aid in troubleshooting problems encountered during the ASST operation.
4. To establish a basis from which the configuration of new test runs can be developed.

How the ASST configuration database/drawing system has been set up to meet these goals is described below.

Data Entry User Interface: The purpose of the data entry user interface is to allow easy entry of configuration information into the database and drawing files. Database information includes that necessary to adequately define the subsystem to subsystem interconnections, required for approved tests. Examples of this information include sensor details and connections to connector pins, cable numbers and channel numbers, signal definition, signal conditioning and recording. Drawing files include those generated by the ASST test group in support of system integration. They provide the information required to put each subsystem component into the overall configuration of the ASST.

The layout of the data entry user interface is shown in Figure 1. Within it, the Oracle forms function supports data entry and the performance of data queries. The Graphical User Interface (GUI) routing display function, is used to generate a graphic representation of the subsystem interconnections defined within the ASST database. The drawing display function,

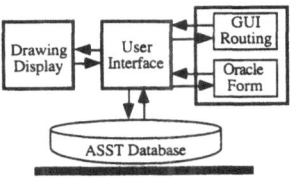

Figure 1. Data Entry User Interface.

supported by a commercially available product from Auto-Scan Systems Inc. called RETREEVE, allows the user to view a raster file of CAD generated drawings.

General User Interface: The functionality of the ASST database is shown in Figure 2. As shown here, the Collider component database top level will be the gateway to the ASST configuration database. Entering a test run number, the user will have access to all data captured on system integration drawings and the configuration database for the specified run.

The ASST database/drawing system allows free access to the configuration drawings and data at any time of the day or night, from any lab location. It also provides control of the configuration and a complete, time tagged record of all proposed changes. Although all of the data and drawings can be printed at anyone time, the system is set up to avoid the need to do so. Database and drawing files can be reviewed, searched, redlined and sorted as well as printed. The database/drawing system is menu based and allows the user to review data at a high level or to locate low level information quickly. Standard configuration data reports can be generated by the user from the perspective of individual magnet string sensors or magnet string connectors. Reports can also be generated to fully define the magnet string sensors and their characteristics. Scanned manufacturer data sheets on sensors, cables, connectors or other pieces of hardware can be viewed and printed. A graphic display of the interconnections from any ASST string sensor to the control and instrumentation subsystem can also be generated automatically by simply identifying a sensor, connector or cable.

SUMMARY

The ASST System Database is expected to have a major impact on the ASST program. Some of the expected results are:

1. A reduction in configuration definition turn around time
2. A reduction in the effort required to define and maintain the ASST configuration
3. Better communication of information to all requiring it
4. Better control of the ASST configuration
5. A reduced need for interconnection drawings
6. A reduction in the paper required to define the ASST configuration.

After implementing and evaluating this approach on ASST, it is our intent to use the lessons learned to support the Collider program. It is recognized, however, that there are significant differences between the ASST program and the Collider. The ASST system configuration is constantly undergoing modification. The Collider configuration will not. Changes will be made to the Collider over time, and will be well documented before they occur. Nevertheless, a database/drawing system like that being set up on the ASST program can be of great value to the Collider program primarily during the system integration and system maintenance phases.

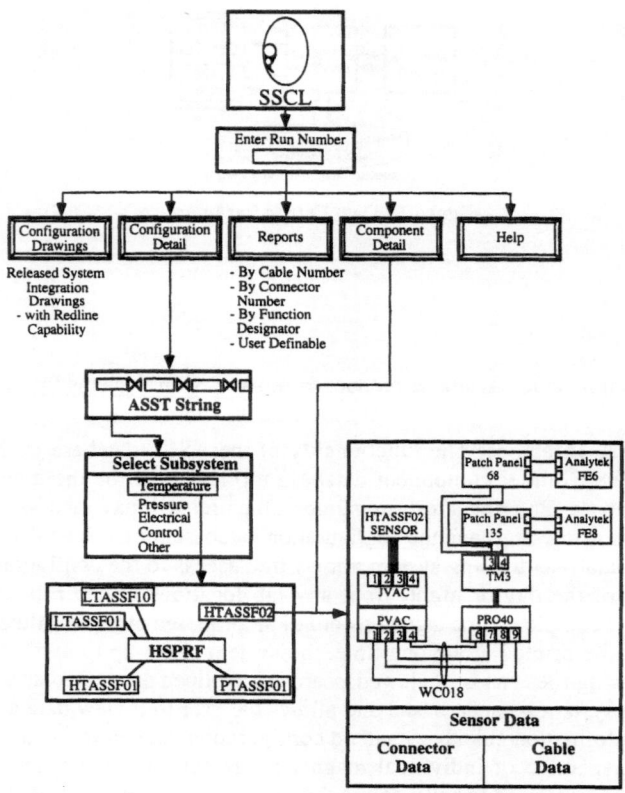

Figure 2. ASST Configuration Database User Interface.

REFERENCES

1. Phil Kraushaar, Presentation III-B-2, Proceedings of the Fifth Annual 1993 International Industrial Symposium on the Supercollider
2. Accelerator Systems String Test Request Procedure, SSCL Document No. E10-0000103.

FABRICATION OF A PROTOTYPE THIN SUPERCONDUCTING SOLENOID COIL FOR THE SDC DETECTOR MAGNET

A. Yamamoto[1], Y. Doi[1], T. Kondo[1], Y. Makida[1], K. Tanaka[1]
H. Mukai[2], H. Hirata[2] and S. Mine[2]

[1] National Laboratory for High Energy Physics(KEK)
1-1 Oho, Tukuba, Ibaraki, 305, Japan
[2] Toshiba Corporation
2-4 Suehiro-cho, Tsurumi-ku, Yokohama, 230, Japan

INTRODUCTION

The SDC thin superconducting solenoid magnet has been designed to provide an axial magnetic field of 2T in a particle-tracking volume of ϕ 3.4m x 8.8m with a transparency of 1.2 radiation length(X_0) in the SDC detector, which is one of the major colliding particle detectors in the SSC accelerator.[1-4] To verify various engineering parameters and to establish technology for the SDC magnet fabrication, a prototype thin superconducting solenoid magnet has been developed in a joint R&D effort amongst KEK in Japan and Fermilab and SSCL in the United States, since 1991. KEK has been responsible to develop the coil and Fermilab and SSCL have been responsible to develop "isogrid" vacuum vessel for the cryostat.[5] The prototype solenoid magnet has a full size in diameter and a quarter length of the detector magnet. Main parameters of the prototype solenoid coil are given in Table 1.

Table 1. Main parameters of the prototype solenoid magnet

Dimensions		
Coil	Mean diameter	3.68 m
	length	1.9 m
Cryostat	Inner diameter	3.4 m
	Outer diameter	4.12 m
	Total length	2.4 m
Central magnetic field		1.5 T
Peak field at conductor		3.8 T
Current		12222 A
Number of effective turns		405
Stored energy		48 MJ

The prototype coil development has been carried out with joint effort between KEK and TOSHIBA CORPORATION in Japan. Progress of the prototype coil fabrication will be described in the following sections.

FABRICATION OF THE PROTOTYPE COIL

General Structure

The SDC prototype solenoid coil consists of a single layer coil and an outer support cylinder placed outside the coil. The superconducting coil is wound with high-strength aluminum stabilized superconductor[6] on inner surface of the support cylinder. The coil is cooled indirectly with forced flow of two-phase helium. A cooling pipe is placed on outer surface of the outer support cylinder with serpentine path having 28 axial passes.

Fabrication of Support Cylinder and Helium Cooling Tube

The outer support cylinder (t=31mm) was made of high strength aluminum alloy(A5083P-H32). In order to use the outer support cylinder as a coil-winding mandrel according to "inner winding concept", the inner surface of the outer support cylinder was accurately machined and finally coated with double layers of glass-epoxy sheets for electrical ground-insulation(>2KV DC).

In order to maximize mechanical shear strength at bonding with epoxy-resin, the epoxy-resin was cured at a temperature of 130℃ for about 10 hours.

Helium cooling tube of 25 mm in inner diameter was pre-assembled by welding to serpentine configuration on a cylindrical template-fixture. After inspection of the cooling tube according to the High Pressure Regulation in Japan, the cooling tube with fin was attached on outer surface of the outer support cylinder by welding as shown in Fig. 1.

Figure 1. Picture of the outer support cylinder with helium cooling tube

Coil Winding

The coil was wound on inner surface of the outer support cylinder by using an automatic coil winding machine specially developed. It was based on "inner winding concept" and a major improvement was made to handle much stiffer and stronger aluminum-stabilized superconductor.[3] Figure 2 shows a bird's eye view of the coil winding machine. At first, the superconductor was insulated by using half-over-lapped Glass/Kapton tapes with B-staged epoxy-resin, and was pre-wound on temporary inner mandrel to have appropriate curvature for the "inner winding". In the inner winding process, the curved conductor was automatically transferred to the inner surface of the outer support cylinder by using a longitudinal compression supplier along the coil circumferential direction, as shown in Fig. 3, to fit the conductor tightly on inner surface of the support cylinder. During the coil winding, wet (A-staged) epoxy-resin was painted on the outer edge of the conductor to be filled between the conductor and the inner wall of the support cylinder to maximize mechanical shear strength at bonding with epoxy-resin.

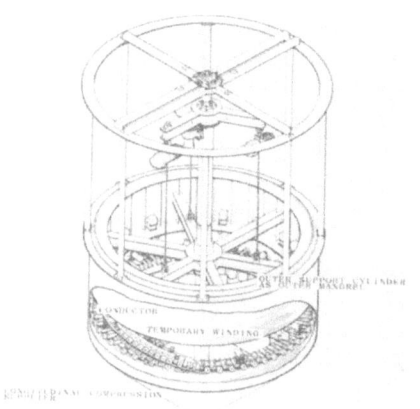

Figure 2. Bird's eye view of winding machine

Figure 3. Picture of winding machine from inside of support cylinder

After the inner ground-insulation, pure aluminum strips with a residual resistivity ratio (RRR) of >2000 were glued in axial direction on the inner surface of the coil. Those are very effective to equalize the coil temperature in steady state operation and also after quench because of those very high thermal conductivity at low temperature.

The coil was finally cured at 130℃ for about 10 hours with the axial and radial compression. Figure 4 and 5 show pictures of prototype coil completed.

Figure 4. Picture of prototype coil after pure Al gluing and curing

Figure 5. Picture of Al gluing and curing from inside

SHEAR STRENGTH OF BONDING WITH EPOXY-RESIN

In advance to prototype coil winding, various basic R&D efforts have been made to understand basic characteristic of each component and to optimize the coil winding condition. A test coil-winding was carried out with the full diameter of 3.7m and with axial length of 30cm to verify the winding machine performance with the inner technique. After completion of the test coil winding and curing, the test coil was cut, as shown in Fig.6, and the cross sectional samples were inspected to verify the characteristics of coil winding and bonding with epoxy-resin. Specially, shear strength of the bonding with epoxy-resin between the coil and support cylinder was carefully measured by using the samples cut from various location. Figure 7 shows the measured results of shear strength test at room temperature and LN_2 temperature. All results have satisfied a design requirement on the shear strength of 15 MPa.

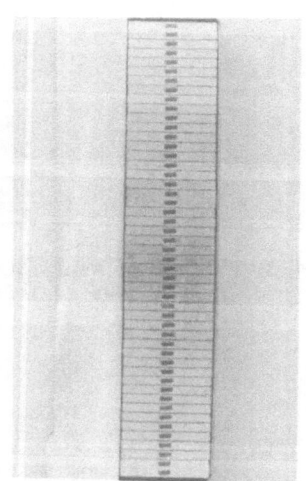

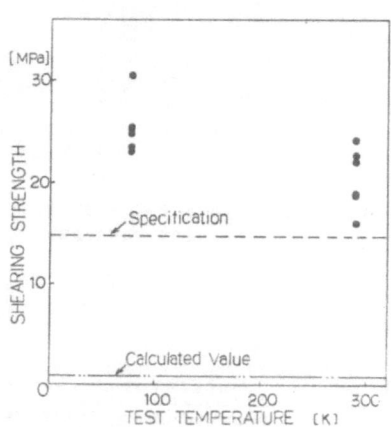

Figure 6. Picture of the cross section of test winding coil

Figure 7. Result of epoxy bonding shearing test

STATUS AND FUTURE PLAN

The prototype coil fabrication has been successfully completed. The coil will be assembled with the vacuum vessel by using "isogrid"technique provided by Fermilab/SSCL and with other cryostat component. The general assembly is planned to be completed in 1993 and the cooling and excitation test will be carried out in Japan to verify the engineering design parameters for the SDC detector magnet.

REFERENCES

1. G. Trilling et. al., Solenoid Detector Collaboration Technical Design Report, SDC-92-201, (1992)
2. A. Yamamoto et. al., Proc. of the Int. workshop on Solenoidal Detector for SSC, KEK, Tukuba, (1990), KEK-Report 90-1- (1990) p142
3. A. Yamamoto et. al., DESIGN STUDY OF A THIN SUPERCONDUCTING SOLENOID MAGNET FOR THE SDC DETECTOR, Proc. of the Applied Superconductivity Conference (ASC-92), in Chicago, (1992), and to be published in IEEE Trans. on Applied Superconductivity.
4. R. Fast, SDC Solenoid Design Note No. 1-179, Fermilab. (1988-1992)
5. L. W. Swenson, C Grozis et. al., Fermilab SDC Solenoid Design Note No. 154, (1991)
6. I. Inoue et al.. Proc. of IISSC-92, Supercollider 4, Edited by J. Nonte (Prenum Press, New York, 1992) p. 943

ELECTRICAL PERFORMANCE CHARACTERISTICS OF THE SSC
ACCELERATOR SYSTEM STRING TEST

W. Robinson,[1] W. Burgett,[1] J. Gannon,[2] P. Kraushaar,[1]
A. McInturff,[1] R. Nehring,[2] V. Saladin,[1] T. Savord,[2]
G. Sorrensen,[2] R. Smellie,[2] G. Tool,[2] and D. Voy[2]

[1]Project Management Office
[2]Electrical Engineering Department
Superconducting Super Collider*
2550 Beckleymeade Avenue
Dallas, Texas 75237-3997

INTRODUCTION

The intent of the Accelerator System String Test (ASST) is to obtain data for model verification and information on the magnitudes of pressures and voltages encountered in an accelerator environment. The ASST milestone run was achieved during July and August, 1992 and consisted of demonstrating the accelerator components could be configured together as a system operating at full current.[1] Following the milestone run, the string was warmed to counteract some design flaws that impeded the operational range. The string was again cooled to cryogenic temperatures in October, and a comprehensive power testing program was conducted through the end of January, 1993. This paper describes how the collider arc components operate in an accelerator environment during quenches induced by firing both strip heaters and spot heaters. Evaluation of the data illustrates how variations in the design parameters on magnets used in a string environment can impact system performance.

CONFIGURATION

The ASST is composed of five 50 mm aperture dipole magnets, a 40 mm aperture quadrupole magnet, a corrector spool, and feed and end spools. The dipole magnets used in the string were industrial prototypes constructed by General Dynamics personnel utilizing facilities at Fermi Lab. Those magnets included DCA313, DCA314, DCA319, DCA315, and DCA316. The quadrupole magnet was built at LBL. The three spool pieces were built to SSCL specification by Meyer Tool, Cryenco, and Consolidated Vacuum Industries. A DC current power supply was used to provide a maximum current of 6500 amps. An energy dump was used with the string to evaluate the energy extraction system.

Each dipole magnet consists of four superconducting coils. The differential voltage across each coil is monitored by the quench protection monitor system and a data acquisition system. There are four strip heaters for each magnet that are positioned along the length of the outer coils in a quadrant configuration. The strip heaters in opposing quadrants are electrically connected together in parallel. Each set of strip heaters is independently controlled by a heater firing unit. This configuration protects the magnet by providing a level of redundancy that ensures the outer coils quench despite a failure that may occur in one of the heater firing units.

Figure 1 illustrates the equivalent electrical circuit of the string. During a quench, the resistive voltage developed by the quenching component causes the bypass diodes to conduct. This allows energy stored in other parts of the string to bypass the quenching region while the energy is extracted from the non quenching portion of the system through the energy dump. Note that the SPR is located between DCA316 and the quadrupole. It was determined that the quadrupole would not withstand the resultant heating that would occur under quenching conditions with two dipoles connected in series with it. In order to protect the quadrupole, an additional bypass diode was added to the system. Also note that only one of the bypass leads is connected. The second bypass lead is only required to support a full cell so it was not necessary to risk using the second bypass lead during initial operation. Both bypass leads were later tested well beyond specification.

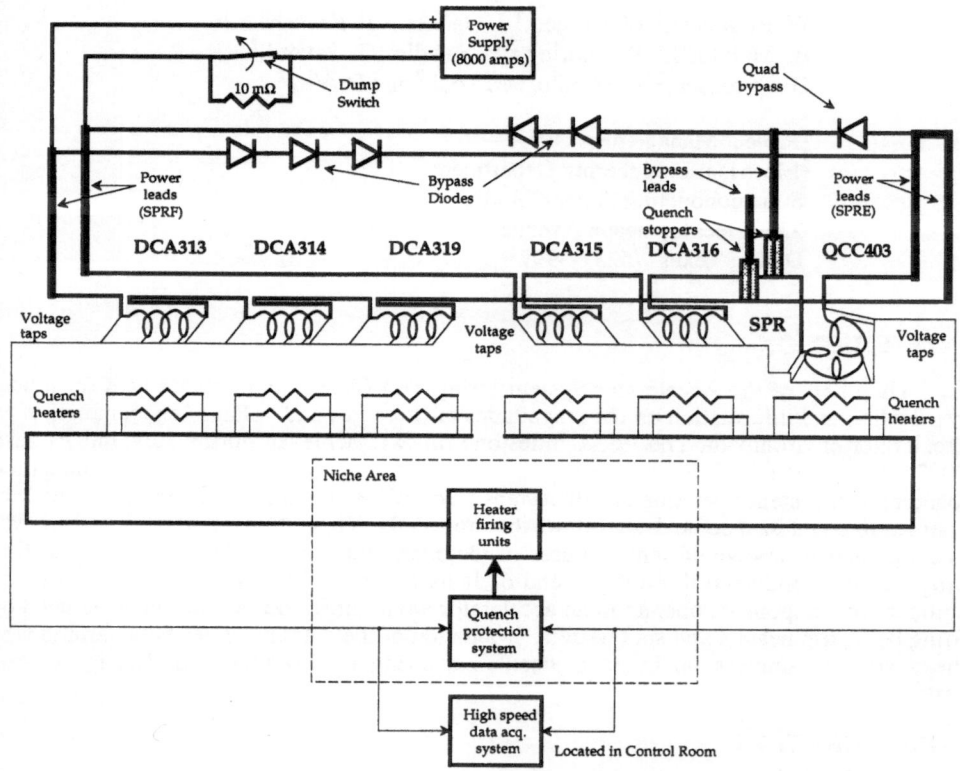

Figure 1. Electrical configuration of the ASST.

QUENCH ANALYSIS

During the initial run, the string was operated at T = 4.65 K partly because of the limitations of the cryogenics plant, and also to keep the amount of operating margin small. One of the unexpected results from testing was the effect differences in residual resistivity ratio (RRR, which is defined as the 300 K resistance divided by the 10 K resistance) between magnets had on the system quench response. Problems were encountered during string testing from using magnets with different RRRs in the same circuit. Table 1 outlines the RRR characteristics of the dipole magnets used in the string. From Figure 1, DCA313, DCA314, and DCA319 are grouped together electrically, and DCA315 and DCA316 are wired together. Some magnets in the circuit were able to dissipate the stored energy at a high rate while other magnets transferred much of their stored energy to the magnets with lower RRR. Figure 2 illustrates the MIITs and peak voltage to ground that was attained during strip heater induced quenches on DCA319. The test was terminated at 6000 amps because the voltage was projected to reach 2300 volts at 6500 amps.

As the MIIts integral demonstrates, RRR has an important role in determining the response of the system. The MIITs is determined by:

$$\text{MIITs} = 10^{-6} \int_0^\infty i^2(t)\, dt = 10^{-6} A^2 d \int_{T_o}^T \frac{C(T)}{\rho(T,RRR,B)}\, dT, \tag{1}$$

where i is the current, A is the cross-sectional area, d is the density, C is the heat capacity of the conductor, B is the magnetic field term for the magneto-resistance, and ρ is the electrical copper resistivity. An approximation for ρ is:

$$\rho(T,RRR) = \frac{1.545}{RRR} + \left(\frac{2.32547 \times 10^9}{T} + \frac{9.57137 \times 10^5}{T} + \frac{.62735 \times 10^{-2}}{T} \right)^{-1} \mu\Omega\text{-}cm. \tag{2}$$

The parenthesis on the right is an approximation to the Grüneisen integral formula for the phonon scattering resistivity.[2] As the RRR of the material is uniformly reduced, the copper resistivity is increased, and MIITs are reduced because the effective time constant of the system is reduced. Stored energy in the magnets is dissipated more rapidly when the RRR of the material is lowered. Since resistivity is increased, peak voltage attained during quench is also increased. In a string environment, quenching magnets with lower RRR experienced higher MIITs, voltages, and temperatures than expected due to additional energy dissipated in those magnets provided by magnets in the system with higher RRR.

Table 1. UI is the upper inner coil, UO is the upper outer coil, LO is the lower outer coil, and LI is the lower inner coil.

RRR VALUES of ASST MAGNETS

Magnet	UI RRR	UO RRR	LO RRR	LI RRR
DCA313	170	174	171	173
DCA314	174	177	174	171
DCA319	105	96	97	108
DCA315	162	173	174	177
DCA316	67	109	109	78

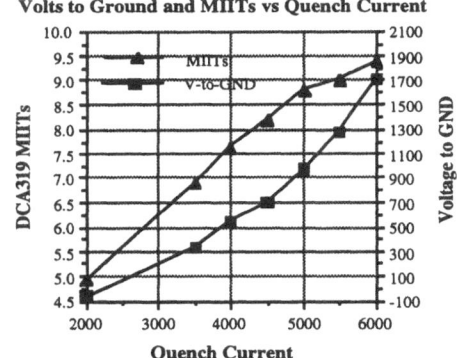

Figure 2. MIITs and voltage to ground as a function current when initiating a string quench by firing the strip heater in DCA319 (t = 4.65 K).

Energy deposition estimates were made using the coil voltage $V_c(t)$, which is a combination of the inductive voltage that results from change in current in a coil of inductance L_c, and a resistive voltage resulting from current passing through the copper in the superconductor wire. The developed coil resistance $R_c(t)$ is given by:

$$R_c(t) = \frac{\left[V_c(t) - L_c \dfrac{di}{dt} \right]}{i(t)}, \tag{3}$$

where di/dt is calculated from the current decay of i(t). Given the resistance of each coil, an estimate is made on how the energy is being dissipated in the string.

The total energy stored in the string is given by $W_L = 1/2\, L\, i_o^2$, where L is the string inductance (approximately 75 mH/dipole and 7.5 mH/quadrupole), and i_o is the string current before a quench occurs. For i_o = 6500 amps, the energy storage in the string is $W_L \approx$ 8 MJoules. The energy deposition for each coil is determined by:

$$W_{R_c} = \int_{t_o}^{\infty} R_c(t)\, i^2(t)\, dt,\qquad(4)$$

where t_o represents when resistance in the coil is detected.

Figure 3 shows how the string energy is dissipated in the string during Event #285 when all strip heaters were fired simultaneously. DCA319 dissipated the most energy because it had the lowest RRR of the three dipoles in its circuit. Although the RRR of DCA316 is similar to DCA319, there was only one other dipole in the DCA316 circuit. The total energy dissipated by Event 285 was 7.79 MJoules. The 3.68% difference from the expected energy of 8.08 MJoules is due to the change in the dipole inductance from the high field iron core saturation.

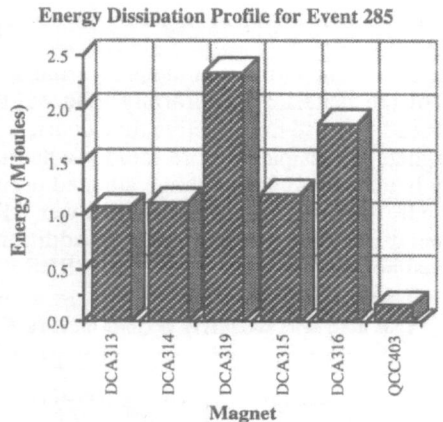

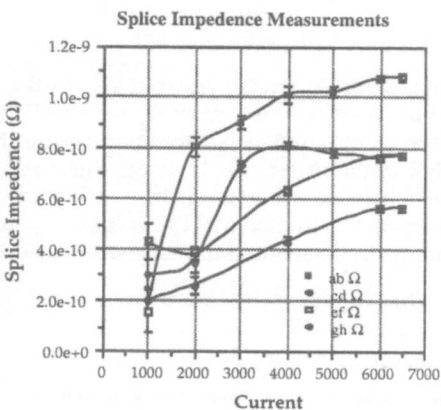

Figure 3. Profile of energy dissipation for Event 285 where all strip heaters were fired simultaneously. Peak current for this event was 6475 amps.

Figure 4. Splice joint impedance as a function of current. The "hump" in the gh Ω curve is probably due to measurement error.

Splice Joint Resistance

Voltage taps were placed on each side of four splice joints in the string. The voltages across the joints were monitored while ramping the string to full current. Figure 4 illustrates the splice joint resistances from 2000 amps to the full current of 6500 amps. These were also the first joints fabricated in the field so it is expected that the impedance will become lower as our processes continue to improve and experience is gained in joint fabrication. The change in resistance as a function of current is due to the superconducting properties of solder at low currents. The differences between splice impedance at high current is probably due to differences in the solder composition or thickness between joints.

FULL CELL RUN

We are currently reconfiguring the string as a full cell that is scheduled for cool down sometime in June. We have grouped magnets together based on RRR into four families. We expect the energy dissipation of the system to become more balanced as each RRR family is configured together through one bypass circuit. We plan to continue monitoring splice joint resistance and collect data at a lower operating temperature.

References

* Operated by the Universities Research Association, Inc., for the U. S. Department of Energy under Contract No. DE-AC35-89ER40486.

[1] Burgett, et al., "Full-Power Test of a String of Magnets Comprising a Half-Cell of the Superconducting Super Collider," SSCL-Preprint-162, October, 1992. (To be published in the journal *Particle Accelerators*.)

[2] M. McAshan, "MIITS Integrals for Copper and for Nb - 46.5 wt% Ti," SSC-N-468, February, 1988.

BRAZED HONEYCOMB VESSEL R&D FOR THE SDC SOLENOID
MAGNET

I. Ohno,[1] H. Kakui,[1] H. Takeda,[1] H. Maehara,[1] K. Nishida,[1]
A. Yamamoto,[2] H. Yamaoka,[2] and Y. Makida[2]

[1]Ishikawajima-Harima Heavy Industries Co., Ltd.(IHI),
  1, Shin-Nakahara-cho, Isogo-ku, Yokohama, 235, Japan
[2]National Laboratory for High Energy Physics(KEK),
  1-1, Oho, Tukuba, Ibaraki, 305, Japan

ABSTRACT

We designed and fabricated a prototype vacuum vessel of ϕ4.1m in diameter and 2.4m in length for the SDC solenoid by using brazed honeycomb techniques. Honeycomb panels have large stiffness in light weight, so the effective thickness of the outer shell composed of honeycomb panels needs only about 1/4 weight of that in comparison with the solid plate shell. Further technical merits in "brazed honeycomb" are its weldability and good performance in bending process. We developed a new bending technique for the brazed flat honeycomb panels to be curved shape. Assembling these curved panels by welding, we have been successful to fabricate a very reliable vessel in vacuum performance. This honeycomb vessel has been verified to have stiffness as designed and appropriate buckling strength through performance tests.

INTRODUCTION

The SDC solenoid is designed to provide an axial magnetic field of 2T over the charged particle tracking volume of ϕ3.4 x 8.8m in the SDC detector.[1,2] The vacuum vessel for the SDC solenoid consists of inner and outer coaxial cylinders with flat annular bulkheads at both axial ends. Since all wide angle particles must pass through the superconducting coil and cryostat before impacting the calorimeter, the desired calorimeter performance requires the material in the coil to be minimized both in terms of radiation length (X_o) and absorption length (λ_o). For the vacuum vessel wall, it is necessary to use aluminum which radiation length is relatively long, and further effort to minimize the material quantity is required.

In Japan recently, fabrication technique for large brazed aluminum honeycomb panels has been developed for industrial uses such as building siding and transportation vehicle walls.[3] Such a material should be very reliable, because the vacuum seal to joint between panels can be welded. To make sure that brazed honeycomb is a possibility for the SDC solenoid, we have developed a prototype vacuum vessel using brazed honeycomb.

DESIGN OF HONEYCOMB VESSEL

The outer cylinder of vacuum vessel for the SDC solenoid has a diameter of ϕ4.1m and a length of 8.73m. Since an external pressure of 0.1MPa due to evacuation and axial decentering force of about 40 tonnes are loaded on the vacuum vessel as the operational loads, we need to design it so as to have sufficient strength against such buckling loads with a safety factor of 2. We have taken a lateral buckling pressure of > 0.2MPa and an axial buckling force of 160 tonnes in the vacuum vessel. The honeycomb vessel design was made according to a reference titled as "Structural Analysis of Shells".[4] As a result, we obtained the design parameters of honeycomb panel as listed in Table 1. The effective thickness is just about 1/4 of that in comparison with the solid aluminum plate.

Table 1. Design parameters of honeycomb outer cylinder in comparison with solid shell.

		Honeycomb	Solid
Aluminum alloy		6951/4045	5083
Total thickness	(mm)	45	27
Skin thickness	(mm)	3.0+3.0	---
Core thickness	(mm)	0.2	---
Cell size	(mm)	19	---
Effect. thick.	(mm)	7.1	27
Weight reduct. ratio		0.26	1
Radiation thick.	(X0)	0.079	0.303

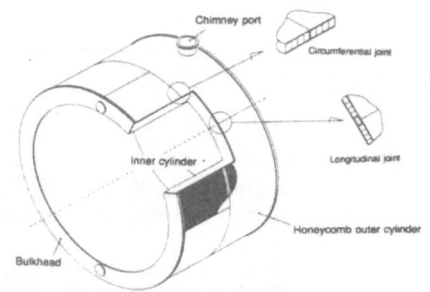

Figure 1. Prototype honeycomb vacuum vessel.

The prototype vacuum vessel fabricated has a full diameter of φ4.1m and a length of 2.4m, corresponding to about 1/4 of that for the main SDC vessel. Because of its much shorter length, the axial buckling strength in the prototype may be much higher than that of the main SDC vessel , as verified in a test described below. Figure 1 shows a bird's-eye-view of the prototype vacuum vessel.

FABRICATION

Figure 2 shows schematic diagram for fabrication of the brazed aluminum honeycomb cylinder. The process was carried out as follows.

----Assembly of flat surface panel with corrugated core (material : A6951 and BA4045).
----Brazing in vacuum furnace to make a flat honeycomb panel.
----"4-point bending" to make curved honeycomb panel. (see Fig.3)
----Jointing four curved panels by welding to make a cylinder unit.

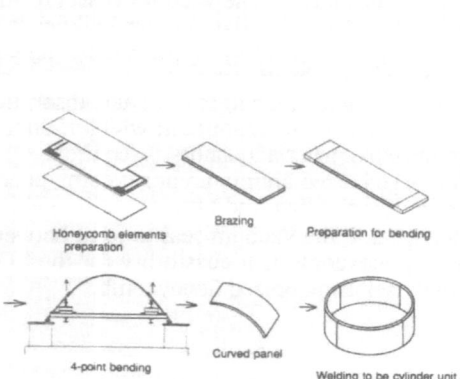

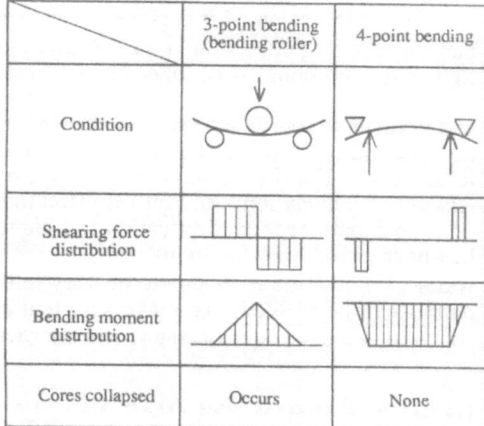

		3-point bending (bending roller)	4-point bending
Condition			
Shearing force distribution			
Bending moment distribution			
Cores collapsed		Occurs	None

Figure 2. Fabrication process of brazed honeycomb cylinder unit.

Figure 3. Concepts of bending honeycomb panels.

The "4-point bending" to make the curved panel was the heart of the process, which was successful after many R&D efforts. Figure 4 shows a picture of bended honeycomb panel.

An inner cylinder of 6mm in thickness and ϕ3.4m in diameter was installed inside the honeycomb outer cylinder and then two bulkheads were attached at both axial ends of the inner and outer cylinders. Figure 5 shows a picture of the completed honeycomb vacuum vessel.

Figure 4. Bended honeycomb panel.

Figure 5. Completed honeycomb vacuum vessel.

TESTS

Two mechanical tests have been carried out to verify the mechanical characteristics of the honeycomb vessel : (1) stiffness test and (2) buckling strength test with lateral and axial load. Before and after the mechanical tests, vacuum leak tests were performed and no leak was found with a sensitivity of 1×10^{-9}Torrl/s.

Figure 6 shows a set up of the stiffness test. The honeycomb cylinder was deformed by applying line forces at top and bottom of the cylinder to measure deviation of the diameter. Figure 7 shows the measured results in comparison with theoretical calculation and analytical computation. The theoretical calculation was made by replacing the honeycomb plate with a solid aluminum plate which has equivalent stiffness of the honeycomb plate. The analytical computation was made by a FEM method. The FEM model of honeycomb cylinder is composed of isotropic solid elements for face sheets and orthotropic solid elements for core part. In this figure, it is clear that the measured values have good agreement with analytical computation and were mostly agreed with theoretical calculation.

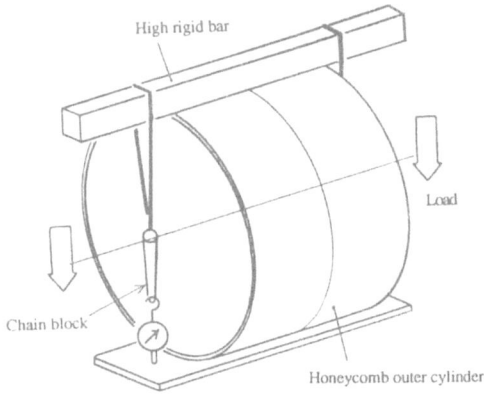

Figure 6. Set up of stiffness test.

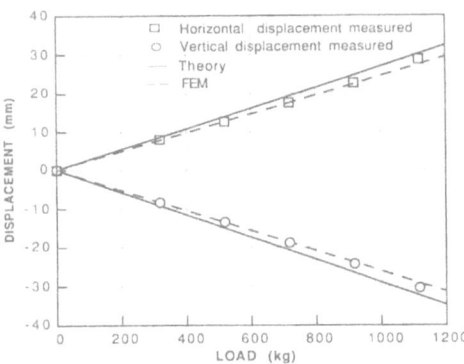

Figure 7. Displacement in diameter in stiffness test.

The operational safety of the honeycomb vacuum vessel was verified by simulating operation of the vessel under a relatively same boundary condition against buckling. A combined load of 0.1MPa in external pressure and 630 tonnes in axial force was loaded to the prototype vessel in order to simulate the operational condition of 0.1MPa in external pressure and 80 tonnes in axial force with a safety factor of 2 in the SDC main magnet. Figure 8 shows a set up of buckling strength test. In the test, at first, the vacuum vessel was evacuated below a pressure of 1Torr and then the axial force was applied with a large press machine, step by step, to reach the maximum value of 626 tonnes. As results of this test, the stress on the outer surface of honeycomb cylinder and the displacement in radial direction of the cylinder are shown in Figure 9. It is well known that stress and displacement change suddenly to be unstable just before buckling occurs. But such a phenomenon was not observed and the vessel still remained under a safe condition.

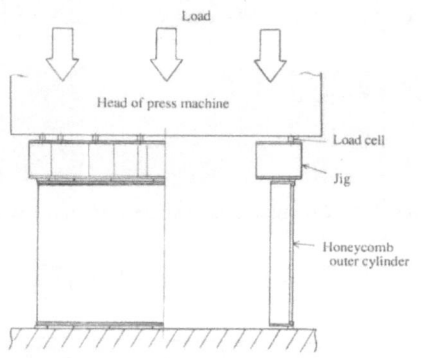

Figure 8. Set up of buckling strength test.

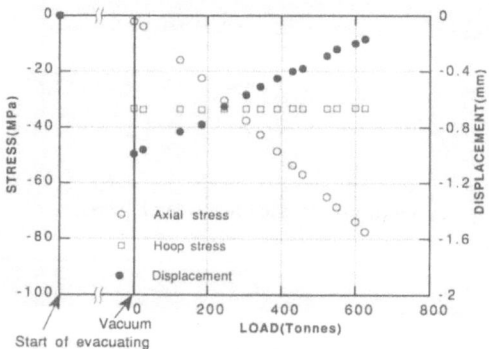

Figure 9. Measured stress of the outer surface and displacement in radial direction of the honeycomb cylinder.

CONCLUSION

As a result of this R&D effort, we have been successful to :

-----design a very thin and light vacuum vessel using brazed honeycomb panels,
-----develop a bending method of flat honeycomb panels to be curved shape without collapsing or buckling, and
-----fabricate a very reliable honeycomb vessel of ϕ4.1m in diameter and 2.4m in length, which has just sufficient stiffness as designed, appropriate safe condition against buckling, and good vacuum performance.

We conclude that the brazed honeycomb panel is very desirable material for the vacuum vessel of the SDC solenoid magnet.

REFERENCES

1. G. Trilling et. al.;"Solenoidal Detector Collaboration: Technical Design Report," SDC-92-201, (1992).
2. A. Yamamoto et. al.;"Design study of a thin superconducting solenoid magnet for the SDC detector," to be published in Proc. of Applied Superconductivity Conf. (ASC-92), Chicago, (1992), IEEE, Trans, Applied Superconductivity.
3. K. Okuto et. al.;"The analysis and design of honeycomb welded structures," Keikinzoku-Yosetsu, Vol. 29, No. 8, (1991), p13 (in Japanese).
4. E. H. Baker et. al.;"Structural Analysis of Shells," McGraw-Hill Book Company, (1972).

ADAPTATION OF LESSONS LEARNED FROM THE EUROTUNNEL PROJECT AND CDM MAGNET PRODUCTION TO SUPER COLLIDER MAIN RING INSTALLATION

J. Belding[2], P. Di Domenico[1], J. Gillin[1], W. Hahn[1],
J. Hopkins[3], J. McCollum[3], R. Naventi[1], M. Nielsen[1],
L.R. Patterson[3], M. Seely[1], and J. Smrha[1]

[1] Bechtel National, Inc.
50 Beale Street
San Francisco, CA 94119-3965

[2] Belding Corporation
130 W. Grand Lake Blvd.
Chicago, IL 60185

[3] General Dynamics Space Systems Division
P.O. Box 85990
San Diego, CA 92186-5990

INTRODUCTION

This paper will present preliminary findings from the Phase I Collider Installation contract studies performed by the Bechtel/General Dynamics/Belding Team related to the installation of technical systems for the SSC main ring north and south arcs. Specific focus is given to the adaptation of lessons learned during construction of the Eurotunnel, including equipment and personnel logistics and transportation. The incorporation of Collider Dipole Magnet manufacturing techniques and process methodologies as related to the handling and interconnection of main ring components is also discussed.

EUROTUNNEL EXPERIENCE AND LESSONS LEARNED

As part of the Collider Installation study, evaluation of past experience with similar projects and adaptation of the lessons learned from these efforts was considered one of the areas key to developing innovative, cost-saving solutions to the Collider Installation work. In addition to using lessons learned from previous experience building similar research machines, the Phase I study effort included evaluating experiences and adapting lessons learned from other projects involving similar technologies and facing similar execution requirements.

One example of such a project is the Eurotunnel. As a key member of the Eurotunnel management team, Bechtel was extensively involved in the overall coordination of the construction and installation planning activities. Based on our experience on the Eurotunnel, several interesting and unique possibilities for adapting and developing new methods for the Collider technical systems installation activity were identified. The Collider main ring and the Eurotunnel are similar in overall size. The installation of the technical systems in the Eurotunnel involved execution of repetitive tasks. Both activities are bound by configurations that limit and restrict access. For both projects special attention to the logistics issues associated with safely moving large numbers of workers and equipment into and out of the tunnel on a daily basis presents a key challenge to a successful installation.

For the Collider Installation study activity, our efforts focused on evaluation and adapation of lessons learned in the areas of safety and loss prevention, personnel control for access to the tunnel and work spaces, communications, lighting and temporary power and control of logistics within the tunnel and each work area. Special equipment potentially adaptable to the Collider Installation was also evaluated.

This paper's focus is on the logistics planning issues associated with developing the installation master plan including scheduling, manpower, productivity and equipment utilization. In examining the Eurotunnel planning activities, planning and monitoring the flow of personnel and material was identified as one of the key areas for improvement. In particular, the ability to rapidly evaluate the impacts of changes to the baseline schedules including delivery of equipment and the ability to develop and evaluate alternate work plans were identified as potential areas for adaptation of lessons learned and for improvement. One of the recommendations was to adapt the simulation modeling used to assess Eurotunnel traffic through-put for use during the installation program. For the Eurotunnel, this traffic simulation was developed using the MODSIM SIMGRAPHICS II language.

For the Collider Installation, MODSIM was used to model the delivery of utilities and ring components to the head houses, the pre-installation preparations done before transporting to the tunnel, transportation down the shaft and in the tunnel, and the amount of time and the number of workers required to install the delivered equipment. The purpose of the simulation is to determine the length of time for delivery and installation of material inside the tunnel given certain constraints. These constraints include the availability of material in the warehouses and head house, the distance from the shafts to the installed locations, the number of delivery vehicles in the tunnel, installation time, number of workers, worker productivity and learning-curve projections.

The scheduled delivery of material, the shaft elevator and the transporter speeds, and the manhours needed to perform each task are modelled over a period of time and for a specific length of tunnel. Statistics are kept for the time that the shaft is idle, for the number of trips, the distance that each transport travels, and for the completion of each task.

At the start of the simulation the model randomly generates trucks which, based on the required installation materials, deliver material to the head house. The simulation time is checked against the elevator and transporter speeds, number of transports and worker shifts. The model schedules the workers to start their shifts at the appropriate times.

For the initial installation planning effort, approximately 150 variables were incorporated into the model. Scenarios were simulated to evaluate the utilization of magnet shafts, to confirm the number and utilization of transport vehicles and installation equipment, and to calculate the utilization and type of crafts required to complete the installation of the technical systems in the north and south Collider arcs. Based on the result of the simulations, an optimized plan for the installation was developed. One of the key results of the simulation was that the installation could be completed using only four of the five magnet installation shafts. Differences between magnet production plans and installation rates were evaluated and the amount and duration of magnet warehousing was assessed.

Two of the key features of the MODSIM software are the graphics interface which allows visual assessments of individual scenarios and the ability to easily change scenario parameters. Typical displays show installation progress of each of the major installation components by niche. The number of active workers in the tunnel and the number of transport carts in use are shown. Productivity and learning curve factors can also be evaluated.

Based on the results of the Phase 1 effort, it is intended that the model will be expanded to cover the entire installation program. Thus the model can serve as a management tool throughout the installation work to determine that installation dates are reasonable and that stated schedule objectives are achievable. Utilization of personnel and materials can also be assessed and optimized. Potential bottle neck areas can be readily identified and corrective actions associated with changing delivery patterns, worker productivity, and availability of transport equipment assessed. Using the simulation model in a dynamic, on-going fashion, as part of the evaluation of actual data based on experience will assist in the risk management and mitigation efforts.

CDM MANUFACTURING LESSONS LEARNED

General Dynamics Space Systems Division's (GDSS) responsibilities as a member of the Bechtel Team for the SSCL's Collider Installation Phase I Program were in the areas of ring component handling and transportation, ring component preparation, and ring component alignment and interconnection. One of the methods taken in addressing solutions to these task areas was evaluation of manufacturing methods and equipment utilized for the Collider Dipole Magnet developmental program for use in installation. The sections that follow discuss some of the considerations given to adapting CDM manufacturing methods to Collider Installation.

Ring Component Transportation and Handling

For the purpose of the Phase I Installation study, component handling and transportation tasks were divided into two areas; above ground and underground. The requirement exists for the movement of main ring components such as CDMs, CQMs and spool pieces from the various above-ground storage and test facilities to the head house locations at given magnet installation shafts. Various concepts for moving these ring components safely and efficiently were assessed. The current concept for the transportation of Collider Dipole Magnets from GD's Hammond Facility to the SSCL site utilizes an environmentally controlled container mounted on a modified air-ride trailer. The container has been designed to allow for either vertical loading of CDMs (by means of removable top sections) or by horizontal loading (by means of a sliding tray system). The horizontal loading concept was designed to work in conjunction with manufacturing process equipment. By eliminating the need for vertical lifting, overall risk of damage to CDM's due to excessive acceleration and tilt is reduced. Accordingly, one option studied for the transfer of ring components from the storage and test facilities to transporter vehicles and then from the transporter vehicles to the head house processing areas was a similar shuttle-like approach which would allow horizontal movement throughout the surface handling process. Additionally, various truck-mounted container configurations were studied for suitability to the various configurations and loads associated with main ring components.

Component Alignment and Interconnection

The steps defining the process required for the installation and alignment of ring components was developed during the Phase I effort. Where possible, processes and equipment being used in the manufacture of CDMs were assessed for use in the alignment and installation tasks.

In order to align the cold mass of the CDM, General Dynamics has developed a laser tracker-based system that utilizes an optical target mounted on the end of the magnetic measuring device (mole). The laser tracker is being used because of its ability to accurately transfer measured magnetic field data to exterior fiducials mounted on the cryostat. For the purpose of aligning ring components in the tunnel, a laser tracker system was considered. This laser system would have the capability of relating position of a given ring component to the tunnel's geologic monuments. Software allowing operators to relate measured fiducial marks to adjustments required on a proposed 6-strut mounting system could be developed.

Operations associated with the interconnection of two CDMs or other main ring components were divided into several major categories, including: electrical connections, welding, and assembly/fit-up. In addition, in-process tests required to ensure integrity of each interconnect were identified. In-process testing includes electrical continuity tests, helium leak tests, and vacuum leak tests. A critical procedure associated with the interconnect region is the splicing of the superconducting cable power bus. A similar splice is required to be performed during the assembly of the CDM. In order to support this manufacturing requirement, GD has developed a specialized splice fixture that applies compression and localized heat to the splice region. High-pot testing of the insulated splice ensures integrity of the splice region. For the purpose of splicing the component-to-component power bus, a system similar to that developed for the Hammond manufacturing operation was considered.

CONCLUSION

The conclusions drawn from the initial Collider Phase I Installation Planning Study activity have identified a number of areas which may provide opportunities for adapting lessons learned to provide innovative solutions and reduction of risks for the Collider Installation. In addition, several operations required to be performed during the installation process are similar in nature to those processes required for the manufacture of main ring components.

Implementing the Eurotunnel lessons learned as well as those associated with building recent large accelerators, combined with adapting similar manufacturing processes and methods where applicable will result in significant savings to the overall installation effort, both in terms of development costs and in the labor required to perform the installation. Safety, quality and reliability will also be enhanced. Opportunities for extending these innovations to produce cost savings as well as quality and safety enhancing methods and equipment for use in other large construction programs also exists.

ACKNOWLEDGMENT

The work described herein was accomplished in conjunction with a contract to the Universities Research Association in support of the Superconducting Super Collider project for the U.S. Department of Energy.

AN ASD PHYSICS EDUCATION PROGRAM

H. R. Barton, Jr.

Superconducting Super Collider Laboratory[*]
Accelerator Systems Division
Dallas, TX 75237

INTRODUCTION

I want to thank the Organizing Committee and Program Chairman of the 1993 IISSC for providing this forum to discuss science education issues here in San Francisco. By all accounts, science education in the United States is in trouble. Clearly, teaching science to the young people of our country is an important goal shared by all of us involved in scientific research. The Director, Dr. Roy Schwitters, has made science education a Laboratory goal for the Supercollider. Those of us who share this goal only can have an impact if we become involved actively in teaching science to the future engineers and scientists enrolled at our colleges and universities. The commitment of IISSC to science education is welcomed by everyone of us who wishes to improve the technological base of the nation for the next generation.

PROGRAM DESCRIPTION

At the request of the Cedar Valley College science department, the Accelerator Systems Division offered to develop an introductory physics course for their undergraduate students. The two-semester course was given by a Supercollider staff scientist for the first time at the College during the 1992–93 academic year. The course includes lectures, laboratory experiments, problem sessions, and tests, of course. The material covered consists of Newtonian mechanics and electromagnetic field theory. The students purchase a popular textbook[1] which contains numerous problems from which homework assignments are made; however, hundreds of transparencies specially prepared for use during course lectures are handed out to the students to provide them with study material.

COURSE PREREQUISITES

Over the last several years, Cedar Valley College has made considerable progress implementing advanced level mathematics courses in the College curriculum. These courses include two semesters of calculus, a semester of advanced calculus, and a semester of

---

[*]Operated by the Universities Research Association, Inc., for the U.S. Department of Energy, under Contract No. DE-AC35-89ER40486.

differential equations. This gives the Cedar Valley students an excellent preparation to learn physics. The serious study of physics can begin only after the students have acquired the mathematical background to understand the complexities of the mathematical formalism used to describe the physical phenomena.

Cedar Valley College, Home to the ASD Super Collider Physics Program.

UNIFYING CONCEPTS OF THE COURSE

In this course for science and engineering students, an attempt is made to adopt some recommendations of science educators who are seeking to make introductory physics courses more effective.[2] The course is designed to emphasize a small number of fundamental concepts that are universal to a wide range of physics beyond the material covered in the first two semesters. These concepts are selected first of all because of their basic scientific importance, but also because they can be illustrated by examples that fall naturally within the course material.

SPACETIME

The symmetry of space and time itself is fundamental to all physical theory. Due to many years of indoctrination using plane geometry, the students accept this symmetry uncritically without considering the consequences of assuming this concept axiomatically. The students resist attempts to work in non-Euclidean geometries. All the students know the Pythagorean theorem, but the concepts of 3-dimensional and 4-dimensional vector spaces have to be introduced and practiced.

PRINCIPLE OF RELATIVITY

The principle of relativity was a proposal made by Galileo, who presented the idea using the example of relative motion within the cabin of a ship which was under way.[3] Newton accepted Galileo's proposal and clearly attempted to incorporate it into his mechanics, at least for the range of velocities within his realm of experience.[4] Newton failed to arrive at what we now would call a relativistic theory of mechanics, perhaps because he had no inkling of the correct transformation between moving coordinate systems. The Newtonian understanding of the principle of relativity is enough to force the description of physics to take the form of second-order differential equations.

LORENTZ TRANSFORMATIONS

One pedagogical benefit of teaching a classical theory of electromagnetic fields is that electromagnetic phenomena naturally are explained on a macroscopic scale by a completely relativistic theory. A discussion of the Lorentz transformation logically comes out of the study of electromagnetism. A covariant formulation of electromagnetic theory is difficult for the students to grasp because the differential operators in 4-dimensional space are unfamiliar to the students, who in fact are still struggling with partial derivatives.

CONSERVATION PRINCIPLES

The solution of specific problems in a mechanics course logically results in constants of the motion in particular cases so that the students can be introduced to the conservation of energy, momentum, and angular momentum in a mathematically rigorous presentation. The conservation of electric charge is introduced as an ad hoc postulate, and the quantization of charge, as well as other quantum effects, are outside the scope of this course.

FROM CONCEPTS TO COURSE LECTURES

During the first semester of the course, two prototypical problems are used to cover the essentials of Newtonian mechanics; however, a considerable amount of lecture time is required before the students are in a position to fully understand these in complete detail. The first problem is depicted in the fresco by Giuseppe Bezzuoli which shows Galileo in Pisa rolling a ball down an inclined plane to experimentally determine the ball's acceleration. A complete understanding of this problem requires a knowledge of the vectorial nature of the gravitational force, a representation of frictional force, and the mathematical description of rotational dynamics. The second problem is the motion of the planets in the solar system. The solution of this problem provides a perfect opportunity to emphasize conservation of both energy and angular momentum in a system constrained by a central force. The integration to obtain the path of motion for a planet is difficult for the students at this level.

Building on the advancing abilities of the students, the second semester challenges them with electromagnetic field theory. The students' background in vector calculus is extended by introducing them to Gauss' (Green's) theorem and Stokes' theorem. A complete presentation of Maxwell's equations in differential form requires numerous hours of lecture. Recasting these equations covariantly depends on the introduction of a 4-potential which has the transformation properties of a vector in spacetime. The 4-potential is an abstract concept which is difficult for the students to comprehend. Several previous lectures on the equation describing wave motion prepare the way for the presentation of transverse waves as the solution of Maxwell's equations in free space.

EDUCATIONAL OUTCOMES

Sixteen students enrolled in the first semester of the course. Their final grades ranged from C to A. Of these students, half continued to the second semester and with much hard work are doing an excellent job of mastering the course material. Certainly, these students benefit from a sound background in mathematics and the low student-to-instructor ratio at Cedar Valley College.

FUTURE—RIGHT ON THE LIGHT CONE

Cedar Valley College has expressed its interest in continuing to offer this program of physics courses in future semesters. A need appears to exist for new textbook material which could be used to teach introductory, calculus-based physics at the college level. The lecture notes from this project could serve as a starting point for such a textbook; however, an extensive editing effort also would be required.

Science education has received prominent recognition in the Director's statement of the Laboratory's goals. Each week, course preparation for this project averages 20 hours and meetings with the students involve 6 contact hours. This expenditure of time represents a significant commitment to this educational activity by the Accelerator Systems Division.

ACKNOWLEDGEMENTS

The suggestion for this program originally came from Dr. Ted Kozman, Head of the Accelerator Systems Division. The success of this program is due in large part to the support of Dr. David Heitman, the Dean of Science at Cedar Valley College in Lancaster, Texas.

REFERENCES

1. David Halliday and Robert Resnick, *Fundamentals of Physics*, 3rd Ed., Wiley, New York, 1988.
2. Alan Lightman, *Great Ideas in Physics*, McGraw-Hill, New York, 1992.
3. Galileo Galilei, *Dialogue Concerning the Two Chief World Systems—Ptolemaic and Copernican*, University of California Press, Berkeley, 1962, first published 1632.
4. Isaac Newton, *Mathematical Principles of Natural Philosophy and his System of the World*, University of California Press, Berkeley, 1962, first published 1686.

COIL WINDING FOR THE SSC GEM DETECTOR MAGNET

Rui Vieira[2], B. Smith[2], P. Marston[2], M. Olmstead[2], Z. Piec[2], P. Titus[3], and G. Naumovich[1]

[1]Everson Electric Company
[2]Massachusetts Institute of Technology Plasma Fusion Center
[3]Stone and Webster Engineering Corporation

INTRODUCTION

The MIT Plasma Fusion Center, as a collaboration member with the Superconducting SuperCollider Laboratory, is currently engaged in the design of the coil modules for the GEM Detector Magnet. All of the winding R&D work to date has been performed at Everson Electric Co. This paper focuses on the components and winding process for each coil module. The magnet is a single layer solenoid composed of two 14.25 m long halves. Each coil half is made up of 12 identical modules. A module consists of 19 turns wound (roll-formed) on the inside diameter of the aluminum bobbin. The bobbin is a cylindrical 6061-T651 aluminum shell, 76.2 mm thick by 1070.0 mm long, composed of four segments (quadrants), joined every 90°. In addition each module has a total of 5 top and 5 bottom compression flanges with which the winding precompression is applied. Fig. 1 illustrates a cross-sectional view of a coil module. The compression flanges also serve to stiffen the module structure and as the means of mechanically attaching one coil module to another.

The GEM conductor design is a cable-in-conduit design, a concept which has been shown to be very stable by the international fusion community and most recently successfully tested in the "Demonstration Poloidal Coil" . The conductor is composed of 450 NbTi superconductor strands cabled, and encased in a 304L stainless steel conduit with a 20 mm inner diameter and a wall thickness of approximately of 2 mm. The cable-in-conduit is encased by a rectangular 1100 "0" temper aluminum sheath which provides protection during the unlikely event of a quench. Fig. 2 illustrates a cross-sectional view of the conductor. Fig. 3 illustrates the option II aluminum sheath which is also being evaluated.

The coil modules are to be wound on site at the South Assembly Building. The building is composed of three stations dedicated to the winding process. Two are winding stations and the third is a pre-assembly station. The current winding process is one in which the conductor is roll-formed on the inner diameter of the bobbin. This roll-forming (winding) process requires the use of two roll-forming machines, a turn table, a pay-out reel (conductor transportation spool) a cleaning station, an insulation station and a conductor measuring device. In addition there are a series of roller stands which will guide and support the conductor between roll-formers and bobbin.

The goal of the current winding R&D program is to design and evaluate the winding equipment and to establish and finalize the winding parameters.

RESULTS AND DISCUSSION

The current winding R&D program is divided into two Phases. Phase I focused on evaluating the roll-forming equipment and in establishing the conductor behavior. The results on the keystone of the conductor indicate that at a radius of 1.825 m its no more than 0.6 mm and at its final radius (9.5 m) approximately 0.05 mm. A final keystone of the magnitude may not be a problem, However this problem can be controlled by extruding the sheath with a counter keystone. To evaluate roll-forming characteristics a set of 7 conductor samples were roll-formed with different roller settings. Fig. 4 summarizes the data. Visual inspection of the aluminum sheath welds before and after roll-forming indicated no evidence of damage even for radii of approximately 500 mm. The behavior is true for both sheath options. The option II sheath was roll-formed to a 3.65 m diameter with the welds in the ID and OD. Visual inspection of these welds also

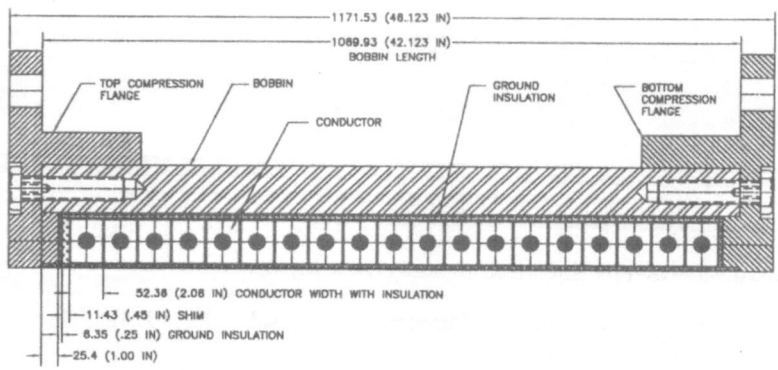

Fig. 1. Cross sectional view of coil module.

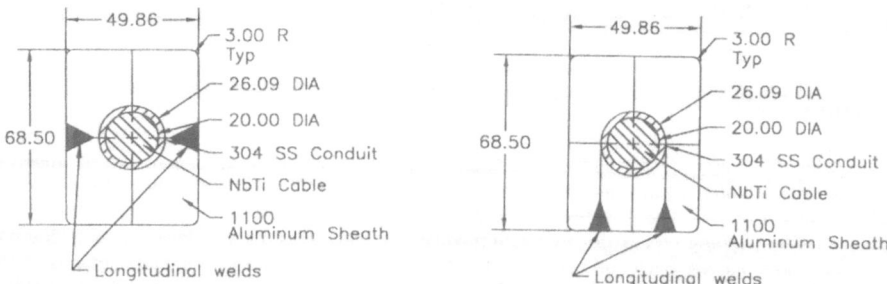

Fig. 2. Option I Conductor Cross–Section without insulation.

Fig. 3. Option II Conductor Cross–Section without insulation.

showed no evidence of damage. Phase II is designed to evaluate and establish the parameter for roll-forming a minimum of two full size turns (19 m dia.). To this end the following steps are currently in progress: (1)assembly of a minimum of 120 m of conductor comprised of conduit and sheath only. (2) Roll-forming conductor onto transportation spool. (3) Development of computer control roll-former. (4) Evaluation of conductor length measuring device. (5) Evaluation of winding parameters. (6) Evaluation of insulation process. The key components of the conductor assembly are: The 3.65 m spool; the longitudinal sheath (MIG) welding; both the conduit and sheath butt (TIG) welding; and the length measuring device. A total of three turns are currently wound. Most of the difficulties with the process have being associated with the welding and with the length measuring device. The welding problem was associated with the gas mixture and has been solved. The length measuring device problem was traced to is readout setup as well as the holding fixture and both problems are currently being work on. Based on the data from phase I, it was evident that a computer controlled roll-former was necessary. A unit has been designed with a curvature measuring system. The unit is designed to maintain the final radius to within +- 25.4 mm. It is schedule to be operational in May 1993. The current length measuring device being evaluated is a wheel with an encoder attached to it. Its circumference is machined to within 0.01 mm accuracy. However the problem with this system is wheel slippage. A calibration process of the wheel is currently under way. Alternative measuring systems are also being evaluated. One that appears very promising is to use a time of travel approach signal system with optic fibers. The fiber can be embedded in the aluminum sheath. During the final roll-forming process (winding) the following parameters will be evaluated: (a) roll-forming rate, (b) curvature consistency, (c) conductor flatness & twist, (d) final conductor dimensions (Average keystone), (e) conductor stretching ,(f) Roll-former performance. In addition the test will evaluate whether there is an advantage to roll-forming to a final radius that is greater than the bobbin radius. This is important because it may help ensure intimate contact between the conductor and the bobbin.

COIL MODULE ASSEMBLY

Bobbin Preassembly

Adjacent to the two winding stations there will be an assembly station to pre-assemble the bobbin. The assembly station is composed of a disc foundation plate fastened to the floor. A ring with a hole pattern

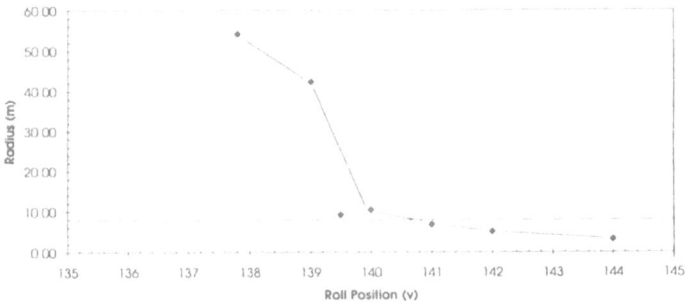

Radius -vs- Roll Position

Fig. 4. Plot of conductor radius as a function of roll position. The final conductor radius is 9.5 m.

congruent with that in the bottom flange, and tapered in thickness at the winding helix angle serves as the interface plate for the bottom bobbin flanges. The bobbins are made up of four quadrants which are held together with an insulated mechanical joint. Each quadrant is made up of a bobbin shell, and an upper and lower set of bobbin flanges. The lower or bottom compression flange arrives from the manufacturer already assembled with its corresponding bobbin shell, cleaned and ground insulated. In addition each bobbin shell arrives with its cooling tubes welded to the OD. Assembly starts by matching and/or aligning the flange's circular bolt hole pattern with the identical bolt pattern of the helical base-plate alignment ring. The same task is repeated on all four bobbin segments. The foundation plate will also have vertical support and alignment posts with adjustable jack screws, Fig. 6, which will hold and position the bobbin shell. The bobbin segment (quadrant) joints are now secured and precise measurements of the coil module circumference are confirmed . These measurements are made to an accuracy of +- 2.54 mm. In the event that adjustments need to be made to the circumference, insulating shims at the segment joints will be used. Once the coil module is assembled, the factory installed bobbin cooling tubes are butt welded at each segment joint. The bolts securing the coil module to the helical ring are removed and the assembly is transported to one of the winding stations. A lifting spider will be used. The spider is attached to the bobbin bolt holes.

BOBBIN WINDING PROCESS

Winding Process

The winding station is composed of a box beam ring turntable driven by a chain and powered by a 1 HP motor see Fig. 5. Secured to the top plate of the turntable is a helical ring, identical to the one in the pre-assembly station. The bolt hole pattern on the bottom bobbin flange mates with the helical ring identical bolt hole pattern and aligns the preassembled coil module on the turntable. The turntable also has alignment support posts with adjustable jack-screws which will hold and position the bobbin. At this point the bobbin is aligned and secured to the turntable, the conductor transport spool is in place and secured, plus all of the necessary equipment such as the roll forming, cleaning station, insulation wrap, and length measuring machines are in place and secure. To insure that the roll forming machines are set correctly, a conductor sample is roll formed and checked against the bobbin for final radius. A curved draw bar is bolted to the conductor end termination (half joint) to begin to pull the conductor to the 1st roll-former. This allows the conductor to be fed through the winding mechanism. The joint half is attached to the bottom of the bobbin at the flange joint opening, and the winding process begins. To monitor the winding process, a measuring device is placed just before the insulation station. Through this length measuring device, a record of the amount of wound conductor will be kept. Adjustment of the conductor mean radius is made by adjusting the thickness of the extra insulation located on the conductor surface facing the bobbin inner diameter. Fig. 7 illustrates four conductors already wound with difference mean radii. Note that neither the bobbin ground insulation nor the conductor wrap insulation are compromised. The bobbin support posts will also contain compression screws which will maintain a maximum load of 10,000 pounds axially on the winding. The axial load will help minimize turn-to-turn gaps as well as maintain the conductor flat. At the end of the winding process, the joint half is placed and secured to the bobbin at the top joint opening. The conductor build or height is measured to identify the minimum shim thickness required to assure proper precompression. A HiPoT and a turn-to-turn electrical insulation test is performed. The top flanges are then installed along with the joint compression plates. The inner radiation shield is bolted to the bottom and top flange. The bolt holes at the top of the shield are slotted to compensate for movement of the top flange. The shield is installed with

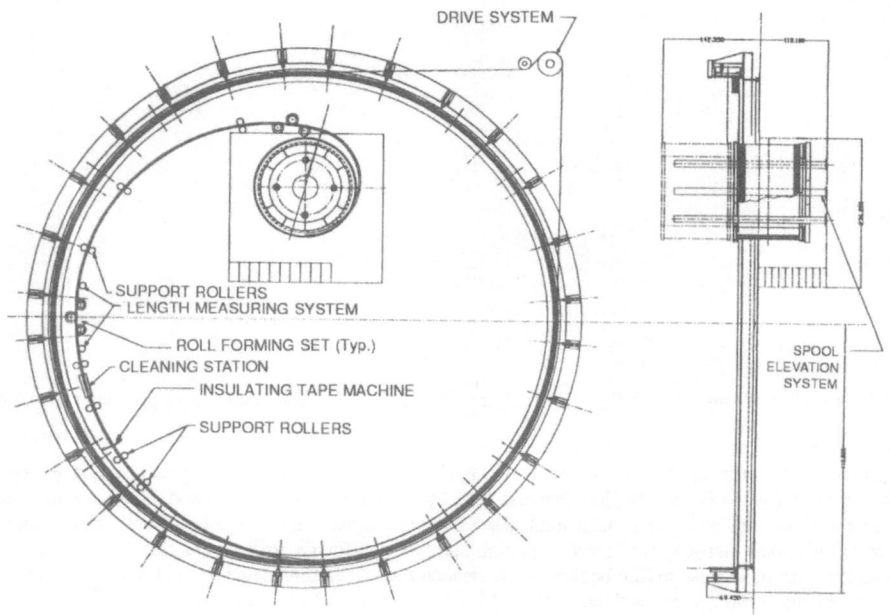

Fig. 5. GEM Conductor Winding Machine.

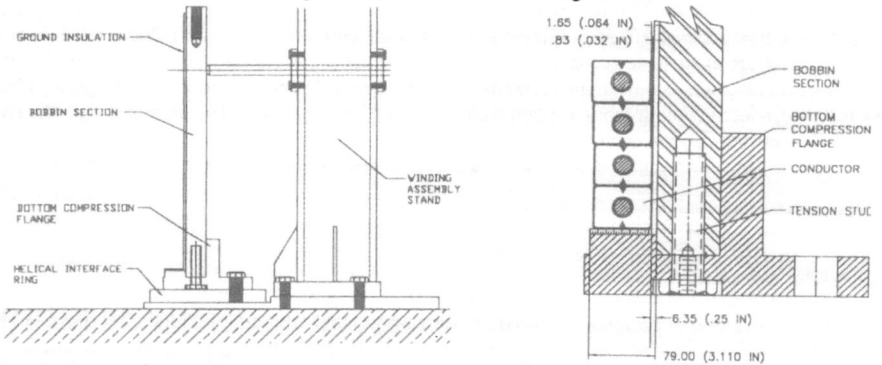

Fig. 6. Pre–assembly Bobbin support alignment post. **Fig. 7.** Adjustment of conductor mean radius.

its insulation pre-assembled. The pre-load is achieved by a set of stud tensioners. The studs that secure the top compression flange to the bobbin are also used to pre-load the winding. The pre-load is monitored by measuring the stud tensioner hydraulic pressure. An alternative approach that is currently being evaluated is the use of one or two large hydraulic presses that are sequentially place along the circumference of the winding table to pre-load the winding. The winding is first compressed with an oversize shim. Once the required load is achieved, the gap between the bobbin and flange is measured, compression is then relaxed and the correctly sized shim is inserted. The winding is re-compressed. Once the winding is pre-loaded the HiPot is performed again. The last step is to install the lifting spider to transport the wound coil module to the magnet assembly building.

CONCLUSIONS

a) The Phase I conductor data indicates that at its final radius (9.5 m) the keystone is approximately 0.05 mm. This level of keystone may require a very miner counter keystone extrusion.

b) Visual inspection of both Option I & II aluminum sheath longitudinal and butt welds show no evidence of damage after being roll-formed to a radius of 1.825 m. In addition the option I conductor also shows no damage at a radius of approximately 500 mm.

c) A computer control roll-former is needed to provide a consistent and uniform final radius due to the variation in properties and dimensions of the conductor.

SYSTEM-LEVEL DESCRIPTION OF THE SDC DETECTOR

Norman E. Wells

Physics Research Division/Solenoidal Detector Collaboration
Superconducting Super Collider Laboratory*
2550 Beckleymeade Ave., MS 2006
Dallas, TX 75237-3997

INTRODUCTION

The detector being designed by the Solenoidal Detector Collaboration (SDC) is an eight story structure with 8.5 million sensor elements[1]. The positions of these elements must be known to less than one millimeter accuracy to meet the physics requirements[2]. However, this 75-million pound experimental device could settle up to 20 millimeters and the entire detector, as well as its components, must be moveable to align with the proton beam and to accommodate detector and floor deflections.

Interesting engineering challenges are presented as the detector is designed. The management challenge of integrating the work of about 1000 physicists and engineers at institutions around the world is also significant. The coordination process is being facilitated by the publishing of a reference book for detector designers, the SDC Detector Parameters Book (Ref. 1), that contains the latest design information. It was used as the source of detector data contained in this paper (unless otherwise indicated).

This paper describes some of the systems issues that are being faced as the detector is designed. After the detector is described, the foundation requirements are reviewed and the method of keeping the detector aligned with the proton beam described. A discussion of detector operations and maintenance completes the paper.

DETECTOR DESCRIPTION

The detector is housed in a hall that is about the size of a football field and is 213 feet underground[3]. The hall is 182 feet high, including the roof structure.

Two counter-rotating beams of protons will be accelerated in a 87-kilometer beamline. A 50-meter section of the beampipe is inside of the detector. The detector beampipe subsystem consists of the beampipe and associated vacuum pumps that maintain a very high vacuum within the pipe to keep air molecules from interfering with the protons. The beampipe is approximately 80 mm in diameter and one millimeter thick. The beam is approximately 10 microns across[4] and its position can vary by up to 100 microns between daily fills of protons[5]. The beampipe position must be accurate to within one millimeter of the beam at the Interaction Point (IP)[2].

Several types of tracking subsystems surround the IP. These contain some seven million sensor elements designed to record the tracks of the particles. The tracker must be independently adjustable and be positioned to within half a millimeter of the beam in radius[2].

---

* Operated by the Universities Research Association, Inc., for the U. S. Department of Energy under Contract No. DE-AC35-89ER40486.

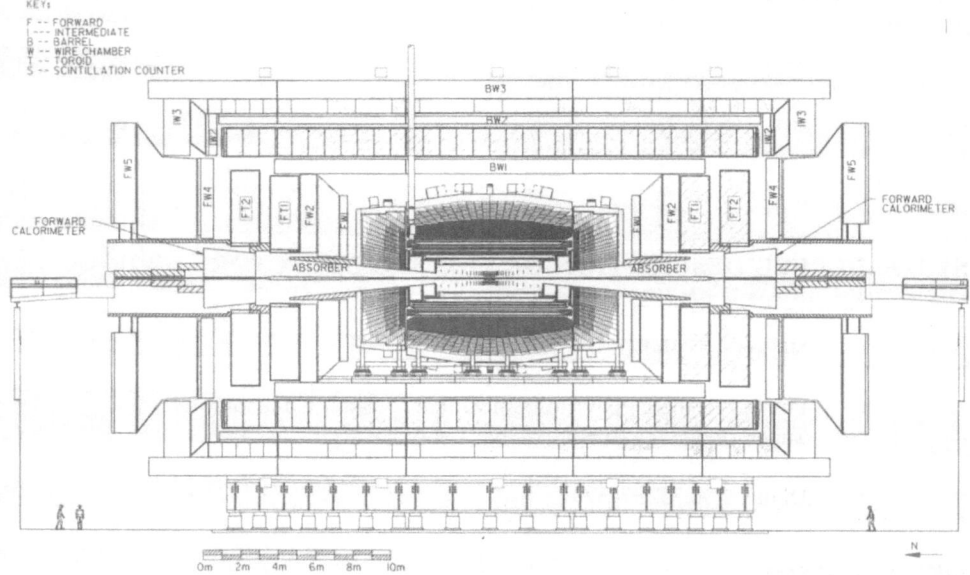

Figure 1. SDC Detector

A superconducting magnet surrounds the tracking system and provides the magnetic field that causes the charged particles to bend as they traverse the detector. The amount of bend in a track provides an indication of the momentum of the particle. A large cryogenic system on the surface provides the coolant to make the magnet superconducting.

The barrel calorimeter supports the magnet and tracker, and provides the ability to measure the energy of the particles emanating from the collisions. A forward calorimeter 12 1/2 meters from the IP provides energy measurement of particles close to the beam line. The four pieces of the 3638-metric tonne (eight million pound) central calorimeter must be rolled precisely into the center of the detector and be held in that position during operation. The 3.2 million pound endcap calorimeter pieces must have the capability to be rolled back to provide access space for maintaining the tracker and magnet. The many crates of electronic equipment that support the data acquisition from the tracker and calorimeter are mounted on the outside of the calorimeter.

The muon subsystem on the outside of the detector provides the capability to record the high-energy muon particles emanating from the IP. The measurements are done in muon chambers that are on either side of another magnet system. The 36 million-pound muon magnet system forms the outer structure of the detector and is 1.5 meter thick steel, 16.5 meters wide and high, and is 28 meters long. The design considered construction time (which favors few pieces), and transportation and handling limitations. The final design consists of 196 blocks weighing about 85 metric tonnes each. Additional muon magnets and measurement chambers are located at the ends of the detector. These must be removable to provide access to the inner components for major maintenance. The entire detector is supported by a large steel structure which rest on 76 jacks.

An extensive electronics system provides the ability to capture and record the data on the collisions. These physics interactions occur at a rate of almost 10^8 per second and each requires about a million bytes of data to record the 8.5 million data channels[6]. Such a data rate would swamp a processing system and yield more data than can be handled. A series of filters, called triggers, are used to capture only the physics events of interest and reduce the amount of data to 10-100 events per second. Even with the large reduction in data, over 10^{15} bytes of data per year are generated because the detector is scheduled to operate over 10^7 seconds per year. However, only a very few data points per year will be seen that lead to new discoveries. An off-line computing system provides long-term storage of the data and will provide the ability for physicists around the world to analyze the results--from their home institution.

The detector is supported by utilities such as cooling systems to remove the heat generated by the electronics, electrical power supplies, safety systems and other support systems.

DETECTOR FOUNDATION

A major requirement of the SSCL was that the physics experiments be located on a foundation that provides low risk of short and long-term deflections. A study[7] in 1991 reviewed the soil/rock conditions in which the detector halls would be constructed. It was found that the East Interaction Regions provided the best foundation for the large detectors (like SDC) because the detector hall floor would be on Austin Chalk, which was the most stable rock in the area. Large detectors on the West Campus would have been on Eagle Ford Shale which is unstable and could have caused unpredictable movements of tens of centimeters. This risk was unacceptable for the large detectors and for the focusing magnets of the collider (who have even tighter positioning and stability requirements). The result of the study was that the large detectors were moved to the Austin Chalk foundations on the East Campus and the smaller detectors (which are not as deep) were located in the Austin Chalk on the West Campus. There was also a cost savings in the process.

The current SDC detector hall floor is ten-foot-thick reinforced concrete[3]. Data indicates that the floor could deflect approximately 5 mm under the load of the detector[8].

ALIGNMENT TO BEAMLINE

The proton beams will collide at a point in space called the Interaction Point (IP). This point (actually, a region about 7 cm long) is only a surveyed point in space when the detector foundation is started. The beampipe for the protons will not be installed until three years after the start of detector construction. A survey accuracy of 2 mm is expected[9] which means that the entire detector, or at least one of its subsystems, may have to move up/down and/or sideways to align with the beam.

The beamline slopes downward 2.16 mm per meter from North to South[10]. This means that the beam drops 86 mm across the detector. The entire detector must also be sloped to maintain the proper positions with respect to the beam. A study[11] was performed to determine if the detector should be assembled on the slope, or if it should be built level and jacked into position. It was found that the detector and floor will deflect during the construction process so a final alignment would be necessary in any case. The decision was made to let the muon subsystem choose the best way to assemble the muon toroid magnet (ie: build on slope, or build level and jack). The entire detector will be aligned with its final position on the slope before the muon system is complete. The inner components will be installed on the slope. This was the least-risk option because alignment of components can take place for several years with the detector in its operating position.

The detector and floor are expected to deflect up to 25 mm as the detector is constructed[12]. Additional deflection over time may also occur. The jacks supporting the detector are being specified to be able to raise and lower the detector through a range of 120 mm [13]. Differential motion between the jacks on either side will be able to translate the detector centerline sideways if necessary. Current design calls for the jacks to keep the detector positioned within 3 mm of the correct position. The calorimeter, tracker and/or beampipe may have to move independently to achieve their tighter positioning accuracies.

DETECTOR OPERATIONS AND MAINTENANCE

The detectors are planning for the proton beam to operate an average of 5631 hours per year (after the initial checkout process)[14]. The remainder of the year the beam and detectors will be scheduled to be off for planned maintenance. The collider will fill the beam with protons about once a day and accelerate them to collision energies. The daily fill is necessary to replenish the protons that are lost to collisions. A stable beam should be available to provide physics data an average of 3754 hours per year[14].

The detector and collider organizations are working to have their systems available for operation 80% of the time. This means that they should operate 80% of the time they are scheduled to operate. During the remaining 20% of the scheduled time they would be down for unscheduled maintenance. This unscheduled maintenance time is being minimized by the use of more reliable components that fail less often (either by their intrinsic design or by designing an installation that will reduce failures), and by designs that reduce repair time.

Much of the detector electronics will be on the surface in the Detector Operations Building where they will be accessible during collider operation. No one is permitted in the detector hall during operation because of the radiation from the beam. This beam radiation is contained by the underground tunnels and halls and does not affect people on the surface. The beam radiation stops when the beam is turned off. The major utility systems that support the detector are also mostly located on the surface where they can be quickly repaired when the beam is on.

SUMMARY

The major system-level focus of the last year has been that of getting a very large steel structure to have the precise adjustability needed to meet the physics requirements. The detector halls have their foundations in the most stable rock available and the hall floor has been specified to minimize deflection. Plans have been made to align the detector and its subsystems to the proton beam when it is assembled in the hall, and to keep them aligned. Efforts are continuing that help define the reliability, availability and maintainability requirements to ensure that this large data acquisition system functions as needed to capture the data the should lead to new physics discoveries. The goal is to ensure that this huge instrument will be able to continue the advancement of science that started thousands of years ago.

REFERENCES

1. SDT-00010, Rev F, SDC Detector Parameters, 14 Sep 92.
2. SDT-000082, Specification for the Alignment Systems of the SDC Detector, 29 Mar 93.
3. CPB-100282, IR-8 SDC Experimental Hall Basic Construction Drawings, 5 Mar 93
4. Memorandum, IR Optics Data for Experimenters, Rae Steining, 26 Aug 92.
5. SDC Note SDC-92-259, "A Summary of Some of the Accelerator Interface Issues of the Past Year," Byron Roe, 14 May 92.
6. SDC-92-201, SDC Technical Design Report, 1 Apr 92.
7. R20-0000450, Study Report on the Location of the Large General Purpose Detectors, Oct 91, J. Western and N. Wells.
8. CPB-100175, Experimental Hall at IR-8 for SDC Detector, Title I Design Submittal, 25 Sep 92, fig 22.
9. SDT-000068, Minutes of SDC Alignment Working Group, 12 Nov 92, pg. 1.
10. ADOD_AGG_91_038, Memo, Subject: Beamline "Drop" Across the Eastern Interaction Regions, David Veal, 21 Nov 91.
11. SDT-000053, Study on the Detector Construction Scheme, V. Kopytoff and N. Wells, 2 Sep 92.
12. SCT-000001, Rev F, SDC Experimental Facilities User Requirements, 10 Feb 93, pg 4-7.
13. SDC-92-342, Muon Barrel Toroid Preliminary Design Report, 7 Oct 92, Appendix G, pg 2.
14. Project Management Memorandum, Availability--Mission Profile--Collider Operations, H. Edwards and R. Stefanski, 21 Sep 90.

PROPOSAL FOR DOUBLE-SIDED SILICON MICROSTRIP DETECTOR

D. Lazarovici,[1] C. Lazarovici,[1] Gh. Caragheorgheopol,[1] R. Ruscu,[1]
V. Cimpoca,[1] E. Muntiu,[2] I. Cernica,[2] and H. Giuroiu[2]

[1]Institute for Atomic Physics, Bucharest, P.O. Box MG-6, Romania
[2]Microelectronica S.A. Bucharest, Romania

INTRODUCTION

Many experiments in high energy research physics make use of silicon detector for trajectory finding. These detectors, realized in different geometries offer the advantage of good assembly compactness around the target, high efficiency for all incident angles and satisfactory position resolution.

We presume that in the future these types of detectors will keep their position in experimental physics, but in the same time, an extension of application in different other fields will take place.

The performance as position detection and the different geometries of silicon structures in accordance with the experimental requirements, are limited only by technological means and the cost of detector process.

This paper considers the experience of detector process, described in literature, in order to design the simplest and less expensive technology.

We think at a double-sided micro strip silicon detector built on his own chip, not including the associated electronics.

The separation of detector lay-out from electronics simplifies the manufacturing technology. Also the sorting and replacement of faulty detector is easier; the number of rejected pieces lowers.

The proposed detector fits not a special experiment ; it was thought as a first step in a system for trajectory finding, having an usual value of position resolution and reliability.

FUNDAMENTALS OF POSITION DETECTORS

To find the position in a plane and in space two types of silicon detectors geometry are used: micro strip and pixel. Stacks of these detectors in layers of perfect parallelism and alignment with thicknesses pending on the energy of studied particles and interaction, are built to compute the trajectories from the successive impact points with the detector planes.

Double-sided detectors are able to give complete information about position in the same manner as pixels do.

In high energy physics experiments both types of detectors are employed, each of them having advantages versus disadvantages. For micro strip detector the electronic circuitry is gathered to the both ends of detector.

The number of readout active channels is at maximum 2N (N is the number of detector strips on a chip). The pixel detector, according to a plane matrix configuration needs an electronic system with considerably more channels: N X M (N,M being the row and column number respectively).

Therefore it is useful to build the associated electronics for pixel detectors on the same detector chip. But, this apparent simplification leads to more complex structures with sophisticated and expensive technology.

The layers containing detectors for trajectory determination are realized from chips of different shapes and value of position resolution must lay in 8...10μm range. At a greater distance from axis the high energy beam is scattered, the detectors need a poorer position resolution (the strip pitch is larger).

For first layers of micro strip detectors, optimal parameters of the structure are: $P = 20\mu$m (strip width, $s = 10\mu$m and gap, $g = 10\mu$m, or $s = g$), for junction type strips. On back side of the detector pitches of 40μm can be provided ($s = g = 20\mu$m or $s < g$).

The readout is made by "clusters" of strips with a number of two-three strips per cluster. The position is determined as the mass center of a gaussian distribution of electrical charge collected from the cluster strips. The structure is p^+nn^+ with implanted junctions and contacts.

Planar technology is used, the same as by MOS structures. The main oxidation for masking to open windows for the p^+ strips is done by thermal growth -MOS oxide (after previous careful cleaning).

The MOS oxidation methods are different ; the most advisable is the one in which an oxidation up to 1000 °C takes place in a gaseous mixture of oxygen, hydrogen and chloride (HCl or trichlorethane). The formed oxide is covered with a protecting layer of Si_3N_4 or Al_2O_2. This layer assures a more stable Si/SiO_2 interface, reduces the positive charge in the oxide layer and hardens the structure to radiation. The number of thermal treatments done at high temperatures must be as low as possible.

For annealing, the following order of processing is preferable: 15min in oxygen atmosphere and two hours in nitrogen atmosphere; this process leads to the lowest value of reverse current.

Boron implantation for obtaining the p^+ strips is preferential to be done from BF_3^+ or BF_2^+ in doses up to 10^{15} /cm² and at low energies (to get superficial junctions), or through oxide layer.

For the ohmic contact n^+ on back side, phosphorus is preferable (implanted or diffused). Implantation is done till 10^{15} /cm⁻ dose and energies chosen pending on the structure (the implantation is done directly or through oxide layer).

Contacts-built on polysilicon layer doped n^+ and thermally treated is advisable because a getterisation of impurities and defects from the detector bulk volume takes place. Consequently, the bulk generation component of reverse current lowers.

The metallisation with Ti + Al + Si is preferred for the sake of increasing the radiation hardness of structure.

A thermal treatment after metallisation is done in $N_2 + H_2$ (10:1) gaseous mixture, at temperature of 400...450 °C (pending of crystal orientation).

The signal charge is fed from the strip to the preamplifier by a capacitor built on the detector chip, by deposition of metal on an insulating layer, grown under controlled process. If for example a common bias line is used, the p^+ strips on front side are connected to ground and the bias for total depletion is applied on back side strips. Also the p^+ strips on the back side, can be connected to a line inserted between the n^+ strips (provided to reduce the effect of space charge accumulation under silicon oxide).

The above considerations are fundamentals of the structures and optimal process of strip (micro strip) detectors. The performances of a manufactured lot of detectors are strongly pending on how rigorous the above technological conditions are respected.

PROPOSAL OF A DETECTOR PROJECT [1-10]

The detector which we proposed (got from N type silicon with 4KΩ.cm resistivity, (111) orientation, EPD < 500/cm , $\tau > =$ 1ms) is a simple p^+nn^+ structure with a pitch of 20μm, $s = g = 10\mu$m on the front side and $p = 40\mu$m, $s = g = 20\mu$m on the back side, the back side strips rectangular to those on front side.

For minimum of rejected chips, the detector system is built without any electronic circuitry. The connection to the electronics is made in a mounting including the detector chip and a chip with 128 channels of preamplifiers. The number of strips is a multiple of 128; a more convenient detector structure is a square one.

We preferred short strips (¯25mm) for the same reason of reducing the number of rejected (faulty) pieces. If the reverse current of strips is low, detector chips can be connected in series forming chains of suitable lengths for each experiment.

Not having pretention to make a standard of the detector chip, the modular chip can be helpful for the experimenter and allow efficiency in manufacture of chips at large scale.

The technological process proposed by us has the goal to apply generally basic principles; for some operations (steps) we described our own methods which gave us satisfactory results.

The detector structure, in spite of having p^+nn^+ configuration has an innovation: the n^+ strips (on the back side) got by doping (implantation) phosphorus in polisilicon layer deposited uniformly on the silicon.

The main purpose was the getterisation of impurities and defects from the volume of silicon, but hardening to radiation also occurs. Also we got rid of the interface Si/SiO_2 from the back side and all the unwanted effects of the space charge accumulation.

Between the strips on backside (n^+) we introduced metallic strips of $6\mu m$ connected to a common bias line; the system with low positive bias (few volts) constitutes "guard intermediate rings" (GIR). These can be polarized or not, pending on the value of leakage current.

In experiments, for the planes further form the target, the detectors are produced by the same process, but with suitable masks. Because capacitive coupling is used for strip readout (through SiO_2 layer) the metallisation is only $Al + Si$.

The electrical design parameters for the micro strip structure proposed are:

-reverse current per strip : $< 1nA/cm^2$

-resistance between two neighboring strips: $>$ than $100k\Omega$(on the front side)

-capacitance between one strip and his neighbour(all other floating) -on the front side: 0,9pF/cm, on the back side : about the same value (the capacitance is practical constant with the s/p ratio, which is the same on both sides)

-capacitance between one strip (with all the others connected together on the front side) and to all the strips on the back side: 2.7pF/cm. This capacitance is important as contribution to electronic noise on the strip channel;

- signal/noise ratio: > 10

The dimensions of the detector chip are function of the length chosen and the number of strips. The dimensions could be 26.6 x 26.6 mm (with 1280 strips on the front and 640 on the back).

TECHNOLOGICAL PROCESS

The proposed detector follows generally the MOS technology.

Grouping some thermal treatments we have reduced the overall number of steps and have optimized some of them, which are to be done at the same temperature.

The cleaning of silicon wafer surface is considered the main condition of success in our technology - We insist strongly on that choosing the solutions which give the best results.

Because the silicon wafers on which we tested the technology had the same dimensions as the masks (4"), we provided an operation of front to back side prealignement.

These masks (two identical pieces) were labeled "O"; there is a mask "1" (for GR on the back side) and two "2" and "3" (for opening in the same time the windows for GR and strips on front and back side); two others, masks "4" and "5" are used for metallisation on front and back.

The sketch of the process is as following:

1. Acidous cleaning (on both side) MOS type in $H_2SO_4 + H_2O_2$ (3:1) solution during 5 min. at 150°C. After that, washing in DIP solution: $HF:H_2O$ (1:10) 20s.
Washing in deionized water and drying in warm nitrogen (RD).

2. Acidous etching in $HNO_3:HF:CH_3COOH$ (5:2:2) at about 0°C, washing drying.

3. Photoresist deposition (on both sides) and prealignement with a pair of "0" masks. Mark and alignments by corrosion on silicon wafer margin.

4. Cleaning before MOS oxidation. After removing the photoresist layer an etching as at point 1 is applied with the solution used but in a reverse succession. The DIP solution is 1:50 ratio.

5. MOS oxidation at 950°C in gaseous mixture $H_2 + O_2 + HCl$ (2,6%). The oxide thickness is 900Å. Annealing in nitrogen atmosphere at 900°C.

6. Deposition of Si_3N_4 in a LPCHD (horizontal reactor) at 630°C. Protection with photoresist. Photolithography for GR on back side.

7. Removing nitride from back side in plasma (Barell method) on the GR window.

8. Implantation in GR back of phosphorus, 100keV, $10^{15}/cm^2$ (through oxide layer).

9. Removing the mask and total oxide on the back side (till silicon surface). Washing , drying.

10. Polisilicon deposition by LPCVD at 660°C. Thermal treatment in nitrogen at 600°C. Thickness of layer~$1\mu m$. Deposition of photoresist on both sides.

11. Applying the mask for strips p^+ and GR on front side and for strips n^+ GR and GIR on the back side. Opening the windows for implantation . Washing the window till oxide layer.

12. Implantation of BF^+, 60keV, $10^{15}/cm^2$ for p^+ junctions.

13. Implantation of phosphorus (as at point 7) in polisilicon on back side after turning the wafer.

14. Removal of photoresist and etching by MOS process (as at point 3).

15. Thermal treatment for both annealing and gettering phosphorus (on the back side). Thermal cycle : 600°C-850°C (5min. in atmosphere as at point 4). Maintain at 850°C 1hour in nitrogen atmosphere. Grow an oxide layer by wet oxidation . Annealing at 850°C during 1hour in nitrogen. Lowering the temperature (slowly, with 1°/min) till 600°C; maintain 3hours. Remark. After this treatment a 500 Å a layer of polisilicon is oxidized due to the oxide growth on the implants. Deposition of photoresist on front and back sides.

16. Removing the layer of oxide on polisilicon from back side with DIP and BHF solutions. Removing of the photoresist.

17. Cleaning for metallisation, washing, drying (RD).

18. Al + Si deposition on both sides. Thickness $0,8\mu m$.

19. Deposition of photoresist on both sides. Applying the masks for metallisation ("4" and "5").

20. Wet etching of metal in standard Al etch solution; remove of SI points in plasma.

21. Removing the photoresist and then wash wafers in cold sulfuric acid.

22. Thermal treatment of metallisation at 450°C in gaseous mixture of $N_2 + H_2$. (10:1)

23. Check for electrical and optical characteristics, contact welding.

24. Final passivation .

The operations underlined in the technological process are considered principal for success of detectors.

PRELIMINARY RESULTS. DISCUSSIONS.

The silicon position detector structure proposed by us needed a technological process fitted and completed for a simple and low cost manufacturing. If the detector structure it is realized as separated chip (not including front end electronics) lower costs, facility for selection, assembling and replacement of pieces in the SSC system, results.The results obtained till now are satisfactory and demonstrate the assumptions and principles used in the project.The quality of technological process experienced by us is evident and is pending on photolithography operation and cleaning.

We hope to demonstrate the feasibility of the proposed detector. On the detector structure already manufactured and measured we want to do radiation hardness and reliability tests.

The authors wish to thank and are grateful to Prof. G. Bellettini and Dr. F. Zetti from INFN, Sezione di Pisa, Italy, for very helpful discussions and encouragement.

REFERENCES

1.J.T.Walton *NIM* A 226, 1(1984), E.H.M. Heijne, ibid 63, H.Boettcher,ibid, 72, J.Kemmer, ibid, 89, E.Gatti,ibid,163, G.Zimmer, ibid, 175.

2.J.Kemmer *NIM* A 253, 365(1987), E.H.M. Heijne, ibid, 325, W.R.Th. Ten Kate, ibid, 333. A.H.Walente, ibid, 558.

3. M.Caccia *NIM* A 260, 124(1987).

4. S.Holland *NIM* A 275, 537(1989), *IEEE Trans. on NS*, 36/1, 283(1989).

5. J.Kemmer *NIM* A 288, 262(1990), E.Fretwurst, ibid, 1, A.Longoni, ibid, 35, T. Maish, ibid, 89.

6. J.T.Walton, F.S. Goulding, *IEEE Trans. on NS* 34/1(1987).

7. H.Dijkstra *IEEE Trans. on NS* 36/1, (1989), P.Holl, ibid, 251, Z.Li,...,ibid,290.

8. H.Becker *IEEE Trans. on NS* 37/2, 101(1990), F.Fujieda, ibid, 124.

9. K.Misra *J.Electroch. Soc.*,137/5 1559(1990), W.Kern,ibid/6,1888

10.M.Pentia, to be published

PHOTON-COUNTING MONOLITHIC AVALANCHE PHOTODIODE ARRAYS FOR THE SUPER COLLIDER

A. Nadeem Ishaque, Donald E. Castleberry, and Henri M. Rougeot

General Electric Corporate Research and Development
Schenectady, NY 12301

INTRODUCTION

In fiber tracking, calorimetry, and other high energy and nuclear physics experiments, the need arises to detect an optical signal consisting of a few photons (in some cases a single photoelectron) with a detector insensitive to magnetic fields. Previous attempts to detect a single photoelectron have involved avalanche photodiodes (APDs) operated in the Geiger mode [1], the visible light photon counter [2], and a photomultiplier tube with an APD as the anode [3]. In this paper it is demonstrated that silicon APDs, biased below the breakdown voltage, can be used to detect a signal of a few photons with conventional pulse counting circuitry at room temperature. Moderate cooling, it is further argued, could make it possible to detect a single photoelectron.

Monolithic arrays of silicon avalanche photodiodes fabricated by Radiation Monitoring Devices, Inc. (RMD) were evaluated for possible use in the Super Collider detector systems. Measurements on 3 element x 3 element (2 mm pitch) APD arrays, using pulse counting circuitry with a charge sensitive amplifier (CSA) and a Gaussian filter, are reported and found to conform to a simple noise model. The model is used to obtain the optimal operating point. Experimental results are described in Section II, modeling results in Section III, and the conclusions are summarized in Section IV.

EXPERIMENTAL CHARACTERIZATION

The tested APD arrays were fabricated by RMD using deep diffusion technology; the following characteristics were measured:

Pitch	= 2 mm
Dead Space	= 300 mm
Breakdown Voltage	= 1390 volts (typical) at 25 C
Cross Talk	< 1% over entire range of voltage
Gain Variation (element-to-element)	< 5%
Total Dark Current	= 3 to 20 nA at 25 C
Bulk Dark Current	= 10 pA at 25 C
Capacitance	= 5 pF at 1300 volts

Fig.1 shows gain of a typical APD on the array as a function of voltage. It is possible to achieve large values of gain (approaching 1000) if the array is operated close to the breakdown voltage. Operating the APD array at 1300 volts (which corresponds to a gain of

81) and using a filter shaping time (t) of 6ms, the signal-to-noise ratio in the detector was measured as a function of the number of incident photons (Fig.2). By extrapolating the data, it is found that the signal-to-noise ratio becomes greater than unity for a signal consisting of 38 or more photoelectrons. However, for an APD gain (M) of 81, the optimum filter shaping time is not 6ms. Fig.3 plots the measured resolution of the pulse height spectrum, at full width half maximum (FWHM) of the peak, as a function of the filter shaping time for an APD gain of 81. The optimum shaping time is found to be 3ms. For a 3ms shaping time, measurements show that the signal-to-noise ratio becomes greater than unity for a signal of 34 photoelectrons.

Due to the excess noise resulting from the statistical nature of the avalanche process in an APD, the signal-to-noise ratio does not improve monotonically with increasing APD gain. Fig.4 plots the measured resolution as a function of voltage for a filter shaping time of 6ms. As expected from APD noise theory, the signal-to-noise ratio is found to be a non-monotonic function of voltage (hence APD gain), with a single extremum at 1330 volts, corresponding to a gain of 150 (the minimum in resolution corresponds to the maximum in signal-to-noise ratio).

The optimum gain is generally not very large, unless a small filter shaping time is employed. In Figs.2&3, one of the parameters of the set (M,t) was held constant at an arbitrary value and the other was varied to find its optimum value. However, there should exist a unique point (M_{opt}, t_{opt}) for which the SNR is globally maximized. Provided the detection system does not impose its own constraints (the scintillation time of the detecting fiber, for example, can set a lower limit on the filter shaping time), the device should be operated near the optimum point for best results.

These experiments were performed to quickly assess the potential of APDs for detecting a small optical signal; no attempt was made to optimize the electronics, to discover the optimum operating point, or to identify the limiting features of the APD arrays experimentally. Instead, a model was developed which was found to agree with the measured experimental data. The model, described in Sec. III, was used to obtain the optimum operating conditions and to identify (possible) device-level improvements.

MODELING AND OPTIMIZATION

If N photons are absorbed by an APD having a gain M and the filter shaping time is t, the resulting signal-to-noise ratio may be expressed as [4]:

$$(\text{SNR})^2 = f(M, \tau) = \frac{q N^2 M^2}{K_s/\tau + K_{p1} M^2 F(M)\tau + K_{p2}\tau + K_f} \tag{1}$$

where K_S, K_{p1}, K_{p2}, and K_f are noise coefficients that depend on circuit and device constants [4]; F(M) is the excess noise factor. A procedure has been developed to calculate SNR for any APD operating voltage and circuit parameters, which employs a theoretical analysis of the APD operation. The electric field profile of the deep diffused APD at a given voltage is obtained from the measured doping profile [5] and is used to calculate the ionization rates using a closed-form analytical expression which approximates Baraff's model [6,7,8]. The ionization rates are used to calculate M and F(M) numerically from McIntyre's formulation [9], which are then used in Eq.1 to obtain SNR as a function of M, or, for a given value of M – and consequently F(M) – as a function of either N or t. Owing to the theoretical calculation of ionization rates, the approach described above allows for an explicit analysis of the temperature dependence of APD performance and is valid over a wide range of electric field; this provides a significant advantage over methods employing empirical descriptions of ionization rates that constrain the analysis to a single temperature and, generally, to a narrow range of electric field.

Table 1 depicts the signal-to-noise ratio for various cases calculated using the above procedure. The third column shows the signal-to-noise ratio for a signal of ten photoelectrons

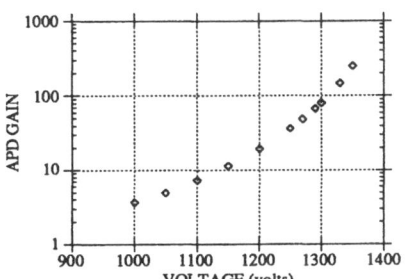

Fig.1 APD gain as a function of voltage.

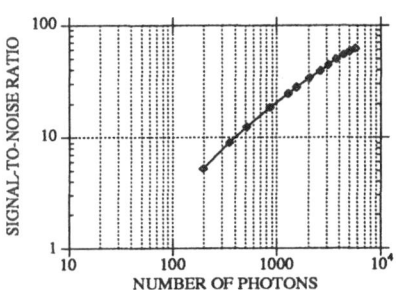

Fig.2 SNR as function of incident photons.

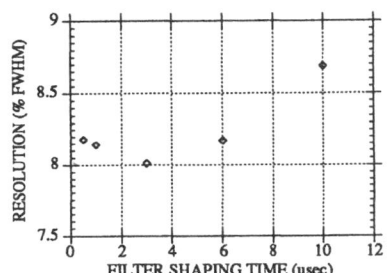

Fig.3 Measured resolution as a function of shaping time for a signal of 2000 photons and an APD gain of 81.

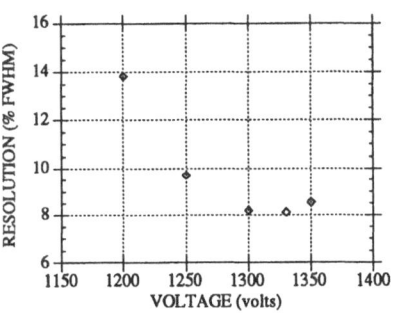

Fig.4 Measured resolution as a function of APD gain for a signal of 2000 photons and shaping time of 6μsec.

Table 1. Signal required to obtain a signal-to-noise ratio greater than unity and the signal-to-noise ratio for a signal of ten photoelectrons for several cases.

	N FOR SNR>1	SNR FOR N=10
M=81, τ=6μs, R_s=400KΩ	30	0.35
M=81, τ=3μs, R_s=400KΩ	27	0.39
M=81, R_s=1KΩ, τ=215ns	7	1.46
M=630, R_s=1KΩ, τ=22ns	3	3.62
M=150, R_s=1KΩ, τ=115ns	5	1.98
1x1mm^2, C_a=C_D, R_s=100Ω M=150, τ=115ns	3	4.29
1x1mm^2, C_a=C_D, M=160 τ=437ns, R_s=100Ω, T=0°C	1	10.16

and the second column gives the number of photoelectrons required to get a signal-to-noise ratio of greater than unity.

Row 1 models the conditions under which experimental data were taken; a reasonable agreement is found. A detailed analysis of the various noise sources shows that the largest contribution comes from the bulk dark current shot noise; reducing the filter shaping time should therefore improve the signal-to-noise ratio. For APD gain of 81, Eq.1 predicts an optimum shaping time (t_{opt}) of 3ms, which agrees with experimental results (see Fig.3). Table 1 (row 2) shows that the signal-to-noise ratio becomes greater than unity for a signal of 27 photoelectrons for this case.

The reason for the relatively large value of optimal shaping time is the large sheet resistance of the APD front contact. The large resistance (400KW) can be reduced to tens of ohms by improved processing techniques. If the series resistance is reduced to 1KW, Eq.1 gives t_{opt}=215ns. Table 1, row 3 depicts that a signal-to-noise ratio of greater than unity is now achievable with as small a signal as seven photoelectrons.

APD gain of 81 was chosen in the preceding calculations to reproduce the experimental conditions. SNR in Eq.17 is a simultaneous function of both M and t. There should therefore exist a point (M_{opt}, t_{opt}) for which SNR is globally maximized; this optimum can be obtained by solving

$$\frac{\partial f(M,\tau)}{\partial M} = 0 \qquad \text{and} \qquad \frac{\partial f(M,\tau)}{\partial \tau} = 0 \qquad (2)$$

simultaneously. Numerical solution of the above equations (with R_S=1KW) gives

$$M_{opt} = 630 \quad \text{and} \quad \tau_{opt} = 22 \text{ nsec}$$

Calculation of SNR with these optimal values shows that a signal of three photons gives a signal-to-noise ratio of slightly greater than unity and a signal of ten photons can be detected with a signal-to-noise ratio of 3.6 (Table 1, row 4). Further improvement is possible by matching the input FET of the CSA to the detector. However, it may not be practically feasible to operate the APD at a large value of gain, such as 630, because this would necessitate operation very close to the breakdown voltage where the gain sensitivity is large. Practically it is more feasible to operate the APD at a gain of 150 (1330 volts). This does not significantly jeopardize the optimal value of SNR, because the surface represented by $f(M,t)$ has a small gradient near the optimal point in the direction of decreasing M. For a gain of 150, Eq.1 gives t_{opt}=115ns. Table 1, row 5 shows that a signal of five photons gives a SNR of unity and a signal of ten photons gives a signal-to-noise ratio of nearly two.

All the measurements and calculations presented above were for 2 mm pitch devices. If, however, light is incident from thin fibres, it is possible to work with devices of a smaller cross-section; all leakage currents and capacitances then scale down, resulting in less noisy operation. Calculations indicate that if the APD area is scaled down by a factor of four and if the amplifier and detector capacitances are matched then, for a gain of 150 and a shaping time of 115ns, a signal-to-noise ratio of greater than unity is obtained at a signal of two photoelectrons (Table 1, row 6). The last row in Table 1 depicts the results of cooling the APD to 0C; under these conditions, the optimal shaping time comes out to be 437 ns for a gain of 160. It is seen that a signal-to-noise ratio of unity is now obtained for a single photoelectron, and a signal of ten photons can be detected with a signal-to-noise ratio of more than ten. Calculations further indicate that a single photoelectron would be detectable with a signal-to-noise ratio of more than four if the APD is cooled to dry ice temperature.

SUMMARY AND CONCLUSIONS

Measurements have been made on 3x3 silicon APD arrays fabricated by RMD. These measurements are in reasonable agreement with a simple noise model. The model has been employed to identify the limiting factors in the performance of existing devices, to obtain the optimal circuit operating point, and to possibly identify device-level improvements. These can be summarized as follows:

- The major limiting factor in existing devices is the series resistance of the front contact; this can be easily eliminated by improved processing techniques.
- The theoretical optimal set of gain and filter shaping time is $M_{opt}=630$ and $t_{opt}=22ns$ for a 2mm pitch pixel and the chosen circuit parameters (assuming that series resistance is reduced to 1KW).
- Practically, a more feasible set of APD gain and filter shaping time is M=150 (APD operating voltage of 1330 volts) and t=115ns. The signal-to-noise ratio approaches unity for this case for a signal of five photons at room temperature.
- By scaling the APD and the input device of the CSA by a factor of four, it appears possible to obtain a SNR greater than unity for a signal of three photons (for M=150 and t=115ns); by cooling the detector, it appears possible to detect a single photoelectric event inside the APD.

Although important issues such as radiation hardness and large volume manufacturability of APD arrays remain to be addressed, and the capability of APDs to detect a single photoelectron is yet to be demonstrated experimentally, the results presented in this paper make a strong case for further evaluation of APDs for calorimetry, fiber tracking, and low-level photon counting applications at the Super Collider.

REFERENCES

1. M. D. Petroff and M. G. Stapelbroek, "Photon-Counting Solid-State Photomultiplier," *IEEE Trans. Nuc. Sci.*, **NS-36**(1), 158-62, 1989
2. B. F. Levine and C. G. Bethea, "Single Photon Detection at 1.3 mm Using a Gated Avalanche Photodiode," *Appl. Phy. Lett.*, **44**(5), 553-55, 1984
3. P. Cushman and R. Rusack, "A PMT Using Avalanche Photodiodes," Conference Record of the 1992 IEEE Nuclear Science Symposium, Vol.1, p.278, 1992
4. A. N. Ishaque, "Noise Modeling of Deep-Diffused Avalanche Photodiodes," GE-CRD Internal Report, 1992
5. A. N. Ishaque, "Electric Field Profile and Potential Distribution in a Deep-Diffused Avalanche Photodiode," GE-CRD Internal Report, 1992
6. G. A. Baraff, "Distribution Functions and Ionization Rates in Semiconductors," *Phys. Rev.*, **128**(6), 2507-17, 1962
7. J. M. Pimbley, "Avalanche Ionization in the Presence of a Bandgap Discontinuity," GE-CRD Internal Report, 1991.
8. A. N. Ishaque, "Analytical Approximation of the Ionization Rate in Semiconductors," GE-CRD Internal Report, 1991.

9. R. J. McIntyre, "Multiplication Noise in Uniform Avalanche Photodiodes," *IEEE Trans. Electron Devices,* **ED-13**, 703-13, 1972.

THE PROTOTYPING AND EVALUATION OF A LIQUID LEVEL ALIGNMENT
MONITORING SCHEME FOR THE SDC DETECTOR

David P. Eartly[1] and John M. Bliven[2]

1 Fermilab, P.O. 500, M.S.205, Batavia, Illinois 60510
2 Fisher High School, Fisher, Minnesota 56723

INTRODUCTION

To know the real time vertical position and shape of the base, barrel toroid, and detector elements of the SDC detector for the Superconducting Super Collider (SSC), a distributed elevation monitoring system with **long term** stability is essential. We have designed, built, assembled, and tested six point and ten point matrices of interconnected point liquid level sensors and a reference reservoir using precision capacitive proximity sensing. We have tested three commercial sensors and have expanded our initial setup based on best test results. Long term measurements on the fixed position setups indicate long term raw sensor standard deviations of 15-30 μm and intrinsic systematic deviations of 5-10 μm as indicated by sensor differences at a given point. Temperature corrections will improve these numbers and a detailed correction study is in progress. The system appears adequate in all respects.

SENSOR SYSTEM UNDER TEST

The basic element of the system is a local liquid level sensor which we call the "pot". It consists of a small stainless steel vessel partially filled with a conductive fluid. An analog output, precision capacitive proximity sensor projects into it above the fluid. The sensor is positioned and held fixed by a flanged seal. The sensor measures the capacitance of the air gap to the fluid. The pot system is an adaptation of that first developed for low beta quadrupole and D0 detector platform monitoring at the Fermilab Tevatron by Hans Jostlein et al.[1,2]

Our modified pots are mounted in cradles which can be leveled and adjusted with precision bubble levels on a three point suspension relative to a reference surface. The pot assembly is shown in Figure 1. We tested probes and electronics units manufactured by Capacitec,[3] Mechanical Technology Inc. [4] and Lion Precision.[5] Units from the latter two companies had good linear output but the wet stability of the prototype units tested was unacceptable. The Capacitec sensor is a probe (with a guard ring) driven by a constant 18Khz current source. The fluid and pot act as a return by capacitively coupling through the air gap. The voltage on the probe is amplified, demodulated, and filtered into an output amplifier which gives 0-10 V dc nominal output. The output is nonlinear but can be made essentially identical from unit to unit. Nominal output sensitivity is 1.2 mV/μm.

Work supported by the U.S. Department of Energy under contract No. DE-AC02-76CHO3000 and contract No. DE-AC35-89ER40486.

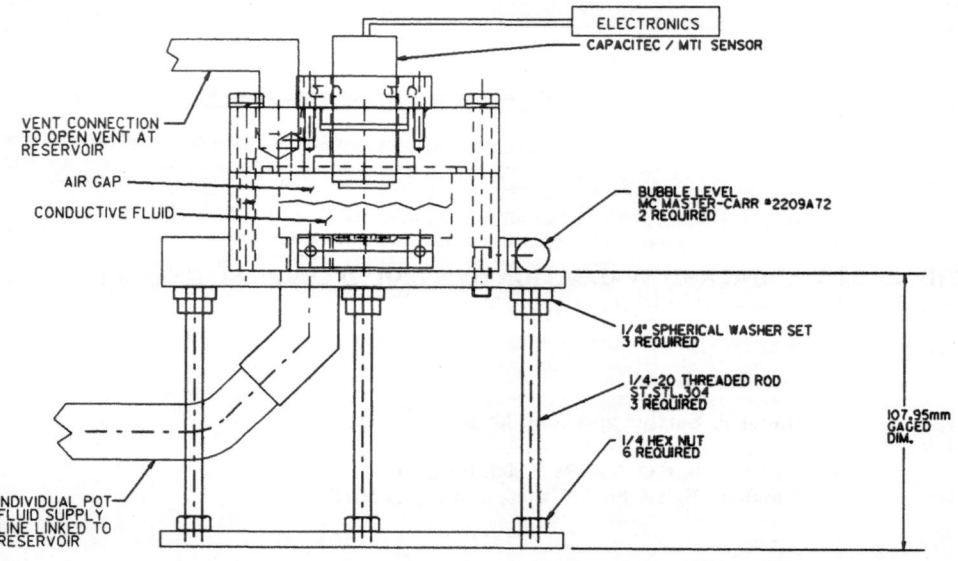

Figure 1. Liquid level pot - capacitive sensor assembly.

A typical sensor dry pot calibration curve is shown in Figure 2. We made a system test of ten channels with Capacitec HPC-500 probes and 410 SC electronics. Each pot uses a self contained single channel amplifier-demodulator card with an attached oscillator-driver. This electronics is local to the pot in a shielded box. DC voltages come from a remote chassis and sensor outputs are bussed there on up to 23m long cables.

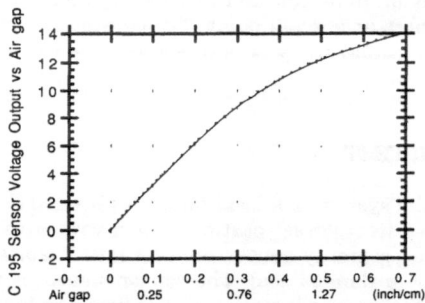

Figure 2. Capacitive sensor C 195 calibration curve.

System pots are interconnected to each other and a supply reservoir by fluid and vent lines. In the detector, the reservoirs and reference pots will sit on vertical monuments. Elevation changes of individual pots will result in a small volume of fluid flowing between pots and the fixed reservoir. RC circuit equivalents modeled in SPICE indicate short time constants (tens of seconds) for settling in the system configuration. The fluid is distilled water made conductive with KNO_3. For solution concentrations of 1-10% by weight the sensor response relative to the solution surface is comparable to the dry pot gap response. All fluid line mechanical connections are inclined so as not to trap air bubbles. All pots are sealed with their vent lines returning to the reservoir. The system is atmospherically vented only at the reservoir. The sensor heads (probe+shield) are humidity sealed with a thin baked on strain gauge epoxy. We embed a heater and a temperature sensor into each probe. The heater prevents condensation which significantly disturbs the gap readout. Sensors dipped into solution with resultant condensation do eventually recover to essentially the original calibrations. Sensor calibration is done by a precision vertical position scan of the sensor in a dry test pot. System calibration is done by varying the fluid level in the system of pots with a vertical position scan of the reservoir without dunking the sensors. The scale uncertainty is resolved from the sensor calibrations. In the test system, the interconnectiong lines are flexible (but hard) nylon tubing to physically decouple the pots. In the detector, lines will be rigid stainless steel with flexible bellows tube connections at the pots. In installation, each pot can be surveyed and adjusted into a defined position and the sensors set to a reference point. The system liquid level can be monitored by a sensor in the reservoir.

In the first version of the test system, six pots with sensors were set up in a small (1.5 x 3m) six point star configuration around the reservoir on a common rigid kinematically supported table. Tests on this configuration extended for six months. We have derived observed resolution, thermal dependences, and stability results; but movements of the system have not been unfolded.

In a second version of the test system, we have interconnected thirteen pots to a reservoir (one detector system equivalent), with ten pots instrumented. We will instrument the others with future test probes. Our first goal is to measure the intrinsic resolution and long term stability of a multi channel system with long cables and many oscillators. Here the pots are grouped by three (plus one reference pot at the reservoir) in a four node star cluster around the reservoir. The whole system is mounted **fixed** on a rigid table sitting on a three point kinematic support. The clusters are at the corners and the reservoir/reference pot at the center. Five precision inclinometers[6] are used to measure the pitch, roll, and yaw of the table. From these, we can determine the relative average elevation displacements of the corner cluster nodes. The system includes air and reservoir temperature monitors to track thermal cycling. Evaporation is compensated for by average system reading shifts. Long term system monitoring is being done via a Keithley multichannel scanning DMM system, IOTech488SCSI controller, and MAC computer. Using the computer clock, a reading is taken every few minutes over a period of ten-fifteen minutes and averaged into a data value. The sensors will be monitored this way for up to six months. Figure 3 illustrates the measured output of a sample fixed pot sensor C195 for a week period as compared to the temperature distribution. The output variation reflects the thermal fluid column height change and temperature dependence of the electronics. Figure 4 illustrates that output with a very simple, non optimized temperature correction. The fluctuations are reduced. The output calibration is 1.2 mV/µm. Residual raw fluctuations are on the order of 30 mV with standard deviations like 10µm as indicated in Table. 1 for different 6 pot runs. Table 2. lists sensor standard deviations for some ten pot runs with long cables. Sensor difference deviations for pairs of adjoining pots at one matrix position are included as estimates of raw systematic resolution.

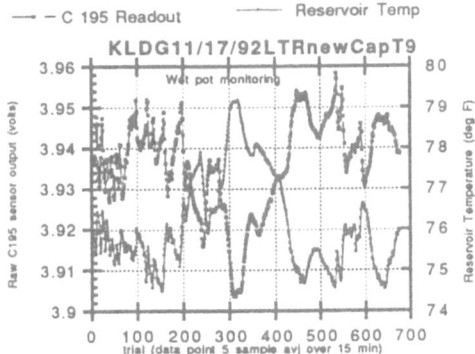

Figure 3. C 195 raw output over 170 hours versus temperature over the same period.

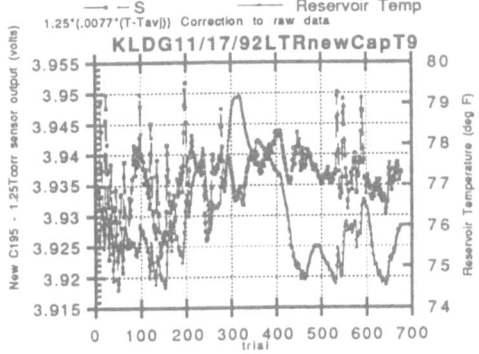

Figure 4. C 195 simply corrected output versus temperature for the same run.

Table 1. Standard deviations of stability measurements on the Capacitec six-pot matrix

Unit	DT=9°F; 2.5 mm gap; 2086 points; 10 minutes each	DT=2°F; 2.5mm gap; 1500 points; 10 minutes each	DT=6°F; 2.5mm gap; 680 points; 15 minutes each	DT=6°F; 4.3mm gap;1074 points; 15 minutes each
C191	10.9 mm	4.7 mm	9.8 mm	14.2 mm
C192	16.1 mm	10.6 mm	8.8 mm	17.8 mm
C196	9.8 mm	1.9 mm	9.2 mm	7.9 mm
C197	14.2 mm	3.9 mm	9.7 mm	15.6 mm
C198	11.4 mm	3.8 mm	8.6 mm	11.2 mm
C195	13.1 mm	6.4 mm	9.5 mm	19.3 mm

Table 2. Standard deviations of stability measurements on the Capacitec ten pot matrix

Unit	DT=7°F; 5mm gap;150 points, 10 minutes each- SHORT RUN		DT=7°F; 5mm gap; 1053 points, 10 minutes each-LONG RUN	
C196	6.1µm	C196-C210= 13.0µm	17.5µm	C196-C210= 5.1µm
C210	14.6µm		14.9µm	
C192	5.4µm	C192-C218= 5.9µm	17.2µm	C192-C218= 6.2µm
C218	3.6µm		14.6µm	
C198	6.3µm		13.4µm	
C191	4.2µm	C191-C209= 6.6µm	19.0µm	C191-C209= 14.1µm
C209	7.0µm		20.4µm	C191-C211= 6.5µm
C211	6.5µm		16.1µm	C209-C211= 9.1µm
C197	13.6µm	C197-C195=12.8µm	26.3µm	C197-C195= 10.5µm
C195	16.6µm		32.5µm	

CONCLUSIONS

The liquid level pot proximity sensors exhibit a good long term resolution (tens µm) and can be used in the SDC conventional magnet fringe fields. We will continue to test the system in a cluster and as a distributed matrix simulating part of the detector base. A thermal correction procedure will be developed to improve the resolutions. We have tested precision inclinometers[6] which have demonstrated the sensitivity and stability to monitor the local orientations of the magnet in sections to insure a 100 µm resolution on any magnet position. These will be coupled with a liquid level system to monitor the detector support system shape.

ACKNOWLEDGEMENTS

We want to thank Hans Jostlein for his valuable comments, review, and guidance. We wish to thank A. Lathrop for his help in getting us set up. The Fermilab Physics Department provided help and support, loan of equipment, and use of facilities for the long test setup.

REFERENCES

1. Hans Jostlein, Unpublished study, Fermilab Physics Department,Batavia, Illinois.
2. Hans Jostlein, The electronic alignment system at the Fermilab D0 detector, D0 Technical note 999, Fermilab, Batavia, Illinois.
3. Capacitec, P.O. Box 819, 87 Fitchburg Road, Ayer, Massachusetts 01432.
4. Mechanical Technology Inc, 968 Albany-Shaker Road, Latham, New York 12110.
5. Lion Precision Div, 563 Shoreview Park Road, St. Paul, Mn 55126.
6. David P. Eartly and Paul T. Johnson, Studies of prototype SDC muon alignment concepts and systems, SDC Technical note SDC-92-0195 Document Office, SSCL, DeSoto, Texas.

MECHANICAL-PROPERTY CHANGES OF STRUCTURAL COMPOSITE MATERIALS AFTER LOW-TEMPERATURE PROTON IRRADIATION: IMPLICATIONS FOR USE IN SSC MAGNET SYSTEMS[*]

John Morena[1], C. Lewis Snead, Jr.[2], C. Czajkowski[3] and John Skaritka[3]

[1]Ace, Inc.
Stuart, FL 34996

[2]Brookhaven National Laboratory
Department of Nuclear Energy
Upton, NY 11973

[3]SSC Laboratory
Dallas, TX 75237

INTRODUCTION

Long-term physical, mechanical, electrical, and other properties of advanced composites, plastics, and other polymer materials are greatly affected by high-energy proton, neutron, electron, and gamma radiation. The effects of high-energy particles on materials is a critical design parameter to consider when choosing polymeric structural, nonstructural, and elastomeric matrix resin systems. Polymer materials used for filled resins, laminates, seals, gaskets, coatings, insulation and other nonmetallic components must be chosen carefully, and reference data viewed with caution. Most reference data collected in the high-energy physics community to date reflects material property degradation using other than proton irradiations. In most instances, the data were collected for room- temperature irradiations, not 4.2 K or other cryogenic temperatures, and at doses less than 10^8-10^9 Rad. Energetic proton (and the accompanying spallation-product particles) provide good simulation fidelity to the expected radiation fields predicted for the cold-mass regions of the SSC magnets, especially the corrector magnets. We present here results for some structural composite materials which were part of a larger irradiation-characterization of polymeric materials for SSC applications.

EXPERIMENTAL

Specimens were supplied to BNL and SSCL by the vendors in either 2.5x.25x.25 or 2.5x.25x0.125 inch lengths. After characterization they were irradiated in liquid helium with 200-MeV protons to nominal doses of either 18^8 or 10^9 rad at the Brookhaven Radiation Effects Facility. The specimen temperatures during radiation did not rise above 20 K. They were then stored in liquid nitrogen until ready for mechanical testing at 4.2 K. Prior to the mechanical tests, the specimens were annealed at room temperature for one week. Standard ASTM short-beam-shear

---

[*]This work was performed under the auspices of the Superconducting Super Collider Laboratory.

tests were employed for the mechanical-property response to the irradiation. For experimental details, dosimetry, references, etc. for these and other materials in the program, see Refs. 1 and 2.

RESULTS

The data for the structural composites are presented in Table 1. The following Results and Observations are based upon these data.

Material #1, Specimen Group #1 (10^8 + 10^9 Rad): This thermoset composite is manufactured by Spaulding Composites. The standard deviation was low and constant, with the specimen color remaining the same after cold irradiation. Failure of the specimens was concentrated in the center with delamination and migration to the ends. Strength after 10^8 Rad increased 5%, and 13.5% after 10^9 Rad. It is reported that this material however is being modified by the original manufacturer as certain constituents of the formulation have been identified as "undesirable" for SSC and prime contractor applications.

Material #2, Specimen Group #2 and #3 (10^8 and 10^9 Rad): This thermoset composite material is manufactured by Allied Signal Corp. The "C" version indicates "cured" and the "P" version "post-cured". The purpose of including two (2) levels of cure was to see if strength would be affected following cold irradiation exposures. As discussed previously, proton irradiation like electron and other forms continue crosslinking of advanced composite materials further than generic materials which lose strength by degradation sooner. All specimens showed a slight darkening with failure concentrated in the center and migrating to ends. All standard deviations remained the same. The "C" version went up 5% after 10^8 Rad, the "P" version down 5% after 10^8 Rad, and up 5% after 10^9 Rad.

Material #3, Specimen Group #4 (10^9 Rad): This thermoset, a vinylester pultrusion is manufactured by Creative Pultrusion Corp. The specimens were cut like Material #3. Dark local discolorations were observed with multiple delaminations across specimens. Standard deviation stayed the same and strength dropped 25% after 10^9 exposures.

Material #4, Specimen Group #5 (10^8 + 10^9 Rad): This thermoset is manufactured by a convertor, Franklin Fiber Lamitex Corp. The "G-10" material specimens were cut from NEMA-grade laminate. As predicted the composite material strength went up 5% after 10^8 Rad, and dropped 58% after 10^9 Rad. All failures (single) were in the center of specimens. Standard deviation of 10^8-Rad specimens was 1/2 of unirradiated and 10^9 standard deviation was X2. Discoloration was localized for 10^8 Rad, and much darker in 10^9 samples.

Material #5, Specimen Group #6 (10^8 Rad): This thermoset is manufactured by Franklin Fiber also. Strength of this "G-11" material was the same after 10^8 Rad. A slight darkening was observed with multiple delaminations along the specimens. Standard deviation stayed the same.

Material #6, Specimen Group #7 (10^9 Rad): This thermoset composite is manufactured by Franklin Fiber also. It is a formulation of a highly crosslinked advanced-composite resin system. The strength went up 7% after testing. Fractures were not visible and standard deviation went up 0.2%. No color changes were recorded.

Material #7, Specimen Group #8 (10^9 Rad): This toughened thermoset composite is manufactured by Bryte Corporation. The mechanical strength decreased by 14% following the irradiation. No color change was recorded. Standard deviation went up .30 and failure was in center of specimens. Specimens were wet indicating a "leaching" of the toughening agent.

Material #8, Specimen Group #9 (10^9 Rad): This thermoset composite submitted as per Material #10. Failure of specimens in center. Standard deviation was up 0.43 without unirradiated test data.

OBSERVATIONS

The short-beam-shear test was used to screen all materials. Materials were selected from very generic thermoset chemistry categories like polyesters, vinylester to epoxies to phenolicotriazine, cyanate ester, polyimide and bismaleimide. Resin injection, resin transfer molding, liquid molding, compression, transfer, bulk molding compound, pultrusion, hand-lamination and other forms of thermoset materials and processes were considered during the selection process. Cost of materials and cost to process was also an important criterion. As always safety and environmental issues entered into the selection process.

The high-energy-physics community industry standard has been "G-10". It is known that because of the curing agent or hardener system used, the radiation resistance at high levels (10^8-10^9 Rads) will be poor. Also the mechanical and physical properties of the material exceed the requirements for some SSC applications. So it is suggested that instead of machining "G-10" for a non-structural application, one might select a more radiation-resistant material that can be "liquid molded", poured, or resin injected. Such applications include stand-offs, insulators, supports, etc.

Table 1. Structural Composites

Specimen Group	Manufacturer of Material	Type	Color Change	Strength After Cold Irradiation[*] (CI)	KSI After (CI)	Sample Condition after (CI)
1	Spaulding	Spaulrad	Same	Up 5% after 10^8	9.2	Failure in center and
			Same	Up 13.5% after 10^9	10.0	Migrates to sample end
2	Allied S.	Cryorad "C"	Slight darken	Up 4% after 10^8	4.8	Migrates to sample end
3	Allied S.	Cryorad "P"	Slight darken	Down 5% after 10^8	3.9	Failure in center and
			Same	Up 5% after 10^9	4.3	Migrates to sample end
4	Creative Pultrusions	1625	Local discolor, dark	Down 25% after 10^9	4.4	Multiple delams
5	Franklin Fiber	G10CR	Local discolor, dark	Down 58% after 10^9	20.8	Single failure in center of all samples
			Slight darken	Same after 10^8	8.3	
6	Franklin Fiber	G11	Slight darken	Same after 10^8	9.7	Multiple delams
7	Franklin Fiber	221CR		Up 7% after 10^{9**}	5.2	Fractures not visible
8	Bryte	EX1524	NA	Down 14% after 10^{9**}	8.1	Single (wet leak) failure in center
9	Permaglas	TE630		10^9 (No unirr.)	13.4	Single failure in center

[*]All mechanical tests are short-beam shear.
[**]Unirradiated baseline data supplied by CTD.

The "structural composites" testing section of the program also proved that most of the more highly crosslinked polymeric and composite materials continued to get stronger (continued crosslinking). The more generic non-aromatic cured (aliphatic) lost strength and this is an indication that molecular bonds were broken or severed. In all cases except "G-10" the loss was less than 50%. If we use the same selection criterion as "adhesive systems" and "filled and unfilled resin systems", we could select the materials tested for applications and radiation zones where we know they will not degrade greater than a certain amount. In other words, if we plan to use a material as a "neat" polymer or composite in a known radiation environment, we can select the appropriate material for that radiation zone.

J. Morena is preparing a design guide[3] that will be available for circulation shortly. The Design Guide will list all common and known nonmetallic materials along with their radiation-resistance levels. This Guide should assist with the selection of composite and polymeric materials for specific radiation-environment applications.

ACKNOWLEDGMENTS

We would like to thank CTD for their participation in providing span widths for the materials in the SBS tests, and for providing data on some of the unirradiated specimens. Special thanks goes to Jack Rothmann who oversaw the irradiations and the accompanying dosimetry, and to Tom Roberts for carrying out the mechanical testing. Assistance from Kim Costin and Amanda Spindel from SSCL is also appreciated.

The information analyzed and reported herein is based upon the estimated performance of a material as it is envisioned for use in actual production applications and environments. It should be noted, however, that ACE, Inc., ACMLC, Inc., John J. Morena, BNL, SSCL and others involved in the subject screening and testing program assume no responsibility for the interpretation or misrepresentation of any data, information, or analysis resulting from the reported results of these tests.

REFERENCES

1. C. Lewis Snead and John J. Morena, "Radiation Testing of Superconducting Corrector-Magnet Organic Polymeric Materials," ATD/SSC-92-2, December 24, 1992.
2. C. Lewis Snead, Jr., John Morena, C. Czajkowski, and John Skaritka, "Mechanical-Property Changes of Polymeric and Composite Materials after Low-Temperature Proton Irradiation," (to be published).
3. "Design Reference Charts for Composites, Plastics, and other Materials Used in High-Energy-Radiation SSC Environments," ACMLC Technical Guide (in preparation).

CORRECTOR/QUADRUPOLE/SEXTUPOLE POWER LEADS FOR THE RELATIVISTIC HEAVY ION COLLIDER AT BROOKHAVEN NATIONAL LABORATORY[1]

R. Shutt, K. Hornik, and M. Rehak

RHIC Department
Magnet Division
Brookhaven National Laboratory
Upton, NY 11973

INTRODUCTION

In RHIC (Relativistic Heavy Ion Collider), there are 492 CQS (Corrector/Quadrupole/Sextupole) assemblies which require leads to carry the current from the power supply to the magnet. The lead assemblies will contain these leads along with instrumentation voltage taps and current carrying wires that are used only for magnet warm-up. These lead assemblies are analyzed for two cooling schemes: 1) gas flow through the lead tube and 2) heat sinking the lead tube along a 40-70 K heat shield (without gas flow). The analysis was extended to include the modeling of the cold and warm ends and effects of superinsulation shielding the lead assembly against radiation (including heat conduction due to residual gas pressure in the surrounding vacuum). Extensive parametric studies of heat exchange areas, specific copper properties, length of the lead, etc. are also included in the analysis.

HEAT SINK vs. HELIUM FLOW

An initial fixed end temperature analysis (cold end set to 4.4 K, warm end set to 293 K) was performed on the lead assembly for both the heat sinking and helium flow cooling schemes. The heat sink temperature was set at 50 K for this initial comparison. The optimal design should have a minimal heat load on the refrigeration system. This means for the gas flow case, the heat leak at the cold end (Q_{in}) should be minimized while the mass flow is also minimized. This is represented by

$$Q_{tot} = Q_{in} + 80 \times M \qquad (1)$$

where 80 is a factor used to translate helium mass flow into heat load on the refrigerator. For the heat sink case, it is necessary to minimize Q_{in} while minimizing the length of the attachment to the shield in order to reduce the load on the refrigerator. Results of the heat sink analysis with the shield temperature set at 50 K show that the heat leak at the cold end is close to, and

---

[1]Work performed under Contract No. DE-AC02-76CH00016 with the U.S. Department of Energy.

in some cases, exceeds the total refrigeration load of the gas flow case. The lead would have to be very long, on the order of 5 or 6 meters, for this type of lead to be considered. When the shield temperature is set to 60 K, the performance is further reduced, and the refrigeration load on the heat shield is increased. Due to the performance indicated by the computer analysis, the difficulties presented with the installation of a 5 or 6 meter long lead, and uncertainty in the precision of the thermal contact that could be achieved between the lead tube and the heat shield, gas flow cooling was chosen.

HELIUM FLOW- FINAL DESIGN

The wires in this lead are such that two carry 100 A each and eight carry 50 A each. An analysis with end temperatures fixed at 4.4 K and 293 K was performed to determine an optimal length within a range that could reasonably fit within the CQS cryostat. According to Table 1, the optimal length would be 110 cm. However, a final lead length of 90 cm was chosen, based on these results and physical restrictions on length due to space considerations. The lead should be long enough for the heat leak (Q_{in}) to be a minimal value, but short enough to fit inside the CQS cryostat assembly.

Table 1. Fixed end temperature analysis of CQS lead (4.4K at cold end, 293 K at warm end)

Length of Lead,cm	M_{crit}, g/s	Q_{in}, W	Q_{tot}, W
80	0.033	1.00	3.65
90	0.034	0.60	3.32
100	0.034	0.48	3.20
110	0.035	0.31	3.11
120	0.036	0.25	3.13

With the length of the lead determined, the end conditions were modeled to see the performance of the complete set-up. The 90 cm length refers to the length of wire enclosed in stainless steel flexible tube of 1.905 cm diameter. At the cold end, a length of wire extends past the end of the flexible tube containing the lead assembly into a large interconnect volume filled with helium. The lead wires are attached to superconducting wires in this volume. The interconnect region is modeled as an infinite volume of helium at 4.4 K that freely convects the heat from the copper wires. The helium gas flow in the flexible tube is modeled as forced convection of helium which removes the heat from the wires. At the warm end, the wires go through an expander and T-joint, where the helium leaves. The copper wires are then spliced to larger copper pieces that are insulated to separate the electric current from ground. Finally, the current passes through large cables that are attached to the power supply. These cables and the insulated copper pieces are exposed to free convection of air. Figure 1 is a schematic of the gas flow case.

RESULTS

Studies were made to determine the effects of material properties, heat exchange area, and sensitivity to exact lengths of each of the sections. Table 2 shows the effect of material properties on lead performance. This shows a great sensitivity of lead performance to the material used, particularly to the relative resistivity ratio, RRR. Q_{in} could be minimized with a very impure copper, but the mass flow would be very high. The mass flow could be minimized with a very pure copper, but Q_{in} would be very high. In most cases, Q_{tot} should be minimized. This is possible over a range of RRR. For this particular case, the RRR should not exceed 120, but cannot be less than 80 due to mass flow constraints on this lead.

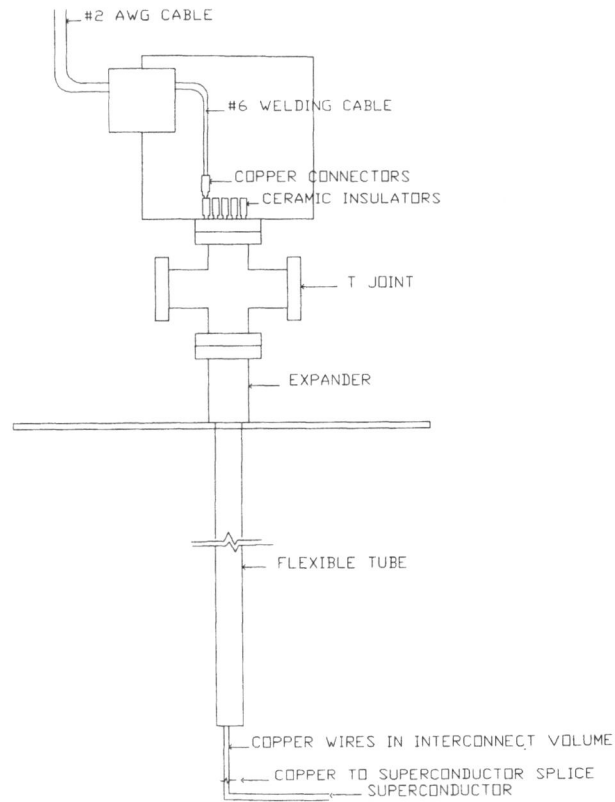

Figure 1. CQS gas flow assembly schematic.

Table 2. Material property study. All results for lead with end conditions included, M=0.036 g/s. Q_{lead} is the heat leak into the interconnect volume, T_{he} is the helium temperature, T_1 is the temperature of the 100 A wires, T_2 is the temperature of the 50 A wires. All temperatures are at the warm end of the flexible tube.

Material	Q_{in},W	Q_{lead},W	Q_{tot},W	T_{he},K	T_1,K	T_2,K
ETP (RRR=100)	0.08	1.00	2.96	213	290	236
OFHC (RRR=106)	0.04	0.56	2.92	230	304	251
RRR=80	0.22	2.27	3.10	225	301	247
RRR=120	0.41	2.80	3.29	202	280	227
RRR=200	0.85	3.72	3.75	182	262	210

A concern in this lead design is the amount of the wire that will actually be exposed to the flowing helium. The wires, when the lead is assembled, are twisted together. Therefore, the entire perimeter is not exposed to cooling helium. Table 3 shows the results as the percentage of the perimeter exposed to flowing helium is varied. the less exposure to flowing helium, the warmer the copper becomes. For this model, a conservative estimate of 75% is applied.

Table 3. Study of percentage of wire perimeter exposed to helium flow. All run at 0.036 g/s with ETP as the material.

% area	Q_{in}, W	Q_{lead}, W	Q_{tot}, W	T_{he}, K	T_1, K	T_2, K
100	0.05	0.63	2.93	194	244	212
90	0.06	0.73	2.94	198	257	218
80	0.07	0.89	2.95	206	276	228
75	0.08	1.00	2.96	213	290	236
70	0.09	1.15	2.97	224	308	247

A study to determine the number of layers of superinsulation needed to keep the effects of radiation on lead performance at a minimum was made. The worst case for radiation would occur if the inside surface of the vacuum chamber were at 293 K. Heat would be radiating into the lead tube along the entire length. With one layer of superinsulation, Q_{lead} is 1.7 W. After fifteen layers, Q_{lead} remains constant at 1.0 W. If the lead was placed in an environment where radiation was not a factor, $Q_{lead} = 0.99$ W. Therefore, it is shown that at least 15 layers of insulation are necessary to keep the effects of radiation on lead performance at a minimum.

A sensitivity study was done in which the length of each section of the lead was varied individually to see the effect on the end lead temperature. The results of these analyses were evaluated and a worst case for burn-out and a worst case for freezing were chosen. Burn-out is defined as the point where the temperature of the lead reaches 450 K. The worst case for freezing would indicate the temperature of the helium and stainless steel tube at the exit of the vacuum chamber and give an idea of the likelihood of condensation or frost forming on the outside of the stainless steel. An analysis where the ends of the leads are fixed at 4.4 K and 293 K does not indicate the presence of condensation on these assemblies. The actual model of the end conditions must be included for this situation to become apparent. Table 4 compares the two worst cases to the proposed design. If the worst case for burnout was actually built, the temperature would still not reach 450 K.

Table 4. Proposed case, worst cases for both freezing and burn-out. All run at M = 0.036 g/s

case	Q_{in}, W	Q_{lead}, W	T_{he}, K	T_1, K	T_2, K
proposed	0.08	1.00	213	290	236
worst/burn	0.01	0.51	265	348	283
worst/freeze	1.61	4.03	180	249	213

CONCLUSIONS

In order to get an accurate picture of lead performance, the entire lead, including specific end conditions, must be modeled. An analysis where the ends are arbitrarily fixed at 4.4 K and 293 K does not indicate the presence of condensation, nor does it indicate the true temperature of the lead as it exits the cryostat. The lead does not end where the helium exits the system. The surrounding equipment must also be modeled in order to find the true temperature of the lead.

The model shows that different RRR and types of coppers can behave very differently. This is an important factor in determining the heat load on the refrigerator and how the heat is dissipated.

5 Parallel Technical Sessions III

DIAMOND DETECTORS FOR THE SSC

K. K. Gan,[1] H. Kagan,[1] R. Kass,[1] R. Malchow,[1] W. Palmer,[1] C. White,[1]
S. Zhao,[1] L. S. Pan,[2] S. Han,[2] D. Kania,[2] M. Lee,[3] S. Kim,[3] F. Sannes,[3]
S. Schnetzer,[3] R. Stone,[3] G. Thomson,[3] Y. Sugimoto,[4] A. Fry,[5] S. Kanda,[6]
and S. Olsen[6]

[1]Department of Physics, The Ohio State University, Columbus, OH 43210
[2]Laser Division, Lawrence Livermore National Laboratory
  Livermore, CA 94550
[3]Department of Physics, Rutgers University, Piscataway, NJ 08854
[4]KEK National Laboratory, Tsukuba-shi, Ibaraki-ken, Japan 305
[5]Physics Division, Superconducting Super Collider Laboratory
  Dallas, TX 75237
[6]Department of Physics, University of Hawaii, Honolulu, HI 96822

ABSTRACT

Diamond is well suited as a particle detector in the high rate and high radiation environment of the SSC. The use of diamond is made possible by recent developments in the chemical vapor deposition (CVD) growth process. CVD diamonds have been studied using radioactive sources and test beams. The measured charge collection distance of CVD diamonds now exceeds that of natural diamond. No degradation of signal is observed up to a rate of 10^4 particles $cm^{-2}s^{-1}$. Exposure to stopping 5 MeV α particles shows no radiation damage with a dose of up to 10^{13} particles cm^{-2}. Prototype diamond/tungsten and silicon/tungsten calorimeters have been constructed and tested in an electron beam at KEK. The energy resolution of the diamond/tungsten detector is comparable to the silicon/tungsten calorimeter.

INTRODUCTION

The Superconducting SuperCollider poses an immense challenge to particle detection technology. At the design luminosity of $10^{34} cm^{-2}s^{-1}$, with two 20 TeV proton beams colliding every 16 ns, there are $\sim 10^4 - 10^6$ particles per second traversing the detector at a distance of 10 cm from the interaction region. Diamond offers a possible solution for this high rate and high radiation environment.

Diamond can operate in this hostile environment due to the strong bonding energy between atoms. This strong bonding results in large band gap which allows the application of a large electric field across the diamond and no p-n junction is necessary to operate the device. The large electric field sweeps the induced charge to the collection surface at high speed. For a detector with a thickness of a few hundred microns, the charge collection time is ~ 1 ns.

The practical use of diamond as a particle detector is made possible by recent developments in the chemical vapor deposition (CVD) process. This process allows diamond to be produced economically over a large area with high purity in comparison to natural diamond. During the last two years, our group has initiated a systematic study of the electronic properties of CVD diamond and correlated them with various growth parameters.

The diamonds samples [1] under study have a cross section area of 1-10 cm^2 with a typical thickness of $\sim 200\mu m$. Electrodes (Ohmic contacts) were placed on both sides of the diamonds by sputtering Ti/Pt/Au [2] or thermally evaporating Cr/Au [3] onto the surfaces. The diamond signal was amplified with a charge sensitive preamplifier [4] followed by a shaping amplifier. A typical signal corresponds to ~ 1000-2000 electrons with noise corresponding to ~ 400-600 electrons equivalent.

CHARGE COLLECTION DISTANCE

The quality of the diamonds can be characterized by the charge collection distance. This is the average drift distance, d, that the electrons and holes move apart before being captured by impurities, defects, or traps. The collection distance is related to the sample thickness t and charge collected Q by,

$$d = \frac{Q}{Q_0}t \quad ,$$

where Q_0 is the charge generated by the ionizing radiation. Q_0 has been calculated using two methods which yield consistent results: (1) by normalizing the diamond pulse height to that of silicon after correction for solid angle and dE/dx, (2) using the EGS Monte Carlo program. In the calculations, 13 eV is used as the energy required to produce an electron-hole pair. Figure 1 shows a comparison of the charge collection distances for CVD and natural diamonds. The samples were measured using a ^{90}Sr radioactive source which produces a β spectrum with a maximum energy of 2.28 MeV. The collection distance of the CVD diamond exceeds that of the natural diamond. In Fig. 2, we chronicle the dramatic improvement in the electronic quality of the CVD diamonds in recent years.

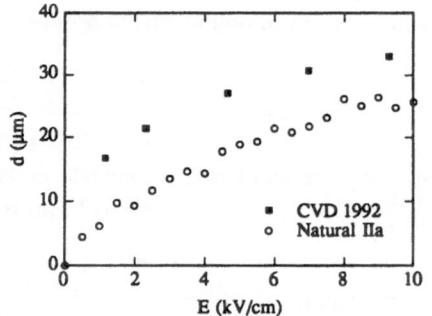

Figure 1. Charge collection distances in CVD and natural diamonds.

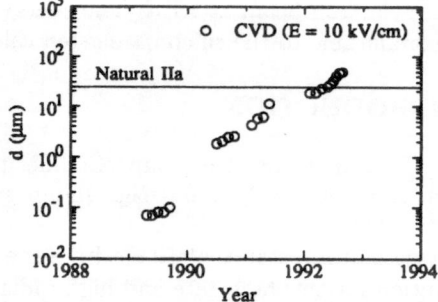

Figure 2. History of charge collection distances in CVD diamonds.

RATE AND RADIATION DAMAGE STUDIES

The rate capability of diamond has been investigated using a test beam at TRIUMF in Canada. The beam line used was M13, a 100 MeV secondary beam line of electrons (minimum ionizing), muons (twice minimum ionizing), and pions (three times minimum ionizing). Figure 3 shows the rate dependence of the pulse height of a CVD diamond. No degradation of signal was observed at a rate of up to 10^4 particles $cm^{-2}s^{-1}$. This is the first experimental indication that diamond will withstand the high rate environment of the SSC.

The radiation hardness of diamond has been studied using the 5 MeV α particle beam at Los Alamos National Laboratory. At this energy, the α particles deposit ~ 1000 times the minimum ionizing energy within the mean range of $12\mu m$. No degradation of pulse height was observed for a dose of up to 8×10^{13} particles cm^{-2}. Figure 4 shows a comparison of the pulse height[6] in a natural diamond before and after stopping 8×10^{15} particles cm^{-2}; the reduction in pulse height is $\sim 1/3$. For doses between 8×10^{13} and 8×10^{15} particles cm^{-2}, the reduction in signal is approximately linear. This study indicates the radiation hardness of the diamond. Further studies of the rate capability and radiation hardness are planned.

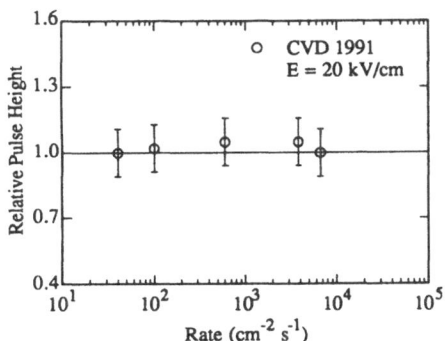
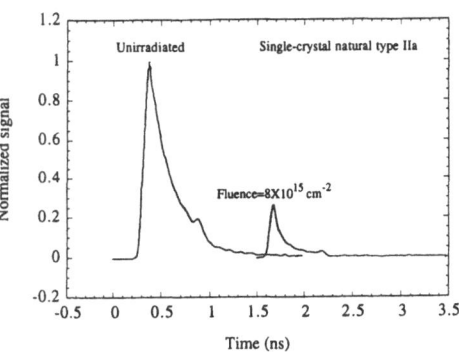

Figure 3. Pulse height dependence on rate, normalized to the lowest rate data point.

Figure 4. Normalized pulse height after high dose α exposure.

ELECTROMAGENTIC CALORIMETER

A prototype electromagnetic calorimeter has been constructed to test the feasibility of diamond as a radiation detector. The calorimeter was designed to use the 5 GeV electron test beam at KEK. Extensive simulation of the electron shower using the EGS Monte Carlo led to the design of a 3 cm $\times$ 3 cm $\times$ 20 X_0 prototype diamond/tungsten calorimeter. A similar calorimeter of silicon/tungsten sandwich was also constructed for a direct comparison. For the tests described below the first 7 layers, which intercept 70% of the shower, were diamond and the remaining layers silicon. The layout of the electrode pattern on the diamonds and silicons is shown in Fig. 5. The electronic instrumentation was identical to that used in the radioactive source test. The calorimeter response was found to be quite linear for both diamond and silicon. A preliminary analysis of the energy resolution of the silicon calorimeter, shown in Fig. 6, yields

$$\frac{\sigma_E}{E} = \frac{20.5\%}{\sqrt{E}} \oplus 2.6\% \quad ,$$

where the constant term is due, in part, to the longitudinal shower leakage. As shown in Fig. 6, the energy resolution of the diamond is comparable to that of the silicon. This is the first demonstration of the feasibility of a diamond calorimeter.

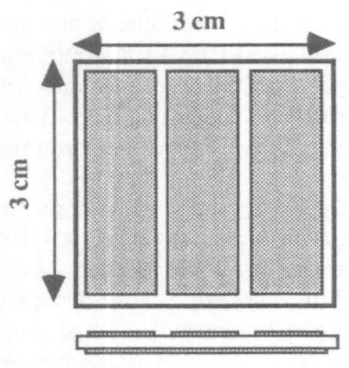

Figure 5. Electrode pattern on the diamond radiator.

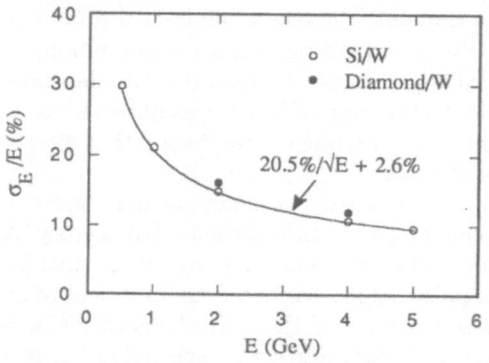

Figure 6. Comparison of the energy resolution of diamond and silicon calorimeters.

SUMMARY

We have made tremendous progress in the last few years in demonstrating the feasibility of using diamond as a particle detector. The quality of CVD diamond now exceeds natural diamond in terms of the collection distance. Preliminary results indicate that diamond can handle the high rate and high radiation environment of the SSC. The energy resolution of the prototype diamond/tungsten calorimeter constructed is found to be comparable to that of a silicon/tungsten calorimeter.

ACKNOWLEDGEMENTS

This work is supported, in part, by grants from the Texas National Research Laboratory Commission (RGFY9164 and RGFY9264), the Department of Energy through the SSC Research Program (92/32100), and the Department of Energy through a grant to Lawrence Livermore National Laboratory (W-7405-ENG-48).

REFERENCES

1. CVD diamond films used for this study were fabricated by Norton Diamond Films and Crystallume Inc.

2. L. S. Pan et al., to be published by J. Appl. Phys. (1993).

3. S. Zhao, Ph.D. Thesis, The Ohio State University (1993).

4. T. Taniguchi, Y. Fukushima, and Y. Yoribayashi, IEEE Trans. Nucl. Sci. **NS-36**, 657 (1989).

5. M. Franklin et al., Nucl. Instr. and Meth. **A315**, 39 (1992).

6. The signal was recorded using a photoconductivity technique: e-h pairs were created by illuminating with a UV laser on a 1 mm space between two electrodes sputtered on the same side of the diamond.

HIGH RATE RESISTIVE PLATE CHAMBERS: AN INEXPENSIVE, FAST, LARGE AREA DETECTOR OF ENERGETIC CHARGED PARTICLES FOR ACCELERATOR AND NON-ACCELERATOR APPLICATIONS

C. R. Wuest, E. Ables, R. M. Bionta, O. Clamp, M. Haro, G. J. Mauger,
K. Miller, H. Olson, P. Ramsey

Lawrence Livermore National Laboratory
P. O. Box 808
Livermore, CA USA 94551

INTRODUCTION

Resistive Plate Chambers, or RPCs, have been used until recently as large area detectors of cosmic ray muons. They are now finding use as fast large-area trigger and muon detection systems for different high energy physics detectors such as the L3 Detector at LEP and future detectors to be built at the Superconducting Super Collider (SSC) and at the Large Hadron Collider (LHC) at CERN.[1,2] RPC systems at these accelerators must operate with high efficiency, providing nanosecond timing resolution in particle fluences up to a few tens of kHz/cm^2 – with thousands of square meters of active area. RPCs are simple and cheap to construct.

We report here recent work on RPCs using new materials that exhibit a combination of desirable RPC features such as low bulk resistivity, high dielectric strength, low mass, and low cost. These new materials were originally developed for use in electronics assembly areas and other applications, where static electric charge buildup can damage sensitive electrical systems. An accompanying paper presented at this conference, "Behavior of Large Resistive Plate Counters at High Rates," describes complementary work carried out at MIT on "standard" RPCs using resistive materials such as Bakelite.[3]

RESISTIVE PLATE CHAMBER OPERATION

A sketch of an RPC is shown in Figure 1. It consists of a 2 mm gas-filled region sandwiched between resistive plates coated with resistive electrodes. An electric potential applied to the electrodes forms a field in the gas of about 4-5 kV/mm. The electrodes typically have surface resistivities of a few 100 kΩ/cm^2. The resistive plates typically have a volume resistivity of 10^{11} Ω-cm and have been made in the past with plasticized PVC, Bakelite (phenolic), or resistive glass.[4-6] The passage of a charged particle, such as a muon, causes ionization to occur in the gas, leading to a rapid electrical discharge in the region around the ionization. The resistivity of the electrodes, the plates, and the properties of the gas, act to rapidly quench the discharge, giving very fast pulses with high amplitude. RPC operation is very dependent on the gas mixture. We use RPC gas mixes of argon, isobutane and Freon 13B1 in a ratio of about 60%/38%/2%, respectively. The "penalty" one pays for desirable RPC behavior is a corresponding depletion of the electric field near the discharge,

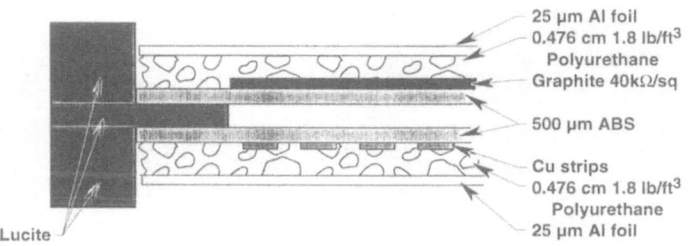

25 μm Al foil
0.476 cm 1.8 lb/ft³
Polyurethane
Graphite 40kΩ/sq

500 μm ABS

Cu strips
0.476 cm 1.8 lb/ft³
Polyurethane
25 μm Al foil

Lucite

Figure 1. Cross-section of an RPC.

leading to inefficient behavior of RPCs at large particle fluxes. The time of the recharge is dependent on the resistivity of the RPC plates.

RPC signals are read out using metal pick-up strips, typically 1 cm wide and a few meters long. Large (~ 500 mV into 50Ω) output pulses are derived from RPCs that can drive standard electronics such as ADCs, TDCs or discriminators *without* any preamplification. Timing resolutions have been measured to be less than a few nanoseconds. Proposed RPC-based trigger systems provide fast, coarse position resolution of muon tracks using the signal and known position of the pick-up strip.

EXPERIMENTAL RESULTS

Reference 3 describes the behavior of "standard" RPCs. The conditions in SSC and LHC environments demand RPCs capable of operating at much higher rates than currently available with standard designs. This is because of the large background of neutrons and charged particles expected at these accelerators. Neutron fluxes of about 10^4–10^5 Hz/cm² are expected, *e.g.*, in the GEM detector at standard luminosity. Bakelite RPCs have a measured efficiency for neutrons of about 0.5%, which gives singles rates due to neutrons of about 50–500 Hz/cm². However, Bakelite RPCs only have a rate capability, at 97% efficiency, of about 50-100 Hz/cm². Therefore, RPCs used in muon detector systems, must have higher rate capability by a factor of 10–100. Our R&D program indicates that this rate capability is now understood to be solely a function of the bulk resistivity of the resistive plate material.

To develop a more reliable detector with the high rate capability necessary for proposed detector systems, we have studied alternative materials for RPCs. We have built a number of RPCs out of specially formulated static-dispersive plastic materials that exhibit a number of desirable RPC properties including large amplitude, nanosecond pulses with very low noise and high rate capability. Since the plastic sheets are thin – 500 μm or less – RPCs using these materials have low mass, with about 0.5% radiation length per RPC.

Figure 2 shows a SPICE model of an RPC. The resistive plates are modeled as a resistor and capacitor in parallel, with the resistor value determined by the measured bulk resistivity and thickness of the material. Table 1 gives parameters for different RPC materials under study. The gas gap is modeled as a simple capacitor. The model is used to predict the dynamic behavior of the RPC when a spark discharge occurs in the gas gap. A characteristic "recovery" time is predicted, associated with the recharging of the electric field in the plate after the discharge occurs. This recharge time varies widely for different materials. Note that the curves in Fig. 2 show the recharge time of the region near the spark discharge and do not represent the signal pulse shape out of the RPC. We define the recovery time as the time between the 10% and 90% amplitude points on the curves of Fig. 2.

The recharge time is directly related to the saturated rate capability of the RPC following a simple model that gives the rate of an RPC at a fixed efficiency given the saturation counting rate. This model usually underestimates the rate for high efficiencies. Let Rs = saturation rate, Re = rate at a fixed efficiency E then, $Re = Rs(1-E)$. This is only an approximation as it assumes a linear relation between saturation counting rate and the period of

Table 1. RPC Resistive Materials Properties

Material	Thickness (cm)	Bulk Resistivity (Ω-cm)	Arc Resistivity (Ω-cm^2)
MIT mirror glass	0.300	$5\ 00\times10^{12}$	1.50×10^{12}
LLNL mirror glass	0.066	4.90×10^{12}	3.23×10^{11}
Kodak glass	0.123	6.42×10^{11}	7.89×10^{10}
Italian RPC Bakelite	0.200	1.00×10^{11}	2.00×10^{10}
LLNL Bakelite	0.161	4.50×10^{9}	7.24×10^{8}
Abstat-M310 plastic	0.072	5.78×10^{9}	4.16×10^{8}
Abstat-M310 plastic	0.060	5.78×10^{9}	3.47×10^{8}
MiTech-411 plastic	0.090	2.03×10^{9}	1.83×10^{8}
MiTech-411 plastic	0.030	2.03×10^{9}	6.19×10^{7}
Corning 0211 glass	0.056	6.70×10^{7}	3.75×10^{6}

inefficiency of the counter. Assuming $E = 0.95$, a typical value, then for $Rs = 20$ kHz we can assume that the counter will operate with full efficiency at 1 kHz. Because of the above approximation this derived rate is actually an underestimate of the real rate.

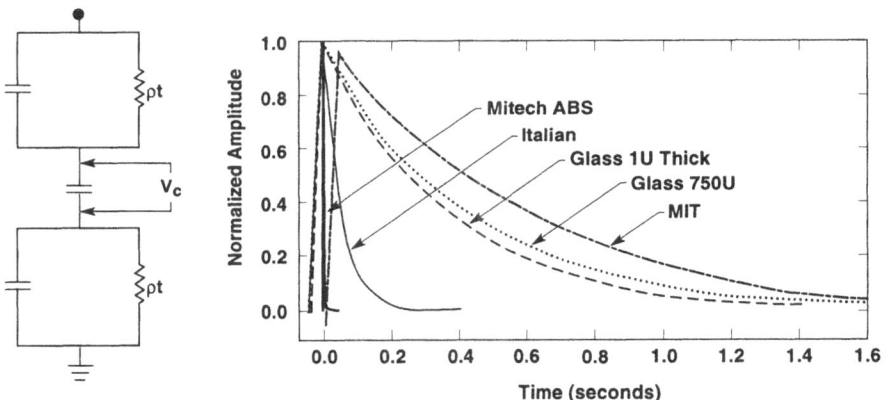

Figure 2. Equivalent circuit and SPICE model predictions of RPC recharge time for various materials. Note that this is *not* the same as the RPC pulse shape, which is much faster.

Figure 3 gives results for saturated rate measurements for different materials compared to the inverse of the predicted recharge time. The measured rates are proportional to the calculated time response of the material, indicating that the operation of the RPC is understood using this model and the material bulk resistivity. From Fig. 3 we see that two plastics formulated with conducting polymer exhibit very high saturated rate capability compared to glasses and Bakelite, with rates of about 15 kHz/cm^2. One plastic is an ABS-based plastic and the other is a PVC-based plastic. We have measured the dielectric strength of the ABS plastic to be superior to the PVC plastic. A number of RPCs constructed from ABS and PVC plastic have been built and tested (the largest being 1.2 m × 2.4 m).

Results from measurements of different RPCs indicate that plastic RPCs operate identically to Bakelite RPCs with regards to pulse width, pulse height, and rise-time jitter. In addition, noise measurements with plastic RPCs give a noise rate of about 0.5 Hz/cm^2 for a

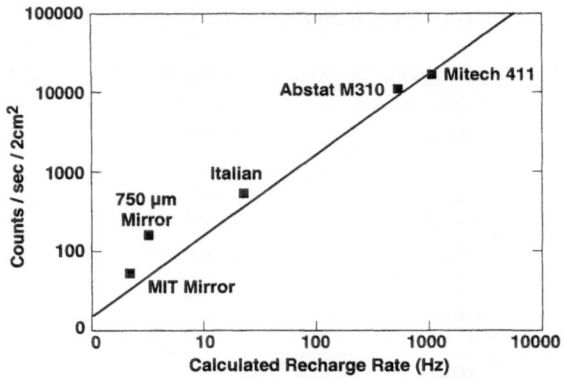

Figure 3. Comparison of SPICE calculated RPC rate capability with measured rates for a number of different RPCs. The RPCs plotted are, in order of increasing rate, MIT glass, LLNL glass, Italian Bakelite, Abstat-M310, and MiTech-411. The linear fit to the points is also indicated.

15 mV discriminator threshold. This is achieved for plastic sheet without any special surface preparation, compared to Bakelite chambers, which require a special coating of linseed oil on the inner surfaces in order to operate properly.

The ABS RPC does exhibit some variability in operation associated with initial turn-on and subsequent voltage changes. This is manifested as a decrease in noise rate over a characteristic settling time of the order of hundreds of minutes. Once the RPC is stabilized the noise rate is the value quoted above. Another characteristic of the ABS plastic is a variability in saturated count rate associated with turn-on and subsequent voltage changes. This appears to be associated with changing bulk resistivity of this particular plastic and is possibly due to polarization, or temperature/humidity effects. We have seen that the saturated rate capability of an ABS RPC is reduced from about 10 kHz/cm^2 to about 1 kHz/cm^2 over a period of a few hours. At this point the RPC is stable, as shown by measurements with constant voltage over a 6-day period.

There are a number of other semiconducting plastics (including epoxy paints) that are being studied and it is expected that the variability discussed above will be solved by the choice of the proper material. Plastics with even lower bulk resistivities (about 10^8 ohm-cm) have recently been identified. Based on our understanding of RPC operation, these lower resistivity plastics will provide even better RPC operation with correspondingly higher saturated rate capability – by as much as a factor of ten over ABS plastic.

ACKNOWLEDGEMENTS

This work was performed under the auspices of the US Department of Energy by the Lawrence Livermore National Laboratory under Contract W-7405-ENG-48.

REFERENCES

1. GEM Collaboration "An Expression of Interest to Construct a Major SSC Detector," GEM-TN-91-1.
2. A. Bohrer, et al., "Status Report of the RD5 Experiment," CERN/DRDC/91-53, January 13, 1992.
3. M. Widgoff, et al., "Behavior of Large Resistive Plate Counters at High Rates," presented at the Fifth International Industrial Symposium on the Super Collider (IISSC), San Francisco, May 7 (1993).
4. G. Battistoni, et al., "Plastic Spark Counters with PVC Electrodes," Nucl. Inst. Meth. **A270**, p. 190, 1988.
5. R. Cardarelli, et al., "Progress in Resistive Plate Counters," Nucl. Inst. Meth. **A263**, p. 20, 1988.
6. M. Anelli, et al., "Glass Electrode Spark Counters," Nucl. Inst. Meth. **A300**, p. 572, 1991.

BEHAVIOR OF LARGE RESISTIVE PLATE COUNTERS AT HIGH RATES

M. Widgoff
Brown University, Providence, RI 02912, USA

E.D. Alyea
Indiana University, Bloomington, IN 47401, USA

E. Ables, R. Bionta, G.J. Mauger, C. Wuest
Lawrence Livermore National Laboratory
Livermore, CA 94550, USA

D. Chen, E.S. Hafen, P. Haridas, I.A. Pless, J. Tomasi
Massachusetts Institute of Technology, Cambridge, MA 02139 USA

R. Santonico
Universita di Roma "Tor Vergata and
INFN Sezione di Roma, Roma, Italy

S. Beridge, W. Bugg, P.Y.C. Du
University of Tennessee, Knoxville, TN 37996, USA

INTRODUCTION

Resistive Plate Counters are being investigated by various experimental groups at laboratories around the world, due to its appeal as a promising, cost effective detector technology for particle, nuclear and cosmic ray physics. Ionizing radiation traversing through these counters give rise to large signals (400 millivolt), with rise times (2-3ns) comparable to scintillator-PMT systems, and time resolution of about 1ns. Furthermore, by using readout strips of optimum width, it might be possible to achieve good spatial resolution, comparable to other standard detectors used to track charged particles.

An important consideration in these RPC investigations is its rate capablility,

Supercollider 5, Edited by P. Hale
Plenum Press, New York, 1994

which is directly related to the local recovery time. Chambers built with resistive plates of volume resistivity $1 \times 10^{11}\Omega$-cm provide a recovery time of about 10ms and can handle rates upto 100 Hz/cm^2 without serious loss of efficiency. However in order to operate RPCs in Multi TeV colliders, like the SSC, it is desirable to have a 1000 Hz/cm^2 rate capability, and consequently a recovery time of about 1ms. Our group has been involved in a search for low resistivity plates (10^8 - 10^{10} Ω-cm), and detailed tests on selected materials are now in progress. Three types of tests are performed to verify if RPCs made of a particular chosen material could provide the required high rate capability. These tests are briefly summarised below.

Resistivity Measurement:

This measurement is made on a small piece of the selected material before the RPC is assembled. A Keithley resistivity cell and a picoammeter is used to make this measurement. Using this setup the resistivity can be measured as a function of the voltage applied to the sample. The resistivity is calculated from a knowledge of the measured current, applied voltage, area of the probe in contact with the sample, and the thickness of the sample.

Measurement of Limiting (Maximum) Rate per unit area:

From a knowledge of the recovery time of the material, and the cross sectional area of the discharge, it is possible to estimate the limiting or maximum rate per unit area (Hz/cm^2) of the chamber. The limiting or maximum rate per unit area is defined as the rate at which the area under consideration is almost totally dead, i.e. a charged particle traversing the region has almost zero efficiency to initiate a signal pulse. The limiting rate can be measured in the laboratory by using a strong radioactive source placed at different heights from the area irradiated, or by using more than one source to demonstrate saturation. In the first case, limiting rate is reached when the counting rate flattens off as a function of distance of the source from the chamber. Precautions have to be taken to ensure that the area irradiated by the source is well defined, and that the source has to be properly shielded so that sensitive regions outside the boundary of the chosen area does not contribute to the counting rate.

Efficiency and Time Resolution:

The final and crucial test of the rate capability of an RPC is to measure the chamber efficiency and time resolution as a function of rate. The measurements can be performed in the laboratory by setting up a cosmic ray telescope. The geometry of the telescope should guarantee that a muon selected by the telescope would also pass through the irradiated region of the RPC. With this setup, the rate dependence of efficiency and time jitter can be studied without the costly and time consuming setup, necessary for a similar study at an accelerator beam.

In the next section we describe our cosmic ray telescope arrangement and in the following section we discuss our results. We find that a simple formula

$$E_r = E_0[1 - \frac{r}{R}]$$

where r is the rate, R is the maximum or limiting rate, E_r is the efficiency at rate r and E_0 is the maximum achievable efficiency in the absence of any high rate inducing source, gives a good paramerization of the efficiency as a function of the rate for both our measurement using a cosmic ray telescope as well the data obtained from efficiency measurements for different beam rates at an accelerator. We also present a measurement of the time resolution (jitter) as a function of rate.

EXPERIMENTAL SETUP

The experimental setup shown in Fig.1 is a cosmic ray telescope using three scintillator-PMT systems to trigger the passage of a charged paricle. The RPC (developed at the University of Rome[1]) is placed 8cm above the second PMT (PMT2), such that three RPC strips (each 3cm wide) run perpendicular to the length of PMT2 (10cm), as shown in the top view of Fig.1.

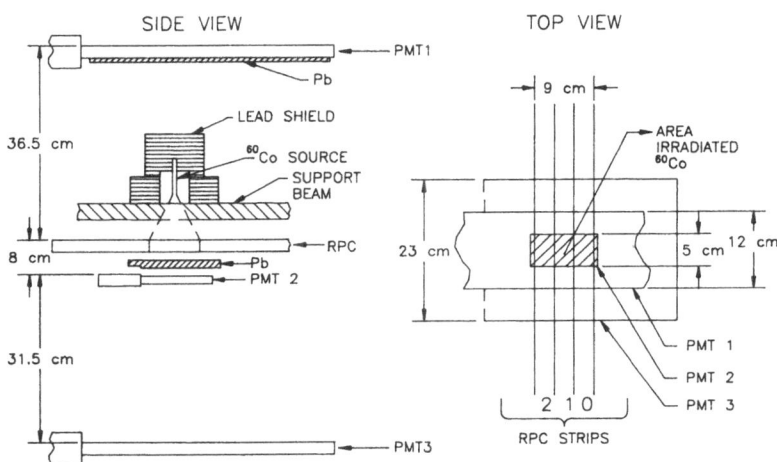

Figure 1.A Cosmic Ray Telescope for High Rate Study

This combination defines an area of 45 cm^2, and all muons triggered by the telescope are constrained to pass through this area. PMT1 and PMT3 have large areas which provide a muon trigger rate of approximately 1 Hz. A 1 millicurie Co^{60} source is placed in a lead enclosure which is supported on iron beams. The readout for the RPC is the ORed output from the 3 RPC strips denoted 0,1,2. The flux rate is varied as described in the previous section. A measurement of the efficiency of the RPC without the Co^{60} source gave of 84%-87% depending on chamber voltage. Since many previous measurements of the efficiency of this chamber using more refined telescopes gave a value of 95%, the smaller value obtained for this system is purely due to edge effects of this particular arrangement. In all results presented in the next section, all measured efficiencies are scaled upto reflect a maximum chamber efficiency of 95%.

RESULTS

The results for the efficiency and time resolution as a function of rate are given in Fig.2 and Fig.3 respectively. In Fig.2 the solid line represents the predicted efficiency as obtained from the formula provided in the first section of this paper, using a value of 0.95 for E_0 and 200 Hz/cm^2 for R. From the limiting rate we can also estimate the resisitivity of the chamber walls of this RPC to be about $5 \times 10^{11} \Omega$-cm. In Fig.3 we show the variation of the time resolution (σ) as a function of rate. The solid line is a fit to the data using a function that is quadratic in r, the rate.

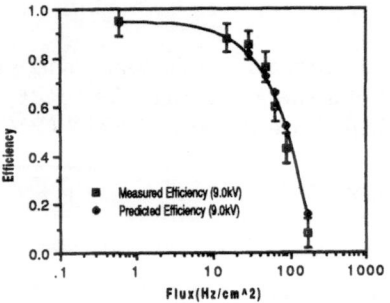

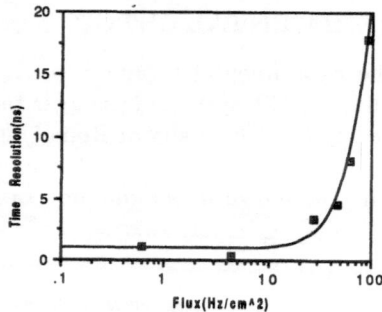

Figure 2. Efficiency vs. Rate.   **Figure 3.**Time Resolution (σ) vs. Rate.

In Fig.4 we provide, for comparison, the efficiency measurement as a function of rate for two types of RPCs in a 50 GeV muon beam at CERN[2]. The dashed curves in both figures (a and b) again represents the predicted behavior from our formula. We have assumed $E_0=1.0$ and $R=1000Hz/cm^2$ for figure 4a (low resistivity, $\rho=1\times10^{11}\Omega$-cm) and $E_0=.97$ and $R=500Hz/cm^2$ for figure 4b (high resistivity, $\rho=2\times10^{11}\Omega$-cm).

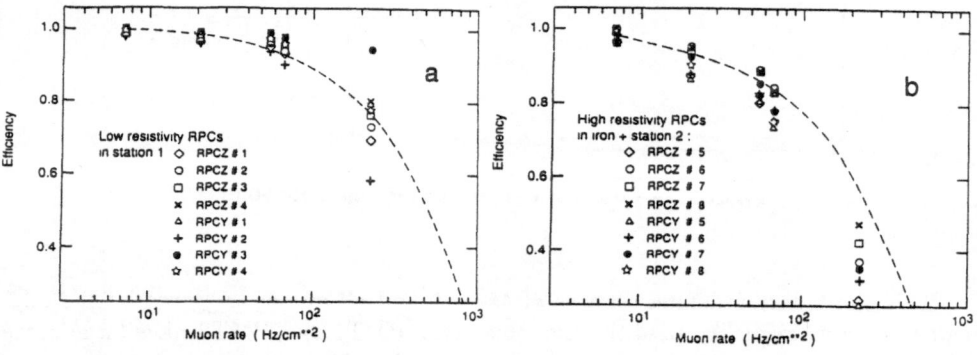

Figure 4. Efficiency vs. Rate measurement in 50 Gev muon beam, (a) for low resistivity RPCs (b) for high Resistivity RPCs.

The results shown in Fig 2 and 3 demonstrate a degradation of both efficiency and time resolution for very high rates. However, for a chamber efficiency higher than 90%, the variation in the time resolution is less than a factor of 2. This conclusion is also demonstrated by an earlier measurement (see Ref.3). In conclusion, we would like to point out that from a knowledge of just the limiting rate value for any RPC chamber we can easily predict its efficiency at any particular rate less than the limiting rate, a useful result for experimentalist developing high rate chambers.

REFERENCES

1. R. Cardarelli et al.,Nucl. Instr. and Meth. A263 (1988) 20.

2. A. Bohrer et al.,CERN/DRDC/91-53 RD-5/Status Report Jan. 1992.

3. M. Bertino et al.,Nucl. Instr. and Meth. A283 (1989) 654.

PRECISION MACHINING AND POLISHING OF SCINTILLATING CRYSTALS FOR LARGE CALORIMETERS AND HODOSCOPES

Craig R. Wuest, Baruch A. Fuchs

Lawrence Livermore National Laboratory
P. O. Box 808
Livermore, CA USA 94551

INTRODUCTION

New machining and polishing techniques have been developed for large barium fluoride scintillating crystals that provide crystalline surfaces without sub-surface damage or deformation as verified by Atomic Force Microscopy (AFM) and Rutherford Back-scattering (RBS) analyses. Surface roughness of about 10–20 angstroms and sub-micron mechanical tolerances have been demonstrated on large crystal samples. Mass production techniques have also been developed for machining and polishing up to five 50 cm long crystals at one time. We present this technology along with surface studies of barium fluoride crystals polished with this technique. This technology is applicable for a number of new crystal detectors proposed at Colliders including the Barium Fluoride Electromagnetic Calorimeter at SSC, the Crystal Clear Collaboration's cerium fluoride calorimeter at LHC, and the KTeV and PHENIX scintillating hodoscopes at Fermilab, and RHIC, respectively.

Lawrence Livermore National Laboratory (LLNL) has an active program of study on barium fluoride scintillating crystals for the Barium Fluoride Electromagnetic Calorimeter Collaboration and cerium fluoride and lead fluoride for the Crystal Clear Collaboration. This program has resulted in a number of significant improvements in the mechanical processing, polishing and coating of fluoride crystals. Techniques have been developed using diamond-loaded pitch lapping that can produce 15 angstrom RMS surface finishes over large areas. Also, special polishing fixtures have been designed based on mounting technology developed for the 1.1 m diameter optics used in LLNL's Nova Laser. These fixtures allow as many as five 25-50 cm long crystals to be polished and lapped at the same time with tolerances satisfying the stringent requirements of crystal calorimeters. We also discuss results on coating barium fluoride with UV reflective layers of magnesium fluoride and aluminum.

BARIUM FLUORIDE SURFACE PREPARATION AND ANALYSIS

Surface preparation is critical to the performance of barium fluoride and other fluoride crystals for a number of reasons. First, an improperly prepared (machined, ground, polished, lapped) crystal suffers from induced stresses and deformations in the first few hundred microns of the surface. These stresses can manifest themselves in the formation of cracks (crazing) over long times, or more quickly when subjected to extremes of heat, radiation, humidity, etc.. Surface stresses can be minimized using well-known polishing and lapping techniques that *gently* bring the surface to a final finish. These techniques have been developed at LLNL for barium fluoride and also applied to cerium fluoride and lead fluoride. Improper surface preparation can also introduce contaminants into the surface of the crystal. Under certain conditions these contaminants can migrate into the bulk of the crystal and cause extended areas of radiation susceptibility. Because these scintillator materials emit their light typically in the UV, surface finish is especially important for good light transport properties.

A number of surface preparation techniques were explored at LLNL, including ion beam milling, diamond turning, and various polishing/lapping techniques. Ion beam milling provides the best crystalline surface, however, the uniformity of the surface, as well as the surface finish is not very good. In terms of surface finish, diamond-turned surfaces are the best with 6 Å RMS demonstrated. However, RBS analysis of diamond-turned surfaces reveal that they are amorphous. Figure 1 shows an example of diamond-turned barium fluoride with a noticeable crystal grain boundary that exhibits different surface roughnesses. Also shown in Fig. 1 is an example of an improperly polished crystal at similar magnification.

A polishing technique – pitch lapping with diamond abrasives – provides the best combination of surface finish (10-20 Å RMS) and surface crystallinity. The technique is applied after more standard polishing techniques and is a simple wheel (lap) prepared with a low melting temperature synthetic pitch. Grooves are formed in the pitch in a pattern to allow cutting fluids, abrasives and ground material to be washed away during the lapping process. The key to the process is a final polish with an abrasive of very uniformly sized diamond, typically 1/2 μm or 1/4 μm diameter, imbedded in the pitch. In addition, a non-aqueous cutting fluid such as low viscosity silicon oil, or ethylene glycol is used to uniformly disperse the diamond and to carry away waste material. Water is not a good fluid for diamond because of the tendency of diamond to agglomerate in water. Water is also not desirable because of the slight solubility of fluoride crystals in water. We have verified the high quality of diamond-lapped surfaces using AFM and RBS. This analysis supports our optical measurements and also provides insights into the mechanics of the polishing technique.

100 µm   100 µm

Figure 1. Photographs (179x) of polished barium fluoride surfaces comparing improper polish techniques (200 Å RMS in regions between the large grooves) to diamond turning (6/80 Å RMS).

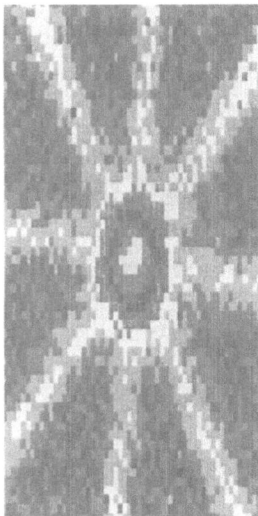

Figure 2. RBS images of polished barium fluoride surfaces showing good (left) and bad (right) crystalline surfaces. The images show the backscatter as a function of x and y tilt angles over a range of $\pm 3°$.

Improperly prepared surfaces are easily identified under optical microscopy, and by using other analysis techniques such as RBS. In the case of RBS, helium ions bombard the surface and can channel into the crystal preferentially along the crystal planes. If the surface of the crystal is amorphous, no preferential backscattering is observed. If the crystal surface is crystalline, the crystal lattice is readily identified as peaks in the backscattering number. Figure 2 shows results for crystals prepared at LLNL using improper and proper polishing methods.

LLNL has designed and fabricated a set of special polishing fixtures that allow up to 5 crystal halves or pairs (50 cm length) to be polished at the same time. These fixtures have been used to demonstrate the technique of multiple polishing. It is expected that flatness can be maintained across the full 25 cm x 25 cm area of grouped crystal halves at the level of a fraction of a wavelength of visible light. Also surface finish can be maintained to about 20 Å. These fixtures are easily adapted to existing techniques and machines in use throughout the world. It is anticipated that these techniques would be very desirable for mass production of crystal segments for calorimeters and hodoscopes. Figure 3 is a photograph of the various polishing fixtures.

The polishing techniques developed at LLNL are simple to implement and are essentially extensions of standard polishing techniques already in practice in the US and elsewhere. We feel that these techniques are easily transferred to industry both in the US and overseas. We expect that our techniques can be utilized in production facilities for large scale production of crystal scintillators with little added cost to the overall production of finished crystals. LLNL engineers and physicists have recently visited China to work with the Chinese to develop this capability.

UV REFLECTIVE SURFACE COATINGS R&D

High quality surface preparation is also important for insuring the proper application of a reflective coating that exhibits good reflectivity in the UV, as well as long term stability. LLNL has experimented with the application of magnesium fluoride and aluminum coatings on barium fluoride. Measurements of front surface reflectance of 500 Å aluminum coatings on barium fluoride have been made along with measurements of reflectance through a thin (2 mm) sample of barium fluoride (back reflectance). Measurements indicate that reflectivity at 220 nm is about 90%.

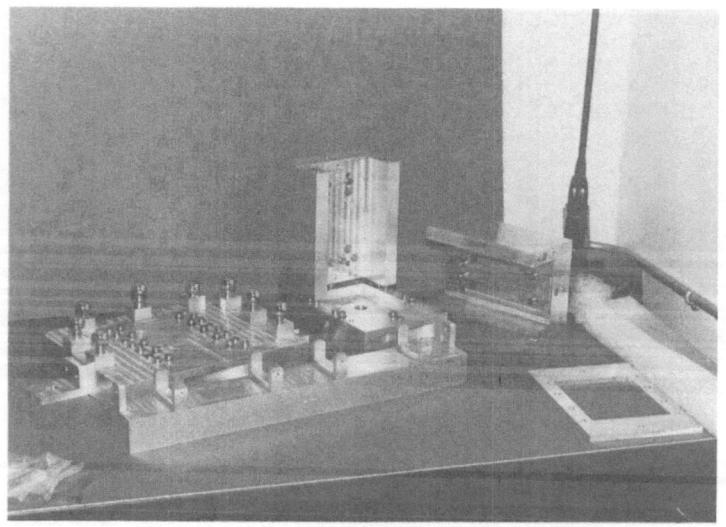

Figure 3. Polishing fixtures developed at LLNL for polishing multiple crystals with trapezoidal shapes. The fixtures allow polishing with a high degree of flatness and excellent surface finish.

Diffuse scattering measurements of the aluminum coating have been made for front surface scattering. It is assumed that this is representative of the diffuse scattering on the back surface into the barium fluoride crystal. Measurements have also been made on magnesium fluoride coatings on barium fluoride to determine the critical angle for total internal reflection.

Additional work is planned to study the long term integrity of coatings. For example, if microscopic pits or pinholes occur, moisture can come into contact with the crystal surface, eventually leading to a degradation of the coating in that region due to chemical reactions that may occur.

LLNL is also helping to provide this data to physicists at Oak Ridge National Laboratory to help model the response of a barium fluoride crystal using a specially written Monte Carlo program. In addition, studies of the response of 50 cm long crystals to cobalt-60 and iron-55 gamma rays and x-rays are being carried out at LLNL, and cosmic ray studies are being made at UC San Diego. These studies are being made for different coating materials and combinations of coatings in an effort to provide uniform collection of scintillation light along the length of the crystal.

CONCLUSIONS

The polishing and coating techniques described here have been shown to produce surfaces with high quality finishes as well as very good macroscopic tolerances. These techniques are extensions and refinements of basic polishing technology and are easily transferred to industry. Other crystals, such as cerium fluoride are becoming increasingly available for scintillation detectors, and we feel that our experiences described here can be applied for precision mechanical processing and coatings of these materials. We have begun a similar program to study cerium fluoride and we have successfully diamond-turned cerium fluoride to surface finishes of the same quality as for barium fluoride. Also, our coating techniques are directly applicable to the somewhat longer wavelength emission of scintillation light in cerium fluoride.

ACKNOWLEDGEMENTS

This work was performed under the auspices of the US Department of Energy by the Lawrence Livermore National Laboratory under Contract W-7405-ENG-48.

BARIUM FLUORIDE CRYSTALS FOR PRECISION EMC AT SSC

Ren-yuan Zhu[1]

Lauritsen Laboratory
California Institute of Technology
Pasadena, CA 91125

ABSTRACT

This report presents experimental facts and color center dynamics related to the optical bleaching of barium fluoride crystals. The optical bleaching is a viable approach to construct a crystal calorimeter at future hadron colliders by using BaF_2 crystals of existing quality.

INTRODUCTION

Main effort in pursuing a barium fluoride (BaF_2) crystal calorimeter for the SSC has been concentrated in improving the radiation resistance of the crystal. This cross disciplinary research has been benefited from a multi institutions collaboration, including crystal manufacturers: Shanghai Institute of Ceramics (SIC) and Beijing Glass Research Institute (BGRI) in China, and material scientists: Tongji University in China and experts in the U.S., especially members of the Expert Panel assigned by the SSC laboratory to review the radiation damage problem of the BaF_2 crystals [1].

It is understood [2] that the radiation damage of BaF_2 (1) is caused by the formation of color centers, (2) saturates at high accumulated dose (~ 100 kRads), (3) is dose rate independent, and (4) recovers extremely slowly under room temperature, but can be fully annealed at 500°C for three hours. There is strong evidence that O^- and OH^- play a key role in forming color centers responsible for BaF_2 radiation damage [3].

The progress of the radiation resistance of BaF_2 crystals is clearly shown in Figure 1 where the transmittances of three 25 cm long crystals produced by the SIC in early 1991 (SIC102), early 1992 (SIC302) and July 1992 (SIC402) are shown as a function of the wavelength. The light attenuation length (LAL) [4] at 220 nm, where fast component of BaF_2 scintillating light resides, after 1 Mrad ^{60}Co γ-ray irradiation is 41 cm for crystal SIC402. The improvement of intrinsic radiation resistance is clearly shown in the increase of the transmittance and relative light output.

---

[1]Work supported in part by U.S. Department of Energy Grant No. DE-FG03-92-ER40701.

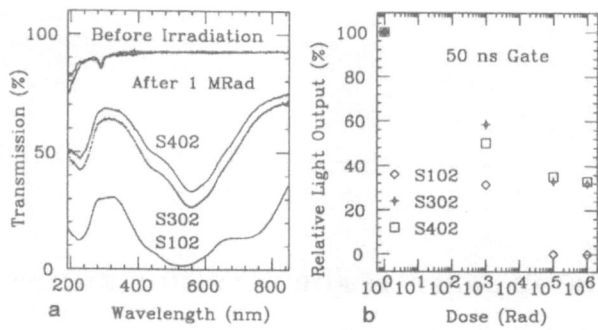

Figure 1. Transmittance before and after 1 MRad γ-ray irradiation (a) and relative light output (b) measured for three 25 cm long BaF₂ crystal produced at SIC in early 1991 (SIC102), early 1992 (SIC302) and July 1992 (SIC402).

OPTICAL BLEACHING

The improvement of intrinsic radiation resistance of BaF_2 crystals, however, does not satisfy the specification of 95 cm LAL under radiation [2]. Optical bleaching *in situ* thus was recommended by the BaF_2 expert panel as an alternative approach. Studies [5] show that optical bleaching with blue light of 400 nm is effective in removing the radiation damage and to reset current production BaF_2 crystals to an LAL of 180 cm. Figure 2a shows the transmittance of SIC302 measured after 1 MRad ^{60}Co irradiation and under illumination with light of different wavelength from a monochromator which has an intensity of 0.85 mW/cm² at 400 nm. Figure 2b, c and d show the transmittance at 220 nm, its corresponding LAL and color center density (1/LAL) as a function of bleaching time.

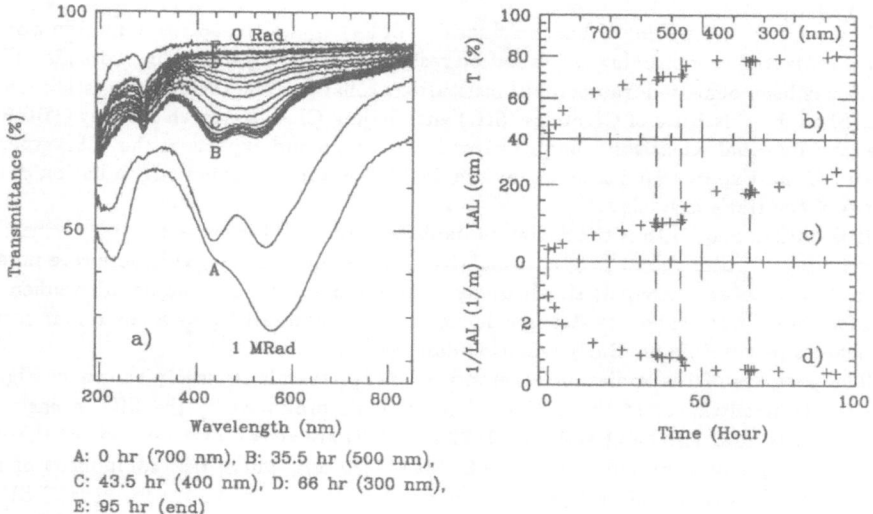

A: 0 hr (700 nm), B: 35.5 hr (500 nm),
C: 43.5 hr (400 nm), D: 66 hr (300 nm),
E: 95 hr (end)

Figure 2. SIC 302 under optical bleaching by light of different wavelengths: (a) the transmittance as a function of wavelength, and (b) transmittance, light attenuation length (c) and color center density (d) at 220 nm as a function of time.

COLOR CENTER DYNAMICS

The dynamics of optical bleaching *in situ* can be explained by a combination of color center annihilation caused by the bleaching and the creation caused by the irradiation [5].

$$dD = -aIDdt + (D_{all} - D)\, bRdt \tag{1}$$

where D is the optically bleachable color center density, a is a constant in units of $cm^2/joul$, I is the light intensity in $joul/cm^2/hr$, D_{all} is the total density of traps related to the optically bleachable color centers in the crystal, b is a constant in units of $1/kRad$, R is the radiation dose rate in units of $kRad/hr$, and t is the time in hours. The solution of Equation 1 is

$$D = D_0 e^{-(aI+bR)t} + \frac{bRD_{all}}{aI + bR}\,[1 - e^{-(aI+bR)t}] \tag{2}$$

where D_0 is the initial value of the bleachable color center density. For each value of I and of R, an equilibrium between annihilation and creation will be established at an optical bleachable color center density (D_w) of

$$D_w = \frac{bRD_{all}}{aI + bR} \tag{3}$$

Table 1. Light Intensity Needed to Maintain BaF$_2$ at LAL = 150 cm

$|\eta|$	0	1	2.5
Dose Rate (kRad/hour)	0.02	0.04	0.4
I (mW/cm^2)	0.21	0.42	4.2

According to this dynamic model, the required light intensity to restore the LAL to a stable value of 150 cm (in a dynamic equilibrium) can be calculated for current production BaF$_2$ crystals to be 10.5 R mW/cm^2, where R is in units of kRad/hr. Table 1 lists the expected radiation dose rate at the front surface of a BaF$_2$ calorimeter, which is much larger than that inside of calorimeter, and the corresponding bleaching light intensities required to set LAL = 150 cm at different rapidities, assuming the BaF$_2$ calorimeter has a barrel of 75 cm radius and two end caps at z = 150 cm. A maximum of 150 W is needed to bleach the entire BaF$_2$ calorimeter at the standard SSC luminosity (10^{33} cm^{-2} s^{-1}). This optical power should be further reduced if taking into account multi-bouncings of the bleaching light inside the crystals and the further improvement of crystal quality.

OPTICAL BLEACHING *in situ*

Experimental tests have been carried out to verify the color center dynamics described in previous section. A total of 15 experiments were performed where a 25 cm long crystal (S402) was illuminated with a 450 nm light with an intensity of I (j/cm^2/hr), and was irradiated under a uniform ^{60}Co γ-ray source with dose rate of R (kRad/hr) at the same time. Both R and I were varied for different experiement, but were fixed during a single irradiation. The transmittance and LAL were measured at the end of each irradiation, and were compared to the expected values calculated according to the color center dynamics. The dose rate R was varied from 0.025 (Run 1—3), 0.1 (Run 4), 0.3 (Run 5) to 0.35 (Run 6—15). The bleaching

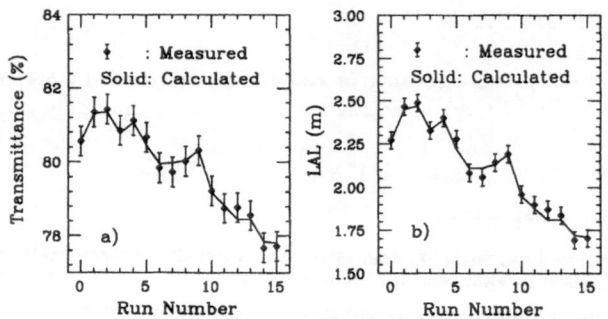

Figure 3. Measured (points) and calculated (solid line) transmittance (a) and LAL (b) of crystal S402 under irradiation and 450 nm light illumination.

light intensity I was varied from 4.4 (Run 1—2), 0 (Run 3), 4 (Run 4—5), 3 (Run 6—7), 3.3 (Run 8—9), 0.85 (Run 10—13) to 0.70 (Run 14—15). The duration of these tests are 1, 4, 2.5, 1, 1, 2.8, 1, 1, 1, 0.75, 1, 1, 1, 3 and 9 hours respectively.

Figures 3 shows the measured transmittance (a) and LAL (b) (data with error bars) and corresponding calculated LAL (solid line) according to Equation 2. The parameters a, b and D_{all} are determined from previous measurement to be: a = 0.68 and 0.95 cm^2/joul for D < and > 0.08 respectively, b = 0.65 kRad^{-1} for accumulated dose < 5 kRad, and D_{all} = 0.73 m^{-1}. The calculated LAL agrees very well with measured data, which indicates the color center dynamics discussed in previous section indeed describes the behavior of the dynamics of radiation damage of barium fluoride crystals.

SUMMARY

Although the intrinsic radiation resistance of production size BaF$_2$ crystals does not yet satisfy the stringent requirement of radiation environment at future hadron colliders, the **optical bleaching** discussed in this report is a viable solution. The proposed model of color center dynamics predicts the optical behavior of BaF$_2$ crystals under irradiation and bleaching very well. It is expected that the LAL of BaF$_2$ crystals can be maintained to a required range *in situ*, so that a precision BaF$_2$ calorimeter can be constructed with production BaF$_2$ crystals of existing quality.

References

[1] S. Majewski et al., BaF$_2$ Expert Panel Report, February 1992 and August 1992.

[2] R.Y. Zhu, **GEM TN-92-48**, January 1992.

[3] L.Y. Chen et al., **GEM TN-92-129**, June 1992; P.J. Li et al., **GEM TN-92-77**, March 1992; and G. Chen et al., **GEM TN-92-78**, March 1992.

[4] D.A. Ma and R.Y. Zhu, **GEM TN-92-148**, August 1992.

[5] D.A. Ma and R.Y. Zhu, **GEM TN-92-149**, August 1992.

OVERVIEW OF THE GEM MUON SYSTEM COSMIC RAY
TEST PROGRAM AT THE SSCL

E. Cas Milner for the GEM Collaboration

Superconducting Super Collider Laboratory[*]
2550 Beckleymeade Ave.
Dallas, TX 75237

INTRODUCTION

Muon track resolution exceeding 75-µm per plane is one of the main strengths of the GEM detector design, and will be crucial in searches for Higgs Bosons, heavy Z-Bosons, technicolor, and supersymmetry. Achieving this resolution goal requires improved precision in muon chambers and their alignment. A cosmic ray test stand known as the Texas Test Rig (TTR) has been created at the SSCL for studying candidate GEM muon chamber technologies. Test results led to selecting Cathode Strip Chambers[1] (CSC) as the GEM muon system baseline chamber technology.

THE TEXAS TEST RIG (TTR)

The triggerable volume of the TTR is large, with a surface area of 1.2 m × 5 m and a height of 3 m, allowing studies of as many as six different chambers simultaneously. All chamber types tested to date have shown excellent performance, with resolutions better than the 75-µm GEM design goal. Comprehensive testing has given information on chamber operation, gas mixtures, calibration, mechanical design, data acquisition, and data analysis. The TTR has become the first user facility at the SSCL and a center for GEM muon system R&D, with more than 100 participating physicists from 19 universities and national laboratories in China, Mexico, Russia, and the United States.

The TTR apparatus has several features contributing to its performance as a powerful test instrument. A 1-m thick stack of steel absorbs cosmic rays with less than about 1.3 GeV/c momentum. Removing the "soft" component of the spectrum makes chamber resolution studies less susceptible to the misleading effects of multiple scattering, and also reduces the trigger rate to about 60 Hz. Scintillator hodoscopes with timing resolution of about 300 ps are positioned above and below the steel to provide the fast trigger. The steel

*Operated by the Universities Research Association, Inc., for the U.S. Department of Energy under Contract No. DE-AC35-89ER40486.

can be magnetized to 15 kG by energizing coils wound in a solenoidal configuration. With the magnet on, a finer position measurement obtained with four planes of 1-cm pitch Iarocci chambers can be used to select the higher momentum component of the muon spectrum, effectively raising the threshold to 10 GeV/c.

The TTR gas system can provide up to five different gas mixtures simultaneously to chambers under test. Since some chambers operated with flammable gases, a gas leak detector system with 14 sensor heads was deployed. This system was sensitive to hydrocarbon gas concentrations as small as 10% of the lower explosive limit. It provided alarms and signals for turning off the chamber gas and high voltage supplies in the event of a leak. TTR safety systems and procedures appeared to serve as a safety "prototype" for the future GEM experiment. The studies reported here were accomplished without injury.

A data acquisition (DAQ) system developed at the SSCL is used at the TTR. It is modular in design, accommodating a wide variety of electronics and software brought to the lab by visiting groups. VME-based processors running the VxWorks real-time operating system are at the heart of the DAQ. They communicate with muon chamber electronics and trigger units residing in CAMAC crates and other equipment such as high voltage supplies. Digitized data are read by the processors from the crates; then events are built and stored in VME memory. A workstation with a dedicated VME link periodically transfers the events from memory to disk and tape. A graphical user interface controls the DAQ. It features a run configuration editor, various run monitors, and an event display. Any computer running UNIX and x-windows and having network access to the TTR can be used to monitor TTR operation and analyze data.

TTR off-line software[2] is a general framework where users place their analysis code. The software automatically fetches the zero-suppressed data file from disk and stores it on an 8-mm tape robot. It also stores on a database (SYBASE) records describing running conditions, the chambers operating, triggers, and other information. The database can return a list of files satisfying queries. All these operations, including running the analysis program, may be controlled through user-friendly pop-up windows. Thus, the user may access the program from any x-terminal, generate a list of data files from the database, edit the list if needed, and activate the offline analysis program to process data stored on the tape robot. One unique feature is a dynamically loaded subroutine-substitution method that allows specifying replacements for default routines in the standard package.

The offline package processes Iarocci chamber and scintillator data, reconstructing tracks using these data both independently and in a combined fit. These tracks may be compared with tracks measured by the test chambers. In addition, there are routines to process data from the technologies. Histograms may be displayed by PAW; the user may easily define new plots. The standard output is used as input to the event display program.

In addition to the TTR cosmic ray test stand, a laser-based test system has been built for small-scale chamber studies, simulating a particle track using a UV-laser beam. The apparatus includes a laser, optical tools, chambers, and a Macintosh-based data acquisition system. The laser has been used to measure the operational speeds of flammable and non-flammable gas mixtures under consideration for the GEM muon system. Recently a magnet was installed for investigating magnetic field effects on CSC performance.

MUON CHAMBER TEST RESULTS

Four types of detectors have been tested at the TTR. Pressurized drift tubes (PDT), limited-streamer drift tubes (LSDT), and cathode strip chambers (CSC) were candidates for muon position measuring detectors, while resistive plate chambers (RPC) could be used for triggering and bunch tagging.

Separate PDT systems were built at Dubna and Michigan State University. They have staggered layers of tubes 3- to 4-cm in diameter, 4-m long, stacked 32 tubes wide. With a flammable gas mixture, resolution below the 100-μm design goal was measured at one atmospheric, improving to 50 μm at 5 atmospheres (Figure 1).[3] A PDT-based GEM muon system[4] appeared to be competitive with the CSC option that was eventually chosen.

Systematic error correction was an important part of the data analysis. An iterative fit to the data yielded a time-to-distance calibration. Muon tracks were used to determine the wire plane relative shifts and rotations, and individual wire displacements. This procedure was demonstrated for drift chambers, and it should apply also to CSC systems.

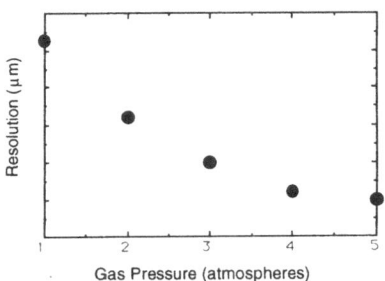

Figure 1. Resolution as a function of gas pressure in the Dubna pressurized drift tube system, measured at the TTR. The gas mixture was Argon:Ethane 50:50.

The MIT-built LSDT system featured precisely machined bridges supporting anode wires inside U-shaped aluminum profiles, and gave resolution below 100 μm A drift tube based trigger concept was tested, suggesting a 94% trigger efficiency.[5]

A 1.2-m × 2.4-m RPC designed and built by an LLNL-MIT group used ABS plastic doped with conducting polymer, and is a low-cost alternative to a scintillator for large area counting. Radioactive source tests indicate rate capability to 1 kHz/cm^2, substantially higher than bakelite RPCs currently in use. PDT and CSC groups were concerned that their chambers would pick up spurious signals from RPCs, but this was not seen in TTR tests.

Three independently developed CSC systems were tested. Designs addressed issues of construction, alignment, and manufacturability. In this type of chamber the cathode is segmented into strips, and the image charge induced on the cathode plane is shared among several adjacent strips. The centroid position is interpolated from the strip charges. All three CSC prototypes have shown spatial resolution better than 70 μm (Figure 2).

The Brookhaven and Dubna CSC designs are proportional chambers with 2.5-mm anode wire pitch. Gas gaps are between paper honeycomb panels with strip boards glued on them. The 4-gap Brookhaven chamber, (0.5-m × 0.5-m), and the Dubna chamber (1.0 m × 1.3 m) were successfully tested at the TTR. A 2-gap trapezoidal chamber, roughly 1.0-m × 2.0-m, with "fanned" strips is currently under test.

The University of Houston CSCs use plastic modules containing eight square tubes enclosing the anode wires. Carbon paint on the wire module interior allows the signals to induce image charges on the cathode strips glued to the wire planes. A chamber with 1-m × 0.5-m sensitive area was tested.

In the baseline CSC design, the strips are made with conventional printed circuit board technology. An alternative being studied is based on package-printing technology. Strip material is made in a subtractive process after the pattern is transferred to a continuous sheet of mylar with a sputtered layer of copper. The strip material appears to be inexpensive and precise. The rms of distribution of measured strip pitches is about 14 μm. Electrical properties were also measured and found to be compatible with CSCs.

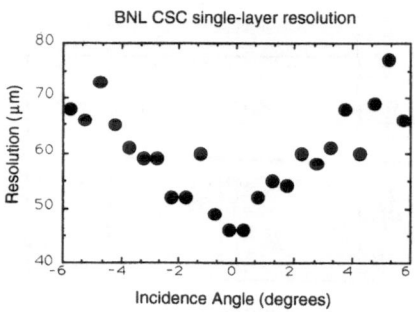

Figure 2. Single-layer resolution of the BNL CSC as a function of incident angle. For the present GEM muon system baseline design, this implies a resolution exceeding 60 μm.

SUMMARY

These results were possible because of advances in chamber design, a sophisticated DAQ system, the size and flexibility of the TTR, and the coherent effort of the GEM muon group. Using cosmic rays at the TTR has been surprisingly fruitful, and the experience has built our confidence in making a high quality muon system for GEM.

REFERENCES

1. "GEM Muon System Based on Cathode Strip Chambers," SSCL TN-92-00199.
2. I. Chow, et al., "Offline Event Reconstruction for TTR (User Guide),"SSCL TN-93-00293.
3. "Muon Technology Choice Performance Comparisons," SSCL TN-93-00282.
4. "The RDT-RPC Technology Option for GEM,"SSCL TN-93-00288.
5. "Muon Trigger Using GEM Drift Tubes,"SSCL TN-92-00161, and SSCL-TN-93-303.

DEVELOPMENT OF HERMETIC ELECTRICAL CONNECTORS FOR SSC SPOOL PIECES

Bill Kountanis[1] and Lou Kalny[2]

[1]Superconducting Super Collider Laboratory*
2550 Beckleymeade Ave.
Dallas, TX 75237-3997

[2]Hi-Rel Connectors, Inc.
760 W. Wharton Drive
Claremont, CA 91711

INTRODUCTION

The Superconducting Super Collider ring is about 54 miles (87 km circumference) and primarily includes a series of magnets. Spool piece assemblies are interspaced in the ring at predetermined intervals to provide specific functions such as cryogenic interfaces, vacuum interface, magnet power, magnet power dump, quench heater power, and special instrumentation. Electrical connectors serve as interfaces for instrumentation and quench heater circuits. These connectors have to meet stringent requirements.

REQUIREMENTS

There are three levels of electrical requirements which are determined by application. These are low voltage instrumentation, voltage taps, and heater circuits. In addition to the electrical requirements, all connectors have to meet unique mechanical and environmental requirements. Common requirements are:

Pressure:	2.5 MPa (375 PSIG)
Vacuum Leak Rate:	2×10^{-10} PaM^3/sec. (2×10^{-9} Std cm^3/sec)
Temperature:	$-65°$C to $+200°$C
Thermal Shock:	$-260°$C to $260°$C
Radiation:	1×10^7 RAD Proton over 25 years

Connectors which are used for low level signals such as temperature sensors, pressure transducers, and accelerometers are required to meet 2000-volt dc high potential with less than 1 micro amp leakage in air before being installed. Connectors for use with voltage taps

---

* Operated by the Universities Research Association, Inc., for the U. S. Department of Energy under Contract No. DE-AC35-89ER40486.

and heaters are required to meet 5000-Vdc high potential with less than one micro amp leakage in air before being installed. After installation, connectors for voltage taps and heaters must meet 3000-Vdc high potential with less than 1 micro amp leakage in helium at 1 atmosphere pressure. Low voltage connectors are not high potted after installation.

DESIGN CONSTRAINTS

Connectors are mounted on the outside of the spool piece as shown on Figure 1 and must be accessible from the aisle side of the spool piece. (The side which is away from the tunnel wall.) Installation must be done from the outside of the spool so that replacement or repair can be easily accomplished. Additionally, a standard mounting pattern would make spool fabrication uniform from a production perspective.

DESIGN SOLUTIONS

In order to meet the pressure and vacuum requirements, a hermetically sealed design is used. The seal has to meet the high potential requirements as well as those of temperature and radiation. The type of seal that met these requirements is the vitreous type which utilized a multi-hole glass preform. Individual seals around each pin did not meet the high pot requirement.

The next obstacle is the high potential requirements in a helium atmosphere. This is accomplished by potting the back side of the receptacle, after the wires are connected, with a suitable epoxy which has to be vacuum degassed to remove air bubbles during encapsulation. A built-in potting ring is used to support the potting process.

Standard box mount receptacles offered the capability of being installed from the outside but could not be adequately sealed for pressure and vacuum. A special flange mount is used with an "o" ring seal. The flange size and hole pattern are the same for all insert arrangements so that any receptacle can be used. Figure 2 shows a typical receptacle.

Shell sizes and insert arrangements correspond to MIL-C-38999 Series III requirements. Any counterpart Series III plug can be used as a mating connector. However, some types of plugs may not meet some of the high potential and environmental requirements. Selection of the plugs must consider the type of material used for contact insulation and the shell materials. Shells for the collider plugs are 300 series stainless steel with a socket insulator material selected to meet the specific performance requirements.

TESTING

Testing, in addition to qualification tests for MIL-C 38999, has included a series of high potential tests. These tests were performed with a high potential tester which limits the breakdown current to 50 micro-amps. The higher voltage type receptacles were subjected to 5000 Vdc in air with less than 1 micro-amp leakage current. All receptacles tested passed this test. The low voltage receptacles were subjected to 2000 Vdc in air with less than 1 micro-amp leakage current and also passed. A limited number of tests using helium atmospheric conditions have been conducted and all receptacles have passed the above required potential values. Additional tests are presently being performed which will include all of the required insert arrangements. A limited number of thermal shock tests have been performed with no failures to date. Hermeticity remains less than 1×10^{-9} Std cm^3/sec.

RELATED DESIGNS

The experience and technology obtained from the development of these receptacles is being extended to developing feedthroughs for superconducting leads. There are two types of superconducting leads; the main magnet bus which is keystone shaped, and the corrector element leads which are round. Both feedthroughs will be in liquid helium.

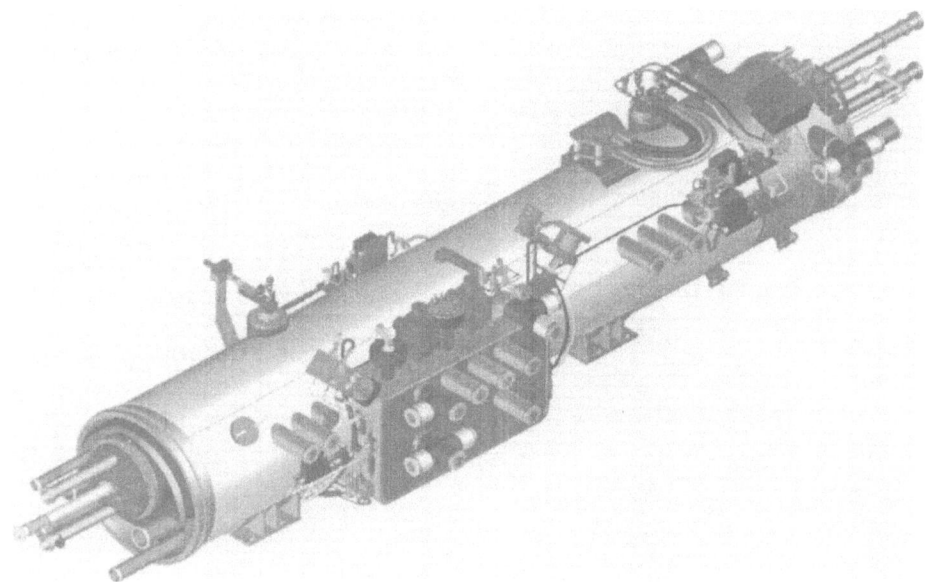

Figure 1. Spool Piece Showing Connectors Installed

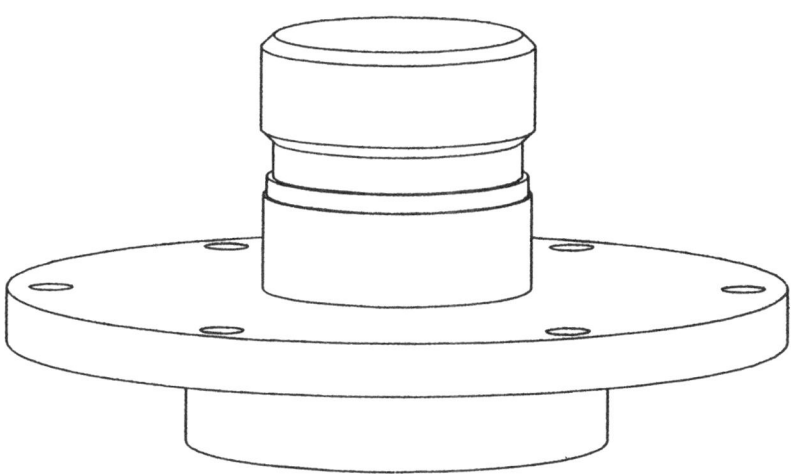

Figure 2. Typical Connector Outline

CONCLUSION

Testing to date has shown that these connectors meet the mechanical and electrical requirements for use in the spool pieces. The only remaining environmental requirement to be accomplished for the collider final design is the selection of an "o" ring material which will have a useful life of 25 years.

DESIGN AND ANALYSIS OF THE COLLIDER
SPXA/SPRA SPOOL PIECE VACUUM BARRIER

Greg Cruse and Gülperi Aksel

Mechanical Engineering Department
Accelerator Systems Division
Superconducting Super Collider Laboratory*
2550 Beckleymeade Avenue
Dallas, TX 75237-3997

ABSTRACT

A design for the Collider SPXA/SPRA spool piece vacuum barrier was developed to meet a variety of thermal and structural performance requirements. Both composite and stainless steel alternatives were investigated using detailed finite-element analysis before selecting an optimized version of the ASST SPR spool vacuum barrier design. This design meets the structural requirements and will be able to meet the thermal performance requirements by using some newer thermal strapping configurations.

INTRODUCTION

Collider Accelerator Arc Sections of the Superconducting Super Collider (SSC) have specific requirements[1] for an insulating vacuum system, one of which is for a vacuum barrier to provide separate vacuum domains per half cell, a 90-m section. There are several other requirements that govern the design of the vacuum barrier,[2] most of which are directly or indirectly generated from the 3B specification. The normal operating loads for the vacuum barrier are the thermal loads imposed from the cryogenic lines that penetrate it. These include the cold mass pipe and the liquid helium (LHe) and gaseous helium (GHe) return lines, all of which operate at approximately 4 K; the 20-K shield helium supply line; and the 80-K shield nitrogen supply and return lines. The vacuum barrier must also withstand a 0.1-MPa pressure at room temperature seen during vacuum pump down. Additionally, the vacuum barrier must be designed for an emergency pressure load of 0.2 MPa at cryogenic temperatures, since the relief valves for the outer cryostat are required to vent at less than 0.2 MPa. The design of the vacuum barrier should also decouple the cold mass axial motion from the vacuum barrier. This should relieve the vacuum barrier from the load resulting from the thermal shrinkage of the cold mass and remove the pressure load on the vacuum barrier from being carried by the fixed post. Overall, the vacuum barrier design should meet a safety factor of 1.5 against yield and 3.0 against critical buckling pressure to insure the safe and successful operation of this component.

---

*Operated by the Universities Research Association, Inc. for the U.S. Department of Energy under Contract No. DE-AC35-89ER40486

In addition to the structural requirements, the vacuum barrier has some thermal constraints on its design. It is required that the heat leaks into the 4-K, 20-K, and 80-K regions for the vacuum barrier must fit within the overall heat budget for the SPXA/SPRA spool pieces. To meet the overall requirements, the goals for the vacuum barrier heat leaks were defined as follows: 130 mW into the 4-K lines, 10.31 W into the 20-K line, and 27.43 W into the 80-K lines

DESIGN CONCEPTS

Several different design ideas for the vacuum barrier were looked at to meet the performance criteria. These included a flat composite bulkhead overlaid with metal, a standard concentric shell design similar to the ASST vacuum barrier (except made out of composite and overlaid with metal), and a stainless steel vacuum barrier that was an optimized version of the ASST design.

Composite Bulkhead Design

This design utilized a 0.125-in.-thick flat composite plate of G-10 with a 2-mil coating of 304 stainless steel. This coating is necessary for the barrier to be helium leak-tight. The 20-K and 80-K thermal shields were integrated into this bulkhead plate, limiting the cryostat to cold mass radiation. Other benefits of this design included space savings in the spool piece as well as simplification in manufacturing. Also, all of the cryogenic lines penetrate this barrier without any bends.

However, there were several shortcomings in this design that removed it from consideration. Foremost were the extremely high stresses in the stainless steel coating and rather high stresses in the composite itself. The stainless steel stresses arise from cooling from 300 K to 4 K with the boundaries restrained, while the pressure loads caused high bending stresses in the composite. Additionally, the heat leak into the 4-K lines was more that six times the goal, as might be expected with such a short conduction path.

Composite Concentric Shell (ASST) Design

This design used the geometry from the ASST vacuum barrier (4-K, 20-K and 80-K plates, each connected by a concentric shell pair) but was made of G-10 composite overlaid by 304 stainless steel. One of the advantages of this design idea was that the use of composites allowed the vacuum barrier to be shortened to about 4 in. and still easily meet the heat-leak budget. Also, geometrically it had more flexibility in allowing the thermal contractions from 300 K down to 4 K. Additionally, it was hoped that this complex geometric piece could be molded as a single composite piece, greatly simplifying the manufacturing process.

High stresses also overshadowed any benefits this design might offer. Although the stresses generated from thermal cooldown were manageable, the stresses from the pressure loading were exceedingly high, especially in the thin stainless steel coating. The highest stresses generally were located in the return end bends of the concentric shell pairs. Other concerns were raised about the radiation resistance of the composites, as well as the ability to make the composite-to-stainless steel connections helium leak-tight. Therefore, it was decided to pursue an optimization of the ASST vacuum barrier design for the final configuration of the collider arc spools.

Optimized SS304 Concentric Shell (ASST) Design

The optimization of the ASST vacuum barrier design into a Collider configuration was begun with goals of decreasing the amount of space the vacuum barrier occupied and simplifying the design from a manufacturing standpoint while still meeting the design requirements. The new Collider vacuum barrier design replaced the three plates and three shell pairs of the ASST design with two plates connected to each other with one concentric shell pair and connected to the outer cryostat with another. (The actual cryostat connection is made through a weld ring.) The inner plate in this design is penetrated by the three 4-K lines: the cold mass pipe, and the LHe and GHe return lines. Additionally, pipe sleeves were added to connect the LHe and GHe lines to the vacuum barrier front plate, increasing the conduction path to the 4-K region and thus decreasing its heat leak. A bellows connection between the cold mass pipe and this front plate also was added, which increased the conduction path, while structurally the connection decoupled the motion of the cold mass pipe from the vacuum barrier, especially in the axial direction.

The outer plate is penetrated by the 20-K and 80-K lines. A pipe sleeve connection between the 20-K line and this plate was needed to keep the 20-K and 80-K regions thermally separated. To keep the heat leak under control, the thermal strapping to the 80-K and 20-K lines was improved. One of the changes was to add copper straps from the 20-K line to a thin copper strip that ran around the circumference of the outer face of the inner shell pair. A similar strapping mechanism was also added between the 80-K lines and the inner face of the outer shell pair. Then the 20-K and 80-K lines intercept more heat than before from the opposite side of the vacuum barrier, reducing the heat into the 4-K region. This allowed the length of the vacuum barrier to be reduced to about 8 in.

Due to high bending stresses at the connection between the front plates and the thin shell pairs, a reinforcing ring was added at these connections, giving a thicker, stronger connection that gradually tapers down to the thin shell.

ANALYSIS

A thermal and structural finite-element analysis was performed on the final design using an established code, ANSYS, version 4.4A. The vacuum barrier is made of 304 stainless steel, and the thermal straps utilize OFHC copper. Temperature-dependent elastic, thermal, and mechanical properties for these materials were used in this analysis. These properties were obtained primarily from "LNG Materials and Fluids"[3] and "Materials at Low Temperatures."[4]

Thermal Analysis

The thermal model of the vacuum barrier (Figure 1) was built almost entirely using ANSYS STIF57 elements, an isoparametric quadrilateral thermal shell element with four nodes in 3D space. The copper thermal strap connections were made using the STIF33 element, a thermal bar with two nodes in 3D space. The weld ring used to connect the concentric shell pairs at the return end was also modelled with the STIF33 element. The bellows connection between the cold mass and the front plate of the vacuum barrier was made using the STIF14 element, a 1D spring-damper element with temperature as its only degree of freedom. The equivalent thermal "spring rate" for the bellows was input as kA/L, where k, A, and L are thermal conductivity, cross-sectional area, and length. The model geometry was defined through the use of parameters for most of the dimensions on the vacuum barrier. This allowed for a more efficient optimization of the design, since new configurations could be quickly implemented and analyzed by changing only a few parameters and real constants.

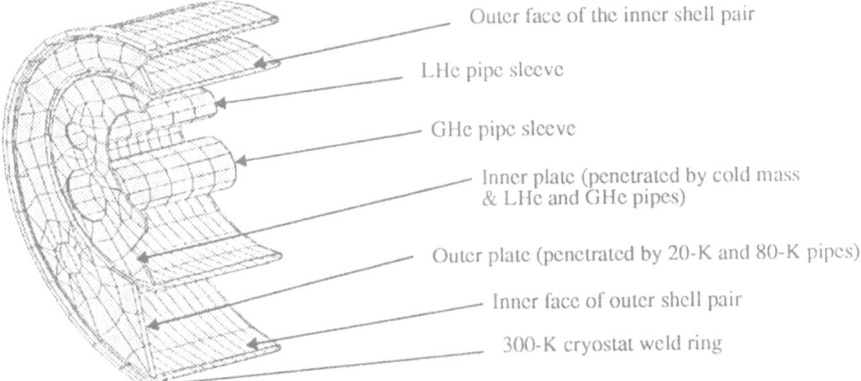

Outer face of the inner shell pair

LHe pipe sleeve

GHe pipe sleeve

Inner plate (penetrated by cold mass & LHe and GHe pipes)

Outer plate (penetrated by 20-K and 80-K pipes)

Inner face of outer shell pair

300-K cryostat weld ring

Figure 1. Finite element model of collider vacuum barrier concept. Cryogenic pipes not shown for clarity.

The boundary conditions for the thermal analysis included convection boundaries on the inner surfaces of all of the pipes. These boundaries were given temperature-dependent convection film coefficients and bulk temperatures of 4.25 K for the cold mass, LHe, and GHe lines, 20 K for the 20-K shield helium supply line, and 80 K for the 80-K shield nitrogen supply and return lines. The outer surface of the cryostat had natural convection with a bulk temperature of 317 K. Due to limitations within ANSYS, radiation was not included in this analysis.

Structural Analysis

After the thermal analysis, the modal was resumed and prepared for a linear static analysis. The STIF57 thermal elements were replaced with STIF63 elastic quadrilateral shell elements, and the STIF33 elements of the weld ring became STIF4 3-D elastic beams. The STIF33 elements that modelled the copper straps were removed, since these were not load-bearing members.

The boundary conditions used both fixed nodes and spring elements. The nodes on the outer cryostat were fixed at a location corresponding to the location of the fixed post, while the lead ends of the 4-K lines were connected to STIF14 longitudinal spring elements that simulated the stiffness of the fixed post. The lead ends of the 20-K and 80-K lines were connected to STIF14 longitudinal and torsional spring elements to simulate the stiffness of the 20-K and 80-K shield connections at the fixed post. The return ends of the cryogenic pipes and cryostat were left free and unconstrained.

This vacuum barrier model was subjected to three loading conditions: thermal loads at operating temperatures, 0.1 MPa pressure at room temperature, and 0.2 MPa pressure at operating temperatures. Additionally, a linear eigenvalue buckling analysis was done on this model to determine the critical buckling pressure. These results were then compared with theoretical calculations.

RESULTS

The results of the thermal analysis show that the final design has a heat leak of 0.131 W into the 4-K region, 12.08 W into the 20-K region, and 36.27 W into the 80-K region. These results are 1%, 17%, and 32% over the previously stated goals, respectively.

The results of the static structural analysis show safety factors with respect to yield of at least 3.7 for the thermal loads, 2.4 for the 0.1 MPa pressure loads, and 1.5 for the 0.2 MPa loads at operating temperatures. For all three load cases the safety factor was greater than 5.0 against ultimate tensile strength. Additionally, the buckling safety factor was 3.07.

CONCLUSION

This proposed design for the collider SPXA/SPRA spool piece vacuum barrier meets or exceeds all of the structural requirements of the applicable specifications, using less space and simpler manufacturing requirements than previous designs. Although the thermal performance of this design is less than the stated goals, this should improve by utilizing newer thermal strapping configurations presently being developed in the design of the vacuum barrier for the High Energy Booster of the Superconducting Super Collider.

REFERENCES

1. Hekking, F.J., "Segment Specification (Level 3B), Collider Accelerator Arc Sections," Superconducting Super Collider Laboratory, Number E10-000027, May 1992.
2. Cleveland, Earl and Webster, Tom, "Performance Specification (Level 4) for the SPXA and SPRA Spool Pieces of the Collider Arc Sections of the Superconducting Super Collider (SSC) Laboratory," Superconducting Super Collider Laboratory, Number AMA-3210005, March 1993.
3. Olien, Neil A., Ed., "LNG Materials and Fluids, Second Supplement," Prepared by Cryogenics Division, Institute for Basic Standards, National Bureau of Standards, Boulder, Colorado, 1979.
4. Reed, Richard P. and Clark, Alan F., eds., "Materials at Low Temperatures," American Society for Metals, Metals Park, Ohio, 1983.

DESIGN AND DEVELOPMENT OF THE SSC COLLIDER SPOOL PIECE

J. Marcus,[1] J. Wollan,[1] J. Cox,[1] R. Riney,[1] R. Johns,[1] E. Rodriguez,[1]
D. Audi,[1] S. Williams,[1] S. Covington,[2] J. Breuel,[2] and G. Brush[2]

[1]Martin Marietta Astronautics Group
P.O. Box 179
Denver, Colorado 80201

[2]Cryenco
5995 North Washington Street
Denver, Colorado 80216

Abstract: Martin Marietta is one of two contractors selected by the SSCL to complete the design and development of the collider standard spool pieces. Spool pieces are complex cryogenic components which provide control of fluid vacuum, power, instrumentation and protection functions for the collider. Requirements development, concurrent design approach, and planned spools test activities will be discussed.

INTRODUCTION

Approximately 1550 standard spool pieces will be required in the SSC collider arc sectors to provide thermal, electrical, and mechanical functions. These integrating functions include 1) routing cryogen fluids to appropriate subsystems and adjacent dipole and quadrupole magnets, 2) limiting heat input to the cryogenic systems via thermal shields, low thermal conductivity materials, multi-layer insulation, and cryostat vacuum, 3) maintaining the temperature of the 4 K helium via a recooler, 4) electrical connection between main magnets, 5) a bypass circuit for main magnet quench protection, 6) connections to external power supplies for corrector magnets, 7) monitoring beam position, 8) instrumentation for system control, 9) isolation of cryostat insulating vacuum, and 10) mounting and interface provisions for corrector magnets. An additional critical function is to assure alignment of the BPM and corrector magnets to the external spool piece fiducials. Also, the prototype spools must contain instrumentation to support performance verification of the spool subsystem.

Each standard spool piece must operate with very high reliability over a twenty-five year life. The spool pieces must be produced at a rate to meet the final collider installation schedules (in excess of forty units per month), and be produced at an average unit cost in "then year" dollars of $70,000.

REQUIREMENT DEVELOPMENT

To assure an orderly development process, Martin Marietta, working with the SSCL and the other contractor, established the spool piece functional, design, acceptance, and test requirements starting in October 1992. Early design work by the SSCL, combined with Martin Marietta pre proposal preparations, reduced the risk associated with simultaneous requirement development and prototype design in support of the aggressive Phase I program schedule.

Spool pieces are complex devices that operate as part of a multi-functional system under various steady state and transient conditions. Since Level 3B collider arc sector requirements are still being developed and revised, completing the Level 4 spool piece requirements has been a key challenge. A summary of Level 4 spool piece requirements is shown in Table I. This table envelopes approximately 300 detail requirements now established for the design of the prototype hardware.

Table 1. Summary Level Spool Piece Requirements

Subject	Operating Requirements		Comments
Cryogenics	• Temperature Range	- 3.97 - 84 K	1) The top level requirements listed are but a portion of 297 total requirements defined to date.
	• Pressure Ranges	- 0.08 - 0.5 MPa	
	• Differential Pressures	- 3-410 Pa	
	• Heat Leak Criteria	- 2.73 W at 4 K	
		16.42 W at 20 K	2) Fifty-four of the 297 requirements involve some verification by test.
		61.2 W at 80 K	
Vacuum	• Beam Tube	- 1.3×10^{-7} Pa	
	• Cryostat	- 1.3×10^{-5} Pa	3) An additional 50 requirements are anticipated from the emerging ICD. Most items are expected to confirm existing designs.
	• Vacuum Barrier	- Leak Tight	
		(2×10^{-11} Pa m3/sec)	
Environments (Induced)	• Shock	- 2.0, 0.5, 0.5 G's Load Diamond	
	• Radiation	- up to 10 MGy	
Quench Protection	• Current/Time	- 7000 A/36 s Time Constant	
Ilities	• Reliability	- 25 year life	
	• Maintainability	- MTTR-227 Hrs Cat 1, 4 Hrs Cat 2, 3	

Typically some requirements will vary inversely as a function of a given parameter and therefore limit design solutions. In an example, shown in Figure 1, the spool piece support posts must be strong enough to withstand a variety of loads (transportation, operation, etc.), but not become a major source of heat leak. A thick wall post would be stronger, whereas a thin wall post would have lower heat leak. Consequently trade studies and detailed analyses are necessary to define the optimal system configuration.

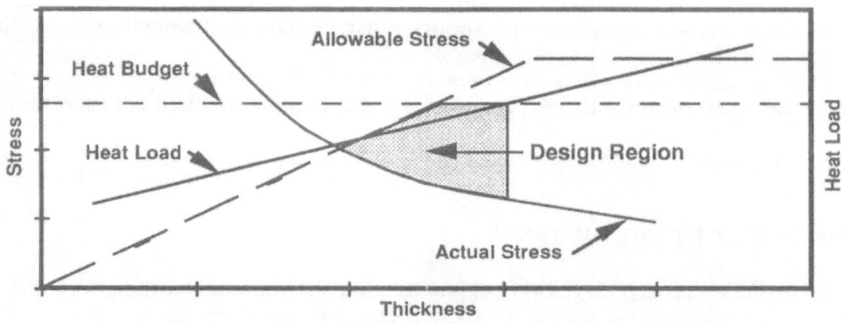

Figure 1 Integrated Thermal and Stress Post Optimization

CONCURRENT DESIGN APPROACH

The Martin Marietta spool piece team is a multi-disciplined, co-located group dedicated to this project. The team includes the SSCL, engineering (systems, analysis, and design), manufacturing, test, procurement, quality, business functions and a key industrial partner - Cryenco of Denver, Colorado. The team's decision-making relies on consensus. The group is dedicated to achieving the best technical and operational decisions on hardware and programmatic issues.

The Martin Marietta team prepared for the June 1992 spool piece solicitation by assimilating early versions of SSCL designs, developing requirements and conducting design trades, and fabricating an engineering development spool (Figure 2). This effort was wholly supported by internal Martin Marietta funds. It enabled a better understanding of SSCL needs and an appreciation of the hardware build process.

Figure 2 Martin Marietta Engineering Development Spool

At contract award, the SSCL provided a top-level "Collider Concept" data base for the standard SPXA (basic unit) and SPRA (basic unit with recooler) to be used as the starting point for industrialization of the spool pieces. Major subsystem elements of the Collider Concept are shown in Figure 3. These spool pieces include cryogenic piping which enables flow and control of cryogenic helium and nitrogen supply and return; access for pump down of insulating and beam tube vacuums; and provisions for beam position monitor, external connections, and corrector magnets. They also incorporate elements of the magnet quench protection system (quench stopper and bypass leads) and vacuum isolation (vacuum barrier). The SPRA includes a liquid helium recooler with a 100W minimum capacity.

The Martin Marietta concurrent design approach starts with trade studies of the major subsystems, evaluating both function and cost. The analytical tools used in the spools concurrent design process are predominantly stress (IDEAS, TK Solver, and ABACUS) and thermal programs (SINDA/FLUENT, TRASYS). Preliminary design efforts and supporting analytical activities proceed in parallel with operations and procurement functions. This iterative process sequentially addresses all subsystems and spool piece final assembly. Cost targets for production options are evaluated and action plans established to assure final production designs are consistent with the requisite $70,000 average unit production cost goal.

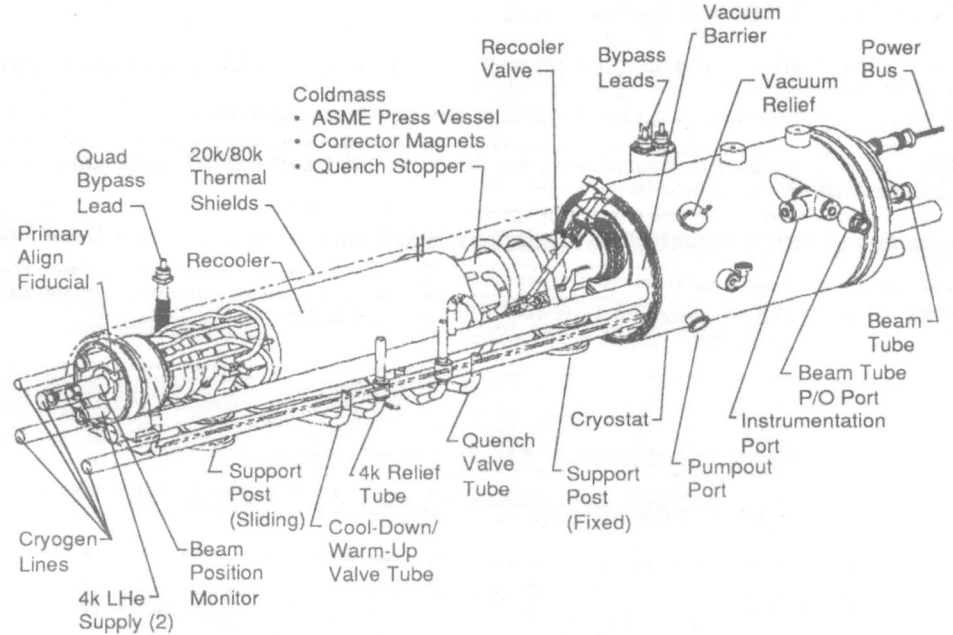

Figure 3 Collider Concept Spool Piece

SPOOL PIECE TEST PROGRAM

The SSCL/Martin Marietta spool piece prototype test program is designed to accomplish numerous objectives. The primary objective is to verify the adequacy of the design. Of the approximately 300 events in the design verification matrix, 54 require testing. Secondary objectives are to obtain data to verify analytical models and performance margins, to support trade studies, to gain experience for production acceptance, and to develop a data base for process control, reliability, and maintainability predictions. In addition to in-process testing accomplished during manufacturing, two major series of tests are planned: a partial functional verification in a stand-alone cold test stand, and a total spool functional verification in the ASST.

The stand-alone cold test in the SSCL Spool Piece Test Facility, will be capable of supplying 50g/s liquid helium and 7000 A through the main power bus. A 30-day test is scheduled for late August and September 1993. After ambient verifications are completed, the tests will include cool-down; cold integrity; steady state fluid, thermal, and electrical; transient fluid, thermal, and electrical; and warm-up. Additional performance verification test sequences will help define operational margins and procedures. These tests will provide confidence in spool piece performance prior to installation in the ASST and reduce the risk of affecting other ASST test objectives.

In the ASST test setup, collider equivalent quench conditions and steady state thermal environments (at 100 g/s) can be achieved. These ASST tests are planned for November and December 1993. The primary objectives of this test series are to verify the spool piece heat leak, quench performance, recooler duty cycle and control method, and to determine the effects of flow induced vibrations. The current program plan includes a CDR in October 1993, with open items closure after the ASST testing is complete.

SUMMARY

Spool pieces are essential elements of the super collider system. Their functionality and reliability must assure meeting collider operating time goals. Martin Marietta has used a multi-disciplined team approach to develop the prototype spool piece. The test series to be conducted at the SSCL in the Fall of 1993 will verify the design and analyses.

CORRECTOR MAGNETS: COMBINED STRUCTURAL ANALYSIS OF
COLLIDER 50MM APERTURE ORDERED WOUND DIPOLES INTERIOR SECTION

Vu H. Tran

Mechanical Engineering Department
Accelerator Systems Division
Superconducting Super Collider Laboratory *
2550 Beckleymeade Avenue, MS 4006
Dallas, TX 75237-3997

ABSTRACT

The 50mm aperture prototype collider ordered wound dipole corrector magnets have been modeled with finite element techniques considering the individual and combined load cases of the preloading from keys, cooldown to 4 K and the effect of magnetic forces during energizing. Results of the analysis are presented as longitudinal, transverse and shear stresses for the ordered wound coils and as maximum von Mises stress for the carbon steel outer laminations, the stainless steel inner lamination, and the carbon steel keys.

INTRODUCTION

The dipole corrector magnets are members of the collider ring correctors which are responsible for providing steering and closed orbit correction of the proton beams. These corrector magnets are located inside of the spool piece. This paper presents results from a finite element stress analysis of the prototype collider 50mm aperture ordered wound dipole corrector magnets. This analysis considers only the interior sections along the length of the structure - which differ greatly from the end sections. The coils for this dipole magnet structure are manufactured using ordered wound techniques. Load cases involved in the analysis include preloading from keys, cooldown to 4 K and magnetic (Lorentz) forces from energizing. The objectives of this stress analysis for the dipole structure are to determine the stresses in the ordered wound coils, the necessary shim thickness around the coils and to evaluate the capability of the outer laminations, inner laminations and keys to resist the stresses induced by reaction with the coils. The results of this analysis will be used as reference for further design of the dipole structure members - including the coils, the inner laminations, the outer laminations and the keys.

The dipole magnet structures consist of magnetic coils held in place by a series of thin, interlocking carbon steel outer laminations. Figure 1 shows two consecutive cross-sections of the dipole. The outer laminations are configured such that each cross-section of the magnetic structure along the length of the coils contains two interlocking outer lamination pieces: one above and one below the

*Operated by the Universities Research Association, Inc., for the U.S. Department of Energy under Contract No. DE-AC35-89ER40486.

magnetic coils. Each outer lamination piece is joined to the pieces immediately in front of and behind it by two continuous pins. Continuous longitudinal carbon steel alignment keys on the right and left sides of the outer laminations provide the "preload" forces necessary to hold the outer laminations in place and to exert pressure on the magnetic coils. The outer lamination sections immediately in front of and behind each individual outer lamination section are identical in geometry but are rotated 180 degrees about the vertical axis (y-axis), yielding the same external shape but a different configuration at the alignment keys. The alternating orientation of the outer lamination pieces permits the "elbow" section of each outer lamination piece keyway to be alternately placed above or below the alignment keys. Layer of Kapton are placed around the coils to act as a shim. Therefore, one of the design criteria in this analysis is to provide enough shim thickness to have the minimum stresses in the coils after cooldown to be at least greater than the maximum stresses in the coils due to magnetic forces only. This criteria should applied to each coil's local longitudinal, transverse and shear direction stresses in the coils. The ordered wound coils structure consists of continual loops of 0.381 mm (15.0 mil)-diameter superconducting Cu:Nb-Ti (2.2:1) wire insulated with a 0.076 mm (3.0 mil)-thick layer of Kapton and Dupont XMPI adhesive.

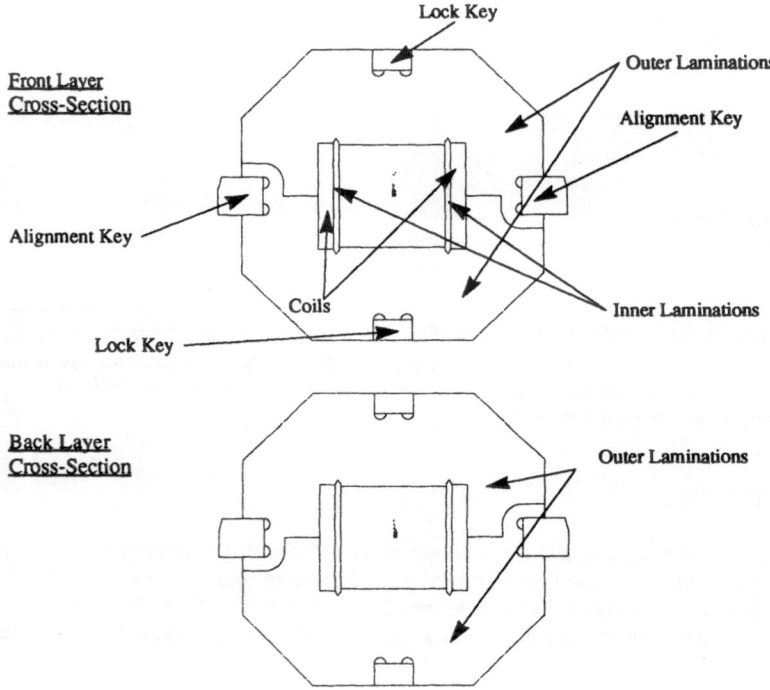

Figure 1. Collider 50mm Aperture Ordered Wound Dipoles - Front and Back Layer Cross-Sections

MODEL CONSTRUCTION

The magnet assembly is modeled as a two-dimensional finite element system. The model consists of two adjacent layers of outer laminations (each consisting of a top and bottom outer lamination piece from the same cross-section) hereafter referred to as the "front" and "back" outer laminations. To get the correct loading from the keys, the two layers of the outer laminations are modeled with two-dimensional plane stress elements. The front and back outer laminations are linked only with coupled nodes at each of the four continuous pin locations. Gap elements having a positive gap distance to act as a interference are used to join the coils to the outer laminations (gap elements only transmit compression load not tension load). These positive gap distances represent the thickness of the shim around the coils. All other gap elements between various parts are given a zero value. The coils, inner lamination and keys are constructed from two-dimensional plane stress elements. Material properties used in this model are shown in Table 1. Material properties for the ordered wound coils are obtained from testing and analysis.

Table 1. Material Properties

Materials		Carbon Steel	304L Stainless Steel [1]	Ordered Wound Coil
Young's Modulus (psi)	T = 293 K	30.0×10^6	27.6×10^6	200,000
	T = 4 K	30.6×10^6	29.2×10^6	300,000
Poisson's Ratio	T = 293 K	0.292	0.29	0.346
	T = 4 K	0.292	0.2788	0.339
Thermal Expansion [2] Coefficients, (K^{-1})		0.685×10^{-5}	1.059×10^{-5}	1.4×10^{-5}
Yield Tensile Strength (psi)	T = 293 K	25,000	58,900	N/A
	T = 4 K	54,000	79,400	N/A
Ultimate Tensile Strength (psi)	T = 293 K	42,000	95,500	N/A
	T = 4 K	65,000	241,000	N/A

LOADING CASES

Preload Case

As mentioned previously, the preloading case experienced upon assembly is simulated by assigning to all the gap elements around the two coils an overlapping gap distance to represent the shim thickness. All material properties for the preload case are referenced to 293 K.

Preload and Cooldown Case

The combined analysis of the preload and the cooldown uses the same model as does the preload alone except that the model is assigned an initial temperature of 293 K and a final temperature of 4 K to represent the cryogenic state after the cooldown has occurred. All material properties for the combined analysis are given at the reference temperature of 4 K.

Magnetic Forces Case

To complete a magnetic stress analysis of the model, a separate magnetic analysis must first be conducted. The magnetic analysis is performed using a somewhat simplified version of the dipole model. The magnetic analysis is concerned only with the location and the magnetic properties of dipole structure components; therefore, the model used in the said analysis contains only the geometric and physical description of the individual components and does not require the use of gap elements. The wire in the coils are given a current density of 4.0297×10^8 A/m^2 at 100% of short sample current, but only 66% is required for operation current. The forces on the nodes of the coils are computed in the magnetic analysis and then inserted into the standard dipole model. A finite element analysis is then performed on the standard model with the magnetic forces alone to determine the influence of the magnetic forces on each component of the dipole structure.

Preload, Cooldown and Magnetic Forces Case

To complete the final combined analysis, the standard dipole model is subjected to the simultaneous loadings of the preload, cooldown and magnetic forces. Once again, a finite element analysis is performed to determine the stresses on the individual components of the coil structure due to the three combined loading cases.

STRESS ANALYSIS RESULTS

In order to have the minimum stresses in the coils after cooldown to be at least greater than the maximum stresses in the coils due to magnetic forces only, the finite element analysis shows that a shim thickness of 4 mils must be used along the two shorter lengths of the two coils. These results are shown on Table 2. As shown in Table 2, for the coils the first value is the minimum and the second is the maximum, thus showing the range of stresses in the coils. The longitudinal and transverse directions are defined along the longer and shorter length of each coil, respectively. The transverse stresses in the coils did not meet the design criteria in this analysis due to the inadequate design of the inner laminations. Even if shims were added to the longer lengths of the coils, this will not produced the necessary transverse stresses in the coils. A new design of the inner laminations should be studied to induce transverse stresses into the coils. Stresses in other parts of the dipole are all less than the material tensile yield strength.

Table 2. Stress Levels for Collider Ordered Wound Dipoles

Loading Cases		Preload of Keys shim = 4 mils *	Preload of Keys & Cool Down to 4 K	Magnetic Forces Only $I = 66\% I_{ss}$ or $110\% I_{op}$	Preload of Keys & Cool Down to 4 K & Magnetic Forces
Coils Min : Max Stress (psi)	$\sigma_{transverse}$	-289 : 4	-5 : 9	-269 : -53	-284 : -53
	$\sigma_{longitudinal}$	-796 : -632	-370 : -343	-200 : -17	-509 : -314
	τ_{shear}	-38 : 38	-4 : 4	-11 : 11	-16 : 16
Carbon Steel Outer Laminations Maximum von Mises Stress (psi)		9,600	4,700	830	4,200
Carbon Steel Keys Maximum von Mises Stress (psi)		2,400	1,200	370	1,100
Stainless Steel Inner Laminations Maximum von Mises Stress (psi)		1,700	50	60	50
* shim thickness between the two shorter lengths of coils and outer laminations					

CONCLUSION

The finite element analysis for the collider ordered wound dipole corrector magnets shows that the stresses induced in the various magnet components under the specified load conditions are well within the acceptable range of allowable material strengths. The coils in this dipole design are loaded only in the longitudinal direction of coils. To help prevent transverse movement of the coils, future designs of the dipoles structure should include inner laminations capable of exerting pressure on the sides of the coils. It should be noted that different values for the material properties of coils will greatly effect the design and analysis. Therefore, accurate values of coils material properties must be obtained for further design and analysis. It should also be remembered that this analysis considers only cross-sections from the interior of the structure in a two-dimensional frame. As the cross-sections at the ends of the dipole structure differ significantly from those in the interior, additional analyses must be performed to evaluate their behavior.

REFERENCES

1. Richard P. Reed and Alan F. Clark, eds., *Materials at Low Temperatures*, American Society for Metals, Metals Park, Ohio, 1983, page 8.1.3-2.
2. Neil A. Olien, Ed., *LNG Materials and Fluids, Second Supplement*, Prepared by Cryogenics Division, Institute for Basic Standards, National Bureau of Standards, Boulder, Colorado, 1979, page 95.

METHOD OF FABRICATING A COMPLETELY ORDERED WOUND COIL

S. Mookerjee, W. Shen,* and B. Yager

Superconducting Super Collider Laboratory†
2550 Beckleymeade Avenue
Dallas, TX 75237-3997

ABSTRACT

An innovative four step coil winding process is under development at the Superconducting Super Collider Laboratory. This method is to be used for correction coils used in magnets for the HEB and Collider Rings of the SSC. High density, orderly wound coils with good mechanical stability and dimensional control are the end products of this technique.

INTRODUCTION

Approximately 10,000 Superconducting Correction Magnets are required for the HEB and Collider Rings of the Superconducting Super Collider. These magnets serve to fine tune the properties of the particle beam in the accelerators. They must operate at cryogenic temperatures for long periods of time while exposed to high radiation. Relatively low operating currents (< 100 A) and tight space constraints result in magnet designs having coils with high conductor densities. The large number of magnets places a premium on methods that can reliably mass produce coils, each with up to hundreds of turns of small (<20 mil) diameter wire. The SSCL has invented a four step coil production process that involves winding multiple layers of flat racetracks without splices and forming them into their final coil shape. Two curing steps are also employed in the process. The paper describes this "flat and form" coil fabrication process that is under development for making "ordered wound" correction coils.

FABRICATION TECHNIQUE

The flat and form coil winding technique consists of four steps. First, a stack of flat "racetrack" layers are simultaneously wound without splices between the layers (see Figure 1a.). The second step is a curing cycle that bonds the wires in each layer together without bonding the layers together. The third step forms the coil into its final shape and the fourth step is a second cure cycle that bonds the layers into a rigid coil (see Figure 1b).

---

* Present address: Texas Accelerator Center, 4802 Research Forrest Dr., The Woodlands, TX 77381.
† Operated by the Universities Research Association, Inc., for the U. S. Department of Energy under Contract No. DE-AC35-89ER40486.

For each cure cycle, the coil is supported by the tooling used in the prior step. The method is dependent on the tooling design as well as the use of thermoplastic bonding materials as part of the wire insulation. To date, the wire used has been 15 mil diameter NbTi superconductor (Cu:Sc = 2.4:1). It is insulated with 0.5 mil Kapton film (with a minimum of 50% overlap). The wrapped wire is overcoated with 0.5 mil layer of XMPI, a radiation resistant thermoplastic made by E.I Dupont Nemours Co. This thermoplastic is used as a bonding agent. The thermoplastic characteristics are integral to the bonding process. The following subsections will describe the bonding and fabrication steps in more detail. Brief descriptions of the tooling will also be included.

Step 1: Flat Winding Process

The flat wound uses a parallel plate mandrel. This is a series of thin plates of alter * Present address Texas Accelerator Center, 4802 Research Forrest Dr., The Woddlands TX, 77381 -nating thickness that are sandwiched between two thicker plates (the base plate and the top plate) Half of these thin plates are referred to as the Mandrel Plates. They are approximately the same thickness, if not thicker than the insulated conductor and, looking at an end view, these plates are of increasing width. This is so that when the coil is formed, the flat ends are all flush in height. The desired thickness of the coil (i.e., how many layers of "racetracks") determine the number of mandrel plates required.The other thin plates are the spacer plates. Their function is to separate and add rigid support to the conductor layers. The important characteristic feature of these spacer plates is that they all have a slot cut in one end. This is where the wires make their transition from layer to layer. (see Figure 2)

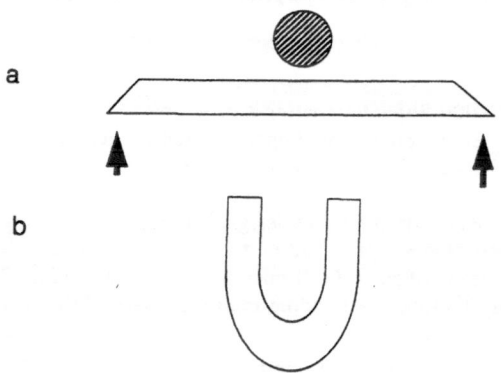

Figure 1. a) Unformed coil end view b) formed coil end view.

When the actual winding procedure is ready to commence, the winding fixture is fastened to a rotating table. The free end of the superconductor is fastened to either the base or top plate. The wire passes through a tensioning system. As the table begins to rotate, the wire will form itself around the bottom-most mandrel plate. After one full revolution, the wire is guided through the slot in the end of the spacer plate outwardly and is wound around the slightly wider mandrel plate on the next layer. Again, after one revolution, the wire filament is guided through the slot in the end of the spacer plate and helped up to the next level. This continues until the wire filament has made it to the topmost layer of the winding fixture. Once this has occurred, the process is repeated, working the wire inward, toward the table. This continues until the desired amount of turns in each layer has been reached.

436

Step 2: First Curing Cycle

Once the coil has been wound, fins are inserted into the slots where the conductor filament has been wound. The transverse pressure exerted by the fins enables the adhesive to extrude evenly, producing a better bond between layers. The winding fixture is then put in an oven preheated to 230° C. Presently, the bake time is approximately ten minutes at temperature. Thermocouples inserted into the winding mandrel monitor its temperature. After the cure time has elapsed, the winding fixture is removed from the oven and allowed to cool. Once this has been completed, the finished coil will resemble a series of Superconducting ribbons that are all joined at one end.

Step 3: Forming Procedure

The next step is to form the coil into its desired shape. The bending tool diagram is shown in figure 3. The coil is aligned properly and fastened to the center block. The center block serves to form the coil. The flat coil and center block assembly are placed into the channel, formed by the base plate and the two side rails. Pressure is applied all around the coil to ensure that the layers bond. Once the coil has been securely captured in the fixture, it is ready for its second cure cycle.

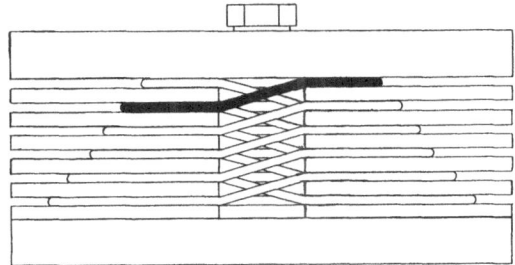

Figure 2. Winding tooling end view (Crossover-slotted end).

Step 4: Final Cure Cycle

Once again, the oven is preheated, this time to 250° C. The bending fixture assembly is placed in the oven for about 15 minutes at temperature. As with the winding fixture, thermocouples are used to monitor the fixture temperature. After time has elapsed, the oven is shut down, the doors opened and the bend fixture is allowed to cool. The coil is then carefully removed from the fixture and is now solidly formed in its desired configuration. For ease of removal, the bending fixture is coated with a Kapton film in areas that contact the coil.

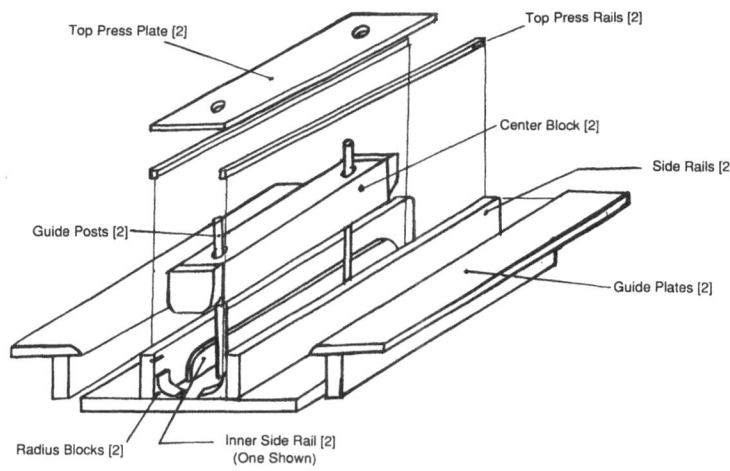

Figure 3. Bending tool.

PERFORMANCE RESULTS

This method of fabricating superconducting coils has been used for the past year-and-a-half by both the SSC Labs and by Everson Electric, a subcontractor to SSC. These coils have been used to manufacture eleven prototype Correction Magnets. Their performance has been inconsistent so far, but they have demonstrated potential. It must be noted that this technique is still under development. The best magnet to date (designated SCOb0.3) had a first quench at 89% design I_{SS}. It performed consistently well in four thermocycles. It was then disassembled for further study. Observation showed that these coils had a very small percentage of voids. One of the goals this technique promises to accomplish is to eliminate voids altogether. These two coils also demonstrated remarkable structural integrity, having an elastic modulus of between 1.3 million psi and 1.6 million psi. It must .be pointed out that there are several factors involved in the performance of the Correction Magnets. Many of these factors, such as coil/yoke interface and coil end /support, are directly related to the assembly procedure. Definitive correlations have not been made but the better magnets have had the best coils. As stated before, the technique is still being researched and observations and suggestions have been proposed to improved performance. These will be addressed in the following section.

Further Research

Most of the problems associated with the Ordered Wound Magnet performance seem to do more with the assembly of the Magnet than the fabrication of the Coils, although significant research has not been done to rule out any possibility. The main problem that occurred during coil fabrication was turn-to-turn shorts occurring in the bending phase of the coil. At first it was believed that reducing the pressure that the coils experience would eliminate this problem. This however, resulted in weaker bonds between the layers of the coil. It is now believed that thicker and more durable insulation around the conductor itself would be the best solution. One magnet made by Everson Electric was made in order to test this theory. It was free of shorts and performed moderately well in its quench test.

It was surmised that another possible reason for the turn-to-turn shorts may have been due to the material that the tooling was fabricated from. It was calculated that aluminum had the same thermal coefficient of expansion as the superconducting coil and hence would be the best material choice. It was later proved that the aluminum expanded much more than the superconductor at the higher range of temperatures it was exposed to. This caused large amounts of shear and tensile stress in the insulation. Currently, the material being used is stainless steel. While switching materials has reduced the amount of turn-to-turn shorts, it has not eliminated them completely.

A third alternative recently was discovered. Recent data has shown that if the superconducting wire is not properly annealed, the wire actually stress-relieves itself during the first bake after winding. The result of this is that the coil appears to shrink. If this is the case, the wire is inducing stresses in itself against the tooling.

Another parameter that is currently being studied is the cure time/temperature/pressure relationship. There appears to be a definite impact between this relationship and the structural integrity of the coil. In the early stages of development, there were some problems with delamination between layers. The problem was researched and ultimately solved by adjusting the temperature and pressure during first and second cure cycles. Further studies must be made to determine a method of controlling coil stiffness by adjusting these parameters.

While there are many variables in the performance of a Correction Magnet, the coils are the most significant part. If they are not properly and rigidly constructed, the magnet will display a poor quench performance. The flat and form technique has the potential to produce strong, densely packed coils with a minimum of voids. It is also a relatively simple alternative to coil fabrication. The potential of adapting it to mass production is currently being explored by industrial collaborators of the SSCL.

MAGNETIC MEASUREMENTS OF THE 5 METER QC SERIES QUADRUPOLES AT LAWRENCE BERKELEY LABORATORY

Paul Barale,[1] B. Benjegerdes,[2] S. Caspi,[2] M.I. Green,[1] A. Lietzke,[2] R. Schermer,[1] C. Taylor,[2] and D. Van Dyke[1]

[1]Magnetic Measurements Engineering Group*
[2]Superconducting Magnet Group
Lawrence Berkeley Laboratory
Berkeley, CA 94720

INTRODUCTION

From May 1991 to September 1992, magnetic measurements were performed on six 5 meter prototype SSC quadrupoles designed and built at Lawrence Berkeley Laboratory (LBL). In addition, one of the quadrupoles was disassembled, reassembled and remeasured. The purpose of this paper is to review the magnetic measurements program and give a summary of some of the results of the magnet testing.

THE MAGNETIC MEASUREMENT PROGRAM

The "MFM" Magnetic Measurement System

The "Magnetic Field Measurement" (MFM) System is a general purpose rotating coil, harmonic analysis magnetic measurement system developed at LBL for measurement of the SSC 5 m prototype quadrupoles. The system has the following significant features:
1. Externally driven tangential measuring coils with quad and dipole bucking coils
2. External optical encoder - angular position to 43.75 µradians absolute
3. Integrated induced voltage measured using digital integrator system
4. Analog bucking to obtain high resolution harmonic content of magnetic field
5. UNIX workstation and VME crate in host-target configuration - all real time data acquisition tasks in dedicated VME crate, operator interface, data analysis and display, and other non-real time tasks in UNIX workstation.

The MFM control, acquisition and analysis software packages were all written in C and developed at LBL. Acquired data is corrected for linear drift, rotated into a 'standard' frame of reference (south pole at 45° when viewed from the non-lead end), and corrected for differences between search coil axis of rotation and the magnetic center of the magnet. In addition, all warm measurements are taken with both positive and negative currents and the results averaged, to remove effects of external fields (e.g. earth's field, iron remnant magnetization in yoke, etc.).

* This work was supported by the Director, Office of Energy Research, Office of High Energy and Nuclear Physics, Division of High Energy Physics, of the U.S. Department of Energy Under Contract No. DEAC03-76SF00098.

The Measurements Program

All magnets were measured both at room and cryogenic temperatures. Our typical test program includes the following:

Room Temperature
1. Uniform field region, ±14 A
2. Axial scan, ±14 A

Cryogenic
1. Uniform field region, 0->6.6->0 kA
2. Axial scan, 3kA
3. Harmonic decay, 640A

Operational details of these tests can be found elsewhere[1].

Both the measurement hardware and software were under development during the measurement of QCC401 and QCC402, consequently the testing of these two magnets was not as extensive as the remainder of the group. In addition, scheduling and research emphasis dictated somewhat the test selection from magnet to magnet.

RESULTS

Axial Scan

The axial scan data provided the most interesting and useful results. Figure 1 illustrates one unexpected effect noted in all cold axial scans. As can be clearly seen, the b5 multipole (first allowed harmonic of b1, the quadrupole) is distorted for positions < 140 cm and > 400 cm, leaving a "uniform field" region of ~260 cm. Centered in these distorted regions are the strain gauge packs, whose influence is integrated over a wide region by the 1 meter long measuring coil. A similar effect can be seen in the b9 (second allowed) harmonic. The effect is not seen in the warm axial scans.

Table 1. Multipole Summary [units]

	a2	a3	a4	a5	a6	a7	a8	a9
QCC402	-0.724	0.553	-0.091	0.222	0.012	0.028	-0.057	0.065
QCC403	1.296	-0.832	0.167	0.010	-0.008	-0.001	0.013	-0.013
QCC404	1.552	0.103	0.126	-0.115	0.015	0.031	0.029	-0.001
QCC405	-0.485	0.280	0.147	-0.046	0.036	0.023	0.005	0.000
QCC405A	-0.456	0.414	0.174	-0.043	0.034	0.017	0.005	0.002
QCC406	-0.662	0.467	-0.236	0.005	0.004	0.025	0.003	0.009
SSC Spec.	2.696	1.550	0.641	0.738	0.209	0.240	0.276	0.317

	b2	b3	b4	b5	b6	b7	b8	b9
QCC402	-0.284	-0.041	-0.195	-1.082	0.086	-0.072	0.002	0.102
QCC403	1.189	-0.155	0.058	-0.850	-0.025	0.041	0.101	0.234
QCC404	-0.694	-1.009	0.023	-0.352	-0.064	-0.014	0.023	0.147
QCC405	-0.184	-0.111	0.195	-0.082	0.044	-0.004	0.008	0.091
QCC405A	-0.151	0.306	0.215	0.004	0.043	-0.016	-0.002	0.087
QCC406	0.148	0.128	0.162	-1.486	0.028	0.006	0.001	0.177
SSC Spec.	2.696	1.550	0.641	1.680	0.209	0.240	0.276	0.776

SSC Spec. in this case is defined as 1 Systematic + 1 RMS for each multipole[3].

Multipole Summary

Table 1 summarizes the average cold axial multipoles for magnets QCC402 through QCC406. The data quoted are the average values over all axial scans and can be considered representative of the "uniform field" region of each magnet at 3000 amps. Detailed explanation of the analysis used to obtain these results are available[2].

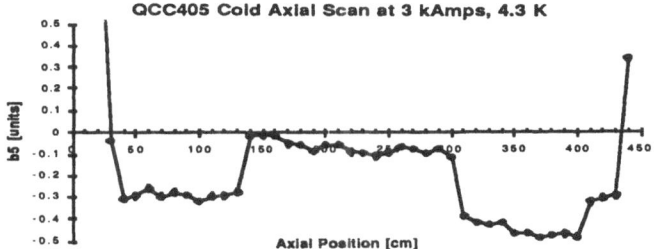

Figure 1. b5 vs Axial Position for QCC405 Cold Axial Scan

Warm-Cold Correlation

Multipole data obtained from warm and cryogenic axial scans were examined for correlation. Axial scan data was used to gain enough points to achieve statistical significance and because some of the multipoles were sensitive to axial positioning. Data distorted by the strain gauges were excluded. Figures 2 below illustrate the correlation for the first allowed multipoles. These multipoles were not corrected for persistent current magnetization effects in the cold measurements (expected to be -0.08 units for b5 and 0.004 units for b9). The data for these multipoles is clustered around points which lie on the diagonal, suggesting little geometric change during cooldown. All the low order multipoles (b2 through a4) exhibit some degree of correlation. In the higher order multipoles, the data lies on a line parallel to the x-axis, suggesting that these are dominated by noise in the warm measurements.[2]

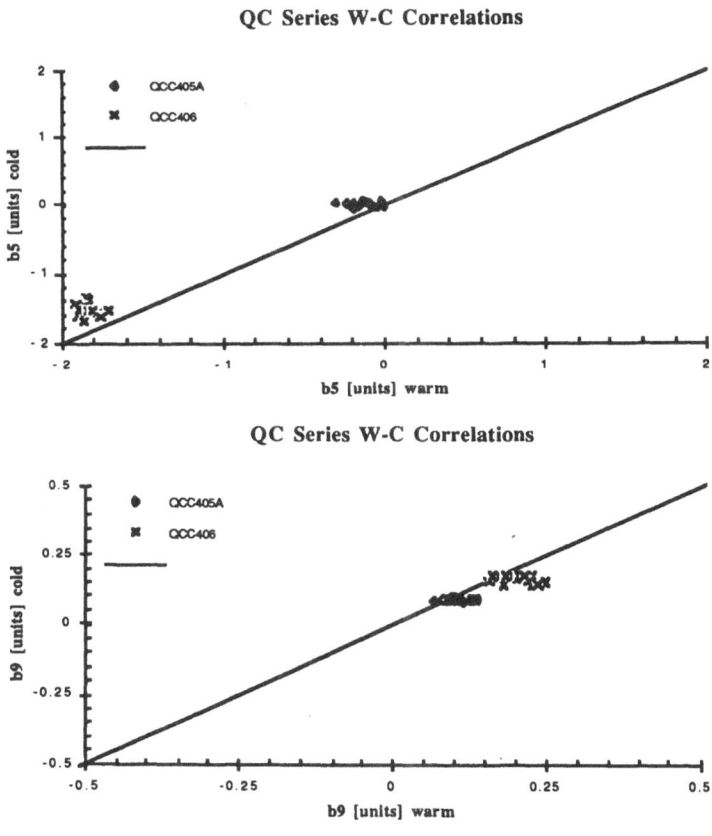

Figure 2. Warm-Cold Correlation of Allowed Multipoles

Injection Decay

None of the QC series quadrupoles exhibited significant injection current first allowed harmonic decay, in part due to small b5 values for the series (except for QCC406, where a significant b5 was deliberately designed in). However, even in QCC406, the decay is on the order of a tenth of a unit over an hour - considerably smaller than decays seen in the SSC dipole first harmonic. Decay for magnets QCC401 through QCC404 are not included in Figure 3 (below) as they are dominated by power supply instabilities. These instabilities were reduced for QCC405 and QCC406, and further corrected prior to QCC405A.

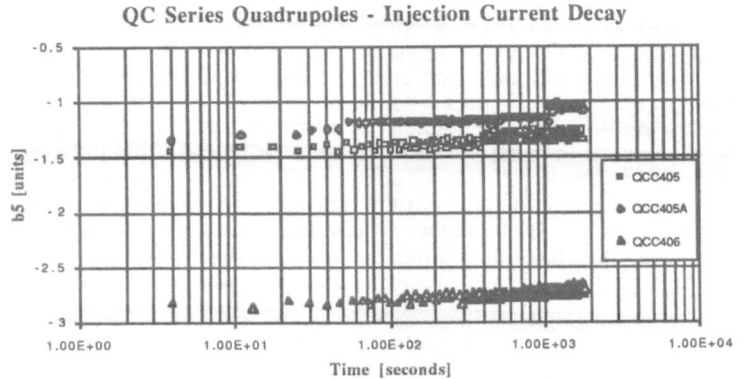

Figure 3. b5 Decay at Injection Current

SUMMARY

From the magnetic measurements performed on the QC series quadrupoles at LBL, it is clear that this design easily meets the SSC specifications on field quality. In addition, multipole decay during injection should not be a significant problem. Finally, it appears that there may be sufficient warm-cold correlation of the lower multipoles to allow for acceptance/rejection decisions to be based on warm measurements, particularly if a more sensitive coil, or higher currents are used for the warm measurements. On the down side, there can be significant local variations in field and field quality that will show up only on axial scans. These are, however, local variations and the effects are minimized when one considers the magnet integral field.

Overall, the QC series of quadrupoles and the related magnet measurement program have to be considered a success, from the standpoint of field quality.

REFERENCES

1. P.J. Barale. "Typical Magnetic Measurements of 5 Meter SSC Quadrupoles for the LBL Superconducting Magnet Group," Lawrence Berkeley Lab LBID 1838, SC-MAG-358, MT-441 (March 1992).
2. Norihito Ohuchi. "Z-scan Magnetic Measurements of LBL Quadrupole Magnets (QCC402-406)," SSC Laboratory, Test and Data Management MD-TA-251 (March 1993).
3. SSC Internal Note Document Number M80-000007, SSC Laboratory (May 1991).

CURRENT LOOP DECAY IN RUTHERFORD-TYPE CABLES

A. A. Akhmetov,[1] A. Devred,[1] R. G. Mints, [2] and R. I. Schermer[1]

[1]Superconducting Super Collider Laboratory*
2550 Beckleymeade Avenue
Dallas, TX 75237, USA

[2]School of Physics and Astronomy
Tel-Aviv University
Tel Aviv, Israel

INTRODUCTION

Recent measurements of superconducting particle accelerator magnets made of multi-strand Rutherford-type cable have shown that the magnetic field and its main harmonics oscillate along the magnet axis with a wavelength nearly equal to the cable transposition pitch length.[1,2] It was also observed that, at low transport current, the periodic magnetic field patterns can persist without any significant decay for more than 12 hours.[1]

The coincidence of the wavelength of the magnetic field oscillations with the cable transposition pitch suggests that slowly decaying current loops exist in the cable even at zero transport current.[3] These loops consist of currents flowing along the cable through one set of strands and returning through another set of strands. In this paper, we consider the process of current loop decay in a Rutherford-type cable.

MODELLING OF A RUTHERFORD-TYPE CABLE

The inner (outer) cables used in SSC 5-cm aperture dipole magnets consist of 30 (36) strands, twisted together, and shaped into a flat slightly keystoned cable. Due to the twisting, each strand goes successively from the inner edge of the cable to the outer edge, and back to the inner edge, over a distance, l_p, given by

$$l_p = \sqrt{4w^2 + p_c^2} \quad , \tag{1}$$

where w and p_c designate the width and the transposition pitch length of the cable.

*Operated by the Universities Research Association, Inc., for the U.S. Department of Energy under Contract No. DE-AC35-89ER40486.

Over the distance l_p, the strand crosses over and has electrical contact with all the other cable strands. Estimate of the cross-over resistance between two cable strands, R_c, varies from 1 to 100 $\mu\Omega$.[4] The electrical contact between a given strand and the rest of the cable can thus be characterized by a transverse conductance per unit length, G_t, given by

$$G_t = \frac{2(N-1)}{R_c\, l_p} \quad , \tag{2}$$

where N designates the number of cable strands. For $l_p \approx 100$ mm, $R_c \approx 5\ \mu\Omega$ and $N = 30$, we have $G_t \approx 10^8\ \Omega^{-1}\mathrm{m}^{-1}$.

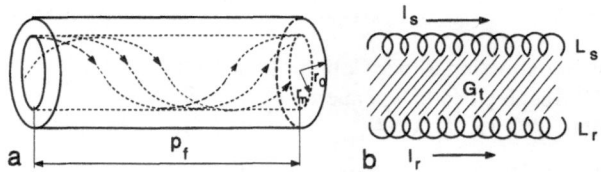

Figure 1. Rutherford-type cable: (a) strand geometry, (b) model describing the process of current redistribution between a given cable strand and the rest of the cable.

As shown in Fig. 1(a), the SSC strands consist of an inner core and an outer sheath of pure copper, surrounding an annular superconducting multifilamentary composite. The filaments of the composite are twisted, with a twist pitch, p_f, different from p_c. While in the superconducting state, the current is confined within the helicoidal filaments. Such a current distribution produces a longitudinal field, H_{sl}, in the strand inner core, and a circular field, H_{sc}, in the strand outer sheath.[5] Calculating the magnetic energy associated with H_{sl} and H_{sc} allows one to estimate their contributions, L_{sl} and L_{sc}, to the strand self-inductance. These contributions can readily be expressed per strand unit length as

$$L_{sl} = \mu_0\, \frac{\pi\, r_m^2}{p_f^2} \quad , \tag{3a} \qquad\qquad L_{sc} = \frac{\mu_0}{2\pi} \ln\!\left(\frac{r_0}{r_m}\right) \quad , \tag{3b}$$

where μ_0 is the magnetic permeability of vacuum, r_m is the average radius of the strand multifilamentary area and r_0 is the strand outer radius. For $r_0 \approx 0.4$ mm, $r_m \approx 0.3$ mm, and $p_f \approx 15$ mm, we have $L_{sl} \approx 2\ 10^{-9}$ H/m and $L_{sc} \approx 6\ 10^{-8}$ H/m.

In this paper, we shall assume that the process of current redistribution between a given strand of a Rutherford-type cable and the remaining $(N-1)$ strands of the cable can be described by the model given in Fig. 1(b). Here, G_t is the transverse conductance calculated above, and L_s and L_r are two effective inductances per unit length defined as

$$L_s = L_{sl} + L_{sc} \quad , \tag{4a} \qquad\qquad L_r = \frac{L_s}{N-1} \quad . \tag{4b}$$

CURRENT REDISTRIBUTION AT LOW TRANSPORT CURRENT

Applying Faraday's law to the circuit of Fig. 1(b) we get

$$L_s \frac{dI_s}{dt} + E_s - L_r \frac{dI_r}{dt} - E_r = \frac{1}{G_t} \frac{d^2 I_s}{dx^2} \quad , \tag{5}$$

where I_s (resp., I_r) and E_s (resp., E_r) are the current and the electric field in the given strand (resp., the remaining $(N-1)$ strands). Let us now assume that I_s and I_r can be written

$$I_s = I^* + \delta I \quad , \qquad (6a) \qquad\qquad I_r = (N-1) I^* - \delta I \quad , \qquad (6b)$$

where I^* is constant and uniform and $\delta I \ll I^*$. If I^* is much less than the strand critical current, I_c, we have: $E_s = E_r = 0$, and Eq. (5) becomes

$$(L_s + L_r)\frac{d(\delta I)}{dt} = \frac{1}{G_t}\frac{d^2(\delta I)}{dx^2} \quad . \qquad (7)$$

With the boundary conditions $\delta I(x=0,t) = \delta I(x=l,t) = 0$, where l is the strand length, the solution of Eq. (7) can be written as a sum of time-decaying waves

$$\delta I(x,t) = \sum_{k=1,2,...} I_{k0} \sin\left(\frac{2\pi x}{l_k}\right)\exp\left(-\frac{t}{\tau_k}\right) \quad , \qquad (8)$$

where I_{k0} is the initial wave amplitude, and l_k and τ_k are given by

$$l_k = \frac{2l}{k} \quad , \qquad (9a) \qquad\qquad \tau_k = (L_s + L_r)\, G_t \left(\frac{l_k}{2\pi}\right)^2 \quad . \qquad (9b)$$

It follows from Eqs. (9) that the larger the wavelength, the longer the time constant. This can be understood if one considers that in our model, the inductances are directly proportional to the wavelength, while the transverse resistance is inversely proportional to l_k. The largest possible wave corresponds to $k = 1$. With the values of G_t, L_s and L_r calculated above, and $l \approx 10^3$ m, we have $\tau_1 \approx 7\ 10^5$ s.

CURRENT REDISTRIBUTION AT LARGE TRANSPORT CURRENT

Let us know consider the case when I^* is close to the strand critical current, I_c. In this case, we need to take into account the influence of the longitudinal electrical field. For a superconducting multifilamentary strand, it is well known that

$$E = E_0 \exp\left(\frac{I - I_c}{I_0}\right) \quad , \qquad (10)$$

where E_0 and I_0 are two constant parameters. (For SSC inner strands, $E_0 \approx 10^{-5}$ V/m, $I_0 \approx I_c/30$, and $I_c(4.3$ K,5 T$) \approx 500$ A.) If $\delta I \ll I_0$, it can be shown that

$$E_s = R^* \left(I_0 + \delta I\right) \quad , \qquad (11a) \qquad\qquad E_r = R^* \left(I_0 - \frac{\delta I}{N-1}\right) \quad , \qquad (11b)$$

where R^* is a resistance per unit length given by

$$R^* = \frac{E_0}{I_0} \exp\left(\frac{I^* - I_c}{I_0}\right) \quad . \qquad (12)$$

Introducing the above expressions of E_s and E_r into Eq. (5) yields

$$(L_s + L_r)\frac{d(\delta I)}{dt} = \frac{1}{G_t}\frac{d^2(\delta I)}{dx^2} - \frac{N}{N-1}R^* \delta I \quad . \qquad (13)$$

Once again, the solution of Eq. (13) can be written as a sum of time-decaying waves

$$\delta I(x,t) = \sum_{k=1,2,...} I_{k0} \sin\left(\frac{2\pi x}{l_k}\right) \exp\left(-\frac{t}{\tau_k}\right) \exp\left(-\frac{t}{\tau^*}\right) \quad , \tag{14}$$

where

$$\tau^* = \frac{(N-1)(L_s + L_r)}{N R^*} \quad . \tag{15}$$

If $I^* > 3I_c/4$, then $R^* > 10^{-10}$ Ω/m. With the values of L_s and L_r calculated above, we have $\tau^* \ll \tau_1$. For long waves and at large transport current, the time constant of current redistribution is thus τ^*, which does not depend on the wavelength. Fig. 2 presents a plot of τ^* as a function of the cable transport current, $I_t = N I^*$. For this computation, we assumed $I_c = I_c(4.35 \text{ K},B)$, where $B \approx 10^{-3} I_t$ corresponds to the field on the pole turn of the inner coil of SSC 5-cm-aperture dipole magnet. It appears that, for $I_t > 6500$ A, the characteristic time of the current redistribution process decreases sharply.

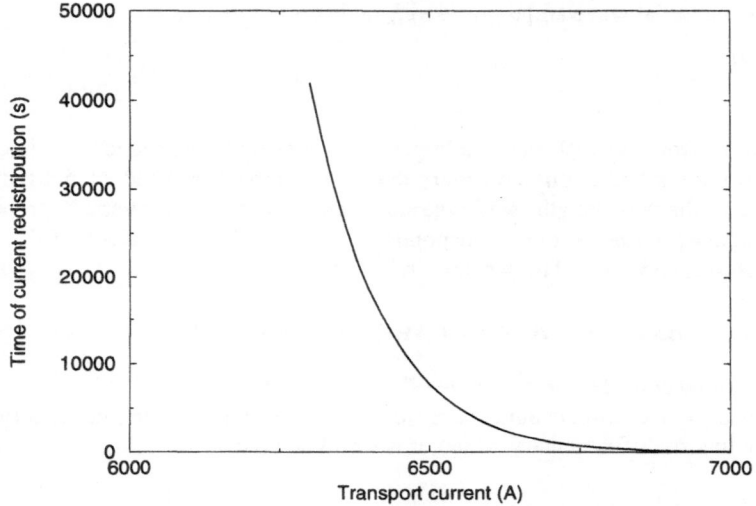

Figure 2. Dependence of τ^* on cable transport current for the pole turn of SSC dipole magnet.

SUMMARY

We have shown that the time constant of current redistribution between one strand of a Rutherford-type cable and the rest of the cable depends on the transport current. At low transport current, the process is dominated by the transverse conductance, resulting in a long time constant, while at large transport current, the series resistance forces the current to redistribute much faster.

REFERENCES

1. P. Schmüser, "Superconducting Magnets for Particle Accelerators," AIP Conference Proceedings, **249**(2), 1992, pp. 1099-1158.
2. A. K. Ghosh, K. E. Robins, and W. B. Sampson, "Axial Variations in the Magnetic Field of Superconducting Dipoles," *Supercollider 4*, J. Nonte, ed., 1992, pp. 765-772.
3. R. Stiening, private communication.
4. A. K. Ghosh, private communication.
5. M. Wake, private communication.

OPTICAL MEASUREMENT OF SSC ARC QUADRUPOLE
COLD MASS LOCATION

C. Corbett,[1] L. Ketcham,[1] R. Pletzer[1] S. Shapiro,[1] R. Viola,[1] Jeff Walton,[1]
R. Wilkins,[1] S. Bagwell,[1] and W. Barnes[2]

[1]Superconducting Super Collider Laboratory[*]
2550 Beckleymeade Ave.
Dallas, TX 75237

[2]Precision Optics Corporation
22 East Broadway
Gardner, MA 01440

INTRODUCTION

The operation of the collider requires that the ARC quadrupoles' beam center and rotational configuration be held to the tight tolerances of 0.2 mm rms and 0.5 mradian rms. The design of the collider superconducting magnets prevents direct access to the cold mass during operation. As a result, cold mass position is inferred from the relationship established by warm measurements relative to a secondary fiducial located on the vacuum vessel and an offset (warm to cold correlation) to account for the position change for a cold magnet. There is concern that the relation of these fiducials to the magnet centerline may change unpredictably due to the temperature drop required for superconducting operation or due to long term creep. This paper reports on a design that will allow visual observation of the cold mass shell position and roll orientation under cold as well as warm conditions.

This paper will present a description of the window configuration and status/results of tests which have been conducted warm and cold in the single magnet configuration. Future testing in a multiple string configuration is being considered.

OPTICAL WINDOW DESIGN

The optical window design was considered as the most straightforward design approach of measuring cold mass position during cold operating conditions. Babcock & Wilcox (B&W) conducted a trade study on various approaches to an "in-situ" cold mass fiducial.[1] The conclusion of the B&W study was that the optical system presented potential reliability problems with vacuum integrity. CERN's approach to the problem[2] is to attach a silicon rod, covered with gold for electrical contact, to the cold mass and determine cold mass position through eight active capacitive sensors which are installed around the rod on two levels and in two directions. One additional sensor is located above the rod and used to determine vertical movements. All nine sensors are attached to

[*] Operated by the Universities Research Association, Inc., for the U.S. Department of Energy under Contract No. DE-AC35-89ER40486.

the flange of the cryostat and there is no physical contact between the rod and the sensors and the sensors work at normal temperatures.

The design goals for our project were to not compromise vacuum integrity, minimize any extra heat leak to the cold mass, and to have enough resolution to make meaningful measurements relative to the tight tolerance for the quadrupole. The design which met these goals is detailed below. The design assumed the position of the magnet field relative to these fiducials on the cold mass would not change from warm to cold.

Mechanical Design

The mechanical design is composed of a clear optical window in the vacuum vessel, an optically coated window in the 80 K shield, a through hole in the 20 K shield, cold mass autocollimation mirrors and targets, and vacuum vessel fiducials. Four locations, at two axial stations on the magnet were instrumented in this manner. A cross section showing the assembly is shown in Figure 1. The tolerances and specifications for the glass, and mechanical parts are contained in Reference 3. A pentaprism was used to make the optical measurements through the top ports.

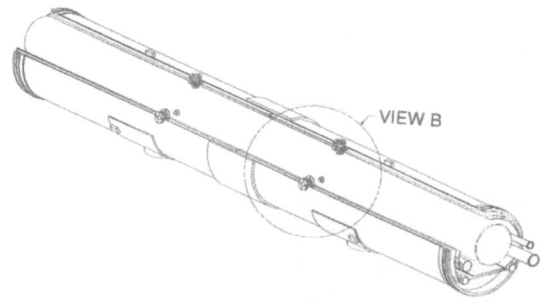

Figure 1. Optical Window Installation.

Thermal Analysis

A thermal analysis was performed in order to determine the effect of the optical windows on the 4 K heat load. The baseline QCC 4 K heat load is predicted to be 0.086 watts (0.232 watt 4 K static budget). Without windows in either the 20 K or 80 K thermal shields the 4 K heat leak would rise to 2.087 watts or a factor of 20. With no 20 K window and an 80 K window with a surface emissivity of 0.15 (actual coating was less) the heat leak was 0.104 watts or a 16% increase over the baseline. This was considered acceptable for these R&D magnets.

Optical Design

A first item of concern was the possibility of apparent mirror displacement caused by warpage in the 80 K shield. A ray tracing analysis performed at Precision Optics for an initial window angle of 20 degrees, yielded an apparent fiducial motion of 6.4 micrometers for a 2 degree change in window tilt. It was judged that this would not invalidate the observations of mirror fiducial motion.

Heating of the window by the 300 K vacuum vessel is of concern since the center of the 80 K window could be at a significantly higher temperature than 80 K. Assuming a good conduction path provided to the aluminum flange by the pressure of the wavy washer, the modified Bessel function solution[4] for this case of heat conduction may be simplified to obtain the center to edge temperature difference in the window as:

$$\Delta T = (300\text{-}80) \times \left[1\text{-}1/ I_o \left(a\sqrt{(h/kd)}\right) \right] .$$

I_o is the modified Bessel function of the first kind, a is the window radius, h is the coefficient of heat transfer, k the window thermal conductivity, and d is the window thickness. The temperature differential is 95 K for an uncoated window. An indium tin-oxide (ITO) coating reduces this temperature differential to 9.3 K. ITO coating both sides of the window would further reduce the temperature difference even further but would also reduce the visible transmission to 65%.

The optical aberration effect of the radial temperature distribution was also examined by Precision Optics. At these low temperatures the index of refraction and thermal length change are not linear functions of temperature. Index values for fused silica are found in Waxler and Cleek[5] and length data as NIST SRM 739. Data for BSC-2 glass appear in Molby.[6] Calculations for the 80 to 175 K temperature differential range give values of optical path difference of 113 nm for BSC-2 and 438 nm (about 0.7 visible wavelengths) for fused silica. The effect of these changes on alignment might be acceptable since most of the temperature difference occurs near the edge of the window; however, the use of the ITO coating substantially eliminates this source of optical aberration.

INSTALLATION

There were no major obstacles to the installation of the optical windows into the QCC magnets. Surveyors were required to get the proper orientation of the cold mass mirrors/targets or else a slight cold mass rotation during cooldown might have resulted in the loss of the ability to autocollimate. Precise placement of the mirrors and targets was observed to make sure that everything would line up and there would be enough room to see the targets warm as well as cold. Worst case axial thermal movement of the cold mass at the axial location selected was predicted to be 7.3 mm. The MLI covering the thermal shields will shrink more than the thermal shields, as a result, it must be cut back slightly larger than the window aperture. In the second magnet, the MLI was taped over to prevent water vapor condensation from the MLI spacer material as discussed below.

TEST PLANS

The assessment of adequate warm/cold alignment correlation and long term stability will be made by optically measuring cold mass position while warm and then again after cooldown to operating temperature to determine the change in position. Periodic measurements will be made over time to determine long term stability. These optical measurements will be cross checked with a beam tube optical target to survey beam tube positional changes between warm and cold states. The use of an optical target requires that a single magnet be tested on a cold test stand in order to make these measurements.

SSCL Applied Geodesy will establish warm and cold mass position relative to a monumented position by both autocollimation and physical measurements of the coordinates on the cold mass mirror. The same procedure will be followed after magnet cooldown, and the change of cold mass position and roll angle will be determined by the difference between the measured values. Additionally, these measurements will be repeated over time to determine long term positional stability. Applied Geodesy has estimated the accuracy of these measurements to be in the range of +/– 50 micron and 20 microradians. The expected results of these tests will be a measurement of the cold mass positional change from warm to cold and a comparison to the overall CQM alignment budget.

This test will be done at the ASST phase II string test; however, prior to installation in the string QCC 406 was cold tested as a single magnet. This provided an opportunity to make an independent assessment of the optical measurement of cold mass position by a second method. The second method requires a survey of cold mass position warm and cold by using a beam tube optical target. A secondary goal of this test is to measure the stability of the cryostat vacuum vessel fiducials which will be mounted in locations that will closely approximate the production CQM positions.

TEST RESULTS TO DATE

Currently, four epochs of measurements have been completed on QCC 406. Unfortunately, difficulties were encountered in the data collection for the first two measurement epochs. Between the first and second epochs of measurement, going from the warm to cold state, the quadrupole was moved along the test bench a number of centimeters, which negated the first epoch of warm measurements. For the second epoch, cold measurements directly to the cold mass were prohibited by a buildup of ice crystals on the 80 K windows and condensation on the mirrors attached to the cold mass. Although three autocollimations and a single vertical translation value were measured, the accuracy of these measurements is dubious at best due to the difficulties created by the limited line of sight access and the possible refraction problems created by the ice particles. Epochs three and four were performed with the quadrupole in its warm and cold states, respectively, and a full set of measurements to the cold mass were successfully gathered for each epoch. These results are summarized in Table 1 with the positions 1 and 2 located at the fixed post end of the quadrupole while positions 3 and 4 are located in the sliding post region.

This single comparison shows the cold mass arching slightly in the horizontal plane (the horizontal directions at positions 1 and 4). The vertical translation is the same at each of the post locations. The magnitude of this "monolithic" translation downward when the magnet is cooled is consistent with warm/cold survey measurements of the center of the beam tube made during an earlier thermal cycle.

The measurement of roll through the side windows, positions 1 and 4, is as much as a factor of six less than that measured through the top windows (positions 2 and 3). The reason for this variation between the top and side positions at each post region is unclear and is still under investigation. More study and/or measurements will be required before any definite conclusions can be drawn from the data.

Table 1. QCC 406 Test Results to Date.

Position	Azimuth Cold – Warm (mrad)	Zenith Angle Cold – Warm (mrad)	Vertical Translation Cold – Warm (mm)	Horizontal Translation Cold – Warm (mm)
Epoch 3 – 4				
Fixed Post				
#1 (side)	0.34	0.13	–0.67	N/A
#2 (top)	N/A	–0.50	N/A	–0.16
Sliding Post				
#3 (top)	N/A	–0.20	N/A	–0.11
#4 (side)	–0.23	0.03	–0.67	N/A

N/A – not applicable, that type of measurement not possible

FUTURE WORK/SUMMARY

The data from the single magnet test stand will continue to be evaluated. Similar measurements will be taken for QCC 405A and QCC 406 as they are installed and operated in the ASST string tests. The ultimate result is to determine the validity of the warm to cold correlation approach and the stability of the correlation over time. The authors acknowledge that two magnets are not statistically significant. Further work might also include an assessment of the physical stability of the mirror fiducial and cold mass shell as they are cooled to liquid helium temperature.

A simple design to measure cold mass position while under operating conditions is now under evaluation. While it is hoped this method will not be required in the tunnel, the authors believe the design is capable of acting as an in-situ fiducial.

REFERENCES

1. Babcock & Wilcox, "Final Report for the Development of the In-situ CQM Alignment Measuring System Proposal," MD-200023, dated 22 September 1992.

2. CERN, Private Correspondence between Jeane-Pierre Quesnel , dated 6/26/92 and 8/24/92.

3. SSCL Document M35-000105, "Installation Procedure for Optical Windows in QCC Quadrupoles," 10-20-92, Charles Corbett.

4. W. P. Barnes, "Some Effects of Aerospace Thermal Environments on High-Acuity Optical Systems," App. Opt., 5, 5, 701–711, (May 1986).

5. R. M. Waxler and G. W. Cleek, "The Effect of Temperature and Pressure on Refractive Index of Some Optical Glasses," J. Res. NBS-A, 77A,6,755-763,(Nov.–Dec. 1963).

6. F. A. Molby, "Index of Refraction and Coefficients of Expansion of Optical Glasses at Low Temperature," JOSA, 39,7, 600–611, (July, 1949).

THE APPLICATION OF MOVING AVERAGE CONTROL CHARTS FOR EVALUATING MAGNETIC FIELD QUALITY ON AN INDIVIDUAL MAGNET BASIS

D. A. Pollock,[1] R. F. Gunst,[2] and W. R. Schucany[2]

[1]Superconducting Super Collider Laboratory, Dallas, Texas 75237*
[2]Southern Methodist University, Dallas, Texas 75275

INTRODUCTION

SSC Collider Dipole Magnet field quality specifications define limits of variation for the population mean (Systematic) and standard deviation (RMS deviation) of allowed and unallowed multipole coefficients generated by the full collection of dipole magnets throughout the Collider operating cycle. A fundamental Quality Control issue is how to determine the acceptability of individual magnets during production, in other words taken one at a time and compared to the population parameters. Provided that the normal distribution assumptions hold, the random variation of multipoles for individual magnets may be evaluated by comparing the measured results to +/- 3 x RMS tolerance, centered on the design nominal. To evaluate the local and cumulative systematic variation of the magnets against the distribution tolerance, individual magnet results need to be combined with others that come before it. This paper demonstrates a Statistical Quality Control method (the *Unweighted Moving Average* control chart) to evaluate individual magnet performance and process stability against population tolerances. The DESY/HERA Dipole cold skew quadrupole measurements for magnets in production order are used to evaluate non-stationarity of the mean over time for the cumulative set of magnets, as well as for a moving sample.

SKEW QUADRUPOLE RESULTS FOR ABB HERA DIPOLE MAGNETS

The HERA tolerance for the skew quadrupole at a reference radius (r_0) of 2.5 cm and 5000 A is +/- 4.0 x 10^{-4} units "maximum variation" and 1 x 10^{-4} units "standard deviation." [1] Skew quadrupole multipole (a_1, USA system) results for 214 HERA dipole magnets produced by Asea Brown Boveri AG, Mannheim, Germany (ABB) are shown in Table 1.

Table 1. Skew Quadrupole * Summary for 214 HERA Dipole Magnets (Producer: ABB)

Minimum	Maximum	Mean	σ	SEM **
-4.55	3.17	-0.0696	1.5897	0.10867

* units x 10^{-4}. ** SEM = Standard Error of the Mean = σ / sqrt (n). Data Source: DESY.

*Operated by the Universities Research Association, Inc., for the U.S. Department of Energy under Contract No. DE-AC35-89ER40486.

ABB was one of two suppliers of dipole magnets for HERA. Due to publication limitations, only the ABB portion of the dipole production will be discussed. It should also be noted that the HERA specification placed no "systematic" tolerance on multipoles.

The distribution of skew quadrupole results for the ABB magnets is assumed to be normal, for demonstration purposes, and is shown in Figure 1. The individual results by magnet production sequence (numbered by coil assembly date) are shown in Figure 2.

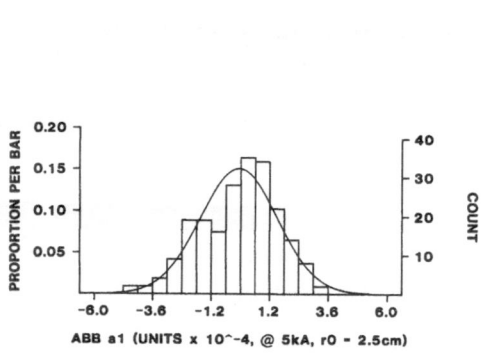

Figure 1. ABB a_1, Frequency Histogram of individual cold magnet results for 214 magnets.

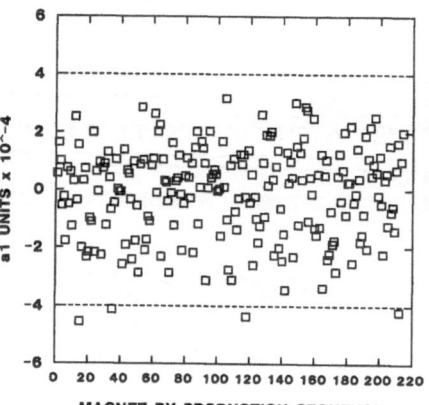

Figure 2. ABB a_1, individual results by magnet production sequence (ordered by coil assembly date).

The cumulative behavior of the sample mean and sample standard deviation every tenth magnet for an increasing size collection of the ABB dipole magnets is shown in Figures 3 and 4. The Standard Error (SEM) is included in Figure 3 to reflect the decreasing uncertainty of the mean with the increasing number of magnets averaged.

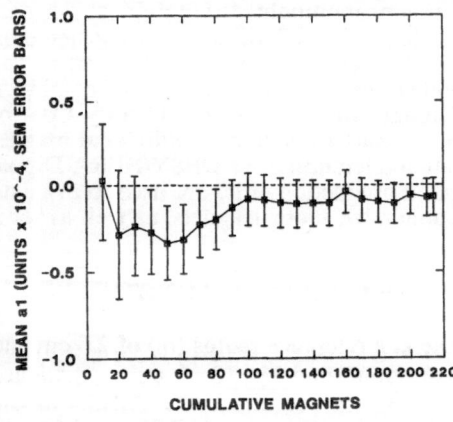

Figure 3. Variation of Mean Over Time, a_1 by cumulative number of magnets, with SEM bars.

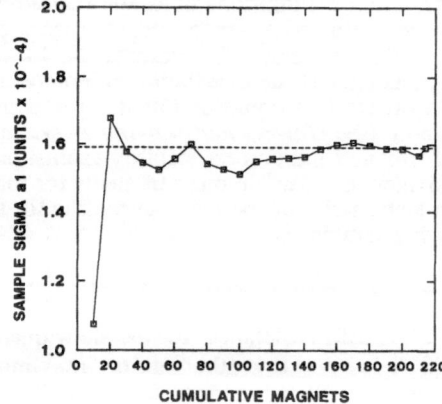

Figure 4. Variation of σ Over Time, standard deviation of a_1 by cumulative magnets.

THE UNWEIGHTED MOVING AVERAGE MODEL (UWMA)

A detailed description of the Unweighted Moving Average may be found in reference 2. The 3σ *Process Limits* for the UWMA are defined as:

$$UCL(M_t) = \mu + \frac{3\sigma}{\sqrt{nw}} \qquad LCL(M_t) = \mu - \frac{3\sigma}{\sqrt{nw}}$$

where M_t = the moving average at time t, w = moving average width or span at time t, n = sample size (for individual magnet evaluation, $n = 1$), and μ = population target mean of all measurements.

For the present application, nw is the number of magnets averaged. During the initial time period (for $t < w$) the control limits are wider than their final steady-state value due to the changing sample size (nw). For a collection of magnets $< w$, the average is computed with the current number of magnets observed. At "steady state" ($t > w$) the oldest magnet observation is dropped and the newest one added to M_t. Since t is incremented for each magnet observed, the influence of each magnet is evaluated as it is observed.

Figures 5 and 6 demonstrate the UWMA for ABB Skew Quadrupole data. In both examples the center is 0 (the target population average), σ (the population standard deviation) is fixed at 1 (the specification limit for σ). The moving average width grows as $w = t$ in Figure 5, and is fixed at $w = 10$ magnets in Figure 6. The data are in time order, based on coil collaring date. Note that the process limits vary with the number of magnets unless $t > w$.

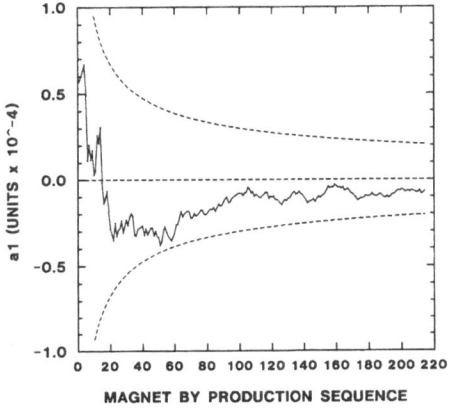

MAGNET BY PRODUCTION SEQUENCE   MAGNET BY PRODUCTION SEQUENCE

Figure 5. HERA a_1 UWMA for ABB Magnets ($w = t$, center = 0, σ = 1), with 3σ process limits for w = the cumulative number of magnets produced up to time t.

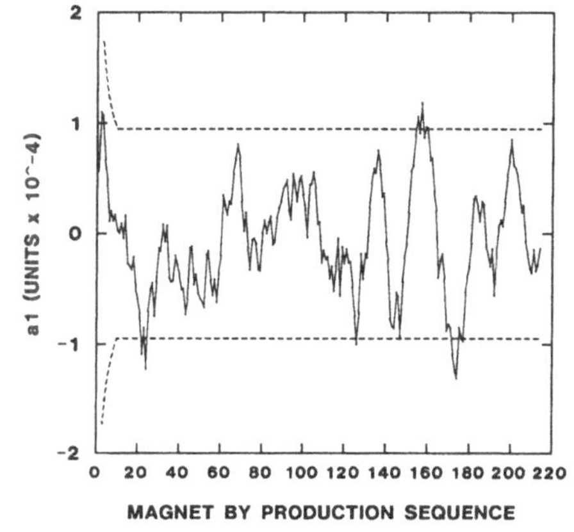

Figure 6. HERA a_1 UWMA for ABB Magnets ($w = 10$, center = 0, σ = 1), with 3σ process limits for w = a moving average of 10 magnets.

The non-stationary behavior of the process mean (systematic multipole) may be observed in the UWMA chart. The choice of the moving average width (w) has a dramatic influence on the output of the model. In Figure 6, relatively stable behavior is observed from magnet 20 to 120. Between magnet 120 and 220 some new source of variation appears to have entered the process as observed in the wide fluctuation of the process mean. Planned research with HERA ABB manufacturing data, which has recently become available to the SSCL MSD, will evaluate possible manufacturing causes for the observed multipole variation.

SENSITIVITY OF THE UWMA MODEL

The ability of a statistical quality control model to detect significant systematic change depends not only on the amount of variation in the process but also on risk probabilities that are tolerated when setting up the control system. The probability that an observation will be outside the statistical control limits and rejected when it is really from the same population and should be accepted is known as alpha risk (α, or producer's risk). For +/- 3σ limits, $\alpha = 0.0027$ or 0.3%. The probability that an observation will be inside the control limits and *accepted* when it is really from another population and should be *rejected* is known as beta risk (β, or consumer's risk). Beta (β) depends on several factors including: sample size, population standard deviation, α, and distance of the alternate mean from the target mean. Table 2 shows various β probabilities for the ABB skew quadrupole data, assuming control limits are based on a target mean of 0, with a population standard deviation of 1.6 or 1, and α of 0.0027 (3σ control) or 0.05 (2σ control) and the process is off center at -0.07 units (the observed mean for ABB dipoles). Though not a requirement for ABB, the -0.07 unit systematic deviation from a center of 0 units is larger than the SSCL systematic a_1 specification allowance of 0.04 units. In order to increase the probability of detecting an alternate mean (with the model control limits) β should be as small as possible. See reference 3 for more discussion of α and β risk and sample size using an SSC example.

Table 2. β Probabilities for Various UWMA Model Conditions.

nw	$\alpha = 0.0027, \ \sigma = 1.6$ β	$\alpha = 0.0027, \ \sigma = 1.0$ β	$\alpha = 0.05, \ \sigma = 1.0$ β
10	0.997	0.997	0.945
50	0.996	0.994	0.924
500	0.980	0.932	0.674
1000	0.951	0.807	0.431
5000	0.500	0.385	0.002

THE EXPONENTIALLY WEIGHTED MOVING AVERAGE

An alternate method for monitoring systematic variation over time is the Exponentially Weighted Moving-Average (EWMA). In the EWMA a smaller weight is applied to older observations in the time series. The advantage of the EWMA is as a *forecasting* tool. The EWMA can be used not only to identify *when* but also by *how much* a process should be adjusted. [4]

CONCLUSION

The Unweighted Moving Average method provides a simple means of monitoring individual magnet multipole performance against population tolerances for systematic multipoles, from the "first" magnet made to the "last." The method also provides an indication of manufacturing process stability over time (for either the "cumulative" observations of all magnets up to the current time, or for a local moving width of observations). The careful selection of the moving average "width" can provide a useful tool for detecting the influence of variation in magnet components and tooling.

Through continued research and analysis of the magnet manufacturing data collected from SSC Prototype Magnets, as well as from the HERA ABB production experience, improved methods for monitoring systematic multipole variation in SSC superconducting magnets may be identified.

ACKNOWLEDGMENTS

S. Wolff, P. Gall, and H. Brueck. DESY, Hamburg, Germany.
Y. Zhao, A. Devred. SSCL, Dallas, Texas.
D. Bonmann. Asea Brown Boveri AG, Mannheim, Germany.
F. Eyβelein. Noell GmbH, Würzburg, Germany.

REFERENCES

1. DESY Technische Spezifikation DES Supraleitungsdipols fur HERA, Teil 1: The superconducting collared coil for dipoles of the proton ring of HERA description and fabrication procedure, S. Wolff, 3 Revision, Feb. 24, 1987.
2. Douglas C. Montgomery, *Introduction to Statistical Quality Control*, John Wiley & Sons, Inc., New York, 1991, pp. 307 ff.
3. D. A. Pollock, et. al., *A Statistical Rationale for Establishing Process Quality Control Limits Using Fixed Sample Size, for Critical Current Verification of SSC Superconducting Wire*, Supercollider 4, Plenum Press, New York, 1992, pp. 491 ff.
4. ibid. Montgomery, pp. 279 ff.

INITIAL OPERATION AND PERFORMANCE TEST RESULTS OF THE
ACCELERATOR SYSTEM STRING TEST (ASST) CRYOGENIC SYSTEM

Ted Kobel and Roberto Than

Process Systems International, Inc.
20 Walkup Drive
Westborough, MA 01581-5003

INTRODUCTION

PSI has supplied three equal capacity helium cryogenic plants[1] (dubbed ASST, MTL, and N15B). The first two plants will provide the helium refrigeration and liquefaction required for magnet testing in the Accelerator System String Test (ASST) facility and the Magnet Test Laboratory (MTL). The third plant (N15B) will supplement the ASST plant and will eventually be utilized for the N15 Sector Refrigeration System (SRS) requirements.

This paper describes the results of the ASST plant initial operational commissioning and performance testing. Operating modes of the MTL/ASST system have been previously described.[2]

COMMISSIONING STEPS

The ASST plant includes the following major components which received I/O, mechanical and electrical functional checkout (pretest operation) followed by performance testing.

1. Computer Control System
2. Instrument Air System
3. LN_2 Dewar and GN_2 System
4. R/L Compressor System
5. CCWP Compressor System
6. CCWP Coldbox
7. CCWP Regeneration System
8. CCWP Dehydration System
9. R/L Regeneration System
10. R/L Coldbox
11. Distribution Box
12. Cold Compressor
13. LHe Dewar
14. Integrated Systems - Plant Final Acceptance Test Run

PERFORMANCE TEST RESULTS

LN₂ Dewar Boil-off Test

The LN_2 dewar boil-off test was successfully completed in November 1992.

Following initial cooldown and fill, the 75,000 liter horizontal LN_2 dewar was allowed to soak and thermally stabilize. Dewar pressure was kept constant by allowing boil-off gas to vent through a back pressure regulator. The discharge vent gas was metered through a positive displacement totalizing gas meter. The total measured boil-off was within the specified maximum allowable equivalent heat leak of 0.3% per day (231 SCFH) at 1.1 Bar.

R/L Compressor Performance Test

Compressor Specs	First Stage (CM11, CM12)	Second Stage (CM13, CM14)
Quantity:	Two (2)	Two (2)
Compressor Type:	Oil Flooded Screw	Oil Flooded Screw
Manufacturer:	Sullair	Sullair
Model Number:	C25LB	C25MA
Rotor Diameter:	255 mm	255 mm
Rotor Length/Diameter:	1.7	1.25
Electric Motor:	250 HP	700 HP
	1.2 Service Factor	1.2 Service Factor
	3 ph/60 Hz/460 V	3 ph/60 Hz/4000 V
	3600 RPM	3600 RPM
	F.L. eff 94.5%	F.L. eff 95.4%

The R/L compressors performance testing began with a 168 hour endurance run at the following nominal design conditions. This run was successfully completed on November 16, 1992.

	CM11	CM12	CM13	CM14
Pd (bar)	3.6	3.3	18.2	17.9
Ps (bar)	1.17	1.11	3.13	3.18
Ts (K)	293	295	299	299
Motor Current (Amps)	200.7	184.2	86.0	92.4
Max. Allowable Amps	212	212	108	108

Following the 168 hour endurance run, the R/L compressors underwent additional testing runs to provide a complete performance map of each first-stage and each second-stage machine.

R/L Compressor Performance Map

The complete performance map consists of 25 test points for each compressor (CM11, CM12, CM13, CM14). The range of test points were selected to cover the extreme limits of suction and discharge pressures.

The following test data are representative samples of the performance map. This data is presented for reference only. The complete data package has been reduced to verify the test results. (The reduced data package with performance curves is too extensive to be included in this report.)

	Design Point	Test date: November 16, 1992 1st Stg Compressors - Sampling of Test Points CM11				CM12			
Flow (g/s)	108.1	116	118	129	118	124	123	139	124
Pd/Ps	3.069	2.904	3.346	2.471	4.705	2.942	3.385	2.538	4.283
Pd (bar)	3.1	3.02	3.48	2.99	4.94	3.06	3.52	3.02	4.54
Ps (bar)	1.01	1.04	1.04	1.21	1.05	1.04	1.04	1.19	1.06
Ts (K)	304.7	298	296	299	295	299	303	304	303
Pwr-shaft (kW)	136.5	135.8	152.8	129.5	205.3	139.9	156.5	138.4	195.3
Isoth Eff (%)	56	56.37	57.37	56.0	54.55	59.41	60.32	59.05	58.15
Vol Eff (%)	-----	85.34	86.32	81.82	85.49	91.55	92.12	91.19	91.34

Note: Shaft power assumes 0.945 motor efficiency

	Design Point	Test date: January 8, 1993 2nd Stg Compressors - Sampling of Test Points CM13				CM14			
Flow (g/s)	211.3	253.8	298.2	297.9	250.8	250.5	291.0	292.0	244.3
Pd/Ps	5.993	5.915	5.146	5.589	6.457	5.847	5.059	5.427	6.428
Pd (bar)	18.1	18.04	18.01	19.56	19.50	18.01	18.01	19.43	19.54
Ps (bar)	3.02	3.05	3.50	3.50	3.02	3.08	3.56	3.58	3.04
Ts (K)	306.4	299	294	293	298	295	297	296	296
Pwr-shaft (kW)	454.3	531.5	538.5	578.3	562.9	539.8	548.3	587.1	570.8
Isoth Eff (%)	53	52.71	55.40	53.95	51.44	50.22	53.08	51.73	48.96
Vol Eff (%)	-----	86.34	86.96	86.63	85.91	83.26	84.28	83.85	82.58

Note: Shaft power assumes 0.945 motor efficiency

CCWP Initial Performance Run

The initial testing of the CCWP system consisted of regenerating the Mol sieve dehydration bed, regenerating the 80K adsorber bed, operation of LN_2 circuits, cooldown to 80K, and a 72 hour CCWP compressor operation. The 72 hour run was successfully completed November 21, 1992. Final testing will occur during the plant integrated systems test (final acceptance testing).

CCWP Compressor Specs

Quantity:	One (1)	Design Flow:	55 g/s
Compressor Type:	Oil Flooded Screw	Design Ps:	2.0 bar
Manufacturer:	Sullair	Design Pd:	8.0 bar
Model Number:	C16LA	Design Ts:	310 K
Rotor Diameter:	163 mm		
Rotor Length/Dia.:	1.7		
Electric Motor:	125 HP, 3 ph/60 Hz/460 V 3600 RPM, F.L. eff 95.2%		

R/L Performance Tests

The R/L performance tests were completed April 16, 1993. The following test data for tests 1, 2, 3 and 4 are representative samples taken from the hourly computer data acquisition. This data is presented for reference only and is currently being analyzed and reduced to verify the test results. The data (*) presented for tests 5 and 6 has been reduced from the hourly computer data acquisition.

Test No.	Oper Mode	100% Design Specs Refrig	Liq	Test Results	Motor[1] In Power	Ave. LN$_2$ Consumption
1	1/2 Liq (T3 off)	0	15.6 g/s	20.4 g/s	664 kW	(tbd) g/s
2	1/2 Ref (T3 off)	1320 W	0	1390 W	649 kW	16 g/s (2.6 bar supply)
3	1/2 Mix (T3 off)	700 W (Guar.)	7.0 g/s (Guar.)	799 W 7.4 g/s	656 kW	30 g/s (2.6 bar supply)
4	100% Liq	0	37.0 g/s	38.9 g/s	1381 kW	113 g/s (2.6 bar supply)
5	100% Ref	3530 W	0	*3900 W	1375 kW	*27 g/s (3 bar supply)
6	100% Mix	2000 W (Guar.)	20.0 g/s (Guar.)	*2225 W *22.7 g/s	*1370 kW	*89.8 g/s (2.6 bar supply)

Note (1): Motor power includes compressor excess flow bypass.

Remaining Performance Testing

The distribution box and cold compressor modules and the LHe dewar will be tested in conjunction with the integrated systems test scheduled for the week of May 3, 1993.

ACKNOWLEDGEMENTS

The author expresses many thanks to all the PSI and SSCL personnel conducting the tests, collecting data and analyzing data.

REFERENCES

1.	U. Wagner and W. Keyer, "Process Design Features of the SSC MTL Cryogenic System," Supercollider 3, Proceedings of the Third International Industrial Symposium on the Super Collider, 1991, Plenum Press.

2.	V. Ganni, R. Than, M. Thirumaleshwar, "Operational Modes and Control Philosophy of the SSCL Magnet Test Lab (MTL) Cryogenic System," Supercollider 4, Proceedings of the Fourth International Industrial Symposium on the Super Collider, 1992, Plenum Press.

OPERATING MODES OF THE SSC SECTOR STATION CRYOGENIC SYSTEM

V. Ganni, S. Abramovich, and T. V. V. R. Apparao

SSC Laboratory,* ASD, Cryogenics Department, Dallas, TX 75237-3997

INTRODUCTION

The magnets in the Collider rings and in the HEB (High Energy Booster) ring are refrigerated by single phase helium flow controlled at a temperature of 4 K to maintain the magnet windings in the superconductive state. To minimize the heat load into the 4 K loop, the magnet cryostats are designed to provide high quality thermal insulation achieved by a high vacuum with multilayer insulation (MLI) and by two thermal shields at a nominal temperatures of 84 K and 20 K, respectively.

An extensive cryogenic system capable of normal operation of the superconducting magnets and handling transient conditions such as quenches, beam activation, ring filling, and beam ramping is required. The cryogenic system must also have the capability to perform other services such as cleanup, cooldown, and warmup, to allow for maintenance and repair of the superconductive components throughout the ring.

The refrigeration system is designed with a specific capacity to meet each of the various loads viz. 4 K refrigeration, 4 K liquefaction and 20 K shield cooling, to meet all the loads during normal and upset conditions and possess some extra capacity. The static heat leak and dynamic load to the 4 K components are the 4 K refrigeration load. It is handled by the latent heat of vaporization in the helium recoolers. The 4 K helium used for cooling the electrical leads returned at 300 K to the refrigerator, is the liquefaction load. The 20 K shield that absorbs part of the radiation and the static heat leak above 20 K is the 20 K refrigeration load. This load is handled by supplying 200 g/s, 3.0 bar helium at 14 K and returns the helium to the refrigerator at a temperature depending on the load. The refrigeration system has one spare compressor. It is capable of delivering 4 K and 20 K refrigeration and part of the liquefaction load in the nominal mode of operation given in Table 1 even when any one of the expanders fail.

Figure (1) is a block diagram showing the major components and their relative locations in the Sector Station Cryogenics System (SCS) which are the Sector Refrigeration Surface System (SRS), Sector Refrigeration Tunnel System (SRT) and the Sector Refrigeration Control System (SRC). There are many operating modes for the SCS and for each mode of operation there is a different scheme for the cryogen flows. The SCS is reconfigured each time by switching and by activation/deactivation of specific sets of equipment that allow for adjusting the process conditions, capacity and the throughput. The line numbers shown in Figure 1 as well as a detailed description of the plant operation is given in Reference.[1]

This paper summarizes the operating modes of the SCS, the various states and processes.

THE OPERATING MODES OF THE SCS

The various modes of operation are defined as follows: Design Mode, Nominal Mode, Standby Mode, Assist Mode, and Utility Modes. Figure 2 shows the various system states and plant operational modes.

The refrigeration capacity required for a sector in four different modes of operation is given in Table 1. The design mode loads have a design margin of 25% for the 4 K loads and 50% margin for the 20 K load compared to the nominal mode. These margins account for the variation in the loads from sector to sector due to the difference in the lengths of the actual sectors and provide extra capacity required in the Assist Mode. The SCS is operated in the Nominal Mode when the collider is at full normal operation. This means the system is cold and the beam is on. The dynamic heat load on the system depends on the beam intensity. The 20 K shield cooling load remains unchanged in the Nominal, Standby and Assist modes of operation. The heat load budgets for each individual sector in the collider as well as the HEB sectors are given in Reference.[1]

The SCS is operated in the Standby Mode when the collider is cold at nominal temperatures, with no beam operating. Hence the refrigeration and liquefaction loads are lower than in the nominal mode. In this mode of operation the refrigeration load depends only on the "static load." The liquefaction load used for the electrical leads may be reduced since there is no electrical current. The excess liquid produced in the assist mode is transferred to a neighboring sector where the refrigeration plant is operating below nominal capacity.

During a quench, a large amount of energy (~7 MJ) is transformed into heat and deposited in the cold mass. The magnets are protected by quickly aborting the beam and ramping down the electric current. The quench valves are opened and helium is vented from the cold mass into the 20 K line. The cold mass is then gradually brought back to nominal operating temperature. Operations such as cleanup, cooldown, warmup and maintenance are part of utility mode operations. The heat capacity of a sector is highly dependent on the temperature. The process times for each of these utility modes depends on the temperature level and the changes in the fluid inventory. Table 2 shows the fluid inventory in each of the cryogenic lines and the process times for each of the utility mode operations in a sector.

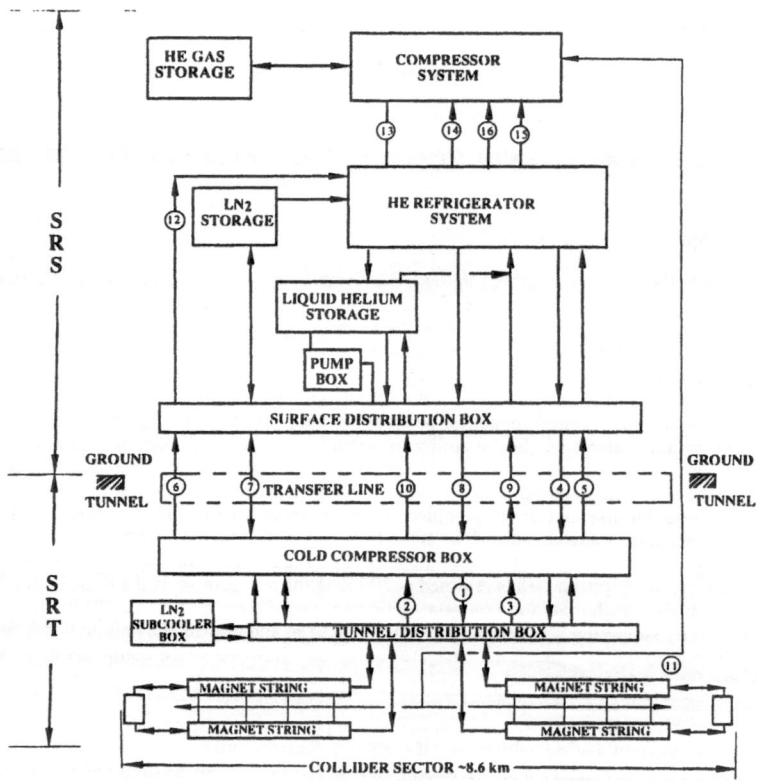

Figure 1. Sector Station Cryogenic System Block Diagram.

Table 1. SCS Capacity for Different Modes of Operation.

	DESIGN MODE							NOMINAL MODE						
	PI	TI	PO	TO	Flow	Load	W(Car)	PI	TI	PO	TO	Flow	Load	W(Car)
	MPa	K	MPa	K	g/s	Watts	KW	MPa	K	MPa	K	g/s	Watts	KW
Refrigeration	0.400	4.0	0.077	3.95	330	6750	574	0.400	4.0	0.077	3.95	264	5400	461
Liquefaction	0.400	4.0	1.05	305	45		320	0.400	4.0	1.05	305	36		256
20 K system	0.400	14.0	0.2	27.8	200	15000	302	0.300	14.0	0.2	23.2	200	10000	208
Cold Compressor	0.075	4.3	0.145	6.07	330	2745	48	0.075	4.3	0.14	6.06	264	2235	45
Total Capacity							1244							970
	STANDBY MODE							ASSIST MODE						
Refrigeration	0.400	4.0	0.077	3.95	138	2800	238	0.400	4.0	0.077	3.95	264	5400	461
Liquefaction	0.400	4.0	1.05	305	25		178	0.400	4.0	1.05	305	54		384
20 K system	0.300	14.0	0.2	27.8	200	10000	208	0.300	14.0	0.2	23.2	200	10000	208
Cold Compressor	0.075	4.3	0.135	6.05	138	1191	27	0.075	4.3	0.14	6.06	264	2235	45
Total Capacity							651							1098

460

The plant is operated in special modes for other types of collider configurations such as single ring cryogenic operation. The plant is said to operate in maximum capacity mode when the sector loads are high, or when assisting a neighboring sector, or when handling additional liquefaction load. It is operated in minimum capacity mode when the sector heat loads are low such as in a standby condition of the ring.

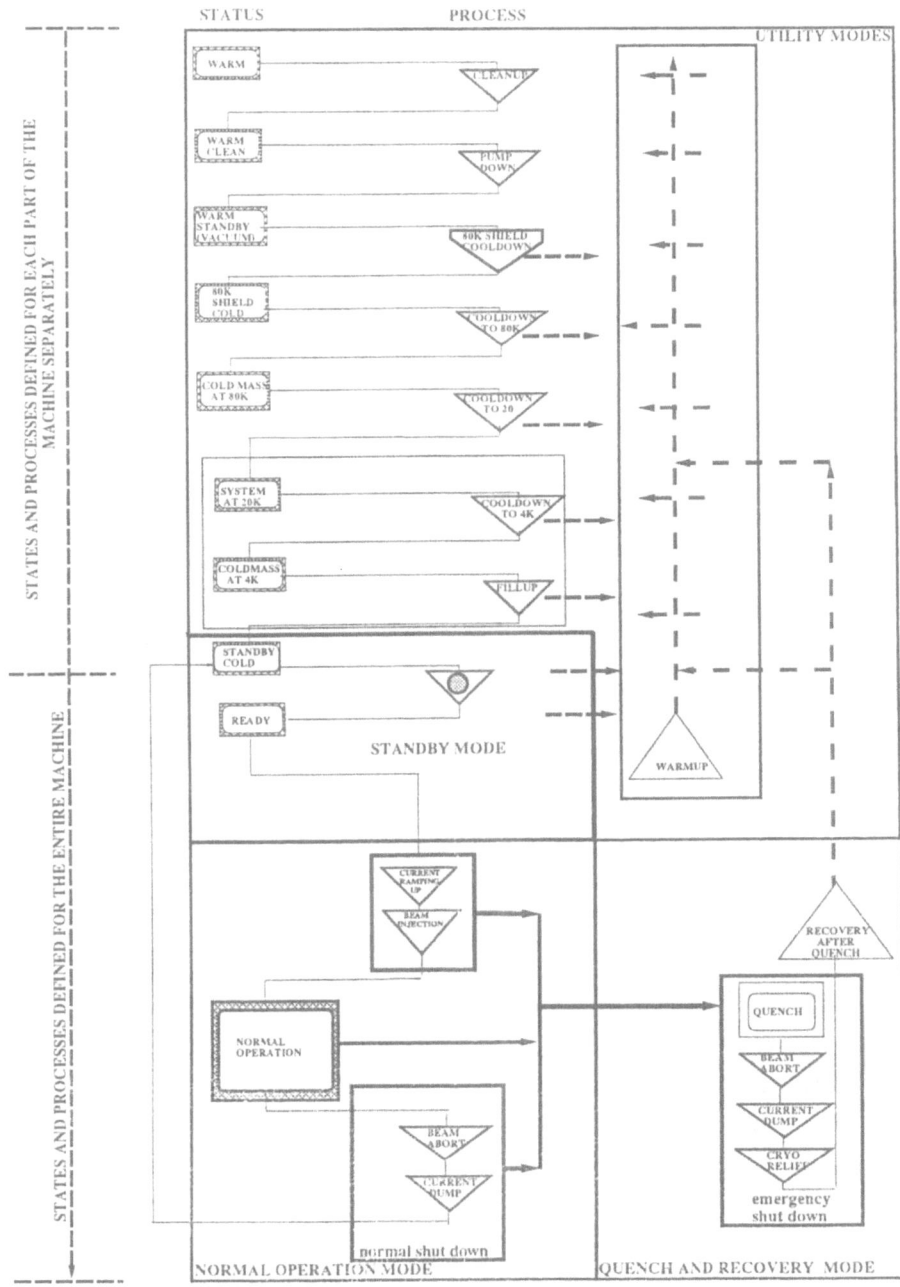

Figure 2. The System States and the Operational Modes.

Table 2. Collider Sector Fluid Inventory.

	I.D. mm	Pressure MPa	Temperature K	Volume liters	Warm clean kg	Warm standby kg	80K shield cold kg	20K shield at 80 K kg	Coldmass at 80 K kg	Coldmass at 20 K kg	Nominal mode kg	Quench and recovery kg	Gas desorption kg
LINE1-4K He feed	45.2	0.4	4.05	133702	21.73	21.73	21.73	21.73	240.13	966.40	18758	-100	966.40
LINE2-4K He Return	45.2	0.35	4.3	27720	4.50	4.50	4.50	4.50	49.79	200.36	3690		200.36
LINE3-GHe return	86.5	0.08	4.05	101500	16.49	16.49	16.49	16.49	182.29	733.64	1250		733.64
Recooler (shell)		0.08	4.05	2592	0.42	0.42	0.42	0.42	4.66	18.73	365		18.73
LINE4-20 K shield	82.6	0.3	14-28	92549	44.52	44.52	44.52	167	167	669	669	100	669
LINE5-80k LN2	57.2	0.7	84	44380	49.84	49.84	34620	34620	34620	34620	34620		34620
LINE6-80k GN2	57.2	0.14	84	44380	49.84	49.84	245	245	245	245	245		245
Warm He gas return	127	0.11	300	218787	35.55	35.55	35.55	35.55	35.55	35.55	35.55		35.55
total helium					123.22	123.22	123.22	245.70	679.41	2623.63	24778		2623.63
total nitrogen					99.68	99.68	34865	34865	34865	34865	34865		34865

HELIUM PLANT AND VERTICAL TRANSFER LINES NOT INCLUDED

Pumpout 192 half cells 22,000 liters each to rough vacuum

insulating vacuum after pumpdown ~0.013 Pa. Expected pump down time including leak check ~48 hours.

Cooldown the 80 K shield from 300 to 80 K (LN2 flow=20 g/s up to nominal flow), 3-4 days/km minimum.

44,530 MJ (2,577kJ/m) to be removed from the shield : 197,760kg LN2/sector (12,360 kg LN2/section , 16 sections/sector)

Cooldown the 20 K shield from 300 to 20 K (20 g/s up to nominal flow),

27,270 MJ (1,578 kJ/m) to be removed from the shield : First cooldown is performed together with the cooldown of the cold mass , approximately 22 days to cooldown to 80 K. Additional 24 hours to cooldown to 20 K.

Coldmass cooldown from 300 to 80 K (400 g/s divided into 4 strings), 22 to 23 days

8.55 E08 KJ to be removed from the cold mass :

Coldmass cooldown from 80 to 20 K (140 g/s divided into 4 strings) 3 to 4 days

4.46 E07 KJ to be removed from the cold mass:

Coldmass cooldown from 20 to 4 K (400 g/s divided into 4 strings) 1 to 2 days

4. E05 KJ to be removed from the cold mass:

Up to 7 MJ transferred into heat. 100 kg , 4 K helium vented into the 20 K shield line. Half cell warmed up to 7 to 10 K. Time ~20 minutes.

Cooldown 1/2 cell to 4 K : 4.5MJoules(heat)+45 minutes static load /sector(3.2MJ). Refill 100kg with 100 g/s ~20 minutes. Control loop activation (the whole sector) 15-20 minutes

Warmup the sector to 20 K by pumping 100g/s 20 K helium (total heat = 4. E05 MJ). 10-15 hours.

Cooldown the sector to 4 K (total heat = 4. E05MJ) increase inventory by 22,000 kg , fillup with 400 g/s ~15 hours.

REFERENCES

1. R. Than, S. Abramovich, V. Ganni. "The SSC Cryogenic System Design and Operating Modes."

* Operated by the Universities Research Association, Inc., for the U. S. Department of Energy under Contract No. DE-AC35-89ER40486.

THERMAL AND FLOW CONSIDERATIONS FOR THE
80 K SHIELD OF THE SSC MAGNET CRYOSTATS

S. Abramovich, A. Yücel, M. Thirumaleshwar[†], J. Demko

Superconducting Super Collider Laboratory[*], Dallas, Texas, USA
[†]Guest Collaborator from Centre for Advanced Technology, Indore, India

INTRODUCTION

The nominal temperatures in the SSC magnets range between 4.2 K in the superconducting coils and 300 K on the cryostat outer wall. To minimize the 4 K heat load, one thermal shield cooled by liquid and vapor nitrogen flows at 84 K, and another cooled by helium flow at 20 K are incorporated in the cryostat.[1] Tubes attached to the shields serve as conduits for the cryogens. The liquid nitrogen tube in the cryostat is used for shield refrigeration and also for liquid distribution around the SSC rings.[2-4] The second nitrogen line is used to return the vapor to the helium refrigerators for helium precooling. The nominal GN2 flow from a 4.3 km long cryogenic string (4 sections) to the surface is 64 g/s. The total liquid nitrogen consumption of approximately 5000 g/s will be supplied at one, two or more locations on the surface. The supply, distribution, circulation and recooling schemes for the LN2 system are described in detail by McAshan et al.[3]

The total heat load of the 80 K shield is estimated as 3.2 W/m: about 50% is composed of infrared radiation; the remaining 50% is by heat conduction through supports, vacuum barriers and other thermal connections between the shield and the 300 K outer wall. The required LN2 flow rate depends on the distribution and circulation schemes. The LN2 temperature will in turn vary depending on the flow rate and on the recooling method used. For example, with a massflow of 400 g/s of LN2 the temperature rises from 82 K to 86 K between two compact recoolers 1 km apart. This temperature is higher than desired. The temperature can be reduced by increasing the flow rate of the liquid or by using the continuous recooling scheme.

This paper discusses some thermal problems caused by certain mechanical designs of the 80 K shield and the possible improvement by using continuous recooling. In the following, we present results of the 80 K shield temperature distribution analysis, the 20 K shield heat load augmentation resulting from the increased 80 K shield temperatures, the continuous nitrogen recooling scheme and some flow timing related analysis.

THE SYSTEM GEOMETRY AND OPERATION

The two parallel rings of the collider and the HEB ring are constructed in parallel tunnels at inclination of 0.2 degrees with a local minimum of 0 degrees. The superconductive magnets strings are divided into sections ranging in length from 350 m to 1350 m with U-tube connections in-between. This arrangement allows for connecting compact nitrogen recoolers which is the basic LN2 recooling system. Vapor generated in the recoolers is returned through the GN2 line back to the helium refrigerator for helium precooling.

A possible alternative to nitrogen recooling uses controlled injection of LN2 into the GN2 lines. The LN2 control valve may be installed in the cryogenic isolation boxes (SPRI) on the uphill side of each section. The liquid flows downhill and by boiling it performs an additional recooling function. In sections of the tunnel where the inclination is too small this method can not be implemented and compact recoolers are still needed. The inlet GN2 pressure to the helium plant (on the surface) is 0.13 MPa. The lowest pressure in the GN2 line in the tunnel in the sector Feed Box –SPRE– is expected to be 0.14 MPa. In this place the ideal recooling temperature is 80.2 K. Due to the vapor flow pressure drop in the GN2 tube, the pressure in the sector End Box –SPRE, 4.3 km upstream– is 0.17 MPa with an ideal recooling temperature 82 K. If the string is longer than nominal, the pressures and the recooling temperatures will be higher.

[*] Operated by the Universities Research Association, Inc., for the U.S. Department of Energy under Contract No. DE-AC35-89ER40486.

Due to the temperature approach in the LN2 recoolers, the thermal resistance between the nitrogen fluids and the tube walls, the resistance of the straps connecting the tubes to the shield, and the thermal resistance of the shield, the shield temperatures will range between 84 K and 98 K. This may cause the 20 K shield heat load to increase beyond the baseline value computed using an average shield temperature of 84 K.

STEADY STATE CONDITIONS FOR NOMINAL STRINGS

Figure 1 shows a sketch of the thermal shield design and the temperature changes in the LN2 line for a flow rate of 400 g/s and a recooling system based on compact recoolers located at different intervals. For ideal recooling (the recooler output temperature lies on line a-i), the highest LN2 temperature is 85K for a recooler every 1 km (point h1), 88.2 K if the recoolers are 2 km apart (point h2), and 95 K if the recoolers are 4 km apart (point h3). In well-designed practical recoolers a temperature approach 2 K may be expected. The 80 K shield temperature distributions and the corresponding 20 K heat load augmentation were calculated[5] for various LN2 recooling temperature between 80 K and 92 K. The vapor line temperature was fixed at 80 K. The following cases were analyzed: Case 1, in which the heat is removed by both the liquid and vapor LN2 lines (with liquid injection), for conservative parameters of heat transfer. These include the effects of the relative low conductivity of the stainless steel, and also the discontinuous or discrete nature of the connections between the shields and the pipes. Case 2 similar to Case 1 but with good thermal conductance between the straps and the nitrogen pipes, neglecting the resistance of the pipe. Case 3 assumes the heat load is removed by the liquid nitrogen line only. Representative temperature distributions for half of a shield section between two posts are presented in Figure 2.

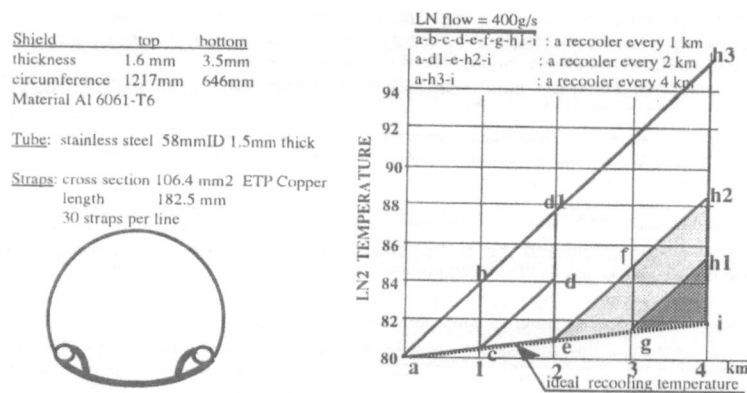

Figure 1. 80 K Shield and the temperature change in the liquid line.

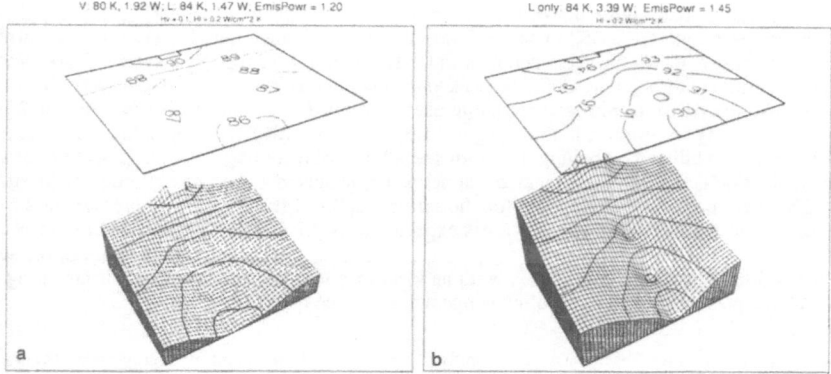

Figure 2. Sample temperature profiles for the 80 K shield.
a) LN2 and GN2 lines active (Case 1), b) LN2 line active only (Case 3).

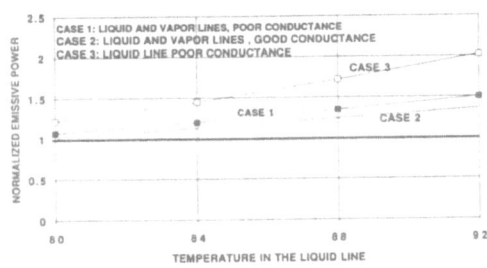

Figure 3. Emissive power as function of temperature in the liquid line.

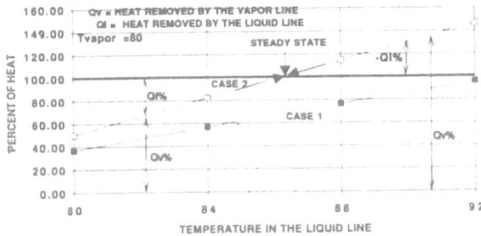

Figure 4. Percent of heat absorbed by each line.

The shield emissive power (normalized to 84 K) variations for each of these cases are shown in Figure 3. The amounts of heat removed by the vapor and liquid lines are given in Figure 4. For Case 2, the liquid temperature increases asymptotically to a steady state point if the initial liquid temperature is low, and decreases asymptotically to the same point if the initial liquid temperature is high.

THE CONTINUOUS RECOOLING MODEL AND INITIAL TRANSIENTS

Liquid nitrogen injected into the GN2 line flows downhill. The pressure in the line is maintained as low as possible due to other system constraints, to enable adequate boiling conditions in the recooling system. Figure 5 represents a typical 1080 meter long section in the collider. The inlet vapor flow to the section is given by $Mv(0)$ and varies from 0 to 48 g/s. The vapor flow rate at the end of each section, $Mv(l)$, can be 16, 32, 48, or 64 g/s as it flows back to the feed box. The flow rate of the injected liquid at the upstream location, Ml is nominally taken to be 16 g/s. Also shown in Figure 5 is a typical control volume used for analysis purposes with a length dx, that has $q(x)$ of nominal heat transfer and qwall of transient heat transfer from cooling down of the tube wall in the full transient situation, and liquid evaporating at a rate of mlv. The flow direction of liquid in the line around the collider ring may be clockwise, counterclockwise, uphill or downhill depending on the N2 delivery and distribution system in use, which may change occasionally during normal operation.

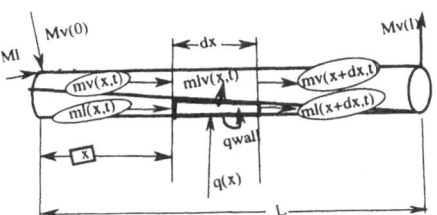

Figure 5. Flow in the vapor line.

Figure 6. Flow development in the inclined plane.

THE FLOW MODEL

Under ideal continuous recooling conditions the whole heat load is absorbed by the latent heat in the vapor line. If the injection rate is smaller than ideal and sections of the tube are dry, the parallel liquid line is assumed to absorb the heat load and the vapor line remains isothermal. Also for ideal continuous recooling there should be no accumulation of the liquid at the end of the nitrogen line to prevent flooding. For the analysis of the continuous recooling scheme, it was assumed that the continuous recooling liquid flow is driven primarily by gravity, the flow of liquid was assumed to be stratified, and that the transient channel flow equations of Manning[6] would apply for the liquid. Additional assumptions made are that the liquid-vapor interface is frictionless, there is perfect heat transfer between the liquid and the shield, and the shield has a high thermal conductance and may be considered isothermal. The mass of liquid evaporated in a control volume is taken as the 80 K heat load divided by the latent heat at the line pressure.

STEADY STATE FLOW

The steady state temperatures are evaluated for different liquid injection rates. The local liquid flow cross section, mass flow rate, velocity and integral liquid inventory are calculated. In Figure 7, the steady state inventory of liquid nitrogen for the first 400 m is shown for different injection rates. Figure 8 shows the time required for the liquid to reach a certain location for the same parameters.

TRANSIENT FLOW

Calculations of transients were performed to determine the time required to reach a steady state liquid flow in the nitrogen line. The analysis and the numerical solution procedure used is described by Abramovich et al.[7] The transient distribution of liquid nitrogen mass flow for a constant initial injection rate of 16 g/s with a tube inclination angle of 0.2 degrees is shown in Figure 9. The liquid front is seen to be very steep, and due to evaporation of the liquid along the tube, the flow linearly decreases proceeding down the tube. The predictions indicate that at these conditions it takes 16,000 seconds for the liquid flow to reach a steady state condition at this inclination angle.

One purpose of this investigation is to provide some analysis that supports the development of a control strategy for continuous injection of liquid into the nitrogen vapor line of the 80 K shield. The previous cases provide some fundamental solutions that provide time scales for filling the tube at a constant mass flow rate of 16 g/s, and the times for some perturbations to propagate along the liquid stream. Simulations were carried out for various step changes in the injection rates.

Another important result is for a case where the initial injection rate of 30 g/s was applied and after some period reduced to a nominal value of 16 g/s This would be done to reduce the time required to supply liquid along the whole length of the nitrogen vapor line. Figure 10 shows results for an inclination of 0.2 degrees with an initial liquid injection rate of 30 g/s which is stepped down after 2000 seconds to a mass flow of 16 g/s. It is seen that the liquid pulse travels the whole 1000 m length without completely disappearing. For this case the tube contains liquid along its entire length after 12000 seconds which is about 4000 seconds less than the constant injection case.

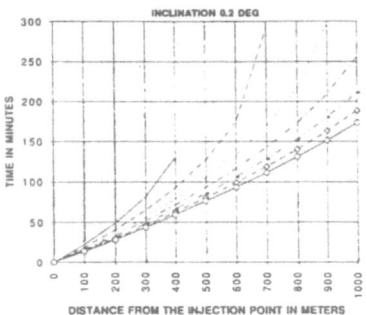

Figure 7. Inventory of LN2 for different injection rates.

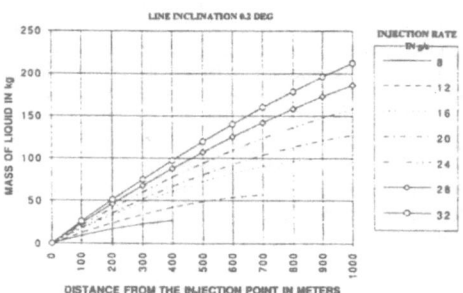

Figure 8. Time of motion of the liquid wave front for different injection rates.

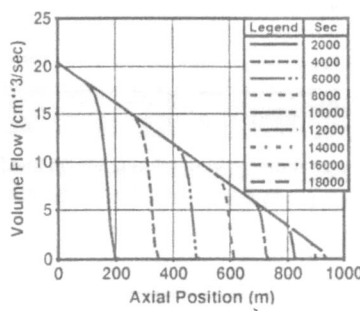

Figure 9. LN2 transient mass flow distributions constant 16 g/s injection.

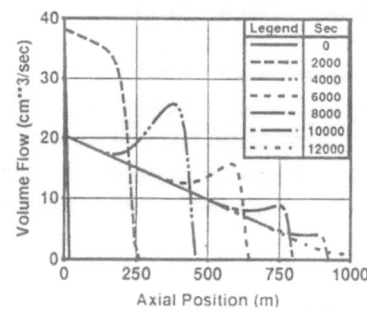

Figure 10. LN2 transient mass flow for distribution for 30 to 16 g/s step.

References

1. R. Than, S. Abramovich and V. Ganni, "The SSC Cryogenic System Design and Operating Modes," Cryo Note 92-12 (1992).
2. S. Abramovich and A. Scheidemantle, "84 K Nitrogen System for the SSC," SSCL-N-712 (1990).
3. M. McAshan, M. Tbirumaleshwar, S. Abramovich, V.Ganni, "Nitrogen System for the SSC," SSCL-592.
4. M. McAshan, M. Thirumaleshwar, S. Abramovich, V.Ganni, and A Scheidemantle, Nitrogen system for the SSC, in: "Supercollider 4," J. Nonte, ed., Plenum Press, New York (1992).
5. A. Yücel and S. Abramovich, "Study of 80 K Shield Temperature Distributions," SSC Cryo Note 92-9.
6. E. B. Wylie and V. L. "Streeter, "Fluid Transients," FEB Press, Ann Arbor, MI, 1982.
7. S Abramovich, J Demko, and M Thirumaleshwar, "Nitrogen System for the SSC: Continous Recooling by Injecting Liquid into the Vapor Line," SSC Cryo Note 94-30 (1994).

THE CONTROL AND OPERATION OF A TEST STAND FOR THE
CRYOGENIC TESTING OF SUPERCONDUCTING MAGNETS

William E. Garber[1] and Rich Fagan[2]

[1]Process Systems Group
Air Products and Chemicals Inc.
7201 Hamilton Boulevard
Allentown, Pa. 18195-1501

[2]General Dynamics Space Systems Division
Space Magnetics
P.O. Box 85990
San Diego, Ca. 92138

ABSTRACT

The design, implementation and use of a control system for the cryogenic magnet test stands at General Dynamics' Hammond, Louisiana facility is discussed. Batch control techniques are used to automate and assist the testing of magnets in a production environment. The test stand control system is integrated into the overall facility control and information systems.

INTRODUCTION

The Hammond facility produces superconducting dipole magnets for the main ring of the SSC. Four cold test stands, a helium refrigerator and the control systems were supplied to the General Dynamics' Hammond Facility by Air Products and CVI. This paper covers the control system for the test stands and how it relates to the operation of the test stands.

KEY OBJECTIVES

The cryogenic test stands at Hammond are an integral part of a magnet production facility. The four test stands will test hundreds of magnets. The control system will record and save about 1 Mbyte[1] of data for each magnet tested. Each magnet will require several tests. During each test the process flows, pressures and temperatures will require precise control in order to make each test as repeatable as possible. The entire test procedure must be completely automated but must permit operator intervention when required.

---

[1] The amount of data is dependent on which tests are run on the magnet. Tests with a large number of quenches or life cycle testing will generate much more data.

CONTROL SYSTEM COMPONENTS

The control system for the test stands is composed of multiple computers and processors. Figure 1 shows the components for one of the test stands[2]. The HP9000 computer and the test program computer were supplied by General Dynamics. The test stand computer and the Bailey INFI90 were supplied by Air Products. The HP9000 contains the complete history of the manufacture and testing of each magnet. The test program computer is similar to other computers that control each tool in the Hammond facility. The test program computer controls and records data on the electrical and magnetic properties of the magnet. In addition, it starts each of the majors steps in the cryogenic process. The test stand computer records the data on the cryogenic process and controls the sequencing of the test procedures. The Bailey INFI90 is a digital control system. It performs the regulatory control, data acquisition, alarms and equipment shutdown. At the conclusion of a test both the test program and test stand computers send a set of data to the HP9000 computer.

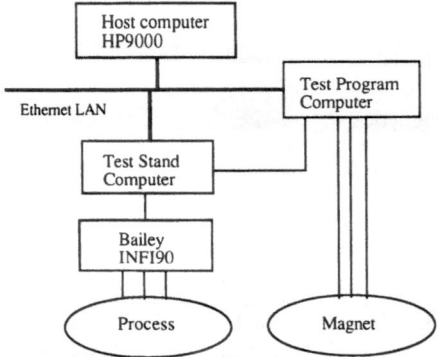

Figure 1. Major components of the control system for a test stand.

HARDWARE AND SOFTWARE SELECTION

The bid schedule and project execution time for the test stands were very tight. No time was available to fully evaluate alternate hardware and software vendors. For the digital control system Air Products already had substantial successful experience with Bailey hardware and software. The Bailey INFI90 system was found suitable for this project and was quickly selected. This system is widely used in chemical, power and the pulp and paper industries for both continuous and batch processes. Similarly, for the test stand computer, a DEC VAX was selected because of successful experience and the availability of an extensive Air Products library of control software written for the DEC VAX. This library met many of the functional requirements. New routines were written in FORTRAN or Batch90.

---

2   There is one HP9000 for the facility. The Bailey INFI90 is common to all four test stands and the He refrigerator.

INTEGRATION OF TEST STANDS INTO FACILITY

The test stand control system has a large number of individual computer systems each performing a set of tasks. All systems must communicate properly with each other. Several meetings were held between Air Products and General Dynamics to agree on the communications protocols. These meetings clearly defined the control responsibilities of each system. General Dynamics had previously developed a standard communications protocol for their computers to communicate with each tool in the Hammond facility. This standard with some minor modifications was used to communicate between the test program and the test stand computers. Test engineers at General Dynamics specified the data recording requirements and the format for transmitting the data to the HP9000. The process data is reported in a tabular data file. The sequence of events in the test stand is reported a time stamped data log file.

CONTROL STRATEGY

The general operating strategy was developed by General Dynamics. It described the necessary tests, the major steps in the test process and the data recording requirements. To develop the control strategy Air Products put together a multidisipline team of controls, process, startup and instrument engineers. The control strategy was developed and refined in parallel with the process design work. The entire batch logic was done in a flow chart format before coding began. This flow chart was reviewed by the Air Products team. The completed logic was reviewed at the critical design review with General Dynamics.

During the development and review process several issues were discussed and resolved. One issue was the protection of personnel during magnet connection and removal, for which a safety system using key locks was designed to prevent accidental flow of process fluids to the magnet. Another issue concerned the quench relief system. During a quench the vaporized helium must be sent back to the refrigerator. This flow is sent to the helium dewar to absorb the contained heat and recondense the vapor. It was determined that the quench valve would have to open within 200 msec in order for this system to function properly. CVI had the quench valve tested for opening speed. As a result their standard valve was modified with an additional vent port on the diaphragm to allow the valve to open in the required time. Also during the development and review process the logic of several regulatory control loops were modified. Most of these changes were for proper control when two streams of helium at different temperatures are blended to control the magnet inlet temperature.

While cryogenic processes are normally continuous, the test stands run in a batch like mode. The test procedure can be divided into 11 major steps as shown in Figure 2. Each step can be initiated by the test program computer. The quench recovery step is automatically initiated by the quench detection system. Each of the steps requires a specific set of control sequences to be performed. The control logic uses the batch control concept of a recipe. A recipe is a set of parameters and events needed to complete a step in the process. For example, the parameters of helium flow to magnet, helium flow through the 20K shield, temperature ramp rate, the maximum temperature difference across the magnet and final magnet temperature are part of the recipe for the cool down step.

TESTING OF THE CONTROL SYSTEM

The testing of the control system has occurred in many phases. First the Bailey INFI90 system was tested at Bailey Controls in New Jersey. Voltage inputs were applied to the input channels. Outputs were measured on volt meters. The shutdowns, regulatory controls and control sequences were tested. The communications between the Bailey INFI90 and the test stand computer were also tested. A simple dynamic model of the process and regulatory controls was used to check the logic of the software on the test stand computer. Testing at site will be first done without flowing process fluid and then with process fluid. A spool piece will be used to replace the magnet for the initial testing.

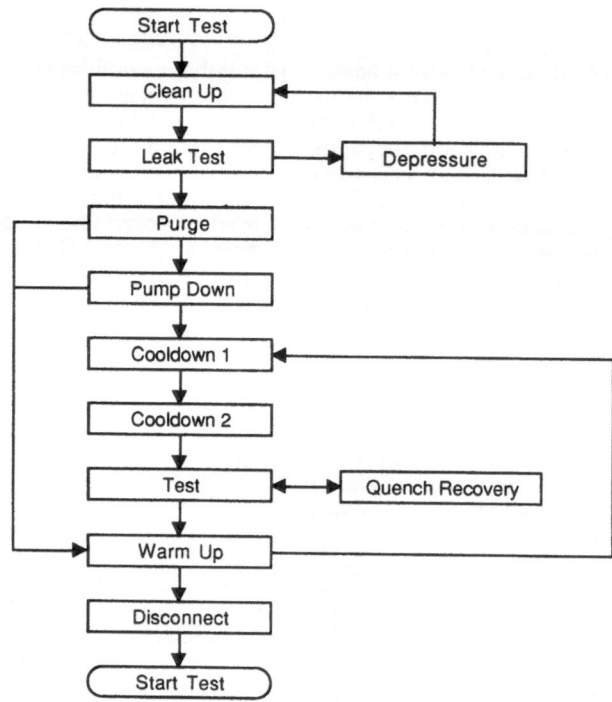

Figure 2. Major steps in the control sequence.

SUCCESS FACTORS

Several factors are key to making the operation of a control system of this complexity a success. One is to utilize a multidiscipline team to develop the control strategy. This team should include process, control, instrument and startup engineers. In addition, a close working relationship between supplier and customer engineers needs to be maintained throughout all project phases. Another key is to develop the process design and control strategy in parallel.

DEVELOPMENT OF CRYOGENIC INSTRUMENTS AND EQUIPMENT FOR SSC MAGNET CRYOGENIC TESTS AT THE MTL

Q. S. Shu, M. Coles, R. Dorman, C. Franclin, R. Fuzesy, G. Gabert, D. Hatfield, I. Syromyatnikov, J. Tompkins, R. Trekell, J. Weisend, and A. Zolotov

Magnet Systems Division
Superconducting Super Collider Laboratory[*]
2550 Beckleymeade Ave., Dallas, TX 75237-3997

INTRODUCTION

The Magnet Test Laboratory (MTL) will test a considerable portion of the total SSC superconducting magnet production in order to control the manufacturing process and verify magnet performance requirements. With ten cryogenic test stands, MTL is capable of housing tests for 30 dipoles and 5 quadrupoles per month. For further understanding and improving the performance of the SSC magnets, there will be two R&D test stands for extensively instrumented magnets, and there will also be three-magnet string test facilities.[1] A large number of instruments were allocated and installed inside the prototype and first production magnets, as well as in the feed and end cans. A data acquisition and control system is developed. A comprehensive cryogenic system (including refrigerator, cryogenic distribution box and, feed/end cans), vapor-cooled power leads, anti-cryostats (warm bore), and other associated systems, have been designed, developed and tested. This paper will briefly discuss the progress to date.

THE CRYOGENIC INSTRUMENTATION SYSTEM

The cryogenic instrumentation to be placed in the MTL Cold Test Stand provides cryogenic parameters – temperature, pressure, flow, cold mass stress (strain gauges) and voltage taps – monitoring and feedback information for the control systems during test operations. The location of gauges on the MTL Test Stand is shown on Figure 1.[2]

Electrical heaters, shown on the Figure 1, together with temperature controllers, will be used for fine temperature control of helium and nitrogen flow (1, 2), warming up part of single-phase helium flow to operate the 20 K shield (3), heating helium gas up to 300 K before returning to the compressor, and the remainder (5, 6) will be used for calibration purposes. All heating elements must be installed in stream.

CRYOGENIC DATA ACQUISITION SYSTEM

The front-end part of the cryogenic data acquisition and control system, shown in Figure 2, has a VME-based architecture, driven by a Motorola MV147 card, containing a 25 MHz 68030 microprocessor. A real time operation system, VxWorks, runs the acquisition and control software as a set of separate tasks, serving different types of instrumentation (temperature, pressure, etc.), that run independently and asynchronous. The necessary timing is provided by the test specific application software. This approach provides the necessary flexibility for the whole system to accommodate various types of tests.

Digital Voltmeters (DVMs) HP 3458A together with 16-channel multiplexer cards form scanning voltmeters used for monitoring signals from temperature, pressure, strain gauges and voltage taps. Using one power line cycle, the voltmeter integration time (which provides 7.5 digit resolution) gives a practical

* Operated by the Universities Research Association, Inc., for the U. S. Department of Energy under Contract No. DE-AC35-89ER40486.

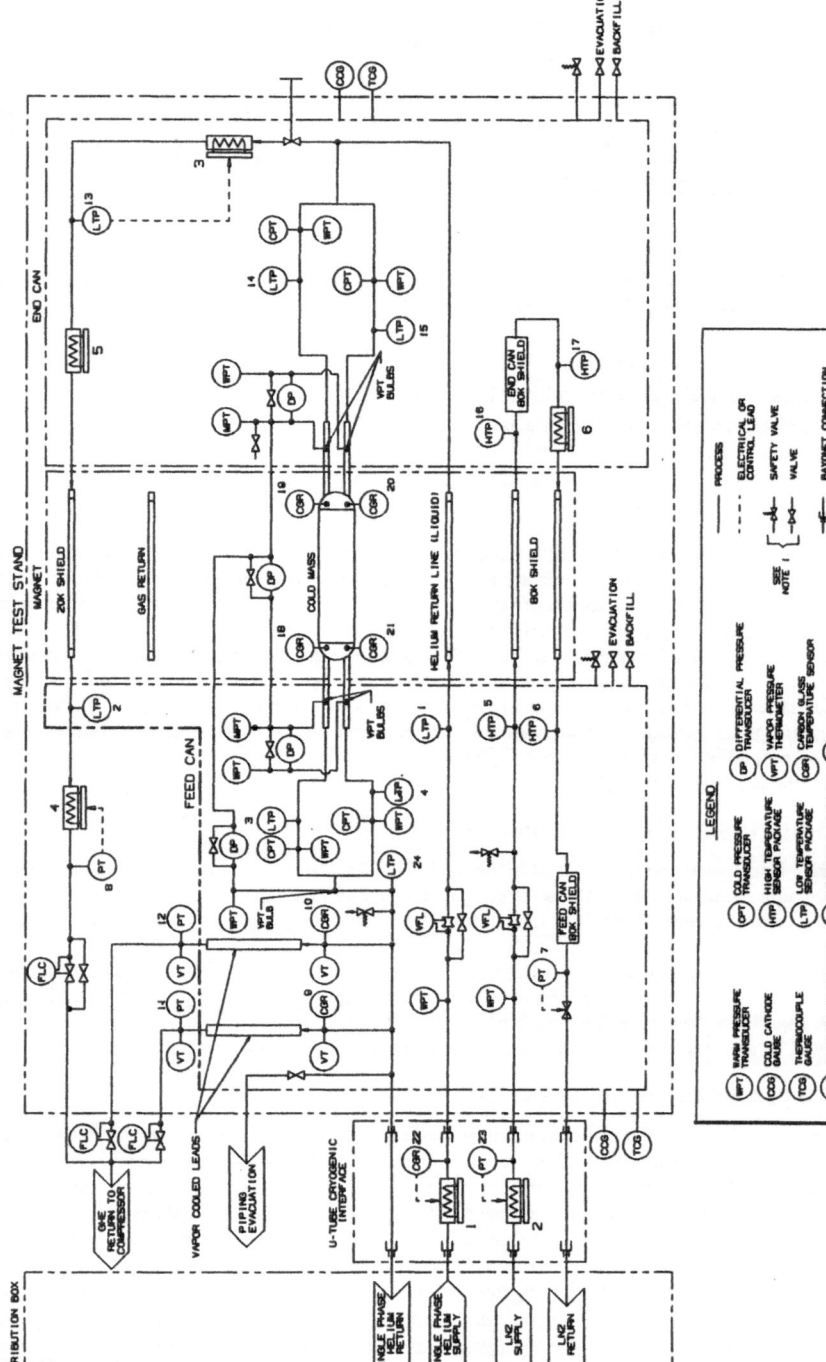

Figure 1. The flow chart of the cryogenic instruments for the MTL Superconducting Magnet Test Stand.

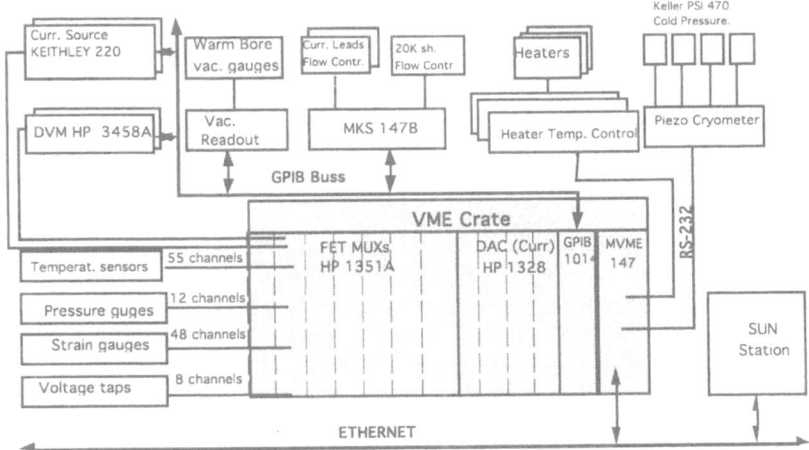

Figure 2. MTL Cold Test Stand Instrumentation Hardware Schematic.

scanning rate of 25 ch/sec. Solid-state FET multiplexers in a HP 1351A provide virtually an unlimited number of switches that increases system reliability. Programmable current sources are used for gauge excitation.

The electrical heater temperature controllers and flow controllers are used as stand-alone devices, with set points downloaded via RS232 and GPIB lines to off-load the VME computer.

THE FEED AND END CAN STATUS

The concurrent design effort with Meyer Tool for the feed and end can is now complete. The fabrication of the first pair of feed and end cans is nearing completion. SSCL confidence in the design has enabled the placement of an option for four additional pairs of feed and end cans from Meyer Tool. Delivery of the first pair of feed and end cans is scheduled to allow cold testing of ASST dipole magnet DCA-207 in August of this year. The end cans are fully instrumented to be able to accurately measure temperature, pressure, and flow rate of helium through the cold mass, 20 K shield and 80 K shield lines. The 20 K shield can be operated from 2.5 to 40 K at a flow rate of 0 to 10 g/s. The flow path for the cold test stand takes the helium through the feed can and into the magnet liquid helium return line to the end can. From the end can the helium flows through the magnet cold mass and into the feed can. As the helium flows through the feed can, it passes through the vapor cooled power lead pot where a portion of the helium is taken through the leads for cooling. The remainder of the helium returns from the feed can back to the distribution box.

THE REFRIGERATION SYSTEM AND DISTRIBUTION BOXES

The refrigeration system for the MTL is in the early stages of commissioning. The compressors will be tested first, followed by the refrigerator liquefier. Figure 3 shows the first cold test stand with the distribution box skid behind. The cans shown on the ends of the magnet are for vacuum system testing which is currently taking place. Cryogenic transfer lines will connect the distribution box to the feed can that will be mounted on the cold test stand. The transfer lines will provide helium and nitrogen supply and return to the feed can. A small heater is included in the helium supply line to provide fine temperature control to the test stand. The

Figure 3. Cryogenic distribution boxes and magnet test stand.

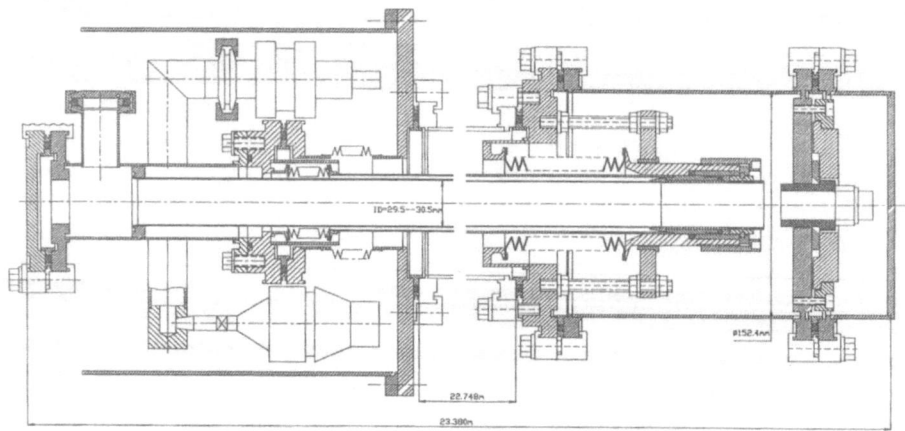

Figure 4. One of the warm bore designs.

nitrogen supply line includes a heater for heat leak testing purposes. The distribution box functions as the interface to the main and CCWP (clean-up, cooldown, warm-up, and purification) cold boxes. It also contains the subcooler required for operating temperatures below 4.5 K. Magnet quench process flow is handled by the distribution box valves. The normal supply and return valves are closed and a quench return valve opens allowing the quench flow to return to the liquid helium dewar and/or the refrigerator cold box.

DESIGN AND TEST OF THE WARM BORE AND FINGER

The anti-cryostats, so called warm bore and warm fingers, for the magnetic field measurements during magnet cryogenic testing have been designed and tested. The warm bore inserts into the magnet beam tube, which is at a temperature around 4.2 K, and accommodates a magnetic measuring field rotating coil, NMR probe and Hall probe within its warm space. Contrary to a normal dewar, the temperature of the outer wall of the warm bore is 4.2 K, whereas the inner space of the bore is maintained above 273 K. Several short warm fingers (3–4 m) have been successfully tested and used in SSC short magnet vertical tests. The thermal performance agrees with design specifications of 2.5 W. A warm bore for SSC full size dipole cryogenic tests has been designed as shown in Figure 4.

DESIGN AND TEST OF 10 kA VAPOR COOLED POWER LEADS

The spiral fin 10 kA helium vapor cooled power leads have been designed for SSC superconducting magnet tests at the Magnet Test Laboratory (MTL). Two different fin geometries and three RRR lead material values were developed to thermally optimize the power lead parameters, including lead diameters, that minimize Carnot work for different lead lengths. In the design, a new thermal barrier device to reduce heat conduction from the vacuum and gas seal area was employed. Therefore, the electric insulation assembly, which isolates the ground potential parts of the lead from the high power parts, was moved into a warm region in order to prevent vacuum and helium leakage in the o-ring seals due to transient cold temperature. The first pair of the power leads were cryogenically tested up to 10 kA.

ACKNOWLEDGEMENTS

The author wishes to thank all colleagues involved in the MTL development.

REFERENCES

1. Q. S. Shu, et. al., "The Cryogenic System for the MTL Magnet Test Stands," IISSC, New Orleans, LA., March 4–6 1992.
2. Q. S. Shu, et.al., "MTL Cold Test Stand Instrumentation Design Specification," M62-000068, MSD, SSCL, 1992.

3. D. Hatfield, "MTL Feed Can and End Can Specification," M62-000025 & M62-000034, MSD, SSCL, 1992.
4. Q. S. Shu, "Thermal Optimum Analyses and Mechanical Design of 10 kA Vapor Cooled Power Leads for SSC Superconducting Magnet Tests at MTL," Applied Superconducting Conference, Chicago, IL. Sept. 1992.

PRELIMINARY DESIGN OF SECTOR CRYOGENIC SHAFT
TRANSFER LINE SYSTEM

B. Zhang and K. Stifle

Accelerator Systems Division
Superconducting Super Collider Laboratory*
2550 Beckleymeade Avenue
Dallas, Texas 75237

INTRODUCTION

The SSC cryogenic system has an unprecedented capacity to supply cryogens at 4 K, 20 K (helium) and 80 K (nitrogen) for both normal operation and transient conditions of the collider magnets and thermal shields.[1] At each sector, the terminal element of the surface refrigeration plant is the surface distribution box (SDB). On the collider side, four half-feed spool pieces (SPRF) interface with the underground cryogenic facilities. The shaft transfer line connects the surface distribution box and the helium cold compressor box in the underground cryogenic chamber. Its reliability is vital to the operation of the SSC cryogenic system. The shaft transfer line consists of 7 cryogenic circuits enclosed in a common vacuum jacket (24 in O.D., schedule 10 carbon steel pipe). The 4 K gas helium, 20 K helium return, and 80 K gas nitrogen circuits use 4" stainless steel tubes, and the other circuits use 3" tubes. The wall thickness for all circuit tubes is 1.651 mm. The length of the shaft transfer line varies from sector to sector. It ranges from 30 to 80 meters. Each shaft transfer line is to be built in basic construction units (straight module and elbow piece) and assembled on-site. The maximum length of the straight module is 12 meters (40 ft). An isometric view of a transfer line module is shown in Figures 1. The thermal shield, covered by two MLI

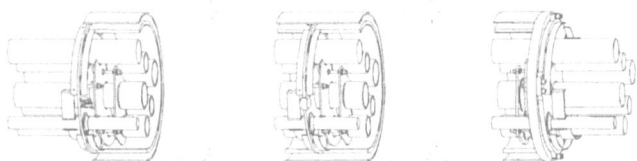

Figure 1. Isometric view of transfer line module

blankets, encloses all the circuit tubes. It is split in halves circumferentially proportional to the locations of the two nitrogen circuit tubes, and soldered to the tubes at a 0.4 m interval. Each helium circuit tube is also covered by an MLI blanket. A shaft transfer line is a single vacuum unit. A vacuum barrier is attached to each end of a transfer line to isolate the vacuum space of the cryostat from neighboring equipment. The enclosed space will be evacuated after the on-site final assembly. A vacuum pumping port is installed on every construction unit, and two pressure relief devices are installed on each shaft transfer line to protect the cryostat from being internally pressurized when a circuit leak occurs. Bellows are used at every interconnect region to accommodate thermal movement for each circuit tube. The elbow piece completes a 90 degree bend. The circuit tubes

*Operated by Universities Research Association, Inc., for the U.S. Department of Energy under Contract No. DE-AC35-89ER40486

are bent twice within the elbow cryostat to complete the 90 degree bend. Straight tube sections between the bends make the tubes parallel to the vacuum jacket. Mittered sections compose the vacuum jacket of the elbow piece.

This work summarizes the preliminary design of the shaft transfer line. Operating modes dictating the design criteria are addressed. System requirements, design constraints, and results of design analysis conducted on key elements are also discussed.

DESIGN

The design and nominal pressure, temperature, and mass flow rates for the seven cryogenic circuits of the shaft transfer line are presented in Table 1.[†] The design also needs to meet certain

Table 1. Shaft transfer line system requirements

Cryo. Circuits	P		T	Flow Rate	
	Design (bar)	Nominal (bar)	(K)	Design (g/s)	Nominal (g/s)
4 K He supply	20	4	4.5	400	400
4 K He return	20	3.5	4.3	400	100
4 K gas He	10	1	4.0	400	300
20 K He supply	20	3	16.5	200	200
20 K He return	20	2	24.0	200	200
80 K liquid nitrogen	20	10	80.0	400	80
80 K gas nitrogen	10	1	80.0	400	300

strength and heat leak requirements. The longitudinal support is designed to withstand a 2 g loading in all directions. The elbow support is designed for hydrostatic, hydrodynamic, and thermal loadings. Design pressures are used to calculate the hydrostatic loading. The estimated heat leaks into the cryogenic circuits per unit length are shown in Table 2. Pressure relief devices protect the vacuum

Table 2. Estimated heat leak per unit length

Heat Leak Budget (W/m)		
4 K circuits	20 K circuits	80 K circuits
0.0984	0.328	3.94

jacket against excessive internal pressure in the vacuum space of the cryostat. Temporary covers are required to protect the interior of the cryostat from damage and contamination at both sides of the modules and elbow pieces during transport. Lugs for handling are installed on the vacuum jacket before a module can be shipped. Appropriate procedures are to be developed for proper transport, handling, and assembly.

The straight module internal suspension system holds the internal elements (circuit tubes, thermal shield, and MLI blankets) and transmits the loading to the vacuum jacket. The suspension system consists of one longitudinal and four radial supports. The longitudinal support anchors the internal elements to the vacuum jacket at one end of the module. Each circuit tube is welded to a stainless steel collar which is bolted to the longitudinal support. One of the radial supports is installed at the opposite end to the longitudinal support, the rest are located at even intervals in between. They restrict the radial movement of the circuit tubes. Three caster modules are attached to the free end radial support to allow longitudinal movements. The internal suspension system for the elbow piece consists of two supports with one at each end of the elbow piece. Both supports are anchored to the vacuum jacket. Due to the 90 degree bend and the use of bellows for circuit tube interconnections, the supports will experience both hydrostatic and hydrodynamic loadings in various operating modes. Within an elbow piece, the thermal movements of the circuit tubes are accommodated by their flexures. Stainless steel brackets are attached to all internal supports to support the MLI blankets. The blankets are secured to the brackets using barbed snap-on type

---

[†] The actual mass flow rates of the nitrogen circuits could deviated from these figures substantially, depending on the final design of the nitrogen system.

nylon fasteners at designated locations. The stress analysis of the internal suspension system is performed using a finite element analysis package ViziCad by Algor, and the results are presented elsewhere[4]. The external suspension system holds the shaft transfer line at two locations on each module. It is to be designed for the operating mode which produces the most severe loading.[‡] All external supports will provide means for three-dimensional adjustment.

The temperature in the surface building (utility shaft head house) is in the range of $0°$ to $50°C$. The controlled collider tunnel temperature is at $80° \pm 5°F$. The assessment of the thermal stress level in the vacuum jacket is based on the the differential of thermal expansion between the vacuum jacket material (carbon steel) and the utility shaft wall (concrete). The thermal stress will be below 1,500 psi for a temperature difference of $50°C$, which falls well below the allowable level. Appropriate spring factors need to be incorporated in the external suspension system design to reduce thermal loading. Attachment to the vacuum jacket is required for external support and handling. The surface area of the attachment to the module vacuum jacket for external support and handling is sized in such a way that a 2 g loading of the module can be supported at one location.

The cryostat serves for the purpose of reducing the heat leaks into the cryogenic circuits. Refrigeration for the thermal protection is available at 20 K and 80 K. Considering the structural simplicity and reliability, only one thermal shield (gauge 19 copper sheet) at 80 K is used. The thermal contact is established by soldering the shield to the 80 K tubes. Numerical analysis indicates that the circumferential temperature variation in the thermal shield is less than 1 K. Two MLI blankets are used over the thermal shield. The blanket design is similar to that for the dipole magnet cryostat.[5] Each blanket has 20 layers of alternating double-sided aluminized mylar and spunbonded polyester spacer. Velcro strips are attached to the blanket to secure the blanket over the thermal shield. An MLI blanket of 20 layers is applied to each helium circuit tube. The internal suspension system interrupts the continuity of the MLI system at each support assembly. To prevent any see-through between the vacuum jacket and the helium circuit tubes, a smaller MLI blanket is used to bridge the gap between two neighboring MLI blankets at these locations. Each shaft transfer line is a single vacuum unit. It is isolated from the neighboring equipment by two vacuum barriers at the ends. A vacuum pumping/leak detection port is provided at each interconnect region. The design vacuum level of the cryostat is 1×10^{-6} torr measured at the inner surface of the vacuum jacket. This vacuum level is to be maintained even at the presence of a minor leak (a helium leak rate less than 10^{-6} std cm^3/s). The vacuum space is rough-pumped to 1.0×10^{-4} torr and then cryopumped. The cryopumping essentially involves solidification and adsorption of residual gas particles on surfaces at cryogenic temperature by means of van der Waals' forces. Both the thermal shield and the 80 K circuit tubes will be cryopumping after the system cool-down. When a specific pumping speed (pumping speed per unit area) $S/A = 14$ l/s-cm^2 is used for the 80 K surfaces, sufficient cryopumping power will be available for the condensible gas constituents due to the outgasing of the surfaces in the cryostat. Cryosorption, a physical adsorption process which also relies on van der Waals' forces, is adopted for vacuum retention purpose. Since the equilibrium pressure of adsorbed gas particles is far lower than the corresponding saturation vapor pressure,[6] particles of the gas constituents difficult to condense (neon, hydrogen, and helium) have to be adsorbed in sub-saturated conditions. Activated charcoal is used as the adsorbent in a cryogenic getter. The getter has an enclosure (shroud) to block the condensable gas especially water vapor particles. This enclosure also serves as a thermal shield to prevent direct radiative thermal communication between the getter adsorbent and warmer surfaces in the cryostat. The getter module, bolted to the 4 K gas helium circuit tube to obtain mechanical support and refrigeration, is located in the interconnect region on the radial support side. The quantity of charcoal to be used in each getter is calculated based on the assumption that the average thermal cycle of the collider is one year. The getters are to be regenerated every time the system is warmed up. In the event when substantial internal leak occurs, a turbo pump will be connected to the cryostat for continuous pumping to maintain the vacuum level at 1.0×10^{-4} torr.

Copper thermal shorting straps are used to maintain temperature boundary conditions of 20 and 80 K at various locations for both longitudinal and radial supports. The maximum temperature drop through one strap is less than 2 K. The vacuum barrier primarily consists of three groups of concentric shells and plates, each holds the circuit tubes at the same temperature level(4, 20, or 80 K). The length of the shells is around 300 mm. Copper plates are soldered to the 20 and 80 K plates to intercept heat leaks into the helium circuits. Table 3 lists the heat leaks into the cryogenic circuits through the composite design option of the internal suspension system.

Two reclosing type pressure relief devices (one at each end of a transfer line) are installed on the vacuum jacket to protect the cryostat against accidental pressure rise. The cracking pressure

---

[‡]Depending on the final design of the bellows, this operating mode could be either the design operating or the CCWP mode.

Table 3. Heat leaks into cryogenic circuits through supports and vacuum barrier

Support/ Barrier	Heat Leak (W)		
	4 K	20 K	80 K
Radial	0.038	0.39	2.6
Longitudinal	0.047	0.62	6.3
Bend	0.062	0.89	9.1
Vacuum barrier	0.023	0.85	10.7

of the relief devices and the maximum allowable pressure in the cryostat are set at 5 and 15 psig respectively. The relief devices are sized for 400 g/s of liquid helium totally vaporized and warmed up to ambient temperature. Each pressure relief device is sized adequately for venting alone. The second relief device is for redundancy purpose. Each pressure relief device is equipped with a protective device for accidental blockage at the inlet to ensure uninterrupted venting.

At the final assembly, a number of construction units (straight modules and elbow pieces) are to be field-connected to form a complete shaft transfer line at each sector. The cryogenic circuits are to be joined by bellows. The alignment tolerance of the circuit tubes is 3 mm between center lines. Before the installation of the thermal shield and MLI system, all cryogenic circuits are to be helium leak-checked using a mass spectrometer. The vacuum jacket interconnection is accomplished using a section of split shells (clam shells). The structure and installation methods of the interconnect region determine the required clearance between the shaft transfer line and the utility shaft wall. This clearance will be kept under 6 inches in order to reserve an adequate drop zone area in the utility shaft. Orbital welder heads will be used whenever possible for the interconnection of the circuit tubes. A longitudinal cutter will be used for the removal of the interconnect section of the vacuum jacket for repair.

SUMMARY

Since the conceptual design of the shaft transfer line was presented in 1992, design effort has been concentrated on narrowing down the design alternatives. After evaluating and comparing various design options, a preliminary design of the sector cryogenic shaft transfer line is concluded. The design requirements and the results of design analysis are quantitatively presented. Open issues are addressed, and further improvement is expected and will be integrated into the final design of the sector shaft transfer line.

REFERENCES

1. M. McAshan, V. Ganni, R. Than, and T. Niehaus, *Refrigeration Plants for the SSCL*, IISSC, Supercollider 3, J. Nonte ed., Plenum Press, New York (1991) p.809.

2. M. McAshan, S. Abramovich, V. Ganni, A. Scheidemantle, and M. Thirumaleshwar, *84 K Nitrogen System for the SSC*, IISSC, Supercollider 4, J. Nonte, ed., Plenum Press, New York (1992) p. 207.

3. B. Zhang and V. Ganni, *Conceptual Design of The SSC Cryogenic Transfer Lines*, IISSC, Supercollider 4, J. Nonte, ed., Plenum Press, New York (1992) p. 453.

4. B. Zhang, *Stress Analysis on SSC Cryogenic Transfer Line Internal Suspension System*, to be presented at this conference.

5. W.N. Boroski, T.H. Nicol, and C.J. Schoo, *Design of the Multilayer Insulation System for the Superconducting Super Collider 50 mm Dipole Cryostat*, IISSC, Supercollider 3, J. Nonte, ed., Plenum Press, New York (1991) p.849.

6. R.A. Haeffer, *Cryopumping Theory and Practice*, J. Shipwright and R. G. Scurlock, trans., Clarendon Press, Oxford (1989).

PHYSICS DETECTOR SIMULATION FACILITY (PDSF) ARCHITECTURE/UTILIZATION

B. Scipioni

Physics Computing Department
Superconducting Super Collider Laboratory*
2550 Beckleymeade Avenue
Dallas, TX 75237

ABSTRACT

The current systems architecture for the SSCL's Physics Detector Simulation Facility (PDSF) is presented. Systems analysis data is presented and discussed. In particular, these data disclose the effectiveness of utilization of the facility for meeting the needs of physics computing, especially as concerns parallel architecture and processing. Detailed design plans for the highly networked, symmetric, parallel, UNIX workstation-based facility are given and discussed in light of the design philosophy. Included are network, CPU, disk, router, concentrator, tape, user and job capacities and throughput.

INTRODUCTION

The SSCL Physics Detector Simulation Facility (PDSF) is currently in Phase II of a multi-phase growth plan. The initial plans called for 500 MIPS of computing in Phase I, 1000 MIPS for Phase II and 4000 MIPS for Phase III. However, due to technology advances, Phase II has 3000 MIPS (or 2000 VAX 11/780 equivalents). The original requirements[1, 2, 3] have recently been re-evaluated to determine whether or how the facility will continue to evolve. The original requirements led to system specification,[4] which was implemented in March of 1991. This initial, Phase I, implementation was a prototype. Lessons learned resulted in Phase II which was implemented in March of 1992. This paper is concerned with the architecture and performance of Phase II.

ARCHITECTURE

After having studied the performance characteristics of Phase I PDSF several conclusions were reached. For parallel processing of GEANT detector simulations it was clear that there was no advantage (from a price/performance standpoint) to the parallel SMP computers (SGI 4D/380) over the single CPU workstation (SPARCstation 2). This was no great surprise based upon the loosely coupled nature of the parallelism and the less than 1/1 speedup of the SMP machines and their premium price. (This situation has changed since and will be discussed below). However, data throughput, especially parallel I/O achieved through disk striping and multithreaded raw disk I/O, showed the SMPs to be the better data machines. In addition, threading the Remote Procedure Call (RPC) mechanism turns a (free) client/server technology into a simple-to-use parallel programming paradigm for networked computers. This led to the selection of the SMP computer as both data server machine and launchpads

---

*Operated by the Universities Research Association, Inc., for the U.S. Department of Energy under Contract No. DE-AC35-89ER40486.

for parallel programs. The ability to have two FDDI interfaces on the SGIs allows for multiple FDDI networks tied together to extend the parallelism to network I/O. This combination of parallel processing, parallel disk I/O and parallel networking has given the PDSF the capability of true overall supercomputer throughput at a fraction of the cost. Another lesson learned might be called spontaneous symmetry breaking. Lack of symmetry in the distributed computing system seems to cause bottlenecks as exemplified by uneven router loads due to upstream/downstream flow of data on an FDDI ring. Sociological factors seem to cause this as well (i.e., familiar computers). Also the multiplexing of Ethernets onto FDDI greatly reduced router packet performance due to excessive fragmentation between dissimilar MTUs. The incorporation of these considerations along with additional acquisitions produced the highly symmetric, multi-networked configfiguration show in Figure 1. It consists of the following:

- 2 clusters of 15 each headless SPARCstation 2s, each with 32 MB RAM and 1 GB SCSI disk
- 2 clusters of 16 each of headless HP 9000/720s, each with 32 MB RAM and 1 GB SCSI disk
- 4 SGI 4D/360s, each with 128 MB RAM and 24 GB SCSI disk
- 2 Summus 8 mm tape robots, each with two drives and 125 GB storage
- 6 SPARCstation 2s, (two for systems database, 4 for multiplexing of consoles)
- Network equipment: 1 HP 9000/720 network management station, 3 Cisco AGS+ routers, 6 FDDI networks with 4 ODS concentrators, and 6 Ethernet networks with 5 ODS concentrators all linking the multiple networks.

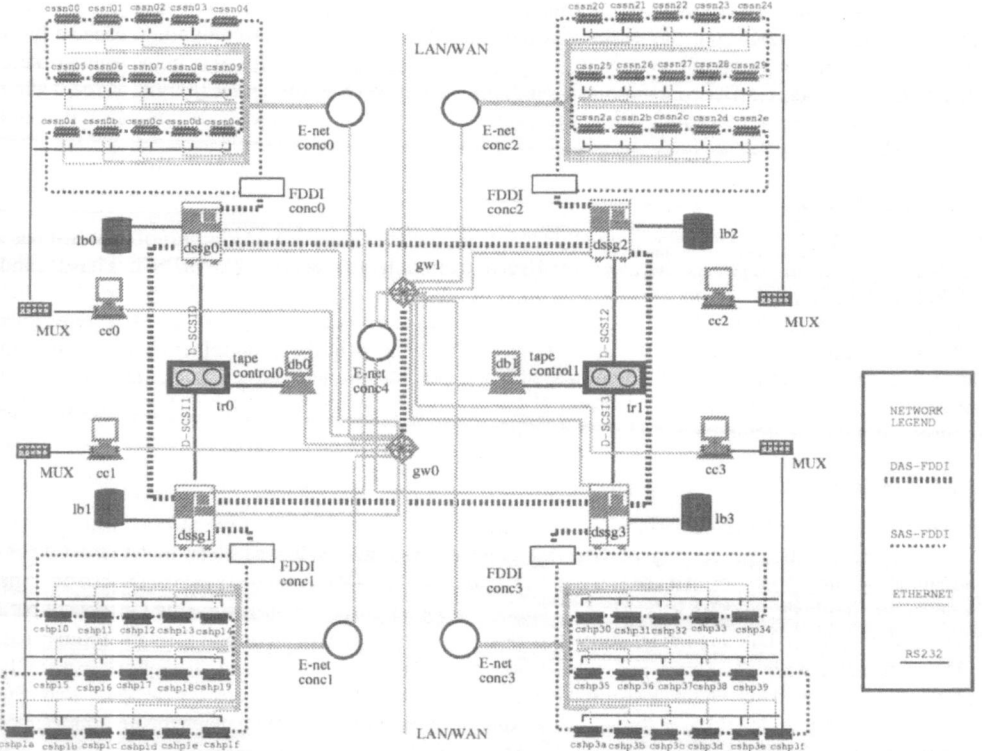

Figure 1. Physics Detector Simulation Facility (PDSF) Architecture.

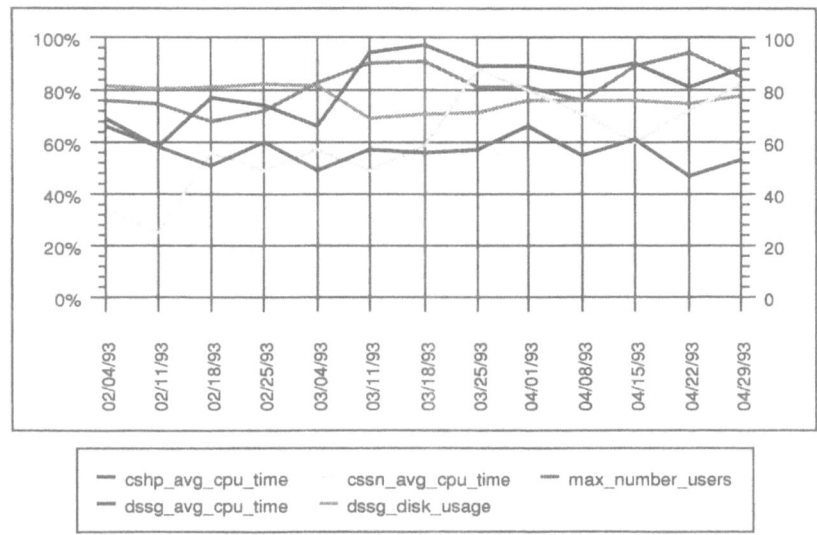

Figure 2. PDSF Utilization History.

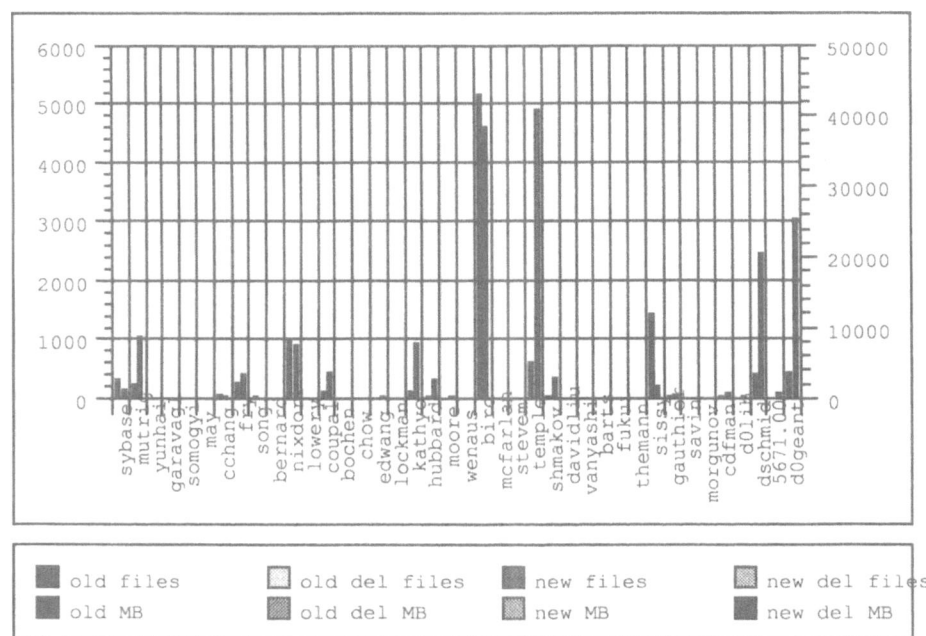

Figure 3. DMS Storage Utilization.

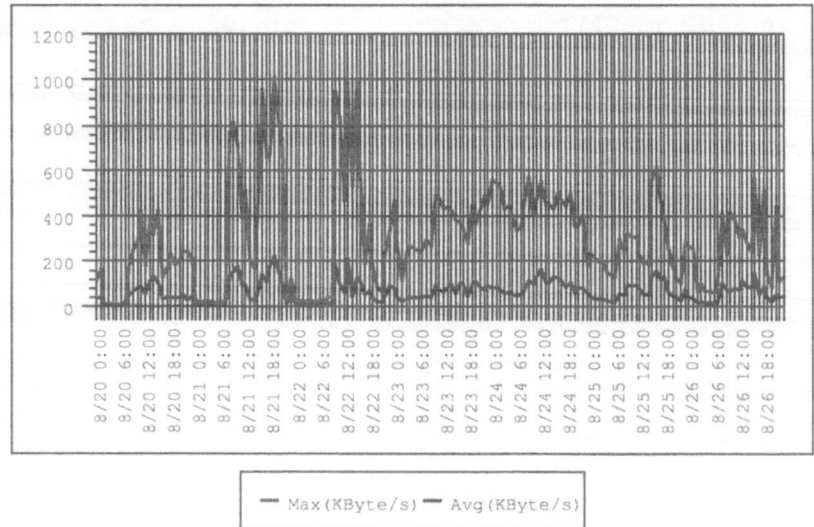

Figure 4. Network Utilization Data.

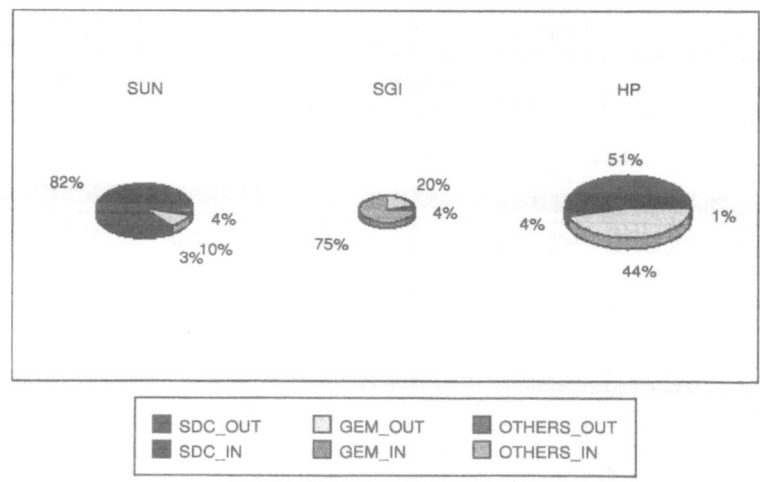

Figure 5. PDSF User Profile.

UTILIZATION

The data contained in Figures 2–4 will be summarized here. There appears to be little true synchronous parallel processing taking place on the PDSF with the exception of the systems analysis software (SISSY).[5] However, there is an event-parallel GEANT which has been run by some SDC members and at least one other application using RPC with the client running outside the PDSF. This, of course, can technically be done over Internet. The parallelism of the architecture, is however, well exploited by the asynchronous (multi-job, multi-user) computing performed on the facility. Some of the more salient points of interest follow:

- **Utilization history**: Over the 1 year Phase II has been in operation certain trends are clear. Figure 2 shows some overall utilization numbers for the last 3 months. The 24 hour CPU utilization of both the SS-2s and the HP-9000/720s has increased from about 25% to approximately 90% during a typical week. The SGIs have risen from 20% to 60%. Higher utilization for SGIs has not been encouraged since they are serving dual roles as both file servers and compute servers. This situation will be remedied in the next Phase of PDSF as discussed below. These numbers fluctuate weekly and are sometimes correlated with collaboration meetings. The two main user groups and three architectures tend to load balance automatically to some extent. The maximum number of concurrent users is now fairly stable at about 80–90. Data server disks has gone from 45% capacity to 80% and has been steady for the last month or two. Among that 100 GB, some file systems remain at capacity most of the time while others fluctuate or remain lower. At the 80% level it is difficult to find large, single pieces of a filesystem for storage of single large files. While there are not many major users of the DMS,[6] there are about 180 GB backed up to 8 mm at this time as seen in Figure 3.

- **Weekly trends**: CPU utilization typically varies widely daily, sometimes reaching close to 100% for entire clusters. This means that weekly averages could only increase by balancing daily loads. However, this in undesirable due to both the low availability and the difficulty of interactive use of the system at higher utilization levels. The current weekly averages are thus considered to be near optimal.

- **Networks**: Hourly snapshots of both the maximum (5 minute average) and average (1 hour average) KB/s are plotted for the week for each computer type. It is readily seen that at times during the week the network traffic exceeds Ethernet maximum bandwidth and at many more times during the week exceeds the practical maximum. This means Ethernets used here would have been a serious bottleneck shutting down the network for many minutes at a time. However, since FFDI is used here there is still good residual capacity on the networks. More tightly coupled parallel applications could easily be run without causing problems.

- **User trends**: Most (>70%) users are from outside the lab (Figure 5). Analysis shows user groups tend to prefer certain architectures. However, during weeks where one of the collaborations is meeting, the freed CPUs tend to be picked up by other groups, although this is not always the case.

PDSF III

Based upon our experience with PDSF II and changes in the computer market several changes in strategy have adopted for the next phase of PDSF. Two of the more notable changes in philosophy of its use, and in the price/performance of SMP computers. The next phase will concentrate primarily on providing batch services, with most interactive and general purpose use being on workgroup clusters apart from but attached to the PDSF. Also the price/performance of parallel machines is now superior to single CPU workstations. Computers like the Sun SPARCstation 10 and the SGI Challenge L will be introduced in the next phase.

CONCLUSION

The overall utilization of the PDSF in terms of resource consumption is close to ideal. This indicates that the concurrency inherent in the facility is both an efficient and effective architecture.

There is however a comfortable amount of network bandwidth which can be used for synchronous, network-based parallel processing. Since this is more technique-oriented it may take more time and/or encouragement to maximize its utilization as a synchronous parallel processing facility.

REFERENCES

1. M. G. D. Gilchriese, editor, "Report of the Task Force on Computing for the Superconducting Super Collider," SSC-N-579, December (1988).
2. L. Price, editor, "Report of the SSC Computer Planning Committee," SSC-N-691, January (1990).
3. L. Cormell, editor, "Physics and Detector Simulation Requirements," SSCL-259, March (1990).
4. G. Chartrand, L. Cormell, R. Hahn, D. Jacobson, H. Johnstad, P. Leibold, M. Marquez, B. Ramsey, L. Roberts, B. Scipioni, N. Shivapuja, and G. Yost, "Physics and Detector Simulation Facility Specifications," SSCL-275, Attachment A, July (1990).
5. B. L. Scipioni, D. Liu, and T. Song, "SISSY: A Multi-threaded, Networked, Object-Oriented Databased Example," IISSC 5 (these proceedings).
6. J. Allen, et. al., "SSCL - PDSF Phase II Software Support Description," IISSC 5 (these proceedings).

PHYSICS DETECTOR SIMULATION FACILITY PHASE II SYSTEM SOFTWARE DESCRIPTION

B. Scipioni, J. Allen, C. Chang, J. Huang, J. Liu, S. Mestad, J. Pan
M. Marquez, and P. Estep

Physics Computing Department
Superconducting Super Collider Laboratory*
2550 Beckleymeade Avenue
Dallas, TX 75237

ABSTRACT

This paper presents the Physics Detector Simulation Facility (PDSF) Phase II system software. A key element in the design of a distributed computing environment for the PDSF has been the separation and distribution of the major functions. The facility has been designed to support batch and interactive processing, and to incorporate the file and tape storage systems. By distributing these functions, it is often possible to provide higher throughput and resource availability. Similarly, the design is intended to exploit event-level parallelism in an open distributed environment.

INTRODUCTION

The Superconducting Super Collider Laboratory (SSCL) has adopted a computing strategy that is intended to provide the greatest amount of low cost computing power for as many users as possible.[1-5] By acquiring open systems and conforming to industry standards, the SSCL has been successful in acquiring and integrating heterogeneous networks of commercially available computers. As a result, we are able to integrate multi-vendor solutions by requiring industry standard interfaces, communication, formats, protocols, and the commonality of UNIX.

MOTIVATION

Detector simulations are characterized by loosely coupled, event parallel data runs. In order to accommodate this, the PDSF has been reconfigured. High speed FDDI networks and high disk I/O have been integrated into a highly heterogenous network of RISC based workstations. The system is composed of four groups of computers call corrals. Each corral contains an SGI 4D/360 data server, 15-16 SUN Sparc2 or HP 9000/720 compute servers with a total of approximately 140 GB of disk space, 40 GB on the corrals and 100 GB on the data servers. All machines within the PDSF are connected via FDDI rings. In addition, two database machines and two 8-mm tape robots are installed to perform database and data management functions. The architecture is represented by Figure 1.

---

*Operated by the Universities Research Association, Inc., for the U.S. Department of Energy under Contract No. DE-AC35-89ER40486.

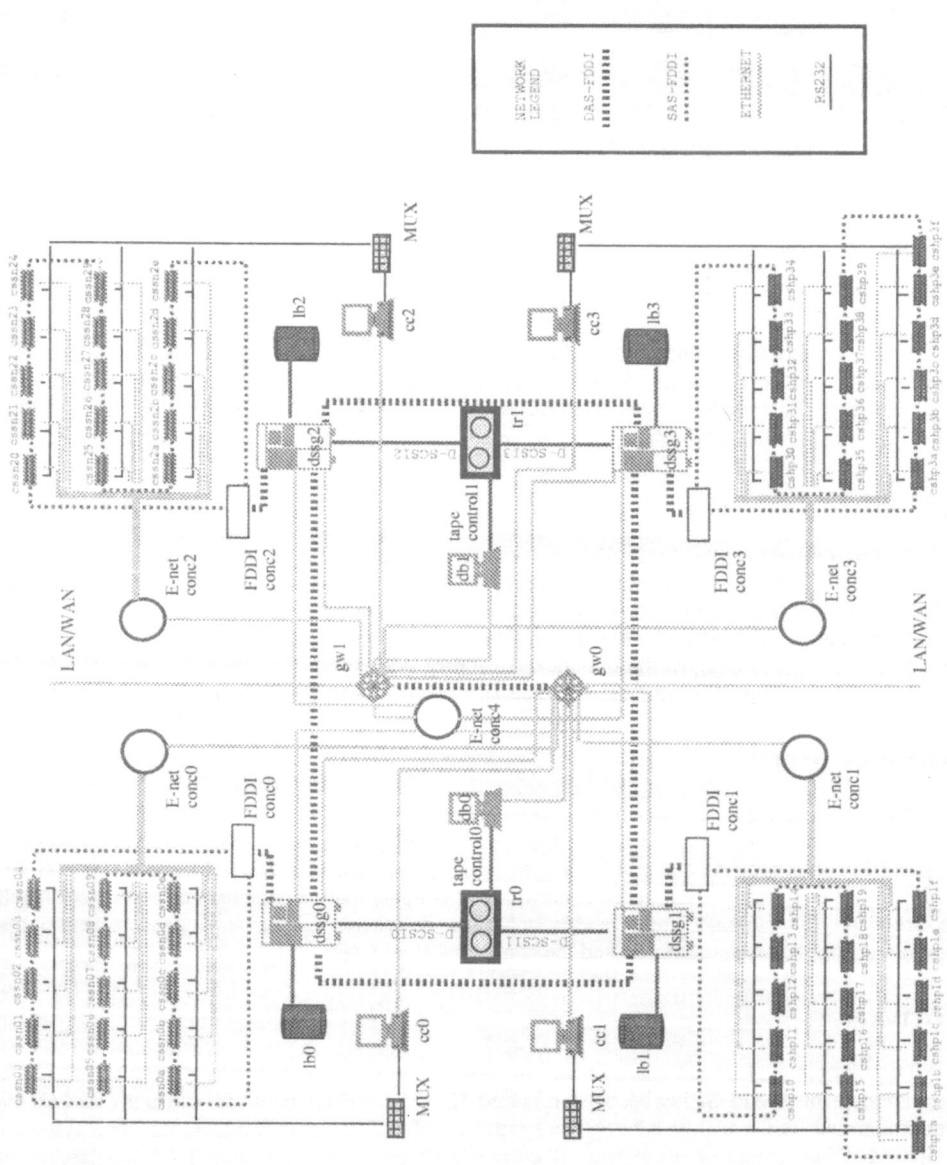

Figure 1. Physics Detector Simulation Facility (PDSF) Architecture.

With multiple gateways available, the PDSF user is allowed to choose between a SUN Sparc2 or HP 9000/720 computing corral. File systems are cross-mounted on each data server thus allowing the sharing of data. A simulation program can be executed on a single computer or fanned out to multiple compute servers.

REQUIREMENTS

The facility operational requirements were broken down into two major functional subsystems:

- Compute servers for interactive and batch usages
- Data servers containing user filesystems and supporting parallel processing

DESIGN

The system software provides the support for the various functions of the PDSF. It is the foundation upon which the interactive and batch processing is built. In addition, it has some responsibility for managing the resources of the PDSF. The system software consists of several subsystems which include the Workstation Allocation System, Console Concentrator, System Database and Polling System, System Mapping Utility, Data Management System, Robotape System, Network Queueing System, Operator Message System and Cooperative Processes Software.

Workstation Allocation System (WASH)

The function of the compute servers is to provide the user with a dedicated resource in order to give instantaneous (or near instantaneous) response to facilitate interactive use. Unfortunately, there are only a finite number of compute servers. Furthermore, there are more users than workstations. The goal of WASH is to intelligently choose a workstation for each login request in order to provide the best possible interactive environment to each user.

Users gain access to the PDSF via *rlogin*, *telnet*, or *dlogin* to one of the four Corrals. The WASH process, in turn, queries the database to determine the best machine for assignment. The database contains information concerning number of users on each workstation as well as system load per box obtained from the polling daemons. After the best machine is chosen, WASH then performs an *rlogin* to that machine on behalf of the user.

The Console Concentrator (CONCH)

In a distributed environment such as PDSF, there are many workstations, most of them headless (without monitor and keyboard). To manage such a configuration of headless workstations all their consoles must be directed to a single workstation with a reasonably large bitmapped display. CONCH was originally written by Neal Ziring of Washington University to run on VAX computers running 4.2 BSD flavor of UNIX. It was extensively modified at SSC. CONCH has a client-server architecture.

The server implements the hardware specific sections of the system. It listens for a connection request over RS232 lines. It also maintains a log file for each managed workstation. Any message that arrives is logged into the log-file. The server also provides a call-up service. Clients use this service to connect to a specific managed workstation. A client is a user's agent. It provides a terminal session with a managed workstation. A client is a separate process; several of them can run simultaneously.

We have combined this program with X Windows to form a more attractive operator interface. Each managed workstation is represented as an icon on a canvas. An alert feature is also provided. When any managed workstation is isolated or down, the icon turns into a different color indicating an alarmed condition.

System Database and the Polling System

The PDSF system database and the polling subsystem are responsible for providing information that is crucial for PDSF subsystems to run on the network. Systems information is gathered by the polling subsystem (SYSPOLL) and sent to the database via a SQL executor service. Information collected by the polling daemons includes running processes, user logins, disk file system usage, and system load for each workstation on the network.

PDSF subsystems, those that take care of network batch job queuing, workstation allocation, and data management, are integrated together through the system database. Each subsystem obtains a certain kind of system information from the system database: WASH monitors the system load to make login assignment decision; SYSMAP makes queries for information regarding running processes, user logins, workstation inet address, and file system usage; NQS uses workstation ID to keep track of job queues for each workstation; DMS queries the database for the status of processes.

The SYBASE database was chosen for its client/server architecture, query performance and availability on major hardware platforms. In addition to the database server software provided by Sybase Inc., a layer of in-house developed SQL executor services is added to the PDSF database system. This structure is devised to extend database service to all workstations on the network.

The System Mapping Utility (SYSMAP)

The SYSMAP utility takes advantage of the system database to assist system administrative personnel in maintaining the system and monitoring its activities. The availability of system information in the database makes the monitoring task easier.

To facilitate administrative functionality, SYSMAP executes procedures to add/delete/ change users and groups on the PDSF system. SYSMAP modifies the UNIX system files, updating system database, broadcasting network information, and executing script to setup user directory along with initial user files.

SYSMAP offers user the option of checking on processes, user logins, system loads, and file system usage of all workstations on the network. Users may also look at the picture of a particular object in a selected range of workstations, such as which workstations a particular user has logged into. Illegitimate background jobs running on a workstation designated for interactive use can be easily spotted by using SYSMAP. SYSMAP has a X window based GUI interface designed to provide graphical display of system usage.

Data Management System (DMS)

Physics processing can be characterized by large amounts of data consumption and data generation. With a number of applications of this type running concurrently, it is clear that on-line disk storage must be supplemented by off-line tape storage. The DMS software is designed to de-emphasize the importance of tape access in physics processing. Therefore the user may concentrate on the data and not the particular storage medium on which it resides. DMS serves a two-fold purpose: to manage the means of data transfer and to provide a catalog service for the data sets transferred between disk and tape.

Robotape

The Robotape subsystem has a series of library routines, a central robotape daemon, and a daemon for each system which has juke boxes connected for use by robotape. The library routines make calls on the robotape daemon for service. The robotape daemon then passes that request to the appropriate system for execution on the proper juke box. Requests are handled concurrently and executed in parallel to the greatest extent possible without collisions.

The Network Queuing System (NQS)

NQS is a package written by Sterling Software for NASA and available from COSMIC. It is a package designed to handle batch processing on a wide variety of UNIX systems. The SSCL has made some enhancements to the software in addition to some minor bug fixes. One modification was to make use of SYBASE to store workstation information instead of NQS's NMAP facility. The other major modification made was to support load balanced queues across multiple execution systems. Some routines were also added to generate reports on the NQS accounting information.

The PDSF consists of four corrals, each of which also has a server. Each corral is divided into systems used for interactive work and systems used to run batch jobs. On each corral, there is a set of short, medium, and long queues, along with a queue to the server. These queues move a job from the interactive system to a batch system or the server for execution. When the job finishes executing, its results are moved back to the system from which it was submitted and the user can be notified of the job's completion if desired.

The Operator Message System

Because of the distributed nature of the PDSF, a simple method of sending messages and soliciting responses from the operators was needed. Programs need only make a call to the message system and indicate if a response is desired and the message system will handle the details of sending the message and returning the response string.

The message system is configured with a list of destinations to route messages. Valid destinations include printing the message, displaying the message, or E-Mailing the message. When a message enters the system, it is broadcast to all the listed destinations. If a message requires a response, the message is assigned a number. A list of unanswered messages can be requested by the operator. Responses to any outstanding message can be sent from any system in the PDSF.

Cooperative Processes Software (CPS)

The Cooperative Processing Software(CPS) is a package of tools that makes it easy to split a computational task among a set of processes distributed over one or more computers. Apart from considerations of speed, the set of processes will operate identically whether on a single computer or spread across multiple computers. Each process runs a program written by the user.

CPS is developed by Fermilab and currently supported by SSCL. The PDSF has contributed many enhancements and modifications to this package through close cooperation with Fermilab.

CONCLUSION

The system software has been fully functional on the PDSF since March, 1991 and is an effective way of integrating distributed computers. It is planned to be at full computing capacity (4000 SSCUPs) during the year of 1993. The facility will continue its primary role to support physics and detector simulations to test and verify SDC and GEM detector designs. The upgrades for PDSF in the near future will consist of adding additional data servers to provide access to tertiary storage devices as well as providing increased on-line storage and memory capacity as required by user demand.

REFERENCES

1. "Report of the task force on computing for the superconducting super collider," SSC-N-579, M. G. D. Gilchriese (editor), Dec. (1988).
2. "Report of the SSC computer planning committee," SSC-N-691. L. Price (editor), Dec. (1989).
3. L. R. Cormell, "High energy physics computing at the SSCL," presented at the *9th International Conference on Computing in High Energy Physics (CHEP91)*, Tsukuba. Japan, Mar. (1991).
4. B. A. Kinsbury, "The network queuing system," Sterling Software, 1121 San Antonio Road, Palo Alto, CA, (1986).
5. Manlio Marquez, "Physics detector simulation facility system software description," SSCL-SR-1182. Dec. (1991).

FDDI EXPERIENCE AT THE SSCL

Mike Jaffe

Physics Research Division
Computing Department, MS-2003
Superconducting Super Collider Laboratory*
2550 Beckleymeade Ave.
Dallas, TX 75237-3997
jaffe@sscvx1.ssc.gov

ABSTRACT

The Physics Detector Simulation Facility (PDSF) is an assemblage of UNIX/RISC workstations and servers which use LAN networking components and standards in a unique way. The PDSF is configured using FDDI much like an internal system bus to a computer system and thus serves as the foundation for the entire PDSF system. This paper will describe the utilization of FDDI in the system, system monitoring, and the interfacing of the PDSF to the local site LAN and WAN environment.

INTRODUCTION

The Physics Detector Simulation Facility (PDSF) is a distributed loosely coupled parallel computing environment made up of groups of workstations and servers functioning together as a single large system. Figure 1 shows the physical and logical layout of the PDSF components.

The compute systems are RISC based architectures running the UNIX operating system and each is equipped with multiple network interfaces. They are interconnected using off-the-shelf LAN technologies in a unique way which utilizes the network as a sort of external/extended system bus or channel. All operations between machines occur via 6 internal FDDI networks which provide the primary communications pathways between systems.

COMPUTING COMPONENTS

There are two types of systems in the PDSF which are referred to as data servers and compute servers. Data servers consist of 4 Silicone Graphics Inc. (SGI) 4D/360 systems each equipped with 6 processors. The compute servers consist of 62 systems arranged in 4

---

*Operated by the Universities Research Association, Inc., for the U.S. Department of Energy under Contract No. DE-AC35-89ER40486.

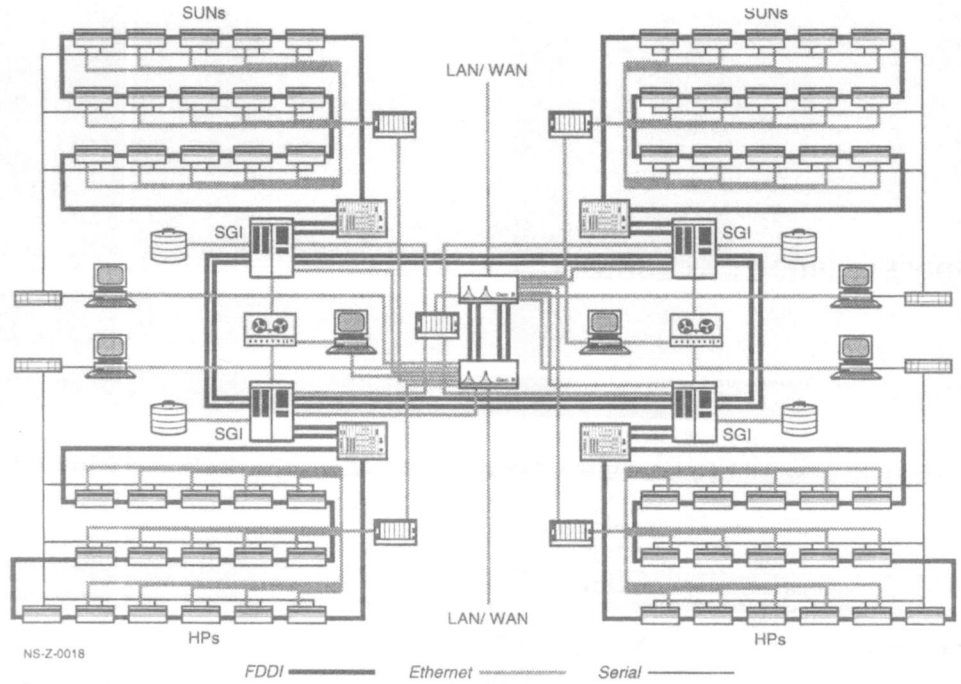

Figure 1. The FDSF Layout

groups, one group per data server, and are made up of two groups of Sun Sparc 2 and two groups of Hewlett Packard (HP) 9000/720 keyboardless and monitorless workstations. There are a total of 30 Sun and 32 HP compute servers with an additional 6 Suns and 1 HP designated for control systems. The total capacity of the PDSF is approximately 3000 MIPs or 2000 VAX equivalents. Secondary storage for the system consists of 160 Gbytes of hard disks and 250 Gbytes of 8mm tape storage arranged in two tape robot systems.

NETWORK COMPONENTS

The strength of the PDSF is its network architecture which allows communication between the many systems comprising the environment. Each compute server is equipped with 1 ethernet and 1 FDDI Single Attach Station (SAS) network interface while each data server is equipped with 2 ethernet and 2 FDDI Dual Attach Station (DAS) network interfaces. The ethernets are connected together into segments using 5 Optical Data Systems (ODS) ethernet concentrators. The FDDI systems are slightly more complex in arrangement and consist of 6 separate FDDI rings as follows: one ring connects the 4 data servers, one ring connects the two routers, four separate rings connect each of the 4 groups of compute servers to their respective data server using 4 ODS FDDI Dual Attach Concentrators (DAC). The entire system is tied together and to the outside world using 2 Cisco Systems AGS+ FDDI/ethernet routers.

FDDI'S ROLE

The foundation of the PDSF is the FDDI which connects its data servers to data servers, data servers to compute servers, compute servers to compute servers, and finally router to router. These FDDI systems make up the backbone of the system and provide a

496

high speed pathway similar to that of a computer system bus or channel over which the activities of the PDSF can occur. These range from resource sharing activities such as Network File System (NFS) and Yellow Pages, to operational utilities used to manage, monitor, and maintain the environment, to user's software utilizing UNIX sockets, RPC, and other network oriented access facilities. All activities internal to the PDSF occur using the FDDI rings and in fact the default names of the systems are the FDDI interfaces which insures that the majority of the network load occurs using these high speed pathways. All of these activities have the ability to access any of the individual network interfaces, and thus pathways, by specifying the IP subnet associated with the interface desired.

IMPLEMENTING FDDI

FDDI has played an instrumental role in the PDSF from its inception and it continues to challenge products, their manufacturers, and the standard itself. Implementing it was no small chore and a number of interesting problems had to be overcome in making it all work. The first difficulty encountered related to Station Management (SMT) and had to do with differences in how the product manufacturers had interpreted "proper" responses to informational request frames from stations on the rings. This didn't pose an operational problem really but did make diagnosing other problems very difficult since it wasn't clear to all stations on the ring who actually existed and who didn't. These problems were worked out with the involved vendors who have since modified their individual implementations of SMT.

Another problem experienced had to do with high signal attenuation (> 6 db) in some of the optical bypass switches which were included in the first phase of the PDSF. This attenuation was causing the systems on the ring to intermittently wrap in response to what appeared to them to be a failed fiber or system. This was the result of the 11 db FDDI defined link budget being exceeded. These particular switches were found to be defective and replacing them corrected the problem. Having experienced this problem, the whole issue of the usefulness of these devices was brought up and the jury is still out as to whether they should be required in an FDDI implementation. Although they can be useful in maintaining and diagnosing problems on a ring by allowing stations to be removed more gracefully (manually because most vendors don't yet provide software bypass control). They can also be troublesome as in the case of more than one station in a row going into bypass mode in which case the ring will probably wrap due to link budgets being exceeded. Their worth is probably best stated as being "system dependent."

Two related issues occurred and centered on the difference between packet sizes allowed on ethernet (1500 byte) and FDDI (4500 byte). The first involved packets transitioning between the FDDI rings and the ethernet segments. In order for the larger FDDI packets to fit onto the ethernet segments either the FDDI hosts or the routers had to be able to fragment them into appropriately sized units. Two solutions were available one of which restricted the FDDI hosts to using the smaller ethernet packets in which case the router passed its packets onto the ethernet segments unmodified. The other solution, and the one that proved most efficient, was to set the FDDI hosts to utilize full size FDDI packets and let the router fragment those needing to transition to an ethernet segment. The related issue dealt with the Maximum Transmission Unit (MTU) size on the FDDI interfaces for the three types of systems in the PDSF (SGI, Sun, and HP). These were found to be set differently by default but fortunately each system allowed this value to be reset to the maximum 4500 byte value.

The problems encountered in implementing FDDI in the PDSF were more of a nuisance than a show stopper. They were for the most part the result of a lack of experience and immaturity in the products, the market, and the standard itself. Additionally, problem resolution was somewhat compounded by a lack of available diagnostic tools specific to

FDDI. It was evident by the time the second phase of the system was implemented that FDDI marketplace was maturing.

ACCESSING THE SYSTEM

Of course the true power of the PDSF can only be realized through its availability to users both local and around the world. This availability is facilitated by its connectivity to the SSCL LAN/WAN which is connected to the Internet through ESNET and BITNET connections. Both of these networks are well developed providing potential T1/DS1 connection speeds to most of the remote users depending on their location and site. A typical access to the PDSF either local or remote is through an X-windows session requiring a significant amount of network bandwidth over a less typical terminal access session. Based on user feedback, response to the system is good which is significant since over 70% of the PDSF users are remote to the SSCL.

MANAGING THE NETWORK

Although the system functions like a single computer system, the PDSF incorporates as much networking as many small computing sites. Thus monitoring and managing its network is crucial to its continued operation and success. A combination of tools are used for this chore including several home-grown UNIX based utilities, Cisco System's NetCentral Simple Network Management Protocol (SNMP) software which includes the Sybase SQL database, and HP's OpenView Network Node Manager SNMP software in addition to a little creativity. These systems run continuously and are used to monitor and collect both instantaneous and trend data for all of the networks in the environment along with assisting in the isolation, diagnosis, and repair of network problems when they occur.

Most PDSF network monitoring and management is performed using the two SNMP tools mentioned but it's important to note that SNMP itself is not without its limitations. It's functionality and usefulness centers on Management Information Bases (MIBs) which amount to lists of all possible values that can be examined on a device. For example MIB entries might be defined to examine total traffic in and out of an interface on a system. In the case of FDDI specifically there are currently no MIB values available to examine its SMT components. A new MIB addressing this limitation has been proposed (RFC1285) and hopefully will be adopted and implemented quickly by the vendor community.

SUMMARY

The PDSF has been in operation in its current configuration for approximately a year now and its network design has proven sound. It just plain works and has required very little modification other than the ongoing occasional tuning of system activities and applications which is performed by its administrators. It is clear from the initial phase of the PDSF and from data gathered in its second phase that the multiple FDDI's are both key and crucial to its operation. The strength of the system is the availability of all of its networks in parallel allowing both high speed and simultaneous access for the systems and user level tasks being performed. Remarkably all of the systems in the PDSF deal very well with the multiple network interfaces and other than a few initial setup issues relating to what used what pathway and who would handle which routing etc. things work very well. All of the networks are being utilized to at least a moderate level and future expansions of the PDSF will certainly add even more FDDI to this unique computing/network environment.

SYSTEMATIC ERROR ANALYSIS OF ROTATING COIL
USING COMPUTER SIMULATION

Wei-chuan Li and Mark Coles

SSC Laboratory[*]
2550 Beckleymeade Avenue
Dallas, TX 75237

INTRODUCTION

This report describes a study of the systematic and random measurement uncertainties of magnetic multipoles which are due to construction errors, rotational speed variation, and electronic noise in a digitally bucked tangential coil assembly with dipole bucking windings. The sensitivities of the systematic multipole uncertainty to construction errors are estimated analytically and using a computer simulation program.

SIMULATION PROGRAM

The simulation program reproduces the operation of the magnetic field quality measurement apparatus or "mole"[1] by generating a sequence of voltage samples and a sequence of intersample times. This data is then input to the standard mole data analysis program and the resulting estimates of magnetic multipole field content are compared to the input field model. This is accomplished by integrating the $(\vec{V} \times \vec{B}) \bullet d\vec{L}$ component of voltage (where $\vec{V}$ is the velocity of the element $d\vec{L}$ of the wire filament and $\vec{B}$ is the local magnetic field) along all of the coil filaments of each coil. The voltage series generated are then analyzed using the standard analysis program in order to compute the harmonic content of the measured magnetic field. Arbitrary magnetic field configurations can be input and the construction details of the mole can be varied in order to estimate the sensitivity of the calculated multipole outputs to these changes. F-mole configuration[2] parameters are used in the simulation program.

---

[*]Operated by the Universities Research Association, Inc., for the U. S. Department of Energy under Contract number DE-AC35-89ER40486.

It was determined that most of the error sensitivity is proportional to the amplitude of the input field harmonics, so a "worst case" field was used for this study which was obtained by using the upper limit of the systematic multipole specification plus three times the standard deviation specification.[3] For this study, various construction parameters, such as coil bowing, misalignment, etc., were varied independently. Sensitivities were estimated by first simulating the ideal F mole design and then comparing the results of that simulation to the results obtained using non-ideal construction parameters. The input magnetic field is shown in Table 1.

Table 1. Comparison of the input and output multipole coefficients.

Multipole index	Input skew a (units)	Input normal b (units)	Output skew a (units) ideal case	Output normal b (units) ideal case	Output skew a (units) nominal case	Output normal b (units) nominal case
1	3.79	1.54	3.79	1.54	3.804	1.545
2	1.082	5.45	1.082	5.45	1.089	5.484
3	0.986	0.506	0.986	0.506	0.9958	0.511
4	0.17	0.74	0.1699	0.7401	0.1723	0.7496
5	0.166	0.076	0.166	0.07599	0.1685	0.07716
6	0.043	0.08	0.04299	0.08	0.04379	0.08144
7	0.04	0.04	0.04001	0.04	0.04077	0.04076
8	0.0305	0.0425	0.0305	0.0425	0.0311	0.04335

In order to check the correctness of the simulation and analysis program the nominal geometry of the coil was simulated. Also, the individual filamentary wires of the cable were modelled as infinitely thin wires and all wires within a bundle were placed at the same angle and radial position on the coil. Table 1 shows the normal and skew harmonics input to the simulation and the computed results obtained after simulation and analysis using this idealized geometry. They are in good agreement. This simulation program was also run using the nominal coil geometry but with a 0.006 inch space between conductors within a wire bundle. These results are shown in Table 1 as "nominal harmonics". Note that the nominal harmonics contain some small errors due to the finite angle subtended by the wire bundles of each coil. Hereafter, unless specified, all the harmonics will be compared to these nominal harmonics.

A typical configuration of dipole bucking coils has one tangential coil and two bucking coils. The amplitude of the dipole field in a typical SSC magnet is approximately ten thousand times greater than the amplitude of the next larger harmonic coefficient. The influence of the dipole term is removed using a technique called "bucking". Digital bucking, the technique considered in this simulation, was developed by the BNL magnetic measurement group.[4] The induced voltage on the tangential coil at time t can be expressed as

$$V_T(t) = -2B_0 N_T r_T L_T \omega \sum_{k=1} (r_T)^{k-1} \sin\left(k\frac{\Delta}{2}\right) \times$$
$$\{ b_{k-1} \cos[k(\omega t - \beta)] + a_{k-1} \sin[k(\omega t - \beta)] \}$$

(EQ 1)

where B_0 is the fundamental dipole field strength, N_T is the number of turns of the tangential coil, r_T is the radius, L_T is the length, ω is the average rotating speed, Δ is the opening angle, β is the initial offset angle, and a_k and b_k are the kth skew and normal multipole, respectively. The induced voltages on the two bucking coils, v_1 and v_2, are given by similar expressions. From these three voltages we can compute a_k and b_k.

As an example of the sensitivity to a coil construction error, consider what happens when the radius of the dipole bucking coil is changed to $r+dr$ from r. The measured values of the a_k and b_k harmonics will be changed by the amounts

$$da_k = \frac{\partial a_k}{\partial r}dr + \frac{\partial a_k}{\partial \Delta}\frac{\partial \Delta}{\partial r}dr \quad \text{and} \quad db_k = \frac{\partial b_k}{\partial r}dr + \frac{\partial b_k}{\partial \Delta}\frac{\partial \Delta}{\partial r}dr. \quad \text{(EQ 2)}$$

The second term in each sum is necessary because the tangential opening angle Δ is computed from the measured ratio of the tangential and dipole bucking coil signals. This effect was also simulated using the computer simulation program, setting dr to 5, 10 and 20 mils. The largest changes occurred for the sextupole components; $da_2/a_2 = -0.082$ and $db_2/b_2 = -0.069$ when $dr=20$ mils. We estimated these changes analytically using the first order approximation (EQ 2), which resulted in $da_2/a_2 = -0.069$ and $db_2/b_2 = -0.069$ for $dr=20$ mils. The two methods are consistent. Since the errors scale approximately linearly with dr, we can extrapolate that $|dr|$ should be less than 0.53 mils to keep $|db_2| \leq 0.01$ units.

This kind of first order estimation can be used for other construction error analyses. We verified that all the analytical estimations are consistent with our simulation program. (For a detailed discussion of error analysis and simulation, see the references 4 and 5.)

Table 2. The tolerance requirements for the mole needed in order to have systematic error of less than 0.01 units with a "worst case" magnetic field.

Parameter	Nominal value	Tolerance
bow	0.0	1.6 mils or 0.041 mm
sag	0.0	40 mils or 0.1 mm
twist	0.0	40 mrad
bearing misalignment	0.0	0.24 mils or 0.0061 mm
radial error of tangential winding	12.3063 mm	0.24 mils or 0.061 mm
radial error of bucking winding	12.3063 mm	0.53 mils or 0.041 mm
angular error of tangential winding	0.262 mm	130 mrad
angular error of bucking winding	3.141 rad	7.1 mrad
parallelism error of tangential winding	0.0	0.48 mils or 0.012 mm
parallelism error of bucking winding	0.0	2.0 mils or 0.052 mm
angular velocity of coil	2.0943	0.021 rad/s
random electrical noise	0.0	2.8 nv(rms)

MECHANICAL TOLERANCE OF THE COIL CONSTRUCTION

We have studied various sensitivities of a rotating coil to construction errors, rotational speed variation and electronic noise. The construction errors include: (1) the radial position of the coil winding may be offset parallel to the rotating axis; (2) the coil winding may have a cone shape; (3) the coil may be bowed during the manufacturing process; and (4) the bearing of the coil may be offset from the geometrical center of the coil. The harmonics are quite sensitive to these errors. Other errors, such as the tangential coil angle offset and the coil twist are less problematic and can be easily compensated if digital bucking is used as a measurement technique. Sagging can be removed provided that a reliable centering correction method is used. If the required maximum tolerable uncertainty on any harmonic coefficient is less than 0.01 units, the coil must satisfy all the tolerance requirements summarized in Table 2. While demanding, these requirements are not impossible to satisfy if careful consideration is given to the fabrication methods to be employed. It was also found that the ideal centering correction not only corrects the center offset, but also fully corrects the sag effect and partially corrects the bearing center offset.

ACKNOWLEDGMENTS

The authors wish to thank Dr. Kiseon Kim and Yu Ping Zhao for many useful discussions regarding magnetic measurements and Fourier spectral analysis.

REFERENCES

1. P. Wanderer, "Production Techniques For Measuring SSC Accelerator Magnets," ICFA Workshop on Superconducting Magnets and Cryogenics, BNL, May 1986.
2. One of a series of moles designed by Brookhaven National Lab. Engineering Drawing Number 25-1105-02.
3. "SSCL Site Specific Conceptual Design," SSCL-SR-1056, July 1990.
4. G. Ganetis, J. Herrera, R. Hogue, J. Skaritka, P.Wanderer, and E.Willen, "Field Measuring Probe for SSC Magnets," Particle Accelerator Conference, March 19, 1987.
5. W.C. Li and Mark Coles, "Rotating Coil Simulation Study," MD-TA-201, May 1992.

STATISTICAL FACTORS TO QUALIFY THE SUPERCONDUCTING MAGNETS FOR THE SSC BASED ON WARM/COLD CORRELATIONS

K. Kim, A. Devred, M. Coles, and J. Tompkins

Magnet Systems Division
Superconducting Super Collider Laboratory*
Dallas, Texas 75237-3997

INTRODUCTION

All of the SSC production magnets will be measured at room temperature (warm), but only a fraction of these will be measured at liquid helium temperature (cold). The fractional information will then be analyzed to determine warm acceptance criteria for the field quality of the SSC magnets. Regarding predictors of the field quality based on partial information, there are several observations and studies based on the warm/cold correlation [Berk 88, Ferb 89, WaEt 92, PoEt 93]. A different facet of the acceptance test is production control, which interprets the warm/cold correlation to adjust the process parameters [Shap 92]. For these applications, we are evaluating statistical techniques relying on asymptotic estimators of the systematic errors and random errors, and their respective confidence intervals. The estimators are useful to qualify the population magnets based on a subset of sample magnets. We present the status of our work, including: i) a recapitulation of analytic formulas, ii) a justification based on HERA magnet experience, and iii) a practical interpretation of these estimators.

NOTATIONS

The requirements for the dipole field quality of an ensemble of magnets building the collider rings are given by i) the systematic multipole errors (SE) and ii) the random multipole errors (RE). The systematic errors are confined mainly to control the tune shift component of the linear aperture criterion, while the random errors are to fit the smear component of the linear aperture inside the physical aperture. To estimate the two errors of a full set of magnets to build a ring, we are using measurements of a subset of magnets (measured respectively at a magnet, at a z-scan position, at a repetition, either warm or cold), using a harmonics-measurement instrumentation named "Mole." Let's introduce some letters and indexes for the description of the quantities:

- χ: a nominal multipole coefficient, i.e., $\sim \{ a_n, b_n ; n = 1, .. \infty \}$
- x : a measurement of χ
- Subscript w (or c, or s) : refers to warm (or cold, or subset) measurements
- M (or K) : total number of magnets measured warm (or cold), $K \leq M$
- P: total number of positions within a magnet
- L: total number of repetitions at a given position within a magnet
- Superscript m (or p, or l) : magnet (or position, or repetition) index, $1 \leq m \leq M$, $1 \leq p \leq P$, $1 \leq l \leq L$. According to the above definitions, the l-th warm (cold) measurement, at the

* Operated by the Universities Research Association, Inc., for the U. S. Department of Energy under Contract No. DE-AC35-89ER40486.

p-th position of the m-th magnet is written: $x_w^{l,p,m}$ ($x_c^{l,p,m}$). Regarding the usage of the indexes, it is noteworthy that i) a missing superscript is after processing on the corresponding index and ii) a missing subscript implies non-specification of the condition [KiE 93]. The SE is defined as $E[\chi_c^m]$ an ensemble average of the magnet mean (χ_c^m), and the (RE)^2 as $E[\{\chi_c^m - E[\chi_c^m]\}^2]$.

ESTIMATION PROBLEMS

The first question was raised when we had successfully manufactured several full size R&D dipole magnets; do these magnets promise that <u>consistent</u> reproduction of the assembly processes will meet the requirements? Or, what kind of additional information do we need to promise that magnets will meet these requirements? The same question could be asked while the production is proceeding (especially when the consistency assumptions are not valid). We are following the framework of statistics and estimation theory to conclude the answers analytically. Specifically, we set three distinct questions:

- Q1: Based on K magnets measured warm (or, cold), how can we estimate the mean (or, variance) of M warm (or, cold) measurements to be produced later?
- Q2: Based on K magnets measured warm and cold, and M measured warm, how can we estimate the mean (variance) of a cold measurement to be produced later?
- Q3: Based on K magnets measured warm and cold, and M warm how can we estimate the mean (variance) of M cold measurements to be produced later?

In short, the information to predict a cold measurement consists of three features coming out of: i) inter-relation of the subset cold measurements, ii) inter-relation of the ensemble warm measurements, and iii) co-relation of the cold and the warm measurements. A simple case assumes that the warm (or cold) measurements quantities are statistically independent and they are linearly co-related, then the relations are well described by the statistical averages. For the case, we can derive i) asymptotic estimators of the systematic errors and random errors, and their respective confidence intervals. In reality, considering that the manufacturing procedures keep the time-history, e.g., aging of the tools and continuity of the processing, we recommend incorporating the dependencies among the warm (or cold) measurements. For the second case, we are considering auto-regressive-moving-average (ARMA) estimators of the cold measurements, and their respective confidence intervals, which are more tightly bound and useful for production control because they utilize the on-going history of the production [BoJe75].

ASYMPTOTIC ESTIMATORS

The aforementioned questions on the estimation problems can be answered by applying the results to the prediction model selection and several estimators corresponding to each parameter, respectively, derived in [KiEt93]. We use a model including the ratio (λ) of measurement noise variances, and the linear prediction coefficients μ and β.

I. Based on a sample of K magnets measured cold; estimate the mean value of the measurements for the ensemble of M magnets

i) calculate the mean $x_{c,K}$ and variance $\sigma_{c,K}$, ii) then, an α-confident estimate of the cold value χ_c is confined: $|\chi_c - x_{c,K}| < Q(\alpha/2)\ \sigma_{c,K}\{\frac{M-K}{MK}\}^{1/2}$

where $Q(\alpha/2)$ is the Gaussian percentile, and the probability that the nominal value χ_c satisfies the inequality is $(1-\alpha)$. A reasonable number of $Q(\alpha/2)$ is 4, for $\alpha < 10^{-4}$.

II. Based on M warm and K cold measurements, estimate the cold multipoles of a given magnet

i) calculate x_W and σ_W, ii) calculate $x_{c,K}$, $\sigma_{c,K}$, iii) based on the ratio λ, calculate the μ and β, iv) then, for the LSE type, an α-confident estimate of the cold value χ_c^m is

$$|\chi_c^m - \mu\, x_W^m - \beta| < Q(\alpha/2) \left\{ \frac{V_s[x_c^m - \mu x_W^m - \beta]}{K} + \frac{(x_W^m - x_{W,K})^2\, V_s[x_c^m - \mu x_W^m - \beta]}{\sum_{m=1}^{K}(x_W^m - x_{W,K})^2} \right\}^{1/2}$$

III. Based on M warm and K cold measurements, estimate the mean of M cold measurements

i) calculate x_W and σ_W, ii) calculate $x_{(c-w),K}$, iii) calculate $\sigma_{(c-w),K}$, iv) based on λ, calculate μ, v) then, an α-confident estimate of the cold mean χ_c is

$$|\chi_c - \mu\,(x_W + x_{(c-w),K}) - (1-\mu)x_{c,K}| < Q(\alpha/2) \left\{ \frac{\sigma_W^2}{M} + \frac{\sigma_{(c-w),K}^2}{K} \right\}^{1/2}$$

Implications: Simple factors to accept 10,000 magnets based on K cold and M ($K \le M \le$ 10,000) warm measurements are: $\left\{ \sigma_W^2 + \sigma_{(c-w),K}^2 \right\}^{1/2} < RE$, and

$$|\mu\,(x_W + x_{(c-w),K}) + (1-\mu)x_{c,K}| + 4 \left\{ \frac{\sigma_W^2}{M} + \frac{\sigma_{(c-w),K}^2}{K} \right\}^{1/2} < SE$$

where the inter-relation information is in x_W, $x_{c,K}$ and σ_W, while the correlation information is in $\sigma_{(c-w)}$, K and μ.

Examples: We have tested the above rules for 5 Fermilab-General Dynamics Magnets. To compare the relative significance of each multipole, normalized numbers of the estimated RE are plotted in Figure 1. The sliced bars represent the estimated RE normalized by the required RE, and the plain bars the contribution from the correlation uncertainty.

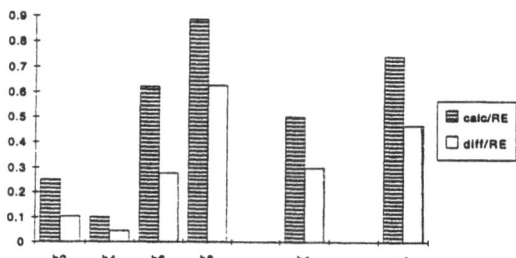

Figure 1. The ratio of the estimated and the required random errors, based on DCA311-315 (sliced bars), and the plain bars for the contribution of the correlation uncertainty.

Regarding SE computations, we do not have enough samples to draw any conclusion, even after the allowed multipoles are assumed to be significantly reduced by a modest redesign [WaEt 92]. To check further the validity of the proposed formulas, we have run simulations on the normal sextupole coefficient b_2 measurements of 220 HERA magnets produced by ANSALDO. Among the 220 Ansaldo magnets, 40 collared-coil assemblies were taken apart and re-assembled after the first set of warm magnetic measurements in order to correct some of their multipole coefficients by shimming the coils. These corrections constitute a deviation in the assembly process, and these 40 magnets are excluded from the study. Also, in order to retain information about equipment wear or personnel training, the magnet index, m, follows the manufacturing sequence. Figure 2 shows the cold and warm b_2 data for the ANSALDO magnets. Figure 2.a shows b_2 values as a function of the sequence. It appears that b_2 varies widely (the mean and standard deviation is about 2 units both for cold and warm measurements). Figure 2.b shows b_2 cold as a function of b_2 warm; note that the Pearson correlation coefficient is 0.961, $\mu = 0.761$ and $\beta = 0.199$ based on the Least-Squared estimator (LSE), or $\mu = 0.824$ and $\beta = 0.069$ based on the Maximum-likelihood estimator (MLE). Let's look at the case III, as an acceptance tool.

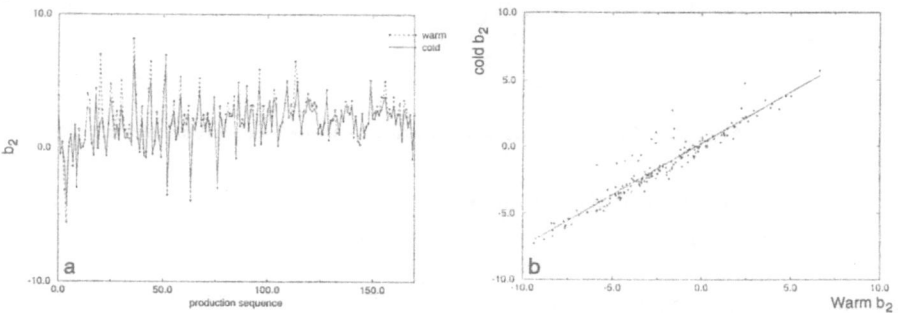

Figure 2. Cold and warm measurement data of b_2 for the HERA/ANSALDO magnets; a) b_2 as a function of product sequence; b) b_2 cold versus b_2 warm (an offset of 12.897 units)

IV. Based on M warm and 20 cold measurements, estimate the mean of M cold measurements

We have 20 magnets measured both warm and cold, and all the magnets produced subsequently were only measured warm. Based on this information, we can estimate the mean of the cold measurements of M magnets. Figure 3 shows the cold estimate of the ensemble mean of M magnets. Two different estimators are shown: the circle for the simplified (asymptotic) and the X for the proposed. It appears that the proposed estimator is closer to the asterisk which is the mean valued calculated from the cold measurements. It is noticeable here that the confidence width of about 1 unit is not significant where the HERA magnets' reference radius is 2.5 cm and the maximum deviation of b_2 error is given by 10 units.

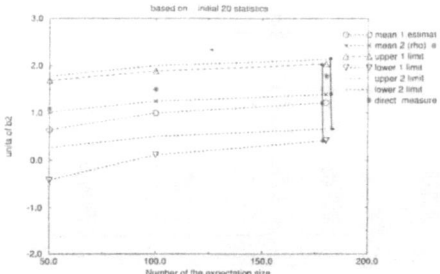

Figure 3. b_2 cold estimate of the ensemble mean of M magnet. The circle for the simple estimator and the X for the proposed estimator.

REFERENCES

[Berk 88] B. Berkes, "Degree of Confidence in Predictions of COLD Multipoles Based on WARM Field Measurements on SSC Dipole Magnets," SSC-N-543, 1988.
[BoJe 75] G. Box and G. Jenkins, *Time Series Analysis: Forecasting and Control*, Holden-Day, 1976.
[Ferb 89] T. Ferbel, "Comments on Prediction of Cold Multipoles from Their Warm Counterparts," SSC-N-578, SSC, 1989.
[KiEt 93] K. Kim et al., "Magnet Indexes and Notations: Correlation Analysis," MTL-draft, 1993.
[PoEt 93] D. Pollock et al., "A Comparison of Least Squares Linear Regression and Structural Relation Models of Warm/Cold Multipole Correlation in SSC Prototype Dipole Magnets," *Supercollider V*, J. Nonte ed., Plenum, 1993.
[Shap 92] S. Shapiro, "Issues in Development of CDM Magnets," unpublished presentation, SSC/MTL, 1992.
[WaEt 92] P. Wanderer et al., "A Summary of SSC Dipole Magnet Field Quality Measurements," *Supercollider IV*, J. Nonte ed., Plenum, 1992.

A PROGRESS REPORT ON THE PRODUCTION OF 1.8 MILLION FEET OF SUPERCONDUCTING CABLE FOR THE RELATIVISTIC HEAVY ION COLLIDER PROJECT AT BNL[1]

S. DelRe, G. Epstein, S. Hong, J. Lichtenwalner
Oxford Superconducting Technology, Carteret, NJ 07008

P. O'Larey, D. Smathers
Teledyne Wah Chang Albany, Albany, OR 97321

M. Boivin, R. Meserve
New England Electric Wire Corp., Lisbon, NH 03585

M. Garber, A. Ghosh, A. Greene, D. McChesney, A. Morgillo, R. Shah
Brookhaven National Laboratory, Upton, NY 11973

INTRODUCTION

The Relativistic Heavy Ion Collider (RHIC) Project under construction at BNL will enable collisions between beams of nuclei as heavy as ^{197}Au, accelerated in two quasi-circular rings to a maximum energy of 100 GeV/u.[1] The bending and focusing of the beams is done by dipole and quadrupole magnets constructed primarily from 30-strand Rutherford-type cable. This is a progress report on the manufacture of superconductor which is a key technical component for the accelerator.

A conservative specification was placed on the superconducting properties. Requested minimum J_c is 2600 A/mm^2 at 5T, 4.2 K and the copper-to-non-copper ratio (C/S) is designed to be 2.25. In order to provide a degree of control of the low field magnetization of the superconducting material, the quantity $I_c(3T)/I_c(5T)$ or "3/5 ratio" is specified for the wire. Table I summarizes some parameter requirements. Specifications on other mechanical properties are given in reference [2].

The production contract for RHIC superconducting cable was awarded to Oxford Superconducting Technology (OST) in September 1991 with conductor fabrication beginning in January 1992. Approximately 112,000 pounds (51 tonnes) of superconductor will be required to produce 1.86 M feet (566 km) of 30-strand cable. Approximately eighteen (18) months have been scheduled for conductor manufacturing with an overlapping period for cabling resulting in a total contract duration of two (2) years.

PROGRAM OBJECTIVES

Superconductor produced by OST for the RHIC contract was planned and is being manufactured with four primary objectives: overall process uniformity, proactive quality assurance, schedule adherence and thorough cost control.

---

[1]Work performed under Contract No. DE-AC02-76CH00016 with the U.S. Department of Energy.

Table I. Some Specification Requirements for RHIC Wire and Cable.

Parameter	Wire	Cable
Filament diameter	6 μm; quantity 3510	
Filament spacing	> 1 μm	
Wire diameter	(0.0255 ± 0.0001) inch	
Wire twist pitch	(1.9 ± 0.2) twists/inch	same after cabling
I_c(5T, 4.2 K)	> 264 A	> 7524 A
Range	< ± 10% Average I_c	< ± 6% Average I_c
I_c(3T)/I_c(5T)	< 1.6	
R(295)K	< 765 $\mu\Omega$/cm	< 26.8 $\mu\Omega$/cm
RRR	> 38	> 38

The most critical raw material component which must be controlled to minimize variation in the final product is Nb-Ti alloy. Negotiations with Teledyne Wah Chang Albany (TWCA) resulted in a concentrated "RHIC-only" campaign for Nb-Ti alloy where strict material and process conformance to a mutually agreed specification was maintained. Approximately 18 tonnes of alloy were melted and fabricated for the program. Statistical Process Control (SPC) monitoring was rigorously applied as well as use of dedicated crucibles, furnaces and fabrication equipment. This procurement of alloy was a landmark in communication and exchange of information between TWCA and OST, and resulted in Nb-Ti alloy with exceptionally uniform properties for use in the RHIC contract.

The next critical element for overall conductor uniformity was for OST to control the fabrication process. This was accomplished by specifically defining the critical process variables in advance of manufacture, and by expanding the process definition to include applicable machinery, manufacturing standards, operator instructions, lot sizes and testing procedures.

With the exception of some testing procedures, no process method was mandated by BNL. OST discloses to BNL on a continual basis all details regarding fabrication and testing. This open style of communication helps to identify potential sources of process nonuniformity and to correct them without adverse impact to the program.

Process and procedure uniformity are also maintained at the cabling facility at New England Electric Wire (NEEW) in Lisbon, NH. Detailed manufacturing and quality plans were established and agreed upon between OST and NEEW prior to commencing cabling and are monitored and disclosed on a regular basis to BNL. An OST representative is in "permanent" residence at NEEW to liaise among the three parties and maintain OST's substantial inventory on-site at NEEW.

OST's approach to Q.A. is intrinsically linked to ensuring process uniformity as well as providing a cornerstone for a productionized superconducting wire manufacturing facility. Detailed definition of Q.A. procedures, audits, testing methods and report formats were established prior to commencing manufacture. Both in-process and final superconductor test results are openly reviewed on a periodic basis with BNL to ensure that specified processes achieve anticipated results.

OST is achieving the RHIC program schedules that were originally defined; these schedules are being reported to BNL on a regular basis. Schedule conformance is important in terms of maintaining a continuous flow of conductor to the cabling facility, efficient lot sizes and achieving contractual commitments.

OST accepted the RHIC contract on a fixed price basis; therefore, control and monitoring of cost are rigorously reviewed. The demonstrated success in terms of uniformity, overall quality, schedule and yield has resulted in a program that has met all measured cost criteria.

WIRE MANUFACTURING

Conductor produced under this contract has been standardized as a double extrusion, 4% barrier, high-J_c optimized fabrication process. RHIC conductor is actually very similar to SSC Outer Dipole conductor. First extrusion monofilament materials are identical to the SSC conductor and differ only in procurement specifications and the cut lengths of hexed monofilament rods. Second extrusion multifilament materials are fundamentally similar with necessary modifications for specified Cu center core size and C/S. Superconductor manufacturing steps for RHIC parallel very closely those of SSC outer conductor and are identical in terms of reaction heat treatments and fabrication equipment.

Piece length performance is important for monitoring process, yield and cost impact. Figure 1 represents actual piece length distribution for about 30 tonnes of superconductor completed. Average piece length is 10,127 meters (33,277 feet).

The distribution is significantly skewed towards longer piece lengths; pieces in excess of 5,000 meters must be cut for loading onto cable reels. Greater than 90% of all wire produced is in excess of 5,000 meters. Less than 0.3% of all conductor produced has been below the minimum length criteria of 670 meters. Because of this performance, the BNL requirement for no cold welds in cable was easily met.

Uniform wire diameter is necessary to enable production of uniform, mechanically stable cables; also, control of superconductor content, or C/S ratio, depends on keeping both the wire resistance and diameter constant.[2] Wire diameters are measured and accurately controlled by OST on line during the final draw using laser micrometers with internal SPC software. Correlation of measurements between OST and BNL data is within about $\pm$ .00002 inches. Figure 2 shows OST data for all of the wire spools produced to date. All wire is eddy current tested at final size to check for inclusions which could disrupt performance uniformity or cause wire breakage during cabling.

Figure 3 shows the distribution of critical currents for 486 wire samples. Measurements were made at BNL on half of the samples to check these results. Agreement with OST data is within 1.5%. The standard deviation is $\pm 2.3\%$, well within the specification requirement of $\pm$ 10% of the running average. Small variation from a Gaussian distribution is caused by improved uniformity during production start-up and adjustments in wire cropping. The 3/5 ratio shown in Figure 4 has remained extremely uniform indicating no process variation.

The superconductor volume fraction is confirmed by monitoring R(295) and wire diameter. For specification purposes an upper limit is set upon R(295) so that the copper fraction may not get too low. The I_c specification ensures that copper fraction does not get too high. BNL and OST have worked to ensure agreement and accuracy of R(295) measurements to a level of about 0.3%. For 486 samples R(295) values have a standard deviation of 0.65% of the mean. Implied C/S variations are within 2.5% of the mean.

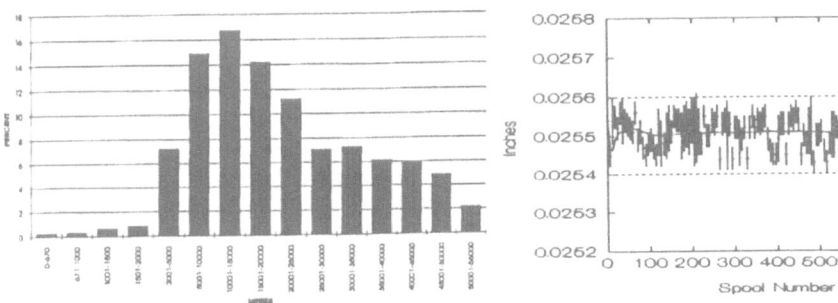

Figure 1. Wire piece length distribution.

Figure 2. Wire diameter for every sixth spool with $\pm 3\sigma$ bars, running average line and specified range.

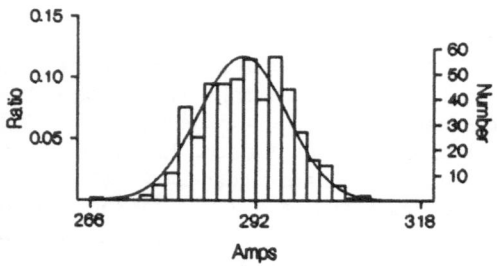

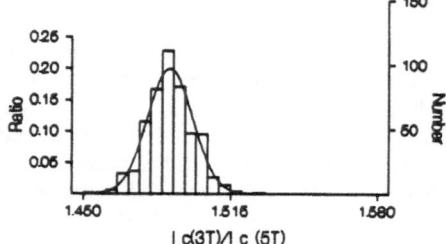

Figure 3. Wire I_c(5T, 4.2 K) distribution. **Figure 4.** Wire "3/5 ratio" distribution.

CABLE FABRICATION

About 650,000 ft. of cable have been produced; the remainder will be completed within seven months. Coil pre-stress and ease of assembly are critically dependent on cable mid-thickness. Keystone angle and width are also important, but generally fall well within tolerances. Figure 5 shows mid-thickness data for cable produced so far at NEEW. The measured variation of mid-thickness along produced cables is about 50% of specified limits.

Critical current data for production cables are shown in Figure 6. The cable I_c average is 10% above the minimum critical current requirement and well within the allowed variation of ± 6%. The slight upward trend in Fig. 6 is due both to smoother operation of the cabling facility giving reduced degradation, and to a small increase in wire performance mentioned above. The standard deviation in I_c for all cables produced is about half of that seen in the wire, and can be attributed to selection of wire for use in cable. The degradation of recent cables is about 1%.

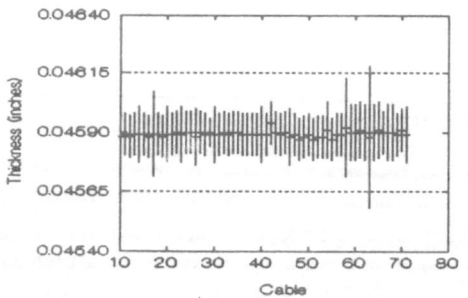

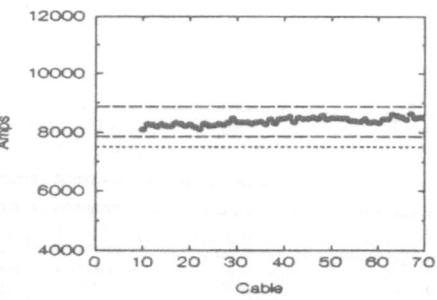

Figure 5. Cable mid-thickness with $\pm 3\sigma$ bars and specified range. **Figure 6.** Cable I_c(5T, 4.2 K) with lower limit and specified range.

SUMMARY

The RHIC production contract is in the final stages of strand manufacture at OST; cabling will be completed, on schedule, by December 1993. This contract has offered a unique opportunity to demonstrate that a significant quantity of SSC type superconductor can be manufactured with high uniformity, reliable schedules and effective cost control. OST would like to acknowledge the efforts of many individuals at BNL, TWCA, NEEW and other contributors and subcontractors who are helping to make this program a success.

REFERENCES

[1]   RHIC Conceptual Design; BNL publication 52195, May 1989.

[2]   RHIC Specification No. RHIC-MAG-M-4141.

RECENT RESULTS IN THE MANUFACTURE OF SSC OUTER WIRE AND CABLE AT AISA WITHIN THE FRAMEWORK OF THE SSC VQP

Hoang Gia Ky, Christian Bruzek, Gérard Grunblatt,
Pascal Mougenot, François Peltier and Philippe Sulten

Alsthom Intermagnetics S.A.
3 Avenue des Trois Chênes
90018 Belfort Cedex
France

INTRODUCTION

Selected for the SSC Vendor Qualification Program, AISA started to develop SSC outer wire and cable in October 1991 with the Phase I Program which consisted of R&D efforts and an initial production of 3400 kg of outer cable. This phase which was achieved in February 1992 resulted in some modifications allowing improvements of the manufacturing process used in the Phase II production of 6300 kg of outer cable. This production, concerning the wire, was achieved four months after receiving the starting NbTi alloy. And 5500 kg of outer cable were delivered to SSCL six weeks later well within the May 1st 1993, date given by SSCL as a target.

This paper reports relevant results of the Phase II production and presents performances of the Phase II wire as well as those of cables manufactured during Phase I and Phase II.

PHASE II WIRE MANUFACTURING RESULTS AND PERFORMANCES

Piece lengths

About 96 % of accepted lengths are greater than 2 km, 65 % of them longer than 10 km and 16 % above 50 km with a longest length of 67 km. This result was obtained in spite of breaks occurred during manufacturing. Many breaks were still caused by Fe rich inclusions. Investigation on the causes of these breaks allowed identification of sources that will be eliminated in the future.

Production yields

Tightening controls as well as process modifications resulting from Phase I manufacture considerations have allowed some improvements in production yields, for both monofilament and multifilament billets. By comparison with those of phase IB: a gain of about 3% is expected for the final wire production.

Critical current IC

An average value of 298 A at 5.6 T and 4.2 K was achieved for Phase II billets, with a dispersion characterized by a CV = 1.57%. This result which allows a margin of about 4% with regard to the specified value, marks a net progress in terms of average value and variation reduction, compared to phase IB critical currents. The variation of Phase II billet mean Ic is illustrated in Fig. 1.

Over the whole phase II billets, the billet average Jc ranged from 2470 to 2570 at 5.6 T and 4.2 K. The cumulative average of Jc at 5.6 T for all Phase II accepted lengths is 2530 A/mm² with a coefficient of variation of 1.31 %.

Concerning n value evaluations, the results obtained are not always consistent. Early results showed some out-of-spec values, the corresponding samples, however, contain intermetallic nodule free filaments. Retests lately performed give n values all above the specification limit. Wrong values previously obtained were due to measuring errors.

Cu/Sc

Variations of Ic's are tightly related to those of Cu/Sc ratios whose distribution is characterized by a mean value : 1.80 and coefficient of variation of 0.94 % (Fig. 2). Again this result indicates an improvement on mean value centering and Cu/Sc variations.

Remaining main wire characteristics as RRR, wire diameter and spring-back values all meet the SSC specifications. Summary of statistics of Phase II wire mechanical and electrical properties are given in table 1.

Table 1. Electrical and Mechanical properties of Phase II outer wire.

	Mean Value	St. Deviation	CV %	Cpk
Strand Diameter (mm)	0.6479	0.0004	0.06	1.90
Cu/Sc	1.795	0.017	0.94	1.89
Ic 5.6 T, 4.2K (A)	298	5	1.57	
n-value 5.6 T	47	7	14	
RRR	186	7	3.57	
Springback	888	31	3.52	

OUTER CABLE MANUFACTURE

During Phase I, 35696 m of dipole outer cable was manufactured, 26257 m of which were made from phase IB in-specification wires to provide about 30 dipole lengths and the remaining materials were "practice" cables. Twenty quadrupole lengths (4698 m) were also produced in three runs with conforming material. Runs of four dipole unit lengths were regularly made at a speed of 3 m/mn. The last run including five dipole unit length was fully satisfactory.

Phase II outer cable were manufactured at a second vendor. Twenty-four cable lengths ranged from 880 to 4880 m were produced in six weeks with a cabling speed of about 6 m/mn. Fig. 3 presents micrographs of the cross sections of cables manufactured. These images show deformations of strands located at the small edge of the cable.

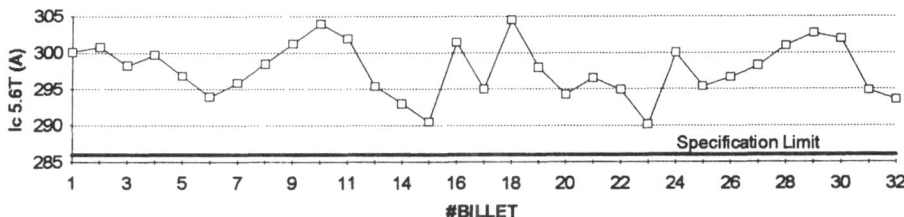

Fig.1 Phase II Ic

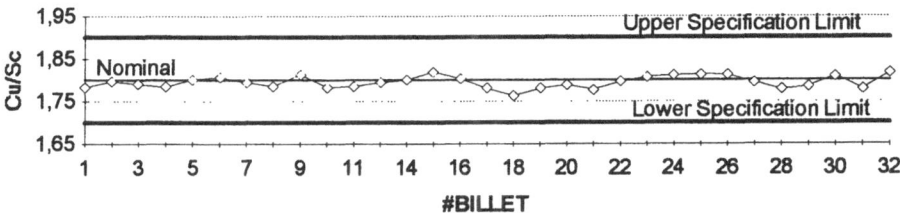

Fig.2 Phase II Cu/Sc ratio

Cable parameters

During cabling operations, main geometrical characteristics: width, mid-thickness, angle were statistically controlled by a cable measuring machine (CMM). All the cables manufactured, Phase I quadrupole and dipole cables, and Phase II dipole cables as well, meet the SSCL dimension specifications.

Statistical data of the 20 quadrupole unit lengths produced are:

	Width (mm)	Mid thickness (mm)	Keystone angle (degree)
Average	9.76	1.166	1.24
Standard Deviation	0.01 %	0.05 %	0.06 %
Spec. Value	9.73	1.166 ± 0.006 mm	1.2 ± 0.1

These cables present excellent aspect : no sharp edges nor poping strand were found.

Variations of Phase II cable parameters are more important, though fall within specification limits, than those of Phase I ones (Fig 4). These differences are probably related to the cabling speed used : higher speed might involve larger variations.

Cable Ic

Cable critical currents were evaluated by measuring critical currents of six extracted

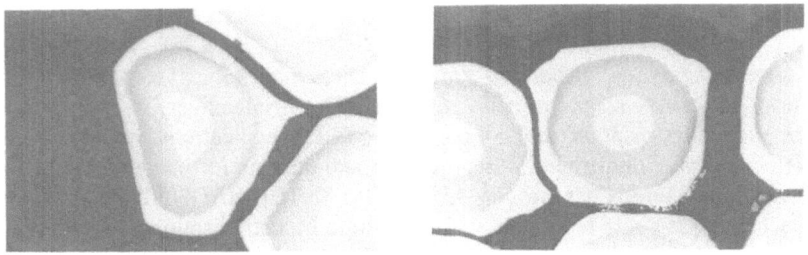

Fig. 3 Micrographs of phase II outer cable cross section.

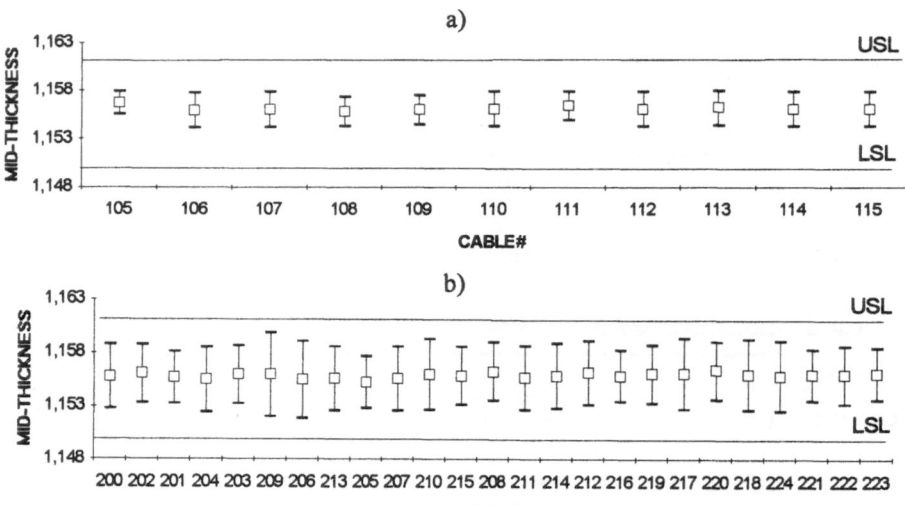

Fig 4. Varition of cable mid thickness: a) Phase Ib   b)Phase II

strand and Ic degradations calculated by comparison with Cable Ic mean values deduced from virgin wire Ic.

Phase I cable Ic tests have been performed by SSCL. Early results are given below :

	Cable Ic (A)	Degradation %	Slope (A/T)	Slope (A/mm²)/T
Quad Cable # SSC-2-A-102	8593	2.61	- 65	- 558
Dipole Cable # 4-A-105	10372	- 0.59	- 67	- 573

Phase II cable Ic's estimated from extracted strand Ic -6 strands per cable- are given below :

Mean Value	10658 A
Sigma	103 A
C.V	0.97 %
Mini/Maxi	10397 A/ 10859A

Compared with the minimum Ic value accepted, 9780 A, Phase II outer cables, exhibited a mean margin of about 9 %. The estimation of Ic degradations with measurements made by AISA shows a maximum value of 2.6 %

CONCLUSIONS

We have successfully produced Phase II wires and cables well within the target manufacturing schedule, in spite of delays on starting material delivery.

Wires and cables manufactured all meet the SSCL specifications. Improvements on the wire quality in terms of uniformity of the products and their performances were achieved.

Phase II manufacture has proven ALSTHOM INTERMAGNETICS ability to easily produce such cable at a rate consistent with Low Rate Industrialisation Program (LRIP).

RECENT PROGRESS AT OXFORD SUPERCONDUCTING TECHNOLOGY
DURING PHASE II OF THE SSC VENDOR QUALIFICATION PROGRAM

Joseph J. Lichtenwalner and John D. Scudiere

Oxford Superconducting Technology
600 Milik Street
Carteret, NJ 07008

INTRODUCTION

Oxford Superconducting Technology (OST) is participating in the Vendor Qualification Program (VQP) for producing superconducting cable for the Outer Dipole Magnets. After a successful completion of Phase I of this VQP, OST began Phase II of the VQP.

Phase II consisted of multifilament billets that utilized the process optimization improvements demonstrated during Phase I.

Along with SSC Phase II, OST is currently producing wire for the Relativistic Heavy Ion Collider (RHIC) Dipole Magnets at Brookhaven National Laboratory. Thirty tonnes are completed at this time.

This paper presents results from Phase II strand processing with correlations to RHIC Dipole processing.

MONOFILAMENT BILLET

Monofilament billets processed for Phase II used NbTi alloy from Teledyne Wah-Chang Albany. The compaction percentage for these billets averaged 3.75% with a coefficient of variation of 5.0%. The percent compaction was calculated using Equation 1.

The average extrusion force was 3392 tons with a coefficient of variation of 3.7%. These values correlate closely with previous SSC Phase I data.

Figure 1 is a graph of k-factor vs. temperature for the monofilament billets.[1] In the temperature range observed, there seems to be no noticeable change in k-factor per change in temperature. This indicates that the slight variations in temperature seen are probably just surface temperature changes, not indicative of the entire billet. This is understandable since the thermal conductivity of copper is so much higher than that of Niobium/Titanium.

After extrusion, bonding was checked between the copper, niobium and NbTi. There was no evidence of un-bonded areas.

The monofilaments were processed down to hex shape and prepared for multifilament billet packing.

Monofilament billets for Phase II met the objective yield. This was expected because for the RHIC Dipole project, over 25 tonnes of monofilament wire have been produced with extremely consistent yield results. (At this point, it should be mentioned that the RHIC Dipole and SSC products parallel each other in terms of process routings and process equipment. The main difference between the two products is the number of filaments in the multifilament billets which, in turn, affects the Cu/Sc ratio.) Because of the similarities between these products, we feel the process uniformity results for SSC and RHIC can be compared to each other.

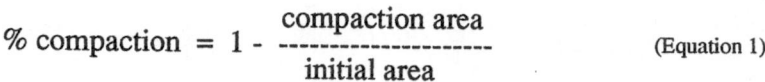

$$\% \text{ compaction } = 1 - \frac{\text{compaction area}}{\text{initial area}} \qquad \text{(Equation 1)}$$

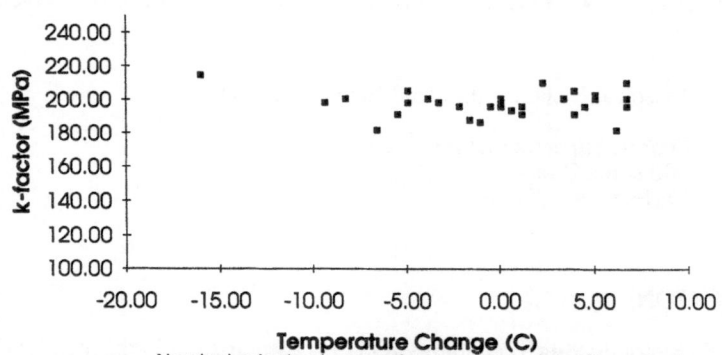

Nominal extrusion temperature corresponds to "0" on chart.

Figure 1. K-factor vs. temperature for monofilament billets.

MULTIFILAMENT PROCESSING

Multifilament billets were packed routinely. They were compacted and extruded in two batches due to time constraints. Compaction results for both batches are very similar, hovering around 5%. Coefficients of variation are also very consistent for both batches. The variation increases a little, to 6.1%, for the two batches combined.

The extrusion results, within each batch, are consistent. The coefficients of variation of extrusion force for each batch are 0.98% and 1.39%. Both batches, together, give a higher coefficient of variation which indicates a difference in averages between the two batches. Extrusion pressure vs. J_c was examined and there is a negligible difference in J_c between the two extrusion batches, indicating that the effects causing the extrusion variation have little or no affect on the final product.

After large-rod draw, bonding was checked between the filaments. Two methods were used: a destructive test and an etch test. Both tests showed acceptable bonding. From here, processing was routine down to small diameter wire.

In Phase I, at small wire diameter, OST experienced numerous wire breaks. Extensive wire-break analysis was performed and the results from this analysis were implemented initially into RHIC Dipole production and then into SSC Phase II production. Thirty tonnes of wire have been completed for RHIC and pieces-per-billet are averaging slightly under 15. SSC Phase II has averaged 13 pieces-per-billet for its 6.5 tonnes of wire produced. Some of these pieces were cut at final size because they were too long to fit on a shipment spool.

OST has demonstrated for over 35 tonnes of final strand that the current production process is capable of producing billets with consistently long lengths. Figures 2a, 2b, and 2c graphically show the trend of piece lengths from SSC Phase IB, RHIC Dipole, and SSC Phase II. As is obvious, much improvement has been made over the past year.

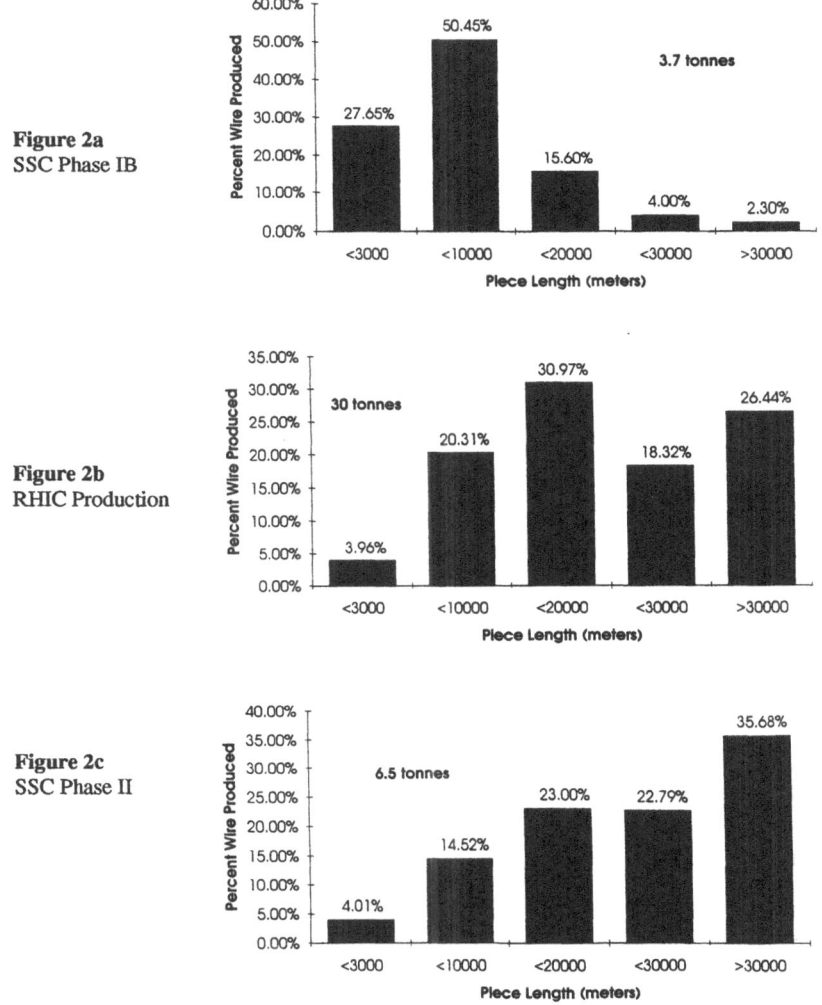

Figure 2a
SSC Phase IB

Figure 2b
RHIC Production

Figure 2c
SSC Phase II

Figure 2a, 2b, 2c. Piece length data for SSC Phase I, RHIC Dipole, and SSC Phase II, respectively.

Table 1. Statistics on critical variables for SSC Phase II strand.

	Ic	Diameter	Cu/Sc*	RRR
	(A)	(mm)		
Ave.	309.8	0.6479	1.779	252.06
St. Dev.	4.28	0.00028	0.0266	18.47
Coef. Var.	1.38%	0.04%	1.50%	7.33%
Cpk		2.86	0.99	
* Cu/Sc data has end effects removed. If the average was at nominal, 1.80, the Cpk value increases to 1.252.				

Table 2. Statistics on critical variables for RHIC Dipole strand.

	Ic	Diameter	Cu/Sc
	(A)	(in.)	
Ave.	290	0.02551	2.21
St. Dev.	6.8	0.0000128	0.055
Coef. Var.	2.34%	0.05%	2.49%
Note** Due to production-unit testing requirements for Ic and Cu/Sc, over 60% of the data points include end-effects. With end-effects removed, the coefficient of variation decreases by almost 1%.			

PERFORMANCE RESULTS

The SSCL has defined four "critical" characteristics for the Phase II qualification evaluation. They are critical current (I_c), wire diameter, copper-to-superconductor ratio (Cu/Sc) and residual resistivity ratio (RRR). Table 1 shows the cumulative statistical results for the 6.5 tonnes of strand produced. Results are excellent, and the low coefficients of variation for all categories is of utmost importance. Looking at Cu/Sc ratio, the Cpk value is .99 (with the non-steady-state end effects removed). The Phase II conductor was shaved using the Phase I process criteria which resulted in an average Cu/Sc ratio of 1.78:1, slightly below the nominal of 1.80:1. If the shave manufacturing step had been adjusted to provide strand with an average Cu/Sc ratio of 1.80:1, the Cpk value would have increased to 1.252.

Table 2 shows the same results for 30 tonnes of RHIC Dipole billets.[2] Again, uniformity is excellent. This data has all end effects included due to the testing requirements. If the end effects are removed, the coefficient of variation decreases by almost 1%, down to levels that are comparable to what the SSC wire achieved.

In Phase II other less critical variables, such as "n"-value and springback, had results averaging 51 and 764, respectively, well within specifications.

CONCLUSIONS

Oxford Superconducting Technology performed very well in Phase II of the SSC VQP. Strand uniformity was consistent as was billet yield. A major step forward was the improvement in piece length, with about 82% of the total wire produced in lengths greater than 10,000 meters. Room for improvement still exists and efforts to further reduce wire breaks are continuously in effect.

Of equal importance to the Phase II results is OST's performance on the RHIC Dipole project to date. With 30 tonnes of strand produced, performance results are excellent and piece length is excellent. With RHIC Dipole production, OST has demonstrated the SSC CDM conductor manufacture rate for LRIP (low rate initial production).

ACKNOWLEDGEMENTS

The authors wish to thank the SSCL for the opportunity to participate in the SSC VQP and to congratulate them for their efforts to keep the Superconducting Super Collider project on schedule and on budget. OST also wishes to thank its employees for their hard work and dedication throughout the SSC VQP.

REFERENCES

1. J.M. Seuntjens, et. al., "Raw Materials and Early Monofilament Analysis from the Vendor Qualification Program," IISSC, New Orleans, March 4-6, 1992.
2. G.H. Epstein, A.F. Greene, et. al., "A Progress Report on the Production of 1.8 Million Feet of Superconducting Cable for the RHIC Project at BNL," IISSC, San Francisco, May 6-8, 1993.

DEVELOPMENT OF NbTi SUPERCONDUCTOR FOR THE SSC AT SUMITOMO ELECTRIC INDUSTRIES

S. Saito, Y. Yamada, T. Sashida, A. Mikumo, M. Koganeya, and M. Nagata

SSC Project Team
Sumitomo Electric Industries, Ltd.
1-1-3, Shimaya, Konohana-ku, Osaka, 554 Japan

INTRODUCTION

Sumitomo Electric Industries, Ltd. (SEI) has been developing NbTi superconductor to be used for the inner coil of the Collider Dipole Magnet for the SSC project. We completed Phase I of the SSC VQP program, and clarified the dependency of some important factors; namely NbTi alloy source, hot extrusion condition, heat treatment condition, and so on, on the superconductive properties, on the productivity, and on the drawability. Furthermore, we developed an unique quality system for the SSC conductor cable.

In Phase II; reflecting the results of Phase I, SEI has been improving the production variations and conductor properties. In addition, SEI has been utilizing our unique quality system in Phase II production.

In this paper, we shall report on the status of improvement on the important properties; that is drawability and critical current. We shall also report on the utilization of our quality system.

PRODUCTION PROCESS

Table 1 shows the main specification. The inner conductor is Rutherford type cable constructed of 30 strands of wire. The diameter of each strand (wire) is ϕ 0.808 mm and consists of approximately 8,000 NbTi filaments with a diameter of 6 μm.

Table 1. Main specification of inner conductor.

Wire (Strand)		Cable	
Wire Diameter	ϕ 0.808 $^+_-$ 0.0025 mm	Mid Thickness	1.458$^+_-$ 0.006 mm
Cu Ratio	1.30 $^+_-$ 0.10	Width	12.34$^{+0.05}_{-0.00}$ mm
NbTi Fila. Size	6μm	Key Stone Angle	1.2^0 $^+_-$ 0.1^0
NbTi Fila. Number	approximately 8,000	Strand Number	30
Critical Current (at 7T)	>368 A	Lay Pitch	86$^+_-$ 5 mm

SEI produced approximately 6,000 kg of superconductor in Phase I, and is producing about the same volume for PhaseII. A 12 inch composite billet with a production unit weight over 300 kg is utilized for the multifilament. The production process has already been reported elsewhere [1,2]. This large billet is designed to improve yield of extrusion. The multifilament billet is constructed of approximately 8,000 hexagonal pieces. From our experience, we have found that drawability depends upon the quantity of surface area between construction materials (Cu, NbTi), and that contamination in the multifilament billet is one of the main causes of wire breakage. Therefore, we assemble the billet in a class 10,000 clean room using an unique single stacking technique under servere cleanliness control. The multifilament billet has a packing factor of 95%. The packing factor approaches 100% after HIPing. Subsequently, the extruded multifilament rod undergoes drawing, heat treatment, twisting, and cabling.

DRAWABILITY

Figure 1 shows the distribution of wire piece length for all 9 Phase IB billets and 13 Phase II billets. In Phase IB, the average length of total finished wire was 80 km; which corresponds to 314 kg per one billet. The average number of wire breakage is 13 per one billet. The percentage of wire less than 1,500 m is 4.2%. This result is within SSC's requirements, and reasonable to prove that SEI's base line process is not far from the optimum process. Therefore, from the view point of drawability, SEI selected almost the same process for Phase II as was used for the Phase I base line.

In Phase II, the average length of total finished wire is about 90 km. This improvement of yield was the result of corrective action SEI undertook based upon the analysis of Phase I results. The long piece length with no breakage is increased as compared with Phase IB; with 25% of the total wire having a piece length greater than 25 km, and the average piece length becoming slightly longer than Phase IB. As there was no significant change in the process except our worker's experience, this results seem reasonable.

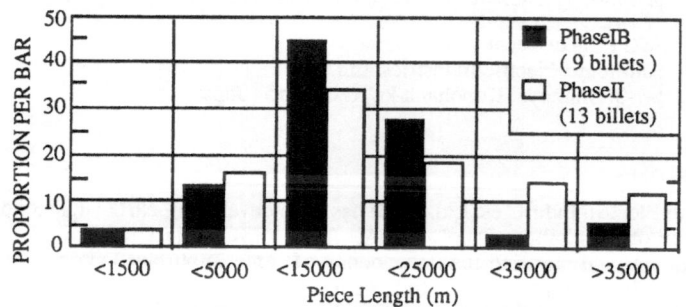

Figure 1. Distribution of wire piece length.

CRITICAL CURRENT

Figure 2 shows the critical current (Ic) variation at 7 Tesla in all the Phase IB billets. The average Ic at 7 Tesla for all the Phase IB billets is 371.7A; the standard deviation (σ) is 7.3 A; and the CV value (<u>C</u>oefficient of <u>V</u>ariation: [σ/Average]) is 1.96%. The CV value is sufficiently small, and satisfies SSC requirements. However, the average Ic does not have sufficient margin as compared with the specification value of 368 A.

From our Phase I investigations on the relation between extrusion condition and critical current density (Jc), we clarified that the Jc had a strong dependency on extrusion temperature; as shown in the following formula:

$$(dJc/(dT) = -1.1$$

Jc: critical current density (A/mm^2 at 7 Tesla)
T: temperature (^{o}C)

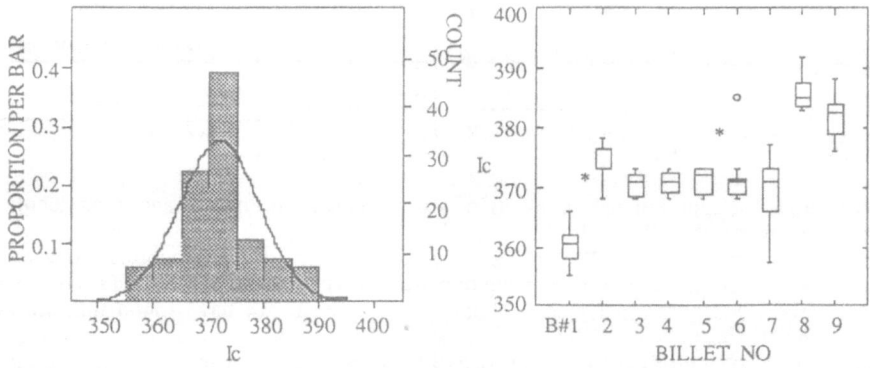

Figure 2. Critical current variation in Phase IB.

In Phase II, the average Ic of 380 A (7 T) was designed to have sufficient margin. To fulfill this requirement, the extrusion temperature is set lower than the Phase I baseline condition. Figure 3 shows the Ic variation of 10 Phase II billets. The average of Ic is 382.4 A (7 T); the standard deviation (σ) is 6.3 A, and the CV value is 1.7%.

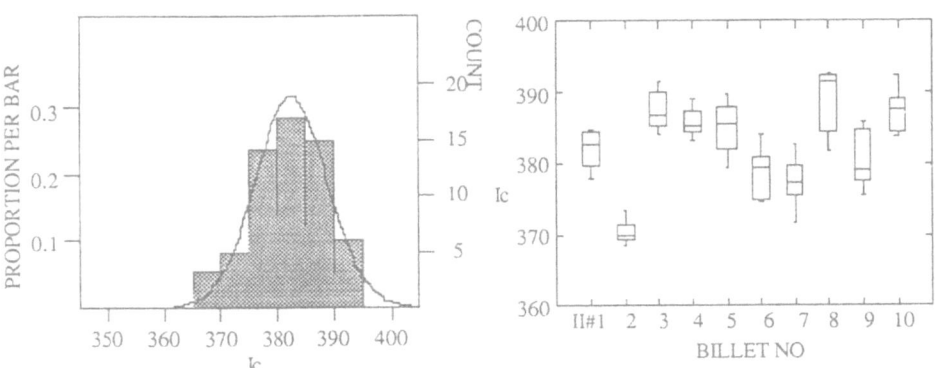

Figure 3. Critical current variation in Phase II.

QUALITY CONTROL SYSTEM

During Phase I, SEI began the establishment of the quality control system, that is so called **TRACE** (Traceability & Reliability Assuring Computerized Environment) system. This system was especially designed for the production of SSC type conductor in accordance to the systematic procedure shown in Figure 4. This system has been utilized in Phase II.

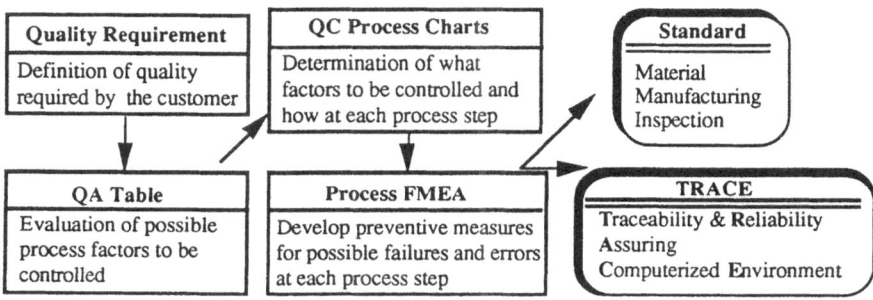

Figure 4. Procedure for establishing of QA system.

The data flow on the **TRACE** system is shown in Figure 5. The important point of the production is to make an instruction and reporting system between the engineers and the operators. This system consists of two engineering work stations and sixteen personal computers on the shop floor. The operator follows the instruction displayed on the CRT, and he measures the properties for each production process according to the standard. The operator enters the data into the PC just aside of machine. In order to prevent any accident, the data entry is on-line and in real time, and the data is checked on the computer display to confirm the input value is within the standard. All data is saved into a main data-base so that it is accessible at any time and everywhere in the factory. The statistical analysis is easily performed to evaluate the process ability such as Cp value. Also, any of the data can be transferred to another computer and analyzed by using software such as "SYSTAT".

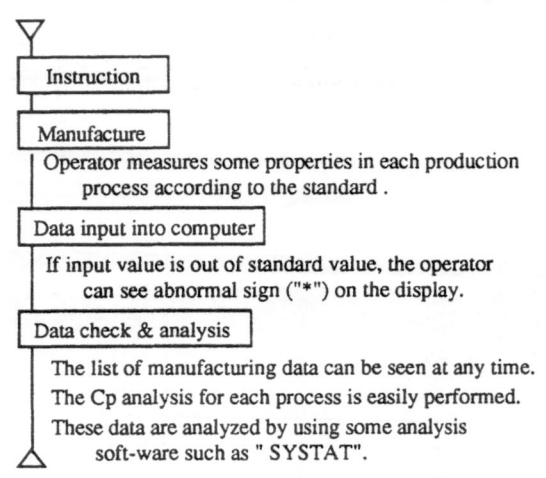

Figure 5. Data flow on the TRACE system.

This QA system has been operating well in the Phase II; especially from the point of instruction and as a reporting system between the engineers and operators. Also, it has been very useful in confirming every process has been done properly. We are confident this system will be useful in the mass production stage of the SSC project.

CONCLUSIONS

1. SEI has been producing Phase II conductor based on Phase IB base-line process. The drawability in Phase II becomes slightly better and the long piece length with no breakage is increased as compared with Phase IB.
2. SEI has been improving the Jc in Phase II by examining the extrusion condition of Phase I. The average Ic of Phase II becomes approximately 10 A higher than that of Phase IB.
3. SEI established a unique QA system in Phase I and confirmed its usefulness in Phase II.

REFERENCES

[1] T. Sashida et al., "Development of Superconducting Wire and Cable for the SSC Project in Sumitomo Electric Industries," Proc. of 3rd IISSC, Atlanta, U.S.A., 1991.
[2] S. Saito et al., "Development of NbTi Superconductor for the SSC by Sumitomo Electric Industries," Proc. of 4th IISSC, New Orleans, U.S.A., 1992.
[3] M. Fuse et al., "Quality System Design and Development for SSC Superconductor Cable," Proc. of 4th IISSC, New Orleans, U.S.A., 1992.

CURRENT DENSITY AND MAGNETIZATION OF FINE FILAMENTARY NbTi SUPERCONDUCTORS

H. Kanithi,[1] W. Wiegert,[1] P. Valaris,[2] M. D. Sumption,[3] and E. W. Collings[3]

[1]IGC Advanced Superconductors, Inc.
Waterbury, CT 06704

[2]Westinghouse
Round Rock, TX 78680

[3]Battelle
Columbus, OH 43201

ABSTRACT

Superconductors containing nominally 7,000 and 39,000 NbTi filaments have been produced by a billet design involving hexagonal cells. The interfilamentary matrices were pure Cu (7K fils) and Cu-0.5 wt%Mn (39K fils). Using the hexagonal cell method, single-stack billets were drawn to filament sizes ranging from 6 μm to 1 μm. The results of scanning electron microscopy, magnetization, and critical current density measurements are described. Filament quality in response to cold work and heat treatment was evaluated using image analysis. It is found that high quality, single stack, fine filamentary strands can be economically fabricated using the hexagonal tube approach.

INTRODUCTION

Multifilamentary superconductive wires with small magnetization and hysteretic loss are desirable both in AC- and DC-dipole magnet applications. These can be achieved by reducing the filamentary diameter (d_f). For a given I_c, it is necessary to increase the number of filaments as d_f is reduced. The standard way of fabricating wire with d_f less than 6 μm is to use a double stacking procedure. However, this approach: (1) produces somewhat lower n values (the exponent in the I-V curve), (2) allows filament distortion at the perimeter of the restacking bundle (because of the large amount of soft surrounding copper), (3) forfeits an additional 10-20 % of the wire because of restacking, and (4) is more time consuming, and thus more expensive, because of the additional extrusion step.

In order to overcome these difficulties IGC Advanced Superconductors has developed a hexagonal cell approach[1]. In this approach, round filaments are inserted into hexagonal copper cells to form the subunits which are then assembled into the final billet. Using this approach, single stack billets fabricated with up to 40,000 filaments can be extruded and drawn down to a give a d_f as small as 1 μm. This paper describes the fabrication and processing of several billets of this type -- see also Refs. 1 and 2.

STRAND MANUFACTURE AND PROCESSING

Billets were made by stacking the extruded monofilaments into Cu hexagonal tube subunits, which were then assembled into the billet can[2]. Discussed here are two of the hexagonal-tube billets previously described, viz ASI08 and ASI40, see Table 1. ASI08 has 7,251 filaments assembled from hexes of 127 filaments each. The interfilamentary matrix as well as the hex material was copper. ASI40 has 38,520 filaments assembled from hexes with 169 filaments each, Cu-0.5%wt Mn interfilamentary matrices and Cu hexagonal tubes. In both cases, round rods were used rather than hexes because they were easier to pack uniformly. Nb sheaths of 4 vol.% (ASI08) and 7 vol.% (ASI40) were used to suppress intermetallic formation. The filamentary region was annular and the interfilament-to-filament (s/d) ratio was 0.22.

Time for billet assembly for ASI08 was about 1/3 of the time necessary for stacking a single stack billet with hexagonal monofilament elements. Assembly time for ASI40 was comparable to the time required for an SSC inner single-stack billet (with $d_f = 6$ μm).

SEM Micrographs and Image Analysis

Figures 1(a) and 1b) show closeups of a number of filaments of strands ASI40HT-25 and ASI08HT-25. Comparing 1(a) and 1(b) we can see that the filaments of ASI40HT are more distorted than those of ASI08HT. This correlates with the percent standard deviation in filament area, CV, extracted from image analysis, Table 1. This difference was evident after extrusion, suggesting that the degree of roundness of the filaments may be inversely proportional to the number of filaments in the stack. Additionally, we can see

Table 1. Strand Specifications

Strand Code[†]	Fil. Diam., μm	CV[††] %
ASI40HT-35	3.43	--
ASI40HT-25	2.45	9.6
ASI40HT-20	1.72	--
ASI40HT-15	1.42	--
ASI40HT-10	0.97	13
ASI40NHT-35	3.44	--
ASI40NHT-25	2.47	3.9
ASI40NHT-20	1.77	--
ASI40NHT-15	1.46	--
ASI40NHT-10	1.00	8.7
ASI08HT-60	5.70	7.6
ASI08HT-45	4.56	--
ASI08HT-35	3.42	--
ASI08HT-25	2.35	7.2

† HT = heat treated, NHT = not heat treated.
†† Percent std. dev. of the filamentary area.

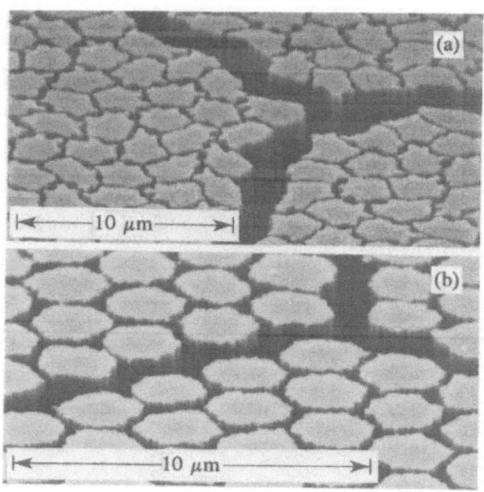

Figure 1. SEM micrographs of filament close-ups for (a) ASI40HT-25 and (b) ASI08HT-25.

that for smaller wires (filaments) CV increases, and that it is greater for the HT samples. These results are to be expected since the filaments deform more for greater reductions due to texturing. The annealing of the matrix material increases the disparity of the flow strength between the matrix and the filaments, which is a known cause of filament non-uniformity.

ELECTRICAL AND MAGNETIC PROPERTIES

Transport and Magnetic J_c

Critical current density (J_c) and n values at 70 kOe are given in Fig. 2 for the heat treated wires as a function of d_f. Thermomechanical processing was adjusted to optimize J_c at $d_f = 6.0~\mu m$ for ASI08HT and at $d_f = 2.5~\mu m$ for ASI40HT. $J_c = 1823~A/mm^2$ for ASI08HT at $d_f = 5.70~\mu m$, which is 10 percent above the SSC specification, and is one of the best values measured for SSC type wire to date. The n-value is better than the typical values which are in the low 30's. J_c for ASI40HT at $d_f = 2.45~\mu m$ is 1613 A/mm^2, just slightly below specification, and the n value is 38. This is, however, at a filamentary diameter less than half of SSC wire. At smaller filamentary diameters both J_c and n drop rapidly, but this is simply because of the chosen optimization point.

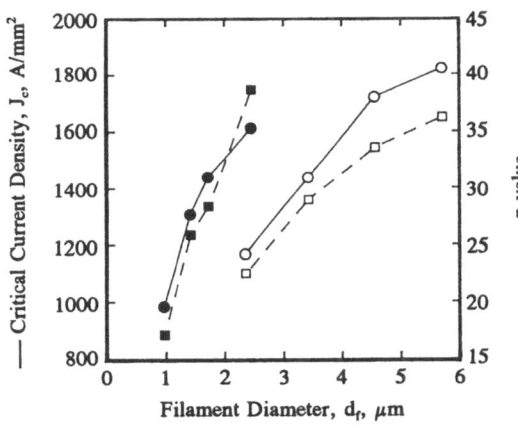

Figure 2. J_c (($\bullet$)) for ASI40HT and ($\bigcirc$) for ASI08HT as well as n (($\blacksquare$) for ASI40HT and ($\square$) for ASI08HT) versus d_f for the heat treated samples at 70 kOe.

Proximity Effect Onset

The magnetization, M, of small bundles of strand was measured as a function of magnetic field strength, H, by vibrating sample magnetometry. As usual[3,4], the full height of the M(H) loop (either ΔM_{clad} or ΔM_{bare}) was controlled by individual-filament bulk- and surface pinning (both clad and etched (bare) strands) and by proximity effect between filaments (clad strands only). Figure 3(a) compares the $\Delta M_{clad}/\Delta M_{bare}$ ratio of the ASI40 series (HT and NHT) with those of a series of Cu-Mn based NHT research alloys (NTCM, 5355 fils) previously measured in [3,4]; Fig. 3(b) is a comparable plot for Cu-matrix strands. PE coupling onset for ASI40HT strands occurs for d_f (d_s) between 2.0 (0.44) μm to 2.5 (0.55) μm, while for ASI40NHT, it occurs for d_f (d_s) between 1.5 (0.33) μm and 2.0 (0.44) μm. These are somewhat earlier onset values than those of the NTCM strands. For ASI08HT, the d_s (d_f) for PE onset seems to fall between 3.5 (0.77) μm and 4.5 (0.99) μm. These values are also somewhat greater than those of the NTCU (4395 fils) strands. The greater value of PE onset diameter in the HT samples is of particular interest, since this also correlates with an increase in CV. This observation invites consideration of the connection between filament quality and PE losses.

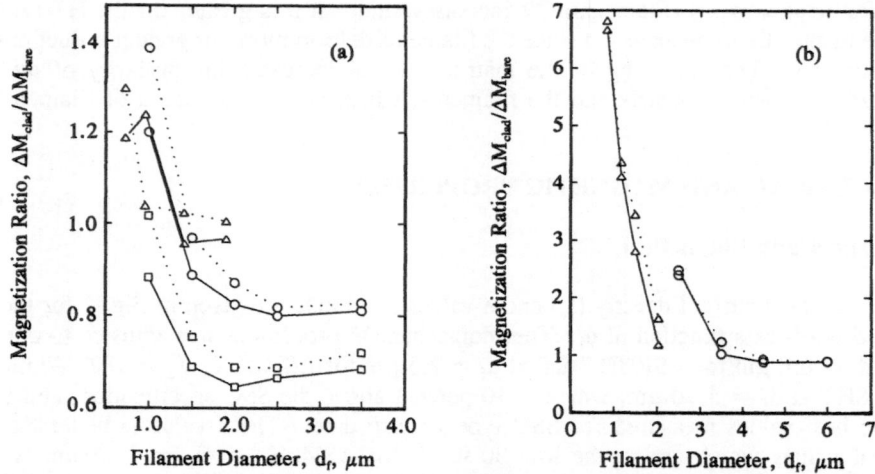

Figure 3. $\Delta M_{clad}/\Delta M_{bare}$ for short magnetization samples of: (a) the CuMn matrix strands; ASI40HT (39K fils ○), ASI40NHT (□), and NTCM (5K fils △), and (b) the Cu matrix strands; ASI08HT (7K fils ○) and NTCU (4K fils △).

DISCUSSION AND CONCLUSIONS

The hexagonal tube method of stacking billets is useful for the manufacture of single (not counting monofilament) extrusion fine filamentary wire. We have fabricated single stack billets with up to 38,520 filaments, and it may be possible reach 100,000 filaments. J_c values and n-values for ASI08HT are among the best of typical SSC wires, and those of ASI40HT compare very well, given the factor of two reduction in d_f. Cabling experiments performed on these wires confirm good mechanical properties, and negligible degradation of electrical properties. Additionally, we find that the hexagonal tube fabrication method is less time-consuming than the comparable double extrusion route, at least for filament numbers greater than 8,000. However, it seems that the filament quality, as measured by area distribution, and the degree of filament roughness, increases with the number of filaments, although the filament quality is better than that typically obtained by double extrusions. The hex cell method allows the fabrication of single-stack billets with large numbers of filaments, useful for low-magnetization- and low-loss applications.

ACKNOWLEDGEMENTS

This research was funded by the U.S. Department of Energy, under contract DE-FG02-90ER81068.

REFERENCES

1. United States Patent No. 5,088,183, Inventor H.Kanithi, filed May 1, 1990, issued Feb. 18,1992.

2. H.C. Kanithi, P. Valaris, and B.A. Zeitlin, Supercollider 3, p. 689-693, 1991.

3. M.D. Sumption, K.R. Marken Jr., and E.W. Collings, IEEE Trans. Magn. **27**, 2166-2169 (1991).

4. M.D. Sumption and E.W. Collings, Adv. in Cryo. Eng. **38**, 783-790 (1992).

6 Next Generation Science

COSMOLOGY AND THE COSMIC BACKGROUND EXPLORER (COBE)

George F. Smoot

Lawrence Berkeley Laboratory,
Space Sciences Laboratory, and
Center for Particle Astrophysics
University of California, Berkeley, CA 94720

The Cosmic Background Explorer (COBE) satellite is famous around the world for measuring the largest and oldest structures ever discovered in the universe. Human beings, living on a small planet orbiting a small star in an ordinary galaxy, have now reached back in time to try to understand the origin of the universe. We have launched a little space probe, orbiting not far above the surface of the Earth, to receive the faint whispers of the cosmic explosion, the Big Bang, which started the expansion of our whole universe 15 billion years ago. The new data have greatly strengthened our conviction that the Big Bang picture is valid, and have begun to fill in some of the details. We have measured tiny differences in brightness of the heat radiation from the Big Bang, only a few parts per million. These differences reveal the structure of the Big Bang itself, as it was only a sub-microsecond after the explosion.

The COBE satellite was built at NASA's Goddard Space Flight Center, located near Washington D.C. A team of hundreds of engineers, scientists, technicians, managers, computer programmers, and many others worked together from 1976 to 1989 to design, build, test, and launch it. The project grew from proposals by three original teams of scientists from universities – the University of California-Berkeley, MIT, and Princeton University – and NASA Centers – Goddard and JPL. Many obstacles were overcome on the way to launch. One of the largest was that the project needed new and difficult technology, such as superfluid liquid helium to cool two of the instruments. An even larger challenge was created when the Space Shuttle exploded in 1986 and launch opportunities on the Shuttle were greatly restricted. The COBE had to be almost completely rebuilt to fit onto an expendable rocket. Nevertheless, this was accomplished and COBE was launched on a small Delta rocket on 18 November 1989.

To study the beginning of the universe, we use the fact that light travels very fast, 300,000 km per second, but not infinitely fast. We look back in time to the beginning simply by looking at things very very far away. One can see one's hand, only 1 meter away, as it was about 0.000000003 seconds ago, which does not make much difference to daily life. However, astronomers study things much farther away. It takes light about 0.02 sec to travel across the Earth, which is 7000 km across, and this time is quite important to radio and computer communications. We receive light from the Sun about 8 minutes after it is emitted, but the nearest star beyond the Sun is so far away that light actually takes 4 years to arrive. When we look at the center of our own Galaxy we see it as it was 25,000 years ago. These vast distances are already a problem for space travel: even if we could travel at the speed of light, a round trip to the center of the galaxy would bring us back to Earth 50,000 years later, and our friends would be long gone. The light from the nearest galaxies takes more than a million years to get here and for the distant galaxies more like one billion to 10 billion years. Looking as far away as we can, to the most distant reaches of the sky, we look back in time about 15 billion years.

How can one possibly measure the distance to something that far away? The simplest way is by triangulation, the same method used by highway surveyors on the Earth. This system only works for fairly nearby stars in our own Galaxy, because to measure a triangle we have to move from one place to another, and we don't move very far - we only get a ride around the Sun once a year as the Earth moves in its orbit. To measure things farther away, we have to use the fact that more distant stars look fainter. This can be made into a reliable, quantitative method if we can convince ourselves that we know how bright each star really is, but that has taken generations of astronomers working with the biggest telescopes, and it still doesn't work perfectly. One of the big uncertainties is that stars and galaxies have evolved over the long times it takes light to travel these distances. The distant stars and galaxies had less time to evolve before the light left them than the ones near us.

Except for the planets and the nearest stars, astronomers can't see things move across the sky, even with telescopes. However, we can measure motion using the Doppler effect. If you listen to a train or car pass as you stand by the highway, you may notice that the pitch of the sounds it makes suddenly becomes lower after it passes. When an object emitting sound moves towards you, the wavelengths of the sound is compressed. When the same object moves away from you, the sound's wavelengths are stretched. A very similar effect occurs for light waves, so one can measure the changes in the wavelengths of light to determine whether a luminous object is coming towards us or going away from us. Visible light is shifted towards the blue when the emitter is approaching us and shifted towards the red when the emitting object is receding. The red- or blueshift is proportional to the speed of motion on the line of sight. To determine a light-emitting object's velocity relative to us, one needs to know the length of the light waves at the time they are emitted, and compare them with the ones we receive. This is possible because the present universe is composed of atoms, and each kind of atom (at least in a gaseous phase) emits light only at certain wavelengths which have been measured in ground-based laboratories.

In the 1920's, Edwin Hubble made an extraordinary discovery, which overthrew mankind's prior concept of the universe. First V.M. Slipher discovered that nebulae have large frequency shifts – nearly all redshifts indicating they are receding from us. Hubble showed that the nebulae are distant galaxies, which are clouds of hundreds of millions of stars. Hubble found the very simple relationship, now called Hubble's Law, between a galaxy's redshift and distance – a galaxy's redshift is proportional to its distance. Hubble's law, called at the time "The General Recession of the Galaxies," shows that the universe is expanding. All the galaxies are moving apart. A naive extrapolation back in time would conclude the galaxies have started their motions at a single spot and a single time. The remarkable implication is that there was an explosion of the whole observable part of the universe! This is one of the greatest mysteries we can imagine - what could have possibly caused such an event? Scientists are of course working on the question and have some possible explanations, but it will be many years before there is agreement about them.

One remarkable aspect of this explosion is that we cannot see either a center or an edge to the Universe. It seems to us that everything is rushing away from us, but since Copernicus we are not so proud as to think that we are in the middle. Mathematically we can show that people living on other planets in other galaxies would have to draw the same conclusions; they would also see all the galaxies rushing away from them too. As a result, we don't even know if there is a center to the Universe at all. The Universe is expanding uniformly everywhere, like a loaf of bread that rises as the yeast grows within it. Scientists think that it is space itself that is expanding rather than that the galaxies are moving out into previously existing space.

The expansion of space produces a redshift by stretching the wavelength of light. The light from more distant objects takes longer to arrive and thus spends more time being stretched than light from nearerby galaxies. A uniformly expanding space automatically give the Hubble law. It give the same Hubble law for an observer located anywhere in space. It is not necessary for a galaxy to move compared to its local region of space; the general expansion of space produces a redshift proportional to distance between emitter and observer. If the galaxy is moving, then there is Doppler shift adding to cosmological redshift caused by expanding space.

Albert Einstein had in the previous decade developed his theory of General Relativity to describe the effect of gravity on space and time. Belgian cleric Georges Lamaitre, Russian Alexandre Friedmann, Albert Einstein, and others had considered what the Universe would do under the influence of its own gravity. They soon realized that the Universe would not be motionless - it would be expanding or contracting. Also, it would look as though we were at the very center of the expansion, as Hubble observed. Even more astounding for the non-mathematician, there need not be a center - the whole universe could be infinite and still have a cosmic explosion. Einstein himself did not like this Big Bang idea at first, but Hubble's experimental data ended the uncertainty. Cosmologists quickly understood that there were two possible histories of the Universe - it could expand forever, or it could expand for a while and then collapse back on itself.

The Big Bang also explains why the night sky is dark. In the 1820's German Heinrich Olbers realized that an infinite, homogeneous, and static universe would look very different from ours. Since every line of sight would eventually end on the surface of some star, the whole sky should have the brightness of the surface of the Sun, and we would be so hot we would have evaporated. The Big Bang says that the sky is dark because the expanding universe dilutes and cools the radiation from distant stars. Moreover, not every line of sight ends on a star because stars have not always existed.

About 20 years after Hubble's discovery, during and after the Second World War, three scientists (George Gamow, Ralph Alpher, and Robert Herman) were thinking about that cosmic explosion, and what it might have done. At that time nuclear reactions were just being understood and it was realized that the atomic element's nuclei were built up from protons and neutrons. At first, they guessed that the Big Bang, starting just with a primordial soup of neutrons and protons, might have manufactured all the chemical elements, with about the relative abundances we have today. They soon found that this idea was wrong. In fact, only the lightest elements, hydrogen, helium, deuterium, and lithium could have been made in the Big Bang, and all the rest must have been made later. They made one key prediction: about 1/4 of all the matter in the universe must be helium, and the other 3/4 is almost all hydrogen. Their numbers are in extremely good agreement with the numbers we get from measuring the composition of stars, so the idea of the Big Bang received a strong confirmation. On the other hand, proponents of opposing theories did not give up until the next major discovery in 1964.

Gamow, Alpher, and Herman in 1948 made one other extremely important prediction: the Universe must be filled with the remains of the heat radiation that existed in the Big Bang. They calculated that it should be very cold, only about 5 degrees above absolute zero. Expansion cooled off the Universe from its blazing hot temperature of billions of degrees minutes from the Big Bang's start. Amazingly enough, the temperature of outer space was not measured precisely until 1964, and by then many scientists had forgotten that the radiation had already been predicted, although it was mentioned in popular astronomy books by Gamow. In that year, Arno Penzias and Robert Wilson discovered the radiation while they were testing an antenna for satellite communications and radio astronomy. Their result was confirmed within a few months by Dicke, Peebles, Roll, and Wilkinson, who had already built an apparatus especially for the purpose, and so the third prediction of the Big Bang model was strongly confirmed. The radiation, which astronomers call the Cosmic Microwave Background Radiation, is a little cooler than the old prediction, only 2.73 degrees Kelvin (Kelvin is unit of temperature starting at absolute zero – about -455F or -273C), but it is there and it is universal. We know it comes from the Big Bang because it comes equally from all directions, completely unrelated to any of the objects known to traditional astronomers.

By the time the COBE satellite was being prepared, astronomers had discovered a new mystery. It had been expected that the Big Bang, being so truly gigantic, must also have been quite uniform when small parts of it were compared to each other. However, maps of the locations of distant galaxies were becoming precise enough to show that the Big Bang was definitely not a uniform, smooth, featureless explosion. Instead, the galaxies are distributed in a very irregular way, with huge groups of them clustered together, and huge empty spaces hundreds of millions of light years across. (A light year is the distance that light travels in a year.) For this to happen, it seems that there must have been something special in the structure of the Big Bang itself that caused this irregular distribution of matter throughout space and that there should be some trace of this in the Big Bang radiation. These are the traces that were discovered by COBE.

COBE carries three instruments, all designed to look at the relic radiation from the Big Bang and the events that happened soon afterward. COBE is in orbit 900 kilometers above the Earth, moving in a circle that is above the sunrise - sunset line on the Earth, and it goes around 14 times each day. In the winter it can even be seen, going from south to north a little after sunset, or from north to south a little before dawn, and it can be recognized because its brightness changes as it spins around once every 72 seconds. We chose this orbit because it lets us protect the instruments - the Sun is always off to the side of the spacecraft and the Earth is always below, so the intense heat (infrared radiation) from both of them does not affect the sensitive instruments. The Big Bang radiation comprises 99% of all the radiant energy in the Universe, but here it is still 100 million times fainter than the heat emitted by the Earth.

The first of the COBE instruments to test the Big Bang theory was one designed to measure the spectrum of the Big Bang radiation. The spectrum describes the color of the radiation – how much energy there is at each different wavelength. This instrument measures wavelengths from 0.1 millimeter to 1 cm. Within two months of the launch of the COBE in November 1989, the team working on this instrument, led by John Mather at NASA - Goddard, had determined that the spectrum was in excellent agreement with the Big Bang predictions. As physicists describe the radiation, it is called "blackbody" radiation, which means that it behaves like heat emitted by an object that has no color at all and is perfectly black, absorbing all light that falls on it. Astronomers breathed an enormous sigh of relief, because only a year before the COBE launch another experiment on a small rocket had made observations that did not agree with the Big Bang prediction. The COBE data agree essentially perfectly with the prediction of a blackbody radiation spectrum. Any real discrepancies must be less than one part in 400, a wonderfully precise verification of the Big Bang model.

The second instrument, the DMR (Differential Microwave Radiometer), was designed to look for those primordial hot and cold spots in the Big Bang radiation. These spots reflect the density variations that would eventually grow into huge clouds of galaxies and huge empty spaces. The microwave receivers, not entirely different from the circuits in a television receiver, compare the brightness of one part of the sky with another. Every year the DMR instrument produces over 300 million measurements of the sky brightness at wavelengths of 3.3, 5.7, and 9.6 millimeters. We have spent over two years combining these measurements in the best way to make maps of the sky.

After extremely careful rechecking of the work, the team announced the results on April 23, 1992 at the American Physical Society meeting in Washington D.C. The team leader, George Smoot from the Lawrence Berkeley Laboratory of the University of California, described the main results and presented the new maps of the early universe. Team member Gary Hinshaw described the data processing. Deputy team leader, Charles Bennett of NASA - Goddard, showed how we separated the signals into cosmic and Galactic components. DMR Instrument team member, Alan Kogut at USRA-Goddard, showed how we tested our equipment and our computer programs to show that the signals were not the result of our own mistakes. COBE team member, Edward Wright of the University of California at Los Angeles, described how we tested various theories of the Big Bang with our data.

The COBE maps show several things. They are maps of the full sky, stretched and distorted a little for presentation on a flat piece of paper. They are oriented so the Milky Way galaxy in which we live runs across the middle of the maps from left to right, and the center of the Milky Way (in the constellation of Sagittarius) is in the middle. The top picture shows how the sky looks when we take the maps as they come from the computer. There is one dominant feature: the top right part of the map is hotter by about 1 part in 1000, (printed as red), and the lower left part is colder by the same amount (printed as blue). This feature, the dipole anisotropy, is caused mostly by the motion of the Earth towards the constellation of Leo. The second (middle) map shows what is left after we subtract the effects of our own motion. In this picture, the main thing we see is the Milky Way galaxy stretched from left to right, with a few streamers sticking up and down. However, away from the galactic plane we see warm and cool regions of various sizes. These are the primordial variations created in the birth of the universe which produce the galaxies, and larger scale structure observed today. The last (bottom) picture shows what we really want to produce – an image of the early universe with the effects of the Milky Way galaxy removed as well. This was the most difficult subtraction, but was possible because we mapped the sky at three different wavelengths, and the Milky Way does not affect them all equally. This map shows randomly-located hot and cold spots, most of them only about 1 part in 100,000 hotter or colder than the average. Unfortunately, it must be admitted that some of the hot and cold spots are also the result of signals produced in our receivers and our removal of the galactic signal, but when we analyze the new data we are now receiving these extra signals will essentially disappear.

Statistical analysis reveals real cosmic patterns hidden in these maps. The details agree very well with the basic idea of the Big Bang, but taken in combination with observations of the galaxies, they also suggest something very strange and exotic. Apparently about 90 to 99% of all the material in the universe is unlike ordinary matter,

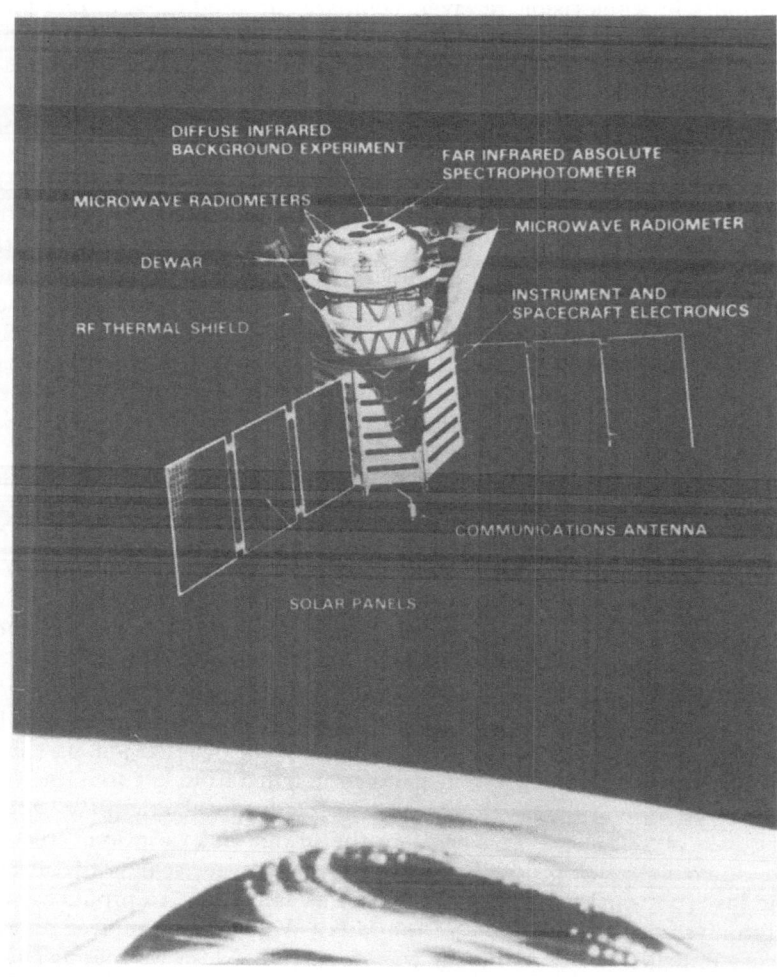

Figure 1. The Cosmic Background Explorer satellite (COBE), in orbit 900 km above the Earth.

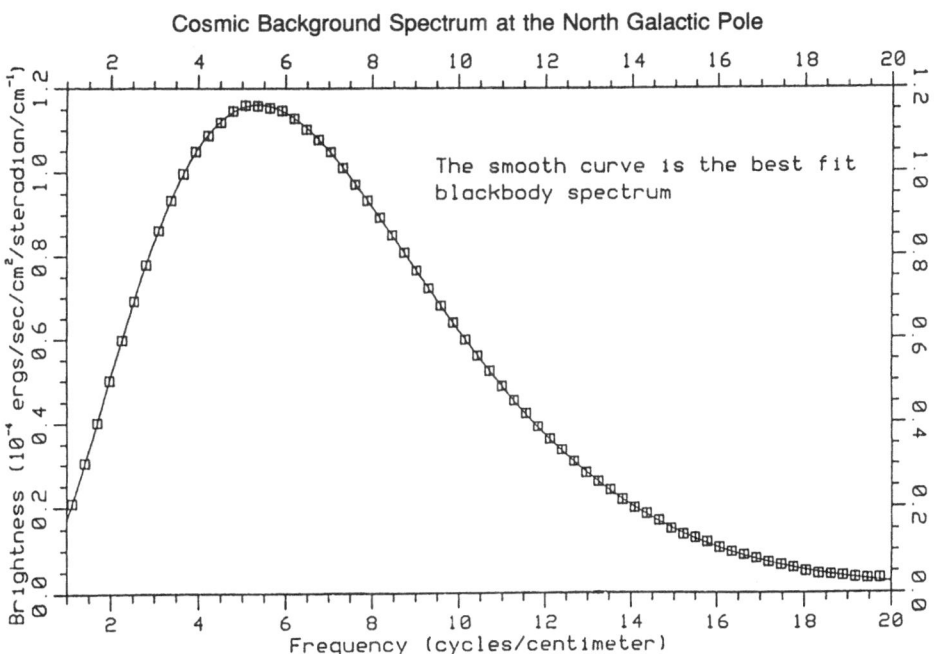

Figure 2. The spectrum of the heat radiation from the Big Bang. The measurements fit the theoretical predictions very well.

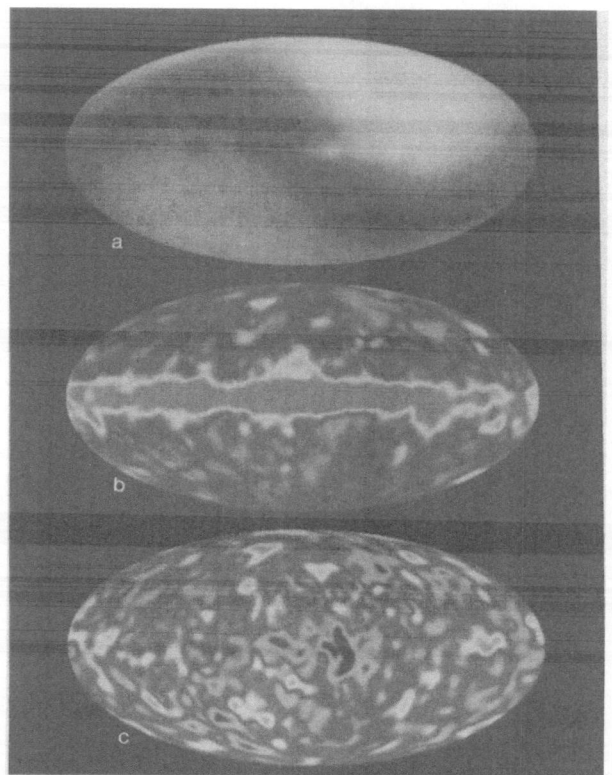

Figure 3. a (top). The microwave radiation from the sky, as seen by the COBE. The radiation is at a temperature of 2.73 degrees Centigrade above absolute zero, and the upper right portion of the map is about 0.12% warmer than the average and lower left portion symmetrically cooler. This tiny difference is caused by the motion of the Earth through the cosmic microwave radiation. b (middle). The same information as in the top panel, but with the effects of the Earth's motion removed. The Milky Way galaxy is the horizontal band through the middle of the map. c (bottom). The same as in the middle, but with the effects of the Milky Way removed as well. The blue and red spots are only 1 part in 100,000 warmer or colder than the rest but show gigantic structures stretching across enormous regions of space.

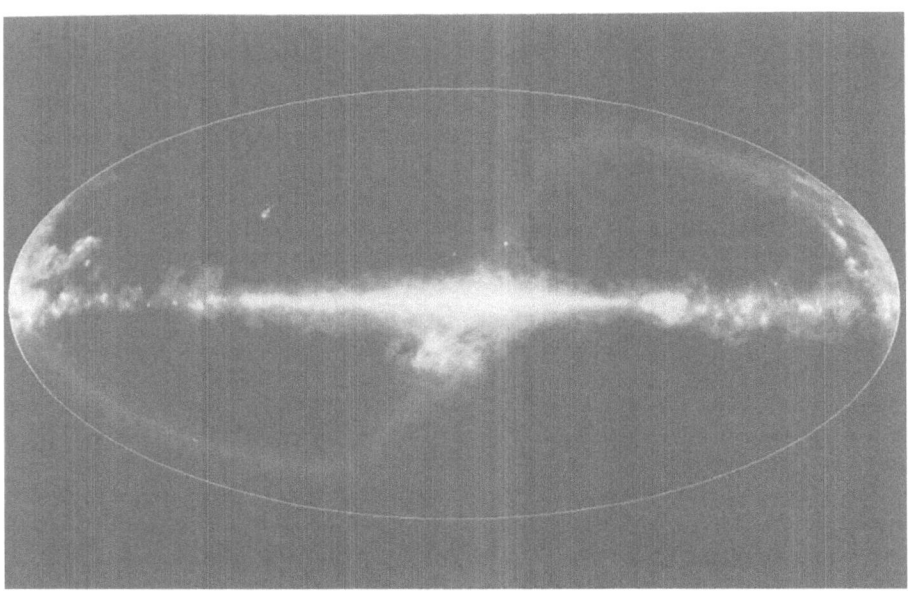

Figure 4. A map of the sky using infrared radiation to show interplanetary and interstellar dust particles. When these are understood we may deduce the existence of radiation from the first generation of stars and galaxies after the Big Bang.

and is instead quite invisible! This interpretation is controversial. We can deduce the existence of this dark matter from the gravity that it produces, but at present we have no other way to detect it. Hundreds of people are building experiments in laboratories around the world to search for this material, called "dark matter," but so far there is no sign of success. Cosmologists have been using this idea for many years, and have grown accustomed to it so it no longer seems so bizarre, but in truth we do not know what constitutes the dark matter. Most of the public debates on cosmology now center on deducing the properties and effects of this dark matter.

The third experiment on the COBE has not yet achieved its primary cosmological objectives. Michael Hauser at NASA - Goddard leads the team in a search for the infrared light from the first galaxies to form after the Big Bang. The light from the first galaxies was originally much like that from stars today but the cosmological expansion has shifted it to the red. For the very earliest galaxies the redshift is expected to be so substantial that it shifts the light past the red to the infrared. It is possible that some of this light was absorbed by dust and re-emitted at even more shifted frequencies. This happens to a portion of the light from our own sun and galaxy.

This search is difficult because there are many other sources of infrared radiation. This instrument has so far made maps of the sky using infrared light at wavelengths ranging from 0.001 millimeter to 0.3 millimeters. They have shown us the shape of the Milky Way galaxy in a striking new way, as shown in the picture at the top. The Earth is not located on this map because it is a map of the sky which surrounds us. We have also made maps that show the interplanetary dust, as shown on the bottom picture in blue. This dust is very interesting, as it is the debris from the asteroids that formed and then collided with one another between the orbits of Mars and Jupiter. Some of it also comes from the remains of comets which remain from the formation of the Solar System 4.5 billion years ago. Both the Milky Way and the interplanetary dust are emitting light at the same wavelengths where we think we may detect the first galaxies. We are now busily calculating how much comes from each source so we may deduce, like Sherlock Holmes, which and how much light must have come from those most distant early galaxies.

What does all this mean for science as a whole? It is of fundamental importance to understand the nature of matter, and we have strong evidence from the COBE that something is missing from our understanding - the invisible matter. One of the great hopes of science is that someday we will understand the interrelation of the four kinds of forces known to physicists. These four are the strong nuclear force (which holds protons and neutrons together in atomic nuclei), the weak nuclear force (which can break up a neutron into a proton, an electron, and a neutrino), the electromagnetic force (which governs the structure of atoms and molecules), and finally the gravitational force. It is already known that the electromagnetic and weak nuclear forces are really aspects of a single more basic force, and there is excellent progress toward unifying them with the strong nuclear force. Gravitational force is different from the others, if we believe Einstein - who said that gravity actually distorts space and time in a definite way, and that all mass and energy cause this distortion. The Big Bang should be an event in which all the forces acted under extreme temperatures and pressures which provide the well-spring of everything in the universe, so it is extremely important to understand it.

The Superconducting Supercollider (SSC) being built in Texas will probe these theories and events more strongly and deeply than any laboratory. The design of the SSC will take us to an energy scale that corresponds to conditions in the universe when it was about one millionth of a millionth of a second old. On a logarithmic scale this is the halfway point from the present universe to the highest energy and shortest length of time we can imagine. Halfway is always an important milestone, but the SSC also reaches into a qualitatively different region. The SSC experimenters will test the unification of the weak and electromagnetic forces and probe how particles acquire mass. I also believe they will find the key to why we have the small (one part in a billion) excess of matter over antimatter in the universe that makes our present universe possible. This will be a giant step forward and point the way towards the ultimate origin of space-time and unifications of forces. I suspect that the unification of forces is what provides the motive force for inflation - the engine that drove the formation of space-time and caused the universal expansion.

The structures discovered by the COBE DMR have much to tell us about this singular event forming our Universe. The DMR maps show us the imprints of structure on the cosmic background radiation when the universe was about 300,000 years old. The structures shown in the DMR maps were much larger than 300,000 light years across at that time. Thus even movement of matter and energy at the speed of light could not form or change these structures significantly in the 300,000 years since the origin of the Universe. The only way we know to account for such large structures is through the process of inflation. We can hypothesize that when the forces were united the universe existed in a different state where space-time was very substantial – unlike its condition today. If space-time has significant energy density itself, then the pressure of space is negative and the universe must expand at an exponentially increasing rate. A small region (less than a millionth of a millionth of a proton in size) could have expanded in a tiny fraction of a second to much larger than a 100 meters in size. This expansion factor is much more than the expansion factor in the 15 billion years since. Thus very small particle (quantum mechanical) fluctuations would have been stretched to macroscopic sizes of cosmological consequence. These originally small fluctuations from the origin of the universe are what have now grown to be the galaxies, clusters of galaxies, and larger scale structure observed today. This is a transcendant concept linking together the microscopic world of particle physics (e.g. the SSC) and the macroscopic world of astronomy and cosmology – the unification of forces and the beginning of space-time and its contents.

This is an important philosophical question and one reason cosmology is so enchanting today to scientists and the public alike around the world.

7 Technical Poster Session II

HEAT LEAK MEASUREMENTS AND THERMAL MODELING OF A MECHANICAL SUPPORT FOR A SSC BEAM TUBE LINER

J. Maddocks, J. Zbasnik, A. Yücel and R. Spidle

Superconducting Super Collider Laboratory[*]
2550 Beckleymeade Ave.
Dallas, TX 75237

INTRODUCTION

Hydrogen desorbed from the beam tube of the super collider by synchrotron radiation may adversely affect the luminosity lifetime of the proton beam.[1] One solution to this problem[2] is to place a distributed cryopump within the beam tube which will trap desorbed gasses.

Such a cryopump can be effected by attaching cryosorber to the cold (4 K) magnet bore tube. A concentric tube, or liner, centered within the magnet bore tube shields the cryosorber from the synchrotron radiation, and becomes the beam tube. By perforating a fraction of the liner surface with small (on the order of 1-3 mm) holes, the liner/cryosorber assembly becomes a distributed pump. The liner temperature may be allowed to equilibrate at a temperature close to that of the 4 K bore tube. However, actively stationing the liner at 80 K is of interest because the synchrotron radiation heat can then be deposited in the liquid nitrogen system. This, at least partially, decouples the allowable beam current from the helium cryogenic system. Active control is accomplished by means of 80 K helium flowing through a trace tube attached to the outside of the liner. A cross section of the magnet bore tube with an 80 K liner is shown in Figure 1.

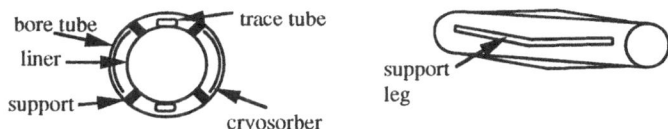

Figure 1. Cross section of magnet bore tube with liner, and schematic of support leg attached to liner.

DESCRIPTION OF HEAT LOADS

The SSC is the first proton machine in which the heat load of the synchrotron radiation will be significant. At baseline operation the synchrotron load is 0.14 W/m, all of which is deposited in the single phase 4 K helium. This represents about 40% of the total 4 K heat load. With an 80 K liner, however, the 4 K synchrotron radiation load is replaced by the static heat load of the liner, while the intercepted synchrotron load is transferred to the LN2 system.

*Operated by the Universities Research Association, Inc., for the U.S. Department of Energy under Contract No. DE-AC35-89ER40486

For an 80 K liner to be practical, it must not impose a heat load on the single phase helium that is greater than the baseline dynamic heat load of the synchrotron radiation. A conservative budget for the static heat load has been set at 0.07 W/m. Non-negligible contributions to the static heat load arise from conduction through mechanical supports, blackbody radiation, end conduction through interconnect pieces where the trace tube penetrates the 4 K bore tube, and conduction through the beam position monitor. By far the largest heat load arises from the support system, so that half of the static heat load budget (0.035 W/m) is allocated to it.

Mechanical supports must be optimized to provide accurate centering of the liner within the bore tube while having minimal cross-section for heat transport. Because the supports are located in the beam tube they must also be radiation resistant. This requirement precludes the use of insulating composites like G-10. Other material candidates, such as polyimides, have been shunned because relatively little is known of their low temperature radiation resistance, thermal conductivity, or mechanical properties. Thus a prototype support consists of four stainless steel legs, each with a rectangular cross-section, approximately 6 mm x 1.2 mm thick. The legs are bent slightly in the middle, as shown in Figure 1, in order to position the liner concentric to the bore tube. To provide the necessary rigidity, support legs must be less than 17.2 cm long, have both ends welded to the liner, and be spaced at 2 m intervals.

The resistance to heat flow of each leg is the sum of the stainless steel resistance and the contact resistance between the support and bore tube. Neglecting contact resistance for the moment, the heat leak through a single leg is given by

$$Q = 2A/l \int kdT \qquad (1)$$

where A is the cross sectional area of a support leg, l is half the leg length and k is the thermal conductivity of stainless steel. The integral is evaluated from 4 K to 80 K, and predicts a heat load of 0.06 W per leg. Assuming that all four legs are in contact with the bore tube, this results in a heat load of 0.12 W/m, which is less than the baseline synchrotron radiation load but more than the static heat load budget.

Since the material and geometry of the support are more or less fixed by other considerations, the contact resistance is the only remaining design parameter with which to reduce the heat load. Contact resistance can be expressed as,

$$R_{contact} = f(\Delta T, k, F, G) \qquad (2)$$

where ΔT is the temperature difference across the contact, k is the mean thermal conductivity of the materials in contact, F is the force with which the contacts are pressed together, and G is a geometric factor related to surface roughness. In general, $R_{contact}$ is not well known. For this reason, tests were conducted to measure both the heat leak of a prototypic support and the average resistance of a stainless to stainless contact.

DESCRIPTION OF EXPERIMENT

Apparatus

For reference, a schematic of the test apparatus is included in Figure 2. The support used is an early, prototype, which has been machined from a piece of Nitronic 40 stainless steel tubing. As indicated in Figure 2, it has two end rings which hold the four legs together in a single unit. The end rings fit tightly over a short piece of liner tubing and are thermally bonded to it with copper impregnated grease. The liner tube (no holes) is plated with 50 μm of copper on the inner surface. Four 2.2 kΩ metal film resistors serve as a heater. They are varnished into brass sleeves that are soldered to the Cu surface of the liner tube, and wired in series. Thermometers are located on the inside surface of each leg directly behind each contact point. In addition, one thermometer is located on the liner tube and another is located on one end ring near a support leg. Three thermometers are carbon ceramic resistors,[3] and three are standard silicon diodes.

The whole assembly is placed in a stainless steel vacuum container and immersed in a saturated helium bath at 4.2 K. Upon insertion of the support assembly into the vacuum container, the contact points of each leg are deflected by 0.75 mm. An independent measurement of the spring constant indicates that this deflection corresponds to an applied force per contact of approximately 12 to 14 N, about the same as that expected to arise from the weight of a liner.

Procedure

The insulating vacuum is pumped to approximately 1×10^{-5} Torr at room temperature using a diffusion pump. During the transfer of liquid, a helium leak detector remains open to the vacuum space. Once the apparatus is immersed in LHe and no leaks have been observed, the detector is valved off.

When the heater is turned on, temperatures slowly rise to steady state values. The time constant for steady state is on the order of a few hours for liner temperatures above 20 K. As a matter of practice, then, the heater is turned up to full capacity and the liner temperature monitored until it is close to the desired value. The heat is then slowly reduced until a steady temperature is achieved. This process takes approximately one half hour. Accuracy of the method was verified on several occasions by allowing the perceived steady state to remain for a few hours. In all cases, the temperatures changed by less than a few percent.

The heater is powered with a dc voltage supply. A precision resistor in series with the heater provides a means of measuring the current, which together with the applied voltage provides a measure of the heat input. When the steady state is achieved, the heat input as well as the temperatures are recorded. In this way, the heat leak of the support is measured for liner temperatures ranging from 10 K to 100 K.

Heat leak measurements are repeated, over the entire temperature range, for a series of contact surface preparations. The first data were obtained with the support and inner vacuum container wall in "as delivered" condition. This means simply that the contact surfaces were oxidized and the contact area undefined. Later, the inner wall of the vacuum container was polished to remove the oxide layer, and the contact area defined. The definition was accomplished by welding beads of stainless steel rod to the contact points, filing them to a specified 2 mm x 3 mm area with a jeweler's file, and lapping them slightly to seat well against the wall of the vacuum can. Since highly polished surfaces (submicron roughness) present the least resistance to heat flow, our treatment was intended to produce a uniformly rough surface.

DATA ANALYSIS

The total heat leak as a function of liner temperature for a number of cases is shown in Figure 3. Scatter in the data indicates the contact resistance is reproducible to approximately ± 15%. In addition, oxide on the "as delivered" contact surfaces seems to have little effect on the heat leak, as expected for rough surfaces. For a liner temperature of 80 K, the measured heat leak is 0.04 W per leg, a reduction of 30% from the estimate obtained by neglecting contact resistance.

As a check of the total heat leak measurement, the heat leak through each leg is computed from ΔT across the leg and the thermal conductivity of stainless steel. The sum over all four legs agrees reasonably well with the total heat leak measured at all temperatures. This indicates that there are no significant parallel heat paths, as for instance through gas conduction, radiation, or instrumentation leads.

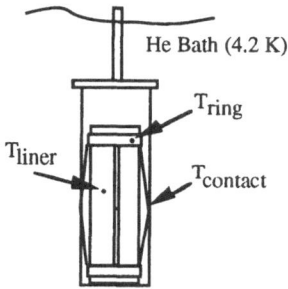

Figure 2. Schematic of apparatus.

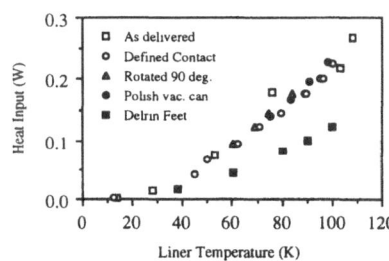

Figure 3. Measured heat leak of a dipole liner support.

The same method can be used to determine an average value of contact conductance (h_c), measured in units of W/m^{2}K. Assume the heat leak through each leg (Q_i) is one fourth of the measured total. Then Q_i is given by equation (1), with the integral evaluated from $T_{contact}$ to T_{liner}.

In general, contact conductance is found to observe a power law dependence on temperature of the form

$$h_c = \alpha T^n, \tag{3}$$

where α is a function of applied force and surface roughness. In addition,

$$Q_i = A_c \int h_c \, dT, \tag{4}$$

with A_c the nominal contact area. Substituting eq. (3) into eq. (4) and integrating, results in a expression for $T_{contact}$.

$$T_{contact} = [\,(n+1)Q_i/\alpha A_c + T_{bath}^{n+1}\,]^{1/(n+1)} \tag{5}$$

Practically speaking, T_{bath}^{n+1} is much less than the leading term and can be neglected. Average values of α and n are extracted from a log-log plot of $T_{contact}$ versus Q_i. The result indicates n = 1.5 and α= 0.75. The value of n compares favorably with stainless to stainless conductances published in the literature.[4] The value of α was determined from the largest values of Q_i so the data could be used to predict an upper bound for the heat leak. The measured value of 0.75 is about two orders of magnitude lower than published data[4] for smooth stainless to stainless surfaces under similar applied load. This indicates the potential sensitivity of the heat load to surface roughness.

In one final run, each contact point of a second support was fitted with a Delrin button. The buttons were attached by press fitting into holes drilled at the points of contact. Only the total heat leak and liner temperature were measured, so that no conductance can be extracted from the data. The data are included in Figure 3. Although Delrin is an unacceptable material for use in the bore tube, the data give an indication of the effect of attaching some sort of plastic buttons to the supports should an acceptable material be identified.

DISCUSSION

The applied contact force in the collider will be determined by three factors: preloading by compression of the supports at the time of insertion in the bore tube, further loading or unloading of support legs due to differential contraction during cooling, and compression of the lower legs and unloading of the upper legs due to the weight of the liner. A simple model of the differential contraction predicts a net reduction in the applied contact force after cooling to 4 K. Thus, if no preloading is required only the lower two support legs will be in contact with the bore tube when the collider is in operation. Under these circumstances, the support heat leak will only be a few percent over budget.

If preloading of the supports turns out to be necessary, the heat leak will be more than double that measured here, depending on the amount of preloading. It appears, however, that the contact resistance can also be increased by the addition of insulating buttons. More work is required, though, to identify an acceptable material and confirm that such is the case.

ACKNOWLEDGEMENTS

The authors would like to acknowledge the assistance of James Morris, Jim Briggs, and Larry Richards in acquiring equipment and collecting data.

REFERENCES

1. I. Maslennikov, et al., Photodesorption Experiments on SSC Collider Beam Tube Configurations, Proceedings of the fifteenth biennial Particle Accelerator Conference held in Washington D.C., May 17-20, 1993.
2. Q.S. Shu, et al., Prototype Liner System for the Interception of Synchrotron Radiation in a Half Cell of the SSCL Collider, Proceedings of the fifteenth biennial Particle Accelerator Conference held in Washington D.C., May 17-20, 1993.
3. V.I. Datskov, Technical Cryogenic Thermometers on the Basis of Commercial TVO Resistors, Communication 8-83-717 of the Joint Institute for Nuclear Research, Dubna (1983).
4. D.N. Lyon, and W.R. Parrish, Low Temperature Thermal Conductivities of Two High Compressive Strength Materials, *Cryogenics*, 7:21 (1967).

AN ENGINEERING DESIGN AND PRODUCTION
OF ELEMENTS OF RF POWER FEEDING AND
DISTRIBUTING SYSTEMS AT IHEP

V.V. Katalev, S.S. Kovalev,
I.I. Sulygin and I.M. Shalashov

Institute for High Energy Physics
142284, Protvino, Russia

INTRODUCTION

The 200 MHz accelerating structure for UNK [1] consists of identical RF units, each one containing of two cylinder-shaped cavities, a 3-dB hybrid, coupled with the cavities by coax lines, a rectangular waveguide and 850 kW power amplifier [2]. For a beam loading varying strongly, a part of the RF power reflected from the cavities is directed by the 3-dB hybrid to the dummy load, so a power amplifier sees actually the mathed load and may be connected to the accelerating structure by a long feeder.

Tne 1st ring of the UNK contains 8, and the 2nd one – 16 RF units. All 24 power amplifiers are placed in the RF-gallery by total length of 90 meters, running on the surface parallel to the UNK tunnel. Waveguide feeders going undeground to the depth of 25-30 m in 4 vertical shafts and running along the satellite tunnel parallel to the main ring tunnel at a distance of 5 m. Thus one waveguide feeder is 40-60 meters long, and a total length of the waveguide is more than 1 kilometer.

RECTANGULAR WAVEGUIDE FEEDER

The waveguide feeder of a cross-section of 860×300 mm^2 is assembled from 2-m straiht sections, a few shorter sections and matched E-plane and H-plane symmetrically truncated right-angle corners. Special compensators are inserted into the waveguide after each 4-5 linear sections to expedite the feeder mounting and to make up thermal expansion. Sections with reflectometers are placed in the input and in the output of each feeder for incident and reflected power control. The waveguide elements are bolted together in an appropriate sequence when a feeder is mounted; RF contacts between the elements are ensured by compressing their plane faces in flange joints.

Straight sections 2 m long (see fig.1) are used for mounting of long straight pieces of the feeders. The tolerances for dimensions specified on the drawing ensure the reflection coefficient value Γ_i of one waveguide joint not more than 0.01. For accelerating systems of the 1st and the 2nd ring of the UNK it is necessary to manufacture about 500 2-m straight sections.

Each section is welded from fours 5 mm thick aluminium alloy plates. The plates are given a preheat-treatment for relieving theirs internal stresses and assembled at a special jig into the waveguide body of nominal dimensions. The plates joint at the jig under welding is shown in fig.2. The mechanical assembly of the waveguide body at the jig ensures high-precicion dimensions, therefore the main problem is to avoid deformations of the plates under welding. Electron-beam welding is not available at IHEP, so helium gas arc welding is used. Experimental optimization of the welding regime and the weld configuration had carried out.

Massive rectangular flanges welded from aluminium alloy bar of cross-section 40×40 mm^2 are pulled from both ends over the ready-made waveguide body. The flanges are joined to the body by short staggered welds. Then the both ends of the sections are faced by machining to get quality contact plane surfaces.

Supercollider 5, Edited by P. Hale
Plenum Press, New York, 1994

Right-angle corners, shown in fig.3, are produced in the same technological process, the feature is their assembling from figured plates at the special jigs. Reflection coefficient of corners is defined by truncation dimensions. For the dimensions, shown in fig.3 it does not exeed the maximum reflection coefficient of straight sections joint: $\Gamma_{E,H} \simeq \Gamma_i \leq 0.01$.

Compensator – short flexible waveguide section (see fig.4) , containing the two short rigid waveguide sections of identical characteristic impedance, which put one into another and connected side to side by bended thin flexible metal plates. The plate bends "roll" to new place freely, without any tensions, when a shift or misalighmeht of the rigid sections occurs.

Reflectometer – the waveguide section 0.5 m long with two loop-type directional couplers, installed on a wide side of the waveguide. Coupling value is 53 or 57 dB, and directivity more then 33 dB.

100 2-m long straight sections, 35 E-corners, 40 H-corners for the 1st ring of the UNK have been produced up to now.

A special automated plant with inductive sensors has been designed for measuring of cross-sectional dimensions of the waveguide sections with an accuracy ±0.2 mm. The measurements data can be represented after computer procesing in a colour "map", on which areas out of tolerance are easy visible.

The waveguide measuring line and a movable short-circuit plunger both 2 m long were constructed at RF lab for matching adjustment of the waveguide elements. Random testing of produced waveguide elements by Weissfloch's method showed for all corners and bended compensators the value of voltage standing wave ratio VSWR = $(1.01 \div 1.02) \pm 0.004$, i.e. ideal matching within an accuracy of measurements, which is defined by measuring line own VSWR value.

The waveguide trace 25 m long with 20 elements was constructed at the RF lab for measurements of parameters of long feeders. The measured values of VSWR = 1.15 at frequency 200 MHz is in agreement with statistical estimates of a reflection coefficient of the waveguide trace for magnitude of reflection coefficient of each discontinuity at the waveguide joints $\Gamma_i \leq 0.01$.

Attenuation constant for the waveguide had been defined by resonance method in a short quarter wave resonator and in a long resonator with many joints; the measured value is in a good agreement with calculated attenuation constant and it is the same in both cases: $\alpha = 1.57 \cdot 10^{-4} \ m^{-1}$, that presents an evidence of high-quality RF contacts in the waveguide joints. For such a value of attenuation constant the waveguide loss per meter is 315 W at an RF power flow of 1 MW. A power 500 W was transfered into a quarter wave resonator by length of about 1.5 m to imitate the power flow 1 MW regime; the temperature of narrow walls was risen up to 30-32°C, while the outside temperature was 19-20°C.

A like rectangular waveguide feeder 140 m long mounted from more then 100 elements is in operation at the particle rebunching station at accelerator U-70 since 1987 [3]. There was no damages or deterioration of the waveguide RF characteristics during the all six years of its usage.

3-DB HYBRID AND WAVEGUIDE-TO-COAX JUNCTIONS

The 3-dB short-slot hybrids are used as combiners of an output RF power in the RF power amplifiers [2], and as dividers and matching devices in the accelerating structure. The hybrid is the two waveguides linked together by a narrow wall. A scattering matrix of a short-slot hybrid and its dimensions required for the hybrid balance had been calculated precisely with Dorfman analysis. A hybrid was choosen without matching reactive elements, and thereby for the frequency 200 MHz a waveguide wide wall dimension $a = 0.86$ m was defined.

The two alternatives of the short-slot hybrid design have been developed: the 1st construction is welded from 10 mm thick aluminium alloy plates by wide of about 1.75 m, and another one constructed from two waveguides welded from 5 mm thick aluminium alloy plates, without one of its narrow walls, bolted together through lengthwise bars welded to the waveguides. The first design calls for a special production procedure, but the second can be produced at the procedure for manufacturing of the waveguide sections. Both the alternatives have been produced and have identical RF characteristics.

The hybrid has one rectangular waveguide port and the three coax ports: there are two waveguide-to-coax junctions for coupling with the cavities (or with an amplifier outputs) and another one for dummy load. The averall dimensions of the short-slot hybrid with junctions are about 2m×2.5m×0.5m, the slot length is 1120 mm. The hybrid used for combining RF power has simple in design junctions with a probe antenna. For hybrids at accelerating units the special intricate (a loop with a gap and a probe) waveguide-to-coax junction had been developed [4]; one of the two identical junctions in the hybrid ports may be used as 180° phase switch to change the direction of particle acceleration in the first ring of the UNK.

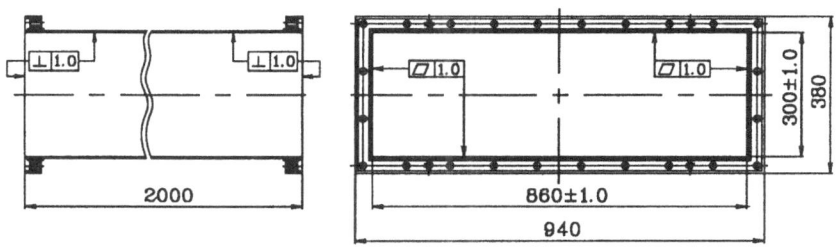

Figure 1. Linear section.

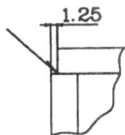

Figure 2. Plates joint for welding.

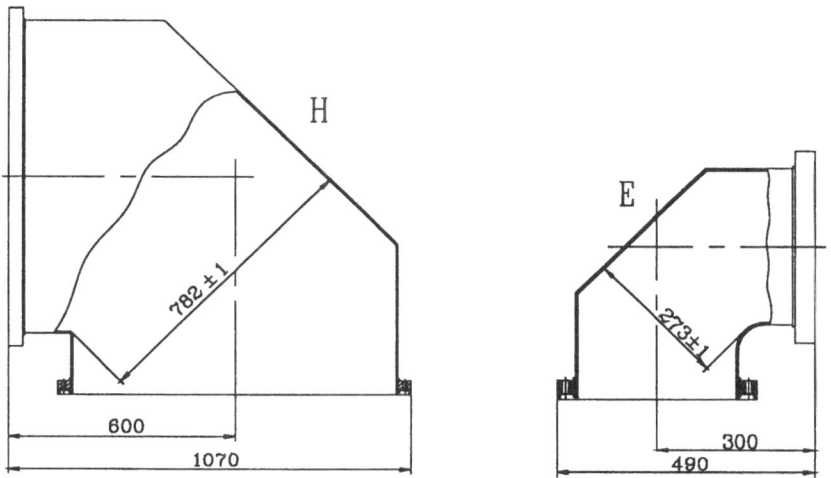

Figure 3. Matched H–plane and E–plane symmetrically truncated right–angle corners.

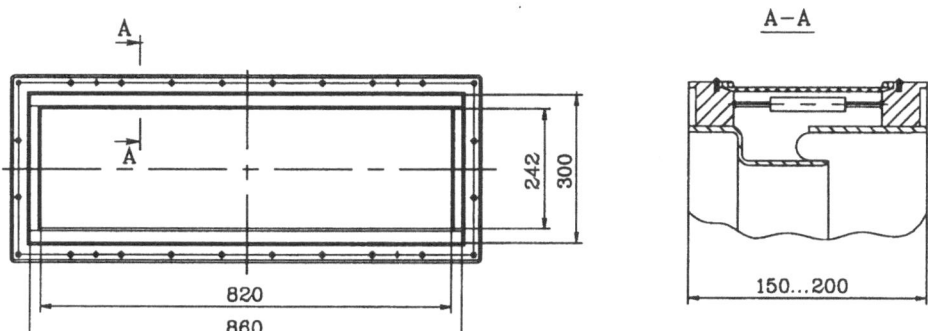

Figure 4. Compensator.

As measured, the hybrid with junctions has a coupling of (3±0.05) dB and insulation more than 32 dB, VSWR is less than 1.05 in all ports at the frequency 200 MHz.

COAX LINES

Copper coax lines having a diameter of the outer conductor of 200 mm and that of the inner one of 85 mm are used to couple the hybrid with the cavities. The coupling coax line is about 2 m long for the accelerating unit of the first ring and about 1.5 m long for that of the second one.

A flange joint design of the copper coax line elements has been developed [5], which does not requires for producing of the coax line such a complicate and expensive procedures as brazing and welding of copper alloys parts; it lowered comprehensively a cost of manufacturing of the coax line. The flanges are screwed on the outer conductor of the coax and copper tube ends are expanded, furnishing RF contact surfaces. The inner conductors of the coax are hollow copper tubes with machined ends. A metal ring spacer with dielectric insulator supporting the inner conductor connectors is inserted into recesses on flange faces. Berillium bronze helical spring laing in ring grooves is used for RF contacts between the coax conductors.

Stright coax sections, matched corners, loop-type reflectometers have been developed and are available for manufactoring.

DUMMY LOAD

The dummy load must dissipate the average power reflected from the cavities during the acceleration cycle, i.e. less than 300 KW. Two cylindrical 50 Ω commercial water-cooled resistors, each designed to dissipate 150 KW on the average, in exponential-shaped screen are used as coaxial matched loads. The power is directed to the loads via two matched probe waveguide-to-coax junctions in the insulated port of the hybrid.

The alternative design is water-coloumn load, in which distillate water, used conventionaly at cooling systems , is used for dissipation of RF power. The problem is that dielectric loss tangent is low in water at frequency 200 MHz, and therefore a load must be very long (10 meters for ensuring VSWR≤1.05.) The load with stright stainless steel coax line by length of 8 m , of cross-section of 64/16 mm, and with quarter wave transformer in the input was examined and worked without problems at the average RF power up to 100 kW, ensuring VSWR = 1.1÷1.15.

Our tentative results in modeling of folded water-coloumn loads of the lower averall dimensions encourage us to continue this work. We are going to construct in future a water-coloumn load, which able dissipate an average RF power 500 -1000 kW.

SUMMARY

As far as our knowledge goes, the frequency 360 MHz has been choosen for the SSC as a guide, but "other multiple of 60 MHz, between 240 and 480 MHz, may also be considered if they offer technical advantages" [6].

In the case of the choice of 240 MHz or 300 MHz the waveguide having been designed at IHEP is applicable directly. Copper coax line elements with cross section of 200/85 mm are available for application up to almost 700 MHz. It is possible for us to design and produce a waveguide and coax elements of the lower dimensions using the similar technical procedures.

A water-coloumn load is a good choice for the frequency 360 MHz: only of 3 m stright length coloumn is needed to obtain VSWR = 1.05, and the load may have an average RF power dissipation ability up to 1 MW.

We would adopt waveguide and coax high power RF components design according to each user requirements.

References
[1] V.V.Katalev et al. Proceedings of the EPAC-92, Berlin, 1992, vol.2, pp.1197-1199.
[2] I.A.Kvashonkin et al. EPAC-92, vol.2, pp.1200-1202.
[3] E.N.Butriakov et al. Proceedings of the X All-Union Particle Accelerator Conference, Dubna, 1987, vol.1, p.193.
[4] S.S.Kovalev, S.A.Kuznetsov. Author's Certificate 1125678, USSR.
[5] I.G.Maltsev et al. Author's Certificate 1479974, USSR.
[6] Georg Schaffer. Proceedings of the EPAC-92, Berlin, 1992, vol.2, p.1251.

SIMULATION AS A TOOL FOR PLANNING THE SSC COLLIDER SPOOL PIECE PRODUCTION

E. Rodriguez[1] and J. Everton[2]

[1]Martin Marietta Astronautics Group
P.O. Box 179
Denver, Colorado 80201

[2]AutoSimulations
P.O. Box 307
Bountiful, Utah 84011-0307

Abstract: Design of the SSC collider ring standard spool piece nears completion, and two production prototype units are being fabricated. Spool pieces are complex cryogenic components of the SSC collider ring which provide interface and/or control of cryogens, vacuum, power, instrumentation, and quench protection and also house the corrector magnets and the beam position monitor. The use of 3D animation in conjunction with simulation in capacity planning, scheduling, and resource utilization and management for production will be discussed.

INTRODUCTION

The producer of the spool pieces for the SSC collider ring is faced with the problem of needing to setup a moderate rate production operation to build and assemble approximately 1400 standard spool pieces at a rate of 42 units per month. Spool pieces are composed of approximately 1700 individual parts and 16 major systems. This problem in itself is a challenge, but as illustrated in Figure 1 the producer must within 3 months of the first preproduction unit delivery increase spool piece deliveries to 5 units per month before shutting down the line prior to production build. For production the producer must increase deliveries to 15 units per month 6 months after start of production deliveries. Subsequently the rate increases to 42 units per month by month 17 for full rate production..

These deliveries can only be accomplished if a well thought out operations plan of the production system can be tested ahead of time to assure all resources (fixed, variable and consumable) are available at the right time and in the proper quantities to meet program requirements. It would be nice to have the facilities, equipment, material, components and personnel available to perform "what if" scenarios to ascertain the proper mix of resources, but this is impractical from a cost standpoint. Simulation is one of the most powerful and cost effective tools available today to perform the analysis required to assure the production system and operations plan is achievable.

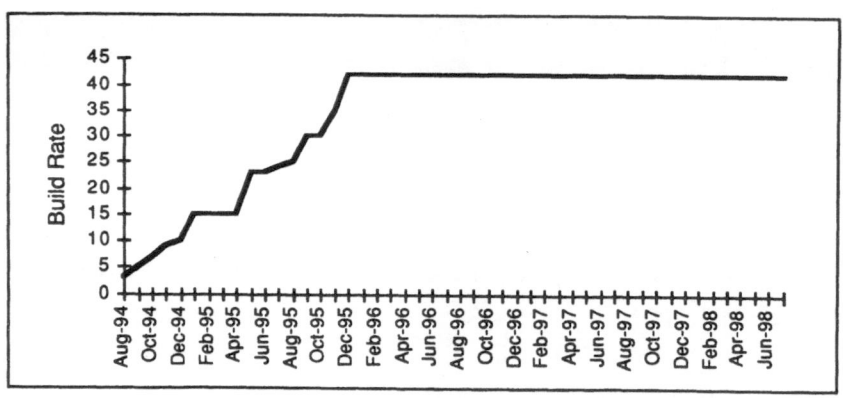

Figure 1. Potential build and delivery rates for SSC collider ring spool pieces

Computer simulations are not new. They have been used for over 20 years to help make these types of planning decisions. Martin Marietta's past experience with other program startups has proven that discrete computer simulations are invaluable. Computer simulations are discrete mathematical models of the manufacturing systems. The model is used to perform "what ifs" and is a tool to understand the interaction of the various components of the manufacturing system. In the past Martin Marietta has used GPSS, SLAM and SIMAN simulation software. Today Martin Marietta is using AutoSimulations' AutoMod software to develop simulations for the spool piece production system because of its versatility and the ability of showing the running of the simulation in a 3-D animation.

SIMULATION

The simulation is composed of the following three parts: the experiment, the model and the animation. Since simulation is the ability of being able to perform "what ifs" or experiments by the use of a model it is essential that the first step in the process is to design the experiment[1]. Our experiment was designed to perform the following analyses:

1. Prediction - Estimating the performance of the proposed manufacturing system under various conditions (schedule, learning curve and performance index).
2. Optimization of the System - Determining how many of the defined factors (resources) of the manufacturing system are required to produce the best system response.
3. Functional Relationships - Estimating the relationship between significant factors (resources) and the system's response.
4. Bottleneck Analysis - Ascertaining the location of bottlenecks within the system which restrict entity (spool pieces) flow through the system.

Table 1 is an example of the bottlenecks that were identified by the simulation as well as the time the entities spent at the bottleneck. This type of analysis helps identify how many and where additional resources are required to make the manufacturing operations run smoothly and without interruptions.

552

Table 1. Time spent at convergence bottlenecks

Station	Mean (Hrs.)	Std. Deviation (Hrs.)	Minimum (Hrs.)	Maximum (Hrs.)
Cold Mass - Instrumentation	9.59341	2.31730	6.67995	17.41982
Cold Mass - Final Assembly	19.90232	6.11239	13.39491	40.58859
Cold Mass - Lead End Sub Assembly	11.23086	1.72791	9.00156	15.08516
Cold Mass - Return End Sub Assembly	9.94380	1.76867	7.55996	13.47517

Model building with AutoSched is simply a matter of providing the definition of the factory, products and desired production requirements in data files; and providing a set of rules for operating the factory. The following data had to be supplied to define the system being modeled:

1. Production Resources - stations, storage operators and tools
2. Products - parts and routings
3. Production Requirements - orders and lots.
4. Operating Rules - task selection rules, batching rules and calendars.

The aforementioned data was developed and input using a Microsoft® Excel spread sheet. Table 2 is an example of some of the data that was input to develop the model.

Table 2. Typical spread sheet input used to develop the model

ROUTE	STEP	DESCRIPTION	STATION	PTIME	UNITS	TOOL	OPERATOR
He_Return	1	Receive Tube Matl	receiving	2.0	hrs		rec_inspect
	2	Cut Tube to Length	tf_bench	0.1	hrs	tube_cutter_1	tube_fabricator
	3	Bend Tube	tf_bench	0.2	hrs	tube_bender_1	tube_fabricator
	4	Weld Tube Section	tf_bench	0.2	hrs	orbital_tube_tf	tube_fabricator
	5	Cold Shock	cryo_coffin_1	0.1	hrs		tube_fabricator
	6	Pressure Test	pressure_coffin	0.2	hrs		tube_fabricator
	7	Clean and Identify	pressure_coffin	0.1	hrs		tube_fabricator
	8	Inspect	pressure_coffin	0.1	hrs		inspector
	9	Move to next assy	delay	0.0	hrs		expeditor

By using the AutoSched software package we also have the ability to perform finite capacity planning and scheduling. Since the model was developed to mimic our real life manufacturing system, data from the MRP and shop floor control systems can be input back into AutoSched. AutoSched can then provide schedule performance data such as production forecasts, lead times and alternate schedules and operating strategies.[2]

The final part of the simulation is the animation. AutoMod uses true 3-D graphic animation that reflects our 3-D world. The animation moves simulation from a list of numbers on a printout that only an operations research person can understand to a full coloar graphic image which can be understool by engineers and managers. This makes the model easy to understand because it looks like and operates like the planned manufacturing system. In the case of the spool piece production system the animation had the layout of the planned facility and the entities comprising the spool piece were routed though the system (Figure 2). The other advantage of using animation is the ease of debugging a simulation. Instead of having to go through a tab run tracking the statistics of a trace entity, the modeler can view the entity in the animation and determine if the model is correct.

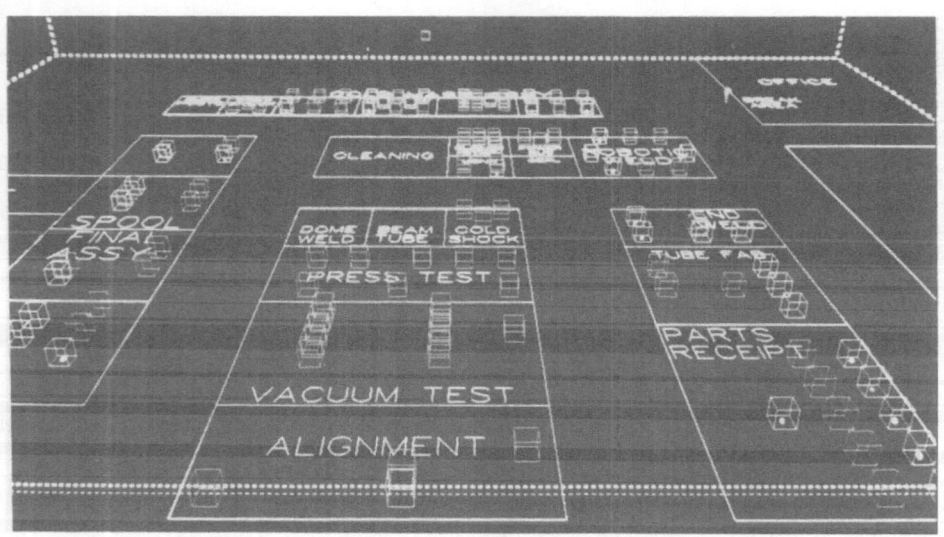

Figure 2. 3-D animation of the spool piece manufacturing system

SUMMARY

The Spool Piece Program is a good example of a situation where simulation is a valuable tool for evaluating and testing out a manufacturing system and operating plan prior to committing capital and tooling dollars. Resource requirements whether they be in the form of equipment, tools, personnel or materials can be identified in relationship to the SSC's desired delivery schedule. The manufacturing system can also be optimized by performing experiments with the simulation model, thus eliminating the need to experiment with the real system when it is up and operational.

Simulation data can be fed back into the management systems used to control the spool piece manufacturing system.

The use of 3D animation in conjunction with simulation is of great benefit in being able to view how the simulation model and experiment are interacting.

REFERENCES
1. C. D. Pegden, R. R. Shannon, and R. P. Sadowski, "Introduction to Simulation Using SIMAN," McGraw-Hill, Inc., New York (1990)
2. M. B. Thompson, "Simulaiton Based Scheduling" Technical Report #MS90-348. Society of Manufactuirng Engineers, presented at the International Manufacturing Technology Conference (Chicago, Illinois, Sept 7)

554

CRYOGENIC CHARACTERISTICS OF THE SSC ACCELERATOR SYSTEMS STRING TEST (ASST)

W. Burgett, D. Franks, P. Kraushaar, M. Levin, M. McAshan,
A. McInturff, R. Pletzer, D. Revell, W. Robinson, V. Saladin,
G. Shuy, R. Smellie, and J. G. Weisend II

Superconducting Super Collider Laboratory*
2550 Beckleymeade Ave.
Dallas, Texas, 75237

ABSTRACT

A series of static and dynamic tests of the ASST were conducted from July 1992 to the present. These tests included heat leak measurements, as well as, temperature and pressure profiles obtained during quench testing of the string. An accurate assessment of heat leak measurements of collider components requires a stable thermal environment with the minimization of end effects. The string test provides the ideal static environment necessary to conduct these measurements. This report summarizes the results of the heat leak measurements conducted on the cold mass, 20K, and 80K shields of the dipoles used in the ASST assembly. We also report on the rapidly changing temperatures and pressures recorded during the string quench tests.

I. DESCRIPTION OF THE EXPERIMENT

For this evaluation the ASST was a half-cell of a regular structure of the collider and consisted of five full length 50 mm aperture dipoles, one full length 40 mm aperture quadrupole and three spool pieces.[1] The dipole magnets in order of their location in the string were DCA313, DCA314, DCA319, DCA315, DCA316, the quadrupole was QC403. The spool pieces installed in the string were: one-half of the standard collider feed spool (HSPRF), a standard recooler spool (SPR), and a special end spool (SPE).[2] All three spools were equipped with quench valves.

The sensor outputs are monitored by the Research Instrumentation Data Acquisition System (RIDAS). This system consists of amplifiers, multiplexers and high precision digital multimeters. All resistance measurements were made by simultaneously measuring the sensor voltage and current. Polarity switching of the thermometer excitation current was employed to remove thermal emfs and amplifier offsets. All sensors were read once every 5 minutes. During quench this rate was increased to 2 kHz for certain sensors. Approximately 300 RIDAS channels were used to support temperature/pressure monitoring.

The arrangements of sensors used in the ASST thermal measurements is shown on Fig.1. Carbon glass resistors with an accuracy of 2.5 mK were used at the lead end of each dipole's cold mass. At the same locations, the cold pressure transducers (Siemens KPY46A) were installed as well. Germanium resistors with an accuracy of 15 mK were installed at the inlet and outlet of each dipole's 20 K shield cooling line. The 80 K shield cooling line contained platinum resistors with an accuracy of 0.5 K at the inlet and outlet of each dipole. In the single phase (cold mass), 20 K and 80 K circuits of each dipole are resistive heaters with which a known

* Operated by the Universities Research Association, Inc., for the U. S. Department of Energy under Contract No. DE-AC35-89ER40486.

amount of heat may be placed into the dipole as a test of the heat leak measuring technique. Silicon diodes and warm pressure transducers are placed in each of the cryogenic pipes at the feed and end spools to determine the inlet and outlet conditions of the cryogenic flows. A Venturi flow rate meter is installed in the single phase line and room temperature volumetric gas meters are installed in the 20 K and 80 K shield lines.

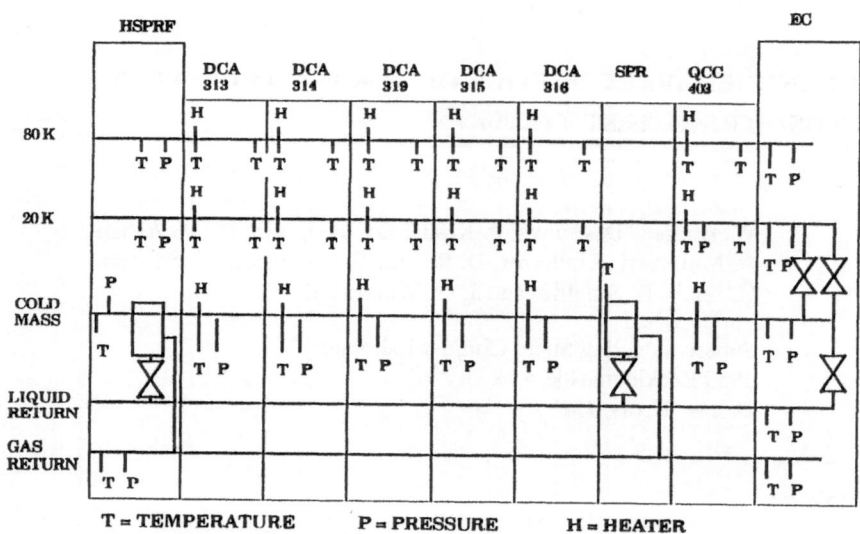

Fig.1. ASST Reduced Instrumentation Schematic.

II. EXPERIMENTAL RESULTS

1. CRYOSTAT HEAT LEAK

The heat leak measurements were performed during two experimental runs on the string (July-Sept. 1992, Nov.-Dec. 1992). The system was allowed to come to thermal equilibrium and after that the inlet and outlet temperatures, flow rates and pressures of the cryogens flowing through the cold mass, 20 K and 80 K shields were measuring. The heat leak was calculated with help of the energy balance equation:

$$Q=G*(h_2-h_1),$$

where: Q - heat leak;
  G - the mass flow rate of cryogen through the magnet;
  h_1 and h_2 - enthalpies of the cryogens at the inlet and outlet points of the magnets.
In Table 1 the dipole plus an interconnect heat leak measurement results are presented.

Table 1. The dipole plus an interconnect heat leak results. (All values are in Watts.)

		D313	D314	D319	D315	D316
Cold Mass						
	RUN 1	9.47	2.40	0.42	1.66	4.07
	RUN 2	9.00	2.40	----	1.46	2.15
20K*	RUN 2	9.58	4.83	----	5.18	3.8
80K	RUN 1	26.5	29.0	26.4	16.0	20.3
	RUN 2	22.6	24.9	22.7	12.5	19.0

* 2 phase flow in 80 K shield

Due to operational problems there were no 20 K heat leak data taken in Run #1. As is clear from the cold mass results, the first and fifth dipoles have much higher heat leak than middle magnets. This may be explained by end effects, i.e. heat leaking from the end devices. These devices are spools and their functional requirements result in significantly higher heat leaks than the dipoles. Additional evidence of the presence of end effects is the significant reduction of heat leak in DCA316 between two runs. A thermal short in the end spool was fixed between the two runs.

It's important to realize that the temperature rise across a single dipole in the case of the cold mass measurements is approximately the same size as the absolute accuracy of the thermometers used in the measurement.

The results shown are the average of 2-6 different data sets which were taken during the experimental runs. Mass flow rates used in the experiments were typically as follows: cold mass - 50 g/s, 20K shield - 1 g/s, 80K shield - 15 g/s.

In order to reduce the resulting uncertainty in these measurements as well as the end effects we excluded from consideration the dipoles at either end of the string and looked at the temperature rise across the middle 3 dipoles. From this data the average heat leaks for a dipole plus an interconnect was calculated. Table 2 shows a comparison of these results with the budgeted values.

Table 2. Average heat leaks for the dipoles plus interconnects.

	RUN 1	RUN 2	BUDGET
Cold Mass	1.3 W	1.4 W	0.36 W
	+/- 33%	+/- 28 %	
20K shield*	N/A	5.59 W	5.06 W
		+/- 1 %	
80K shield	28.0 W	24.5 W	37.0 W
	+/- 14%	+/- 16 %	

* 2 phase flow in 80K shield

The uncertainties shown with these results in Table 2 are calculated by comparing the temperature rise across the middle three dipoles with the absolute accuracy of the temperature sensors. The results in Table 2 show that the average value of cold mass heat leak is significantly over budget. The reason for this is not yet clear. It might be problems with the support posts or thermal shorts between the 20 K shield and the cold mass. Indications are that such a thermal short may exist at the beginning of DCA319.

Inaccuracy of the flow rate measurements and other systematic problems may be estimated by placing a known amount of heat into the cryostat and comparing this to the additional heat measured by the increase of the temperature rise across the dipoles. Several heater tests were conducted and they have showed that these errors are less than 10 % in the cold mass and 80 K cases and less than 30 % in the 20 K case.

2. QUENCH RESULTS

The power test program included a number of quenches at different currents. At each current level, quenches were induced in particular magnets to study the performance of the system and to monitor string parameters.

One of the important goals of the power tests is to measure the magnitude and the time profile of the pressure wave caused by the quenching magnets. In Fig.2 the typical pressure profile is shown as a function of time after quench at 6.5 kA induced by a strip heater in DCA316. The quench protection system then fired the strip heaters in the DCA315 dipole and in the quadrupole. The second pressure peak at 0.9 sec occured when the other three magnets thermally quenched. The maximum pressure of about 205 psia occured during that quench and was observed in dipole DCA313, as shown in Fig.2. The peak quench pressure observed during full current quenches ranged between 180 and 205 psia. The magnets are currently certified to 310 psia and the maximum allowable working pressure is 250 psia. Thus, the observed pressures are within the design limits. Moreover, the typical full current quench pressure was around 180 psia, and the 205 psia is the highest observed to date.

The cold mass temperature during the same quench is shown in Fig.3. The highest temperature is about 6.5 K, and was observed in DCA313, DCA319 and DCA316 dipoles. The temperature sensor in the dipole DCA314 didn't work at that time.

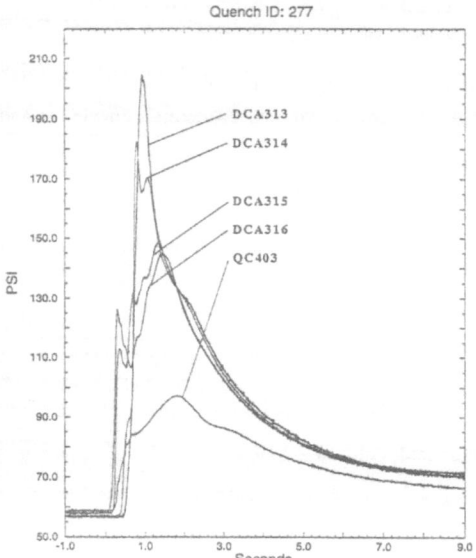

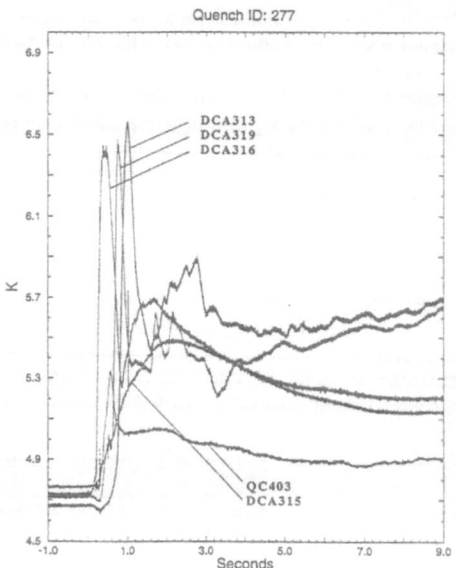

Fig.2 Cold Mass Pressure During Quench.　　　**Fig.3** Cold Mass Temperatures During Quench

SUMMARY

During the assembly of these prototypical dipole magnets considerable difficulty was experienced with the positioning of the thermal shields due to the lack of the adequate tooling. It is not known if this has affected the cold mass heat leak, which is significantly over budget. In the planned third run of the ASST experiment, the fourth dipole will be replaced with DCA323, which is highly instrumented. This additional instrumentation and modeling that is currently under way should help us understand and fix the cause of the higher than expected cold mass heat leak. It should be pointed out that small changes in cryostat systems can have large impacts on heat leaks of this magnitude. In the case of the 40 mm dipoles at the FNAL string test facility the cold mass heat leak was within budget.[3]

ACKNOWLEDGEMENTS

The authors would like to thank the refrigerator crew and many technicians that worked under direction of H.Carter, M.Hentges and C.White.

REFERENCES

1. W. Burgett et al, Full power Test of a String of Magnets Comprising a Half-Cell of the Superconducting Super Collider, SSCL-Preprint-162 (1992), accepted for publication in "Particle Accelerators".
2. D.Clark et al, SSC Spool Piece Design, "Proc of the Third International Industrial Symposium on the Super Collider", Plenum Press, N.Y. (1991)
3. A.D. McInturff et al, Measured Control Characteristics of the Half-Cell 40 mm Aperture Magnet String, "Proc.of the Fourth International Industrial Symposium on the Super Collider", Plenum Press, N.Y. (1992)

DEVELOPMENT OF A CORRECTOR ELEMENT POWER LEAD

John J. Wollan,[1] Sam Lucas,[1] Tom Byram,[2] and Ken Efferson[2]

[1]Martin Marietta Astronautics Group
P.O. Box 179, Denver, Colorado 80201

[2]American Magnetics, Inc.
112 Flint Road, Oak Ridge, Tennessee 37831

INTRODUCTION

In the SSC main collider up to five orders of superconducting corrector magnets will be incorporated into the spool pieces to correct for component and machine aberrations. Each order of corrector magnet will be separately powered. This will require a corrector element power lead (CEPL) with a single, room temperature-to-4K interface with multiple leads. CEPLs represent a potentially significant heat load to the 4K helium and must be vapor-cooled to meet heat leak requirements. Analysis and design of vapor-cooled power leads for cryogenic systems have been extensively documented in the literature [1,2]. The purpose of this paper is to focus on the requirements and associated issues for this particular application.

Two design alternatives exist for this application. One utilizes high temperature superconductor (HTS) for the low temperature end of the leads; the other utilizes standard materials throughout. Leads utilizing HTS have heat loads at 4K which are 25 to 50% lower than conventional lead technology, which makes them attractive. However, the technology is still under development, long-term reliability is yet to be proven, and radiation effects may not be thoroughly understood. Standard technology leads, on the other hand, are well proven, highly reliable, and considerably less expensive. Although refrigeration operating costs can be reduced appreciably with HTS leads, it is prudent to develop an alternate CEPL based on conventional technology. This paper will discuss a CEPL based on standard lead technology. Other papers at this conference[3,4] discuss the HTS option.

Performance requirements for the CEPL have been developed by the SSCL and the Phase I spool piece contractor teams. The requirements are:

Number of Leads	12
Maximum Current/Lead	100A
Maximum Current/All Leads	1000A
Maximum Ramp Rate	1A/s
Fault Pressure	3.0MPa
Voltage Standoff In Air at 300K	3000VDC, <1µA

In addition to these explicit requirements from the Level 4 spool piece performance specifications, there are several derived or implied requirements. First, there is a 4K heat budget for the spool. A related requirement is allowed venting of 4K helium to 300K of 74 mgm/s. The CEPL allocation is approximately 60 mgm/s. This flow must support operation with different currents in different leads. There are also other potential requirements which could become part of the CEPL specification. Operation under fault - the loss of cooling due to either internal flow blockages or a failure of an external flow control. Testing corrector magnets at currents above the 100A operating limit, possibly as high as 140A. There may also be the need to assure no build up of ice or frost on the top of the CEPL.

The combination of these requirements impose constraints on the design.

DESIGN

Many design options exist upon which to base a conventional technology CEPL. Our design is based on commercially available, standard, vapor-cooled leads developed and manufactured by American Magnetics, Inc. [1] Leads based on this technology have been proven in many applications, including 36, 2KA leads and 74, 5KA leads operating in the Fermi Tevatron for the last six years.[5]

For this application design challenges include: 1) the need for very high heat transfer efficiency between the cooling gas and multiple leads to minimize heat leak to the 4K system, and 2) the need to have adequate cooling flow for each lead and to maintain it under conditions of different currents in different leads. Uniform cooling could be achieved by either cross flow of all vent gas across all leads or by venting gas up each individual lead, with flows controlled by maintaining a constant pressure drop across each lead. The latter approach was the one chosen.

A prototype lead (Fig. 1) based on this technology has been developed, fabricated and tested. The active lead length is 51.7 cm, which is driven by the spool piece internal and external envelope constraints. Each lead has a total copper cross-sectional area of 4.08 x 10^{-2} cm^2. made from multiple, #38 silver-coated copper wires. The basic lead design was modified to establish the necessary pressure drops. Provision was also made to assure that individual leads could not be plugged due to possible contamination in the 4K helium supply.

Voltage breakdown is a strong function of configuration. From the Paschen curve for air at ambient temperature and pressure,[6] a 3KV standoff voltage accounting for geometry effects yields a minimum required gap of about 3.0 mm which is accounted for in our design.

Design concepts to address the potential of frost or icing on the top of the lead assembly have been developed but were not implemented on this prototype.

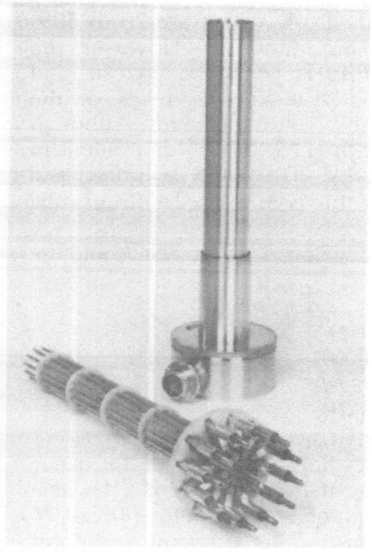

Figure 1. Prototype CEPL.

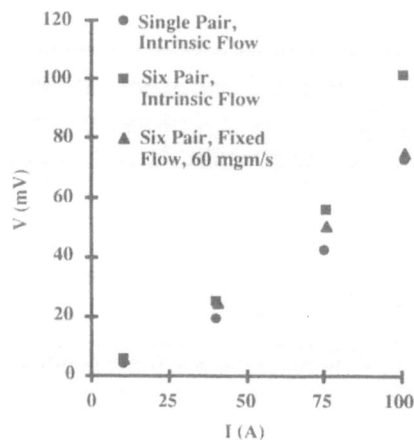

Figure 2. Voltage vs. current, three operating conditions.

ANALYSIS

A thermal/fluid model of a CEPL has been developed based upon SINDA/FLUINT, a thermal/fluid code developed by Martin Marietta under contract to NASA. The model includes temperature and pressure dependent properties of helium,[9] and temperature dependent thermal properties of copper and stainless steel. It calculates the copper electrical resistivity using a polynomial expression $P = f(\sum A_n T^n)$ from McAshan.[8]

The code output includes: 1) temperature along each lead length, 2) helium temperature and pressure as a function of position, and 3) voltages across each lead. Analyses simulating the various test conditions have been made.

TEST

The prototype CEPL was tested in a 142 cm high, 25.3 cm diameter dewar in one atmosphere, two-phase helium. Pairs of leads were connected at the cold end with 61 cm long, stabilized NbTi superconducting bus. At the top, lead pairs were connected in series with pairs of 1-0 cables. Voltage taps were connected to the top of each lead and to the power bus joints at the bottom. Boil off gas could be vented through the CEPL and/or through a vent line in the dewar lid. These flows were measured with a Sierra Instruments, Top Trak model 821-2-V1 flow meter. Dewar pressure was measured with a Dwyer Series 475 digital manometer.

Six series of tests were run:
- Baseline heat leak for the dewar, the housing, and the full CEPL assembly
- V(I) for a single lead
- V(I) for all leads for intrinsic and for fixed flow (60 mgm/s)
- V(I) for unequal currents for fixed flow (60 mgm/s)
- V(I) for I > 100A
- V(I) for no flow

Figure 2 shows voltage vs. current for a typical lead pair. Maximum variations from lead to lead were +/- 10%. The 75mV test result for 100A is in good agreement with the calculated 80 mV for an optimized lead.[2] With current in all 12 leads, results show an increase in voltage at 100A but well within normal limits. It should be noted, however, that this operating point, 1200A, exceeds the requirement by 20%.

Figure 3 shows voltages for unequal currents; leads not carrying the indicated currents carried only 10A. Flows were fixed at 60 mgm/s for all cases. These results show that at the fixed flow rate there was no inbalance in cooling flow with unequal currents.

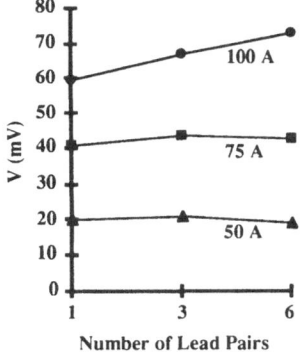

Figure 3. Imbalanced currents.

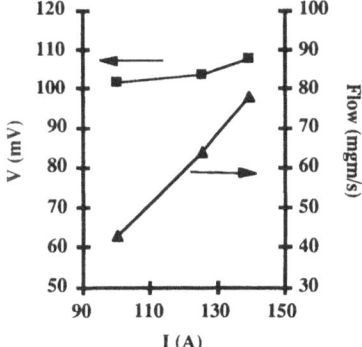

Figure 4. Voltage and flow vs. overcurrent.

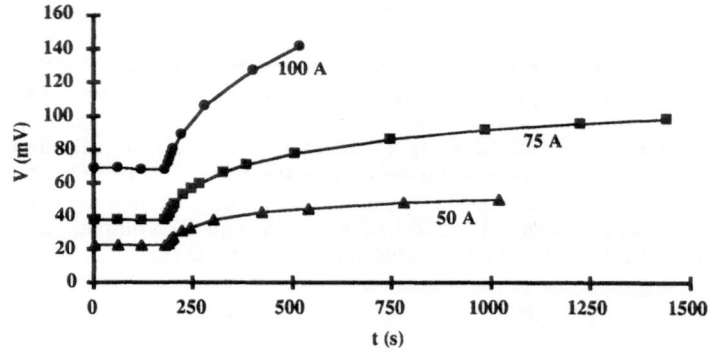

Figure 5. Voltage as a function of time for no cooling flow.

To evaluate the design margin all 12 leads were powered to 125A and 140A. Initially cooling flow was fixed at 60 mgm/s. At 125A after 13 minutes the flow had to be increased to 62 mgm/s to achieve equilibrium. At 140A the flow had to be increased to 78 mgm/s to achieve equilibrium. Voltages and flows are plotted in Figure 4.

To test for fault operation, currents of 50A, 75A and 100A were applied to 5 lead pairs with a flow of 60 mgm/s. For each case, once the leads equilibrated the CEPL flow was reduced to zero. The increase of voltage with time is shown in Figure 5. At 50A the leads reached equilibrium in about 17 minutes. At 75A the test was terminated at 20 minutes. It appears that the leads would have eventually reached equilibrium. At 100A the leads reached 150 mV in about 6 minutes and the test was terminated.

The CEPL was successfully hipot tested in ambient air to 4KV between adjacent leads and between all leads and the housing following these tests.

SUMMARY

Test results confirm that the prototype CEPL meets the performance requirements, has significant margin for overcurrent, and can operate without flow to at least 75% of operating current. Gas flow and boil off measurements confirmed that the net heat load to the bath, accounting for the required flow, is negligibly small.

REFERENCES

1. K. R. Efferson, Helium Vapor Cooled Current Leads, Rev. of Sci. Inst. 38:1776 (1967).
2. M. N. Wilson, "Superconducting Magnets," Clarendon Press, Oxford (1983).
3. J. Wu, J. Dederer, and S. Singh, Development of a Multiple HTS Current Lead Assembly for Corrector Magnets Application, to be published in: "Supercollider V," Plenum Press.
4. K. Ueda, K. Takita, T. Uede, M. Mimura, N. Uno and Y. Tanaka, Design Study and Model Test of High T_c Superconductor Current Leads, to be published in: "Supercollider V," Plenum Press.
5. Dr. J. Theilacker, FNAL, private communication.
6. S. W. Schwenterly, Design and Testing of Electrical Insulation for Superconducting Coils, "Adv. Cryo. Eng." V33, Plenum Press, New York (1987).
7. V. Arp and R. D. McCarty, "Thermophysical Properties of Helium-4 from 0.8 to 1500K with Pressures to 2000 MPa," NIST Tech Note 1334, U.S. Government Printing Office, Washington (1989).
8. M. McAshan, "MIITS Integrals for Copper and for N6-46.5 wt% Ti," SSC-N-3468 (1988).

AN EXPERIMENTAL EVALUATION OF JOINT ELECTRICAL RESISTANCE ON POWER LEAD THERMAL PERFORMANCE

V. I. Datskov[1], J. A. Demko, S. D. Augustynowicz, and R. D. Hutton

Superconducting Supercollider Laboratory[2]
2550 Beckleymeade Ave., MS 4003
Dallas, Texas 75237-3946

INTRODUCTION

The amount of electrical resistance in braze joints is not known for certain. In addition the annealing processes that occurs during a braze or solder operation can change the residual resistivity ratio (RRR) of the copper [1,2]. The change in the electrical resistivity of samples of copper because of exposure to conditions that a high current lead would see during a brazing operation were experimentally investigated. A sample was taken from a manufacturing and brazing trial of the high current power leads for the Superconducting Super Collider (SSC), and from oxygen free high conductivity copper (OFHC) 101 rod similar to that used in the trial. The samples were heated under conditions that a current lead would undergo during the brazing process. Measurements were made of the electrical resistance of the copper specimens and across a braze joint in the manufacturing trial sample for temperatures ranging from liquid helium to room temperature.

A prototype of the SSC high current lead is shown in figure 1. This lead was fabricated from 5 sections that were brazed together. Some results for the measured residual resistivity ratio (RRR) along this lead are given.

EXPERIMENTAL APPARATUS AND PROCEDURE

One set of samples were cut from a brazing trial of a power lead head section which included a braze joint. The second set of copper specimens were cut from a rod of OFHC copper alloy 101 with a diameter of 3.175 mm. The samples were mounted in a fixture that provided electrical current to the ends of the specimen. Two additional electrical contacts were installed that were used in the measurement of the voltage drop across a fixed distance

---

1. Guest Collaborator from JINR/Laboratory of High Energy, Dubna, Russia
2. Operated by the Universities Research Association, Inc., for the U.S. Department of Energy under Contract No. DE-AC35-89ER40486

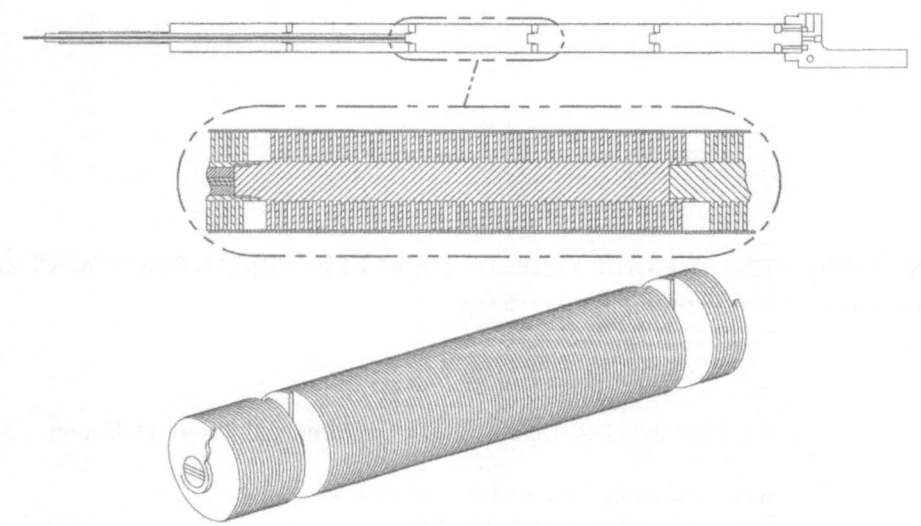

Figure 1. Prototype SSC power lead and typical section from which lead is constructed

along the specimen. The temperature was measured with a RTD that was calibrated for temperatures from 4 K to over 310 K.

The electrical resistivity can be calculated from the measured quantities using equation 1

$$\rho = \frac{VA}{IL} \tag{1}$$

where ρ is the electrical resistivity (ohm-m), V is the voltage drop, L (m) is the distance between voltage taps, A (m^2) is the cross sectional area of the specimen and I (amperes) is the current displayed on the power supply. A HP6011A (0-120A) power supply was used to provide a constant current source for the experiments. The voltage drop along the specimen was measured with a Keithley 182 sensitive digital nanovoltmeter.

To determine the RRR distribution along the prototype 6.5 kA lead, measurements of the voltage drop for a current of 100 A along the lead were made at several locations across two fins at room temperature (300K) and liquid nitrogen temperature (77 K). Using the ratio of the measured voltage drops at these two temperatures and the resistivity correlation from Devred[3,] the RRR could be determined.

EXPERIMENTAL RESULTS

Results for the sample cut from a brazing test of a current lead flag onto the conductor shaft are shown in figure 2. The braze (anneal) temperature for all of the samples is 920 K. Data from the sample provide measurements of the resistivity of copper in an annealed state and, in the same sample, the resistivity of a solder joint. Figure 2 also shows the measured electrical resistivity for the annealed sample and a geometrically similar annealed sample with a braze joint. The RRR of the annealed sample falls between 60 and 80, whereas the annealed sample with a joint has an apparent RRR less than 40. Table 1 contains the specific contact resistance that was determined from this data as well as the resistive heat generation for a joint in a 1.6 cm diameter power lead core carrying 6500 amps.

The measurements show that the specific joint resistance increases with temperature.

The last test performed on the second test specimen, thermally cycled it between room temperature and liquid nitrogen temperature 21 times. Then a measurement of the resistance from liquid helium to room temperature was made. The result, was that there was no measurable change in the electrical resistance of the specimen or the joint.

Table 1. Specific resistance and resistive heating of braze joint in annealed sample

Temperature (K)	Specific Contact Resistance (Ohm-cm^2)	Joule Heating at 6.5 kA(W)
4.2	4.725(10^{-8})	1.009
77.3	5.175(10^{-8})	1.105
296.	7.200(10^{-8})	1.537

The changes in RRR for different anneal times are shown in figure 3. Initially the copper sample had a RRR of about 77. The RRR increased to 275-280 after 1 to 2 minutes in the furnace at braze temperatures. Additional annealing decreased the RRR to approximately 190. This result indicates that brazing and soldering operations must be minimized if the properties of a current lead are to be maintained. This also suggests that it may be possible to tailor electrical properties in a current lead by heat treating.

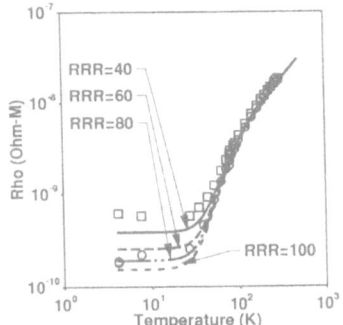 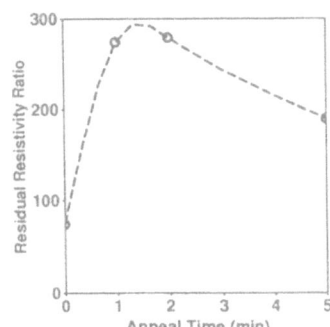

Figure 2. Electrical resistivity of annealed sample (O) and joint (□)

Figure 3. Change in RRR with anneal time

Figure 4 shows measurements of the resistivity from some of the OFHC samples. The procedure to determine RRR from measurements of the ratio $\rho(300K)/\rho(77K)$ is illustrated graphically in figure 4. The figure shows that there is a unique functional relationship between this ratio and the RRR of copper. These data indicate that the RRR values measured using the ratios between liquid nitrogen and room temperatures are approximately 55, 150, and 170. It is evident from the shape of these curves that for accurate measurement of RRR, especially at higher values, that the room temperature must be precisely measured.

Measurements of the RRR distribution along a prototype lead are shown in figure 5 where the lead flag (warm end) is at 835mm. It is clear that there are substantial variations in the RRR along the lead and that it would not be correct to characterize the lead assuming a uniform value. The large drops in RRR at axial locations of 0.160m, 0.325m, 0.475m, 0.660m, and 0.830m occur because sections of the lead are brazed together at these locations as shown in figure 1. The large drop in RRR at these locations reflects the joint contact resistance present. On either side of these low RRR values are high peaks in the RRR due to localized annealing during the brazing operation

SUMMARY AND CONCLUSION

The results from the annealing tests indicate that the RRR for OFHC 101 copper will increase for a small amount of annealing, and then with more annealing the RRR will decrease from the peak value to a RRR that is higher than its initial condition. As a result any brazing or annealing operation to OFHC copper will increase the local RRR. Results from a prototype brazing trial tested indicate that there can be a significant contact resistance due to the brazed or soldered connection. The amount of electrical dissipation was estimated for some of these values. Additional work is required to characterize the contact resistance for other brazing and soldering applications.

A measured distribution of the RRR along a prototype power lead shows that there can be significant variation in properties. The results indicate that there is a reduction in the electrical conductivity at joints but that the surrounding region has an increased conductivity due to localized annealing that takes place during the brazing procedure.

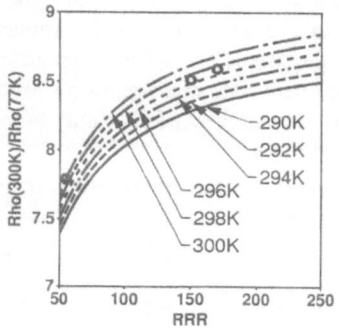

Figure 4. Relationship between RRR and resistivity ratio $\rho(300K)/\rho(77K)$

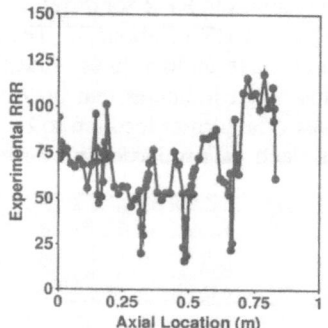

Figure 5. Measured RRR profile along a prototype 6.5kA power lead

ACKNOWLEGEMENTS

The authors would like to acknowledge Michael McAshan, Bill Fietz, Bob Smellie, and Ruben Carcagno for providing technical advice and support, as well as Paul Turley and Darren Cox for assisting in the preparation of the tests and the necessary drawings for the program.

REFERENCES

1. B. V. Elkonin, and J. S. Sokolowski,"Simple Technique for Increasing the Conductivity of Copper for Current Lead Conductors", Cryogenics, Vol 31, December 1991
2. J. Cl. Puippe and W. Saxer, "Electrodeposition of Copper on the Internal Walls of Colliders in Beam Tubes", Supercollider 4, Plenum Press, 1992
3. A. Devred, SSC Laboratory, Internal Communication

DEVELOPMENT OF A MULTIPLE HTS CURRENT LEAD ASSEMBLY
FOR CORRECTOR MAGNETS APPLICATION

J. L. Wu, J. T. Dederer, and S. K. Singh

Westinghouse Science & Technology Center
Pittsburgh, PA 15235

INTRODUCTION

Vapor-cooled current leads used for transmitting power to superconducting power equipment such as the corrector magnets in the SSC spools can introduce a significant heat leak into the cryostat which results in cryogen boil-off. Replenishing the boil-off or refrigerating and liquefying the vapors associated with the cooling of these leads may constitute a significant portion of the operating cost and/or the capital investment of the power equipment. Theoretical studies and experiments have demonstrated that the heat leak introduced by a current lead can be significantly reduced by using ceramic high temperature superconductor (HTSC) as part of the conductor in the current leads.[1-11]

A HTSC reduces heat leak in a current lead by being superconducting in the temperature range below its critical temperature and by having a low temperature thermal conductivity which is generally orders of magnitude lower than the copper alloys commonly used as the current lead conductors. This combination reduces Joule heating and heat conduction, resulting in lower heat leak to the cryostat. The advantage has been clearly demonstrated by a HTSC lead tested up to 2 kA and operating in a self-cooled mode.[2] Application of HTSC to current lead design also permits efficient interception of heat leak at a temperature substantially higher than 4.2K and therefore further reduces the power requirement associated with cooling of current leads. In addition, due to low thermal conductivity of the HTSC material, the overall lead length can be greatly reduced, which could be a significant advantage for application in confined environments such as in the SSC spools.

To demonstrate the advantages and large scale application of this technology, Westinghouse Science & Technology Center has continued its efforts in High Temperature Superconducting (HTS) current lead development. The efforts include qualification testing and selection of commercial sources of HTSC for current leads and the successful development of a 12 x 100 A multiple HTS current lead assembly prototype for SSC Corrector Element Power Lead (CEPL) application. The work of the HTSC testing was reported earlier,[12] the efforts on the design, fabrication and testing of the multiple HTS lead assembly is reported below.

DESIGN OF 12 X 100 A MULTIPLE HTS LEAD ASSEMBLY

The 12 x 100 A multiple HTS lead assembly consists of six pairs of mechanically integrated but electrically isolated helium vapor-cooled current lead elements. Each lead element conductor is made of a short length of cylindrical HTSC rod and two parallelly connected copper wires. A common helium vapor chamber encloses all the HTSC rods portion of the lead elements. In the copper wire portion of the lead assembly, each lead element is housed separately within individual fiberglass tube. This arrangement provides good electrical insulation between the lead elements in the region where helium vapors are at higher temperature and thus has a lower dielectric withstand capability. To eliminate parallel flow instability problems which might exist when only part of the twelve lead elements are carrying currents, a helium flow intermixing arrangement is incorporated in the mid-section of the lead.

Depending upon the desired operation mode of the current lead and the HTSC material, different cross-sectional area and length of the HTSC rods can be selected. For the prototype multiple HTS lead assembly, we have selected a length which permits us to test the lead assembly in both self-cooled (i.e., all helium vapors are generated by the heat leaks through the lead conductors) and with a heat leak intercept with an externally introduced helium gas at temperatures of higher than 4.2 K.

Fig. 1 shows the 12 x 100 A multiple HTS lead assembly prototype. This lead assembly has an outside diameter of approximately 7 cm and a length of 50 cm, from terminal to terminal.

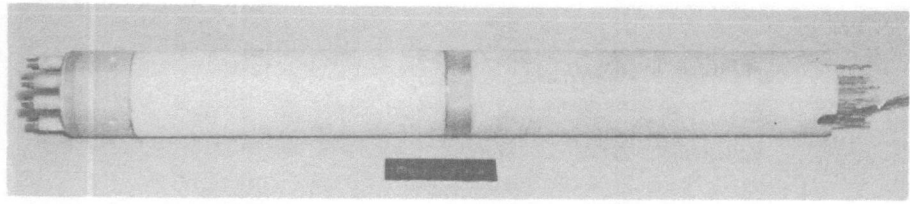

Figure 1. Westinghouse 12 x 100 A HTS Lead Assembly.

PERFORMANCE TESTING

The performance of the 12 x 100 A multiple HTS lead assembly has been extensively tested at the Cryogenic Laboratory of Westinghouse Science & Technology Center. The testing included: steady state helium boiloff measurements in the self-cooled mode of operation and with heat leak intercept with externally introduced gaseous helium at a temperature of 13 K; transient operation testing in current ramping and zero flow stability, and dielectric tests in air and in gaseous helium. Table 1 lists test parameters, lead element connection configuration and the test run time.

Table 1. Tests performed on the 12 x 100 A HTS lead assembly.

I. Steady State Helium Boiloff
- Self-Cooled Mode Operation:
    - All Six Pairs of Leads Connected in Series and Carrying 75, 100, and 125 A.
      Total Run Time for These Tests: 4 Hours and 46 Minutes.
    - Three Pairs of Leads Connected and Carrying 100 A.
      Run Time: 3 Hours and 12 Minutes.
    - One Pair of Leads in Circuit and Carrying 100 A.
      Run Time: 1 Hour 30 Minutes.
- Test with Externally Introduced 13 K GHe:
      All Leads Connected in Series and Carrying 100 A.
      Run Time: About 3 Hours.
II. Transient Operation Tests, Self-Cooled Mode:
    - Current Ramping Tests.
      Current Increased to or Decreased from 100 A with a Ramp Rate of Higher than ± 5 A/s.
      Lead Connection Configurations: One Pair and Six Pairs in the Circuit.
    - Zero Flow Stability Test.
      Cooling Helium Vapor Flow Shut Off for 20 Minutes from the Down-Stream Side. All Leads
      Connected in Series and Carrying 100 A.
III. DC Hi-Pot Tests
- In Room Temperature Atmospheric Air.
- In Room Temperature Gaseous Helium at P = 0.1, 0.2 and 0.3 MPa.

In the helium boiloff measurement and in the transient operation testing, the lead elements were paired by connections with Nb_3Sn wires at the lower terminals and the lead assembly was installed vertically within a helium dewar. Liquid helium level was maintained at a level well above the lower opening end of the lead assembly housing such that all gaseous helium boiloffs generated by the lead conductors have to exit through the flow passages in the upper terminals of the leads and be collected and measured by flow meters either individually for each lead or in combination of all twelve lead elements. Sensors were located at various locations for measuring temperatures on the lead conductor surface as well as of the gas helium. Voltage drop across the lead terminals and across the HTSC portion of the lead were recorded for each of the twelve lead elements. Steady state helium boiloff rates were determined through adjusting the gaseous helium flow rate through the lead elements and the liquid helium level such that a steady-state condition of temperatures and lead voltage drops are maintained for a sufficiently long time. All tests were conducted with helium pressure maintained at near 1 atmospheric pressure.

TEST RESULTS ANALYSIS AND DISCUSSION

Fig. 2 shows the plots of data from the steady-state helium boiloff test with all leads carrying 100 A, or for a total current of 1200 A. Total gaseous helium flow rate ($\dot{m}$) started at 32.02 mg/s but was subsequently reduced to 28.79 mg/s. After about 120 minutes, an essentially steady-state condition was reached as evident

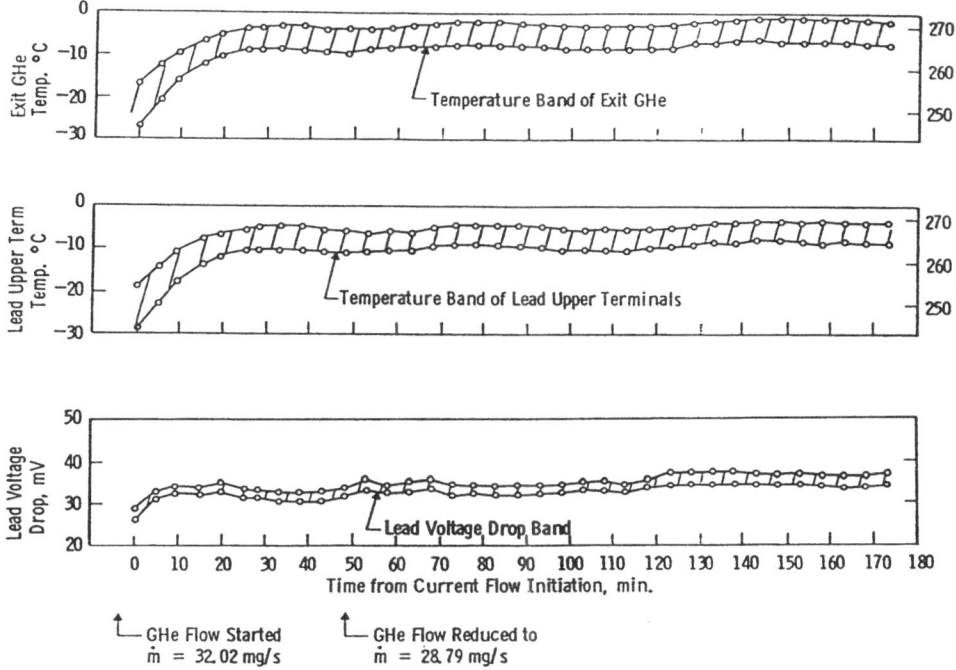

Figure 2. Steady-state data of 12 x 100 A HTS led assembly, I = 100 A for all lead elements.

by the constant values of temperatures and lead voltage drop. Expressing the boiloff measurement in terms of specific boiloff rate, i.e., boiloff rate per KA of current, we have rates of between 23.99 to 26.68 mg/s·KA. Adjusting this rate upwards to account for the additional heat conductions into the leads when six pairs of power supply cables are used, we obtained a specific boiloff rate of about 31 mg/s·KA. This rate is about half of the rate of an optimum designed conventional vapor-cooled lead (~60 mg/s·KA).

Table 2. Test results summary of the 12 x 100 A HTS current lead assembly.

I. Steady State Helium Boiloff:
- Self-Cooled Operation
  - $I = 12 \times 100$ A
    $\dot{m}/I$: 23.99 to 26.68 mg/s $\cdot$ kA
    Temp. at Upper Lead Terminals: -9 to -3 °C
  - $I = 12 \times 125$ A and $I = 12 \times 75$ A Have Comparable $\dot{m}/I$ Values.
  - Adjusted to Six Pairs of Cables to power Supplies:
    $\dot{m}/I \approx 31$ mg/s $\cdot$ kA (0.63 W/kA)
- Tests with Heat Leak Intercept with 13 K GHe:
  - 60 to 90% of LHe Boil-Off Replaced by 13 K GHe.
  - At Near Steady-State (12 x 100 A):
    $\dot{m}/I = 24.0$ mg/s $\cdot$ kA
    $\dot{m}/I$ from LHe Boil-Off: ~ 4.2 mg/s $\cdot$ kA
    $\dot{m}/I$ from 13 K GHe: ~ 19.8 mg/s $\cdot$ kA
- No Parallel Gas Flow Instability Problems.

II. Transient Operation Tests, Self-Cooled Mode
- No Problem with $\dot{i}$ Tested (~ ± 2 to 6.6 A/s).
- Zero Flow Stability of Longer than 20 Minutes.

III. DC Hi-Pot Tests
- Withstands 5 kV in Air, with Leakage Currents of ≤ 0.3 μA.
- Withstands 3.5 kV in 0.3 MPa GHe, Leakage Current ≤ 0.5 μA.

Table 2 summarizes the results of all the tests performed.

CONCLUSIONS

We have successfully developed a 12 x 100 A vapor-cooled HTS current lead assembly for SSC corrector magnet application. The lead assembly has demonstrated a specific boiloff rate which is only half the rate of an optimally design conventional vapor-cool lead. The HTS lead assembly has also successfully withstood the higher than required current ramp rate and the dielectric breakdown voltage. It has demonstrated a zero flow stability of more than 20 minutes and is stable for various operation configurations. Employing these HTS lead assemblies in the SSC spools can produce a significant saving in refrigeration cost.

REFERENCES

1. J.L. Wu, J.T. Dederer, O.R. Christianson and S.K. Singh, "Testing and Performance of High Temperature Superconducting Current Leads," Supercollider 4, Plenum Press, New York(1992).
2. J.L. Wu, J.T. Dederer, P.W. Eckels, S.K. Singh, J.R. Hull, R.B. Poeppel, C.A. Youngdahl, J.P. Singh, M.T. Lanagan, and J. Balachandran, "Design and Testing of a High Temperature Superconducting Current Lead," IEEE Trans. Mag., 27:1861(1991).
3. K. Ueda, T. Bokno, K. Takita, K. Mukae, T. Uede, I. Itoh, M. Mimura, U. Uno and T. Tanaka, "Design and Testing of a Pair of Current Leads Using Bismuth Compound Superconductor," 1992 Applied Superconductivity Conf., Chicago, IL(1992).
4. K. Tasaki, D. Ito, H. Ogiwara, K. Numata, A. Hane and K. Hoshino, "Development of Oxide Superconducting Current Lead," 1992 Applied Superconductivity Conf., Chicago, IL(1992).
5. Y. Yamada, T. Yanagiya, T. Hasebe, K. Jikihara, M. Ishizuka, S. Yasuhara and M. Ishihara, "Superconducting Current Leads of Bi-Based Oxide," 1992 Applied Superconductivity Conf., Chicago, IL(1992).
6. P.F. Herrmann, C. Albrecht, J. Bock, C. Cottevieille, S. Elschner, W. Herbert, M-O. Lafon, H. Lauvray, A. Leriche, W. Nick, E. Preisler, H. Salzburger, J-M. Tourre and T. Verhaege, "European Project for the Development of High Tc Current Leads," 1992 Applied Superconductivity Conf., Chicago, IL(1992).
7. J.L. Wu, J.T. Dederer, P.W. Eckels, S.K. Singh, J.R. Hull, "High Temperature Superconducting Current Leads for Fusion Magnets Systems," 14th IEEE Symposium on Fusion Engineers, San Diego, CA(1991).
8. J.R. Hull, A. Unal, and M. Chyu, "Analysis of Self-Cooled Binary Current Leads Containing High Temperature Superconductors," Cryogenics, 32:822(1992).
9. J.R. Hull, "High Temperature Superconducting Current Leads for Cryogenic Apparatus," Cryogenics, 29:1116(1989).
10. F. Grivon, A. Leriche, C. Cottevieille, J.C. Kermarrec, A. Petitbon, A. Fevrier and Y. Laumond, "YBaCuO Current Lead for Liquid Helium Temperature Applications," IEEE Trans. on Mag. 27:1866(1991).
11. A. Matrone, G. Rosatelli, and R. Vaccarone, "Current Leads with High T_c Superconductor Bus Bars," IEEE Trans. Mag., 25:1742(1989).
12. J.L. Wu, "Testing of High Temperature Superconductors for Cryogenic Current Lead Applications," 1992 Applied Superconductivity Conf., Chicago, IL(1992).

DESIGN AND APPLICATION CONSIDERATION OF HIGH TEMPERATURE
SUPERCONDUCTING CURRENT LEADS

J. L. Wu

Westinghouse Science & Technology Center
Pittsburgh, PA 15235

INTRODUCTION

As a potential major source of heat leak and the resultant cryogen boiloff, cryogenic current leads can significantly affect the refrigeration power requirement of cryogenic power equipment. Reduction of the heat leak associated with current leads can therefore contribute to the development and application of this equipment. Recent studies and tests have demonstrated that, due to their superconducting and low thermal conductivity properties, ceramic high temperature superconductor (HTSC) can be employed in current leads to significantly reduce the heat leak.[1-8] However, realization of this benefit requires special design considerations pertaining to the properties and the fabrication technology of the relatively new ceramic superconductor materials. Since processing and fabrication technology are continuously being developed in the laboratories, data on material properties unrelated to critical states are quite limited. Therefore, design analysis and experiments have to be conducted in tandem to achieve a successful development.

Due to the rather unique combination of superconducting and thermal conductivities which are orders of magnitude lower than copper, ceramic superconductors allow expansion of the operating scenarios of current leads. In addition to the conventional vapor-cooled lead type application, low heat leak conduction-cooled type current leads may be practical and are being developed.[4,9] Furthermore, a current lead with an intermediate heat leak intercept has been successfully demonstrated in a multiple current lead assembly employing HTSC.[1]

These design and application considerations of high temperature superconducting (HTS) current leads are addressed below. The discussion is focused on the areas related to the use of ceramic superconductors.

DESIGN CONSIDERATIONS

For development of a device employing HTSC, the first thing which needs to be considered is the status of HTSC fabrication technology with respect to production of device-size HTSC components having satisfactory properties and an affordable price. For current lead application, the HTSC fabrication technology has advanced enough that components can be produced with adequate properties, as demonstrated in several prototype devices. Other important design considerations for current leads are techniques for joining HTSC and normal metal (copper); effects of thermal cycling to the HTSC and the HTSC/normal metal joint; issue of superconductor quench and its protection, and effects of the operating environment, such as radiation, upon the properties of the HTSC material. These design considerations are discussed below.

Supercollider 5, Edited by P. Hale
Plenum Press, New York, 1994

571

HTSC Component Size and Properties

To fully utilize the benefits of HTSC in a current lead application, it is best to employ the HTSC without a metallic stabilizer so that the low thermal conductivity advantage of HTSC does not diminish. This generally is not a problem since in a current lead application, the required lengths of the HTSC components are typically very short and the operating current densities are low enough that, with adequate design, quench of superconductor can be avoided. Furthermore, it is desirable to operate at low current densities from a thermal capacity standpoint. Table 1 gives general requirements of HTSC component size and critical superconducting properties, as well as the thermal and mechanical property requirements for low currents and high currents applications. These specifications are essentially within the reach of today's HTSC fabrication technology.

Table 1. General specifications of high temperature superconductor for current lead application.

SUPERCONDUCTOR GEOMETRY	For Low Current Application (e.g., 100-200 A)	For High Current Application (e.g., 1-2 KA)
Cross-sectional area:	0.3 to 1.0 cm^2	3.0 to 7.0 cm^2
Length	4.5 - 15.0 cm	7.0 - 18.0 cm
SUPERCONDUCTOR PROPERTIES		
Critical temperature (B=0):	$\geq$ 85K	$\geq$ 85K
Critical current density:		
@ T = 77K:	$J_c \geq 300$ A/cm^2 for B = 0	$J_c \geq 800$ A/cm^2 for B = 0
	$J_c \geq 150$ A/cm^2 for B = 100G	$J_c \geq 500$ A/cm^2 for B = 1000G
@ T = 4.2K:	$J_c \geq 750$ A/cm^2 for B = 100G	$J_c \geq 1000$ A/cm^2 for B = 1000 G
Thermal conductivity:	$\leq$ 2 w/mK at 4.2K	$\leq$ 2 w/mK at 4.2K
Mechanical properties (at room temperature):		
Flexural strength:	$\geq$ 65 MPa	$\geq$ 65 MPa
Elastic modulus:	$\geq$ 80 GPa	$\geq$ 80 GPa
Hardness:	$\geq$ 1.5 GPa	$\geq$ 1.5 GPa
Fracture toughness:	$\geq$ 1.7 MPa$\sqrt{m}$	$\leq$ 1.7 MPa$\sqrt{m}$

HTSC/Normal Metal Joint

Contact interface joint between a HTSC and a normal metal such as copper is a critical design area. The joint is a source of undesirable Joule heating, which increases cryogen boiloff, and a potential problem area mechanically due to the significant difference in mechanical and thermal properties between the ceramic and the metallic materials.

For a current lead application, contact surface resistivity of the order of 1 $\mu\Omega \cdot cm^2$ at 4.2K is desired. This will allow a design of a contact joint with a resistance lower than 1 $\mu\Omega$ at 4.2K when carrying a current of 100A. A proportionally lower resistance at higher currents can be achieved by increasing the contact surface area. The Joule heating at the contact joint will be less than 10 mW for a current of 100A, which is insignificant as compared to the overall lead cooling requirement. Contact joints of low resistance has been accomplished in the 12 x 100A HTS lead assembly developed by Westinghouse using BSCCO bars.[1] Contact resistance of $\mu\Omega$ magnitude at 4.2K was obtained for a current of 100A.[10] In addition, the contact joint shows a metallic behavior with respect to temperature, which is a desirable characteristic for a current lead application.

Effect of Thermal Cycling on HTSC and HTSC/Normal Metal Joint

Thermal cycling may degrade HTSC materials mechanically due to the possibility that residual stress and microcracks may be present in the material. It may also create degradation in the HTSC/normal metal joint as a result of a difference in thermal expansion of the two materials. However, with proper design the thermal cycling effects can be minimized. In a previous paper,[10] it was shown experimentally that a contact joint can withstand 60 cycles of cool-down/warm-up cycles between room temperature and liquid nitrogen temperature with no degradation of critical current density in HTSC and having only a 20% increase in contact joint resistance. Based on these results, the thermal cycling effects on HTSC components appears manageable.

Superconductor Quench and Protection

Due to its lower operating current densities, higher heat capacity at the operating temperature and higher operating temperature margin, the ceramic superconductors used in current leads have significantly higher stability against quench as compared to the conventional low temperature superconductor (LTS). We have estimated that for the HTSC bars used in the Westinghouse 12 x 100A HTS lead assembly, the minimum MPZ (minimum propagation zone) triggering energy is about 8 mJ at 4.2K and 3.16J at 50K. These values are orders of magnitude higher than typical LTS wire used in magnets.

It was also commonly agreed that, due to its low thermal conductivity and high heat capacity, the quench propagation velocity of ceramic superconductor is rather slow. Quench protection of HTSC may be necessary to prevent a burnout. For current lead application, due to low operating current density and short length of conductor, there is generally sufficient time for removal of power before burnout occurs. Our estimate shows that, for the case of our 12 x 100A HTS lead assembly, the time to heat HTSC bars from 84K to 300K adiabatically is about 24 s, sufficiently long for power removal. Long burnout times have also been obtained by Hull for YBCO operating at low current densities.[11]

Effects of Radiation on HTSC

For applications such as in a particle accelerator system, the effects of radiation on the HTSC material properties needs to be considered. Based on published results, no appreciable adverse effects are expected for HTSC employed in current leads. Radiation has been used to increase the flux pinning capability of ceramic superconductors for increasing the critical current density in the presence of magnetic field. The results[12-14] suggest that radiation tends to increase the critical current density in the high temperature region and decrease the critical current density in the lower temperature region, and it only marginally affects the critical temperature of the superconductors. The decrease of critical current density in the low temperature region should pose no problem since, at low temperature, the ceramic superconductors have much higher critical current density than when they are in high temperature. Currently, data of the effects of radiation on the mechanical properties of ceramic superconductors are very limited.

APPLICATION CONSIDERATIONS

Various schemes of applying a HTS current lead to reduce refrigeration power requirement have been addressed by Hull.[11] Selection of an operating scheme of current lead depends, of course, not only on the lead design itself but also on the characteristics of the cryogenic sub-system, as well as the overall system. The benefit of various operating schemes of HTS current leads has to be evaluated against the mechanical complexity and the cost associated with it. For example, conduction-cooled leads (Fig. 1(c), with $\dot{m}_2 = 0$) are simple in mechanical design, but tend to have a higher heat leak as compared to vapor-cooled leads. Conduction-cooled HTS leads are currently under development for use, in conjunction with cryocoolers, in MRI magnet systems.[4,9]

Use of ceramic HTSC can enhance the practicality of vapor-cooled lead designs having conduction or helium vapor cooling in the low temperature region and nitrogen vapor cooling in the high temperature region (Fig. 1(c) and Fig. 1(b), $\dot{m}_2$ is nitrogen flow). This may produce a substantial reduction in refrigeration requirement as compared to the self-cooled type application using helium vapor for the entire lead length. The reduction in refrigeration requirement is accomplished with an additional design complexity for allowing a second cryogen (nitrogen) in the system, requiring cryogenic seals for separating the two cryogens and requiring means to ensure that the temperature in the nitrogen region does not drop below its solidifying temperature.

The low thermal conductivity of the ceramic HTSC allows the intercept, at a temperature significantly higher than 4.2K, of the heat leaks coming from the copper portion of the lead. This operation scheme has been successfully demonstrated[1] in an experimental prototype using externally introduced gaseous helium at a temperature of 13K. The scheme of operation is shown in Fig. 1(b), where $\dot{m}_1$ is the boiloff and $\dot{m}_2$ is the externally introduced 13K gaseous helium. Test results indicated that introducing $\dot{m}_2$ can reduce the boiloff rate by as much as 90% from the case of self-cooled operation mode (Fig. 1(a), i.e., the operation mode in which $\dot{m}_2 = 0$. Operation with a heat leak intecept can therefore further reduce the total refrigeration power requirement for lead cooling by more than a factor of two. Since the self-cooled type operation of a current lead employing HTSC was found to reduce boiloff by a factor of two from the conventional all copper lead,[1] the employment of heat leak intercept in HTS leads could therefore reduce the boiloff to only 25% of the conventional leads.

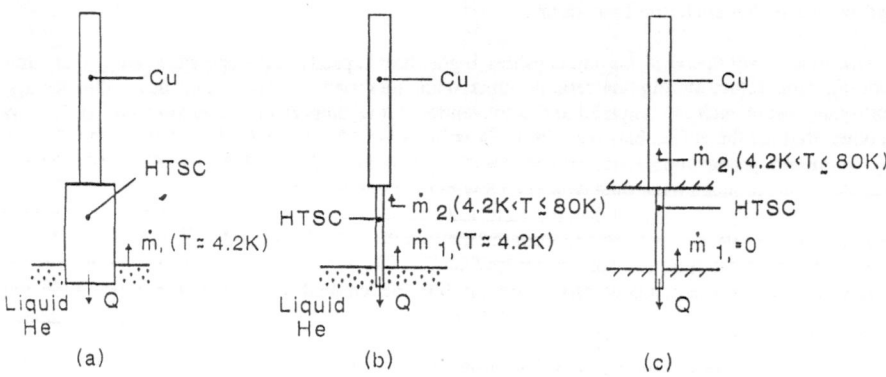

Figure 1. Operation schemes of HTS current leads.

CONCLUSION

Design and application issues of current leads employing ceramic high temperature superconductors were evaluated from a viewpoint of realizing the maximum benefits in practical device application. These evaluations have paved the way for successful developments of HTS current leads by Westinghouse.

REFERENCES

1. J.L. Wu, J.T. Dederer and S.K. Singh, "Development of a Multiple HTS Current Lead Assembly for Corrector Magnets Application," 5th IISSC, San Francisco, CA, May 1993.
2. J.L. Wu, J.T. Dederer, P.W. Eckels, S.K. Singh, J.R. Hull, R.B. Poeppel, C.A. Youngdahl, J.P. Singh, M.T. Lanagan, and J. Balachandran, "Design and Testing of a High Temperature Superconducting Current Lead," IEEE Trans. Mag., 27:1861(1991).
3. K. Ueda, T. Bokno, K. Takita, K. Mukae, T. Uede, I. Itok, M. Mimura, U. Uno and T. Tanaka, "Design and Testing of a Pair of Current Leads Using Bismuth Compound Superconductor," 1992 ASC, Chicago, IL.
4. K. Tasaki, D. Ito, H. Ogiwara, K. Numata, A. Hane and K. Hoshino, "Development of Oxide Superconducting Current Lead," 1992 ASC, Chicago, IL.
5. Y. Yamada, T. Yanagiya, T. Hasebe, K. Jikihara, M. Ishizuka, S. Yasuhara and M. Ishihara, "Superconducting Current Leads of Bi-Based Oxide," 1992 ASC, Chicago, IL.
6. P.F. Herrmann, C. Albrecht, J. Bock, C. Cottevieille, S. Elschner, W. Herbert, M-O. Lafon, H. Lauvray, A. Leriche, W. Nick, E. Preisler, H. Salzburger, J-M. Tourre and T. Verhaege, "European Project for the Development of High Tc Current Leads," 1992 ASC, Chicago, IL.
7. J.R. Hull, A. Unal, and M. Chyu, "Analysis of Self-Cooled Binary Current Leads Containing High Temperature Superconductors," Cryogenics, 32:822(1992).
8. J.R. Hull, "High Temperature Superconducting Current Leads for Cryogenic Apparatus," Cryogenics, 29:1116(1989).
9. K.G. Herd, B. Dorri, E.T. Laskaris, J.E. Tkaczyk and K.W. Lay, "Grain-Aligned YBCO Superconducting Current Leads for Conduction-Cooled Applications," 1992 ASC, Chicago, IL.
10. J.L. Wu, "Testing of High Temperature Superconductors for Cryogenic Current Lead Applications," 1992 ASC, Chicago, IL.
11. J.R. Hull, "High Temperature Superconducting Current Leads," 1992 ASC, Chicago, IL.
12. K. Shiraish, Y. Yano and Y. Otoguro, "Gamma Irradiation Effect on a $Bi_{1.5}Pb_{0.5}Sr_2Ca_2Cu_3O_{10}$ Superconductor," Japanese Jour. of Appl. Phys., 30:L120(1991).
13. L. Luo, Y.H. Zhang, S.H. Hu, W.H. Liu, G.L. Zhang and W.X. Hu, "Gamma Radiation Effects on Some Properties of YBCO," Physica C, 178:11(1991).
14. H.W. Weber, H.P. Wiesinger, W. Krischa, F.M. Sauerzopf, G.W. Crabtree, J.Z. Liu, Y.C. Chang, and P.Z. Jiang, "Critical Currents in Neutron Irradiated YBCO and BiSCCO Single Crystals," Supercon. Sci. & Tech., 4:103(1991).

DESIGN STUDY AND MODEL TESTS OF HIGH Tc SUPERCONDUCTOR CURRENT LEADS

K. Ueda[1], K. Takita[1], T. Uede[1],
M. Mimura[2], T. Kinoshita[2], and Y. Tanaka[2]

[1]Fuji Electric Corporate Research and Development, Ltd.,
    2-2-1 Nagasaka, Yokosuka, 240-01 Japan
[2]Furukawa Electric Co.,
    2-4-3 Okano, Nishi-ku, Yokohama, 220 Japan

INTRODUCTION

Current leads are the dominant source of the heat load in a common superconducting magnet system. Therefore, many theoretical and experimental studies have been made[1-6] to apply high temperature superconductors(HTS) to the current leads. Westinghouse and Argonne National Laboratory[1], and we[5,6] have demonstrated 25–40% reduction in required liquid helium consumption. Theoretical consideration showed that refrigerator input power can be decreased to 1/3 that of the conventional lead if a higher temperature coolant is utilized[5,6].

In this paper, thermal performance calculations for a 6.6 kA current lead cooled by intermediate temperature coolant are given. Performances and applicability of the HTS lead depend on the material development. Some of our work concerning HTS material is presented. Our 1 kA model lead system was modified in order to test the cooling scheme using intermediate temperature coolant. Test results are presented.

THERMAL PERFORMANCE CALCULATION OF 6.6 kA HTS CURRENT LEAD

Design calculations have been done for a current lead with an operating current of 6.6 kA[7] and effective length of 0.7 m. We evaluated the input power to the refrigerators for 4 cooling schemes (Figure 1); (a) conventional copper lead cooled by helium gas supplied at 4.2 K into the cold end (b) copper plus HTS with the same cooling scheme as in (a), (c) copper cooled by helium gas supplied at an intermediate temperature into the bottom of the copper portion, and the HTS uncooled but cold end kept at 4.2 K, (d) same as (c) but liquid nitrogen cooled. Main design parameters and thermal performances obtained are given in Table 1.

Supercollider 5, Edited by P. Hale
Plenum Press, New York, 1994

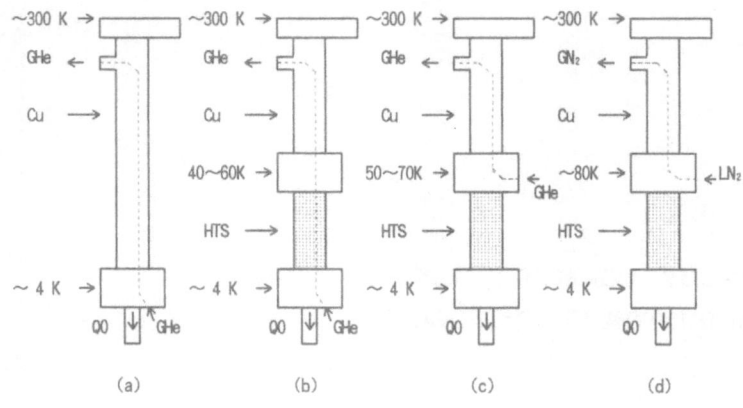

Figure 1. Cooling scheme for conventional and HTS current leads.

Table 1. Main design parameters and thermal performances in 4 cooling schemes for 6.6 kA current lead.

case	(a)	(b)	(c)	(d)
cooling scheme	Cu 4K GHe (conventional)	Cu 4K GHe HTS 4K GHe	Cu 55K GHe HTS uncooled	Cu LN$_2$ HTS uncooled
length/material/	0.15/Cu/380	0.15/HTS/825	0.15/HTS/1100	0.15/HTS/2200
(m) /cross section (mm^2)	0.55/Cu/380	0.55/Cu /550	0.55/Cu / 380	0.55/Cu / 900
coolant	LHe	LHe	55K GHe	LN$_2$
flow rate [g/s]	0.390	0.260	0.457	0.758
heat flow [W] at temp. [K]	8.0 at 4.2	0.0 at 4.2	0.37 at 4.2 +23.4 at 65	1.04 at 4.2 +151 at 78
temp[K] at 0.15m	16.0	56.1	65	78
ideal input power [kW]	2.67	1.77	0.626+0.026 =0.652	0.585+0.073 =0.658
relative to (a)	1.00	0.66	0.24	0.25

The ideal input power can be reduced in cases (c) and (d) to 1/4 that of the conventional lead. One can select either scheme depending on whole refrigeration system construction. In SSC, 80K level cooling is accomplished by liquid nitrogen[7]. Overall dimensions of the 6.6 kA current lead in case (d) will be 150 mm in diameter (in low temperature portion) and 1.0 m in total length (length can be adjusted to fit to the device to be incorporated).

PROPERTIES OF Bi–BASED SUPERCONDUCTOR FOR CURRENT LEAD

In the prototype 1 kA current lead, Bi(2223) sintered bulk rods measuring approximately 9.5 mm in diameter and 145 mm in length were used.

Three–point bending strength was measured with HTS rods for current leads. The strengths were over 80 MPa whether or not they had been immersed in liquid nitrogen.

576

The maximum load itself was over 800 N and ample to withstand handling in current lead assembling and use.

Contact resistivity Rc between the silver electrode and the oxide superconductor rod is related to Jc of the HTS and has a strong dependence on temperatures. Measurement has been made on a small sample(ca. 1x3x30 mm) of sintered bulk of Bi(2223) having silver foils on the both ends, with current below 20 A. Results are shown in Figure 2.

Thermal conductivity measurement were conducted on the sample cut from the Bi (2223) sintered rod. Thermal conductivity integral were 70 W/m for 77–4 K. Details will be published elsewhere.

PERFORMANCE TESTS OF 1 kA PROTOTYPE LEAD PAIR

A pair of 1 kA current leads was fabricated and tested[5,6].   The lead was originally designed to be cooled by helium gas over its total length.   The testing showed that helium flow rate required to keep steady state was about 3/4 that for conventional copper lead.   The prototype lead pair has been modified to test in the intermediate temperature helium gas cooling scheme(Figure 1 (c)).   Helium gas at an intermediate temperature, e.g. 40–70 K, is supplied at the joint part between HTS and copper.   In practical applications, the gas will be supplied by a refrigerator.   In our test model,   vaporized gas from the cryostat bath was gathered in a hood and conducted through a duct to the joint part. A heater and a temperature sensor are arranged in the duct and used to control the entering gas temperature.   Figure 3. shows the test set–up of the 1 kA HTS model leads. Few examples of the test data in stable operations are given in Table 2 as well as the ideal refrigerator power estimated based on the test data.   Ideal input power of a refrigerator for the HTS current lead estimated from the experimental data was around 30 % of conventional copper lead.   The value is near the theoretical value.

Another modification has been done to the 1kA prototype to test the liquid nitrogen cooling scheme(Figure 1(d)).   In the preliminary test at 500 A, measured temperature distribution was near the calculated one.

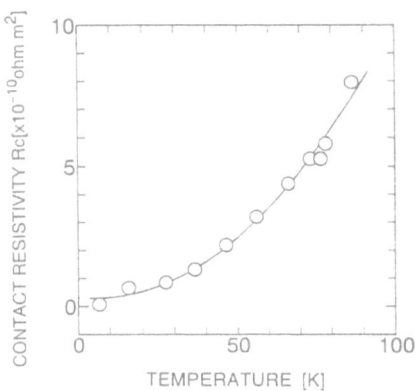

Figure 2. Temperature dependence
of contact resistivity Rc

Figure 3. Test set–up of the 1 kA
HTS model leads

Table 2. Examples of the test data for intermediate temperature helium gas cooling and refrigerator input power estimated from the test data.

RUN NO	I	Tog	T4	F	Wi	Pi (Tog)	Qo HTS	Qo FRP	Pi (4K)	Pi tot	Pi HTS ──────── Pi Cu
	A	K	K	mg/s	J/g	W	W	W	W	W	
500A-2	506	42.9	50.6	39.6	1697	67.2	0.136	0.123	18.2	85	0.44
750A-4	759	44.5	54.2	41.4	1648	68.2	0.153	0.132	20.1	88	0.30
1000A-1	1016	23.0	35.5	42.8	2566	109.8	0.072	0.087	11.2	121	0.31

Tog : He gas temperature supplied to the joint part
T4 : Cu block temperature at the joint part
Pi(Tog) : ideal input power for supplying helium gas at Tog and F
Qo : estimated from the T4 data and thermal conductivity integral
Pi(4K) : ideal input power for refrigeration Qo at 4 K
Pi(HTS)=Pi tot, Pi(Cu):386 W/kA=1.15 W/kA * 4.77 * 70 W/W

But, because the model lead is optimized for cooling with low temperature helium, it is not suitable to measure thermal performances accurately with nitrogen cooling scheme.

CONCLUSIONS

1. Design calculations of 6.6 kA current lead incorporating HTS indicated that the refrigerator power requirement can be reduced to as low as 1/4 of conventional lead.
2. Testing of 1 kA model lead with intermediate temperature helium gas cooling has been conducted. Ideal refrigerator power estimated from that testing was about 30 % that for conventional copper lead.
3. Items for further development will be,
−safety in the event of quenching of HTS.
−durability of the HTS material and joint.

REFERENCES

1. J.L.Wu, J.T.Dederer, P.W.Eckels, S.K.Singh, J.R.Hull, R.B.Poeppel, C.A.Youngdahl, J.P.Singh, M.T.Lanagan, and U.Balachandran, "Design and testing of a high temperature superconducting current lead", IEEE Trans. Mag. 27:1861(1991)
2. F.Grivon, A.Leriche, C.Cottevieille, J.C.Kermarec, A.Petitbon, A.Fevrier, and Y.Laumond, "YBaCuO current lead for liquid helium temperature applications", IEEE Trans. Mag. 27:1866(1991)
3. J.R.Hull, "High temperature superconducting current lead", 1992 Applied Superconductivity Conf., August 1992, Chicago, IL
4. P.F.Herrmann, C.Albrecht, J.Bock, C.Cottevieille, S.Elschner, W.Herkert, M-O.Lafon, H.Lauvray, A. Leriche, W.Nick, E.Preisler, H.Salzburger, J-M.Tourre, and T.Verhage, "European project for the development of high Tc current leads", 1992 Applied Superconductivity Conf., August 1992, Chicago, IL
5. K.Ueda, T.Bohno, K.Takita, K.Mukae, T.Uede, I.Itoh, M.Mimura, N.Uno, T.Tanaka, "Design and testing of a pair of current leads using bismuth compound superconductor", 1992 Applied Superconductivity Conf., August 1992, Chicago, IL
6. K.Ueda, K.Takita, T.Uede, M.Mimura, N.Uno, Y.Tanaka, "Thermal performance of a pair of current leads incorporating bismuth compound superconductor", Fifth International Symp. on Superconductivity, November 1992, Kobe, Japan
7. J.R.Sanford and D.M.Matthews, eds. "Site-Specific Conceptual Design of the Superconducting Super Collider", SSCL-SR-1056, SSC Lab., July 1990

DATABASES FOR ANALYSIS OF SUPERCONDUCTING
CABLE MANUFACTURING

V. A. Bardos, E. S. Coleman, M. J. Erdmann, B. A. Jones
K. S. Kozman, D. J. Little, and J. M. Seuntjens

Superconducting Super Collider Laboratory*
2550 Beckleymeade Ave.
Dallas, TX 75237

INTRODUCTION

Starting in September 1991, eight cable vendors began fabricating approximately 10,000 kg each of Inner or Outer superconducting cable for the Superconducting Super Collider's (SSC) cable Vendor Qualification Program (VQP). This program, designed to identify vendor's for competition for the supplying of superconducting cable for the manufacture of SSC magnet systems, will conclude in June, 1993.

The conductor database was developed as an integral part of the VQP in order to analyze the origins of variation within the conductor fabrication processes, and develop and implement control procedures to minimize such variations. In addition, the database development effort will provide a direct link to the MAGCOM database system being implemented by the Test and Data Management Department of the Magnet Systems Division of the SSCL.

DATA ANALYSIS

At no time in the past has such extensive access to superconducting wire/cable processes been available to a superconductor development program. The database development effort is designed to contain every critical process variable for the production process of each of the vendors in the VQP. These parameters are presently stored and analyzed using FoxBASE+/Mac, a Macintosh based database management program which is dBase compatible.

Data has been analyzed to compare the chemical composition of NbTi and Nb sheet, the metallurgical parameters of both the raw materials and completed monofilament, the extrusion parameters of monofilament and multifilament billets, yield vs. aspect ratio (of the mono and multifilament billets), heat treatment studies, electrical and metallurgical properties of final size wire, electrical and metallurgical properties of finished cable, and

*Operated by the Universities Research Association, Inc., for the U.S. Department of Energy under Contract No. DE-AC35-89ER40486.

cable degradation. In Phase II of this program, planned to test variability in each vendor's optimized process, all cable vendors happen to use the same alloy source. As a result, the program provides a uniquely broad perspective of the processing effects on the electrical, mechanical, and superconducting properties of NbTi. The VQP database will be a vital tool in the analysis of this effort.

Database Structure

At the start of the VQP, each vendor was given a disk containing the 16 database files listed in Table 1. Each database is created of fields which contain the pieces of information needed to examine a particular component, such as the billet number or the I_c of wire. FoxBASE+ allows for six field types: **character** which includes all keyboard characters; **numeric** includes numbers, decimal point and a leading plus or minus sign, and will accommodate limited mathematical analysis; **date** in a MM/DD/YY format; **logical** for true or false conditions; **memo** for explanatory text having no size limitations, and **picture** for information such as radiographs which would have to be scanned into the database.

Table 1. VQP Database Files

Database Name	Description	Number of Fields
CU_PROD	Raw Material Copper Data	11
NbTi_ig	NbTi Ingot Chemical Analysis	55
NbTi_LT	NbTi Lot Analysis	31
NbSHEET	Nb Sheet Analysis	47
MonoProd	Monofilament Billet Assembly Data	18
MonoWC	Monofilament Billet Weld and Compaction Data	17
MonoExt	Monofilament Billet Extrusion Data	32
MultiProd	Multifilament Billet Assembly Data	22
MultiWC	Multifilament Billet Weld and Compaction Data	17
MultiExt	Multifilament Billet Extrusion Data	32
HEATHIST	Wire Heat Treatment Data	21
STNDPROD	Strand Production Data	11
STNDTEST	Strand Test Data	27
STNDMAP	Strand Map Data	10
CblProd	Cable Production Data	16
ExtStnd	Extracted Strand Data	27

The SSCL is primarily a Macintosh environment but several of the vendors participating in the VQP are collecting data on DOS equipment. FoxBASE+/Mac is a Macintosh based data management system chosen for its compatibility to PC/MS-DOS versions of FoxBase+ and dBase III.[1] This facilitates the transfer of data without the need for reentry, therefore reducing the possibility of human error.

In addition to analyzing data entered in Numeric Fields, FoxBASE+ can generate reports that allow the operator to filter fields which contain information not suitable for wide distribution, such as proprietary process data.

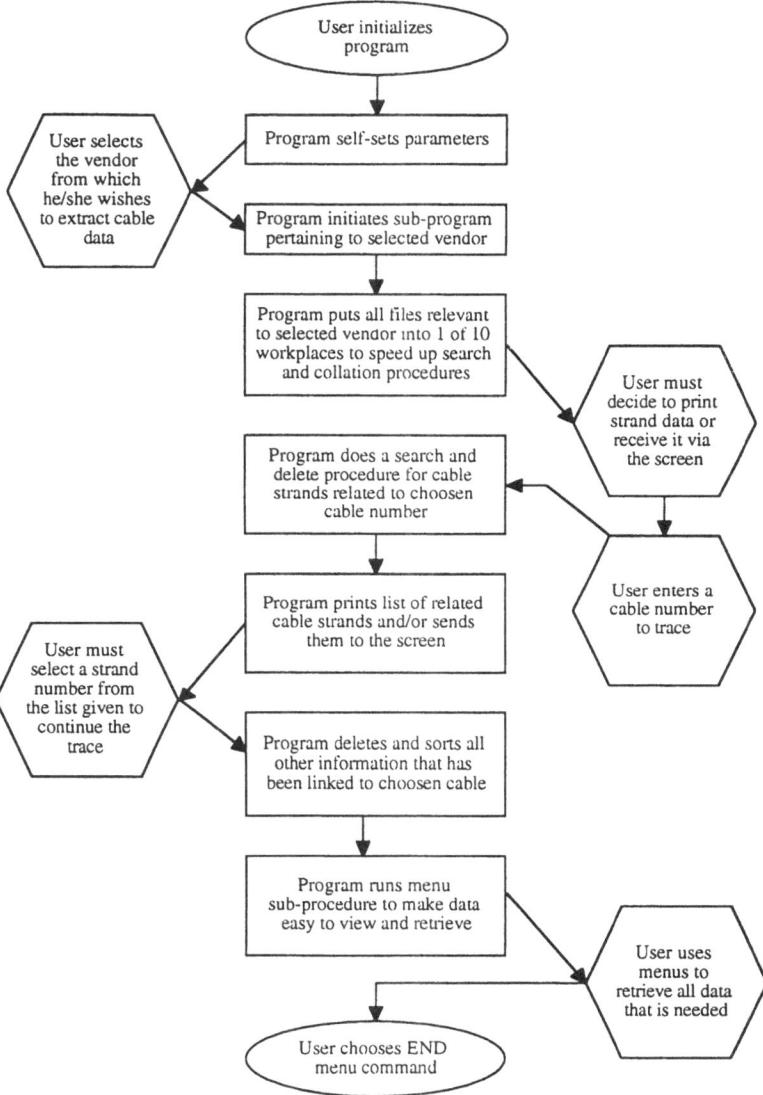

Figure 1. Flow chart of "Cable Info-Link"

Programming

FoxBASE+ uses the industry-standard dBase programming language. A program has been written to create a relational database system. The goal is to enter a cable number and trace process and raw materials data for the individual strands of superconductor (Inner CDM cable, 30 strands; Outer CDM cable, 36 strands; CQM cable, 30 strands). The system goes about this process by performing various search and delete sub-procedures which enable for quick and easy data retrieval. Figure 1 is a flow diagram of the program which has been titled "Cable Info-Link".

The major design setback for the program to work is it requires data from the later stages of cable development. Because of the differences in data entry each vendor file requires a customized program.

FUTURE PLANS

When the VQP started in September 1991, FoxBASE+ appeared to be the best database package available. Further investigation proved that 4th Dimension (4-D) might actually be more suitable. The importing of Phase I data into 4-D (which is also IBM compatible) is in process and there will soon be a running version of this program. At the time of completion, a comparison of the two database packages containing identical data will be conducted to determine the most appropriate package for data collection during full rate production.

The major difference between the two applications is that to obtain the desired end result, FoxBASE+/Mac has to do numerous search, sort, and delete routines. With 4-D the procedure is much simpler. By using a process known as singular relation[2], it allows users to relate, view, and filter the fields without actually tampering with the data. A runtime only version of 4-D is also available at a reduced cost which makes distribution to vendors easy and cost effective.

REFERENCES

1. "FoxBASE+/Mac Users Guide" Fox Software, Inc., Perrysburg, OH (1991).
2. "4th Dimension User Reference" ACI US, Inc., Cupertino, CA (1992.)

RESULTS OF OUTER WIRES AND CABLE FOR THE SSC VQP
AT HITACHI CABLE, LTD.

S. Sakai[1], G. Iwaki[1], Y. Suzuki[2], M. Seidoh[2], H. Moriai[2], T. Kamiya[2], T. Ohta[3], H. Noda[4], and K. Kamata[5]

[1]System's Material Laboratory, [2]Tsuchiura Works, [3]Toyoura Works
[4]Hitaka Works, [5]Tokyo Head Office
Hitachi Cable, Ltd.
3550 Kidamari-cho, Tsuchiura-shi, Ibaraki-ken 300, Japan

ABSTRACT

Hitachi Cable, Ltd. has finished phase IB practice for the outer strands and cables. We were required to process of 20 pieces of the multifilamentary single stacked outer billets extruded using our present extrusion facility and to practice cabling operation. In phase IB, the electro-magnetic, mechanical, and dimensional properties of wires and cables were excellent meeting required specifications. However, wire breakage often occurred and average wire piece length was 3,340 m.

At present, we move into phase II practice. The 0.648 mm diameter outer strands fabricated from the first 11 pieces of the multi billets among the 36 pieces of those required for us. In these 11 pieces, the average wire piece length is dramatically improved to 12,100 m because of the application of the process to increase bonding force between monofilamentary wires.

INTRODUCTION

Low initial and operating costs, high stability and reliability are required for the superconducting magnets of the particle accelerator. To meet these requirements, superconducting cables should play an important and leading role. On this point of view, Hitachi Cable has developed various types of superconducting wires and cables. In these few years, large keystone angle cables, the high current density wires over 3,200 A/mm^2 at 5 T, and low AC loss cable with Cu-Mn alloy matrix had been developed.[1] Although these wires and cables achieved the excellent superconducting electro-magnetic properties, the yield on the wire fabrication has not been sufficiently discussed. For the establishment of the superconducting wire and cable production process to get high yield and reliability, we, Hitachi Cable attend the SSC vender qualification program (VQP).[2]

In this paper, points of our fabrication procedure are described first, and piece length distribution and the STRAND REST results of dimensional, mechanical and electrical testings for the phase IB and the first 11 pieces of the multi billets for phase II are reported.

FABRICATION PROCEDURE

The monofilament billets and the multifilament billets are extruded by the hydrostatic extrusion press. The hydrostatic extrusion method is the most suitable for getting extruded composite materials with high uniformity in the cross-sectional and longitudinal direction. On the phase IB, the 20 pieces of multifilament billets were to be assembled using monofilament rods processed from 15 pieces of the monofilament billets. The 36 multifilament billets are to be assembled from the 30 monofilament billets for phase II.

We procure NbTi bars and Nb sheets meeting the requirements in the specifications No. SSC-Mag-M-4000A[3] and SSC-Mag-M-4001[4] each from Teledyne Wah Chang Albany. The volume fraction of the Nb barrier in a filament is designed to 4.8 %. The Cu/SC(Nb,NbTi) volume ratio for the monofilament rod is designed 0.44 corresponding to the S/D (Spacing/Diameter) ratio of 0.20 for phase IB, 0.52 corresponding to the S/D ratio 0.23 for phase II. The billet diameters are 166 mm for multifilamentary strand, and the total length is 1,180 mm. The extruded monofilament rod of 50 mm in diameter is drawn and finally straightened and cut a specified length after forming it to a hexagonal shape with a flat to flat width of 1.77 mm.

The assembly of a multifilament billet is made by inserting 3,966 pieces for phase IB and 4,164 pieces for phase II of straightened and cut hexagonal monofilament rods into an Oxygen Free Copper (OFC) can around a OFC central rod. Designed values for the packing factor inside the OFC can and Cu/SC ratios are 93.2 % and 1.73 in phase IB, 94.4 % and 1.80 in phase II. The multifilament extrusions of 44 mm in diameter are drawn to the final size 0.648 mm in diameter through 3 times aging heat treatments at 405 °C for 50 hours.[1]

Table 1 shows the specified values and the Hitachi Cable design values for the major parameters of the strands and cables. In phase IB, we used 2 stage of non powered type Turk's head rollers designed by ourselves for stabilizing the dimensions of 36-strand outer cable along the length during operation. In phase II, we procure the motor powered type Turk's head rollers to decrease the stage and improve the dimensions of cable. The dimensions for keystone angle, width and thickness are monitored at every 5 feet interval along whole the length during cabling operations using the CMM calibrated by the SSCL.

Table 1. Specifications and Hitachi Cable Design for the SSC outer strand and cable.

Item		unit	specification	design phase IB	phase II
Strand:	Diameter	mm	0.648±0.0025	0.648	0.648
	Cu/SC ratio		1.8±0.1	1.73	1.80
	Filament Diameter	μm	6±0.5	6.2	6.0
	Twist Pitch	mm	13±1.5	13	13
	Ic at 5.6T	A	≥286	as	as
Cable:	No. of Strands		36	as	as
	Mid Thickness	mm	1.156±0.006	as	as
	Width	mm	11.68+0.05,-0.00	as	as
	Keystone Angle	deg.	1.02±0.06	as	as
	Ic at 5.6T*	A	≥9,780	as	as

* Cable Ic including self-field correction

THE RESULTS FOR STRAND PRODUCTION, STRAND TEST, AND CABLES

Figure 1 shows the cross-sections for the obtained strand and cable. The composites have been worked very uniform in the cross-section. The average total length, average piece length and piece number for 0.648 mm diameter outer strands obtained finally for phase IB are 45,700 m/billet, 3,340 m and 14, and 56,200 m/billet, 12,100 m and 4.6 for phase II. The average yields are approximately 80 % for phase IB and 94 % for the first 11 billets of phase II. In the first billets of phase II, average piece length is dramatically improved to 12,100 m because of the applications of the process to increase bonding force between monofilamentary wires. The cold weld-free cable yield is expected as 97 %.[5]

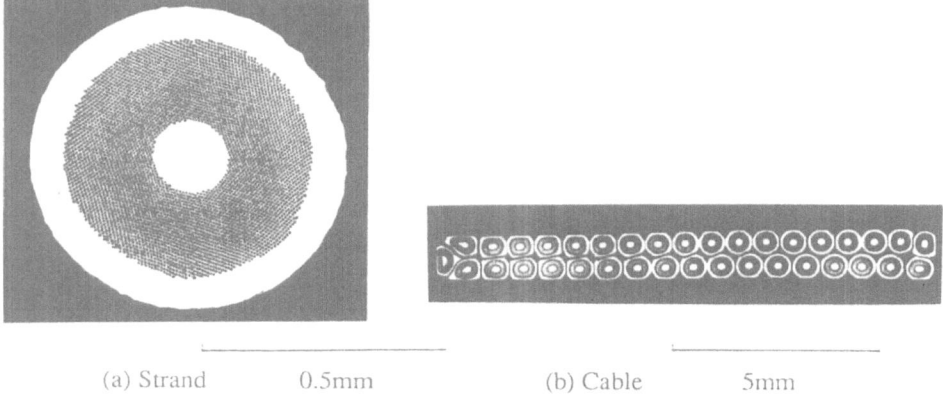

(a) Strand   0.5mm   (b) Cable   5mm

Figure 1. Cross-sections for the outer strand and cable.

Figure 2 shows the piece length distribution for phase IB and the first 11 pieces of multifilament billets of phase II. Phase II' shows the calculated result of the piece length distribution in the case of no cutting after extrusion for phase II. We have planned the installation of combine type drawing facility to avoid the cutting extrusions.

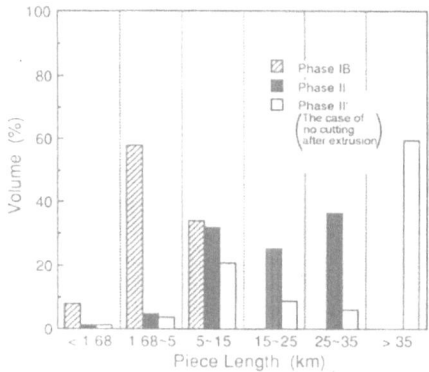

Figure 2. The piece length distribution for phase IB and phase II.

Table 2 shows the STRAND TEST results. The testings for these properties are practiced in accordance with the specification No. SSC-Mag-M-4146[6], APPENDIX B SUPERCONDUCTING WIRE TEST METHOD, SSC-Mag-T-9001~9006. The Cu/SC ratios of strands in phase II are increased to meet the specification, sufficiently. In the properties of spring back test, values in phase II is smaller than those in phase IB. The final annealing temperature is elevated from 240 °C to 270 °C. In this range of temperature, as the flexibility of the OFC material is rapidly changed[7], the spring back test values in phase II are decreased. The critical currents Ic are around 300 A in phase IB and 295 A in phase II. Ic values of phase II strands are decreased by the increase of Cu/SC ratio. The critical current densities of strands in phase II are increased from 2,430 A/mm^2 at 5.6 T to 2,490 A/mm^2, contrary to the Ic values. Ic's of the cable in phase IB are in the range of 10,500 A at 5.6 T to 10,700 A, to the specified values over 9,780 A. Ic degradation estimated by comparing the Ic's for strands before and after cabling is approximately 1.5 % at 5.6 T.

Table 2. STRAND TEST results.

	Diameter (mm)	Cu/SC	Spring Back (deg.)	Sharp Bend (%)	Ic at 5.6T (A)	n Value	RRR
Phase IB	0.647~ 0.649	1.70~ 1.72	655~ 690	0	294~ 306	50~ 55	148~ 166
Phase II First 11 billets	0.647~ 0.649	1.78~ 1.84	540~ 620	0	287~ 302	45~ 49	127~ 145
Specification	0.648±0.0025	1.8±0.1	< 1,090	No Crack on Cu Surface < 1% filament damage	≥ 286	≥ 35	≥ 70

SUMMARY

We are required 20 pieces of multi billets for the outer strands and cables in phase IB, 36 pieces in phase II. Phase IB practice has been finished, and at present the first 11 pieces of multi billets for phase II were fabricated as the final outer strands.

The average piece length was 3,340 m in phase IB. In the first 11 pieces for phase II, the average piece length is dramatically improved to 12,100 m because of the application of the process increase bonding force between monofilamentary wires. The average yields on multi billet process are improved from 80 % for phase IB to 94 % for the first 11 billets of phase II, too. The STRAND TEST results of dimensional, mechanical and electrical testings for the phase IB and the first 11 pieces of the multi billets satisfied the specified requirements.

REFERENCES

1. S. Sakai, et al., "Supercollider 4," Plenum, 1081(1992)
2. K. Kamata, et al., "Applied Superconductivity Conference in Chicago (August 1992)"
3. Superconducting Super Collider Lab. Magnet System Division Specification, No. SSC-Mag-M-4000A, "NIOBIUM TITANIUM ALLOY BARS AND RODS"
4. Ibid., No. SSC-Mag-M-4001, "NIOBIUM BARRIER MATERIAL"
5. J. M. Seuntjens, et al., "Supercollider 4", Plenum, 685(1992)
6. Superconducting Super Collider Lab. Magnet System Division Specification, No. SSC-Mag-M-4146, "NbTi SUPERCONDUCTING WIRE FOR SSC DIPOLE MAGNETS (OUTER)"
7. S. Sakai, et al., "Hitachi Cable Review," 81(1989)

STUDY ON THE AC MAGNETIZATION OF LHC TYPE OF RUTHERFORD CABLES

A. P. Verweij,* L. E. Eriksson, and H. H. J. ten Kate

University of Twente
Applied Superconductivity Centre
Enschede, The Netherlands
*presently at CERN, Geneva

INTRODUCTION

In accelerator magnets, and especially at low excitation values, the field errors are mainly determined by the filament hysteresis. To obtain an understanding of this hysteresis, measurements are normally performed on strands under the assumption that the magnetization of the cable is equal to the magnetization of the strands. In this paper a new test arrangement is described to measure the AC and DC magnetization of flat cables up to lengths of about 13 cm. Magnetization measurements on various NbTi keystoned Rutherford type of cables are presented. In order to investigate the influence of the geometry of the cable in terms of the strand and cable pitches, some strands are also separately tested in both their original (straight) and "cable" shapes. The field is applied parallel to the wide side of the cable. One can calculate for this case that the magnetization is mainly determined by the hysteresis of the filaments and the inter-filament coupling. For strands with a copper matrix the inter-strand coupling in the parallel field is negligible, even for relatively low inter-strand contact resistances. An experimental study of the inter-strand magnetization is only relevant if the transverse pressure on the cable under test conditions is about the same as the pressure in the coils in which the cable is used.

THEORETICAL MODELS FOR THE MAGNETIZATION

A numerical model is used to calculate the filament hysteresis, analogous to the model described by Hartmann[1]. The main advantages are that the self field is included, that there is no assumption made on the shape of the boundaries between saturated and non-saturated areas and that any $J_c(B)$ relation can be implemented easily.

A general description of the inter-filament coupling loss is given by Campbell[2]. Here the main formulas are briefly noted. In a strand with a twist pitch l_p, an electromotive force will be induced between the filaments due to the internal field variation dB/dt. The induced voltage will give rise to a transverse electric field $E_x = -l_p \, dB/dt/(2*\pi)$. The total field in the strand $B = B_a - \tau*(dB/dt)$, with the time constant of the system and defined by $\tau = \mu_{eff} \tau_0$ with $\tau_0 = \mu_0 l_p^2/(8\pi^2 \rho_e)$ for round wires. The effective resistivity ρ_e depends on the SC filling factor η as $\rho_e = \rho_0*(1\pm\eta)/(1\mp\eta)$. The signs depend on whether the filaments do or do not contribute to the transverse conductivity. The μ_{eff} depends on the shape of the sample: $\mu_{eff} = \mu/(1+N(\mu-1))$ with N the demagnetizing factor. For a sinusoidal variation of the applied magnetic field B with an amplitude B_a^m, the loss per cycle Q per unit volume is determined by the integral:

$$Q = \int_0^T M dB_a = \frac{n\pi\mu_{eff}\omega\tau\left(B_a^m\right)}{\mu_0\left(1+(\omega\tau)^2\right)},$$

with $n=(1-N)^{-1}$ the shape factor of the sample.

MEASURING EQUIPMENT AND SAMPLES

The cold part of the measuring equipment is shown in Figure 1. It consists of a concentric set of four

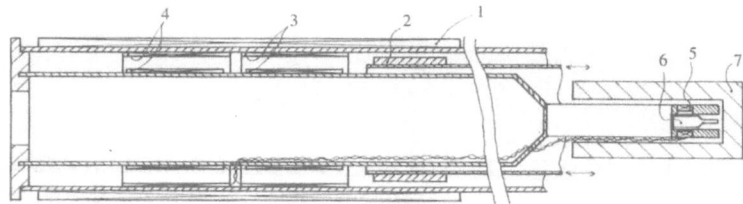

Figure 1. The cold part of the measuring equipment: 1: AC solenoid, 2: Sample, 3: Upper pick-up coils, 4: Lower pick-up coils, 5: Sensing coil, 6: Hall probe, 7: PbBi shield.

pick-up coils that can be inserted in a superconducting solenoid providing an AC field of 1T. The magnetization is measured on a ring shaped sample placed between the two pick-up coils of the upper half of the set. By this way both the inductive voltage and the empty coil effect are reduced to a minimum. The circumference of the sample is about 130mm. The pick-up coils are wound from a superconducting wire and are connected to a small superconducting sensing coil. The current in this circuit is proportional to the time integral of the pick-up voltage. The circuit can be regarded as a flux transformer, where the flux change in the pick-up set, due to the magnetization of the sample, is transferred to the sensing coil. It is therefore also possible to perform accurate measurements of the magnetization of a sample under DC conditions. The flux in the coil is measured using a Hall probe, located inside the sensing coil. The calibration factor relating the Hall probe voltage to the magnetic moment is calculated and measured using a Nb foil and is accurate within 1%. In order to eliminate the stray field of the solenoid and other magnetic disturbances both the sensing coil and the Hall probe are placed inside a superconducting shield made of PbBi. Two heaters are present to enable a reset of the magnetic shield and the flux transformer in the case flux has penetrated through the shield or is trapped in the flux transformer circuit.

Five different cables are measured of which the main characteristics are listed in Table 1. The samples S5, S6 and S7 are used in several 1m LHC dipole model magnets; S5 and S7 for the inner layer and S6 for the outer layer. The samples S1 and S2 are envisaged to be used for the inner layer of the LHC itself. Additionally there are two samples S1so and S1ss which consist of 4 strands of S1 in their original (twisted) shape and in a straightened shape respectively. For S1ss the strands are completely separated from each other while for S1so the strands cross each other several times.

Table 1. Overview of the main sample parameters.

	S1	S2	S5	S6	S7
Width (mm):	17.0	17.0	17.0	17.0	17.0
Thickness (mm):	2.04/2.50	2.04/2.50	2.04/2.50	1.30/1.67	2.04/2.50
# of strands:	26	26	26	40	26
$\varnothing$ strand (mm):	1.29	1.29	1.29	0.825	1.29
Lp strand (mm):	117	130	130	95	130
Cu/SC ratio:	1.6	1.6	1.9	1.8	1.735
# of fil. per strand:	27954	21780	900	2550	3060
Lp fil. (mm):	25	25	25	25	25
Nominal $\varnothing$ fil. (μm):	4.8	5.4	25	9.5	14
J_c at 8T, 4.2K (A/mm^2)	938	1087	986	1136	1084

RESULTS AND DISCUSSION

All the results are presented per volume sample; to obtain the results per volume superconductor the magnetization has to be multiplied by $(1+\lambda)$ with λ the Cu/SC ratio (see Table 2). All measurements are performed using a sinusoidal applied field with peak to peak value of B_{pp}.

The magnetization loops for the samples 1 to 5 for $B_{pp}=1.2T$ are shown in Figure 2. At the frequency of 0.02Hz only a very small fraction of the magnetization is due to the coupling currents. It can be easily seen that the magnetization curve of the various samples have rather different shapes. Due to the proximity effect, especially S1 and S2 show a relatively large increase of the magnetization for small values of B. For S2 this effect is even more pronounced which could be due to the fact that the distance between the bundles of filaments is smaller than for S1 resulting in an inter-bundle coupling. The peak in the magnetization curve is always displaced slightly from the origin as the internal field is not zero due to the screening currents. As

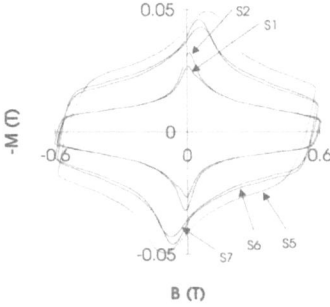

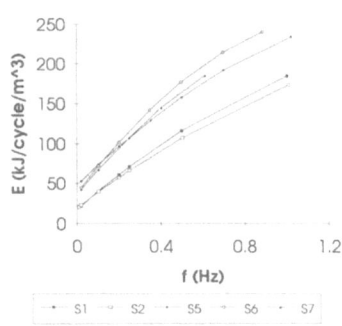

Figure 2. Magnetization loops for f=0.02Hz and B_{pp}=1.2T for S1, S2, S7, S6 and S5 (in order of increasing magnetization at B=0T).

Figure 3. The energy loss per cycle as a function of the frequency for B_{pp}=1.2T.

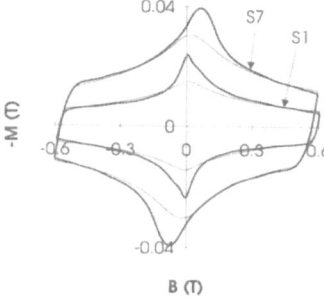

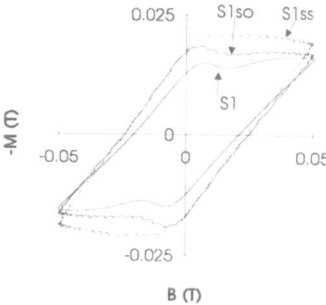

Figure 4. Comparison between the experimental (bold) and the theoretical (thin) M(B) curve for S1 and S7 for f=0.02Hz and B_{pp}=1.2T.

Figure 5. Magnetization loops for S1, S1so and S1ss for f=0.02Hz and B_{pp}=0.1T.

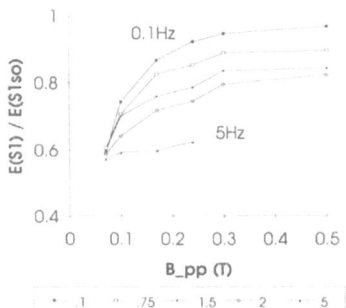

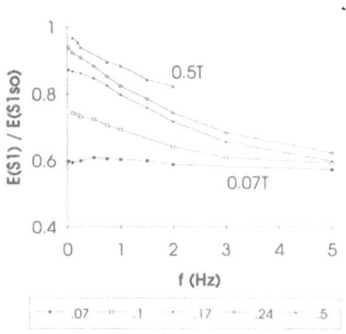

Figure 6. E (S1)/E(S1so) as a function of the peak-peak value of the applied field for several frequencies between 0.1Hz and 5Hz.

Figure 7. E (S1)/E(S1so) as a function of the frequency for several values of B_{pp} between 0.07T and 0.5T.

expected this displacement increases for increasing filament diameter and critical current density. After subtraction of the coupling current part the obtained curves are used to determine the nCu $J_c(B)$ relation by fitting the Kim relation $J_c(B)=J_0/(1+|B|/B_0)$. The results are presented in Table 2. The fit is at least valid in the field range from 0.2 to 0.5T.

Table 2. Determined parameters of the Kim relation that describes $J_c(B)$ and of the effective resistivity.

	S1	S2	S5	S6	S7
B_0 (T)	0.30	0.32	0.30	0.35	0.31
J_0 (10^{10} A/m2)	3.7	3.1	2.3	4.5	3.0
ρ_{eff} (10^{-10} Ωm)	1.1	1.1	1.1	0.9	0.9

The energy loss, defined as the area of the $M(B)$ curve, is shown in Figure 3. The inter-filament loss of the samples differs by about a factor of 1.5. S1so has the same loss as S1 for low frequencies and high field amplitudes while the coupling loss of S1ss is about a factor $cos^{-1}(\theta)$ higher than S1 with θ the cable twist angle. Table 2 presents the effective resistivities as can be determined from the slope of the curves at f=0Hz. These effective resistivities are approximate values since there is no unique time constant due to the non-uniformity of the local field.

Figure 4 gives the comparison between two calculated and measured hysteresis loops. As proximity effects are not included in the model there is a discrepancy between the curves for small field values. For higher fields the model and the measurements are in good agreement. To obtain a better agreement at low fields one should suppress the proximity effect by simply etching away the copper matrix.

Figure 5 shows the magnetization loops of S1, S1so and S1ss for a small B_{pp} and a low frequency. Two bumps are seen in the curve of S1 and S1so while only one is observed for S1ss. The second bump is no longer present for higher B_{pp} nor for S5, S6 and S7. It is therefore likely that this effect is due to the magnetization caused by the field change longitudinal to the strands.

Figures 6 and 7 show the ratio of the energy loss per cycle per volume cable between S1 and S1so as a function of B_{pp} and the frequency respectively. It can be clearly seen that for small B_{pp} ($B_{pp}<2B_p$ with B_p the penetration depth) or a high frequency the demagnetization of the cable becomes relevant. This demagnetization effect is qualitatively in good agreement with calculations of Eikelboom on flat cables[3]. For low frequencies and high B_{pp} the coupling current part of the magnetization of S1 and S1so is decreased by a factor $cos(\theta)$ compared to S1ss (with θ the twist angle of the strands). These figures show that it is possible to estimate the losses of a cable at low frequency and high field amplitudes by measuring a single strand.

CONCLUSIONS

A new test arrangement has been developed to measure the AC and DC magnetization of Rutherford type of cables up to lengths of one cable twist pitch which is about 13 cm.

A good understanding of the magnetization of a Rutherford cable can be obtained by simply measuring the magnetization of a strand in a certain range of field and frequency. Only for small field values or high frequencies the demagnetization of the cable will cause a considerable decrease of the magnetization. This difference can be reduced by winding the strand into a coil with a similar geometry as the cable.

For high field variations parallel to the wide side of a cable and for low frequencies the inter-filament coupling loss of a cable is a factor $cos(\theta)$ smaller than that of straight strands with θ the cable twist angle.

ACKNOWLEDGMENTS

This work is part of a collaboration between the Applied Superconductivity Centre of the University of Twente and CERN in the framework of the LHC magnet development program and supported in part by FOM, the Netherlands Foundation for Fundamental Research on Matter.

The authors would like to thank Dr. L. Oberli for providing the cable samples of various manufacturers.

REFERENCES

1. R.A. Hartmann, A contribution to the understanding of AC losses in composite superconductors, PhD thesis, University of Twente, The Netherlands (1989).
2. A.M. Campbell, A general treatment of losses in multifilamentary superconductors, *Cryogenics*, Vol. 22, 3-16 (1982).
3. J.A. Eikelboom, AC losses in prototype conductors for the NET toroidal field coils, PhD thesis, University of Twente, The Netherlands, (1991).

THE EFFECT OF SILICON ADDITION TO THE INTERFILAMENTARY
COPPER ON J$_C$, COMPOUND FORMATION AND INTERDIFFUSION

H. Liu,[1] K. J. Faase,[2] E. Gregory,[1] and B. A. Zeitlin[1]

[1]IGC Advanced Superconductors, Inc.
Waterbury, CT 06704

[2]Department of Mechanical Engineering
Oregon State University, Corvallis, OR 97331

INTRODUCTION

One of the reasons why high critical current density is difficult to achieve in fine filament Nb-Ti superconducting wire is that a reaction occurs between the copper matrix and Nb-Ti filaments. A diffusion barrier around each filament was introduced in the processing of fine filamentary wire in order to achieve J$_c$ values close to the intrinsic ones.[1] One study of diffusional reaction rates through the Nb barrier[2] has indicated that, for typical SSC composites, a barrier area of 4% and 9% is necessary for producing 6μm and 2.5μm diameter filaments respectively. Consequently, if diffusional interactions can be eliminated without adding a large volume of barrier material, it is possible to achieve higher J$_c$'s at lower cost. Another limitation on the J$_c$ in fine filament Nb-Ti superconducting wire results from the mismatch in mechanical properties of Nb-Ti filaments and copper matrix at high wire strains. The hardness and ultimate tensile strength (UTS) of Nb-Ti filaments increase with increasing amount of the cold work and no UTS saturation has been seen[3], whereas the UTS of copper saturates. An improper filament array also adversely affects J$_c$, but this can be resolved by changing the filament distribution geometry, i.e. by reducing the interfilamentary spacing.[4] Improving mechanical strength of copper matrix is important for reducing the amount of fine filament sausaging. Recently, in work that was primarily directed towards the development of material for ac applications, it was reported that, when silicon is added to the copper matrix, the formation of intermetallic compounds can be greatly reduced[5,6]. Cu-Si alloy also has mechanical properties more compatible with NbTi than copper. If the above results can be verified, the technique can probably be applied to the manufacture of high J$_c$ SSC type conductors and large filamentary NbTi superconductor materials for general use. The work described here was to investigate these effects of silicon additions to the Cu matrix.

EXPERIMENTAL PROCEDURES AND RESULTS

Samples

Two 152 mm diameter, 229 mm long castings were made in a vacuum melting furnace, one high purity Cu2.5 wt.% Si and the other high purity Cu3.5 wt.% of Si. Four monofilamentary cans, 99.7 mm O.D., 76.9 mm I.D. and 114.3 mm long, were made. The characteristics of the monofilamentary billets are shown in Table 1. These four billets were processed in a same way, HIP'd and extruded into 25.4 mm diameter monofilamentary rods (monos 1, 2, 3 and 4). These extruded rods were then drawn down. The 3.5 wt.%Si matrix work hardened more rapidly than did with 2.5 wt.%Si matrix, but, they were both vacuum annealed followed by a fast cool; mono 6 at a strain following extrusion of 2.2 and mono 4 at a strain of 2.9.

Table 1. Characteristics of monofilamentary billets.

Billet #	Can Matls	Filamentary Alloy	Nb Barrier
1	C101	Nb-46.5wt.%Ti	None
2	C101	Nb-46.5wt.%Ti	8%
3	Cu-2.5wt.%Si	Nb-46.5wt.%Ti	None
4	Cu-3.5wt.%Si	Nb-46.5wt.%Ti	None

Monofilamentary wires

In the filament / matrix interdiffusion experiments we have concentrated on three main factors: anneal, cold work and precipitation heat treatment. Mono 1 and 3 were drawn down from extrusion size to 1.83 mm and at this size samples were annealed. Scanning Electron Microscope Back Scatter Electron Images (SEM BEI) and Electron Microporbe (EMP) traces of the interfaces, before and after annealing, were taken. An SEM BEI of this annealed material showed there to be much less interaction in the Cu2.5wt.%Si / Nb-Ti interface than in the Cu / Nb-Ti interface. A silicon concentration peak of 20 at. % appears in the interface of the Cu2.5wt.%Si / Nb-Ti, in the EMP trace shown in Figure 1. It is interesting to note that the Cu penetrates a considerable distance into the filaments. The same type of penetration is noted, however in the Cu matrix material with an 8 % barrier. This fact is of some significance and is being investigated further [7].

Mono 1 and 3 samples were also drawn down from the extrusion size to 11.68 mm and then annealed. These wires were drawn down further to the diameters of 5.8 mm and 0.68 mm, imparting additional strains (ε) of 1.4 and 5.7 respectively. The 0.68 mm diameter samples were then given a heat treatment of 420°C/120h. Samples for SEM BEI and EMP analysis were taken in the above conditions. Table 2 shows the widths of the interdiffusion layers, measured from the EMP traces. It can be seen that Cu2.5wt.%Si reduces compound formation as well as interdiffusion that results from the anneal. Under all other conditions, Cu2.5wt.%Si reduces compound formation. Figure 2 shows the SEM BEI of the Cu2.5wt.%Si / Nb-Ti interface at a wire diameter of 5.8 mm after the anneal at 11.68 mm. The discontinuous phase at the Cu2.5wt.%Si / Nb-Ti interface is silicon rich. It has a limited ductility, as do the intermetallics formed in the Cu / Nb-Ti interface, and it fractures during the cold work.

Multifilamentary wires

Monos 1, 2 and 3 were drawn down to 1.47 mm from the extrusion size and assembled into thin walled hexagonal tubes [8] which were packed and sealed in 50.8 mm O.D., OFHC copper cans, and designated as multis 1, 2, and 3. They were HIP'd and extruded into 12.7 mm diameter rods. Three heat treatments of 375°C/40h were given at wire diameters of 4.57 mm, 2.89 mm and 1.83 mm respectively, during wire drawing. No attempt was made to process mono 4 into multifilamentary material at this time because ductility problems were anticipated. Sample wires after the last heat treatment were taken for EMP (with a resolution of 0.5 μm) analysis on matrix / filament interfaces. The EMP traces show that, for the Cu2.5 wt.% sample while there does not appear to be a concentration of Si in the interface, there is no evidence of the compound which is shown in the Cu matrix material. Metallography of wire cross sections are shown in Figures 3 and 4. These show the Cu2.5wt.%Si material to have more uniform and regularly spaced filaments than the Cu material, due to the higher mechanical strength of the copper-silicon alloy matrix.

Multis 1 and 3 wires were drawn down further to 0.46 mm after last heat treatment. The filaments as extracted from the matrices, are shown in Figures 5a & b.

J_c's were measured at 5T and 6T and the results from multis 2 & 3 wires, compared, Figure 6. Multi 1 material could not be made into J_c samples because of wire breakage due to compound formation. The J_c's of multi 3 wire , which peaked at the strain of 4.5 after the

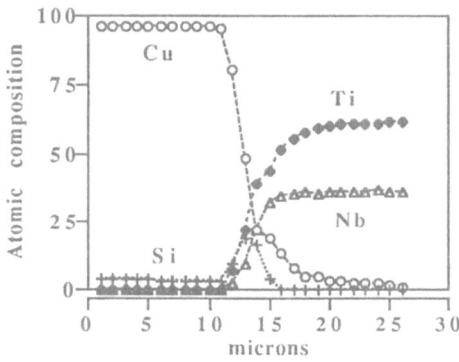
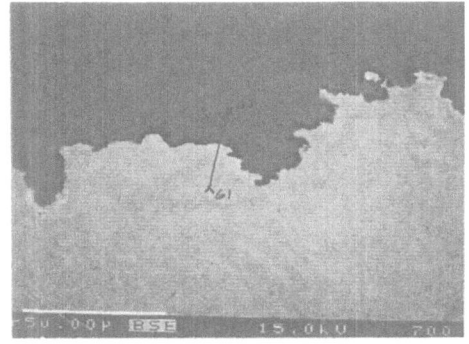

Figure 1. EMP trace across Cu2.5wt.%Si / Nb-Ti interface, 1.83mm dia.wire, annealed at 1.83 mm.

Figure 2. SEM BEI of Cu2.5wt.%Si / Nb-Ti interface, 5.8 mm dia. wire, annealed at 11.68 mm.

last HT, is at least 16% higher than the J_c's of multi 2 wire, which peaked at a strain of 4 after last HT.

Table 2. Width of the interdiffusion layers[§].

Dia.(mm)	Conditions	D_1 (μm)	D_2 (μm)
1.8	Before anneal	3* (6)	3* (6)
1.8	After anneal	>24 (42)	7 (16)
5.8	Anneal & ε of 1.4		6.4 (18) (across Si rich zone)
			4* (12) (across free Si rich zone)
0.68	Anneal & ε of 5.7	5.2* (11) (across free remnant intermetallics)	3.5* (11) (across free Si rich zone)
0.68	Anneal, ε of 5.7 & 420°C/120hr.	4.5* (11) (across free remnant intermetallics)	3.7* (10) (across free Si rich zone)
		19 (24) (across a remnant intermetallic)	

§: The bracketed values under D_1, D_2 are the interdiffussion widths between Cu/Nb-Ti, and Cu2.5wt.%Si/Nb-Ti, respectively, and are defined as the distance between the Cu and Ti traces at one atomic percent level. The values not in the brackets are defined as the distance between the normal Cu level in the matrix and the normal Ti level in the filaments *: It should be pointed out that the EMP we used in this analysis has a resolution of 3 μm. Therefore, these data are in a range approaching the limitation of EMP resolution. The locations where the EMP traces were taken are shown in the brackets following the measured values.

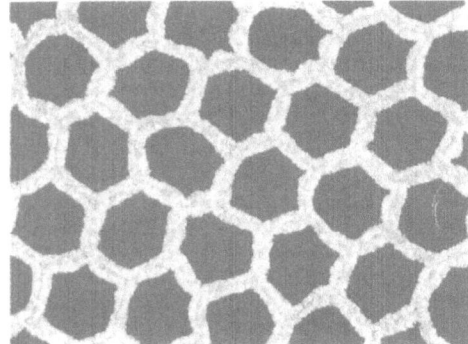
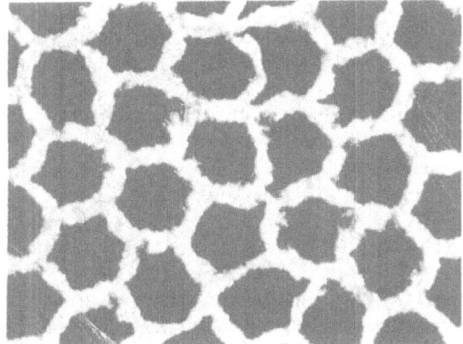

Figure 3. Cross section of multi 3 wire, at a dia. of 1.83 mm after last HT. 260X

Figure 4. Cross section of multi 1 wire at a dia. of 1.83 mm after last HT. 260X

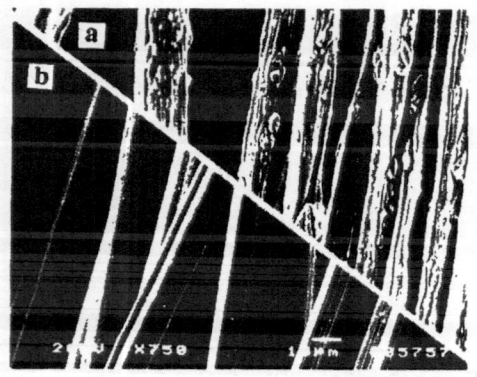

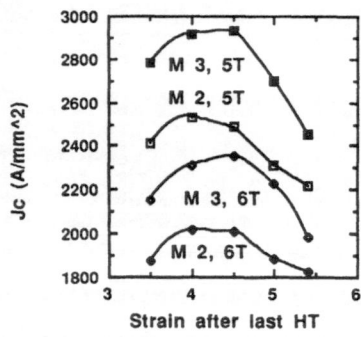

Figure 5. The surface of the filaments extracted from **a**, multi 1, **b**, multi 3, after the last HT.

Figure 6. J_c vs. ε at 5T and 6T for multi 2, M2, and multi 3, M3 wires.

SUMMARY

Interdiffusion between Nb-Ti filaments and a Cu2.5wt.%Si matrix has been studied and J_c data on multifilamentary materials obtained. The results show that the addition of 2.5wt.%Si to a copper matrix significantly reduces compound formation that takes place at the matrix / Nb-Ti interface, compared with that which occurs when OFHC copper is the matrix. A 50.8 mm diameter multifilamentary billet with Cu2.5wt.%Si as the matrix between the filaments has been successfully processed into fine wire. Enhanced matrix strength and resulting improved filament quality have been shown in the material with Cu2.5wt.%Si. Material with Cu2.5wt.% Si as the interfilamentary matrix also exhibits J_c's superior to those shown by Nb shielded Nb-Ti filaments in an OFHC copper matrix.

ACKNOWLEDGEMENTS

The authors would like to thank J.M.Seuntjens of SSCL for his helpful discussions, and EMP analysis results and Brian Boyle of IGC-ASI for his assistance with the experiments. This work was carried out under a Phase I grant, (No. DE-FG02-92ER81332) from the Small Business Innovative Research Office of DOE/SSC.

REFERENCES

1. T.S. Kreilick, E. Gregory, and J. Wong, "Fine Filamentary NbTi Superconducting Wires", Adv. in Cryo Eng.,A.F. Clark & R.P. Clark, 32 pp 739-745, (1986).
2. K.J. Faase, P.J. Lee, J.C. McKinnell, and D.C. Larbalestier, "Diffusional Reaction Rates Through the Nb Wrap In SSC and Other Advanced Multifilamentary Nb46.5wt.%Ti Composites," *Adv. in Cryo Eng.*, 38B:723 (1991).
3. Z.Guo, and W.H. Warnes, "Mechanical Behavior of Fine Filament Nb-Ti as a Function of Processing," *IEEE Trens actions on Applied Superconductivity*, 3,1: 1022, (1993).
4. E. Gregory, H. Liu, G.M. Ozeryansky, M.D. Sumption, K.R. Marken Jr., and E.W.Collings, "Experiments To Improve Materials For SSC Magnets," "Supercollider 4," John Nonte, ed., Plenum Press, New York, pp.923-930, (1992).
5. K. Tachikawa, J. Ninomiya, T.Ajioka, M. Terada, and K. Sakinada, "Recent Studies On Composite Superconductors," 7th U.S.-Japan Workshop on High Field Superconductors, Fukuoka, Japan, Oct. 22-24, 1991.
6. S. Akita, S. Torii, H. Kasahara, K. Matsumoto, Y. Tanaka, T. Ajioka, and K. Tachikawa, "Ultrafine Multifilamentary Nb-Ti Wires with Cu-Si Alloy Matrix," *Cryogenics*, 33,2:199 (1993).
7. J.M. Seuntjens, SSCL, Private communication.
8. H. Kanithi, P. Valaris, and B.A. Zeitlin, "A Novel Approach to Make Fine Filament Superconductors," "Supercollider 4," John Nonte, ed., Plenum Press, New York, pp. 41-47, (1992).

CONDUCTOR DESIGN FOR THE GEM DETECTOR MAGNET

J. V. Minervini, P. G.Marston, B. A. Smith, R. J. Camille,
Z. S. Piek, R. F. Vieira, and P. Titus

MIT Plasma Fusion Center, 167 Albany Street, Cambridge, MA 02139

G. Deis, N. Martovetsky, and P. J. Reardon

SSC Laboratory,* 2550 Beckleymeade Avenue, Dallas, TX 75237

R. Stroynowski

Southern Methodist University, Dallas, TX

INTRODUCTION

The Gammas, Electrons, Muons (GEM) Detector, one of the two large detectors planned to be built at the SSC, features high muon momentum resolution. This is achieved by magnetization (by a huge magnet, about 20 m in diameter and 31 m long) of roughly 10,000 m^3 of space within the muon chambers. The GEM Detector Magnet[1] should be designed to operate with highest possible reliability level to ensure maximum availability of the Detector Systems. That means that the magnet and the conductor should be as stable as practically possible. The conductor should be reliably protected against overheating and electrical breakdown in the case of a quench or fast discharge. For reliability reasons, the magnet dump voltage is relatively low, 500 V to ground, which implies low current density in the conductor. Table 1 lists general requirements for the conductor.

There are several conductor options for such a big magnet and the final choice of the conductor is always a tradeoff between the performance, cost, reliability, quality assurance, R&D efforts, and readiness of industry to manufacture the conductor.

Table 1. General requirements for GEM conductor

Central field	T	0.8
Peak field in winding	T	1.6
Operating temperature	K	4.5 – 4.8
Maximum hot spot temperature in a quench event	K	100
Maximum Dump Voltage to ground	V	500
Stored Energy	GJ	2.5
Charging time	hr	8

Indirectly Cooled Conductor (ICC) has been used in most detector magnets for high energy physics so far.[2,3] Forced Flow Cooled conductors (FFC), with smooth channels for helium circulation showed good stability and reliability in magnets for fusion research.[4,5] The third option is the Cable-in-Conduit Conductor (CICC), where strands are in intimate contact with helium for stabilization. CICC was tested in several magnets for fusion[6,7] and energy storage[8] and showed very

* Operated by the Universities Research Association, Inc., for the U. S. Department of Energy under Contract No. DE-AC35-89ER40486.

high stability and low ac losses. Although ac loss is not an issue in dc magnets, high conductor stability is a very attractive feature in a big magnet such as GEM. Table 2 shows comparative energy margins for these three conductor types.

Table 2. Energy margins for different conductor options for GEM.

	ICC (20 kA)	FFC (20 kA)	CICC (50 kA)
Energy margin, J/m	7	50	315
Equivalent displacement, mm	0.35	2.5	7

It is worth noting that for a monolith conductor with high current and, consequently, with a big cross section of stabilizer, the stability is affected by slow diffusion of the current from the superconducting cable to the stabilizer.[9] Proper design of the conductor can relax this phenomenum, especially in FFC, but slow current diffusion still substantially affects stability. In CIC conductor, the cable is evenly distributed and this effect does not take place if distribution of the current over the cross section is uniform.

These numbers show that for the CIC conductor the tolerable disturbances are so large that coil design can eliminate the very laborious and delicate operation of impregnation, which saves a lot of winding time and effort which is why cable-in-conduit was chosen for the GEM Magnet. Although the winding seems to be easier to make with CIC, essential efforts should be applied on quality assurance/quality control during CIC manufacture and verification tests. Figure 1 shows the cross section of the GEM conductor.

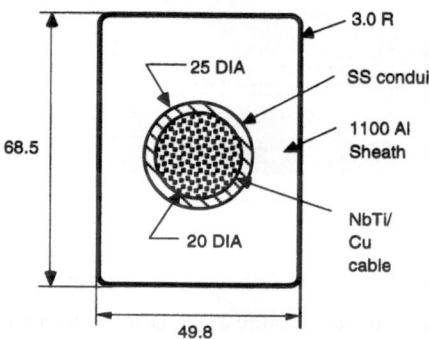

Figure 1. Cross section of the GEM Conductor. Dimensions are in mm.

RATIONALE AND CONDUCTOR DESIGN

Operating Current: Higher current is desirable from protection considerations, smaller number of lengths and joints, smaller number of turns in the magnet, smaller amount of insulation and better mechanical integrity. Lower current offers lower cryogenic loads due to the current leads, easier manufacture and transportation, less expensive current supply, distribution and protection equipment. The 50 kA operating current is chosen as a reasonable compromise within industry manufacturing capability.

Conductor Length: A conductor length of 1140 m is chosen to be the maximum practicable from the manufacturing point of view. The coil design uses one length of the conductor for one coil segment. There are 24 coil segments altogether for a total conductor length of 27.4 km.

Parameters of the Strand: To take advantage of the high heat capacity of the helium in the conduit, one has to provide the conditions, corresponding to the "well cooled regime," where heat transfer from the strand in the normal state should be higher than Joule heat generation. The criterion describing those conditions is :

$$I_{op} \leq (hPA_{Cu}(T_c - T_b) / \rho_{Cu})^{0.5} \quad \text{where:}$$

I_{op} – operating current, A; h – heat transfer coefficient (W/m^{2}K); P – wetted perimeter (m); A_{Cu} – area of copper stabilizer in cable (m^2); ρ_{Cu} – copper stabilizer resistivity (Ohm*m); T_c – critical temperature (K); and T_b – initial helium temperature (K). On the other hand, the enthalpy of the helium available in the event of a disturbance is driven by the temperature of the current sharing between the SC filaments and copper matrix, so the lower the ratio I_{op}/I_c, the more the energy margin is. For GEM this ratio is chosen conservatively at 0.25. Table 3 lists the strand parameters for GEM CICC.

Table 3. Main parameters of the strand

Parameter	Value
Total wire length in the magnet (km)	13,041
Total weight of the wire in a magnet (kg)	46,250
Wire diameter (mm)	0.73
Cu:SC ratio	3.6:1
NbTi filament size (μm)	<40
I_c @ 2 T & 4.2 K (at 10^{-5} V/m) (A)	>400
Matrix material	Type 101 Copper, (Oxygen Free)
RRR in the wire	>150
Twist pitch (mm)	20

Cable: The cable is designed to be fully transposed, 4-stage right-hand twisted, made of 450 strands. The cabling pattern is 3x5x5x6. The final cable is wrapped with 304 SS tape 0.05 mm thick with a 40-50% overlap to keep tight tolerances on the cable diameter and to eliminate appearance of "bird caging" during cable spooling on shipping spools. The total cable length is – 24 x 1150 m piece lengths = 27.6 km. Cable weight/piece is about 1920 kg.

Cable in Conduit: In the event of a quench, the conduit will experience high pressure[10] up to 340 ATM. To ensure high quality of the conduit, 304L steel is chosen because of its excellent weldability. To date, at least two different manufacturing methods have been used for CIC Conductor. Both methods start by cabling superconducting strand, and then encasing the cable in stainless-steel conduit. The first manufacturing method consists of feeding flat stainless steel strip along with the superconducting cable through a five-stage tube mill machine followed by TIG seam weld. In the second approach, the required length of cable is pulled through an oversized conduit which is assembled in full length out of butt welded seamless (i.e., extruded) tubes.

An advantage of the second approach is that the length of the welds is much shorter, and butt welds are much easier to inspect. However, there is no experience with pulling a cable through 1 km of conduit and this issue is currently under study. After reduction, in both options, the ID of the conduit will be 20 mm, the OD will be – 25 mm and the void fraction of the cable will be 37%, which will minimize internal disturbances inside the conduit.

Aluminum Sheath: Aluminum sheath is added around the conduit and serves as a shunt in the event of a quench, keeping the hot spot temperature below the design value of 100 K. For economy Al 1100 is chosen as a material for the sheath with RRR = 17. Total weight of the sheath in the GEM Magnet is about 190 tons.

Electrical joints: Electrical connections will be done between those coil segments using specially designed joints. To provide maximum reliability, joints will have a built-in cooling path (secondary loop) for heat removal which is separate from the conduit. The joint design and length of the joint should accommodate tolerances of the conductor length and uncertainties arising during the winding process (expected to be in the 30–50 cm range). For the redundancy, there will be secondary leak-tight containment added around the joints when assembled in the field. Electrical resistance of the joint is expected to be less than 0.5 nOhm at 50 kA.

PREPRODUCTION EVALUATION AND VERIFICATION PROGRAM

Although the objective of the GEM conductor design is to use existing technology well within the state of the art, reliability concerns imply that some verification and justification steps should be taken.

Strand: No Verification is required. This is within normal industry capability.

Cable: There are several issues which should be addressed in this program, such as: what should be diameter of the take-up spool to eliminate excessive deformations of the cable and SS tape on the last stage, whether the last stage of cabling will give a mechanically stable configuration, and whether annealing of the cable is necessary after the last stage. Manufacturing experiments showed that the 3x5x5x6 pattern is possible without substantial problems, though the take-up reel should have a diameter more than 1 m. Additional efforts are required to understand the RRR value during the cabling process.

Conduit: The primary issues for the conductor are establishment of the QA/QC procedure and verification of at least one technique for assembling the cable-in-conduit. Design of the cryogenic leak test facility is now in progress. Some preliminary estimates carried out for evaluation of the friction force for pulling the cable in the conduit indicates that it can cause a yield of the copper in the cable.[11] A pulling experiment on 60–100 m conduit is planned in the near future to obtain experimental data.

Aluminum Sheath: Different types of Al sheath design were evaluated. Bending experiments have been conducted to understand the behavior of the sheath-conduit interface, keystoning effect, and prospective effects for the winding procedure. The effect of welding on the temperature inside the conduit was also studied. As a result, the preferred design of the sheath is established, which is two symmetric profiles with two longitudinal welds. This minimizes conductor distortion and eases the manufacture and transportation of the aluminum sheath. The shipping spool diameter is specified as well.

Joints: Significant efforts were put to develop stable and reproducible technology for the joint assembly and also to increase joint reliability. High conductivity aluminum was introduced in the design to limit heat generation in the joint at temperatures above the critical down to 10–12 W, which can be easily taken by mass flow of only 0.5-1 g/s. Subscale joints were manufactured and tested. These tests confirmed that the target value of the joint resistance of 0.3–0.5 nOhm is achievable.

Conductor Test Experiment: To demonstrate the conductor performance in conditions close to the GEM magnet requirements before the beginning of full production, we plan to carry out a conductor test experiment with the main objectives: verification of the conductor performance at full and higher current; measure the operating margin; measure stability to external and internal disturbances; study propagation of the normal zone; study current transfer to the sheath at a quench; simulate heating the conductor as a result of the quench and study hot spot temperature; verify joint design and mass flow rate through the joint; and study stability of the joints against the external disturbances. This test is planned to be performed on about 60 m of the GEM conductor wound on a mandrel 1.1 m in diameter and about 1 m high. It is expected that this test coil could be charged up to 100 kA, then stresses on the conductor will be comparable with stresses on the conductor in GEM. Table 4 lists the main parameters of the test coil in comparison with GEM.

Table 4. Comparison of the Test coil parameters with GEM.

Parameter	Test coil at 50 kA	at 100 kA	in GEM
Stored Energy, kJ	400	1600	2.5e6
Maximum Field in Winding, T	1.7	3.4	1.6
Max Hoop Stress, MPa	4	16	24.7
Compressive Axial Stress, MPa	1.33	5.3	5.7

CONCLUSION

Rationale and design parameters of the cable-in-conduit conductor for GEM have been presented. The current evaluation and verification program results in clarification and improvements of some design features and manufacturing approaches. The planned conductor test experiment should demonstrate performance of the conductor and joints, and verify design parameters.

REFERENCES

1. B.A. Smith, P.G. Marston, et al., "Design concept for the GEM detector magnet," presented at the 1992 ASC, Chicago, IL, Aug. 23-28, 1992.
2. J.M. Baze, et al. "Design, construction and test of the large superconducting solenoid Aleph," *IEEE Trans.on Mag.,* Vol. 24, No. 2, p. 1260, March 1988.
3. P.T.M. Clee and D.E. Baynham, "Towards the realisation of two 1.2 Tesla superconducting solenoids for particle physics experiments," *Proc. Int. Conf. MT-11,* Tsukuba, Japan, Vol.1, p. 206, 1988.
4. D.P. Ivanov et al., "SC toroidal field coils of Tokamak-7," *Atomnaya Energia,* 45, (1979) 3, p. 171.
5. V.A. Alkhimovich, I.O. Anashkin, et al. "The current capacity tests of the Tokamak T-15 Nb_3Sn toroidal coil assembly," *IEEE Transactions on Magnetics,* Vol. 27, No. 2, 1991, p. 2057-2059.
6. M.S.Lubell, J.A. Clinard et al. "The IEA Large Coil Task Test Results in ISMTF," *IEEE Transactions on Magnetics,* Vol. 24, No. 2, 1988.
7. M.M.Steevs, M.O.Hoenig, et al., "Progress in the manufacture of the US-DPC test coil," *IEEE Transactions on Magnetics,* Vol. 25, No. 2, pp. 1738-1741, March 1989.
8. S.D. Peck and P.H. Michels, "Test results from the 200kA SMES/ETM conductor," Presented at ASC-90, Snowmass, Colorado, 24-28 Sept., 1990.
9. D.E. Baynham, N.V. Fetisov, N.N. Martovetsky, "Stability of indirectly cooled conductors with large cross section," Presented at ASC-92, Chicago, August 23-28, 1992.
10. Minervini, J.V., et al., "Cable-in-Conduit Conductor Concept for the GEM Detector Magnet," presented at the *1992 Applied Superconductivity Conference,* Chicago, IL, Aug. 23-28, 1992.
11. R.L. Huddleston, "Summary of GEM Magnet Cable/Conduit Assembly Analysis," Feb. 10. 1993, private communication.

AN OVERVIEW OF MAGNETIC, MECHANICAL, THERMAL, AND ELECTRICAL ANALYSES PERFORMED ON THE HDM MODEL MAGNETS

J. F. Lowry,[1] O. R. Christianson,[1] H. L. Chuboy,[1] E. F. Daly,[1] D. J. Hall,[1] D. C. Johnson,[1] M. P. Krefta,[1] G. T. Mallick,[1] J. F. Roach,[1] S. K. Singh,[1] J. R. Snyder,[1] and Consultants W. J. Carr, Jr. and J. H. Parker, Jr.

[1]Westinghouse Science and Technology Center
1310 Beulah Road
Pittsburgh, PA 15235

INTRODUCTION

Westinghouse Electric Corporation (WEC) is under contract to the SSCL[1] to design, develop, fabricate, and deliver superconducting dipole magnets for the High Energy Booster. The first phase of this program involves the analysis, fabrication, assembly, and testing of 1.8 m model magnets designed by the SSCL to operate in a vertical dewar. They have many of the operating characteristics (central field, transfer function sag, dipole field purity, temperature and field margins, insulation stress margin, etc.) required of full length High Energy Booster Dipole Magnets (HDM).

WEC has performed coordinated magnetic, mechanical, thermal, and electrical analyses of the HDM model magnets, with these objectives: 1) perform a thorough technical review and analysis of the SSCL model magnet design; 2) develop a thorough understanding of the design; 3) verify through analytical and numerical analyses that the SSCL model magnet design, when built, will satisfy the requirements of the HDM Design Requirements Document;[2] and 4) identify any deficiencies in the SSCL design and work with the SSCL to resolve such deficiencies. The design analysis process began with release of the model magnet drawings and process specifications by the SSCL. WEC reviewed these drawings and data for dimensions, tolerances, manufacturability, and assembly fit and passed them to the analysis groups for development of analytical and finite element (FE) models of various magnet components and assemblies.

The various analyses were integrated and proceeded in parallel, using a limited number of models for a wide range of calculations, with the results of calculations and analysis in one area (or on one model) serving as input to other areas. For example, in order to calculate the model magnet field and temperature margins under operating conditions, it is first necessary to calculate the peak magnetic field in each cable. As a second example, the integrity of the electrical insulation system is closely related to such factors (among many others) as the MIITS and the magnet hot spot temperature during a quench, which in turn depend on the interaction of the magnet with its quench protection circuit.

MAGNETIC ANALYSIS

The initial step in developing detailed mathematical models of the model magnet is to generate a geometric representation of the two-dimensional (2D) coil cross section. The computer program STACKUP was written for this purpose. Given the compressed insulation thicknesses, dimensions of the inner and outer cables and wedges, and appropriate stacking radii, this program "stacks" the cables and wedges against perfect arcs and then, for a given current, uses a closed-form field solution to calculate the spatial harmonics. Several stacking options are incorporated, including stacking against either an inside or outside radius, and positioning cables and wedges at the first point of contact with the radius, or with the midpoint of the cable (or wedge) inner or outer edge attached to the radius. The stackup can be initiated from either the pole or midplane and the program has an option for constraining either or both while varying the insulation thicknesses to adjust the coils to fill the available windows. Several alternatives for representing the current distribution internal to each cable are available: a thin rectanglar region of uniform current density; two thin parallel lines as in the computer program SSCMAG (a coil cross section optimization program written by SSCL); or 30 (or 36) discrete, square current elements each containing equal current. It was found that the harmonics calculated using any of these methods did not vary greatly, and also compared favorably with the output of the SSCL program SSCMAG.[3]

STACKUP outputs a complete set of geometric coordinates for the corner points of all conductors and wedges. This file is used as a basis for developing a FE model of the magnet 2D cross section, including the coil elements, collars, and iron yoke. The FE model was created in FIGURES II, a preprocessor for HOTMAP (a version of WEMAP—Westinghouse Electromagnetic Analysis Program— which performs high accuracy calculations of spatial field harmonics). Cable currents were distributed throughout thin rectangular regions of uniform current density. Collar details were included so that effects of a small but nonzero collar susceptibility could be examined. The geometry was meshed using 12,175 triangular elements of third order. The magnetization characteristics used to represent the yoke iron at 4 K were originally supplied by Ramesh Gupta from Brookhaven National Laboratory (BNL). Yoke packing factors of 0.975, 0.99, and 1.0 were examined. The transfer function relating magnet current to central field (B_0) is one of the analysis results. From this, the transfer function sag was computed and compared with SSCL results. The operating current (6640 A) required to produce a central field of 6.67 T was interpolated from the transfer function data. The effects of magnetic saturation and collar susceptibility on spatial harmonics (b_n) were investigated by calculating the harmonic transfer functions over a range of currents. Other information extracted from the FE model included the magnet inductance, stored magnetic energy, and Lorentz forces; the latter were required as input to structural analysis to determine cable displacements resulting from magnet operation. Finally, the skew quadrupole (a_1) due to top half to bottom half yoke iron packing factor mismatch was calculated for the predicted worst case weight imbalance in an actual model magnet. The FE analysis is described in detail by Krefta et al.[4]

One result of this FE analysis is the magnetic field throughout the 2D cross section, including the peak and average fields in each cable, which are needed as input to the calculation of quench current and field and temperature margins. To accurately calculate the peak magnetic field in a cable, the current representation must be refined, since the field in the immediate vicinity of a current element is sensitive to its precise geometric representation. The current is generally represented as thin rectangular regions of uniform current density, which yields accurate spatial harmonics at r = 1 cm, far enough away from the current sources that the precise details of the current distribution are less important. Ideally, each filament in each superconducting strand would be represented as a current source, which is impractical because of the large number of filaments and the inability to know the precise location of each filament in the strand cross section. A practical alternative is a discretization or strand-by-strand representation of the current. The current is subdivided evenly among the strands in a cable, the current density is taken to be uniform within each strand, and the individual strands are meshed in the FE model. This refined model is still only an approximation to the actual current distribution; nevertheless, it results in a considerable increase in the complexity of the model. To minimize this increase in complexity, only a few cables are subdivided— those in the critical regions (i. e., near the pole) where the highest fields are known to occur.

The effect of superconductor magnetization currents on field uniformity (spatial harmonics) was also calculated. Persistent currents are induced in the superconducting filaments everywhere they are subjected to a transverse (nonaxial) component of magnetic field. Expressions were developed for the magnetic field generated by persistent currents, and these relationships were used to calculate the resulting field angular harmonics. The Fortran code PERSIST was written to numerically calculate the persistent current field.

STRUCTURAL ANALYSIS

A second quarter-section FE model of the magnet 2D cross section was developed in FIGURES II for structural analysis. The model consists entirely of 2D plane stress elements with gap elements placed between the bearing surfaces. All gap elements were specified without friction and all material properties were taken to be linear elastic and isotropic with the exception of the coils: the model incorporates the differing radial and azimuthal coil properties. Structural details of the collar, yoke, and shell, and interactions (interferences) among them are critical in this model. Nominal dimensions were assumed except for deviations in pole face shim size and the amount of stress-free interference between the collar and yoke. Calculations were performed for both Nitronic 40 and high manganese steel collar materials.

Structural analysis was performed using the ANSYS code. Four steps covering magnet assembly and operation were simulated; conditions imposed at each step were carried through the subsequent steps. The first step simulated coil collaring with an imposed azimuthal prestress of 70 MPa on the inner coil and 56 MPa on the outer coil. The second step simulated yoking and skinning with an assumed azimuthal shell tension (after welding) of 205 MPa. Cool down from ambient to 4.2 K was simulated in the third step. Finally, Lorentz forces were calculated in the magnetics FE model and applied to the structural model to simulate magnet operation at fields to 6.67 T. Details of these calculations are discussed by Snyder.[5]

The simulation of actual magnet operation demonstrates the integration required among the various models and types of analysis. Lorentz force loading causes small changes in conductor positions; these displacements can cause small changes in magnet field uniformity. In the initial magnet cross section simulation the superconducting cables and wedges are stacked against ideal arcs formed by the collars. The structural analysis steps described above show that the collar arcs are distorted by strains induced during collaring, yoking, shell welding, magnet cooldown, and operation. In order to determine the resulting conductor displacements, an initial displacement solution vector was established using a structural FE model in which the collars are assumed to be rigid, i. e., the collar modulus is taken to be so large that collar deformation is negligible. Conductor displacements resulting from any of the other analysis steps (including Lorentz force loading due to magnet operation) are expressed relative to the initial (rigid collar) position by subtracting the final displacement solution vector from the vector resulting from a coil compressed in a rigid collar. Subsequently, the conductor displacement file containing deflections of conductor blocks for each load case is supplied to magnetic analysis, interpolated from blocks to individual conductors, and used for calculation of the changes in spatial harmonics.

THERMAL ANALYSIS

Thermal analyses of the model magnet design were performed using both analytical and numerical methods, including: cool-down performance, including elapsed time and quantity of cryogen required for cool-down and operation; field and temperature margins for both inner and outer coils; dependency of quench current on ramp rate; evaluation of the capability of the model magnet design to meet specified dump conditions; evaluation of the model magnet stability against quench (i. e., calculation of minimum quench energy); adiabatic quench propagation velocity; quench pressure rise, quantity of cryogen vaporized during a quench, and quench recovery time; and magnet current versus time after quench, MIITS, and magnet temperature rise and hot spot temperature.

Calculations of quench current and field margin require as input the peak and average magnetic fields in each conductor; fields were calculated by discretizing certain conductors and using the FE code HOTMAP. The field margin is then found from the load line of the magnet and the magnetic field dependent critical current of the superconductor. The temperature margin is found from the magnetic field dependent critical current, critical temperature, and the operating temperature of the superconductor. Details and results of these calculations are discussed by Christianson et al.[6]

Calculations of quench pressure rise and temperature rise as functions of time after quench were performed for the HDM model magnet for several values of porosity, void fraction, and test circuit parameters, by obtaining solutions to the time-dependent mass and energy equations using an empirical form of the momentum equation. This analysis includes both boiling LHe and stagnant GHe heat transfer, phase change effects, the adiabatic quench propagation velocity, and magnetoresistivity of the cable copper matrix. Sensitivity analyses were performed to determine bounds for coil porosity, void fraction, and test circuit parameters. Details and results are discussed by Daly and Christianson.[7]

Magnet current decay and temperature rise following quench, and MIITS were calculated by analyzing the interaction of the magnet with its quench protection circuit. A coupled numerical procedure and MIITS calculation was used to estimate temperature rise in the HDM model magnets. The increasing size of the normal zone was based upon measured propagation velocities; the temperature was calculated from the MIITS integrals. Magnet current decay depends upon the temperature and the time-varying resistance of the normal zone. These calculations are summarized by Christianson et al.[8]

Calculations of AC losses during HDM model magnet ramping were performed using methods similar to those of W. J. Carr, Jr.[9] and G. Snitchler et al.[10] AC losses include superconductor and iron hysteresis losses and eddy current losses in all conductors, including the superconductor. Also included are dynamic resistance losses. The FE calculations provide the peak and average field in each conductor. The computer code HEBLOSS was written to input this field data and calculate the eddy current, hysteresis, and dynamic resistance losses. Hysteresis power losses in the superconductor depend linearly on dB/dt. Dynamic resistance losses also depend on dB/dt, and on a "g function" which is itself dependent on the operating and critical currents. There are both **intrastrand** and **interstrand** eddy current losses in the superconducting cable; both depend on $[dI/dt]^2$. Hysteresis losses in the iron depend on the product of the coercive force H_c and the magnetic induction B_0 near the knee of the B-H curve. HEBLOSS sums the losses over each cable and outputs the energy loss per unit magnet length for each cable. Summing over all cables, all four quadrants, and along the length of the magnet, produces the total loss. Losses that are accurately predictable are within the proposed heat budget; however, cable coupling losses resulting from interstrand resistance could easily exceed the predictable losses, since the interstrand resistance is not yet fully specified and controlled. AC loss calculations are discussed by G. T. Mallick et al.[11]

ELECTRICAL ANALYSIS

Electrical analyses of the model magnet design were performed using both analytical and numerical methods. Many analyses involve the interaction of the magnet with its quench protection circuitry, including the circuitry for dumping the magnet energy following quench. The integrity of the model magnet electrical insulation system was analyzed, particularly those features unique to the coil-on-coil cure design. The analyses included an assessment of electrical insulation creep paths as well as evaluation of the electrical stress in gas voids (air or GHe). Various magnet insulation tests (Hipot and ringer circuit) were evaluated to determine peak electrical stresses on the insulation system.

Insulation ratings were reviewed and compared to insulation strengths for the various dielectric materials in the magnet: polyimide films, G-10, epoxy/fiberglass cable wrap, LHe, GHe, and air (magnet testing during assembly). Safety margins were determined under worst-case conditions, i. e., Hipot testing in air at room temperature and pressure, and the lowest density GHe state during a quench cycle. Peak voltage during quench was determined to be less than 1 kV, while 3 kV peak voltage (coil-to-coil) and 5 kV (coil-to-ground) will be applied during full coil Hipot testing. Under the latter condition air void partial discharges may be expected between the inner coil and pole ground for high void fraction. Such discharges may result in leakage currents exceeding the maximum value allowed by the specifications (25 µA).

The code PSPICE was used to analyze the voltage distribution in the magnet during impulse voltage conditions. The magnet was modeled as four coils, with each coil treated as a Π section with distributed winding capacitances lumped on each end of the inductance. The FE code HOTMAP generated the self and mutual inductance matrix required as input to PSPICE. Inner and outer coil resistances were taken to be 108 mΩ/m and 172 mΩ/m, respectively. Simulations were carried out to calculate the ringer circuit voltage at line and splice terminals for a 20 µF ringer capacitor at 1500 V, and to predict the effect of a single shorted turn on ringdown frequency. These calculations are described by Roach et al.[12]

A lumped parameter model of the quench protection (energy dump) circuit was analyzed to predict quench voltages under expected model magnet test conditions. Terminal voltages, internal and external splice voltages, magnet resistance, hot spot temperature, and MIITS were predicted as functions of dump circuit resistance and firing delay. In all cases the peak quench terminal voltage was less than 650 V.[12]

CONCLUSIONS

Magnetic, structural, thermal, and electrical analyses of the HDM model magnet design have been performed. The various analyses were integrated and proceeded in parallel, using a limited number of models for a wide range of calculations, with the results of calculations and analysis in one area (or on one model) serving as input to other areas. In general, the results of our analyses agree closely with those of the SSCL.[3] Discussions of analysis details and results can be found in references three through twelve.

REFERENCES

1. High Energy Booster Dipole Magnet Program, Contract No. SSC-91-B-01705, to WEC.
2. Design Requirements Document for Technology Model Magnet (DSB-101), Document M80-000048, Rev. A, 01-13-92.
3. Interim Design Review Number 1 Data Package, Vol. 1, CDRL-16 (Rev. A), 11-04-92.
4. M. P. Krefta, H. L. Chuboy, and J. H. Parker, Jr., Magnetic field calculations for the HDM model magnets, Paper VII-25, 5th IISSC.
5. J. R. Snyder, HDM model magnet mechanical behavior with high manganese steel collars, Paper IV-16, 5th IISSC.
6. O. R. Christianson et al., HDM model magnet margin, Paper IV-12, 5th IISSC.
7. E. F. Daly and O. R. Christianson, HDM model magnet quench pressure rise estimates, Paper VII-38, 5th IISSC.
8. O. R. Christianson et al., HDM model magnet quench performance, Paper IV-11, 5th IISSC.
9. W. J. Carr, Jr., "AC Loss and Macroscopic Theory of Superconductors," Ch. 6, Gordon and Breach, NY (1983).
10. G. Snitchler et al., Design and AC loss considerations for the 60 mm dipole magnet in the High Energy Booster, Supercollider 3, J.Nonte Ed., Plenum Press, NY (1991), p. 625.
11. G. T. Mallick et al., Improved loss calculations for the HDM magnets, Paper VII-35, 5th IISSC.
12. J. F. Roach et al., Design and issues associated with the HDM electrical insulation system, Paper IV-9, 5th IISSC.

INCREASED CRITICAL CURRENT AND IMPROVED MAGNETIC FIELD RESPONSE OF BSCCO MATERIAL BY SURFACE DIFFUSION OF SILVER

Y. Z. Negm,* G.O. Zimmerman,* K. A. Eckhardt, and R. E. Powers*

Boston University, Physics Department, Boston, MA 02215
*Also ZerRes Corporation, Boston, MA 02215

ABSTRACT

We have developed a procedure of increasing the critical current of **BSCCO** ceramic superconducting material , the value of the critical current is increased by 30%. Moreover the degradation of the critical current with the applied magnetic field had been decreased. The procedure consists of applying a thin layer of silver to the surface of the conductor. The details of the procedure and the improved performance are discussed . This procedure has great significance for any future application of HTSC materials where high current carrying capacity is necessary. It will therefore be important in the application of HTSC materials to SSC high current leads.

INTRODUCTION

The application of High Transition Temperature Superconductors (HTSC), for practical purposes has been limited by their low critical currents and the deterioration of this critical current on the application of a magnetic field. Although some HTSC materials have been produced which have critical currents of the order of $10^5 A/cm^2$, they can only be made in films or small laboratory quantities, and can not be produced in moderate quantities because of the slow rate and cumbersome and unreliable process by which they are produced. The forms of HTSC materials which can be produced in quantity are bulk sintered components which generally are produced from a powder. In general, the powder is mixed with an organic binder which is then carefully removed. The surface tension of the binder provides the necessary force to bring the powder grains together so that they can be sintered once the binder is eliminated. The ultimate limitation on J_c is thought to be the weak link behavior which develops between the powder grains during sintering[1]. This behavior is enhanced, or may be a result of the extremely short coherence length in these materials. In long length of the material, the critical current of the HTSC is limited by micro cracks, and the alignment of the grains since their conductivity is anisotropic. To a certain degree those micro cracks can be filled by the addition of silver powder to the HTSC powder before the binder is added and before sintering[2-4]. In addition, the silver may provide pinning centers for quantized magnetic flux lines and thus diminish flux motion which results in resistivity in superconducting materials. The addition of silver, however, increases the thermal conductivity of the material which in some applications is an undesirable effect.

We have found a method which enhances the critical current density in HTSC sintered material both at zero and finite magnetic field. It consists of an application of a small quantity of silver to the surface of the HTSC bulk material. This method does not significantly increase the thermal conductivity of the material.

Supercollider 5, Edited by P. Hale
Plenum Press, New York, 1994

Our aim is to build electrical power leads to conduct current between 70 K and 4.5 K. For this high critical currents and low thermal conductivity are of importance. In our application the HTSC material is in the form of long rods of either rectangular or circular cross section. Those rods are reinforced for mechanical stability and provided with low electrical resistance contacts which on one end connect to copper leads and on the other to conventional low temperature superconducting leads such as copper clad niobium titanium.

EXPERIMENTAL PROCEDURE

A long rectangular rod of BSCCO material, characterized as 95% 2223 phase, of length 8 cm and 0.3 cm x 0.3 cm is cut into two halves. Thus the rods are identical except for the treatment of one of the rods as described below.

The surfaces of each rod are treated by using very fine sand paper to introduce an uncontaminated surface. Then one of the rods is coated with a thin layer of silver . During the heat treatment, the silver layer diffuses into the HTSC grains.

We then apply contact pads for current and voltage leads to both rods in order to measure their critical currents by means of the conventional four point method. Several methods are conventionally used for this purpose[5-9]. Those are pressed indium contacts, ultrasonically soldered bonds, silver paste which is then baked in and soldered to, sputtering, metal sprayed, and silver vapor deposition. In practical application the contact resistance should be minimal to avoid heating of the sample and consequently quenching it.

The area of each silver pad on any rod was approximately 0.3 cm x 0.3 cm. To make the contact between the metallic conducting wire and the High T_c material we apply Supersolder[10], a low temperature soldering alloy developed by the ZerRes Corporation. We found this alloy to have a contact resistance which is lower by a factor of three from that of conventional solder contacts between the HTSC deposited silver and copper.

RESULT AND DISCUSSION

We measured the critical current of both the silver surface coated and the uncoated rod by means of the conventional four probe technique at 77 K. The results are shown in Fig. 1. The critical current of sample #1, without silver coating, as seen from Fig.1, is 20 A, i.e. the critical current density is 250 A/cm^2. The value of the critical current of the second rod, the silver coated one, is about 30 A, i.e., the current density is 370 A/cm^2. The coating had the effect of increasing Jc by 30%.

Fig. 2 shows the current-voltage characteristics of the silvered rod in various magnetic fields.

Fig. 3 shows the comparisons of the critical currents of both the silver coated and the uncoated rod in a magnetic field at 77 K. At 100 G the critical current density decreases from 370 A/cm^2 to 250 A/cm^2. Its behavior is greatly superior to that of an uncoated rod where the critical current decreases from 250 A/cm^2 to 144 A/cm^2 in a magnetic field of 100 G.

Several attempts have been made to increase the critical current density of the HTSC material by mixing different materials with the HTSC powder. The mixing of the two phases of BSCCO, the 2212 and 2223[11] , or the addition of silver powder to the BSCCO powder before sintering. Those methods have had both desirable and undesirable results on the critical current and its behavior in a magnetic field.

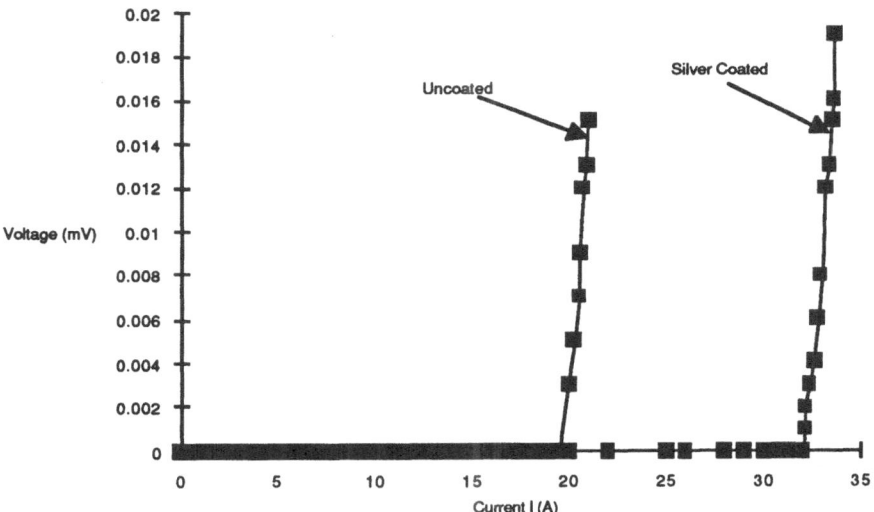

Figure 1. Current voltage characteristics of silver coated and uncoated rods.

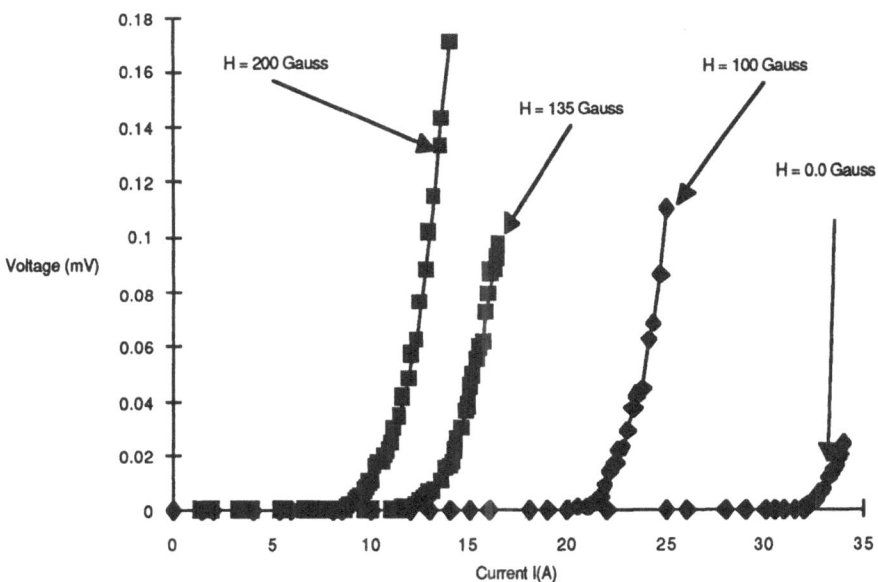

Figure 2. Current voltage characteristics of the silvered rod in various magnetic fields.

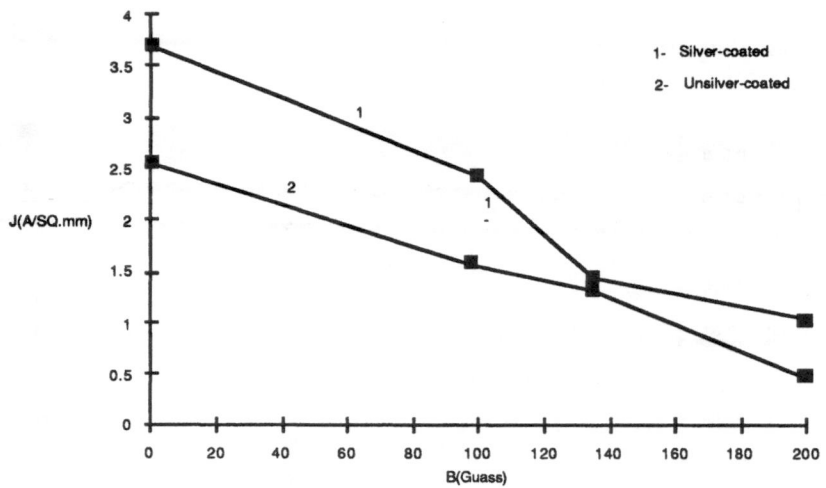

Figure 3. Critical current density vs. Magnetic field for silvercoated and uncoated rods.

Here, we demonstrate for the first time, that the critical current characteristics of an HTSC material can be significantly improved by the application of a thin surface layer of silver which does not significantly add to the thermal conductivity of the rod. In addition, the above results suggest that even in granular HTSC material, the bulk of the current is carried near the surface.

REFERENCES

1. S. Gotoh, M. Murakami, H. Fujimoto, and N. Koshizuka, J. Appl. Phys. 72, (6) 15 , Sept. (1992).
2. M.J.Neal, D.B. Chandler, L.J. Klempetner, and M.V. Parish, IEEE   Transactions, Applied Superconductivity 1, No. 4, (December, 1991).
3. G. G Peterson, B.R. Weinberger, L. Lynds, and H.A.Krasinski, J. Mater. Res. 4, 605, (1988).
4. M. Itoh and H. Ishigaki, J. Mater. Res. 6, 2272, (1991)
5. J. W. Ekin, A. J.Panson, and B. A. Blankenship, Appl. Phys. Lett. 52, 331, (1988).
6. J. W. Ekin, T. M. Larson, and N. F. Bergen, Appl. Phys. Lett. 52, 1819, (1988). Lett. 54, 666, (1989).
8. S. Jin, M. Davis, T. Tiefel, R. Van Dover, R. Sherwood, H. O'Bryan, G.Kammlott, and R. Fastracht, Appl.Phys. Lett..54, 2605, (1989).
9. J.Katz, J.Willis, M. Maley, and R.Castro, J. Appl. Phys. 65, 1792, (1989).
10. G.O. Zimmerman and A. Kaplan, Patent Number 4,966,142, (1990).
11. Y.C. Guo, H. K. Liu, and S. X. Dou, Physica C 200, 147, (1992).

CONCEPTUAL DESIGN STUDIES FOR HE II COOLED LOW BETA
QUAD IR MAGNETS

D. Baritchi , C. Corbett , A. Jalloh, N. V. Mokhov,
R. Schermer, and R. Stiening

Superconducting Super Collider Laboratory*
2550 Beckleymeade Ave.
Dallas, TX 75237

INTRODUCTION

This paper discusses the results of the Cryostat Design Group's effort on the Conceptual Studies of the Low Beta Quad Magnets (LBQ) cooled by superfluid Helium (He II). The conclusion of this study is that the "adapted CERN" He II cooling design does present a satisfactory system for cooling the LBQ magnet interaction region (IR) radiation heat loads; however, for the increased luminosity heat load some research on the cable insulator is required to achieve an effective thermal connection between the outer coil and the He II bath. The LBQ magnets are the final four magnets on either side of the IR detector. These four magnets are named Q1, Q2A, Q2B, and Q3, reaching outward from the interaction point (IP). All magnets have a 50-mm magnet bore and produce a gradient of 193 T/m. Magnetic lengths were all considered to be 15 meters for the purposes of this study.

Radiation from the proton-proton collisions results in a significant heat deposition on the LBQs which is most severe for Q1 and least severe for Q2 and Q3. This is reduced with a 25-mm aperature, 3 meter long steel collimator installed upstream of Q1. Total heat to be removed from Q1 is 47 watts at the baseline luminosity (10^{33} proton/cm2 –sec) which presents the most severe design challenge.[1] Maximum local heat deposition is 0.25 mw/g. The average value for the inner coil compared to the worst 2.5-meter length is 1.01 w/m average compared to 2.24 w/m maximum. The thermal model results were generated for both the maximum heat loads and the average heat load. The design loads for the 10^{34} luminosity design can be obtained by multiplying the above values by ten. The assumption has been made that a cooling collimator is installed inside the beam tube along the entire length of the magnets to absorb one-half of the total heat load for the increased luminosity design.

"ADAPTED CERN" CONCEPT

The CERN superfluid cooled magnet design[2] was adapted to our requirements to come up with the "adapted CERN design" shown in Figure 1. Heat is removed from the cold mass

---

*Operated by the Universities Research Association, Inc., for the U.S. Department of Energy under Contract No. DE-AC35-89ER40486.

by a 1 atm He II bath at approximately 1.9 K. The heat sink for the 1 atm bath is two-phase He II at 1.8 K flowing inside a heat exchanger(s) that runs the entire length of the cold mass. Cryogenics would supply 2.2 K liquid He II which is then expanded through a Joule-Thompson (JT) valve into two-phase flow at 1.8 K saturation conditions (1.8 K, 0.001638 MPa) and flows to a concentric tube heat exchanger. It was concluded from a mechanical design viewpoint that routing the liquid supply portion (or low quality two-phase flow) outside the cold mass (inside the 4.2-K shield) would be preferable to making concentric tube connections at the interconnects. The recirculating pump shown in Figure 1 is not required. By design the pressure drop in the system is very low and the required head pressure can be obtained from pressure downstream of the JT valve. The 1.8-K heat exchanger is pumped to keep constant saturation conditions. The intermediate shield is a 4.2-K cooled shield that plays a part in the cooldown and quench protection procedures. Magnet cooldown begins by diverting some of the 4.2-K supercritical He and circulating this fluid through the magnets. The safety relief valves (SRV) are open during this cooldown period. When the magnet has been cooled to 4.2 K, the SRV and the cooldown and fill valves are closed. The magnets are now filled and pressurized to 1 atm. The He II heat exchangers are now activated to bring the magnets down to their operating temperature of 1.9 K. During a quench the SRV opens at 20 atm and the He II bath is relieved into the 4-K cooling duct. All cryogenic resources are concentrated in the spool piece which is sufficiently far from the IR not to impose any significant maintenance problems. The advantages of the "adapted CERN" design are that it requires no re-invention of the wheel, the design makes efficient use of cryogenic resources, and has planned cooldown and quench recovery procedures. Also heat exchangers can be sized for either baseline luminosity or for increased luminosity levels.

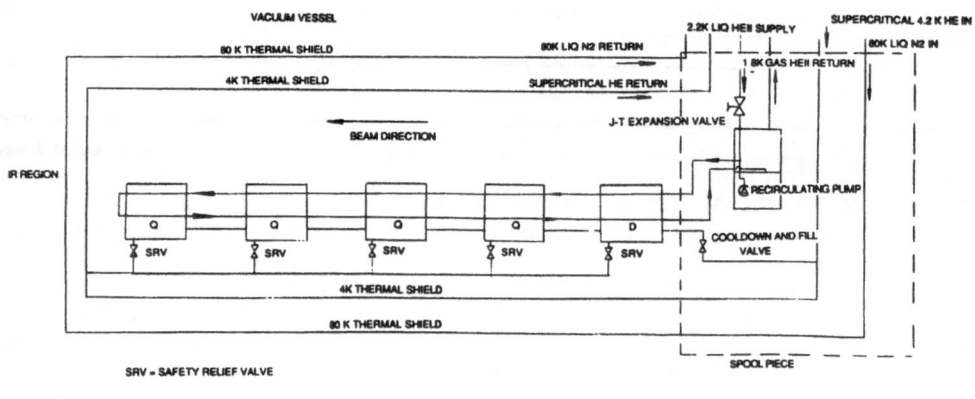

Figure 1. Cooling Flow Schematic for the "Adapted CERN" Concept.

He II Heat Exchanger Duct Size Calculations

The number of heat exchangers and their dimensions were determined by the need to limit pressure drop and the need to provide enough area to limit heat flux and the associated temperature drop. The design assumption was to limit pressure drop to 0.14 kPa. This was done to keep the two-phase He II as close as practical to 1.8-K saturated conditions and to minimize the "reverse" Joule-Thompson effect.[3] The total heat load was assumed to be applied uniformly over the heat exchanger length and all heat was absorbed by the latent heat of He II (20 J/g). The two-phase pressure drop was calculated by homogeneous model.[4] The assumption was made that initial quality was zero and the quality rose toward

0.95 at heat exchanger exit as it absorbed heat. The pressure drop predicted by the homogeneous model was close to a gas-only pressure drop.

Heat Transfer Design

The heat transfer analysis approach was to evaluate the heat flux that the LBQ cooling system is required to remove against the available driving delta temperature. It is a design ground rule that the maximum temperature of the superconducting material is 2.0 K. This was chosen in part to provide a 150 mK margin above the Lambda temperature at 1 atm. The ultimate sink temperature is saturated He II at 1.8 K. Consequently, in our design we have, without margin, a driving temperature of 200 mK to remove heat deposited by IR radiation and static heat leaks. The available driving temperature is the sum of five primary components: (1) ΔT_1 – Temperature difference required to remove the IR radiation energy deposited on the inner/outer coils (also collars/yoke) to the He II bath at 1 atm, (2) ΔT_2 – Temperature difference across the He II bath to carry heat to the outside of the heat exchanger duct, (3) ΔT_3 – Kaptiza Resistance at heat exchanger duct – He II interface, (4) ΔT_4 – Conduction resistance across heat exchanger wall, and (5) ΔT_5 – Kaptiza resistance at He II/inner wall interface. The first and last of these are discussed below. The assessment resulted in the conclusion that the 200 mK limit could be obtained.

ΔT_1. CERN conducted a test of the heat transfer of heated conductors in a superfluid bath (1.7 – 1.9K).[5] The test utilized stainless steel as a pseudoconductor and electrically heated a central turn and two of its neighbors and measured the temperature difference between the sample and the surrounding bath. The conductor was prepared similarly to the way that SSC fiberglass epoxy conductors are prepared. A 2-mm channel between rows of fiberglass tape was left for He II flow. CERN varied the heat input from 0 to 56 mW which corresponds to 0 to 8.78 mW/cm^3 using their cable dimensions. The 10^{33} luminosity maximum heat load value of 0.896 mw/cm^3 corresponds to CERN test data at 5.7 mw. The 10^{34} luminosity maximum heat load of 2.816 mw/cm^3 corresponds to CERN data at 17.98 mw. From CERN data with heat transfer at all four faces we would expect only a 14 mK delta temperature rise at the increased luminosity heat load. The CERN test investigated the effect of one of the small faces insulated from transferring heat to the bath by thermal grease. The increased luminosity delta temperature rises to to 25 mK and the overall heat transfer coefficient is backed out to be 0.0132 W/cm^2 –K. The power removed by a small face is shown in the CERN paper[5] and from this data a heat transfer coefficient can be calculated separately for the large face and the small face. The small face heat transfer coefficient is 0.140 W/cm^2 –K and the large face coefficient is 0.00462 W/cm^2 –K.The CERN data did not address the isolated outer layer of conductors which are effectively isolated from the bath by the inter-layer Kapton sheets and the Kapton sheets at the collar interface.

ΔT_5. The temperature difference between the inner wall and the He II depends on test data. Normal engineering boiling correlations are not applicable here because He II does not boil (nucleate bubble growth, etc.) like normal fluids. Because of the very high thermal conductivity, He II boiling takes place on the liquid surface and not the heat transfer surface. The most applicable test data is a heated two-phase flow test of He II conducted at CERN.[6] The CERN test heat exchanger tube was 25.6 mm, and the heated test section was 2 m (the actual flow length was 24 m). The heat flux input was 0–3.5 W/m (0.0044 W/m^2), the flow rate 0.83 g/sec, and the saturation temperature was 1.9 K. The input quality was 0.64 and exit quality ranged from 0.68 to 1.0. The test results indicate that the heat transfer coefficient (ΔT_{3-5}) was 0.059 W/cm^2 –K and was independent of quality up to dryout. The He II liquid covered 35% of the tube perimeter independent of local heat flux. The results for a heat flux of 0.00164 W/cm^2 is ΔT_{3-5} =27.8 mK. CERN test results are independent of quality and heat flux within the limits tested. Heat flux is linearly proportional to ΔT. The CERN heat transfer coefficient includes the effect of the inside/outside Kapitza resistance, conduction resistance across the heat exchanger duct, and the effect of 35% liquid coverage of the inner perimeter.

2D Thermal Model

A two-dimensional thermal model of the LBQ cross section was created[7] using SINDA[8] to investigate the effectiveness of the He II cooling and to conduct whatever trade studies (such as insulation thickness) might be required.The thermal model was generated on the following assumptions: He II bath radial temperature drop was neglected, He II/ inner coil heat transfer based on the convection coefficients backed out of the CERN experiment, outer coil heat transfer by conduction only, no He II heat transfer between laminations, and Kaptiza resistance at all He II/ solid boundary interfaces. Kaptiza resistance and gas conduction SINDA subroutines were created to include these effects. Model results for baseline luminosity show inner conductor maximum temperature of 1.808 K and outer conductor maximum temperatures of 2.07 K. The increased luminosity maximum conductor temperatures were 1.860 K and 2.97 K for the inner and outer conductors respectively. The thermal model results at increased luminosity heat load show some outer conductor layers at greater than 2.0 K temperature due to the assumption of heat transfer by conduction only. Some future experimental work is recommended to measure the extent of outer row thermal coupling with the static superfluid bath.

SUMMARY

As a result of the above, our study recommended that the "Adapted CERN" design should be selected over other investigated designs because of: workable concept for increased luminosity design, i.e., there is room for growth, greatest design margin, design concentrated cryogenic resources at one end, baseline design would require only two heat exchangers. The recommended design parameters are: four 88-mm cooling channels in yoke, liquid supply line routed outside of the cold mass, 78-mm diameter corrugated cooling ducts, two heat exchangers for baseline heat load; the design is upgradable with the addition of a beam cooling insert and two additional heat exchangers.

REFERENCES

1. N. V. Mokhov, R. Schermer, and R. Stiening, "The SSC Low_Beta Quads Design Study," SSC Report, dated 4 June, 1991.

2. CERN, "Design Study of Large Hadron Collider," CERN, July 1991.

3. B. J. Huang, "Joule-Thompson Effect in Liquid He II," *Cryogenics* **V 26**, pages 457–477, (1986).

4. S. V. VanSciver, "Helium Cryogenics," Plenum Press, New York (1986).

5. Meuris, "Heat Transport in Insulation of Cables Cooled by Superfluid Helium," *Cryogenics* **V 31** (1991).

6. Cyvocy, LeBrun et al, "Heated Two-Phase Flow of Saturated He II over Length of 24 m," CERN-AT/91-28, LHC Note 169 (1991).

7. C. Corbett, D. Baritchi, and A. Jalloh, "Documentation of Conceptual Design Studies for He II Cooled Low Beta Quad Magnets," SSC Memo 92 6420 R ME04, dated 14 August 1992.

8. SINDA '85/FLUINT, Version 2.3, COSMIC Program MSC-21528, COSMIC, The University of Georgia, Athens, GA.

REDUCING THE ENERGY REQUIREMENTS OF QUENCH PROTECTION
HEATERS FOR SSC DIPOLES - TEST RESULTS

C. Haddock and J. Kuzminski

Superconducting Super Collider Laboratory*
2550 Beckleymeade Ave.
Dallas, TX 75237

D. Orris and P. Mazur

Fermi National Accelerator Laboratory
P.O. Box 500
Batavia, IL 60510

INTRODUCTION

Design considerations and first test results of quench protection heaters for Superconducting Super Collider (SSC) collider dipole magnets have been presented in earlier papers. [1,2]

The heaters have been shown to fully protect the magnet against excessive peak temperatures which would represent damage to the superconducting coil. Installation and operation of the heaters does not place the magnet at any increased risk of failure, since the energy densities applied are relatively low ($\sim 1 \text{J/cm}^2$) and the construction technique was made as simple as possible.

The energy required by the heaters in order to protect the magnet is considerably larger than that amount estimated during the planning of the collider ring protection scheme. Therefore, three long magnets following the Accelerator Systems String Test (ASST) construction series at Fermi National Accelerator Laboratory (FNAL) were made available for quench protection heater "R&D" studies. All of the ASST series magnets deliberately kept the high energy requirement heaters for the purpose of commonality for the string test.

This paper describes the results of the "R&D" heater tests and the amount of energy reduction achieved. It is shown that it has been possible to reduce the heater energy requirement to a value below the original estimate, and to therefore potentially save collider cost.

SOURCES OF ENERGY INCREASE

The first full length ASST 50 mm aperture dipole magnet built and tested at FNAL (DCA311) incorporated a number of changes from the 40 mm program which affected quench heater energy requirements. First, a second layer of Kapton 5 mil (0.127 mm) in thickness, was added between the quench heater and the superconducting coil, intended to reduce the risk of a heater to coil short. In order that the required quantity of heat to quench the magnet be conducted across the insulation thickness in the required time, (200ms @ 2,000A) a larger energy pulse was now needed at the quench heater terminals.

* Operated by the Universities Research Association, Inc., for the U. S. Department of Energy under
  Contract No. DE-AC35-89ER40486.

Supercollider 5, Edited by P. Hale
Plenum Press, New York, 1994

Secondly, the operating margin of 50 mm dipoles was designed higher than that for the 40 mm dipole program. In order to quench the magnet in a given time interval, an extra amount of energy was required.

Thirdly, it was required that the quench protection heaters should have a cold resistance value of the order of 10 times that of a previous spot like heater design. This is required such that in the collider ring most of the energy from the heater firing unit (HFU) is deposited into the quench heater and not into the connecting leads between the HFU and the magnet. These connecting leads can be quite long ($\sim$ 300m). This in turn required that the quench heater should be essentially a long stainless steel strip along the length of the magnet. The strip has been partially copper plated in order to meet the cold resistance requirements. The active area of the strip is approximately 50 times the spot heater type design; its area was chosen to be conservative since it was hypothesized that the quench propagation velocity in 50 mm magnets might be slower than that for the 40 mm case and, therefore, that a larger coil volume would need to be quenched by heaters in order to protect the magnet. This in turn would require a higher heater energy.

R&D MAGNET HEATERS TEST RESULTS - HEATER ENERGY REDUCTION

T_fn is the characteristic response time interval of the quench heaters. It is the time interval between firing of the quench heater unit and the development of a normal resistive region in the coils. It has been typically shown that T_fn = 200/100ms is sufficient to protect a magnet when operating at a current of 2000A/5000A.

The heater energy increase due to the larger operating margin of 50 mm dipoles was measured in the following way. The energy/unit area of quench heater required to produce a T_fn=200ms @ 2000A was measured on short 40 mm dipole DSO313. All of the quench heaters in this study have the same thickness (0.025mm) and width (12.7mm). The experiment was repeated on short 50 mm magnet DSA328 which had been built with one set of heaters placed one layer of 0.127 mm Kapton from the coil. The results are shown in Table 1.

It was also possible to determine how much extra energy was required due to the second layer of Kapton in DSA328, where a second set of heaters was placed two layers of 0.127mm Kapton from the coil. The results are also shown in Table 1 and Figure 1.

Table 1. Increased heater energy requirement due to increased margin and increased insulation thickness.

Magnet Current (A)	T_fn (ms)	DS0313 (40mm) 1 Layer Kapton	DSA328 (50mm) 2 layers Kapton	DSA328 (50mm) 1 layer Kapton
2000	200	0.43 J/cm^2	1.47 J/cm^2	0.77 J/cm^2

The required energy increase due to the increased operating margin is 80%. It can be seen also from Table 1 that the required energy increase due to the second layer of Kapton represents a factor of 1.9 over the case for one layer of Kapton. Since the risk of heater-to-coil shorts had been mitigated by improvements in the heater design itself, it was not necessary to also include an extra layer of Kapton in the design, after the ASST (DCA311- DCA 320) series was completed. The first full-length magnet "R&D" test was to place a so-called "baseline" heater set one layer of 0.127mm Kapton from the coil, and compare it with another set placed two layers from the coil., i.e., to repeat the DSA328 experiment on a full length magnet. The reduction in heater energy was by a factor of 2.3 times. This energy reduction has been since incorporated into the baseline design of the collider ring dipoles.

On the succeeding two R&D full-length magnets (DCA322 AND DCA323), all quench heaters were placed one layer of 0.127mm Kapton from the coil. These two magnets studied the effect of reducing the "active area" i.e., the unplated length of the quench heater.

Magnet DCA322 contained one set of baseline heaters and one "R&D" set on which the active area had been reduced by a factor of 6 compared to the baseline. On magnet DCA323 the active are was reduced by a factor of 12. The results for magnet DCA311, which contained two sets of baseline heaters placed two layers from the magnet coil, are compared to the results for magnets DCA322 and DCA323 in Figure 2.

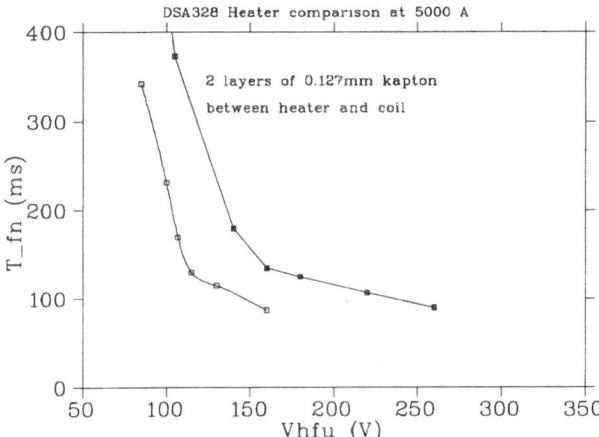

Figure 1. T_fn vs HFU Voltage comparison for quench heaters placed one and two layers of 0.127mm Kapton from coil. The reduction in HFU voltage represents a factor of 1.9 in Energy.

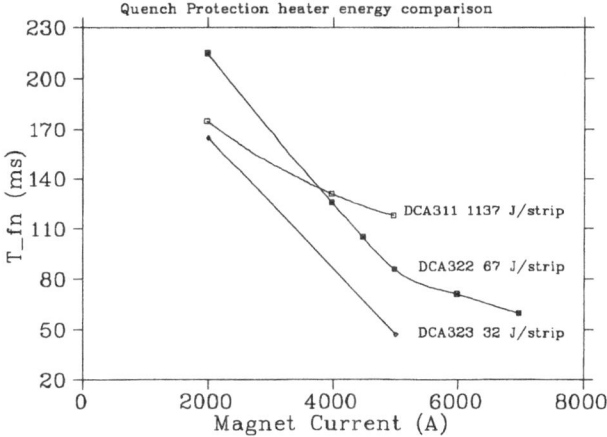

Figure 2. Quench Protection heater energy reduction on three full-length dipoles.

The performance is similar in each case. The "R&D" heaters protect the magnet with 1/17th and 1/35th the value of energy of the DCA311 tests for magnets DCA322 and DCA323, respectively. The energy required (67 J or 32 J for DCA322 and DCA323 respectively) is now considerably less than the original estimate used for evaluating HFU costs for the collider.

FUTURE STUDIES

The reduction in quench protection heater energy by reducing the active length of the heaters also involves decreasing the resistance of the heater. This in turn results in a higher loss in the connecting leads to the HFU. However the reduction in energy achieved is such that the energy loss increase in the leads is tolerable. The cost of the quench heaters is essentially the same regardless of their plated length since they are simply plated strips of stainless steel in a Kapton insulation. However, the cost of the connecting cable, capacitor cost, cabling tray cost, HFU niche size and number, may greatly depend upon the value of the quench heater resistance, and further may experience sharp trends as one finds it necessary to move to non-standard sizes of capacitor, cable tray, etc. Since the performance of the quench heaters of different active length and resistance is now well known, it is possible to optimize the rest of the components of the collider protection electronics with respect to overall cost. Based on the provided quench heater performance data, this task will be performed by the Accelerator Division, Electrical Department of SSCL [1]. This will, in turn, allow a statement of the necessary quench heater resistance to minimize the cost of the collider protection scheme.

SUMMARY

Through a series of tests on three dipole magnets, it has been possible to reduce the energy required by quench protection heaters by a factor of 17, while maintaining the same magnet protection.

REFERENCES

1. C. Haddock, R. Jayakumar, F. Meyer, G. Tool, J. Kuzminski, J. DiMArco, M. Lamm, T. Jaffery, D. Orris, P. Mazur, R. Bossert and J.Strait,``SSC Dipole quench protection heater development," Proceedings of the 12th International Conference on Magnet Technology, June 24-28th 1991, Leningrad, USSR.
2. C. Haddock, B. Aksel, F. Meyer, J. Jayakumar, and G. Tool, "SSC Dipole quench protection heater test results" Proceedings of the 1991 IEEE Particle Accelerator Conference, May 6-10th 1991, San Francisco, CA., pp 2215-2217.

DEVELOPMENT OF SUPERCONDUCTING CORRECTION MAGNETS FOR THE SSC

J. Mitchell[1], T. Adams[1], B. Clark[1], L. Clark[1], N. DiGiacomo[1], M. Fitzgibbon[2], D. Hansen[1], D. Hirschfield[1], R. Keogh[2], D. Ladner[1], R. Peterson[1], L. Pokorny[2], B. Wilson[1]

(1) Martin Marietta Astronautics Group, Denver, CO 80201
(2) Advanced Interconnection Technology, Islip, NY 11751

ABSTRACT

Martin Marietta has completed the development of a number of prototype corrector magnets based on the "Jelly Roll" fabrication technique. Tooling and fabrication and test methodology will be discussed, and magnet performance data will be presented.

INTRODUCTION

Approximately 7,000 dipole, quadrupole, and sextupole correction magnets will be needed within the collider ring to help confine and steer the proton beams through the 54 mile long ring. As the name implies, correction magnets are used to make very fine adjustments to the beam, thus ensuring that the proton bunches remain precisely centered within the 41.4 mm diameter beam tubes as well as properly grouped and phased for acceleration. This fine tuning is essential for the collider to achieve it's ambitious energy and luminosity performance goals. The large number of magnets and the precision performance represents a considerable challenge to both design and manufacturing. Indeed, the SSCL is funding numerous development studies within industry to determine which magnet fabrication techniques meet the performance requirements, and are best suited for cost effective large scale production. The work presented here was done during Phase 1 of the Corrector Magnet Development Program accomplished under contract to the Universities Research Association in support of the Superconducting Super Collider Project for the U.S. Department of Energy.

"JELLY ROLL" FABRICATION TECHNIQUE

The Jelly Roll fabrication technique was developed by Advanced Interconnection Technologies, Inc. (AIT) with support from the U.S. Department of Energy. This automated method of manufacturing the correction coils is described in Reference 1. Briefly, the superconducting wire is precisely laid on a adhesive backed, flat sheet of Kapton using the Multiwire™ technique of bonding the wire to the substrate using an ultrasonic stylus. This method can be used for all multipole types. The geometric configuration of the flat coil is fed into the wiring machine software. The ultrasonic head places the wire within 0.001 inch in a quadrupole pattern of two layers thick and 37 racetracks long.

During Phase 1 of the SSCL Corrector Magnet Program, Martin Marietta made eight quadrupoles using these coils. After we receive the flat pattern coil from AIT, it is readied for roll-up by removing the adhesive from the center of the coil racetracks. The coil is then rolled around an octagon shaped beam tube in the fashion of a "jelly roll" cake. The coil is insulated and shimmed with Kapton sheet for the correct preload and superferric iron laminations are placed around the coil to clamp the coil in place. The magnet is compressed to a specific preload (2,000 psi) and keys are inserted to hold the assembly together (see Figure 1).

TOOLING AND FABRICATION

Tooling for the "Jelly Roll" technology can be divided into the two main categories of winding tooling and collaring and final assembly tooling. The winding tooling has been described in Reference 1. The assembly tooling consists of a coil roll-up tool, a lamination stacking tool, and a collar/key tool.

Figure 1. Phase 1 Quadrupole Correction Magnet - "Jelly Roll" Style.

The coil roll-up tool is a simple device that locates the coil layers into the quadrupole's four poles with wheeled spokes to simulate each pole face. Each double layer is rolled onto the tool successively to form the quadrupole shape. The final configuration yields a matrix of 18 layers by 34 to 51 rows (increasing by one turn per layer inside to outside) for a total of 765 turns. After the coil is rolled-up, it is secured to set the shape. Then insulation is wrapped around the coil to both insulate and shim the coil to provide the proper preload. This was a hand operation during this development phase, but shows promise for a more automated approach during large scale production.

Now that the quadrupole coil has been formed, we surround it with collar quadrants that were previously stacked and assembled. The lamination stacking tool indexed the laminations on their pole face edge. The laminations are stacked in a right hand/left hand sequence and positioned in the stacking tool to a preset quantity. Two stainless steel rods are inserted axially and nuts and lock washers are attached to secure the quadrants. After four quadrants are assembled, they are positioned around the coil and zippered together. This forms a octagonal shaped outer collar around the coil that will preload the coil.

The magnet is now placed in the collar/key tool (Fig. 2) for compression and keying. Four hydraulic rams compress the coil so that the laminations are closed tight. Once they are closed around the coil, the key way is opened to it's widest configuration. At this point, the keys are pressed into the laminations using eight cylinders to hold the preload on the magnet.

TEST METHODOLOGY

Testing of the Phase 1 corrector magnets consisted of electrical property tests (warm and cold), training behavior, and quench level stability tests as a function of ramp rate and current polarity. The electrical property tests included coil resistance, inductance, and high-pot measurements to verify the insulation. Resistance measurements were taken at various places during the assembly process to ensure no turn-to-turn or coil-to-ground shorts appeared. Inductance measurements were performed at ambient temperature and again at LHe temperature after the magnet was assembled.

After the warm tests were completed, the magnet was installed in the test dewar and connected to the power supply and instrumentation. Figure 3 shows our test set-up. The magnet was then cooled to LN_2 temperature, the dewar purged, and then the magnet was cooled to LHe temperature. The primary test requirements for Phase 1 testing were to determine the training behavior and quench level stability of the magnets. To establish the training curve of the magnets, we tested the magnet current at 2 amps per second. After each quench, we allowed the magnet to stabilize for 5 to 8 minutes and re-tested. Upon completion of a sequence of runs in a given polarity, each resulting in a quench, we determined (1) the magnet training curve, (2) whether the magnet had plateaued after five quenches, and (3) whether the required plateau stability of 1% of the short sample limit was reached. We then repeated the testing with the opposite current polarity. In both polarities the sensitivity to ramp rate changes was investigated on all magnets. We tested the magnets at ramp rates of 0.1, 0.5, 1.0, 3.0, 5.0 and 10.0 A/s with no degradation in quench performance.

Figure 2. This Tool Compresses the Collar Around the Coil and Presses in the Four Keys Simultaneously.

MAGNET PERFORMANCE

In general, all the magnets trained quickly and remembered their training after thermal cycling. Figure 4 shows the first cycle training of several magnets. As shown, the early magnets trained progressively (20 - 25 quenches to plateau) with first quench approximately at 50% of Ic. After some adjustments to the assembly process; specifically, adding a longitudinal Teflon slip plane and increasing the compression preload, the later magnets improved dramatically. For example, magnet JR-005's first quench was at 71% of Ic and plateau was reached in seven quenches. Magnets JR-007 and JR-008 also showed a steeper training curve but with a slightly lower first quench.

Once the magnets were trained, they remembered their training very well with thermal cycling. A typical training curve showing the first and second thermal cycles is shown in Figure 5 for magnet JR-005.

During the summer of 1992, C. Lewis Snead of Brookhaven National Laboratory (BNL) and John J. Morena of Ace, Inc. conducted "Radiation Testing of Superconducting Corrector Magnet Organic Polymeric Materials" at BNL. Preliminary results from these tests indicate that two of the adhesive materials used in the coil winding process for "Jelly Roll" magnets (bondall and XP17) did not perform well in the 10^9-Rad irradiation. This data, and the low packing fraction of superconducting wire when the coil is rolled into the magnet geometry using the "Jelly Roll" technique, has caused the SSCL to eliminate this technique from further consideration in the SSCL Corrector Magnet Program.

However, the "Jelly Roll" fabrication technique shows a lot of promise and with further development I believe this technology could prove itself to be a simple method of assembling corrector magnets. It can be used where field requirements and packing fraction are less stringent and the irradiation levels are lower over the life of the system. For example, the Relativistic Heavy Ion Collider (RHIC) is using this technique for the fabrication of its corrector magnets. There the multipole layers are radially nested on the beam tube.

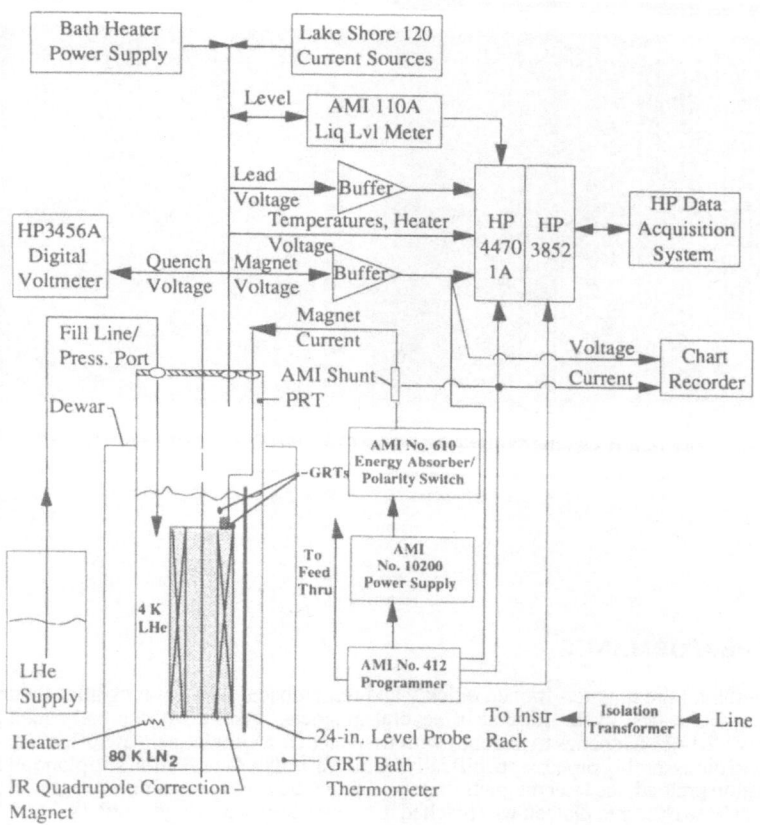

Figure 3. Test Instrumentation for Phase 1 Testing.

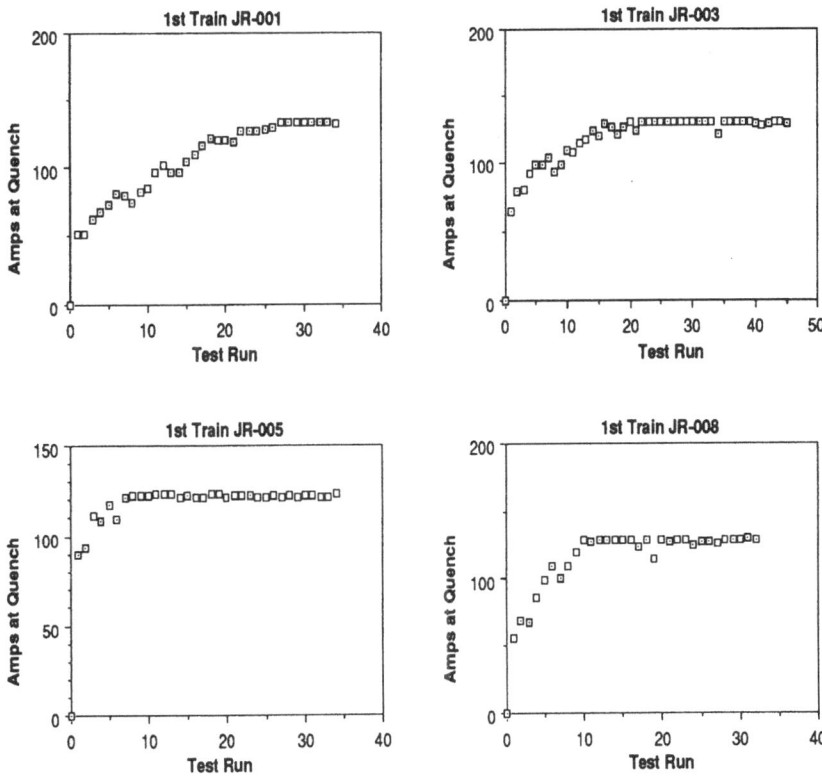

Figure 4. Typical Correction Magnet Training Histories.

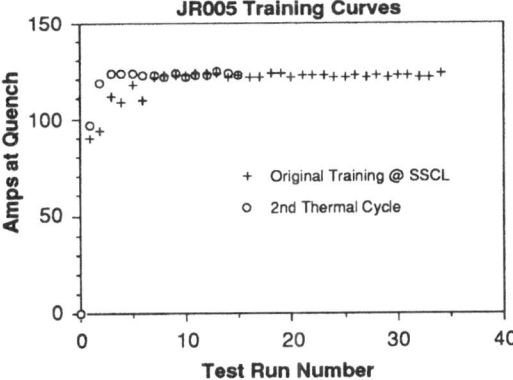

Figure 5. Typical Training Curve Showing Results for the First and Second Thermal Cycles.

REFERENCES

1. Schieber, Leonard, A Precise Technique for Manufacturing Correction Coils, Super Collider 4, 411, (1992) J. Nonte, ed.

CORRECTOR MAGNETS: COMBINED STRUCTURAL ANALYSIS OF COLLIDER 50MM APERTURE ORDERED WOUND QUADRUPOLES INTERIOR SECTION

Vu H. Tran

Mechanical Engineering Department
Accelerator Systems Division
Superconducting Super Collider Laboratory *
2550 Beckleymeade Avenue, MS 4006
Dallas, TX 75237-3997

ABSTRACT

The 50mm aperture prototype collider ordered wound quadrupole corrector magnets have been modeled with finite element techniques considering the individual and combined load cases of the preloading from keys, cooldown to 4 K and the effect of magnetic forces during energizing. Results of the analysis are presented as longitudinal, transverse and shear stresses for the ordered wound coils and as maximum von Mises stress for the carbon steel outer laminations, the stainless steel inner lamination, and the carbon steel keys.

INTRODUCTION

The quadrupole corrector magnets are members of the collider ring correctors which are responsible for tune correction and control of the proton beams and augment the main quadrupoles. These corrector magnets are located inside of the spool piece. This paper presents results from a finite element stress analysis of the 50mm aperture prototype collider ordered wound quadrupole corrector magnets. This analysis considers only the interior cross-sections along the length of the structure - which differ greatly from the end sections. The coils for this quadrupole magnet structure are manufactured using ordered wound techniques. Load cases involved in the analysis include preloading from keys, cooldown to 4 K and magnetic (Lorentz) forces from energizing. The objectives of this stress analysis for the corrector magnet quadrupole structure are to determine the stresses in the ordered wound coils, the necessary shim thickness around the coils and to evaluate the capability of the outer laminations, inner laminations and keys to resist the stresses induced by reaction with the coils. The results of this analysis will be used as reference for further design of the quadrupole structure members - including the coils, inner laminations, outer laminations and keys.

The quadrupole magnet structures consist of magnetic coils held in place by a series of thin, interlocking carbon steel outer laminations. Figure 1 shows two consecutive cross-sections of the quadrupole. The outer laminations are configured such that each cross-section of the magnetic structure along the length of the coils contains four interlocking outer lamination pieces. Each outer lamination piece is joined to the pieces immediately in front of and behind it by two continuous pins.

*Operated by the Universities Research Association, Inc., for the U.S. Department of Energy under Contract No. DE-AC35-89ER40486.

Continuous longitudinal carbon steel alignment keys (on the right and left sides) and lock keys (on the top and bottom sides) provide the "preload" forces necessary to hold the outer laminations in place and to exert pressure on the coils. The outer laminations sections immediately in front of and behind each individual outer laminations sections are identical in geometry but are rotated 180 degrees about the vertical axis (y-axis), yielding the same external shape but a different configuration at the alignment and lock keys. The alternating orientation of the outer lamination pieces permits the "elbow" section of each outer lamination piece keyway to be alternately placed above or below the alignment keys and placed on the left or right side of the lock keys. Layer of Kapton are placed around the coils to act as a shim. Therefore, one of the design criteria in this analysis is to provide enough shim thickness to have the minimum stresses in the coils after cooldown to be at least greater than the maximum stresses in the coils due to magnetic forces only. This criteria should applied to each coil's local longitudinal, transverse and shear direction stresses in the coils. The ordered wound coil structure consists of continual loops of 0.381 mm (15.0 mil)-diameter superconducting Cu:Nb-Ti (2.2:1) wire insulated with a 0.076 mm (3.0 mil)-thick layer of Kapton and Dupont XMPI adhesive.

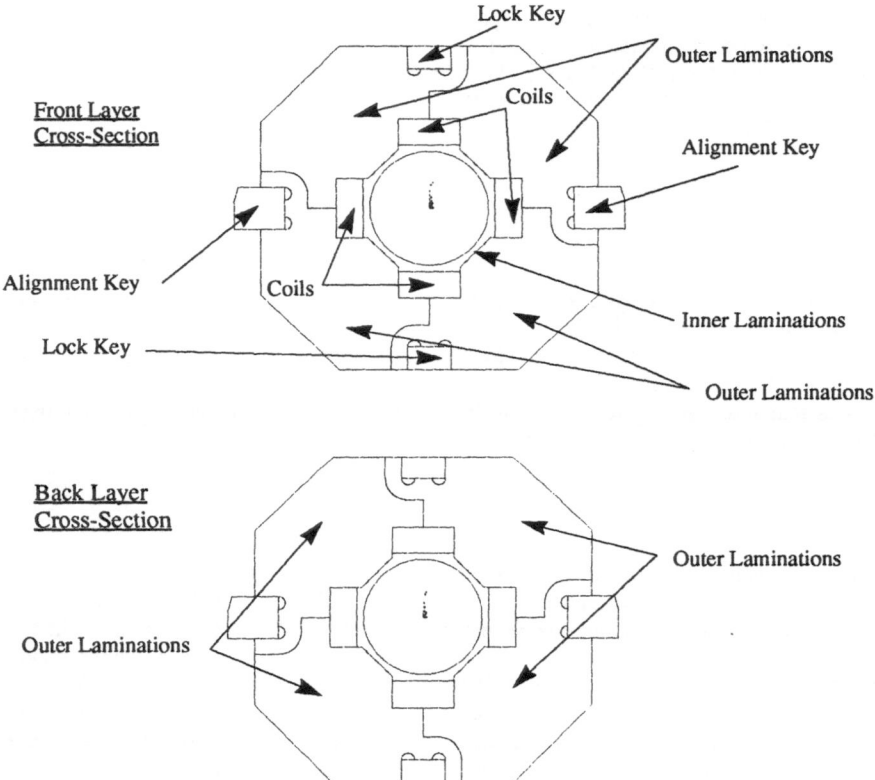

Figure 1. Collider 50mm Aperture Ordered Wound Quadrupoles - Front and Back Layer Cross-Sections

MODEL CONSTRUCTION

The magnet assembly is modeled as a two-dimensional finite element system using ANSYS finite element code. The model consists of two adjacent layers of outer laminations (each consisting of four outer lamination pieces from the same cross-section) hereafter referred to as the "front" and "back" outer laminations. To get the correct loading from the keys, the two layers of the outer laminations are modeled with two-dimensional plane stress elements. The front and back outer laminations are linked together by the coupled nodes at each of the eight continuous pin locations. Gap elements having a positive gap distance to act as a interference are used to join the coils to the outer laminations (gap elements only transmit compression load not tension load). These positive gap distance represent the thickness of the shim around the coils. All other gap elements between various parts are given a zero value. The coils, inner lamination and keys are constructed from two-dimensional plane stress elements. Material properties used in this model are shown in Table 1. Material properties for the ordered wound coils are obtained from testing and analysis.

Table 1. Material Properties

Materials		Carbon Steel	304L Stainless Steel [1]	Ordered Wound Coil
Young's Modulus (psi)	T = 293 K	30.0×10^6	27.6×10^6	200,000
	T = 4 K	30.6×10^6	29.2×10^6	300,000
Poisson's Ratio	T = 293 K	0.292	0.29	0.346
	T = 4 K	0.292	0.2788	0.339
Thermal Expansion [2] Coefficients, (K^{-1})		0.685×10^{-5}	1.059×10^{-5}	1.4×10^{-5}
Yield Tensile Strength (psi)	T = 293 K	25,000	58,900	N/A
	T = 4 K	54,000	79,400	N/A
Ultimate Tensile Strength (psi)	T = 293 K	42,000	95,500	N/A
	T = 4 K	65,000	241,000	N/A

LOADING CASES

Preload Case

As mentioned previously, the preloading case experienced upon assembly is simulated by assigning to all the gap elements around the four coils an overlapping gap distance to represent the shim thickness. All material properties for the preload case are referenced to 293 K.

Preload and Cooldown Case

The combined analysis of the preload and the cooldown uses the same model as does the preload alone except that the model is assigned an initial temperature of 293 K and a final temperature of 4 K to represent the cryogenic state after the cooldown has occurred. All material properties for the combined analysis are given at the reference temperature of 4 K.

Magnetic Forces Case

To complete a magnetic stress analysis of the model, a separate magnetic analysis must first be conducted. The magnetic analysis is performed using a somewhat simplified version of the quadrupole model. The magnetic analysis is concerned only with the location and the magnetic properties of quadrupole structure components; therefore, the model used in the said analysis contains only the geometric and physical description of the individual components and does not require the use of gap elements. The wire in the coils are given a current density of 4.0596×10^8 A/m^2 at 100% of short sample current, but only 66% is required for operation current. The forces on the nodes of the coils are computed in the magnetic analysis and then inserted into the standard quadrupole model. A finite element analysis is then performed on the standard model with the magnetic forces alone to determine the influence of the magnetic forces on the quadrupole structure.

Preload, Cooldown and Magnetic Forces Case

To complete the final combined analysis, the standard quadrupole model is subjected to the simultaneous loadings of the preload, cooldown and magnetic forces. Once again, a finite element analysis is performed to determine the stresses on the individual components of the coil structure due to the three combined loading cases.

STRESS ANALYSIS RESULTS

In order to have the minimum stresses in the coils after cooldown to be at least greater than the maximum stresses in the coils due to magnetic forces only, the finite element analysis shows that shim thickness of 3.6 and 5.5 mils must be used along the shorter and longer length of all the coils, respectively. These results are shown on Table 2. As shown in Table 2, for the coils the first value is the minimum and the second is the maximum, thus showing the range of stresses in the coils. The longitudinal and transverse directions are defined along the longer and shorter length of the coils, respectively.

Table 2. Stress Levels for Collider Ordered Wound Quadrupoles

Loading Cases		Preload of Keys shim1 =3.6 mils * shim2 = 5.5 mils **	Preload of Keys & Cool Down to 4 K	Magnetic Forces Only $I = 66\% I_{ss}$ or $110\% I_{op}$	Preload of Keys & Cool Down & Magnetic Forces
Coils Min : Max Stress (psi)	$\sigma_{transverse}$	-1,733 : -881	-1,364 : -569	-568 : -92	-1,214 : -505
	$\sigma_{longitudinal}$	-1,078 : -509	-918 : -383	-381 : -56	-840 : -396
	τ_{shear}	-109 : 109	-104 : 104	-19 : 22	-70 : 53
C. S. Outer Laminations Max. von Mises Stress (psi)		43,000	32,100	12,900	33,100
C. S. Keys Max. von Mises Stress (psi)		30,100	22,200	8,900	22,900
S. S. Inner Laminations Max. von Mises Stress (psi)		30,100	20,600	240	13,600
* shim1 = shim thickness between the two shorter lengths of coils and outer laminations ** shim2 = shim thickness between the longer length of coils and outer laminations					

CONCLUSION

The finite element analysis for the 50mm aperture collider ordered wound quadrupole corrector magnets shows that at shim thickness of 3.6 and 5.5 mils, there are sufficient stresses in the coils to restrain movement. The stresses in the coils have peak values of maximum and minimum, therefore, further design of the quadrupoles should try to reduce these peak values and get an evenly distributed stresses in the coils. It should be noted that different values for the material properties of coils will greatly effect the design and analysis. Therefore, accurate values of coils material properties must be obtained for further design and analysis. It should also be remembered that this analysis considers only cross-sections from the interior of the structure in a two-dimensional frame. As the cross-sections at the ends of the quadrupole structure differ significantly from those in the interior, additional analyses must be performed to evaluate their behavior.

REFERENCES

1. Richard P. Reed and Alan F. Clark, eds., *Materials at Low Temperatures*, American Society for Metals, Metals Park, Ohio, 1983, page 8.1.3-2.
2. Neil A. Olien, Ed., *LNG Materials and Fluids, Second Supplement*, Prepared by Cryogenics Division, Institute for Basic Standards, National Bureau of Standards, Boulder, Colorado, 1979, page 95.

CORRECTOR MAGNETS: COMBINED STRUCTURAL ANALYSIS OF COLLIDER 50MM APERTURE ORDERED WOUND SEXTUPOLES INTERIOR SECTION

Vu H. Tran

Mechanical Engineering Department
Accelerator Systems Division
Superconducting Super Collider Laboratory *
2550 Beckleymeade Avenue, MS 4006
Dallas, TX 75237-3997

ABSTRACT

The 50mm aperture prototype collider ordered wound sextupole corrector magnets have been modeled with finite element techniques considering the individual and combined load cases of the preloading from keys, cooldown to 4 K and the effect of magnetic forces during energizing. Results of the analysis are presented as longitudinal, transverse and shear stresses for the ordered wound coils and as maximum von Mises stress for the carbon steel outer laminations, the stainless steel inner lamination, the carbon steel floating pole tips and the carbon steel keys.

INTRODUCTION

The sextupole corrector magnets are members of the collider ring correctors which are responsible for linear chromaticity control of the proton beams. These corrector magnets are located inside of the spool piece. This paper presents results from a finite element stress analysis of the 50mm aperture prototype collider ordered wound sextupole corrector magnets. This analysis considers only the interior cross-sections along the length of the structure - which differ greatly from the end sections. The coils for this sextupole magnet structure are manufactured using ordered wound techniques. Load cases involved in the analysis include preloading from keys, cooldown to 4 K and magnetic (Lorentz) forces from energizing. The objectives of this stress analysis for the corrector magnet sextupole structure are to determine the stress in the ordered wound coils, the necessary shim thickness around the coils and to evaluate the capability of the outer laminations, inner laminations, floating pole tips and keys to resist the stress induced by reaction with the coils. The results of this analysis will be used as reference for further design of the sextupole structure members - including the coils, the inner laminations, the outer laminations and the keys.

The sextupole magnet structures consist of magnetic coils held in place by a series of thin, interlocking carbon steel outer laminations. Figure 1 shows two consecutive cross-sections of the sextupole. The outer laminations are configured such that each cross-section of the magnetic structure along the longitudinal axis of the coils contains four interlocking outer lamination pieces. Each outer lamination piece is joined to the pieces immediately in front of and behind it by two continuous pins. Continuous longitudinal carbon steel alignment keys (on the right and left sides) and lock keys

*Operated by the Universities Research Association, Inc., for the U.S. Department of Energy under Contract No. DE-AC35-89ER40486.

(on the top and bottom sides) provide the "preload" forces necessary to hold the outer laminations in place and to exert pressure on the coils. The outer laminations sections immediately in front of and behind each individual outer laminations section are identical in geometry but are rotated 180 degrees about the vertical axis (y-axis), yielding the same external shape but a different configuration at the alignment and lock keys. The alternating orientation of the outer lamination pieces permits the "elbow" section of each outer lamination piece keyway to be alternately placed above or below the alignment keys and placed on the left or right side of the lock keys. There are two floating pole tips located on the left and right sides of the inner laminations. Layer of Kapton are placed around the coil to act as a shim. Therefore, one of the design criteria in this analysis is to provide enough shim thickness to have the minimum stress in the coil after cooldown to be at least greater than the maximum stress in the coil due to magnetic forces only. This criteria should applied to each coil's local longitudinal, transverse and shear direction stresses in the coils. The ordered wound coil structure consists of continual loops of 0.381 mm (15.0 mil)-diameter superconducting Cu:Nb-Ti (2.2:1) wire insulated with a 0.076 mm (3.0 mil)-thick layer of Kapton and Dupont XMPI adhesive.

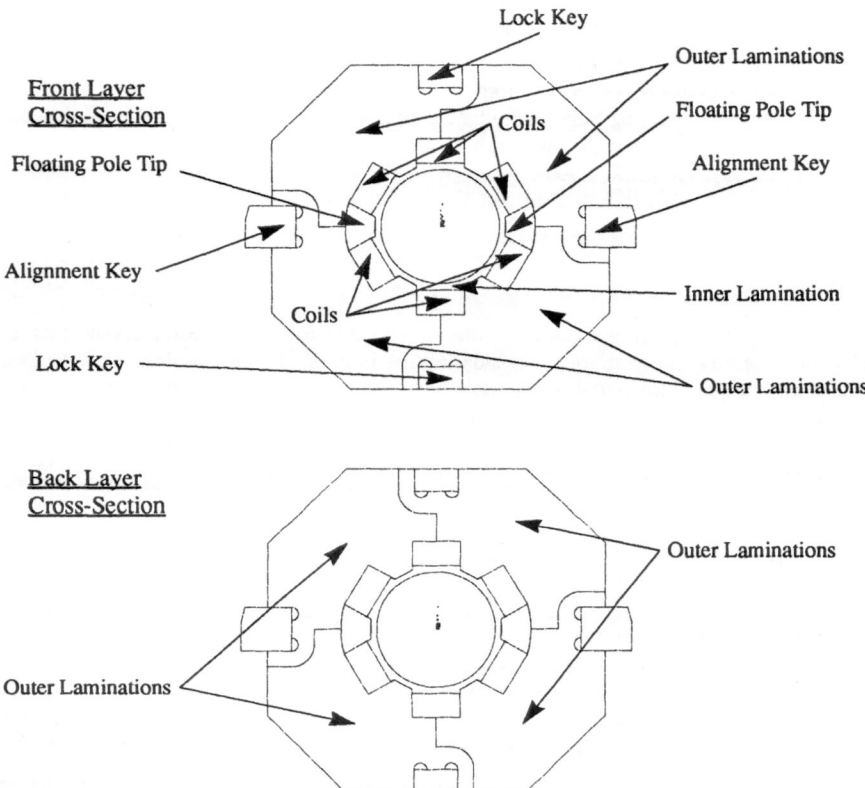

Figure 1. Collider 50mm Aperture Ordered Wound Sextupoles - Front and Back Layer Cross-Sections

MODEL CONSTRUCTION

The magnet assembly is modeled as a two-dimensional finite element system using ANSYS finite element code. The model consists of two adjacent layers of outer laminations (each consisting of four outer lamination pieces from the same cross-section) hereafter referred to as the "front" and "back" outer laminations. To get the correct loading from the keys, the two layers of the outer laminations are modeled with two-dimensional plane stress elements. The front and back outer laminations are linked together by the coupled nodes at each of the eight continuous pin locations. Gap elements having a positive gap distance to act as a interference are used to join the coils to the outer laminations (gap elements only transmit compression load not tension load). These positive gap distance represent the thickness of the shim around the coils. All other gap elements between various parts are given a zero value. The coils, inner lamination, floating pole tips and keys are constructed from two-dimensional plane stress elements. Material properties used in this model are shown in Table 1. Material properties for the ordered wound coils are obtained from testing and analysis.

Table 1. Material Properties

Materials		Carbon Steel	304L Stainless Steel [1]	Ordered Wound Coil
Young's Modulus (psi)	T = 293 K	30.0×10^6	27.6×10^6	200,000
	T = 4 K	30.6×10^6	29.2×10^6	300,000
Poisson's Ratio	T = 293 K	0.292	0.29	0.346
	T = 4 K	0.292	0.2788	0.339
Thermal Expansion [2] Coefficients, (K^{-1})		0.685×10^{-5}	1.059×10^{-5}	1.4×10^{-5}
Yield Tensile Strength (psi)	T = 293 K	25,000	58,900	N/A
	T = 4 K	54,000	79,400	N/A
Ultimate Tensile Strength (psi)	T = 293 K	42,000	95,500	N/A
	T = 4 K	65,000	241,000	N/A

LOADING CASES

Preload Case

As mentioned previously, the preloading case experienced upon assembly is simulated by assigning to all the gap elements around the four coils an overlapping gap distance to represent the shim thickness. All material properties for the preload case are referenced to 293 K.

Preload and Cooldown Case

The combined analysis of the preload and the cooldown uses the same model as does the preload alone except that the model is assigned an initial temperature of 293 K and a final temperature of 4 K to represent the cryogenic state after the cooldown has occurred. All material properties for the combined analysis are given at the reference temperature of 4 K.

Magnetic Forces Case

To complete a magnetic stress analysis of the model, a separate magnetic analysis must first be conducted. The magnetic analysis is performed using a somewhat simplified version of the sextupole model. The magnetic analysis is concerned only with the location and the magnetic properties of sextupole structure components; therefore, the model used in the said analysis contains only the geometric and physical description of the individual components and does not require the use of gap elements. The wire in the coils are given a current density of 4.3884×10^8 A/m^2 at 100% of short sample current, but only 66% is required for operation current. The forces on the nodes of the coils are computed in the magnetic analysis and then inserted into the standard sextupole model. A finite element analysis is then performed on the standard model with the magnetic forces alone to determine the influence of the magnetic forces on the sextupole structure.

Preload, Cooldown and Magnetic Forces Case

To complete the final combined analysis, the standard sextupole model is subjected to the simultaneous loadings of the preload, cooldown and magnetic forces. Once again, a finite element analysis is performed to determine the stresses on the individual components of the coil structure due to the three combined loading cases.

627

STRESS ANALYSIS RESULTS

In order to have the minimum stresses in the coils after cooldown to be at least greater than the maximum stresses in the coils due to magnetic forces only, the finite element analysis shows that shim thickness of 3 and 6 mils must be used along the two shorter and one longer lengths of all the coils, respectively. These results are shown on Table 2. As shown in Table 2 for the coils the first value is the minimum and the second is the maximum, thus showing the range of stresses in the coils. The longitudinal and transverse directions are defined along the longer and shorter length of the coils, respectively.

Table 2. Stress Levels for Collider Ordered Wound Sextupoles

Loading Cases		Preload of Keys shim1 = 3 mils * shim2 = 6 mils **	Preload of Keys & Cool Down to 4 K	Magnetic Forces Only $I = 66\% I_{ss}$ or $110\% I_{op}$	Preload of Keys & Cool Down & Magnetic Forces
Coils Min : Max Stress (psi)	$\sigma_{transverse}$	-1,187 : -954	-877 : -657	-505 : -80	-927 : -478
	$\sigma_{longitudinal}$	-672 : -515	-586 : -426	-372 : -17	-618 : -282
	τ_{shear}	-34 : 34	-33 : 33	-38 : 20	-14 : 21
C.S. Outer Laminations Max. von Mises Stress (psi)		57,200	44,000	18,000	45,300
C.S. Floating Pole Tips Max. von Mises Stress (psi)		4,500	4,000	50	3,000
C.S. Keys Max. von Mises Stress (psi)		31,300	23,100	10,800	23,700
S.S. Inner Laminations Max. von Mises Stress (psi)		15,500	10,900	700	6,800
* shim1 = shim thickness between the two shorter lengths of coils and outer laminations ** shim2 = shim thickness between the longer length of coils and outer laminations					

CONCLUSION

The finite element analysis for the 50mm aperture collider ordered wound sextupole corrector magnets shows that at shim thickness of 3 and 6 mils, there are sufficient stresses in the coils to restrain movement. The stresses in the coils have peak values of maximum and minimum, therefore, further design of the sextupoles should try to reduce these peak values and get an evenly distributed stresses in the coils. It should be noted that different values for the material properties of coils will greatly effect the design and analysis. Therefore, accurate values of coils material properties must be obtained for further design and analysis. It should also be remembered that this analysis considers only cross-sections from the interior of the structure in a two-dimensional frame. As the cross-sections at the ends of the sextupole structure differ significantly from those in the interior, additional analyses must be performed to evaluate their behavior.

REFERENCES

1. Richard P. Reed and Alan F. Clark, eds., *Materials at Low Temperatures*, American Society for Metals, Metals Park, Ohio, 1983, page 8.1.3-2.
2. Neil A. Olien, Ed., *LNG Materials and Fluids, Second Supplement*, Prepared by Cryogenics Division, Institute for Basic Standards, National Bureau of Standards, Boulder, Colorado, 1979, page 95.

ENGINEERING DESIGN OF A HIGH-TEMPERATURE
SUPERCONDUCTOR CURRENT LEAD

R. C. Niemann,[1] Y. S. Cha,[1] J. R. Hull,[2] M. A. Daugherty,[2]
and W. E. Buckles[2]

[1]Argonne National Laboratory, Argonne, IL 60439
[2]Superconductivity, Inc., Madison, WI 53705

INTRODUCTION

As part of the U.S. Department of Energy's Superconductivity Pilot Center Program, Argonne National Laboratory and Superconductivity, Inc., are developing high–temperature superconductor (HTS) current leads suitable for application to superconducting magnetic energy storage systems.

The principal objective of the development program is to design, construct, and evaluate the performance of HTS current leads suitable for near–term applications. Supporting objectives are to (1) develop performance criteria; (2) develop a detailed design; (3) analyze performance; (4) gain manufacturing experience in the areas of materials and components procurement, fabrication and assembly, quality assurance, and cost; (5) measure performance of critical components and the overall assembly; (6) identify design uncertainties and develop a program for their study; and (7) develop application–acceptance criteria.

DESIGN CRITERIA

A current–lead design was generated for a near–term energy storage system. Significant elements of the criteria are given in Table 1.

HTS LOWER STAGE

Geometry

Based on the current–carrying characteristics of available, well–characterized HTS conductors and the geometrical requirements of the magnet cryostat, the lower stage was designed to consist of six parallel current–carrying HTS conductor rods. The rods are connected to the transition at their warm ends and to the bus collector assembly at their cold ends. Each rod is contained in an epoxy fiberglass tube that channels the helium vapor coolant and provides structural support for the rod. A safety lead is located central to the HTS rod array. The geometry offers flexibility that can readily accommodate conductors of different materials and/or geometries.

Table 1. HTS current–lead design criteria.

Parameter	Design value
Current (A)	1500
SSD background field (T) at HTS conductor	
* Radial	≥0.050
* Axial	≥0.100
Temperature (K)	
* Upper–stage warm end	300
* Transition	50 ≤ T ≤ 70
* Lower–stage cold end	≤10
Heat leak per lead to 4 K (W/lead at current)	≤0.7

Table 2. Outline specifications for HTS current–lead conductor.

Material	YBCO w/15 vol.% Ag
Critical current	$J_c \geq 200$ A/cm^2 @ 77 K @ 5 mT
Thermal conductivity	$\lambda \leq 12$ W/mK @ T ≤ 77 K
Mechanical integrity	Flexural strength ≥ 180 MPa @ 300 K
	Fracture toughness ≥ 3.2 MPa $\sqrt{m}$ @ 300 K
Geometry	Cylindrical rod (O.D.) = 1.3 cm ± 0.03 cm
	Length = 20 cm ± 0.15 cm
	Straightness = ± 0.04 cm about a centerline joining the centers of the two ends

Conductor

YBCO with 15 vol.% silver was selected as the baseline HTS rod material. Conductor selection was based on near–term material availability, balanced electrical and mechanical properties, and existence of an extensive data base on this material's performance characterization. Significant elements of the HTS conductor specifications are given in Table 2.

Thermal Performance

The thermal performance estimate of the overall lower stage assembly is shown in Table 3.

Structural Performance

The objective of the structural design and performance analyses was to ensure the long–term effective and reliable performance of the lower stage assembly. The loading included (1) static (gravity and fabrication), (2) dynamic (handling, transportation, seismic, and magnet charge and discharge), and (3) thermal cycles. Lateral and axial loading of the conductor rods is controlled by assembly geometry and materials.

Magnetic Shielding

The conductor rods are in the presence of applied magnetic fields from the magnet solenoid and the other five rods in the lower stage assembly. The resulting applied radial field at the cold end of a conductor is approximately 0.03 T. Magnetic shielding employing conventional material and geometry is employed to attenuate the field.

Safety Lead

A safety lead is incorporated in the lower stage assembly as an alternate current path in the event of HTS conductor malfunction. The safety lead consists of a stainless steel tube connected in parallel to the conductor elements. The safety lead contributes 0.11 W to the lead assembly's 0.87 W cold–end heat leak. Lead temperature rises to 500 K when the safety lead is employed to discharge the system's stored energy.

Table 3. Thermal performance estimate of lower stage assembly.

Operating conditions	• $I = 1500$ A • $T_{WARM} = 60$ K • $T_{COLD} = 10$ K	
Heat leak per lead		
	Component	Heat leak to 10 K (W)
	HTS elements	0.70
	Element shroud tubes	0.01
	Safety lead	0.11
	Voltage isolation tubes	0.01
	Helium gas	0.04
	Total	0.87

TRANSITION

The transition's functions are to (1) provide electrical and mechanical connections between the lower and upper stages, (2) provide flow paths for the helium–vapor cooling stream, (3) provide a heat–intercept connection to a cryocooler, (4) provide electrical isolation between the lead and the adjacent cryostat components, and (5) provide pressure-vessel continuity within the cryostat.

The transition consists of (1) a central copper disk, (2) an insulating tube, (3) an outer copper ring, (4) twelve copper cables connecting the ring and the cryocooler heat intercept, and (5) stainless steel tubular sections for welding to the cryostat's helium vessel lead-penetration tubes. The lower and upper stages are connected to the disk by screw–joint connections. The connections provide low electrical resistance, low thermal resistance, structural integrity, and long–term reliability.

The transition is configured to minimize the temperature difference between the disk and the cryocooler intercept; this is achieved through geometry, materials selection, and fabrication methods. The transition must also provide effective and reliable electrical isolation.

CONVENTIONAL UPPER STAGE

Configuration

The upper stage requirements are to (1) provide an efficient and reliable electrical connection between the ambient environment and the transition, (2) provide a low heat load to the transition, and (3) be compatible with the magnet cryostat geometry.

Thermal Performance

The thermal performance estimate for the overall upper stage assembly is given in Table 4.

PERFORMANCE EVALUATIONS

Component

A component performance evaluation program has been developed to confirm the analytical performance predictions and/or qualify the design features for construction.

Table 4. Thermal performance estimate of upper stage assembly.

Operating conditions	• $I = 1500$ A
	• $T_{WARM} = 300$ K
	• $T_{COLD} = 60$ K

Heat leak per lead

Component	Heat leak to 60 K (W)
Conventional current lead	45.00
Helium vessel pressure tube	1.85
Voltage isolation tubes	0.09
Helium gas	0.13
Total	47.07

Components and/or design features considered by the program include (1) HTS conductor characterization, (2) lower stage element and assembly structure, (3) screwed–joint connections, (4) transition heat transfer, (5) magnetic shield attenuation, and (6) voltage isolation.

System

A system performance evaluation program has been developed to evaluate the performance of current lead assemblies. This program consists of three separate phases: the zero–field test, energy storage operations test, and life–cycle test.

CONCLUSIONS

The engineering design of a HTS current lead assembly suitable for SSD applications has been completed. The general current–lead design approach is applicable to superconducting magnets for the SSC and other applications. Readily adaptable design features include (1) modular lower stage, transition, and upper stage; (2) lower stage structure and connections; (3) transition heat intercept–voltage isolation configuration; and (4) upper stage structure and connections. The design lends itself readily to alternate HTS conductor materials and geometries, system–specific installation geometries, and system–specific operational modes. A component performance evaluation program is underway.

Lead assemblies are being constructed for system performance evaluations that include zero–field operation, energy storage operation, and life–cycle simulations. The knowledge gained by the performance evaluations will further the understanding of the application of HTS current leads in a true engineering sense to actual superconducting magnetic systems. The resulting information will contribute to the design data base and will identify areas for further development.

ACKNOWLEDGMENTS

The authors gratefully acknowledge the engineering contributions of J. D. Gonczy, T. H. Nicol, and B. R. Weber; the technical expertise of M. Bordson and A. S. Wantroba;

and the manuscript preparation skills of J. A. Stephens and the editorial contributions of C. A. Malefyt. Work at ANL was supported by the U.S. Department of Energy, Energy Efficiency and Renewable Energy, as part of a program to develop electric power technology, under Contract W-31-109-Eng-38. The work at Superconductivity, Inc., was conducted as a part of the corporation's internal systems development program.

MAGNETIC FIELD CALCULATIONS FOR THE HDM MODEL MAGNETS

M. P. Krefta, H. L. Chuboy and J. H. Parker, Jr.

Westinghouse Science and Technology Center
1310 Beulah Road
Pittsburgh, PA 15235

INTRODUCTION

Westinghouse Electric Corporation (WEC) is presently under contract to the SSCL to design and build the dipole magnets for the High Energy Booster (HEB).[1] As a starting point in the magnet design process, a baseline design was provided by SSCL, identified as DSB-701, to be used for the analysis and construction of three 1.7 m model magnets. The magnet drawings and associated documentation were used by WEC to perform a thorough analysis and develop an understanding of the design. The analysis work performed included magnetic, mechanical, thermal, and electrical modeling of the baseline design DSB-701. The particular focus of this paper is to summarize the magnetic analyses that were performed.

CABLE STACKUP

The starting point of the analysis is to determine the locations of the current carrying elements throughout the magnet cross section. The computer program STACKUP[3] was developed for stacking cables and wedges on perfect arcs so their locations and those of the current elements they contain can be accurately determined. The inputs to the program include: the dimensions of the cables and wedges under azimuthal pressure; the pole locations as fixed by the collar, nominal pole shim and ground wall insulation; the midplane gap as fixed by the ground wall insulation; and, a set of stacking rules which are dictated by tooling considerations. Some of the stacking rule options include stacking against either an inside or outside radius, and positioning cables and wedges either at the first point of contact with the radius, or with the midpoint of the cable's (or wedge's) inner or outer edge attached to the radius. The program permits initiating the stackup from either the pole or midplane and also has an option for constraining both while varying the insulation thicknesses to fit the available windows. Several alternatives for representing the current distribution internal to each cable exist. These include representing the current with a thin rectangle, using two thin parallel lines as in the computer program SSCMAG,[2] or using discrete current elements (30 inner, 36 outer cable) to represent individual strands in the cable.

The cross section resulting from the cable stackup is shown in fig.1. In this case, the cables and wedges were stacked against an outer radius. The stacking arc contacts the cable outer edges at their midpoint and the wedges are slid radially outward until they first touch the arc. The harmonics for this configuration are given in Table 1 along with results obtained using SSCMAG. The harmonics for the cases of rectangular, two-lines and strand current representation are, for all practical purposes, the same. However, the harmonics resulting from the program SSCMAG are somewhat different. These differences are attributed to the stacking method. In the case of SSCMAG, the cables are stacked on an inner radius intersecting the uninsulated cable edge with the stacking arc. The wedges are not considered in this program, since the purpose of the program is to optimize the location of current elements without regard to wedges. The wedges are then designed from the output of the SSCMAG program. Correspondingly, STACKUP is an analysis program which uses both the cable and wedge geometries to stack. This program allows for stacking the cables using a variety of methods. In the WEC analysis, the cables and wedges were stacked against an outer radius, using a midpoint method for cables and first point of contact method for the wedges. The different methods of positioning the cables result in differences in the harmonics shown in Table 1. This was verified by setting the options in the STACKUP program to be identical to those in SSCMAG. The same harmonics are obtained using either the STACKUP or SSCMAG.[3]

Table 1. Calculated field harmonics for DSB-701 using the computer programs STACKUP and SSCMAG.

	STACKUP RECTANGULAR	STACKUP TWO-LINES	STACKUP STRANDS	SSCMAG
Transfer Function (T/kA)	1.027706	1.02701	1.02700	1.0276
b2 (units)	-0.6115	-0.5662	-0.5685	-0.1372
b4 (units)	0.0188	0.0179	0.0185	-0.0188
b6 (units)	0.0121	0.0128	0.0129	0.0056
b8 (units)	0.0031	0.0033	0.0033	0.00106
b10 (units)	0.0076	0.0076	0.0076	0.00698
b12 (units)	0.0021	0.0021	0.0021	0.00195
b14 (units)	-0.00096	-0.00095	-0.00095	-0.000105

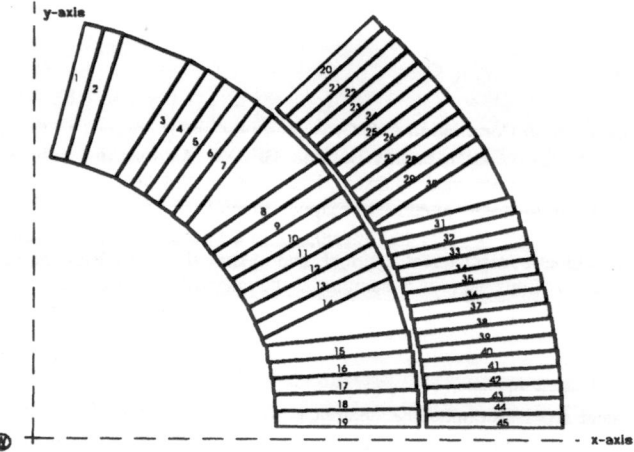

Figure 1. Stackup of cables and wedges for HDM model magnet cross section DSB-701.

MAGNETIC SATURATION EFFECTS

In addition to the geometric component of the field harmonics, there is an additional contribution from the yoke iron as it begins to saturate. Magnetic saturation can be accurately predicted using a detailed finite element (FE) model of the magnet cross section. A complete set of geometric coordinates for all conductor and wedge corner points is generated by the program STACKUP and used as a basis for developing a FE model of the magnet 2D cross section. The complete FE model includes the coil current elements, collars, and iron yoke. The model was created in FIGURES II, a preprocessor for HOTMAP.[4] Currents were distributed in each cable as thin rectangles of uniform current density. Collar details were included so effects of a small but nonzero collar susceptibility could be examined. The FE mesh contains 12,175 third-order triangular elements. Yoke packing factors of 0.975, 0.99, and 1.0 were examined.

Resulting from this analysis is the transfer function relating the transport current to the central field. The calculated transfer function is shown in fig. 2 for cases of yoke packing factors of 1.0 and 0.975. At low fields, the transfer functions are nearly identical since the relative permeability of the iron is high. At higher fields however, the yoke packing plays a more significant role since the net magnetization associated with the iron becomes limited by saturation. A packing factor of 1.0 results in a larger transfer function when compared with the case of 0.975. A third curve showing calculations from SSCL[5] is also superimposed on the plot. This data corresponds to a packing factor of 0.975. The results are similar to those of WEC (0.975 pf) except for a slight offset in the curve. The slight offset is attributed to small differences in the inner yoke radius assumed in the SSCL (70.0 mm) and WEC (70.2 mm) analyses.

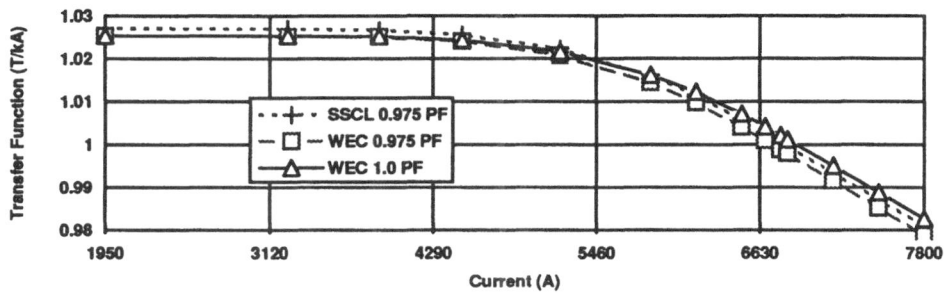

Figure 2 . Transfer function comparison for magnet design DSB-701 for cases of WEC 1.0 packing factor, WEC 0.975 packing factor and SSCL analyses.

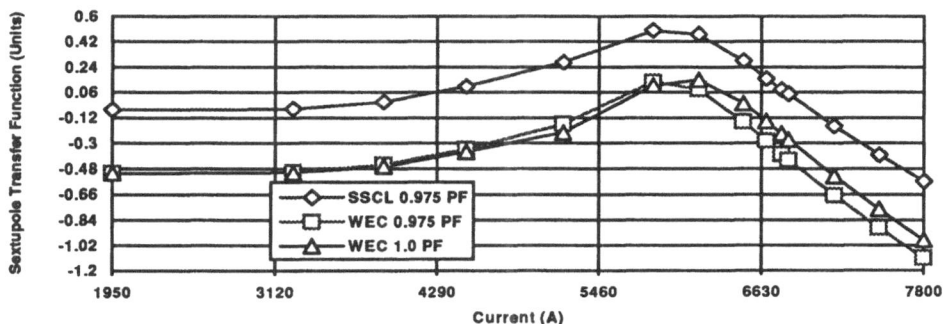

Figure 3 . Comparison of b_2 harmonic for cases of WEC 1.0 and 0.985 packing factor and SSCL analyses.

The finite element model also predicts the effects of saturation on the field harmonics. For example, fig. 3 shows the sextupole harmonic (b_2) as a function of current for the cases of 0.975 and 1.0 packing factor. The case of 1.0 packing factor begins to show signs of saturation at slightly lower currents when compared with the case of 0.975 packing factor. This trend is expected since less total iron is available using the 0.975 packing factor. Also superimposed on this plot are calculations made by SSCL[5] using computer program PE2D. The SSCL results are offset from the Westinghouse calculations at all current levels. This can be explained by the differences in the stacking algorithms used to locate the cables. As described in the previous section, the SSCL model used SSCMAG to establish the cable locations, while the WEC model used STACKUP. The different stacking algorithms result in slightly different geometric components in the field harmonics. The offset in the b_2 harmonic is apparent in fig. 3. Perhaps of greater importance is the change in b2 from its low current value to its peak at roughly 6 kA. This represents the harmonic component due to magnetic saturation. It is reassuring that a similar Δb_2 is predicted for both SSCL and WEC analyses.

The FE model was also used for calculating several other key magnet parameters.[6] These results are not described in detail due to the space limitations of this paper, but are briefly summarized in Table 2.

Table 2 . Summary of results from finite element analysis of magnet design DSB-701.

DSB-701 Magnet Parameter	Current at 6.67T	TF Sag	Stored Energy	Inductance
Calculated Value	6640 A	2.08 percent	108 kJ/m	4.9 mH/m

FIELD HARMONICS FROM OTHER SOURCES

In addition to magnetic saturation and the geometric multipole components, several other factors are known to influence the field harmonics. These effects include: a slight collar susceptibility; the component strain due to collaring, yoking, cooldown and Lorentz forces; and, the superconductor magnetization currents . Each of these effects has been accounted for using analytic models in an attempt to predict the net or integrated field harmonics as a function of current. The contribution to the b_2, b_4, b_6 and b_8 harmonics from these components has been calculated at several different current levels[7,8]. These components were then superimposed to produce plots of the integrated field harmonics.

An example of this is shown in fig. 4 for the case of the sextupole harmonic. The initial data for the sextupole harmonic was obtained at several current levels from a finite element solution which included geometric and saturation components. A third component was added to account for effects of the collar susceptibility. Next, the effects of component strain, cooldown and Lorentz forces were superimposed to give the curve identified with the triangular symbols. Lastly, the persistent current harmonics were superimposed (filled diamonds) to give the net or integrated field harmonic associated with the magnet cross section. Since the persistent current component is multivalued, only the case of increasing current is considered. Also shown on the plot are the systematic multipole limits at injection and high fields that the magnet design must satisfy. It may be noted that at injection the calculated level of the sextupole harmonic (-2.54 units) is greater than the systematic limit (2.0 units). This will be corrected in the HDM design.

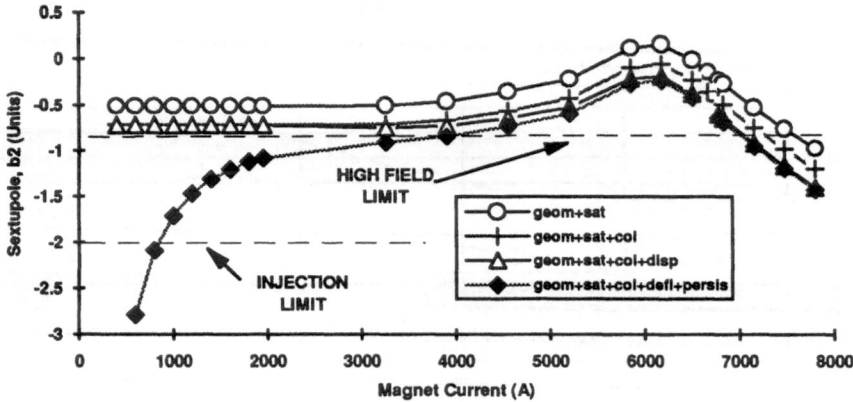

Figure 4. Sextupole harmonic vs magnet current including geometric, saturation, collar susceptibility, structural deformations, and persistent current components.

MAGNETIC FIELD IN THE VICINITY OF THE CABLES

It is important to know the peak and average magnetic field throughout the cross section as input to the calculation of quench current and field margin. To accurately calculate the peak magnetic fields in the cables, the current representation must be refined since the field in the immediate vicinity of a current element is sensitive to precise geometric representation. Generally the current is represented as thin rectangular regions of uniform current density. However, when determining the peak magnetic field in the superconductor, a strand model for the cables is used in regions where the highest fields are known to occur. The calculated peak fields at 6640A are 6.93 T (inner coil) and 5.74 T (outer coil), respectively.

CONCLUSIONS

A thorough magnetic analysis of the DSB-701 design has been performed and is summarized in this paper. The results have been compared with SSCL calculations where appropriate and are currently being used for the analysis of magnet test data from the first model magnet, DSB-701.

REFERENCES

1. High Energy Booster Dipole Magnet Program, Westinghouse contract No. SSC-91-B-01705.
2. Computer Program SSCMAG, a coil cross section optimization program written by SSCL.
3. Interim Design Review Number 1 Data Package, Vol. 1, CDRL - 16 (Rev. A), Nov. 1992.
4. Computer Program HOTMAP is a specialized version of WEMAP--Westinghouse Electromagnetic Analysis Program--developed to perform high accuracy calculations of spatial field harmonics.
5. V. Venkatraman, "Magnetic analysis of the HDM Dipole," presented 5/92 and personal correspondences with G. Snitchler of SSCL, 1992.
6. Chuboy, H. L., and Krefta, M. P. ,"HDM Model Magnet Magnetic Analyses -- Two-Dimensional Finite Element Model," Westinghouse STC Rpt. No. 92-8TM4-HEBCG-R4, Jan. 1993.
7. Krefta, M. P. ,"HDM Model Magnet Magnetic Analyses -- Effect of Mechanical Deformations on Field Strength and Harmonics," Westinghouse STC Rpt. No. 92-8TM4-HEBCG-R7, 1993.
8. Parker, J. H., Jr.,"HDM Model Magnet Magnetic Analyses -- The Effect of Superconductor Magnetization Currents on Field Uniformity in SSC Magnets," Rpt. 92-8TM4-HEBCG-R5,1993.

PRELIMINARY MULTIPOLE CALCULATIONS OF THE SSC VERTICAL BENDING MAGNETS

L. V. Nguyen, H. Gurol, and K. Gudimetta

General Dynamics Space Systems Division
Space Magnetics
P. O. Box 85990
San Diego, CA 92186-5990

INTRODUCTION

The purpose of this paper is to present the results of the magnetics calculations performed on the vertical bending magnets of the Superconducting Super Collider ring. These magnets are among the special dipole magnets needed in the interaction regions of the collider ring. Each magnet consists of two Collider Dipole Magnet (CDM) cold masses rotated by 90 degrees and placed in a single cryostat. The distance between the beam tube centerlines varies between 0.45 and 0.606 m. The cold mass used for the calculations is the General Dynamics version B coil cross-section. The main concern is whether the CDM cold masses are able to shield the field sufficiently to prevent significant "cross-talk" between the cold masses as well as magnetization of the asymmetrically placed cryostat. The calculations were performed with a 180 degree finite element model of the magnet cross-section. The calculation approach and the results will be discussed in this paper.

CALCULATION APPROACH

Finite Element Model and Definition of Multipoles

A 180 degree PE2D[1] model of the magnet cross-section was generated to perform the calculations. This model consists of 36,903 quadratic elements. Figure 1 shows the model with the field lines. The spacing between the cold masses is variable within a given magnet. The spacing between the cold mass centerlines of this model was chosen to be the minimum value of 0.45 m. This was initially deemed the most conservative case for the analysis. The collars were modeled as free space. The off-set due to the permeable Nitronic-40 collars for the CDM[2] is about -0.5 units of b2 and about 0.05 units of b4. These off-sets can be added to the resulting multipoles.

The multipoles of the vertical bending magnets are defined with respect to their local dipole field, which is along the Y_t and Y_b axes (see Figure 1). The field expansion used to calculate the multipoles is the same as that of the CDM:

$$iB_x + B_y = 10^{-4}B_0 \sum_{n=0}^{\infty} (ia_n + b_n)(x + iy)^n$$

Where B_0 is the dipole field strength, a_n and b_n are the normal and skew components of the multipoles. For a perfectly constructed magnet, only the normal components exist. The odd normal multipoles ($b1$, $b3$, $b5$, etc.) of each cold mass are induced by the asymmetry about the Y axis.

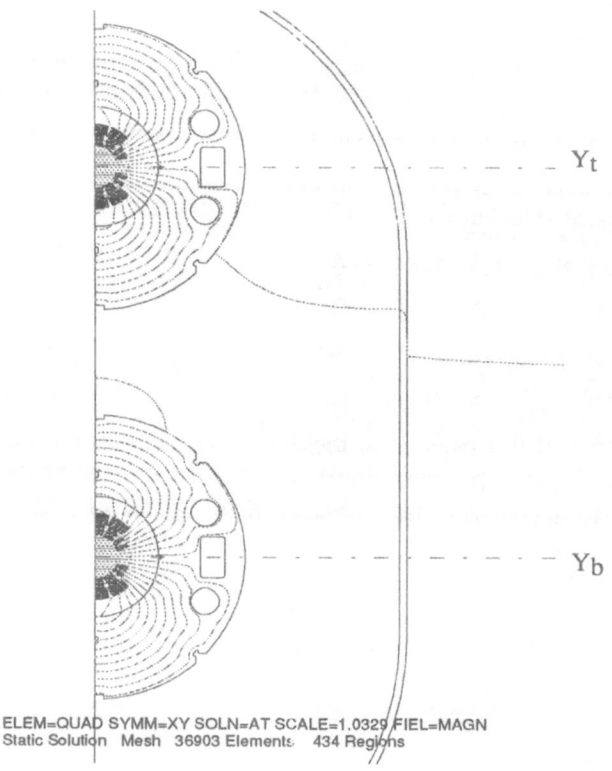

ELEM=QUAD SYMM=XY SOLN=AT SCALE=1.0329 FIEL=MAGN
Static Solution Mesh 36903 Elements 434 Regions

Figure 1. The 180 degree finite element model of the vertical bending magnet cross section. This model has been verified against a 180 degree model of the version B cross-section.

Numerical Accuracy Verification

The model was first verified for numerical accuracy by comparing with a much smaller 180 degree model of a single cold mass. The verification was performed by setting the lower cold mass of the vertical bending magnet model to air. In addition, the permeabilities of the cryostats in both models were set equal to unity. The resulting values of the allowed multipoles $b2$, $b4$ and the normal and skew quadrupoles $b1$ and $a1$ calculated by the two models are compared in Table 1. It is seen that there is good agreement between the two models for the multipoles shown. Since these test cases are physically symmetric, both $b1$ and $a1$ must be zero. The results indicate that both models give values close to zero. Some numerical inaccuracies (not shown) were found for higher order multipoles, since the mesh itself is not perfectly symmetric. These higher multipoles can be calculated with either a higher mesh density, or a symmetric mesh.

Table 1. The numerical accuracy of the vertical bending magnet model is acceptable for the multipoles that were calculated.

	Vertical Bending Magnet (top cold mass only)	180 degree model of the CDM
b2	0.60	0.61
b4	-0.06	-0.06
b1	0.0003	0 (by symmetry)
a1	0 (by symmetry)	-0.0004

RESULTS

The resulting multipoles b1 through b4 with and without cryostat are shown in Figure 2 for the top cold mass, and Figure 3 for the bottom cold mass. The multipoles are plotted for currents ranging from 650 A to 6760 A. Note that the operating 10F lattice current is 6714 A. A number of observations can be made. The induced b1 in the cold masses is comparable to the a1 induced in the CDM by off-centered cryostat. At low field, the effects of the "cross-talk" between the two masses are virtually zero. At high field without cryostat, "cross-talk" produces b1 of about 0.05 units with opposite signs for the top and bottom cold masses. The transfer functions of b2 and b4 without cryostat are seen to be very comparable for the top and bottom cold masses and also almost identical to the CDM version B transfer functions (once the effects of the permeable collar, the persistent currents and the off-centered cryostat are subtracted). The presence of the cryostat has significant effects on b1 and b2. It induces about -0.04 units of b1 and 0.3 units of b2 for the top cold mass, and -0.01 units of b1 and 0.2 units of b2 for the bottom cold mass. These values are comparable to those of the CDM.

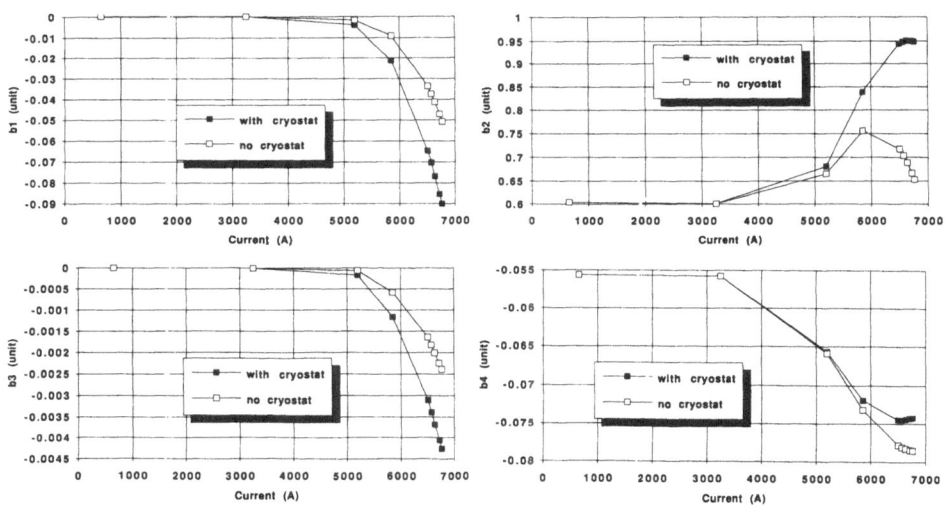

Figure 2. Resulting multipoles b1 through b4 for the top cold mass with and without cryostat.

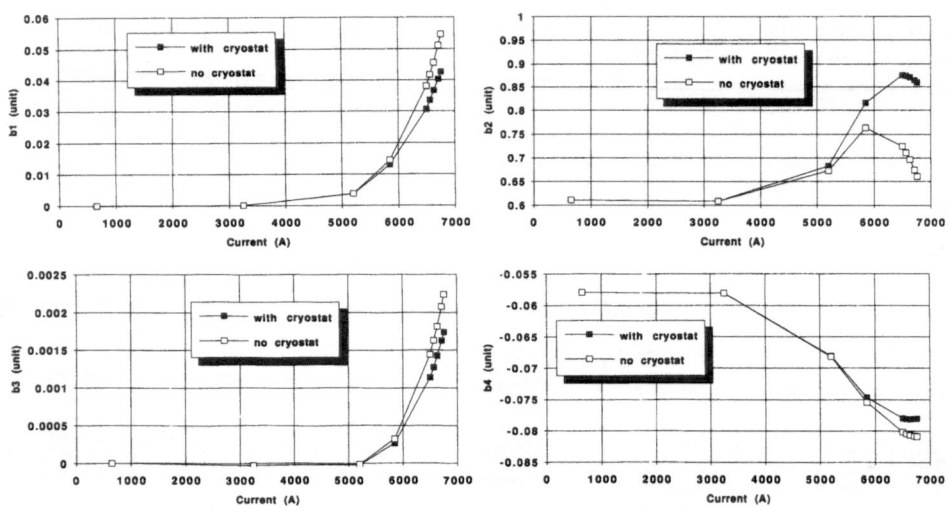

Figure 3. Resulting multipoles b1 through b4 for the bottom cold mass with and without cryostat.

CONCLUSIONS

Preliminary multipole calculations on the vertical bending magnets were performed using a 180 degree finite element model. Based on the results, the main conclusion is that the same cold mass used in the 15 m CDM magnets can be used in the vertical bending magnets. This can be achieved without any need to re-optimize the coil cross-section. The induced b1 is comparable to a1 of the CDM, and can probably be corrected in a similar fashion using iron bars in the yoke tooling slots[2]. Other correction approaches, such as optimization of the cryostat geometry to off-set the b1, may be feasible for these magnets. The resulting multipoles are comparable to the CDM magnets, based on the present 2 dimensional calculations. However, the 3 dimensional end effects of these magnets are expected to have a greater impact on the length averaged multipoles since some of the magnets are shorter, and they are closer together and without a cryostat separating them. A number of future studies must be undertaken to understand the magnetic design of these magnets. These include: three dimensional calculations of the multipole contribution of the coil ends, refinement of the two dimensional finite element model to reduce numerical error in the multipole calculations, sensitivity study for the cold mass spacing, investigation of the correction approaches for the odd multipoles, and understanding the beam accelerator optics requirements and its impact on the multipoles.

Acknowledgments

The authors would like to thank Chuck Gibson of General Dynamics for the useful information on the vertical bending magnets.

Portions of the data used to perform the work described herein was accomplished under contract to the Universities Research Association in support of the Superconducting Super Collider project for the U.S. Department of Energy.

REFERENCES

1. PE2D, 2 Dimensional Magnetics Software by Vector Fields.
2. H. Gurol, et al., "Status of the Magnetic Design of the SSC Dipole Magnets", IISSC, New Orleans, LA, March 4-6, 1991.

DESIGN OF A MODEL DIPOLE MAGNET FOR
THE SSC HIGH ENERGY BOOSTER

N. Hassan, K. Couzens, S. Dwyer, A. Jaisle, R. Jayakumar,
S. Krishnamurthy, R. Mihelic, S. Phillips, R. K. Puri, K. Sarna,
G. Snitchler, and V. Venkatraman

Superconducting Super Collider Laboratory*
2550 Beckleymeade Ave.
Dallas, Texas 75237

INTRODUCTION

A superconducting model dipole magnet has been designed to serve as a vehicle in an R&D program to develop a dipole magnet for potential use in the SSC High Energy Booster. The objective has been to use the Brookhaven National Laboratory (BNL) and Fermi National Accelerator Laboratory (FNAL) 50 mm aperture dipole designs to the maximum possible extent for design of a dipole magnet with the same size aperture and a field intensity of 6.67 T. Objectives of this program have also included an evaluation of magnet cross section designs which provides increased margin and includes a field quality iteration on BNL and FNAL dipole designs. The salient parameters of this magnet are listed in Table 1. In this paper the 2D magnetic and mechanical design of the cold mass in conceptual and detail form is presented.

MAGNETIC DESIGN

The field quality specification for the HEB model dipole magnet is presented in Table 2. The SSCMAG[1] code was used to determine the optimized coil cross section design and PE2D®,[2], a finite element analysis code, was used to study the effects of iron saturation on transfer function and field harmonics. The 2D cross section of the magnet, which consists of a two-layer coil assembly and mechanical components, is illustrated in Figure 1. Magnetic field harmonics for the optimized cross section are presented in Table 2. Physical parameters of the the inner and outer coils as well as other cold mass dimensional parameters are outlined in Table 1. In Figure 2, variation of transfer function as a function of current is illustrated. The most notable difference between the magnetic design of the HEB model dipole and those of the other SSC 50 mm aperture dipoles is that the conductor block located in the pole region of the inner coil consists of two turns. All other 50 mm aperture SSC dipole magnets are designed with three turns for the same conductor block. This variation from other designs has provided for an improvement in field quality for the HEB dipole. In studies carried out for design of the magnetic cross section every attempt was made to arrive at a symmetric wedge design for the inner and outer coils. However, use of symmetric wedges would result in unacceptably steep pole angles or field harmonic values higher than those obtained with the final cross section design.

---

* Operated by the Universities Reasearch Association, Inc., for the U.S. Department of Energy under Contract No. DE-AC35-89ER40486.

Table 1. Major parameters of HEB model dipole magnet cross section.

Current	6.63 kA	Inner Layer	–
Central Field	6.67 T	Number of turns	19
Non-Linearity	2.07 %	Conductor blocks	5, 7, 5, 2
Margin	13.3 %	Number of strands	30
Operating Temperature	4.25 K	Wire diameter	0.81 mm
Coil Inner Diameter	49.7 mm	Cu/Sc ratio	1.3
Coil Outer Diameter	99.97 mm	Pole angle	76.1 °
Intercoil Spacing	0.47 mm	Outer Layer	–
Collar Width	19 mm	Number of turns	26
Yoke Inner Diameter	140.40 mm	Conductor blocks	15, 11
Yoke Outer Diameter	330.10 mm	Number of strands	36
Cold Mass Diameter	340.00 mm	Wire diameter	0.65 mm
Coil End-to-End Length	1678 mm	Cu/Sc ratio	1.8
Coil Straight Length	1350 mm	Pole angle	44.5 °
Cold Mass Length	1.8 m	NbTi Filament Diameter	6 μm

Table 2. HEB model dipole magnet field quality terms expressed in units of 10^{-4} of the main field evaluated at a radius of 1 cm from the center.

Multipole	b_2	b_4	b_6	b_8
Specification	0 ± 0.8	0 ± 0.08	0 ± 0.013	0 ± 0.01
Optimized	0.000	0.000	0.004	0.002

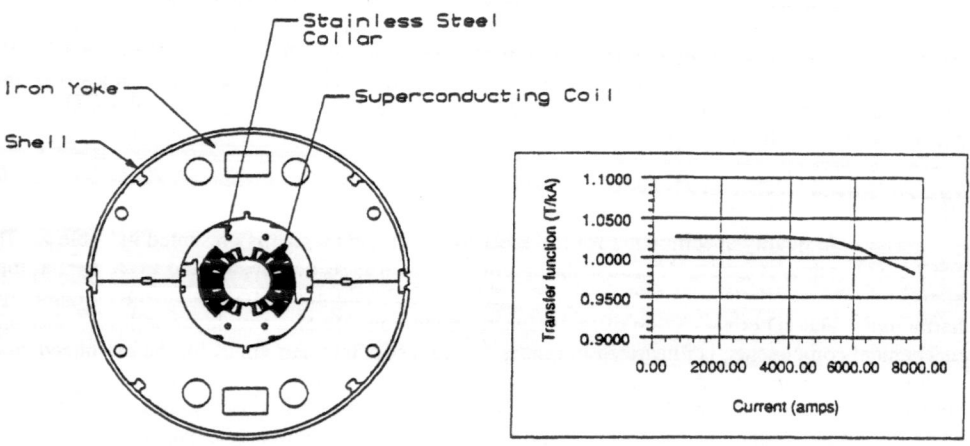

Figure 1. HEB model dipole magnet cross section.

Figure 2. HEB model dipole.magnet transfer vs. current.

MECHANICAL DESIGN

In this section, design of various components of the cold mass is described. The coil assembly consists of two layers. The superconducting cable is insulated with 0.03 mm thick Kapton®,[3] film with 50 % overlap configuration. The insulated cable is then wrapped with 0.11 mm thick fiberglass tape impregnated with B-stage epoxy. The same configuration is used for insulating and wrapping of the copper wedges. BEND,[4] TOSCA,[5] and ENDS3D[6] codes were used to determine the optimum design for the coil ends. The "grouped end" configuration was selected for coil end design. G-11 CR design for the coil ends. The "grouped end" configuration was selected for coil end design. G-11 CR is the selected material for the coil end parts. Unlike the BNL and FNAL dipole magnets, in which the inner and outer coils are wound and cured separately, the outer coil in the HEB model dipole is wound on the already cured inner coil. This process offers several advantages, the most important of which involves the internal ramp splice[7] designed for the HEB model dipole. With two separately wound coils, as it is done in other SSC dipoles, the splice is made after the coils are assembled together. Making the ramp splice on "separate wound" coils requires breaking the bond between the splice lead and the adjacent turns. That portion of these splice leads which is not contained in the splice housing has to be epoxied back onto the adjacent turns. This is a time consuming operation with potential reliability problems. With the "wind on coil" technique, the lead of the outer cable is soldered to the lead of the cured inner coil prior to winding of the outer coil. This is a simple and quick process. Perhaps, the most important advantage of the "wind on coil" technique is that as the first turn of the outer coil is wound, the splice will be tensioned and will be cured with the outer coil in a tensioned condition. This is considered to be an improvement with regard to ramp splice mechanics, particularly for magnets which have to withstand high ramp rates. Furthermore, the "wind on coil" method allows curing the coil end keys/ramp splice housing with the coil and the ramp. This allows the splice and the ramp to perfectly conform to their mating parts. Using the inner coil as the winding mandrel also allows outer coil to conform to the inner during the curing process. The "wind on coil" approach requires using a spacer between the inner and outer coils. The spacer produces a needed smooth and rigid surface for winding the outer coil. The rigid spacer between the coils offers flexibility to electromagnetic design by eliminating quench concerns with regard to locating the pole of the outer coil on an inner coil turn, instead of an inner coil wedge. The spacer creates a barrier which separates the inner coil conductor from the outer coil pole shim and collar interface. It also provides for a more rigid coil structure as it is bonded to the outer coil during the curing cycle. Another advantage of this winding technique is the cost savings of a die for outer mandrel lamination. Calculations have indicated that the interlayer spacing in the HEB dipole does not adversely effect transfer of heat between the coil layers.[8] Among disadvantages of this coil winding technique are the added cost of the spacer and the slightly more complex ground insulation configuration. The "wind on coil" technique puts the inner coil through two curing cycles, which may increase AC losses. The second cure cycle changes the inner coil size also, resulting in a slight increase in modulus of elasticity of the inner coil. Since the soldered splice between the inner and outer coils will be exposed to high temperatures with the outer coil curing cycle, "silver solder" is selected for joining the coil leads to avoid solder softening and loss of splice tension. In the HEB model dipole several layers of 0.13 mm thick Kapton are used to completely insulate the coils from the collars. A radial space of 0.15 mm is included for installing quench protection heaters that will be needed in full size prototypes of this design. Both the collaring shims and the 0.50 mm thick collaring shoe are made of brass. The collaring shims were designed to clip onto the collars for ease of assembly. The collaring shoe is used to avoid tearing of the ground insulation during collaring, and extrusion of the insulation into the voids of the collar assemblies. A beam tube design was not necessary for the model magnet since it will be cold tested under pool boiling conditions.

Mechanical behavior of the cold mass components and assemblies at various stages such as after collaring, shell welding, cooldown, and excitation were analyzed using ANSYS®,[8,9,10]. Results of the performed structural analysis are presented in Table 3. Design of the 1.5-mm thick collar laminations for the HEB dipole provides an option to make these laminations from Nitronic-40 or high manganese stainless steel. The intent of the program was to build two magnets at the SSC, using Nitronic-40 in one and high manganese in the other. The objective was to evaluate the performance of both materials for use in the future magnets. It is thought that high manganese steel may be more desirable due to its lower coefficient of thermal expansion. The low coefficient of thermal expansion of this material will maintain the collar-yoke contact at 4.3 K and helps to minimize the possibility of premature quenches due to loss of collar-yoke contact. The collar-yoke interface for the HEB dipole has been designed for a line-to-line fit. These collars are circular, i.e., not "antiovalized".[8] This configuration also helps to maintain a vertical collar-yoke contact with cooldown to 4.3 K. The collar laminations are spot welded in left-handed and right-handed pairs to avoid coil twist during assembly. The collar laminations are assembled in packs using swaged tubes and are interlocked using phosphor bronze keys.

Table 3. Structural analysis results for HEB model dipole magnet.

Condition	Inner / Outer Coil Azimuthal Stress (Mpa)	Vertical / Horizontal Collar Deflection (mm)	Yoke Mid-Plane Gap (mm)
After Collaring	69.7 / 54.8	0.10 / -0.03	—
After Shell Welding	85.9 / 68.0	0.06 / 0.007	0.04
After Cooldown	56.5 / 51.20	0.05 / 0.03	0.00
At 6500 A	23.1 / 28.0	0.03 / 0.06	0.00

The HEB model dipole uses horizontally-split yoke laminations made of 6.35 mm thick low carbon steel for the middle and 3.2 mm thick stainless steel for the coil ends. The smoother lamination edge produced in fineblanking is intended to reduce friction at the collar-yoke-shell interfaces and eliminate the "ratcheting" effect observed in BNL dipole magnets.[8] This is an important consideration since these parts shrink at different rates with cooldown. Use of thick laminations on the ends is also intended to eliminate the need to epoxy them in order to avoid buckling under press and welding loads. The iron yoke laminations are coated with manganese phosphate which acts as a rust inhibitor and provides an electrically resistive barrier between adjacent laminations, reducing losses due to yoke eddy currents. The half shells and end plates for the HEB dipole are made of 304 stainless steel. The shell behavior under magnetic, thermal, and mechanical stresses was also analyzed using ANSYS. The cold mass end plates contain the axial force developed with magnet excitation and are welded to the half shells during assembly.

ACKNOWLEDGEMENTS

The authors would like to express their appreciation to B. Archer, A. Fluhmann, Y. Chen, F. Nobrega, D. Orrell, S. Smith, and E. Vrsansky of SSC Laboratory and G. Spigo of CERN for their valuable suggestions and support throughout the design effort.

REFERENCES

1. B. Archer, D. Orrell, and G. Snitchler, SSCMAG users manual, MD-ENG-92-A-005, (1992).
2. PE2D is trademark of Vector Fields, Inc., Oxford, England.
3. Kapton is a registered trademark of DuPont, Inc.
4. J. M. Cook, An application of differential geometry to SSC magnet end winding, SSCL-N-720, (1990).
5. TOSCA is trademark of Vector Fields, Inc., Oxford, England.
6. ENDS3D, Lilly, D. Orrell, G. Snitchler, private communication.
7. N. Hassan and R. Jayakumar, A trade study of using cured inner coil as a mandrel for winding outer coil, (1992).
8. N. Hassan, Design of SSC high energy booster model dipole magnet, (1993).
9. Y. Chen, Mechanical analysis of DSB cross section, (1992).
10. ANSYS is trademark of Swanson Associates, Inc., Houston, PA.

DESIGN AND PERFORMANCE OF A NEW 50 MM
QUADRUPOLE MAGNET FOR THE SSC

G. Spigo,[1,2] G. Cunningham,[1] C. Goodzeit,[1] D. Orrell,[1,3]
J. Turner,[1,3] R. Jayakumar[1]

[1]SSC Laboratory,* 2550 Beckleymeade, Dallas, TX 75237
[2]Present Address CERN CH-1211, Genève 23, Switzerland
[3]Present Address CEN-Saclay, F-91191, Gif-sur-Yvette CEDEX, France

INTRODUCTION

A superconducting quadrupole model magnet with a 50 mm aperture and a gradient of 190 T/m, in operation at 4.35 K and 6500 A, has been designed, built and tested at the SSC. This accelerator magnet is expected to have application in the interaction regions of the collider main rings. Its dipole-type stainless steel collars with mated self-aligning pole spacers were a major innovation in design. The model had stringent requirements on field quality and a conservative 21% current margin.[1] The first two articles have now demonstrated satisfactory quench performance over several thermal cycles, reaching plateau at approximately 8660 A with minimal training. This paper is a brief sketch of the design and preliminary results on the first model. Fabrication[2] and testing[3] are described in other papers of this conference.

ORIGINAL SPECIFICATIONS AND REQUIREMENTS

A project began in mid-1990 at the SSC to design and build a 50 mm superconducting quadrupole, soon called the QSE magnet. Central gradient was set at 180±10 T/m at the operating current of collider dipole magnets, then 6500 A. Operating current has since been revised to 6714 A. The magnet was to use an available cable, and the 36-strand outer cable of the CDM was chosen. J_c and other characteristics of this conductor permitted the QSE design to attain its conservative margin. Magnetic field harmonics were specified as in Table 1, in order to permit several possible applications in the machine lattice at the time.

Preliminary requirements on coil and yoke were determined by traditional approximate methods for the critical current[4] and short sample characteristics of NbTi, and infinite iron assumption for the field calculation. Electromagnetic design began from a coil having an inner layer of eleven turns and outer layer of sixteen turns to achieve the gradient transfer. Yoke outer radius was selected to fit the existing CQM cryostat. Rough dimensions of conductor and yoke thus set, stainless steel collars of Nitronic 40 were chosen for their high strength early in the process, and a number of collar and yoke layouts were analyzed with ANSYS[5] programs for (2-d) mechanical structures. While Lorentz forces[1] which these structures counteract are somewhat less than for dipoles, the quadrupole coil assemblies have more parts and are prone to shift under assembly stresses, cooldown and excitation. Therefore great effort was considered necessary to perfect the mechanical "packaging" of this device.

---

* Operated by the Universities Research Association, Inc., for the U. S. Department of Energy under Contract No. DE-AC35-89ER40486.

Table 1. Field harmonic specifications in "units". Note 200 units of dipole error is equivalent to 200 μm of centering error in (x,y), and that 5 units of skew quadrupole error is equivalent to a magnet rotation of 0.5 mrad. Allowed terms marked (*).

n	Normal Systematic	Normal Random	Observed Warm Test	Skew Systematic	Skew Random	Observed Warm Test	Multipole
0	20.000	200.000	——	20.000	200.000	——	dipole
1	10000. *	——	10000. ± 0	2.000	4.796	0.0 ± 0	quadrupole
2	0.184	1.765	-2.589 ± .038	0.184	1.765	1.883 ±.038	sextupole
3	0.085	0.812	0.531 ± .013	0.085	0.812	-1.945 ±.013	octupole
4	0.078	0.224	-0.067 ± .011	0.078	0.224	0.013 ±.011	decapole
5	0.573 *	0.103	0.551 ± .011	0.072	0.206	-0.102 ±.011	dodecapole
6	0.033	0.032	-0.022 ± .004	0.033	0.032	0.026 ±.004	quadecapole
7	0.030	0.029	0.006 ± .004	0.030	0.029	-0.005 ±.004	sexadecapole
8	0.028	0.027	0.005 ± .004	0.028	0.027	-0.005 ±.004	octadecapole
9	0.103 *	0.025	0.020 ± .004	0.026	0.025	-0.009 ±.003	icosapole

MAGNET DESIGN OVERVIEW

Through a series of reviews from November, 1990 to February, 1991, the conceptual approach to the magnet was refined. Notable was a decision to use dipole-type collars, confining and positioning the quadrants with self-aligning brass pole spacers. This would permit horizontal assembly in an available dipole collaring press and permit extension of the straight section to 15 m length in prototype. First known application of this technique was soon to be announced[6] at MT-12. A dipole structure was thought to increase stiffness and strength of the coil assembly, though that was further enhanced by a line-to-line fit at the collar-yoke interface at the cold operating point. The adopted mechanical cross section design is shown in Fig. 1, with some other quadrupoles for comparison. Yokes, stainless steel collars and brass spacers are all laminated structures, formed into packs of various convenient lengths. Collar laminations are spot welded in alternating left and right pairs, and these weldments are themselves interleaved with left and right hand types to form standard collar packs of 158 mm length. Spacers are assembled in 30 mm packs with roll pins at three points. Finally the iron yokes are pressed into long half-packs held together by flared stacking tubes. Stainless steel collars without spacers extend over coil end sections, and yoke packs are terminated at each end of the magnet by monolithic blocks of stainless steel, cemented with epoxy and baked. This latter feature helps to prevent "chevron" collapse of yoke packs in the yoking press. Laminated structures are confined by a stainless steel shell.

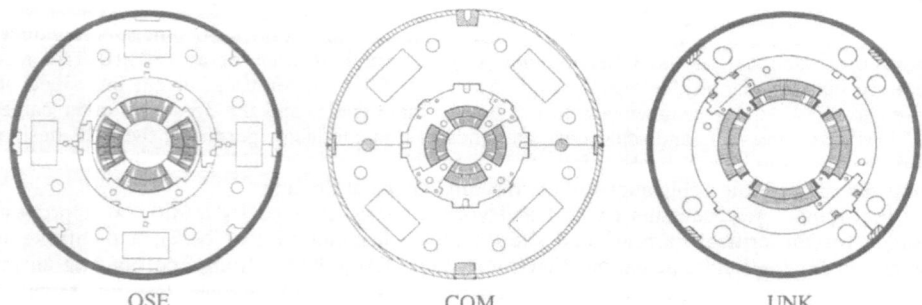

QSE   CQM   UNK

Figure 1. Cross section of the QSE magnet with comparison to other recent quadrupoles—the SSC (LBL) 40 mm main ring quadrupole CQM, and the IHEP 80 mm UNK quadrupole. Drawings not to scale.

ELECTROMAGNETIC APPROACH

It was stipulated that inner and outer coils would have the same pole angle in order that spacer elements have the simplest shape and to permit easier winding of the layered coil with less chance of damage to pole turns. This also avoids pole wedges which can complicate problems of tolerance and bonding. PE2d[7] codes and other special programs were used to optimize the 2-d cross section and obtain an estimate of the allowed harmonics. Unallowed harmonics are essentially constrained only by the very close tolerances on the collared coil and by process control specifications. These harmonics were estimated by simulation of small errors in a coil without symmetry. Experience and simulation runs set the design tolerances. A solution was found to approximate the cos(2θ) current distribution of a perfect quadrupole. It required one asymmetric copper wedge for the inner coil, and another for the outer coil.

Studies of the existing DESY quadrupole, and those in development for LHC and the SSC main ring all used the same cable for both inner and outer coils. Preliminary calculations had shown that an 11.5 mm-wide insulated cable would be suitable here. It was important for production and design scheduling to use an existing cable. The outer CDM conductor was chosen for QSE because of its width and low rigidity. This choice made possible winding the new magnet coil-on-coil without a splice between inner and outer coils. Such technique had been used to advantage in HERA quadrupoles and reduces the number of soldered splice joints from 7 to 3. In some magnet designs, such cable splices had been found to have an adverse effect on quench performance. Salient parameters of the QSE cable are in Table 2.

Table 2. Conductor parameter specifications and *estimates.

Parameter	Bare Cable	Insulated	Final Cured*
Width	11.680 mm	11.982 mm	11.960 mm
Thin Edge	1.054	1.356	1.222
Thick Edge	1.260	1.562	1.426
Keystone Angle	1.05°		
Cable Pitch Length	89 mm with left-hand lay		
Critical Current	9870 A at 5.6 T and 4.2 K		
RRR	> 30 Ratio R_{295}/R_{10} (cold worked condition)		
Cu: SC Ratio	1.8 : 1		
Insulation	50% overlapped Kapton plus fiberglass-epoxy tape		
Twist Pitch	15 mm with right-hand twist		
Strand Diameter	0.648 mm		
Number Strands	36		

Coil design was optimized after basic dimensions—insulation cured thickness, intercoil spacing, ground insulation, collar width, yoke inner radius—were all set and cross-checked. Ground insulation is determined by maximum voltage at quench. Two 125 µm layers of Kapton[8] were judged sufficient, but doubled because they must be cut, and also in order to provide good force transmission during collaring and yoking. Pole-to-pole voltage estimates required two such Kapton layers at the parting planes. Turn-to-turn voltage was estimated at 25-30 V, and a cable insulation composed of a 25 µm thick Kapton layer 50% overlapped, and a 0.10-0.11 mm-thick fiberglass-epoxy tape spaced by 2 mm was judged adequate.

A key dimension to be fixed was the yoke inner radius. Preliminary mechanical calculations were done with the conservative condition that magnet collars had to take the entire Lorentz force at full excitation. It was found that a 20 mm-thick stainless steel collar was suitable for the self-supporting assumption. Hence inner yoke radius was set to 70 mm. With these main parameters, conductor cross section was optimized for the precise position of each turn, and thence the dimensions of all wedges. Thus there were to be four conductor blocks per quadrant, and the QSE cross section is denoted 8-3 and 13-3 by SSC convention. The computed field was within original requirements. The final conductor geometry is input to the PE2d finite element programs, which refine and complete the field calculations, given a hysteresis curve for the iron and precise definition of yoke geometry. PE2d will produce plots of the field, flux and force. The force estimate on conductors is an important input to refine the cross section, in an iterative procedure[9] which aims to guarantee that the final design places every turn at its precise position at cold condition and under full excitation.

Field was optimized in the end region with 3-d modeling by arranging conductor block U-bends of both coil layers along the z-axis, in order to reduce peak field and minimize the integral of allowed harmonics b_5 and b_9. The end parts are called keys, spacers and fillers, and saddles, according to their placement inside a pole turn, between blocks, or outside the final turn. Fig. 2 shows a quadrant coil. Direct numerical specification of these end parts comes out of this optimization process. Design techniques for the coil end parts have been published[10] in a previous conference. Rapid manufacture and cost containment of such small, rigid, and irregularly shaped plastic (G-11CR) parts remains a challenge. Major design issue is to reconcile cured and uncured dimensions for these items.

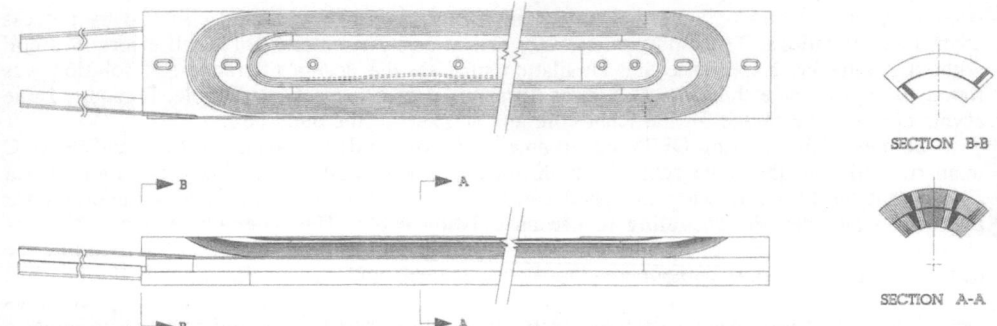

SECTION B-B

SECTION A-A

Figure 2. Quadrant coil mechanical design of QSE. Inner and outer coils are wound without a splice.

ITERATIONS TO FINAL MECHANICAL DESIGN

Mechanical analysis was performed by hand and with the finite element code ANSYS. In the calculations plane stress is assumed, creep characteristics of coils are ignored, and coil properties are assumed linear (program limitation). Desired coil prestress after collaring was estimated to be about 43 MPa on average, based on prior experience and preliminary hand calculations. Main objective was to have the coils always in contact with the pole faces of the collars. To limit creep effects in the coils (flow of insulating materials) that mainly occur at room temperature, it is important to have coil prestress as small as possible at warm condition, consistent with cold prestress needs. This prestress was effected in the model by setting a 0.094 mm interference between coils and collars. The finite element calculation predicted that a small yoke gap would exist after the yoking and welding operation (0.021 mm at the inner yoke radius). The yoke tends to force the collars back to their original round shape because of the line-to-line fit between yoke and collars. Like the collaring process, this increases stress in the coils—by an average of about 14 MPa over the inner and outer layers.

Since the Nitronic 40 collars have a higher coefficient of thermal expansion than the iron yoke, cooldown results in a decrease in contact force between the yoke and collars. Also coils have a higher coefficient of thermal expansion than collars, so they shrink away from the collars during cooldown. Combining these effects means that coils lose a substantial amount of prestress on cooldown. These effects aid in closing the yoke midplane gap which remained after welding. As the magnet is energized, Lorentz forces act to separate the coils from the collars at the poles and compress the coils at the parting planes. However model calculations and strain gage data showed the coils still have a safe prestress remaining at full excitation. More complete QSE mechanical analysis will appear in a forthcoming MT-13 paper.

SUMMARY OF MAGNET FEATURES

- Extensive finite element modeling of magnetic field in 2-d and 3-d
- Iterative design achieves mechanical / magnetic specification at cold state
- Conservative 21% current margin for acceptable quench performance
- Outer coils wound on inner coils without a conductor splice
- Strong dipole-type collars with self-aligning pole spacer inserts
- Requires only dipole presses and permits industrialization to 15 m length
- Coil prestress holds at full excitation, aided by collar-yoke interference
- Inter-quadrant splices of doubled cable in copper stabilizer external to shell

CONCLUSIONS FROM THE FIRST TESTS

Production tests and quench performance runs have shown the QSE101 mechanical design to be very satisfactory. Quench data are summarized in Fig. 3 for the first model, which had its first quench at 7120 A, and plateau around 8660 A or slightly above short sample estimate, following 3 or 4 training quenches. Slight improvement was noted in a 2nd thermal cycle. Warm magnetic field tests are summarized by "observed" entries of Table 1. These measurements were made with a 0.25 m rotating coil device (mole) and represent average multipole estimates in the straight section of the magnet, ±0.375 m from the center. Cold measurements so far cannot be directly compared to the warm data, and are continuing. There appears to be a significant amount of sextupole, octupole and decapole in the magnet. This could occur from parts out of tolerance or process variations which alter coil position—100 μm midplane shift in a quadrant can generate ~3 units of sextupole, 1 unit of octupole and 0.5 unit of decapole. The new model may require some tuning of its conductor cross section.

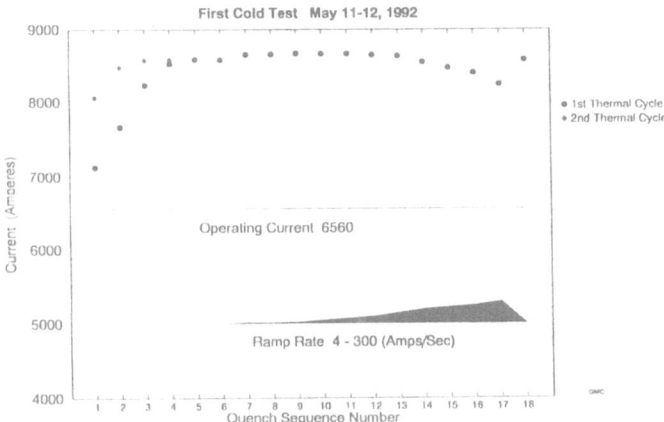

Figure 3. Current at quench for two thermal cycles. First quench occurred at 7120 A, 4.22 K.

ACKNOWLEDGEMENTS

The authors are indebted to many SSC colleagues who have contributed their ideas and experience during the conceptual and detailed design phase, as well as fabrication and final test. Special appreciation is expressed to S. Smith, F. Nobrega, R. Puri, Y. Chen, V. Venkatraman and G. Snitchler of the Cold Mass Design Group; and to N. Hassan who guided the magnet through manufacture; to S. Stromberg, P. Nema, R. Wood, S. Dwyer, E. Cox, D. Bonifert, E. Vrsansky, M. Caracciolo, and W. Stone, who developed all the tooling and production procedures to make the magnet; to D. Albone, A. Fluhmann, D. Covington, D. Block, M. Lewis, A. Jaisle, S. Allen, G. Boswell, R. Williams, L. Brooks, E. McGuire, A. Brown, D. Hart, R. Clerihew, J. Lovell, R. Quick, J. Stephens, R. Rudzinski, L. Thorne, R. Webb, S. Zacharias, A. Leighton and K. Bigert who built the first models with care and diligence; to B. England, R. Lathrop, V. Li and G. McAdams who produced the many detail drawings; to S. Bagwell and F. Ott who found the vendors and made the drawings real; to D. Bailey, R. Fitzgerald, J. Thorne and B. Malnar of the Production Management team for their dedicated leadership and support; to A. Devred, T. Ogitsu, D. Bein, R. Zeigler, J. DiMarco, Z. Wolf and to the entire test laboratory staff, for design and assistance with instrumentation and testing. Without the strong efforts of all these colleagues, the QSE magnet and this paper would not have been possible.

REFERENCES

1. Martin N. Wilson, Superconducting Magnets, Monographs on Cryogenics, Clarendon Press Oxford, 1983.
2. D. Albone et al., "Modifications and Fabrication of the 50 mm Aperture...," III-D-6 in these Proceedings.
3. B. Archer et al., "Test Performance of the 50 mm Aperture SSC Model Quadrupole Magnets," VII-39 in these Proceedings.
4. M. Green, "Calculating the J_c,B,T Surface for Niobium Titanium Using a Reduced-State Model," IEEE Transactions on Magnetics, v. 25, no. 2, 2119-2122, March 1989.
5. ANSYS is trademark of Swanson Associates, Inc., Houston, TX.
6. A. Ageev et al., "Development and Study of the Superconducting Magnets (for the UNK)," IHEP Preprint 90-158, Protvino (Russia), 1990.
7. PE2d is trademark of Vector Fields, Inc., Oxford, England.
8. Kapton is a registered trademark of DuPont, Inc.
9. N. Kallas, C. Haddock, D. Orrell, G. Snitchler and R. Jayakumar, "A Software Package Linking PE2D and ANSYS for SSC Magnet Design," Particle Accelerator Conference, San Francisco, 1991.
10. D. Orrell et al., "Mechanical and Electromagnetic Design of the SSC QSE101 Quadrupole Ends," IEEE Transactions on Magnetics, v. 28, no. 1, January 1992.

IMPROVED LOSS CALCULATIONS FOR THE HDM MAGNETS

G. T. Mallick, Jr., W. J. Carr, Jr., M. P. Krefta, and D. Johnson

Westinghouse Science and Technology Center
Pittsburgh, PA 15235

INTRODUCTION

Losses due to ramped fields and currents, quite adequate for the initial design, were calculated previously by Snitchler, Jayakumar, Kovachev, and Orrell[1] for the high energy booster magnets to be used in the SSC. The present analysis considers the loss problem in more detail.

Hysteresis loss in the superconductor

Hysteresis loss in a superconductor results from irreversible motion of boundaries between regions of critical current density, computed here as the hysteresis in magnetic moment plotted against magnetic field. The magnetic moment is proportional to the critical current density j_c, and the dependence of j_c on the magnetic field is introduced into the calculation with the use of the analysis of Morgan and Sampson[3]

Under some conditions the hysteresis can be significantly affected by the transport current. To avoid an elaborate calculation, the effect of transport current on the hysteresis is examined for a dc transport current, and the result is then applied to a ramped current. The dc approximation is equivalent to assuming that each filament in the strand carries the same current. The magnetic moment of a strand per unit volume of superconductor becomes approximately[4]

$$\frac{m}{V} = \frac{2}{3\pi} \lambda j_c d \left[1 - \frac{I^2}{I_c^2} \right]^{3/2} \tag{1}$$

where d is the filament diameter, $I_c = I_c(B)$ is the critical current at field B, and λ is the fraction of superconductor in the strand. The instantaneous power per unit volume in a large magnetic field (large compared with the penetration field) is

$$\frac{P}{V} = \left| \frac{dB_a}{dt} \right| \left(\frac{m}{V} \right) \tag{2}$$

where dB_a/dt is the time derivative of the applied magnetic field. The power depends on the position of the wire within the magnet, through both dB_a/dt and I_c, since I_c varies from point to point in the magnet with the magnetic field. The loss per cycle at a point in the magnet is obtained by integrating the power over time.

Dynamic Resistance Loss

The total power loss in a filament carrying a dc current I and acted upon by a large changing transverse applied magnetic field is given by[4]

$$[\frac{P}{V}]_{fil} = \frac{2}{3\pi} j_c d |\dot{B}_a| g(\frac{I}{I_c})$$ (3)

where $g(I/I_c)$ is a function defined by Murphy and Walker (see reference 4). The function g can be written approximately as[4]

$$g(\frac{I}{I_c}) \approx (1 - \frac{I^2}{I^2_c})^{3/2} + \frac{3\pi}{4} \frac{I^2}{I^2_c}$$ (4)

where the first term gives rise to the hysteresis loss and the second term an additional loss which is the dynamic resistance loss. This loss can be quite large near $I = I_c$ but is found to be only a minor term in the present analysis after an average over the cycle and over the magnet.

Intrastrand Eddy Current Loss

Eddy current loss which occurs within a strand (per unit strand volume) is given by

$$\frac{P_{intrastrand}}{V} \approx \lambda' \sigma_\perp \dot{B}^2_a [\frac{L}{2\pi}]^2$$ (5)

where $\sigma_\perp$ is the transverse conductivity of the matrix, L the filament twist pitch, and λ' is the fraction of total area of the strand which is occupied by the filamentary region (which differs from the fraction of superconducting volume λ). As a result of magnetoresistance it is assumed that

$$\sigma_\perp = \frac{\sigma_\perp(0)}{(1 + kB_a)}$$ (6)

where k is constant, taken to be 0.6. Since $|dB_a/dt|$ is constant in time for a constant ramp, $\sigma_\perp$ is the only term which depends on time, and the average over a cycle with peak field B_m is

$$\langle \sigma_\perp(B) \rangle = \sigma_\perp(0) \frac{\ln(1+kB_m)}{kB_m}$$ (7)

The value of $\sigma_\perp(0)$ is proportional to the residual resistance ratio of the matrix and to geometrical factors. We have used for $\lambda' \sigma_\perp(0)$ a typical value deduced from previous measurements[5] at three tesla to give for the loss per (HDM) cycle

$$\frac{Q_e}{V} = 0.0086 \frac{B_m \ln(1+kB_m)}{3\ln(1+3k)} \quad (J/cm^3)$$ (8)

which is the loss per unit strand volume near a point in the magnet with maximum field B_m. To obtain a value for the magnet, $B_m \ln(1+kB_m)$ is averaged over the magnet cross section, using another computer program.

Interstrand Eddy Current Loss

The most uncertain term in the calculation of losses comes from eddy currents in the conductor cable due to relatively low interstrand resistance. As estimated by Nah et. al.[6] with the use of Morgan's analysis[7], the loss in joules per cycle is 340 (dI/dt)/r_c for a cycle of 500-5000-500 amperes, where r_c is the crossover resistance between strands in micro-ohms. For a full bipolar cycle, this result must be multiplied by two, and also by 6000/4500 to account for a slightly larger current in our case. Thus, in joules per full cycle, the interstrand coupling loss is estimated to be

$$Q_{interstrand}(J/cycle) = 907 \frac{\dot{I}}{r_c} \qquad (9)$$

and is found to become excessive for r_c less than about 20 $\mu\Omega$.

Iron Hysteresis Loss

An approximation which is frequently used to estimate the hysteresis loss per unit volume in a ferromagnetic material is

$$\frac{Q_{iron}}{V} \approx \frac{4 H_c B_0}{4\pi} *10^{-7} \left(\frac{J}{cm^3}\right) \qquad (10)$$

where the magnetic quantities are in emu. H_c is the coercive force in oersted and B_0 is the induction somewhere near the knee of the B-H curve. For $H_c = 1.25$ oersted and $B_0 = 12,000$ gauss, the result is 4.8×10^{-4} joules/cm^3.

Sum of Losses

Since the various components vary with position in the magnet, the field dependent equations were solved by computer utilizing the field map for the cross section (SSCL DSB-701 design) of the magnet[8].

Table 1. Summary of significant losses for the HDM. Loss is in joule/cm. Results are averaged over the HDM cycle, which is 500 seconds with peak field of 6.7 tesla and a peak current of 6700 amperes.

Item	Inner Coil	Outer Coil	Total	Snitchler
Q_{hyst}	0.867	0.566	1.433	
$Q_{dynamic\ resistance}$	0.033	0.023	0.056	
$Q_{hyst} + Q_{dyn}$	0.900	0.589	1.49	1.24
$Q_{strand\ eddy\ currents}$ [a]	0.25	0.10	0.35	0.174
$Q_{strand\ total\ loss}$	1.16	0.679	1.84	1.41
$Q_{cable\ eddy\ currents}$ [a] (calculated $20\mu\Omega$)[b]	0.11	0.04	0.15 (1.6)	0.08
Total SC losses (Upper limit r=20$\mu\Omega$)	1.27	0.719	1.99 (3.44)	1.49
$Q_{iron\ losses}$			0.30	0.31
GRAND TOTAL (Upper limit r=20$\mu\Omega$)			2.29 (3.74)	1.80

[a] Eddy current losses are extrapolated from experimental measurements of strand loss[5].
[b] Items in parentheses are calculated using Morgan's equation[7] with interstrand resistance of 20$\mu\Omega$.

The results are plotted in Figure 1 as a function of cable number, showing the relative magnitude of the various components across the magnet cross section. The cable numbering system starts at the inner coil midplane (1) and goes to the inner pole (19), the outer midplane is (20), and the outer coil farthest from the midplane is (45). The cable eddy current losses are highest where the entire cable face is exposed to the most flux.

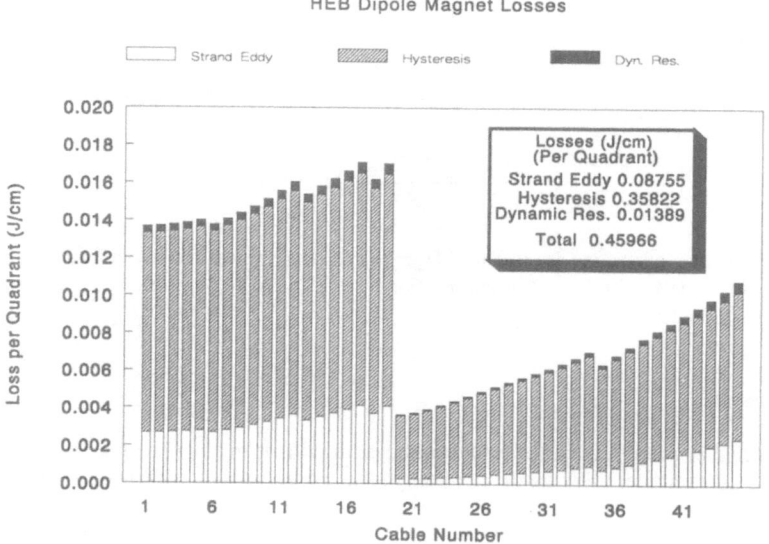

Figure 1. Losses in the HDM for various cable positions.

REFERENCES

1. G. Snitchler, R. Jayakumar, V. Kovachev, and D. Orrell, "Design and A.C. Loss Considerations for the 60mm. Dipole Magnet in the High Energy Booster." Supercollider 3, edited by J. Nonte, Plenum Press N.Y. 1991, p. 625.
2. W.J. Carr, Jr., "AC Loss and Macroscopic Theory of Superconductors," Gordon and Breach, N.Y., (1983) Ch. 6.
3. G. Morgan and W.B. Sampson, SSC Tech Note #76, SSC-N-519, June 10, 1988.
4. W.J. Carr, Jr., op. cit. p. 73. The factor y_1/R_0 in this reference can to good approximation be replaced by I/I_c.
5. G.T. Mallick, Jr., J.M. Toms, W.J. Carr, Jr., G. Snitchler, V. Kovachev, and R. Jayakumar, "Results of AC Loss Measurements on SSC Conductors," Supercollider 3, edited by J. Nonte, Plenum Press, N.Y. 1991, p. 695.
6. W. Nah et al., "Quench Characteristics of 5-cm-Aperture, 15-m-long SSC Dipole Magnet Prototypes."
7. G. Morgan, "Eddy Currents in Flat Metal-Filled Conducting Braids," JAP 44, No. 7, July 1973, p. 3319.
8. H.L. Chuboy and M.P. Krefta, "HDM Model Magnet Magnetic Analyses and Performance Prediction, Two Dimensional Finite Element Model," Westinghouse Report 92-8TM4-HEBCG-R4.

DESIGN AND ANALYSIS OF THE CRYOSTAT SYSTEM FOR A 15 m LONG PROTOTYPE QUADRUPOLE MAGNET

A. R. Jalloh, C. Corbett, C. J. Chang, D. Baritchi,
S. Bagwell, S. Chen, and J. Riddle

Magnet Systems Division
Superconducting Super Collider Laboratory*
2550 Beckleymeade Avenue
Dallas, TX 75237

E. Daly

Westinghouse Science and Technology Center
1310 Beulah Road
Pittsburgh, PA 15235

INTRODUCTION

Quadrupole magnets with a gradient of 190 T/m are required in several sections of the interaction region of the SSC collider ring. These magnets have an aperture of 50-mm and greater and they are expected to operate at 4.3 K. They vary in length from approximately 6 m to about 15 m. Superconducting magnets with a 50-mm aperture and 15-m length have never been built before. It is necessary to develop and prove out the design and technology for long quadrupole magnets before embarking on the design of tunnel magnets of similar design. A research and development program was initiated in January 1992 in the Magnet Systems Division of the SSC Laboratory to develop such a magnet. The goal of this R&D program was to develop a producible and cost effective magnet design and fabrication process that can satisfy performance needs.

The baseline design is the 50-mm Dipole Cryostat developed at Fermi National Accelerator Laboratory.[1] Attempts were made to improve upon baseline design wherever this was feasible. This paper presents the design that was developed for the cryostat system of the prototype quadrupole magnet. A cross-section view of the magnet, showing the cryostat system components is shown in Figure 1.

---

* Operated by Universities Research Association, Inc., for the U.S. Department of Energy under Contract No. DE-AC35-89ER40486

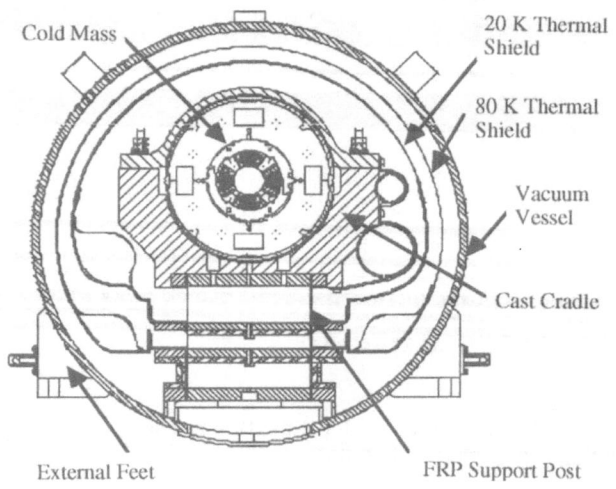

Figure 1. QCE Magnet cross-section.

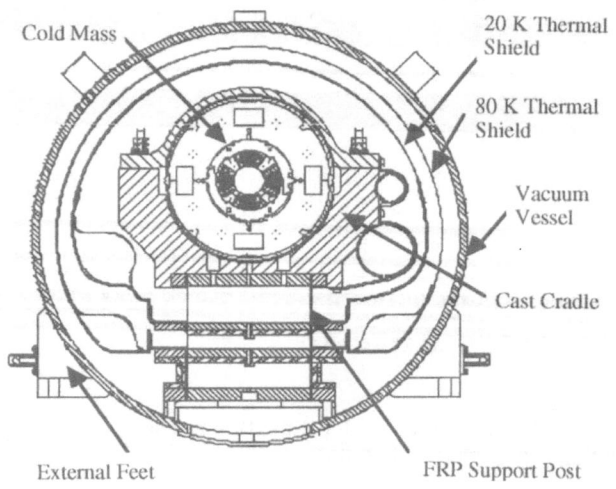

Labels on figure: Cold Mass, 20 K Thermal Shield, 80 K Thermal Shield, Vacuum Vessel, Cast Cradle, External Feet, FRP Support Post

DESIGN REQUIREMENTS

The design requirements for the quadrupole magnets in the interaction region had not been developed at the inception of the program. As such, the cryostat system design requirements were derived from available requirements for magnets with similar functions or physical characteristics. or physical characteristics. Individual interaction region quadrupole magnets will probably have requirements that are different from these derived requirements. However, this serves as a reasonable starting point. The structural requirements were derived from collider magnet specifications. The collider dipole static heat leak budget was the specification used since both magnets have approximately the same physical length. The alignment requirement was derived from the collider quadrupole specifications. The following is a summary of the requirements:

Structural Loads:	0.5 G lateral
	0.5 G axial
	3 G vertical
Static Heat Leak Budget	0.363 watts at 4 K
	5.055 watts at 20 K
	37 watts at 80 K
Alignment: Cold Mass Sag:	< 0.06 mm

In addition to these requirements, the magnet was also required to satisfy ASME pressure vessel and piping code requirements (Section VIII and B31.3). The magnet will be tested in the Magnet Test Laboratory and as such its external supports and interfaces must be compatible with the test stands.

CURRENT CONFIGURATION

The physical characteristics of the magnet are summarized below. The cold mass parameters were derived from the 40 mm collider dipole magnet design.

Cold Mass OD	=	276.78 mm (Maximum)
Cold Mass Length	=	15123.7 mm (end plate to end plate)
Shell Thickness	=	4.93 mm (nominal)
Weight Intensity	=	3650 N/m
Cold Mass Stiffness	=	4.7×10^7 mm^4 (125% skin stiffness – theory)
(second moment of area)	=	6.25×10^7 mm^4 (measured from 40 mm collider dipole magnet cold mass DD0018)

VACUUM VESSEL

The vacuum vessel design is simpler than the Accelerator System String Test (ASST) dipole design. The present design has a wall thickness of 12.7 mm, compared to the 8.3 mm wall thickness of the ASST design. The added wall thickness provides enough structural strength to eliminate the reinforcing rings at the support post locations. The vessel wall is uniform throughout. The length of the vacuum vessel is 14 960 mm. It has four circumferential welds and three longitudinal welds. The outside diameter of the vessel is 711.2 mm. The vacuum vessel is 74% stiffer than the ASST vessel. The fundamental frequency of the entire magnet assembly increases about 20% as a result of this increased stiffness and will therefore improve dynamic response to ground motion and therefore stability. The material is structural steel ASTM A516 Grade 70, consistent with the material selected for the collider dipole and quadrupole magnets.

One of the design requirements is that the magnet has to interface with the test facilities in the SSCL Magnet Test Laboratory. To accomplish this, the ASST dipole feet and end rings are integrated into the design. The end rings are modified on the faces that mate with the vacuum vessel since the inner and outer diameters of the vacuum vessel are greater than those of the ASST vacuum vessel. The side that mates with the test facility is not modified.

SUSPENSION SYSTEM

The suspension system is comprised of the cold mass support posts and the sliding cradles. The magnet has seven posts compared with five for the ASST Design. The number of posts was driven by the cold mass sag requirements. The support posts are of single tube design. Figure 2 shows the support post with the rings and disks. The material used is fiber reinforced plastic (FRP), the same material used in the outer tube of the re-entrant posts of the ASST design. Extensive development of both the FRP material and shrink fit concepts are documented in the literature.[2,3] The 4 K and 300 K disks and rings, fastened by shrink fits, are stainless steel while the 20K and 80K disks and rings are aluminum 6061-T6. The outer dimensions of the rings were modified to match the corresponding dimensions of the support post being designed by Genaral Dynamics for collider dipole

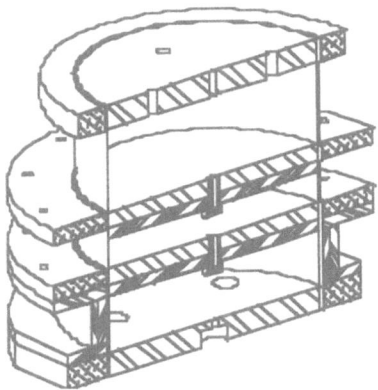

Figure 2. Fiber Reinforced Plastic Single Tube Support Post.

magnets. This was done to assure compatibility between the two post designs if the need came up to use the CDM support post in the magnet. The post design is much simpler than the re-entrant post design and it has far fewer components than the re-entrant post. The seven posts meet the static heat leak budget of the five re-entrant posts, the structural design loads for the collider magnets of 0.5 G axially, 0.5 G laterally and 3 G vertically. There is a factor of safety of 2 on buckling loads.

The cast cradle is a modification of the ASST design. It is made up of two parts, a lower section and an upper steel strap. The lower section is made of cast steel and machined. It is dimensioned to accommodate the support post which is about 25 mm longer than the re-entrant post.

The bolted split ring that holds the cradle in place was eliminated for ease of assembly. The center cradle is attached to the cold mass by welding eight 1-inch gussets to the cradle on one side and the cold mass on the other side. This technique eliminates ASME qualifications of the cast cradle but does not compromise structural integrity.

INSULATION SYSTEM

The insulation system for this magnet consists of thermal shields at the 80 K and 20 K levels, two multi layer insulation blankets wrapped around the 80K shield and cryogenic pipes to carry the cryogens. The 4 K and 20 K blankets of the baseline ASST design were eliminated.[4] The major requirement for the shields was that they had to conform to ASST interface requirements at the ends of the magnet. The QCE shield design eliminated the transition joints and used extruded 304 L stainless steel pipes to improve reliability. The design provided more room to adjust the shields during assembly. The thermal shields are connected thermally to the cryogenic pipes by thermal straps. They are restrained in troughs in the shields with structural aluminum straps.

SUMMARY

The cryostat system for a 15-m long quadrupole magnet with a 50-mm aperture has been designed. The magnet has a uniform walled vacuum vessel with dipole external feet and ASST end rings to ensure compatibility with Magnet Test Laboratory test facilities. The port is a low profile platform that is welded onto the inside of the vacuum vessel. The support post is a fiberglass reinforced plastic single tube post with shrink fit rings and disks at 4 K, 20 K, 80 K and 300 K.

REFERENCES

1. T. H. Nicol, "Design Development for The 50 mm Superconducting Super Collider Dipole Cryostat," Supercollider 3, Edited by J. Nonte, Plenum Press, New York, 1991, pp. 1029–1036.
2. T. H. Nicol, "SSC 50 mm Collider Dipole Cryostat Single Tube Support Post Conceptual Design and Analysis," Supercollider 4, Edited by J. Nonte, Plenum Press, New York, 1992, pp. 747–755.
3. T. H. Nicol, J. D. Gonczy, R. C. Niemann, "Design and Analysis of the SSC Dipole Magnet Suspension System," Supercollider 1, Edited by M. McAshan, Plenum Press, New York, 1989, pp. 637–649.
4. R. Pletzer, "Results of Preliminary Analysis of 50 mm Dipole Magnet Without 4 K and 20 K MLI Blanket," MSD Memorandum, 1991.

CONCEPTUAL DESIGN OF THE CRYOSTAT SYSTEM
FOR THE 2:1 VERTICAL BENDING DIPOLE MAGNET
OF THE SSC COLLIDER RING

A. R. Jalloh, C. Corbett, C. J. Chang, D. Baritchi, S. Chen,
J. Riddle, and Z. Gruzdeva

Magnet Systems Division
Superconducting Super Collider Laboratory[*]
2550 Beckleymeade Avenue
Dallas, TX 75237

E. Daly

Westinghouse Science and Technology Center
1310 Beulah Road
Pittsburgh, PA 15235

INTRODUCTION

The beam separation is 90 cm throughout most of the SSC collider ring. As the beams approach the interaction points, the vertical beam separation is reduced from 90 cm down to 45 cm. The beams then travel through several quadrupole magnets and empty cryostats where the beam separation is 45 cm. Finally the separation is reduced from 45 cm down to zero, essentially, as they head for the interaction point. This final reduction takes place in a vertical bending dipole magnet, designated as BV1. The beam separation in this magnet starts out at 45 cm and goes down to 44 cm. A schematic of the beam lines in the interaction region is shown in Figure 1. This magnet is two coillider dipole cold masses that are rotated 90° to achieve vertical beam bending. If there is enough room, the cold masses could be contained in separate cryostat systems. However, if the available room is not large enough, then they will have to be contained in one cryostat system. Also, the normous weight of the two cold masses will impose severe transportation and handling structural loads on the cryostat system. The design must be robust enough to withstand these loads. Another prime design constraint of the cryostat is that it should fit in the unnel space without requiring local tunnel enlargements or encroaching on aisle space. Consequently, a vacuum vessel design that makes optimum use of tunnel space is required.

[*] Operated by Universities Research Association, Inc., for the U.S. Department of energy under Contract No. DE-AC35-89ER40486

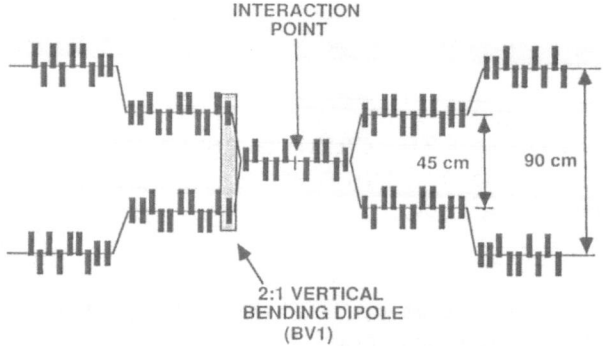

Figure 1. Layout of Magnets in Interaction Region.

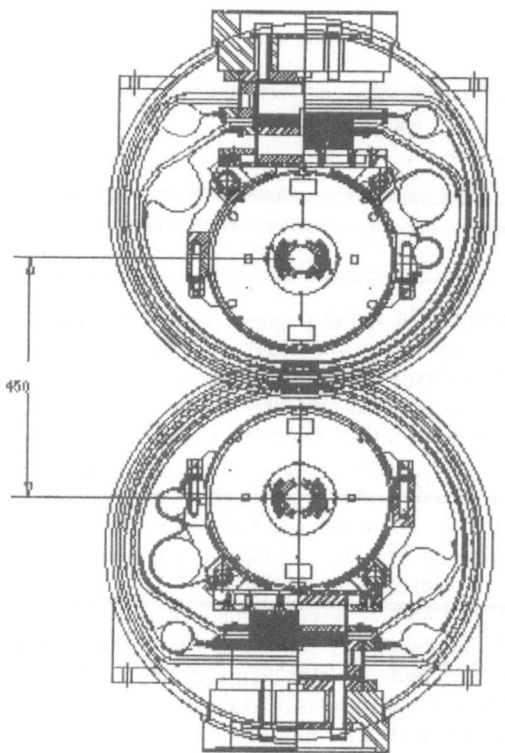

Figure 2. Separate Cryostat for Each Cold Mass and Rotated Upper Magnet.

This paper presents the various conceptual designs that were investigated to determine a feasible cryostat system for BV1. The three concepts investigated for this work were:

1. Two separate collider dipole magnets with a rotated upper magnet
2. Ellipto-cylindrical vacuum vessel
3. Cylindrical vacuum vessel

Each of these concepts is discussed in the following sections. It should be pointed out that detailed engineering analyses to study the thermal and structural performance characteristics of the various concepts have not been carried out. These analysis will be done when the design requirements are available. A cylindrical vacuum vessel design was the recommended design for the 2:1 cryostat system.

ROTATED UPPER MAGNET

The first concept considered was installing one collider dipole magnet (CDM) directly above another while maintaining the 44–45-cm beam separation. This option is shown in Figure 2. The outside diameter of the CDM vacuum vessel is 686 mm. The clearance between the cold mass shells is 11 cm. This concept was not feasible because the vacuum vessels would interfere with each other. Also, it is desirable to couple the cold masses through the suspension system to facilitate alignment. This concept clearly does not provide this feature.

ELLIPTO-CYLINDRICAL VACUUM VESSEL

The ellipto-cylindrical vacuum vessel concept is a true 2 in 1 cryostat. A schematic of this concept is shown in Figure 3. The vacuum vessel is made up of cylindrical and elliptical sections that are welded together longitudinally. Here the two cold masses are coupled through a super cradle. The cradle consists three sections that bolt together to secure the cold masses. The lower section bolts onto the top of the support post and supports the lower cold mass. The middle section bolts onto the lower section, secures the lower cold mass and acts as a saddle to support the upper cold mass. Finally, the upper section bolts onto the middle saddle section and secures the upper cold mass. The vacuum vessel is very susceptible to buckling when it is evacuated. Also, the design poses fabrication and reliability issues that would drive up the cost of the magnet.

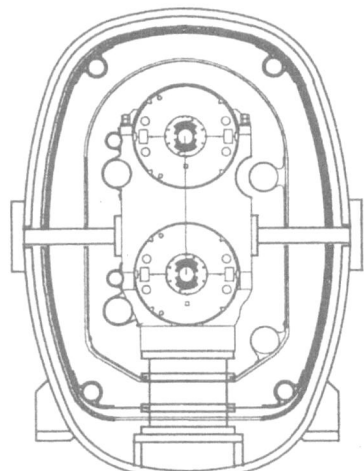

Figure 3. Ellipto-Cylindrical Cryostat - Two Cold Masses in One Cryostat.

CYLINDRICAL VACUUM VESSEL

The cylindrical vacuum vessel concept is similar to the cylindro-elliptical design. A drawing showing the concept is presented in Figure 4. The major difference between the wo is the shape of the vacuum vessel. It is a cylinder. with an outer diameter of 1364 mm. The wall thickness is 25 mm. The problems of buckling stability, fabrication and reliability associated with the cylindro-elliptical design are minimal. The cradle design is similar in the two concepts. Also, lateral struts are incorporated in the design to address the bending load effect of the cold mass assembly on the support posts. These struts would be attached to the cradles on one end and the vacuum vessel wall on the other end. The one big disadvantage this design has over the cylindro-elliptical design is that it takes up more aisle space in the tunnel.

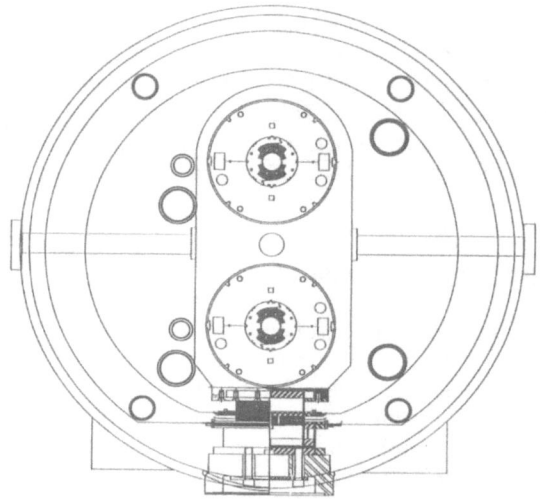

Figure 4. Cylindrical Cryostat - Two Cold Masses in One Cryostat.

SUMMARY

Three concepts were investigated for the 2-in-1 cryostat design of the vertical bending dipole magnets. One of the concepts involved the installation of one complete magnet vertically above another. This concept was not deemed feasible because there was not enough clearance to prevent the vacuum vessels from interfering with each other. The other two concepts both involved the use of a single vacuum vessel to contain both cold masses. These concepts were referred to as the 2-in-1 designs. One of these 2-in-1 designs used a vacuum vessel made up of cylindrical and elliptical sections that were welded together longitudinally. The complexity of the shape and the amount of welding required to fabricate this vacuum vessel would introduce fabrication and reliability issues that will drive up the cost of the system. Also, the shape of the vacuum vessel makes it very susceptible to buckling. The other 2-in-1 concept used a cylindrical vacuum vessel. This design minimizes the problems of buckling, fabrication and reliability. It is the preferred design concept; however, detailed engineering analyses will have to be carried out to validate the concept.

REFERENCES

1. Superconducting Super Collider Laboratory, "Site-Specific Conceptual Design," SSCL-SR-1056, July 1990.

HDM MODEL MAGNET QUENCH PRESSURE RISE ESTIMATES

E. F. Daly and O. R. Christianson

Advanced Electromechanical Systems
Westinghouse Science & Technology Center
1310 Beulah Road
Pittsburgh, PA 15235

INTRODUCTION

The DSB model magnet design is used to verify manufacturing tooling and processes as well as quench performance and, to some extent, magnetic field quality. The testing scenario for a model magnet involves a vertical dewar that can accommodate a magnet that is 1.8 meters long x 0.34 m in diameter. The cooling scheme for the model during test is pool boiling atmospheric helium liquid, which is markedly different from the cooling scheme, 0.4 MPa single phase gaseous helium, provided for the prototype magnets. While in test, the models will be tested for plateau current, field quality and ramp rate dependence, and may be subjected to many quenches.

During quench, the magnet is subjected to a high rate of Joule heating, yielding potentially high local coil temperature. Some heat is transferred to the liquid helium, and as a result, the liquid helium contained in the windings vaporizes and rushes out of the coils through gaps and spaces in the insulation. The rapid vaporization of helium can lead to potentially damaging high pressure within the windings, adversely affecting the insulation system and consequently further magnet operation.

The calculations and modeling described in this report predict the quench pressure rise within the DSB model magnet design under a variety of testing conditions. This evaluation includes consideration of test scenarios and two design parameters, coil porosity and coil void fraction.

MATHEMATICAL MODEL

The quench event and ensuing thermal hydraulics combine to form a complex set of processes that are not well understood and difficult to accurately model. A large number of assumptions, discussed below, are used to either help to simplify the analysis or render the problem tractable.

The quench modeled occurs in the outer coil at the axial midpoint of the magnet. Adiabatic temperature rise predictions for the outer coil are higher than the inner, thus a worst case event is modeled. The axial midpoint is expected to have the highest pressure rise along the length of the magnet. The helium is treated as a simple compressible substance. A phase transition occurs in the windings on a time scale much shorter than any other thermodynamic or hydraulic processes. Essentially, the helium flashes to vapor in the first several milliseconds after the coils become normal. This process is termed "dryout". After dryout, the helium expands, its pressure and temperature increase rapidly and it flows from the coil void.

One dimensional helium flow is modeled in all flow passages. Helium flows radially inward from the coils to the aperture of the magnet. From there it flows axially to the ends of the magnet where a fixed atmospheric pressure is assumed. There are no cross flow cooling vents or bore tube in this model.

The Joule heat generated is redistributed as energy stored in the temperature rise of the coil, heat transferred by conduction through the Kapton to the collars, and heat transferred to the helium. The energy stored in the temperature rise of the coil is much greater than heat transferred to either the helium or the collars. For this analysis, the collars are assumed isothermal and recovery is not included.

It is assumed that the quench propagates along the coil at a calculated adiabatic quench propagation velocity which is a function of coil temperature. Since the quench velocity is included, it is assumed that no axial thermal conduction in the coil occurs during the quench event.

The insulation porosity is assumed to be a known function of pressure. The voids in the coil are filled with liquid helium prior to the initiation of the quench. A nominal void fraction of 3% is used as a baseline, while a range of void fractions are evaluated.

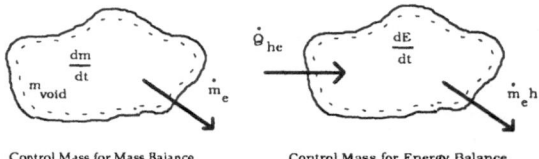

Control Mass for Mass Balance   Control Mass for Energy Balance

Fig. 1 Control Masses for Derivation of Equations

$$\frac{dm}{dt} + \dot{m}_e = 0 \qquad\qquad \frac{dE}{dt} + \dot{m}_e\, h = \dot{Q}_{he} \qquad\qquad (1,2)$$

The dump circuit consists only of the magnet and dump resistor. The dump resistor is used to extract energy from magnet following the quench. There is a delay time included for sensing a quench and switching off the current supply to the magnet. This is important since the delay time will affect the amount of current flowing through the magnet. Using the assumptions explained above, the mass and energy equations are derived from the rate-basis control mass shown in Figure 1.

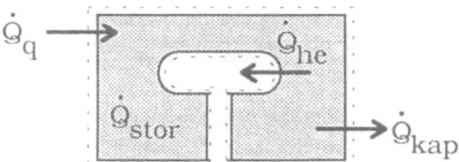

Fig. 2 Energy Balance for a single coil element.
The symmetric model contains 10 elements over the half length of the magnet.

An additional energy balance for each time step on the coil is required to determine what portion of the total energy is conducted to the helium. The resulting equation, based on the schematic in Figure 2 is:

$$\dot{Q}_q = \dot{Q}_{he} + \dot{Q}_{kap} + \dot{Q}_{stor} \qquad\qquad (3)$$

where the term on the left is the Joule heating produced by the quench. The three terms on the right are, from left to right, the heat transferred to the helium, the heat conducted to the collars through the Kapton and the energy stored in the temperature rise of the coil. The energy stored in the magnet during operation is 108 kJ/meter and the inductance is 4.90 mH/meter as determined from magnetics calculations[1] for the DSB cross section. The operating current which yields the desired magnetic field is calculated from these two parameters.

The momentum equation for a viscous fluid, which relates the mass flow and pressure drop, usually contains the appropriate friction factor, cross sectional flow area and flow path length. Solutions to date rely on some broad assumptions with regard to these parameters. The approach adopted here is to use an empirical correlation derived from flow tests conducted at Brookhaven National Laboratory (BNL) by Sampson [2] on a 40mm 16-turn inner coil. The tests compare the Kapton/fiberglass insulation scheme to an all Kapton scheme to determine relative porosity of each. Sensitivity analyses are conducted to determine how the porosity affects the peak pressure developed in the windings. It is interesting to note that the

Kapton/fiberglass scheme is roughly seven times more porous than the all Kapton system. The resulting momentum equation for flow between the void and the bore is

$$m_e = K(P_{void} - P_{bore})$$ (4)

where K is the porosity of the coil in (kg/s/Pa/m). Hilal et al [3] employed the parameter β, which is essentially K, to investigate sensitivity of maximum pressure to porosity. The range of values Hilal used for β were 10^{-13} to 10^{-17} kg/s/Pa/m while actual measurement yields β or K in the range of 10^{-10} kg/s/Pa/m. This suggests that the coils are much more porous than early estimates.

The equations above are discretized, coded in Sun FORTRAN, using an HEPAK subroutine for the helium properties [4], and solved explicitly to determine temperature and pressure rise within the coil.

RESULTS

There are two classes of quenches that can occur, one related to conductor material limitations and the other related to mechanical stability of the strands within the coil. At a maximum current of 7400 A, which is assumed a conductor limited quench, the temperature is almost 120 K and the peak pressure is about 3.3 MPa. The second class of quench, due to conductor motion, would most likely occur at or near the operating current, and the resulting temperature rise would be roughly 100 K, although it is not shown in the figure below. Dryout, circled in the figure, occurs in the first several milliseconds while the liquid helium within the void is increasing in temperature and pressure.

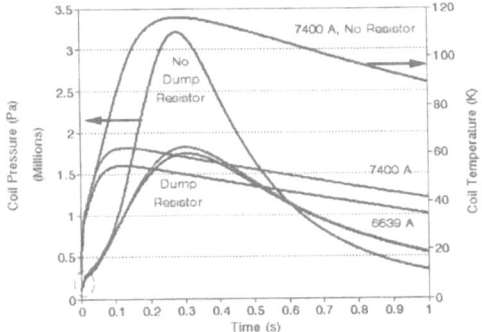

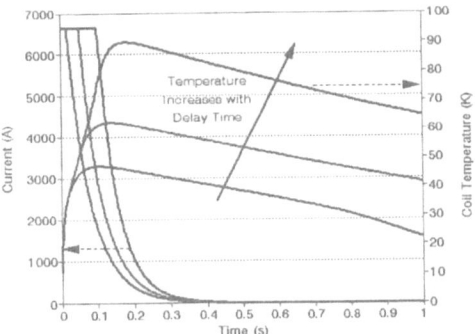

Fig. 3 Effect of Operating Current and Test Set up on Maximum Pressure and Temperature Rise

Fig. 4 Effect of Delay Time on Current I and Maximum Temperature

By adding a 100 mΩ dump resistor to the circuit, the peak pressure and temperature are reduced by a factor of two. The maximum temperature rise is 62 K at the operating current and the corresponding pressure rise within the coil is 1.8 MPa. Since the current decay time is dominated by the dump resistance, the initial current has a much smaller effect on the coil temperature and subsequent pressure rise. The dump resistor will absorb more energy, which results in less helium vaporized, lower maximum coil temperatures and quicker recovery times.

A second set of analysis cases were run to determine the effect of delay time on the amount of energy stored in the coil. The void fraction is 3% for this set of cases and the operating current is 6639 A. The current supplied to the magnet is constant until the quench is sensed. In a real system, after the quench is detected, the power supply is "crowbarred", and then the dump resistor is brought into the test circuit. The total delay is the sum of the individual switching times. The net effect is that longer delay times for sensing the quench result in more energy deposited in the magnet prior to "crowbarring" the power supply.

The temperature of the coil and pressure within the coil void is shown in Figure 5 as a function of time after the quench occurs. Longer delay times lead to greater pressure rise within the coil. Since more energy is dumped into the magnet resulting in higher coil temperatures, again more helium is vaporized.

The next several runs investigate the sensitivity of the helium pressure rise with respect to coil porosity and void fraction. The delay time is fixed at 34 milliseconds and the initial operating current selected is 6639 A. Increasing porosity tends to reduce the maximum pressure rise as well as reduce the duration of the pressure pulse. In Figure 6, the porosity is varied by orders of magnitude, holding the void fraction constant at 3%. Increasing the porosity by factors of two or three from the experimental value yield pressures ranging from 1.5 to 2.0 MPa. Next, the void fraction is varied over a wide range while holding the porosity constant at the experimental value obtained from the BNL data. The operating current is 6639A, and the delay time is 34 milliseconds. As void fraction increases, the peak pressure decreases while the duration of the pulse increases.

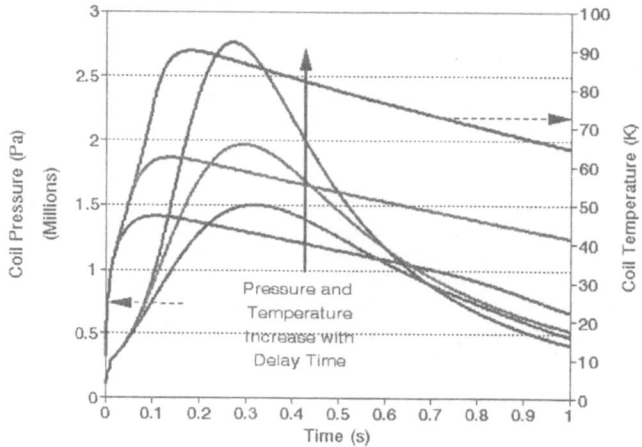

Fig. 5 Effect of Increased Delay Time on Maximum Pressure and Temperature

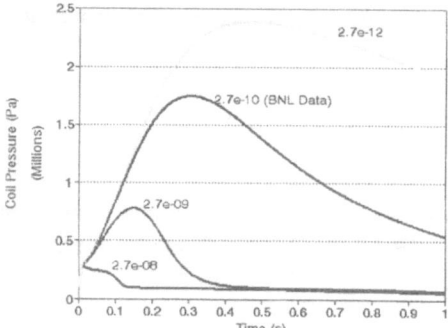

Fig. 6 Sensitivity of Maximum Pressure to Coil Porosity

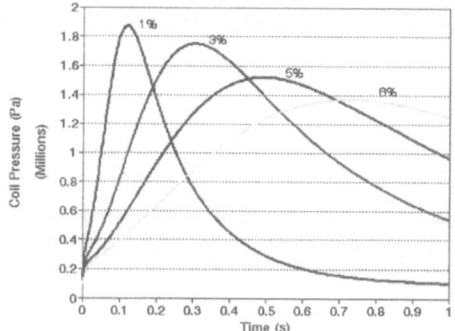

Fig. 7 Sensitivity of Maximum Pressure to Coil Void Fraction

Table 1. Comparison of quench pressure rise calculations*.

Design & Reference	Void Fraction (%)	Pmax (MPa)	Tcoil (K)	Time to peak (ms)	Cross Flow (?)	2-Phase Helium (?)
DSB Model	3	1.8	62	354	no	yes
HDM Prototype [5]	3	2.9	77	0.6	yes	no
CDM SSCL 5 cm Prototype [6]	3	2.3	70	13	yes	no
CDM GD 4 cm Prototype [7]	1	4.7	205	700	no	no
CDM GD 4 cm Prototype [3]	4	3.2	NR	50	yes	no
CDM BNL 4 cm Prototype [8]	NR	1.6	86	>1000	yes	no

*NR = Value not reported.

CONCLUSIONS

The quench pressure analysis yields realistic values for the maximum pressure rise within the windings when compared with calculations of other workers. Several parameters can significantly impact the peak pressure within the coil winding. For delay times less than 50 ms, coil porosity less than 2.7×10^{-10} kg/s/Pa/m, coil void fraction 3% or greater and a dump resistor of roughly 100 mΩ, the pressure rise is 1.8 MPa. At maximum operating current, it is recommended that a dump circuit, whose quench sensing delay time is less than roughly 50 ms, be employed. Also, a dump resistor is recommended to minimize quench pressure rise, minimize coil temperature rise, optimize use of helium inventory, and minimize recovery time.

REFERENCES

1 Chuboy, H. L. and M. P. Krefta, "HDM model magnet magnetic analysis-- two-dimensional finite element model", Westinghouse STC Report N0. 92-TM4-HEBCG-R4, January (1993).

2 W. B. Sampson, Data from BNL porosity testing on All-Kapton and Kapton/Fiberglass Insulation systems, Private communication, August (1992).

3 Hilal, M. A.et al, "Helium pressure rise of superconducting super collider dipole magnets following a quench", *Advances in Cryogenic Engineering*, Vol. 33, pp. 143-148 (1988).

4 HEPAK v. 2.3, Copyright (1990) by CRYODATA, P. O. Box 558, Niwot, CO 80544.

5 Fagan, T. J. and P. W. Eckels, "Process control oriented quench analysis of an SSC magnet" *SuperCollider 2*, pp. 501-517 (1989).

6 Carcagno, R. H. and W. E. Scheisser, "Helium venting computer simulation during an SSC dipole quench", *Advances in Cryogenic Engineering*, Vol. 37, Part A, pp. 709-718 (1992).

7 Ibrahim, E. A., M. A. Hilal, and S. D. Peck, "Quench pressure analysis of adiabatically stable magnets", *IEEE Transactions on Magnetics*, Vol. MAG-23, No. 2 (1987).

8 R. P. Shutt and M. L. Rehak, "Cross-Flow Cooling", Pres. by M. L. Rehak at MSIM on 4-7-1992.

TEST PERFORMANCE OF THE QSE SERIES OF 5cm APERTURE QUADRUPOLE MODEL MAGNETS

B. Archer, D. Bein, G. Cunningham, J. DiMarco,
T. Gathright, J. Jayakumar, A. LaBarge, W. Li,
D. Lambert, M. Scott, G. Snitchler, and R. Zeigler

SSC Laboratory[*]
2550 Beckleymeade Ave.
Dallas, TX 75237

INTRODUCTION

A 5 cm aperture quadrupole design, the QSE series of magnets were the first to be tested in the Short Magnet and Cable Test Laboratory (SMCTL) at the SSCL. Test performance of the first two magnets of the series are presented, including quench performance, quench localization, strain gage readings, and magnetic measurements.

Both magnets behaved reasonably well with no quenches below the collider operating current, four training quenches to plateau, and good training memory between thermal cycles. Future magnets in the QSE series will be used to reduce the initial training and to tune out unwanted magnetic harmonics.

QUENCH PERFORMANCE

Training

The training behavior of QSE101 is shown in Figure 1. Four training quenches were needed to reach a plateau current of about 8660 A at 16 A/s during the first thermal cycle. The first training quench is at 7120 A, 6% above the collider operating current of 6714 A. All four training quenches occurred in the end parts of the magnet, with plateau in the inner coil, pole turn, straight section. Because of various problems that occurred during the testing of QSE101, the first magnet tested at the SMCTL, the quenches in the second and third thermal cycles could not be localized. The magnet shows good retraining behavior with one, or perhaps two, training quenches during the second thermal cycle. Retraining behavior during the third thermal cycle was obscured by trouble with the power supply

* Operated by the Universities Research Association, Inc., for the U.S. Department of Energy under Contract No. DE-AC35-89ER40486.

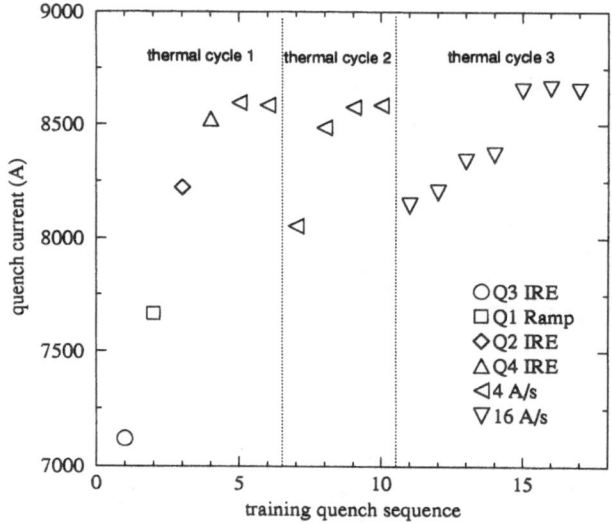

Figure 1. QSE101 training history for each of the three thermal cycles. IRE stands for inner coil, return end.

and a heat leak caused by the magnetic measurement warm bore, thus, the retraining is not thought to be representative.

QSE102 training behavior is shown in Figure 2. There are four training quenches during the first thermal cycle, all occurring in the end parts. The plateau current is about 8670 A at 16 A/s. As the magnet was being warmed up at the end of the first thermal cycle it was accidentally warmed to at least 358 K, or 85° C. The resultant thermal stresses are believed to have returned the magnet to an essentially virgin state, resulting in the complete lack of training memory exhibited by thermal cycle two. Thermal cycle three has only one training quench, demonstrating that the magnet does retain training memory between thermal cycles.

The training performance is good for both magnets, especially the training memory between thermal cycles. Almost all the training quenches occurred in the ends of the magnets, pointing to a need to refine the end parts, which were modified by hand.

Ramp Rate Sensitivity

Figure 3 shows that QSE101 has little ramp rate dependence, whereas, QSE102 exhibits a strong dependence at ramp rates above 50 A/s. At ramp rates of 75 A/s and above the quench location for QSE102 is in the ramp between the inner and outer coils in quadrant 3. The tooling for making the ramp was changed between QSE101 and QSE102 to prevent tubing of the cable in the ramp, which apparently succeeded in keeping the strands in contact with a resultant reduction of interstrand resistance in QSE102. QSE101 style ramp tooling is being used for all further QSE magnets.

QSE102 had no quenches when ramped down from 6500 A to 4000 A at ramp rates of 100 A/s, 200 A/s, 300 A/s, and 400 A/s.

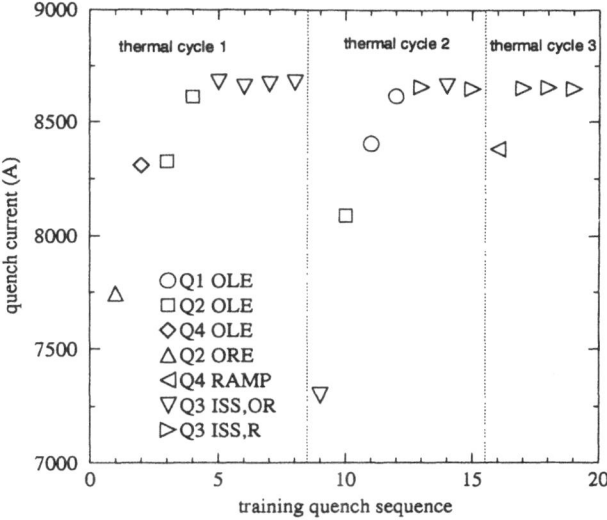

Figure 2. QSE102 training history for each thermal cycle. OLE stands for outer coil, lead end, ORE stands for outer coil, return end; ISS,OR stands for inner coil, straight section pole turn, opposite ramp; ISS,R is inner coil, straight section pole turn, ramp side.

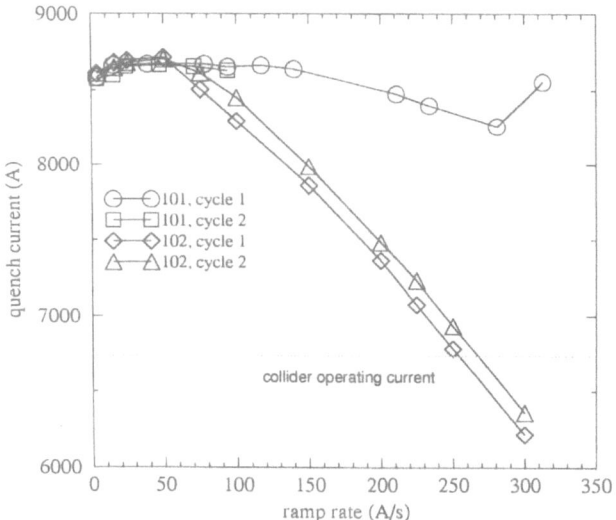

Figure 3. Ramp rate dependence of QSE101 and QSE102.

STRAIN GAGES

Average prestress measurements for QSE101 and QSE102 are shown in Table 1. The

Table 1. Average prestress measurements for QSE101 and QSE102.

	QSE101		QSE102	
	avg inner (MPa)	avg outer (MPa)	avg inner (MPa)	avg outer (MPa)
After collaring and yoking	89	95	85	86
After collaring and yoking, theory	63	45	63	45
After cool down	39	35	48	57
After cool down, theory	30	18	30	18
Energize loss (6500 A)	8	11	13	8
Energize loss, theory	10	8	10	8
Prestress loss from thermal cycle 1 to 2	-	-	26	23

prestress after collaring and yoking, and after cooldown, are greater than expected perhaps because the outer coils are oversize by about 0.2 mm. Prestress loss due to energization is about as expected. The temperature excursion experienced by QSE102 caused a loss of about 25 MPa, both warm and cold, due to either ground plane insulation creep or cable deformation.

MAGNETIC MEASUREMENTS

Magnetic harmonics measured for QSE101 and QSE102 using a 25 cm rotating coil are shown in Table 2. Coil size mismatch causes the a_2 and b_2 components, whereas, the

Table 2. QSE101and QSE102 warm magnetic harmonics in "units" at 10 A.

n	QSE101		QSE102	
	a_n	b_n	a_n	b_n
2	1.88 ± 0.04	-2.59 ± 0.04	-1.42 ± 0.03	2.27 ± 0.03
3	-1.95 ± 0.01	0.53 ± 0.01	-0.28 ± 0.01	0.52 ± 0.01
4	-0.01 ± 0.01	-0.07 ± 0.01	-0.07 ± 0.01	-0.07 ± 0.01
5	-0.10 ± 0.01	0.55 ± 0.01	0.15 ± 0.01	0.02 ± 0.01

significant a_3 and b_3 indicate that the collars have become slightly oval due to the dipole collaring technique used for the QSE magnets. Reliable cold magnetic measurements are not available.

ACKNOWLEDGEMENTS

The authors wish to acknowledge the valuable contributions of S. Averitte, M. Deak, R. Garlow, K. Rust, G. Williamson, and R. Wood to this work.

A QUASI-3D MODEL TO PREDICT TEMPERATURE DISTRIBUTIONS IN SSC MAGNETS DUE TO 3D HEAT LOADS

B. Archer, R. Schermer, and G. Snitchler

SSC Laboratory[*]
2550 Beckleymeade Ave.
Dallas, TX 75237

H. Kohli

Dept. of Engineering Mechanics
The University of Texas at Austin
Austin, TX

INTRODUCTION

Hadronic and electromagnetic cascades in the vicinity of the interaction regions deposit energy into the final focusing superconducting magnets as a heat load that varies radially, azimuthally, and axially. The helium used to cool the magnets convects heat axially as it flows, thus, the temperature distribution at any point in the magnet depends on the previous history of the helium. A detailed map of the temperature distribution of the superconducting coils is required to determine if the cooling is sufficient to prevent the magnet from quenching.

A three-dimensional model is required to analyze the magnet temperature distribution due to the 3D heat load and the axial helium flow. A 3D finite element thermal model of a 15 m magnet is not practical due to the large amount of CPU time that would be required. Therefore, a quasi-3D model has been developed that divides the magnet into a large number of segments axially, then calculates the temperature distribution for each segment using a 2D finite element model with boundary conditions derived from the previous segment by conservation of energy.

---

* Operated by the Universities Research Association, Inc., for the U.S. Department of Energy under Contract No. DE-AC35-89ER40486.

FORMULATION

Neglecting the end regions, the mechanical cross-section of the magnet is constant axi-ally whereas the material properties and heat loads can vary radially, azimuthally, and axi-ally. Axial conduction through the coils, collars, and yokes is assumed to be negligible. As currently implemented, the mass flow rates, or helium velocities, differ between the annulus and the bypasses, but have a constant axial flow rate. The basis of the method is to subdivide the magnet into short, and perhaps non-uniform, lengths over which the material properties can be taken as a constant. At the center of each segment a steady state 2D solution can be obtained using a finite element model if the helium temperatures in the annulus and bypass are known, as well as the interpolated heat generation rates. The helium temperature at the helium inlets ($z=0$) is known and the 2D steady state solution can be found for this section. An estimate of the helium temperature at the center of the next section can be found by com-puting the amount of heat absorbed by the helium in each flow path of length δz, assuming that the heat load conditions are constant for that length. By repeating this process the tem-perature distribution can be calculated for the entire length of the magnet.

If T_z and $T_{z+\delta z}$ are the absolute temperatures of the helium in a flow passage at two planes located at z and $z + \delta z$, the rate of change of enthalpy between the two planes is

$$\dot{H} = \rho A v C_p (T_z - T_{z+\delta z}) \tag{1}$$

where A is the cross-sectional area of the flow passage, v is the helium axial velocity, ρ is the helium density, and C_p is the specific heat of the helium. The distance δz between the planes is chosen so that the change in density, pressure, and specific heat is small between the planes.

In any longitudinal section of a helium flow passage energy is conserved. Neglecting the change in other forms of energy, such as kinetic and potential energies, the energy removed by the helium must equal the energy transferred into the helium across the flow passage walls. For a yoke bypass the power going through the walls for a section of length δz is

$$\dot{U} = h_{by} \pi D_{by} \delta z (T_y - T_z) \tag{2}$$

where T_z is the bypass helium temperature at z, T_y is the average wall temperature of the yoke in contact with the helium at z, D_{by} is the diameter of the bypass, and h_{by} is the heat transfer coefficient between the bypass and the yoke. Setting the sum of the input energy and the output energy to zero, then rearranging terms gives the bypass helium temperature at cross section $z + \delta z$

$$T_{z+\delta z} = T_z + \frac{h_{by} \pi D_{by} \delta z (T_y - T_z)}{\rho A_{by} v_{by} C_p}. \tag{3}$$

For the annulus the geometry is slightly more complicated because there are three heat sources: the beam tube, the coil, and the collar. Setting the sum of heat transfer into the helium and the heat transported out by the helium to zero and rearranging terms gives the annular helium temperature at the $z + \delta z$ cross section

$$T_{z+\delta z} = T_z + (\rho A_a v_a C_p)^{-1} [\alpha h_{coil} \pi D_{coil} \delta z (T_{coil} - T_z)$$

$$+ (1 - \alpha) h_{coll} \pi D_{coil} \delta z (T_{coll} - T_z) + h_{bt} \pi D_{bt} \delta z (T_{bt} - T_z)] \tag{4}$$

where the first term in the brackets is the coil to annulus heat transfer, the second term is for the collar to annulus heat transfer, and the third term is the beam tube to annulus heat transfer. For the coil term, α is the fraction of the circumference of the annulus that is superconducting coil, h_{coil} is the heat transfer coefficient between the coil and annulus, D_{coil} is the inner diameter of the coil (and collar), and T_{coil} is the average temperature of the coil in contact. Similarly, h_{coll} is the heat transfer coefficient between the collar and annulus, and T_{coll} is the average temperature of the collar in contact. For the beam tube term h_{bt} is the heat transfer coefficient between the beam tube and annulus, D_{bt} is the outer diameter of the beam tube, and T_{bt} is the average temperature of the beam tube in contact with the annular helium.

RESULTS

This algorithm has been used to predict the steady state temperature profile of a hypothetical QL1 quadrupole, the final focusing quadrupole nearest the interation point. The heat load due to the hadronic and electromagnetic cascade on the 15 m long QL1 is approximately 50 W, which is the highest heat load of any accelerator magnet in the SSC. Preliminary heat load data for QL1 is available[1] that varies both radially and azimuthally as shown in Table 1. The heat load is strongly peaked at the lead end of the magnet, farthermost from

Table 1. Heat deposition data for the QL1 magnet (W/kg).

axial distance from lead end	0-3 m	3-6 m	6-9 m	9-12 m	12-15 m
beam tube	1.798×10^{-1}	1.728×10^{-1}	1.486×10^{-1}	1.228×10^{-1}	1.050×10^{-1}
inner coil/collar	1.290×10^{-1}	1.240×10^{-1}	1.066×10^{-1}	8.810×10^{-2}	7.536×10^{-2}
outer coil/collar	4.050×10^{-2}	3.894×10^{-2}	3.348×10^{-2}	2.767×10^{-2}	2.366×10^{-2}
collar/yoke	6.306×10^{-3}	6.064×10^{-3}	5.213×10^{-3}	4.308×10^{-3}	3.684×10^{-4}

the interaction point, which is also the helium inlet end. A helium mass flow rate of 0.1 kg/s is used.

Figure 1 shows the annular helium temperature and the average inner coil temperature for two cases in which the bypass diameter is either 3.2 cm or 2.25 cm. The primary effect of changing the bypass diameter is to redistribute the mass flow so that more helium goes down the annular space. For the 3.2 cm bypass the annular mass flow rate was 0.0056 kg/s, with a total pressure drop of 42 MPa, whereas the 2.25 cm bypass resulted in an annular mass flow rate of 0.0129 kg/s and a total pressure drop of 206 MPa.

The model predicts that for a 3.2 cm bypass the maximum temperature rise in the inner coil is 0.70 K, which leaves the magnet with an unacceptably low quench margin, 0.3 K to 0.7 K, depending on the inlet helium temperature supplied by the cryogenic system. Increasing the annular helium flow by decreasing the bypass to 2.25 cm results in a peak inner coil temperature rise of 0.51 K, a significant improvement in performance. Better performance is achieved principally because of the larger mass flow rate which keeps the annular helium temperature below the temperature of the inner coil.

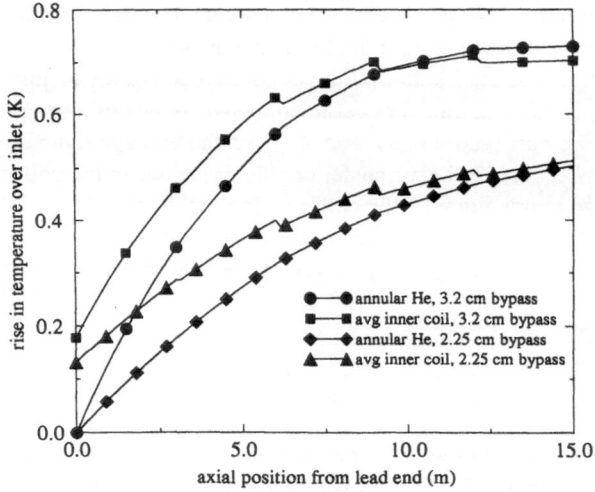

Figure 1. Inner coil average temperature versus axial position for a 3.2 cm diameter bypass and a 2.25 cm diameter bypass. Also shown is the annular helium temperature versus axial position for each case.

This method is being extended to allow analysis of alternative cooling designs such as cross-flow cooling and multiple bypass channels for use in QL1.

ACKNOWLEDGEMENTS

The authors wish to acknowledge the valuable contribution of Dr. Bulent Aksel to this work.

REFERENCES

1. N.V. Mokhov, R. Schermer, and R. Stiening, private communication, June 4, 1991.

CALORIMETRIC AC LOSS MEASUREMENT OF 1.3-M MODEL DIPOLE MAGNETS BY USING A 1.8 K CRYOSTAT

Y. Morita,[1] K. Hara,[1] N. Higashi,[1] A. Kabe,[1] H. Kawamata,[1] Y. Kojima,[1] H. Nakai,[1] T. Ohmori,[3] T. Takahashi,[2] S. Taneda,[4] A. Terashima,[1] A. Yamanishi,[3] and K. Hosoyama[1]

[1] National Laboratory for High Energy Physics, 1-1, Oho, Tsukuba, Ibaraki 305
[2] Hitachi Works, Hitachi Ltd., 3-1-1, Saiwai-cho, Hitachi-shi, Ibaraki 317
[3] Ishikawajima-Harima Heavy Industries Co., Ltd., 1,shin-nakahara-cho, isogo-ku, yokohama 235
[4] Kobe Steel Co., Ltd., Takatsukadai, Nishi-ku, Kobe, Hyogo 500

INTRODUCTION

To measure AC losses of dipole magnet by the calorimetric method, the magnets were cooled to 1.8 K with 120 liters of pressurized superfluid helium, using a 1.8 K cryostat. Heat dissipated in the magnet during a ramp cycle was measured by observing the temperature of the superfluid helium. AC losses were evaluated by the production of the heat capacity and temperature rise of helium. The 1.8 K cryostat consisted of a 4.2 K helium vessel, a 1.8 K pressurized superfluid helium vessel, and heat exchangers (1.8 K saturated helium pot and precooler). Liquid helium from the 4.2 K vessel, cooled to 2.5 K in the pre-cooler, expanded in the 1.8 K saturated helium pot. It cooled the magnet and pressurized superfluid helium to 1.8 K. Evaporated helium gas in the pot was evacuated by a pumping unit. The magnets were model dipole magnets fabricated at KEK for SSC dipole magnet R&D. The overall length is 1.3 m and inner aperture of the coil is 50 mm. Two types of magnet were made to study the AC properties: Type 1 with 6 μm NbTi filament cables and Type 2 with 2.5 μm filament cables. AC losses were measured in a ramp cycle from 500 to 5000 A with 5 second duration at the flat top. Ramp rates were varied from 70 to 400 A/s. Results show that the Type 2 magnet had smaller AC loss than the Type 1.

MODEL DIPOLE MAGNET

The magnets used were model dipole magnets fabricated at KEK for SSC dipole magnet R&D.[1] This is a two-layer and cosθ-type magnet, originally designed by the SSC Laboratory. The inner layer coil has a 50 mm aperture, 19 turns of cables and three copper wedges. The outer has 26 turns and one copper wedge. The coils are clamped by stainless steel collars. Brass shims are inserted between collars and coils to adjust stress on the coil. A low carbon iron yoke and a stainless steel shell mechanically support the collars and coils. The overall length of the assembled magnet is 1.3 m.

The superconducting cables are Rutherford-type 6 μm NbTi filament cables. The inner cable has 30 strands and the Cu to SC ratio of 1.3. The outer has 36 strands and the Cu to SC ratio of 1.8. Coils of the Type 1 magnet were wound with this cable. For better AC characteristics, 2.5 μm NbTi filament cables were specially made. These have a sub-bundled structure with Cu and Mn (0.5% by weight) matrix. Cu to NbTi ratios are 1.5 and 1.8 for inner and outer cables, respectively. The Type 2 magnet coils were wound with this cable.

1.8 K CRYOSTAT

The cryostat consists of a vacuum vessel, a 80 K radiation shield, a 4.2 K helium vessel, a 4.2 K radiation shield, a 1.8 K helium vessel, a precooler, a J-T valve, a 1.8 K saturated helium pot, and a pumping unit. The 4.2 K and 1.8 K helium vessels and 80 K and 4.2 K radiation shields were wrapped with 40 layers of the superinsulation. The 1.8 K vessel is connected at the bottom of the 4.2 K vessel by stainless steel pipes, through which liquid helium is supplied to the 1.8K vessel. When the cool-down from 4.2 K to 1.8 K begins, the pipes are closed by glass fiber-reinforced plastic "pop-up" valves. They also function as a safety valve when magnet quench occurs. The model dipole magnet is suspended in the 1.8 K vessel by 4 support rods. The vessel is sealed with indium. The 10,000 A current leads are set in the 4.2 K vessel and connected to the superconducting bus bars. The bus bar feeds the transport current to the magnet through the connection pipe. A schematic diagram of the cryostat is shown in Fig. 1.

Heat exchangers cool the magnet from 4.2 K to 1.8 K by the J-T expansion of 2.5 K liquid helium. They consist of a precooler and a 1.8 K saturated helium pot. The precooler cools 4.2K liquid helium to 2.5K before the J-T expansion . It has a lamination structure of 58 copper disk fins (thickness : 1 mm), 30 stainless steel rings (thickness : 2 mm), and 29 stainless steel spacers (thickness : 1 mm). Each copper disk has 12 tear-drop shaped paths for helium vapor, while the stainless steel rings provide a path for liquid helium. The stainless steel rings and spacers also suppress thermal conduction along the precooler, so the cooler has relatively small dimensions (diameter : 59 mm, length : 152 mm). The discs, rings and

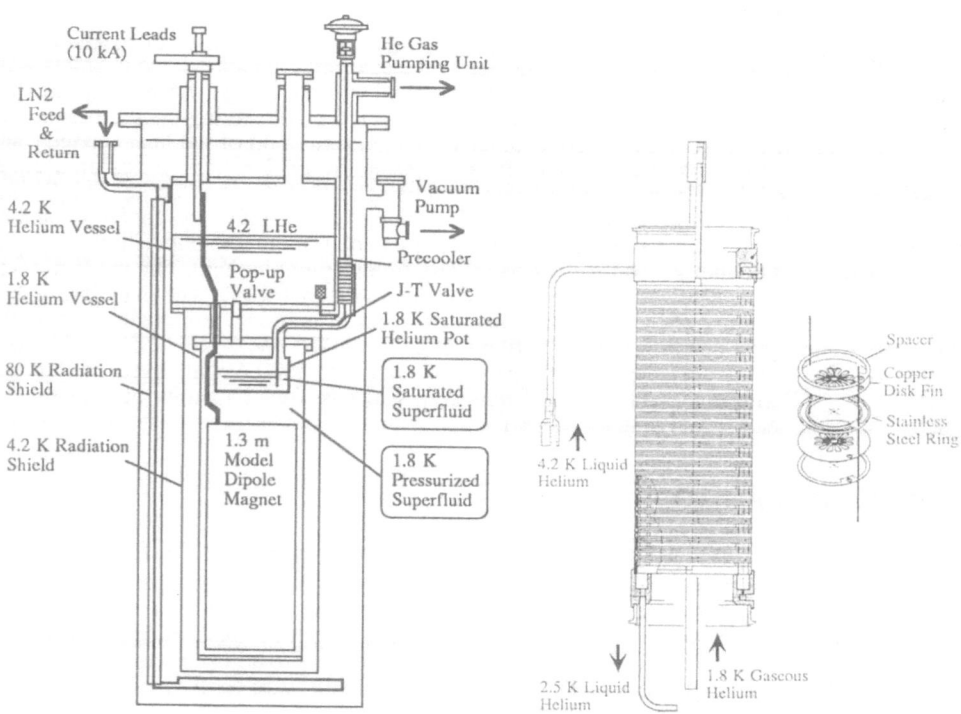

Figure 1. 1.8 K cryostat.   **Figure 2.** Pre-cooler

spacers were nickel-brazed in a vacuum furnace at 1000 °C. A schematic diagram of the precooler is shown in Fig. 2. The saturated 1.8 K helium pot is a donut-shaped stainless steel vessel. It has 333 copper tubes brazed at the bottom to increase the heat exchanging surface. The pot contains 12 liters of saturated helium under 17 hPa, and cools the magnet. Boil-off helium is evacuated by the pumping unit. The unit consists of two mechanical boosters and two rotary pumps. The pumping rate of the unit is 700 m³/Hr at 17 hPa. The pressure of the 1.8 K saturated helium pot is monitored by an absolute pressure manometer.

By feeding liquid nitrogen into the 1.8 K saturated helium pot, the magnet was first cooled to 80 K with pressurized helium gas of about 500 hPa. The magnet was then cooled to 4.2 K by feeding the liquid helium into the 1.8 K vessel through the "pop-up" valves. The cool-down to 1.8 K begins by opening the J-T valve and pumping boil-off helium gas from the 1.8 K saturated helium pot. The temperature in the 1.8 K vessel was monitored with two carbon glass resistance temperature sensors set at the top and the bottom of the magnet. The cooling power of the heat exchangers was 7.5 W at 1.8 K, and 3.5 W at 1.5 K. The magnet was cooled from 4.2 K to 1.8 K for 5 hours. The heat leakage of the cryostat was 1.6 W. The lowest temperature attained was 1.45 K.

AC LOSS MEASUREMENT

After the magnet was cooled to 1.8 K, the J-T valve was closed and superfluid helium in the 1.8 K saturated helium pot was evaporated using a heater. The heat dissipated by the AC loss of the magnet raised the temperature of the superfluid helium. The dissipated heat was evaluated by the integration of a product of the heat capacity and the mass of superfluid helium;

$$Q = \int_{T_1}^{T_2} m \, C_p \, dT \quad ,$$

where Q is heat, m mass, C_p heat capacity, T temperature, T_1 initial temperature, and T_2 final temperature of helium during the ramp cycle. Prior to the AC loss measurement, the measurement system was calibrated by a 50 Ω heater attached at the bottom of the magnet. The heat leakage of the cryostat was measured several times during the AC loss measurement and was subtracted from the measured heat.

The magnet was ramped-up from 500 to 5000 A, and after 5 seconds ramped-down to 500 A. Another cycle started after 5 seconds. The ramp rates were varied from 70 to 400 A/s. As an example, time variation of the temperature measured by the carbon glass resistance temperature sensor and current pattern at a ramp rate of 294 A/s are shown in Fig. 3. Change of temperature rise rate can be clearly observed. AC losses were measured for three magnets (KEK#3, KEK#4 and KEK#6). Magnets KEK#4 and KEK#6 are Type 1 and KEK#3 is Type 2. Figure 4 shows a result of the calorimetric AC loss measurement as a function of ramp rate. Solid lines are linear fits for measured data. The coefficients of the linear fits are 0.08, 0.22, and 0.33 J/(A/s), and intercepts at zero ramp rate are 127, 147, and 122 J/cycle for KEK#3, KEK#4 and KEK#6, respectively.

CONCLUSIONS

A 1.8 K cryostat was made and used successfully to cool the model dipole magnets. The cryostat had a cooling power of 7.5 W at 1.8 K with heat leakage of 1.6 W. The AC losses of the magnet were measured by a calorimetric method. The ramp rate ranged from 70 to 400 A/s. Two types of magnet were measured: Type 1 made from the 6 μm NbTi filament cables and Type 2 from the 2.5 μm filament cables. Results show that the Type 2 magnet had smaller AC losses and a smaller coefficient of ramp rate dependence than did the Type 1 magnet.

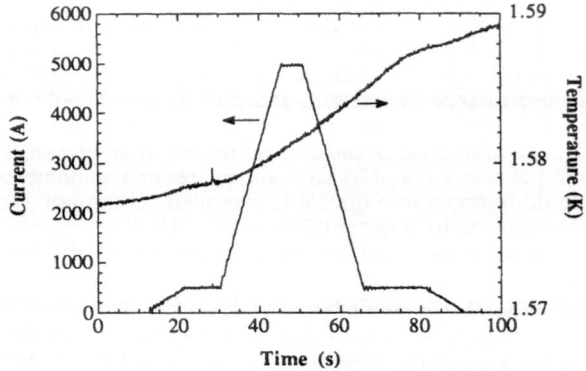

Figure 3. Time variation of transport current and temperature.

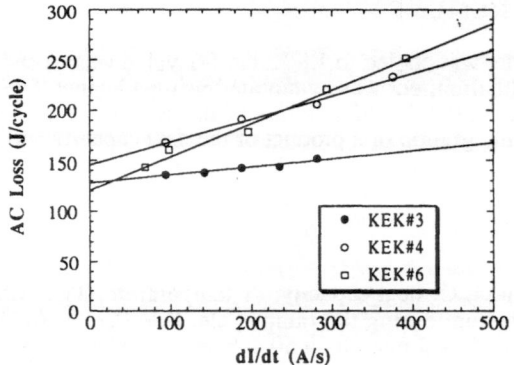

Figure 4. AC losses in a ramp cycle.

ACKNOWLEDGMENTS

The authors express their gratitude to Professors T. Shintomi, H. Hirabayashi and Y. Kimura for their continuous support and encouragement. They also wish to thank Dr. A. Sato for useful discussions.

REFERENCES

1. K. Hosoyama, "Fabrication and Test of a 5-cm Aperture, 1-m Long SSC Collider Dipole Magnet," Proc. Third International Symposium on the Super Collider, 1992.

THE LIQUID HELIUM THERMOSYPHON FOR THE GEM DETECTOR MAGNET

R. P. Warren

Lawrence Livermore National Laboratory*
P. O. Box 808
Livermore, CA 94550

INTRODUCTION

The GEM detector magnet, a horizontal solenoid 19.5 m in diameter and wound with a niobium-titanium cable in conduit, will be located with it's axis 19.5 m below grade. The conductor is wound on the inside of an aluminum bobbin which is cooled by liquid helium which flows by natural convection in a thermosyphon loop from a large storage dewar located at the ground surface. The function of the thermosyphon system is to absorb the environmental heat load as well as any internally generated heat. In the first category is included that heat which is transfered to the magnet by way of the mechanical supports, the insulation and the current leads. The internally generated heat includes the resistive heating within the normally conducting conductor splices and the inductive heating of the bobbin during current transients. Though similar systems[1] have been employed elsewhere, there are some unique aspects to the present design. By taking advantage of the large vertical head available, the parallel heat exchanger passes within the magnet remain sub-cooled, thus insuring single phase coolant within the magnet. It is believed that this will be the first instance of such a large vertical head being used to this advantage in a helium system.

SYSTEM DESCRIPTION

A flow schematic for the entire helium system, of which the thermosyphon is part, is shown on Figure 1. The refrigeration of the current leads is shown on Figure 1 as being provided by a separate forced flow circuit which also provides supercritical coolant to the conductor conduit to enhance conductor stability. An option to provide the current lead refrigeration with the thermosyphon is presently under consideration.

Because of the need for an extreme level of reliability, the system has been configured to minimize both the total number of actively controlled cold valves and the number of active control elements in the experimental hall.

---

* This work was performed under the auspices of the U.S. Department of Energy by Lawrence Livermore National Laboratory under Contract No. W-7405-Eng-48.

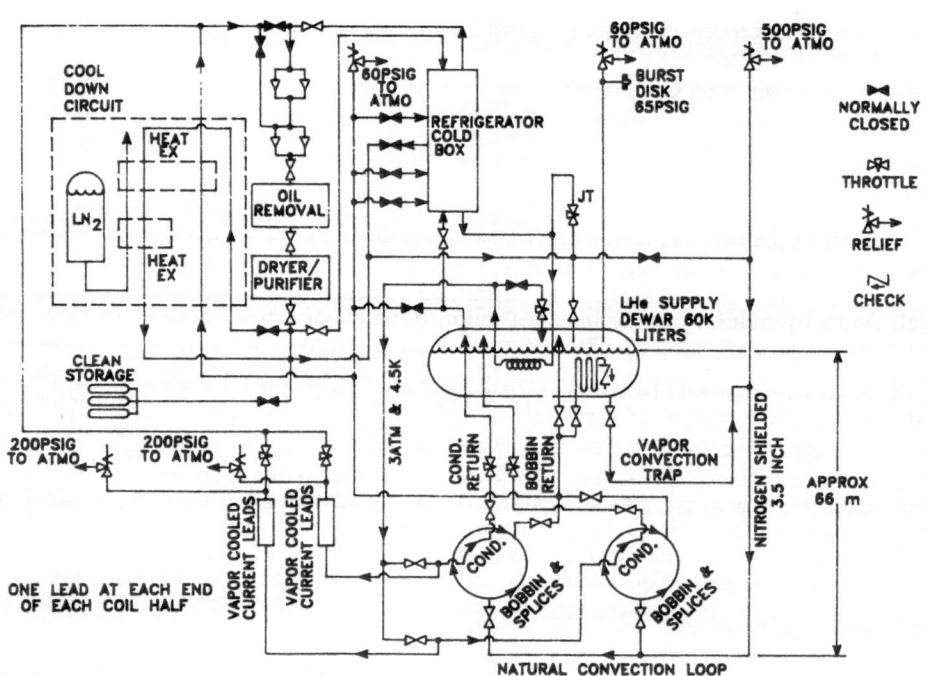

Figure 1. GEM Magnet, Helium Flow Schematic.

HEAT LOADS

The steady heat loads are estimated on what we believe is a conservative basis at 0.8 kW. The inductive heating of the 7.6 cm thick bobbin wall has been analyzed by Radovinsky[2]. The bobbin wall is segmented in both the axial and circumferential directions to reduce the heating during current transients, and as a result, the heat generation on charging which is done at a slow rate (eight hours) is low. The heating on emergency discharge which occurs exponentially with a time constant of 100 s is another matter, however. The bobbin is assembled from twenty four spools, each with a 1.2m axial dimension, which are electrically, and therefore, to some extent thermally isolated from each other. During emergency discharge, Radovinsky calculates a peak heating in the end spools of 1.8 kW at 9 s which is followed with an exponential decay.

THERMOSYPHON THERMODYNAMICS

There is a slight warming of the liquid helium as it descends from the storage reservoir at the surface[3] and then a much more significant warming as it is heated in the bobbin heat exchanger and then finally some vaporization as the hydrostatic head is reduced at the upper end of the return leg. The heating on the return side of the thermosyphon gives rise to the bouyancy which drives the natural convection flow. For 1.3 atmospheres pressure in the surface reservoir, the pressure at the bottom of the thermosyphon is 2.1 atm. and 1.8 atm. at the top of the magnet. With a worst case storage temperature of 4.5 K at the surface (saturated at 1.3 atmospheres), the temperature at the bottom of the thermosyphon is 4.6 K and if we specify a maximum temperature of 4.75 K at the top of the magnet, for the assumed heat load, a flow of about 0.7 kg/s is needed. The 4.75 K is well below the current sharing temperature of 7.5 K. The quality of the return flow at the surface is about 7%.

PIPING

The supply side, or down comer, piping is liquid nitrogen shielded to minimize the inlet temperature to the magnet. The supply and return piping is sized based on a balance between the frictional head loss and the differential of the hydrostatic head between the two the two sides of the flow loop. On this basis, for a mass flow of about 700 g/s, 3.5 inch pipe is chosen for the supply side and 4 inch pipe on the return. For the bobbin heat exchanger, a single 3/4 inch tube on each bobbin segment looks about right when considering the frictional pressure drop, however, a somewhat larger size, or perhaps multiple tubes, would be desirable from the standpoint of increasing the bobbin heat capacity to stabilize the temperature during rapid turn down of the magnet current. The liquid helium contained within a one inch ID heat exchanger tube has twice the heat capacity of the 3 inch thick aluminum bobbin at 4.5 K. These sizes appear to be more than adequate for the cool down flow which is limited by compressor and nitrogen heat exchanger capacity.

References

[1]Lottin, J.T. and Duthil, R., Aleph Solenoid Cryogenic System, Proc. ICEC 12, Butterworth, 1988.
[2]Radovinsky, A., Joule Heating of GEM Detector Bobbins, MIT-GEM-EM-014, 16 Dec 93.
[3]Because the energy equation requires the conversation of the sum of the enthalpy and elevation, a warming results from a decrease in elevation.

THE THERMAL BEHAVIOR OF AN SSC MAGNET IN AN UNCONDITIONED
OUTDOOR ENVIRONMENT

Randy K. Pletzer

Magnet Development Group, Analysis Section
Superconducting Super Collider Laboratory*
2550 Beckleymeade Ave., Ste. 125
Dallas, TX 75237-3997

SUMMARY

The superconducting coils of the Superconducting Super Collider (SSC) magnets are subjected to potentially harmful warm temperatures during transportation and storage operations in certain outdoor environments. A realistic, harmful operating environment to which a magnet could be subjected at some point in the SSC construction would occur in the scenario in which an uncovered magnet remains outdoors during a hot, bright summer day, without environmental conditioning or protection for an extended period. Such a scenario could occur during a transportation truck breakdown, or if a magnet were left outside awaiting entry to the construction shaft over a weekend, for example. The magnet would be subjected to elevated ambient temperatures and solar radiation which could heat the magnet and raise the coil and bus assembly at some location to a temperature greater than the 32.2 °C (90 °F), at which the B-stage epoxy used in the coil ground plane insulation is thought to weaken. A thermal analysis of an Fermi National Accelerator Laboratory (FNAL) 50mm Dipole style magnet under such a hypothetical operating condition was conducted to predict the maximum coil temperature rise and rate of rise during such a scenario.

The dipole magnet thermal model was reconfigured to account for the lack of cryogens and presence of atmospheric pressure dry Nitrogen gas in the cryostat for its transportation configuration. The magnet model was subjected to a 48 hour continuous outdoor soak scenario, running from midnight at the beginning of the 1st day to midnight at the end of the 2nd day. The initial temperature for the entire magnet was 26 °C (78.8 °F). A daily schedule of ambient air temperature and of solar heat flux was imposed on the magnet. The maximum ambient air temperature to which the magnet was exposed was 37.8 °C (100 °F). The heat flux schedule was generated according to ASHRAE (American Society of Heating, Refrigeration, and Air Conditioning Engineers) standard methods for June 21 at the SSC N-15 location, for a clear sky. The results of this analysis showed that the maximum coil temperature rise of 4.4 °C over 48 hours, for an average of 0.092 °C/hour, occurred at the

---

* Operated by the Universities Research Association, Inc., for the U. S. Department of Energy under Contract No. DE-AC35-89ER40486.

coil ends on the outer coil. The maximum coil temperature after 48 hours was 30.4 °C (86.7 °F), which is 7.4 °C (13.3 °F) less than the maximum ambient temperature of 37.8 °C (100 °F) to which the magnet is exposed, and 1.8 °C (3.3 °F) less than the coil temperature limit of 32.2 °C (90 °F). Based on the average coil temperature rise rate, it would require an additional 19.6 hours for the maximum coil temperature to reach the coil temperature limit of 32.2 °C (90 °F).

Based on this result, the coils seem to be adequately buffered against exposure to undesirable temperatures in the event of an exposure not exceeding 67.6 hours to an unconditioned outdoor environment. Since the predicted coil temperature rise rate is low, it is recommended to consider an actual outdoor soak test of a dipole magnet for the purpose of measuring coil temperatures and environmental conditions, to verify the analysis results.

ANALYSIS

The 50mm FNAL Dipole thermal model[1] was used as the basis for the study. In the magnet's transportation configuration, the cryogen lines are dry and capped at the ends, the interconnect shield bridges are not present, the ends are covered with the shipping bell, and the magnet cryostat is filled with dry Nitrogen gas at atmospheric pressure. The standard Dipole thermal model was modified to account for these differences.

A magnet in an outdoor environment will be heated at the vacuum vessel surface through natural convection from ambient air which is initially at a higher temperature than the vacuum vessel surface and by the total solar thermal radiation which falls on the unshaded magnet exterior surface. A worst case scenario for outdoor heating of the magnet was developed. The magnet was assumed to be outdoors at the SSC N-15 site location during a clear, hot summer day typical for this climatic region. The daily ambient air temperature profile to which the magnet was exposed is shown in Figure 1 (results section).

The total solar heat flux incident on an unshaded surface is comprised of three components. The direct normal solar flux is that which is received by an unshaded surface directly from the solar disk. The diffuse solar flux is the direct solar flux which is scattered and reflected by the atmosphere and reaches the exposed surface indirectly from all portions of the sky visible to the surface. Reflected solar flux is the direct and diffuse solar flux which is reflected from the ground and other surfaces and reaches the exposed surface indirectly from these other surfaces. The worst case daily total solar heat flux incident on the magnet was calculated using the ASHRAE method[2] for a horizontal magnet oriented with its long axis North to South on June 21 at the N-15 site, for a clear sky condition. The surface over which the magnet rested was assumed to be light colored concrete with a nominal solar reflectivity of 0.3[3]. Since the ASHRAE method applies to flat surfaces, the circular magnet vacuum vessel exterior surface was treated as a 12 sided regular polygon, and the solar heat flux incident on each facet calculated separately and averaged over the circumference to obtain the hourly average. For purposes of the diffuse and reflected solar components, the possible presence of other surfaces near the magnets was ignored, and the magnet was assumed to be exposed to the sky and ground only. The resulting total solar heat flux generated for the assumed worst case magnet outdoor environment is also shown in Figure 1.

In the outdoor environment, the magnet will exchange heat with it's surroundings via convection with the ambient air and thermal radiation to surrounding surfaces. In the worst case, the magnet would not have the benefit of being cooled by a breeze, so that it was assumed that the magnet resides in still, windless air and exchanges heat with the atmosphere by free, rather than forced, convection. The solar thermal radiation is treated as a net heat load on the magnet surface, and is equal to the incident total solar heat flux multiplied by the solar absorptivity of the magnet surface. A solar absorptivity of 0.9 was used as a worst case for gloss red paint on metal surface.[3] However the magnet surface will still exchange heat by thermal radiation with surrounding surfaces such as the ground and sky, since these surfaces will emit and absorb thermal radiation apart from the solar heat flux. The worst case (highest) apparent sky temperature occurs for the case of blackbody sky radiation, in which case the sky temperature is equal to the local atmospheric ambient.[4] The temperature of the ground surface, assumed to be concrete, is a complex function of absorbed solar flux, ground conductivity and initial temperature, and heat transfer coefficient with the ambient air. As a worst case, approximation the ground temperature was assumed to follow the ambient air temperature profile.

The magnet thermal response to the estimated hot soak environment was simulated over a 48 hour period. The magnet was assumed to be exposed to the outdoor environment starting at midnight of the 1st day. An initial temperature of 26 °C (79 °F) was used to represent the magnet coming from a benign, conditioned environment such as a building interior. A 48 hour soak would adequately characterize the daily coil temperature rise and reveal any hysteresis in the day to day thermal behavior of the coil temperatures.

RESULTS

The predicted temperatures of the exposed magnet over the duration of the outdoor soak scenario are shown in Figure 1. The vacuum vessel, exposed directly to the ambient air and solar radiation, tracks the ambient temperature until the sun rises the 1st day, at which point it heats rapidly to a maximum temperature of 75.8 °C (168 °F) between 12 P.M. each day. The results show that the vacuum vessel peak daily temperature does not vary significantly during the 2nd day. Figure 1 shows that the 80K shield peak temperature is 62.9 °C (145 °F) on the 2nd day, and that the 20K shield peak temperature is 55.7 °C (132 °F) also on the 2nd day. The 20K shield peak temperature does show some hysteresis from the 1st to the 2nd day. The results also show the advance of the heat input by the shift in the peak temperature times as one goes inward from the vacuum vessel to the shields and finally to the Cold Mass. The large difference in the magnet cryostat components daily peak temperatures and the daily peak ambient temperature indicates that the component temperatures are primarily driven by the solar heat input rather than heat input from the ambient air.

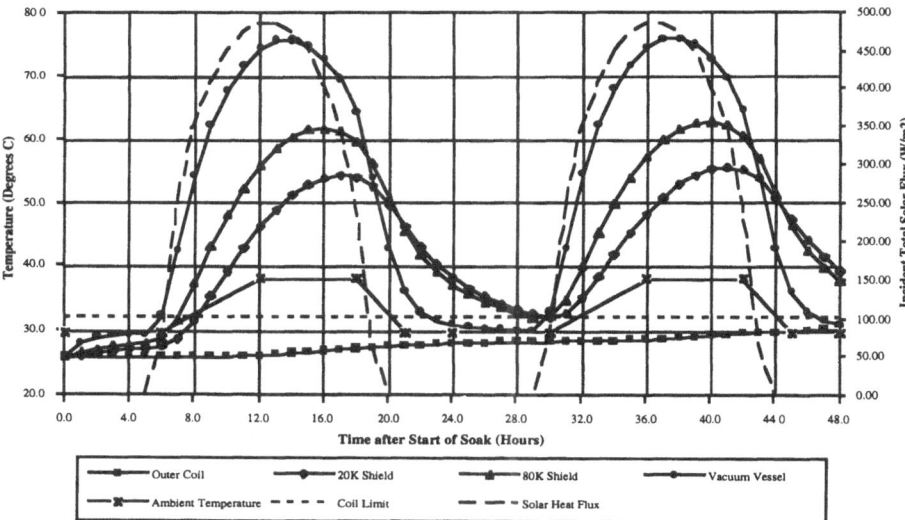

Figure 1. Dipole Magnet 48 Hour Unshaded Outdoor Hot Soak Results.

The maximum coil temperature occurs at the lead and return ends of the magnet. It shows a steady, nearly linear rise at a rate of 0.15 °C/hour during a period of duration nearly equal the daylight period of 14 hours, but offset forward from the daylight period by about 4 hours. The average maximum coil temperature rise rate over the 48 hours was 0.092 °C/hour, resulting in a maximum coil temperature of 30.4 °C (86.7 °F). The maximum coil temperature falls below the nominal coil temperature limit of 32.2 °C (90 °F). Based on this 48 hour average rise rate, the coil temperature would not reach the limit for an additional 19.6 hours. This would be till the end of daylight on the 3rd day of soak, based on a June 21 solar heat profile. The reason that the Cold Mass responds very slowly to the exterior conditions is due primarily to its very large thermal capacitance compared to that of the cryostat components.

Although the coils appear to be protected against any excursions above the nominal high temperature limit of 32.2 °C during the 48 hour outdoor unshaded hot soak scenario, several other failure modes due to high temperature may also exist. Exceeding the char/ignition point temperature of the MLI PET layers could create a fire hazard, and excess temperatures could be detrimental to the structural integrity of the support post tubes. It is recommended that a review of the maximum temperature limits of these and other non-metallic magnet components be conducted to determine of any additional failure modes due to high temperatures exist.

If a scrap dipole magnet was or becomes available, an outdoor soak test could be conducted to experimentally determine the coil region temperature and verify the analytical predictions. Such a test would involve the measurement of temperatures of the coils, shields, vacuum vessel, and ambient air, and the measurement of direct and diffuse solar flux and wind speed over the duration of the test. In addition, or as an alternative, additional analysis may be performed to improve the modeling. There was insufficient time to remove the shields from the interconnect region, and the ground temperature behavior and free convection effects inside the cryostat need to be more adequately modeled. An immediate solution would be to require that all transported magnets be equipped with an independently powered air conditioning unit and that all magnets which must be placed outdoors for any reason be completely shaded. Otherwise, it is recommended that the magnets be tested in outdoor conditions to measure their response to the environment, so that adequate handling requirements may be drafted.

ACKNOWLEDGMENTS

The author wishes to thank Ed Daly, formerly of the SSCL Magnet Division and presently at Westinghouse Science and Technology Center, for his contributions to this study. Mr. Daly initiated the study and performed the initial analytical work until his departure from the SSCL.

REFERENCES

1. Pletzer, Randy K., "Results of Revised Analysis of 50mm Dipole Magnet With and Without 4K and 20K MLI Blankets," SSCL Magnet Engineering Memorandum 92 5830 H M ****, dated 3-23-92.
2. ASHRAE Handbook of Fundamentals 1989, Chapter 26 "Fenestration," p.26.1.
3. Seigel, Robert and Howell, John: "Thermal Radiation Heat Transfer," 2nd edition, McGraw-Hill Book Co., 1981, p. 836.
4. Bliss, Raymond W., "Atmospheric Radiation Near the Surface of the Ground: A Summary for Engineers," no publication source available.

COLLIDER QUADRUPOLE MAGNET (CQM) OPTICAL ALIGNMENT WINDOW HEAT LEAK PREDICTION

Randy K. Pletzer

Magnet Development Group, Analysis Section
Superconducting Super Collider Laboratory*
2550 Beckleymeade Ave., Ste. 125
Dallas, TX 75237

SUMMARY

The additional static heat leak to the cryogenic refrigeration systems of an SSC Collider Quadrupole Magnet (CQM) due to the Optical Alignment Window apparatus was calculated for three configurations: (1) panes in the 20K shield, 80K shield, and vacuum vessel, (2) empty aperture in the 20K shield and panes in the 80K shield and vacuum vessel, and (3) empty apertures in both the 20K and 80K shields and a pane in the vacuum vessel only. In addition, two cases for pane infrared optical coating configuration were simulated: (1) coating on both sides of a pane and (2) coating on the top side of a pane only. For a four windows per magnet design, the results of the analysis show that the 4K and the 20K total static heat leak budgets will be exceeded only for the empty apertures in both of the shields configuration. This result applies for a single side or double sided coated pane. The existing baseline configuration (no 20K shield pane, pane in 80K shield, pane top side coated) is sufficient to meet the CQM total static heat leak budget.

There is not enough difference between the thermal performance of the windows with both sides or only one side of the panes coated to affect whether or not the heat leak exceeds budget due to the windows, therefore it will not necessary to coat both sides of the panes to meet the existing CQM static heat leak budget. Since the 4K static heat leak for the configuration without a pane in the 80K shield grossly exceeds budget, the optical window assembly will have to include a pane in the 80K shield.

ANALYSIS

The CQM Optical Alignment Window configuration consists of a polished stainless steel alignment mirror affixed to the Cold Mass, above which is located a single circular aperture in each of the 20K shield, 80K shield, and vacuum vessel cryostat components. The apertures allow the mirror to be viewed by the alignment instrument outside of the vacuum vessel. The baseline design calls for a total of four such windows in each CQM. The shields do not have to contain a pane to meet any vacuum quality or integrity requirements, thus a

---

*Operated by the Universities Research Association, Inc., for the U.S. Department of Energy under Contract No. DE-AC35-89ER40486.

simple empty aperture cut through the shields would suffice, unless required to minimize heat leak. The current baseline design [1] specifies a pane in the 80K shield but not in the 20K shield. Where a pane is present, it will consist of ordinary optical quartz window glass with an infrared reflecting, visibly transmitting optical coating of indium-tin oxide (ITO) applied over the entire top surface of the pane. The total hemispherical emissivity of the coated surface of the glass used by the optics consultant, Precision Optics, in their analysis is 0.15, and of the uncoated surface is 0.5. These emissivity values were used in this heat leak analysis. The vacuum vessel aperture is 4 cm diameter, the shield apertures are 5 cm diameter, and the Cold Mass mirror is 4 cm diameter. The cryostat configuration for the LBL 40mm Quadrupole was used as the magnet baseline for this analysis.

A diagram of the thermal model of a single window is shown in Figure 1. The additional heat leak due to the presence of the optical alignment window is due primarily to the increase in thermal radiation heat transfer from glass panes in a higher temperature cryostat component to the adjacent lower temperature cryostat components and pane, and from the direct thermal radiation between non-adjacent components (such as the 80K shield and Cold Mass) possible where an empty aperture permits such surfaces to see each other. The thermal radiation between adjacent components, such as the vacuum vessel and the 80K shield, increases over that for a configuration without windows since the emissivity of the coated or uncoated glass is higher than that of the metal surfaces (aluminum or stainless steel) which it replaces at an aperture. Ordinary optical window glass has a narrow band of high (0.9) transmittance from 0.2 μm to 3 μm which covers the visible and near infrared bands. The transmittance of glass in the far-infrared beyond 3 μm, where 99.99% of the total thermal radiation energy for bodies below 310 K is emitted, is essentially zero, so that the glass panes simply appear as higher emissivity patches on the shields for purposes of thermal radiation modeling. However, where an empty aperture appears, thermal radiation from a higher temperature cryostat component beyond the aperture is allowed to pass through the hole and fall on a surface which would normally be protected by the adjacent apertureless surface, in the windowless configuration.

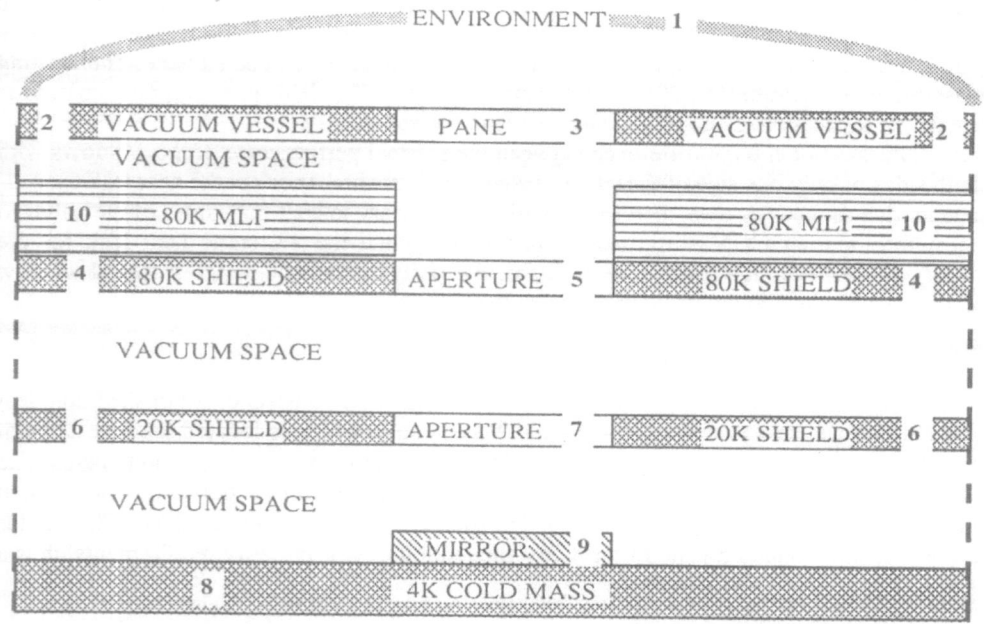

Figure 1. CQM Optical Alignment Window Thermal Model Configuration

The increase in heat leak due to the presence of the window from residual gas conduction is insignificant above 20K since residual gas conduction at collider vacuum design levels (1E-6 torr) is insignificant compared to radiation. The increase in the 4K total static heat leak due to residual gas conduction between the 80K shield and the Cold Mass through the empty aperture in the 20K shield was ignored in this analysis since it is also not significant compared to the thermal radiation.

The additional heat leak for each of the three cryogenic systems due to the optical alignment window from thermal radiation was calculated on a per window and per CQM basis for Collider operational conditions using an Excel spreadsheet. The Quadrupole Thermal Model was not needed for the calculation. The temperature of the vacuum vessel, 80K shield, 20K shield, and 4K Cold Mass were fixed at 310, 84, 20, and 4.2 K for purposes of this analysis, based on previous results from the full Quadrupole Thermal Model [2]. The temperature of the glass panes, where present, was assumed to be the same as that of the cryostat component in which they were emplaced, based on results of the pane thermal analysis [1] which showed that the maximum difference in pane to structure temperature would only be 6 K where the ITO coating was used.

RESULTS

The predicted total static heat leaks per CQM with 4 windows (not per window) due to the Optical Alignment Windows are shown in Table 1. Table 1 pertains to the baseline case of panes coated on the top side only, except for the entries showing the heat leak due to the windows for the both sides coated case as identified. Table 1 displays the design heat leak for each system both with and without the windows and compares the design with the budget heat leak. The comparison values apply only to the top side coated case only.

The results presented in Table 1 for the CQM with four Windows show that none of the heat leaks will exceed budget for either the baseline configuration (no pane in 20K shield) or the configuration with a pane in both shields, for either the top or both sides of the panes coated. There is no significant difference in the design to budget heat leak margin between these two acceptable configurations. The heat leak for both pane sides coated is always less than that for the top pane side only coated as expected. There is no need for further consideration of a pane in the 20K shield to meet heat leak requirements, as the current baseline is sufficient. In fact, the number of windows per magnet could be increased. In addition, the ITO coating on the top surface of the panes is also sufficient, and coating of both sides of the panes will not be necessary to meet static heat leak requirements for either of the acceptable configurations.

Table 1 shows that the 20K heat leak and the 4K heat leak will be significantly, exceeded for the configuration with no pane in both of the shields, for either the top or both sides of the vacuum vessel pane coated. The large increase in 4K heat leak is due entirely to direct exposure to the 310 K vacuum vessel and pane when the 80K shield pane is removed, as a consequence of the absolute temperature to the 4th power nature of thermal radiation heat transfer. The heat leak per window (not shown) also exceeds the 4K total static budget significantly, so that a reduction in the number of windows per CQM would not allow the removal of the 80K shield pane from the design. The configuration without an 80K shield pane is unacceptable since it exceeds the 4K heat leak requirements and should not be considered further.

The small net decrease in 80K heat leak for the configuration without an 80K pane is attributable to several phenomena. The increase in 80K heat leak for the other two configurations was entirely due to the higher emissivity of the pane than the shield which it replaced, coupled with the fact that the 80K shield is covered with MLI. The increase in 80K heat leak with an 80K pane was entirely localized in the pane, rather than over the entire shield surface, since it was covered in MLI. In contrast, the increase in 20K and 4K heat leak per window is not localized to the 20K shield pane or 4K Cold Mass and mirror since all of these surfaces can exchange heat with the higher temperature surfaces above, and not just the window components. When the 80K pane is not present, there is no 80K surface to absorb thermal radiation from the vacuum vessel wall and pane, and this heat is transferred entirely to the 20K and 4K surfaces below, decreasing the 80K heat leak in comparison with no window present.

Table 1. Predicted CQM Heat Leaks with Optical Alignment Windows

Heat Leak Parameter [All in Watts except Relative Difference]	Pane in 20K and 80K Shields	Baseline: No Pane in 20K Shield	No Pane in 20K or 80K Shield
80K Total Static Design Heat Leak w/o Windows	13.705	13.705	13.705
80K Static Heat Leak due to Windows, Top Coated	0.866	0.854	-0.016
80K Static Heat Leak due to Windows, Both Coated	0.721	0.711	-0.020
80K Total Static Design Heat Leak With Windows	14.571	14.559	13.689
80K Total Static Budget Heat Leak	14.970	14.970	14.970
80K Total Static Heat Leak Design - Budget	-0.399	-0.412	-1.281
Relative Difference, Design vs. Budget	-2.7%	-2.7%	-8.6%
80K Heat Leak Budget Status	**UNDER**	**UNDER**	**UNDER**
20K Total Static Design Heat Leak w/o Windows	1.068	1.068	1.068
20K Static Heat Leak due to Windows, Top Coated	0.012	0.006	1.794
20K Static Heat Leak due to Windows, Both Coated	0.006	0.002	1.633
20K Total Static Design Heat Leak With Windows	1.080	1.074	2.862
20K Total Static Budget Heat Leak	2.081	2.081	2.081
20K Total Static Heat Leak Design - Budget	-1.001	-1.007	0.781
Relative Difference, Design vs. Budget	-48.1%	-48.4%	37.5%
20K Heat Leak Budget Status	**UNDER**	**UNDER**	**OVER**
4K Total Static Design Heat Leak w/o Windows	0.086	0.086	0.086
4K Static Heat Leak due to Windows, Top Coated	0.000	0.018	2.001
4K Static Heat Leak due to Windows, Both Coated	0.000	0.014	1.804
4K Total Static Design Heat Leak With Windows	0.086	0.104	2.087
4K Total Static Budget Heat Leak	0.232	0.232	0.232
4K Total Static Heat Leak Design - Budget	-0.146	-0.128	1.855
Relative Difference, Design vs. Budget	-62.9%	-55.0%	798.9%
4K Heat Leak Budget Status	**UNDER**	**UNDER**	**OVER**

In conclusion, the results of the thermal analysis showed that the baseline configuration of a pane in the 80K shield and no pane in the 20K shield with top side of the pane only coated would not introduce sufficient additional heat leak into the cryostat to exceed the present CQM static heat leak budgets.

ACKNOWLEDGMENTS

The author wishes to thank Charles Corbett, Bob Viola, and Sol Shapiro of the Magnet Division for their help in providing the configuration data upon which this analysis depended.

REFERENCES

1. "Positioning Monitoring Requirements for the Cold Mass Quadrupole Magnet Subassembly", a Design Report prepared for URA, Precision Optics, Inc., 10-August-92.
2. Pletzer, Randy K., "Results of Revised Analysis of 40mm Quadrupole Magnet With and Without 4K and 20K MLI Blankets.", SSC Magnet Division memorandum M E 04, 92 5830 H M ****, 23-March-92.

SUMMARY OF DIPOLE FIELD ANGLE MEASUREMENTS ON 50MM-APERTURE SSC COLLIDER DIPOLE MAGNET PROTOTYPES

J. Marks,[1] M. Bleadon,[2] J. DiMarco,[1] M. Kuchnir,[2] J. Kuzminski,[1]
T. Ogitsu,[1] Ed. E. Schmidt,[2] Y.Yu,[1,3] and H. Zheng[1]

[1]Superconducting Super Collider Laboratory,* Dallas, TX 75237
[2]Fermi National Accelerator Laboratory, Batavia, IL 60510
[3]Institute of Electrical Engineering, Bejing 100080, China

INTRODUCTION

At several stages in the production of the SSC collider dipole magnets and their final installation (as well as for beam orbit calculations) the magnetic field angle needs to be known. A simple device using a permanent magnet which aligns itself with the magnetic field (the Vertical Field Angle Probe, FAP) had been developed at FNAL to survey the direction of the magnetic dipole field with respect to the vertical (as determined by gravity) along the magnet axis.[1] The determination of the dipole field angle was part of the field quality characterization of a series of thirteen full-length 50mm-aperture SSC Collider Dipole Magnet Prototypes (DCA311-DCA323) which were built for R&D purposes at FNAL (for design specification).[5,6]

Measurements with the first developed FAP system (FAP1) were performed on a regular basis through several stages of the magnet production process with the intention of fabrication quality control.[2] Part of these included measurements performed before and after cryogenic testing: these data are summarized here. The performance of a second system (FAP2) with an improved probe and data aquisition system was tested on part of the DCA series as well. This paper includes a presentation of time stability, noise and angular resolution data of this second probe. Another alternative instrument to determine the dipole field angle is the "mole" rotating coil system developed at BNL used mainly to measure the multipole components of the magnetic field[4]. In the case of magnet DCA320, a comparison is made between the field angle as determined by the mole and those determined by both of the FAPs.

SETUP AND MEASUREMENTS

The vertical field angle measurement systems consist of a small permanent magnet which is housed in a jewelled gimbal system to which an electrolytic bubble level sensor is connected. The bubble level sensor determines the angle of the permanent magnet with respect to the vertical. The probe is positioned inside the beam tube along the magnet axis by a set of interconnecting rods. Spacers attached to the rods keep the probe in the center of the beam tube. A detailed description of the FAP system can be found in reference.[1,2]

For the FAP2 system, the probe and data aquisiton were redesigned. The main modifications of the probe were placing the amplifier close to the actual measuring device (inside the G10 cover). This should result in lower electronics noise (better angular resolution) and better signal stability. The heat dissipation from the amplifier should also maintain a more constant temperature on the electrolytic bubble level sensor. In addition, improve-ments on the balance of the gimbal system should give faster stabilization of the signals.

The dipole field angle is measured at various locations (z positions) along the magnet axis (every 0.0762 m – resulting in 180 measurements for the 15-m prototypes tested) once with the probe head pointing to the lead end of the magnet and then again with the head pointing towards the

* Operated by the Universities Research Association, Inc., for the U. S. Department of Energy under Contract No. DE-AC35-89ER40486.

non-lead end (both measurments are taken so the zero level can later be removed). The magnet current was 8 A which results in a magnetic dipole field of the order of 0.008 Tesla.

The "FA1 mole" with an air motor and a 1m coil was used to measure the field angle for comparison with the FAPs. Mole measurements were made every meter (since the mole integrates the field over its 1m coil length), but only in one direction axially; hence the zero level of the mole is a matter of calibration. The mole data in this paper were taken at 10 A with the same polarity of magnet current as for the FAPs.

DATA PROCESSING AND RESULTS

We let $\vartheta^+(z)$ represent the measured dipole angle as a function of the z position in the case where the probe head points towards the lead end of the magnet, and $\vartheta^-(z)$ the case where the probe head points to the non-lead end. Since both of these axial scans contain mechanical and electrical contributions to the zero level, $\vartheta^+(z)$ and $\vartheta^-(z)$ must be combined according to

$$\vartheta\,(z) = 0.5 \cdot (\,\vartheta^+(z) - \vartheta^-(z)\,) \tag{1}$$

to obtain the true dipole field angle $\vartheta(z)$. This method cancels constant effects on the zero level, but still contains contributions from iron yoke magnetization[3] because the FAP system operates only with one polarity of magnet current. The offset $o(z)$ defined by

$$o\,(z) = 0.5 \cdot (\,\vartheta^+(z) + \vartheta^-(z)\,) \tag{2}$$

contains information about the zero level. The variation of $o(z)$ (noise) is a measure for the angular resolution of the measurement system. Because $o(z)$ shows, in some cases, a systematic behavior as a function of z, we evaluate this portion of the error by finding a 3rd order polynomial fit, $o^f(z)$, to the offset $o(z)$. The systematic portion is thus defined as

$$\sigma_{sys} = 0.5 \cdot |o^f(z)_{max} - o^f(z)_{min}| \quad, \tag{3}$$

where $o^f(z)_{max}$ and $o^f(z)_{min}$ are the maximal and minimal value of $o^f(z)$. The random portion is determined by the deviations of $o^f(z)$ from $o(z)$

$$\sigma_{noise} = \sqrt{\frac{1}{n-1}\sum_z [o^f(z) - o(z)]^2} \quad, \tag{4}$$

where n is the number of z positions.

Table 1. Summary of dipole field angle measurements using FAP1 before and after cryogenic testing.

Magnet	Δ [mrad]	<o(z)> [mrad]	σ_{noise} [mrad]	σ_{sys} [mrad]
DCA311	6.46±0.71	1.15	0.43	0.63
DCA312	8.33±0.46	1.00	0.38	0.30
DCA312	5.54±0.56	1.96	0.56	0.30
DCA313	8.26±0.34	0.02	0.34	0.19
DCA313	7.79±0.40	0.14	0.33	0.45
DCA314	8.35±0.46	-0.18	0.45	0.11
DCA314	7.48±0.78	-0.48	0.37	1.17
DCA315	6.11±0.39	0.33	0.39	0.52
DCA315	5.08±0.65	0.56	0.34	1.00
DCA316	7.65±0.86	-1.26	0.33	1.24
DCA316	8.73±0.62	-0.87	0.61	1.34
DCA317	7.36±0.43	0.50	0.43	0.25
DCA317	6.73±0.36	-1.07	0.32	0.40
DCA318	5.21±0.43	-1.06	0.38	0.52
DCA318	5.14±0.35	-0.65	0.34	0.55
DCA318	5.37±0.50	-2.14	0.40	0.59
DCA319	6.40±0.35	-0.61	0.35	0.35
DCA319	5.91±0.49	-1.33	0.43	1.46
DCA320	6.65±0.29	-3.05	0.29	0.09
DCA320	9.40±0.49	2.00	0.48	0.31
DCA321	7.75±0.31	-2.21	0.28	0.40
DCA321	8.39±0.63	0.31	0.41	0.70
DCA322	7.14±0.42	2.18	0.32	0.41
DCA322	8.22±0.34	2.99	0.32	0.24
DCA323	7.99±0.32	2.08	0.30	0.23
DCA323	8.98±0.36	1.14	0.34	0.41

SUMMARY OF FAP1 DATA

Table 1 summarizes field angle probe data of the 50 mm aperture magnet prototype series taken with the FAP1 system. In principle both the average dipole field angle and the difference in its maximum and minimum values,

$$\Delta = \vartheta(z)_{max} - \vartheta(z)_{min} \; , \tag{5}$$

are of interest. But because the average depends on the mounting of the magnet on the test stand we present only Δ in the summary table. The error on Δ consists of two contributions, a random part represented by σ_{noise} (col. #4 Table 1) and a systematic part defined as being

$$\delta = 0.5 \cdot (o^f(z_{max}) - o^f(z_{min})) \; , \tag{6}$$

where z_{max} and z_{min} are the z positions of the maximum and minimum field angle. Finally the error quoted in column 1 of Table 1. is obtained by adding σ_{noise} and δ in quadrature.

Changes in Δ as high as 2.79 mrad are observed in the measurements before and after cryogenic testing. These are not real changes but rather reflect differences in the magnetization of the iron yoke (discussed further by M. Kuchnir et al.[3]). The average Δ for the thirteen magnets is about 7 ± 1 mrad. This has to be compared with the collider dipole specification for allowed variation in $\vartheta(z)$ which is ± 2.5 mrad from the average.

Comparison of FAP1 and FAP2 Performance

For three magnets the dipole field angle has been measured using both systems FAP1 and FAP2. A summary directly comparing the $\vartheta(z)$ measured by these two devices is shown in Table 2. The average values of the dipole field angle agree very well. The point by point signatures are also in good agreement, though these are not presented in this paper. The errors quoted on the average are calculated by

$$\delta <\vartheta(z)> = \sqrt{\frac{\sigma^2_{noise}}{n} + \sigma^2_{sys}} \; , \tag{7}$$

where the factor $\frac{1}{\sqrt{n}}$ on σ_{noise} is gained from the statistics on the average. The point by point angular resolution of the new FAP2 system is 0.16 mrad and about a factor 2 smaller than in the case of FAP1. The systematic contribution for FAP2 measurements are a factor 2 smaller than the random variation except for magnet DCA320 where they are measured to be of the same order of magnitude as the random variation.

Table 2. Comparison of the measured dipole field angle for FAP1 and FAP2.

Magnet	FAP1		FAP2	
	$<\vartheta(z)>$ [mrad]	σ_{noise} [mrad]	$<\vartheta(z)>$ [mrad]	σ_{noise} [mrad]
DCA320	2.04±0.31	0.48	2.16±0.15	0.16
DCA322	-0.24±0.25	0.32	-0.01±0.09	0.16
DCA323	-0.96±0.23	0.30	-0.72±0.05	0.17

Stability and Reproducibility of the FAP2

In order to evaluate stability and reproducibility of the FAP2 system, seven measurements (21 z positions each) were made on magnet DCA320 in the region of z = -6.89 m to z = -5.89 m over a period of six days. For data recording the aquisition system of FAP1 was used, except for measurement #8 which was done using the complete FAP2 system. As shown in Table 3. the average dipole angle is measured to be stable within ± 0.1 mrad. The noise varies between 0.39 mrad and 0.1 mrad; a systematic contribution was not observed. The offset varies between ± 0.15 mrad, except for measurement #5 where the probe was cooled down to a temperature of about 5 °C (beginning of the measurement) in order to evaluate temperature dependence. While the average field angle is not influenced by the temperature change it seems that the offset increased. Neither in $<\vartheta(z)>$ nor in the point by point signature $\vartheta(z)$ is a systematic drift in time observed. Standard deviations among the seven measurements at each of the 21 z positions were calculated. These point by point standard deviations are bounded by 0.16 mrad and 0.65 mrad with an average of 0.35 ± 0.02 mrad.

Table 3. Stability check of the FAP2 system for magnet DCA320.

Measurement	$<\vartheta(z)>$ [mrad]	$<o(z)>$ [mrad]	σ_{noise} [mrad]
1	-2.95±0.07	-0.10	0.30
2	-2.79±0.08	-0.15	0.36
3	-2.86±0.09	-0.04	0.39
4	-2.82±0.03	0.15	0.14
5	-2.77±0.05	0.40	0.22
6	-2.96±0.04	0.01	0.19
7	-2.85±0.02	-0.09	0.10
8	-2.92±0.02	-0.14	0.08

Comparison of FAP and Mole Measurements

The mole system determines the magnitudes and phases of multipole field components (including the phase of the dipole). Since gravity sensors are also mounted inside the mole, the phase of the dipole with respect to gravity can be measured. To accommodate the coil length (1 m) of the mole and the different positioning of the two devices during measurements, the FAP dipole field angle data $\vartheta(z)$ have been averaged along z accordingly. The field angle measurements presented in Figure 1 show a comparison among the three probes for magnet DCA320 after cryogenic testing. The mole data contains a zero level of 3.5 mrad (removed in Figure 1) probably reflecting a miscalibration between the gravity sensors and the angular encoder of the mole (this calibration was not repeated on site). The errors presented for the FAPs have been calculated according to equation (7); the dipole field angle patterns measured with the three devices agree well within these errors.

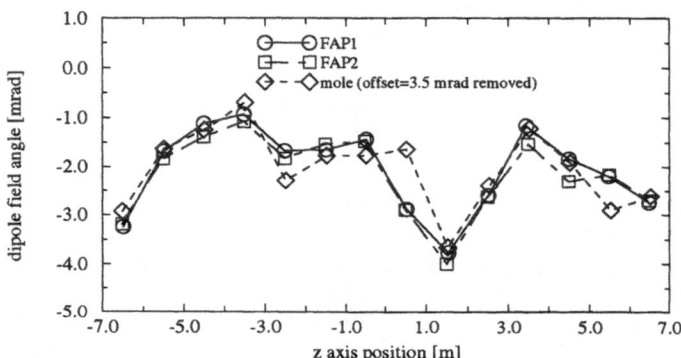

Figure 1. Comparison of dipole field angle pattern for the mole and the FAP systems (after cryogenic testing).

SUMMARY

Measurements of the dipole field angle of thirteen FNAL-built, full-length, 50-mm aperture SSC Collider Dipole Magnet Prototypes using a Vertical Field Angle Probe were presented. The average of the differences between the maxima and minima of the axial dipole angle profiles for the thirteen magnets is about 7 ± 1 mrad. Using a redesigned system the point-by-point angular resolution improved by about a factor 2 and was measured to be 0.16 mrad. A stability study shows no systematic drift of the measured field angle and the average value over 1m length is stable within ±0.1 mrad. The axial dipole angle variations as measured by the mole and the FAP systems are in good agreement.

REFERENCES

1. M. Kuchnir and Ed. E. Schmidt, "Measurements of Magnetic Field Angle Alignment," IEEE Trans. Mag., Vol.24, p 950 (1988).
2. M. Kuchnir et al., "SSC Collider Dipole Magnets Field Angle Data," *Proceedings of the XVth International Conference on High Energy Accelerators*, Hamburg, Germany.
3. M. Kuchnir et al., "Magnetic Field Angle Changes during Manufacture and Testing of Collider Dipoles," Proceedings of the 1992 Applied Superconductivity Conference, Chicago.
4. G. Ganetis et al., "Field Measuring Probe for SSC Magnets," *Proceedings of the 1987 IEEE Particle Accelerator Conference*, Washington, D.C. p 1393.
5. J. Strait et al., "Mechanical Design of the 2D Cross-Section of the SSC Collider Dipole Magnets," *Proceedings of the 1991 IEEE Particle Accelerator Conference*, page 2185.
6. J.S. Brandt et al., "Coil End Design for the SSC Collider Dipole Magnet," ibid, page 2182.

TEST OF FERMILAB BUILT, POST-ASST, 50-MM-APERTURE, FULL LENGTH SSC DIPOLE MAGNETS

J. Kuzminski,[1] A. Akhmetov,[1] R. Bossert,[2] T. O. Bush,[1] D. W. Capone II,[1]
J. Carson,[2] R. Coombes,[1] S. W. Delchamps,[2] A. Devred,[1] J. DiMarco,[1]
C. Goodzeit,[1] S. Gourlay,[2] C. Haddock,[1] R. Hanft,[2] W. Koska,[2]
M. Kuchnir,[2] M. J. Lamm,[2] P. Mantsch,[2] J. Marks,[1] P. O. Mazur,[2]
W. Nah,[1,3] T. Ogitsu,[1,4] D. Orris,[2] J. Ozelis,[2] T. Peterson,[2] E. G. Pewitt,[2]
P. Sanger,[1] R. Schermer,[1] R. Stiening,[1] J. Strait,[2] J. C. Tompkins,[1]
J. Turner,[1] M. Wake,[1,4] J. Zbasnik,[1] Y. Zhao,[1] and H. Zheng[1]

[1]SSC Laboratory,* 2550 Beckleymeade Ave., Dallas, TX 75237
[2]Fermilab, Batavia, IL 60510
[3]Korea Electrotechnology Research Institute, Changwon,
   Kyongnam, 641-120, Korea
[4]KEK, National Laboratory for High Energy Physics,
   Tsukuba, Ibaraki 305, Japan

INTRODUCTION

During 1992 at Fermilab, a series of nine 50-mm-aperture, 15-m-long, SSC superconducting dipole magnets, designed jointly by Fermilab, Brookhaven National Laboratory, and the SSC Laboratory, have been built and successfully cold tested. Seven of these dipole magnets, designated for the Accelerator System String Test (ASST) carried out at SSCL in Dallas, were assembled at Fermilab by General Dynamics personnel, and have achieved the nominal operating current level without significant training.[1,2] In addition, a series of four R&D magnets (DCA320 323) were manufactured at Fermilab to test an alternative insulation schemes. In this paper we present the quench performance of these four R&D magnets, which were cold tested at the Fermilab Magnet Test Facility at nominal temperatures of 4.35 K, 3.85 K, and 3.50 K. An extended characterization test was performed on one of these magnets (DCA322). During this test the magnet was successfully cooled down to superfluid He temperature (1.8 K) and reached a field B ≥ 9.5 T.

MAGNET CONSTRUCTION

The design of the full-length, 50-mm-aperture SSC dipole magnets has been previously described.[3,4] Here we note only the main features. The magnetic field is generated by a two layer, cos(q) type coil clamped by stainless steel collars. The collars serve to position the conductor as specified by the magnetic design and to restrain conductor motion under excitation. The upper and lower collars are locked together by tapered keys and left-right pairs of collars are spot welded to provide greater horizontal stiffness. In the Fermilab design, a vertically split yoke is employed to

* Operated by the Universities Research Association, Inc., for the U. S. Department of Energy under Contract No. DE-AC35-89ER40486.

provide mechanical support to the collars near the horizontal mid-plane and thus to limit deflections under Lorentz force. The 4.95 mm thick, 340 mm O.D. stainless steel shell, made from two half-cylinders welded at the vertical parting plane of the yoke, serves as a helium containment vessel and as a structure clamping together the two halves of the vertically split yoke. To provide axial restraint under excitation, a 38-mm thick end plate is welded to each end of the cold mass shell and the collared coil is preloaded axially against these plates by means of four set screws at each end.

The inner coil of each magnet is instrumented with 53 voltage taps located in the six turns nearest the poles. These voltage taps allow for a quench origin determination with a resolution of a few cm for quenches occurring in the instrumented turns.

All magnets are equipped with two collar packs instrumented with beam-type strain gauge transducers for azimuthal coil stress measurements.[5] These packs are located at positions corresponding to the minimum and maximum of inner coil size. In addition, each magnet has one assembly of load cells,[5] mounted on the non-lead end of the magnet, to measure the forces between the coil and the end plate during excitation, and gauges on the cold mass shell.

Table 1 presents the details of insulations and adhesives employed in various magnets. There are 5 mil shims in the outer coils of magnet DCA320-321 to account for change in the insulation thickness. Note that both sides of the insulation had adhesive coating in DCA322-323 to increase resistance to conductor motion.[6]

Table 1. Cable insulation employed in magnets DCA320-323.

MAGNET	INSULATION	ADHESIVE
DCA320	DuPont Kapton 2H+2LT	3M 2290 epoxy one side
DCA321	DuPont Kapton 2H+2LT	3M 2290 epoxy one side
DCA322	Allied Signal Apical 2NP+2NP	Allied Signal Cryorad both sides
DCA323	Allied Signal Apical 2NP+2NP	Allied Signal Cryorad both sides

EXPERIMENTAL PROCEDURE

The magnet cold testing was carried out at nominal temperatures of 4.35 K, 3.85 K and 3.50 K. The mass flow of supercritical helium at 4 atm was ~50 g/s.

The generic test sequence was essentially the same used during testing of ASST magnets, with cool down without a restriction on the temperature difference between the helium inlet and outlet ends of the magnet, and includes two test cycles, separated by warm-up to room temperature. Both testing cycles started at T=4.35 K with a strain gauge run to quench with subsequent ramps followed to establish a quench current plateau. During the second testing cycle, after re-establishing a quench plateau at 4.35 K, additional tests at 3.85 K and 3.50 K were performed to determine quench performance at these temperatures. Strain gauge ramps to currents 100 A below $I_{plateau}$ were taken at each test temperature.

TEST RESULTS

Figure 1(a) shows the spontaneous quench performance of the four magnets tested, ordered in the sequence they were tested. All quenches displayed in this plot occurred at a ramp rate ≤ 4 A/s. The horizontal dashed line shows the design SSC operating current of 6, 600 A (corresponding to 6.7 T). The overall quench performance of these magnets is slightly below that obtained in the DCA311-319 series. Among the four, DCA322 showed the poorest quench performance, reaching operating current after the second quench and plateau only on the fifth. As in the ASST series of magnets, no training was observed during the second Testing Cycle (TC).

Figure 1(b) shows quench performance of the magnets at low temperatures. Magnets DCA320 and 321 show little or no training and reach the short-sample limit at 3.85 K and 3.50 K. There is a one training quench at 3.85 K in DCA322, and DCA323 exhibits substantial training at all tested temperatures with several quenches originating in the upper outer coil.

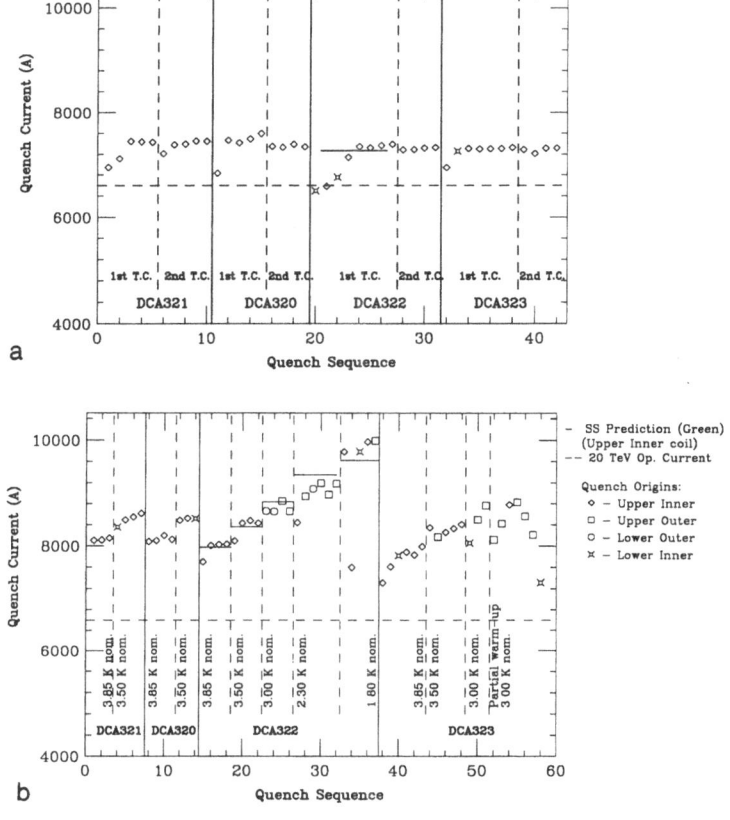

Figure 1. Spontaneous quench performance at 4.35 K (a) on the first and second cooldown and at lower temperature (b) during the second cooldown.

An extended characterization test was performed on magnet DCA322. It consisted of cooling-down the magnet to 3.0 K, 2.3 K and then to the super fluid He temperature of 1.80 K. At each step a strain gauge run to quench was performed to check for possible coil unloading during the magnet excitation. Figure 2 shows an example of the azimuthal inner coil stress (a) and end forces (b) as a function of I^2. The average stress loss between 0 and 6,500 A is about 20 MPa. The stress is linear with I^2, (i.e. with force during excitation) and the prestress is still positive even at the highest current at which strain gauge data were taken (8,894 A or 8.8 T.) At this field the magnetic forces are almost twice as large as at the SSC operating field. In all the 50 mm magnets tested to date, there is no indication of coil unloading. After careful evaluation of strain gauges data, an attempt was made to establish a quench plateau at each temperature. Figure 1(b) illustrates magnet DCA322 quench performance; at 1.8 K the magnet reached a limit of 10,000 A (B ≥ 9.5 T) after several training quenches at 2.3 K.

SUMMARY AND CONCLUSIONS

Four 50-mm-aperture, full-length SSC R&D magnets, in addition to nine ASST magnet prototypes, have been built at Fermilab, following the baseline design but with a different cable insulation scheme. These magnets showed more training quenches then previously tested ASST magnets. However, a conclusive correlation between insulation material used and the spontaneous quench performance could not be established. One of these magnets (DCA322) was tested at 1.8 K and reached a quench plateau at nearly 10,000 A (B ≥ 9.5 T). This extended characterization test confirmed a confidence in the mechanical design of 50-mm aperture magnet prototype.

701

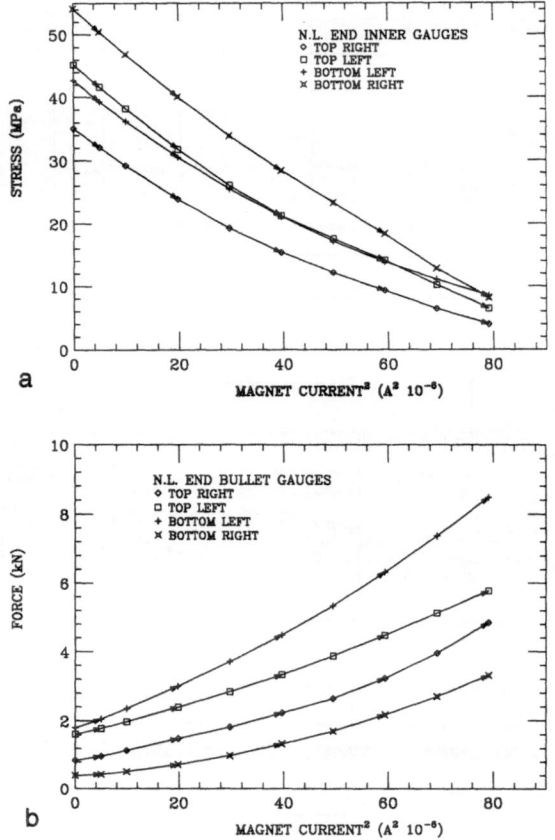

Figure 2. Azimuthal coil stress (a) and end-forces (b) as a function of I^2 for the magnet DCA322 Data taken at 1.8 K during the strain gauge run to quench.

ACKNOWLEDGMENTS

We are grateful to the Fermilab and the SSC Laboratory design team and the Fermilab production staff for their excellent work during this project. We also wish to thank the staff of the Fermilab Magnet Test Facility for their dedication and support of the test program there.

REFERENCES

1. A. Devred, et al., "Review of SSC Dipole Magnet Mechanics and Quench Performance," *Supercollider 4,* J. Nonte ed., 1992, pp. 113-136..
2. J. Kuzminski, et al., "Quench Performance of 50-mm Aperture, 15-m-Long SSC Dipole Magnets Built at Fermilab", Proc. of the XV-th International Conference on High Energy Accelerators, Hamburg, 1992, *Int. J. Mod. Phys. A (Proc. Suppl.)* **2B** (1993) pp. 588-591.
3. J. Strait, et al, "Mechanical Design of the 2D Cross-Section of the SSC Collider Dipole Magnets", *Proceedings of the 1991 IEEE Particle Accelerator Conf.,*1991, p 2127.
4. J. S. Brandt, et al., Coil End Design for the SSC Collider Dipole Magnet, ibid, p.2182.
5. C. L. Goodzeit, M. D. Anarella, and G. L. Genetis, "Measurement of Internal Forces in Superconducting Accelerator Magnets with Strain gauge Transducers", *IEEE Trans. Mag.* **25, No. 2**, 1989, pp 1455-1458.
6. R. Sims, et al., "A Study of Variation in Cable Insulation Systems and their Effect on Creep", presented at the Fiftht International Industrial Symposium on the Supercollider, San Francisco, May, 1993. To be published in *Supercollider 5.*

STRAIN GAUGE PACK IN PROCESS MEASUREMENT SYSTEM FOR THE COLLIDER QUADRUPOLE, DESIGN AND TECHNOLOGY TRANSFER

Richard E. Zeigler,[1] D. Bein,[1] L. Brooks,[1] M. Garza,[1]
B. Rodgers,[1] and G. Novak[2]

[1]Superconducting Super Collider Laboratory*
2550 Beckleymeade Ave.
Dallas, TX 75237-3997

[2]Babcock & Wilcox Magnet Systems Division
200 Lang Horne Rd.
Lynchburg, VA 24503

INTRODUCTION

The in process measurement system that is described below was designed to monitor the strain gauge instrumentation collar pack during all phases of quadrupole magnet manufacture. The primary goal was to implement a monitoring system that incorporated standard measurement techniques while maintaining signal integrity. The flexibility and robustness required in a production environment were taken into account throughout the design and implementation. All major components were chosen from commercially available supplies. The driving software was chosen to provide a user friendly interface while keeping software development efforts to a minimum without sacrificing adaptability.

MEASUREMENT CRITERIA AND TECHNIQUE

The gauges that are monitored with this system consist of sixteen (16) series connected 350Ω strain gages. An excitation current of 2.5mA is applied to the series circuit and the resulting voltage from the individual gauges is monitored using a digital multimeter. The goal of the measurement was to accurately resolve the measurement of the strain gauge to 10mΩ (0.010Ω). This level of accuracy corresponds to approximately 14μ-strain (or ≈20 PSI coil pre-stress). There were no significant data collection rates required for this system.

There are many sources of error that effect this level of accuracy. Some sources of error are digital multimeter (DMM) accuracy, current source noise, reference resistor accuracy, and thermoelectric effects. The DMM chosen for this system has a 10.5 part per million (ppm) accuracy which equates to approximately 4mΩ. The current source has a

* Operated by the Universities Research Association, Inc., for the U. S. Department of Energy under Contract No. DE-AC35-89ER40486.

noise specification of 0.8μA which equates to approximately 2mΩ. The technique used required measurement of the current through a precision shunt resistor. The measurement of a shunt resistor was chosen because of the inaccuracies of the DMM involved in directly measuring the current, in addition to presenting problems in the implementation. The precision resistor chosen was a 350Ω (±0.001%) resistor with a 5ppm long term accuracy and a temperature coefficient of 5ppm/°C, which corresponds to ≈2mΩ (neglecting any change in temperature). Due to the number of connections and relays involved in this implementation there was an attempt to reduce any thermoelectric effects that may have been present. This was done by reversing the current in the series circuit and averaging the reading during the measurement. It should be noted that the inputs to the DMM were also switched so as to always be in the same region of the analog to digital converter therefore reducing any non-linearity effects of the DMM. This list of possible errors is not meant to be exhaustive but should reflect the primary sources of error that were addressed in the implementation.

HARDWARE

The hardware chosen for this system was commercially available (see Table 1). Each component was chosen for its flexibility and/or accuracy. The majority of the components were available with an IEEE-488 option which made automation more convenient. A large and varied choice of plug in cards made the scanner an optimal choice for this implementation. Due to the characteristics of the low voltage scanner card, low contact resistance and long life, the card was selected for the primary multiplexing equipment. The DMM was chosen for its stability and long term (1 year) accuracy. Although the current source was not IEEE-488 compatible, the stability and noise specifications made this device a satisfactory choice. The computer chosen was based on an earlier implementation of the measurement system that utilized a similar computing platform.

Table 1. Major Hardware Components.

NOMENCLATURE	MANUFACTURER	MODEL
Scanner	Keithley	706
DMM	Hewlett Packard	3458
Current Supply	Hewlett Packard	6181
Computer	Apple	Macintosh ci

SOFTWARE

The software was chosen for several reasons. Limited development time, no special programming expertise by designers, adaptability, excellent user interface, and flexible output formats made LabVIEW 2 by National Instruments a good choice. Many of the drivers for the hardware instruments existed and only minor modifications were required. The environment promoted the program adaptability that was needed. In addition, the superior user interface provided by this program allowed for reduced operator training and familiarity. The output was readily acceptable to many other presentation and analysis programs that are widely distributed in the Laboratory.

PACKAGING

Due to the environment that this system was expected to be operated in, several steps were taken to shield and protect the system. A robust upright 19" rack was chosen to mount all equipment of the system. This, in combination with large casters, provided convenient maneuverability throughout the production floor. The cabinet was equipped with a large ventilation system to maintain temperate conditions. A multiple standard input uninterruptable power supply (UPS) was also installed to provide power conditioning and backup. All connections were made within an electromagnetic interference (EMI) shielded enclosure while maintaining good cable shielding and grounding techniques. Special attention was given to equipment supporting brackets. Maintenance and repair required accessibility and was provided for by slide mounts and hinged mounting where appropriate.

CONCLUSION

While maintaining basic signal integrity, this implementation allowed for flexible use on the magnet production floor. It was easily transportable by technicians performing the monitoring function while providing easy, accurate data collection. The basic technique for measurement of the strain gauges was adhered to throughout the system implementation. The attention to detail, documentation, and readily available equipment selections has led to a highly maintainable system that has proven to be extremely flexible. This design has led to an implementation that monitors dipole strain gauge collar packs, end force transducers (bullets), and shell gauges in addition to quadrupole strain gauge collar packs.

ADVANCES IN THE USE OF TOMOGRAPHIC INSPECTION TECHNIQUES FOR NON-DESTRUCTIVE ANALYSIS OF GEOMETRIC CONDUCTOR POSITION AND CORRELATION WITH MAGNETIC CROSS-SECTION MODELING

D. Bein,[1] G. Snitchler,[1] G. F. Rabaey,[2] J. Bolger,[2] R. Crane,[2] I. L. Morgan,[2] and M. Vinson[2]

[1]Superconducting Super Collider Laboratory*
2550 Beckleymeade Ave.
Dallas, TX 75237-3997

[2]International Digital Modeling Corp.
1901 Rutland Dr., Austin, TX 78758

ABSTRACT

Industrial Computerized Tomography has been applied to magnet components in various stages of the manufacturing process. These Computerized Tomographic images can be analyzed to infer detailed dimensional information about magnet component positions (conductor, wedges, collars, etc.) throughout the magnet manufacturing process (cable winding, collaring, yoked/skinned). An analysis technique will be presented and measurement accuracies will be discussed.

INTRODUCTION

The impact of gamma ray Computerized Tomographic (CT) imaging as a medical diagnostic tool has been revolutionary. This diagnostic tool has also been applied (with much less notoriety) to industrial applications such as steel tubing (and I-beams), turbine blades, rocket motors, toxic waste drums, concrete piers, electric power poles, and electronic components to name just a few[1].

Under the direction of the Superconducting Super Collider Laboratory (SSCL) Magnet Systems Division,† International Digital Modeling Corp. (IDM) has developed CT inspection and analysis techniques for use on magnets and magnet components. These inspection techniques have been utilized to inspect a broad range of samples, which include NbTi billets, cured un-collared winding sections, 50–mm collared dipole sections, and 40–mm coldmass quadrupole sections (cf. Figure 1). The CT image shown in Figure 1 was obtained using IDM's IRIS™ VARIScan Laboratory System. The VARIScan system

---

* Operated by the Universities Research Association, Inc., for the U. S. Department of Energy under Contract No. DE-AC35-89ER40486.
† This work is supported by Superconducting Super Collider Laboratory Magnet Systems Division under Contract No. 92-Z-06849.

consists of a 7 Curie CO-60 isotopic source and a collimated detector aperture of 2 mm by 5 mm. Data was acquired at 1/8 degree increments with a maximum detector spacing of 0.16 mm. The nominal tomographic resolution of the VariScan Laboratory System is approximately 1.0 line pair per mm.

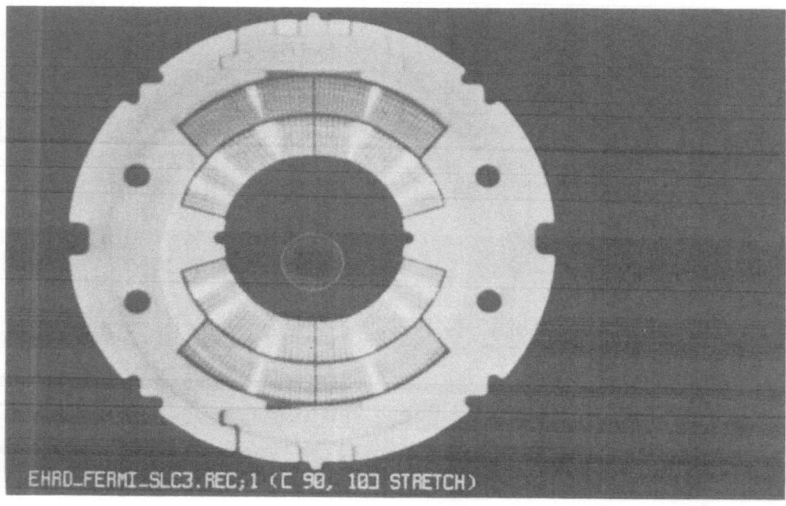

Figure 1. A CT image of a 50 mm collared magnet section. Notice the density contrast between the copper wedges (8.96 g/cc) and the cable insulation material (< 2.00 g/cc).

Results such as Figure 1 can provide a significant amount of useful information on aspects such as missing components, gross manufacturing defects, and improper assembly. Besides the large scale information "visually" obtainable from CT images, detailed dimensional information can be inferred from the full precision CT data (floating point, not 8 bit integer image data) using appropriate analysis techniques.

GENERAL DATA ANALYSIS CONSIDERATIONS

The location of features (conductors, copper wedges, collars, etc.) within magnet samples have been determined from the CT image data using the following analysis procedure. Suppose one has a group of cables which have been resolved in a CT image (cf. the sketch in Figure 2). If one selects a series of data profiles which intersect these resolved cables, then by using sensitivity analysis techniques[2] and a knowledge of the system transfer function (STF), a database of observed conductor edge locations can be generated. Knowledge of the STF is necessary because the STF has the effect of "smearing or smoothing" features within a CT data set, that is the observed CT data set is the convolution of the STF and the ideal CT data. The accumulated information in the database can be processed to determine dimensional parameters such as average insulation gap widths, average conductor thicknesses, conductor corner positions, and conductor centroid locations.

The sensitivity analysis techniques used to determine the location of conductor edges (or other magnet components) can be described quantitatively by recognizing that an observed data profile is a function of the edge locations, the densities of the conductor and insulation materials, and the STF parameter (x_k, c, i, and , respectively); $I(x_{k,c,i,})$. Using the chain rule for differentiation gives:

$$dI_{(x_k, \rho_c, \rho_i, \sigma)} = \frac{\partial I}{\partial \rho_c} d\rho_c + \frac{\partial I}{\partial \rho_i} d\rho_i + \frac{\partial I}{\partial \sigma} d\sigma + \sum_{k=1}^{n} \frac{\partial I}{\partial x_k} dx_k \quad , \tag{1}$$

where the summation is over all edges of interest. For the simplest case, the densities of the conductor and insulation materials are constant for a given profile, and the system transfer function is constant for a given CT image, hence d_c, d_i, and d are equal to zero. If one assumes the conductor edges are defined by step functions and the system transfer function is well represented by a Guassian, then the sensitivity functionals, $\partial I/\partial x_k$, are Gaussians located at the positions x_k, with width . Expressing $dI(x_{k,c,i,})$ as the finite difference between the observed data profile and a theoretical (as designed) profile, and discretizing the sensitivity functionals allows one to recast equation (1) into a matrix equation. This matrix equation can be solved for the dx_k's using standard matrix inversion techniques. The dx_k's give the deviations (or correction factors) from the theoretical (as designed) locations. By adding the correction factors to the theoretical edge locations the algorithm described above can be iterated until convergence is achieved.

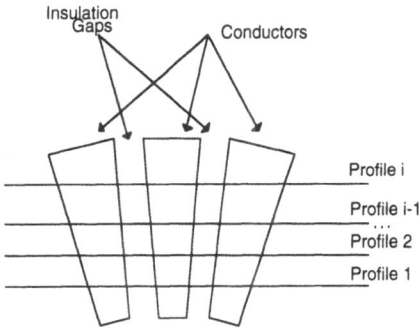

Figure 2. This sketch shows the general placement of data profiles on a set of cables (cable => conductor and insulation).

CHARACTERIZATION OF THE SYSTEM TRANSFER FUNCTION

As pointed out above, the STF acts to "smear or smooth" features within a CT data set. In order to obtain accurate and reproducible information from CT data sets, one must have an understanding of the general shape and the effective width of this function. This was accomplished by designing and fabricating the calibration/gage block.[3] The calibration/gage block was designed to simulate a 50–mm dipole coldmass. Design features include simulated windings with insulation gaps of 0.15, 0.20, 0.25, and 0.31 mm in width and separation gaps between collar/yoke and yoke/skin pieces. An effort was made to maintain the density contrast of an actual 50–mm dipole magnet.

CT data was acquired from the calibration/gage block. This information was used to determine the effective width of the STF and to verify that the STF is well represented by a Gaussian.

RESULTS AND DISCUSSION

Using the analysis techniques outlined above, preliminary results have been obtained for parameters such as average insulation gap widths, average conductor thicknesses, conductor corner positions, and conductor centroid locations. Average insulation gap widths and average conductor thicknesses have been measured with accuracies of ± 0.05 mm. Conductor positions have been determined with accuracies of ± 0.13 mm. The need for dimensional and positional information on magnet components has been widely discussed and demonstrated.[4,5,6] This need can be met with the application of industrial CT through out the magnet manufacturing process.

REFERENCES

1. H. Ellinger, I. L. Morgan, R. Klinksiek, F. Hopkins and J. Neils Thompson, Tomographic analysis of structural materials, *in:* "Imaging Applications for Automated Industrial Inspection and Assembly," *SPIE vol. 182,* p. 179, (1979).

2. R. D. Rosenwald and G. F. Rabaey, "Application of the continuous orthonormalization and adjoint methods to the computation of solar eigenfrequencies and eigenfrequency sensitivities," *Ap. J. Supplement Series*, 77:97-117, (1991).

3. D. Bein, G. Snitchler, G. F. Rabaey, J. Bolger, R. Crane, I. L. Morgan, and M. Vinson, SSCL Engineering Note in process.

4. L. Nguyen and H. Gurol, "The Effects of Manufacturing Tolerances on the Multipoles of the SSC Dipole Magnet," *Supercollider 3*, J. Nonte, ed., Plennum Press, New York, (1991).

5. R.C. Bossert, J.S. Brandt, J.A. Carson, K. Coulter, S. Delchamps, K.D. Ewald, H. Fulton, I. Gonczy, S.A. Gourlay, T.S. Jaffery, W. Kinney, W. Koska, M.J. Lamm, J.B. Strait, M. Wake, M. Gordon, N. Hassan, R. Sims, and M. Winters, "Initial results from 50mm short SSC dipoles at FermiLab," *Supercollider 3*, J. Nonte, ed., Plenum Press, New York, (1991).

6. D. Bein, J. Zbasnik, "Utilization of Gamma Ray Inspection System for Tomographic Imaging and Dimensional Analysis of Complete Model Magnet Cold Masses and Collar Coil Sections," *Supercollider 4*, J. Nonte, ed., Plenum Press, New York, (1992).

INDUSTRIAL HARMONIC ANALYSIS SYSTEM FOR MAGNETIC
MEASUREMENTS OF SSC COLLIDER ARC AND HIGH ENERGY BOOSTER
CORRECTOR MAGNETS

M. I. Green, R. Sponsel, and C. Sylvester

Superconducting Super Collider Laboratory*
2550 Beckleymeade Ave.
Dallas, TX 75237-3997

INTRODUCTION

The SSCL collider arc and high energy booster corrector magnets are 50 mm bore cryogenic magnets. The integral strength and harmonics will be measured by industry at full current at 4.2 K and at plus and minus 400 mA at room temperature. Dipoles, quadrupoles, and sextupoles have error tolerances of a few tens of units. A prototype harmonic analysis system for magnetic measurements of production and prototype dipole, quadrupole and sextupole magnets has been designed and is being fabricated. We describe the criteria for search coil designs, data acquisition system hardware and software. Radial search coil arrays are being fabricated utilizing multifilar[1] wire. Two digital integrators will allow simultaneous accumulation of unbucked and bucked configurations.

DESCRIPTION OF MAGNET MEASUREMENT REQUIREMENTS

Field strength and harmonics of dipole, quadrupole, sextupole, octupole and decapole corrector magnets need to measured. The bore is 50 mm and magnet lengths vary from about 0.2 to 1.6 meters. The required field quality of the error harmonics is 10 to 30 units. A unit is defined as 10^{-4} of the fundamental field evaluated at a reference radius of 1 cm. The system resolution will be better than 1 unit.

INSTRUMENTATION

Figure 1 is a block diagram of the instrumentation. The voltage from rotating radial coil arrays is fed to digital integrators which are latched by an incremental optical encoder. Using integrators rather than reading the voltage directly eliminates the need to correct the coil voltages for rotational speed variations.

---

* Operated by the Universities Research Association, Inc., for the U. S. Department of Energy under Contract No. DE-AC35-89ER40486.

Measurements of the error harmonics will be facilitated by utilizing analog bucking, i.e., connecting coils in series opposing to cancel out the fundamental, and one lower harmonic. As the digital integrators have a high input impedance, the system will be able to simultaneously obtain data from the fundamental coil by itself and the fundamental coil output bucked by the output of the compensating coils.

The data acquisition instrumentation is VME based. A Motorola MVME-167 will control the VME bus. The digital integrators are Metrolab[2] PDI 5035 VME bus models which are based upon a voltage-to-frequency converter developed at CERN.

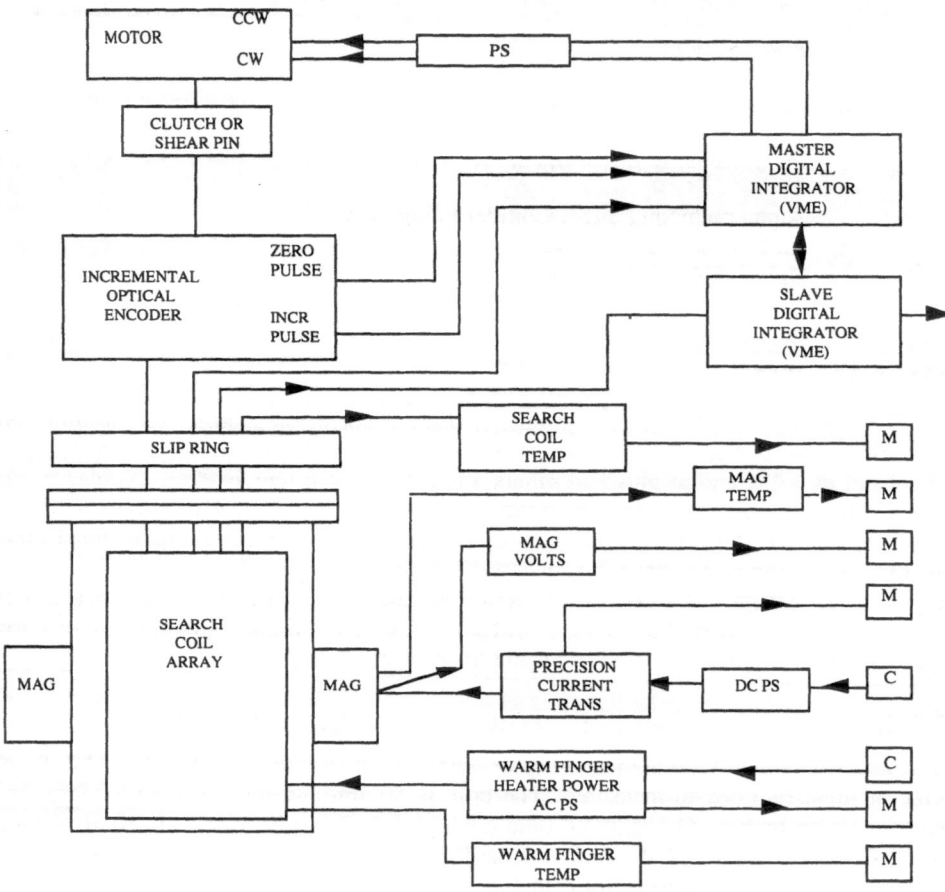

Figure 1. Instrumentation Block Diagram.

SOFTWARE

Figure 2 is a global structure chart of the software. The operating system will be UNIX based. At the time this publication is being written, we are re-evaluating the computer system architecture and the real time kernel. The *initialize & test hardware* module for example, will open, control, monitor and log windows and the log file. It will also check whether the VME crate is on line and whether the instrumentation required for the test is powered and initialized.

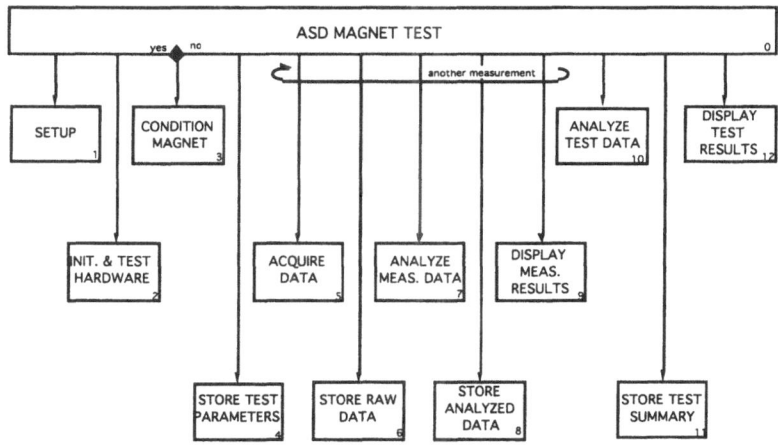

Figure 2. Global structure chart of software.

ELECTRONIC DESIGN OF SEARCH COIL ARRAYS

A Lotus 123 spreadsheet has been developed for designing search coil arrays for dipoles, quadrupoles and sextupoles. One enters the outside diameter of the rotating coil array, some wire dimensions, a few magnet parameters, and the spread sheet calculates the dimensions of the coil forms, the sensitivity of the search coil array to the fundamental and harmonics and also the expected signal to noise ratio.

The coils must satisfy $\sum p_i N_i r_i^n = 0$ where p_i is the parity of the coil bundle (+1 or −1), N_i is the number of turns in each coil bundle, and r_i is the mean location of the coil bundle, and n is the harmonic number that is being bucked. We are using the notation that $n = 1$ indicates the dipole harmonic. For dipole search coil arrays, only one equation needs to be satisfied. For quadrupole search coil arrays, two equations, $n = 1$, and $n = 2$ must be satisfied.

COIL FORMS

The dipole search coil array consists of three coplanar "identical" coils that extend completely through the magnet with sufficient length to include end effects. Figure 3 depicts the three identical coil forms. The fundamental coil is on the right, the bucking coil is in the center, and the coil on the left is a dummy coil for mechanical symmetry. The axis of revolution passes through the center of the center coil.

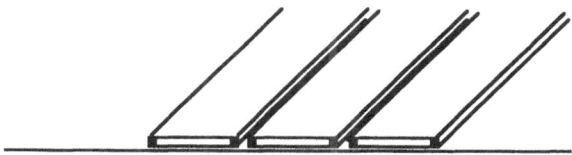

Figure 3. Cross section of dipole search coil array.

The quadrupole search coil array consists of five "identical" radial coils.

COIL FABRICATION

The coils are wound around a flat Nema G-10 mandrel which has been machined to the required width, thickness, and flatness.

713

In the coil winding operation, the mandrel is supported between two flat plates. The multifilar wire is then wound around the mandrel, one layer at a time, with each layer bonded to the previous layer. Completed coils are then calibrated and attached to the support structure. The specified number of coils for a particular assembly, are supported on a Nema G-10 rod and are aligned with the use of two alignment pins, located on the semicircular support rod. A matching semicircular cover, also machined from Nema G-10, is then fastened to the winding support rod to complete this assembly, after the electrical connection for the signal leads have been made.

COIL SUPPORT STRUCTURE

The coils on the support structure, together with bearings and drive shaft are housed in a brass support tube, which in turn is supported inside a warm finger during use for cold magnetic measurements. The warm finger assembly is being fabricated using type 316-L seamless stainless steel tubing. A heater is installed in the annular space between these two tubes to facilitate vacuum conditioning.

STATUS

The mechanical design of the system is completed and many of the fabricated components are on hand at SSCL. The majority of the purchased electro-mechanical components such as the optical encoder, stepper motor, slip rings etc., are also on hand. Coil winding is ongoing at SSCL and it is expected that the five coils which are required for the quadrupole system will be completed in approximately two weeks. The instrumentation has been specified and the majority of the components are on hand. The Data Acquisition System software is being designed. Figure 4 shows the coil assembly for the prototype system. Except for the slip ring assembly, this is the drive system design which will be delivered to industrial magnet manufacturers, for measurement of SSCL superconducting corrector magnets.

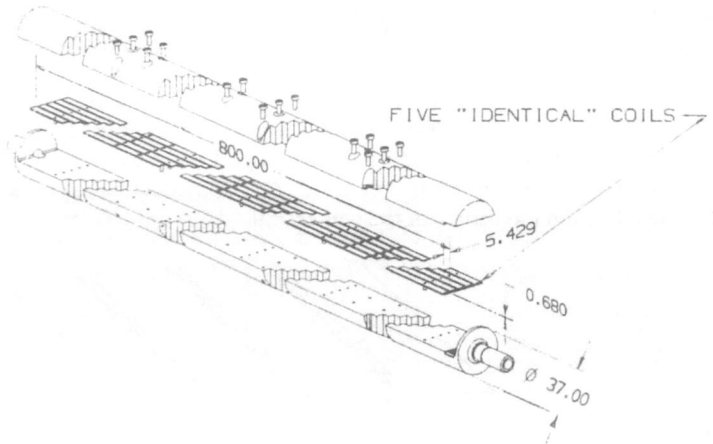

Figure 4. Quadrupole Coil Assembly.

REFERENCES

1. MWS Wire Industries, 312 Cedar Valley Drive, Westlake Village, CA 91362.
2. Metrolab Instruments SA, 110 ch. du Pont-du Centenaire, CH-1228 Geneva, Switzerland (US rep: GMW, P.O. Box 2578, Redwood City, CA 94064).

PERFORMANCE OF FIELD MEASURING PROBES FOR SSC MAGNETS[*]

R. Thomas, G. Ganetis, J. Herrera, R. Hogue, A. Jain, W. Louie,
A. Marone, and P. Wanderer

RHIC Project
Magnet Division
Brookhaven National Laboratory
Upton, New York 11973-5000

ABSTRACT

Several years of experience have been acquired on the operation of probes ("moles") constructed for the measurement of the multipole components of the magnetic fields of SSC magnets. The field is measured by rotating coils contained in a 2.4-m long tube that is pulled through the aperture of the magnet by an external device—the transporter. In addition to the measuring coils, the tube contains motors for rotating the coil and a system for sensing local vertical using gravity sensors to provide an absolute reference for the field measurements.

We describe the steps that must be taken in order to ensure accurate, repeatable measurements; the design changes that have been motivated by difficulties encountered (noise, vibration, variations in temperature); and other performance issues. The mechanical interface between the probe and the beam tube of the magnet is also described.

MAGNETIC MEASURING SYSTEM

Components

The magnetic field in the aperture of an SSC magnet is determined by measuring the voltages induced on a number of rotating windings. These voltages are then Fourier analyzed to obtain the multipole coefficients. The system at BNL measures the voltages at 128 positions over one complete rotation of the coil. The field angles are determined relative to an absolute reference, the direction of the gravitational field. The part of the magnetic measurement system that is inserted into the aperture of the magnet (the "mole") is made up of the following components:
- three or more coil windings
- a hollow, cylindrical, coil-form made of epoxy-fiberglass or ceramic
- slip rings to bring out the electrical signals
- an electric or air motor for rotating the windings that is connected to the coil-form through gear reducers, torsional flex joints, and viscous dampers
- bearing supports
- precision and coarse-resolution gravity sensors mounted on a rigid, but rotatable, platform

---

[*]Work supported by the U. S. Department of Energy.

o a precision optical encoder, mounted to the gravity sensor platform, for determining the angular position of the coil-form relative to the platform
o an air or electric motor for rotating the gravity sensor platform and encoder
o a tachometer for the gravity sensor drive
o a metal (brass) shell that encloses all the elements listed above
o electrical connectors (and, in the case of air motors, connectors for the air hoses).

External to the mole itself are:
o the mole control chassis and its remote interfaces
o cables for carrying the electrical and air signals ("tethers")
o the mole transporter which moves the mole through the magnet under computer control, the transporter control system, and its connecting cables
o a digital volt meter for each winding
o for cold measurements, a warm bore tube to maintain the mole at the desired temperature
o a temperature controller and heater cable for the warm bore tube
o a computer for controlling the equipment, performing the measurements, and storing the results.

Construction, Assembly and Calibration

Coil and Gravity Sensor Motors: In order to avoid damage and excessive wear, it is important that the motor, gear reducer and other mechanical parts of the mole should be operated only within the desired speed range. After assembly, each motor is tuned by finding the voltage values that give the desired fixed speed or speeds. The nominal rotation period of the coils is 3.5 ± 0.1 s (17.14 ± 0.5 rpm). The gravity sensor platform has three nominal rotation speeds as well as a pulse mode in order to allow the platform and encoder to be accurately zeroed with respect to gravity under program control.

Gravity Sensors: The mole incorporates electrolytic gravity sensors (Figure 1) in which the resistance of a conducting fluid in a curved glass tube is sensed and converted into an angle relative to gravity. The sensor has three terminals and acts as a potentiometer. Two sensors (one serves as a back-up) have a resolution of about 2 millidegrees and a range of $\pm 2°$. The third sensor has a resolution of about 30 millidegrees and a range of $\pm 60°$.

Each gravity sensor is adjusted and then fixed to the platform so as to minimize its sensitivity to a tilt of the longitudinal axis of the platform. The sensors are then rotated to specified angles using an indexing head, and a linear fit of angle vs. sensor output is produced for each sensor.

An angle of 0° is defined as a potentiometric voltage of 0 across the primary gravity sensor.

Coil Windings: The principal winding for obtaining the harmonics is the *tangential* winding and has a small opening angle (Figure 2). Other windings are provided to measure the fundamental component of the field and are wound on the same coil-form as the tangential winding. For example, to measure a dipole field, two dipole bucking windings are provided. The digitized voltages from these two windings are analyzed by the computer and digitally combined so as to produce a dipole component of the same phase and amplitude as that sensed by the tangential winding. This combined signal is then subtracted from the digitized voltages of the tangential

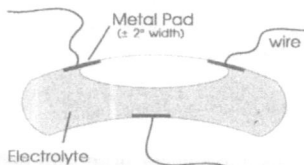

Figure 1. The high resolution electrolytic gravity sensor acts as a potentiometer whose setting is determined by the position of the sensor relative to vertical.

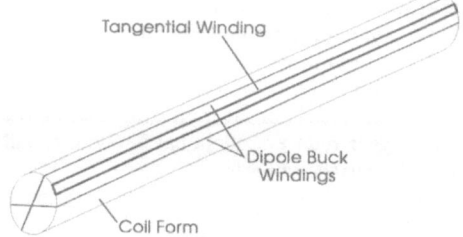

Figure 2. The coil form and three typical windings.

winding. The process is called "digital bucking." For the measurement of quadrupole fields, the coil-form has two bucking windings sensitive to the quadrupole field plus a dipole winding for removing the dipole component that is present when the coil-form is not collinear with the magnetic axis.

The amplitude of the field is not adjusted in the mole calibration procedure. Instead, the measured opening angle, radius, and number of turns of each coil is used to convert the voltage measurements into a field amplitude. For air moles, a dipole field of about 1.37 T can be applied, so it is possible to check the agreement with an NMR measurement of the same field.

The angle of the field is calibrated by determining an offset angle for each winding. This offset angle is the angle of the dipole field as measured by a given winding in the reference dipole magnet. This procedure assumes that: 1) the dipole component of the field produced by the reference dipole is precisely vertical, 2) the position of the index pulse from the encoder relative to vertical (gravity) is correctly reported by the primary gravity sensor, and 3) all 128 encoder positions used for acquiring field measurements during a single rotation are precisely spaced relative to the index pulse.

DIFFICULTIES AND SOLUTIONS

Vibration

Rhythmic vibration of the gravity sensor platform causes the liquid in the sensor to develop standing waves, and the resulting reading deviates actual angular position of the platform. New, larger gravity sensors are being incorporated in the moles that contain an electrolyte with a higher viscosity.

When the mole is calibrated in the reference dipole, it is placed in a supporting tube. It was observed that this support tube would vibrate when the coil motor was turned on. The support tube was stiffened up and the mole support modified to reduce these vibrations.

During the measurement of magnets at cryogenic temperatures, warm nitrogen gas flows through the bore tube. It was found necessary to shut off the gas during the time that a measurement was being acquired in order to eliminate vibrations produced by the gas flow.

Temperature

The gravity sensor reading is temperature dependent. To keep the sensor at a constant temperature during measurements, a small heater on the platform is used to maintain the temperature at $\sim 32 \pm 0.5$ °C.

This heater is sufficient for measurement of magnets at ambient temperature. For the cold measurements, a warm bore tube is inserted in the magnet. Both a single-wall and a double-walled bore tube have been tried with the double-walled giving more satisfactory results. The warm bore tube has an electric heater arranged in four strips along the length of the tube, spaced 90° apart. The current flow is such as to cancel the torque on the tube when the magnet is powered. The cancellation is not perfect however, so it has been found necessary to turn off the heaters during the time the measurement is being taken.

In the double-wall design, an insulating vacuum exists between the inner and outer tubes. This vacuum space is filled with 15 layers of superinsulation. The stainless-steel heater strips are mounted on the outer surface of the inner tube. Thermocouples are located in the vacuum space to monitor and regulate the temperature.

Magnetic Field Gradients

To verify that the gravity sensor readings do not change significantly in a dipole field, gravity sensors were rigidly mounted and inserted into a SSC dipole which was then ramped from 0 to 6600 A. No significant field dependence was observed. A magnetic field *gradient*, however, strongly affects the gravity sensor readings. The dependence is a strong function of the magnitude of the field gradient, so at low field gradients (< 65 T/m), the effect is negligible, but as the quadrupole current is increased, the deviation from the expected reading increases as $I^{-4.5}$. The electrolytic fluid in the gravity sensors is, apparently, paramagnetic, and moves when subjected to magnetic field gradients, thus producing an incorrect reading.

Impact Forces

When the calibration of a mole is rechecked by placing it in the reference dipole, the reported phase angle will sometimes be shown to be different from 0.0° by as much as 0.1°. The shifts never occur gradually, but instead appear suddenly. The shift has been observed to alternate in sign.

The most likely sources of calibration shifts are 1) the encoder, along with the shaft and couplings connecting it to the coil form and the encoder connection to the gravity sensor platform, and 2) the gravity sensors themselves or the method used to affix them rigidly to the gravity sensor platform.

It is known that application of any dc voltage to a gravity sensor will plate out material on the metal pads inside the sensor and cause the readings to shift. However, shifts have now been observed in moles where the gravity sensor itself is protected by capacitors from ever having a dc voltage applied to it.

Even though moles are handled carefully, they can not be entirely protected from mechanical shock. To investigate the effects of impact forces on the mole, one mole was repeatedly struck, lightly but firmly, with a hammer in the region of the gravity sensor platform. This caused the phase angle to shift by $\sim 0.05°$, and it did not return to the original value.

The shifts may therefore principally result from jarring, mechanical shock, and bending of the mole when moving it from place to place, inserting it into a magnet, and pulling it through the magnet.

Some moles (FA3 and RA1), while showing shifts initially, are not observed to change subsequently over periods of months (two and three months, respectively).

To prevent errors from calibration shifts, a short permanent magnet that can be precisely leveled will be used. Before the mole enters or exits a magnet, it will pass through the permanent magnet where a measurement will be taken to verify that the phase angle calibration has not changed.

FUTURE WORK

Field Gradients

For short quadrupole magnets, it may be possible simply to locate an electrolytic sensor beyond the influence of the field. Otherwise, it will be necessary to develop or use a new type of gravity sensor. For example, a pendulum with some optical means of determining the pendulum's angle relative to some reference.

Calibration Shifts

In order to investigate the sudden small shifts in the measured phase angle of the reference dipole, two mole subsystems will be constructed. One will examine the output of the small Teledyne-Gurley optical encoder used with the moles relative to that of a larger (and presumably more stable) BEI encoder on the same shaft. The other will investigate the effects of shock and movement on the gravity sensors. Each subsystem will be portable and will be subjected to the same kinds of impact forces that these devices would be expected to encounter when used in a mole.

CONCLUSIONS

The moles that have been developed are capable of highly precise measurements of the multipole components of the magnetic fields of SSC magnets. Small shifts in the reference phase angle are observed to occur. A permanent magnet will be used to detect these shifts to insure reliable absolute phase angle measurements. Other difficulties resulting from rhythmic vibration of the mole and temperature fluctuations have been largely overcome.

A COMPARISON OF LEAST SQUARES LINEAR REGRESSION AND
MEASUREMENT ERROR MODELING OF WARM / COLD MULTIPOLE
CORRELATION IN SSC PROTOTYPE DIPOLE MAGNETS

D. A. Pollock,[1] R. F. Gunst,[2] K. Kim,[1] and W. R. Schucany[2]

[1]Superconducting Super Collider Laboratory, Dallas, Texas 75237*
[2]Southern Methodist University, Dallas, Texas 75275

INTRODUCTION

Linear estimation of cold magnetic field quality based on warm multipole measurements is being considered as a quality control method for SSC production magnet acceptance. To investigate prediction uncertainties associated with such an approach, axial-scan (Z-scan) magnetic measurements from SSC Prototype Collider Dipole Magnets (CDM's) have been studied. This paper presents a preliminary evaluation of the explanatory ability of warm measurement multipole variation on the prediction of cold magnet multipoles. Two linear estimation methods are presented: *least-squares regression,* which uses the assumption of *fixed* independent variable (x_i) observations, and the *measurement error model,* which includes measurement error in the x_i's. The influence of warm multipole measurement errors on predicted cold magnet multipole averages is considered.

MSD QA is studying warm/cold correlation to answer several magnet quality control questions. How well do warm measurements predict cold (2kA) multipoles? Does *sampling error* significantly influence estimates of the linear coefficients (slope, intercept and residual standard error)? Is estimation error for the predicted cold magnet average small compared to typical variation along the Z-Axis? What fraction of the multipole RMS tolerance is accounted for by individual magnet prediction uncertainty?

THE DATA

To compare the two linear estimation models, the joint behavior of warm/cold multipole pairs have been studied. The data selected is from six CDM's (DCA311, 312, 314, 315, 317, and 319) which were measured at FNAL using mole B2. The data includes multipoles (a_n and b_n, $n = 1,11$) which were measured at 24 Z-scan positions (between -7.01 and 7.01 m) for each magnet under both cold (2kA) and warm (+/- 10 A) conditions. The reported values are the position average of repetitive coil rotations. Also, the measurements have been centering corrected. Traditional linear regression assumptions regarding independence, constant variance and normality have been verified for the source data using graphical residual analysis techniques.[1] Research results for the b_2 (normal sextupole) multipole will be demonstrated in this paper.

---

*Operated by the Universities Research Association, Inc., for the U.S. Department of Energy under Contract No. DE-AC35-89ER40486.

COMPARING ORDINARY LEAST SQUARES AND MEASUREMENT ERROR MODEL FITS

Ordinary Least Squares (OLS) is typically used to estimate the relationship between paired observations for which y is assumed to be linearly dependent on or explained by x. For the Warm/Cold multipole study, y_i is a cold multipole observation (mole-position average), x_i is a warm multipole observation. The dependent variable is traditionally assumed to be "fixed", (i.e., observed without error).

Since the Warm observations are not fixed (i.e., each individual coil rotation at a given mole-position generates a different result), the OLS prediction model is subject to both random and measurement errors. According to linear estimation theory, the predicted slope, intercept and residual standard error using the OLS model under such conditions will be biased.[2, 3] Without adjusting for the measurement error, the cold prediction model will *underestimate* the slope. In the case that x and y are themselves random variables (i.e., error-free cold and warm multipole measurements are each random variables based on samples from several coil-rotations) the traditional linear model assumptions should be replaced by a Measurement Error Model.[3]

The key to applying the measurement error model (MEM) to cold multipole predictions is determining the value of the *ratio of error variances* (λ) which for our example is defined as the ratio of (variance cold) / (variance warm). To estimate λ, sample variance for both warm and cold multipoles by position (using individual coil rotations) within individual magnets have been calculated. The position variance estimate has been calculated as the average of individual position variances for all Z-scan positions and available magnets. Outliers were removed by using the Shewhart Variance Control Chart. Extreme values due to magnet end effects and strain gauge packs were also removed. The estimated λ for the b_2 data is 0.000392 units. It should be noted that a recent upgrade to the SSCL magnetic measuring system has reduced warm position variance significantly. [4] The data in this study are for measurements taken before the improvement was introduced.

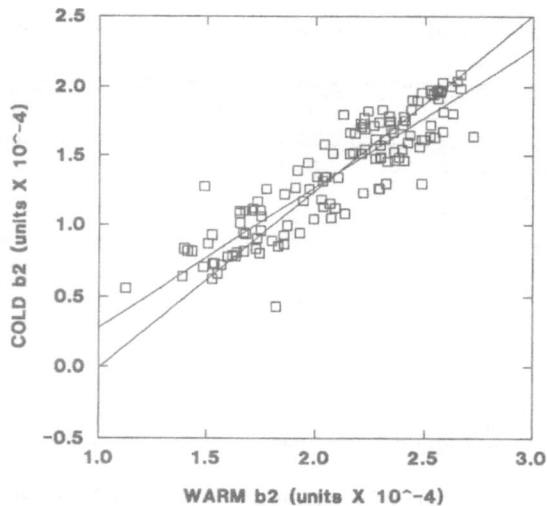

Figure 1. Warm vs. Cold b_2. The linear fits using OLS (shallower slope) and MEM (steeper slope) methods. Data does not include strain gauge position values.

A comparison of OLS and MEM prediction equations for b_2 is shown in Figure 1. The figure shows how the OLS estimate tends to under-estimate the slope. The estimated coefficients (slope, intercept and residual standard error) and a 95% prediction interval for both OLS and MEM models of individual mole-position estimates are listed in Table 1. The wider 95% prediction interval for the MEM estimate shown in Table 1 is due to including measurement error (λ).

Estimate	Slope	Intercept	Std. Error of Residuals	95% Prediction Interval *** (1 sided, at mean warm value)
OLS *	0.976	-0.717	0.23319	0.46023
OLS **	0.997	-0.718	0.18397	0.36194
MEM *	1.397	-1.600	0.27892	0.54875
MEM **	1.256	-1.263	0.20762	0.40847

*	with "strain gauge" effects.	***	95% prediction interval
**	without "strain gauge" effects.		= 1.96 x Std. Error x sqrt(1/n + 1)

Both the OLS and MEM prediction methods have been used to estimate cold multipoles by Z-position. See Figure 2 "Warm/Cold b_2 Estimates by Mole-Position (DCA311)" for an example of the predictions by position. Based on One Way Analysis of Variance (ANOVA) of DCA311 warm multipole coil-rotation data, the average variance "within" position (i.e., due to measurement error) is 0.01482 units and the "between" postion variance is 0.12842 units. Differences due to position (i.e. manufacturing variation along the length of the magnet) explain approximately 92% of the variance in DCA311 warm b_2. Similarly, differences due to position explain approximately 99% of the variance in DCA311 cold b_2. In both cases, measurement error is very small compared to variation along the magnet.

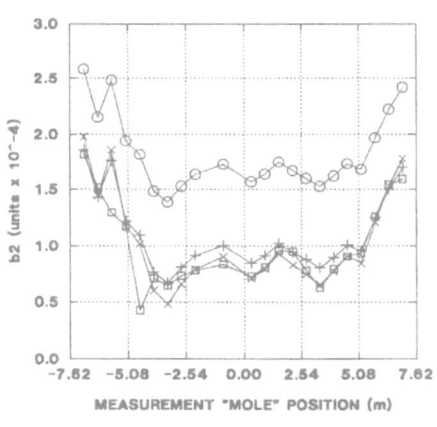

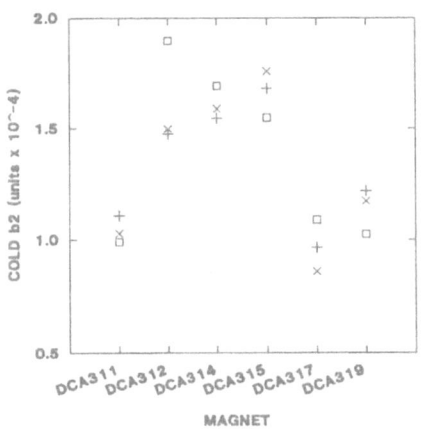

Figure 2. Warm/Cold b_2 Estimate by Mole-Position (DCA311). Strain gauge effets are not shown. Predicted cold multipole values by mole-position are plotted using OLS and MEM methods. Actual Warm and Cold Measurements are shown for comparison.

Figure 3. Warm/Cold b2 Estimate of Magnet Averages. Mole-position values have been averaged over the length of the magnet, without positions -0.31, and -1.52 (strain gauge effects).

PREDICTING COLD MAGNET AVERAGE MULTIPOLE VALUES

For magnet acceptance purposes, the magnet mean of the multipoles is of primary interest. The magnet mean is estimated as an average of the position measurements (or predictions). A summary of magnet mean predictions for b_2 is presented in Figure 3 and Table 2. From these summaries, one may compare the relative difference between the Cold measurements and the OLS or MEM estimates. The MEM prediction was closer than the OLS prediction to the measured cold mean in 4/6 of the cases studied.

Table 2. Magnet Mean Results for b_2 (normal sextupole).

Note: Position <> -0.31 m, -1.52 m (strain gauge effect removed), n = 22 positions per magnet.
Bold indicates minimum prediction delta (absolute deifference from Cold Mean).

Magnet	DCA311	DCA312	DCA314	DCA315	DCA317	DCA319
Cold Mean	0.99052	1.89581	1.69006	1.54929	1.08904	1.02556
OLS Mean	1.10590	1.47492	1.54623	1.67855	0.96566	1.21579
Delta OLS	0.11538	0.42089	0.14383	*0.12926*	*0.12338*	0.19023
MEM Mean	1.02881	1.49925	1.58910	1.75583	0.85753	1.17272
Delta MEM	*0.03829*	*0.39656*	*0.10096*	0.20654	0.23151	*0.14716*

DISCUSSION OF QUALITY CONTROL QUESTIONS PRESENTED IN THE INTRODUCTION

1. Using a linear model, how well do warm measurements predict cold (2kA) multipoles? For the example presented, the average absolute difference between six cold magnet averages of b_2 and predicted magnet averages is: 0.18716 units (OLS), and 0.18684 units (ML), see Table 2.

2. Does warm measurement sampling error significantly influence estimates of the linear coefficients (slope, intercept and residual standard error)? Yes, at the "mole-position" level as demonstrated, see Table 1.

3. Is estimation error small compared to typical variation along the Z-Axis? Using the difference between the observed cold magnet average and the predicted magnet average, estimation error appears to be *small*. In this example the predicted magnet mean differed from the COLD measurement mean by 0.18716 units (OLS average absolute difference) and 0.18684 units (MEM), see Table 2. Using ANOVA of individual coil-rotation data, measurement error was shown to be very small compared to variation along the magnet.

4. What fraction of the multipole tolerance (b_2 RMS 1.15, Systematic 2.0 at injection, 0.8 at high current) is accounted for by individual magnet prediction uncertatinty? Based on the average absolute difference between cold and predicted magnet means, OLS estimation error accounts for approximately 16.3 % of the $b2$ RMS tolerance and 23.40% of the high field b_2 systematic tolerance, while MEM estimation error accounts for 16.2% of the RMS tolerance and 23.36% of the high field systematic tolerance.

CONCLUSION

This study has demonstrated two methods which may be considered for making cold magnet predictions from warm magnetic measurements. The study is based on a very small number of magnets. For the b_2 example presented the two estimation methods produced similar predictions of the cold mean. Recent improvements to the measurement system may make the issue of "measurement error" insignificant. Uncertainty of the predicted mean due to magnet axial variation will continue to be studied as more magnets and data become available.

ACKNOWLEDGMENTS

A. Devred, J. DiMarco (SSCL, Dallas, Texas). K. Kussmaul (Westinghouse Electric Corporation, Round Rock, Texas).

REFERENCES

1. D. C. Montgomery (1991). *Introduction to Statistical Quality Control*, Second Edition, John Wiley & Sons, New York. pp. 465 - 468.
2. J.D. Jobson (1991). *Applied Multivariate Data Analysis. Volume I: Regression and Experimental Design*, Springer-Verlag, New York. pp. 121, 122.
3. M. Reilman, R. Gunst, and M. Lakshminarayana (1986). Stochastic Regression with Errors in Both Variables, *Journal of Quality Technology*, Vol. 18, No.3, pp. 162 ff.
4. J. DiMarco, et. al., *Magnetic Measurements, DCA320-322*, MSIM Report, September 22, 1992.

QUENCH PERFORMANCE OF SEXTUPOLE CORRECTOR MAGNETS

C.M. Rey and E.T. Gossler

Babcock & Wilcox-Accelerator & Magnet Systems
P.O. Box 785 MS 80
Lynchburg, VA 24505

INTRODUCTION

We report the quench performance of the first four sextupole corrector magnets built by B&W. Three of the four magnets were cold tested at the Central Facility in Waxahachie, TX and are by denoted the nomenclature BWDb2.1[1]- BWDb2.4.

Babcock and Wilcox (B&W) and Advanced Interconnect Technology (AIT) have joined in a cooperative effort to develop and build corrector magnets using the Direct Wire technique for the Superconducting Super Collider (SSC). The Direct Wire process uses an automated wiring machine consisting of an ultrasonic head, a precision positioning system, and computerized control. The superconducting wire is fed through the wiring head and under a precision grooved stylus which is excited by ultrasonic transducers. As the wire is guided along a computer controlled path the ultrasonically excited stylus imparts energy to the wire which locally melts an adhesive coating, bonding the wire in place. A more detailed description of the Direct Wire technique has been given elsewhere.[2]

TEST PROCEDURES

Each magnet was mounted in a vertical cryostat and tested in pool-boiling liquid helium at one atmosphere. Magnets typically had two thermal cycles of quench testing with typically 30 to 40 quenches per thermal cycle. A testing cycle consisted of performing magnet training quenches until a plateau was reached or a maximum of 50 quenches. No sub-cooling or any other method was used to accelerate training.

Magnet training quenches were produced by increasing the current at a steady rate until a quench was detected. No energy extraction operations were initiated after the detection of a quench. The current ramp rates varied for each individual magnet, from a minimum of 50 mA/s to a maximum of 10 A/s, depending on the magnet.

TEST RESULTS

BWDb2.2 and BWDb2.3

BWDb2.2 and BWDb2.3 were produced during the early stages of winding development. These magnets all experienced problems with improper wire nesting at the coil ends (in the turn around section). Proper wire nesting occurs when the wires stack in a hexagonal packed structure. In these coils, problems were encountered in the end regions where improper nesting of the wires at times resulted in placement of wires directly on top of each other. As a result, a substantial build up of wire (~ several mm in the radial direction) was occurring at the coil ends. This gave the coils an over-all "dog-bone" shape (see Figure 1). In addition, wire insulation was scraped away in the end regions where the machine manipulates its turn around. As a result, the coils experienced several electrical shorts when the end clamps were pressed around the coil form. The electrical shorts were detected by both dc resistance and ac inductance measurements. Thus, only very low ramp rates ranging from 50 mA/s to 200 mA/s were used during quench testing. Ramp rates higher than 200 mA/s showed appreciable rate dependence in the quench current.

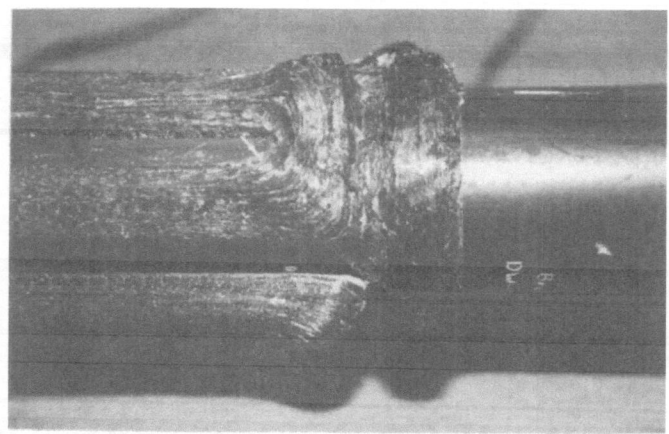

Figure 1. Coil BWDb2.2 (45° point to point turns).

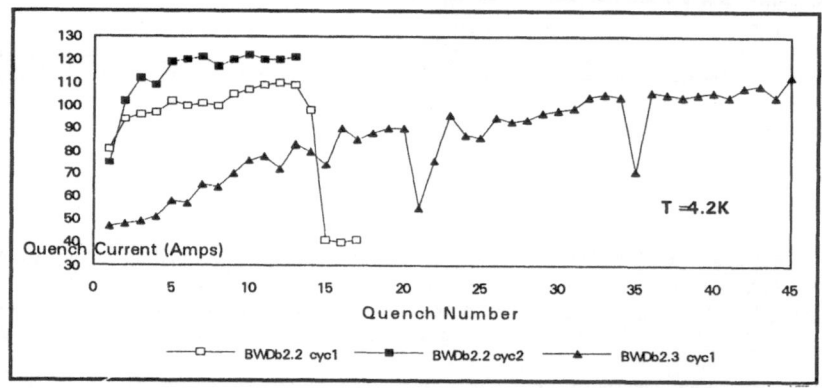

Figure 2. Quench summary of BWDb2.2 (two thermal cycles) and BWDb2.3.
Note the degradation in quench performance of BWDb2.3 resulting from the loss of end restraint.

The quench summary for BWDb2.2 is shown in Figure 2. On the first thermal cycle, the first quench occurred at a current level of 81 A.[3] Thirteen quenches were performed but the magnet never reached a plateau current. During the second thermal cycle, the first quench occurred at a current level of 75 A. The magnet reached a plateau current level of approximately 120 A in 5 quenches.

The quench summary for BWDb2.3 is also shown in Figure 2. During the first cycle of quench testing, the coil end clamps which offer restraint in the end regions fell off during testing. As a result, the coil end regions were left unsupported during quench testing. In addition, electrical shorts were detected by dc resistance and ac inductance measurements. Thus, only ramp rates <200 mA/s were used. Test results show that the first quench occurred at a current level of 42 A. Forty-five training quenches were performed but the magnet never reached a plateau current. A second thermal cycle was not performed on this magnet.

BWDb2.4

BWDb2.4 was fabricated after a modification was made on the wiring process. This modification was made in an effort to eliminate electrical shorts and reduce the radial build-up of wire at the coil ends. The modification consisted of replacing the four abrupt 45° turns with a series of twelve 15° turns, which

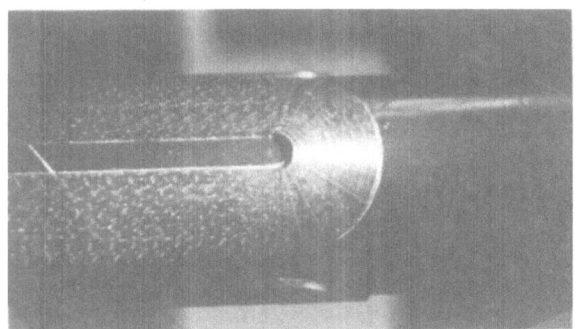

Figure 3. Coil BWDb2.4 (15° multiple vector turns).

closer approximates continuous motion. In addition, an encoder was added to the wire feed mechanism to provide feedback to the ultrasonic transducer (UT) energy controller. Using the feedback controller, UT energy was delivered to the stylus in proportion to the rate of wire feed. This provided a constant energy input per unit length of wire. These modifications improved the bonding and reduced the insulation damage in the turns (see Figure 3). In fact, BWDb2.4 was the first magnet built in which no detectable electrical shorts were encountered.

A reduced preload configuration was chosen to further minimize the risk of electrical shorts during collaring. This was accomplished by reducing the thickness of shimming material between the coil and the iron laminations. However, this left the conductor unsupported after accounting for differential thermal contraction. The lack of preload most likely led to the poor training behavior on the first thermal cycle.

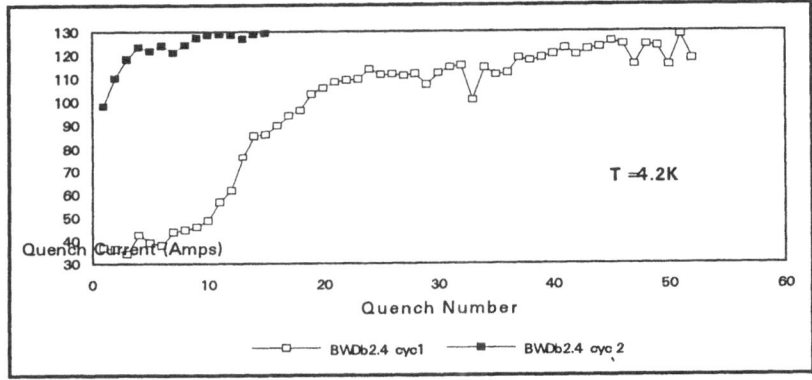

Figure 4. Quench summary of BWDb2.4 (both thermal cycles).
Note the substantial improvement of the quench performance on the second thermal cycle.

On the first thermal cycle, the first quench occurred at a current level of 37 A. It took approximately 40 quenches to reach a near plateau current of 120 A. The second thermal cycle showed marked improvement. The first quench occurred at a current level of 98 A. It took four quenches to reach a plateau current level of 124 A. After 15 quenches, the current level increased to 129 A. Ramp rates up to 10 A/s were used with no apparent degradation in quench performance.

This improved behavior would suggest that considerable conductor movement was occurring during training on the first thermal cycle caused by lack of support from the iron laminations. The data from the second thermal cycle suggests that these magnets do indeed remember their training.

SUMMARY

B&W and AIT have fabricated four sextupole corrector magnets for the SSC. Three of the four magnets have been quench tested at the Central Facility in Waxahachie, TX. A modification was made to the winding process after the completion of the third coil. All subsequent coils show considerable improvement in coil construction. Two of the three magnets which have been cold tested have shown poor quench performance on the first thermal cycle, but improved behavior on the second thermal cycle. The poor quench behavior on the first thermal cycle is believed to be caused by inadequate coil support (caused by the large differential thermal contraction) between the superconducting wire and the iron laminations. Various methods to improve coil support at operating temperature are being investigated. These methods include enclosing the coil in an encapsulated tube and potting the coil with an alumina filled cryogenic epoxy, and investigating ways to increase the prestress in the laminations, without inducing electrical shorts.

REFERENCES

2. E.T. Gossler and C.C. Coghill, Supercollider 4, ed. by J. Nonte, Plenum Press, New York 1992.

END NOTES

1. Coil BWDb2.1 was only used as a manufacturing mock-up and was not intended for cold testing.
3. The collider operating current for these magnets is approximately 78 A, or 60 % of the short sample value. Short sample current is estimated at approximately 130 A.

QUENCH PERFORMANCE OF B&W-SIEMENS 1 m QUADRUPOLE MAGNETS

C. M. Rey, J. A. Waynert, G. Dun, M. Xu, J. Savignano, J. P. Kelley,
K. Dixon, J. Maloney, P. Hlasnicek, A. L. Billingsly, and B. I. Cantor

Babcock & Wilcox
Accelerator and Magnets Section
P.O. Box 785 MS 80
Lynchburg, VA 24505-0785

INTRODUCTION

We report the quench performance of the first three short quadrupole model magnets designed and built by Babcock & Wilcox (B&W) and Siemens. These magnets are the first magnets designed and fabricated entirely by industry for the SSC. This series of short model magnets is denoted by the nomenclature QSH-801 to QSH-803.

MODEL MAGNET CONSTRUCTION

The four coils in each collared coil are arranged in a two layer cosine 2Θ pattern around a 40 mm diameter circular bore shown in Figure 1. This design is based on the original Lawrence Berkeley Laboratory (LBL) 40 mm collared coil cross section.[1] A new feature of the design is the spliceless transition from the inner to the outer coil layer. The magnet conductor is a 30 strand NbTi Rutherford type cable with a 1.2 degree keystone angle. Each strand has a diameter of 0.648 mm with a Cu/Superconductor ratio of 1.8 to 1. The cable is insulated with 2 layers of HN-50 kapton (12.5 mm thick, 50 % overlap) and fiber-glass cloth (0.1 mm). The fiber-glass cloth is impregnated with B-stage epoxy for later curing of the coil into a rigid form. The magnet cross section consists of a two piece collar assembly in which the stainless steel collar laminations are made of nitronic-40. The yoke laminations (~6 mm thick) surrounding the collared coil are made of ultra-low carbon steel. The collar-yoke interface uses a sliding design in which the collared coil is allowed to "breath" within the yoke.[2] The collared coil and cold mass assembly process has been described elsewhere.[3]

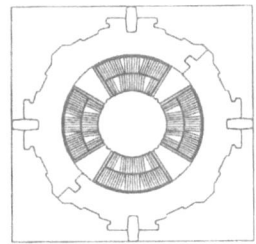

Figure 1. B&W - Siemens 40mm Quadrupole Magnet Collared Coil cross section

TEST PROCEDURES

Despite its nomenclature, model magnet QSH-803 was the first magnet built in the QSH series and was tested at the SSCL in Dallas, TX. Model magnets QSH-802 and QSH-801 were tested at the B&W Vertical Test Facility (VTF). QSH-801 through QSH-803 experienced three thermal cycles of testing.

Quench testing was performed by ramping the magnets at 16 A/s to 6500 A, waiting for 5 minutes and then ramping to quench at 4 A/s. No energy extraction methods were initiated after the quench. The stored energy being dissipated as heat into the helium bath. Magnet training quenches were performed until the magnet reached plateau. No sub-cooling was necessary to accelerate training on any of these short model magnets. Ramp rate studies were then performed to determine their effect on the plateau current. Typical ramp rates that were used varied from 4 A/s to 500 A/s.[4] The mechanical behavior of these magnets has been described in another paper.[5]

TEST RESULTS

QSH-803

The quench current summary for all thermal cycles of QSH-803 is shown in Figure 2. The first quench on each of the three successive thermal cycles occurred at the current levels 7557A, 7749 A and

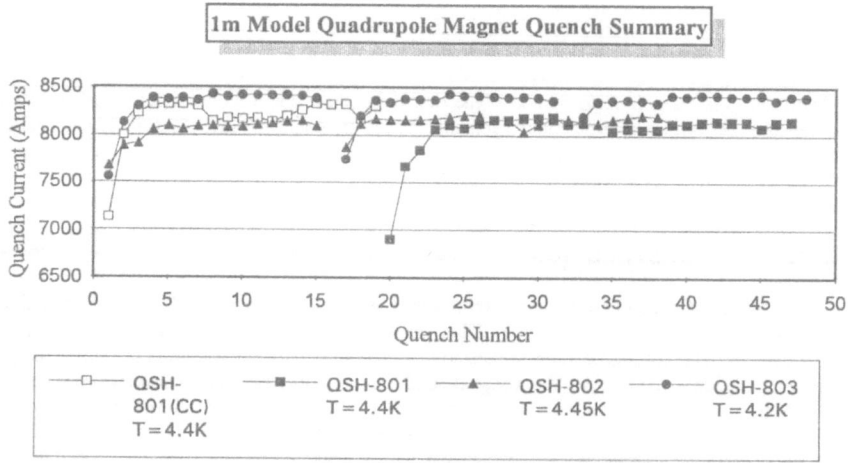

Figure 2. Quench current summary for B&W - Siemens Short Quadrupole Model Magnets.

8198 A respectively.[6] The plateau current for each thermal cycle was approximately 8360 A at 4.23 K. Short sample critical current for this cable (all Oxford) is estimated at ~8084 A at 4.23K.

The ramp rate behavior of QSH-803 is shown in Figure 3. No appreciable ramp rate dependence was observed for ramp rates up to 300 A/s (the highest rate studied).

Nearly all training quenches (88 %) originated in the inner- outer coil ramp region. Of these training quenches 50 % originated in the ramp of quadrant 4. In the ramp rate studies, all of the quenches occurred in the outer pole turn straight section.

QSH-802

The first quench on each of the three successive thermal cycles occurred at the current levels of 7680 A, 7882 A, and 8040 A, respectively. The plateau current for each thermal cycle was approximately 8160 A at 4.40 K.[7] After the second thermal cycle, axial end restraint was imposed on the magnet. In this configuration, the magnet still had a sliding yoke-collar interface but the magnet could not grow in the axial direction during energization.

Over 70 % of the training quenches on QSH-802 occurred in the ramp region which bridges the inner layer to the outer. For ramp rates between 16 and 300 A/s, all quench origins appear in the outer pole turn straight section of quadrant one.

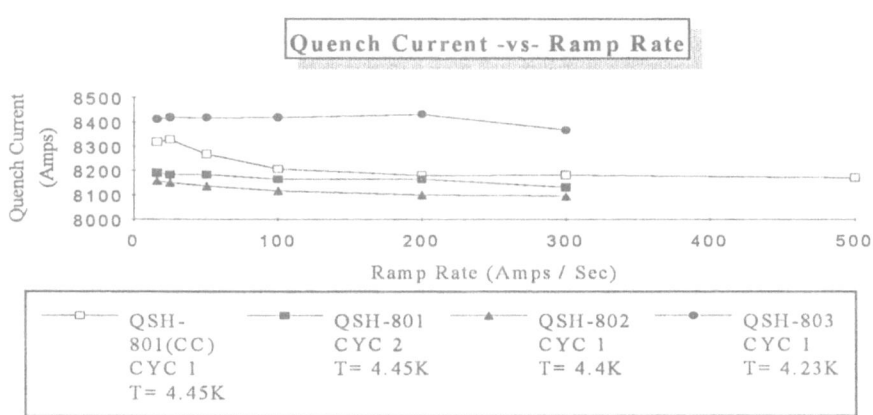

Figure 3. Ramp rate summary for B&W - Siemens Short Quadrupole Model Magnets.

QSH-801

The collared coil for QSH-801 was cold tested prior to its final assembly into a yoked and shelled cold mass. Results show that the first quench occurred at a current level of 7130 A at 4.45 K. It took four quenches to reach a plateau current level of 8320 A. Ramp rate studies show no appreciable rate dependence up to 500 A/s (the highest rate studied).

Once the collared coil was yoked and shelled into its cold mass assembly (QSH-801) it was again tested. The first quench occurred at a current level of 6905 A. It took four quenches to reach a plateau current of 8060 A at 4.45 K. The magnet again showed no appreciable ramp rate dependence up to 500 A/s. After this thermal cycle, axial end restraint was imposed on the magnet. In this configuration, the magnet still had a sliding yoke-collar interface but the magnet could not grow in the axial direction during energization. The magnet was retested, with the end restraints, to determine its quench performance. The magnet had no training and went right to plateau current (8060 A).

Based upon the previous quench performance of QSH-803 and QSH-802, we were somewhat surprised that on the first thermal cycle of QSH-801 (the second thermal cycle experienced by its collared coil), that the magnet took four quenches to reach plateau. QSH-803 and QSH-802 all had less training on their second thermal cycle. QSH-801 on its second thermal cycle (the third thermal cycle for its collared coil) showed no training behavior. This behavior would suggest that either these collared coils do not remember their training, or that the training behavior may be influenced by the yoke-collar interface or the yoke-shelling process itself, or both. However, without a more comprehensive study in which several thermal cycles are performed on collared coils, this is only speculation.

There is one interesting result from the QSH-801 testing which our thermal models do not adequately explain. When the collared coil was cold tested, ramp rate studies showed a slight degradation (<2 %) from plateau current for ramp rates between 16 A/s and 200 A/s. This is not unusual behavior, and is usually attributed to additional eddy current heating in the cable at these higher ramp rates. After the magnet had been yoked and shelled into the cold mass assembly; however, ramp rate studies showed an increase (~ 1 %) from plateau current. This type of behavior has also been observed before on short magnets tested in pool boiling liquid helium. What we cannot explain, is why both types of behavior were measured on the same magnet, and furthermore we do not understand what characteristic in the yoke-shell process/interface causes this change in behavior.

SUMMARY

In summary, B&W-Siemens has built and tested three short model quadrupole magnets. Two of the short model magnets were tested at the B&W Vertical Test Facility. All magnets meet and exceed all quench requirements for the operation of the collider quadrupoles. The model magnets show current margins in excess of 20 % above operating current (6714 A) and all three magnets have taken only four training quenches to reach plateau current on the first thermal cycle. Subsequent thermal cycles show little or no training behavior. Training, plateau, and ramp rate quench origins all appear to be reproducible within the magnet.

REFERENCES

[1]. A.F. Lietzke,et al., "Quadrupole Magnets for the SSC," presented at MSIM meeting, Sept. 1992.
[3]. M.W.Hiller, et al., "Advances in Cryogenic Engineering," ed. by R.W.Fast, Plenum Press, 1991.
[5]. C.M. Rey, et al., "Mechanical Behavior of B&W Short Quadrupole Magnets," Submitted to IISSC 5.

NOTES

[2]. For these 1 m model magnets, at room temperature, typical forces required to move the collared coil within the yoke were ~200-300 N.
[4]. Note, all ramp rate studies were performed by ramping from 0 A to quench.
[6]. The collider operating current is 6714 A.

AXIAL DECONVOLUTION OF MEASURED MAGNET FIELD HARMONICS

Daniel W. Bliss and Husam F. Gurol

General Dynamics Space Systems Division
Space Magnetics
P.O. Box 85990
San Diego, CA 92186

INTRODUCTION

The measurement of magnetic field harmonics for the Super Collider magnets is currently performed using a rotating one meter long coil referred to as the "mole". The fact that the mole is 1 meter long hides the fine structure of the harmonics as a function of axial position. Along the length of the magnet, particularly in the ends, there are axial variations in harmonics that are much shorter in wavelength than the mole search coil. This is not important to accelerator operation since only the integrated harmonics have a significant effect on the beam physics. However, in trying to understand manufacturing effects and end design, the fine axial harmonic structure is important. In addition, the mole length places significant limitations on the usefulness of discrete Fourier transforms in studying production effects on magnetic field harmonics. This paper details a method of determining the fine axial harmonic structure using the present mole design.

PHYSICAL AND MEASURED HARMONICS DEFINITION

The mole measurement of the magnetic field can be expressed in terms of harmonics using a series expansion of the magnetic field integrated over the mole length. Equations 1 and 2 define physical harmonics and measured harmonics. In general, the coefficients are modified by various scaling terms such as central field or an arbitrary value of 10,000. However, the particular scaling is not important to this analysis, and can be ignored. It is interesting to note that the series expansion is in two dimensions, in the presence of fields with gradients in all three dimensions. Consequently, a single such expansion is insufficient to completely describe the field.[1] Nonetheless, the two dimensional local field harmonic coefficients can still be useful in understanding the local manufacturing and end design effects.

$$\mathbf{B} = \sum_{n=0}^{\infty} [b_n(s) + i\, a_n(s)]\, \mathbf{z}^n \tag{1}$$

$\mathbf{B}$ is the complex representation of the magnetic flux density, $B_y + i\, B_x$,
$\mathbf{z}$ is the complex representation of transverse displacement, $x + i\, y$,
n is the harmonic number, where $n=0$ is a dipole,
$b_n(s)$ and $a_n(s)$ are the normal and skew harmonics or multipoles,
s is the longitudinal displacement

Supercollider 5, Edited by P. Hale
Plenum Press, New York, 1994

$$D_n(s) = \int_{s-L/2}^{s+L/2} C_n(s') \ ds' = \int_{-\infty}^{\infty} m(s-\lambda) \ C_n(\lambda) \ d\lambda \qquad (2)$$

$D_n(s)$ is the measured harmonics,
$C_n(s)$ is either a_n or b_n,
L is the length of the mole,
s' is a variable of integration,
m(s) is gate function with a value of one within the limits of -L/2 and L/2, zero
      otherwise,
λ is a variable of integration

The function m(s) is introduced to represent the mole. The mole function is a gate function with a width of L and a height of unity. Taking the Fourier transform of equation 2 produces equation 3.

$$D_n(s) <=> D_n(\omega) = M(\omega)C_n(\omega) \qquad (3)$$

M(ω) is the Fourier transform of m(s), L $\mathrm{sinc}\left(\dfrac{\omega L}{2\pi}\right)$,

$C_n(\omega)$ is the Fourier transform of $C_n(s)$

From equation 3, it is clear that the frequency distribution of the measured harmonics is perturbed by the frequency distribution of the mole function. One might be tempted to try to convolve the measured harmonics with some inverse function of the mole, or divide the frequency distribution by M(ω), producing the physical harmonics. However, these methods tend to be cumbersome. The following is a simpler alternative.

RECURSION METHOD DEFINITION

By differentiating equation 2, and rewriting this in a more convenient form, the recursive relation, equation 4, is found.

$$C_n(s) = \frac{dD_n(s-L/2)}{ds} + C_n(s-L) \qquad (4)$$

The relation is applied recursively to determine the harmonic at some arbitrary value of s. To reconstruct the harmonics as a function of s, the relation is applied until an axial location of (s-kL) is reached, where the value of k is some integer such that C_n(s-kL) is known to be essentially zero. This occurs at some small distance from the end of the magnet.

HARMONIC RECONSTRUCTION EXAMPLE

The following is an example of axial harmonic deconvolution applied to a calculated b_2 distribution for the model Super Collider, collider dipole magnets. In the following figures b_2 is referred to as C_2 for the sake of consistency with the above equations. The model magnets are slightly less than two meters long. The mole was taken to be 1m in length. Both the physical and the reconstructed harmonic distributions are displayed in Figure 1. Figure 1 also contains an expanded view of the comparison at the far end of the magnet; the end of the magnet that required multiple iterations. The original and reconstructed harmonics are well matched, even at the far end of the magnet.

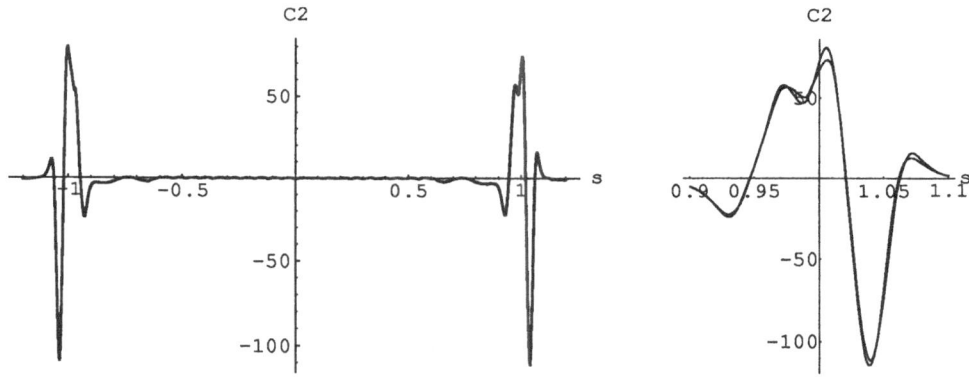

Figure 1. Overlay of physical and reconstructed harmonic distributions, $C_2(s)$

The measured harmonics are displayed in Figure 2. The distribution was produced by simulating mole measurements at 0.01m axial intervals. This distance is a reasonably course sampling interval. It is interesting to compare the measured to the physical harmonics. They bare little resemblance to each other. The derivative of the measured function was performed by taking a numerical derivative of a third order interpolation of the mole measured harmonics, see Figure 3.

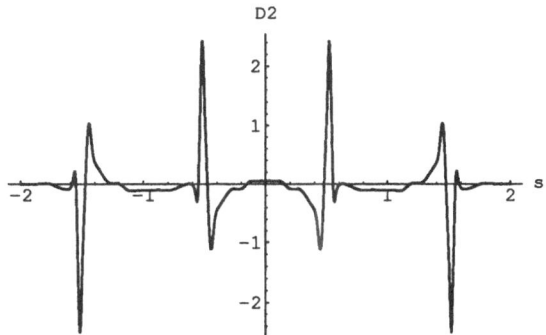

Figure 2. The harmonic distribution measured by the mole, $D_2(s)$

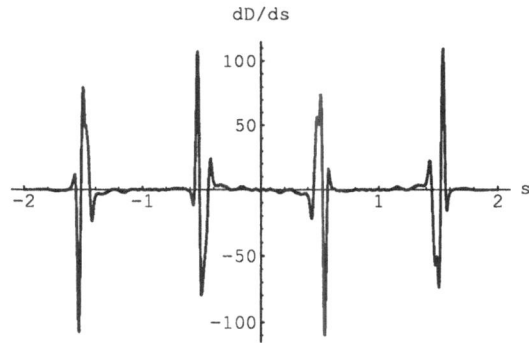

Figure 3. Derivative of the mole measured harmonics, $\dfrac{dD_2(s)}{ds}$

DISCUSSION

In order to use this method, the measurement of the harmonics using the mole must be of sufficient axial resolution that the derivative can be accurately calculated. Viewing the calculated harmonic distributions for Super Collider CDM's, an acceptable resolution would be a spacing of less than 1cm. This high resolution requires that a large number of measurements must be taken. It will require a significant investment in time to completely characterize a full length magnet. Clearly, this would not be desirable for production, but during prototyping when we are trying understand the field harmonics of the ends and manufacturing effects, the additional time spent measuring the magnet harmonics is eminently reasonable. In addition, it may be possible to reduce the total measurement time by varying the sampling resolution as a function of axial position. Due to the large axial variations of harmonics in the magnet ends, it may be desirable to take measurements at very high resolution in that region, on the order of 5mm or less. Conversely, in the central region, it may be found that a few centimeters is a reasonable spacing. This variation in sampling spacing would not significantly complicate the calculations.

There are some limitations to this method. The recursion method amplifies the measurement error of the mole in a couple of ways. First, taking the derivative of the experimental data causes the error to be exacerbated. This effect is mitigated by using a higher order interpolation of the measured data, essentially filtering the signal. Second, the harmonics of the central region of the magnet accumulate the error from several reconstructions further out axially. This will place an upper limit on the accuracy of the measurements. However, this should be compared to the current situation where there is very limited information is obtained on the axial variation in harmonics.

CONCLUSION

This paper presented a method of determining the fine axial structure of magnetic field harmonics for accelerator magnets using existing test equipment. The method allows investigation into production and end design effects on axial variations of magnetic field harmonics. An example was provided to demonstrate the method. The example displayed that physical magnetic field harmonics could be accurately reconstructed from mole measurements.

ACKNOWLEDGMENTS

We would like to thank Roger Simon of General Dynamics, Peter Wanderer, and Animesh Jain of Brookhaven National Laboratory for their comments.

Portions of the work described herein are being accomplished under contract to the University Research Association in support of the Superconducting Super Collider project for the U.S. Department of Energy.

REFERENCES

1. D. W. Bliss, M. P. Krefta, and H. F. Gurol, Magnetic and mechanical considerations in the design of the SSC dipole magnet end region, Applied Superconductivity Conference, 1992.

APPLICATIONS INTERFACES TO THE MAGNET DATABASE SYSTEM

M. J. Ball, N. Delagi, B. Horton, J. C. Ivey, L. Jones, R. Leedy, X. Li,
B. Marshall, S. L. Robinson, and J. C. Tompkins

Magnet Systems Division
Superconducting Super Collider Laboratory[*]
2550 Beckleymeade Ave., MS 1002
Dallas, TX 75237-3997

ABSTRACT

The Test and Data Management Department of the Magnet Systems Division (MSD) of the Superconducting Super Collider Laboratory (SSCL) maintains a central database of information that is relevant for analysis and comparison of test results for all SSC superconducting magnets, including those from all the magnet subcontractors and those built in-house. Interactive interfaces have been developed to allow the SSCL magnet scientists to easily extract desired data in the format required by their analysis programs, so that the database can serve as the sole source for all magnet studies. Examples are given, including applications for strain gauge and quench voltage tap data analysis and magnetic field harmonics measurement.

DATABASE FUNCTION

Over 10,000 superconducting magnets will be required for the SSC collider and High Energy Booster rings. Some of these magnets will be built 'in house', but most will be built by industrial vendors under contract to the SSCL Magnet Systems Division. All magnets will be tested warm at the vendor site, and many (including all of the early units) will be tested cold at the vendor site or the SSCL Magnet Test Laboratory (MTL).

The Data Management Group of the MSD Test and Data Management Department is integrating all the data that is relevant for analysis of test results on all SSC superconducting magnets into a network-accessible data management system and developing a user interface that facilitates access to this data. The focal point of the data management system is a central database which we have named MagCom. This database will include test results for critical component materials, configuration information (serial or batch numbers) for critical components in all magnets, data collected at major production steps, summaries of test

---

[*] Operated by the Universities Research Association, Inc, for the U.S. Department of Energy under Contract No. DE-AC35-89ER40486.

results on all magnets, and reference 'pointers' to the computer files that have the full data. It will serve as the source for the standard information needed by the magnet analysis programs and any ad hoc requests for information. It will provide summary information for each magnet to the magnet installation contractor and the Accelerator Division controls databases. We are designing the user interfaces to allow simple access and manipulation of any magnet data, whether it is in the database or the external files.

The MagCom database is implemented using the Sybase relational database management system in a UNIX X11 environment on a SUN 4/280 server connected to Internet (grumpy.ssc.gov) and DECnet (SSCSUN)[1]. We recently acquired a faster server (SUN SPARCserver 630MP) with 24 GigaBytes of magnetic disk storage and are transferring the database operation to it.

Past effort [2,3,4] has focused on development of database structures and data import procedures for R&D data from magnets built at Brookhaven National Lab, Fermilab, and Lawrence Berkeley Lab. An automatic input processor has been developed that detects the presence of new files and runs the import procedures for known types of input files. A text menu system provides access to standard reports. Specialized user interfaces have been written for strain gauge, quench, and harmonics test runs, to assist the analysis staff in selecting data from the database for input to their application programs.

QUENCH ANALYSIS INTERFACE

For quench analysis, we created an interactive interface which allowed us to build an integrated application from an existing FORTRAN quench analysis program (MPLOT) and an existing PV-Wave graphical channel analysis program (PLVT). The interface provides a means by which an external program can interact with the user via a graphical user interface (GUI) using windows called widgets.

The quench test data is stored in the SDS (Self-Describing Data Standard) format developed by the SSC Central Design Group with the ISTK project [5]. The MPLOT program reads a quench SDS file, computes certain additional information such as resistances on coils, allows the user to select channels to plot, and produces an output file suitable for PLVT. The PLVT program plots the channels in the output file, allows the user to zoom in on specific parts of the plot, and provides curve-fitting options.

The quench analysis interface was written in IMSL/IDL command language. IMSL/IDL and PV-Wave have a common core of commands; the 3200 line quench PLVT runs under IMSL/IDL with no code changes. The interface is actually a more general tool which possesses the following capabilities:

1) The user can execute a FORTRAN program from within the interface and communicate with the program using IDL widgets or the text window.

2) The user or an external program can invoke an IDL (PV-Wave) procedure and communicate with it via UNIX system pipes.

This approach is faster and more intuitive than a text-based interface. In this application, the interface speeds up the operation of selecting a magnet and test, selecting channels to plot (within MPLOT) and then displaying them (using PLVT). A specialized widget procedure facilitates the selection of a magnet and test by displaying a list of existing data files for user selection. The user selects channels within a test by another specialized widget.

The interface has been used to analyze quench test data from FNAL and BNL. We have also created a program which imports quench and ramp rate test results from the Short Magnet and Cable Testing Laboratory (SMCTL) at the SSCL so that it can also be analyzed with this interface.

STRAIN GAUGE INTERFACE

The strain gauge interface links data in the database with the user's PV-Wave graphical interface programs. Full advantage of the Sybase utilities was taken to provide a file in the format needed by the PV-Wave programs. Sybase utilities use 4th generation tools for faster development and easy maintenance.

The user runs a Unix script which provides calls to C code and the database. The C software requests choices of magnet, test run, etc. and sends these choices as parameters to a Sybase Report. Some of these reports will also prompt the user for further information. This use of parameters within the reports makes the reports more general purpose. However, as the user's needs change (for example, based on experience gained as the prototype magnets are built), the reports can be easily changed within Sybase's Report Workbench utility without having to recompile the C code that calls the report or accesses the report output.

Most of the file formatting is done with the Report Workbench utility. However, some reports require additional massaging of their output files that is more efficiently done with Unix utilites such as "grep". After this last step, the PV-Wave software grabs the generated file as input to its graphical user interface (GUI) environment from which the user generates plots interactively.

The set of reports currently available provides various views of the data: measurement data from a specific strain gauge and run, or summary data over a run (with averages, maximum/minimums and standard deviations), or listings of channels for non-zero or zero amp data by test run, or listings of timestamped user comments entered during the test runs. Existing plots display the stress values from eight strain gauges for a given strain gauge run vs. current squared.

Future plans include developing a more general application program interface to directly link the database to the PV-Wave programs, thus eliminating the intermediate formatted file.

HARMONICS INTERFACE

A prototype harmonics database, graphical user interface (GUI), and data loaders are now available for use by the MSD Test Department. The interface allows analysts to select individual magnet test runs, and then specify the dataset (subset or entirety of the full harmonics test data) to download to files that may be used as input into existing analysis programs.

Data for input to the database came from R&D magnet tests at BNL, FNAL, and LBL, and the new magnetic field measurement system (mfm) developed by the MSD Test Department. All of these sources use different data formats and methods of organizing the data in subdirectories. A two-step loader was developed to load all the data for a given magnet from its directory structure. One module (common to all data sources) interpreted the subdirectory hierachy "database" which contained information about the test conditions (Warm/Cold), type of test, etc. and in turn called the second module (format specific) that actually read and loaded the data. The first module was written using Perl and SybPerl while the module that actually loaded the data was written in C++.

Two approaches to the graphical user interface were implemented; one downloads data to an ASCII file and the other creates SDS format files. Both use the ISTK tools (glistk, sds, glish) and C++ to provide a 'point and click' access to the database, allowing the user to specify the data source by working through a menu of selections for the magnet type, magnet-id, test type, and test number. The user then specifies the dataset content (full test results or selected subsets) to be returned for that selected test. One interface allows the user to output the requested dataset to an ASCII file which can be used as input to existing analysis routines. The other interface creates output in an SDS file format which can be used as input to an existing analysis program or to a display program. The display program allows the user to select subsets of the SDS file to send to a graphics server which was written using PV-Wave in server mode. Thus the user could choose to look at one harmonic plotted along the length of the magnet, or several harmonics at a specific z-position.

Several lessons have been learned from experience with this prototype.

1) File system databases (which rely on directory structure for certain identifying information) are very error prone and probably not a good way to store important experimental data. This descriptive information should be included in the files themselves.

2) It is difficult to deal with multiple types of data inputs that were not designed with consideration of data management plans.

3) The user interface needs to provide links to analysis programs but also needs to provide more ad hoc ways to specify contents of datasets.

FUTURE PLANS

The next step is to standardize formats and procedures based on our past experience. We are developing data content and format specifications for data to be submitted by the magnet vendors and are collaborating with the MTL Software Group on development of test data file structures and general data processing software that is needed for their initial operations. This general software, which will be usable with the MagCom database also, includes an application program interface and an interactive browser that will allow ad hoc selection of data to be displayed in table or plot format. Our goal is to have a library of procedures that will facilitate study of data, from fabrication through testing steps, on one magnet or comparison of data across a series of magnets.

REFERENCES

1. James Ivey, Xiaoyu Li, Joe Garbarini, Penny Ball, "System Design Overview for the Magnet Test Database System," Supercollider 4, M. McAshan, ed., Plenum Press, New York (1992).
2. M.J. Baggett et al., "The Magnet Components Database System," Supercollider 2, M. McAshan, ed., Plenum Press, New York (1990).
3. P. Baggett et al., "The Magnet Database System," Supercollider 3, M. McAshan, ed., Plenum Press, New York (1991).
4. M.J. Ball et al., "The Magnet Database System," Supercollider 4, M. McAshan, ed., Plenum Press, New York (1992).
5. Erika Lutz, "The ISTK Overview Document," Lawrence Berkeley Laboratory (1991).

MTL DISTRIBUTED MAGNET MEASUREMENT SYSTEM

Jerzy M. Nogiec, Paul A. Craker, Joe P. Garbarini, Vladimir Ilushin,
James C. Ivey, J. David Lambert, and Wei-chuan Li

Magnet Systems Division
Superconducting Super Collider Laboratory[*]
2550 Beckleymeade Avenue
Dallas, TX 75237-3946

INTRODUCTION

The Magnet Test Laboratory (MTL) at the Superconducting Super Collider Laboratory
will be required to precisely and reliably measure properties of magnets in a production
environment. The extensive testing of the superconducting magnets comprises several types
of measurements whose main purpose is to evaluate some basic parameters characterizing
magnetic, mechanic and cryogenic properties of magnets.

The measurement process will produce a significant amount of data which will be subjected to complex analysis. Such massive measurements require a careful design of both the
hardware and software of computer systems, having in mind a reliable, maximally automated system. In order to fulfill this requirement a dedicated Distributed Magnet Measurement System (DMMS) is being developed.

HARDWARE CONFIGURATION

The existence of multiple test stands for testing magnets, the inherent parallelism of the
testing process and the distribution of the measurement system led naturally to the concept
of applying multiple computers connected via a local area network. Such a computer network will serve two purposes: (1) monitoring and control of measurement test stands, and
(2) operation of the system, data storage and analysis of measurement results. This requires
the computer system to possess real-time system features as well as user-computer dialog
oriented features. In order to fulfill these different requirements, two categories of computers are used. VME-based Motorola 68030 computers running the VxWorks real-time operating system are used for data acquisition and monitoring. Analysis of results and operation of

---

*Operated by the Universities Research Association, Inc., for the U.S. Department of Energy under Contract
No. DE-AC35-89ER40486.

the measurement systems are implemented on Sun Sparc workstations. Some of the measurement instrumentation is connected to the computers using the IEEE-488 interface system. A dedicated Sun server is used for long-term data storage purposes. All the computers are connected, and form a subnet, separated from the outside traffic via a router (see Figure 1).

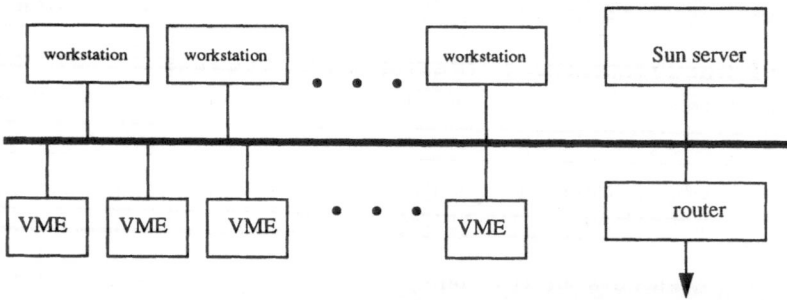

Figure 1. DMMS computer configuration.

SYSTEM ORGANIZATION

One can distinguish the following elements which compose the DMMS: measurement subsystems (called also data acquisition subsystems), services, and a data handling subsystem.

The core of the DMMS is composed of measurement and quench characterization subsystems. The measurement subsystems consist of communicating processes running on two computers: (1) a workstation that hosts the user interface and stores data, and (2) a VME based computer that controls the measurement process by directly interacting with hardware modules.

The distributed character of the system introduces problems associated with managing resources and synchronization in a network of computers. Our approach to these problems is to view the distributed system not only as a collection of multiple local systems but also as a single coherent system. There will be no central authority in the DMMS to resolve problems of several activities (applications) competing for limited resources (test stands, data acquisition computers, power supplies etc.). Instead, the need for synchronization in a system sharing resources will be satisfied in a distributed fashion. Resources are available in the DMMS through services. Each service is accessible and fully controlled by a single manager process. A manager decides which requested operations to perform and in which order. The decision is made based on the current resource status and the client authorization to request service.

There will be a number of services that differ in complexity, reliability requirements, and implementation. A partial list of anticipated services follows:

1. *Protection service*: The DMMS will be protected against the use by non-authorized people by using passwords for authentication purposes and access privileges for authorization definition purposes.

2. *Interlock service*: The interlock service is an interface to the interlock system that will enable users to obtain current interlock status, reset an interlock status, and access the interlock log file.

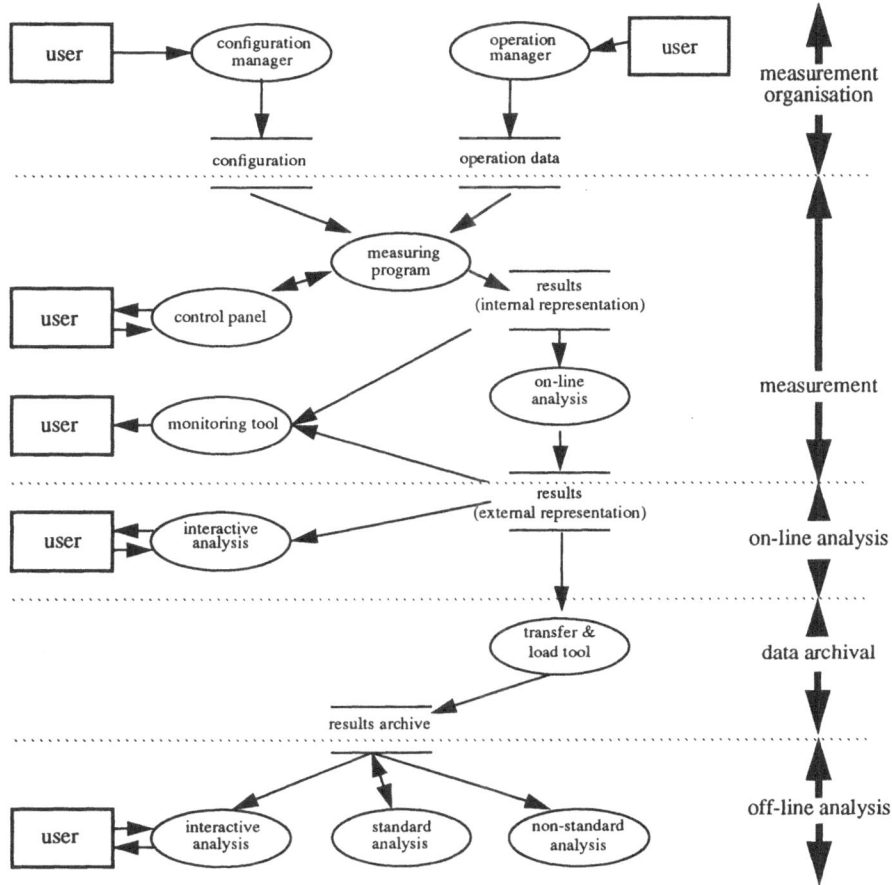

Figure 2. Simplified data flow diagram.

The described system is still under development and it will be achieving its final version gradually. The methodology applied to development of the system is based on both the classical software engineering approach called waterfall lifecycle, and on prototyping. The classical software development methodology emphasizes through identification of software development phases (requirements analysis, system design, implementation, testing, operation and maintenance), whereas prototyping methodology manifests itself in the fact that the development process modifies the specification as the project progresses. The users are provided with prototype systems to allow them to check their ideas and acquire practical measurement experience.

3. *Power supply service*: The power supply service will be responsible for communication between the magnet power supply control hardware and software processes requesting changes in magnet current values. It will also coordinate requests coming from the different client processes and maintain a current change history log.

4. *Transporter service*: The transporter service is responsible for positioning probes inside magnets.

5. *Electronic logbook*: The logbook service will enable operators to write all the comments,

741

notes, problems, questions etc. to the common log. All entries will have assigned origin data (operator name, computer) and timestamps.

6. *Exception and event logger:* The exception and event logger servers use a log technique to permanently store information about exceptions and important events that take place during operation of the system. The logs can be examined later for debugging and maintenance purposes. The present version of the service is logically centralized, implemented as a central multiple process server and distributed sub-servers, running on every computer which is under control of the system.

7. *Checklist service:* The progress of testing of a particular magnet will be reflected in the magnet checklist.

8. *System news service:* The system news service will be responsible for storing and making available to operators information which concern the recent changes in operation of the system.

DATA MANAGEMENT

The storage and retrieval of various data produced by the DMMS is based on the centralized Sybase data base system. Three logically separated data bases have been defined: magnet measurement, operation, and configuration data bases. The magnet measurement data base will contain data collected during measurements and data produced as a result of analysis. The operation data base will contain data necessary to facilitate and automate the operation of the MTL, whereas the configuration data base will maintain data describing hardware and software configuration of the DMMS.

Data produced by magnet measurement subsystems of the DMMS will be passed in files to the data handling subsystem that will be responsible for loading data to the data base and archiving original files. To facilitate the archival part of the system, the dedicated catalog data base is under construction. The purpose of this data base is mainly to maintain information about location of all original data files stored in the archives.

A simplified data flow diagram is shown in Figure 2. Dotted lines mark boundaries between logical phases in the process of measuring magnets and analyzing measurement results.

SUMMARY

The DMMS is a distributed computer system devoted to measuring properties of accelerator magnets. In this system several applications can run simultaneously and distributed resources are allocated dynamically using a non-centralized solution. In contrast to this, the data management part of the system is based on a central data base server. The data acquisition parts of the system and the data handling part are loosely coupled that eliminates the possibly negative influence of a non-real-time data base on the performance of real-time constrained measurement subsystems.

A QUENCH DETECTION/LOGGING SYSTEM FOR THE SSCL MAGNET TEST LABORATORY

K. Kim, M. Coles, J. Dryer, and D. Lambert

Magnet Systems Division
Superconducting Super Collider Laboratory*
Dallas, Texas 75237-3997

INTRODUCTION

The quench in a magnet describes a process which occurs while the superconductivity state goes to the normal resistive state. The consequence of a quench is the conversion of the stored electromagnetic energy into heat. During this process the initiating point will reach a high temperature, which will char the insulation or melt the conductor and thereby destroy the magnet. To prevent the magnet from being lost, it is standard practice to observe several resistance and/or inductance voltages across the magnet as quench signatures — Detection. When a quench symptom is detected, protection operations are initiated: proper shutdown of the magnet excitation systems and treatment to dilute the heat energy at a spot — Protection. The temperature rise is diluted by firing heaters along the length of the magnet to insure that the dissipated energy is spread. It is interesting that there is not a significant amount of published research on detection. One reason is that the detection procedure is believed to be a transparent operation based on properly balanced voltage tap signals. Another reason is that there are working detection systems at other laboratory test facilities, which perform fine. Subsequently, for a new test facility, it is customary to adopt an existing system with minor modifications to retain the working experience of the existing system. In reality, the noise environment varies, facility to facility, and it is necessary to understand the existing systems quite well for the specific modification. For example, digitally controlled power supply systems, adopted for the precise control of the excitation current, can produce strong spike noises and consequently increase the false quench detection rate. The addition of simple low pass filters can sometimes reduce the spike heights, but at the expense of a broadened plateau. Power line noise (60 Hz) and its harmonics, communications interference and ground noises can also vary at new installations. To afford a more reliable quench detection system, two distinct approaches have been tried in the past: i) Understanding of the Noise Mechanism [ZhEt92] and Sub-system Optimization [ShEt85], and ii) Escaping from the Known Electromagnetic Noises by Observing Optical Waves [TsEt87] or Acoustic Waves [Dres88]. The MTL of SSCL confronts a mass-measurement of about 10,000 production magnets [Cole92]. To meet the testing schedule, the false quench detection rate needs to be further optimized while the true quench detection rate remains secure for the magnet measurement safety. To meet these requirements, we followed an iterative top-down approach. First we defined the signal and noise characteristics of the quench phenomena by using existing software tools to build a rapid prototype system incorporating all proven functionality of the existing system. Then we further optimize the system through iterative upgrading based on our signal and noise character findings.

* Operated by the Universities Research Association, Inc., for the U. S. Department of Energy under Contract No. DE-AC35-89ER40486.

OVERALL ARCHITECTURE

The general design goal of the new system is four-fold: i) reliability, the system will be equipped with more powerful tools such as digital signal processing techniques, ii) flexibility, the system will be modularized for further expansion and external operator interfacing will be enhanced to control the operation of the system, iii) ruggedness, the system will depend on the well-proven standard protocol and unit-tested off-the-shelf sub-systems with minimum re-invention, and iv) adaptivity, each quench event will record the detection history and enable us to later evaluate the algorithms and parameters for future upgrading.

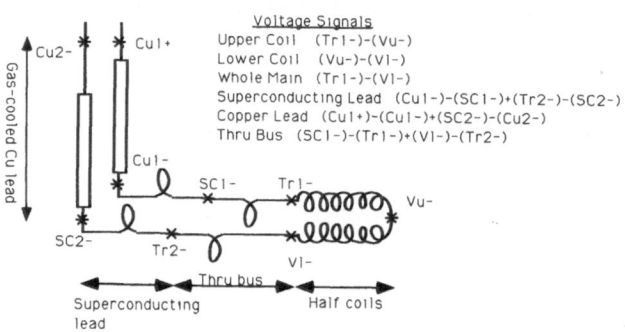

Figure 1. Voltage tap convention.

Hardware: For the SSCL magnet test facility, the overall instrumentation system is modularized with the quench detection system being one module [LaEt92]. The VME readout instrumentation system for the quench detection system is an Analog-to-Digital Converter (ADC), which takes voltage inputs from the isolation amplifiers directly tapping the voltages inside the measurement magnet, the cryostat feed-can interconnection, the transductor signal measuring the power bus current, and the TTL-compatible triggering signal from the backup card. Figure 1 depicts voltage taps from the magnet and the feed can. Isolation amplifiers for the quench detection system have fixed gains to swing the signal of interest within 10% of the full scale of the ADC. Traditionally, the most reliable input voltages are across the magnet coil. To use the prevailing concept of the simple bridge comparison, the whole magnet coil voltage is bisected into half cell voltages of the upper and lower half coil. The well-balanced bridge comparator cancels the inductive voltages of the half coils, and any output of the bridge is attributed to the normal resistive component in one of the coils. To minimize the number of channels for the sampling, symmetric tapped voltages of the positive and negative power buses are added. The power bus-related inputs are the Cu power bus voltage drop, the superconducting (SC) power-lead voltage and the SC through-bus voltage. The isolation amplifiers use Analog Devices AD210 amplifiers, which isolate input, output and power via the transformer couplings which are internal to the modules. The ADC is a DVX-2502 from Analogic accepting 8 differential inputs within +/- 10-V range. A digital output module VMIC/VME2170 is used to output the TTL compatible levels to control the protection system. For the fail-safe operation, an Uninterrupted Power Supply (UPS) backs up all critical electronics, including the quench detection crate and isolation amplifier cards. Another fail-safe card is the backup card, whose main function is to compare the voltage drops across the leads against a fixed threshold, the resistive voltage across the coil against a threshold, and the half-coils voltage difference against a threshold. In the event that the VME crate malfunctions, (for example, due to an unexpected software hangup or a broken hardware card) the analog quench detection system on the backup system will trigger the relevant quench protection systems. The backup card has redundant circuits with separate power supplies. Isolation amplifiers for the voltage taps and the quench detection system are all on the backup card, hence the backup card was designed with the high voltage failures consideration. Also, an analog differentiator is implemented on the backup card, which differentiates the input up to 3 Hz and has a −20 dB/dec roll-off sufficient enough to remove the high frequency noise.

Application Software: The application software has been divided into a set of 3 concurrent modules: i) acquisition, ii) processing and iii) action modules. The acquisition module handles tasks related to the ADC H/W, and continuously manages several sets of ring buffers. The ring buffer size can be set either short or long. The short size is limited to 20 words, in which the worst system latency is less than the stacking time of the buffer (about 20 ms) for detection and protection. The long size is 1000 words to record the quench signature signals for about 1 second both before and after processing. The processing module implements the core algorithm of the application, to enhance the received signal and test the object measurement quantities. The action module is to interface to the surrounding systems including power supply controller, cryogenic controller, quench protection system and operator interfacing, to broadcast the processing results and gather necessary information. The main processing module is composed of 4 stages:

1. DATA ACQUISITION: The data acquisition is the highest priority task except for the system tasks, and affords a uniform sampling timing. Every millisecond the ADC is triggered at the maximum sampling rate of 200 Hz. The first set of 8–channel data is captured and transferred via Direct Memory Access (DMA) to a corresponding ring buffer on MVME147, which results in about 5 ns delay between adjacent samples. Once a quench the event is detected, the contents of the ring buffers are written on the SUN side, via NFS. The data acquisition works like a logger for the quench characterization. With the normal configuration, a dump-out data file keeps the history of 7 buffers of sampled values of the raw signals, 9 buffers of processed signals, and 6 buffers of the quench signature signals.

2. FILTERING PROCESSING: Some typical examples of quench induced signals show very strong spike noises over the half cell voltages and the significant 60 Hz component. Several filtering methods used to control these noises include i) the nonlinear spike filtering, ii) linear low pass filtering, iii) differentiation filtering, and iv) 60 Hz harmonics filtering. Since a spike interval is much longer than a single sampling time, the 3-input median filter provides a sufficient spike rejection. Some remaining spikes and resultant false decision will be taken care of by the 2-stage variable threshold decision. A linear low pass filter is implemented by a simple moving average, which is especially useful for low level signature processing with differentiation. Differentiation filtering, giving a 90 degree phase shift while keeping the same magnitude of the input, is implemented in a bi-quad rational form covering all the frequency range up to one half of the sampling frequency [SpBe79].

3. BALANCING: Practical data from the DCA 300 series SSC collider dipoles show that the offset level is about 10–30 mV and it becomes a harmful bias to the QDC1 decision threshold when combined with the 100–mV level noise fluctuation. The signature signal offsets are adjusted by initially applying a known signal, such as a periodic saw-tooth, and numerically adjusting the multiplying coefficients.

4. THRESHOLDING: To decrease the false decision rate, most probably due to residual spike components, we confirm that the quench decision remains valid through an additional fixed interval. Once a quench signature is detected, the next 10 consecutive samples are tested, and all must exceed the threshold before a quench is declared. If the detected signature value is much bigger than the threshold values, the decision will be shortened proportionally.

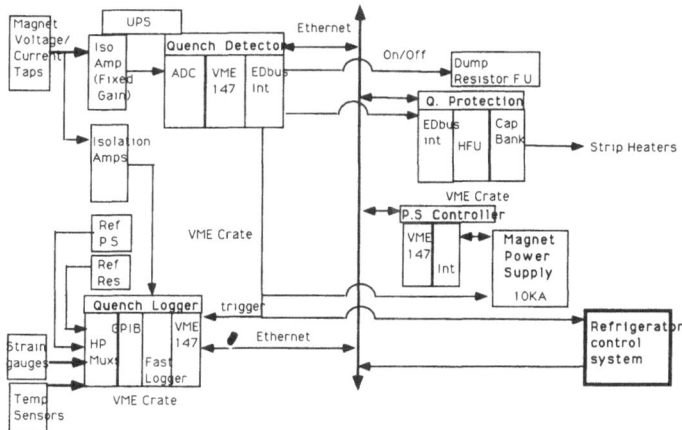

Figure 2. Quench detection system and neighbors.

INTERFACING AND TESTING SETUP

Some of the interacting systems around the quench detection system include the power supply controller, refrigerator control system, and quench protection systems as depicted in Figure 2. The quench detection interface to the power supply puts it into the bypass (short circuit) mode of operation. Interfacing to the quench protection system includes a system to discharge capacitors into heater strips built into the magnet in order to induce quenching over the entire length of the magnet, and de-energization of the magnet through switching in an external dump resistor system. There is additional interfacing to the host for the purpose of operation. Three X-windows are assigned: System-monitor, Quench-monitor and Interactive input. After completing the extensive test of functionality based on the simulation inputs, field experiments were performed at the Fermi National Accelerator Laboratory (FNAL) magnet test facility (MTF), while they were testing super-conducting dipole magnets designed for SSC (series number DCA 321, DCA 322 and DCA 323). The quench detection system implemented originally for MTF has run simultaneously with the SSC design. The trip signal from the MTF quench detection system was also recorded and compared to the buffered values of the SSC system. One sample of the results is displayed in Figure 3, where the decision delay between the FNAL and SSC systems is about 6 ms including the software and hardware delays, as expected in the simulations.

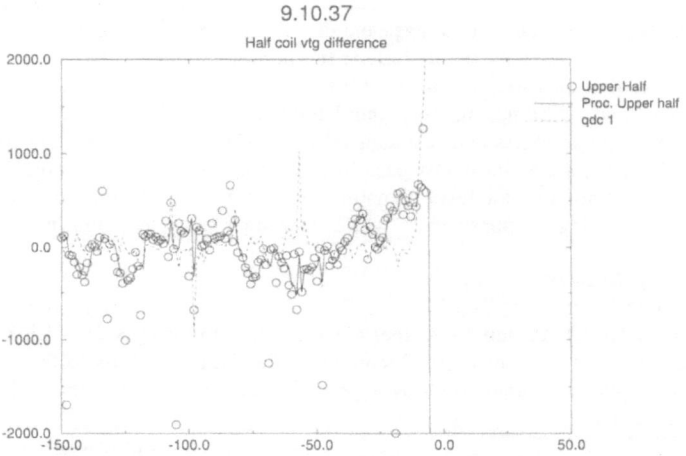

Figure 3. A typical test result: the time scale for the x-axis is 1 ms per tick and the instant t = 0 corresponds to the time at which a quench was declared by the SSC system, and shows about 6 ms latency after the time tripped by the FNAL (see the abrupt change at t = –6 ms), including the software processing part and the overall hardware.

REFERENCES

[AuEt92] J. Augueres et al., "Quench Detector and Analyzer for a UNK Superconducting String," *IEEE Trans. on Magnetics*, pp.178-181, Jan. 1992.

[Col292] M. Coles, "A Facility Description of the SSCL MTL," *IISSC 92*, 1992.

[Dres88] L. Dresner, "Quench Detection by Fluid Dynamic Means in Cable-in-conduit Superconductor," *Adv. Cryo. Eng.*, pp.167-174, 1988.

[Gane91] G. Ganetis, "Equipment Used in Testing SSC Magnets at BNL," BNL note,.

[LaEt92] D. Lambert et al., "Software Design Philosophy for the SSCL MTL," SSCL, MTL-92-009, 1992.

[SpBe79] J. Spriet and J. Bens, "Optimal Design and Comparison of Wide-Band Digital On-line Differentiators," *IEEE Trans. on ASSP*, pp. 46-52, Feb. 1979.

[ShEt85] S. Shen et al., "The Design and Preliminary Test Results of the Quench Detection System for IFSMTF," *Proc. of the 9th Int. Conf on MT* Sept. 1985.

[StEt89] J. Strait et al., "FERMILAB R & D Test Facility for SSC Magnets," FERMILAB-TM-1563, 1989.

[TsEt87] O. Tsukamato et al., "Quench Detection of Superconducting Magnet by Dual-core Optical fiber," *IEEE Transactions on Magnetics*, pp.1572-5, March 1982.

[VoSa91] D. Voy, T. Savord et al., "ASST Electrical System Final Design Review : CECAR/QPM Software and Hardware," unpublished note, Oct. 1991.

[ZhEt92] M. Zhelamsky et al., "Problems of Quench Detection in the ITER Magnet System," *IEEE Transactions on Magnetics*, pp.244-6, Jan. 1992.

DIGITAL ELECTRONICS FOR THE INCLUSION OF SHOWER MAX
AND PRESHOWER WIRE DATA IN THE CDF SECOND-LEVEL TRIGGER

J.W. Dawson[1], K.L. Byrum[1], W.N. Haberichter[1],
L.J. Nodulman[1], A.B. Wicklund[1], K.J. Turner[2], D.W. Gerdes[3]

[1]Argonne National Laboratory
Argonne, IL 60439
[2]Fermi National Accelerator Laboratory
Batavia, IL 60510
[3]University of Michigan
Ann Arbor, MI 48109

ABSTRACT

As part of the upgrade program at CDF, electronics has been built to bring the shower max (CES) and preshower (CPR) data into the trigger at level 2. After each crossing, 384 bits from shower max and 192 from the preshower wires are latched. Data from tracks are bussed to this module to provide the wire address and momentum which are then successively compared to the wire data in large look-up tables. Approximately 50 nanoseconds is required to determine a match, write the results in FIFO, and make the results available to track memory. Monte Carlo analysis has indicated that an increase in efficiency of a factor of three in triggering on b decays will be achieved with this hardware.

PHYSICS CONSIDERATIONS

Hadron colliders have an enormous potential for the study of b-physics, provided that efficient triggers can be devised for the B-decay products. For example, the $B\bar{B}$ production rates measured at CDF, in the central region alone (e.g., abs (Y) < 1, P_t > 6 GeV/c), correspond to 100 Hz at Tevatron luminosities of 10^{31}. This rate may be contrasted with that expected for an e^+e^- B factory, which would be around 3 Hz at design luminosity of $3 \cdot 10^{33}$. To compare with LEP, the total $B\bar{B}$ production in the CDF central detector, 10^9 pairs in a 100 pb^{-1} run, would be equivalent to $5 \cdot 10^9$ hadronic Z decays. To date, the most useful B-triggers in the CDF experiment are the single lepton (electron or muon) triggers, and the J/ϕ dimuon trigger. Of course, the additional requirement of a second lepton in conjunction with these basic triggers is useful for studies that require b-tagging. While the J/ϕ modes are very useful for reconstructing exclusive B decays (eg, B $\rightarrow$J/ϕ +K, K*, ϕ, Λ) and are the most likely vehicle for the observation of CP violation in the B sector (using B $\rightarrow$J/ϕ K$_S$), the semileptonic decay triggers are needed if one is to identify the charge of the B. For example, the detection of B_s-$\bar{B}_s$ mixing, an important measurement for the Standard Model, could not be accomplished with J/ϕ modes, but would likely be based on observation of electron-D$_s$ correlations.

The CDF trigger employs a "level-3" processor farm, which does offline reconstruction to select highly enriched event samples for physics analysis. The input to this farm is provided by the "level-1" and "level-2" triggers, which use fast analogue signals from the calorimeter and raw TDC information from the central tracker and muon chambers, to define electron, muon, jet and missing E$_t$ trigger objects. To match the bandwidth of the hardware Event Builder and the level-3 farm, the level-2 accept rate is constrained to be less than around 35 Hz, or 3000 nb at 10^{31}. The CDF single electron trigger, with a 9 GeV threshold, has a level-2 accept rate of 500 nb, or about 18% of the total bandwidth. The actual electron purity is around 7%, or 30 nb after level-3 and offline processing. For central muons, with 1992 central upgrade chambers, the sample purity and level-2 rates are comparable. However, the muon fiducial coverage is less than that for electrons, and with a fixed P_t threshold, the electron trigger provides about four times the B-physics rate of the muon trigger.

There are several handles for further improving the purity of the level-2 electron trigger, and any improvements translate into a corresponding increase in the rate of B-events written to tape, since the overall trigger is bandwidth-limited. The offline identification of electrons relies heavily on the detailed information on shower development provided by the central strip and preradiator chambers, which are located at depths of $6X_0$ and $1X_0$ respectively. The baseline level-2 trigger associates a high-P_t charged track with an electromagnetic calorimeter cell. The strip and preradiator chambers would allow a much finer-grained match between the track position and the shower energy, and this tighter matching can reject the main backgrounds in the baseline trigger - namely "overlap" events with charged tracks plus gamma rays in the same calorimeter cell. Figure 1 shows the expected level-2 rates for electrons as a function of the calorimeter E_t threshold, for the baseline system (top curve), and with a strip-chamber pulse height requirement (bottom curve). The overall rejection provided by the strip chamber alone is about a factor of three. This would represent a significant increase in the purity of B's written to tape. Numerically, with the baseline trigger and electron bandwidth, CDF can write approximately 10^6 B $\rightarrow$ electron events to tape in 100 pb^{-1} (all backgrounds subtracted); the strip chamber trigger would allow this rate to increase to approximately $3 \cdot 10^6$ per 100 pb^{-1}. That may be compared with B $\rightarrow$ electron or muon rates at LEP, around (45,000 efficiency) per 10^6 Z decays.

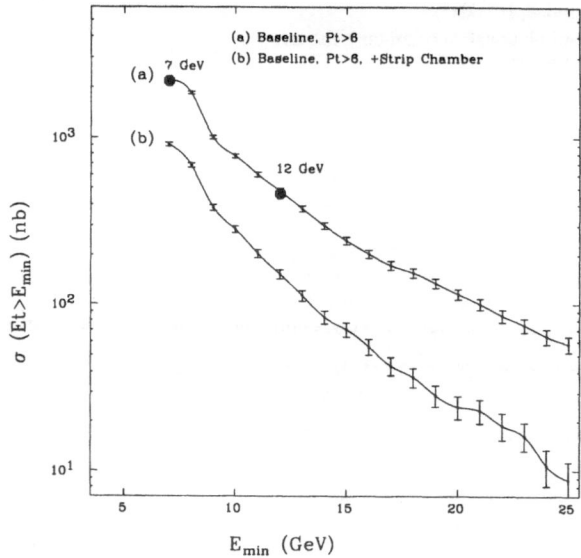

Fig. 1 Level-2 trigger rates in CDF as a function of the calorimeter E_t threshold. The top curve ("a") shows the baseline CDF trigger used in 1989, with a track P_t threshold set at 6 GeV/c (3.8 GeV/c for E_t<9 GeV). The bottom curve ("b") shows the same rates with an additional strip-chamber pulse height requirement, set to be 90% efficient for electrons for each E_t threshold. The solid points are the measured level-2 cross sections for the 1989 CDF run.

HARDWARE DESCRIPTION

The digital electronics to incorporate CES and CPR data in the CDF second level trigger is built in FASTBUS protocol. Two identical cards, one for CES and one for CPR reside in FASTBUS adjacent to the CTCX card. The electronics is built on two cards because it is extremely difficult to provide space for chips and the necessary connector space required for the 384 CES and 192 CPR bits. The two physically identical cards are configured under software to accommodate either CES or CPR information. The cards are built using multi-layer technique as 12 layer cards with essentially all devices implemented in surface mount, and with integrated circuits mounted on both sides of the card. Layout and routing were done at Argonne using the Telesis system.

In order to deal with the connector problem the 384 CES bits and 192 CPR bits are brought to an active patch panel near the digital electronics. The bits are carried from the detector on approximately 50 meters of twisted pair as differential TTL and translated to single-ended TTL on this patch panel. They go from there as single-ended TTL signals on approximately 2 meters of flat ribbon cable to the digital electronics. Honda high density connectors with .025" pitch are used to bring the single-ended bits from the active patch panel to the digital module. FASTBUS interfaces are provided on both cards since it is necessary to program both cards and execute diagnostics on both individually, and since it was desired to make the cards identical as far as layout and routing. After a crossing, a sequence of sets of track data is provided by the Central Fast Track Processor (CFT). Track data provided by the Central Fast Track Processor is latched by strobes from the CTC. Data and strobes are bussed from the adjacent CTCX card.

The logic is shown in Fig. 2. After a crossing, the 384 bits from the CES and the 192 bits from the CPR are received from the front end electronics, and after an appropriate time for data to become valid, are latched. The CES bits are partitioned as 48 busses, each containing 8 bits from the East or West wedges, and numbered from 0 to 23, E or W. The CPR bits are partitioned in like fashion as 48 busses, each containing 4 bits from the East or West wedges, and numbered from 0 to 23, E or W. Each of these busses forms part of the address word which addresses an SRAM look-up table. Following the latches, the bits are masked by a set of mask words previously written into on-card registers from FASTBUS. Mask bits may be used to remove noisy wires and for diagnostic purposes. The latched and masked bits are logically OR'ed to proved CES and a CPR bits for each wedge, which are translated to differential ECL and provided at a connector.

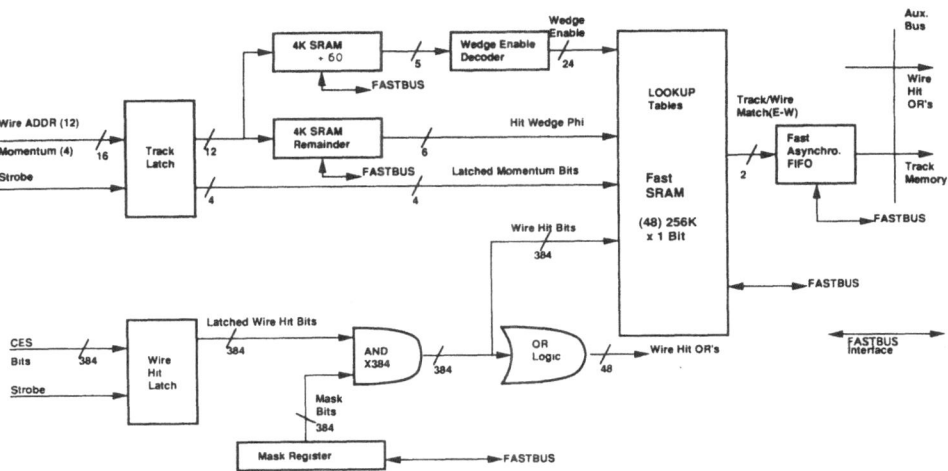

Fig. 2 Simplified Block Diagram: Digital Electronics for the Inclusion of Shower Max and Preshower Wire Data in the CDF Second-Level Trigger

When an 12 bit WIRE_ADDR word and 4 bit MOMENTUM word are received from the CFT, they are latched by RD_STR which is also received from the CTCX. The 12 bit WIRE_ADDR word constitutes the address word for two 4k SRAM's. The first SRAM performs an algorithm which divides the value represented by the 12 bit word by 60, and produces a 5 bit output word. Since there are 1440 wires, the quotient is a wedge number going from 0 to 23. This 5 bit wedge number can then be decoded to yield 24 enable signals, which can be used to enable the appropriate SRAM's, each of which is associated with one wedge. The other 4k SRAM performs an algorithm which yields the remainder when the wire address is divided by 60, which is a representation of the hit position in phi across the wedge which has been selected by the other SRAM. It is necessary only that there be a defined one-to-one relationship between WIRE_ADDR and wedge number and phi position across a wedge. There is no requirement for exact monotonicity.

In the case of the CES, the 6 bits from the second SRAM, the 4 MOMENTUM bits, and the 8 CES bits constitute the address supplied to the SRAM which is enabled by the decoded wedge number. In the case of the CPR, the 6 bits from the second SRAM, the 4 MOMENTUM bits, and the 4 CPR bits constitute the address supplied to the SRAM which is enabled by the decoded wedge number. The outputs of these SRAM's are tied together and since 23 of the 24 chips are not enabled, the one which is enabled will produce output data on the CESE, CESW, CPRE, and CPRW line. This output data is strobed into an asynchronous FIFO, where it can then be strobed into track memory in time with the muon data by a strobe supplied by the muon hardware. The logic is implemented in high-speed CMOS SRAM's and ACT logic and data is valid at the output FIFO in less than 50 ns. from the time WIRE_ADR is latched by RE_STR. The card has logic to allow all SRAM's to be read/written from FASTBUS to permit initialization and execution of diagnostics.

All memories and registers can be written/read from FASTBUS, and for diagnostic purposes it is possible to set bit patterns of the CES and CPR hits. This is accomplished by permitting all the input latches to be set, and then the mask words are used to create the bit patterns which can be run through memory, and the output data tested. It is necessary that test bits be latched in the WIRE_ADR latch so that complete address information is supplied to the 256k SRAM's.

TESTING AND SOFTWARE

The first tests of the digital electronics fastbus card were performed at Argonne in a test setup which consisted of an IBM PC interfaced to a fastbus crate. A programmable LeCroy 1821 Segment Manager provided the interface between the fastbus board and the IBM PC, the latter from which diagnostic commands were issued. We used a Single Platform Diagnostic Package, called SPUDS[1] both to troubleshoot the board as well as to read and write to the various board registers and RAMs which resided in both data and CSR space. In addition, we built a test input cable which allowed us to fake data to the input latches to test data flow through the board.

The second phase of testing occurred at Fermi National Laboratory. We used a mini fastbus crate which housed a Kinetic Systems Q-bus Processor (QPI). The QPI was interfaced to a mini VAX computer which was clustered with the CDF online computers. Several diagnostic software packages were developed.

The first diagnostic program performed tests similar to those already conducted at Argonne. Namely, the program performed read and writes to the various registers and RAMs. Also available in this program are more extensive tests which march 0's and 1's through the registers and rams. In addition, the program tests the performance of the asynchronous FIFO. A second program allowed the user to download specific look-up tables to the various RAMs. This second program was designed to test software algorithms which matched electron pulses measured in the shower max detector with tracks measured in the CFT.

The final stage of testing the fastbus board consisted of integrating it into the L2 trigger in a parasitic mode. This was done by installing it into its designated L2 fastbus crate and connecting it to electronic members both upstream and downstream of itself. A schematic of the output signals received by the track list card is shown in Fig. 3.

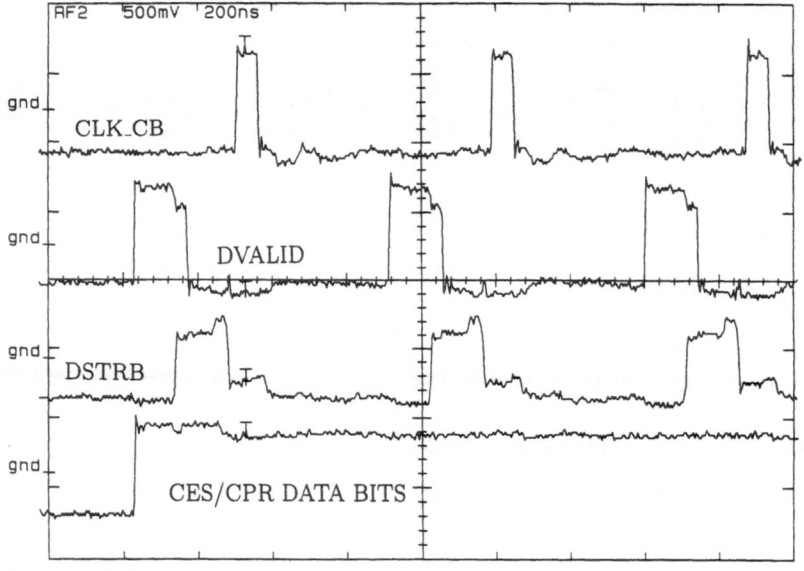

Fig. 3 The FIFO generates the signal DVALID and transmits it to the track list card along with the CES/CPR data bits for every track. When the track list card has received the DVALID signal, it latches the CES/CPR data bits and responds to the our card with the handshake signal DSTRB. When the signal DSTRB is received by the FIFO, it causes the FIFO to upload the next pattern of CES/CPR data bits and generates another DVALID.

ACKNOWLEDGEMENTS

Work supported in part by the U.S. Department of Energy, Division of High Energy Physics, under contract W-31-109-ENG-38.

REFERENCES

1.   J. Anderson and J. Franzen, Data Acquisition Hardware Group, SPUDS User Manual, FNAL (1992).

DESIGN AND IMPLEMENTATION OF MULTI CHANNEL READ-OUT
AND LEVEL 2 STORAGE FOR TIME MEASUREMENT CHIP

T. Ekenberg[1], E. Gerds[2], R. Van Berg[1], H. H. Williams[1]

[1]Department of Physics
[2]Department of Electrical Engineering
University of Pennsylvania
Philadelphia, PA 19104

ABSTRACT

This paper describes the design, implementation, and testing of a multi-channel read-out and level 2 buffering circuit for a time measurement chip, L2I. L2I version 1.0 interfaces a four-channel version of the Time Measurement Cell (TMC) with the next level in the front-end electronics chain, the Data Collection Chip (DCC) for read-out of the SDC straw tracking detector. The chip was implemented as a standard cell design, except for a full-custom SRAM block, and fabricated in a 1.2 μm n-well CMOS process. The total area of the chip including pads and off-chip driving circuitry is 12 mm^2. Power consumption during full-speed operation is 12 mW. High level simulation of the circuitry was performed using Verilog HDL, and detailed timing simulations were carried out with Hspice. The chip was tested on an HP82000 IC test-station and was found functional with a minor design error in the control path.

SYSTEM DESIGN

The L2I [2] was developed to provide read-out capability in a "full-density" test of the SDC straw tracker. This is a test where several hundred straw tubes are instrumented at, or close to, the design density and read out under realistic conditions in a beam line. A test of this type will provide information of the signal to noise ratio in a experiment-like environment with high-speed digital signals close to sensitive analog front-end devices and other possible interference. The L2I and the DCC [3] provide the interface between the TMC [1] and the rest of the data acquisition system. See Fig. 1 for signals interfacing the L2I to the TMC and the DCC.

The TMC implements the time measuring functions in an all digital fashion. The chip is organized in a 32 by 32 bit memory array, with one 32 bit row at a time selected for reading and another one for writing. The write operation is carried out by presenting the discriminator signal (indicating a track in the chamber) from the front-end device to the data input of a memory row. A write signal is then propagated

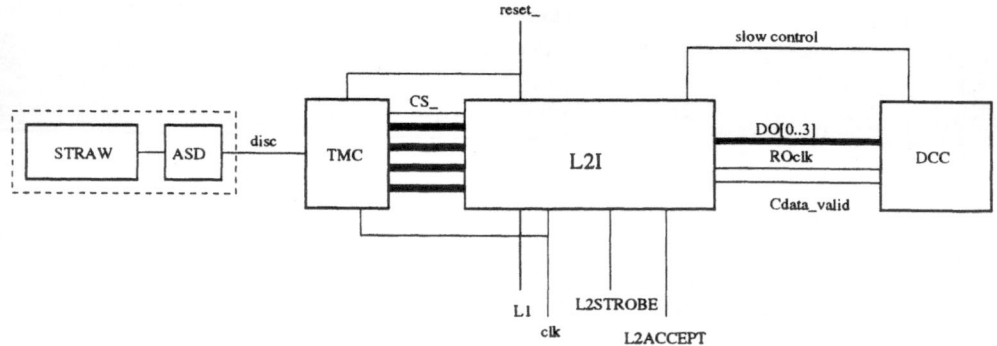

Figure 1. Signals interfacing L2I with TMC and DCC. Included are also signals from trigger and clock distribution network: $L1$, CLK, RESET_, L2ACCEPT, L2STROBE.

down a 32 stage delay chain with 1 ns delay per stage. Each stage drives a cell in the selected memory row, allowing writing every ns in successive storage locations. The delay of a delay element is controlled with a phase-locked loop like structure. The write and read pointer cycles through the memory in a synchronous fashion, allowing a fixed delay before a level 1 trigger decision must be made for the data in a memory row. If no level 1 trigger accept is received, the memory row is over-written the next time the write pointer arrives.

If a level 1 trigger accept is received, the 32 bits of the memory row is encoded into 6 bits and the output driver is enabled. The six bits are presented to the L2I data input. They are latched into the level 2 buffer together with a tag generated from counting the number of global level 1 accepts and storing this level 1 id. The storage time required for a level 2 trigger is of variable length and the level 2 buffer's SRAM cells can hold the data for an indefinite amount of time.

The level 1 id is used to recognize proper data for read-out. A system requirement is that every level 1 accept must be followed by a level 2 decision. This is accomplished by strobing the L2STROBE for every decision and activating the L2ACCEPT for valid events. The L2I will raise the Cdata_valid flag if a level 2 accept with data on-chip is found. The L2I will then wait for the ROclk to start strobing the data out. Each channel datum is 24 bits long and the ROclk must be strobed 6 times to read out a channel completely. After the channels with data have been read-out an End of Datum (EoD) marker is read out. The EoD marker is also 24 bits long but contain a certain error code to flag its special function.

A level 2 accept without data will also raise the Cdata_valid flag, but the only datum available on-chip will be the EoD flag. This is done in order to simplify the book-keeping of events in the DCC.

CIRCUIT DESIGN AND SIMULATIONS

In order to minimize the design and development time of the L2I the chip was implemented as a standard cell design with a few full-custom blocks. The actual level 2 buffer is a dual port SRAM which because of its regular structure and high requirements on write access speed was chosen for a full-custom implementation. A dynamic counter that have to operate at the full system clock frequency of 60 MHz is also full-custom. The standard cells are designed by the AμE center at the University of

Mississippi for the scalable MOSIS design rules and are available in the public domain.

The L2interface is the core of the L2I. This is the block where the storage of the level 2 data and data flags is done. It is also where the data is selected for read-out if associated with a valid level 2 trigger. All level 1 trigger accepts of the detector system are received at the L2I and counted in an eight bit counter. This is the tag associated with all data present at the time of a level 1 accept. The six bits of data and the eight bit tag is stored in the level 2 buffer. The level 2 buffer is a SRAM connected as a FIFO. The write pointer is advanced every time there is data present in connection with a level 1 accept. If data is present at the input without a level 1 accept, or there is a level 1 accept without data, the write pointer is not incremented.

The read pointer in the level 2 buffer is advanced every time data in the buffer is accepted or rejected by the level 2 trigger. A system requirement is that every level 1 accept is followed by a level 2 decision. This is either an accept — L2STROBE activated when L2ACCEPT is already active — or a reject — L2STROBE active without L2ACCEPT. The L2STROBEs increment a control counter that tracks the level 1 accept counter with an average delay equal to the average time for a level 2 decision to be made. The value of the level 2 decision counter is compared to the value of the data tag at the head of the level 2 FIFO. If they match and the data is rejected, the read pointer is advanced in the FIFO. If there is a match and the data is accepted, the datum is latched in temporary registers. An internal line from the L2interface to the readoutMux is strobed to indicate valid data in the channel.

The readoutMux is the circuitry interfacing the four channels and the End of Datum-generator with the read-out bus. If level 2 trigger accept is received, channels with data are multiplexed onto the read-out bus followed by the End of Datum. Since the width of the read-out bus is four bits and the datum length is 24 bits, six strobes of the ROclk are necessary to read out one channel. If there is no data available on the L2I the EoD is read out by itself.

Extensive simulations with Verilog HDL was carried out during the design phase of the L2I. All blocks up to the channel level were simulated for correct logical functionality with mixed behavioral and structural models. Detailed timings simulations were done with Hspice on critical paths. Finally an Hspice simulation was carried out on a partial extracted model of the chip with 7,000 transistors out of the total 17,000 transistors on-chip. This was done in order to verify the timing characteristics of certain critical global interconnects. After the chip was submitted the entire design was simulated with Verilog to verify read-out characteristics and the interface to other parts of the system.

Version 1.0 of the L2I provides a four location deep level 2 storage buffer. The read-out bus is four bits (one "nibble") wide and two hand-shake line synchronizes the read-out. There are 39 primary I/O pins, 8 power and ground pins, and 13 pins dedicated to test functions.

MEASUREMENT RESULTS

The L2I was tested on an HP82000 IC test stations at the University of Pennsylvania — for initial verification of functionality — and on another HP82000 at the SSCL for more complete characterization. Initial test indicated full functionality, but further testing at speed (60 MHz) revealed a design error in a counter in the control path. In most instances this error have no effect on the circuit functionality, but the ordering of the bits out of the EoD-generator is only correct every fourth event, when the counter wraps by itself. This problem can in a fairly simple manner be worked around in the read-out software since the information provided by the EoD marker is redundant to the information on the Cdata_valid line. This signal also drops when the chip is completely read out. One solution would be to read out the chip until the Cdata_valid drops, and then discard the six last "nibbles" coming from the corrupt EoD.

The read and write operations of the level 2 buffer were tested together with the over-write protection and the wrapping of the read and write pointers. The buffer was found fully functional and operated with a maximum tested input frequency 5 MHz.

The maximum ROclk frequency is 2.5 MHz. This is mainly limited by the rise and fall times of the data on the output bus. The intrinsic delay involved in shifting between "nibbles" is about 50 ns. With output pin drivers designed for resistive load, a maximum read-out frequency of more than 10 MHz should be achievable.

Total power consumption during full-speed operation was measured at 12 mW, 3 mW per channel.

SUMMARY

L2I, a level 2 buffering and read-out chip has been designed for the straw-tracking detector of the SDC. This chip will read out a four-channel TMC when a level 1 trigger accept is received. It will further provide the level 2 storage pending the level 2 trigger decision, and multiplex the four channels together into the output stream. A four bit bus with two synchronizing hand-shake lines provides the interface to the next chip in the front-end electronics chain, the DCC. The L2I CMOS ASIC is fabricated in a 1.2 μm n-well process and covers an area of 12 mm^2 including pads and output drivers. Version 1.0 was found functional with a minor design error in the control path that re-ordered the output bits of the End of Datum marker during read-out. This design error can be worked around in software if desired, since the information of in the End of Datum is redundant with respect to other hand-shake signals.

Version 1.1 of the L2I have been submitted. This version has the design error in the control path corrected, as well as the order of the data in the datum changed to accommodate the design of the DCC.

REFERENCES

[1] Y. Arai, et al. *A CMOS Four-Channel × 1 K Time Memory LSI with 1-ns/b Resolution,* Journal of Solid State Circuits, **27**, No. 3, p. 359 – 65 (1992).

[2] T. Ekenberg, et al. *Design and Implementation of Level 2 Buffering and Read-out for Multi-Channel TMC and TCC/AMU,* internal University of Pennsylvania HEP Engineering report, (1992).

[3] G. Stairs, et al. *Data Collection Circuit for the SDC Straw Tube Tracker,* Proc. IEEE Nuclear Science Symp. 1992, p. 520 – 22 (1992).

[4] A. E. Stevens, et al. *A fast Low-Power Time-to-Voltage Converter for High Luminosity Collider Detectors,* IEEE Trans. Nucl. Sci. (USA), **36**, No. 1, p. 517 – 21 (1989).

AN EXAMPLE OF COMBINING SYBASE WITH PAW TO PERFORM
GRAPHICAL DATA DISPLAY AND ANALYSIS

A. Fry and I. Chow

Superconducting Super Collider Laboratory*
2550 Beckleymeade Avenue
Dallas, TX 75237-3997

INTRODUCTION

The program PAW (Physics Analysis Workstation) [1] enjoys tremendous popularity within the high energy physics community. It is implemented on a large number of platforms and is available to the high energy physics community free of charge from the CERN computing division. PAW combines extensive graphical display capability (HPLOT/HIGZ), with histogramming (HBOOK4), file and data handling (ZEBRA), vector arithmetic manipulation (SIGMA), user defined functions (COMIS), powerful function minimization (MINUIT), and a command interpreter (KUIP).

To facilitate the possibility of using relational databases in physics analysis, we have added an SQL interface to PAW. This interface allows users to create PAW N-tuples from Sybase tables and vice versa. We discuss the implementations below.

METHODS AND RESULTS

An "isql" interface was constructed within PAW using KUIP commands [2] so that users can type in SQL commands at the PAW prompt. The interface takes the SQL statements and sends them to the Sybase DB-library interface [3]. Therefore, all the SQL computations are done at the SQL server. Users can also issue commands to transfer the contents of their Sybase tables to and from PAW n-tuples. Results can be displayed using the HIGZ graphics library which is contained in PAW. The commands that we added to PAW for processing SQL commands are listed in Figure 1.

---

*Operated by the Universities Research Association, Inc., for the U.S. Department of Energy under Contract No. DE-AC35-89ER40486.

1: * ISQL	SQL Command Interface	
2: * MVEC	Retrieve values from a column to a vector in PAW	
3: * SQL	Single line SQL input processor	
4: * SNTUPLE	Move values from tables in database to N-tuples in PAW	
5: * SYBTAB	Move N-tuple values from memory back to Sybase	

Figure 1. Commands added to PAW for SQL interface to Sybase

After creating these commands within PAW, we performed a test using data from the CDF experiment at Fermi lab. The data are organized into named structures called YBOS banks [4]. The YBOS banks were extracted into corresponding Sybase tables. The schema of the tables are based the original YBOS tree structures. Unique IDs are generated to specify each records. We performed a Z-mass analysis using pure SQL statements. This analysis selected 243 events from a loose sample of 5007 Z candidates. The mass plot (fig. 2) of the selected events shows the Z mass peak near 92GeV.

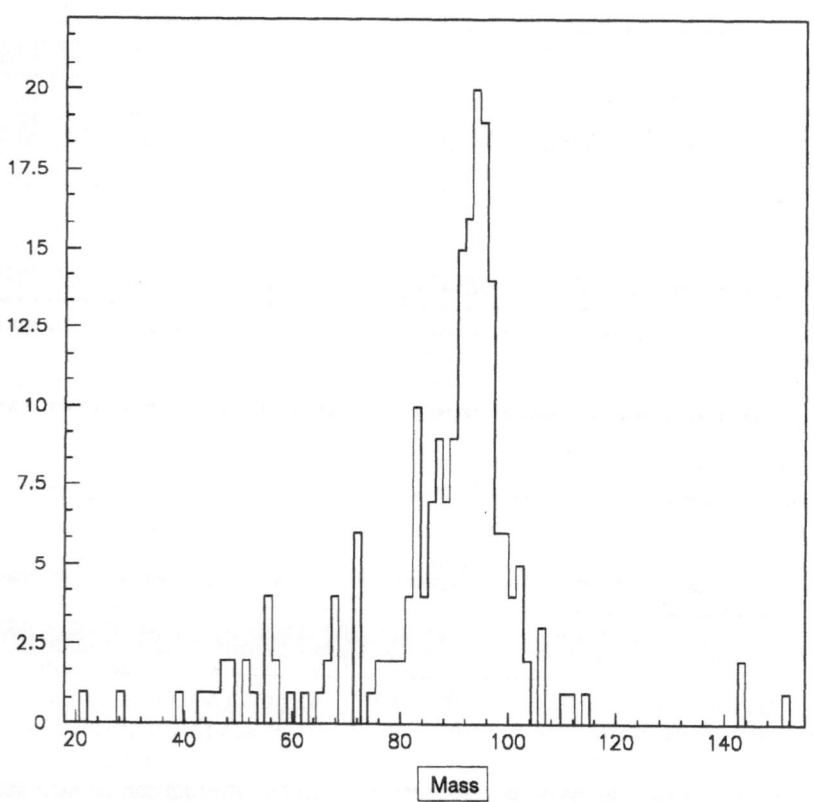

Figure 2. The display of the results of our Z-mass SQL analysis.

FUTURE RESEARCH

When performing this test, we found that the YBOS data structures as used by the CDF experiment were not optimal for a relational model. A better data model of High Energy Physics data can be constructed. We are continuing the effort to define a more appropriate model for HEP data within the relational and Object-Oriented database paradigms [5][6]. Hopefully, such effort will benefit both the database and HEP communities. We also believe that the graphical interface to Sybase using PAW can be found useful in many other application which do not as yet have a graphical interface.

ACKNOWLEDGMENTS

We wish to thank Craig Blocker for his assistance in our understanding of the CDF YBOS structure.

REFERENCES

1. R. Brun, O. Couet, C. Vandoni and P. Zanarini, "Physics Analysis Workstation -- PAW", CERN Computing, CERN (1989).
2. R. Brun and P. Zanarini, "KUIP -- Kit for an User Interface Package", CERN Computing, CERN (1988).
3. "DB-Library Programming Guide", Notes from Sybase Educational courses.
4. "CDF notes", Fermi Lab, Chicago, (1988)
5. A. Baden, C. Day, R. Grossman, D. Lifka, E. Lusk, E. May, S. Mehta, L. Price and X. Qin, "A Data Model for Scientific Databases Supporting Objects, Attributes, Methods and Collections: Preliminary Report", PASS project, (1991)
6. R. Grossman, X. Qin, C. Day, S. Loken, F. MacFarlane, E. May, D. Lifka, E. Lusk, L. Price, A. Baden, L. Cormell, P. Leibold, D. Liu, U. Nixdorf, B. Scipioni, T. Song, "Database Challenges: Analyzing High Energy Physics Data Using Database Computing", PASS project, (1992)

WHAT THE QUALITY PHILOSOPHY BRINGS TO A RESEARCH AND DEVELOPMENT ENVIRONMENT LIKE THE SSC

Steve Davis and John L. Wentz

Quality Assurance
Superconducting Super Collider Laboratory*
2550 Beckleymeade Avenue
Dallas, TX 75237-3946

ABSTRACT

In achieving major schedule and performance milestones with a project as technologically advanced as the Superconducting Super Collider, many activities must be coordinated simultaneously without the luxury of a conventional design review process. Because the design may change several times prior to the delivery of a one-of-a-kind or prototype component or subsystem, close verification and monitoring of design, manufacturing and test processes are needed on a real-time basis. This verification and monitoring is performed on two levels by Quality Assurance at the SSC Laboratory; Division and General Management. The Division level is involved in day-to-day activities at the Laboratory and the Suppliers; the General Management level performs the independent oversight function for all the Laboratory quality processes. In the divisions, continuous monitoring of design, procurement, manufacturing, installation, and testing activities is performed. At the General Management level, quality program development and implementation is evaluated within each division. Critical suppliers involved in system design, manufacturing and testing are evaluated against contract and program requirements to assure systems safely perform their intended functions.

Responsibilities for quality are extended by the participation of the SSCL Quality Assurance Office in Accelerator Readiness Reviews (ARR) previously known as Operational Readiness Reviews (ORR) for each major machine developed and designed at the Laboratory. The ARR process for the Accelerator Systems String Test (ASST) provided a real test. Quality Assurance promoted continuous awareness of DOE contract requirements, SSC Laboratory Quality Assurance requirements and the Safety Analysis Report requirements which is not an easy task in the scientific community. A result has

---

* Operated by the Universities Research Association, Inc., for the U.S. Department of Energy under Contract No. DE-AC35-89ER40486.

been a new cooperative attitude in which physicists, scientists, engineers, safety and quality professionals can work together towards a common goal .

INTRODUCTION

Designing and building the World's largest and most powerful "anything" draws considerable attention from skeptics in each part of the World, but to design and build the World's largest and most powerful scientific instrument draws significantly more attention. This increased attention not only comes from average people populating the earth's four corners, but specifically World leaders currently involved in scientific and technological research and development. The Superconducting Super Collider Laboratory (SSCL) when completed will be the World's largest and most powerful particle accelerator. A project of such magnitude and complexity requires tremendous technical and professional expertise, planning, coordination, and management controls to be a success.

Because this project is being built in the United States, there is increased attention given by the rest of the World on the project's ultimate success or failure. This will influence the World's view of the United States' commitment to quality in education, science and technology. Along with the enormous financial considerations for a project of this size, high energy physics expertise from the world's sharpest minds is needed to collectively tackle new challenges in accelerator and detector design and construction techniques. This is necessary to improve the envelope required to expand the knowledge of what we know today about the make-up of matter.

In an effort to integrate research and development practices, quality principles and DOE contract requirements, a cooperative effort needs to be exerted. To accomplish this task without limiting the creative nature of the scientists involved, the Program Management, Safety, and Quality organizations focus on the Laboratory program and contract compliance issues. Quality involvement helps provide consistency in the Laboratory's interpretation of administrative and contract requirements for the project.

QUALITY INVOLVEMENT IN THE FIRST MAJOR MILESTONE

One of the most significant achievements at the SSC Laboratory to date in terms of milestones was the installation, testing and operation of a string of superconducting magnets and inter-connecting spool pieces. This specific milestone was mandated by Congress to prove that a string of industrially manufactured magnets would function prior to being placed inside a tunnel some two-hundred feet below ground. This milestone was completed in August 1992, two months ahead of schedule. Because of the tight schedule to complete this milestone, most activities had to be monitored on a real-time basis. Verification of installation and testing activities was critical to safety, quality and operability of the String components. These steps determined the readiness to proceed with energizing the two primary hazardous energy sources, electrical power and cryogenics.

THE ACCELERATOR SYSTEM STRING TEST (ASST)

The specific major components of the ASST included five dipole magnets, one quadrupole magnet, and three spool pieces. These major components are supported by other subsystems which provide the required 6800 amperes(A) of electricity and the liquid helium temperature environment.

The milestone objectives were to cool down the String components and apply 6800A to the electrical bus and cables safely. Upon performing this test, a manually induced quench would be initiated to test all related safety systems for de-energizing the String and compensating for the increase in heat and pressure produced during the quench event. Data is collected from various monitoring points throughout the String for input into production design and for determining component characteristics when operated as a String in tunnel conditions.

All of the activities which take place during the development, design, installation, testing and operation of the ASST must be monitored and verified in accordance with contractual and administrative requirements. The quality monitoring, verification and oversight functions are performed by two different organizations within the Laboratory; the Accelerator Systems Division (ASD) and the General Management Office (GMO).

ASD QUALITY ASSURANCE DEPARTMENT

The primary responsibilities of the ASD QA/QC include reviewing, monitoring, inspecting, and surveying all quality related activities of contractors for major components and support systems as well as division activities associated with design, construction, installation, testing and maintenance of test facility components and systems for the ASST. The design of the Superconducting Magnets is the responsibility of Magnet Systems Division (MSD).

ASD also provides technical support for the Project Management Office (PMO) who has responsibilities for technical direction, accelerator physics design, and oversight of the construction for bringing the ASST and other research and test facilities to operating status. This close interface helps provide awareness to scientists and engineers on critical items which might require independent verification to assure that we consistently demonstrate how we achieved expected or unexpected results and to document how we accomplished these activities.

GENERAL MANAGEMENT OFFICE QUALITY ASSURANCE

The responsibilities and functions of the GMO QA Office are currently twofold. The primary function of GMO QA Office is to evaluate whether the SSC Laboratory has adequately established and effectively implemented a quality assurance program which meets the requirements set forth in Department of Energy (DOE) Orders 4700.1 "Project Management System," and 5700.6C "Quality Assurance." These documents provide the primary requirements for managing and controlling administrative and technical activities of the SSC Laboratory. This evaluation is accomplished by performing system and activity audits and surveillances of activities considered important to safety of personnel and equipment and reliable operation of equipment. Audits are also performed at the facilities of key contractors responsible for providing critical services and equipment such as superconducting magnet components, spool pieces and support systems.

The other responsibilities of the GMO QA organization include developing and conducting training programs for QA awareness at the Laboratory. GMO develops upper level QA procedures which give direction to each division within the Laboratory on program activities that need to be addressed and controlled procedurally. In addition, GMO provides oversight support for Accelerator Readiness Reviews (ARR) for each major research and test facility (i.e. ASST, LINAC). In the case of the ASST, GMO QA has the organizational independence and insight to program and contractual compliance issues requiring consideration in this process.

ACCELERATOR READINESS REVIEWS (ARR)

The ARR process has been very useful in providing a forum for scientists, engineers, professionals and management personnel to collectively address technical, environmental, safety, quality and reliability issues for SSCL facilities including the ASST. The ARR process is conducted by a committee appointed by the PMO and comprised of scientists, engineers, environmental, health, safety and quality professionals from different organizations within the Laboratory. Because this committee is diverse and independent of activities necessary to make the facility physically ready for operation, members are free to identify issues which concern the readiness of the facility to be operated and shutdown safely. This may include issues identified in the Safety Analysis Report (SAR) which is the primary facilities and systems description and function document. This document also identifies and addresses mitigation actions necessary for all known facility and system safety and environmental hazards. Issues identified by the committee must be satisfactorily resolved and verified prior to the committee recommending a permit to operate the facility. The permit to operate will then be issued by PMO to the Director of the Laboratory. In turn, concurrence by DOE is required prior to conduct of operations.

The ARR Committee reviews documentation and interviews subject matter experts throughout the final readiness stages of the system and facility testing to determine if any additional hazards exist and whether the facility can be safely operated as required by DOE without undue risk to personnel or equipment. Any concerns raised by the committee are documented as "Action Items" and issued to the group leaders for the subject matter experts to resolve. After actions have been completed for all Action Items issued during this process, a final presentation is conducted to confirm facility readiness.

The quality assurance representative on the committee, just as other members of the committee, provides a different perspective and consciousness to the ARR process. Where technical representatives focus on design, operability and reliability of the systems to perform their function, the quality assurance representative focuses on processes established to assist in accomplishing the desired results safely and systematically as defined by DOE requirements and Laboratory plans. The scientists and engineers may develop, design and build a machine that works, but may not provide the evidence needed to prove the machine works the way it was intended. This information is captured in program plans, specifications, procedures, records, and analysis and test results. This evidence is a big part of what the quality assurance representative emphasizes during the process.

CONCLUSION

In a research and development environment like the SSC Laboratory, the present and future role of quality assurance will ultimately be proactive versus reactive. The Laboratory continues to establish plans and procedures to comply with DOE, industry and administrative requirements. These plans and procedures provide consistency and continuity within organizations throughout the Laboratory. By establishing performance criteria and measuring performance results, meaningful changes can be made to programs which add value to the existing processes without overburdening those processes with controls that do not affect the quality of the final product. Therefore, the goals of the SSC Laboratory Quality Assurance Organization are to promote continuous improvement by developing process controls where needed and enhance those that already exist where possible.

STRUCTURAL BEHAVIOR OF THE SIX-STRUT SUPPORT SYSTEM

Bui Dao

Mechanical Engineering Department
Accelerator Systems Division
Superconducting Super Collider Laboratory*
2550 Beckleymeade Avenue, MS 4006
Dallas, TX 75237-3997

ABSTRACT

A simple mathematical model was used to understand the structural behavior of the six-strut support system. By breaking the system into symmetry and antisymmetry configurations, the vertical and torsional displacements were decoupled, thus making simple calculations possible.

INTRODUCTION

As its name implies, the six-strut system supports the accelerator component with six struts arranged in various ways. The arrangement considered in this paper has three vertical, two transverse, and one axial strut as graphically shown in Figure 1. This support system will be used to support many components for the Superconducting Super Collider. Figure 2 shows a photograph of the SSC Linear Accelerator (LINAC) Radio Frequency Quadrupole (RFQ) utilizing this support system. By using differential threads on both ends, the six-strut system increases movement resolution, ease of installation, and alignment efforts. It is of interest to determine the effect of the support system on the deformation of the components.

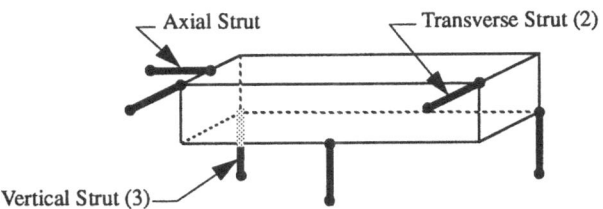

Figure 1. Six-Strut Support Arrangement with 3 Verticals.

---

* Operated by the Universities Research Association, Inc., for the U.S. Department of Energy under Contract No. DE-AC35-89ER40486.

STRUCTURAL BEHAVIOR

The six-strut system puts the minimum constraint on the component. The system is structurally determinate. Because of the bearing mounts, the struts only resist axial loads. The reactions of the six struts can be computed using the static equations:

$$\Sigma F = 0 \quad \text{for } x, y, z$$

$$\Sigma M = 0 \quad \text{for } x, y, z$$

Any solution of the displacements will be a combination of elastic deformation and rigid body displacements. The latter one is due to and depends upon the boundary conditions applied at the support end of the struts. Because the structure is determinate, from the small displacement point of view, rigid body displacement is not a primary concern. It does not generate any additional stresses

Figure 2. The SSC LINAC RFQ with Six-Strut Support System.

nor affect the deformed shape. This is not completely true if large displacements are imposed. Geometric effects will take place. In the normal case, only the deformed shape due to gravity is of interest. The transverse and axial struts provide the movements for alignment but they do not participate in resisting the gravity loads. The primary deformed shape can be understood by decomposing the problem into symmetry and an antisymmetry components as shown in Figure 3. The former provides a solution for the vertical deformation. The latter provides a solution for the torsional deformation.

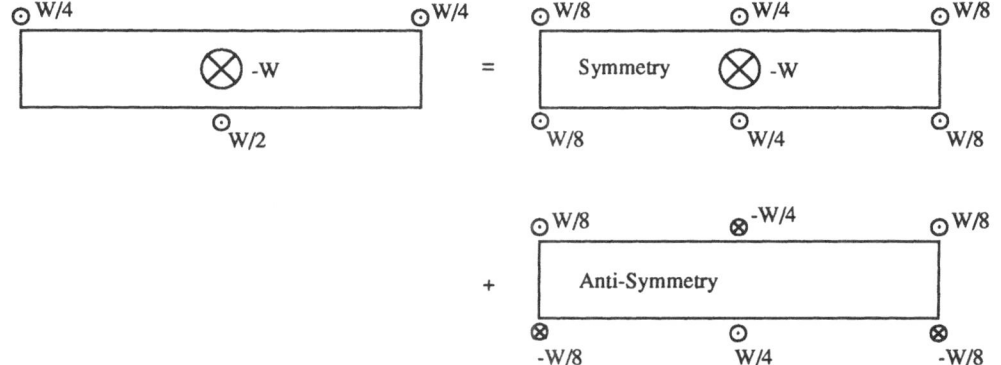

Figure 3. Problem Decomposed into Symmetry and Antisymmetry Solutions.

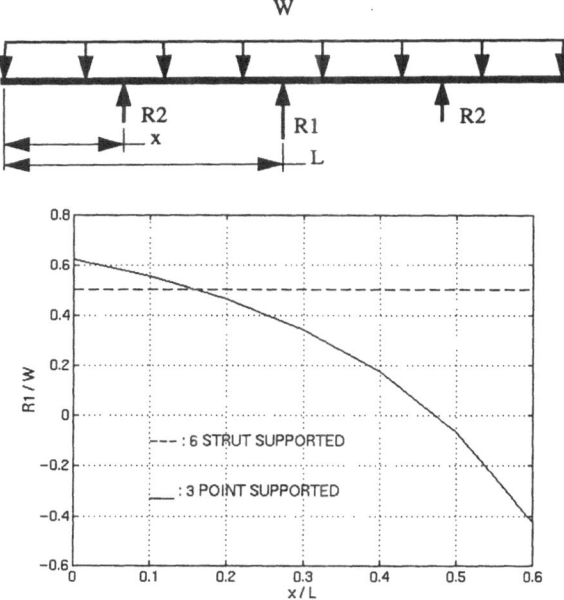

Figure 4. Middle Reaction is a Function of Support Location.

VERTICAL DEFORMATION

For resisting the vertical loads, the six-strut system leads to many misconceptions. This system is not similar to a three point supported beam. Since the six-strut is determinate, the reactions of the vertical struts are independent of their axial position along the beam, unlike the three point support beam. Figure 4 shows the middle reaction for the two different support cases. The vertical deflection depends on the location of the supporting strut. Figure 5 shows the effect on vertical displacement as the location of the outside strut varies. Since the outside strut carries 1/4 of the weight, the deflection can be minimized by optimizing the location of the support between the quarter point and the end of the beam. The lowest maximum vertical displacement for the six-strut system is 0.28mm compared to 0.68mm for a 2-point simple support system.

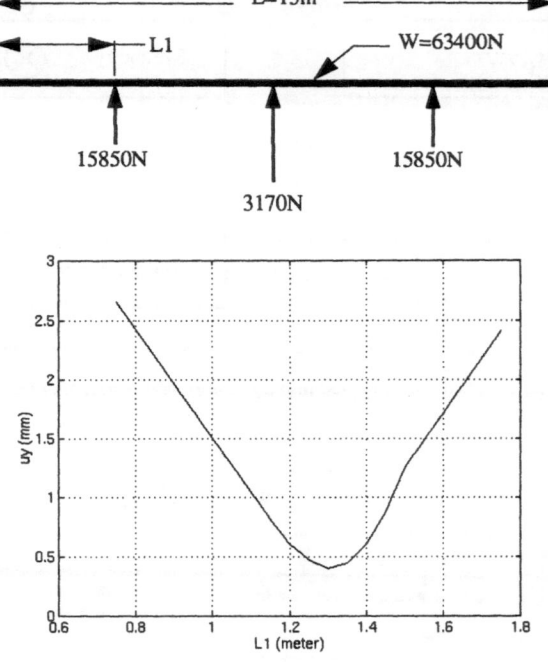

Figure 5. Six-Strut Support Maximum Vertical Deflection is a Function of Outside Support Location.

TORSIONAL DEFORMATION

This arrangement of the six struts will inherently generate a torsional deformation in the structure. This is due to the asymmetry of the arrangement as shown in Figure 3. The torsional angle can be computed as

$$\theta = \frac{TL}{JG}$$

where

W = component weight
T = Wd/8
J = torsional moment of inertia
G = shear modulus
d = transverse distance between support
L = axial distance between outer and middle supports

CONCLUSION

Superimposing the two solutions gives the total deformation. The method above can give a quick assessment on the deformed shape. Local deformation due to the strut is not considered in the solution. This methodology applies for arbitrary loading and non-symmetry location of the struts as well. In such a case, the reaction on a strut is not trivial but can be obtained from the static equations.

THERMAL ANALYSIS OF THE SSC BEAM SCRAPER

N. Tran and B. Dao

Mechanical Engineering Department
Accelerator Systems Division
Superconducting Super Collider Laboratory[*]
2550 Beckleymeade Avenue, MS 4006
Dallas, TX 75237-3997

ABSTRACT

When a particle beam impacts a beam scraper, heat is generated resulting in a rise in the temperature of the material. The maximum temperature rise should be kept to a minimum in order to maintain scraper efficiency and performance. In this paper the results of a thermal analysis of a scraper are presented.

INTRODUCTION

Beam scrapers are devices, located in the West Utility region of the collider rings, for removing the unwanted, diffused halo of particles that surrounds the dense central core of the accelerated beam[1]. It is important to minimize the peak scraper temperature to avoid excessive thermal deflection in a lateral direction on scraping surface. As indicated in Figure 1, three main scraper designs, using copper material, were evaluated from a thermal viewpoint. The first design involves no effort to cool the scraper. The second involves solid cooling while the third design adds water cooling.

ANALYSIS

A transient, three dimensional finite element model was developed using ANSYS[2]. The temperature dependent thermo-physical properties of copper were accounted for. The scraper initial temperature was assumed to be 305 K for all the analyses. Load boundary conditions included a constant scraping rate of about 1.45×10^9 protons per second[3] resulting in about 1284 watts applied for 20 minutes. Figures 2 and 3 show the energy deposition density per proton along the longitudinal and lateral directions of a copper scraper[4]. Radiation effects were omitted and an adiabatic condition was assumed on the scraper outside surfaces. Based on these analyses, an optimum scraper design can be chosen.

---

* Operated by the Universities Research Association, Inc., for the U.S. Department of Energy under Contract No. DE-AC35-89ER40486.

Table 1. Peak scraper temperatures for different design concepts (after 20 minutes).

Case No.	Description		ΔTmax (K)
1	Baseline design with 1" x 2" x 47" copper rod		456
2	Solid cooling with 6" Dia. x 47" copper block, 2" scraping surfaces		54
3	Solid and water cooling with two 1/2" Dia. water channels spacing 1" from scraping surface	2 liters/min. per channel	27
4		5 liters/min. per channel	20
5		10 liters/min. per channel	17

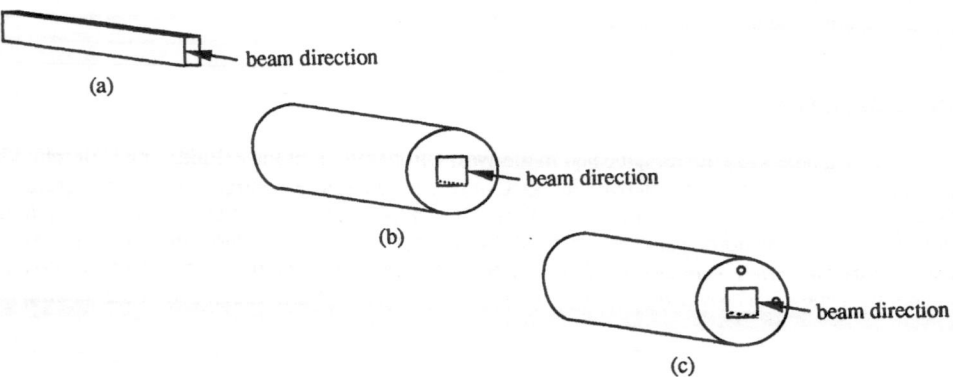

Figure 1. Schematic of the SSC beam scraper with different design concepts: (a) baseline design, (b) solid cooling, and (c) solid and water cooling.

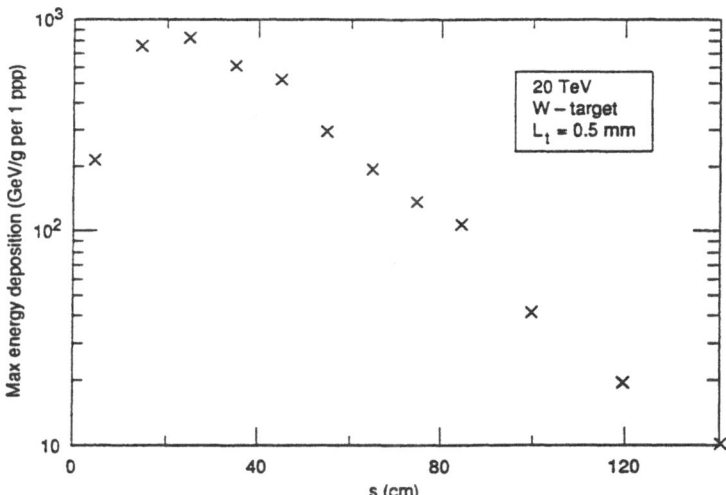

Figure 2. Longitudinal distribution of maximum energy deposition density in the copper scraper for chosen parameters of the scraper system.

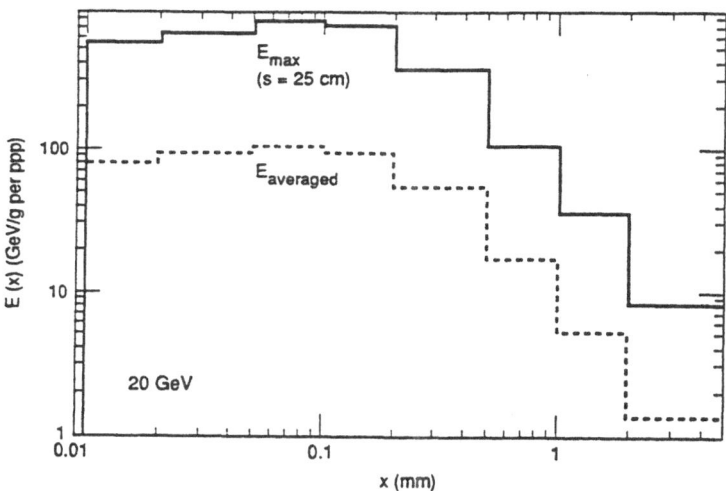

Figure 3. Lateral distribution of energy deposition in the copper scraper at shower maximum and averaged over 120 cm of length.

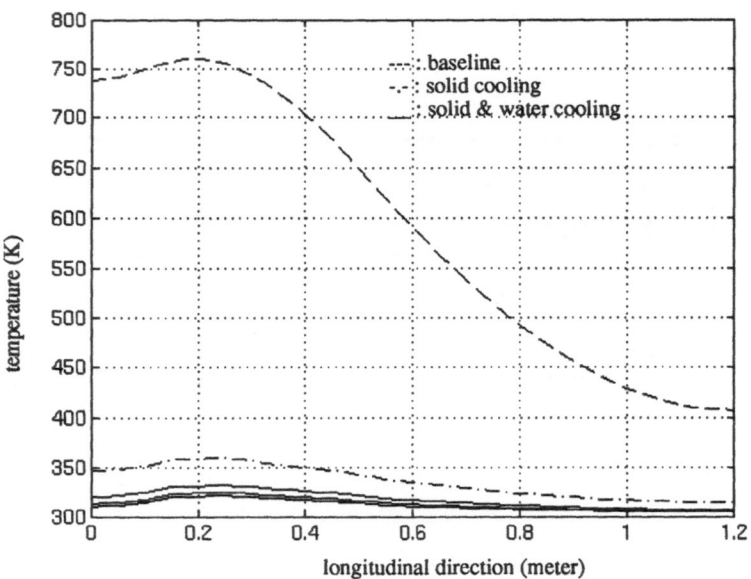

Figure 4. Variation of temperature on scraping surface along scraper length for the three design concepts.

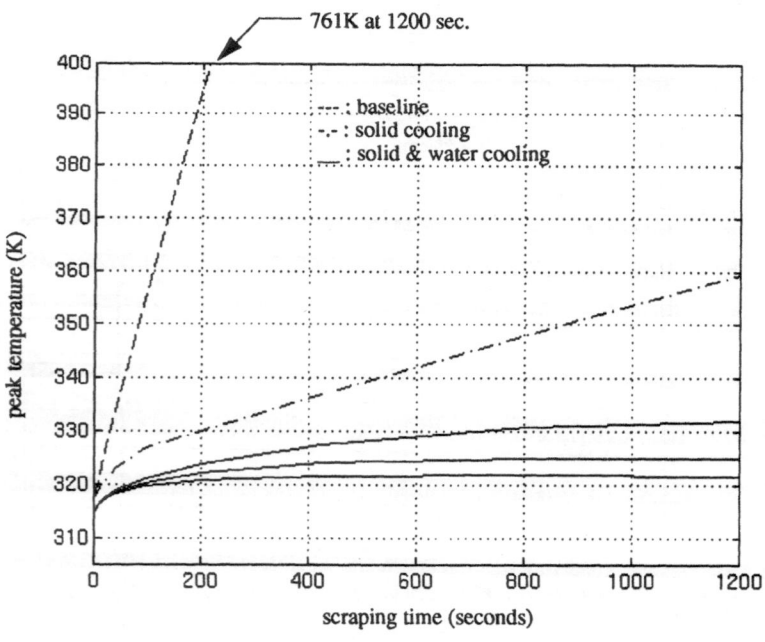

Figure 5. Variation of scraper peak temperature with time for the three design concepts.

RESULTS

Table 1 shows the thermal analyses results associated with three different scraper design concepts. A temperature rise was significantly reduced almost 400 K from a baseline (case 1) to a solid cooling (case 2) and improved even further with a water cooling design (cases 3, 4 and 5).

CONCLUSION

Based on the present thermal analysis, a water cooled scraper gives the lowest peak temperature after 20 minutes at full scraping rate.

REFERENCES

1. Element Specification (Level 3B), Collider Accelerator Arc Sections, Superconducting Super Collider Laboratory (SSCL), Number E10-000027, August, (1992).
2. ANSYS Engineering Analysis System, Revision 4.4, Swanson Analysis Systems, Inc., Houston, Pennsylvania, (1989).
3. Discussion with R. Soundranayagam.
4. M. Maslov, N. Mokhov, and I. Yazynin, "The SSC Beam Scraper System," SSCL-484, June, (1991).

REFERENCES

TECHNICAL STATUS OF SSC RF AMPLIFIER AND ACCELERATING CAVITY SYSTEMS

J. Ferrell, J. Rogers, P. Coleman, J. Curbow, and J. Mynk

Superconducting Super Collider Laboratory*
2550 Beckleymeade Ave.
Dallas, TX 75237

ABSTRACT

An overview of the design status of each RF amplifier system is presented and potential opportunities for industry involvement are identified. Specifically, brief descriptions of the LINAC, LEB, MEB, HEB, and Collider RF Amplifier systems are presented, as well as the LEB and MEB cavities. Status in the LINAC ranges from an RFQ Amplifier being operational at the SSC to procurement of subsystems still pending. The LEB prototype system is in tests at SSC, and the MEB prototype system is in construction. Preliminary design of the Collider RF system has begun, and design of the HEB RF system will begin in 1994.

LINAC RF SYSTEMS

The SSC LINAC is comprised of an ion source, Radio Frequency Quadrapole (RFQ), two bunchers, a four tank Drift Tube LINAC (DTL), two more bunchers (driven by a single amplifier), a nine-module Coupled Cell LINAC (CCL), and a compressor (located in the transfer line between the LINAC and Low Energy Booster).

The RFQ system for the LINAC operates at a frequency of 427.617 MHz through the DTL and at 1282.51 MHz for the following bunchers, CCL, and compressor. The LINAC pulse repetition rate is 10 Hz with RF pulse lengths from 65 to 90 μs depending on cavity fill time.

The 600-kW, 427-MHz RFQ amplifier utilizes a tetrode final amplifier and is presently in test at the SSC. Bunchers #1 and #2 will utilize identical 50-kW amplifiers that have triode amplifiers. These units are presently being manufactured. The 4-MW, 427-MHz DTL amplifiers are cathode-modulated klystrons. The klystrons have been manufactured, and the modulators are being manufactured now. The 20-MW, 1282-MHz CCL amplifiers are also cathode-modulated klystrons. Several klystrons have been manufactured, and the modulators are in fabrication. Bunchers #3 and #4 will be driven by a single CCL klystron operated at reduced cathode voltage to achieve 2.5-MW output.

---

* Operated by the Universities Research Association, Inc., for the U.S. Department of Energy under Contract No. DE-AC35-89ER40486.

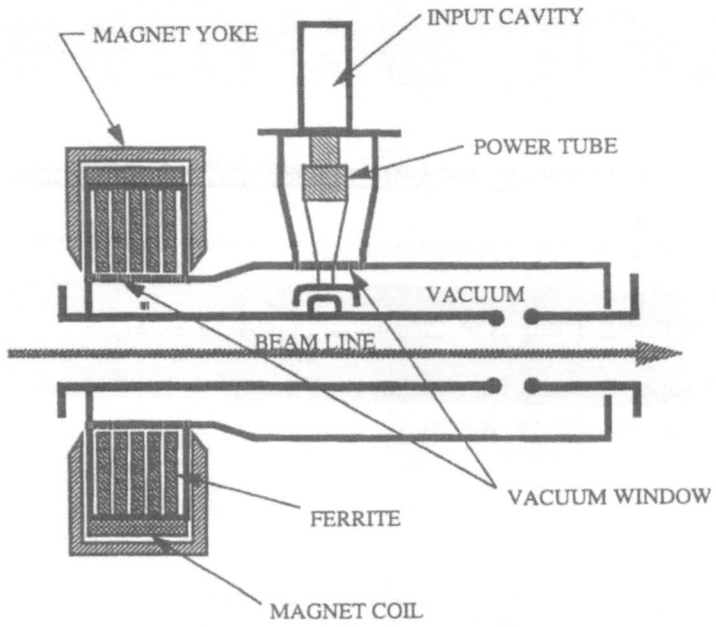

Figure 1. LEB cavity simplified diagram.

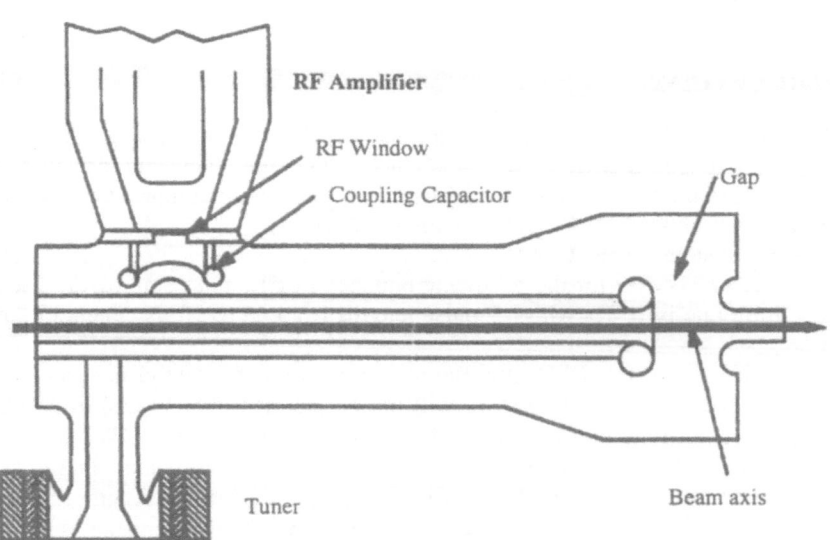

Figure 2. MEB cavity simplified diagram.

The 100-kW, 1282-MHz compressor amplifier will be a cathode-modulated klystron as well. Procurement documents for this unit are in work.

LOW ENERGY BOOSTER (LEB) RF SYSTEM

The LEB RF system is required to provide a circumferential accelerating peak voltage of 765 KV. The RF frequency varies from 47.51 to 59.78 MHz over the acceleration cycle. The accelerating cycle is 50 ms long at a 10-Hz repetition frequency. The accelerating voltage will be provided by at least 8 and no more than 12, quarter-wave, ferrite-tuned cavities. Figure 1 is a simplified diagram of the LEB cavity design. Two prototype cavity designs have been fabricated that differ primarily in the ferrite-cooling technique. One cavity utilizes conduction cooling of the ferrites, while the other uses direct liquid cooling of the ferrites. The conduction-cooled design has been tested with mixed results to date. The frequency range was found to be adequate, but the prototype design suffered ferrite damage due to arcing. The liquid-cooled version is presently in test.

MEDIUM ENERGY BOOSTER (MEB) RF SYSTEM

The MEB RF system is required to provide a circumferential accelerating peak voltage of 1600 KV. The RF frequency varies from 59.78 to 59.96 MHz over the accelerating cycle of 4.5. The accelerating voltage will be provided by at least 8 and no more than 12 quarter-wave, ferrite-tuned cavities. Figure 2 is a simplified diagram of the MEB cavity design. The tuner design utilizes conduction cooling of the ferrites. The prototype cavity is presently being fabricated.

HIGH ENERGY BOOSTER (HEB) RF SYSTEM

The HEB RF system is required to provide a circumferential accelerating peak voltage of 1250 KV. The RF frequency is 59.96 MHz and varies only a few hundred Hz over the accelerating cycle of 100. Design of the HEB RF system is in the very preliminary stages and will commence in earnest in 1994. As presently conceived, the cavity will be similar to the Fermilab Tevatron cavity design. Seven cavities are planned to achieve the accelerating voltage. The power amplifier will be located at the surface.

COLLIDER

The RF systems for the SSC Collider are required to furnish a peak circumferential voltage of 20 MV per ring at a frequency of 360 MHz. The RF system design has concentrated so far on the use of superconducting versus normal-conducting cavities and whether to locate the amplifiers on the surface or at the beam-line level. A decision has been reached to locate the amplifiers at the surface. The power required per klystron is dependent on the cavity choice. Possible cavity choices include normal-conducting single cell, normal-conducting multi-cell, and superconducting single cell cavities. The cavity choice is still being studied, but is expected to be made shortly.

COMMON RF SYSTEMS

The SSC RF systems are planned to have a high degree of commonality. The low-level RF and control systems will be very similar for all machines and will be based on VME/VXI technology. The booster systems will all utilize the same power supplies, driver amplifier, and power amplifier tube.

THE LOW ENERGY BOOSTER PROJECT STATUS

G.W. Tuttle

Superconducting Super Collider Laboratory*
2550 Beckleymeade Ave.
Dallas, TX 75237

INTRODUCTION

In order to achieve the required injection momentum, the Superconducting Super Collider (SSC) has an accelerator chain comprised of a Linear Accelerator and three synchrotrons. The Low Energy Booster (LEB) is the first synchrotron in this chain. The LEB project has made significant progress in the development of major subsystems and conventional construction. This paper briefly reviews the performance requirements of the LEB and describes significant achievements in each of the major subsystem areas. Highlighted among these achievements are the LEB foreign collaborations with the Budker Institute of Nuclear Physics (BINP) located in Novosibirsk, Russia.

LEB PERFORMANCE REQUIREMENTS

The LEB is a rapid cycling (10 Hz) synchrotron and has a unique triangular shaped, three-fold symmetric lattice. The primary design challenge for the LEB is the preservation of the transverse beam emittance at low energy, the difficulty being a substantial space-charge tune spread. The lattice was designed to avoid transition, $\gamma_t = 22.14$, in an effort to mitigate the emittance problem. The performance requirements are outlined in Table 1.

Table 1. Low Energy Booster Performance Requirements.

Parameter	Requirement	Units
Injection Momentum	1.22	GeV/c
Extraction Momentum	12.0	GeV/c
Particles/Bunch	1×10^{10}	
Input $\varepsilon_{rms-normalized}$	0.3	πmm-mrad
Output $\varepsilon_{rms-normalized}$	0.6	πmm-mrad

* Operated by the Universities Research Association, Inc., for the U. S. Department of Energy under Contract No. DE-AC35-89ER40486.

MAGNET STATUS

The LEB includes a complex lattice of 186 main magnets, 90 two meter dipoles and 96 quadrupoles of varying lengths, and 244 corrector magnets including: 90 dipoles, 94 quadrupoles (90 trim and 4 skew), 48 sextupoles, and 12 skew sextupoles, and specialty magnets used for the injection, the extraction, and the transport of the beam. Although the magnets were primarily designed by the SSC, some aspects of the design are the result of a BINP collaboration .

The SSC signed an agreement with BINP to build and test a prototype of the main dipole and quadrupole magnet. BINP has built the main dipole and will begin testing in May 1993. BINP has also built and tested a prototype main quadrupole. The SSC also commissioned Lawrence Berkeley Laboratory (LBL) to build a prototype quadrupole and Stanford University to build a prototype main dipole magnet, all based on the same design.

Lawrence Berkeley Laboratory (LBL) has built and tested the prototype main quadrupole magnet and Stanford University is expected to complete the prototype main dipole in May 1993. The test results on both main quadrupole prototypes indicate that the design will meet or exceed specification and only minor structural support changes are contemplated. BINP and their subcontractors are scheduled to start full-rate production of the main dipole and quadrupole magnets in late 1993.

A prototype of the two basic types of quadrupole corrector magnets (high field and low field) were built by the SSC. The quadrupole correctors design makes use of the same lamination as the main quadrupole magnet. A sextupole corrector magnet prototype and a dipole corrector magnet prototype were also built by the SSC. Some of the early testing was accomplished at Los Alamos National Lab and Lawrence Berkeley Laboratory. Subsequently all corrector magnets have been tested at the SSC. The test results to date reveal a promising design and required only minor modifications the production design.

BINP is currently building another set of prototype corrector magnets based on the modified SSC design, under an agreement similar to the main magnet agreement. BINP and their subcontractors are scheduled to start full-rate production of the corrector magnets in early 1994.

POWER SYSTEM STATUS

The SSC has completed specifying the major components of the Ring Magnet Power System. This system will be capable of delivering 4000 amperes of current to a resonant circuit comprised of the main magnets, twelve 40 mH Energy Storage Inductors (ESI), and twelve 18 mF Capacitor Banks, tuned to resonate at 10 Hz.

Scheduled to begin delivery in May 1993, the ESIs will be the first production LEB components to arrive on site. BINP designed, and arranged for the fabrication of the ESIs in Ekaterinberg, (Siberia) Russia The ESI design has been verified and two production units have been tested. The other component of the LC circuit, the Capacitor Bank Assembly, is in the source selection phase and several outstanding proposals have been received from Industry.

The remaining portion of the LEB power system are elements of a larger acquisition strategy to buy similar power system components for the entire injector chain. These include the 15 kV switch gear and transformers, the rectifier converters cabinets, thyristors, and passive filters. The LEB phase of these procurements will be out for competition to industry in the next 12–18 months.

The corrector power supplies, also operating at 10 Hz, are lower capacity, operating at load currents as low as 3.5 Amperes for the skew sextupole, and as high as 100 Amperes for the high-field quadrupole. The corrector power supplies are of the rack mounted variety and the present design includes powering a series of similar amplifiers from a single bulk power supply. This procurement is planned for competition in late 1993.

RF SYSTEM STATUS

The RF System is the most technically challenging LEB system, the principal challenge being the RF tuner/cavity design. The LEB cavity is a quarter wave coaxial resonator with the accelerating gap at the high-voltage end and a ferrite tuner at the high current end. Tuning is accomplished by varying the permeability of the ferrite. The LEB RF system is required to provide a circumferential voltage of 765 kV and tune over a substantial range (47.5 – 59.8 MHz). Although the baseline design is 8 cavities, lattice space is available for a maximum of 16 cavities, should the required gap voltages and the associated heating problems preclude the 8 cavity design.

The perpendicularly biased field design for the LEB tuner results in a higher magnetic Q in the ferrite, thereby allowing higher gap voltages than using the conventional parallel biased design. Even so, substantial heat is developed due to the losses. This has lead to two designs: a liquid-cooled tuner which was designed and built by the SSC and a conduction-cooled design which is a modification of the SSC design produced by BINP.

The conduction-cooled design has been tested at the SSC and has demonstrated good performance. However a failure recently occurred at higher gap voltages, apparently due to a void in the elastoner used to bond the ferrites, resulting in arcing. A failure analysis is being conducted by the SSC and BINP to determine the viability of the design.

The liquid-cooled tuner is currently being assembled and testing should begin in May 1993. Upon completion of the prototype testing, a final design will be selected and a build-to-print procurement will be initiated. The tuner/cavity assembly also includes a 4CW150,000 tetrode which is mounted directly on the accelerating cavity.

The RF tuner/cavity is supported by an array of power supplies. The largest being the Anode Power supply which is an oil filled 400 KVA rectifier/transformer. The tuner is biased by another power supply which is tap selectable (150Vdc – 300 Vdc) in 50-V increments, and capable of providing 1500 A of output current. These RF power supplies will be put out for competition in the next 12 months. A complex low-level RF and supervisory control system is also being developed at the SSC in cooperation with Los Alamos National Laboratory.

VACUUM SYSTEM STATUS

The LEB beam line is defined by 198 vacuum chambers which fit into the aperture of the main and corrector magnet packages. A prototype of each major chamber was been built by industry and has been tested at the SSC. A build-to-print procurement for 198 chambers and the associated bellows will be issued within the next 12 months.

The 10^{-7} Torr vacuum environment within the chamber will be maintained by 90 noble diode ion pumps. These pumps will be supported by a number of power supplies and controllers located in racks in the LEB surface buildings. The ion pumps and associated support equipment include: thermocouple gauges, cold cathode gauges, and Turbo Molecular Pump carts obtained in a build-to-specification procurement within the next 12 months.

INSTRUMENTATION SYSTEM STATUS

Another prominent feature of the beam line is the more than 100 monitors that are located in strategic locations throughout the LEB lattice. These monitors and their associated electronics characterize the beam position, beam profile, beam current, bunch length and other vital beam information. The instrumentation system also includes a number of beam loss monitors, located on the tunnel walls, which detect any substantial beam losses. All monitors and their associated electronics are in various stages of design and development. The majority of the LEB instrumentation system will be procured, at the component level, via build-to-print specifications in the next 18–24 months.

CONTROL SYSTEM STATUS

The LEB, like all other machines, will be centrally controlled from the Accelerator Main Control Room (AMCR). The LEB will communicate with the AMCR via a SONET fiber optic backbone that supports a minimum OC-3 bandwidth. The LEB interfaces to the SONET backbone via Add Drop Multiplexers (ADM) located in each of the LEB surface buildings. The ADMs will interface directly to the individual subsystems located within the building. The LEB also includes a precision timing system which will be the basis for synchronizing all LEB subsystems. Most of these systems will be procured in a build-to-specification procurement in the next 18–24 months.

CONVENTIONAL CONSTRUCTION STATUS

The most visibly exciting LEB progress to date is that of the $570 \times 3.7 \times 3$ meter tunnel . The construction began in January 1993 and is scheduled to be completed by mid 1994. The construction of the six primary surface buildings, one located at each arc and one located at each straight, and the installation building will begin in the next few months.

SUMMARY

Although technical challenges remain in areas such as the RF System, the LEB design and development is progressing exceeding well. Most major subsystems will be procured in the next year and the tunnel will be ready for installation activities in early 1994. The LEB is currently scheduled to be operational in 1996.

REFERENCES

1. E. D. Courant, A. A. Garren., and U. Wienands, "Low Momentum Compaction Lattice Study for the SSC Low Energy Booster," *Proceedings of the IEEE Particle Accelerator Conference*, San Francisco, CA (1991).
2. R. C. York *et al.*, "Low Energy Booster: A Status Report," *Proceedings of the 1991 IEEE Particle Accelerator Conference*, San Francisco, CA (1991).
3. R. L. Poirier, T. Enegren and C. Haddock, "Perpendicular Biased Ferrite Tuned RF Cavity for the TRIUMF KAON Factory Booster Ring," *IEEE Particle Accelerator Conference*, Chicago, IL (1989).
4. C. Jach, "Switchable 10 Hz/1 Hz LEB Magnet Power Supply System," IEEE Particle Accelerator Conference, San Francisco, CA (1991)
5. R. Winje, "Equipment Acquisition Plans or the SSCL Magnet Excitation Power System," This Conference, San Francisco, CA (1993).

8 Parallel Technical Sessions IV

AN ENGINEERING DESIGN STUDY OF THE SUPPORT PLATFORM
ASSEMBLY FOR THE SSC SDC DETECTOR

H. J. Krebs,[1] R. H. Wands,[2] and J. L. Western[1]

[1]Superconducting Super Collider Laboratory*
2550 Beckleymeade Ave.
Dallas, TX 75237-3997

[2]Fermi National Accelerator Laboratory
Wilson Road, P. O. Box 500
Batavia, IL 60510*

INTRODUCTION

A large angular acceptance high energy physics particle detector is presently being designed by the Solenoidal Detector Collaboration (SDC) for the purposes of doing high p_t physics at the Superconducting Super Collider Laboratory (SSC Laboratory). The support platform assembly is the structural device which transfers the 30,000 tonne gravitational load of the octagonally shaped muon barrel toroid (MBT) and the other detector components to the foundation below. The detector components are very sensitive to differential deflection and rely on the barrel toroid and support platform for stability. Figure 1 shows a front elevation view of the support looking in the direction of the beam axis. A side elevation view of the support looking perpendicular to the beam axis from the collider ring center (looking east) is shown in Figure 2.

The operational load path is provided by two pairs of inclined longitudinal plates resting at 67.5° on three pairs of plate girders that are positioned in-line in Z. The plate girders are held together laterally with 38 tie bars and supported vertically by the vertical adjustment system. The lateral stability of the inclined plates is provided by 22 stabilizer beams with cross bracing between each beam. The Z location of each split in the plate girder is coincident with the Z location of the gap in the calorimeter (4428 mm from the detector center.) The width of each split is 155 mm to allow installation of the alignment reference system.

The collider beam line in the IR-8 underground experimental hall is oriented at a 2.16 mm/m slope from south to north as shown in Figure 2. The support is designed and installed to provide this slope at the top surface of the inclined plates. The assembled support rests on a ten foot thick steel reinforced 8000 psi concrete slab. The slab has a 2 mm differential deflection criteria under normal gravitational loading.

---

*Operated by the Universities Research Association, Inc., for the U.S. Department of Energy under Contract No. DE-AC35-89ER40486.

The total length of the support platform after assembly is 28,060 mm. The X width of the assembly is 12,430 mm. The height of the assembly is 4650 mm measured perpendicular to the beam line at the detector interaction point. The height of the vertical adjustment system is 1000 mm. The weight of the support platform assembly and the vertical adjustment system is approximately 2000 tonnes. The physical size and weight of the individual components are driven by a 18 meter x 11 meter equipment access shaft and the 100 tonne capacity crane in the underground experimental hall.

OVERVIEW OF PLATFORM PARTS AND THEIR FUNCTIONS

INCLINED PLATE

The inclined plates provide the load path from the detector to the plate girder assemblies below. Precision 100 mm radii are provided top and bottom to provide a moment free connection between the MBT and the inclined plates and also between the plate girder assemblies and the inclined plates. These edges are flame hardened and will be coated with a molybdenum disulfide based lubricant prior to installation. There are vertical grooves machined in the outside surfaces of the plates at Z=4428 mm and Z=11453 mm to provide clearance for the alignment reference system. The grooves are 155 mm wide in Z and 65 mm deep. There are two inclined plates on each side of the support assembly. Each pair of plates is connected tongue-in-groove so that they may act as a single mechanical unit. Each inclined plate is a 275 mm thick by 3220 mm wide by 14,030 mm long stress relieved steel forging. The material will be selected on the basis of weldability and mechanical properties of 40,000 psi minimum yield and 70,000 psi minimum tensile strengths. The weight of each inclined plate is 99 tonnes.

PLATE GIRDER ASSEMBLIES

The plate girder assemblies are ASTM A36 structural steel weldments that provide the load path from the inclined plates down to the vertical adjustment system. The girders have enough axial bending stiffness to allow vertical adjustments by the vertical adjustment system without over stressing the inclined plates. The girders are designed in such a way that the pivot groove centerline and the bottom machined surfaces will provide the 2.16 mm/m angle to allow assembly of the upper detector components parallel to the collider beam.

The girder assemblies consist of dual 50mm thick vertical trapezoidal plates spaced in Z to coincide with the spacing of the base plates that are embedded in the pit floor. The center to center spacing between these plates is 230mm to maximize the effective area of the vertical adjustment system. The trapezoidal plates are connected in Z with 75mm thick vertical webs. The vertical height of the girder is 1585 mm, 1565 mm and 1545 mm for the south, mid and north girder assemblies respectively. The girders also contain 200 mm × 613 mm steel blocks with three 78 mm diameter through holes at each tie bar location. The blocks provide buckling resistance in the vertical webs from bolt preload and aid in the insertion of the three M72 bolts that connect the tie bar yoke to the inner web of the plate girder. The bolts are spaced vertically 219 mm to standardize the assembly tooling with the MBT. The inner and outer surfaces of the plate girder webs are reinforced with a 25 mm thick plate in the area of this bolted connection. The webs and trapezoidal plates are sandwiched between two 75 mm thick plates that run the entire length of the assemblies. The top plate is 700 mm wide and the bottom plate is 3000 mm wide. The Z length of the plate girders is 9525 mm, 8700 mm and 9525 mm for the south, mid and north girder assemblies respectively. The interface between the trapezoidal plates and the base plate is braced with 500 mm tall by 400 mm wide by 50 mm thick gussets.

Two 50 mm thick pads straddle each pair of trapezoidal plates to provide mounting surfaces for the fixtures that are necessary to install and position the inclined plates and the stabilizer beams. The lower pads are horizontal and provide a pushing surface in the vertical direction for positioning of the stabilizer beams during assembly. The upper pads are located at 67.5° from horizontal to provide a pushing surface that is perpendicular to the inclined plate on the opposite side.

The bottom plate of the plate girder is reinforced locally at the interface with the vertical adjustment system with 900 mm wide by 3000 mm long plates. The stock thickness of these plates varies from 75 mm to 100 mm. The bottom surface of these plates is planed to provide the 2.16 mm/m slope of the detector.

The interface between the plate girder and the inclined plate is a 300 mm x 200 mm saddle block which has been machined to 22.5° with a 100 mm radius semi-cylindrical notch machined to fit the radiused end of the inclined plates. The 22.5° angle limits loading of the inclined plate to the axial direction. The block is inserted in a notch that is provided in the trapezoidal plates thus enabling the trapezoidal plates to resist the shear load that is transmitted from the inclined plate. The shear load that would develop from lateral loading in the Z direction is resisted by mechanical stops machined into the inclined plate/plate girder interface. These stops are located at Z=±2850 mm and ±8815 mm. Lateral loads in the X direction are resisted by the pivot sockets.

The weight of the plate girder assembly is 84 tonnes, 71 tonnes and 83 tonnes for the south, mid and north girders respectively.

STABILIZER BEAMS

The 22 stabilizer beams are W36x210 ASTM A36 hot rolled sections that provide the connection between the inclined plates and are also used to stabilize the support and resist lateral loads. Each connection to the inclined plate is designed as a moment carrying connection with dual L8x8x1 ASTM A36 hot rolled sections with twenty 1.5 inch diameter A325 bolts in shear and forty four 1.5 inch diameter A325 bolts in tension and bending. Each beam has six 500 mm wide x 300 mm tall reinforced openings for cable tray access to the inner muon chambers of the bottom octant. The weight of each beam is 2.5 tonnes.

TIE BARS

38 tie bars run laterally between the plate girder assemblies to react the horizontal component of the detector load on the inclined plate. The tie bars are ASTM A36 steel that are 3700 mm long and have a cross section of 200 mm wide x 500 mm high. The large cross section is necessary to limit axial elongation of the bars (separation of the plate girders) to 1 mm during normal detector operational loading. The connection to the plate girder uses a 175 mm diameter x 225 mm long hardened steel pin in double shear to minimize bending moments when alignment corrections are made with the vertical adjustment system. The weight of each tie bar is 6 tonnes.

UPPER SADDLE BLOCKS AND SHIMS

An upper saddle block will be incorporated as the interface between the support platform assembly and the muon barrel toroid. 2 mm thick shim stock shall be positioned on the top of the blocks to provide the Y (vertical) alignment capability for the barrel floor. The alignment of the barrel in the X direction is accomplished using a two degree stainless steel wedge shaped key between the saddle block and the inner surface of the notch in the MBT bottom block. The wedge key will enable a 5 mm correction in X during installation of the MBT. A 25 mm thick hardened steel sleeve is positioned between the radiused end of the inclined plate and the block, and resides in a semi-cylindrical pocket that has been machined in the block. A 25 mm x 25 mm steel key prevents rotation of the sleeve with respect to the saddle. The saddle blocks are 874 mm long and 300 mm wide x 195 mm tall. Some blocks must have machined notches to accommodate clearance for the alignment reference system. The material of the upper saddle block is 6061-T6 aluminum in order to provide magnetic resistance between the support platform and the muon barrel toroid.

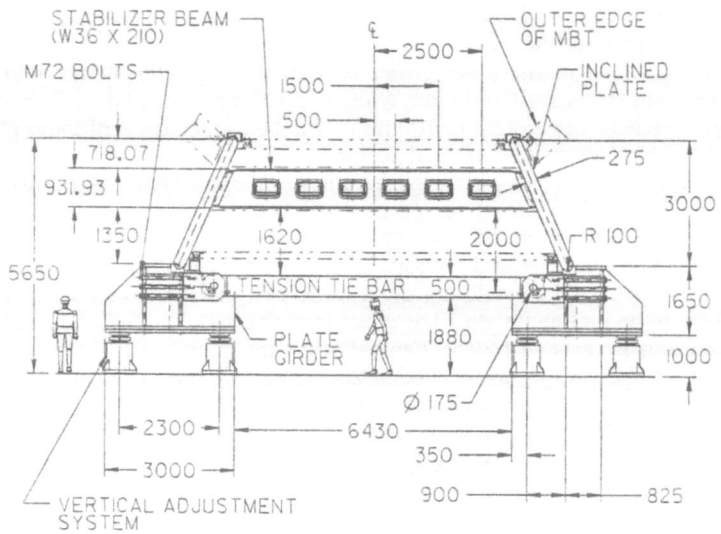

Figure 1. Front Elevation View.

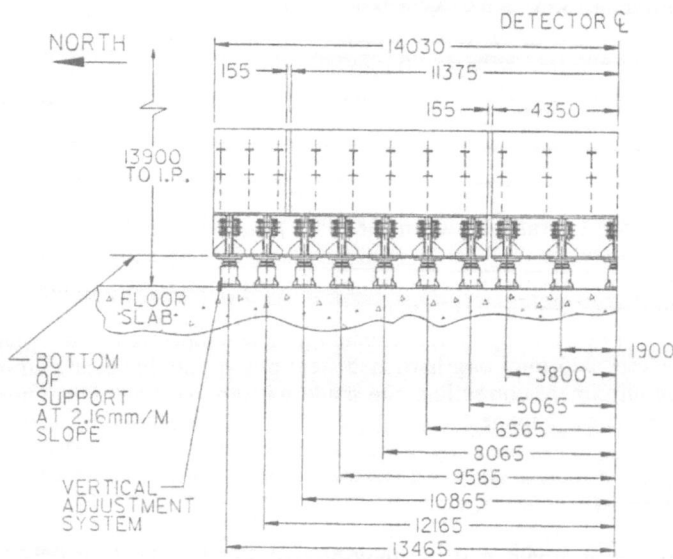

Figure 2. Side Elevation View.

CONCLUSION

The support platform assembly is presently in the preliminary design phase and is being optimized for function, material usage, fabricability and ease of installation. A detailed finite element model is being created that will encompass the muon barrel toroid, support platform and foundation as a single mechanical unit. The model will be used to investigate the effects of slab warping as well as failure modes and effects analyses of the vertical adjustment system.

AN ENGINEERING DESIGN STUDY OF ALTERNATIVE DESIGN OPTIONS FOR THE FORWARD TOROIDS OF THE SSC SDC DETECTOR

L.I. Dittert and J.L. Western

[1]Principal Engineer
ICF Kaiser Engineers, Inc.
1800 Harrison Street
Oakland, CA 94612

[2]Engineering and Design Group
Conventional Construction Division
Superconducting Supercollider Laboratory*
2550 Beckleymeade Avenue
Dallas, TX 75237-3997

ABSTRACT

The Muon Forward Toroids (MFTs) are 3-meter-thick sections of magnetized iron located at both ends of the detector proposed by the Solenoidal Detector Collaboration (SDC). The forward calorimeter and the outer muon chamber layers are supported by the MFT. This paper investigates viable and cost effective construction alternatives for the MFTs.

INTRODUCTION

The SDC detector is a major SSC experiment located at the IR8 site of the east campus. The total weight of the detector is approximately 32,000 metric tons.

The MFTs are located at both ends of the Muon Barrel Toroid (MBT) and are designed to be moved out of the MBT to facilitate access to the detector components located inside the MBT. The shape, dimensions, magnetic field distribution, and the overall tolerances of the MFTs must satisfy the physics goals of the experiments and other requirements of the detector.

The principal objective of this work was to develop cost-effective methods of fabrication and assembly of the MFT steel and its support system that will meet these requirements.

GENERAL DESCRIPTION

Each MFT is made up of two 1.5-meter-thick steel sections (MFT1 and MFT2), as shown in Figure 1. The forward detector system, including the Forward Calorimeter, is supported by the MFT. Each forward toroid assembly weighs about 3075 metric tons, the forward detection system weighs about 825 metric tons, and MFT1 and MFT2 weigh about 1075 and 1165 metric tons, respectively. The toroid

---

*Operated by the Universities Research Association, Inc., for the U.S. Department of Energy under Contract No. DE-AC35-89ER40486.

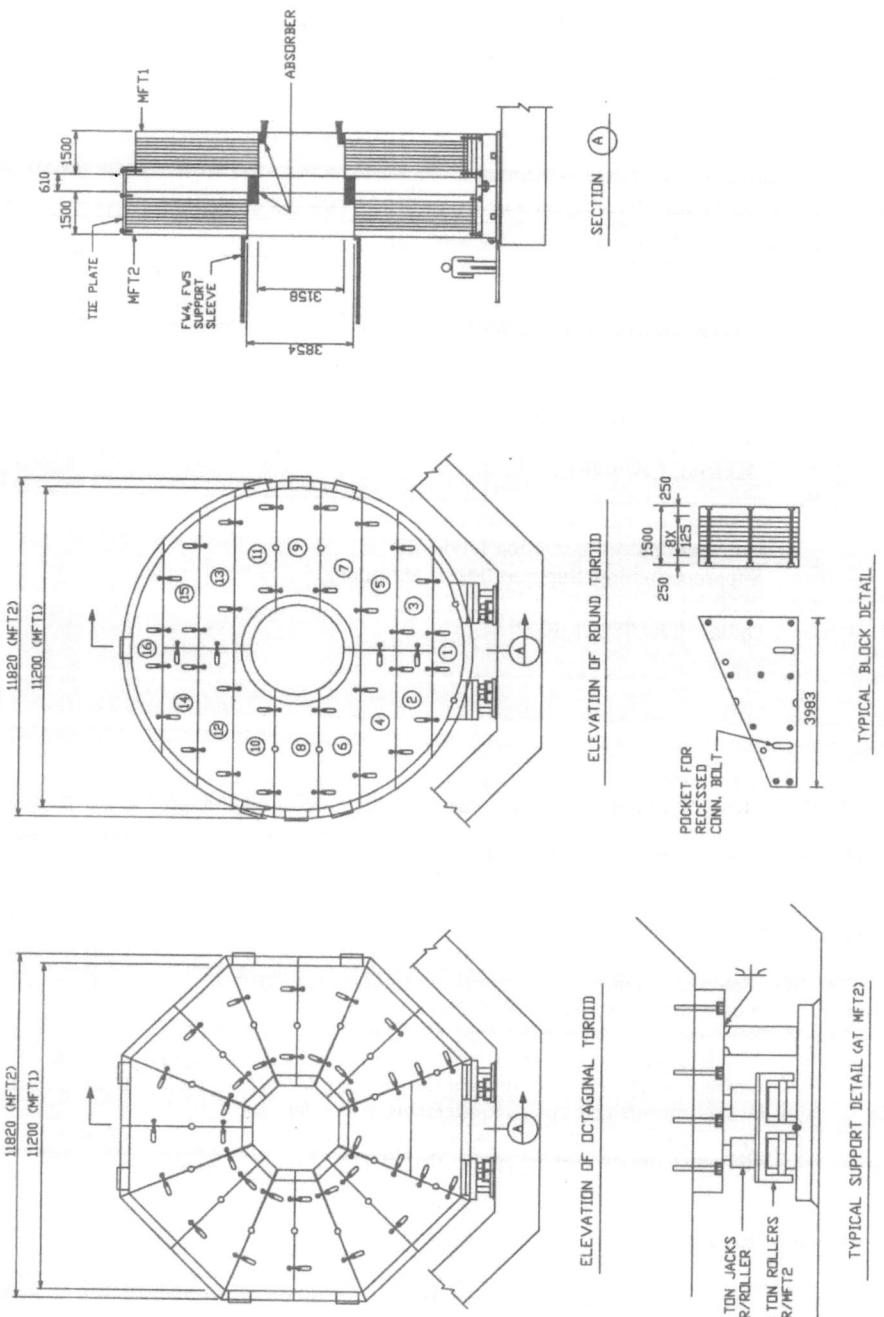

Figure 1. Forward Toroid Alternatives.

assembly is supported on two T-shaped weldments bolted to the underside of the MFT, as shown at the lower left of Figure 1. The supports are equipped with hydraulic jacks and rollers to permit adjustment and movement of the MFT. To reduce costs and construction time in the hall, the steel of the MFT is assembled from blocks of plates bolted together in the fabrication shop.

ALTERNATIVE SHAPES INVESTIGATED

Two toroid shapes, one round, one octagonal, and two block shapes, one pie-shaped, one rectangular, were investigated. Each toroid section was assembled from 16 blocks weighing from 60 to 75 metric tons. The octagonal toroid was built from 14 pie-shaped and two trapezoidal blocks, and the round toroid from 16 rectangular blocks. A round toroid assembled from pie-shaped blocks was also investigated. A comparison of the three types of toroid designs investigated is presented at the conclusion of this paper.

BLOCK CONSTRUCTION AND ASSEMBLY

Each block is built in the fabrication shop from vertical plates bolted together by longitudinal high-strength bolts placed along the edges of the block, as shown in Figure 1, Typical Block Detail. To accommodate the recessed holes of the longitudinal bolts and the pockets of the block connection bolts, the outer plates of the blocks are 250 mm thick, while the inner plates are thinner to facilitate flattening the plates by the longitudinal bolts. Because of this arrangement, the flatness tolerance of the inner plates is not critical; however, the outer plates must be rolled to within the required flatness of the finished block.

After tightening the longitudinal bolts, the contact surfaces of each block are machined for proper magnetic continuity and fit. Since the machined surfaces of the rectangular block round toroid alternative are parallel to each other, the angular milling accuracy and resulting flatness of the completed toroid should be easier to achieve.

For uniform flux distribution the eight radial surfaces of the octagonal toroid alternative are separated by nonmagnetic plates tapering from 26 mm to 2.6 mm (approximately), from the inner to the outer corners of the toroid.

The blocks are held together by 64 mm diameter high-strength stud-bolts recessed into the outer block plates. There are four bolts between adjacent blocks. After the bolts are tightened, the slots are plugged with steel inserts to satisfy the MFT ray penetration requirements. To prevent sliding, some of the blocks are tied to each other by 200 mm diameter fitted shear pins.

To provide sufficient stability against overturning in the "Z" direction, and a more uniform distribution of the eccentric weight of the forward detector system to the rollers, MFT1 and MFT2 are tied together with stiff shear transfer plates located on the sides of the toroid.

To ensure the required flatness tolerance of the completed toroid and proper fit of the blocks during assembly in the experimental hall, the toroid blocks are pre-assembled in the fabrication shop. Any tolerance problems and misfits discovered during pre-assembly can be corrected before blocks leave the fabrication shop. After pre-assembly, the blocks are prime-painted and match-marked before shipment to the SSCL. The machined areas of the blocks are not painted.

MATERIAL SPECIFICATIONS

Materials for the MFT and its supports are as follows:

Component	Material Specifications
Plates for block construction	SAE 1010, or equal
Toroid shear keys and bolt pocket inserts	SAE 1010, or equal
Block connecting studs and Z bolts	ASTM F568
Plates for MFT support frame and tie plates	ASTM A36, or equal

FABRICATION TOLERANCES

To meet the physics goals of the experiments and the specification requirements, the fabrication tolerances of the blocks and of the completed MFT must be as follows:

- Tolerance for the total thickness of all plates in a block must be +/- 5 mm.
- Flatness tolerance (maximum deviation from a true flat plane) of the outer vertical block plates must be 10 mm at any location of the plate.
- Angular deviation of the machined contact surfaces of the block must be such that the deviation of any location of the vertical surface of the completed toroid from the true plane is not more than 15 mm.

SUMMARY OF FINDINGS

The structural integrity, method of fabrication, and assembly of octagonal and round-shaped toroids assembled from large blocks were investigated. We found that in comparison with octagonal or round toroids built from pie-shaped blocks, the structural integrity of a round shaped toroid built from rectangular blocks is less dependant on the block-to-block connection bolts, and the fabrication and machining of the blocks is less complicated and should result in better tolerances. The number and total length of the shear pins is also less, however, the total area of the machined surfaces is about 24 per cent more than what is required for toroids assembled from pie-shaped blocks.

Table 1 summarizes the findings on the principal features of the three design alternatives investigated.

Table 1. Comparison of Forward Toroid Shapes.

FEATURES EVALUATED	OCTAGONAL TOROID	ROUND TOROID	
	Pie-Shaped Blocks	Pie-Shaped Blocks	Rectangular Blocks
Plate Material Waste	Moderate	Moderate	Minimum
Toroid Support	Flat plate	Machined plates and shear keys	
Flux Uniformity MBT to MFT	Good	Not as good	Not as good
Flux Uniformity of MFT	Requires tapered spacers	Excellent	Excellent
Block Fabrication	Machining tolerances are critical and more difficult to maintain		Machining tolerances are less critical, easier to maintain
Block Assembly	Not difficult, but requires extra step to install tapered spacers	Not difficult	Easy, fewer shear pins to install

ENGINEERING DESIGN OF THE GEM NOBLE LIQUID CALORIMETER

T. Adams, K. Barnstable, N. DiGiacomo, B. Easom,
R. Hund, M. Lajczok, L. Mason and G. Velasquez

Martin Marietta Astronautics (MS DC6069)
P.O. Box 179
Denver, CO 80201

ABSTRACT

The GEM calorimeter includes a large cryogenic noble liquid system with both electromagnetic and hadronic sections. The electromagnetic section employs accordion technology, while the hadronic sections are constructed from parallel plate absorber and readout boards. The system is divided into a barrel, where the use of liquid krypton is envisaged, and two endcaps containing liquid argon. The status of the design is presented.

INTRODUCTION/GENERAL DESCRIPTION

The GEM detector employs several unique design approaches involving both noble liquid and scintillating fiber technology to achieve its ambitious goals of precision measurement and resolution. These approaches, in turn, demand equally unique and challenging engineering solutions to prove them feasible. The design of the calorimeter, which is central to the overall performance of the detector, has evolved remarkably since the experiment officially received approval to proceed in early 1992.[1] Indeed it has matured to the point where a great deal of detailed engineering design work has already begun. It is no surprise that the areas of greatest activity are also those presenting the greatest engineering challenges, namely the accordion electromagnetic sections.[2] The following paragraphs will present an overview of the noble liquid calorimeter design emphasizing the barrel section, particularly the barrel electromagnetic section. The current baseline configuration for the GEM calorimeter is presented in Figure 1. Table I provides a summary of important physical parameters. Note that both the noble liquid barrel and endcap sections will reside inside of the scintillating fiber calorimeter which is an interesting engineering challenge in itself.[3] The complete

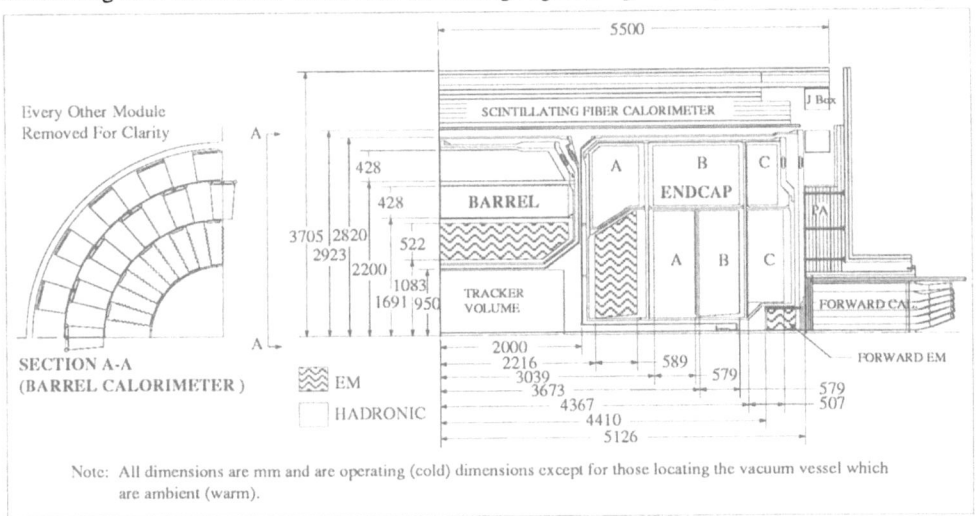

Figure 1. Cross Sectional Views of the GEM Calorimeter (1/4 Model).

Table 1. General Parameters for the GEM Noble Liquid Calorimeter.

COMPONENT	SEGMENTATION η x φ (x longitudinal)	No. OF MODULES in φ	MODULE MASS Mg
BARREL			
EM	.026 x .026 (x 3)	40	1.6
INNER HADRONIC	.078 x .078 (x 1)	80	2.0
OUTER HADRONIC	.078 x .078 (x 1)	80	2.2
ENDCAP			
EM	.026 x .026, 1.23 < η < 1.83	N/A	23.0
	.0522 x .0522, 1.83 < η < 2.46	N/A	
	.0783 x .0783, 2.46 < η < 2.95	N/A	
INNER HADRONIC A	.039 X .157, 1.31 < η < 1.73	20	2.4
	.078 X .157, 1.73 < η < 3.20		
INNER HADRONIC B	.039 X .157, 1.48 < η < 1.89	20	2.4
	.078 X .157, 1.89 < η < 3.39		
INNER HADRONIC C	.039 X .157, 1.64 < η < 2.04	20	2.1
	.078 X .157, 2.04 < η < 3.09		
OUTER HADRONIC A	.078 x .078	40	1.4
OUTER HADRONIC B	.078 x .078	40	2.2
OUTER HADRONIC C	.078 x .078	40	0.6

noble liquid calorimeter (barrel and both endcaps) weighs about 1240 tonnes, is roughly 10 meters long and is about 5.5 meters in outside diameter. The barrel and endcaps are each enclosed in their own independent cryostats consisting of inner liquid vessels, surrounded by insulating vacuum vessels. This allows for the removal of each approximately 390 tonne endcap so that access may be gained to the barrel calorimeter and tracker. The method of support for the three large cryostats is similar to that developed for the DØ detector at Fermi National Lab which uses special legs, designed to flex during cool down, to carry the dead weight of the coldmass through the vacuum vessels.[4] The major vessel joints, which will be bolted and seal welded, are designed to simplify the assembly process while providing shear and moment load capability in corner transition areas. The liquid vessels will be ASME code stamped while the vacuum vessels will be designed to the intent of ASME.

The sensing elements of the calorimeter are divided into the two categories of electromagnetic (EM) representing approximately 105,000 channels of data, and hadronic, representing roughly another 19,400 channels. The EM sections, lie closest to the interaction point, provide the finest granularity of measurements and constitute the greatest engineering challenge of the two. We begin, however, by briefly describing the hadronic section which uses the more conventional parallel plate design approach.

BARREL HADRONIC CALORIMETRY

The hadronic section of the GEM barrel calorimeter consists of 160 wedge shaped modules weighing an average of 2.1 tonnes each, arranged in two radially concentric and two axial sections. The two radial layers are staggered 4.5 degrees in φ in order to minimize the effects of projective cracks between adjacent modules. A typical module, shown in Figure 2, consists of a brass box equipped with a brass strongback on the outer periphery. The outside of the strongback provides a convenient space to mount electronic circuit boards and cooling loops. The sides of the module act as shear panels to limit deflection and are perforated with holes to promote the free circulation of krypton between modules. The modules will be supported from the 12.7 mm thick end plates which are constructed with keyways designed to mate with support and alignment rails located on the liquid vessel center and end washers. After a module is slid into position along the rails a locking block is inserted through a machined hole in the outer end plate to secure it against radial movement. Retaining bolts are used to rigidly fasten the opposite end to the central washer. This approach allows the module structure, which is mostly brass, to expand and contract during thermal cycling independently of the aluminum cold vessel while also remaining symmetrical with respect to the calorimeter center.

The contents of each module are comprised of parallel layers of lead absorber, and signal electrode plates and brass ground plates, separated by 2mm thick liquid krypton gaps. The absorber and signal electrode plates, shown in Figure 2, are complex laminates of lead, prepreg cloth, and a resistive coating divided into discrete pads corresponding to tower segmentations of 0.08 x 0.08 in η and φ. Individual cell configurations are illustrated in Figure 2. The ground plate is a much simpler continuous sheet which adds to overall module stability. The ground plates through which particles from the interaction region enter and leave a module, i.e. the entrance and exit plates, serve double duty as the inner and outer structural panels. Acetal (DELRIN) spacer buttons that maintain a uniform liquid krypton gap, also position the absorber and electrode pads relative to one another and provide a load path between ground plates. Axial movement of plates is controlled through the use of notches and matching machined bosses. The plate ends located farthest from the interaction region have a rectangular notch machined into them, while the ends nearest the central washer have "tee" shaped notches. These slide over matching bosses which are part of the module end plates. Both notches keep

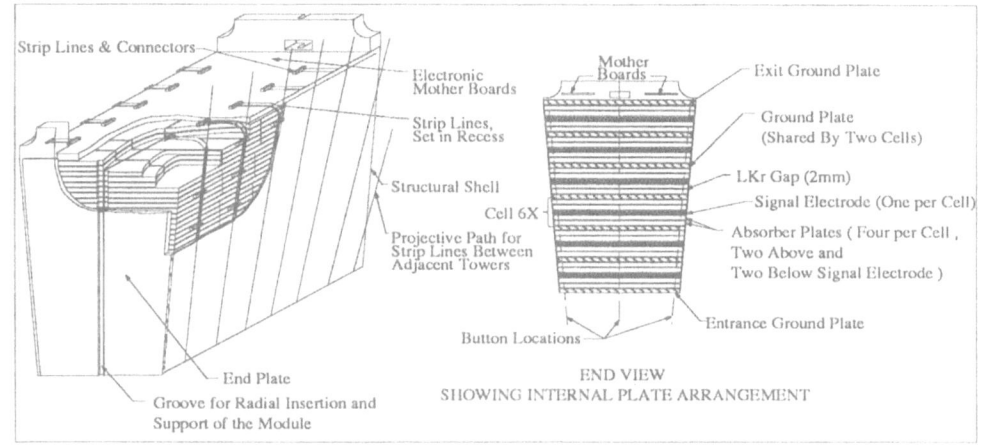

Figure 2. Typical GEM Calorimeter Barrel Hadronic Module.

the plates centered within the module, while the rectangular notch also allows the plates to expand and contract independent of the surrounding module structure. The assembled parallel plate and button stack is confined radially by compressing it between the entrance and exit ground plates, with the module side panels providing the tension.

ENDCAP HADRONIC CALORIMETRY

There are 180 endcap hadronic modules distributed as shown in Figure 1 and Table 1. They are significantly different from the barrel hadronic modules in both shape and construction. For example, the ground, absorber, and signal electrode plates are copper instead of lead and, they are arranged vertically instead of horizontally. Tension straps, rather than side panels, compress the stack between the entrance and exit ground plates which form the for and aft ends of a module. The endcap segmentation is shown in Table 1, the shifts that occur within the inner modules are necessary to insure that the minimum absorber and signal electrode pad dimensions are greater than the 2.0 cm manufacturing limit.

BARREL ELECTROMAGNETIC CALORIMETRY

The accordion technology used in the electromagnetic sections of the GEM calorimeter combines the advantages of high sampling fraction with excellent signal to noise characteristics[2].

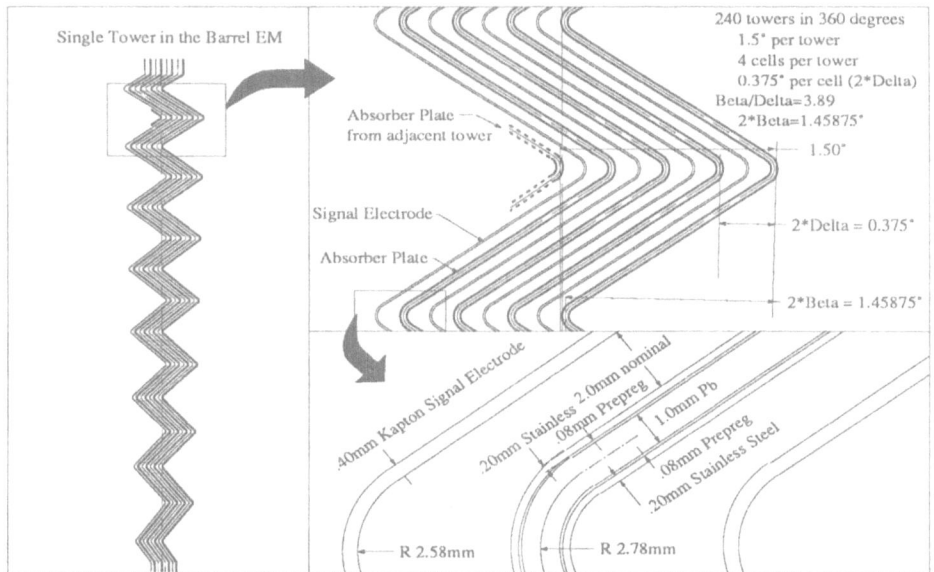

Figure 3. Accordion Geometry for the GEM Barrel Electromagnetic Calorimeter.

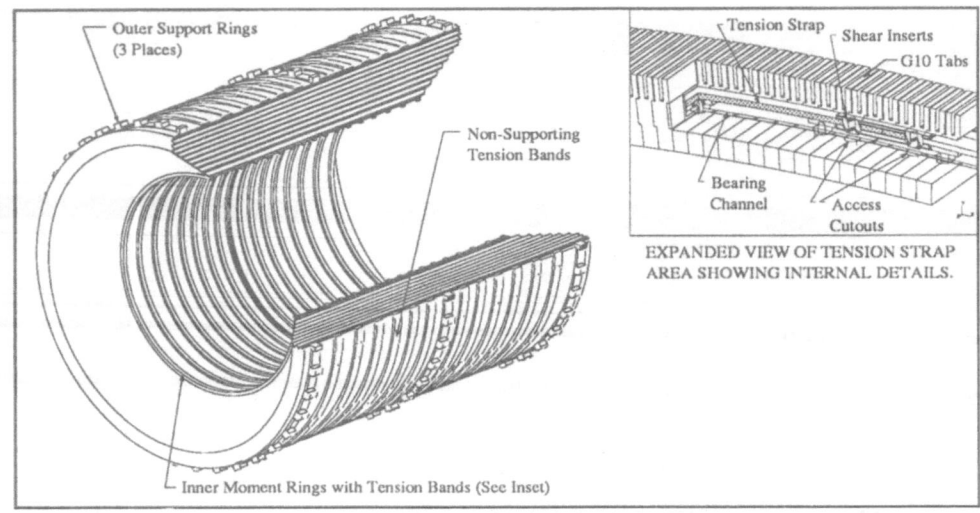

Figure 4. GEM Barrel Electromagnetic Calorimeter.

The name derives from the unusual geometry which bends the absorber and signal electrode plates into a shape resembling the bellows of an accordion as illustrated in Figure 3. The bends of individual accordion sheets are bounded by projective lines separated by 1.45875° in φ. Note that the accordion bend angle must be adjusted as a function of radial location in order to maintain a constant thickness liquid krypton gap between adjoining plates. Towers, which are defined by 1.5° increments in φ, consist of 4 intermeshed sets of steel clad lead absorber and Kapton signal electrode accordions. An absorber shared with the adjacent tower completes the last cell. Figure 4 shows a cut away view of the completed EM barrel constructed of 40 longitudinal modules. Each module is defined by 6 towers in φ and weighs 1.6 tonnes. The ends of the modules are beveled inwards along a projective line in order to minimize the effects of particle showers that begin within the cryostat walls near the inside corners. Structural integrity is maintained via tabs located at the inner and outer radii. These tabs employ integral G10 spacers designed to accept stainless steel circumferential tension bands (see Figure 4 inset). The outer G10 spacers are extended radially at the module center and end locations to provide support load paths to the surrounding aluminum shell. Radial fasteners installed through the shell and into the center section of the EM provide axial alignment and stability. This arrangement allows the EM section to contract and expand about the calorimeter centerline.

ENDCAP ELECTROMAGNETIC CALORIMETRY

The 18 tonne endcap EM section also uses accordion technology, only here the plates form vertical curtains that hang parallel to the beam line.[5] "L" section stiffeners located between adjacent towers at the front and back of the EM section provide uniform load paths to a surrounding steel shell. The load path from the stiffeners to the outer shell is via the front and back walls which are formed from large circular G10 plates that are perforated to promote liquid argon circulation. This design allows the entire assembly to be handled as a monolithic structure.

FUTURE WORK

The complex geometry of the accordion design makes it difficult to analyze as a structure. Future work will, therefore, concentrate on the thermo structural characteristics and behavior of individual EM modules as well as the completed EM assemblies. Load conditions will include initial assembly, cool down, and operation. It will be necessary to conduct analyses in three dimensions in order to fully understand the interplay between the individual accordion plates, G10 end tabs, tension straps, and the surrounding aluminum shells. Material characterization tests are also planned for the composite accordion plates.

REFERENCES

1. B. Barish *et al.*, GEM Letter Of Intent, SSCL-SR-1184, GEM TN-92-49, (30 Nov. 1991).
2. B. Aubert *et al.*, "Performance of a Liquid Argon Electromagnetic Calorimeter with an "Accordion" Geometry", CERN-PPE-91 - 73, (Submitted to Nucl. Instr. and Meth.) (1991)
3. The GEM Technical Design Report, GEM-TN-93-262, (1993).
4. The DØ Revised Design Report, DØ Note 512, Fermilab, (1985).
5. B. Easom *et al.*, Structure of the GEM Endcap Electromagnetic Calorimeter, GEM TN-903-321, (April 1993).

OVERVIEW OF THE SUPERCONDUCTING MAGNET SUBSYSTEM FOR THE GEM DETECTOR AT THE SSC

G. Deis*, J. Bowers, A. Chargin, J. Heim, A. House, C. Johnson,
G. Oberst, L. Pedrotti, J. Swan, R. Warren, S. Wineman, R. Yamamoto

Lawrence Livermore National Laboratory
P.O. Box 808, Livermore, CA 94550

R. Camille, G. East, P. Marston, J. Minervini, R. Myatt, S. Myatt,
R. Pillsbury, Z. Piek, B. Smith, J. Sullivan, P. Titus, R. Vieira

Massachusetts Institute of Technology, Plasma Fusion Center
185 Albany Street, Cambridge, MA 02139

J. Krupczak, N. Martovetsky, P. Reardon, R. Richardson, D. Richied

Superconducting Super Collider Laboratory†
2550 Beckleymeade Ave., Dallas, TX 75237

R. Stroynowski

Southern Methodist University, Dallas, TX

INTRODUCTION

The SSC Laboratory plans to deploy two "large" detectors for the essential high-energy physics experiments at the initial startup of the collider. The GEM detector is optimized to emphasize precise measurement of photons and electrons, as well as precise tracking of high-energy muons. An essential part of the GEM detector is the magnet subsystem, which provides the magnetic field necessary for identification and high-resolution tracking of charged particles. This large superconducting magnet system, with ferromagnetic field-shapers, presents a variety of engineering challenges in superconductor technology, in magnet-winding technology, fabrication, assembly and installation of large and heavy components, and in ensuring the required high operating availability.

DESIGN OVERVIEW

A basic philosophy of the GEM Detector is to provide a large magnetized solenoidal volume, in which the rest of the detectors are installed[1]. Within this basic concept, the GEM magnet is required to provide a combination of size and magnetic field sufficient for high-

---

*Present address is Superconducting Super Collider, Dallas, TX
†Operated by the Universities Research Association, Inc., for the U. S. Department of Energy under Contract No. DE-AC35-89ER40486.

resolution measurement of charged-particle momentum. Though better performance is achieved with larger, higher field magnets, certain detectors, which will be installed inside the solenoids, limit the magnitude of the allowable field, and the overall size of the underground detector hall restricts the magnet size. The overall cost of the detector favors both lower field and smaller radius. Trade studies indicated that the best choice was a large solenoid, 18 m inner diameter and 31 m overall length with modest magnetic field, 0.8T on center, and with no flux return. This approach satisfies the requirements at minimum cost, and avoids higher-risk magnet designs, such as multi-layer solenoids. In order to meet our muon-resolution requirements for trajectories which are close to the solenoid axis, each end of the solenoid is occupied by a large, conical, ferromagnetic steel structure, the "Forward Field Shaper".

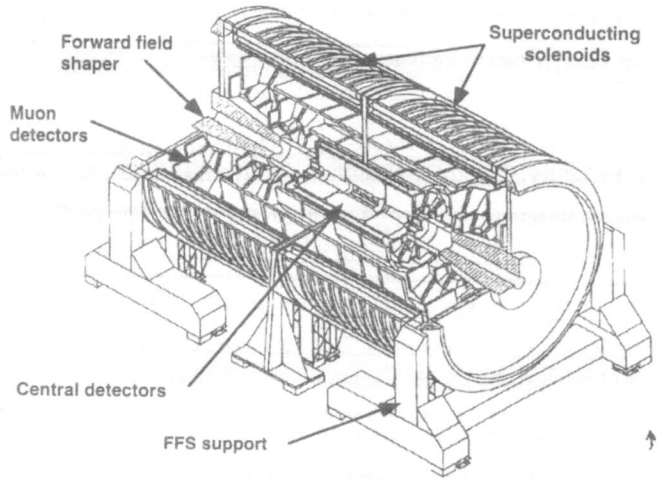

Figure 1. GEM Detector

The GEM Magnet, shown in Figure 1, comprises five main elements, two superconducting solenoids, two Forward Field Shapers, and the Central Detector Support, along with the required auxiliary systems, such as vacuum, cryogenics, power/protection, and control. The main system characteristics are listed in Table 1. The primary engineering challenges are associated with the large size and weight of the system. All major components must be field-assembled, and this often dictates different design approaches than used for previous large detector magnets. For example, the difficulties inherent in winding such a large coil in the field led, in part, to our choice of a conductor which was less sensitive to minor winding imperfections. Practical aspects of fabrication have been considered at each step, and will be brought into even clearer focus when an industrial contractor is retained, starting in summer of 1993, to complete the design, fabrication, and installation of the magnet subsystem. The design details are outlined below.

Cold mass

We chose to use a cable-in-conduit (CIC) superconductor design for GEM, because it offered excellent stability against quench, as well as enhanced opportunities for performing

Table 1. GEM Magnet System Parameters

Central induction	0.8	T	Stored energy	2.5	GJ
Mean winding radius	9.5	m	Peak field at conductor	1.6	T
Vessel inner diameter	18.0	m	Operating temperature	4.5	K
Vessel outer diameter	21.8	m	Total conductor length	27.2	km
Overall coil length	31.0	m	Charging time	8	hr
Number of turns	456		Total cold mass	1050	Mg
Operating current	50.2	kA	Total coil ass'y mass	1500	Mg
Inductance	1.98	H	Mass of FFS + support	2000	Mg

key quality-assurance tests prior to installation[2,3]. This conductor is formed in a single layer against the inside of a 76mm thick cylindrical aluminum coil form, which provides mechanical support and cooling (by conduction to helium tubes on the outside of the coil form) for the windings. The coil form also provides an axial mechanical prestress to the conductor to minimize slipping (and potential for quench) during charging. Each of the two separate cold mass halves consists of twelve coil segments, where each coil segment is a 19-turn single-layer coil, roughly 19m in diameter and 1.2m long, containing a single 1.2 km length of conductor. The twelve segments are connected mechanically with flanges, and electrically with low-resistance, large-area lap-type joints[1]. A completed cold-mass half contains a total of 228 turns and weighs 525 metric tons.

Vacuum Vessel

Each of the two cold-mass halves is enclosed in a separate annular vessel, which provides insulating vacuum and physical support[4]. The two vacuum vessels each consist of a simple membrane-like inner vessel, a reinforced outer vessel, and two thick end rings which add stiffness to the two vessels as well as provide stable attachment points for the cold mass supports. The vessels will be constructed of a combination of low-carbon steel and 300-series stainless steel, to satisfy the requirements on mechanical performance, magnetic field, cryogenic performance, and cost. These large structures are field-fabricated because of their large size and weight (nearly 22m OD, 15m long, and 900 metric tons). Assembly of the cold mass and the vacuum vessels is accomplished with their axes vertical, since this is the preferred orientation for fabricating vessels, and eases the problem of ensuring roundness. The cold mass and the vessels are assembled together by consecutively nesting the vessel shells with the cold mass, and then completing two final close-out welds. This assembly scenario requires, for each magnet half, at least two high, heavy lifts, with limited clearances (e.g., lift an outer vessel over and then onto the cold mass); a final lift to rotate the magnet axis to horizontal is also needed. The vessel design has been chosen to simplify these assembly operations.

Cold-mass supports and thermal shields

Two different sets of titanium rods are used to support the cold mass within the vacuum vessel. These rods are designed to support the weight and the considerable magnetic force on the coil (52MN), while preserving the roundness of the coil ends, allowing for dimensional changes during cooldown and charging, and minimizing the heat conducted to the 4.5K structure. A set of 32 rods lying in the plane of the coil ends, and oriented tangentially to the coil, support the coil weight. A system of 8 axial supports connect to the cold mass at roughly the middle of the coil, and to the outer end ring of the vacuum vessel; these axial supports resist the attractive force between the coil halves.

The 4.5K structure is completely surrounded by LN-cooled thermal radiation shields, consisting of lightweight aluminum panels with welded-on cooling tubes. These shields are surrounded on both sides by multilayer insulation, which reduces the consumption of LN and the heat load to the cold mass. The thermal shields also provide conductive thermal intercepts to reduce the heat conducted by the cold mass supports to the cold mass.

Forward field shapers

The two forward field shapers each consist of approximately 1300 metric tons of ferromagnetic steel, supported by a welded-steel support structure. The structure must resist both the magnetic force of 12MN pulling toward the center of the solenoids, and the large moment resulting from the cantilevered weight. The cantilevered design is required by the design of the detectors which fit around it. The FFS cones are constructed out of a number of disks, which interlock to support the large shear load from the cantilevered weight. The entire set of disks is held together with a number of pre-tensioned rods which penetrate the whole group. The FFS support is fabricated off-site in approximately 10 large pieces, which are then shipped to the detector site, where they are bolted together. The FFS is bolted on last, just prior to installation.

Auxiliary Systems

Cooling for the cold mass is provided by a passive thermosiphon system, with the LHe in tubes welded to the outside of the coil forms[5]. Supercritical helium (3 atm, 4.5K) is also supplied to the inside of the CIC, the joints between coil segments, and the vapor-cooled leads. Helium refrigeration requirements are comparable to those for the SSC's Accelerator System String Test, approximately 2kW refrigeration, plus 20g/s liquifaction. To improve availability, we plan to ensure that the magnet can operate even during refrigerator outage, by using the inventory of stored liquid.

The power, protection, and control system includes the main 50kA, 20V DC power supply, which can charge or discharge the magnet in 8 hours. The DC supply is connected to the magnet with forced-air cooled aluminum busses. A dump resistor and redundant circuit interrupters are included in the circuit to provide a means for quickly discharging the magnet in the event of an emergency such as a fire in the underground hall or a magnet quench. Finally, sensors and controls are provided to constantly monitor the condition of the magnet, and control all system simultaneously.

SUMMARY

The GEM magnet will be one of the largest superconducting magnets to be constructed. Although its size presents a number of engineering challenges, we have adopted a very conservative design in order to ensure the high operating availability required for a key system a major SSC experiment.

REFERENCES

1. R. Stroynowski, et.al., Magnet Technical Design Report, SSCL report number GEM-TN-92-116, 1992.
2. N. Martovetsky, et. al., Conductor Design for the GEM Magnet, *Supercollider 5/IISSC Proceedings,* (1993)
3. B. Smith, et.al., Design Concept for the GEM Detector Magnet, *Applied Superconductivity Conference, Chicago, Ill,* [1992]
4. J. Bowers, et. al., Design of the Vacuum Vessel Subsystem for the GEM Detector at the SSC, *Supercollider 5/IISSC Proceedings,* (1993)
5. R. Warren, et. al., The Cryogenic System for the GEM Detector Magnet, *Supercollider 5/IISSC Proceedings,* (1993)

LOW ENERGY BOOSTER RADIO FREQUENCY CAVITY STRUCTURAL ANALYSIS

Kennedy Jones

Mechanical Engineering Department
Accelerator Systems Division
Superconducting Super Collider Laboratory*
2550 Beckleymeade Avenue, MS 4006
Dallas, TX 75237-3997

ABSTRACT

The structural design of the Superconducting Super Collider Low Energy Booster (LEB) Radio Frequency (RF) Cavity is very unique. The cavity is made of three different materials which all contribute to its structural strength while at the same time providing a good medium for magnetic properties. Its outer conductor is made of thin walled stainless steel which is later copper plated to reduce the electrical losses. Its tuner housing is made of a fiber reinforced composite laminate, similar to G10, glued to stainless steel plating. The stainless steel of the tuner is slotted to significantly diminish the magnetically-induced eddy currents. The composite laminate is bonded to the stainless steel to restore the structural strength that was lost in slotting. The composite laminate is also a barrier against leakage of the pressurized internal ferrite coolant fluid. The cavity's inner conductor, made of copper and stainless steel, is subjected to high heat loads and must be liquid cooled.

The requirements of the Cavity are very stringent and driven primarily by deflection, natural frequency and temperature. Therefore, very intricate finite element analysis was used to complement conventional hand analysis in the design of the cavity. Structural testing of the assembled prototype cavity is planned to demonstrate the compliance of the cavity design to all of its requirements.

INTRODUCTION

This paper presents the stress, displacement, dynamic, random vibration, and thermal analysis performed on the Low Energy Booster (LEB) Radio Frequency (RF) Cavity. The Cavity is unique in that it is constructed of 304 stainless steel, SS, copper, and a fiber reinforced composite laminate (similar to G10). The SS provides structural strength, the copper is for good electrical conductivity, and the composite is to reinforce the SS in the

*Operated by the Universities Research Association, Inc., for the U.S. Department of Energy under Contract No. DE-AC35-89ER40486.

Supercollider 5, Edited by P. Hale
Plenum Press, New York, 1994

areas where the SS housing is slotted to reduce eddy currents. The components of the cavity, figure 1, are the tuner housing, ferrite disks, outer conductor, inner conductor, tetrode, higher order mode damper, and ion pump. The tuner housing, outer conductor, and inner conductor are presented in this paper to show that their prescribed design specifications were met. These components were analyzed with ANSYS revision 44a FE software and conventional hand analyses. The following sections discuss the results that were obtained in conducting these analyses.

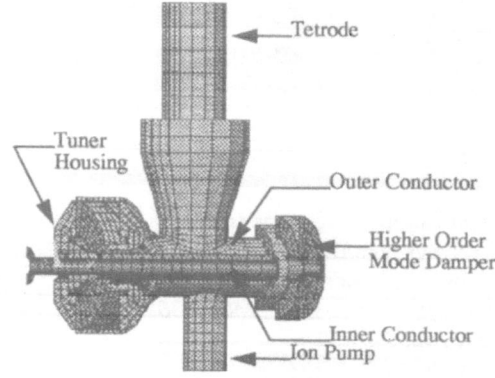

Figure 1. RF Cavity FEM

TUNER HOUSING

The tuner housing serves as a containment vessel for the ferrite and cooling fluid while at the same time providing stiffness for the inner conductor. The tuner has SS fins on both sides, figure 2, which are welded to the SS housing. The surface and perimeter of the

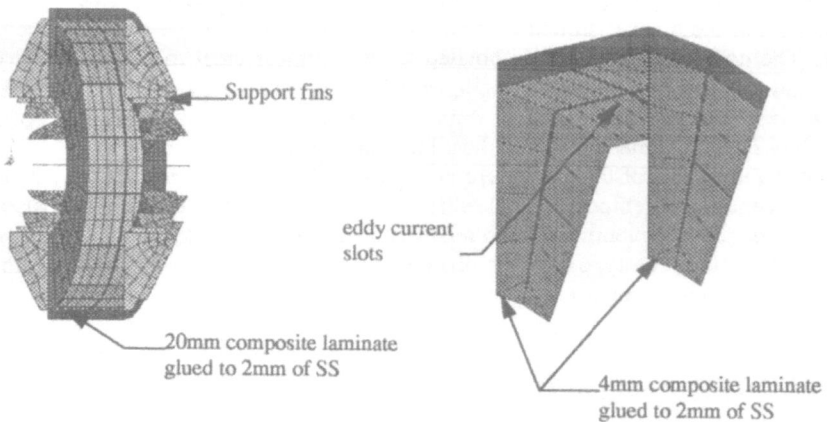

Figure 2. Tuner Housing FEM

housing are slotted to reduce the eddy currents but this also reduces the stiffness. To regain this stiffness a 4-mm composite laminate is glued to the surface and a 20-mm composite laminate is glued to the outer perimeter of the housing. The FEM represents the 20-mm laminate as solid elements layered on plate elements representing the SS housing outer perimeter. The surface of the housing is modelled with plate elements having combined properties of SS and a 4-mm laminate of the composite. The major loading on the housing is due to the 875 N weight of the ferrite disks and a 34.5 MPa internal coolant pressure load. The maximum stress levels in the housing are very low with $\sigma_{1\text{-max}}$ less than 7.5 MPa and τ_{max} less than 4.2 MPa. This indicates that there will be low deflection levels in the inner conductor. From hand analysis the maximum stresses in the bolted joints are also less than 20% of the yield stress of 304 SS (205 MPa).

OUTER CONDUCTOR

The outer conductor, figure 3, is made of 8-mm 304 SS with 400 microns of copper plating. It is subjected to structural and thermal loadings. Structurally, it sees a vacuum load and also supports the entire weight of the cavity, including the tetrode and the ion pump. It is also attached to a massive stand by support brackets through which random vibrations are

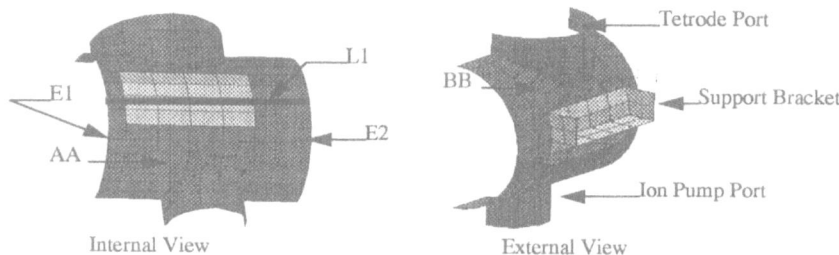

Figure 3. Outer Conductor FEM

introduced. The outer conductor, OC, is copper plated to reduce the RF losses from 3550 watts to 288 watts. Requirements specify that the stress levels in the OC must be less than 136 MPa and the maximum operating temperature be less than 200°C. The thermal boundary conditions are temperatures along the line L1 and at the edges, E1 and E2 are maintained at 45°C. The maximum predicted principle stress due to structural and thermal loading is 81 MPa at point BB. The maximum predicted operating temperature is 126°C at point AA.

INNER CONDUCTOR

The Inner Conductor, figure 4, is the single most important component of the Cavity. It is constructed of a 2-mm copper outer shell brazed to a 2-mm SS inner shell with water

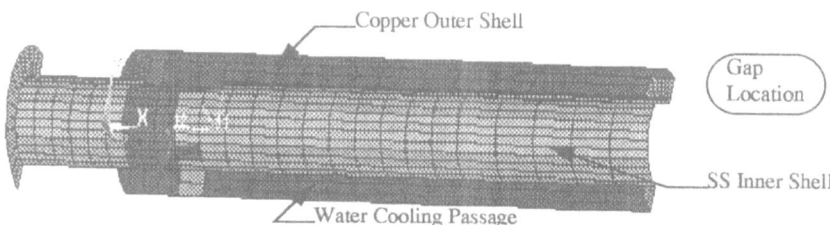

Figure 4. Inner Conductor FEM

cooling channels in between. The maximum allowable static or dynamic vertical deflection at the gap is 25 microns. The maximum allowable static or dynamic horizontal deflection is also 25 microns. The finite element analysis predicts that the maximum static vertical deflection will be 13 microns and the maximum horizontal deflection will be 19 microns. Once the static analysis was completed a modal analysis was performed to determine if there were any natural frequencies below 100 hz. The first three frequencies were 61 hz, 85 hz, and 123 hz. Since the general requirement specified that all natural frequencies be above 100 hz, it was necessary to perform a random vibration analysis to demonstrate that the excitation of the IC at the gap would be less than 25 microns. The random vibration analysis would allow determination of the mean output displacement generated from random displacements imposed at the support bracket. The input power spectral density, PSD, data that was used in the analysis was obtained from several different laboratory in the United States. The output response of the RF Cavity to this input PSD was a mean frequency, ω_o, of 27 hz and a mean displacement, σ_m, of 0.104 microns. Next a statistical analysis on the probability of the peaks from the output response[1] was done using ω_o, σ_m, and a 25 year time period as input. This yielded a maximum peak response of 1 micron.

CONCLUSION

In conclusion, these analyses demonstrate that the tuner housing, outer conductor, and inner conductor of the LEB RF Cavity will indeed meet and exceed their design requirements. If the components of the Cavity are manufactured within the tolerances called out in the machine drawings and assembled properly, the Cavity should perform structurally and thermally as expected without any problems.

REFERENCES

1. M.G. Hallam, N. J. Heaf, L. R. Wootton, "Dynamics of Marine Structures," CIRIA Underwater Engineering Group, Report UR 8, 2nd Edition, pp 57.

LOW POWER RF BEAM CONTROL ELECTRONICS FOR THE LEB

L.K. Mestha, J. Mangino, V. Brouk, T. Uher and R. C. Webber

Superconducting Super Collider Laboratory*
2550 Beckleymeade Ave., Dallas, TX 75237

ABSTRACT

Beam Control Electronics for the Low Energy Booster (LEB) should provide a fine reference phase and frequency for the High Power RF System. Corrections applied on the frequency of the rf signal will reduce dipole synchrotron oscillations due to power supply regulation errors, errors in frequency source or errors in the cavity voltage. It will allow programmed beam radial position control throughout the LEB acceleration cycle. Furthermore the rf signal provides necessary corrections during adiabatic capture of the beam as injected into the LEB by the Linac and will guarantee LEB rf phase synchronism with the Medium Energy Booster (MEB) rf at a programmed time in the LEB cycle between a unique LEB bucket and a unique MEB bucket. We show in this paper a matured design and possible interfaces with other subsystems of the LEB such as the beam instrumentation, High Power RF Stations, global accelerator controls and the precision timing system. The outline of various components of the beam control system is also presented followed by some test results.

INTRODUCTION

Some early thoughts on the overall design and development of the Low Power RF control electronics were presented in earlier conferences on the Super Collider.[1,2] Particularly in Reference 1 the new beam transfer synchronization was discussed without any experimental results. Upon thorough investigation of the machine requirements and due to considerable progress in the design of other LEB subsystems we were able to design a more matured system. Some of the redundant hardware was taken out from the design shown in Reference 2. Due to the clear picture of the link with the SSC precision timing system we are now able to show for the first time the beam synchronization system working together with the essential components of the timing system. Among the test results shown at the end of this paper, some were done with beam on Fermilab booster.

ESSENTIAL PARTS OF BEAM CONTROL LOOP HARDWARE

Generally, loops associated with the beam control are (1) Beam Phase Loop, (2) Radial Loop, and (3) Synchronization Loop. We have shown schematically in Figure 1 a top level block diagram of digital implementation. The Direct Digital Synthesizer (DDS) uses 1-Ghz clock to produce the frequency between 47 Mhz to 60 Mhz and a strobe signal to latch the data. Since the Synthesizer needs data in the form of a 32-bit binary word the hardware is digital. Digital Signal Processor (DSP) #1 is used to produce the basic frequency ramp.

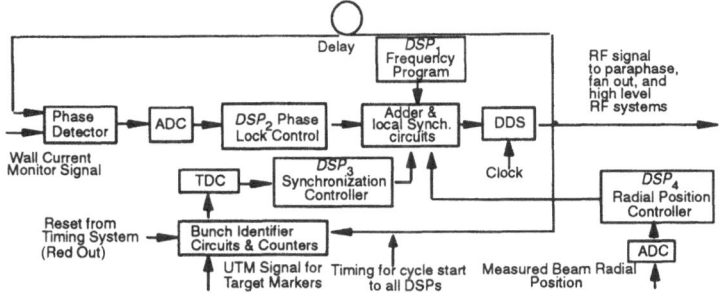

Figure 1. Schematic diagram of the Synchronization and Beam Control System.

*Operated by the Universities Research Association, Inc., for the U.S. Department of Energy under Contract No. DE-AC35-89ER40486.

Beam Phase Loop

The beam phase loop has a Phase Detector, a 12-bit ADC, a Digital Signal Processor #2 and additional hardware with external adder circuits. The beam signal is compared with the Synthesizer rf signal through a delay cable in a beam phase detector. The beam phase is then digitized using the 12-bit ADC and the data is read into the DSP. The beam phase data is multiplied inside the Signal Processor by a time varying gain (if needed). Additional filters or non-linear controllers (if found useful) are used in the Processor on real time. The ADC is sampled at DDS strobe signal period. With the hardware we have used in the beam phase loop, a sampling period of less than 3 μs has been achieved in Laboratory tests. For the total phase lag to be less than 45 degrees introduced in the beam phase loop at twice the maximum synchrotron frequency (the power amplifier to cavity, beam, the sensor such as a wall current monitor, cables, phase detector and digital processing), a stringent requirement is placed on the sampling period. In case the delay turns out to be larger than expected then the plan is to replace the phase loop DSP with analog processing and do the analog to digital conversion afterwards.

Synchronization Loop

The synchronization loop shown in Figure 1 has been designed to phase lock the LEB rf signal to the MEB rf signal with the MEB running at its injection frequency. LEB revolution markers and the MEB target markers are used for locking the frequencies. The target markers are derived from the MEB rf signal by dividing its frequency by the harmonic number (792) and displacing the marker train appropriately for multiple transfer. This is done in a timing module and is discussed later in this document. The revolution markers for the LEB rf signal are derived by dividing the LEB rf signal by the harmonic number (114) after identifying a particular bucket to track. This is done in 'bunch identifying circuits' of Figure 1. Two revolution markers are used as 'Ref' and 'Hit' for the Time to Digital Converter (TDC). A 20-ps TDC is used in the synchronization loop. Marking the bucket from the LEB rf signal is shown in the timing diagram of Figure 2. The 'reset' pulse in this figure is generated from MCLK (MEB beam synchronous clock – same as MEB rf signal, but with coded messages sent over the timing system) and is shown in the Interface section below. The 'reset' pulse indicates the arrival of the first target marker. After the arrival of the first target marker, count the pulses from the LEB rf signal. After a known number of count (say 40 or any bucket we want to identify) the rf signal is divided by the harmonic number, 114, to generate the revolution marker. This means we have identified a 40th bucket and are tracking at each revolution in the MEB. The TDC will output the time interval between two markers. The Signal Processor #3 will read the TDC data each time a new data appears and the new data appears each time the target marker appears. In the synchronization DSP, the time interval is processed and then compared to the 'trip-plan' values. The trip–plan values indicate the position of the LEB reference bunch in the ring when the MEB target marker arrives. These values can be calculated (theoretical) or measured. The data representing the error between the trip-plan values and the processed TDC values are then multiplied by a constant to produce the frequency shift on the adder interface circuits at the input end of the DDS. In this way, the hardware provides the ability to phase lock the MEB rf signal to the LEB rf signal. Time for closing the synchronization loop can be done at a predetermined count of the MEB target marker. In this way when the synchronization loop is closed the LEB rf signals are phase matched to the trip-plan data.

Since the trip-plan values are known in advance, we can select a particular count on the MEB target markers after reaching the required momentum to fire the kickers. Say for 3770 counts of the target markers we would have reached the target momentum (indication of the momentum match with the MEB can come from a gauss clock looking at the LEB main magnetic field) and are within the momentum tolerance of the MEB. Then, the LEB beam can be extracted any time. However, if there is a requirement on firing the kickers to phase match to the gap in the LEB, then there is no change required for the synchronization loop because the reference bucket created in the bunch identifier circuits can be made to phase match with the gap.

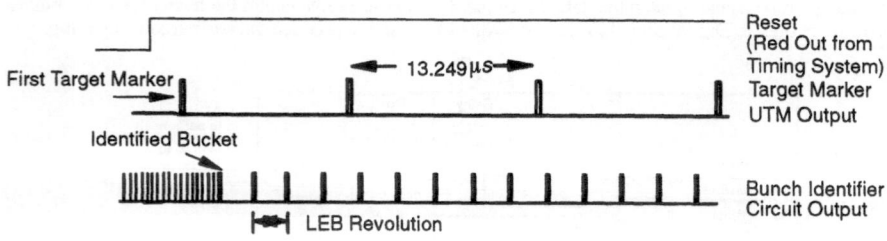

Figure 2. Basic timing diagram of the synchronization loop.

Radial Loop

With increased understanding of the feedback loops the radial loop is planned to configure differently from that of the Fermilab booster. The radial position signal is digitized using a 12-bit ADC and is read into DSP #4. The radial position values are compared with the desired reference data inside the signal processor. Resulting data is multiplied by the time varying gain and then converted to frequency shift and applied to the adder circuits. In this way the radial loop is producing additional frequency modulation on the DDS signal.

INTERFACES WITH OTHER SUBSYSTEMS

Interface with Beam Instrumentation

Beam instrumentation interface is needed to acquire on real time the longitudinal and radial position information of the beam. Unprocessed resistive wall current monitor is used to detect the longitudinal position. The wall current monitor is a device measuring the beam current almost virtually with a flat amplitude response. It is a 'current transformer' responding to currents in the vacuum chamber which were created largely due to the beam travelling down the pipe. The shape and the size of the wall current monitor signal depends on the particle distribution in the beam tube and the number of particles.[3] For the LEB, a signal dynamic range of 48 dB is expected for operations with particles per bunch of 2.0×10^8 in commissioning mode to 5.0×10^{10} in test beam mode. The signal from the wall current monitor is connected to the beam phase detector as shown in Figure 1. For operations with such a wide dynamic range the signal processing electronics in the beam phase detector needs additional circuits which could be made switchable depending on the beam intensity.

A VXI card (block diagram is shown in Figure 3) which gives the processed 'sum' and 'difference' signals of the horizontal beam position monitor electrodes will be used to measure the radial position. This card will be located in the LEB RF building (Building S2) along with the Synchronization and Beam Control electronics. As shown in Figure 3 the output of two horizontal electrodes placed at the opposite sides of the vacuum chamber (Electrodes A & B) of the Beam Position Monitor (BPM) are used to measure the radial position. The beam's radio frequency component is selected and logarithmically demodulated. The difference of the output of the logarithmic amplifiers of the signal processing circuits produces a voltage proportional to the beam position. A prototype board was developed and tested which contained a special circuit topology and modern analog circuit components to obtain a low noise, high bandwidth and wide dynamic range in position measurement.[4] Particularly for the radial loop a bandwidth of few kilohertz is found sufficient. The BPM closest to arc A1 is designated for the radial feedback which has a highest dispersion of 2.7 in the ring.

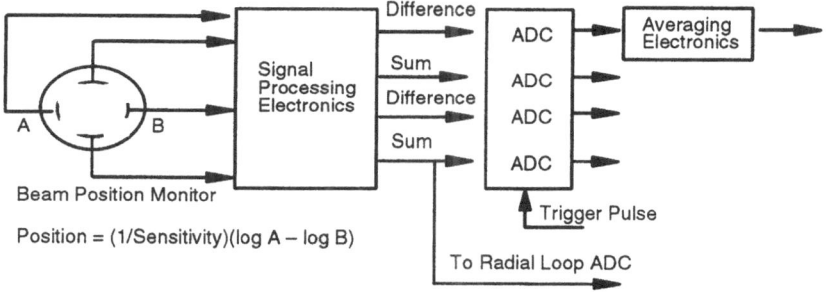

Figure 3. Radial position detector and processing circuits.

Interface with the Timing System

The LEB Low Power RF and other electronics for the rf systems are located in LEB RF building. Similarly the MEB RF electronics are located in the MEB RF building. For phase matching the MEB bucket with the LEB bucket the MEB rf signal is brought over a special timing cable from the MEB RF building to the LEB RF building. This signal is called MCLK – the MEB beam synchronous clocks of the timing system. The electronics required for firing the kicker system are placed in the LEB extraction building. The codes on global precision timing system carry information required to fire the LEB and MEB kickers. In Figure 4 we show a typical top level timing diagram of generating the target markers from the codes. Codes such as 'black' appear on the 'MCLK' timing signal at the end of every MEB revolution marker. This code is associated with a delay of say 'd_1' in the target markers. Since the target markers happen to be the destination bucket for the LEB reference bunch, the delays 'd_1', 'd_2', 'd_3', etc., will be needed to phase shift them for multiple transfer from a common reference. Each delay is associated with a particular color code on the MCLK timing signal. The black code will get changed to red code when LEB reset is detected which is associated with a state change on the signal line 'red out' to high. At the same time the 'black out' signal goes low to indicate the elapse of black code. The encoder in the timing system will detect the red code and delay the target markers by 'd_2' as indicated by the 'Target Markers' signal. The first target pulse after the 'red out' signal goes active high. This will be used to identify the first target marker for beam transfer synchronization. Since the synchronization is guaranteed after a definite number of MEB rf counts after the first target marker, all the power supplies associated with the kicker charging and firing system can be timed out of this pulse to extract the beam from the LEB.

The 'LEB Reset' is coded on GCLK timing signal. The reset appears at 10 Hz, which is derived from the main magnet field or current measurement system to time the injection of the Linac beam before the field minimum on the LEB magnet system. The 'LEB Reset' signal is used to ramp the frequency in the DDS and various other cavity tuning functions in the High Power RF System. This signal is asynchronous to the target markers. Hence an uncertainty of one MEB turn with respect to the 'trip-plan' is expected while identifying the target markers after the LEB reset has gone active high. Since all the kicker system including the bunch identifier circuits in the synchronization system (Figure 1) will see the same first target marker, there will be no mismatch of the target marker counts. However, the synchronization loop will cure any phase errors to maintain the required phase relationship with the target markers by maneuvering the radial orbit.

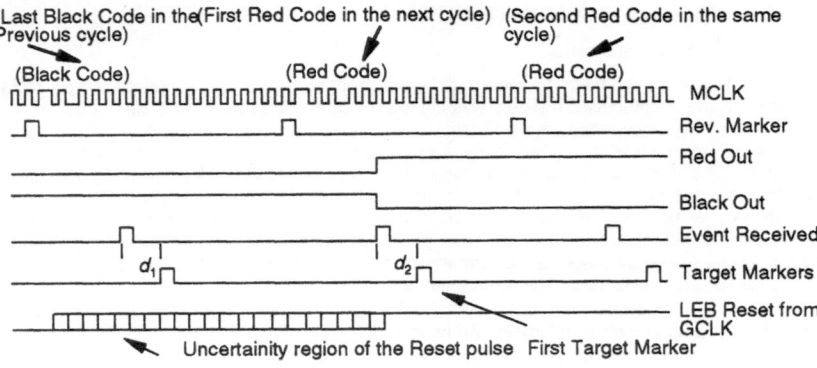

Figure 4. Top level timing diagram of the target marker generator.

Interface with High Power RF System

Analog Interface: The DDS signal is fanned out to all the LEB cavities through a delay unit or a phase shifter as shown in Figure 5. The programmable phase shifters are used to generate required counterphasing between the selected pair of cavities. At this stage it is unclear as to how many pairs of cavities are involved during counterphasing. Also the feedback loops local to the rf cavities are expected to maintain the phase constant between the rf signal and the gap voltage during acceleration.

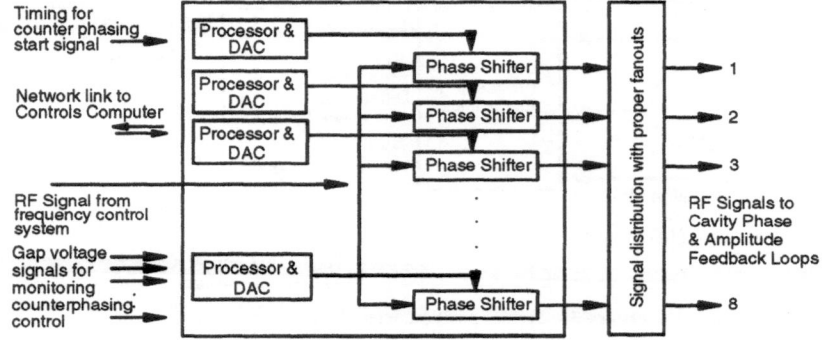

Figure 5. Schematic diagram of the counter phase control units.

Digital/Computer Interface: A schematic layout of the planned data and control flow is compiled in Figure 6. The synchronization and beam control hardware is resident in rf group control area. In addition to generating an accurate rf signal with appropriate counterphasing, the voltage profiles, cavity tuner function profiles and power amplifier turn-on/off sequences are also supplied from the rf group control area. The computer link between main control consoles to the group control area provides necessary functions (listed in Figure 6). For diagnostic purposes beam phase error, radial position error and synchronization phase errors are stored by the DSPs in a separate memory. Values are read by the main ring control computers. The multiple transfer sequence is supplied to the timing modules to generate delays 'd_1', 'd_2', .. *etc.*

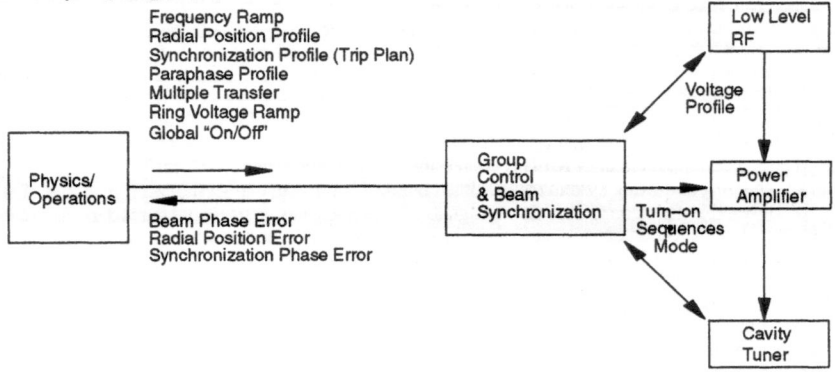

Figure 6. Data flow diagram from main controls computers.

SOME EXPERIMENTAL RESULTS

Several parasitic beam tests were done at Fermilab to study the feasibility of the digital hardware. Also, we have demonstrated the phase-locking capabilities of the beam phase loop by accelerating the booster beam. We achieved an efficiency indistinguishable from that of the existing booster system.[5] In addition to this in Figure 7 the bench test results of the synchronization loop are shown. The experiment was carried out with the hardware configuration discussed in Figure 6 of Reference 2 for phase locking two fixed frequencies (60 MHz). The x-axis shows the phase error in degrees of the rf wave after subtracting the phase between the DDS rf signal and the fixed frequency with the trip-plan. The y-axis shows the number of samples. Clearly, on most occasions the loop was able to control the phase to within 3 degrees (equivalent to about 4 cms wave length). A 20 ps TDC was used for the time interval measurement.

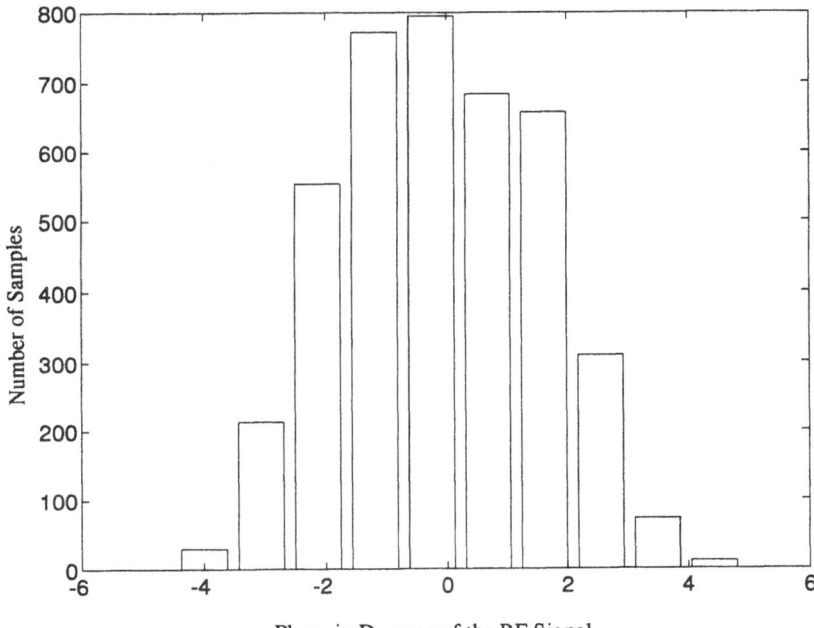

Figure 7. Histogram of the phase values between two RF frequencies when synchronized using trip-plan approach.

CONCLUSIONS

We believe the system proposed in Figure 1 can provide the precise reference phase and frequency needed for the LEB RF System which is necessary to accelerate the beam without beam loss, allow programmed beam radial position control, capture the linac beam adiabatically and guarantee synchronism with the MEB between a selected LEB bunch and a selected MEB bucket. We have also shown briefly how the beam control system can function with the precision timing system to provide synchronous triggers for extraction, how the rf signals are distributed to each of the LEB RF stations and a pathway to interface with the controls computers.

REFERENCES

1. D.J. Martin et al., "Early Instrumentation Projects at the SSC," 2nd International Industrial Symposium on the Super Collider, Miami, March 14–16, 1990.
2. L.K. Mestha et al., "Early Stages in the Development of the Global RF Feedback for the SSC Low Energy Booster," 3rd International Industrial Symposium on the Super Collider, Atlanta, Georgia, March 13–15, 1991.
3. R.C. Webber, "Longitudinal Emittance, An Introduction to the Concept and Survey of Measurement Techniques Including Design of a Wall Current Monitor," *AIP Conference Proceedings, 212, Accelerator Instrumentation*, Upton, NY 1989.
4. G. Roberto et al., "Log–Ratio Technique for Beam Position Monitor Systems," EPAC, Berlin, 1991.
5. L.K. Mestha et al., "A Digital Beam Phase Loop for the Low Energy Booster," *Proceedings of the IEEE Particle Accelerator Conference*, Washington, D.C., May 17–20, 1993.

THREE-DIMENSIONAL MODEL OF A LIQUID-COOLED, LOW ENERGY BOOSTER, RADIO-FREQUENCY CAVITY TUNER AT THE SUPERCONDUCTING SUPER COLLIDER

R. Ranganathan, A. Propp, B. Campbell, and B. Dao

Mechanical Engineering Department*
Accelerator Systems Division,
SSC Laboratory, Dallas, TX 75237

ABSTRACT

A three-dimensional computational heat transfer and fluid flow model was developed to analyze a forced-flow, liquid-cooled, low energy booster (LEB), radio-frequency (RF) cavity, tuner concept. The results for a commercial dielectric heat transfer fluid[1] indicated safe temperatures in the ferrite.

INTRODUCTION

The RF cavity tuner (modeled) consisted of four ferrite rings encased in a copper-plated titanium housing (Figure 1). A ceramic vacuum window is present at the inside radius of the housing. The perpendicular biasing of the ferrites used for tuning the LEB RF cavity results in heat generation in the ferrites, copper and the coolant. A cooling system is needed to remove this heat[2,3] and ensure that the peak ferrite temperature is maintained at safe levels. Therefore, a model was developed to analyze liquid cooling of the tuner. The results for a commercially available dielectric coolant[1] (chosen based on RF considerations) are reported here. Details of this work are documented elsewhere.[2]

ANALYSIS

Due to symmetry, only one-fourth of the tuner was modeled. The flow and heat transfer were assumed to be steady, three-dimensional, turbulent and incompressible. Buoyancy effects were included and the thermo-physical properties assumed to be constant. The problem was solved using PHOENICS.[4] The heat generated in the ferrites, housing and the coolant were estimated to be 1150, 5000 and 300 W, respectively. Row 1, Table 1 specifies the baseline. Grid dependence (rows 1, 2, Table 1) and other studies[2] were made to obtain reliable results. The uncertainty of the results should be of the order of $10°C$.

RESULTS

Qualitative Results for the Baseline (Figure 1)

The velocities show the path of least resistance to be the annular space between the ferrites and the housing. The isotherms (in the ferrite) indicate recirculation cells, where peak temperatures are present. The isobars show a large pressure drop at the exit.

---

*Operated by the Universities Research Association, Inc., for the U.S. Department of Energy under Contract No. DE-AC35-89ER40486.

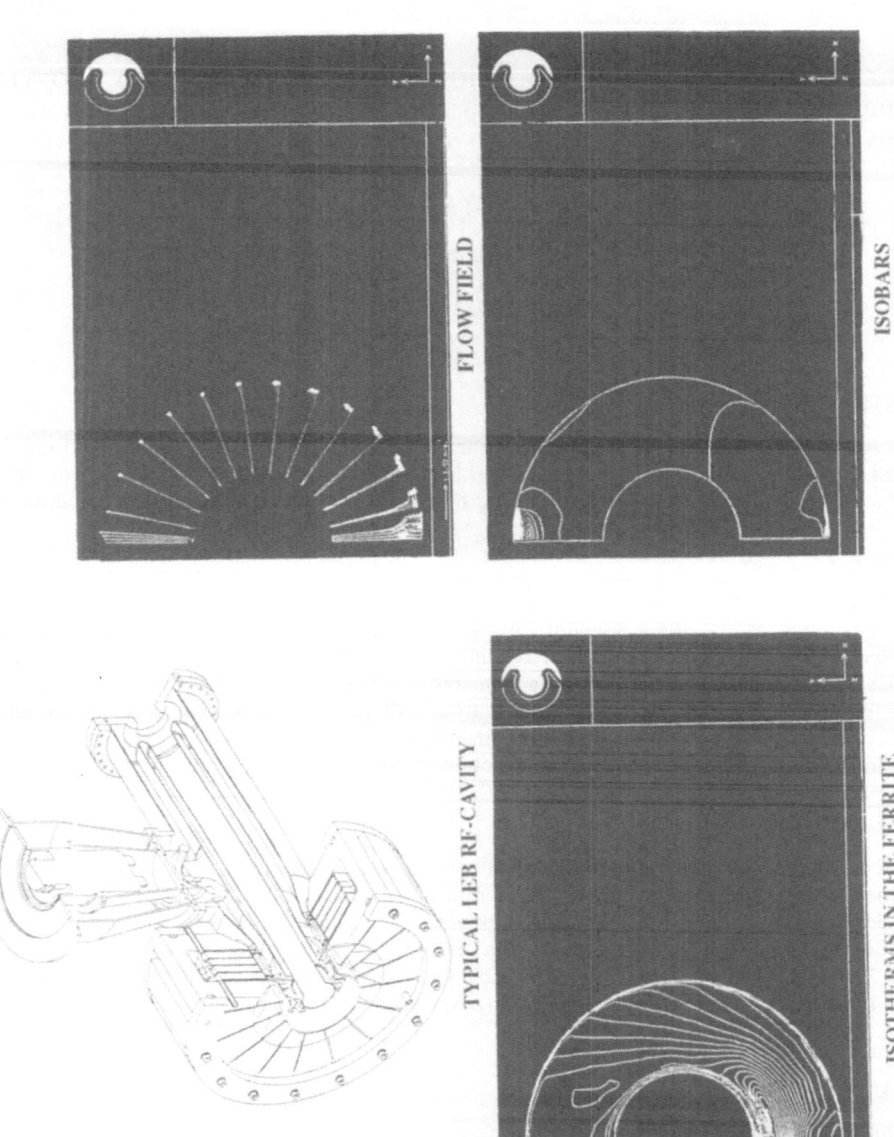

Figure 1. Qualitative results for the baseline.

FLOW FIELD

ISOBARS

TYPICAL LEB RF-CAVITY

ISOTHERMS IN THE FERRITE

Table 1. Sensitivity Results

Row	Grid	T_{ci}	ΔT	N_i	N_e	S_a	a_i	a_e	m	B ?	C ?	S_w	ΔP	T_{pc}	T_{pf}
1	25000	5	5	2	2	6.5	585	585	80	Y	Y	3.8	4800	21	28
2	50000	5	5	2	2	6.5	585	585	80	Y	Y	3.8	4800	29	21
3	25000	-30	5	2	2	6.5	585	585	80	Y	Y	3.8	5200	-14	-9
4	25000	5	5	6	6	6.5	585	585	80	Y	Y	3.8	4600	24	23
5	25000	5	5	2	2	2.5	585	585	80	Y	Y	3.8	3850	18	26
6	25000	5	5	2	2	6.5	1170	585	80	Y	Y	3.8	4800	18	23
7	25000	5	5	2	2	6.5	585	585	20	Y	Y	3.8	560	41	39
8	25000	5	5	2	2	6.5	585	585	80	N	Y	3.8	4900	38	33
9	25000	-30	5	2	2	6.5	585	585	80	N	Y	3.8	4900	3	-4
10	25000	5	5	6	6	6.5	585	585	80	N	Y	3.8	4700	28	23
11	25000	5	5	2	2	2.5	585	585	80	N	Y	3.8	3850	24	25
12	25000	5	5	2	2	6.5	2377	2377	325	Y	N	10.0	-	41	23

Quantitative Sensitivity Results (Table 1)

A 35°C reduction in the coolant inlet temperature, lowers the peak temperatures by a similar amount (row 3), indicating a linear dependence. Multiple inlets and exits lower the peak temperatures due to better coolant distribution (row 4) but do not affect the pressure drop (rows 1 and 4). This may be because the total exit flow area was a held constant. A smaller annular space lowers the peak temperatures by inducing coolant into the interior (row 5). The peak temperatures are sensitive to the inlet cross-section area (row 6) since they are located near the ceramic window across from the inlet. A lower coolant flow rate results in higher peak temperatures and a lower pressure drop (row 7) as expected. Further, the location of the peaks for this case shifted to the top of the ceramic window (not shown in this paper), since at a lower flow rate buoyancy effects are relatively stronger. Buoyancy effects were found to be important (rows 8–11), consistent with other findings.[5,6] The only exception is row 10 where a more optimal coolant distribution due to multiple inlets and exits reduces the relative influence of buoyancy.

Absence of Copper Plating

Without the copper plating, the heat generation in the housing increases five–fold. Therefore, the space between the ferrites and the side walls of the housing was increased to increase coolant circulation there. The overall coolant flow rate was also increased in proportion to the total heat load. The results indicate safe temperatures (row 12).

SUMMARY

A three-dimensional, numerical model was developed to evaluate a liquid-cooled LEB RF cavity tuner. The model was used to evaluate the performance of a commercially available dielectric coolant.[1] Results show that the peak temperatures are sensitive to: the coolant flow rate and inlet temperature, the number of inlets and exits, the space between the ferrites and the housing, and the inlet area. The pressure drop across the tuner was dependent on the coolant flow rate and exit area. Buoyancy effects were found to be important. When the tuner walls are not copper plated, an increased coolant flow together with increased space near the housing can ensure safe temperatures.

REFERENCES

1. Galden Heat Transfer Fluid, AUSIMONT, Morristown, New Jersey.
2. R. Ranganathan, "Three dimensional numerical analysis of a liquid-cooled LEB RF cavity tuner," AMC-2210001, SSC Laboratory, (1992).
3. R. Ranganathan, R., "LEB RF cavity tuner solid cooling," AMC-2210005, SSC Laboratory, (1993).
4. PHOENICS, an acronym for Parabolic, Hyperbolic or Elliptic Numerical Integration Code Series, software developed by Concentration Heat and Momentum Ltd., United Kingdom.
5. B. Campbell, "SSC LEB cavity mechanical design considerations," presented at the RF Workshop, TRIUMPF, Vancouver, B.C., October 25–26, (1990).
6. B. Campbell and R. Ranganathan, "An experimental and analytical study of a buoyancy driven cooling system for a particle accelerator," 1993 Particle Accelerator Conference, May (1993).

NOMENCLATURE AND UNITS

a_e (a_i)	total cross–section area of all the exits (and inlets), mm^2
B	buoyancy effects included or not ? yes (Y) or no (N)
C	is a copper coating present on the housing or not ? yes (Y) or no (N)
ΔP	pressure drop across the tuner cavity, Pa
ΔT	average temperature rise of the coolant across the tuner cavity, °C
Grid	number of cells used in computations
m	coolant flow rate, kg/minute
N_e (N_i)	number of coolant exits (and inlets)
S_a	annular space between the ferrites and housing, mm
S_w	space between the ferrites and the side walls of the housing, mm
T_{ci}	average coolant inlet temperature, °C
T_{pc} (T_{pf})	peak coolant (and ferrite) temperature, °C

MULTI-MeV ELECTRON COOLING -- A TOOL FOR INCREASING THE PERFORMANCE OF HIGH ENERGY HADRON COLLIDERS? *

Tim Ellison[1], Jim Adney[2], Dan Anderson[1], Mark Ball[1], Dave Caussyn[1], Mike Ellison, Jim Ferry[2], Brett Hamilton[1], Sergei Nagaitsev[1], Dag Reistad[3], Mark Sundquist[2], Peter Schwandt[1], and Miro Sedlacek[4]

[1]Ind. Univ. Cyclotron Fac., 2401 Milo Sampson Ln., Bloomington-IN 47405
[2]The Nat. Electrostatics Corp., Graber Rd. Box 310 Middleton, WI 53562
[3]The Svedberg Laboratory, Box 533 S-751 21 Uppsala-Sweden
[4]Alfvén Lab, Royal Institute of Technology, S-100 44 Stockholm-Sweden

SUMMARY

An electron cooling system installed in an available straight section of the SSC MEB could reduce the beam emittance by an order of magnitude without significantly increasing the SSC fill time. Consequently, the SSC could operate at a higher luminosity with reduced emittance beams, or at the same luminosity with beams of both reduced current and emittance. Since luminosity is a direct measurement of the SSC physics research potential, the 6.5 − 10 M$ system is very economical in light of the SSC capital and daily operating costs. A further economical benefit for future colliders is the reduced aperture requirement in the following higher energy synchrotrons. A system[1] is being assembled to demonstrate the necessary technology, leading to its commercial availability. This paper summarizes the application and feasibility of electron cooling 2 − 20 GeV/nucleon ion beams.

INTRODUCTION

The luminosity of a bunched-beam collider depends upon the ion beam current and emittance as I^2/ε; the space charge tune shift (ΔQ_{SC}) limit, however, varies as I/ε [n.b. ε *always* refers to the *rms normalized* emittance]. Consequently, when pushing colliders to higher luminosities, I is usually increased, while keeping the quantity I/ε constant once the tune shift limit has been reached. Another possible approach is to reduce the beam emittance after acceleration in the first (or second) ring. Since ΔQ_{SC} is usually only

---

*This work was supported by the Texas National Research Laboratory Commission under grants RGFY9158 and RGFY9258, Indiana University, IUCF, and the National Electrostatic Corporation.

important in the lowest energy synchrotron, smaller values of ε may be allowed later in the chain of accelerators providing the beam–beam tune shift is not a limitation ($\Delta Q_{SC} \sim C/\gamma^2$ where $C \sim \gamma$ is the ring circumference; thus $\Delta Q_{SC} \sim 1/\gamma$). Such an approach could provide increased luminosity by allowing operation with the same beam current with reduced emittance, or provide the same luminosity by allowing operation with a reduction in the beam current and a greater reduction in emittance. There are advantages resulting from both reductions: the lower beam current reduces radiation damage and synchrotron radiation at very high energies; the lower beam emittance reduces the aperture requirements in all the following machines -- a *very* economical benefit.

ELECTRON COOLING RATE

Electron cooling is accomplished by merging an ion beam with a co-moving electron beam in a short region of a storage ring. Ions moving in the accompanying electron beam rest frame lose energy by coulomb interactions. The action damping time, τ, of an ion with action ε in both planes and momentum amplitude error $\Delta p/p$ is well approximated as

$$\tau \approx \frac{A}{Z^2}\,\frac{\gamma^2}{4\pi\eta\Lambda}\,\frac{M}{m}\left(\frac{kT_e}{mc^2}\right)^{3/2}\frac{1}{nr_e^2 c}\left[\sqrt{\frac{8}{\pi}}+\left(u_\perp+\sqrt{u_\parallel}\right)^3\right]. \tag{1}$$

Equation 1 includes the effects of the "flattened" electron velocity distribution and the ion betatron and synchrotron motion. A and Z are the ion atomic number and charge state; γ is the usual relativistic parameter; Λ is the Coulomb logarithm (≈ 10); k is the Boltzmann constant; m and M are the electron and proton masses; c is the speed of light; n is the electron density; and r_e the classical electron radius; all other symbols are defined in Table I. The quantities $u_\perp$ and $u_\parallel$ are the nondimensional ion transverse and longitudinal velocities normalized to the electron beam rms transverse velocity:

$$u_\perp = \sqrt{\frac{\beta\gamma\epsilon_i mc^2}{\beta' kT_e}}; \qquad u_\parallel = \beta\,\frac{\Delta p}{p}\sqrt{\frac{mc^2}{kT_e}} \tag{2}$$

where β is the usual relativistic parameter and β' is the radial aperture function in the cooling region. For electron and ion beams of equal radii, $u_\perp$ corresponds to $(\epsilon_i/\epsilon_e = mT_i/MT_e)^{1/2}$. Electron cooling is most effective when $u \leq 1$; in this regime, corresponding to the dominance of the first addend in the square brackets of Eq. 1, $T_i \approx A\cdot 2{,}000\cdot T_e$ and the ion beam emittance will damp exponentially until thermal equilibrium is reached, or until the cooling force becomes balanced by an opposing heating force (e.g. IBS, multiple scattering, diffusion from nonlinear resonances, etc.).

ELECTRON GUN AND CONFINEMENT SYSTEMS

For the fastest possible cooling, we thus require $\epsilon_e \leq \epsilon_i$. At the cathode $\epsilon_e = \pi(r_c/2)\cdot(kT_c/mc^2)^{1/2}$. A standard tungsten dispenser cathode with $r_c = 3$ mm and $T_c = 1300$ K would thus produce an electron beam with $\epsilon_e < 0.75\pi$ μm, less than or equal to the emittance of the brightest non-cooled ion beams. Such cathodes can also provide the necessary current, > 2 A. This cathode emittance must then be preserved in the electron gun, during electrostatic acceleration in the Pelletron, and during transport through the cooling region. An electron gun, modeled using *EGUN* and *SAM*, behaves nearly as well as a model with perfect Pierce geometry. The aberrations in the electrostatic column lenses and magnetic focussing solenoids have also been shown to be insignificant. We believe, however, that experimental verification of adiabatic acceleration is necessary (though difficult since the optics are dominated by space charge rather than emittance).

The simplest electron focussing channel in the cooling region is a series of very weak solenoids with focal length f_{sol} spaced by the distance L_{sol}. From here on we assume the radius of the uniform electron distribution is $\sqrt{2}$ times the rms radius of the normal ion distribution, and $\varepsilon_e \lesssim \varepsilon_i$ before cooling. The magnetic confinement system maintains the space charge (rather than emittance) dominated beam divergence less than the incoherent divergence due to the beam temperature. This condition then determines both f_{sol} and L_{sol}:

$$f_{sol} > \frac{\beta'}{\sqrt{2}} \gg L_{sol} = \frac{2\bar{\lambda}^2}{f_{sol}} = \sqrt{2}\,(\beta\gamma)^2 \frac{\varepsilon_i}{r_e} \frac{ec}{I_e} \qquad (3)$$

where λ is the electron beam plasma wavelength divided by 2π (i.e., $\beta c/\omega_p$, where ω_p is the plasma frequency). Note that L_{sol} is independent of β'.

The electron and ion beam temperatures can be reduced by adiabatic expansion which increases the cooling rates even while reducing the electron beam density (T/n remains constant at a focus whereas $\tau \sim T^{3/2}/n \sim 1/\sqrt{\beta'}$). These larger beams have correspondingly smaller angular divergences and consequently more stringent alignment tolerances in order to preclude *effective* temperatures in excess of the cathode temperature, $T_c \approx 0.12$ eV/k. Neglecting obvious limits to size of β', such as magnet apertures, β' is limited by the precision of the beam position monitors used for the relative alignment of the electron and ion beams. Assuming the position electrodes are located inside the solenoids and have an rms resolution δ, the angular misalignment can be limited to $2\delta/L_{sol}$ thus limiting β' as:

$$\beta' < \beta^3 \gamma^3 \left(\frac{ec}{I_e r_e}\right)^2 \frac{\varepsilon_i^3}{\delta^2}. \qquad (4)$$

Scaling β' as $(\beta\gamma)^3$ and using Eqs. 1 and 2, we find for constant I_e and ε_i, that $\tau \sim \gamma$, a far cry from the γ^5 scaling, based upon different assumptions, often seen in the literature.

A three dimensional particle tracking study showed that electron beam, with a uniform spatial distribution and Gaussian angular distribution, develops small tails and a slight amount of hollowness in this transport system; the emittance growth, however, is negligible. Since solenoids are second order focussing element, the focal length for ions is $(AM/m)^2$ times greater than for the electrons. The ion beam tune shift and betatron plane coupling is consequently negligible. The ion beam space charge can, however, adversely affect the less rigid electron beam and provisions must be made to prevent space charge neutralization of the electron beam.

EXAMPLES OF ELECTRON COOLING FOR COLLIDES AND RINGS

Table I summarizes 5 possible applications of intermediate energy electron cooling. None are optimized, but rather designed to fit into existing lattices.

The system for CoSy[2], a ring being commissioned at KFA-Jülich, would fit into 1 of the 4 straight sections currently reserved for stochastic cooling. The electron system would cool beams approximately 10 times faster to much smaller equilibria than the stochastic system. It is necessary to reduce $I_e \sim (\beta\gamma)^2$ as the proton beam momentum is reduced from its rigidity limit of 3.3 GeV/c due to fixed L_{sol}; τ, however, remains almost constant.

Cooling would be extremely fast in the proposed KEK-PS[3] heavy ion collider due to the A/Z^2 scaling of the cooling time and relatively low beam energy. This system would counteract intrabeam scattering consequently increasing the luminosity by a factor of ≈ 10.

The emittance of proton beams in HERA is limited by space charge effects in the first

Table I. Proposed intermediate energy electron cooling systems.

Parameter	Symbol	CoSy	KEK-PS	Petra II	MEB-1	MEB-2	Units
Ring Properties							
Circumference	C	0.1835	0.283	2.3	3.96	3.96	km
Cooling region length	L_c	4	3.5	50	25	40	m
Fraction (L_c/C)	η	2.2×10^{-2}	1.2×10^{-2}	2.2×10^{-2}	6.3×10^{-3}	1×10^{-2}	
Cool region beta funct.	β'	8	10	40	50_x; 20_y	100	m
Ion Beam Properties							
Ion species		H^+	H^+	$^{197}Au^{79+}$	H^+	H^+	
Momentum/nucleon	p	3.2	5; 8	8.4	13		GeV/c
Norm rms emittance	ε_i	2.5	2	4	0.70		$\pi\ \mu m$
Rms mom. sprd.	$\Delta p/p$	1×10^{-3}	4×10^{-4}	5×10^{-4}	2×10^{-4}		
Peak ion current	$I_{i,\ peak}$		$Q \times 1.6$	1,000	800		mA
Lasslet tune shift	ΔQ_{SC}		8; 3×10^{-4}	0.03	0.07		
Beam-beam tune shift	ΔQ_{b-b}		1×10^{-5}		8.6×10^{-4}		
Electron Cooling System Parameters							
Electron current	I_e	2	2	2	2	2	A
Electron kinetic energy	U	1.3	2.2; 3.8	4.1	6	6	MeV
Cathode radius	r_c	3.2	3.2	3.2	3.2	3.2	mm
Electron beam radius	r_b	3.2	3.2	4.2	3.2	4.5	mm
Electron temperature	T_e	0.12	0.12	0.07	0.12	0.06	eV/k
Solenoid focal length	f_{sol}	5	none	28.4	19	91	m
Solenoid spacing	L_{sol}	1	none	2.5	6	2	m
Emittance damp. time	τ						s
initial		6	0.4; 1.2	35	100	30	
for cold beams		0.5	0.05; 0.13	5	35	20	
Approximate cost		1	2	4	6.5	6.5	M$

synchrotron, DESY III. In the following ring, PETRA II, $\Delta Q_{SC} \approx 0.03$; consequently the emittance could be reduced by $\approx \times 10$ in about 1/3 the ring ramp time, 108 s.

Cooling in the SSC MEB has been investigated in great detail[4]. MEB-I is a design which leaves the ring lattice entirely unmodified; MEB-II re-arranges the ordering of a few elements in the long straight insertion to increase β' and L_C.

PROJECT STATUS

We proposed a 3 year study to (1) prepare a design report on the feasibility of cooling beams in the SSC MEB and (2) to demonstrate the necessary technology. The technical design report showed that all the necessary subsystems are technically feasible; most of the equipment needed to demonstrate that the system as a whole will function reliably in an accelerator has been procured . Our major funding source, the TNRLC, however, has discontinued support. An international collaboration of interested laboratories is being formed to complete this work. Electron recirculation test should begin within one year.

REFERENCES

1. D. Anderson, M. Ball, D. Caussyn, T. Ellison, B. Hamilton, S. Nagaitsev, P Schwandt, J. Adney, J. Ferry, M. Sundquist, D Reistad and M Sedlacek, "The development of a prototype multi-MeV electron cooling system", in proc. 1993 Part. Accel. Conf. (17-20 May 1993, Washington-DC).
2. R. Maier, U. Pfister and J. Range for the COSY-Team, "The COSY-Jülich project April 1991 status", in *Proc. 1991 IEEE Part. Acc. Conf.* (IEEE cat. no. 91CH3038-7, IEEE Piscataway NJ)2808-2900.
3. J. Chiba et al., "The PS Collider design report 2 (in English)", KEK Report 90-13 (September 1990).
4. The MEBEC group, *MEBEC E-Cool Design Report 1992*, (IUCF publication, October 1992).

DESIGN OF AN 80 K LINER PROTOTYPE IN SSCL ASST
FOR SYNCHROTRON LIGHT INTERCEPTION

Q. S. Shu, W. Chou, D. Clark, W. Clay, Y. Goren, R. Kersevan,
V. Kovachev, P. Kraushaar, K. Leung, J. Maddocks, D. Martin,
D. Meyer, R. Mihelic, G. Morales, J. Simmons, G. Snitchler,
M. Tuli, W. Turner, L. Walling, K. Yu, and J. Zbasnik

Superconducting Super Collider Laboratory*
2550 Beckleymeade Avenue
Dallas, TX 75237-3997

INTRODUCTION

The Superconducting Super Collider (SSCL) is the first proton superconducting accelerator designed to operate at 20 TeV (each beam) with beam current of 72 mA in which synchrotron radiation is a significant design factor. The Collider will produce a synchrotron power of 0.14 W/m and 18 kW total at 4.2 K. This synchrotron light will produce considerable photodesorbed gases in the beam vacuum. The photodesorbed gases may greatly reduce the beam lifetime and scattered beam power may lead to quenching of superconducting magnets. The Collider availability may be unacceptable without properly addressing this concern. A liner is one method under consideration to minimize the presence of photo-desorbed gases. Such a liner may improve Collider cryogenic thermal efficiency that would allow a potential luminosity upgrade.

A liner prototype has been developed for prototype testing at the Accelerator System String Test (ASST) facility since the half cell (five dipoles, one quadrupole and one spool piece with a beam position monitor) is an existing basic unit of the Collider. The liner operational temperature was required to be 80 K based on photodesorption data available from the CDG and SCDG measurements.[1] Those data showed that liners at lower temperatures 20 K or 4.2 K had either unacceptable impedance margins or excessively long conditioning periods. An 80 K liner also replaces the 4 K dynamic heat load of the synchrotron radiation with a static heat load, independent of the beam intensity, and transfers the intercepted heat to the liquid nitrogen system. In January of 1993 new photodesorption data indicated the viability of 4 K and 20 K systems.[1]

Developing a liner system presents scientific and technical challenges. The system design addresses photodesorption, particle beam stability, magnetic field quality, beam induced wake fields, rf impedance, cryogenics, magnet quenching (especially quenches induced Lorentz pressure), and many other interdisciplinary technical problems. This paper

* Operated by the Universities Research Association, Inc., for the U. S. Department of Energy under Contract No. DE-AC35-89ER40486.

presents the results of trade studies, analyses and engineering design of an 80 K liner and also briefly discusses the preliminary consideration of lower temperature liners.

80 K ASST LINER CONFIGURATION

As shown in Figure 1, an 80 K Liner System consists of an 80 K perforated tube (the liner tube) located coaxially inside the 4.2 K magnet bore tube. The liner temperature is maintained at 80 K by high pressure GHe loops of 0.25 g/s. The GHe flow is recooled by LN_2 in the cooling pipe of a magnet cryostat. A special end conducting cooling structure is designed to cool the quadrupole magnet liner since the ASST quadrupole beam tube ID (32.3 mm) is much smaller than an ASST dipole ID (42.3 mm). Low heat leak supports hold the liner in the center of the beam tube. A thin layer of cryosorber (0.5 mm) on the inner surface of the 4.2 K beam tube pumps the photodesorbed gases through the holes on the liner tube. The rf joint, a good thermal contact joint, assures the continuity of the image current and aids assembly and maintenance. The magnet interconnect with these joints, and a compact heat exchanger are included in a cryogenic box. The cryogenic box also allows each liner to have up to a 54-mm thermal contraction during cooldown and warmup. However, the liner is discontinuous when routed through the Beam Position Monitor (BPM).

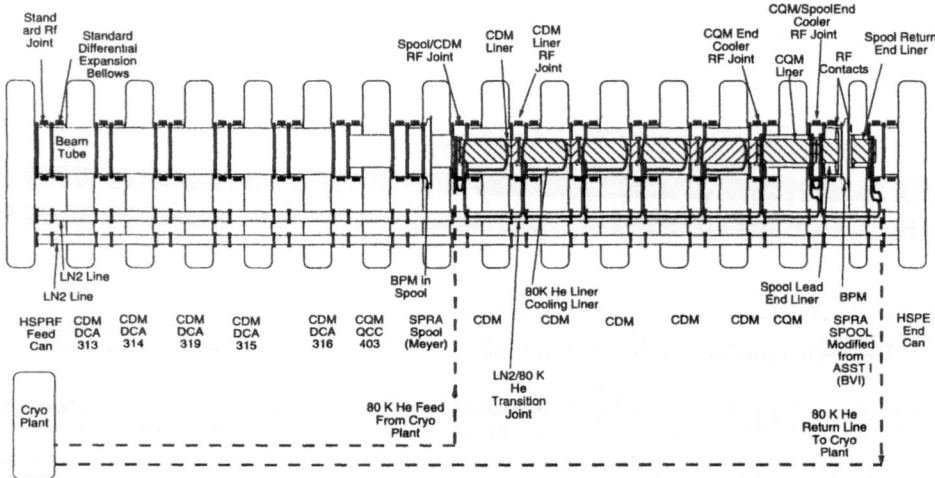

Figure 1. A schematic layout of the ASST 80 K liner system.

80 K LINER TUBE DESIGN

Figure 2 shows the tube cross section of the Collider Dipole Magnet (CDM) liner and Collider Quadrupole Magnet (CQM) liner. A uniform and maximum possible liner inner diameter (ID) is needed due to: (1) particle beam commissioning, (2) particle beam dynamic stability, and (3) safety margin. However, the maximum liner ID is constrained by: (1) the available magnet beam tube inner diameter (ID), and (2) the minimum liner radial space. Using parallel loop cooling, the minimum liner radial space is 6 mm and using end conducting cooling, the radial space needs to be 3.5 mm. Tables 1 and 2 show the maximum liner ID and the liner impedances in various options, respectively.

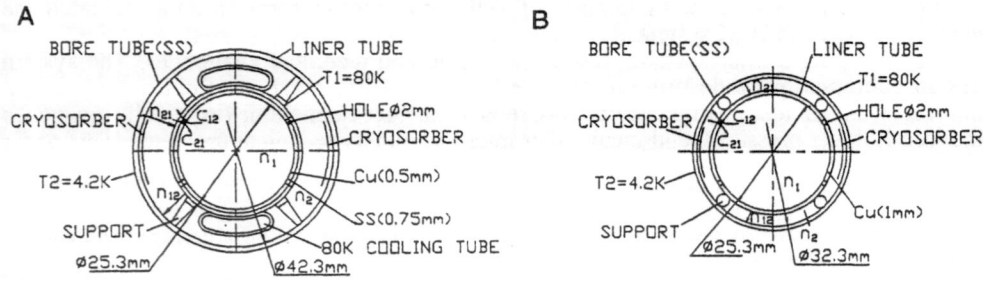

Figure 2. The cross sections of the (A) CDM liner and (B) CQM liner.

Table 1. Possible maximum liner ID in various cases.

Object	Dipole Beam tube ID, mm	Magnet Liner ID, mm	Quad Beam tube ID, mm	Magnet Liner ID, mm	Spool Beam tube ID, mm	Piece Liner ID, mm
ASST	42.3	25.3	32.3	25.3	32.3	25.3
GD, B & W	32.3	20.2	32.3	20.2	32.3	20.2
Desired	42.3	31	42.3	31	42.3	31

Table 2. Comparison of impedances.

Case	Liner ID, mm	Hole/Slots Coverage	Z (liner) M ohm/m	Z (other) M ohm/m	Z (total) M ohm/m	Safety margin
Baseline	32.3			40	40	6.7
With liner	25.3	2 mm, 2%	22	112	133	2
With liner	25.3	2 mm, 4%	44	112	156	1.7
With liner	25.3	2 × 6, 2%	8	112	120	2.2
With liner	33	2 mm, 4%	15	40	55	4.9

The liner tube design also must meet the following requirements:

1. Inner wall conductivity and thickness $\sigma * \delta > 2 \times 10^5$ Ohms^{-1}
2. Liner impedance $Z\,L/n < 0.34$ Ohm
   $Z\,T < 20$ M Ohm/m
3. Inner wall photodesorption coefficient $\eta = 0.02$
   $\alpha = 0.3$ for H_2
4. Liner pump speed — 600 l/m/s for H_2
5. Total liner heat leak to 4 K — < 1 W for dipole
6. Cryosorber pump speed — 1200-3000 l/m/s for H_2
7. Cryosorber pumping capacity — 30 Torr l/m at 294K
8. Cryosorber activation temperature — 294 K; regeneration < 80 K H_2, <294 K all Gases; recovery fraction regeneration > 98%
9. Liner quench survivability and ASME code — 100 quenches in 25 yrs
10. Radiation dose tolerance — 1400 MRad in 25 yrs

RETROFIT ASSEMBLY, END COOLING AND HEAT LOAD BUDGET

An extensive trade study has been performed to develop a retrofit 80 K liner structure and flow return cooling loop, Figure 3A. If an 80 K liner is chosen for collider upgrade, the structure will help its insertion into magnets to be retrofitted in the tunnel. As shown in Figure 3B, the CQM liner is refrigerated through thermal conduction by 80 K GHe in a compact heat exchanger at the end of the liner tube outside the CQD cold mass. The maximum temperature increase ΔT could be less than 5 K for a Spool Piece liner, and 10 K for a CQM. Table 3 shows the heat load budget. Radiation is difficult to reduce to budgeted levels because the cryosorber likely to have an emissivity near unity.

Table 3. Fixed or static 80 K liner heat budget.

	Dipole (w)	Quadrupole (w)	Spool (w)	Half Cell (w)
Supports	0.5	0.5	0.3	3.30
I R Radiation	0.2	0.06	0.04	1.1
Interconnects	0.05	0.05	0.05	0.35
BPM	–	–	0.26	0.26
Total	0.75	0.61	0.73	5.01

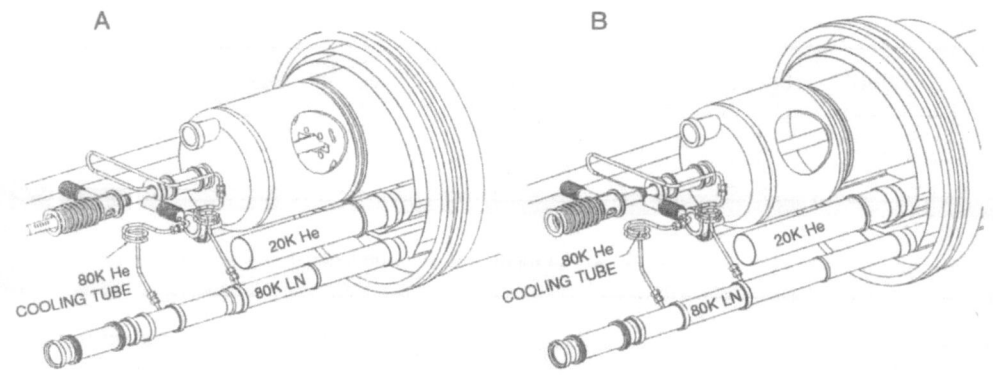

Figure 3. (A) Assembling a CDM liner, (B) Assembling a CQM liner.

INTERFACE BETWEEN CQM, BPM, AND SPOOL PIECE

Figure 4 shows the liner interfaces with the CQM, BPM and the Spool Piece. Details were discussed in the liner design report.

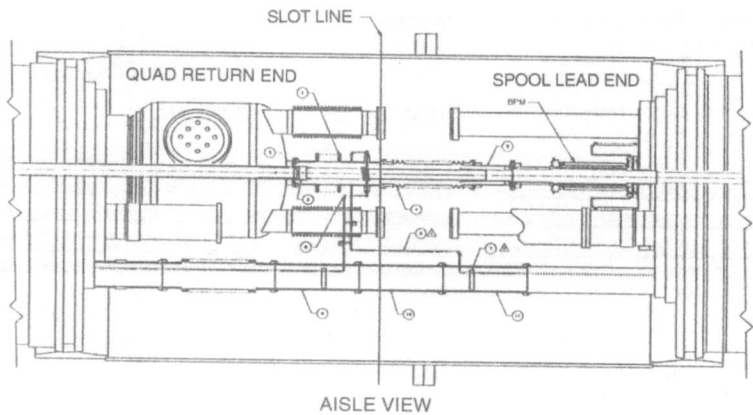

Figure 4. Liner interfaces between the CQM, BPM and the Spool Piece.

LOWER TEMPERATURE LINERS

Several concepts of 4.2 K and 20 K Liners have been studied as a result of the new photodesorption tests. Concept A had a complex extruded shape with three supports integral to the liner tube. Concept B showed a circular beam tube with three brazed hat shaped supports running the full length of the tube (to ensure even thermal distribution between liner and bore tube). Concept C showed the same support system as proposed for 80 K liner, i.e., discrete supports located every 1 m. All three concepts would be optimized for good thermal contact between the liner and beam tube. In the three concepts, the cryosorber is located on the liner outer surface. If the photodesorption coefficient at lower temperature is much less than at higher temperature, the number of the holes on the liner and the rf transverse impedance in the structure will be further reduced. Concept D considered addition of a 4.2 K channel on the liner tube to boost cooling capacity. The 20 K liner system concept is very similar to the 80 K system, but with the option of cryosorber on liner outer surface.

REFERENCES

1. 80 K ASST Liner Design Report, edited by Q. S. Shu, in preparation.

PROGRESS IN UNDERSTANDING RAMP RATE SENSITIVITY
IN HIGH ENERGY BOOSTER DIPOLE MAGNETS

J. M. Butler,[1] C. A. Swenson,[1] and W. J. Carr[2]

[1]Westinghouse Magnet Systems Division
I-H 35 North and Westinghouse Road
Round Rock, TX 78680

[2]Westinghouse Science & Technology Center
1310 Beulah Road
Pittsburgh, PA 15235

INTRODUCTION

Quench current on some string test magnets decreased significantly with increased ramp rate. As subcontractor for the design and development of the HDMs, Westinghouse is responsible for ensuring that HDMs reach full field during the bipolar operating cycle which ramps at 62 A/s, compared with only 4 A/s for CDMs. Our objective is to specify and control design and processes to consistently produce magnets that meet ramp rate requirements. Understanding observed ramp sensitivity is necessary, along with aggressive testing of candidate design and process improvements.

DIAGNOSIS

Magnets have been characterized as "Type A" (I_q vs. dI/dt concave downward) or "Type B" (concave upward), with some in between. To explain type B magnets, consider the time constants of various interstrand eddy currents induced in pairs of strands with finite crossover resistance at various contact points as in Figure 1. In this model, it is assumed that most contacts have high resistance. A circuit element formed by a string of diamonds that has a *net* voltage between the points of low resistance is called an *active circuit element*. The *net* voltage is one "diamond" in Figure 1b because the number of diamonds is odd. Strings with an even number of diamonds are *passive circuit elements* since they have no net voltage. Voltage is assumed constant and proportional to ramp rate. The eddy current in a single active circuit element can be calculated by treating all other circuit elements as passive. The total eddy current flow is then obtained from superposition of the solutions for each active element. Currents for long active elements separated by passive elements of low resistance will not strongly overlap.

Supercollider 5, Edited by P. Hale
Plenum Press, New York, 1994

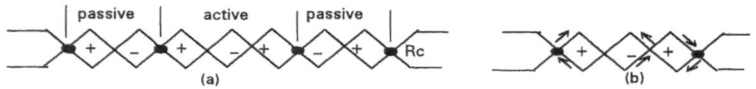

Figure 1. (a) Two strands in the cable which form a diamond pattern between crossovers. Dots indicate low crossover resistance. The remaining points are assumed to have infinite resistance. (b) Eddy current in an *active circuit element*. The net voltage of one "diamond" causes current to flow in both the active element and outside of it, but the latter falls off with distance, due to inductance and resistance. The + and - indicate the polarity of the voltage induced in the each diamond by the ramping field.

The single active circuit in Figure 2 shows end effects which can be important because of the discontinuous field in the ends of the magnet. Up to about ±2 "diamonds" of voltage can be obtained from one end. Quenches near the ends of the magnet may be the result of currents due to these higher end voltages. The end voltage is highly dependent on the end geometry and cable twist pitch. In long elements voltage from consecutive ends can add if there is an odd number of diamonds in both the straight part *and* each end.

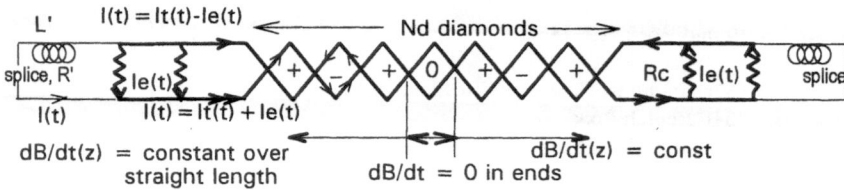

Figure 2. Darker lines indicate possible eddy current I_e in an active circuit element. I_e is superimposed on the transport current. Interstrand resistance is high except where indicated by resistors. Discontinuous field in the end turn results in a net voltage of two "diamonds" in the active element shown. Inductance L depends on the number of diamonds in the active element, N_d. Inductance outside the active element L' helps force current through resistors shown.

$I(t)$ = the total current in a strand in the active element.
$I_t(t)$ = the transport part = I_{cable}/N_w for N_w identical strands.
$I_e(t)$ = the eddy current part, which adds to I_t in one strand and subtracts from I_t in the other.

Then in one strand, $I = I_e + I_t$ and at a constant ramp rate, a quench will occur when $I = I_c$, the critical current. Let t_q = time to quench and let I_q = strand transport current I_t at quench time t_q. Then

$$I_q = I_c - I_e(t_q) \quad \text{where } t_q = I_q/\dot{I}_t \tag{1}$$

The quench current I_q is determined by evaluating I_e and I_c. For $L' \gg L$ and $L'/R' \gg L/R$ in Figure 3, an approximate solution for I_e is:

$$I_e \approx \frac{V}{R}(1 - e^{-Rt/L}) + \frac{V}{R'}(1 - e^{-R't/L'}) \tag{2}$$

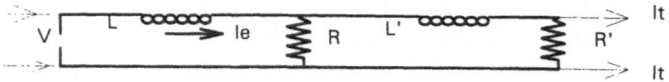

Figure 3. Simplified circuit. L is the inductance of the active circuit, R comes from a *group* of nearby crossover resistances in parallel, and L' and R' represent the remainder of the circuit.

822

Equation (2) describes build up of current in the active element, and a slower build up of current in the remainder of the circuit. If the ramp is terminated before appreciable eddy current is established in the remainder of the circuit, (2) reduces to just the first term:

$$I_q = I_c - \frac{V}{R}(1 - e^{-Rt_q/L}) \quad \Rightarrow \quad t_q = \frac{c_1 - \frac{V}{R}(1 - e^{-Rt_q/L})}{\dot{I}_t(1 + c_2)} \tag{3}$$

where the wire critical current I_c is assumed to vary linearly with field B in the range of interest (3 - 8 Tesla) and B, in turn, varies linearly with wire transport current I_t so that $I_c = c_1 - c_2 I_t$. The effect of temperature on I_c is neglected because it is small in type B magnets. Figures 4 through 7 show calculation results solving Equation (3) for the case listed below. Other realistic combinations of input variables give the same results.

$V = N_v k_v \dot{I}_t$	Voltage in the active eddy circuit element where:
$N_v = 4$	Number of diamonds of voltage ($\approx$ two consecutive ends)
$k_v = 8.3~\mu V/(A/s)$	Voltage per diamond (at the midplane) at $\dot{I}_t = 1$ A/s
$L = N_d L_d$	Inductance of the eddy circuit where:
$N_d = 580$	Number of diamonds of inductance ($\approx$ two straight lengths)
$L_d = 35~\eta H$	Inductance contributed by each diamond length
$R = 0.102~\mu\Omega$	10 parallel resistances of $1\mu\Omega$ each or 100 of $10\mu\Omega$
$N_w = 30$	Number of wires in a cable
$I_c = c_1 - c_2 I_t$	Wire critical current at the quench ($I_t = 1000B/N_w$)
	$c_1 = 1140$, $c_2 = 3.63$ for wire in DCA318

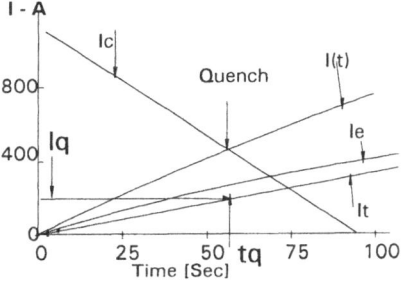

Figure 4. Components of strand current at $\dot{I}_t$ = 3.3 A/s ($\dot{I}_{cable}$ = 100 A/s). Quench occurs at $t_q = I_q/\dot{I}_t$ when $I = I_c$. $I_q = I_q$ cable/N_w

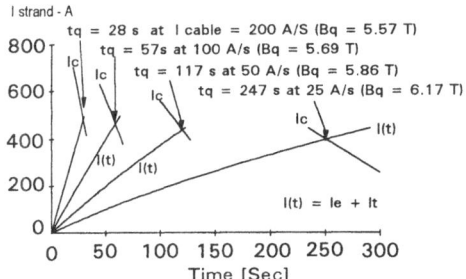

Figure 5. The intersection of I(t) and $I_c(t)$ gives t_q for different ramp rates.

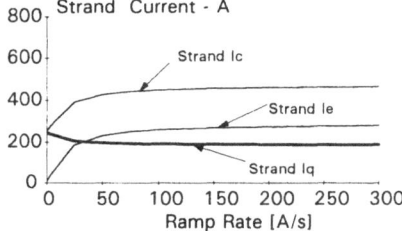

Figure 6. Calculated strand currents vs. ramp rate. $I_q = t_q * \dot{I}_t$ from Figure 5.

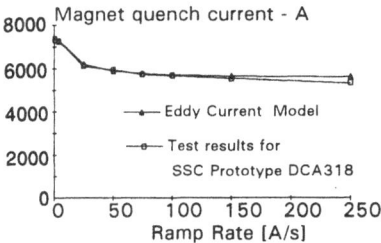

Figure 7. Comparison to test. I_q magnet = 30 I_q from Figure 6. The small negative slope at large ramp rate is attributed to thermal effects.

The I_q vs. $\dot{I}_t$ curve begins to flatten when the exponent of e equals unity, i.e., when ramp rate is $\dot{I}_t = I_q R/L = I_q/\tau$ where τ = the time constant. At low ramp rate ($t_q \gg \tau$ and $\dot{I}_t \rightarrow 0$,), equation (3) reduces to:

$$I_q = I_c - V/R \qquad \text{or} \qquad I_q = \frac{c_1}{1+c_2} - \frac{N_v k_v \dot{I}_t}{R(1+c_2)} \qquad (4)$$

where $-N_v k_v/[R(1+c_2)]$ is the initial slope of the quench curve and $c_1/(1+c_2) = I_q(0)$. At high ramp rate ($t_q \ll \tau$), the right hand term in Equation (3) can be expanded to give:

$$I_q \approx \frac{I_c}{1+N_v k_v/L} \qquad \text{or for } \dot{I}_t \rightarrow \infty \qquad \frac{I_q(\infty)}{I_c(0)} = \frac{1}{1+N_v k_v/L(1+c_2)} \qquad (5)$$

Combining (4) and (5) to eliminate $N_v k_v/(1+c_2)$ gives the time constant in terms of quench plot characteristics:

$$\tau = \frac{L}{R} = -\left.\frac{dI_q}{d\dot{I}_t}\right|_{\dot{I}_t=0} \left/ \left(\frac{I_q(0)}{I_q(\infty)} - 1\right)\right. = \frac{\text{initial slope}}{\text{fractional decrease in plateau height}} \qquad (6)$$

To explain type A behavior in the present model, a larger value of R is required, along with a small value of N_v, to make the initial slope small. Figure 8 shows the calculation result when $N_v = N_d = 1$, and $R = 1.2\mu\Omega$. Although the agreement with experiment for high ramp rates is not good, much of this discrepancy can be explained by eddy current heating not considered in the present model.

POSSIBLE DESIGN SOLUTIONS

One candidate solution for ramp rate sensitivity this analysis suggests is to fully insulate the strands. For perfect insulation with all crossover resistances infinite, the voltage coming from the end effects can add to a rather large value, tens of times that for a single diamond. However, the inductance L would then be that for the entire winding, or

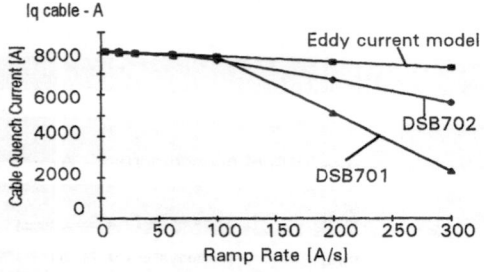

Figure 8. For Type A magnets, the calculated eddy current model curve is nearly linear for $N_v=1$, $N_d=1$, and $R=1.2\mu\Omega$. ($\tau=0.03$ s) Increased temperature from eddy current heating may explain the downturn at 100 A/s in test curves.

the order of one mH. Taking the splice resistance to be 1 to 10 $\eta\Omega$ leads to an enormous time constant, measured in days, and solution of (3) for any time t_q of practical interest is simply $I_q = I_c$. Coating all strands with ebonol is one possible way to insulate. No coldwelds could be allowed, which appears possible without significant cost, based on piece lengths now being produced.

QUENCH PROPAGATION STUDY FOR BNL-BUILT, FULL-LENGTH, 50mm APERTURE SSC MODEL DIPOLES

J. Muratore, M. Anerella, G. Cottingham, M. Garber, G. Ganetis, A. Ghosh,
A. Greene, R. Gupta, S. Kahn, E. Kelly, G. Morgan, A. Prodell, W. Sampson,
R. Shutt, P. Thompson, P. Wanderer and E. Willen

Brookhaven National Laboratory[1], Upton, NY 11973

INTRODUCTION

As part of the program to build and test SSC 50mm aperture prototype dipole magnets, a series of seven full-length (15m-long) dipoles were built and tested at BNL. The design of these magnets and their differences from the earlier 40mm aperture dipole magnets have been discussed elsewhere[1]. In order to thoroughly evaluate the performance of the new design, an extensive and varied schedule of tests were done, and these included quench testing and stress measurements to exhibit mechanical and thermal behavior, a ramp rate quench program to study eddy current heating effects in the conductors, and a multitude of magnetic field measurements to characterize field quality.

An important part of the testing program was the study of quench propagation velocity and hot spot temperature over a range of experimental conditions in order to characterize the safety of the conductor during quenches experienced under different circumstances. Such studies are important tools in the design, implementation, and verification of quench protection strategies in superconducting accelerator magnets. This investigation was facilitated by artificially inducing quenches under controlled experimental conditions with spot heaters placed at carefully chosen locations on the magnet coils. Such studies were done as part of the 15m-long magnet test program and were performed on five of the magnets in the series. All were equipped with spot heaters on an inner coil, and two of these also had spot heaters on an outer coil. Therefore, in addition to the usual studies in the inner coils, where most spontaneous quenches originate, it was also possible to study quench propagation in the outer coils, where slower quench velocities and higher conductor temperatures are expected, in comparison to that in the inner coils. In addition, the two magnets with outer coil spot heaters were the only ones in the series which used the new all-Kapton cable insulation scheme, so such a study would provide valuable data about the thermal flow characteristics of this insulation during quenches.

In spontaneous quenches, where there may be no voltage taps, it is not possible to measure the conductor hot spot temperature. It is very straightforward, however, to measure the number of MIITs generated, since only the magnet current and voltage need be measured. The concept of MIITs then becomes a valuable diagnostic tool which can characterize the temperature behavior of a conductor during quench and can be used to determine limits for safe operation of the coil. With spot heaters placed at known locations and closely bracketed by voltage taps, hot spot temperature can be measured. Research such as is described in this paper is therefore important in order to determine the validity of the MIITs approach and to establish a correlation between temperature and MIITs.

[1]Work performed under Contract No. DE-AC02-76CH00016 with the U.S. Department of Energy.

MAGNET DESIGN AND CONSTRUCTION

The BNL-built, full-length 50mm aperture dipole magnets all shared the same design and construction features. They differed only in cable conductor and insulation parameters, the most important of which are listed in Table 1 for selected magnet coils. In all cases, spot heaters were mounted on the midplane turn of the lower coils only, and information for only those coils for which spot heater studies were done are included in the table. As can be seen, the conductor, for which there were three industry suppliers, varied to some extent in critical current, copper-to-superconductor ratio, and RRR. In particular, the lower inner coil cable for DCA207 had a post-cure RRR of 222, compared to the nominal value of about 135 for the other inner coils; and the copper-to-superconductor ratio of the lower inner coil of DCA208 was 1.34, lower than the nominal 1.5 for the other inner coils. Note that the last two magnets in the series, DCA212 and DCA213, had Kapton-Kapton cable wrap, which resulted in higher curing temperatures and pressures.

EXPERIMENTAL PROCEDURES

Each spot heater element is a $50.8\mu m$ thick rectangular strip of stainless steel 38.1mm long and 11.9mm wide, mounted on the wide flat dimension of cable on a section of the midplane turn. The mounting is accomplished by first stripping away a 50.8mm length of cable insulation at the location where the heater is to be attached, then placing two layers of Kapton, the heater element strip, then another layer of Kapton. In addition to this, there is a Kapton cap placed over the midplane of both the lower and upper coils. Projections on each end of the heater element strip are bent away from the cable and brought out into leads, which are routed along the coil inner diameter to the outside of the magnet for connection to a pulser power supply. A quench is initiated by generating a pulse by capacitive discharge through the heating element with enough energy to heat the small section of conductor to its critical temperature. Energies needed to quench depended on the field strength, and therefore magnet current, at the quench location. For the inner coils, spot heater energies ranged up to 17J, while for the outer coils, energies were as high as 80J.

Table 1. Coil cable and cure parameters related to quench performance for those magnet coils involved in the study.

Magnet	Coil	Cable ID	Wire Mfg.	$I_c(A)$ [7.0T]	Cu:SC Ratio	RRR Post Cure	Cable Wrap	Cure T (C)	Cure P (MPa)
DCA207	LI	SSC-3-I-00055	IGC	10836	1.54	222	K/EF	135	73.1
DCA208	LI	SSC-3-I-00067	IGC	11371	1.34	132	K/EF	135	73.1
DCA211	LI	SSC-3-O-00050	OST	11283	1.48		K/EF	135	49.0
DCA212	LI	SSC-3-S-00034	SCN	10368	1.51	136	K/K	225	81.4
	LO	SSC-4-S-00027	SCN	6924	1.77	228	K/K	225	80.7
DCA213	LI	SSC-3-S-00040	SCN	10039	1.54	140	K/K	225	81.4
	LO	SSC-4-S-00028	SCN	6832	1.78	221	K/K	225	82.7

K/EF = Kapton/Epoxy-Fiberglass cable insulation scheme; K/K = Kapton/Kapton cable insulation scheme; LI = Lower inner; LO = Lower outer.

For each of the magnets, two spot heaters were mounted on the midplane turn of the lower inner coil: spot heater #1 was located 68.58cm from the lead end; spot heater #2 was 22.86cm from the nonlead end and was 13.81m from a voltage tap pair on the same straight section of cable. For a quench from spot heater #2, therefore, the time-of-flight of the quench front from the quench origin at the spot heater to the tap pair was used to calculate the average quench velocity along the straight section of cable. In DCA212 and DCA213, the magnets employing the all-Kapton insulation scheme, there were also two spot heaters on the midplane turn of the lower outer coil: spot heaters #3 and #4 were 7.62cm from the lead and nonlead ends, respectively. Both these heaters were each closely bracketed by voltage taps spaced 10.16cm apart with the spot heater at the center.

For the outer coil spot heaters with their bracketing voltage taps, conductor hot spot temperature was measured during a quench by acquiring the voltage signal vs time during the quench event and dividing this signal by the magnet current, point-by point. From the resulting time development of the conductor resistance, the temperature could be determined by using a resistance-temperature calibration curve obtained by measurements made on a short sample of the cable used in the outer coils[2].

MIITs values for each quench were calculated for both inner and outer coil spot heater quenches by integrating the square of the acquired current signal waveform from the quench start time, determined by the onset of resistive voltage, to the time at which the magnet current has decayed to zero. The relationship between MIITs and hot spot temperature is given by the expression

$$MIITs \times 10^6 = \int_0^{\infty} dt \ I^2(t) = S^2 \int_{T_0}^{T_{max}} dT \left(\frac{r}{1+r}\right) \frac{C(T)}{\rho(RRR,T,B)} \qquad (1)$$

which is derived from a heat balance equation[3] that assumes adiabatic and constant field conditions, usually good assumptions during the initial part of the quench propagation. In this equation, T_0 is the initial temperature, T_{max} is the hot spot temperature, S is the conductor cross-section, C is the specific heat per unit volume of conductor, B is the magnetic field at the quench spot, ρ is the copper resistivity, and r is the copper-to-superconductor ratio.

As can be seen, if the conductor properties and the magnetic field at the quench location are known, the temperature can be calculated from MIITs by an implicit solution of Eq. 1. A computer model which does this was used to generate families of MIITs-temperature correlation curves, for both the inner and the outer coils, for a range of magnetic fields[4]. These theoretical calculations have two purposes: to establish upper bounds for MIITs that give the safe conductor temperature limits of, typically, 500K; and to determine if MIITs and temperature measurements during quenches, such as were done on the outer coils, fit the adiabatic model and its assumptions. Doing this may give insight into the actual mechanisms occurring during quench propagation.

In addition to temperature and MIITs measurements, quench propagation velocities were also calculated from the voltage signal data and fitted to an adiabatic model[5] to obtain information on mechanisms at work during quenches. Velocity may be calculated from the quench signals in two ways. If the quench occurs in a section of cable between two voltage taps, as it does for the outer coil spot heaters, or if the quench occurs at a known distance from a single tap, as it does for the inner coil spot heater #2, the time of flight to these taps can be used to calculate velocity and position. Such a calculation results in the average velocity over the length of the section. If the quench front is accelerating, this will be different from the initial velocity of the quench but this is the most important value from the standpoint of temperature and MIITs generated.

If the copper resistivity $\rho(B)$ is known as a function of magnetic field B, the initial velocity can be calculated by measuring the slope dV/dt of the resistive voltage increase with time and using the formula

$$v_{init} = \frac{S}{\rho(B)I} \frac{dV}{dt}, \qquad (2)$$

which can be derived from Ohm's Law and where S is the cross-sectional area of the conductor and I is the quench current.

For spot heater #2 in the inner coils of all five magnets and spot heater #3 in the outer coils of DCA212 and DCA213, quenches were induced at a series of currents and the MIITs values calculated. Then, in each of these cases, a series of quenches were initiated with increasing quench protection heater (strip heater) time delays at a selected current, usually that for which the MIITs number was a maximum. For the outer coil quenches, the hot spot temperature was determined from measurements of resistance vs. time and quench propagation velocities were calculated from time-of-flight measurements of voltage tap signals. For the inner coil quenches, average velocities were calculated by time-of-flight and initial velocities from Eq. 2. Results were compared to theoretical calculations of temperature from MIITs and of the initial quench velocities from adiabatic models to determine the validity of the models and their usefulness as predictive tools in quench testing. All test results reported here were done nominally at the SSC operating temperature of 4.35K.

EXPERIMENTAL RESULTS

1. MIITs

Quenches were induced at a series of magnet currents and, for each magnet, a plot of MIITs vs. I_q was generated. Figure 1 shows these plots for the lower inner coil quenches of all five magnets. As can be seen, the shapes of the plots are similar, and they agree qualitatively with those studies done on earlier magnets in the SSC and RHIC projects. The current for peak MIITs ranged from 54% to 68% of the spontaneous plateau quench current. This behavior is typical and can be understood by a simple argument. Peak MIITs occurs at an intermediate current, at which the quench velocity has become fast enough to transport the heat away from the hot spot and limit the temperature there as stored electromagnetic energy is converted to heat. This current is typically about 2/3 of the quench plateau mean value. Offsets among the MIITs vs. quench current curves reflect differences in conductor properties among the five magnets, as was shown in Table 1. Note that the two all-kapton magnets DCA212 and DCA213 exhibit the highest MIITs curves. It will be seen later that the initial quench propagation velocities are slightly less for these magnets than for the others. Note also that the curve for DCA212 is higher than that for DCA213 even though they use similar conductor. This is due to a higher quench detection threshold used for DCA212 and so resulted in a longer delay between the start of quench and the time at which the power supply is shut off and the strip heaters are fired.

For DCA213, Fig. 1 also shows the MIITs as a function of current for the outer coil spot heater #3, and, as for the inner coil data, MIITs values increase with current and reach a peak value, but much closer to the conductor short sample cable limit, about 93%.

Studies were also done to determine if the magnet coils were self-protecting during quenches under conditions of maximum MIITs and, therefore, maximum hot spot temperature. This was done by inducing a series of quenches at the current of peak MIITs while increasing the time delay for strip heater firing after power supply shutoff. Figure 2 shows the results of this study for four of the magnets and plots MIITs vs strip heater time delay for each of the magnets at the current at which it exhibited peak MIITs during the previous study of MIITs variation with quench current. Except in the case of DCA213 inner coil data, the increasing MIITs numbers approach an asymptotic value and level off, as is expected in a self-protecting system. For DCA213, more data needed to be taken at longer time delays to determine the asymptotic MIITs value. For inner coils, the peak MIITs number reached during these studies corresponds to a temperature no higher than 302K, as calculated from Eq. 1, which provides an upper bound for hot spot temperature as a function of MIITs and magnetic field at the quench origin. These results show that the inner coil quenches generate hot spot temperatures below 500K and are within safe limits. For the outer coil quenches of DCA213, data at 6000A asymptotically approach a limit, as with the inner coil behavior, but this limit corresponds to a measured temperature of about 700K, which is considered to be above the safe limit. These results therefore indicate a need for strip heater protection for the outer coil conductor in DCA213.

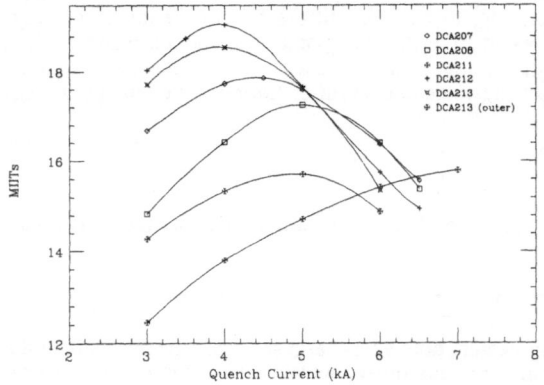

Figure 1. MIITs vs. quench current for magnets in spot heater studies. Quench detector threshold was 0.5V for DCA213 (outer), 4V for DCA212, and 2V for all others.

2. Quench Velocity

For the inner coil quenches with spot heater #2, the average quench velocity through the midplane turn straight section of cable was calculated by time-of-flight for the quench front to reach the tap closest to the heater. This was done for different currents and the results were similar for all five magnets. The average over the five magnets is plotted against current in Fig. 3. It should be noted that the average velocities for the two all-Kapton magnets were similar to those for the other magnets, whereas the initial velocities were less for the all-Kapton magnets.

Initial quench velocities were calculated with Eq. 2, using the slopes of the quench voltage signals. Data for DCA208 is shown in Fig.4, along with a theoretical curve of velocity vs. quench current calculated from a classical model that assumes no energy transfer by cooling and no change in magnetic field during the quench process[5]. The calculation also does not correct for temperature dependence of the specific heat and thermal conductivity of materials. In reality, some or all of these factors may be significant, and, for the all-Kapton magnets, this may be the case. Also, in some cases, the model may be valid only over the time span of the initial quench process. As with DCA208, the data fit the model predictions very closely in the case of the other epoxy-fiberglass wrap magnets; but, for the two all-Kapton magnets, the data fall below the model curves, as shown for DCA212 in Fig. 5. This implies that at least one of the assumptions in the model are not valid. In particular, this could reflect a larger cooling process than expected, and more cooling than in the epoxy-fiberglass wrap magnets.

Initial quench velocities for outer coil quenches were calculated by time of flight to the two closely bracketing voltage taps. Velocities ranged from about 5 m/s at 3000A to about 10 m/s at 7000A.

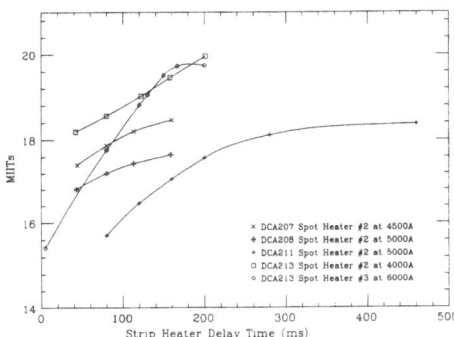

Figure 2. MIITs vs. strip heater delay.

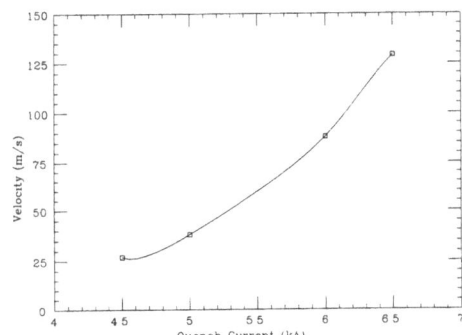

Figure 3. Average quench velocity vs. quench current for inner coil midplane turn.

3. Hot Spot Temperature

Since the outer coil spot heaters were the only ones which were closely bracketed by voltage taps, these were the only spot heaters for which we could directly measure the temperature of the quench hot spot and therefore quench studies with these heaters provided valuable temperature data for the determination of quench safety criteria in the superconductor. Also, the resulting MIITs-temperature correlation provided a check of the adiabatic MIITs model represented by Eq.1.

Due to problems with the quench detector circuitry, outer coil data from DCA212 is difficult to interpret and will not be presented here. For DCA213, hot spot temperature was determined by measuring the increase in resistance of the 10.16cm length of cable between the voltage taps which bracket the spot heater, as was described earlier, and using short sample cable data of resistance vs. temperature. Temperatures can be plotted vs. MIITs values for all quenches performed but quenches were done at different magnet currents and therefore at correspondingly different magnetic fields, which varied from 0.58T to 1.36T at the midplane, as current was varied from 3kA to 7kA, respectively. Since the MIITs-temperature correlation depends on the field at the quench location, all this data cannot be compared to a unique correlation curve generated by Eq. 1, since a different curve will result for each field at quench. However, since seven quenches were done at 6kA (1.16T) for one of the strip heater time delay studies, as was shown in Fig. 2, these data can be plotted against a theoretical MIITs-temperature curve calculated from Eq. 1 for that field. As can be seen in Fig. 6, the measured data points appear to diverge from the calculated curve as MIITs and T increase. This is to be expected in the case of an adiabatic model such as that used here. This curve can thus serve as an upper bound for temperatures if the MIITs values during a quench are known.

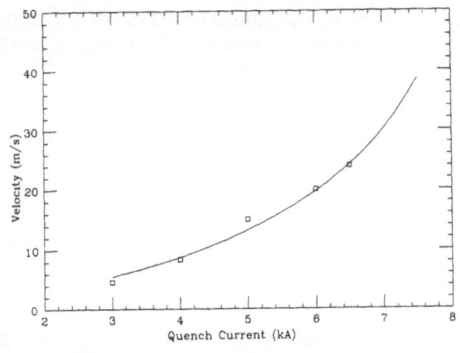

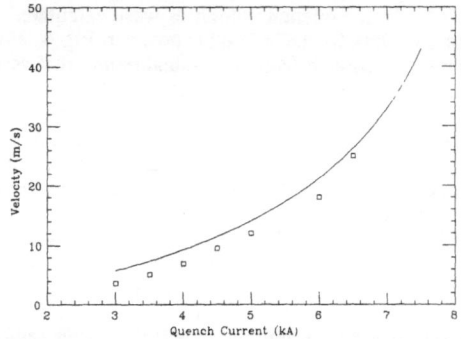

Figure 4. Initial quench velocity vs. quench current in inner coil midplane turn of DCA208. Solid curve is from model calculation.

Figure 5. Initial quench velocity vs. quench current in inner coil midplane turn of DCA212. Solid curve is from model calculation.

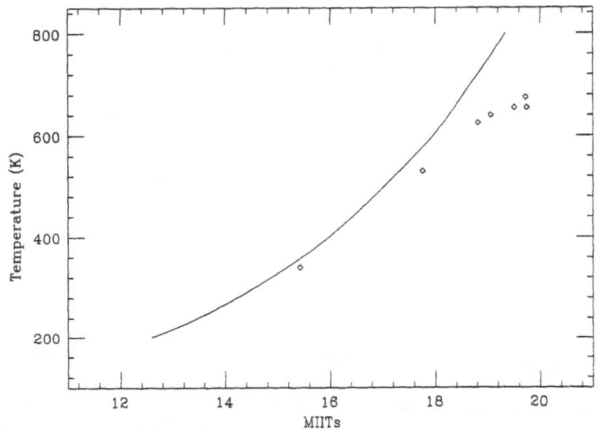

Figure 6. MIITs-temperature correlation curve from adiabatic model, with data from DCA213.

CONCLUSION

Since it is not usually possible to measure hot spot temperature directly but it is straightforward to calculate the MIITs numbers, these results show that correlation curves generated by the adiabatic model are still useful as upper bounds for MIITs values when temperatures cannot be measured. In order to get a more realistic fit to the data and be a better predictor of quench temperature, a model must take into account various effects, which include: 1) cooling by transfer of heat to the helium; and 2) changes in the magnetic field at the quench location during the time scale of the quench.

Also, this study has shown that the inner coils are self-protecting but that the outer coils, with the type of conductor being used, may not be and may require strip heaters or some alternate protection scheme in case of quenches.

REFERENCES

[1]   J.F. Muratore et al., "Construction and Test Results from 1.8 m-Long, 50 mm Aperture SSC Model Collider Dipoles," Supercollider 4, p.559, Plenum Press, 1992.
[2]   M. Garber, private communication.
[3]   N.M. Wilson, Superconducting Magnets, Oxford Clarendon Press, p.200, 1983.
[4]   A. Devred, MAGLIB Scientific Subroutine Library, Superconducting Supercollider Laboratory, 1992.
[5]   A. Devred, "General Formulas for the Adiabatic Propagation Velocity of the Normal Zone," IEEE Trans. Magnetics, 25(2), 1698 (1989).

PARAMETERS AND CONCEPTUAL DESIGN OF A VERTICAL BENDING MAGNET FOR THE SUPER COLLIDER

C. Goodzeit, R. Jayakumar, F. Nobrega, G. Snitchler, and R. Stiening

Superconducting Super Collider Laboratory*
2550 Beckleymeade Avenue
Dallas, Texas 75237-3997

INTRODUCTION

Special dipole magnets are required near the intersection regions of the Supercollider to bend the beams in the vertical plane so that they can be made to converge for collision. This paper presents the design requirements and parameters for one of these magnets (the BV1C splitter) along with the magnetic design and conceptual mechanical design for the cold mass.

FOREWORD

A high field (6.5 T), large aperture (87mm) dipole magnet is required to bend the Collider beams in the vertical plane to cause them to converge and then separate in the SSC Interaction Regions. The location of these magnets in the lattice is shown in Figure 1 and eight such magnets are required for the Collider. Presently the magnet is in the conceptual design stage and a program has been planned for its development. This paper presents: (a) the design requirements, (b) the parameters and conceptual design, and (c) outlines the development plans for the cold masses for these magnets.

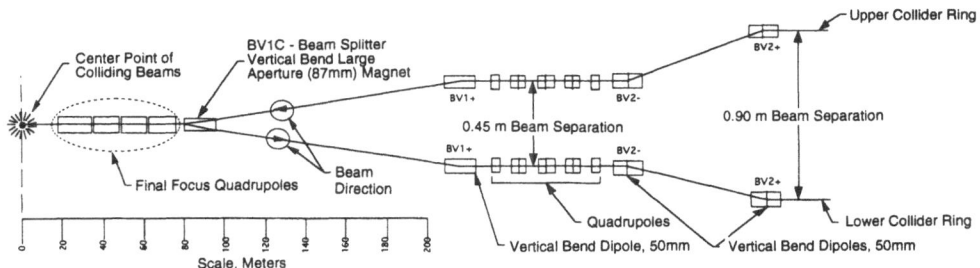

Figure 1. Position of BV1C- in the Lattice

* Operated by the Universities Research Association, Inc., for the U. S. Department of Energy under Contract No. DE-AC35-89ER40486.

DESIGN REQUIREMENTS

These magnets are related in performance to the Collider dipole magnets; however, they have to achieve this performance at the large aperture of 87mm. In general the requirements are as follows in Table 1.

Table 1. General Magnet Requirements.

Parameter	Requirements
1. Aperture	Between 85 and 100mm (87mm selected)
2. Magnetic Length	To match that of the CDM. Two versions may be required, one at 15m and the other @ 13m.
3. Central Field	6.5±5% at 6714A
4. Operating Temperature	4.2K
5. Multipole Tolerances	High field only. Obtained by scaling the high field CDM multipole tolerance to 2.5 cm radius from 1 cm radius. No specification for persistent current multipoles.
6. Energy Deposition in Coils	Synchrotron radiation: ≤ 0.69 W/m Beam Scattering: 2.90 W per magnet
7. Operating Margin	≥10% at high field in the presence of the above heat load.
8. Training Performance	No training quenches to reach full field. No more than 2 to reach 105% of full field.
9. Coil End Design	Design objective is to have zero net multipole content when averaged over the individual coil end regions.
10. Helium Containment	The helium containment assembly is to required to be inspected and stamped for conformance to the ASME Unfired Pressure Vessel Code, Div. 1.
11. Radiation Hardness	Coil and components ≥10^9 Rads without significant deterioration.

PARAMETERS AND CONCEPTUAL DESIGN

The cold mass uses a two-layer cosine theta coil design with an 87mm aperture. A detailed discussion of the magnetic design for this magnet along with the multipole requirements is given by Venkatraman, et. al[1]. Table 2 lists cold mass parameters while Table 3 shows the conductor characteristics used in this design. In order to satisfy the performance requirements for this magnet, the superconducting cable that is used for the inner coils of the CDM has been selected for both the inner and outer coils.

Table 2. Magnet Parameters

	Inner Coil	Outer Coil
Number of Turns	32	32
Number of Wedges	4	1
Inner Diameter, mm	87	112.7
Outer Diameter, mm	111.7	137.4
Calculated Critical Current, A	7522	8847
Operating Margin, B_0/B quench	11%	25%
Other dimensions:		
Width of Collar @ Outer Coil, mm		30
Inner Diameter of Yoke, mm		197
Outer Diameter of Yoke, mm		470

Table 3. Cable Characteristics

	Units	Value
NbTi, High Homogeneity, %Ti	%	47
Strand Diameter	mm	0.808
Twist Pitch of Strand	mm	13±.15
Copper: S. C.		1.3
Number of Strands per Cable		30
Width of Cable (Bare)	mm	12.34
Mid-Thickness of Cable (Bare)	mm	1.458
Keystone Angle	Deg.	1.2
Lay Pitch of Strands in Cable	mm	86
Filament Diameter	μ	6
Number of Filaments/Strand		8000
Critical Current Density in NbTi @ 4.2K and 7T	A/mm^2	1700
Critical Current in Cable @ 4.2K and 7T	A	10000

The two dimensional cross section of BV1C- is shown in Figure 2. In addition to the above features, the baseline design concept uses:

a. A horizontally split yoke with notches at the midplane for a weld backup strip. The yoke contains features for stacking and alignment at the 45° position which are used to obtain a "twist free" cold mass during the shell welding process. Standard bus slots and helium by pass holes are also present in the yoke. It is intended that these yoke lamination will be fine blanked from 6.35mm thick ultra low carbon steel.

b. High manganese steel collars which index in precision slots at the poles and have a line to line fit with the yoke. These collars will be stamped from 1.5 mm thick material and spot welded into left/right alternating pairs. Hard phosphor bronze keys having a taper angle of 3° are used to assemble the pre-stressed coils into the collars.

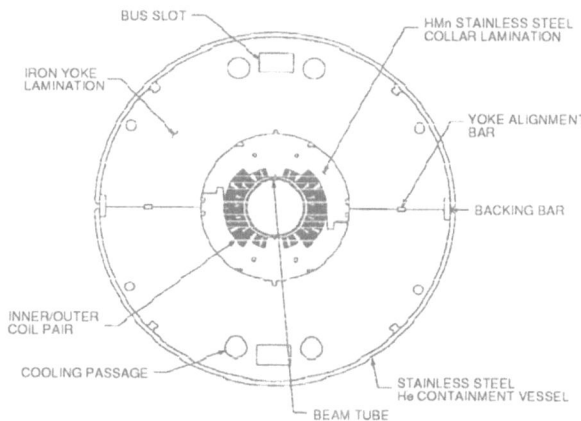

Figure 2. Two Dimensional Cross Section of the 87mm BV1C- Dipole

The baseline cable insulation is to double wrap the cable first with a 48-50% overlap layer of a 1-mil thick polyimide insulation such as Kapton® or Apical® without adhesive. The second layer is a butt wrap of the same type of film as used in the first wrap covered with a thin film of thermosetting adhesive such as Sheldahl 3PM® or Cryorad®. This system has been tested for punch through resistance and compressed insulation thickness indicating that it has about twice the punch through rating for the previously used insulation with fiberglass-epoxy outer wrap and the compressed insulation thickness is 4.25 mils per side.

The construction of the cold mass at the (lead) end of the magnet is shown in Figure 3. The end plate is welded to the inside of the shell and supports the ends of the coils. In the non-lead end of the magnet, coil ends are loaded with four end force transducers which are adjusted to the required end load of about 20 KN. Four set screws are used in the lead end to apply a similar load. Other details of this construction include a ramp section of the inner coil first turn which is contained in an insulated section clamped in the collars at the end of the straight section as shown. This cable is solder spliced to the outer coil first turn at the radius of the outer coil. The upper and lower coils are solder spliced together using the outer coil leads on the outside of the end plate and encapsulated in the insulated retainer. The inner coil leads continue into the end region where they are connected to the power busses. The 6.35mm thick ultra low carbon laminations terminate at the end of the straight section of the magnet. The collared coil is then supported radially by 3mm thick stainless steel lamination which extend over the saddle end region of the coil to reduce the peak field effect at the ends. Helium containment is provided by the end dished heads which are welded to the shell and by the beam tube which passes through the inner coil aperture and it welded to the end domes. One end of the beam tube is supported from the end dome by a bellows to take up the differential contraction in case of a large transient temperature difference that might occur between the beam tube and the rest of the cold mass.

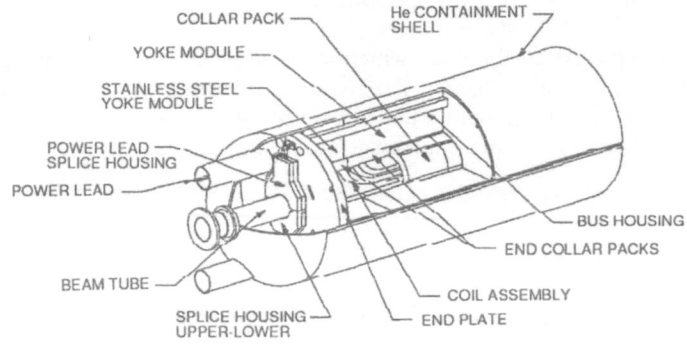

Figure 3. Lead End of the BV1C- Dipole Cold Mass

PROGRAM PLAN

One lot of 4 model magnets, each 1.8 meters long is scheduled to be completed and cold tested by July 1995. The prototype magnet phase begins after the completion of the model phase in July 1995. This phase consists of one prototype cold mass and cryostat and is scheduled to be completed in December 1996. The cold test measurements of the model and prototype phases will be used for the final magnetic iteration if needed in the production phase that will begin in July 1997. The production phase consists of eight magnets.

REFERENCES

1. V. Venkatraman, et, al., "Magnetic Design Considerations for the SSC Vertical Bending (BV1C) Magnet", IISSC 1993, Plenum Publishing Corporation, 1993.

RELIABILITY OF THE ELECTRICAL INSULATION SYSTEM FOR THE HDM

J. F. Roach,[1] S. K. Singh,[1] A. G. McConnon,[2] S. B. Suttle,[2]
and K. L. Kussmaul[2]

[1]Westinghouse Science and Technology Center
1310 Beulah Road
Pittsburgh, PA 15235

[2]Westinghouse Magnet Systems Division
I-H 35 North and Westinghouse Road
Round Rock, TX 78680

INTRODUCTION

The Westinghouse Electric Corporation (WEC) is under contract to design and build the High Energy Booster dipole magnets (HDM's) for the SSCL through low rate initial production (LRIP). The first phase of the HDM program is the fabrication and test of short 1.8 m HDM model magnets designed by the SSCL. This technology transfer phase is well underway with the delivery of the first WEC built HDM model magnet, DSB701 to the SSCL and the completion of the test program conducted at the SSCL Superconducting Cable and Magnet Test Laboratory (SMCTL) in April of this year. Model magnet DSB702 has been delivered to the SSCL for testing in May and June of this year, and DSB703 is nearing completion. This paper presents a summary of lessons learned from electrical insulation problems discovered during the DSB model magnet program. Insulation reliability and risk issues impacting the HDM prototype design being developed by WEC are discussed.

HDM ELECTRICAL INSULATION REQUIREMENTS

The HDM electrical insulation design voltage rating requirements in the SSC HEB Dipole Magnet Specification[1] are presented in Table 1. The voltage ratings are defined for dielectric performance of the HDM in an atmospheric air environment. The insulation safety margins under actual operational conditions in gaseous and liquid helium must be verified for the electrical stresses induced by the design voltages in Table 1 to correctly assess insulation reliability and performance. The maximum leakage current requirement at rated voltage is 25 µA and must be met "during the useful life of the magnet."

The radiation resistance of the insulation materials for the magnet coils and beam tube is a critical design issue. The HDM is required to withstand a radiation dose of 10 MGy at the inner radius of the inner coil in the horizontal plane and 1.67 MGy at the outer radius of the outer coil over the 25 year life of the magnet. The radiation dose at the beam tube insulation and dielectric support structure may be even higher than 10 MGy. The SSCL has an ongoing program to evaluate the radiation resistance of materials under consideration for use in the SSC.[2,3] The evolving SSCL radiation resistance data base is being used by WEC as a guideline for HDM prototype insulation material acceptability with respect to radiation performance. Reliable data on radiation resistance is essential to the sucess of the HEB operation, and WEC is working with the SSCL and key vendors to ensure that dielectrics are adequately tested and evaluated.

Supercollider 5, Edited by P. Hale
Plenum Press, New York, 1994

Table 1. Insulation ratings in air for the HDM.

Component	Reference Component	Minimum Rated Voltage (kV)
Coils	All Others Except Quench Heaters	5
Connected Bus	All Others Except Coils	5
Thru Bus	All Others	5
Coils	Adjacent Coils	3
Coil Turn	Other Coil Turn in Same Coil	3
Coils	Quench Heaters	2 + 0 / -0.5
Quench Heaters	Collars/Yokes/Shell	2

HDM MODEL MAGNET INSULATION EXPERIENCE AND LESSONS LEARNED

The DSB series model magnets were designed by the SSCL and fabricated by WEC. The design is a coil-on-coil configuration which uses a G-10 intercoil spacer and internal ramp splice confirguration. The cable / wedge insulation is Kapton / fiberglas prepreg (Hexel F185) and the groundwall insulation uses 0.127 mm Kapton sheets, including Kapton replacement for the quench heater. The insulation design and verification analysis is discussed in Reference 4. During the fabrication of the model magnets several insulation problems were encountered, solutions implemented, and lessons learned documented. Three such issues are discussed:

Issue 1: Inner coil shorting to mandrel during winding - During winding around tight end turns, usually closest to the pole where bend radius is smallest, insulation on the ID of cable is sometimes torn. Megger testing of coil-to-mandrel detected shorts. Since for model magnets insulation damage in this area is of not consequence, Megger testing was discontinued for coil-to-mandrel. However, this issue must be examined further during prototype development because insulation damage in this area of the coil could lead to faults for fully assembled magnets with beam tubes.

Issue 2: Pealing off of wedge OD insulation during liner removal following cure - A stainless steel liner is placed over coil after winding to provide a protective interface between the coil OD and curing mold. The liner ID was initially sprayed with Frekote 700-NC releasing interface for a non-stick surface for damage-free removal of coil after cure. Although this worked for most of the interface region, portions over the wedge area peeled off in some cases anyway, believed due to high wedge friction. A surface sealant, Frekote B-15 Mold Release, is now used on the ID of the liner to seal any surface imperfuctions prior to spraying with Frekote 700-NC. This has allowed the releasing interface to be more effective and has virtually eliminated insulation tearing.

Issue 3: Hipot failure of coils-to-ground after shell welding - An outer coil lead which exits the magnet at the midplane grounded to the brass collaring shoe which protruded abnormally beyond the end of the collared coil assembly. The OD of the coil lead abraded against the sharp edge of the collaring shoe and shorted. Steps were taken in the manufacturing process to ensure proper axial location of the collaring shoe, preventing its protrusion out of the ground wall insulation.

Minor insulation problems were also encountered in the form of "nicks" in the cable insulation and "flashing", that is surface roughness on midplane and pole surfaces. Small bare conductor regions were repaired with varnish. "Flashing" was sanded off with fine grit paper. The complexity of end turn regions, as well as springback mechanics led to turns separating from end parts and subsequent end turn insulation damage. Minor "nicks" of tears in the groundwall insulation were repaired with adhesived Kapton tape. These insitu insulation repairs and methods, some learned from WEC experience during the CDM technology transfer magnet builds at BNL, must be throughly reviewed during the HDM prototype phase, and strict procedures established for corrective actions with respect to noncompliance of insulation.

MODEL MAGNET IN-PROCESS DIELECTRIC TESTING

In-process dielectric testing is an integral part of the HDM model magnet manufacturing process. The principle electrical tests include: insulation resistance measurements for integrity of groundwall and midplane insulation; coil resistance, inductance, and quality factor for turn-to-turn integrity; coil ringer (impulse) for turn-to-turn integrity; and Hipots for groundwall, midplane, and quench heaters (in prototypes). The dielectric tests and test sequence for in-process insulation verification is based on existing practices at the National Laboratories and lessons learned during the WEC model program. Our test philosophy is to detect any nonconformities as early as possible in the manufacturing cycle to allow for timely and cost effective corrective actions and to evaluate the impact on subsequent operations. However, the frequency of high voltage testing (Hipot and ringer) for both the model magnet program and the prototype phase will be limited to avoid unnecessary magnet damage, without compromising the assessment of insulation reliability. Optimization of in-process high voltage testing strategies will be an important issue to be resolved during HDM prototype development.

Typical DSB model magnet dielectric test results are as follows. The insulation resistance using 1000 VDC is typically > 2000 MΩ and leakage currents < 0.3 μA at 3 kV Hipot for coils and voltage taps to ground as well as for upper to lower coils at the midplane. The HDM model magnet Hipot testing maximum voltage was limited to 3 kV as recommended by WEC and accepted by the SSCL as the result of a potential creep stress problem in the SSCL groundwall insulation design.[4] No dielectric tests were planned for the cold test program of the HDM models by SSCL at the SMCTL In the case of HDM prototypes, WEC plans for cold dielectric testing after cool down prior to magnet energization; dielectric testing will follow thermal cycles as well. The WEC Accelerated Degradation Testing (ADT) plan will include dielectric testing at stress levels sufficient for assessment of long term insulation system reliability as a function of thermal and magnetic cycling.

CABLE AND WEDGE INSULATION FOR HDM PROTOTYPES

The baseline cable and wedge insulation system for the HDM prototype is all polyimide film using thermoset adhesive with optional organic or heavy metal catalyst to reduce cure temperature requirements.[5] This choice was made primarily on the basis of reducing risk to ramp rate sensitivity, multipole tuneability impact, radiation resistance, dielectric punch-through, end part compatibility, and manufacturability. The specific insulation configuration is 0.0254 mm film (Apical or Kapton) halfed lapped on the cable/wedge followed by a halfed lapped 0.019 mm film with 0.005 mm thermoset adhesive on both sides. The baseline adhesive is Cryorad with catalyst and a 135 C cure temperature. Optional adhesives under consideration are uncatalzed Cryorad, Sheldahl 3PM, and CTD-105 all with nominally 185 C cure. The DuPont XCI / XRCI all polyimide system with aliphatic XMPI to reduce the cure temperature to 185 C and reduced base film thicknesses tailored to the HDM design is being considered as an alternative These 185 C cure systems will be evaluated and selected only if they have satisfactory ramp rate dependence, which is a critical reliability and performance issue for HDM magnets.

The HEB cycle is bi-polar with ramping to full field (6700 A) at 62 A/s. Prototype magnets built at FNAL and BNL to date have exhibited significant ramp rate dependence at this ramp rate requirement. WEC is performing a ramp rate sensitivity study to solve the problem for HDM's. One of the variables being studied in this program is the coil insulation curing cycle. Testing performed at KEK[6] has shown a direct correlation between AC losses and coil curing temperature and pressure. Since increased cure temperatures lower cable interstrand resistance, enhancing AC losses and multipoles associated with eddy currents during ramping, the HDM curing temperature was selected to be 135 C with upper limit of 150 C to reduce the risk of poor ramp rate performance. Insulation systems that support this cure temperature decision include Kapton / fiberglass used in the DSB model magnets or all polyimide film with catalyzed thermoset adhesive. The Kapton / fiberglass system was not considered for production HDM magnets because of evidence that its punch-through resistance is marginal to support high dielectric reliability.

The systematic multipole requirements for the HDM[1] will be difficult to meet without tuning the cold mass cross section design before LRIP production. One of our tuning strategies is to optimize the cross section as a function of compressed azimuthal insulation thickness range so that the need for shimming is removed while still achieving an acceptable pre-stress range to be determined. SSC Magnet Design Software, SSCMAG was applied to the baseline HDM cross section with insulation thickness builds (t) varied from 50 to 95 μm on the inner coil with outer coil insulation assumed 5 μm larger due to lower prestress; tuning can be achieved until wedge subtended angles reach the design limit of 0.4° on inner/0.6° on outer, that is wedges are not permitted to have any sharper points. Figure 1 shows the results of this analysis. Multipoles can be tuned for insulation builds less than 87/92 μm inner/outer coils. Based on this analysis, a target build size for the HDM baseline insulation was selected to be 72/77 μm inner/outer coils which allows for tuning margin to thicker insulation. The baseline HDM insulation system described achieves this design target by using reduced base film thickness of 0.019 mm for the outer adhesived layer as described above. The DSB model magnet Kapton/fiberglass system insulation build is 84/89 μm. The all Kapton CI/RCI system for the CDM magnets has a nominal insulation build of 89/94 μm.

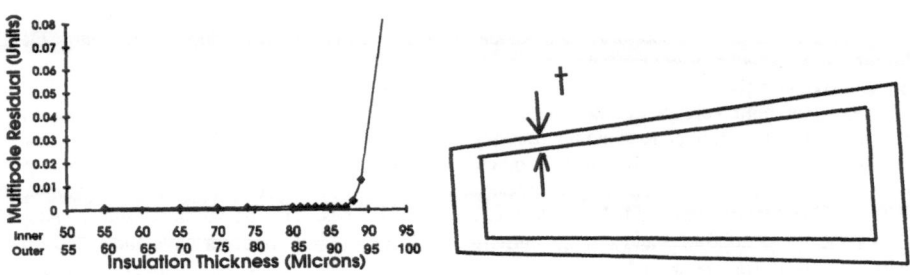

Figure 1. Multipole tuneability as a function of cable insulation thickness.

The selection of all polyimide film with thermoset adhesive for the HDM baseline insulation system is recognized as a risk and a development testing program is being formulated to mitigate these risks. The baseline insulation system has been used in only a few long dipole magnets and therefore, has limited data base for reliability assessment. However, this insulation system is being considered for the SSCL QSE coils and the 85 mm speciality dipoles, and these programs are providing insulation performance characteristics which will be evaluated for applicability to the HDM design.[7] Principle reliability concerns are the mechanical and dielectric characteristics of polyimide films less than 0.0254 mm thick, specifically for the 0.019 mm film with thermoset adhesive baselined. Pinholes in the base film and punch-through resistance are major issues. Also, catalyzed adhesives are expected to have reduced radiation resistance, and must be tested to ensure 10 MGy capability with respect to their function in the coils. There is some concern also with respect to thermoset adhesive flow which could find its way on the the bare cable surface as well as spill over on to curing tooling. For example,the viscosity of Cryorad adhesive drops rapidly at its cure temperature is approached which can lead to undesireable flow properties.[7] An Insulation System Development Test Plan is underway to address manufacturing issues, coil size and modulus, and dielectric reliability of the proposed polyimide/thermoset insulation system. The WEC model magnet test facility will be used to wind 1.8 m inner coils to evaluate the performance of the HDM prototype baseline insulation system as well as the XCI/XRCI system alternate as tailored for the HDM design.

CONCLUSIONS

Insulation reliability of the HDM is critical to the sucessful performance of the HEB ring. DSB model magnets are being limited to 3kV Hipot due to a potential creep path deficiency in the groundwall insulation design. The baseline HDM prototype cable and wedge insulation system is all polyimide film using a thermoset adhesive with low curing temperature to reduce the risk of ramp rate sensitivity. The insulation system targets cable insulation thickness for flexibility in multipole tuneability. Critical reliability issues for performance of the baseline insulation are radiation resistance, impact of curing cycle on ramp rate, and dielectric performance of reduced polyimide films.

REFERENCES

1. SSC HEB Dipole Magnet Specification, M80-000045A, January 15, 1993.
2. Cable Insulation Working Group, SSCL MSD, Dallas, TX, February 5, 1992.
3. A. Spindel, SSCL Radiation Program, MSIM, General Dynamics, Hammond, LA, April 20, 1993.
4. J.F. Roach, S.K. Singh, and A.G. McConnon, Design and issues associated with the HDM electrical insulation system, Paper IV-9, 5th IISSC.
5 Cable and Wedge Insulation Trade Study, M3C-400011, WEC, February, 1993.
6. T. Shintomi and Y.Z. Lei, AC loss measurements of the SSC conductors at KEK.
7. Private communications with F. Nobrega and S. Krishna at SSC MSD.

A STUDY OF VARIATIONS IN DIPOLE CABLE INSULATION SYSTEMS AND THEIR EFFECT ON CREEP

Richard E. Sims,[1] Rodger Bossert,[2] Jay Hoffman,[2] Pamela Schmidt,[2] Amanda Spindel,[1] and James Strait[2]

[1]Superconducting Super Collider Laboratory[*]
2550 Beckleymeade Ave.
Dallas, TX 75237

[2]Fermi National Accelerator Laboratory
Batavia, IL 60510

INTRODUCTION

In late 1991 the Fermilab Manufacturing Group was directed by the Superconducting Super Collider Laboratory (SSCL) to investigate, build and test a series of magnets utilizing a low temperature cured all polyimide insulating system. Since an all polyimide insulated system was suspected to have higher post- collared stress relaxation than the existing glass tape system, a study was initiated to find ways to minimize the stress relaxation.

For the study six different polyimide films possessing different modulus characteristics, coated with either 3M type 2290 Epoxy or Allied Signals Cryorad[†] were wrapped onto five inch pieces of 50 mm inner cable in various combinations and in various build-ups. Ten of these wrapped cable segments were made into vertical stacks and cured with a low temperature cure cycle of approximately 290°F. The curing cycles were carefully reproduced for each sample. Samples were run with constant cure pressures of both 6 K psi and 10 K psi. These stacks were then measured for creep at a constant pressure of 12 K psi for one hour and for modulus of elasticity using the size delta between 8 K psi and 12 K psi. The results of this data as compared to the glass tape base line are presented here.

In the graphs and charts in this paper the legend code is as follows: i.e., HA5, 2/3HA, BLT means a stack named HA5 was composed of a first layer containing a 2/3 overlap (3 layers) of type HA film and a second layer of butt wrapped type LT film. Later in the study the name became the description plus a dash number (i.e., the above stack would be HA3LT-1). When the lower layer contained glue the letter E for epoxy or C for Cryorad was added in.

Figure 1 is a plot of "equivalent coil creep" versus cure pressure. Equivalent coil creep is obtained by multiplying the measured stack creep by 2.033 which adjusts the ten material layers of the stack data to approximate the layers in a curved inner 50 mm dipole coil. (It is derived from the 19 turns of cable and insulation plus the insulation only on the wedges). This scale adjustment allows a rough indication of how the various wrap combinations would perform in a coil configuration.

---

[*] Operated by the Universities Research Association, Inc., for the U.S. Department of Energy under Contract No. DE-AC35-89ER40486.
[†] "Cryorad" and "Apical" are trade names of Allied Signal. "Kapton" is a trade name of Dupont. "3P" is a trade name of Sheldahl.

Supercollider 5, Edited by P. Hale
Plenum Press, New York, 1994

The number of layers of polyimide insulation around each cable is indicated down the middle of the graph. The legend identifies the two stacks cured at 6K psi and 10 K psi used for each graph line as well as the shorthand code for the layer identification.

The following indications can be drawn from Figure 1:

1. The glass tape baseline has a low creep compared to this selection of layer combinations.
2. Lower numbers of layers of polyimide produce less creep.
3. The creep is generally lower when the first layer is composed of a higher tension modulus material such as type H Kapton (Dupont) or type N Apical (Allied Signal).
4. Lower number of layers and higher tensile modulus materials are less effected by the 6K psi to 10K psi curing pressure variation.
5. Increasing the number of layers which contain glue decreased creep greatly (the three glue layer systems are only shown as 10K psi points).
6. Bare cable stacks have less creep when "cured" at 6K psi than at 10K psi, the opposite of insulated stacks. This is not well understood but may be a mechanical effect caused by friction of the polyimide film on the individual cable strands not allowing the strands to move in the length direction. The bare cable was surrounded by steel mold during the cure measure.

The strongest effect found in these studies was that the number of layers of insulation that were glued together had a strong effect on coil equivalent creep and on the stack modulus of elasticity (as shown in Figures 2 and 3, respectively). The scatter in theses graphs is because the number of layers and combinations of materials are varied in each data point. The lines shown are least-squares curve fits.

Some of the scatter was cleared by plotting the first layer grouped by their brand names, Kapton and Apical. Since the two brands of polyimide film are made by different processes, the film characteristics are somewhat different. The author suggests that these differences are measurable by the tensile modulus. Figure 4 displays the equivalent coil creep of a narrow selection of stacks with either one or two glue layers cured at 10K psi mold pressure versus the tensile modulus of the material of the first layer. This display generally shows that the higher tensile modulus films yield lower creep results.

Studies of stack modulus do not predict coil modulus directly because the curved shape of the coil and keystoned shape of the cable distribute the pressure very unevenly compared to the even pressure distribution in a vertical ten stack. However, the stack modulus may be used to rank the effects of the various layer combinations (see Figure 5).

Figure 6 displays a selection of stacks with various film layer and glue layer combinations ranked in order of creep. A legend has been added indicating the number of glue layers within the insulation layers. The lowest creep stacks have glue on the first layer and the second layer. Microscopic studies were made of a dissected stack made with Cryorad on the first and second layers. Because the Cryorad (as well as epoxy) viscosity drops to a watery state before hardening and the stack is under great pressure during cure, the glue was squeezed onto the superconducting cable at the underside of the wrap line. The amount of glue was small (approximately 0.002 to 0.020 inches in diameter puddles and loose spheres) but any glue in contact with superconducting cable is thought to be a possible source of quenching.

CONCLUSION

In conclusion, this study suggests for lowest stress relaxation, the preferred cable insulation system should have the fewest number of layers of insulation (three is practical) which are high tension modulus polyimide films. The system should be cured at the highest practical mold pressure for the longest practical time. In order to reach a compromise in obtaining low creep without using glue on the bottom wrap it is suggested that the top layer have glue on both sides and the bottom layer have no glue. This would constitute a two glue layer system that should come close to the attributes of a three glue layer system. The resultant coil system should have a stress relaxation of only slightly greater than the glass tape system. This choice was pursued by Fermilab on one of the low temperature cured study magnets made in 1992.

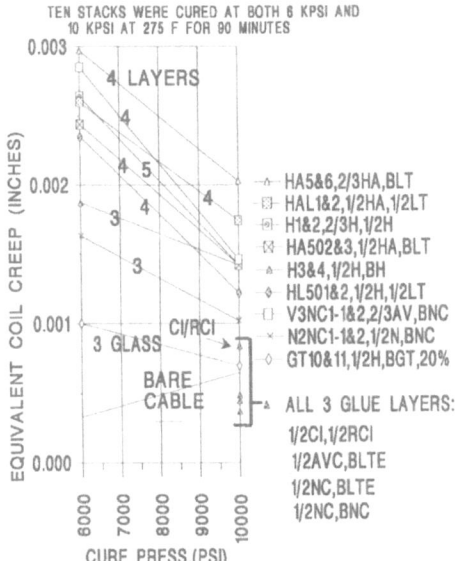

Figure 1. Coil Equivalent Stack Creep vs. Cure Pressure.

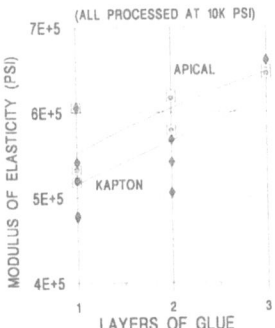

Figure 2. Effect of # of Layers of Glue on Coil Equivalent Creep.

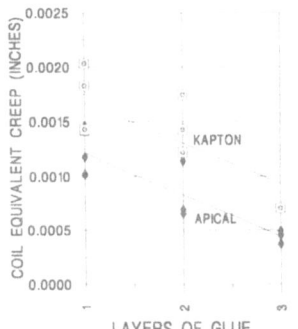

Figure 3. Apical and Kapton Stack Modulus of Elasticity vs # of Glue Layers.

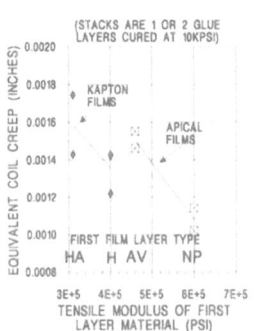

Figure 4. Equivalent Coil Creep vs Film Tensile Modulus.

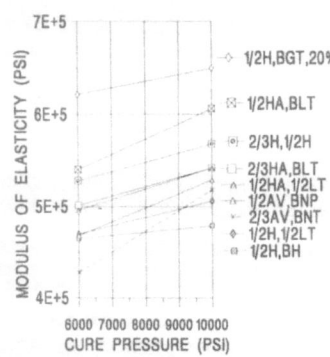

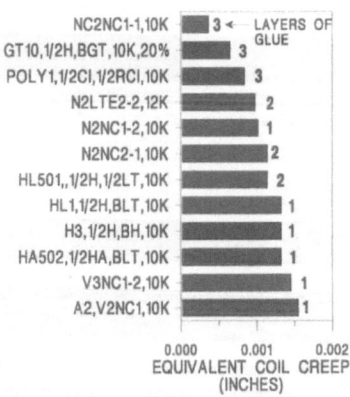

Figure 5. Modulus of Elasticity
vs. Curing Pressure for Various Films.

Figure 6. Comparison of Various
Layering Schemes.

REFERENCES

1. J. Strait, Memo, "Proposed Insulation Systems for Remaining Short and Long SSC Dipoles" FNAL, Technical Document TS-SSC-92-019, Feb. 5, 1992.
2. R.E. Sims, Slides, "Cable Insulation Workshop," Feb. 5, 1992, SSCL, FNAL Technical Document TS-SSC 93-004, Feb. 5, 1992.
3. J. Strait, "Cable-Pair Insulation Breakdown Test Results," FNAL Technical Document TS-SSC 92-065, May 26, 1992.
4. Various Authors, "Review of Fermilab Short Magnet Dipole Test Results," FNAL Technical Document, TS-SSC-92-087.

A STUDY OF STRUCTURAL INTEGRITY OF THE SSC
MAGNET SIX-STRUT SUPPORT SYSTEM

C. J. Chang, S. Chen, A. Jalloh, and D. M. Martin

Magnet Systems Division
Superconducting Super Collider Laboratory
2550 Beckleymeade Avenue
Dallas, TX 75237

INTRODUCTION

In the preliminary SSC design configuration, the overall magnet cryostat assembly is supported externally at three mechanical positioning points by two stands attached to the collider tunnel floor. Differing from the SSC design concept, a well designed and simply operated six-strut support system has been used at the Lawrence Berkeley Laboratory/Advanced Light Source (LBL/ALS) successfully.[1] In brief, the struts are equipped with ball joints at each end, and the struts are so arranged in a manner of three vertical, two lateral, and one longitudinal directions that provide the minimum constraints necessary to support the ALS system in the accelerator tunnel. Because the struts are made of differential screws which are adjustable in length, the six-strut system also provides minute controllability for cryostat alignment purposes. Also, the six-strut support system keeps the magnet from direct exposure to the accelerator tunnel floor which may lower the potential risks for the magnet from the ambient vibration and ground motion.[2] Because of the mentioned merits, the six-strut support configuration is being considered as an alternative to the SSC magnet support system.

This paper presents a qualitative study for the structural integrity of the SSC six-strut support system. Several six-strut design configurations are investigated.

THEORETICAL BACKGROUND

The six force equilibrium equations required for a three dimensional system are

$$\Sigma F_x = 0 \qquad \Sigma F_y = 0 \qquad \Sigma F_z = 0$$

$$\Sigma M_x = 0 \qquad \Sigma M_y = 0 \qquad \Sigma M_z = 0$$

*Operated by the Universities Research Association, Inc., for the U.S. Department of Energy under Contract No. DE-AC35-89ER40486.

where the summations imply the inclusion of all force and moment components in the specified coordinate system. The above equations can be deduced to a fact that at least six reactions are required to maintain overall equilibrium of a body in space, which is the basis for the six-strut support system design.

BASELINE SIX-STRUT DESIGN

The SSC magnet baseline six-strut support system design configuration uses three vertical, two lateral, and one axial struts to maintain the overall cryostat assembly in the collider tunnel space. Two of the three vertical struts are placed on one side near both ends, and the third vertical strut is placed at the midpoint of the opposite side of the cryostat shell, which provide the vertical movement, axial rolling, and pitching controllability for the cryostat. The two lateral struts are connected to the attachment bosses on the bottom shell near both ends of the cryostat to control the lateral movement and yawing of the cryostat assembly; the axial strut is connected to the middle attachment boss on the cryostat bottom shell to control the cryostat axial movement.

Figure 1 shows the cryostat finite element analysis result for a typical SSC 15 meter long dipole magnet for the baseline six-strut design configuration.

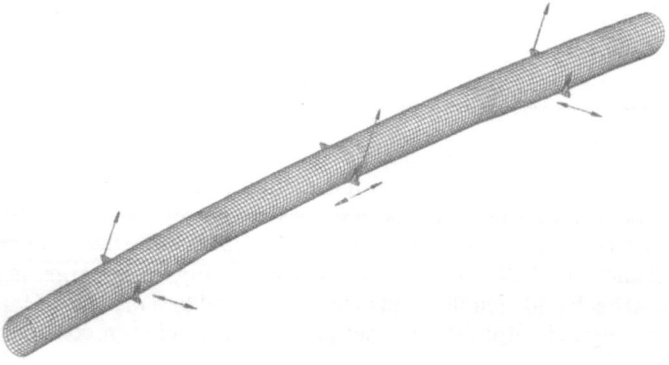

Figure 1. Cryostat deformed shape for the baseline six-strut configuration for a typical SSC 15 meter long dipole magnet.

Note that the cryostat has a twist, toward the opposite direction of the vertical struts, which results from the couples produced at the vertical-strut lug positions. Analysis results also indicate that the cold mass is shifted laterally and has a roll, following the cryostat deformed pattern, which increases the cold mass deflection. Also, the cold mass support posts are no longer in the original upright positions and remain in a bending state, that may cause the support posts to buckle during the operating life of the magnets. Although, by relocating the vertical struts along the cryostat axis it is possible to reduce the cryostat twist, the twist presents a significant cryostat design effort to meet the SSC magnetic alignment requirements.

ALTERNATE SIX-STRUT DESIGNS

Three-Symmetrical-Vertical-Strut System

To avoid the cryostat twist, as seen in the baseline six-strut design, one alternative is to place two of the three vertical struts symmetrically at both sides near one end, and the third

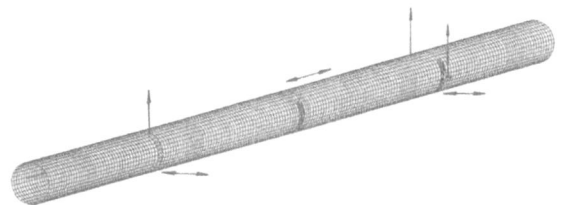

Figure 2. Alternate six-strut configuration with three vertical struts.

vertical in the centerline of the other end of the cryostat, as shown in Figure 2. As in the baseline design, two lateral struts are used to control the lateral movement and yawing, and one axial strut is used to control the axial movement of the cryostat.

Two-Symmetrical-Vertical-Strut System

Another alternate six-strut configuration is to use two vertical, three lateral, and one axial strut to support the cryostat. The two vertical struts are placed in the centerline of the cryostat to control vertical movement and pitching; the three lateral struts are used to control the lateral movement, axial rolling, and yawing; the longitudinal strut is to handle the axial movement of the cryostat.

Cryostat Deformed Shapes

For the two above mentioned alternate six-strut configurations, because the space of the SSC collider tunnel is limited, the upper cryostat assembly must be suspended from the top and the lower cryostat supported at the bottom. Finite element analysis results for the upper and lower cryostats supported with two vertical struts are shown in Figures 3 and 4, respectively.

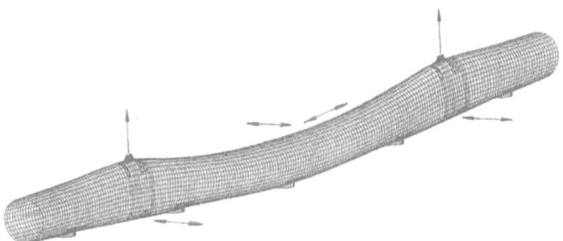

Figure 3. Upper cryostat deformed shape with two vertical struts suspended from the top.

The upper cryostat shell is pulled into a thin oval shape and the lower cryostat is compressed into a fat oval shape at the vertical-strut lug position. Similar deformed shapes were observed for the upper and lower cryostats (not shown in the paper) for the alternate three-vertical-strut design configuration.

Although these two alternate six-strut configurations avoid the cryostat twist, as seen in the baseline six-strut design, the significantly different cryostat deformations for the upper and lower magnets in the collider tunnel still present complexity for alignment purposes.

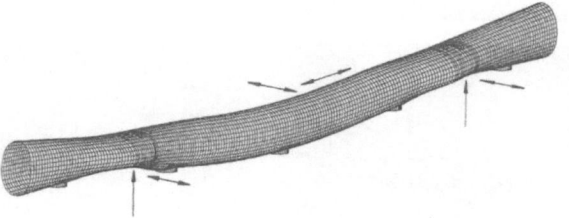

Figure 4. Lower cryostat deformed shape with two vertical struts supported at the bottom.

MODIFIED SINGLE VERTICAL-STRUT ATTACHMENT

To eliminate the cryostat shell deformation difference between the upper and lower magnets in the collider tunnel, the single vertical-strut attachment to cryostat for the above mentioned alternate six-strut configurations may be replaced with an adapter connected to both sides of the cryostat shell as shown in Figure 5.

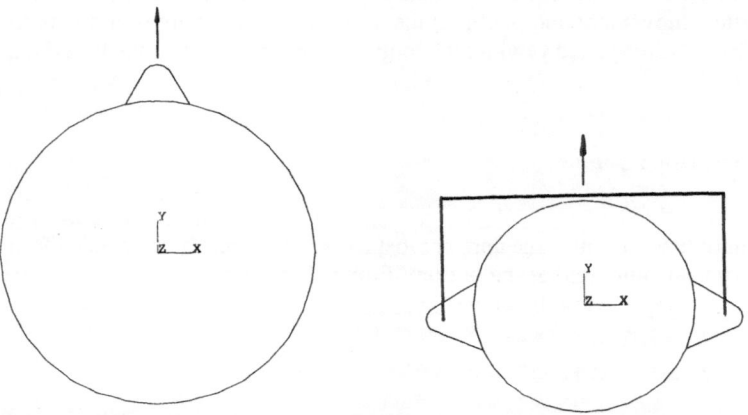

Figure 5. Illustration of a modified attachment for the single vertical-strut to cryostat.

SUMMARY

The SSC magnet baseline six-strut design configuration results in a cryostat twist, which presents significant design efforts and magnetic alignment complexities. To avoid the cryostat twist, two possible alternate six-strut configurations are examined first; after that, a modified attachment concept for the single vertical-strut to cryostat is proposed, which may provide a direction for the SSC magnet six-strut support system design consideration.

REFERENCES

1. T. Lauritzen, SSC Laboratory six strut support system talk, November 12, 1991, presented at the Superconducting Super Collider Laboratory, Dallas, Texas.

2. P. A. Manning, and R. B. Burdick, Dynamic Analysis of a Six Strut System, June 19, 1992, presented at the Superconducting Super Collider Laboratory, Dallas, Texas.

SSC 50mm DIPOLE MAGNET CRYOSTAT THERMAL MEASUREMENT RESULTS

W.N. Boroski, T.H. Nicol, M.K. Ruschman, and C.J. Schoo

Fermi National Accelerator Laboratory
P.O. Box 500
Batavia, IL 60510 USA

INTRODUCTION

A prototype Superconducting Super Collider (SSC) 50 mm dipole magnet cryostat, DCA323, was instrumented at Fermilab and delivered to the SSC Laboratory for installation into the accelerator systems string test facility. In series with other magnets, the instrumented cryostat will be used to quantify and verify cryostat thermal performance with respect to design requirements. Prior to leaving Fermilab, DCA323 was subjected to magnetic testing at the Magnet Test Facility (MTF). This presented an opportunity to obtain preliminary thermal performance data under simulated operating conditions. It should be noted that measurements of overall cryostat thermal performance were not possible during the MTF measurements as the magnet test stands are designed for magnetic rather than thermal testing. They are not designed to limit heat inleak to the ends of the cryostat, which has been shown to have a significant effect on overall measured thermal performance[1]. Nonetheless, these measurements do offer insight into the performance of several of the cryostat components and sub-systems.

CRYOSTAT INSTRUMENTATION

DCA323 was instrumented with 127 transducers of various type to measure temperature, strain, and vacuum. Thermometers were installed on the inner and outer thermal shields, on one support post assembly, internal to the inner and outer 80K MLI blankets, and on the cryostat vacuum shell. Strain gages were installed on each support post to measure axial loads applied during thermal cycling. Vacuum gauges were installed in the vacuum shell to measure system pressure. A comprehensive discussion of the instrumentation is not possible here given space constraints. However, information on exact transducer placement, transducer types, etc., is detailed in an internal Fermilab document[2].

DATA ACQUISITION SYSTEM

A PC-based data acquisition system monitored transducer outputs on a continuous basis, logging readings to file on 30-minute intervals. The stand-alone data acquisition system consisted of an IBM-compatible computer connected to a data acquisition (DA) unit via an IEEE-488 interface. All transducer outputs were connected to the DA unit. Constant current sources provide excitation for the transducers: separate current sources were used for each transducer type. Current polarity was reversed and averaged for each measurement to eliminate thermal EMF effects.

THERMAL PERFORMANCE RESULTS

Thermal Cycling

DCA323 was subjected to two thermal cycles during testing at MTF. A thermal cycle is defined as a cooldown to operating temperature and subsequent warm-up to ambient temperature. Of particular interest is the length of time required for the MLI blankets to equilibrate during transient conditions. Figure 1 shows the change in temperature of an inner layer of the 80K MLI blanket stack from the start of cooldown #1. Layer 32 was selected as it is near the middle of the 64-layer blanket stack. Note that even after 170 hours, the temperature of the layer is changing, albeit at a very slow rate ($\approx$ 70 mK per hour). For these tests, equilibrium was reached after approximately 170 hours, or 150 hours after the 80K shield reached operating temperature. Thermally, the second thermal cycle occurred much like the first. Based upon the results of the two thermal cycles, the 80K blanket assembly equilibrates approximately 6 days after the 80K shield reaches steady-state operating temperature.

Temperature Change in MLI Blanket Layer 32

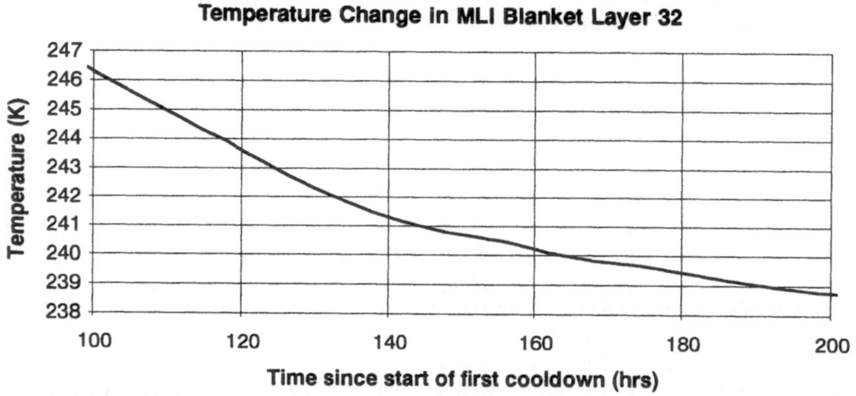

Figure 1. Time required for MLI blanket to reach equilibrium during initial cooldown.

Support Post Performance

The support post located between the center and end posts on the lead end of the magnet was instrumented with thermometers and strain gages positioned along the conductive path. Steady-state support post temperature profiles derived from data recorded on three separate days are presented in Table 1. The values in the table were obtained by averaging data recorded over 10-hour periods.

By design, the top of the support post should operate near 4.2K. However, in operation the post top equilibrates at a higher temperature due to thermal resistance of the cold mass cradle assembly. These measurements indicate a 2.25K temperature gradient across the cradle connection. The table also shows that the 80K intercept operated at a temperature significantly above the design temperature of 80K. This is due in part to the MTF liquid nitrogen supply system which caused the 80K thermal shield to operate near 90K . Notwithstanding, there was a significant thermal gradient between the thermal shield and support post intercept which may suggest some inefficiency in the thermal shorting straps.

Heat load into the 4K ring and the 80K intercept were calculated using physical properties and measured temperatures and are included in Table 1. The average calculated heat load into 4K was 0.041 watts, which is greater than the design heat load of 0.032 watts. This calculation is made using available thermal conductivity values for a graphite reinforced plastic that may be of slightly different composition than the composite tube actually used. Thermal conductivity of the actual material has not been characterized. The average calculated heat load into the 80K intercept is 3.235 watts, slightly greater than the design heat load of 3.160 watts. One would have expected the calculated heat load into the 80K intercept to be slightly lower than design given the higher operating temperature of the intercept. As it was not possible to install instrumentation on the lower end of the support post inner tube, it is not possible to calculate the conductive heat load into the 20K intercept.

Table 1. Steady-state support post temperature profile.

Post Location	Temperature (K) _Data recorded on:_		
	10/24/92	11/02/92	11/10/92
4K ring	6.323	6.234	6.781
GRP tube, near 4K ring	13.037	13.100	13.315
GRP tube, near 20K intercept	16.696	16.815	16.885
20K intercept	18.391	18.623	18.449
80K intercept	103.798	103.715	105.294
FRP tube, near 80K intercept	185.570	184.822	186.165
FRP tube, near 300K ring	234.118	233.065	234.520
300K ring	290.606	289.224	290.707
Heat load into 4K ring	0.041 W	0.042 W	0.040 W
Heat load into 80K intercept	3.250 W	3.220 W	3.234 W

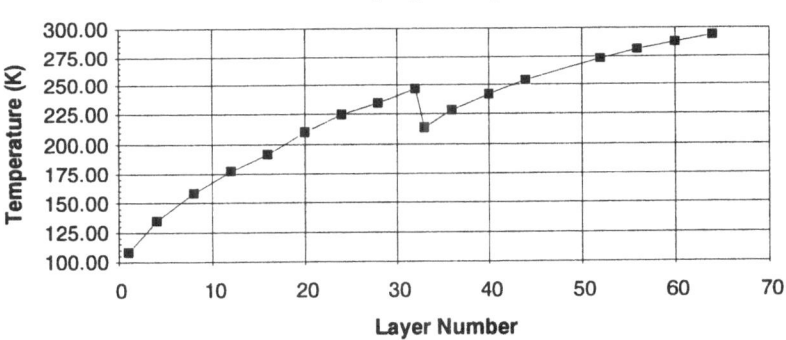

Figure 2. Temperature profile through the 80K MLI blankets.

MLI Blanket Performance

There were a total of 54 thermometers in the 80K MLI blankets. Arrays of 18 thermometers each were located at three positions along the cryostat length. Figure 2 shows the temperature profile through the inner and outer blankets at one location during one steady-state period. Temperatures plotted in the graph were averaged over a 24-hour period. Similar graphs were made using data from thermometers located at the other two thermal "slices" through the blanket; those temperature profiles were nearly identical to that shown in Figure 2.

Analysis of the thermometer locations showed that the temperature measurements were significantly influenced by the sewn seams on the adjacent blanket. When the blankets are installed in the cryostat, the blanket seam/joint areas are offset to preclude excessive build-up due to hook-and-loop closures, additional cover materials, etc. Unfortunately, this offset caused the thermometers in the instrumented blankets to be located immediately above or below the seam/joint area of the adjacent blanket, which caused the step in the thermal profile. The result of this thermometer positioning is that data obtained from these instrumented MLI blankets is of little use in predicting heat flux through the MLI, or for thermal modeling purposes. Nonetheless, the data clearly shows that the seams have a significant impact on the thermal gradient through the adjacent blanket, which equates to a potentially significant impact on overall thermal performance. It is not clear from this data how localized an effect this is.

Data consistently showed that the temperature gradients through the blanket at the three instrumented "slices" were nearly identical. One of the concerns with long cylindrical applications of MLI is the ability to evacuate the interstitial regions of the multilayer assembly. MLI measurements have recorded the temperature profile through an 80K blanket under various insulating vacuum levels[3]. A distinct curve in the temperature profile was observed at pressures near 10^{-5} torr; however, as the pressure increased above 10^{-4} torr the curve became linear due to increased residual gas conduction. Figure 3 shows the curve to be non-linear, suggesting that the interstitial volumes had been adequately evacuated to reduce residual gas conduction effects. And the fact that the profile was the same at several positions along the blanket length suggests that the blanket had been uniformly evacuated over its length.

Conclusions

Measurements made on DCA323 indicate the cryostat instrumentation is functioning properly. Placement of the MLI blanket thermometers caused the data to be of limited use in calculating heat flux through the blanket assembly. However, the thermometers did indicate that the interstitial volume of the blankets was evacuated to sufficiently low levels to preclude excessive heat transfer through residual gas conduction. Calculations of conductive heat transfer through the instrumented support post indicate that heat loads into the 4K and 80K temperature stations may exceed the design budget by a small amount.

REFERENCES

1. Niemann, R.C., et al., "SSC Dipole Magnet Cryostat Thermal Model", in: "Advances in Cryogenic Engineering", Vol. 33, Plenum Press, New York (1988).

2. Boroski, W.N., "Instrumentation of SSC Dipole Cryostat DCA323", Fermilab TSS/Engineering Laboratory Document IGEL-040.

3. Gonczy, J.D., Boroski, W.N., and Niemann, R.C., "Thermal performance measurements of a 100 percent polyester MLI system for the Superconducting Super Collider", in: Advances in Cryogenic Engineering, Vol. 35, Plenum Press, New York (1989).

CQM THERMAL SHIELD DESIGN

K.E. Grut, G.A. Lehmann, X.R. Huang, J. Waynert

Babcock & Wilcox Company
Naval Nuclear Fuel Division
PO Box 785
Lynchburg, Va. 24505

ABSTRACT

The SSCL baseline cryostat design utilizes thermal shields that are a welded assembly of aluminum sheet and extruded aluminum tube sections. In an attempt to reduce the cost of Collider Quadrupole Magnets (CQM) B&W investigated alternatives to the SSCL baseline design. The alternative chosen uses stainless steel tubes that are supported by the lower shield. Choosing an alternate material for the cryogen lines introduced a number of variables that needed to be addressed in order to establish an efficient, cost effective design. This paper presents B&W's baseline thermal shield design and the methodology used to obtain that design.

BACKGROUND

The original design for the CQM utilized all aluminum shield assemblies. These assemblies included extruded aluminum tubes for cryogen fluids. These extrusions were welded to the upper and lower shields that were made of break-formed aluminum sheet. The all aluminum shield required a bi-metallic transition joint to mate up with the stainless steel interconnect hardware. The cost of extrusions and bi-metallic joint hardware is high and drove B&W to investigate alternative configurations to meet the cost goals for the CQM.

Possible Alternatives

B&W proposed to use a stainless steel tube for the CQM cryostat which would be supported by the thermal shield assemblies. The difference in materials posed a potential problem, however. Aluminum has a coefficient of thermal contraction approximately 1/3 greater than that of stainless steel. Consequently, the shield must permit relative motion between the tube and shield while still effectively transferring heat. Several methods of achieving this goal were investigated and the list of candidates was narrowed to two.

The first design (used by General Dynamics) uses a stainless steel tube and metallic straps that are brazed to both the shield and tube. Heat is transferred through the thermal straps directly to the tube at various locations along the tube. This design permits axial motion between components but is expensive due to the high part count and manufacturing cost. Additionally, the tube ends are free to move laterally requiring a positioning device (or spider).

The second design is one in which the shield entraps the tube within its own profile (Fig. 1). The shield not only supports and positions the tube but also enables metal-to-metal contact between the tube and shield for the entire length. This contact is used to transfer heat to the cryogen fluids. This approach is similar to existing cryostat designs. The challenge with this design is to define the amount of contact force to minimize thermal contact resistance yet provide for relative motion.

Trade-Off Study

Each of the two basic concepts for transferring heat can be incorporated into a functional shield design, however, the number of parts, the forming method and the amount of time necessary to manufacture them, all have an impact on the overall cost. This cost impact relative to a baseline needed to be evaluated to determine the most feasible design. Six configurations based on the two concepts were evaluated and the normalized results for three typical configurations are provided in Table 1. The first concept is similar to the SSCL baseline and is the basis for comparison.

Table 1. Piece part and overall costs associated with thermal strap and entrapped tube designs.

Shield Concept	Part Cost	Man-hours	Total Cost
Extruded section with entrapped tube	$1.0	1.0	$1.00
Superplastically formed shield	$0.59	0.54	$0.59
Break formed with transitions and thermal straps	$0.54	3.5	$1.08

Clearly the most feasible design is an entrapped concept with superplastically formed shields. Superplastic forming provides the ability to form complex shapes (i.e. interface features) into the shields and eliminate a number of sub-components from the overall assembly. This also allows B&W to optimize the support structure for thermal efficiency as well as structural integrity.

PRELIMINARY DESIGN

The goal in designing the entrapped concept was to obtain a shield profile that would support and position the cryogen tubes while maximizing the amount of heat transfer between the tube and shield. The shield design must also allow the components to move relative to each other and permit easy installation of the tube.

Bottom Shields

The initial design had approximately 210 degrees of contact around the tube. The intent was to restrain the tube as well as support it. However, with a convoluted profile, tube installation and forming the convolution are difficult. A profile of 180 degrees is easier to form and was therefore chosen.

A thermal model of the CQM was built to establish a minimum amount of contact force between the tube and shield to meet thermal needs. The results show that the minimum force required for the 80K shield is 823 N (443 N for the 20K shield). Since the entrapped concept is a cylinder within a cylinder and the tube profile is all but perfect, a realistic assumption is that the tube is only in contact at high spots in the entrapped region. This means that the entrapped region of the shield needs to be as good as reasonably achievable.

Since the tubes and shields need to move relative to each other and the tubes are not truly entrapped, the method used to contain the tubes needs to resist axial loads and lateral movement of the tube. To do this a series of band clamps is installed through slots along the length of the shield. The tension in these clamps is just enough to restrain the tube and provide sufficient contact force but permit axial movement under thermal contraction. Two of the clamps near the support post are tightened to resist installation loads.

Top Shields

The top shields are thin and prone to distortion by vibration and thermal loads. The top shields will be formed by conventional methods (roll forming) and will be attached to the bottoms by an intermittent stitch weld. This type of weld is expected to minimize distortion upon cool down which has been observed in the ASST magnets. Figure 1 shows a thermal shield using the entrapped concept.

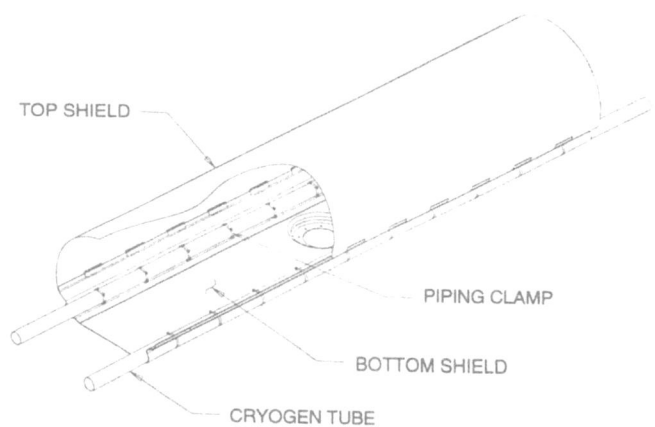

Figure 1. B&W's thermal shield design. The bottom shield supports the tube as well as transfers heat.

ANALYSIS

The 20K tube and shield were modeled to evaluate the loading of this region due to different material properties. The different rates of contraction between the shields and the tubes were expected to produce stress due to friction in the entrapped region. As was expected, the shield and tube did have a slight loading due to friction but this is very small (about 20 MPa). The maximum stress is located around the support post and is due to thermal contraction around the clamped region.

The 80K shield undergoes a rapid cooling where the 20K shield is cooled by slowly ramping the temperature to operating conditions. It is anticipated that this the 80K shield undergoes the worst case thermal loading for the shields. A preliminary thermal analysis of the CQM cryostat was performed to determine the most drastic thermal distribution. The results show that temperatures vary from 185K at the tube interface to 291K at the top of the shield.

Distortion of the shield cross section is also an issue because the 80K shield is cooled with only one cryogen line and therefore asymmetric loading results. A finite element model of the

80K shield was made to evaluate the effects of the thermal gradient on shield stress and distortion. The maximum stress in the 80K shield occurs in the top shield and is approximately 21.5 MPa. The top shield distortion is 5.9 mm. This is primarily a lateral displacement but should not be a problem during cool down since there is approximately 11 mm of clearance between the 20K and 80K shields in this area.

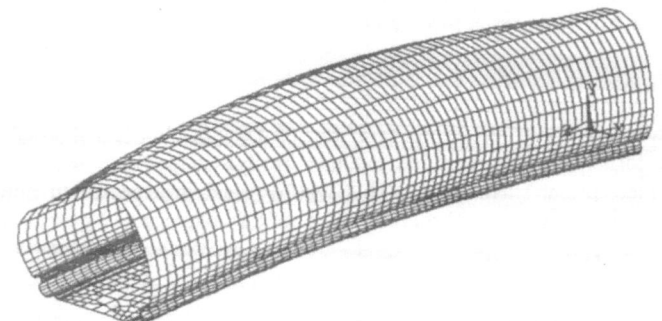

Figure 2. Asymmetric cooling in the 80K shield causes distortion in the top shield.

B&W also performed a modal analysis of both 20K and 80K shield assemblies. The lowest mode of vibration occurs at approximately 7 Hz. This a lateral motion in the 80K top shield and is indicative of the top shield's thinner cross section. The bottom shield is much stiffer because of the complexity of the shield profile and the additional stiffness of the tube.

PLANS

B&W will continue to pursue the entrapped thermal shield design. It is expected to be an efficient method of transferring heat and a low cost method for manufacturing the CQM cryostat. B&W is presently having the aluminum-to-stainless steel contact resistance evaluated to quantify the clamping force necessary to meet the CQM heat budget. Thermally conducting greases are also being investigated as an option to ease sliding. Finally B&W will evaluate the overall shield performance with the first prototype CQM which is expected to be tested by the SSCL.

ACKNOWLEDGMENTS

The work for which paper has been prepared has been supported by the SSCL under contract SSC-91-B-01703.

REFERENCES

1. X.R. Huang, "CQM Thermal Performance Analysis (2) CQM Cryostat 2-Dimensional Thermal Model," M3D-200036, January, 1993.
2. K.E. Grut, Thermal Shield Structural Analysis," M3D-200116,TBD.
3. K.E. Grut, "Modal Analysis of the 80K Thermal Shield Assemblies," M3D-200037, October, 1992.
4. K.E. Grut, "Modal Analysis of the 20K Thermal Shield Assemblies," M3D-200038, October, 1992.
5. K.E. Grut, "Thermal Shield Trade-Off Study," M3D-200039, November, 1992.
6. Personal Communiqué with Glen McIntosh, August, 1992.

SINGLE TUBE SUPPORT POST THERMAL ANALYSIS AND TEST RESULTS

T. H. Nicol, W. N. Boroski, and C. J. Schoo

Fermi National Accelerator Laboratory
P.O. Box 500
Batavia, IL 60510 USA

INTRODUCTION

Cold mass structural supports used in prototype Superconducting Super Collider (SSC) 50 mm dipole magnets built at Fermilab and Brookhaven are adaptations of the design developed during the 40 mm design program at Fermilab.[1,2] The design essentially consists of two composite tubes nested within each other as a means of maximizing the thermal path length. In addition it provides an ideal way to utilize materials best suited for the temperature range over which they must operate. Filament wound S-glass is used between 300K and 80K. Filament wound graphite fiber is used between 80K and 20K and between 20K and 4.5K. An alternate design for supports which uses a single composite tube has been developed at Fermilab and continues to be refined by the industrial contractors.[3] The advantage of the new design is cost reduction due to a significantly simpler assembly and incorporation of many common parts. This report describes the thermal analysis and testing of a single composite tube support post whose function is identical to that of the current reentrant design.

THERMAL ANALYSIS

Characterizing this support design as being a single tube is somewhat misleading. It actually consists of two separate composite tubes joined at the 20K intercept by clamping the tube overlaps between the outer ring and inner disc, both of which are aluminum, and which are machined to effect a shrink fit joint. Single tube in this context refers to the non-reentrant design of the assembly which is what distinguishes it from all SSC prototype assemblies. The support structure used in the thermal analysis and test is shown in Figure 1. It consists of a fiberglass reinforced epoxy tube (FRP) between the 300K connection and 20K intercept and a graphite reinforced epoxy tube (GRP) between the 20K intercept and the top ring.

Equations 1 through 4 are expressions for the heat load to each thermal intercept in terms of the support geometry and material properties. Table 1 lists the relevant geometry and material property values used in the analysis. Note that thermal conductivity integrals are shown only for the case in which the low temperature intercept is at its nominal operating temperature of 20K.

Typically when making thermal conductivity measurements the temperature of the low temperature intercept is varied as a means of documenting support performance at points away from nominal conditions. Data points are taken with this shield at 10K, 20K, 30K, and 40K. The results of the thermal analysis at these four operating conditions are summarized in Table 2.

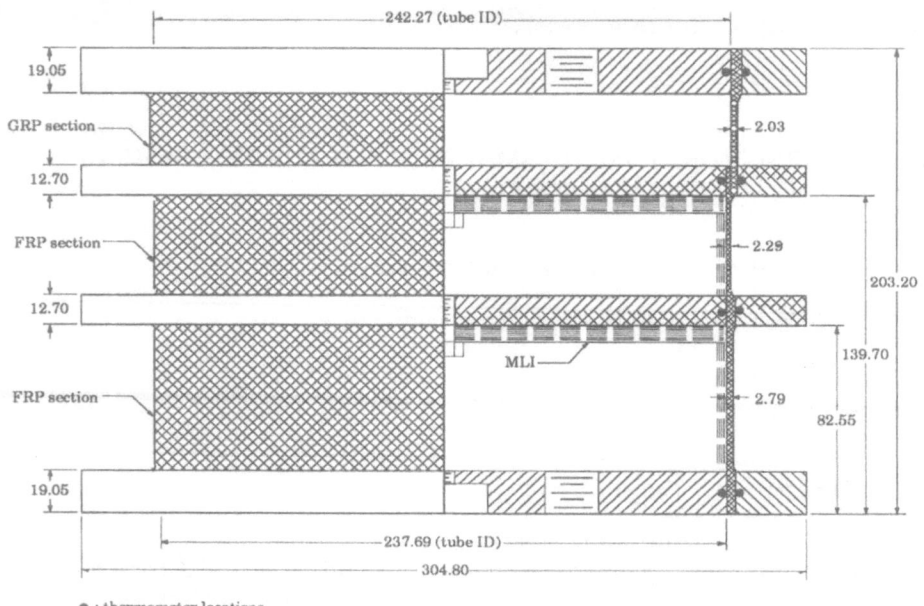

Figure 1. Cross section of the single tube support used in thermal analysis and testing

$$Q_{4.5} = \frac{A_3}{L_3} \int_{T_{top}}^{20} \kappa_{grp} dT \tag{1}$$

and

$$Q_{4.5} = \frac{A_{eq}}{L_{eq}} \int_{4.5}^{T_{top}} \kappa_{ss} dT \tag{2}$$

where

$$\frac{A_{eq}}{L_{eq}} = \frac{1}{\dfrac{\ln\left(r_o / r_i\right)}{2\pi t} + \dfrac{L_{hm}}{A_{hm}}} \tag{3}$$

$$Q_{20} = \frac{A_2}{L_2} \int_{20}^{80} \kappa_{frp} dT - Q_{4.5} \tag{4}$$

$$Q_{80} = \frac{A_1}{L_1} \int_{80}^{300} \kappa_{frp} dT - Q_{20} - Q_{4.5} \tag{5}$$

where: A_1, A_2, A_3 = cross sectional areas of the lower, middle, and upper tube sections
A_{eq} = equivalent cross sectional area of the top post disc and heat meter
A_{hm} = cross sectional area of the heat meter active element
L_1, L_2, L_3 = thermal path length of the lower, middle, and upper tube sections
L_{eq} = equivalent length of the top post disc and heat meter
L_{hm} = length of the heat meter active element
κ_{ss} = thermal conductivity of stainless steel
κ_{grp} = thermal conductivity of graphite reinforced composite
κ_{frp} = thermal conductivity of glass reinforced composite
r_i, r_o, t = top disc inner radius, outer radius, and thickness
T_{top} = temperature at the top support ring

Table 1. Thermal analysis parameters

Parameter	Units	Value
A_1	mm^2	2110.90
A_2	mm^2	1723.48
A_3	mm^2	1559.55
A_{hm}	mm^2	285.02
L_1	mm	63.50
L_2	mm	44.45
L_3	mm	31.75
L_{hm}	mm	8.53
r_o	mm	121.14
r_i	mm	12.70
t	mm	19.05
$\int \kappa_{ss}$ (9K-4.5K)	W/mm-K	0.002
$\int \kappa_{grp}$ (20K-9K)	W/mm-K	0.001
$\int \kappa_{frp}$ (80K-20K)	W/mm-K	0.016
$\int \kappa_{frp}$ (300K-80K)	W/mm-K	0.122

Table 2. Thermal analysis results

Low temp intercept (K)	T_{top} (K)	$Q_{4.5}$ (W)	Q_{20} (W)	Q_{80} (W)
10	6.03	0.012	0.658	3.386
20	8.92	0.043	0.564	3.449
30	12.06	0.096	0.434	3.526
40	15.81	0.189	0.254	3.613

THERMAL TEST

Thermometers were positioned along the conductive path of the support assembly to map the temperature of the various metal components under different operating conditions. 100 ohm platinum RTD's were inserted into the 80K and 300K discs and rings. 100 ohm carbon resistors were used as thermometers on the 20K and 4.5K metal components. Thermal performance measurements were conducted in a Heat Leak Test Facility.[4] The support assembly was installed in a helium dewar in much the same way it is installed in a magnet cryostat. Each shield is cooled by internal cryogen lines and the top of the support is attached to an LHe vessel. Total heat flow to 4.5K through the support assembly was measured by means of a heat meter that measures heat flow as a temperature gradient across a thermal impedance with an accuracy of ± 1 mW at 4.5K.[4] Unfortunately, the current configuration of the measuring system does not permit measuring heat flow to the intermediate shields.

One goal of the measurement program was to measure the heat load to 4.5K at different 20K shield temperatures. This provides not only multiple points at which to confirm the analysis, but also provides information on total magnet system performance at degraded operating conditions. Precise flow control of the cooling gas to the 20K shield allows operation of this shield at any temperature. For this test, we chose to perform measurements with this shield at 10K, 20K, 30K and 40K.

Table 3 presents a tabulated summary of the measurement data. When comparing this with the calculated results in Table 2 it is clear that there is reasonably good agreement between the predicted performance at 20K shield temperatures of 30K and below. Agreement would be improved with thermal conductivity data specific to each composite material used in the support assembly. In addition to heat load, this table provides insight into other support performance characteristics, most notably the thermal efficiency of the shrink fit joints. Small temperature gradients across each joint are indications that these joints provide a good means by which to transfer heat out of the composite tubes to each thermal intercept.

Table 3. Thermal test results

Sensor location	Sensor type	Low Temperature Intercept (K)				
		10	20 (*meas 2*)	20 (*meas 1*)	30	40
4K disc	Carbon resistor	6.567	9.070	9.168	13.659	16.337
4K ring	Carbon resistor	6.435	9.142	8.981	13.815	15.978
20K disc	Carbon resistor	15.779	22.847	22.865	34.595	40.275
20K intercept	Carbon resistor	9.924	19.142	19.162	31.752	37.439
80K disc	Platinum RTD	85.798	85.664	85.859	86.355	86.531
80K intercept	Platinum RTD	83.650	83.519	83.728	84.191	84.312
300K disc	Platinum RTD	282.044	282.223	282.189	282.287	282.891
300K ring	Platinum RTD	281.842	282.022	281.984	282.084	282.689
20K shield	Carbon resistor	8.311	19.118	19.140	33.357	39.852
80K shield	Platinum RTD	81.730	81.611	81.816	82.262	82.340
Vacuum	Glass ion gauge	1.36E-6	1.34E-6	1.67E-6	1.34E-6	1.04E-6
Heat load to 4.5K (W)	Heat meter	0.010	0.030	0.029	0.088	0.130

SUMMARY

Thermal analysis and test capabilities at Fermilab have evolved a great deal over the course of the prototype SSC magnet development program. The results presented here are encouraging in the sense that they provide some assurance that system performance, determined largely on the basis of analysis results, is predictable assuming that care is taken in the magnet cryostat fabrication process. This work also points out areas in which our current analysis and measurement capabilities could be strengthened. First, better correlation between calculated and measured results would be improved with material specific thermal conductivity data. Cost constraints precluded development of a system to accurately measure thermal conductivity from 300K to 4.5K. Second, overall performance predictions would be improved by developing the capability of measuring heat flow to the 80K and 20K shields.

ACKNOWLEDGMENTS

The authors would like to thank Mssrs. Michael Kramer and James Leslie of ACPT, Huntington Beach, CA for their expertise and assistance in design and prototype development and Mr. Richard Kunzelman, formerly of Fermilab, for prototype assembly and testing.

REFERENCES

1.   Nicol, T.H., et al., "SSC Magnet Cryostat Suspension System Design," Advances in Cryogenic Engineering, Vol. 33, Plenum Press, New York, 1987, pp. 227-234.

2.   Nicol, T.H., J.D. Gonczy and R.C. Niemann, "Design and Analysis of the SSC Dipole Magnet Suspension System," Supercollider 1, Plenum Press, New York, 1989, pp. 637-649.

3.   Nicol, T.H., "SSC 50 mm Collider Dipole Cryostat Single Tube Support Post Conceptual Design and Analysis," Supercollider 4, Plenum Press, New York, 1992, pp. 747-755.

4.   Gonczy, J.D., "Heat Leak Test Facility," Advances in Cryogenic Engineering, Vol. 31, Plenum Press, New York, 1986, pp. 1291-1298.

THE DESIGN AND DEVELOPMENT OF A SINGLE PIECE, SANDWICH TUBE COLDMASS SUPPORT POST FOR THE SSC COLLIDER DIPOLE MAGNETS

Aziz Ahmad[1] and Gregory Mehle[2]

[1]Superconducting Super Collider Laboratory*
2550 Beckleymeade Avenue
Dallas, TX 75237

[2]General Dynamics Space Systems Division
PO Box 85990
San Diego, CA 92186

ABSTRACT

This paper presents the design and development of the single piece, carbon/epoxy syntactic core sandwich tube support post concept for the 13-m and 15-m Collider Dipole Magnet (CDM). The engineering studies focus on balancing the structural and thermal requirements for the structure, and most importantly the implicit requirement of "Designing for Producibility." Results from these studies, as well as the development program plan, will be presented.

INTRODUCTION

The SSC CDM coldmass and cryostat thermal shielding is supported within the vacuum vessel by support posts. There are five posts supporting the 15-m dipole; four posts support the 13 meter dipole. The support posts are required to meet very rigorous structural, thermal and geometric requirements. The structural requirements for the support posts include the static and dynamic loadings during transportation and handling of the magnet, as well as the introduction and maintenance of the induced cold mass curvature (sagitta) loads. The four outboard support posts of the 15m CDM accommodate sliding hardware to mechanically isolate the thermal contraction of the coldmass from the vacuum vessel. This sliding hardware is at the (300 K) support post base. The center support post is the geometrically fixed support between the coldmass and vacuum vessel.

The support posts thermally isolate the 4 K coldmass, the 20 K inner thermal shield, the 80 K outer thermal shield, and the 300 K vacuum vessel. The limited space between the coldmass and vacuum vessel requires designing the support post with a short thermal length. The significance of this thermal path is the primary design driver for the configuration. Therefore, optimization of the structure is directly associated with thermal performance of the support post material system.

The initial design and development efforts for the CDM support post were performed at Fermilab[1] in support of the 40 mm CDM development. The design began evolving from the complex re-entrant post configuration to a simpler nested tube concept. The nested tube concept is composed of a carbon/epoxy cylinder existing in the thermal path between the 4 K coldmass and the 20 K inner thermal shield; and a fiberglass/epoxy tube existing between the 20 K and 300 K thermal paths. The two tubes are connected by a single lap shear bonded joint

* Operated by the Universities Research Association, Inc., for the U. S. Department of Energy under Contract No. DE-AC35-89ER40486.

which is loaded between an interference fit external (shield) attachment ring, and an internal support disk located at the 20 K intercept. The support post mechanical attachments at the 4 K, 80 K, and 300 K intercepts are consistent with the 20 K attachment (interference fit metal external-ring to internal-disk). The configuration takes advantage of the low thermal conductivity of the carbon/epoxy material below 20 K, and the low thermal conductivity of the fiberglass/epoxy above 20 K.

The Single Piece Composite Tube Approach

Since the design of the support post is cost driven (8000 dipole magnets with five support posts each for the collider) the implicit design requirement is "Designing for Producibility." This means that the ability to minimize hands-on fabrication processes will reduce the support post assembly costs (since the material costs are relatively fixed). Operations such as fabricating two separate tube structures (fiberglass and carbon/epoxy), machining land areas for dimensionally restrictive interference fit locations on the tube, preparing and bonding the lap shear joint region, etc. contribute significantly to the cost of the support post assembly. Minimizing, or eliminating any of these steps provides a substantial program cost reduction.

Since composite materials provide the thermal and structural capabilities required for the support post, and metal attach hardware is required for assembly, the decision becomes which design features of the Fermi two-piece can be optimized to reduce manufacturing complexity (which translates into lower costs). The chosen optimization locations were the tube composition/configuration, and the interference fit.

The initial design work at General Dynamics Space Magnetics (GDSM) focused upon determining the thermal conductivities of a selection of candidate post materials. Since the thermal budget requirements for the post structure are restrictive, only composite (non-metal material) structures can be considered for use. Epoxies, and more importantly micro-balloon filled epoxies, have lower thermal conductivities than glass or carbon fiber filled epoxies. Intuitively, chopped fiber filled epoxy will have lower conductivities than long fiber filled epoxies, but chopped fiber composites have significantly lower compression strengths, have brittle (failure) behavior at cryogenic conditions, and are more susceptible to creep. Long-fiber reinforced epoxies provide the necessary strength and stiffness to meet the structural requirements of support post while still maintaining thin cross-sections to minimize thermally conductive paths.

Point of Departure Configuration

Figure 1 is a diagram of the Support Post assembly. The support post assembly is composed of a single piece carbon/epoxy, syntactic core sandwich cylinder of a constant thickness (2.5 mm). There are three plies of carbon/epoxy cloth on the Internal Diameter (I.D.) side of the tube, a 0.76 mm ply of glass micro-balloon filled epoxy (syntactic core) center, and two plies of cloth on the tube Outer Diameter (O.D.) Although the constant thickness provides structural overkill for the 4 K to the 80 K regions of the post (lower bending induced loads) the configuration allows simplified fabrication requirements.

The internal support disks, as well as the 4 K and 300 K external rings are made of 304LN CRES. The 20 K and 80 K intercept rings are made of 6061-T6 aluminum. The 20 K and 80 K intercept regions (external rings) are used to support thermal shielding, and the cryogenic return piping (and hardware) for the coldmass. The interference-fit internal disks (304LN) are required to distribute end loads into the post cylinder, and to allow load carrying at the 20 K and 80 K intercepts.

The tube is fabricated by wrapping carbon/epoxy cloth onto a cylindrical mandrel to form the three skin plies at the tube I.D. — the three I.D. plies are applied in a continuous wrap. All skin plies are wrapped under mechanical loading (follower roller) to eliminate entrapped air. Syntactic core film is then wrapped over the I.D. skin build-up. Finally, additional carbon/epoxy cloth is wrapped over the syntactic core to form the two O.D. plies of the tube.

The tube rolling approach to fabrication allows multiple tubes to be fabricated at one time. The cured tube is cut into the 177.8 mm (7 inch) lengths required for the post. Achieving the continuous wrapping of the skin plies allows the possibility of creating a quasi-automated process for tube fabrication. Tube wrapping allows flexibility in material orientation (material direction placed parallel with load direction), and provides a low life cycle cost due to the high rate of material that can be processed.

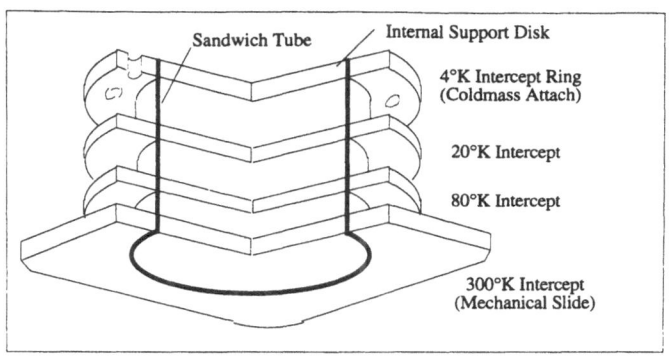

Figure 1. CDM support post assembly.

For the GDSM support post the basic premise is to eliminate all toleranced machining operations for the sandwich tube prior to assembly. The thermal intercept ring inside diameters are machined to slip over the as-fabricated tube O.D. A nominal 0.38 mm (0.015 inch) gap between the tube O.D. and the ring I.D. is filled with an aluminum filled structural adhesive. The adhesive is "buttered" onto the mating surfaces of tube and rings prior to ring installation. The I.D. of the rings are machined with a slight taper to allow the adhesive to fill the radial gap between the tube and ring during ring placement. The adhesive is cured while the rings are constrained in a tool to achieve controlled positioning. Structurally, the cured adhesive taper acts as a wedge lock during loading.

The internal disk outside diameters are dimensioned to be approximately 0.38 mm (0.015 inch) larger than the as-cured tube I.D. The internal support disks are cryo-installed (liquid nitrogen cooled to reduce dimension) into the bore of the ringed tube. The disks are held in place until the disks warm. As the disk warms, the enlarging disk O.D. creates the interference fit for the disk/ring/tube system. The interference fit of this configuration requires the internal disk to be the only precision toleranced (interference) surface. The sandwich tube I.D. remains consistent and repeatable due to the fabrication process selected.

DEMONSTRATION PROGRAM

A structural and thermal demonstration program has been identified to examine critical features of the material and assembly of the support post at the coupon, component and CDM assembly level. The program is streamlined to provide timely information for the optimization of the single piece sandwich tube support post design and fabrication.

The initial structural demonstration for the support post consists of the fabrication and compression test of sandwich cylinder coupons (76.2 mm length) to determine the typical ultimate strengths at room temperature (300 K), and liquid nitrogen (80 K) temperature. The objective of this test is to determine appropriate lamination and cure parameters which will be incorporated into the support post assembly.

A flat panel of the sandwich is used to test the in-plane shear strength of the sandwich cross-section at 300 K and 80 K temperature. The shear test method is the Iosipescu Shear test.

The support post assembly test program will demonstrate the quasi-production configuration support post in bending and shear to limit load at 300 K, followed by compression loading (limit and ramp-up to failure) at 300 K and 80 K.

The bending/shear test utilizes an angle plate fixture to cantilever the support post assembly in the test machine. The post assembly has a load fixture fastened to the top of the post for load application at the coldmass centerline location. The load applied represents the predicted worst-case combined sagitta load and 0.5g lateral acceleration (limit) loading induced in the post. The load will be applied for 5 cycles at 300 K. Assemblies will then be loaded in uniaxial compression to 123.6E6 N (28000 lb) for five cycles at both 300 K, and 80 K; then be loaded uniaxially to compression failure at both 300 K and 80 K respectively.

The 4 K/300 K intercepts rings will be demonstrated during the bending/shear limit loading. However, a separate test configuration is used to demonstrate the typical failure load capacity of the structural-adhesive/shrink-fit interface of the 20 K/80 K thermal intercept attachment. The test is performed at 300 K only.

Thermal properties for the support post tube material are being developed from 300 K to 20 K. Thermal conductivity and expansion testing is to be performed on the carbon/epoxy cloth material, the core material, and the sandwich material. The data will be used to update the thermal analysis for the cryostat system.

Development of the long term creep characteristics of the support post tube will not be performed at the component level. For the specific application of the support post assembly if creep were to occur it would manifest itself as lost sagitta, as measured by the beam tube. Laser alignment measurements will be taken after cryostat assembly on the first prototype magnets to determine if support post creep exists.

CONCLUSIONS

This paper has presented the basic design and development parameters for the single piece sandwich tube support post configuration. The development work is continuing at General Dynamics Space Systems (GDSS). Currently, the structural test results have more than satisfied the requirements for the CDM. However, the final cross-sectional configuration for the support post will be based upon the combined thermal/structural test results. Specific areas for optimization include reducing the number of skin plies, or terminating skin plies at specific thermal intercept zones (the current configuration uses a constant thickness tube along the full length of the support post); or utilization of fiberglass/epoxy material between 20 K and 300 K intercepts. Table 1 lists the preliminary thermal analysis results for heat leak of the 15 m CDM cryostat with a four-ply carbon/epoxy, syntactic core sandwich support post.

Table 1. Preliminary Thermal Analysis

Thermal Intercept	Cryostat Total Thermal Budget (watts)	Cryostat Total Heat Leak (watts)
80 K	37	40
20 K	5.06	4.80
4 K	0.36	0.27

Since the GDSS support post was designed with producibility in mind all optimization concepts do not affect the lamination, cure, or assembly processes/features of this configuration.

ACKNOWLEDGEMENTS

The work described herein is being accomplished under contract to the Universities Research Association in support of the Superconducting Super Collider project for the U.S. Department of Energy, Inc.

REFERENCES

1. T. H. Nicol, "SSC 50 mm Collider Dipole Cryostat Single Tube Support Post Conceptual Design and Analysis," Supercollider 4, Plenum Press, New York, 1992, pp.747–755.

MINIMUM RECOMMENDED
SSC LABORATORY SEISMIC DESIGN REQUIREMENTS

Jeffrey L. Western and Martin W. Butalla

Engineering and Design Group
Conventional Construction Division
Superconducting Super Collider Laboratory
2550 Beckleymeade Avenue
Dallas, Texas 75237

INTRODUCTION

The Superconducting Super Collider (SSC) Laboratory has been strategically located in Ellis County, Texas where design, construction, operation, and maintenance costs would be minimized. One of the parameters that affects costs during the design and construction phases of the laboratory life-cycle is the geological stability of the area with respect to seismic ground motion. Historical records demonstrate that bedrock accelerations expected to occur at any one time during the 20-year operating life of the facility for the project site will generate forces on both surface and subsurface structures of a magnitude less than the minimum forces required by current governing engineering design codes. Specifically, the Uniform Building Code locates the site within the Seismic Zone 0, but remains silent with respect to a Zone 0 minimum lateral loading. The specific magnitude of the minimum applied forces must therefore be determined from an application of both ASCE 7-88 *Minimum Design Loads for Buildings and Other Structures* (formerly ANSI A58.1) and FEMA 222, *Federal Emergency Management Administration*, referenced by and in compliance with DOE 6430.1A, the governing design code for the SSC.

PURPOSE

The purpose of this study was to develop minimum lateral loading requirements for the safest yet least costly design of all enclosed technical equipment as well as submerged conventional facilities, where surface wind does not govern.

SCOPE

The SSC Laboratory minimum lateral loading requirements stem from a sound engineering need to enforce lateral stability through application of a virtual seismic equivalent lateral acceleration. Since the actual seismic activity in a Zone 0 is viewed as being less than adequate for lateral stability, a larger virtual seismic equivalent is used as the minimum. The *General Design Criteria* of the Department of Energy (DOE 6430.1A), which governs the global design of all facilities of the SSC Laboratory, refers to the American Society of Civil Engineers publication ASCE 7-88, *Minimum Design Loads for Buildings and Other Structures* (formerly ANSI A58.1) and FEMA 222, *Federal Emergency Management Administration*. The design of all facilities and technical equipment of the SSC Laboratory shall be subject to this minimum lateral stability constraint unless it can be show that a higher lateral loading would govern.

SEISMICITY UNDER DOE 6430.1A GENERAL DESIGN CRITERIA

The Department of Energy (DOE) has published guidelines in order 6430.1A, *United States Department of Energy General Design Criteria* for the design of structures with respect to earthquake and other seismic activities. The Uniform Building Code (UBC) indicates that the SSC Laboratory is located within the Seismic Zone 0. Since the UBC remains silent regarding seismicity for Zone 0, other codes such as American Society of Civil Engineers (ASCE 7-88), *Minimum Design Loads for Buildings and Other Structures* (formerly ANSI A58.1), and the Federal Emergency Management Administration, (FEMA 222 / 223) shall be applied to determine the minimum requirements for a specific Zone 0 design.

The approach used by ASCE 7-88 and FEMA 222 / 223 is consistent with that depicted in UCRL 53582. Therefore ASCE 7-88 and FEMA 222 / 223 will govern for the SSC, since both of the references give a specific seismic zone coefficient for Zone 0.

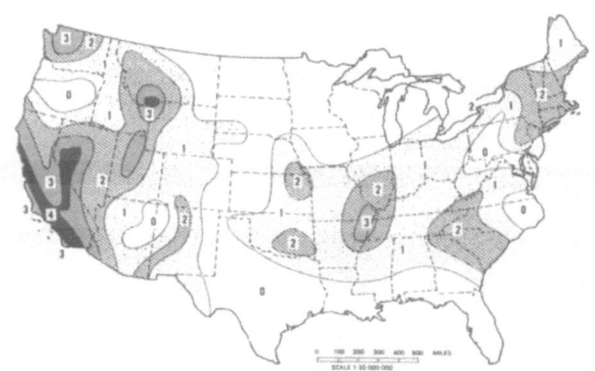

Map for Seismic Zones--Contiguous 48 States (ASCE 7-88; p 33)

SEISMICITY UNDER ASCE 7-88 (FORMERLY ANSI A58.1)

The seismic zone map of ASCE 7-88 reflects a SSC Laboratory locale of Zone 0.[1] For structures in Seismic Zone 0, ASCE 7-88 requires that the minimum lateral seismic force be dissipated through the structural connections into the lateral resistive elements, i.e., floors, roofs, etc.[2]

Anchorage of Concrete and Masonry Walls

A positive direct connection between a laterally-supporting floor/roof and a concrete/masonry wall connected to it shall be capable of resisting a horizontal force (Fp) specified by the equation:

$$Fp = Z*I*Cp*Wp \tag{1}$$

where Z represents the seismic zone numerical coefficient, shown in Table 2 with a value of 0.125 corresponding to Zone 0. All facilities of the SSC Laboratory reside within Zone 0 on the seismic zone map. The Occupancy Importance Factor (I) for buildings that will house more than 300 people (Category II) is 1.25. The remaining SSC buildings and structures fall within Category I, for which I = 1.0. The Horizontal Force Factor for the SSC Laboratory facilities is defined to be 0.3. The weight of the portion or part of the structure (Wp) includes the total dead and applicable live load for that section.

$$Fp = 0.125*1.25*0.3*Wp < 0.05*Wp, \tag{1a}$$

so that Fp = 0.05*Wp will be taken as the minimum required horizontal force.

---

[1] American Society of Civil Engineers, *Minimum Design Loads for Buildings and Other Structures (ASCE Standard 7-88)*, July 1990, p 33.
[2] Ibid., pp 32, 40.

Load Paths

All parts of the building or structure that transmit seismic forces shall be connected through a continuous path to the resisting element. The connection and the elements along the path to the resisting element shall be capable of withstanding a minimum force (Fp), determined to be the greater force from the following equations (2) or (3):

$$Fp = 0.15*Z*I*Wp \tag{2}$$

For Seismic Zone 0, Z = 0.125; for Category II occupancy, I =1.25. Thus

$$Fp = 0.023*Wp$$

for equation (2), or

$$Fp = 0.05*Wp \tag{3}$$

for equation (1a). Therefore, the minimum horizontal force (Fp) shall be determined by:

$$Fp = 0.05*Wp$$

where Wp is the dead load and any applicable live load portion of the structure.

SEISMICITY UNDER FEMA 222

The requirements imposed by FEMA 222 shall be considered the minimum loading requirements based upon an Effective Peak Acceleration (EPA) of less than 0.05g. This EPA has a 0.1 probability of being exceeded in 50 years. Seismic Hazard Exposure Group I and an EPA = 0.05g categorizes SSC Laboratory facilities.[3] SSC Laboratory facilities fall into Group I because the facilities are not emergency stations nor public places. With Group I classification and EPA of 0.05g, the Seismic Performance Category is A, corresponding to a zone of low earthquake risk where the building seismic design need only account for proper connections and anchorage of concrete or masonry walls.[4]

Anchorage of Concrete or Masonry Walls

Concrete or masonry walls shall be anchored to the roof and all floors that provide lateral support for the wall. The positive direct connection between the walls and the floors/roofs shall withstand the greater of the two seismic lateral forces:

$$Fp = 1000*Av \text{ (plf of the wall)} \tag{4}$$

or $$Fp = 0.05*Wp. \tag{5}$$

A_v is the EPV acceleration in units of g, here 0.05. Wp is the live and dead load of the portion of the structure. Substituting we have

$$Fp = 50 \text{ pounds per linear foot of the wall}$$

or $$Fp = 0.05*Wp.$$

The equation applicable to the SSC is the resulting combination of equation (4) and equation (5):

$$Fp = 0.05*Wp \text{ with a minimum of 50 pounds per linear foot.}$$

---

[3] Federal Emergency Management Agency, *NEHRP Recommended Provisions for the Development of Seismic Regulations for New Buildings,* 1991 Edition, Part 1, pp 3-5
[4] Ibid., pp 45-6.

Connections (Load Paths)

All parts of a building between separation joints shall be interconnected. Connections shall be capable of transmitting (resisting) the greater of the following two seismic forces (Fp) induced by the parts being connected:

$$Fp = (A_v/3) * Wp \qquad (6)$$

or $\qquad Fp = 0.05 * Wp. \qquad (7)$

where A_v is the effective peak velocity-related acceleration corresponding to the effective peak velocity (EPV) in units of g, which is less than 0.05. The probability of exceeding this value is less than 10% for any 50-year period. Wp is the applicable live load and dead load of the **smaller** connected part.

By comparison the greater Fp results from equation (7). The SSC Laboratory shall use the following seismic force equation for design purposes:

$$Fp = 0.05 * Wp$$

where Wp is the total live and dead load of the **smaller** connecting part.

Also note, in general, that a positive connection for resisting a horizontal force acting parallel to the member shall be provided for each beam, girder, or truss to its support. The connection shall have a minimum strength so as to transmit (resist) a force Fp defined as follows:

$$Fp = 0.05 * Rp$$

where Rp is the total dead and live load **reaction** of the beam, girder, or truss.[5]

SPECIAL STRUCTURES AND CONSIDERATIONS

The minimum requirements for the SSC Laboratory may not suffice for certain technical components. Special structures that contain large, concentrated masses, high aspect ratios, and unique support requirements, such as large detectors, may exhibit dynamic characteristics that fall outside of this static envelope and so may need to be subjected to additional and more comprehensive response spectrum or time history analysis.

Non-structural components that may be critical to SSC Laboratory operation, such as control units, high voltage electrical equipment, and cryogenic devices may experience secondary damages from seismic loadings due to overturning or failed supporting or anchoring mechanisms. These secondary or non-structural damage potentials may require additional seismic hazard loading considerations.

COMPARISON OF ASCE 7-88 (ANSI A58.1) AND FEMA 222

Both ASCE 7-88 and FEMA 222/223 impose the identical minimum lateral force criteria, with the additional constraint of member reaction minimums on connections mentioned in FEMA 222 noted above. The design requirements and the connection type constraints are identical. The information from the two standards is elementary and readily available. Both standards provide minimum requirements limitations. The strength of the structural connections must exceed the lateral force (F_p).

CONCLUSION

The equation from ANSI A58.1 gave the lateral force (Fp) to be no less than $0.05 * Wp$ (Wp is the total live and dead load). FEMA 222 dictates the lateral force to be $0.05 * Wp$ with a minimum of 50 lb. / ft. Both documents give the same requirements for the lateral force. The guidelines of both documents are in compliance with the UBC. In conclusion, both documents are used as standards for constructing the seismic requirements for the SSC

---

[5]Ibid., p 46.

RECOMMENDATIONS

We recommend that all facilities and structures, including technical systems components, at the SSC Laboratory be designed to withstand a minimum lateral force as specified in ASCE 7-88. We also recommend that a Seismic Design Section, that reflects the herein contained minimum constraints, be incorporated into the SSC Laboratory Facility Design Criteria.

ACKNOWLEDGEMENTS

The Superconducting Super Collider (SSC) Laboratory wishes to recognize and thank the individual members of the Seismic Design Requirements Working Group who collaborated to development the minimum recommended seismic design requirements. They are Joseph E. Keller of the Physics Research Division, H. Joseph Weaver of the Project Management Office Collider Group, David Goss of the Directorate, Jeffrey L. Western and Martin W. Butalla of the Conventional Construction Division.

REFERENCES

1. American National Standard (American Society of Civil Engineers), "Minimum Design Loads for Buildings and other Structures ," ANSI A58.1 (ASCE 7-88). July, 1990.
2. Building Seismic Safety Council, "1991 Edition NEHRP Recommended Provisions for the Development of Seismic Regulations for New Buildings, Part 1, Provisions," FEMA 222. January, 1992.
3. Building Safety Council, "1991 Edition NEHRP Recommended Provisions for the Development of Seismic Regulations for New Buildings, Part 2, Commentary, "FEMA 223. January, 1992.
4. International Conference of Building Officials, "Uniform Building Code 1991 Edition," ISSN 0896-9655. May 1, 1991.
5. United States Department of Energy, "General Design Criteria," DOE 6430.1A. April 6,1989.

ANALYSIS AND DESIGN OF UNDERGROUND DETECTOR HALLS

Birger Schmidt and Dan Yavorsky

The PB/MK Team
5510 South Westmoreland Road
Dallas, Texas 75237

INTRODUCTION

Construction has begun on two large underground halls that will house the detectors of the Superconducting Super Collider (SSC). The detector halls will measure 34 meters wide, 110 meters long, with the floor founded 66 to 69 meters below ground surface. A crane hook clearance of about 31 meters is required. Radiation shielding requires a roof cover equivalent to 6 meters of concrete and an equal distance to the adjacent ancillary shafts.

The halls bottom out in competent Austin Chalk, but the walls are largely composed of lower strength Taylor Marl. While the chalk is a well-behaved, elastic material, the marl is less well understood. Depending on the types of laboratory tests performed on the marl, it was found to behave either as a rock or a soil, and the marl's behavior when exposed in a deep excavation was in doubt, as conventional analyses could not completely be relied on.

Among the many options investigated, a scheme of excavation, permanent ground support, and internal structure was selected combining economy, simplicity of construction, reliability, and verification of long-term performance having the following components:
- Initial excavation to 9 meter depth.
- Vertical cut to 69 meters, with support by soldier piles and permanent tieback anchors.
- Placement of 3 meter thick concrete floor to minimize and distribute detector settlement.
- Construction of 34 perimeter columns, 1.8 meter square, supporting a steel frame roof and about 9 meters of backfill.
- Use of truss roof as a strut for long-term support, accepting part of the lateral wall load.

This scheme permits long-term lateral wall displacements to occur without significant effects on the internal hall structure. Anchor testing and proof-loading, together with instrumentation and monitoring during and after excavation, are intended to confirm design assumptions and ascertain long-term safety and reliability of the structure.

As for other underground structures, the main construction material forming part of the permanent structure is the rock mass presented to us by nature. While steel and concrete behavior can be engineered and is predictable within narrow margins, natural materials must be accepted as they are and do not follow rigorous and predictable behavior patterns. One of the greatest design challenges was to create an economical structure that takes advantage of the ability of the rock mass to support itself, yet accepts the uncertainty and relative unpredictability of exact rock mass behavior.

GEOLOGY

One reason for selecting the North Texas site for the SSC was the simplicity and benign nature of the geologic formations. Only three significant geologic strata are encountered at the site: Taylor Marl outcropping to the east above the Austin Chalk, found at the ground surface over the remainder of the site, and the Eagle Ford Shale below the chalk.

The sites for the two detector halls are similar. Below a 10 m thickness of residual soil and weathered marl, the Taylor Marl reaches a depth below the middle of the wall of the halls. The Austin Chalk forms the foundation of both halls. Two steep faults with a vertical displacement of some 3 m have been detected at one of the halls, but additional faults of small displacements may be encountered during construction. Based on previous experience the faults are not expected to result in significant weakness of the rock formations.

Properties of the two principal geologic materials are presented in Table 1. Although the Austin Chalk is layered, with seams or layers of shaly material, the properties of the chalk are generally isotropic, except that the layering makes the horizontal permeability much greater than the vertical permeability.

Table 1. Rock properties assumed for design.

Parameter	Chalk	Marl	Remarks
Compressive Strength (MPa)	15.2	$1.2 + 0.06z$	z equals depth (m) below top of fresh marl
Modulus (GPa)	3.45	0.83	
Cohesion (MPa) (Drained)	-	0.21	
Friction Angle (Degrees) (Drained)	-	30	
Permeability (cm/sec)			
Vertical	10^{-7} to 10^{-8}	10^{-8} to 10^{-9}	
Horizontal	10^{-6} to 10^{-7}	5×10^{-7} to 5×10^{-8}	

DEVELOPMENT OF STRUCTURAL CONCEPTS

For open-cut construction in sound rock where roof cover is not required, structural support elements to supplement the innate stability of the ground is often not needed for stability of the excavated space. Safety, performance, economy, and serviceability can be achieved with rock reinforcement, surface deterioration protection, grouting, and dewatering. However, if a roof is required, additional structural support elements are needed. When the roof system also supports a great thickness of backfill, the structure becomes a formidable part of the cavern structure. The structural elements must transfer the heavy roof loads into the rock formations without creating rock instability.

During construction and the service life of a cut-and-cover structure, the structural elements are stressed concurrently with the rock reinforcing (relatively flexible) and the rock formations (relatively stiffer). The redundant nature of the problem and the uncertainties in material properties over time result in stress-strain compatibility evaluation difficulties.

870

The design optimization and value engineering resulted in the structural configuration shown in Figure 1. The roof system is a structural steel tied-arch frame constructed as a truss. Many true arch configurations were studied, but the selected solution minimizes the backfill and roof dead weight, and avoids the need for high falsework to support overhead forms for concrete placement.

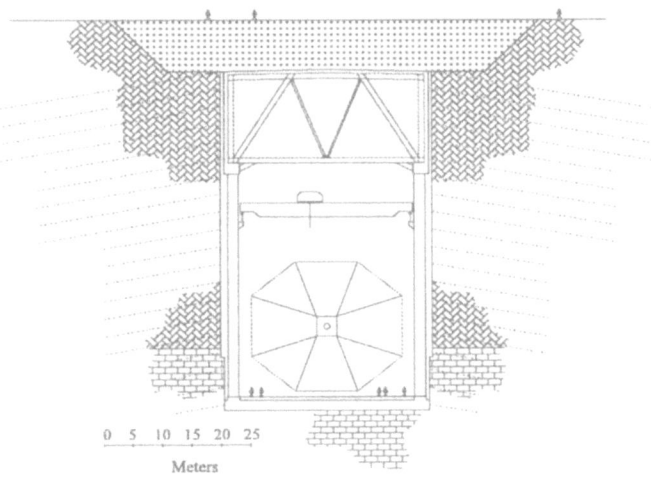

Figure 1. Design Schematic of SSC interaction hall.

The foundation slabs of the detector halls are subject to stringent settlement criteria. Large placements of high-modulus concrete are used to meet these tolerances. The flexural stress in the slabs does not control the design, rather, the elastic, creep, and shrinkage deformations of the concrete under low stress levels control the base slab design.

WALL SUPPORT CONCEPT

The walls of the experimental halls are permanently supported by tieback anchors. The ground support has the following major components, as shown in Figure 1:
• Soldier piles in pairs, embedded in concrete caissons.
• Tieback anchors threaded through the soldier piles.
• Fiber reinforced shotcrete placed in arches between soldier piles.

The tieback anchors consist of seven-strand prestressing wire tendons, placed in boreholes inclined 10 degrees and up to about 40 meters long. At the hole bottom, the tendons are grouted in place, with grout filling the borehole; this is the anchor zone. Between the anchor zone and the stressing head at the soldier piles, no grout surrounds the tendon, and the tendon is free to stretch.

For corrosion protection, the tendon is surrounded by corrugated HDPE tube and individual strands are covered by a plastic sheath in the free zone.

Design of the anchoring system requires the determination of the following:
• Geometric parameters: vertical and horizontal spacing, and anchor inclination (10 degrees from horizontal)
• Design load, anchor length, and free length of individual anchors
• Stressing load, less than design load for individual anchors

- Instrumentation, testing, and monitoring requirements.

Construction will be performed in stages. After excavation to 9 m (the depth to the roof structure, caissons with twin soldier piles will be installed 3 m apart by drilling 300 mm holes, inserting the piles, and filling the holes with weak concrete. The main hall, with vertical walls, will be excavated in stages. After 3 to 5 m are excavated, anchors will be installed in rows, tested, and stressed. The rock exposed between the soldier piles will be covered with shotcrete to form a structural arch and to protect the rock from deterioration.

As excavation proceeds, the walls will tend to move towards the open face, adding load onto the anchors by straining the anchor's free length. For this reason, the initial stressing load of the anchors must be less than the design load. This flexibility will allow the ground to adjust to the removal of the existing horizontal stresses at the wall face.

Total anchor load and individual anchor lengths are determined based on overall wall stability analyses of several kinds, and the initial stressing or lock-off loads are determined from numerical analyses simulating the excavation and anchor installation sequence.

For long term stability, the traditional analysis consists of determining the equilibrium of a wedge cutting into the wall, limited by a vertical tension crack. Observations of both natural and man-made cuts indicate that such vertical tension cracks occur at a distance of 0.4 to 0.5 times the height of the cut. Friction on the bottom of the wedge and the supporting anchor force are the stabilizing components of the analysis, while the weight of the wedge is the driving force. The tension crack is assumed to be partly or completely filled with water, adding a driving force from the water pressure. The geometric parameters of the wedge are varied to find the critical wedge, which forms the basis for the selection of the total design anchor force, with a suitable safety factor, which varies subjectively with the severity and credibility of the assumed circumstances.

Once the total required anchor force has been determined, individual anchor design loads are determined. The free anchor lengths are selected such that the anchor zones are outside the extent of the wedge. As check-analyses, standard slope stability analyses, assuming circular potential failure surfaces, were also performed.

A finite difference numerical analysis, using the FLAC code, was used to simulate the excavation and support sequence. The program simulated the step-wise excavation and the installation and stressing of the anchors and produced final anchor loads at the completion of the excavation, verifying that when anchors were stressed to 80 percent of the design load, the final loads were less than 95 percent of the design loads.

Altogether, nearly 3000 tieback anchors will be used for the two experimental halls having the world's tallest vertical, anchor supported walls. It is of some interest to examine the components of safety for this unprecedented structure.

- The best strength data fit for the marl indicates a phi of 50 degrees and a cohesion of 0.38 MPa. A phi of 30 degrees and a c of 0.21 MPa were used to approximate the lower range of the test data.
- The safety factor used for the wedge analyses was 1.5 or greater for the most credible assumptions, and in excess of 1.2 for the worst assumed case.
- The calculated anchor loads are less than the design loads, which include a safety factor on the tendons and the anchors of at least 1.67.
- The design includes horizontal drain pipes, intended to remove water from any tension cracks and reduce porewater pressures elsewhere.
- In the final condition, the roof truss will act as a compression member, a strut, which will take up load in the event that anchors relax with time.
- All anchors will be proof tested to 1.33 times the design load before being locked off.

An extensive instrumentation and monitoring program will be implemented to verify assumptions regarding loads and displacements, including instrumentation of some of the anchors, permitting corrective action, should it become necessary.

DESIGN OF A TUNNEL COOLING SYSTEM

Allen Cheng,[1] Robert Dawson,[2] and Richard Zhao[3]

[1] D'Escoto Associates (PB/MK)
[2] Conventional Construction Division/SSCL
[3] The PB/MK Team
 5510 S. Westmoreland Rd.
 Dallas, TX 75237

ABSTRACT

Based on a thermal analysis of the SSC main tunnel and requirements for space conditions in the underground electronics niches, thermal cooling systems will be required to reject the heat dissipated underground. Many systems were considered and compared to decide the optimum solution. Construction documents are being prepared and contract administration will proceed as technical and non-technical objectives are pursued.

INTRODUCTION

Like all existing large accelerators, the SSC is being built underground for reasons of radiation safety. All of the magnets, power supplies, control electronics, and instrumentation will be located either in the underground tunnel or the tunnel niches. In order to maintain the alignment of the accelerator magnets and keep the electronics operating reliably, a stable temperature not exceeding 80°F will be critical.

The spacing of the niches will be every 0.3 miles, with the shafts to the surface every 2.7 mile, which limits direct assess to the niches. The following paper will point out the specific needs that demand a controlled environment in the tunnel and niches.[1,2]

REQUIREMENTS FOR THE TUNNEL AND NICHES ENVIRONMENT

Niche Electronics

Within the niches the following electronics transfer heat to air and must not be subjected to temperatures greater than 80°F or temperature variations greater than 5°F per day.

- Corrector power supplies
- Beam position monitors
- Timing system
- Cryogenics controls
- Vacuum systems
- Beam loss monitors
- Quench protection

For each of these systems to consistently perform as designed and calibrated, the environment must be maintained within the prescribed limits. Total heat rejected in each niche will be 12 kW.

Tunnel Environment

To ensure the precision of the alignment of the magnets is not adversely affected, a stable of the tunnel temperature must be maintained. Thermal stability is also a requirement of the bellows in the spool pieces and the scraper alignment. Temperatures must be maintained below 85°F and not exceed a gradient of 7.2°F/half-sector. Heat loads within the tunnel originate from the following sources:

- Power cables
- Corrector cables
- Dump resistor cables
- Transformers

Total heat rejected in a typical sector will be 216 kW.

HEAT TRANSFER PROCESS IN TUNNEL AND EXTERIOR ROCK REGION

In developing the system to extract the heat from the niches and the tunnel, the heat transfer process in the tunnel and the surrounding rock region was analyzed. If the rock could be used as the heat sink and had sufficient capacity to maintain the tunnel and niche temperatures within the allowable limits, the heat extraction system would be greatly simplified.

Generation of Thermal Layer around Tunnel

Transferring heat through the tunnel wall generates a thermal layer from the wall surface out to a distance where, beyond which for practical purpose, there is no heat flow. Within the region of the thermal layer, the heat transfer process can be described by an energy-balance equation and Fourier Equation, i. e., net rate of heat entering through the bounding surfaces is equal to the rate of storage of energy in this region and heat flux is proportional to the temperature gradient.

Based on these principles, there must be a temperature gradient in the rock region to force heat flow to the deep rock. By testing rock samples and taking bottom-hole temperatures at SSC monitoring wells, values were determined for conductivity, specific heat, thermal diffusivity, and undisturbed ambient rock temperatures at the tunnel depth. As heat is continually added in the tunnel and niches, the temperatures of the rock in the thermal layer continues to rise as the energy is stored in an ever increasing volume of rock. To determine tunnel and niche temperatures, a heat transfer model was developed and solved by the integral method. The solution has proved to be a very good approximation for the exact solution.[3,4] For these calculations it was assumed the rock mass was infinite, homo-geneous, initially at 70°F, and the flow of groundwater was so small its effect was insignificant. Using this information it was determined that with only the heat transfer into the rock, within five hours, the niche air temperature will exceed 105°F and the tunnel temperature will exceed 90°F in 200 days.

DEVELOPMENT AND COMPARISON OF TUNNEL COOLING SYSTEM

Many cooling systems were considered during the study. These systems can be classified into three groups based on their transport media: air cooling system, air/refrigerant cooling system, and chilled water cooling system.

Air Cooling System

In this system, cooling air is supplied from one shaft and removed from the next shaft. Fans are used in the niches to transfer the heat to the tunnel. When cooling air is forced through the tunnel, the distributed heat in the tunnel as well as the niche heat load is picked up and warm air will be exhausted at the next shaft.

This is the simplest system because only air is involved in the whole process below grade. The highest tunnel temperature that would occur at the niche closest to the exhaust shaft must still be no greater than 75°F to adequately cool the niche. But to keep the temperature gradient to less than 10°F, the air volume required is so great the size of the shafts would have to be greatly enlarged. At the supply shafts, the rock becomes an additional heat source instead of a heat sink because the supply air temperature is lower than the rock temperature.

Air/Refrigerant Cooling System

This method uses direct-expansion (DX) units to reject heat from the niches to the tunnel. This allows the tunnel temperature to reach the maximum limit of 85°F while cooling is provided for the niches. Cooling air is forced to flow from one shaft to be removed at the next shaft, picking up the heat from the distributed tunnel loads and the niche DX units.

The rock will remove some heat from the tunnel where the temperature is greater than 70°F, however, the work of compression of the DX units increases the load in each sector by approximately 50 kW. This system requires an even greater amount of air to maintain the temperature gradient and the reliability of the DX units represents a very serious maintainability and reliability concern.

Chilled Water Cooling System

Because the greatest concentration of heat load is in the niche, the best approach is to extract the heat as directly as possible from the heat source. This is the basic principle of the chilled water cooling system.

In this system chilled water is distributed through the tunnel in an insulated chilled water supply pipe to each niche. Within the niches, terminal units consisting of coils and fans provide cool air and transfer the heat to the water. The return pipe is left uninsulated so it can absorb the distributed heat loads in the tunnel, thereby creating an environment in the tunnel that is practically isothermal. Chilled water is supplied from a chiller above ground and circulated to the tunnel.

Minimal air flow through the tunnel will increase the heat transfer to the return line and will be utilized to maintain humidity control and ventilation requirements. Estimates of installed and operating costs indicate the chilled water cooling system has the lowest initial and life cycle cost.

CONCLUSION

Due to the geotechnical conditions of this region, the SSC tunnel is basically in a warm, dry, insulating rock that is ideal for tunneling but does not function well as a heat sink. This condition is especially critical in tunnel niches where the heat load is concentrated and exceeds the heat dissipation capability of the surrounding rock. A cooling system had to be designed to extract the heat in order to provide a constant ambient temperature suitable for the operation of the electronic systems. Studies have determined the chilled water cooling system can best meet the criteria.

Ten identical cooling systems will be provided with a sum total capacity of 1500 tons produced by chillers at alternating remote sites. A major element of the initial cost will be the 100 miles of steel pipe required. More than 300 fan coil units, each with a capacity of 4 tons, will be installed in the adits and niches. The operation of these equipment will be controlled by multiple direct digital control panels networked locally into ten systems, which in turn will be networked into a site-wide system .

REFERENCES

1. "Design Requirements Collider Tunnel Cooling And Ventilation System," SSC Document Control Number Y32-00042 Rev. A, Superconducting Super Collider Laboratory, Dallas, TX, November 1992.
2. G. Dugan et al., "Collider Tunnel/Niche Temperature Requirements," SSCL-N-807 Superconducting Super Collider Laboratory, Dallas, TX, January 1993.
3. T. J. Lardner and F. B. Pohle, "Journal Applied Mechanics," pp. 310-312, June 1961.
4. M. Necati Ozisik, "Heat Conduction," 1980.

AN ENGINEERING STUDY OF INSTALLATION METHODS FOR AN
IN-TUNNEL 69 kV CABLE SYSTEM : (A DESIGN STATUS REPORT)

W. D. Smith, P.E.[1] and L. Norman[2]

[1]Chief Electrical Engineer
The PB/MK Team
5510 South Westmoreland Road
Dallas, TX 75237

[2]Superconducting Super Collider Laboratory
2550 Beckleymeade Ave.
Dallas, TX 75237-3997

INTRODUCTION

This paper provides an overview of the concepts and considerations analyzed during the development of the Superconducting Super Collider (SSC) tunnel power distribution system.

SSC Electrical Characteristics

The electrical nature of the SSC process creates a need for close interaction between the power demands of the SSC machine and the design of the power distribution system connecting the machine to the utility. Accelerator facility loads may be described as having typical conventional power requirements except for the power supplies. These accelerator power supplies are furnished with special harmonic filters and provide the drive current which must increase as these machines accelerate their beams to higher energies. It should be noted that for the superconducting machines much of the inductively stored energy is returned to the power distribution system.

POWER DISTRIBUTION SYSTEM CONCEPT

The general concept of providing a power distribution system that accepts high voltage bulk utility power and distributes this power to designated load centers is described in this section. High voltage power will enter the facilities at the West and East Main Substations. These sources will then be transformed and distributed via underground ductbank to the various SSC loads.

The power distribution system concept for the Collider uses a segmented bus arrangement at 69 kV. Ten 69 kV substations ('5' sites) around the Collider are divided into five north substations, N15 through N55, and five south substations, S15 through S55. The distance between each '5' site is approximately 5.4 miles. See Figure 1.

Collider High Voltage Feeder Distribution

Under normal operating conditions, one Collider configuration is to feed two north and three south 69 kV substations from the West Main Substation, and three north and two south 69 kV substations from the East Main Substation. Only one 69 kV tie circuit breaker in each north and south arc shall be left normally open so that all 69 kV feeders are energized. Under emergency

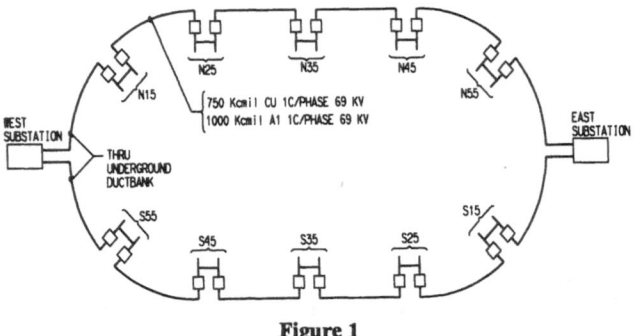

Figure 1

operating conditions, the 69 kV system shall be capable of providing power to five sequentially connected substations.

Four 50/67/83 MVA transformers will provide 69 kV power for distribution to collider surface and tunnel loads. An arrangement of 69 kV switchgear will serve a peak load of 137 MVA, and an average load of 94 MVA in the Collider.

Distribution Options: The two basic distribution options are surface via cable or transmission line or in the tunnel via cable. Using 69 kV for aerial subtransmission lines is a practical method to distribute bulk electric power. However, high levels of lightning activity in the project area and the need for system-wide security make overhead transmission lines less attractive. The cost of right-of-way acquisition, Public Utility Commission certification, and the associated politics further reduce the value of this alternative.

Two locations were investigated for feeder distribution in the tunnel, one in the crown overhead, and the other in the concrete floor (invert). The crown location offered the options of cable in tray, armored cable clamped to the tunnel wall, and cable in conduit. For an invert location, cable in conduit and cable in trench were reviewed as possible options. Both options require the use of solid dielectric, low smoke, non-halogenic jacketed cable designed for radiation levels that reach a maximum of 600 KRAD over 25 years. To reduce the possibility of fire, toxic smoke or explosions only cross-linked polyethylene or rubber-based insulation are being specified for the cable.

Preliminary engineering analysis of the crown location indicates that only a cable-in-tray configuration would be feasible because of space limitations, safety considerations and cost. Cable-in-conduit had higher capital costs, a more difficult installation, and required large spaces in the tunnel for pull boxes and conduits. The armored cable was eliminated for safety and cost reasons.

CABLE INSTALLATION CONSIDERATIONS

During the development of the detailed design of the mechanical and electrical systems, the facilities requiring space in the collider tunnel were located, sized and their configurations identified. The tunnel concept, as outlined in Figure 2 includes technical trays for magnet and cryogenic cable systems, piping, lighting and grounding and space for the magnet transport vehicle power rails.

With space in the crown of the tunnel at a premium, a cost/benefit analysis, was conducted using the following criteria:

- Personnel and equipment safety •
- Available real estate in crown •
- Reduced utility coordination problems

- Ease of inspection and maintenance
- Reduced installation costs

Cable Trench

The proposed trench will be sized to minimize the effect on the available floor space and allow for cable expansion expected during normal usage. Drainage will be provided to remove water removal from the trench, where required. The trench would be constructed as shown in Figure 3.

878

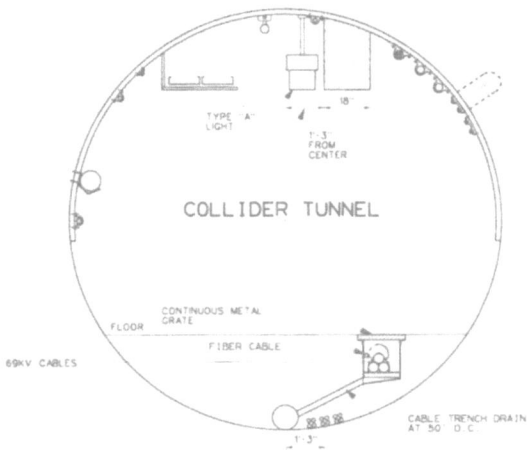

COLLIDER TUNNEL

Figure 2

Physical Protection: The trench installation provides satisfactory support and protection for the cables. A covered trench isolates the cables from all activities in the tunnel during installation of the technical systems and when removing magnets. The cover and the trench location also provide protection for both tunnel equipment and personnel. If a cable fault occurs, the danger would be confined to the trench. The installation is designed to ensure full compliance with the National Electrical Code (NEC).

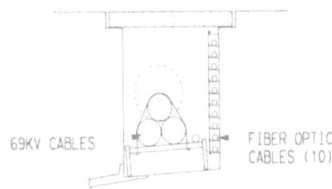

Figure 3

Space Availability: Locating the 69 kV cable in a trench in the tunnel invert affords better space allocation for conventional systems mounted around the tunnel walls and improves access to other technical systems such as light fixtures, the transport power rail system, and technical trays.

Utility Coordination: The trench arrangement for the 69 kV cables mitigates the coordination problems associated with routing the cables through the '5' site adits. The adit crown is occupied by cryogenic piping, HVAC piping and equipment, magnet high current and dump busses, and other technical systems.

Cable Inspection and Maintenance: The use of removable steel grating will allow the cables to be inspected quickly and easily. Any repair or maintenance work, with the removal of a few sections of cover, can be completed efficiently without the need for scaffolding or special supports .

Cable Installation: The floor level location reduces the labor required during initial installation, as the cables can be pulled along rollers at floor level without any longitudinal bracing for the pulling forces that will be generated. The cable can then be spliced and laid into the trench without the additional effort of lifting the cables into an overhead support system. The installation of cables in a trench system allows for maximum construction flexibility for the "conventional systems" portion of the contract. Because of the protection provided by a covered trench, it is possible to install the cables independent of other conventional construction.

Cable Tray Support Systems

Physical Protection: The cable tray provides both physical and mechanical protection for the cable, and physical protection for equipment and personnel in the tunnel. The cables are protected by the cable tray and are located out of the way of other elements in the tunnel. However, the location of the 69 kV cables at only 9.5 ft above the floor may present potential personnel safety or equipment hazards.

Cable Access and Space Availability: The cable tray provides an accessible support system. Cables can be readily inspected visually, and are accessible for maintenance. If a segment of cable or a splice fails and needs to be replaced, the tray and support would allow access to the cables for removal and replacement. However, close proximity to the energized transport rail and light fixtures makes working on or near the cable tray difficult. De-energizing a segment of the transport system power rail and removal of one or more light fixtures may be required. Additionally, temporary supports will be required to prevent damage to the shielding or insulation of the suspended portions of the cable when making repairs out of the trays.

Cable Tray Installation: Installation of the cable tray may be completed in two steps. First the preassembled tray supports would be installed and bolted to the strut in the tunnel crown. Upon completion of the supports installation, the cable tray sections will be fastened into place. These operations would be successive and both would require the use of mobile scaffolding which would preclude any other construction work being performed in the same area simultaneously.

Cable Installation: Installation of the 69 kV cables into the tray would be scheduled to follow the installation of the other conventional systems. The cables could be pulled into place along the tunnel floor, a single reel length at a time, spliced in place and lifted into the tray. Another method would provide rollers in the tray to enable single reel lengths to be pulled into the tray. Splicing would be done on an elevated platform.

Cable Routing: Routing the 69 kV cables in cable tray through the '5' site adits (rooms) to the utility shaft will be difficult due to the number of other utility systems located in the adit such as cryogenic piping, magnet high current bus structures, chilled water piping, technical system cable trays and miscellaneous electrical systems.

COST COMPARISON

Our preliminary review indicates that the 69 kV cable system can be relocated to the invert of the tunnel with minimum cost impact and with maximum overall space and installation/maintenance benefits. The cost comparison is as shown in the table below.

Table 1. Cable Type Cost Comparison

SYSTEM	CABLE TRAY	CABLE TRENCH
Cable Tray	$4,248,000	NA
Supports - Tray	$1,197,000	NA
Forming	NA	$1,040,000
Supports - Trench	NA	$1,682,000
Trench Cover	NA	$6,619,000
Concrete - Savings	NA	($3,700,000)
Labor To Install Cable	$3,314,000	$2,651,000
TOTALS	**$8,759,000**	**$8,292,000**

DATA ACQUISITION SYSTEM FOR THE GEM TTR

Luis Villaseñor for the GEM Collaboration

Institute of Physics
University of Guanajuato
P.O. Box E-143
León, Gto. 37000

INTRODUCTION

The TTR, described elsewhere[1], played an important role in helping select the appropiate muon detector technologies for GEM, i.e., those with muon track resolution better than 75 μm, as well as being the first user's facility at the SSCL. Of great importance for the overall success of the TTR operation was its data acquisition system, which we describe next.

DATA ACQUISITION SYSTEM

At the time of the proposal of the TTR, in November, 1991, it was decided to follow a conservative aproach in the design of the DA system, mainly due to the tight schedule: five months to develop the system starting from scratch. Therefore we decided to base the DA system on the CAMAC standard, using the General Purpose Interface Bus (GPIB) as the link between the CAMAC crates and the computer.

Two DA systems were developed in parallel: a simple one based on a Macintosh fx computer running an acquisition program written in LabView, which is a graphical programming language from National Instruments; and a more powerful one based on a Sun SPARCstation running under UNIX. The data collected by the Macintosh were logged into either a 1.2 GB external hard disk or an 8 mm tape (an Exabyte 8500 tape drive was connected to the SCSI port of the Macintosh computer) or both. In addition to the acquisition program, there was a simple histograming LabView program fetching the data from disk, i.e., running in pseudo on-line mode, to monitor the integrity and quality of the data; on the other side, no effort was done to provide off-line software for the Macintosh: as the computer was on the SSCL computer network, it proved more efficient to transfer (ftp) the data into a more powerful machine like a UNIX workstation and do the analysis there. The Macintosh-based system turned out to be quite easy to develop and it became itself invaluable while debugging the UNIX system.

The UNIX-based DA system[2] consists of a Sun SPARCstation connected to a VME crate by means of two independent links: ethernet and the Multisystem Extension Interface (MXI bus); by default ethernet is used for the command/response link and MXI for the data link. The VME crate contains the Motorola 68030 processor (MVME147 board), a National Instruments GPIB interface, as well as the Ethernet and MXI interfaces. A characteristic of the system is its modularity and expandability, i.e., there can be more modules in the VME crate, such as memory boards or additional CPU's. A 1.2 GB external hard disk and an exabyte EXB8500 tape drive with tape capacity of 2 or 5 GB were connected to SCSI ports of the SPARCstation. Figure 1 shows this hardware configuration.

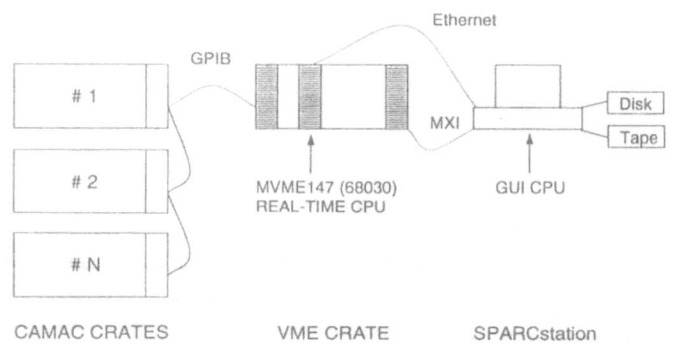

Figure 1. TTR DA system hardware configuration.

The software used for the UNIX-based DA system divides in two: the software that runs on the SPARCstation and the software that runs in the real-time operating system of the 68030 CPU. Figure 2 shows the default software configuration. The TTR server process handles the graphical user interface (GUI), the data collection and the tape control processes running on the SPARCstation. The GUI process was developed using the OPENLOOK GUI application environment of the SPARCstation and it makes extensive use of the windows capabilities of the latter. This process uses various subwindows to control and monitor the DA system as shown in figure 3. The TTR server process handles all the communications among the different processes running on the same SPARCstation by means of the interprocess communication mechanism provided by UNIX; it also handles the command/response communication with the real-time 68030-based system. The data collection process gets the data from the real-time system buffers and writes it to disk; everytime the disk file exceeds a user-selectable size, the file is closed and a new file is opened. The tape control process writes the data to tape (only if the system is instructed to write to tape, by means of a flag in the TTR GUI), this happens when the disk file reaches a predetermined size. If the system is instructed not to write to disk, the disk file is erased after the data has been written to tape. In addition the system can be configured in the TTR GUI not to write to disk nor to tape.

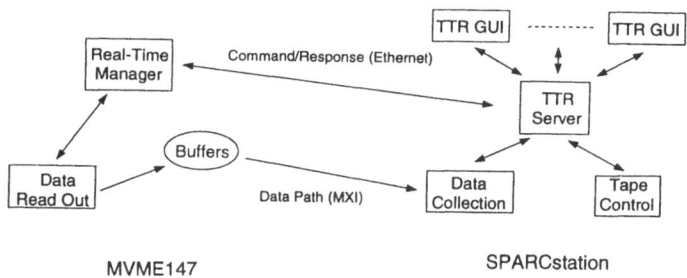

Figure 2. TTR DA system software configuration.

The real-time operating system on the 68030 is VxWorks 5.0.2. The tasks running under VxWorks are the TTR executive task and the read out task. The TTR executive task handles all initialization and acts a the command/response link with the SPARCstation, the commands it responds to are: start, pause, stop, quit, probe, event dump, insert comment event, and status request. The read out task reads the data from the CAMAC crates, formats the event and puts it in the VME buffer. This task is an event driven task in the sense that it starts when there is an external trigger.

The event format contains a header with the event size, DAQ mask, trigger mask, run number, event number, the date and the time; it can accommodate a variable number of data groups, where the data groups are determined by type of read-out module and by type of module owner. The DAQ mask provides the type of event: normal, comment, threshold, special, voltage, calibration, zero-suppressed or Monte Carlo. The trigger mask encodes the type of level 2 trigger in effect for the run, i.e., one of the various hit patterns of the scintillation counters programmable with the LeCroy 4508 Lookup units.

In addition to the event dump process, there are two more on-line monitoring processes: a histograming process based on the CERN library and an event display written in C, using Motif widgets and the X library. The histograming application, written in Fortran, provides a set of predetermined "key" histograms. The event display[3] provides two 2D projections of the TTR; it can display the hit patterns in the different technologies, along with superimposed straight line fits to a number of user-selectable combination of hits. It also provides the option of opening a subwindow for each of the detector prototypes under test. Rather than using plain Motif for writing the event display application, we used the TeleUSE application builder, which provided a significant reduction in the development time.

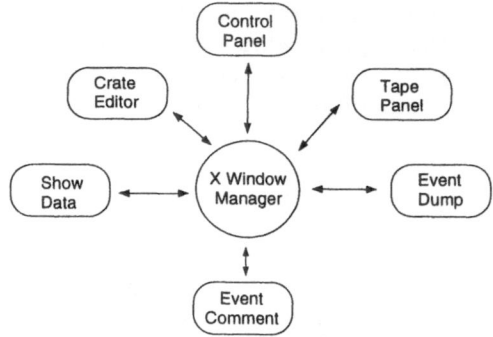

Figure 3. TTR graphical user interface.

An extensive TTR off-line software system was also developed[4]; it automatically fetches the data from disk and stores it on the tape-based data management system of the SSCL and simultaneously saves records describing run conditions on the SYBASE database. The off-line analysis is done through a GUI, written in C using Motif widgets and the X library, that allows the user to retrieve data files on the basis of criteria from the database. The GUI also allows the user to start the off-line analysis application on the selected data files, this application uses dynamically-linked user-selectable Fortran routines, a feature that provides the off-line analysis system with great flexibility.

SUMMARY

The two DA systems developed from scratch for the TTR were very successful, in particular it proved quite useful to have had the simple Macintosh/LabView system during the development of the more powerful SPARCstation/UNIX/VxWorks system. The latter system is now mature enough to be used at test beams.

REFERENCES

1. E.C. Milner, Overview of the GEM Muon System Cosmic Ray Test Program at the SSCL, SSCL-Preprint-256.
2. M. Botlo et al.,TTR Data Acquisition System, GEM TN-93-369.
3. L. Villasenor, TTR Event Display User's Guide, GEM TN-93-364.
4. I. Chow et al., Offline Event Reconstruction for TTR (User Guide), SSCL TN-93-00293.

OFF-LINE ANALYSIS SOFTWARE FOR THE TEXAS TEST RIG

G. P. Yost for the GEM Collaboration

Superconducting Super Collider Laboratory*
2550 Beckleymeade Avenue
Dallas, TX 75237-3997

ABSTRACT

Data analysis for the TTR requires integrating a large number of muon chamber technologies, each with different requirements, into a single analysis chain. Many of these technologies come with their own software, which have different conventions; these packages are grafted on. Data are stored on a tape robot with essential information stored in a database where it may be queried. Operation is done from special-purpose X^{TM} windows designed to facilitate data selection and its subsequent analysis. Program development was done using the Hewlett-Packard Softbench^{TM} product.

INTRODUCTION

The Texas Test Rig (TTR) is a 5 meter tall cosmic-ray telescope used to test and calibrate prototype muon chambers for the GEM Collaboration. Two planes of scintillator hodoscopes at the top and bottom provide a trigger. A one-meter thick steel filter which can be magnetized, located between the trigger planes and below the prototype chambers, restricts valid triggers to muons of energy above about 1.4 GeV/c^2. Four planes of Iarocci chambers of 1 cm pitch provide straight-line determination above (two planes) and below (two planes) the steel. These are used, when the magnet is energized, to measure the bend angle and thereby enable the muon momentum to be estimated. In fact, the bend angle in both the "bend" and the "non-bend" planes (relative to the field) can be measured, allowing the amount of multiple scattering to be used as a second factor in the muon momentum measurement, as described in Vanyashin and Yost.[1]

Further details on the TTR may be found in the paper by E. Cas Milner, submitted to this conference.

The data for the TTR are acquired as described by Luis Villasenor in this conference. The data are then split into two streams. The first stream contains the complete data before zero-suppression; this is stored on an 8-mm tape. The second stream contains that portion of

---

* Operated by the Universities Research Association, Inc., for the U. S. Department of Energy under Contract No. DE-AC35-89ER40486.

[1]A. Vanyashin and G. Yost, "Monte Carlo Studies of the Texas Test Rig Performance," GEM TN-92-99 (1992).

the data surviving zero suppression; this is stored on a disk. There is also an event display which is accessible on-line.

The off-line software is responsible for moving this zero-suppressed data to a tape robot for permanent storage, for entering indicative information regarding this data into a database, and for the analysis of this data. We proceed to describe the philosophy and practice of this software.

RUNNING THE OFF-LINE SOFTWARE

The TTR data flow is shown in Figure 1. The data come from the on-line system to reside temporarily on a disk. At regular intervals (e.g., every 15 minutes) the disk is queried for the presence of new data. The presence of new data is signified by a file inserted into a particular directory with a particular name structure. These files are not written until a run is complete. The query is initiated by a SYBASE™ database server running on the "Physics/ Detector Simulation Facility" (PDSF) at the SSC Laboratory.

When a new data file is present, it is retrieved over the network and stored on a tape robot. Certain indicative records, always present at the beginning and the end of the file, are scanned. These records contain information on the technologies which are taking data, the number of triggers, the temperature and humidity at the start and the end of the run, and other information. This indicative information is stored in the database along with a pointer to the associated data on the robot.

Once the data file has been stored, the filename on the temporary disk is changed. If the network should go down before the transfer is complete, this last step is not accomplished and the data file will be retrieved anew when the network returns. In this case, any portion of the file which may have been accessed prior to the network problem is over-written.

Data may now be analyzed using a package which currently runs on an HP 9000/730 workstation. A pop-up window presents three choices: query the database, run the analysis program, or run the event display.

In inverse order, the event display is identical with the on-line version and will not be discussed further. To run the analysis program a new pop-up window appears. In this window one must specify files to be processed. These may be located on the PDSF robot or on a local disk. The window calls for a file containing a list of files, both for the robot files and for the local files. Each such list of files may be constructed and/or edited from the window. The robot has a typical access time of about five minutes, so one is well-advised to transfer data files to a disk if frequent access is desired. The off-line program will read any number of files, beginning with the tape robot and proceeding to the disk when desired files off the robot are completed.

The off-line program also requires two files containing survey data and calibration data. Each file is in binary format and can be changed only by running a special program. This is to ensure that quick changes don't run the risk of creating a file which can not be read. The names of these files are entered in the off-line window; defaults are provided.

The off-line program has one additional running feature. The user may substitute his or her own special subroutines in place of standard routines. The user must place all such routines in a special archive file which is specified in the window. The program is then loaded dynamically, so that routines are loaded when called. The loader looks in this archive file first, then in the standard files, when loading a routine which has been called. Routines which happen not to be called are not loaded, and need not be present.

This subroutine-substitution feature is essential for a program which is intended to be used by a variety of people from different institutions. Each user will want to install their own code to analyze the data from their own chambers.

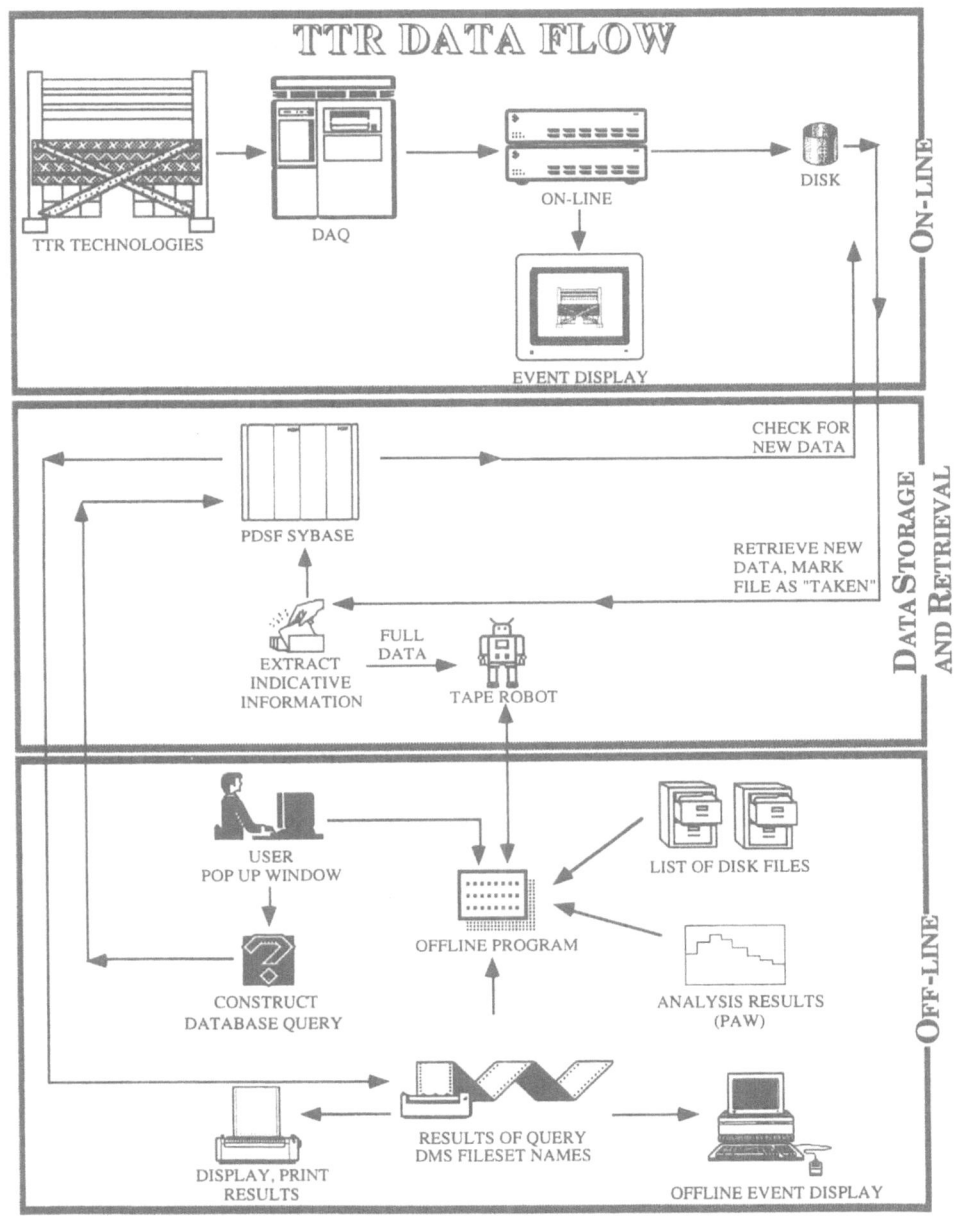

Figure 1. Texas Test Rig (TTR) Data Flow.

The final choice in the primary window is to access the database. This offers in its turn a set of windows with selections which are automatically entered into a database query. One may, for example, select runs featuring a certain set of chambers running with certain gases. One may select temperature and/or humidity ranges, magnet current in a range, and a range of run numbers. One may select a range of trigger counts, e.g., more than 1000 to ensure processing only of runs with adequate statistics.

Following the selections, one may press "fetch." The database typically returns the results in a few seconds. A window displays the filenames on the robot and the associated run numbers and list of the technologies active for that run.

Clicking on a particular run causes the complete set of database information for that run to be displayed in another window. This information may be printed. Finally, the results of the selections are passed to the off-line program window described above.

STRUCTURE OF THE ANALYSIS SOFTWARE

The analysis software is structured to provide "hooks" for user routines which may be substituted for dummy routines provided with the standard package. It also accomplishes a number of other tasks. It will read and unpack the data. It will read the survey file and store the surveyed locations of the chambers for later use (e.g., so that a given track could be followed from one set of chambers to another). It will read the calibration file and apply the calibration constants to, e.g., convert ADC counts to induced charge.

The program then proceeds to process the track. It makes an estimate of the location in the scintillators which was struck. It attempts to correlate this with Iarocci information and construct a track from the combined information both above and below the steel. If, as is sometimes the case, some of the Iarocci data are missing, it uses what information it does have to make an estimate of the trajectory. If a complete trajectory (both above and below the magnet) in either the bend or the non-bend plane, or both, is available, it is then used to estimate the momentum.

When this is done, the program calls standard routines (or their user-supplied replacements) to reconstruct the track in the prototype chambers and calculate a χ^2 for the fit.

This will be done for all tracks in the data files requested. If desired, selected summary data for each track may be written to disk for later input to a global fit. In such a fit, quantities such as survey locations or calibration constants may be varied to find optimized values for the collected set of tracks. This requires rerunning the entire set of tracks for each iteration, of course, and can therefore be very time-consuming.

SUMMARY

An off-line data storage, retrieval, and analysis package with novel features has been constructed for data collected by the Texas Test Rig. This package has successfully stored all the TTR data on a tape robot with indicative data stored in a SYBASE™ database. The program queries this database, retrieves desired data, and processes it in an analysis package.

MEASUREMENT AND ALIGNMENT GOALS FOR THE SDC DETECTOR

Robert J. Hovde

Physics Research Division/Solenoidal Detector Collaboration
Superconducting Super Collider Laboratory*
2550 Beckleymeade Ave, MS 2006
Dallas, TX 75237-3997

INTRODUCTION

There are two major goals driving the requirements for alignment and measurement of the detector being designed by the Solenoidal Detector Collaboration (SDC). First is each subsystem's goal of measuring its particular physics data as accurately as possible. Second is the goal of correlating the data from all subsystems in order to form a complete picture of the event. It is the ability to correlate all of the data that makes the SDC detector so powerful.

The major subsystems of the SDC detector surround the interaction point in a series of concentric layers. The tracker is at the center, surrounded by the superconducting solenoid magnet and the central calorimeter. The magnetic field of the solenoid is along the beamline, with the return path being through the steel of the calorimeter hadron section. The central systems sit within the muon system, which consists of a magnetized-iron barrel toroid, wire chamber modules and scintillation counters for muon identification and momentum measurement. At the ends of the barrel toroid are forward toroids and muon chambers, as well as a forward calorimeter.

The momentum of charged particles can be measured by the bend in their tracks as they pass through the magnetic fields (sideways bend in the solenoidal field, and towards or away from the beamline in the toroids). To complete the event description, the calorimeter system is designed to detect neutrals through the measurement of overall missing transverse energy.

The tolerances for both placement (location) and measurement (knowledge of the location) of components are very stringent. There may be several hardware and software combinations that can achieve the desired results, and the decision as to which route to take will be determined by performance, reliability and cost.

The SDC detector will be largely built on-site at the SSC Laboratory. The requirements for precise measurements, therefore, extend from basic manufacturing tolerance control to the final alignment of the entire detector. Manufacturing precisions are at the tens of microns level, while the overall alignment of the detector is at the hundreds of microns level..

---

* Operated by the Universities Research Association, Inc., for the U. S. Department of Energy under Contract No. DE-AC35-89ER40486.

FACTORS WHICH AFFECT ALIGNMENT

Mid-way through the detector, the two beams will cross at approximately 150 milliradians, resulting in a volume in which interactions will take place between protons of the opposing beams. Those interactions produce particles which will be measured by the detector . In order to relate all of the information, however, a trigger system must first discriminate events of interest from the massive volume of background events, passing only data from events of interest to be collected and stored for later off-line analysis.

Because of the speed of the interactions (10^8 events per second[1]), the first-level trigger must decide which data to save with very little processing time. This Level 1 trigger depends on precise placement of the detector elements, assuming that a particle that strikes a given element has traveled from the origin along a particular trajectory. If that data or combination of data from several detector elements match an expected trigger scenario, the trigger accepts the data.

The Level 2 trigger then takes the data that was accepted by the Level 1 trigger and compares the reactions of more elements before deciding that the combined group of data represents an interesting interaction event. The combined Level 1 and 2 triggers lower the accepted rate of events passed to the Level 3 processing by about five orders of magnitude.[2] Level 3 processing includes the entire saved data set, sending data from approximately 100 events per second to permanent storage.[3]

PLACEMENT ACCURACY

The requirements for positioning of the different detector components are based on the functions of the trigger, as noted above. Since there is essentially no time for computations in the Level 1 trigger, it must depend on the accuracy of a component's placement to compare its event measurement to expected values and to the measurements of other components. The accuracy of the placement is chosen so that the error in placement will not seriously degrade the component's inherent measurement capability. These decision variables are described below, in the discussion of each subsystem's position goals. Other variations, caused by the system-level environment, must also be corrected.

The collider beamline slopes down 2.16 mm per meter through the hall from North to South. This means that the entire detector must be positioned on a 0.12 degree slope, with the north end 60.5 mm higher than the south end. A system of jacks will be used to keep the detector coaxial with the beamline. The detector will be assembled and aligned several years before the beamline magnets are actually installed. The initial survey of the experimental hall, therefore, will be crucial to the interface of the detector and collider.

The 75 million pound weight of the detector will cause deflections in the floor and barrel support structure - which could be as much as 20 mm. These movements will be jacked out by the main support jacks. The weight of the barrel iron itself will cause a change in the cross section from octagonal to ellipsoidal (the top will move down as much as 4.1 mm and the sides will move out 3.6 mm[4]). There will also be movements due to removal and re-installation of the end toroids and calorimeter.

Movement of components due to changes in temperature or the force of magnetic fields must also be taken into account. The diameter change (flat-to-flat) for the barrel toroid, for example, would be approximately 0.9 mm for a 5 degree Celsius temperature rise.[5] The change in the radius of the barrel toroid when the magnet is excited will be of order 70 microns,[6] approximately the same as the measurement precision of the muon alignment system.

MEASUREMENT ACCURACY

The Level 2 trigger and the off-line reconstruction of tracks require a very closely measured location of each component. While the Level 1 trigger divides the detector into 64 bins around the circumference, the Level 2 trigger uses 1024 bins. For off-line track reconstruction, each individual silicon chip and each drift-tube wire must be located precisely in relation both to each other and to a common reference system. The reconstruction of tracks and, therefore, the calculation of particle momenta, will only be as good as the measurement system's output.

Since it is impractical to physically measure each component's actual position on a continual basis, the alignment system will only locate fiducial markings. These locations, along with temperature measurements, inclinometer readings, and the positions calculated from previous data, will be fed into a computer model which will compute the locations of individual detector modules and wires. The computed positions will then be used in track reconstruction calculations.

ALIGNMENT/MEASUREMENT REQUIREMENTS

Central Tracker

The central tracker consists of three major components which must be in close relative alignment to act as one unit for trigger purposes. The central, silicon system must be centered within the outer barrel tracker to 150 microns and with respect to the intermediate tracker to 1 mm.[7] For off-line track reconstruction, these two values must be known to 15 microns and 40 microns, respectively.[8] Since transverse momentum is measured by changes in circumferential position with respect to radius, the rotational alignment of the silicon system vs. the barrel and intermediate systems is held to a placement of 10^{-4} radians, with measured position knowledge of 10^{-5} radians.[9] For Level 2 trigger purposes, which assume that a track starts from the orgin, the tracker needs to be centered on the interaction point to within 500 microns.[10]

Silicon Tracker: The specifications for alignment of the internal silicon tracker components are at the micron level. Manufacturing tolerances will result in a 10 micron local alignment of the silicon substructures, with a build-up of tolerances resulting in the resolution at the external fiducial markings of 135 microns.[11]

Barrel (Straw) Tracker: The barrel tracker is very sensitive to circumferential errors, with circumferential placement and measurement requirements of 140 microns and 35 microns, respectively, while the radial position requirements are only 1,200 microns.[12]

Intermediate Tracker: The intermediate tracker, presently designed as a gas-microstrip detector, has three parallel, circular planes of detectors attached to a cone-shaped support frame. The detector elements are precisely placed on each plane to a local circumferential alignment of 40 microns, while the radial position is only required to be within 1 mm.[13]

Superconducting Solenoid Magnet

The superconducting solenoid magnet will be adjusted during installation to a location precision of 500 microns,[14] with its magnetic axis aligned to within 5 microradians of the beamline.[15] Once in place, there will be no further adjustment of its position.

Calorimeter

The central calorimeter (barrel halves and end caps) rests on a wheeled support system that must transfer its 8 million pound weight through a rail structure ("picnic tables") to the muon barrel toroid steel. The forward calorimeter will be supported by a similar system. The jacks in the central calorimeter support system must be able to move through approximately 5 cm vertically and 2.5 cm horizontally to ensure that the shower maximum sections are within 3 mm of being concentric with the beamline[16] and that the tracker is within its approximately 5 mm adjustment range. Additionally, the tracker and calorimeter are divided into 1024 bins for the Level 2 trigger, and must remain within this six milliradian bin alignment for the trigger to operate properly.

Muon Measurement System

The muon barrel and intermediate modules are supported on rails which are fixed to the muon barrel steel. The rails will have an adjustment capability during assembly of approximately 10 mm with their position and alignment determined by survey. The intermediate modules are cantilevered from the ends of the barrel, and are installed near the end of the assembly sequence, after the detector has undergone its major deformations. The forward modules are attached to the forward toroids and will have some post-installation adjustment capability.

The placement tolerances for each of the muon modules is calculated based on the movement (translation and rotation) that will cause a 10 percent degradation in the measurement resolution of the module.[17] All of the muon modules have relatively loose placement tolerances of several millimeters of translation but the forward modules have a very tight 0.41 milliradian tolerance in pitch.[18]

Since all of the detector components depend on the barrel toroid support system for basic positioning, that system has been designed to be adjustable. The jacking system will have to align the entire detector to the sloping beamline, as well as having to compensate for settling of the floor or of the detector itself. Since the muon alignment system will not have good lines-of-sight through the support structure beneath the barrel, a liquid-level system and temperature monitors will provide data for that part of the system. A combination of data from straight-line monitors, proximity sensors, a range-only measurement network, and the liquid-level, temperature sensors will be integrated via a computer model to determine actual detector element locations.[19]

SUMMARY

A combination of precise manufacturing tolerances and careful assembly will relate each detector component to a fiducial. The placement of the fiducials with respect to the primary detector coordinate system and with respect to each other will be accomplished by surveys and measurements at each successive step of assembly. The precise locations will then be continuously monitored by the alignment system, either by direct measurement or by inference through a computer model. The subsystem fiducial placement tolerances vary from 500 microns to several millimeters in translation and 0.41 milliradians to several milliradians in rotation, while the manufacturing and location measurement tolerances are very tight, at the microns and tens of microns levels. Alignment and measurement equipment and procedures continue to evolve, still remaining flexible enough to accommodate new ideas and components.

REFERENCES

1. A. Lankford et al., *Conceptual Design of the SDC Trigger from the SDC Trigger Group,* SDC-91-00098, September 30, 1991, pg. 1.
2. *Solenoidal Detector Collaboration Technical Design Report,* SDC-92-201, 1 April 1992, pg. 8-3.
3. Ibid., pg. 8-7.
4. H.J. Krebs, Ed., *Preliminary Design Report - Muon Barrel Toroid, Version 1.2,* SDC-92-342, October 7, 1992, pg. 2–8.
5. Ibid., pg. 2-8.
6. R. Bell, Committee Chairman, Ed., *SDC Muon Barrel Toroid Preliminary Design Review Committee Report,* SDT-000069, November 18, 1992, pg. 4.
7. R. J. Hovde, Ed., Minutes, *SDC Alignment Working Group, November 12, 1992,* SDT-000068, November 16, 1992, pg. 61.
8. Ibid., pg. 61.
9. Ibid., pg. 61.
10. Ibid., pg. 61.
11. Ibid., pg. 62.
12. Ibid., pg. 61.
13. Ibid., pg. 61.
14. A. Yamamoto, Telefax to R. Hovde, SSCL, with position and measurement requirements for the superconducting solenoid magnet, 28 Sep 92.
15. R. Little, *Misalignment of Solenoid Field,* PMO Memo 93 4700.52 E B, March 18, 1993.
16. *Solenoidal Detector Collaboration Technical Design Report,* SDC-92-201,1 April 1992, pg. 6-7.
17. S. Behrends, P. Hurst, J. Bensinger, *Performance-Based Tolerances For Muon Supermodule Alignment, Version 3.3,* SDC-92-376, 4 December 1992, pg. 1.
18. Ibid., pgs. 8-9.
19. R. Hovde, Ed., *Minutes, SDC Alignment Working Group, November 12, 1992,* SDT-000068, November 16, 1992, pg. 34.

AN ENGINEERING DESIGN STUDY OF THE TRANSFER BRIDGE FOR THE SSC SDC DETECTOR

J. L. Western, PE, SE,[1] G. W. Tulk, PE,[2] and L. I. Dittert, PE[3]

[1]Conventional Construction Division
[2]Physics Research Division
 Superconducting Super Collider Laboratory*
 2550 Beckleymeade Avenue
 Dallas, Texas 75237

[3]Principal Engineer
 ICF Kaiser Engineers, Inc.
 1800 Harrison Street
 Oakland, California 94612

ABSTRACT

The transfer bridge is a major structural installation component for the large detector proposed by the Solenoidal Detector Collaboration (SDC). The transfer bridge is the structural device that allows for installation of the Muon Forward Toroid (MFT) and calorimeter components into the Muon Barrel Toroid (MBT). This paper summarizes the design and construction of the SDC transfer bridge.

INTRODUCTION

The SDC detector is a major SSC experiment located at the IR8 site of the east campus. The total weight of the detector will be approximately 32,000 metric tons.

The Muon Barrel Toroid (MBT) is the primary support mechanism for most components of the SDC detector such as the Muon Forward Toroid (MFT) and Barrel and End Cap Calorimeters.

Each MFT has a weight of about 3,000 metric tons. One is located at each end of the MBT. The MFT is designed to be moved in and out of the MBT to facilitate access to the detector components located inside the MBT (Figure 1).

Each half of the Barrel Calorimeter weighs 1,200 metric tons and each of the End Cap Calorimeters weigh about 600 metric tons. They are located inside the MBT and are designed to be moved in and out of the MBT for detector assembly, repair and/or modifications (Figure 1).

---

* Operated by the Universities Research Association, Inc., for the U. S. Department of Energy under Contract No. DE-AC35-89ER40486.

The principal objective of the transfer bridge design is to develop a removable, easy to assemble, and structurally safe transfer bridge that will permit the movement of the MFT and the Barrel and End Cap Calorimeters between the operating floor and the MBT.

GENERAL DESCRIPTION

The transfer bridge will consist of two main structures: the support grillage and the bridge girder assembly, and two cantilever walkways weighing a total of about 85 metric tons. Each component of the transfer bridge will be installed by the hall crane between the ends of the MBT and the operating floor.

The columns and bracing of the support grillage will be fabricated from standard structural shapes and the bridge deck will be made up of two braced box girders and heavy floor plates (Figure 1). Two removable cantilevered walkways outboard of the bridge deck will also be provided. The floor plates will be designed to permit small mobile equipment such as fork lifts, welding machines, etc. to access the detector, and the top flange of the box girders will serve as the track plates for rolling the MFT and the calorimeters into the detector.

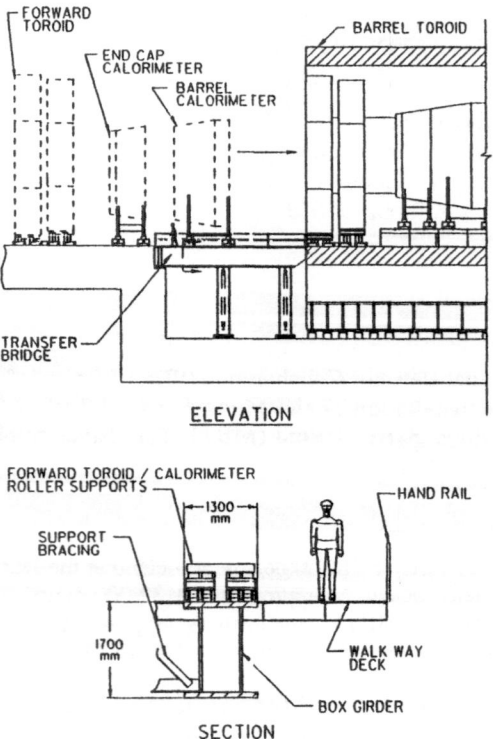

Figure 1. Top figure: illustrates bridge elevation. Bottom figure: illustrates section at box girder.

SPECIAL DESIGN REQUIREMENTS

The SDC Detector, due to MBT plate block construction, will not be able to support large concentrated transfer loads and will be sensitive to local deflections greater than a few millimeters. The transfer bridge interface with the MBT will be designed to limit loads transferred, as well as match stiffness at this interface to limit deflection differentials.

894

BRIDGE GIRDER CONSTRUCTION

As shown in Figure 1, the bridge girders will be fabricated from four thick plates welded together to form a box girder. The top flange of the girder will be grooved to act as roller guides. The main reasons for using a box girder design will be to permit lowering of the MFT onto the transfer bridge, to adjust or repair the rollers during transit of the MFT, and to reduce the transverse bending and twisting of the top flange due to the roller loading. Due to roller orientation, the minimum flange width will be 1074 mm plus the edge distance required.

The two box girders will be tied together with wide flange beams and bracing to prevent rotation and to maintain the gage of the track grooves at all times. A floor plate, level with the top flange of the girders, as well as several lifting lugs for the crane will also be provided. The weight of the completed bridge girder assembly will be about 71 metric tons.

The box girder will have welded stiffeners and bearing plates at the operating floor end, and thick end plates with horizontal shear pins welded to them at the MBT end.

BRIDGE SUPPORT GRILLAGE

The box girders will be supported at the operating floor and at two other locations by a grillage. The grillage will be a simple welded space frame built from standard rolled shapes. The four columns of the grillage will rest on base plates bearing on the detector basement floor. The weight of the grillage will be about 70 metric tons.

DESIGN ANALYSIS

A three-dimensional model was created to determine the stresses, deflections, and support reactions of the transfer bridge subjected by the roller loads of the MFT moving across the bridge.

To find the maximum deflections, stresses, and reactions the model was subjected to several loading positions to represent the MFT moving across the bridge. The results indicated that to reduce girder deflections and loading on the shear pins, the columns of the grillage must be larger than what would be required for stability alone.

Not considering the movement of the concrete floor slab, the maximum deflection of the box girder was 1.5 mm. More importantly, the maximum vertical distortion on a 3.5 meter length of the track (the length of the roller train under the MFT) was calculated to be only 1.0 mm.

INSTALLATION

First, the transfer bridge support grillage will be placed on the basement floor of the detector hall. Then, the girder assembly will be installed on top of the grillage by sliding it forward and dropping the girders into the pockets of the operating floor. In the next step, horizontal hydraulic jacks in the floor pockets will move the girder assembly towards the MBT until the shear pins are fully engaged with the holes in the last plate of the MBT. The collar nuts of the jacks will be locked to ensure that the longitudinal force of the MFT moving jacks will not displace the girders.

In the final step, the support grillage will be jacked up against the box girders for full bearing, and the base plate of the grillage will be grouted to plates embedded in the concrete floor using a quick-setting flowable grout.

The removable walkways outboard of the box girder will be installed to provide access around the MFT moving jacks.

INTERFACE WITH MBT, MFT, CALORIMETERS, AND HALL FLOOR

At the ends of the MBT, two areas will be kept clear of all piping, cables, etc. to permit butting of the bridge girders.

The details and gage of the roller guide grooves in the top flange of the box girders (used as track plates) must be coordinated with the roller designs of the MFT and the calorimeter installation frame. The transition details of the track plates at each end of the bridge must be also coordinated.

The size, elevation, and details of the four plates embedded in the concrete floor for the grillage columns, and the details of the box girder supports at the edge of the operating floor have to be developed for the Architect/Engineer designing the Experimental Hall.

Most importantly, bridge loadings for the MFT, End Cap and Barrel Calorimeter installation frames, and for mobile equipment to be used for installation or maintenance of detector components must be established.

MATERIAL SPECIFICATIONS

Materials Specification for the transfer bridge and support grillage follows:

- Plates and shapes for bridge assembly, except girder top flanges
  - ASTM A36 material.
- Girder top flanges - ASTM A514 material.
- Plates and shapes for grillage - ASTM A36 material.

TOLERANCES

Tolerances for the transfer bridge and support grillage fabrication will be in accordance with the American Institute Steel Construction (AISC) Specifications for the Design, Fabrication, and Erection of Structural Steel for Buildings, except as follows:

- Tolerance for the working lines of individual columns of the grillage will be 1:1000
- Track plate groove gage tolerance will be +/-3 mm
- Track plates, including the top flange of the box girders will have a flatness tolerance of 3 mm on any 5 meter length of track, and will be level across the tracks to a tolerance of 1:1000
- Tolerance for track plate elevations will be +/-3 mm

CONCLUSION

The transfer bridge will be a major structural component for assembly and disassembly of the SDC Detector, with the detector itself sensitive to the transfer bridge performance. The relative stiffness of the SDC Detector and transfer bridge interface needs to be considered in the design of the transfer bridge.

ACKNOWLEDGMENTS

The authors would like to acknowledge the contribution of R. Gates in preparation of this paper.

REFERENCES

Leung, K. K. and Western, J. L., 1992, An Engineering Design Study of Detector Deformation Limits in the SDC - Detector, in: "IISSC Supercollider 4," pp. 167-174 Plenum Press, New York.

Western, J. L. and Butalla, M. W., 1992, SDC Detector Foundation Requirements, in: "IISSC Supercollider 4," pp. 1119-1125 Plenum Press, New York.

Dittert, Les, 1992, "SDC Forward Toroids Transfer Design Conceptual Design," ICF Kaiser Engineers, Inc., California.

SDC MUON BARREL TOROID 1/9.197 SCALE TEST MODEL AT SSC

C.W. Jeff Tseng,[1] Y. Gusakov,[2] J. J. Cherwinka,[3] and R. H. Wands [4]

[1]Superconducting Super Collider Laboratory, Dallas, TX 75237*
[2]Joint Institute for Nuclear Research (Dubna), Moscow, Russia
[3]University of Wisconsin, Madison, WI 53760
[4]Fermi National Accelerator Laboratory, Batavia, IL 60510

INTRODUCTION

The purpose of the scale model as shown in Figure 1 for the muon barrel toroid (MBT) is to discover any problems in the preliminary design[1] associated with the fabrication, assembly and installation.

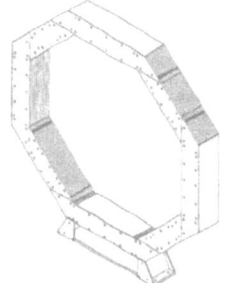

Figure 1. MBT test model ring assembly and support.

The information obtained from the model fabrication, assembly, installation, and testing processes will be used to evaluate the MBT preliminary design and to verify the finite element (FE) analysis. The final design of the MBT will take advantage of the experience gained from the test model to improve the design and engineering, fabrication methods, and assembly methods. The goals that are considered relevant to the test model include the following:

- Verify the in-plane and out-of-plane stiffness of the test model as predicted by the FE analysis.
- Verify the corner joint stiffness of the test model as predicted by the FE analysis.
- Verify the fabrication and assembly methods for the blocks, pins and keys on a small scale.
- Demonstrate the plate flattening scheme on a small scale.
- Insure that the preliminary design of the MBT can be assembled into the correct shape on a small scale.

A magnetic measurement will be performed to record the field density inside an air gap. The main goal is to find out the field distribution for the test model due to geometry variation. The field measurements will also provide the information to verify the magnetic property of the steel material as well as the impact on the field distribution due to material variation.

* Operated by the Universities Research Association, Inc., for the U. S. Department of Energy under Contract No. DE-AC35-89ER40486.

DESCRIPTION OF TEST MODEL

Magnet Toroid

The test model has a scale of 1:9.197 compared to the full MBT. The octagonal ring consists of eight long blocks and four short blocks and represent 1/16 of the MBT. The details of design and dimensions for the test model are provided in Reference 2. Each long block consists of four 5/8" inside plates and two 3/4" outside plates. Each short block consists of ten 5/8" inside plates and two 3/4" outside plates. The long block uses 12 longitudinal bolts to hold the six plates together, while the short block uses eight longitudinal bolts to hold the twelve plates together. After assembly, the blocks are machined to final size. At either side of the block, there are six counterbore bolts for the long block, and twelve counterbore bolts for the short block. These corner bolts are used to connect the short and the long blocks into an octagonal ring.

There are two pins for each of the long and short blocks that are threaded to act as nuts for the corner bolts. There are two taper pins to be installed for each of the long and short blocks after the blocks are assembled together. These pins are intended to carry the shear when the blocks are being machined. There are two square keyways for each long and short block to help locate the blocks and carry the shear at the joint. There are four shear plates in the longitudinal direction for connecting the long blocks together. Since the scale model represents only one ring, the shear plates have much less impact than the shear plates for the full MBT.

Mechanical Tests

(1) Vertical Deflection (x-y plane) – Measure the deflection in x and y at the top, middle and bottom sections of the ring, and the bolt load at the lower corner of the ring under various loads. This tests the stiffness of the ring in x-y plane (Figure 2).

(2) Lockwasher Deflection (y-z plane) – Measure deflections in z at the top portion of the ring and the bolt load at the lower corner when the load is applied in opposite directions above and below the upper block joint. This tests the lockwasher stiffness of the ring (Figure 3).

(3) Vertical Misalignment (x-y plane) – Measure forces in y at the upper corner to create a vertical gap between the two adjacent blocks. The bolts at one corner are removed to allow a vertical gap under a load. This tests the stiffness of the upper block joint and corresponding bolt loads (Figure 4).

(4) Block Stiffness (x-y plane) – Measure deflections and relative displacement between laminated plates for a simply supported long block. This tests the stiffness of the laminated block (Figure 5).

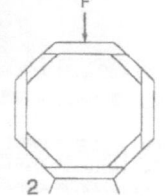 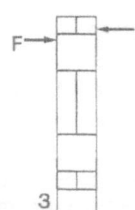 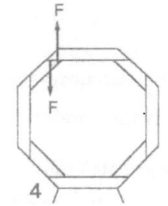 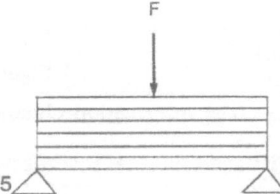

Figures 2 – 5

Magnetic Tests

Magnetic Field (x-z planes) - Use a Gaussmeter and a transverse Hall probe to measure the magnetic field density in y inside the air gap (2 mm) between the adjacent blocks. The coils are placed in the center of each octane of the toroid. This tests the field uniformity and strength. (Figures 6 and 7)

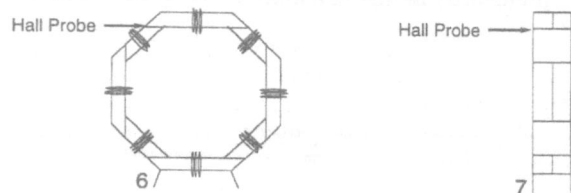

Figures 6 – 7

TEST RESULTS

Mechanical Test Results

The comparison between the test model results and the FE analysis results is only applicable where the test model behavior is fairly linear because the FE models are constrained to behave linearly over the entire load range. The torque for all corner bolts is 90 in-lbs for the scale model and the FE models. The comparison is provided in Table 1. There is good agreement between the stiffness predicted by the FE model and the stiffness measured in the tests.

Table 1. Force/Displacement (kg/mm).

Case	Test Model	FEA	% Difference
1	13,500	13,700	+ 1.5
2	5.5	6.5	+ 18.0
3	356	288	− 19.0
4	3,496	4,130	+ 18.0

Magnetic Test Results

The magnetic field density measured in y for various points inside the air gap (in x-z plane) indicates the maximum field being 1.3 Tesla. The cross sectional area for the air gap is about 1.4 times the cross sectional area in the central region for the rest of the blocks. Since the filed lines are fairly perpendicular through the air gap and the total flux remains constant for each cross section, the magnetic field in the center region of the block is about 1.82 Tesla with the proper geometric correction. This measured value is in close agreement with the design value of 1.8 Tesla which is based on the permeability of the steel and the amp-turn design of the coils. (Figure 8)

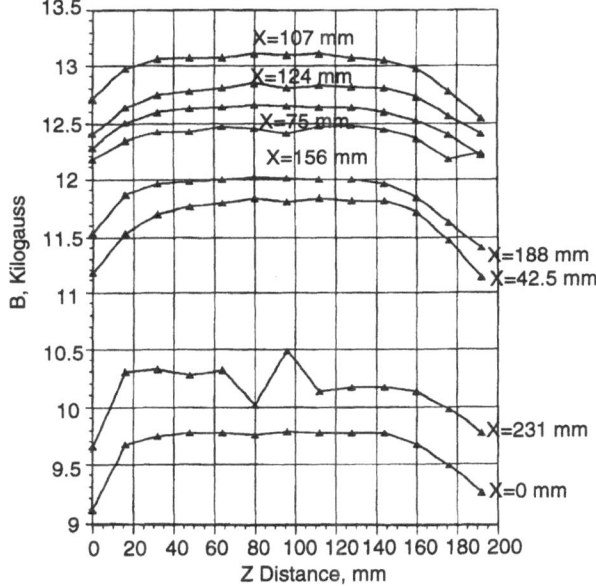

Figure 8. Magnet Field Distribution for Various X Distance inside the Air Gap (X=0 mm to X=231 mm).

CONCLUSION

The scale model does not indicate any difficulties in fabricating the parts, assembling the blocks, and installing the ring to meet the preliminary design of the MBT. All tolerances as required by the design can be achieved by the fabricator. The main deviation from the design is that all blocks except one used all-machined plates before the assembly.

The corner joints are tested within the safety limit for the M6 bolts. The corner joints remained under compression throughout the testing loads. There is no indication of failure under any testing case.

The comparison between the test results and FE results for case 1 are in good agreement as shown in Figure 9. An additional test case was performed to evaluate the corner joint stiffness for a tip-loaded cantilevered long block. The comparison between the test results and FE results is good and the FE model predicted higher (conservative) bolt loads when the external force gets higher (Figure 10). This provides the verification that the FE assumptions and modeling technique are sound and FE results are reliable in predicting the behavior of the full size structure.

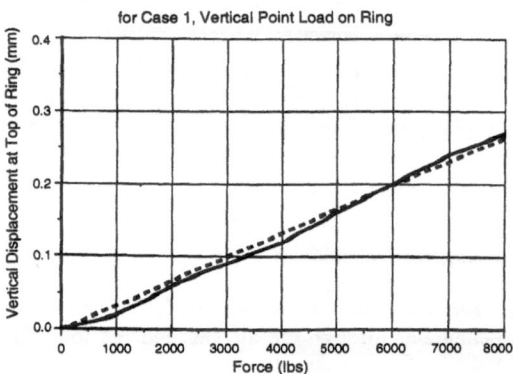

Figure 9. Comparison between test and FE results for case 1.

The magnetic distribution in the test model is quite uniform. The maximum field density of 1.3 Tesla is in agreement with the coil design and the magnetic property of the steel material used for the scale model. The impact from a material variation in one long block is found to be not as significant as predicted for the test model. The magnetic simulation for the scale model is not complete, but the preliminary results from computer simulation are in close agreement with the field measurement.

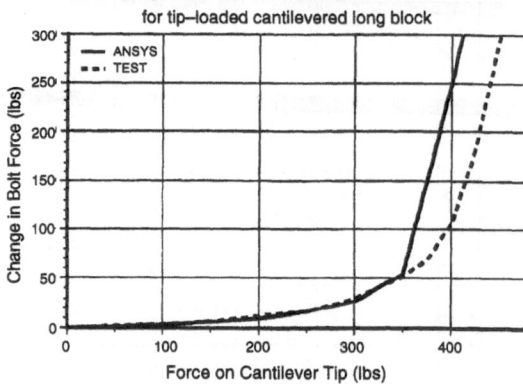

Figure 10. Comparison between test and FE results for tip-loaded case.

REFERENCES

1. J.J. Cherwinka and C. W. Jeff Tseng, SDC Muon Barrel Toroid Specification, *SDC Solenoidal Detector Notes*, SDC-92-284, June 1992.
2. C.W. Jeff Tseng and Y. Gusakov, SDC Muon Barrel Toroid Model, Scoping Document for MBT 1/9.197 Scale Test Model, *SDC Solenoidal Detector Notes*, SDC-92-258, May 1992.
3. M.F. Spotts. "Design of Machine Elements," Prentice-Hall, Inc., Englewood Cliffs, 1978.
4. L.S. Marks. "Standard Handbook for Mechanical Engineers," McGraw-Hill Book Company, New York, 1966.

THE PROPOSED VACUUM SYSTEM FOR THE INTERACTION REGIONS
OF THE SUPERCONDUCTING SUPER COLLIDER

G. R. Chapman

Superconducting Super Collider Laboratory*
2550 Beckleymeade Avenue
Dallas, TX 75237-3997

INTRODUCTION

The design of the beam pipe located in the GEM and SDC detectors situated at interaction regions 5 and 8 of the SSC is influenced by three major considerations. The first is the desired pressure in the vacuum system. Experience has shown that a pressure of 10^{-8} torr is achievable under realistic operating conditions. Preliminary calculations show that at 10-8 torr, the beam gas background will remain acceptably low. In the Central Tracker, the beam gas contribution at a pressure of 10^{-8} torr is 2 orders of magnitude below that due to primary interactions. In the muon system, the neutron fluence due to beam gas is a factor of 20 below that due to particle-to-particle (pp) interactions. However, if the vacuum were degraded to 10^{-7} torr, the neutron fluence in the muon system due to beam gas interactions, would become comparable to the rate of pp interactions. Furthermore, continuing studies will enable us to better shield the pp source rather than the beam gas source. The second major input to the beamline design is driven by the desire to minimize secondary interactions in the beam pipe itself. In the central region, this is achieved by the use of a thin walled Beryllium section of pipe. In the forward regions, the path length of particles traversing the beam pipe is much longer, and it is necessary to avoid a potentially large flux of neutrons which could be created in the beam pipe walls. The GEM vacuum assembly has been designed to provide a conical section of beamline arranged such that it lies outside a cone of 0.5 degrees from the interaction point and is therefore entirely in the shadow of the Forward Calorimeter. Work is ongoing to incorporate appropriate flares in SDC beam pipe. The third consideration is a more practical one. The assembly sequence of both the GEM and SDC Detectors proceeds in such a way that the beam pipe must be designed in discrete sections that can be installed concurrently with the overall installation of the experiment.

BEAM LINE DESCRIPTION FOR THE GEM VACUUM SYSTEM

Length and Diameter

The total length of the beam pipe between the collimators of the final focusing quadrupoles is 62 m: ± 31 m from the Interaction Point (IP). It consists of a central section of Beryllium pipe running through the Central Tracker, coupled to successive lengths of Stainless Steel, or Aluminum alloy pipe which vary in diameter. In addition to the pipe itself, there are associated pumps, bellows, valves and Beam Position Monitors. The beam pipe and its associated hardware from the IP to +31 m is shown in Figures 1 and 2. The current length, diameter and location of each sector are listed in Table 1.

Material and Thickness

The beam pipe passing through the Tracker is Beryllium, nominally 1.5 mm thick. The low density, low out-gassing rate and high rigidity of Beryllium makes it an ideal material to satisfy the design requirements.

* Operated by the Universities Research Association, Inc., for the U. S. Department of Energy under Contract No. DE-AC35-89ER40486.

Stainless Steel and Aluminum alloys have been considered for the subsequent sections of pipe. Stainless Steel is preferred for its ease of fabrication and vacuum conditioning. Aluminum is more difficult to fabricate and condition, and its use could result in a higher average pressure; however, it's residual activation characteristics are better than those of Stainless Steel. Never the less, GEM is designing with stainless steel as the beam pipe in the Calorimeters is so close to the cryostat walls that there will be negligible difference in activation.

The nominal thickness of the Stainless Steel pipe will be not greater than 1 mm for 80 mm diameter sections and not greater than 2 mm for 200 mm diameter sections. The thickness of the conical section can be gradually increased by increasing the pipe diameter, from 1.2 mm, where the diameter of the pipe is 134 mm, to 4.2 mm thick at its largest. Table 2 lists the beampipe by values for the SDC.

Table 1. Current length, diameter and location of eack sector.

PIPE SECTION	NO.	LOCATION (MM)	LENGTH (MM)	DIAMETER (MM)
Central Tracker with Bellows	1	IP to ±2000	4000	80
End Cap Calorimeter with NEG	2	±2000 to ±3500	1500	200
Forward Calorimeter with BPM Assembly	2	±3500 to ±8512	5012	80
Forward Field Shaper conical with Ion Pumps	2	±8512 to ±31000	22488	134 TO 518

Table 2. SDC Configuration.

PIPE SECTION	NO.	LOCATION (mm)	LENGTH (mm)	DIAMETER (mm)	TAPERED ANGLE (deg)
Central Barrel Tracker	1	IP to 4327	8654	80	
End Cap Calorimeter	2	±4327 to ±6837	2510	80 to 126.95	0.53
From Far End of End Cap Calorimeter to Near End of Forward Calorimeter (with Sector Value)	2	±6837 to ±12670	5833	126.95 to 204	0.53
Forward Calorimeter	2	±12670 to ±15420	2750	60 to 90.446	0.140
From Far End of Forward Calorimeter to Collimator with Sector Value	2	±15420 to ±21240.8	5820.8	90.446 to 121.61	0.140

PRESSURE REQUIREMENTS IN THE BEAM PIPE VACUUM SYSTEM

To achieve reliable operating pressure of 10^{-8} torr under realistic conditions, requires a conservative approach to the design of the vacuum system. For example, it is desirable to perform a final *in situ* bake-out of the entire beam pipe although at present this appears to be difficult in certain areas of the experiment and impossible in others. To understand the vacuum profile under these conditions, a more detailed set of vacuum calculations, assuming less than perfect conditions, has been accomplished and is discussed in detail in technical note GEM TN-93-290. This information is summarized in Figures 3 and 4. For GEM and Figure 5 for the SDC vacuum system.

The pumping speed requirement and pumping locations are determined by these calculations, consistent with the detector geometry. This implies that unique pumping systems are required.

GEM Vacuum Assembly

A combination of ion pumps and Non-Evaporable Getter (NEG) pumps are required to reduce pressure in the Beam Pipe to the required ultra high vacuum range (10-9 torr) or a lower parameter will require ion pumps plus sublimation with pumping speeds of about 1000 Liters-per-second (1/s) located at positions shown in Figure 1. The ion pumps must be able to operate in the GEM magnetic fields. These pumps will be specially designed to operate without the use of standard permanent magnets. Pumping will be achieved using the magnetic field of the GEM solenoid when it is on. The pumps will be fitted with a supplementary coil to supply a low magnetic field (1.5 k gauss) for pumping when the GEM magnet is off. To maximize pumping speed, the connections between the ion pumps and the beam pipe must be the same diameter as the ion pump inlets and as short as possible.

In the central region the problems are more acute. The long distance and low conductance of the 80 mm beam pipes in the Central Tracker and Endcap Calorimeter sections require some auxiliary pumping. We plan to use heat activated Non-Evaporable Getter (NEG) pumps with pumping speeds of 300 l/s in front of the End Cap Calorimeters in order to achieve the best possible vacuum without the inclusion of more massive ion pumps. A prototype NEG pump is currently under construction. It uses conduction heat activation rather than a direct electrical connection through the vacuum wall. If successful, this technique will allow for more reliable operation in the inaccessible Central Tracker region.

The SDC detector assembly (due to a smaller solenoid) is arranged so that it is possible to locate all ion pumps outside the influence of the magnetic field of the solenoid, hence commercially available ion/sublimation pumps are proposed to be used positioned as shown in Figure 5.

Calculation Results for GEM and SDC

GEM contains one NEG of 300 l/s and four ion pumps plus sublimation of 1000 l/s, the calculated pressure distribution for the beam pipe from the interaction point to 31 m is shown in Figure 4. The average pressure is 7.06×10^{-9} torr. The calculation in the interaction region, shown in Figure 3, shows that the pipe diameter of the End Cap Calorimeter section is critical to the average pressure in this region. If an 80 mm diameter pipe is used in this section, the average calculated pressure changes from 1.02×10^{-8} to 1.28×10^{-8} torr, remaining below the design figure, no matter where the NEG is placed. However, if a 200 mm diameter pipe is used in the End Cap section, the average calculated pressure drops to 7.66×10^{-9} torr.

SDC consists of the calculated average pressure without the addition of a supplementary NEG pump and with ion pumps positioned as shown in Figure 5 is 4.87E-9 torr with a pressure profile as shown in Figure 6.

Out-gassing Rate for standard calculations for pressure distribution in beam pipes assumes perfect conditions with correctly baked-out pipes. As mentioned above, a final bake-out of the GEM and SDC beam pipes appears to be difficult for most of the pipe sections and impossible for the Central Tracker area. Therefore, less perfect conditions must be assumed in our calculations. The out-gassing rates in these calculations, in contrast with perfect conditions, are listed in Table 3.

Table 3. Out-gassing Rates (torr.liter/sec. cm^2).

MATERIAL	IN THIS CALCULATION	PERFECT CONDITIONS
Beryllium	5.0×10^{-11}	7.0×10^{-14}
Stainless Steel	5.0×10^{-11}	2.0×10^{-13}

INTERFACES

Detector Beamline/Machine Interface

The final focusing quadrupoles are located at ±35 m from the interaction point, with 3 m long protection collimators located in front of the quads, beginning at ±32 m. It is proposed that the warm flange of this collimator should be the connecting interface between the Gem Detector vacuum assembly and the collider vacuum assembly. This and other machine/detector interface issues are being defined and documented in an ongoing series of machine/detector interface meetings.

Beamline/Detector Interface

Assembly of the beam pipe will be in sections as the assembly of the GEM and SDC Detector progresses. This assembly procedure (for GEM) is described in detail in Chapter 9, Facilities, Assembly and Installation. Further documentation is provided in a Gem Technical Note, GEM TN-93-289.

GEM: Beam Position Monitors (BPM) and Pumping Station Provisions are being made for the installation of two Beam Position Monitors (BPM's), one on each side of the interaction point, located immediately after the Forward Calorimeter. Each BPM assembly would consist of a gate valve (about 4" ID), a BPM and two bellows assemblies. A decision on the final location of the BPM's will be made after the details of other detector parameters are frozen.

There are four pumping stations on each side of the interaction point. Each pumping station consists of an ion pump plus sublimation with pumping speeds of 1000 l/s, a 6" diameter gate valve and a cross, which has a port for an ion gauge or a probe of a mass spectrometer and a port for the roughing line. The positions of the pumping stations are shown in Figure 1. A pressure distribution calculation of the beam pipe is shown in Figures 3 and 4. The pumping speed requirements and pumping locations are determined by this calculation. Access must be maintained to all ion pumping stations. To minimize material in the highly radioactive

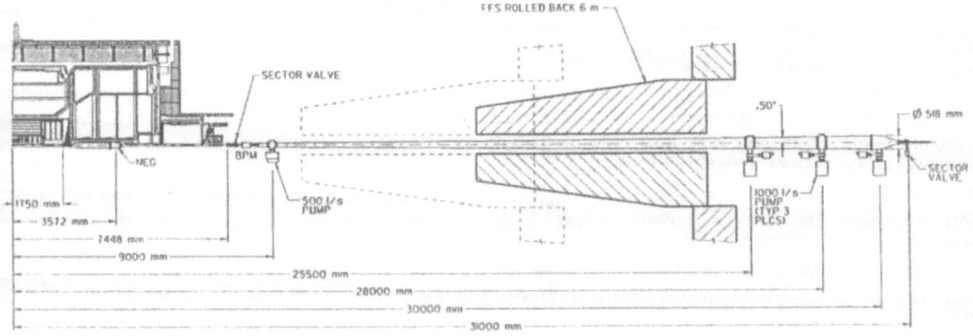

Figure 1. Beam Pipe in the Entire Interaction Region.

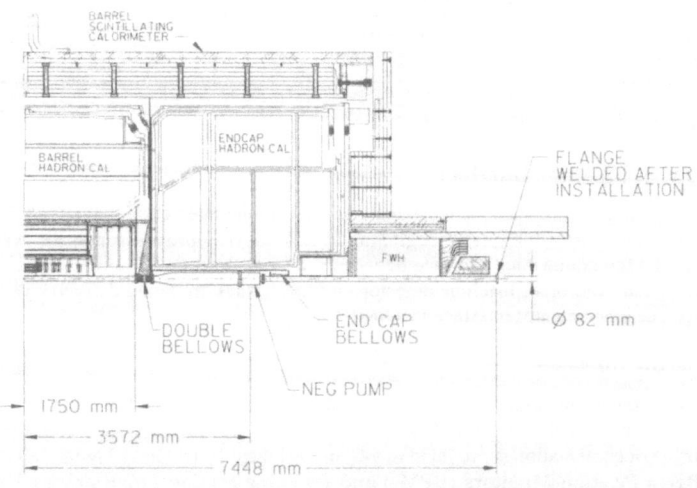

Figure 2. Beam Pipe Installed in the Central Region of GEM.

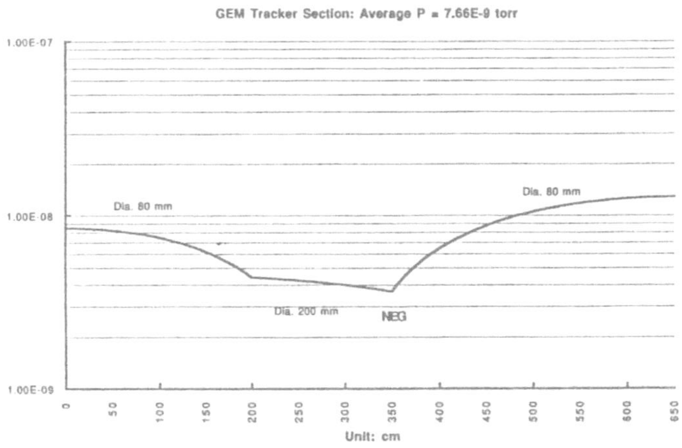

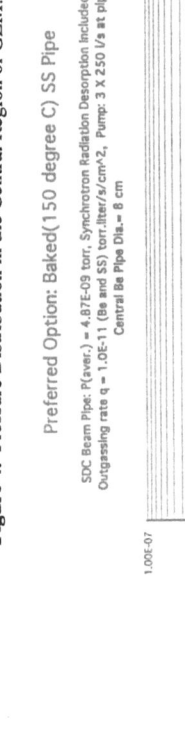

Figure 4. Pressure Distribution in the Central Region of GEM.

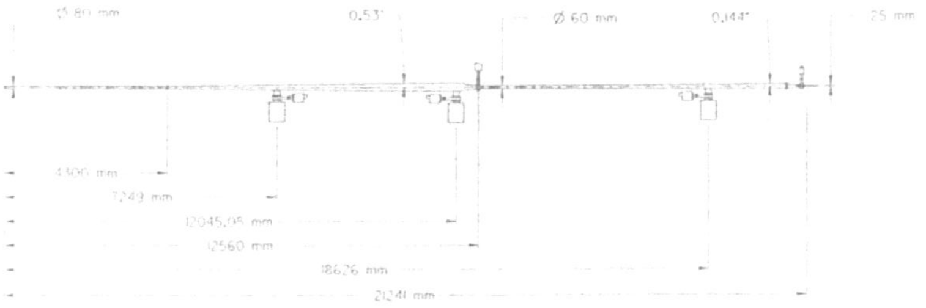

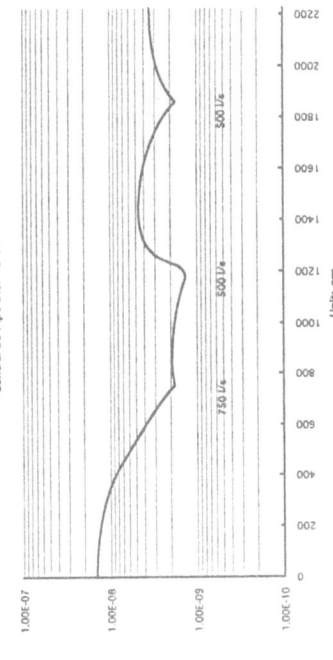

Figure 6. Pressure Distribution in the SDC Interaction Region.

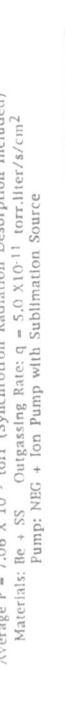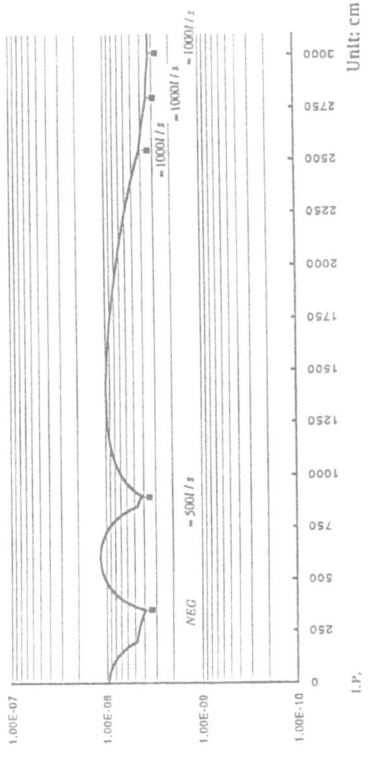

Figure 3. Pressure Distribution in the Entire Interaction Region of GEM.

Figure 5. Beam Pipe in the SDC Interaction Region

environment near the beam line, there are no permanently mounted roughing pumps included in the design. Any routine maintenance or repairs will require access for personnel and equipment to the ion pumping stations.

SDC: No firm decision has yet been taken concerning the installation of beam optic monitors in the beam line of the interaction region of the SDC detector.

Beam Pipe Support and Alignment of the support of the beam pipe and associated hardware must meet two requirements: 1) it must support the weight of the pipe and vacuum related hardware, and 2) it must be adjustable to accommodate deflections. In most of the sections, the beam pipe must also be able to slide horizontally along the beam direction during the installation procedure. A wire or band support will be used between the Central Tracker pipe section and the Barrel Calorimeter. It must be vertically adjustable and should allow a small amount of movement along the pipe direction due to thermal retraction. The beam pipe, in the Endcap Calorimeter section, needs a guiding-slide support mechanism to meet the requirement of installation. The section of conical pipe through Forward Field Shaper (FFS) (unique to GEM)will be supported on the inner wall of the FFS. In front of the collimator, there will be a 4 m outer diameter concrete shield around the beam pipe. The support for the pipe and pumps should sit within the shielding. The detailed design of these supports is on going.

A SYSTEMS APPROACH TO HAZARD AND RISK MANAGEMENT
FOR THE SDC DETECTOR

Robert F. Lavelle

Superconducting Super Collider Laboratory*
2550 Beckleymeade Avenue
Dallas, TX 75237-3997

INTRODUCTION

The realities of the day seem almost insurmountable: Environmental impact statements, public hearings, stringent enforcement of the existing government ES&H regulations together with new and seemingly costly requirements; the list goes on. It is self-evident that the recent years have seen a significant increase in environmental and system safety awareness. We are now beginning to see the effects of this awakening on the Department of Energy (DOE) and its High Energy Physics (HEP) programs.

The days of a laissez-faire and sometimes cavalier approach to safety and environmental issues are over. Environmental awareness and personnel safety are now paramount design criteria. In our environment, the importance of ES&H is now on par with physics performance itself, especially in the eyes of the customer. The predicament is further complicated by budget-driven extensions of development profiles, sometimes elongated out to ten or twelve years. All of these facts combine to render the traditional "post-hoc" safety analysis program obsolete.

The predominant question today centers on how HEP Project Managers can implement a cost and performance effective ES&H program. This program must at once address the needs of the customer, the needs of the research, the needs of the public, and must be dynamic enough to be viable over a ten or twelve year development schedule.

We must begin by coming to an understanding of our working environment, an understanding of the customer's requirements and a utilization of lessons learned from others who have existed in a similar environment. Surprisingly enough, there is real help in answering these questions. Let's characterize the current HEP environment in which such a program must operate.

THE NEW HEP ES&H ENVIRONMENT

Two facts of life conspire to present a complicated development environment. First, is just the nature of the beast; as we look deeper and deeper into the subatomic realm, we are required to use mechanisms of increasing size and complexity. Second, the more we learn

---

*Operated by the Universities Research Association, Inc., for the U.S. Department of Energy under Contract No. DE-AC35-89ER40486.

about the world around us, the more we realize that our performance as global custodians has been, to use a Washington term, dispositive. Thus, the new emphasis on ES&H in the first place.

So, now we find ourselves at the SSC, designing the world's largest and most complex P+P- particle detector, the SDC. It is a detector with significant ES&H challenges, a multi-million dollar ES&H budget, and a design environment that is the most restrictive and most carefully monitored in HEP history. We can anticipate a continuous string of assessment and audits in one form or another. Detailed design reviews, documentation reviews, and quantitative and qualitative ES&H analyses are all built into the development management approach.

The New Detectors

Today's high energy particle detectors are significantly larger and more complex than their predecessors. The SDC Technical Design Report (TDR) and the SDC Conceptual Safety Analysis Report (CSAR) describe in detail both the size and complexity of the SDC as well as the types of hazards and risks indigenous to a detector of this ilk. Of note for this article are the inherent hazards associated with the various detection mechanisms, including flammable gases and liquids, health physics concerns, and hazardous atmospheres. In addition, the detector is honeycombed with maintenance access confined spaces that will have to be managed.

The New Regulatory Environment

The DOE Energy Research Office has had at its disposal a comprehensive set of ES&H requirements and guidance that can be applied to the design and operation of particle detectors. In the past, these regulations, in the form of the nuclear-oriented DOE Orders, have been applied with some degree of uncandor. In today's environment, the DOE is calling for a much more sincere approach as well as attempting to better define the safety criteria for detectors in its new accelerator order, DOE 5480.25, Safety of Accelerator Facilities. A sincere attempt at compliance with this list is a formidable task indeed.

The Systematic Approach

The SDC ES&H program is an attempt to provide a viable approach to ES&H while living within tight budget constraints. To do this, three tasks are paramount. First and foremost, we must design and operate a safe, effective detector. Second, we must comply with the letter of the customer's (DOE's) ES&H requirements. And, third, we must capitalize on existing approaches and lessons learned from others in heavily scrutinized, requirements laden environments. Both the Department of Defense and the Nuclear Power Utility organizations operate in similar environments. Both are environments characterized by protracted development schedules and detailed design and analysis requirements. A number of important aspects dominate a successful program in their environments as well as in ours:

1. A well defined, detailed ES&H Program Plan and Schedule.
2. An austere but adequate funding profile that provides for continued engineering support for the life of the development.
3. Consistent and effective documentation trail.
4. Cost and Schedule Management.
5. Judicious use of quantitative and probabilistic analyses.
6. Gravitation towards design implementations that can be adequately analyzed using qualitative techniques.
7. Standardized analysis techniques.
8. Flowdown of responsibility from ES&H management to the actual design engineers.

The SDC ES&H approach coalesces all of the aforementioned points into an effective, albeit somewhat patchworked, systematic approach to requirements flowdown, design analysis, and analysis documentation that is both safety requirements and cost effective.

THE SDC SYSTEMATIC APPROACH

Figure 1 depicts both the four key elements of an effective program and the schedule flow. Foremost is a well defined program plan. This plan provides a framework for the ES&H organization to function. The majority of effort lies in the area of requirements flowdown into the design activities. The results of the design engineering efforts are subjected to an iterative series of qualitative and quantitative design reviews.

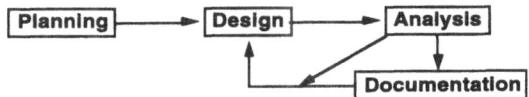

Figure 1. SDC ES&H Program Key Elements.

While ES&H is not the only area addressed at these both formal and informal reviews, it is a key player. The results of both the design work and the subsequent analyses are documented in a sequentially more detailed set of Safety Analysis Documents (SAD's). It is important to note that both the design reviews and the SADs act as a closed loop feedback system allowing both designers and management to assess design and program compliance.

The ES&H Program Plan

The ES&H Program Plan details the design, analysis and management tasks to be performed. It is a description of the methodology that you intend to employ to identify, evaluate and either resolve or manage all hazards and risks associated with the system in question. The SDC plan is modeled after the requirements MIL-STD-882 task 101 and its associated Data Item Description (DID) DI-SAFT-80100. The SDC plan provides the following types of information:

- ES&H Organization: Describes the organizational elements within the SDC collaboration with ES&H responsibility. It defines the responsibilities, authorities, and accountability of both management and engineering personnel.

- ES&H Program Schedule: Identifies integrated system activities and provides milestones for design requirements flowdown, design analyses, design reviews, and documentation deliveries. While the official version of this document does not include manloading for each task, the initial versions were used extensively for just that purpose.

- ES&H Requirements: This section describes three important aspects of the ES&H program: 1) the methods to be used to identify ES&H requirements, including health physics and human factors engineering. These requirements are provided to the engineering community via the SDC Engineering Guidelines. 2) ES&H assessment procedures for characterization of hazards identified in both the initial conceptual qualitative analyses and the more detailed quantitative type analyses, i.e., FMECA. 3) Management controls to be used to ensure compliance with requirements or justify deviations from those requirements

- Hazard Analysis Procedures: Detailed procedures for both qualitative and more detailed quantitative hazard analysis procedures.

This type of plan, with its associated level of detail provides a basis of understanding between the SDC collaboration and their customer, the SSC Laboratory and the DOE. This plan ensures that adequate consideration is given to ES&H issues during the development of the detector system.

Requirements Flowing into Design

The starting point in the implementation of this program is the identification and understanding of the requirements. Requirements can be thought of as consisting of two main categories, programmatic and criteria. Programmatic requirements refer the program precepts mandated by the customer; in this case, the DOE. Criteria refer to design stipulations. These are hard requirements that directly affect the hardware/software design of the system. The requirement for a Critical Design Review (CDR) is a programmatic requirement. Compliance with the National Electrical Code is a design criteria.

Both the programmatic and criteria requirements for the SDC detector are located in several different types of documents, including the DOE orders and their associated references (it's not uncommon to have requirements documents nested four and five deep on a government contract), SSC Laboratory Standards, Physics Research Documents, and even some self-imposed documents. In the interests of design acceptance and simplicity, these requirements are coalesced and flowed down to the engineering community in the form of SDC Engineering Guidelines. It's important to note that verification of compliance with these guidelines is a central task at both internal and external design reviews. These ES&H guidelines are part of a larger set of guidance documents that encompasses all the engineering best practices approach that SDC is taking.

Design Reviews

The SDC program is tied to the overall programmatic requirements in the SDC Program Management Plan. A formal design review process has been implemented consisting of five formal reviews as shown in Table 1.

Table 1. Design Phase, Design Reviews and associated documentation.

CONCEPTUAL DESIGN	DETAIL DESIGN	CONSTRUCTION	COMMISSIONING
PRELIMINARY DESIGN REQUIREMENTS REVIEW	PRELIMINARY DESIGN REVIEW CRITICAL DESIGN REVIEW	ACCEPTANCE TEST PLAN REVIEW	OPERATIONAL READINESS REVIEW
CSAR	dPSAR PSAR	SAR	FSAR

Documentation

Detailed information on the breadth of these reviews is available in the SDC Program Management Plan. The intent is simple: Provide a vehicle for assessing design compliance with criteria at each logical transition point in the engineering development and construction of the SDC detector system.

Each of the above referenced design reviews is associated with an ES&H SAD delivery from the system or subsystem under review. The reviews and the SADs track the various life cycle phases of the system as illustrated in Table 1.

PROGRAM STATUS AND INITIAL RESULTS

The SDC ES&H program was initially implemented in the fall of 1991, subsequent to the approval of the SDC Letter of Intent and receipt of the approval to proceed with a Technical Design Report (TDR). This initial phase of the program has been dubbed the "conceptual phase," and the SAD that resulted from the first round of qualitative analyses is the SDC Conceptual Safety Analysis Report (CSAR). The initial version of the CSAR was provided to the laboratory for review together with TDR on 1 April 1992. The CSAR REV "A" was released in October of that same year. The SDC design and CSAR have been

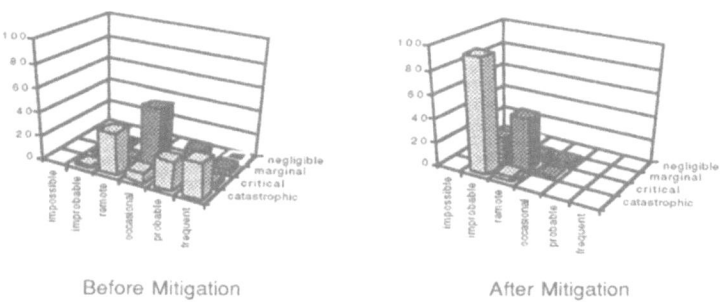

Before Mitigation   After Mitigation

Figure 2. Initial ES&H Program Results.

subjected to expert review by three panels: a Laboratory Review Panel, the Program Advisory Review, and by the DOE. Program approach, conceptual designs, and the CSAR all received positive reviews. The first results from the ES&H program are presented in Figure 2. The initial analysis of the conceptual SDC design yielded 60 catastrophic accidents with a probability of occurrence that was unacceptable. Subsequent to this analysis, and after engineering iterations in light of the analysis, the probability of occurrence of these accidents has been reduced to either improbable or impossible.

9 World Converging on SSC

International Participation

PRC/US COLLABORATION IN SSC CONSTRUCTION

F. Shou-xian

Institute of High Energy Physics
19 Yuquan Road, Shijing Shan District
Beijing 100039
P. R. C.

INTRODUCTION

In the past decade, China has made great progress in developing high energy physics research. Since Beijing Electron Positron Collider (BEPC), the first high energy accelerator in China, realized collision in October 1988, it has experienced constant upgrading in performance. In the second half of 1991, the performance of BEPC reached a record high since its inception. The luminosity of BEPC at the energy of 2 GeV is 6×10 cm^{-2} s^{-1} with a lifetime of 8 to 10 hours. Operation has been very stable, trouble rate is less than 10% of the total operation time.

The above fact shows that Chinese industry is able to meet the requirements of BEPC and BES (Beijing Spectrometer). Many components of BEPC and BES proved to be very reliable after long time operation. By the construction of BEPC, a batch of experienced and high-caliber scientists have been cultivated and an experienced operation team trained. All these have laid a good foundation for today's international cooperation.

PROGRESS ON SSC COLLABORATION

SSC is the largest basic science project that mankind has ever built, it will provide mankind with an opportunity to deepen understanding of the origins and basic constituents of matter. Its potential scientific merits have attracted many scientists from all over the world. We had gotten valuable help and collaboration from our American colleagues during the construction of BEPC/BES to further develop this collaboration and friendship, Chinese scientists are very happy to make contributions to the construction of SSC. But as China is a developing country with limited economic power, we can only make our contributions by fabricating some accelerator and detector components at a no-profit cost and exchanging experienced scientists. This idea has gotten strong support from our government.

To push this collaboration, Prof. Fang Shouxian as the Chinese representative for the collaboration with the delegation of Prof. Zhou Guangzhao, the President of the Chinese Academy of Sciences, had visited SSCL many times. The first time was in March, 1991.

And two months later the same year, Dr. Bob Sheldon and some other SSCL experts, and Prof. Panofsky visited China to explore the possibility and ways of collaboration on the SSC. Through these visits, a mutual understanding was achieved. In later visits to each other, both sides proposed items for collaboration. In October 1991, the PRC Delegation for the 12th Meeting of the US/PRC Joint Committee on High Energy Physics headed by Prof. Zhou Guangzhao visited SSCL and exchanged views with the directors of SSCL on the collaboration. The collaborative program on SSC was first officially discussed at the 12th Meeting of the US/PRC Joint Committee on High Energy Physics held at Lawrence Berkeley Laboratory on October 31–November 1, 1991. It was agreed that the collaboration on the SSC construction be included within the framework of the ongoing agreement on high energy physics and a high level working group be established to draft the official Implementing Accord. It was also agreed that a Symposium on SSC Physics, Experiment and Technology would be held in Beijing in the following May. Since then considerable progress has been made on the collaboration.

On May 18–19, 1992, a working group meeting on SSC collaboration between CAS and DOE was held at IHEP, Beijing. The PRC Delegation was headed by Prof. Zhou Guangzhao and the U.S. Delegation was headed by Dr. W. Hess. At the meeting, an Implementing Accord for the collaboration was drafted. Both sides agreed that the IHEP under the CAS and the SSCL under the DOE would be the executive institutions for the collaboration, and in order to perform this collaboration, the IHEP and the SSCL would sign an Inter-lab Agreement. To this end, the two labs exchanged many visits which resulted in the first inter-lab agreement with nine accelerator related tasks in May, 1992, the second inter-lab agreement with eight detector related tasks. In these two inter-lab agreements, contributions were clearly stipulated. It reads: "the IHEP will perform Tasks under this Agreement as defined in Statements of Technical and Program Requirements which will become a part of this Agreement. The tasks will describe the specific efforts to design, develop, and/or fabricate major components of SSC systems and to supervise the production, test, and delivery of these major components. The IHEP will provide personnel, equipment, facilities, materials, and services as necessary to perform these tasks." Following the stipulations in these Agreements, we have made great efforts in order to fulfill our obligations. Our efforts have met quick responses from our American colleagues. I would specially mention Dr. Bob Sheldon, and Dr. Eddie Duck for his timely information on the status and progress of SSC.

The official Implementing Accord was formally signed by Prof. Zhou Guangzhao and Dr. William Happer in November 1992 at the First Meeting of the PRC/US Joint Committee on SSC Collaboration. The signature of the official agreement put the SSC collaboration under direct governmental coordination. The Chinese government put this collaboration on a very high priority. A special leading group was formed to ensure it a success. Our vice premier Zhu Rongji personally heads this group.

The following brief introductions are on the four tasks we are undertaking under the lab to lab agreement.

Coupled Cavity Linac (CCL)

With rich experiences in the fabrication of electron linac and energy doublers, IHEP agreed to start collaboration on CCL first. When Dr. Bob Sheldon and his colleagues visited us in May 1991, both sides agreed upon the scope of work.

CCL will be composed of 9 modules with 72 tanks. Each tank has 16 cells and there are 63 bridge cavities between the tanks. The IHEP is supposed to undertake the fabrication of all the 9 modules, and in addition, part of the R&D work. This task contains considerable technical difficulties and its schedule is very tight too. To catch up, a lot of preparatory work had been actually done before the Agreement was signed. Prof. Xu Shaowang and several others from IHEP were invited to visit SSCL to get familiar with the

CCL configuration design and prefabrication study. In November 1991, a R&D group on CCL fabrication was set up at IHEP. It is composed of more than 20 experts on accelerator physics, rf system and mechanics. This group first proposed a fabrication plan for the model copper cavity section with an aim to gain experiences and master as quickly as possible the techniques for fine machining, brazing, rf measurement and adjustment. Technical as well as administrative preparations were also made for the organization of the prototype and batch production.

By the time the first Inter-lab Agreement in which the CCL task is included was signed in May, 1992, much progress had been made. In 1992, two review meetings were successfully held at IHEP. Both sides frankly exchanged views on the problems concerned and discussed most of the technical procedures for the fabrication. In general, both sides were satisfied with the progress. These two review meetings served to push the CCL task a step forward. Up to now, the following work has been completed.

Prototype

To obtain experiences, we chose to make a prototype with the parameters of Tank 1 which includes a main cavity with 8 cells and a half-bridge coupler. At the end of last February, we successfully brazed the link cavity and the end cell to the prototype tank. The unbrazed bridge coupler was completed on March 10. The prototype was shipped to SSCL on March 21 for further R&D test.

Tanks 2 and 3 and the Bridge Coupler B12

Tanks 2 and 3 were completed in the middle of March. Fabrication of middle cell, end cell and disk of B12 was finished. The RF tuning method had been developed with the joint efforts of the engineers of both IHEP and SSCL. It took about one month to measure the RF frequency of the cells, field flatness in the cavity.

The most difficult part of this task is the actual fabrication process and RF measurement of the monolithic structure adopted on CCL. In addition, raw material consumption and manpower needs are also very high.

MEB Corrector Magnets

The contract on this task was signed in November 1992. But collaboration on this started in May, 1991. This task includes the fabrication of 168 sextupole magnets, 111 vertical dipole correctors and 130 horizontal correctors. IHEP will fabricate vertical and horizontal correctors, Shanghai Electric Machinery Plant will undertake to fabricate the sextupoles. Recently, the SSCL technical representative, Dr. M. Schulze visited IHEP and the Shanghai Electric Machinery Plant. There has been some revision upon the original lamination drawings of the sextupoles. With the approval of the SSCL technical representative, the machining of the lamination dies for the sextupoles and vertical dipoles has been arranged according to the new design. According to the schedule milestones, these should be completed at the end of 1995. We are confident that this task will be accomplished on time with the efforts of both sides.

SDC Collaboration

We are mainly involved in the following two tasks which are included in the lab to lab agreement on detector collaboration.

Radiation Damage Hardness Test

From the end of 1990, a fruitful collaboration between FNAL and IHEP in the field of radiation damage test was carried out. Till now we have finished testing ten SDC calorimeter modules. Important data and reliable parameters have been provided for selecting materials for SDC Calorimeter. Among these, the radiation hardness test upon the multi-optical fiber structure which is much better than single optical fiber structure was the most important. It provided important experimental data for the final selection of the scheme of the electromagnetic barrel calorimeter. Now the test of new materials 3HF/o2 scintillator tile/fiber has also been finished. It seems that this material is the best candidate for use in the HAD endcap. Both sides think this work was very well done and it is very essential to the success of the SDC Collaboration.

SDC Calorimeter Steel Absorber Wedges

This task includes the fabrication of 64 wedges and a prototype. We signed the contract on the prototype and the intention of undertaking 64 wedges in December 1992. The Xinhe Shipyard is recommended to undertake the fabrication of the SDC prototype hadronic absorber steel structure with the supervision of IHEP. Now the first real prototype is near completion, it will be shipped to SSCL in May. This success has been achieved through the efforts of both sides and due to the hard work of the Xinhe Shipyard workers. Both sides are satisfied with the present progress. A review meeting will be held this coming fall, the fabrication plan of all the 64 barrel wedges will be decided at the meeting. If the actual fabrication can be started at the end of the year, all the components can be delivered in the first season of 1996.

GEM Collaboration

As GEM started later than SDC, up to now we have not signed any formal contract to undertake tasks. But it is the intention of both sides for China to undertake the fabrication of up to 40–50 percent of the barrel part of the muon detector. Assembling and fundamental testing is also planned to be at IHEP. Now the cathode strip chamber has been chosen for the TDR. IHEP has actually started to be involved from this year, we have joined the R&D of the chamber technology and the chamber factory preparations. IHEP will also participate in muon simulation electronics and trigger.

Addendum to Supercollider 4

OPERATIONAL MODES AND CONTROL PHILOSOPHY OF THE SSCL MAGNET TEST LAB (MTL) CRYOGENIC SYSTEM

V. Ganni, R. Than, and M. Thirumaleshwar[*]

Superconducting Super Collider Laboratory[†]
2550 Beckleymeade Avenue
Dallas, TX 75237-3997

INTRODUCTION

The MTL's function is to test prototype and industrially manufactured magnets for the Superconducting Super Collider Laboratory (SSCL). The cryogenic system of the MTL has a main refrigeration system consisting of a two-stage compression system, a refrigerator/liquefier coldbox, a liquid helium dewar, warm gas storage, and a regeneration skid. The MTL cryogenic system also includes the following auxiliary equipment: two *cleaning, cooling, warmup* and *purification* (CCWP) coldbox modules with a regeneration skid for the charcoal beds, two CCWP compressors, a dehydration skid with its own regeneration system, a pump box, a refrigeration recovery unit, and five distribution boxes. Figure 1 describes the general flow of these systems. At any given time, the refrigeration system has the capacity to simultaneously test at least six magnets under normally required testing conditions.

Every magnet will undergo cleaning, cooldown, and filling prior to general testing, conditioning, quench testing, and other experiments. At the completion of general testing, etc., the magnet must be emptied prior to warming it up to ambient temperature. Furthermore, conditioning, training, and testing of the magnets can be carried out at different temperatures between 4.5 K and 2.5 K. The cryogenic system is designed to test multiple magnets, not all of which are necessarily in the same preparational or operational stage. This paper describes the different operational modes and the behavior and control of the total cryogenic system during multiple magnet tests.

EQUIPMENT DESCRIPTION

The refrigerator/liquefier system has two 185 kW (250 HP) first-stage and two 525 kW (700 HP) second-stage compressors. The 4 K refrigeration is accomplished using liquid nitrogen precooling and work extraction by four expanders. The system has a design capacity of 2100 W and 21 g/s at 4.4 K. A 40,000 liter helium dewar serves as the liquid storage and energy manager. The design and basic operation of the refrigerator and its response to varying load conditions has already been explained in a previous paper.[1]

The refrigeration recovery system, shown in Figure 2, has a two-pressure-level vacuum system that provides the vacuum to subcool the supply helium to the magnet. This system has two 110 kW (145 HP) vacuum pump trains. A refrigeration recovery heat exchanger module

---

* Guest collaborator from Center for Advanced Technology, Indore, India.

† Operated by the Universities Research Association, Inc., for the U.S. Department of Energy under Contract No. DE-AC35-89ER40486.

with two Joule-Thompson valves recovers the cooling capacity of the boil-off flow from the subcoolers by cooling high pressure flow and supplying the resulting two-phase helium to the dewar. The two-pressure-level vacuum system allows for efficient, flexible, and stable testing of multiple magnets over the required temperature range of 4.5 K to 2.5 K.

The pump box, shown in Figure 2, contains a make-up pump and two circulation pumps. The make-up pump has a capacity in excess of 200 g/s and the two circulation pumps can provide a circulation rate in excess of 500 g/s of helium flow to the supply header. Two surge vessels in the pump box attenuate pressure pulses generated in the system.

The CCWP system's function is to clean, cool down, and warm up a magnet test stand and purify the helium gas. This system, shown in Figure 3, has its own compressors, oil removal skid, and regeneration skid. It also has a dehydration skid for moisture removal and an 80 K charcoal bed to remove impurities from the helium. The return pressure is maintained at 2 bar and the supply pressure to the magnet test stand can be as high as 6 bar. The CCWP system has two 95 kW (125 HP) compressors. Refrigeration for the CCWP flow is provided by liquid nitrogen.

The distribution box, as shown in Figure 2, serves as the valve box between the magnet test stand and the other equipment (refrigerator, pump box, refrigeration recovery and the CCWP units). It also contains heat exchangers for subcooling the flow of supply helium to the desired temperature. The subcooler bath temperature is maintained by the refrigeration recovery system connected to the phase separator. The distribution boxes are nitrogen heat shielded. There are separate supply and return lines with control valves on each distribution box for each of the following flows:

- Warm helium gas supply from CCWP
- Cold helium gas supply from CCWP
- Return helium line to CCWP
- Liquid helium supply from pump box
- Liquid helium return to pump box
- Quench return line
- Vacuum 1 from phase separator (for 2.5 K)
- Vacuum 2 from phase separator (for 3.6 K)
- Gaseous nitrogen supply
- Liquid nitrogen supply
- Nitrogen vent line

The magnet test stand cryogenic system, shown in Figure 2, consists of a feed can, the magnet, and an end can. These three components are treated as a unit for all operating conditions. The feed can is connected to the distribution box through U tubes. A magnet change does not require warmup of any subsystems upstream of the U tubes.

CLEANING OF A MAGNET TEST STAND

The magnet is cleaned by circulating clean warm helium gas through the magnet test stand from the CCWP unit. The CCWP system return pressure is controlled at 2 bar and the supply pressure can be as high as 6 bar. In the CCWP coldbox, the high pressure helium is cooled to 80 K with the aid of liquid nitrogen. The helium gas exits the second heat exchanger (nitrogen boiler) and enters the charcoal adsorber. A portion of the 80 K clean gas may be sent to the test stands for cooldown purposes while the remainder of this gas is warmed back to room temperature by exchanging energy with the incoming flow from the compressors. This clean warm gas is sent through the warm supply header to the distribution box(es) and on to the magnet test stand(s). The impure exit gas containing moisture and air from the test stand is returned through the CCWP return header and back to the CCWP module. This impure gas is passed through the dehydration skid for moisture removal before it is returned to the CCWP compressors. The nominal flow through the magnet system for cleaning is about 15 g/s. The flow rate to the magnet is controlled by a flow control valve in the distribution box on the warm supply line. The return gas is sampled for moisture content. The cleanup continues until in the return helium reaches the required moisture level.

COOLDOWN AND FILLING OF A MAGNET TEST STAND

Cooldown and filling of the magnet test stand occurs in three phases. Cooling from 300 K to 80 K is accomplished with the CCWP unit, using nitrogen as the refrigerant. Below 80 K, cooling and filling are accomplished with liquid helium supplied from the refrigerator/pump box.

The first phase of cooldown, from 300 K to 80 K (~ 90 percent of cooldown load), is achieved using high pressure gas from the CCWP compressors which has been cooled to 80 K by the nitrogen pre-cooling loop in the CCWP coldbox. The helium gas exits the pre-cooling loop heat exchangers and enters the carbon adsorber. If no cooldown ramp is required for the magnet (to reduce the thermal stress), the 80 K gas is sent through the CCWP cold supply header to the distribution box(es) and on to the magnet test stand(s). If a temperature ramp is required for the cooldown, clean warm helium is mixed with cold helium in the distribution box to generate gas at any required temperature between 300 K and 80 K. The appropriate supply temperature is automatically obtained by controlling cold and warm supply flows with a feedback signal from the load supply gas temperature. The gas from the magnet is returned through the CCWP return header to the CCWP module. Based on its return temperature, the gas either bypasses the heat exchanger or returns through the appropriate section of the heat exchanger to recover refrigeration. The gas is finally returned to the compressors after warmup and moisture removal by the dehydration skid.

The second phase of cooldown, from 80 K to 12 K, is accomplished by supplying 4.5 K helium at 4 bar from the refrigerator/pump box to the test stand via the distribution box. A flow control valve located downstream of the subcooler in the distribution box maintains a flow rate of 20 g/s to the magnet test stand. The helium from the test stand is returned through the quench return header (by resetting the quench return/cooldown valve control to 2.5-3 bar) to the refrigerator where it is selectively admitted to the proper heat exchanger depending upon the helium's temperature.

The third phase of cooldown starts when the return helium temperature drops below 12 K. At this temperature, the return flow in the quench header is diverted to the dewar. As the magnet test stand is further cooled and filled, the supply flow rate is ramped from 20 g/s to the required circulation rate. The normal circulation loop can be established when the magnet return temperature has reached 5 K by opening the return heat exchanger bypass valve to the liquid return line (pump box suction header) and by resetting the quench return valve to its normal control pressure of 4.5 bar. For lower temperature operation, the bypass valve is closed and the flow is returned through the heat exchanger. The time required for each phase of cooldown and warmup is calculated by computer simulation and is given in Table 1. The duration of these transient processes will be longer if control of the temperature gradient in the magnet is required.

Table 1. Magnet test modes

Mode	System	P inlet	T inlet	Flow	Return Temp.	Duration	Remarks
Clean	CCWP	3 bar	300 K	15 g/s	N/A	Depends on moisture	Stop on low dewpoint indication
Cool	CCWP	6-3 bar	300 K- 80 K	80 g/s	300 K-80 K	5 hrs	80 K supply, or > 5 hrs for programmed cooldown
Cool	Main	4 bar	4.5 K	20 g/s	80 K-12 K	2 hrs	Recover in main coldbox
Cool	Main	4 bar	4.5 K	20-100 g/s	12 K-4.5K	0.5 hr	Cool and fill, send return flow to dewar
Test	Main	4 bar	4.5K- 3.6K	100 g/s*	4.6K-3.8K		*Nominal flow rates for these temperature ranges
		4 bar	3.6K- 2.5K	50 g/s*	3.9K-2.9K		
Warm	CCWP & Main	4 bar	80 K	20 g/s	4.5 K-12 K	0.5 hr	Empty inventory into dewar
Warm	CCWP & Main	4 bar	80 K	20 g/s	12 K-60 K	3 hrs	Recovery in main coldbox
Warm	CCWP	6-3 bar	300 K	50 g/s	60 K-300 K	10 hrs	Recover in CCWP coldbox

NORMAL OPERATION OF A MAGNET TEST STAND

A pair of circulation pumps in the pump box maintains the circulation through the distribution boxes and the magnet test stands. The distribution box can deliver 4 bar subcooled helium between 2.5 K and 4.5 K to the magnet test stand with the aid of the refrigeration recovery vacuum system. The normal cooling loop is established once the magnet test stand is cooled and filled. The circulation pumps provide the 4 bar subcooled helium flow for maintaining the magnet temperature during normal testing. The booster pump in the pump box draws liquid from the dewar at a nominal pressure of 1.2 bar and supplies this to the circulation pumps at approximately 3.5 bar. The supply to the circulation pumps comes from the test stand return and makeup flows. The circulation pumps have the capacity to provide flow rate in excess of 500 g/s and at a pressure of 4 bar. This flow is subcooled before leaving the pump box. The supply flow enters the distribution box and is precooled through a heat exchanger by the flow returning from the magnet. However, if this returning flow is above 4 K, then it bypasses the heat exchanger. The supply flow is further cooled in the subcooler and a flow control valve located downstream of the subcooler regulates the flow to the magnet test stand. Part of the supply flow from the pump box is flashed into the subcooler bath to provide cooling to the subcooler flow. Two vacuum pressure lines are connected to the subcooler phase separator. Vac1 provides a maximum vacuum corresponding to a saturation temperature of 2.4 K, and Vac2 produces a saturation temperature of 3.5 K. The helium supply temperature is indirectly controlled by the subcooler phase separator vacuum valves on the vacuum lines in the distribution box that maintain the required vapor pressure.

WARMUP OF A MAGNET TEST STAND

The magnet test stand is warmed up in three phases. The first phase empties the cold helium (4 K) into the dewar. This is accomplished by supplying up to 50 g/s of 80 K helium from the CCWP unit to the test stand with the return flow being directed into the quench header by resetting the quench/cooldown valve control to 2.5-3 bar. The return flow is directed into the dewar so long as the temperature is below 12 K. When the return flow reaches 12 K, the second phase of warmup begins. The CCWP flow is limited to 20 g/s and the magnet return flow is directed to the refrigerator coldbox until the return temperature reaches 50-60 K. In the third phase of warmup, the return flow into the quench header is shut off, and the CCWP takes over and warms the magnet to room temperature. The clean warm gas is supplied through the warm header and the cold gas returns to the CCWP module through the return header. The cooling capacity of the return gas is recovered through the CCWP heat exchanger. The nominal flow rate for warmup of one magnet is approximately 50 g/s. This flow rate depends upon other activities of the CCWP module.

OPERATION OF MULTIPLE TEST STANDS

The system is presently configured with five distribution boxes that are connected to two common headers containing multiple supply and return lines. The refrigeration recovery system is designed to simultaneously handle three test stands at 3.6 K with a 100 g/s circulation rate and two at 2.5 K with a 50 g/s circulation rate. The pump box circulation pumps can provide flow rate in excess of 500 g/s and, along with direct makeup supply from the refrigerator, more than five magnets can be tested at a given time. Various combinations of cleaning, cooling, and warmup functions can be performed on multiple stands up to the CCWP system's capacity of 110 g/s. All the CCWP activities can be executed in parallel on some stands while cold tests are being carried out on other stands. The quench header experiences varying operating conditions depending on whether quenches and cooldowns happen simultaneously or independently. The controls and operation of the quench/cooldown header with multiple test stands will be discussed in a following section. Table 2 shows the static and minimum pump work and Table 3 shows the test stand loads. Table 4 shows the corresponding load on the refrigerator/liquefier. From these tables, one can determine the various possible test stand operating combinations that the system capacity will allow. The system's inventory and energy management philosophy allows transients such as quenches, cooldowns, and quench recoveries to be treated as average loads.

OTHER OPERATIONAL CONFIGURATIONS

The system can be operated without the pump box makeup pump by supplying the makeup flow directly from the refrigerator via the dewar subcooler into the liquid helium supply header (Figure 2). The system can also be operated without either circulation or makeup pumps by supplying all the flow directly from the refrigerator at 4 bar (180 g/s) via the dewar subcooler to the supply header. In this case the flow will return to the dewar via the quench return header. The flow can also be returned to the dewar via the pump bypass if the pump box remains cold, allowing the quench return header to be used for cooldown, warmup, and quenches. The circulating flow is limited to the supply available from the refrigerator. When the system is operated without pumps, the bypass valve between the supply and quench return header will be used for transient operating conditions to minimize the pressure fluctuation to the refrigerator and to the other test stands.

The system can also be operated without the refrigeration recovery unit. A subcooler bath temperature of 4.3 K can be achieved since the vapor return lines from the distribution boxes and pump box are connected to the low pressure side of the refrigerator. If the refrigeration recovery module connection to the low pressure side of the main refrigerator is available, then the subcooler in the distribution box can be used; otherwise, the low temperature is limited to the dewar conditions or pump box subcooler conditions.

QUENCH MANAGEMENT

A pressure rise in the magnet loop or a quench signal initiates a quench sequence for the cryogenic system. This involves the opening of the quench return valve and the closing of the liquid helium supply and return valves in the distribution box, and the opening of the isolation valve between the quench header and the dewar. The dewar valve will remain open for minimum preset time (e.g., 30 seconds) before the temperature controller is allowed to take over. Sudden closing of the distribution box supply valve will cause the pressure to build up in the supply header and will activate the circulation pump bypass or the pressure control valve between the supply header and quench header. When the temperature of the flow in the quench header exceeds 12 K, the isolation valve to the dewar is closed. The flow then goes to the coldbox, where it is admitted to the appropriate heat exchanger. Because the isolation valve to the dewar remains open for a brief preset period before the temperature controller takes over, cooldown gas from another magnet stand may also be admitted to the dewar. The pressure control valve on the coldbox cooldown/quench return line maintains the quench header pressure at 2.1 bar, which is above the dewar maximum operating pressure. The cooldown/quench line in the coldbox and the compressor suction line have independent vents to prevent the compressor suction pressure from rising above the maximum limit. The bypasses and the surge vessels in the pump box will also smooth out the transients caused by the quench sequence.

Table 2. Static heat load and minimum pump work.

Static heat leak		Minimum pump work (idle & bypass control)	
Pump box heat leak	80 W	Circulation pump (Idle 10%)	20 W
Transfer lines heat leak	163 W	Make-up pump (Idle 10%)	90 W
Dewar heat leak	7 W	Total Idle	110 W
Refrigeration recovery	30 W	Circulation pump (Bypass 2%)	4 W
Total	280 W	Make-up pump (Bypass 2%)	18 W
		Total bypass	22 W

Table 3. Test stand loads

Test stand at:	4.4 K	3.6 K	2.5 K
Circulation + make-up flows	100 g/s + 3 g/s	100 g/s + 8.8 g/s	50 g/s + 7.3 g/s
Pump work	86 W	110 W	70 W
Distribution box leak	39 W	39 W	39 W
Test stand heat load	51 W	51 W	51 W
Lead cooling	1 g/s	1 g/s	1 g/s
Refrigeration Recovery	0 g/s + 0 W	1.5 g/s + 50 W	1.5 g/s + 40 W
Total /stand	176 W + 1 g/s	250 W +2.5 g/s	200 W + 2.5 g/s
# of stands	N1	N2	N3

Table 4. Estimation of average load on the refrigerator/liquefier.

Total for stands @4.5K	N1 × (176 W + 1.0 g/s)	Q1
Total for stands @3.6K	N2 × (250 W + 2.5 g/s)	Q2
Total for stands @2.5K	N3 × (200 W + 2.5 g/s)	Q3
Total static heat leak	280 W	Q4
Pump bypass load	22 W	Q5
Average quench load N4 # of quenches/8 hrs*	N4 × 17 W	Q6
Average quench recool/recovery N5 # of recools/8 hrs**	N5 × 0.35 g/s	Q7
Average cooldown load (80-4.5 K) N6 # of cooldowns/week	N6 × 0.18 g/s	Q8
Average leakage losses	0.5 g/s	Q9
Total load ***	Q1+ ... +Q9	Q$_T$
System capacity	2100 W + 21 g/s	Q$_A$

*   Based on 0.5 MJ/quench.
**   Based on 1.0 MJ/quench
*** 1 g/s of liquefaction capacity ≈ 100 W of refrigeration capacity at 4 K
N1, N2...N6 are selected so that Q$_T$ stays within system capacity (Q$_A$).

LHE DEWAR QUENCH MANAGEMENT AND ENERGY CAPACITY

The 40,000 liter liquid helium dewar serves as a liquid inventory and quench energy manager for the refrigerator/liquefier.

The dewar serves as an energy buffer between the refrigerator/liquefier and the magnet quenches. Approximately half a megajoule is expected to be released into the return helium during a single magnet quench. The magnet test stand contains approximately 20 kg of helium inventory and the average enthalpy increases by 25 j/g, which is returned to the dewar. In order to absorb the energy, dewar vapor pressure is allowed to rise to 2.0 bar. A maximum operating dewar pressure of 2 bar is selected to prevent the dewar pressure from reaching the supercritical state. Figure 4 shows the mass that can be added to the dewar when the dewar pressure is limited to 2 bar (or to fill the dewar) as a function of initial dewar liquid volume fraction for various average enthalpy levels of the quench mass flow. For high initial dewar levels, and low enough enthalpy levels of the quench flow, the dewar will fill before 2 bar is reached. Points on the downward trend of these curves correspond to a full dewar. At an initial liquid level of approximately 50 percent volume, the mass absorbed in the dewar at an enthalpy level of 35 j/g is about 400 kg, corresponding to a total energy addition of 14 MJ.

The dewar also serves as a buffer during phase three of cooldown. The flow rate and energy level for this condition is very small and would be a small transient for the dewar. A higher amount of energy may be added to the dewar if a quench should occur during the second cooldown phase of another magnet. In this situation, the warmer gas is also admitted into the dewar until the temperature controller takes over and the valve to the dewar is closed. The additional amount of energy that can be dumped into the dewar, assuming the cooldown gas in the quench header is at 430 j/g, can be as high as 1 MJ. The dewar should be able to absorb over 15 single magnet quenches in a short duration and still maintain a satisfactory operating pressure.

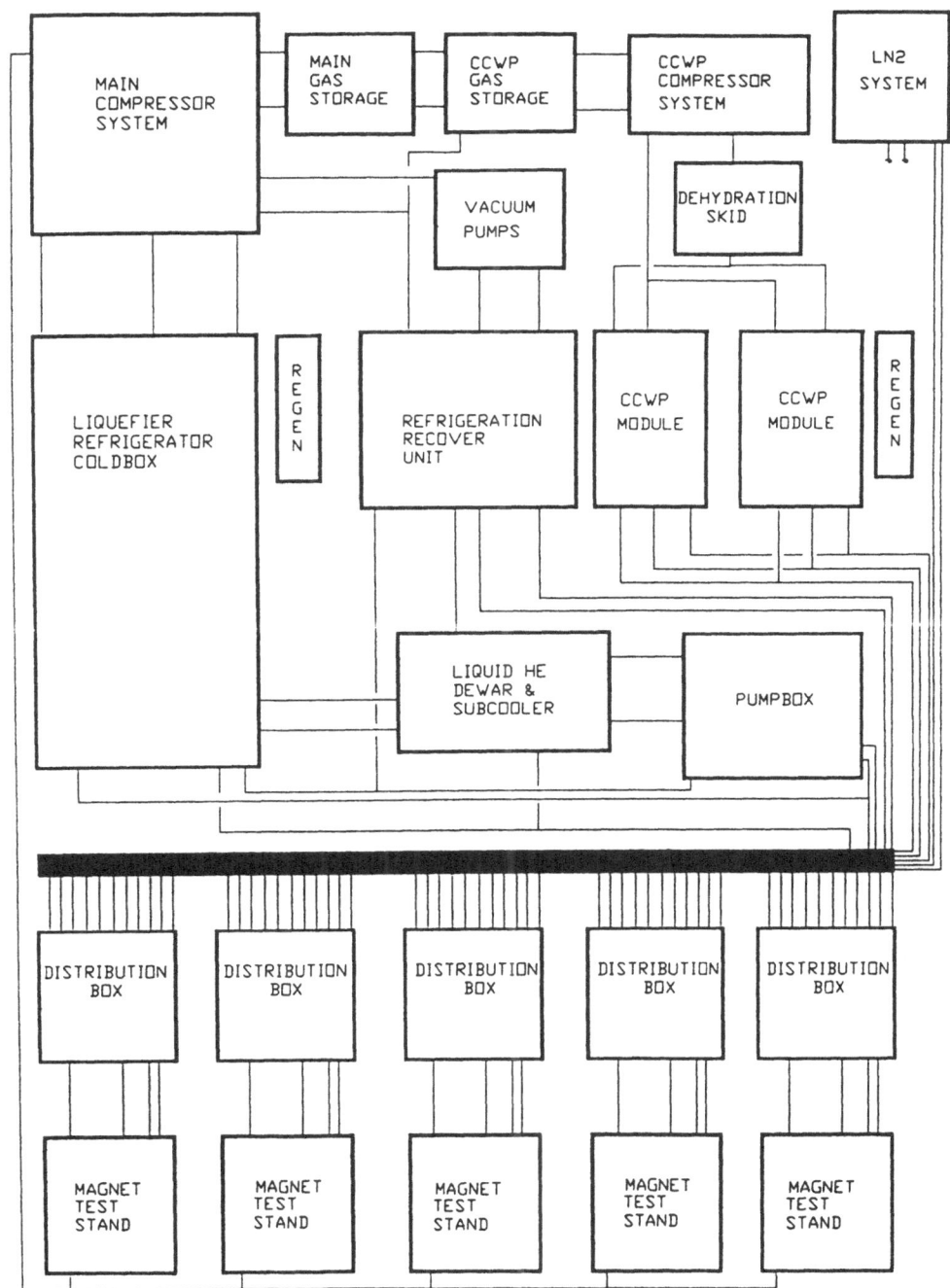

Figure 1. MTL Cryogenic System Block Diagram.

927

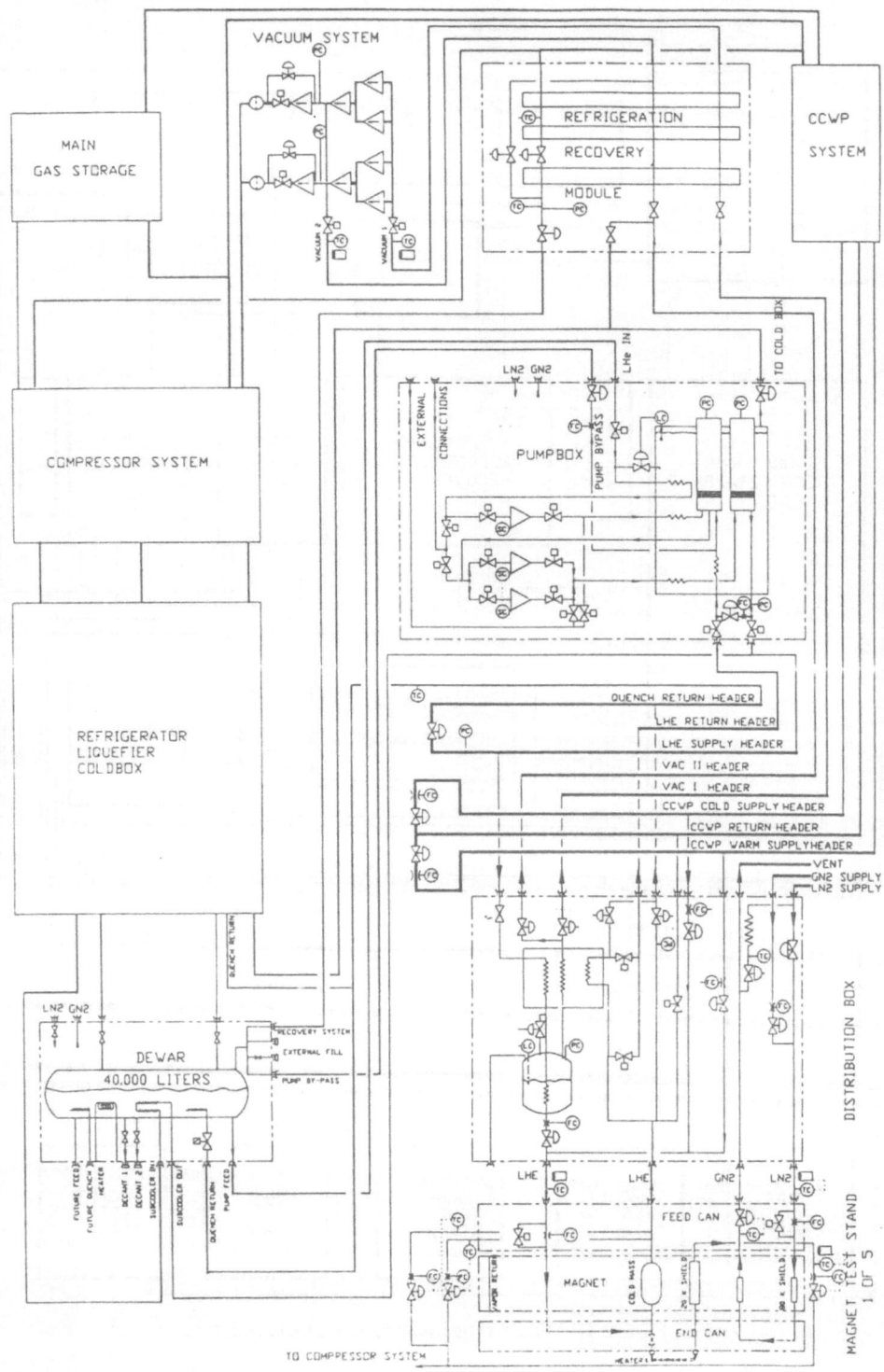

Figure 2. Distribution System, Pumpbox, and Refrigeration Recovery.

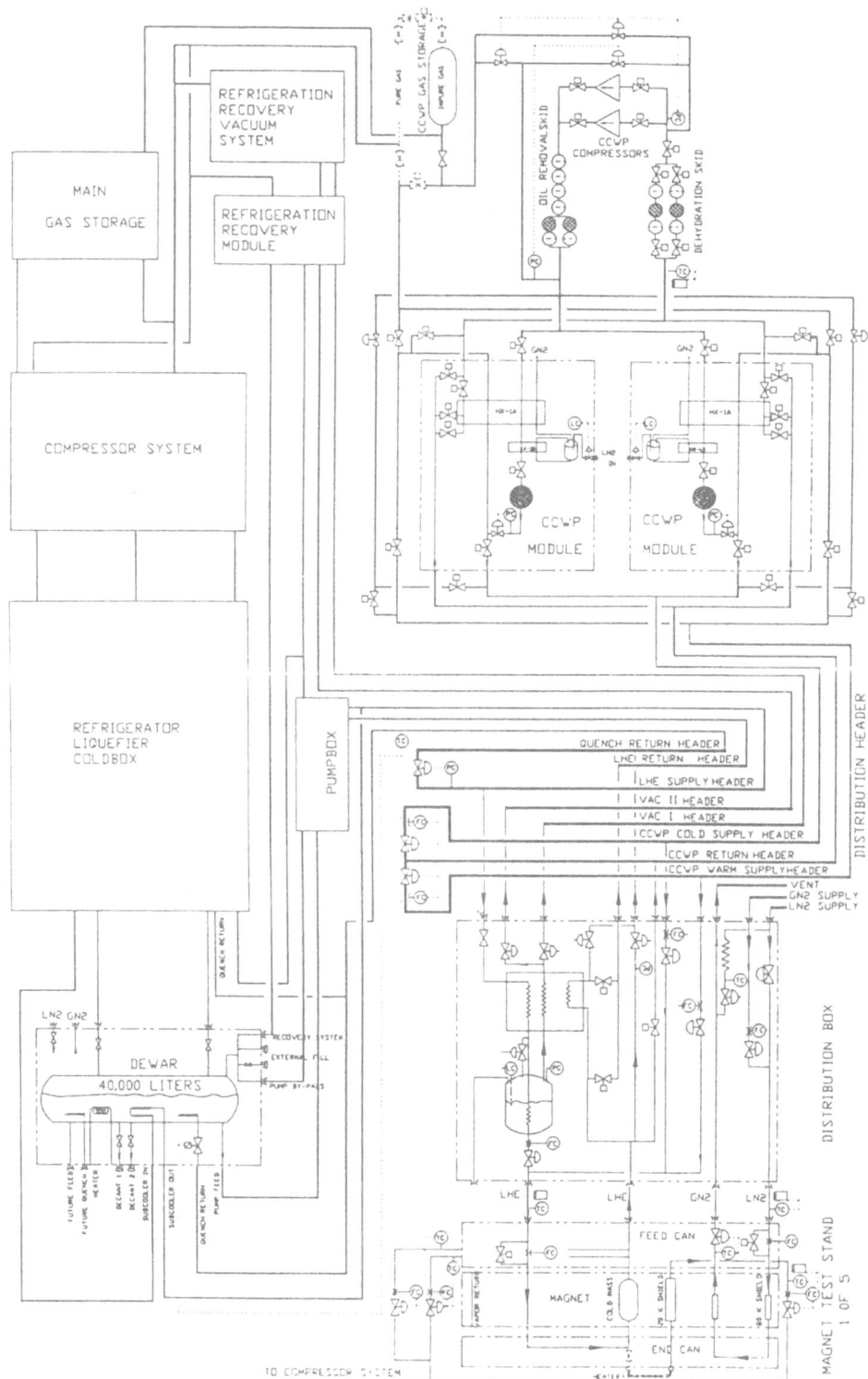

Figure 3. CCWP and Distribution System.

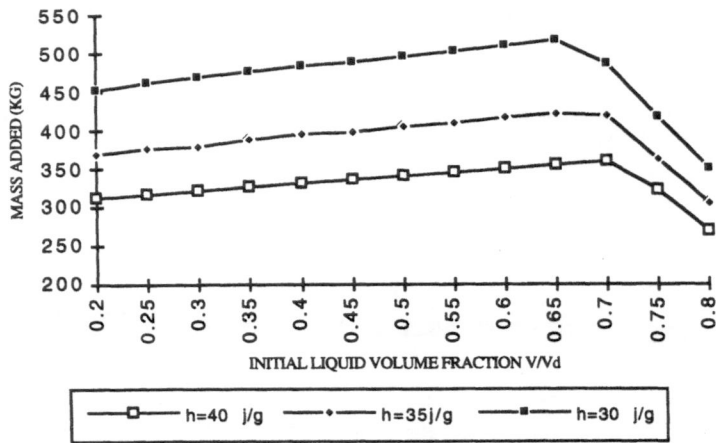

Figure 4. Mass of helium added to dewar vs. initial liquid volume (for three average enthalpies of the quench flow to dewar to reach 2 bar or to fill the dewar)

INVENTORY MANAGEMENT

The total MTL cryogenic system has two helium compressor systems. The gas inventory varies between the two systems depending on the mode of operation of the multiple test stands. The inventory in the CCWP system varies depending on the cooldown and warmup of the magnet test stands with this unit. The net inventory transfer between the main refrigerator /liquefier system and the CCWP system depends on the condition at which the transfer of the cooldown or warmup function occurs between these systems. The excess inventory in the CCWP system is purified and stored in the pure gas tank, and it is returned to the first stage main compressors or to the refrigeration recovery system. Helium gas from the main refrigeration system storage may be supplied to the CCWP gas storage if needed for inventory management.

SUMMARY

The MTL system can simultaneously perform at least six cold magnet tests at collider operating conditions of 100 g/s and 4 K. Magnets can be conditioned at temperatures between 4.5 K and 2.5 K. The average installed compressor power for each operating cold test stand is 270 kW (360 HP); including the vacuum pumps it is 305 kW (410 HP). This overview has presented the various operating modes, features, and flexibility of the Magnet Test Laboratory cryogenic system. The design of the system and the accompanying control philosophy allow the test facility to be operated with maximum flexibility and schedule efficiency. The cryogenic system for the Accelerator System String Test (ASST-A) is similar to this system. Many of the procedures and features described herein also apply to the ASST cryogenic system.

REFERENCES

1. U. Wagner and W. Keyer, Process design features of the SSC MTL cryogenic system, *in:* "Supercollider 3," J. Nonte, ed., Plenum Press (1991).

Attendees

Chuck Abel
General Dynamics Space Systems

Shaul Abramovich
SSC Laboratory

Frank S. Adams
TSA Inc.

Joanne K. Adams
TSA Inc.

Aziz Ahmad
SSC Laboratory

Alexander A. Akhmetov
SSC Laboratory

Toshio Akita
Akita Co. Ltd.

David Alderson
IGC Advanced Superconductors

Bruce Andersen
H&J Tool & Die Co.

Charles E. Anderson
Air Products & Chemicals Inc.

Kathy Anderson
SSC Laboratory

David Andrews
Oxford Instruments

Julie Andrews
General Dynamics

Tom Ankermann
Process Systems International

Tamirisa R. Apparao
SSC Laboratory

William A. Appleton
CVI Incorporated

Bill Archer
SSC Laboratory

Craig S. Arden
SSC Laboratory

Julie B. Armstrong
SSC Laboratory

David N. Arnau
Editorial Services International

Richard Atkinson
Silvex

Al F. Axt
Westinghouse

Paul S. Ayres
Babcock & Wilcox

Kingo Azuma
Mitsubishi Heavy Industries

Robert Baer
Efficient Engineering Co.

Scott Bagwell
SSC Laboratory

David B. Bailey
SSC Laboratory

Rich Bailey
General Dynamics Space Systems

Bob Baldi
General Dynamics Space Systems

M. J. Ball
SSC Laboratory

Wayne Ballard
SSC Laboratory

Antonio Baltramin
INFN-LNL

Paul Barale
Lawrence Berkeley Laboratory

Ronald L. Barber
Los Alamos National Laboratory

Marge Bardeen
SSC Laboratory

Victoria A. Bardos
SSC Laboratory

Dan Baritchi
SSC Laboratory

Liana Baritchi
SSC Laboratory

Paul Barnett
Senate Energy Committee

Vinod Barot
E. I. Du Pont

Henry Barton
SSC Laboratory

Honorable Joe Barton
U.S. House of Representatives

Fred Baumann
Hirel Connectors Inc.

Barbara J. Beam
AT&T Federal Markets Group

James E. Bear
SSC Laboratory

Gerald F. Beatty
LTV Steel Company

William M. Bedell
Votaw Precision Technologies Inc.

Brian L. Beers
SAIC

Jewel C. Bell
Bell Machine Company Inc.

Thomas J. Bell
Bell Machine Company Inc.

Peter F. Bellavigna
Hitachi America Ltd.

Allan Bennett
Allied Signal Aerospace Co.

Richard S. Benson
LeCroy Corporation

Bob Bentley
Integrated Resources Group

William Bernsdorf
U.S. Department of Energy

Steve Berridge
University Of Tennessee

Joginder S. Bhore
The PB/MK Team

Duane Bingaman
Kurt J. Lesker Co.

John Bird
SSC Laboratory

Dan W. Bliss
General Dynamics

Michael C. Boivin
New England Electric Wire

Chris F. Boltz
Metal Surfaces Inc.

John W. Bonn
CVI Incorporated

William Bonow
E. H. Wachs Company

Franz V. Boos
MPI International Inc.

Regina Borchard
Martin Marietta Energy Systems

Rolf G. Bork
SSC Laboratory

William N. Boroski
Fermilab

Bruce Boswell
Westinghouse MSD

Brad Boulais
General Dynamics Space Systems

John Bower
Westinghouse Electric Corp.

Louise Branch
Canadian Consulate Trade Office

Neil Breckenridge
SSC Laboratory

Richard J. Briggs
SSC Laboratory

John E. Brogan
Danco Precision Inc.

Donald P. Brown
Brookhaven National Laboratory

Graham Brown
SSC Laboratory

John C. Bruno
J & L Specialty Products

Warren Buckles
Superconductivity Inc.

Lawrence D. Buhl
Copper & Brass Sales

Alex M. Bulc
Bechtel

Timothy E. Burnett
Westinghouse MSD

Bob Burton
Transpec Fasteners Inc.

Jackie Burton
SSC Laboratory

Thomas O. Bush
SSC Laboratory

Martin W. Butalla
SSC Laboratory

John M. Butler
Westinghouse MSD

Max Buttinger
GEC-Marconi

Yunhai Cai
SSC Laboratory

George E. Callender
Westinghouse MSD

Hugh A. Calvin Jr
Westinghouse Electric Corp.

H. Glenn Campbell
Babcock & Wilcox - NNFD

Ronald A. Campbell
EBASCO Constructors Inc.

Richard A. Cancilla
Westinghouse Electric Corp.

R. Michael Capers
Scientific Instruments Inc.

Donald W. Capone
SSC Laboratory

William M. Caracciolo
SSC Laboratory

Ruben H. Carcagno
SSC Laboratory

J. W. Carey
TRIUMF

Ronnie Carleton
Office Of Congressman Martin Frost

Per Carlson
MSI

Lawrence A. Carlstrom
CVI Incorporated

Clayton H. Carmean
The PB/MK Team

Jim Carney
U.S. Department of Energy

Richard A. Carrigan Jr.
U.S. Department of Energy

John A. Carson
Fermi National Accel. Laboratory

John Carusiello
Meyer Tool & Mfg.

Shlomo Caspi
Lawrence Berkeley Laboratory

Annette Caudiano
Mitsui & Co. (USA) Inc.

William J. Cesarek
Stellar Industries Inc.

Jeff Chang
SSC Laboratory

Gerry Roy Chapman
SSC Laboratory

Honorable Jim Chapman
U.S. House of Representatives

Coby Chase
SSC Laboratory

Yanping Chen
SSC Laboratory

Allen Cheng
The PB/MK Team

Enrico Chesi
CERN

Warren Chester
Crest Products Corp.

Michelle Chin
Office Of Congressman Joe Barton

Ivan K. Chow
SSC Laboratory

Robert W. Dawson
SSC Laboratory

Dean De Benedet
Martin Marietta

Gary A. Deis
SSC Laboratory

Joe G. Deley
Westinghouse MSD

Jonathan Demko
SSC Laboratory

Arnaud Devred
SSC Laboratory

Nicholas Di Giacomo
Martin Marietta

Joseph Di Marco
SSC Laboratory

Les I. Dittert
ICF Kaiser Engineers

Thomas W. Dombeck
SSC Laboratory

Dell' Orco Domenico
Lawrence Berkeley National Laboratory

Paul Domigan
Field Effects

Rene Donaldson
SLAC Beam Line

Guglielmo Donati
Europa Metalli - LMI SPA

Jheroen Dorenbosch
SSC Laboratory

Gene Dretke
U.S. Department of Energy

Gilbert Drouet
CERN

Rens L. Dubbeldam
Holec Ridderkerk

John D. Dubbs
Air Products & Chemicals Inc.

Greg Duchnak
Teledyne Metal Forming

Eddie Duek
SSC Laboratory

Trent Duffy
Parsons Brinckerhoff

Gerald F. Dugan
SSC Laboratory

Gene G. Dun
Babcock & Wilcox - NNFD

John Dykstra
Minnesota Valley Engineering Inc.

David P. Eartly
Fermilab

Brian Easom
Martin Marietta

David O. Eberle
Copper & Brass Sales Inc.

Ronald Egginton
British Embassy

Kevin Einswiler
Lawrence Berkeley Laboratory

Tor L. Ekenberg
University of Pennsylvania

John E. Elias
Fermilab

Amanda Elioff
The PB/MK Team

James L. Elliott
General Dynamics

Tim Ellison
IND W. Cyclotron Fal.

Michael S. Emery
Oak Ridge National Laboratory

Glenn H. Epstein
Oxford Superconducting Tech.

Mark J. Erdmann
SSC Laboratory

Jorge L. Espinosa
SSC Laboratory

David B. Everson
Everson Electric Co.

Yvonne Everson
Everson Electric Co.

Richard W. Fagan
General Dynamics

Shawn Fagen
Advanced Photonix Inc.

Bernard J. Fallon
Verify Inc.

Shou Xian Fang
Inst. of High Energy Physics

Roger A. Farrell
Intermagnetics General Corp.

Chris Farver
Allen-Bradley

Anthony Favale
Grumman Aerospace Corp.

Lee Fennema
Allen-Bradley

Robert J. Ferrari
Marotta Scientific Controls

Jim Ferrell
SSC Laboratory

Robert Ferrell
Airco Gases

Jan A. Fickling
Westinghouse MSD

Arnold Field
AFA Industries

William A. Fietz
SSC Laboratory

Ronald Fincher
AT&T Federal Group

Roderic R. Fink
Acme Cryogenics

Terry Fleener
Ball E-0 And Cryogenics

Gareth D. Flora
Martin Marietta

Carlos M. Flores
Barcel Wire & Cable Corp.

Adrian W. Fluehmann
W. Fluehmann Inc.

Donn Forbes
Supercollider News

Paul T. Forde
H. C. Starck Inc.

Harold K. Forsen
Bechtel Corp

Dave Fortunato
SSC Laboratory

James B. Fraivillig
E. I. Du Pont

Poul Frandsen
CERN

James W. French
SSC Laboratory

Melissa I. Fromm
Air Products & Chemicals Inc.

David G. Frost
Supercon

Honorable Martin Frost
U.S. House of Representatives

Ben Fuchs
Lawrence Livermore National Laboratory

Ken Fukawa
Japan Electrical Manufacturers

Tripp Funderburk
Office of Congressman Bob Livingston

Jim Fusco
ITEN Industries

Tony A. Gabriel
Oak Ridge National Laboratory

Tom Gadsden
SSC Laboratory

Irwin Gaines
Fermilab

Robert W. Galvin
Motorola Inc.

Michael T. Gamble
SSC Laboratory

Henry M. Gandy
TNRLC

Bill Garber
Air Products & Chemicals Inc.

James L. Garner
SMC Technology

Riccardo Garre
Europa Metalli-LMI

Dan Garrett
Furon

Bill Garth
SSC Laboratory

Raj K. Gattu
SSC Laboratory

Alain Gauthier
SSC Laboratory

Eric J. Gerds
University of Pennsylvania

Joel Gerhard
National Mine Service Co.

Rod Giboney
Kaiser Electroprecision

Chuck Gibson
General Dynamics Space Systems

Gary Gigg
General Dynamics Space Systems

Paul H. Gilbert
The PB/MK Team

Gil Gilchriese
Lawrence Berkeley Laboratory

L. Gill
Siemens Industrial Automation

Chris Gillespie
Superconductor Industry

James R. Gillin
Bechtel Construction Inc.

Guy M. Gistau
Air Liquide

Carl L. Goodzeit
SSC Laboratory

Ronald A. Grass
Astroseal Products Mfg.

James A. Gray
SSC Laboratory

Thomas H. Gray
Allegheny Ludlum Corp.

David Green
SAES Getters/USA Inc.

Mike I. Green
Lawrence Berkeley Laboratory

Richard C. Greeno
Teledyne Eng. Services

Eric Gregory
IGC-ASI

Vincent J. Grillo
Cryofab Inc.

Gerard Grunblatt
Alsthom Intermagnetics

Kenneth E. Grut
Babcock & Wilcox - NNFD

John A. Gruver
Westinghouse

Ya- Nan Guo
SSC Laboratory

Fred J. Gurney
Westinghouse MSD

Sam Gurol
General Dynamics Space Systems

Ahmet Gursoy
Parsons Brinckerhoff

Anthony R. Hakl
ICF Kaiser Engineers

Garvie G. Hale
SSC Laboratory

Phyllis J. Hale
SSC Laboratory

Chuck A. Hallquist
Everson Electric Company

Jack Hanlon
Los Alamos National Laboratory

William Happer
U.S. Department of Energy

Stephen J. Harper
Kaiser Marquardt

Alexander Harvey
SSC Laboratory

Nick Hassan
SSC Laboratory

Michael P. Hechlet
SSC Laboratory

Robert V. Hedderick
SSC Laboratory

Jay Heefner
SSC Laboratory

Ezra D. Heitowit
Universities Res. Assn. Inc.

Roland Hellmer
Balzers

Larry Hemmel
Ketema Aerospace & Electric

Steve L. Hensley
CVI Incorporated

Wilmot N. Hess
U.S. Department of Energy

Susan Hickey
Lockheed Missles & Space

Jeanne Hicks
Office of Congressman Jim Chapman

Michael W. Hiller
Babcock & Wilcox - NNFD

Arvle F. Hix
Irving Burton Assoc. Inc.

Gia Ky Hoang
Alsthom Intermagnetics

Steve Hoenig
Lake Shore Cryotronics

Wolfgang Hoernschemeyer
KM-Kabelmetal Ag

Kirsten M. Hofmann
Westinghouse MSD

Matt Hogan
Office of Congressman Pete Geren

Hank Holt
EDS Unigraphics

Seung Hong
Oxford Superconducting Tech.

Charles B. Hood

Richard P. Hora
General Dynamics

Kathryn E. Hornik
Brookhaven National Laboratory

Kenji Hoshoyama
KEK

Robert J. Hovde
SSC Laboratory

Donald Howard
Superconductivity News

Joseph D. Hudspeth
Kerns Mfg. Corp.

James R. Hupton
Westinghouse Marine Div.

Jukka Hurskainen
Outokumpu Superconductors OY

Peter Hwang
Gardner Cryogenics

Hideki Ii
The Furukawa Electric Co. Ltd

Masaru Ikeda
Furukawa Electric Co. Ltd.

Yoshio Imai
Mitsubishi Electric Corp.

Bob Inman
Barcel Wire & Cable Corp.

A. N. Ishaque
GE CRD

Jon R. Ives
SSC Laboratory

Pat Jacobs
Airco Gases

Mike Jaffe
SSC Laboratory

Robert A. Jake
American Magnetics Inc.

Abdul R. Jalloh
SSC Laboratory

Michael A. Janoski
Irving Burton & Assoc.

Robert H. Janowski
The PB/MK Team

Andrew J. Jarabak
Westinghouse Electric

Raghavan Jayakumar
SSC Laboratory

Wendell W. Jesseman
New England Electric Wire

Chuck Johnson
Los Alamos National Laboratory

Dave Johnson
Standard Mfg. Co. Inc.

Duane P. Johnson
SE Systems Inc.

Robert A. Johnson
JBC Associates

Honorable J. Bennett Johnston
Senator D-LA

Kennedy Jones
SSC Laboratory

Leonard F. Jones, Jr.
EBASCO Constructors Inc.

Proctor Jones
Senate Appropriations Committee

Ann G. Jordan
Oak Ridge National Laboratory

N. B. Buck Jordan
Waxahachie Chamber of Commerce

David C. Joyce
UCLA

James T. Joyce
H. C. Starck

Robert J. Jozwiak
Allen-Bradley

Atsushi Kabi
KEK

Vladimir Kadyshevsky
Joint Inst. For Nuclear Research

Yuri A. Kamyshkov
Oak Ridge Nat'l Laboratory

Hem Kanithi
IGC Advanced Superconductors

Raphael G. Kasper
SSC Laboratory

Kenneth Kauffmann
SSC Laboratory

A. Kawai
Nippon Steel Corporation

Sam Kawamata
Hitachi Cable America

B. Kazimierzak
SCE Inc.

Shigeo Kido
Kawasaki Steel Corp.

Ed Killian
Brookhaven National Laboratory

Kiseon Kim
SSC Laboratory

Glenn E. Kinard
Air Products & Chemicals Inc.

Daniel J. Kinzie
General Dynamics

Thomas Kirk
SSC Laboratory

Sherry Kivlighn
SSC Laboratory

Bruce Kling
SSC Laboratory

Larry Knecht
Everson Electric Co.

Ted Kobel
Process Systems International

Rocky Kolb
Univ. of Chicago/Fermilab

Bill Kountanis
SSC Laboratory

Vess Kovachev
SSC Laboratory

Ted A. Kozman
SSC Laboratory

Max Kozyrczak
Process Systems International

David F. Kramer
Inside Energy

Philip F. Kraushaar
SSC Laboratory

H. James Krebs
SSC Laboratory

Mark P. Krefta
Westinghouse STC

T. Scott Kreilick
Hudson International Conductors

Kenneth A. Kreinbrink
CVI Incorporated

Ed Krenik
Office of Congressman Dean Gallo

Detlef Krischel
Siemens AG

S. Krishnamurthy
SSC Laboratory

Bala Krishnan
New Bedford Panoramex

Michael Kropf
General Dynamics Space Systems

Jacek Krzywinski
SSC Laboratory

Ravikumar V. Kudaravalli
UCLA

Jozef Kuzminski
SSC Laboratory

Vince La Bon
U.S. Department of Energy

Richard La Salle
SSC Laboratory

Aaron E. Labarge
SSC Laboratory

Horace N. Lander
Nippon Steel USA, Inc.

John Larkin
Highland Technology Inc.

Buddy Latham
Westinghouse MSD

Glen Lathrop
SSC Laboratory

Robert F. Lavelle
SSC Laboratory

Ruth D. Lawson
Oak Ridge National Laboratory

Daniel R. Lehman
U.S. Department of Energy

Thomas A. Leiser
Trammell Crow Co.

De Lynden R. Lersch
SSC Laboratory

James Leslie
ACPT Inc.

Lou Lestochi
Air Products & Chemical Inc.

Kent K. Leung
SSC Laboratory

Fred Levin
SSC Laboratory

Michael Levin
SSC Laboratory

We- Chuan Li
SSC Laboratory

Joseph J. Lichtenwalner
Oxford Superconducting Tech.

John Liedl
Elin EA Magnet Technology

Alan Lietzke
Lawrence Berkeley Laboratory

Steve Lindmayer
B.C. Trade Development Corp.

He Liu
IGC-ASI

Rosai Livio
SAES Getters

Paula B. Lobe
Hitachi America Ltd.

Jim Lonergan
SAIC

D. K. Longley
AT&T Federal Group

Jerald F. Lowry
Westinghouse STC

Cesar A. Luongo
Bechtel

Charles Lyraud
CEA Saclay

Jeff Mac Kinnon
Office of Congressman Joe Barton

Donald A. Mac Nab
Westinghouse Marine Div.

John Mackay

James Maguire
AET Ltd.

Joe Maguire
U.S. Department of Energy

Michael A. Main
Westinghouse MSD

David Main
H. C. Starck

R. Makino
Hitachi Ltd.

G. T. Mallick
Westinghouse STC

Robert J. Malnar
SSC Laboratory

David W. Mantei
Sullair Corp.

David S. Mapes
Babcock & Wilcox - NNFD

Joe P. Marcus
Martin Marietta

Trent Marion
AT&T Federal Group

Richard H. Marloff
Westinghouse STC

Robert J. Marsh Jr.
Teledyne Wah Chang

Nicolai Martovetsky
SSC Laboratory

Michael Marx
SSC Laboratory

Joseph F. Masso
Maxwell Labs

Toshiaki Masuda
Mitsui & Co. (USA) Inc. - DFW

Peter O. Mazur
Fermilab

Susan Mc Colley
Office of Congressman Ron Packard

Tony Mc Connon
Westinghouse MSD

Ann M. Mc Coy
Oak Ridge National Laboratory

Ellen M. Mc Coy
Sheldahl Inc.

Kim A. Mc Donald
Chronicle of Higher Education

John Mc Donough
EG&G Nuclear Instruments

Mac Mc Ginnis
Allied Signal Aerospace Co.

James R. Mc Harg
Varian Vacuum Products

Glen E. Mc Intosh
Cryogenic Technical Services

Debra L. Mc Keown
TNRLC

Lawrence M. Megan
Teledyne Pittsburgh Tool Steel

Greg Mehle
General Dynamics Space Systems

Robert F. Meserve
New England Electric Wire

Steve Mestad
SSC Laboratory

Frank J. Meyer
Meyer Tool & Mfg.

Bob Michaud
Los Alamos National Laboratory

Brian Miller
Office of Congressman Don Young

Joseph L. Miller
Irving Burton Assoc. Inc.

William O. E. Miller
Los Alamos National Laboratory

Edward C. Milner
SSC Laboratory

John Mitchell
Martin Marietta

T. Miyatake
Furukawa Electric Technologies

Martin W. Moffat
Teledyne Metal Forming

Henry A. Mollicone
Stellar Industries Inc.

James E. Monsees
The PB/MK Team

Anthony J. Montgomery
SSC Laboratory

Suhas Mookerjee
Abb Technology Co.

Sumit Mookerjee
SSC Laboratory

Billy Moore
Office of Congressman Jim Chapman

Allan Mord
Ball E-O & Cryogenics

Stefano F. Moretti
Ansaldo N.A. Inc.

Hiroaki Morita
Mitsubishi Heavy Industries

Yoshiyuki Morita
KEK

Paul F. Morris
Teledyne SC

Vicki Lynn Morris
Structural Composites Industries

Pascal Mougenot
Alsthom Intermagnetics

Hiroshi Mukai
Toshiba Corp.

George T. Mulholland
SSC Laboratory

C. Richard Mullen
Intermagnetics General Corp.

Ray Muller
Hamamatsu Corporation

Timothy K. Muller
Westinghouse

P. J. Mundo
Ellis County Chambers

A. Munro
Lotepro Corporation

Joseph Muratore
Brookhaven National Laboratory

James A. Musser
Indiana University

Honorable John Myers
U.S. House of Representatives

David Nahmias
Air Products & Chemicals Inc.

Uriel Nauenberg
University Of Colorado

Greg J. Naumovich
Everson Electric Co.

Ron Naventi
Bechtel National Inc.

Yehia Z. Negm
Zerkes Corp.

Mitch Newcomer
University of Pennsylvania

Long V. Nguyen
General Dynamics

Renee A. Nichols
Ceramaseal

Wolfgang Nick
Siemens AG

Thomas H. Nicol
Fermilab

Marshall G. Nielsen
Bechtel National Inc.

Ralph C. Nieman
Argonne National Laboratory

Fred Nobrega
SSC Laboratory

Jerzy M. Nogiec
SSC Laboratory

K. Nohara
Kawasaki Steel Corp.

Edd Nolen
Office of Congressman Tom Bevill

James J. Nonnie
Senior Flexonics

John Nonte
SSC Laboratory

Leonard S. Norman
SSC Laboratory

Joyce North
Lancaster Chamber of Commerce

Mary Ann Novak
Parsons Brinckerhoff

David T. O' Brien
Westinghouse MSD

Richard H. Oeler
Air Products & Chemicals Inc.

Michael I. Oged
SSC Laboratory

Toru Ogitsu
SSC Laboratory

Kenji Ohki
Mycom Corp.

Fumio Ohkubo
Mitsubishi Heavy Industries

Isamu Ohno
IHI

Dave W. Ohst
Westinghouse Electric

Masao Oka
Allied Apical Co.

Yoshifumi Okamoto
Kaneka Co.

Minoru Okumura
Toshiba Co.

Kenneth O. Olsen
Martin Marietta Corp.

Mark Olson
Teledyne Rodney Metals

Yas Onel
University of Iowa

Catherine Opila
Sheldahl

David Orrell
SSC Laboratory

M. Oshima
Hitachi Ltd.

Aubie J. Oslin
SSC Laboratory

Gene L. Oster
Oster Eng.

N. Ozaki
Hitachi Ltd.

Honorable Ron Packard
U.S. House of Representatives

Chris P. Palasinski
Gardner Cryogenics

Stephen Palermo
MKS Instruments

Wolfgang Panofsky
SLAC

Joe Parish
E. I. Du Pont

Russ K. Parman
Westinghouse Marine Div.

Kenneth D. Partain
Bechtel

Gary P. Patterson
Martin Marietta

Lee Patterson
General Dynamics Space Systems

Ben Peach
Hackett Pacific Corp.

James B. Peeples
CVI Incorporated

Yuan Ji Pei
National Synchrotron Radiation Lab.

Charles J. Pencinger
Abb Power T&D Co. Inc.

Ruggero Pengo
INFN-LNL

Diego Perini
CERN

Jacques Perot
CEA Saclay

Charles R. Perry
TNRLC

Suzanne Person
General Dynamics Space Systems

Lynn Peugh
North Texas Commission

Karl A. Pfister
MPI International Inc.

James G. Pierce
PHPK Technologies Inc.

Suzanne Pigott
SSC Laboratory

Jon A. Piles
SSC Laboratory

F. Plasil
Oak Ridge National Laboratory

Randy K. Pletzer
SSC Laboratory

Douglas A. Pollock
SSC Laboratory

Alexander S. Popp
Inverpower Controls Ltd

John P. Porter
Brown & Root Inc.

David R. Potter
Westinghouse Marine Div.

Steve Potter
Merrick Engineering

Robert L. Powell
Process Systems International

Ronald P. Pratt
Weldaloy Products Co.

Lawrence E. Price
Argonne National Laboratory

Roy A. Prince
SSC Laboratory

Jean- Claude Puippe
W. Fluhmann AG

Helen Quinn
SLAC

Paul D. Quirk
Allied Signal Aerospace Co.

Greg Ragaey
IDM Corp.

Myrna Ivette Ramos
Hewlett Packard

Raj P. Ranganathan
SSC Laboratory

Gerard R. Rankin
Hoechst Celanese

Larry A. Rehn
Interfet Corp.

James Reidy
SSC Laboratory

Richard L. Rerig
IGC/APD Cryogenics Inc.

Chistopher Rey
Babcock & Wilcox - NNFD

Richard L. Rhodenizer
Intermagnetics General Corp.

James H. Richardson
SSC Laboratory

Robert A. Richardson
SSC Laboratory

Phil Ridley
Silvex Inc.

Edward C. Rinck
Weldaloy Products Co.

Richard L. Riney
Martin Marietta

Michael Riordan
SLAC Beam Line

Chris Rista
SAES Pure Gas Inc.

John A. Ristuccia
Westinghouse Electric

J. Frank Roach
Westinghouse STC

Jack Roberts
Tri-Gas Inc.

Walt Robertson
General Dynamics Space Systems

Will Robinson
SSC Laboratory

Edward G. Rodriguez
Martin Marietta

John A. Romans
Eurus Technologies

Doris C. Rosenbaum
Southern Methodist University

Carl H. Rosner
Intermagnetics General

Jim Rossi
Kyser Company

Henri Rougeot
GE Corporate R&E

John Royet
Lawrence Berkeley Laboratory

Charles E. Rufer
CERN

Robert F. Ryan Jr.
Babcock & Wilcox - NNFD

C. Frank Rydeen
SSC Laboratory

Gary Saenz
New Bedford Panoramex

Brenda Safreed
General Dynamics Space Systems

Toshiyuki Saito
Hitachi Works

Shuji Sakai
Hitachi Cable Ltd

Vincent L. Saladin
SSC Laboratory

Galdino Salvador
General Dynamics

Nicholas P. Samios
Brookhaven National Laboratory

Gary H. Sanders
SSC Laboratory

Michael G. Sandy
Westinghouse Electric Corp.

James R. Sanford
SSC Laboratory

Phillip A. Sanger
Westinghouse STC

Yoshihisa Sato
Toshiba Corporation

Ronald M. Scanlan
Lawrence Berkeley Laboratory

Terry Scanlan
Airco Gases

Giuseppe Scarfi
Ansaldo

Samuel A. Scheer
Westinghouse

John C. Scheider
Lockheed Austin Div.

Robert I. Schermer
SSC Laboratory

David J. Schmid
SSC Laboratory

Birger Schmidt
The PB/MK Team

Donald Schwartz
Cabot Performance Materials

Roy F. Schwitters
SSC Laboratory

Brian L. Scipioni
SSC Laboratory

Frank Scrima
Katie Group

John Scudiere
Oxford Superconducting Tech.

Armin Senn
Linde Kryotechnik

Jeffrey M. Seuntjens
SSC Laboratory

Howard W. Shaffer
Westinghouse Electric

Ren- Chen Shang
SSC Laboratory

Betty J. Shannon
Allied Signal Aerospace Co.

Lawrence E. Shaw
EBASCO Services

Jerry D. Shay
KM-Kabelmetal America Inc.

Bob Sheldon
SSC Laboratory

Philip E. Shelley
SSC Laboratory

Xiaorong Shi
Caltech

N. Shibata
Hitachi Ltd.

Shin Shimada
Sumitomo Electric

Takao Shiota
FTAC

David C. Shoaf
Westinghouse Electric Corp.

Fang Shouxian
IHEP

Mike H. Shrubsole
Murdock Inc.

Quan- Sheng Shu
SSC Laboratory

Charles W. Simpson
Morrison Knudsen Corp.

James E. Simpson
SLAC

Richard E. Sims
SSC Laboratory

Pekka K. Sinervo
University of Toronto

Sharad K. Singh
Westinghouse STC

Mark J. Sisinyak
Brown & Root Inc

Edward J. Siskin
SSC Laboratory

Dave Smathers
Teledyne Wah Chang Albany

David J. Smith
Westinghouse MSD

Kevin D. Smith
Oxford Instruments Ltd

William D. Smith
The PB/MK Team

George F. Smoot
Lawrence Berkeley Laboratory

George W. Snapp Jr.
Bi-Braze Division

Gregory Snitchler
SSC Laboratory

John R. Snyder
Westinghouse Electric

Robert S. Sokolowski
Intermagnetics General Corp.

John H. Sondericker
Brookhaven National Laboratory

945

Kyle Spann
Stellar Industries Inc.

Gene Spence
CVI Incorporated

L. Clay Spencer
Brown & Root Inc.

J. A. Spieckerman
Zerres

Amanda L. Spindel
SSC Laboratory

Frank Spinos
SSC Laboratory

Robert Sponsel
SSC Laboratory

Richard P. Stanek
Fermilab

Michael J. Stanko
Westinghouse MSD

John Stekly
Intermagnetics General Corp.

Kirk Stifle
SSC Laboratory

Jack Story
SSC Laboratory

Robin Stringfellow
TNRLC

Erkki Strom
Outokumpu Copper OY

Takayasu Sugiyama
Nippon Steel Corp.

John Supancic
Von Roll Isola Inc.

Sam B. Suttle
Westinghouse MSD

Toru Suzuki
Nippon Steel Corp.

Chuck Swenson
Westinghouse MSD

Gary Swenson
SSC Laboratory

Walter L. Swift
Creare Inc.

Robert H. Swinderman
Pitt-Des Moines Inc.

Cosmore Sylvester
SSC Laboratory

Balazs Szeless
CERN

Kasuke Takahashi
KEK, Nat'l Lab for H.E. Physics

Greg Tanimoto
C. Itoh Pipe & Tube Inc.

Claude D. Tapley

Clyde Taylor
Lawrence Berkeley Laboratory

Herman Ten Kate
University of Twenty

Vigdor L. Teplitz
SMU

Richard Thomas
Brookhaven National Laboratory

Werner Thomi
CERN

Timothy C. Thompson
Los Alamos National Laboratory

John Thorne
SSC Laboratory

Timothy S. Thurston
SSC Laboratory

John S. Toll
Universities Rs. Assn. Inc.

John C. Tompkins
SSC Laboratory

Gerry Tool
SSC Laboratory

Richard J. Tosetti
Bechtel National Inc.

Junji Toyoda
Nippon Steel USA Inc.

Nghia Dai Tran
SSC Laboratory

Vu H. Tran
SSC Laboratory

Herb Trenham
SSC Laboratory

Kelly Tribble
American BOA

Craig Trimble
SSC Laboratory

Jeff Tseng
SSC Laboratory

Manny Tuli
SSC Laboratory

Gary W. Tulk
SSC Laboratory

Jon Turner
SSC Laboratory

Gary W. Tuttle
SSC Laboratory

Kazuo Ueda
Fuji Electric Corp.

Fred W. Ullrich
Fermilab

Robert Urbach
Ball Aerospace Systems

George J. Urich
Noell Inc.

M. Van Berkum
Cryovat Int. BV

Jeremy E. Van De Bogart
Babcock & Wilcox - NNFD

Johannes M. Van Oort
Lawrence Berkeley Laboratory

Voitto Vanhatalo
Outokumpu Superconductors OY

Alex P. Varghese
Gardner Cryogenics

Dale Vest
Ketema Aerospace & Electronics

Rui Vieira
MIT/PFC

Rui Vieirs
MIT Plasma Fusion Center

Luis Villaseñor
SSCL/IFUG Mexico

James E. Vincent
Teledyne SC

Michael Vitale
Sumitomo Corp. of America

Kotov Vladislav
JINR Dubna

Joann Vogt
AT&T Federal Markets Group

Douglas L. Vollmer
MPI International Inc.

Richard J. Vopelak
Babcock & Wilcox - NNFD

Udo Wagner
Linde Kryotechnik

Ian Walker
GMW Associates

Bill Wallace
Hewlett Packard

Martin Walt

Richard Warren
Lawrence Livermore Nat'l Laboratory

Tom Washenfelder
Siemens IA

Doug Wasitis
Office of Congressman John Myers

Jerry M. Watson
SSC Laboratory

Joseph A. Waynert
Babcock & Wilcox - NNFD

Robert C. Webber
SSC Laboratory

Charles Weber
Babcock & Wilcox

Jim Wei
Westinghouse Electric Corp.

Bernard Weinstein
Los Alamos National Laboratory

Kevin Welch
Martin Marietta

Norman E. Wells
SSC Laboratory

Siegfried Wenger
Elin EA Magnet Technology

Milton J. Werkema
Sheldahl

Jeffrey L. Western
SSC Laboratory

Randy C. White
SSC Laboratory

Walter Wiegert
IGC Advanced Superconductors

Jon Wikstrom
Minnesota Valley Engrg Inc.

Erich H. Willen
Brookhaven National Laboratory

Peter M. Williams
Oxford

Ricky Williams
SSC Laboratory

Ron Williams
House Science Committee

Stephen H. Williams
SLAC

Bob Williamson
SSC Laboratory

Lola Williamson
SSC Laboratory

Kenneth F. Wilson
CVI Incorporated

Thomas R. Winch
SSC Laboratory

Russ Winje
SSC Laboratory

Philip W. Winkler
Air Products & Chemicals Inc.

Randall Withrow
Lockheed Missles & Space

John Wollan
Martin Marietta

Ronn P. Woolley
SSC Laboratory

James V. Worth
Oxford Superconducting Tech.

William C. Worthington
Leybold Inficon Inc.

Scott N. Wright
Allied Signal Aerospace Co.

Jiing-liang Wu
Westinghouse STC

Kuo- Chen Wu
Brookhaven National Laboratory

Craig R. Wuest
Lawrence Livermore National Laboratory

Russ Wylie
SSC Laboratory

Song- Jiang Xia
SSC Laboratory

Dan Yamamura
Teledyne Japan

Victor A. Yarba
SSC Laboratory

James H. Yeck
U.S. Department of Energy

J. Yoshida
Hitachi Ltd.

George Yost
SSC Laboratory

Adnan Yucel
SSC Laboratory

Richard Zeigler
SSC Laboratory

Bruce Zeitlin
IGC-Advanced Superconductors

Burt Zhang
SSC Laboratory

Richard Zhao
The PB/MK Team

Ren-Yuan Zhu
Caltech

Author Index

Subject Index

ABACUS, 429
Absorber wedges, 918
ABS plastic, 401, 402, 417
Accelerator configuration, 13
Accelerator Readiness Reviews (ARR), 759, 761–762
Accelerator Systems Division (ASD)
 education programs, 359–362
 Quality Assurance, 761
Accelerator System String Test (ASST)
 beam tube liner, 817–820
 configuration parameters, 65, 339–342, 347–348
 construction and installation, 63–65
 cryogenic system, 71–74, 82, 455–458, 555–558, 930
 data acquisition and controls, 79–82
 dipole cryostat design, 659, 660
 ER preliminary string test, 323–326
 heat leak tests, 69, 555–558
 objectives, 63
 power systems, 75–78
 program management, 67–68
 quality assurance, 283–286, 760–762
 quench tests, 65–66, 555–558
 schedule of tests, 24, 63–64, 347
 spool pieces for, 423, 424, 430
 superconducting cable for, 148, 149, 150, 156, 157
 thermal insulation for, 221–224
Accordion calorimeters, 793–794
Accounting system, 118
Adapted CERN cooling design, 607–610
Add Drop Multiplexers (ADM), 10, 780
ADM, see Add Drop Multiplexers
Advanced Interconnection Technologies, 615, 723
AFM, see Atomic Force Microscopy
AHARM, 212, 265
Air Products and Chemicals, 83–86, 469
Akita Co., 233
Algor, 317
Alignment
 of beam pipe, 906
 laser tracker for, 358
 optical alignment windows, 447–450, 691–694

Alignment (cont.)
 of SDC detector, 369, 381–384, 889–892
 six-strut support and, 763–766, 843–846
Allied Signal Corp., 179, 386, 387
Alsthom Intermagnetics, 147–158, 511–514
Alumina-reinforced polyimides, 195–196, 255
Amada Metrecs Co., 238
American Magnetics, 560
Amplifier-shaper-discriminator, see ASD-8 signal processor
Analytek, 81, 82
ANL, see Argonne National Laboratory
Annealing of copper, 563–566
ANSALDO, 505
ANSYS Engineering Analysis System
 beam scraper thermal analysis, 767–771
 cryogenic support post stresses, 174
 magnet stresses, 185–188, 217–220, 265, 600–601, 621–624, 626, 645–647, 650
 muon barrel stresses, 900
 RF cavity stresses, 169, 799–802
 vacuum barrier stresses, 425–426
Apical/epoxy insulation
 coil properties and, 205–208, 225–228, 840
 for Collider dipoles, 700
 for HEB dipoles, 837
 for vertical bending magnets, 833
Appletalk, 144
Applied Superconductivity Centre, 587
Arc/Info, 124
ARC quadrupole magnets, see Quadrupole magnets, Collider
Argon, 399, 791, 794
Argonne National Laboratory (ANL), 629, 747
ARR, see Accelerator Readiness Reviews
ASD, see Accelerator Systems Division
ASD-8 signal processor, 43–46, 52–54, 59
Asea Brown Boveri AG, 451
ASST, see Accelerator System String Test
Atomic Force Microscopy (AFM), 407, 408
Audits of contractors, 761
Automation, 279–282, 307, 330
AutoMod, 551–554
Auto-Scan Systems, 341

957